Physical and Historical Geology

Tidewater Community College

James Monroe
Reed Wicander
John Huntsman

CENGAGE Learning™

Australia • Brazil • Japan • Korea • Mexico • Singapore • Spain • United Kingdom • United States

CENGAGE Learning™

Physical and Historical Geology
Tidewater Community College

James Monroe
Reed Wicander
John Huntsman

Executive Editors:
Michele Baird
Maureen Staudt
Michael Stranz

Project Development Manager:
Linda deStefano

Senior Marketing Coordinators:
Sara Mercurio
Lindsay Shapiro

Senior Production / Manufacturing Manager:
Donna M. Brown

PreMedia Services Supervisor:
Rebecca A. Walker

Rights & Permissions Specialist:
Kalina Hintz

Cover Image:
Getty Images*

* Unless otherwise noted, all cover images used by Custom Solutions, a part of Cengage Learning, have been supplied courtesy of Getty Images with the exception of the Earthview cover image, which has been supplied by the National Aeronautics and Space Administration (NASA).

ISBN-13: 978-0-495-43444-3

ISBN-10: 0-495-43444-2

Cengage Learning
5191 Natorp Boulevard
Mason, Ohio 45040
USA

Cengage Learning is a leading provider of customized learning solutions with office locations around the globe, including Singapore, the United Kingdom, Australia, Mexico, Brazil, and Japan. Locate your local office at: **international.cengage.com/region**

Cengage Learning products are represented in Canada by Nelson Education, Ltd.

For your lifelong learning solutions, visit **custom.cengage.com**

Visit our corporate website at **cengage.com**

Printed in the United States of America

Contents

1 Understanding Earth: A Dynamic and Evolving Planet 2

2 Plate Tectonics: A Unifying Theory 26

3 Minerals—The Building Blocks of Rocks 58

4 Igneous Rocks and Intrusive Igneous Activity 86

5 Volcanism and Volcanoes 112

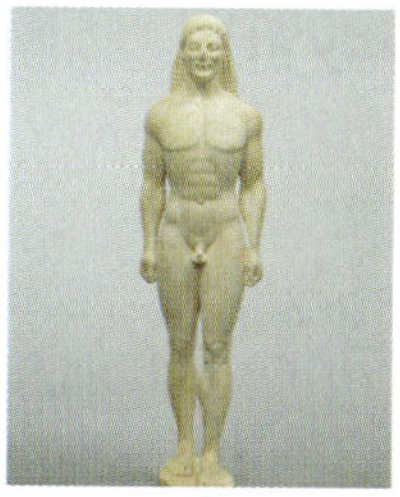

12 Running Water 320

13 Groundwater 352

14 Glaciers and Glaciation 384

15 The Work of Wind and Deserts 412

16 Shorelines and Shoreline Processes 438

17 Geologic Time: Concepts and Principles 470

18 Evolution—The Theory and Its Supporting Evidence 504

19 Precambrian Earth and Life History 530

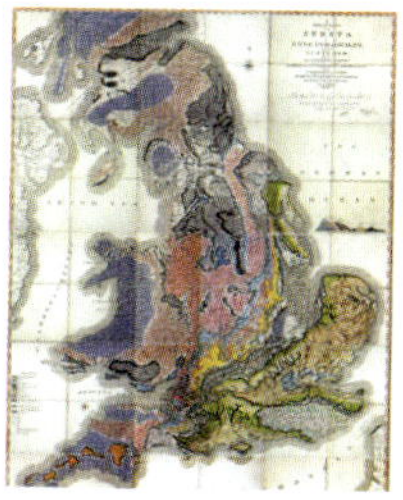

20 Paleozoic Earth History 562

21 Paleozoic Life History 602

22 Mesozoic Earth and Life History 636

23 Cenozoic Earth and Life History 680

Preface

Earth is a dynamic planet that has changed continuously during its 4.6 billion years of existence. The size, shape, and geographic distribution of the continents and ocean basins have changed through time, as have the atmosphere and biota. We have become increasingly aware of how fragile our planet is and, more important, how interdependent all of its various systems are. We have learned that we cannot continually pollute our environment and that our natural resources are limited and, in most cases, nonrenewable. Furthermore, we are coming to realize how central geology is to our everyday lives. For these and other reasons, geology is one of the most important college or university courses a student can take.

The Changing Earth: Exploring Geology and Evolution, Fourth Edition, is designed for an introductory course in geology that can serve both majors and nonmajors in geology and the Earth sciences. One of the problems with any introductory science course is that students are overwhelmed by the amount of material that they must learn. Furthermore, most of the material does not seem to be linked by any unifying theme and does not always appear to be relevant to their lives.

The goals of this book are to provide students with a basic understanding of geology and its processes and, more important, with an understanding of how geology relates to the human experience—that is, how geology affects not only individuals but society in general. It is also our intention to provide students with an overview of the geologic and biologic history of Earth, not as a set of encyclopedic facts to memorize but rather as a continuum of interrelated events that reflect the underlying geologic and biologic principles and processes that have shaped our planet and life upon it. With these goals in mind, we introduce the major themes of the book in the first chapter to provide students with an overview of the subject and enable them to see how Earth's various systems are interrelated. We also discuss the economic and environmental aspects of geology throughout the book rather than treating these topics in separate chapters. In this way, students can see, through relevant and interesting examples, how geology affects our lives.

NEW FEATURES IN THE FOURTH EDITION

The Fourth Edition has undergone considerable rewriting and updating to produce a book that is easier to read, with a great amount of current information, many new photographs, a completely revamped art program, and various new features to help students maximize their learning and understanding of Earth and its systems. Drawing on the comments and suggestions of reviewers, we have incorporated many new features into this edition.

Perhaps the most noticeable change is that Chapter 19 in the Third Edition, "A History of the Universe, Solar System, and Planets," has now been incorporated into Chapter 1 to give students a complete view of Earth's earliest development and relation to the rest of the planets in our solar system. Because plate tectonic theory is such an important theme in geology, it is now covered in Chapter 2 so that students can appreciate its significance to material covered in subsequent chapters. In addition, a new Chapter 24 provides students with an overview of the concepts presented throughout the book and ties together the various themes covered.

The former Prologue and Introduction in each chapter are now combined into a new Introduction. These new sections begin each chapter with a story related to the chapter material and also address why each chapter's material is relevant and important to the student's overall understanding of the topic.

Concept art spreads are found throughout the book. These two-page art pieces are designed to enhance students' interest in the chapter material by visual learning. Some of the topics are "The Burren Area of Ireland," "Rock Art for the Ages," and "The Many Uses of Marble," to name a few.

Another new feature, Geology in Unexpected Places, discusses interesting geology or geologic phenomena in unusual places. We think this feature will be particularly appealing to students because it relates geology to the human experience. "Time Marches On—The Great Wall of China," "Diamonds and Earth's Interior," "Ancient Ruins and Geology," and "Floating Burial Chambers" are just a few of the topics covered.

We also have a powerful new interactive media program called GeologyNow, which has been seamlessly integrated with the text, enhancing students' understanding of important geologic processes. It brings geology alive with animated figures, media-enhanced activities, tutorials, and personalized learning plans. And, like other features in the Fourth Edition, it encourages students to be curious, to think about geology in new ways, and to connect their new-found knowledge of the world around them to their own lives.

Many of the popular What Would You Do? boxes have been rewritten, posing new topics and questions. These boxes are designed to encourage students to think critically about what they're learning. They incorporate material from each chapter and ask open-ended questions to elicit discussion and formulate reasoned responses to particular situations.

The Third Edition's Perspectives have been replaced by Geo-Focus sections on a variety of new topics and updated previous ones.

Many photographs in the Third Edition have been replaced, including most of the chapter opening photographs. In addition, many photographs within the chapters have been enlarged to enhance their visual impact.

The art program has been completely revamped to provide the most accurate and visually stimulating figures possible. In addition, new paleogeographic maps have been commissioned that vividly illustrate in stunning relief the geography during the various geologic periods. Students will find two global views of Earth for each time period.

We think that the rewriting and updating done in the text as well as the addition of new photographs and newly rendered art greatly improve the Fourth Edition by making it easier to read and comprehend as well as a more effective teaching tool. Additionally, improvements have been made in the ancillary package that accompanies the book.

TEXT ORGANIZATION

Plate tectonic theory is the unifying theme of geology and this book. This theory has revolutionized geology because it provides a global perspective of Earth and allows geologists to treat many seemingly unrelated geologic phenomena as part of a total planetary system. Because plate tectonic theory is so important, it has been moved to Chapter 2 and is discussed in most subsequent chapters as it relates to the subject matter of that chapter.

Another theme of this book is that Earth is a complex, dynamic planet that has changed continuously since its origins some 4.6 billion years ago. We can better understand this complexity by using a systems approach to the study of Earth and emphasizing this approach throughout the book.

We have organized *The Changing Earth: Exploring Geology and Evolution,* Fourth Edition, into several informal categories. Chapter 1 is an introduction to geology and Earth systems, geology's relevance to the human experience, the origin of the solar system and Earth's place in it, a brief overview of plate tectonic theory, the rock cycle, organic evolution, and geologic time and uniformitarianism. Chapter 2 deals with plate tectonics in detail, whereas Chapters 3–7 examine Earth's materials (minerals and igneous, sedimentary, and metamorphic rocks) and the geologic processes associated with them, including the role of plate tectonics in their origin and distribution. Chapters 8–10 deal with the related topics of Earth's interior, the seafloor, earthquakes, and deformation and mountain building. Chapters 11–16 cover Earth's surface processes. Chapter 17 discusses geologic time, introduces several dating methods, and explains how geologists correlate rocks. Chapter 18 explores fossils and evolution. Chapters 19–23 constitute our chronological treatment of the geologic and biologic history of Earth. These chapters are arranged so that the geologic history is followed by a discussion of the biologic history during that time interval. We think that this format facilitates easier integration of life history with geologic history. Chapter 24 summarizes and synthesizes the concepts, themes, and major topics covered in this book.

Of particular assistance to students are the end-of-chapter summary tables found in Chapters 20–22. These tables are designed to give an overall perspective of the geologic and biologic events that occurred during a particular time interval and to show how the events are interrelated. The emphasis in these tables is on the geologic evolution of North America. Global tectonic events and sea-level changes are also incorporated into these tables to provide global insights. In the Fourth Edition, we have reduced these tables so that each fits on one page instead of spreading over two pages.

We have found that presenting the material in the order discussed above works well for most students. We know, however, that many instructors prefer an entirely different order of topics, depending on the emphasis in their course. We have therefore written this book so that instructors can present the chapters in any order that suits the needs of a particular course.

CHAPTER ORGANIZATION

All chapters have the same organizational format. Each chapter begins with a photograph that relates to the chapter material, an Outline that engages students by having many of the headings in the form of questions, an Objectives list that alerts students to the learning outcome objectives of the chapter, followed by a new Introduction that is intended to stimulate interest in the chapter by discussing some aspect of the material and showing students how the chapter material fits into the larger geologic perspective.

The text is written in a clear, informal style, making it easy for students to comprehend. Numerous newly rendered color diagrams and photographs complement the text and provide a visual representation of the concepts and information presented. In addition, GeologyNow icons appear throughout the text, indicating opportunities to explore interactive tutorials, animations, or practice problems available on the

GeologyNow website at *http://earthscience.brookscole. com/changingearth4e*.

Each chapter contains one Geo-Focus section that presents a brief discussion of an interesting aspect of geology or geologic research. What Would You Do? boxes, usually two per chapter, are designed to encourage students to think as they attempt to solve a hypothetical problem or issue that relates to the chapter material.

Topics relating to environmental and economic geology are discussed throughout the text. Integrating economic and environmental geology with the chapter material helps students relate the importance and relevance of geology to their lives. Mineral and energy resources are discussed in the final sections of a number of chapters to provide interesting, relevant information in the context of the chapter topics. In addition, each of the chapters on geologic history in the second half of the book contains a final section on mineral resources characteristic of that time period.

Geology in Unexpected Places sections are found in most chapters. This new feature is designed to focus on interesting geology in unusual places or settings you might not have thought about.

The end-of-chapter Geo-Recap begins with a concise review of important concepts and ideas in the Chapter Summary. The Important Terms, which are printed in boldface type in the chapter text, are listed at the end of each chapter for easy review along with the page numbers on which they are first defined. A full Glossary of important terms appears at the end of the text. The Review Questions are another important feature of this book; they include multiple-choice questions with answers as well as short-answer, essay, and thought-provoking and quantitative questions. Many new questions have been added in each chapter of the Fourth Edition. Each chapter concludes with World Wide Web Activities that provide students with the URL for this book. At the Brooks/Cole website, students can assess their understanding of each chapter's topics, take quizzes, and participate in comprehensive interactivities as well as access up-to-date weblinks and find additional readings.

ANCILLARY MATERIALS

FOR INSTRUCTORS

We are pleased to offer a full suite of text and multimedia products to accompany the Fourth Edition of *The Changing Earth*.

GeologyNow GeologyNow is the first assessment-centered student learning tool for a course that combines physical and historical geology. It is tied to your lectures and the text through Living Lecture Tools, which bring geologic processes to life. GeologyNow is web-based and free with every new copy of the text.

The Brooks/Cole Earth Sciences Resource Center
http://earthscience.brookscole.com

Book Companion Website *http://earthscience.brookscole.com/changingearth4e*
The Brooks/Cole Earth Sciences Resource Center and the Book Companion Website feature a rich array of learning resources for your students. The text-specific companion website includes quizzing and other web-based activities that will help students explore the concepts presented in the text.

Multimedia Manager with Living Lecture™ Tools
Create fluid multimedia lectures using your own lecture notes and clips, traditional art from all of our earth sciences texts, and three-dimensional, animated models of *Active Figures* from the text itself.

Instructor's Manual with Test Bank This valuable instructor resource contains approximately 2000 test questions updated for this edition, chapter outlines/overviews, learning objectives, lecture suggestions, important terms, and a list of key resources.
ISBN 0-495-01022-7

ExamView Create, customize, and deliver tests and study guides in minutes with this easy-to-use assessment and tutorial system.
ISBN 0-495-01024-3

Transparencies Full-color transparency acetate collections are available for Physical Geology and Historical Geology.
Physical Geology acetate package: ISBN 0-534-39994-0
Historical Geology acetate package: ISBN 0-534-39297-0

FOR STUDENTS

Study Guide A valuable tool to help you excel in this course! Filled with sample test questions, learning objectives, useful analogies, key terms, drawings and figures, and vocabulary reviews to guide your study.
ISBN 0-495-01397-8

Current Perspectives in Geology Michael McKinney, Kathleen McHugh, and Susan Meadows (University of Tennessee, Knoxville)

This book is designed to supplement any geology textbook and is ideal for instructors who include a writing component in their course. The articles are culled from a number of popular science magazines (such as *American*

Scientist, National Wildlife, Discover, Science, New Scientist, and *Nature*). Available for sale to students or bundled at a discount with any Brooks/Cole geology text.
ISBN 0-534-37213-9

Essential Study Skills for Science Students Daniel Chiras (University of Colorado—Denver)

Designed to accompany any introductory science text. It offers tips on improving your memory, learning more quickly, getting the most out of lectures, preparing for tests, producing first-rate term papers, and improving critical thinking skills.
ISBN 0-534-37595-2

Book Companion Website *http://earthscience.brookscole.com/changingearth4e*

This website features a rich array of learning resources, including quizzing and other web-based activities that will help you explore the concepts that are presented in the text.

GeologyNow GeologyNow is available through the book companion website at *http://earthscience.brookscole.com/changingearth4e*. In additional to GeologyNow, students who use the website have access to maps, weblinks, Internet and InfoTrac® College Edition exercises, learning objectives, discussion questions, chapter outlines, and much more.

ACKNOWLEDGMENTS

As the authors, we are, of course, responsible for the organization, style, and accuracy of the text, and any mistakes, omissions, or errors are our responsibility. The finished product is the culmination of many years of work during which we received numerous comments and advice from many geologists who reviewed parts of the text: Kenneth Beem, Montgomery College; Patricia J. Bush, Dèlgado Community College; Paul J. Bybee, Utah Valley State College; Deborah Caskey, El Paso Community College; William C. Cornell, University of Texas at El Paso; Kathleen Devaney, El Paso Community College; Richard Diecchio, George Mason University; Robert Ewing, Portland Community College; David J. Fitzgerald, St. Mary's University; Dann M. Halverson, University of Southwestern Louisiana; Ray Kenny, New Mexico Highlands University; Glenn B. Stracher, East Georgia College; Monte D. Wilson, Boise State University; and Guy Worthey, St. Ambrose University.

We wish to express our sincere appreciation to the reviewers who reviewed the Third Edition and made many helpful and useful comments that led to the improvements seen in this Fourth Edition: Renee M. Clary, University of Louisiana at Lafayette; Michael Conway, Arizona Western College; Kathleen Devaney, El Paso Community College; Kristi Higginbotham, San Jacinto College; Gary L. Kinsland, University of Louisiana at Lafayette; Bob Mims, Richland College; Michelle Stoklosa, Boise State University; and Azam M. Tabrizi, Tide Water Community College.

We also wish to thank Kathy Benison, Richard V. Dietrich (Professor Emeritus), David J. Matty, Jane M. Matty, Wayne E. Moore (Professor Emeritus), and Sven Morgan of the Geology Department, and Bruce M. C. Pape of the Geography Department of Central Michigan University, as well as Eric Johnson (Hartwick College, New York), and Stephen D. Stahl (St. Bonaventure, New York) for providing us with photographs and answering our questions concerning various topics. We are also grateful for the generosity of the various agencies and individuals from many countries who provided photographs.

Special thanks must go to Keith Dodson, Earth Sciences Editor at Brooks/Cole, who initiated this fourth edition, and to Alyssa White, Development Editor, who saw this edition through to completion. We are equally indebted to our production manager, Nancy Shammas of New Leaf Publishing Services, for all her help. Her attention to detail and consistency as well as general good nature are greatly appreciated. We would also like to thank Carol Reitz for her copyediting skills. We appreciate her help in improving our manuscript. We thank Kathleen Olson for her invaluable help in locating appropriate photographs and checking on permissions. We would also like to recognize Cheryll Linthicum, Thomson Production Project Manager; Carol Benedict, Assistant Editor, who managed the accompanying ancillary package; Ericka Yeoman-Saler, Technology Project Manager, for developing the media program; Jennifer Somerville, Marketing Manager; Kelley McAllister, Marketing Communications Manager; and Megan Asmus, Editorial Assistant. We also extend thanks to the artists at Precision Graphics, who are responsible for updating much of the art program.

As always, our families were very patient and encouraging when much of our spare time and energy were devoted to this book. We again thank them for their continued support and understanding.

James S. Monroe
Reed Wicander

Understanding Earth: A Dynamic and Evolving Planet

CHAPTER

OUTLINE

- Introduction
- What Is Geology?

  GEO-FOCUS 1.1: *Interpreting Earth History*
- Geology and the Formulation of Theories
- How Does Geology Relate to the Human Experience?
- How Does Geology Affect Our Everyday Lives?
- Global Geologic and Environmental Issues Facing Humankind
- Origin of the Universe and Solar System, and Earth's Place in Them
- Why Is Earth a Dynamic and Evolving Planet?
- The Rock Cycle
- Organic Evolution and the History of Life
- Geologic Time and Uniformitarianism
- How Does the Study of Geology Benefit Us?
- Geo-Recap

Geology Now *This icon, which appears throughout the book, indicates an opportunity to explore interactive tutorials, animations, or practice problems available on the GeologyNow website at http://earthscience.brookscole.com/changingearth4e.*

Satellite-based image of Earth. North America is visible in the center of this view as well as Central America and South America. The present locations of continents and ocean basins are the result of plate movements. The interaction of plates through time has affected the physical and biological history of Earth. Source: NASA

OBJECTIVES

At the end of this chapter, you will have learned that:

- Geology is the study of Earth.
- Earth is a complex, integrated system of interconnected components that interact and affect one another in various ways.
- Theories are based on the scientific method and can be tested by observation or experiment.
- Geology plays an important role in the human experience and affects us as both individuals and members of society and nation-states.
- The universe is thought to have originated about 15 billion years ago with a Big Bang. The solar system and planets evolved from a turbulent, rotating cloud of material surrounding the embryonic Sun.
- Earth consists of three concentric layers—core, mantle, and crust—and this orderly division formed during Earth's early history.
- Plate tectonics is the unifying theory of geology and revolutionized the science.
- The rock cycle illustrates the interrelationships between Earth's internal and external processes and shows how and why the three major rock groups are related.
- The theory of organic evolution provides the conceptual framework for understanding the history of life.
- An appreciation of geologic time and the principle of uniformitarianism is central to understanding the evolution of Earth and its biota.
- Geology is an integral part of our lives.

Introduction

A major benefit of the space age has been the ability to look back from space and view our planet in its entirety. Every astronaut has remarked in one way or another on how Earth stands out as an inviting oasis in the otherwise black void of space (see the chapter opening photo). We are able to see not only the beauty of our planet but also its fragility. We can also decipher Earth's long and frequently turbulent history by reading the clues preserved in the geologic record.

A major theme of this book is that Earth is a complex, dynamic planet that has changed continuously since its origin some 4.6 billion years ago. These changes and the present-day features we observe result from the interactions among Earth's internal and external systems, subsystems, and cycles. Earth is unique among the planets of our solar system in that it supports life and has oceans of water, a hospitable atmosphere, and a variety of climates. It is ideally suited for life as we know it because of a combination of factors, including its distance from the Sun and the evolution of its interior, crust, oceans, and atmosphere. Life processes have, over time, influenced the evolution of Earth's atmosphere, oceans, and to some extent its crust. In turn, these physical changes have affected the evolution of life.

By viewing Earth as a whole—that is, thinking of it as a system—we not only see how its various components are interconnected but also better appreciate its complex and dynamic nature. The system concept makes it easier for us to study a complex subject such as Earth because it divides the whole into smaller components we can easily understand, without losing sight of how the components all fit together as a whole.

A **system** is a combination of related parts that interact in an organized fashion (■ Figure 1.1). Information, materials, and energy that enter the system from the outside are *inputs*, whereas information, materials, and energy that leave the system are *outputs*. An automobile is a good example of a system. Its various subsystems include the engine, transmission, steering, and brakes. These subsystems are interconnected in such a way that a change in any one of them affects the others. The main input into the automobile system is gasoline, and its outputs are movement, heat, and pollutants.

We can examine Earth in the same way we view an automobile—that is, as a system of interconnected components that interact and affect each other in many ways. The principal subsystems of Earth are the *atmosphere, biosphere, hydrosphere, lithosphere, mantle,* and *core* (■ Figure 1.2). The complex interactions among these subsystems result in a dynamically changing body that exchanges matter and energy and recycles them into different forms (Table 1.1). The rock cycle is an excellent example of how the interaction between Earth's internal and external processes recycles Earth materials to form the three major rock groups (see Figure 1.12). Likewise, the movement of plates profoundly affects the formation of landscapes, the distribution of mineral resources, and atmospheric and oceanic circulation patterns, which in turn affect global climate changes.

We must also not forget that humans are part of the Earth system; our presence alone affects this system to some extent. Accordingly, we must understand that actions we take can produce changes with wide-ranging consequences that we might not initially be aware of. For this reason, an understanding of geology, and science in general, is of paramount importance. If the human species is to survive, we must understand how the various Earth systems work and interact and, more important, how our actions affect the delicate balance between these systems.

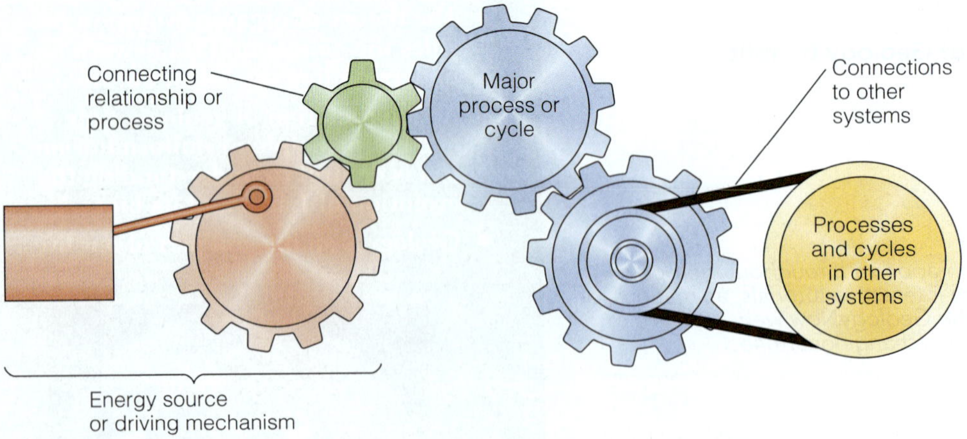

■ **Figure 1.1**

A series of gears illustrates how some of Earth's systems and processes interact. Pistons and driving rods represent energy sources or driving mechanisms, large gears represent important processes or cycles, and small gears represent connecting processes or relationships. Pulleys show relationships to other systems. Although gears are a useful way to represent systems diagrammatically, remember that real Earth systems are far more complex.

When people discuss and debate such environmental issues as acid rain, the greenhouse effect and global warming, and the depleted ozone layer, it is important to remember that these effects are not isolated but part of the larger Earth system. Furthermore, remember that Earth goes through time cycles that are much longer than humans are used to. Although they may have disastrous short-term effects on the human species, global warming and cooling are also part of a longer-term cycle that has resulted in many glacial advances and retreats during the

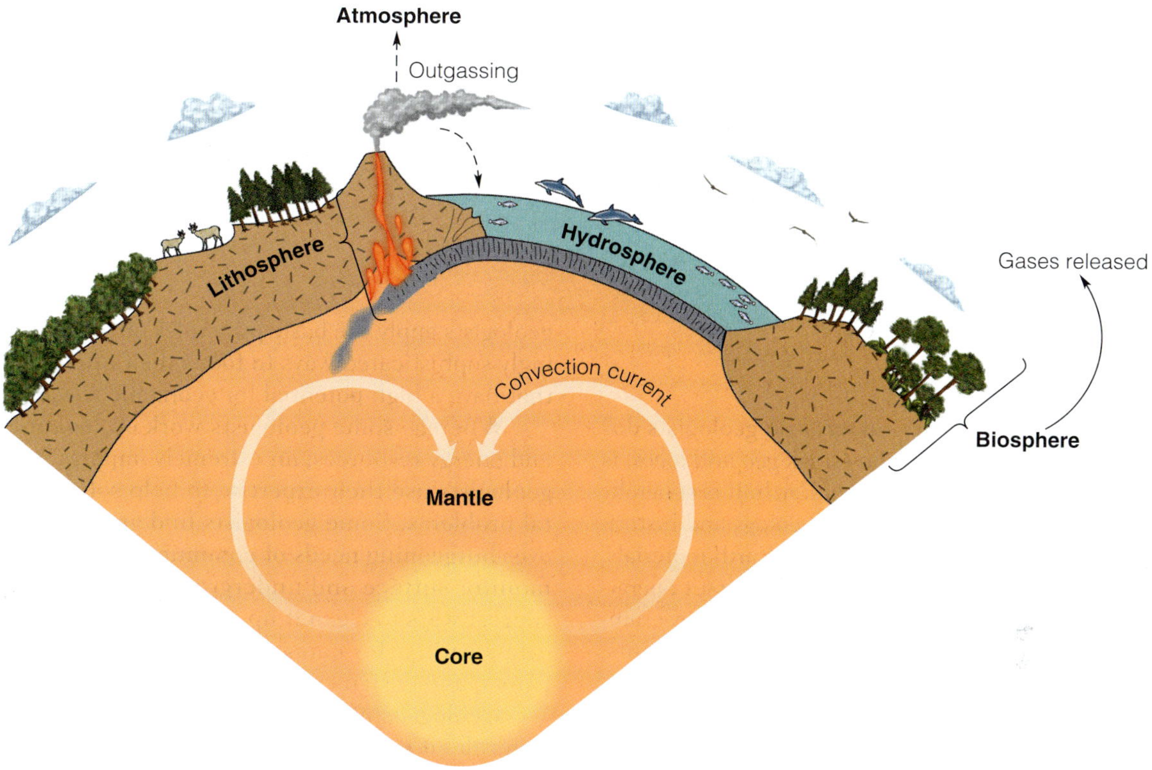

■ **Figure 1.2**

The atmosphere, biosphere, hydrosphere, lithosphere, mantle, and core can all be thought of as subsystems of Earth. The interactions among these subsystems make Earth a dynamic planet that has evolved and changed since its origin 4.6 billion years ago.

Table 1.1

Interactions Among Earth's Principal Subsystems

	Atmosphere	**Hydrosphere**	**Biosphere**	**Lithosphere**
Atmosphere	Interaction among various air masses	Surface currents driven by wind; evaporation	Gases for respiration; dispersal of spores, pollen, and seed by wind	Weathering by wind erosion; transport of water vapor for precipitation of rain and snow
Hydrosphere	Input of water vapor and stored solar heat	Hydrologic cycle	Water for life	Precipitation; weathering and erosion
Biosphere	Gases from respiration	Removal of dissolved materials by organisms	Global ecosystems; food cycles	Modification of weathering and erosion processes; formation of soil
Lithosphere	Input of stored solar heat; landscapes affect air movements	Source of solid and dissolved materials	Source of mineral nutrients; modification of ecosystems by plate movements	Plate tectonics

past 1.6 million years. Because of their geologic perspective, geologists can make vital contributions to the debate on global warming. They can study long-term trends by analyzing deep-sea sediments, ice cores, changes in sea level during the geologic past, and the distribution of plants and animals through time.

As you read this book, keep in mind that the different topics you study are parts of a system of interconnected components and not isolated pieces of information. Examined in this manner, the continuous evolution of Earth and its life is not a series of isolated and unrelated events, but a dynamic interaction among it various subsystems.

WHAT IS GEOLOGY?

What is geology and what do geologists do? **Geology,** from the Greek *geo* and *logos,* is defined as the study of Earth. It is generally divided into two broad areas—physical geology and historical geology. *Physical geology* is the study of Earth materials, such as minerals and rocks, as well as the processes operating within Earth and on its surface. *Historical geology* examines the origin and evolution of Earth, its continents, oceans, atmosphere, and life (see Geo-Focus 1.1).

The discipline of geology is so broad that it is subdivided into many fields or specialties. Table 1.2 shows many of the diverse fields of geology and their relationship to the sciences of astronomy, biology, chemistry, and physics.

Nearly every aspect of geology has some economic or environmental relevance. Many geologists are involved in exploration for mineral and energy resources, using their specialized knowledge to locate the natural resources on which our industrialized society is based. As the demand for these nonrenewable resources increases, geologists apply the basic principles of geology in increasingly sophisticated ways to focus their attention on areas that have a high potential for economic success.

Whereas some geologists work on locating mineral and energy resources, an extremely important role, other geologists use their expertise to help solve environmental problems. Some geologists find groundwater for the ever-burgeoning needs of communities and industries or monitor surface and underground water pollution and suggest ways to clean it up. Geologic engineers help find safe locations for dams, waste-disposal sites, and power plants, and design earthquake-resistant buildings.

Geologists also make short- and long-range predictions about earthquakes and volcanic eruptions and the potential destruction that may result. In addition, they work with civil defense planners to draw up contingency plans should such natural disasters occur.

As this brief survey illustrates, geologists pursue a wide variety of careers and roles. As the world's population increases and makes greater demands on Earth's limited resources, we will depend even more on geologists and their expertise.

Table 1.2

Specialties of Geology and Their Broad Relationship to the Other Sciences

Specialty	Area of Study	Related Science
Geochronology	Time and history of Earth	Astronomy
Planetary geology	Geology of the planets	
Paleontology	Fossils	Biology
Economic geology	Mineral and energy resources	
Environmental geology	Environment	
Geochemistry	Chemistry of Earth	Chemistry
Hydrogeology	Water resources	
Mineralogy	Minerals	
Petrology	Rocks	
Geophysics	Earth's interior	Physics
Structural geology	Rock deformation	
Seismology	Earthquakes	
Geomorphology	Landforms	
Oceanography	Oceans	
Paleogeography	Ancient geographic features and locations	
Stratigraphy/sedimentology	Layered rocks and sediments	

GEOFOCUS 1.1

Interpreting Earth History

Historical geology is the study of the origin and evolution of Earth. Geologists are interested not only in placing events in a chronological sequence but also, and more important, in explaining how and why past events took place. Recently, historical geology has taken on even greater importance because scientists in many disciplines are looking to the past to help explain current events (such as short- and long-term climatic changes) and using this information to try and predict future trends.

We look at Earth as a system consisting of a collection of various subsystems or related parts interacting with each other in complex ways. By using this systems approach, we can see that the evolution of Earth, far from being a series of isolated events, is a continuum in which the different components both affect and are affected by one another. An example is the early history of Earth in which the evolution of the atmosphere, hydrosphere, lithosphere, and biosphere are intimately related. Today, scientists are examining the effect humans have on short-term climate changes and the environment, as well as what a decrease in global biodiversity means for both humans and the planet.

Geologists seek to know not only what happened in the past but also why something happened and what the implications are for Earth today and in the future. Thus it is important to understand present-day processes and to have an accurate means of measuring geologic time so as to appreciate the duration of past events and how these events might affect Earth and its inhabitants today.

An important component of historical geology is understanding how we know what we know. How do we know that dinosaurs became extinct 65 million years ago, or that glacial conditions prevailed over what is now the Sahara Desert during the Carboniferous Period? How can we be so sure that the early atmosphere was devoid of oxygen and evolved over millions of years to one that today has oxygen? Historical geology addresses these questions by seeking answers in rocks and fossils. As more information becomes available from new observations or scientific techniques, geologists become more confident in their interpretations of past events.

One of the many exciting aspects of geology, and of science in general, is that there are still so many unanswered questions. For example, there is still heated debate on what caused the Permian mass extinction. Another exciting area of research is the determination of past environments. New studies indicate that changes in the chemistry of the oceans may have significantly affected the carbon cycle and have important implications in present-day reef ecology and evolution.

What is important to remember is that rocks and fossils provide the clues to Earth's evolution. By applying the various principles of geology, we can interpret Earth's history. It is also equally important to remember that geology is not a static science but one that, like the dynamic Earth it seeks to understand, is constantly evolving as new information becomes available.

GEOLOGY AND THE FORMULATION OF THEORIES

The term **theory** has various meanings. In colloquial usage, it means a speculative or conjectural view of something—hence, the widespread belief that scientific theories are little more than unsubstantiated wild guesses. In scientific usage, however, a theory is a coherent explanation for one or several related natural phenomena supported by a large body of objective evidence. From a theory, scientists derive predictive statements that can be tested by observations and/or experiments so that their validity can be assessed. The law of universal gravitation is an example of a theory that describes the attraction between masses (an apple and Earth in the popularized account of Newton and his discovery).

Theories are formulated through the process known as the **scientific method.** This method is an orderly, logical approach that involves gathering and analyzing facts or data about the problem under consideration. Tentative explanations, or **hypotheses**, are then formulated to explain the observed phenomena. Next, scientists test the hypotheses to see whether what they predicted actually occurs in a given situation. Finally, if one of the hypotheses is found, after repeated tests, to explain the phenomena, then the hypothesis is proposed as a theory. Remember, however, that in science even a theory is still subject to further testing and refinement as new data become available.

The fact that a scientific theory can be tested and is subject to such testing separates it from other forms of human inquiry. Because scientific theories can be tested, they have the potential to be supported or even proved wrong. Accordingly, science must proceed without any appeal to beliefs or supernatural explanations, not because such beliefs or explanations are necessarily untrue but because we have no way to investigate them. For this reason, science makes no claim about the existence or nonexistence of a supernatural or spiritual realm.

Each scientific discipline has certain theories that are of particular importance. In geology, the formulation of plate tectonic theory has changed the way geologists view Earth. Geologists now view Earth from a global perspective in which all its subsystems and cycles are interconnected, and Earth history is seen to be a continuum of interrelated events that are part of a global pattern of change.

Collection of the New York Public Library, Astor

Figure 1.3

Kindred Spirits by Asher Brown Durand (1849) realistically depicts the layered rocks along gorges in the Catskill Mountains of New York State. Durand was one of numerous artists of the 19th-century Hudson River School, which was known for realistic landscapes. This painting was done to show Durand conversing with the recently deceased Thomas Cole, the founding force of the Hudson River School.

HOW DOES GEOLOGY RELATE TO THE HUMAN EXPERIENCE?

Many people are surprised at the extent to which we depend on geology in our everyday lives and also at the numerous references to geology in art, music, and literature. Many sketches and paintings represent rocks and landscapes realistically. Examples by famous artists include Leonardo da Vinci's *Virgin of the Rocks* and *Virgin and Child with Saint Anne,* Giovanni Bellini's *Saint Francis in Ecstasy* and *Saint Jerome,* and Asher Brown Durand's *Kindred Spirits* (■ Figure 1.3).

In the field of music, Ferde Grofé's *Grand Canyon Suite* was, no doubt, inspired by the grandeur and timelessness of Arizona's Grand Canyon and its vast rock exposures. The rocks on the Island of Staffa in the Inner Hebrides provided the inspiration for Felix Mendelssohn's famous *Hebrides* Overture.

References to geology abound in *The German Legends of the Brothers Grimm,* and Jules Verne's *Journey to the Center of the Earth* describes an expedition into Earth's interior. On one level, the poem "Ozymandias" by Percy B. Shelley deals with the fact that nothing lasts forever and even solid rock eventually disintegrates under the ravages of time and weathering. Even comics contain references to geology. Two of the best known are *B.C.* by Johnny Hart and *The Far Side* by Gary Larson (■ Figure 1.4).

Geology has also played an important role in history. Wars have been fought for the control of such natural resources as oil, gas, gold, silver, diamonds, and other valuable minerals. Empires throughout history have risen and fallen on the distribution and exploitation of natural resources. The configuration of Earth's surface, or its topography, which is shaped by geologic agents, plays a critical role in military tactics. Natural barriers such as mountain ranges and rivers have frequently served as political boundaries.

"You know, I used to like this hobby. ... But shoot! Seems like *everybody's* got a rock collection."

Figure 1.4

References to geology are frequently found in comics, as this *Far Side* cartoon by Gary Larson illustrates.

HOW DOES GEOLOGY AFFECT OUR EVERYDAY LIVES?

Most readers of this book will not become professional geologists. Everyone, however, should have a basic understanding of the geologic processes that ultimately affect all of us. We can trace many connections between geology and various aspects of our lives. Natural events or disasters, by their sheer magnitude, provide perhaps the most obvious connection. Less apparent, but equally significant, are the connections between geology and economic, social, and political issues.

Natural Events

Events such as destructive volcanic eruptions, devastating earthquakes, disastrous landslides, gigantic sea waves, floods, and droughts make headlines and affect many people in obvious ways. Although we cannot prevent most of these natural disasters, the more knowledge we have about what causes them, the better we will be able to predict, and possibly control, the severity of their impact.

Economics and Politics

Equally important, but not always as well understood or appreciated, is the connection between geology and economic and political power. Mineral and energy resources are not equally distributed and no country is self-sufficient in all of them. Throughout history, people have fought wars to secure these resources. We need look no further than 1990–1991 to see that the United States was involved in the Gulf War largely because it needed to protect its oil interests in that region. Mineral and energy availability and needs in many cases shape foreign policy. The sanctions imposed by the United States on South Africa in 1986, for example, did not include most of the important minerals we had been importing and needed for our industrialized society, such as platinum-group minerals. Many foreign policies and treaties develop from the need to acquire and maintain adequate supplies of mineral and energy resources.

Our Role as Decision Makers

You may become involved in geologic decisions in various ways—for instance, as a member of a planning board or as a property owner with mineral rights. In such cases, you must have a basic knowledge of geology to make informed decisions. Furthermore, many professionals must deal with geologic issues as part of their jobs. For example, lawyers are becoming more involved in issues ranging from ownership of natural resources to how development activities affect the environment. As government plays a greater role in environmental issues and regulations, members of Congress have increased the number of staff devoted to studying the environment and geology.

Consumers and Citizens

Most people are unaware of the extent to which geology affects their lives. If issues like nonrenewable energy resources, waste disposal, and pollution seem simply too far removed or too complex to be fully appreciated, consider for a moment just how dependent we are on geology in our daily routines.

Much of the electricity for our appliances comes from the burning of coal, oil, or natural gas or from uranium consumed in nuclear-generating plants. It is geologists who locate the coal, petroleum, and uranium. The

What Would You Do?

The concept of sustainable development links satisfying basic human needs with safeguarding our environment to ensure continued economic development. The standard of living we enjoy depends directly on our consumption of geologic materials. You are the president of a large multinational mining company. Discuss how you might balance the need to extract valuable ore deposits and turn a profit for your company with the need to safeguard the environment, particularly if these deposits are located in an undeveloped country with no environmental laws.

copper or other metal wires through which electricity travels are manufactured from materials found as the result of mineral exploration. The buildings we live and work in owe their very existence to geologic resources. Consider the concrete foundation (concrete is a mixture of clay, sand or gravel, and limestone), the drywall (made largely from the mineral gypsum), the windows (the mineral quartz is the principal ingredient in the manufacture of glass), and the metal or plastic plumbing fixtures inside buildings (the metals are from ore deposits, and the plastics are most likely manufactured from petroleum distillates of crude oil).

When we go to work, the car or public transportation we use is powered and lubricated by some type of petroleum by-product and is constructed of metal alloys and plastics. And the roads or rails we ride over come from geologic materials, such as gravel, asphalt, concrete, or steel. All these items are the result of processing geologic resources.

As individuals and societies, we enjoy a standard of living that is obviously directly dependent on the consumption of geologic materials. Therefore, we need to be aware of geology and of how our use and misuse of geologic resources may affect the delicate balance of nature and irrevocably alter our culture as well as our environment.

Sustainable Development

The concept of *sustainable development* has received increasing attention, particularly since the United Nations Conference on Environment and Development met in Rio de Janeiro, Brazil, during the summer of 1992. This important concept puts satisfying basic human needs side by side with safeguarding our environment to ensure continued economic development. By redefining "wealth" to include such natural capital as clean air and water, as well as productive land, we can take appropriate measures to ensure that future generations have sufficient natural resources to maintain and improve their standard of living.

If we are to have a world in which poverty is not widespread, then we must develop policies that encourage management of our natural resources along with continuing economic development. A growing global population will mean increased demand for food, water, and natural resources, particularly nonrenewable mineral and energy resources. Geologists will play an important role in meeting these demands by locating the needed resources and ensuring protection of the environment for the benefit of future generations.

GLOBAL GEOLOGIC AND ENVIRONMENTAL ISSUES FACING HUMANKIND

Most scientists would argue that the greatest environmental problem facing the world today is overpopulation. With the world's population reaching 6.4 billion in 2004, projections indicate that this number will grow by at least another billion people during the next two decades, bringing Earth's human population to more than 7 billion. Although this may not seem to be a geologic problem, remember that these people must be fed, housed, and clothed, and all with a minimal impact on the environment. Some of this population growth will be in areas that are already at risk from such geologic hazards as earthquakes, volcanic eruptions, and landslides. Safe and adequate water supplies must be found and kept from being polluted. More oil, gas, coal, and alternative energy resources must be discovered and utilized to provide the energy to fuel the economies of nations with ever-increasing populations. New mineral resources must be located. In addition, ways to reduce usage and to reuse materials must be devised to decrease dependency on new sources of these materials.

The problems of overpopulation and how it affects the global ecosystem vary from country to country. For many of the poor and non-industrialized countries, the problem is too many people and not enough food. For the more developed and industrialized countries, it is too many people rapidly depleting both the nonrenewable and renewable natural resource base. And in the most industrially developed countries, it is people producing more pollutants than the environment can safely recycle on a human time scale. The common thread tying these varied situations together is environmental imbalance created by a human population exceeding Earth's carrying capacity.

One result of environmental imbalance and an excellent example of the interrelationships among Earth's systems and subsystems is global warming caused by the greenhouse effect. Carbon dioxide is produced as a by-product of respiration and the burning of organic material. As such, it is a component of the global ecosystem and is constantly being recycled as part of the carbon cycle. The concern in recent years over the increased level of atmospheric carbon dioxide relates to its role in the greenhouse effect. The recycling of carbon dioxide between Earth's crust and atmosphere is an important climatic regulator because carbon dioxide, as well as other gases such as methane, nitrous oxide, chlorofluorocarbons, and water vapor, allows sunlight to pass through but traps the heat reflected back from Earth's surface. Heat is thus retained, causing the temperature of Earth's surface and, more important, the atmosphere to increase, producing the greenhouse effect.

With industrialization and its accompanying burning of tremendous amounts of fossil fuels, carbon dioxide levels in the atmosphere have been steadily increasing since about 1880. Many scientists conclude that a global warming trend has already begun and will lead to severe global climatic shifts. Most computer models based on the current rate of increase in greenhouse gases show Earth warming by as much as 5°C during the next hundred years. Such a temperature change will be uneven, however, with the greatest warming occurring in the higher latitudes. As a consequence of this warming, rainfall patterns will shift dramatically, which will have a major effect on the largest grain-producing areas of the world, such as the American Midwest. Drier and hotter conditions will intensify the severity and frequency of droughts, leading to more crop failures and higher food prices. With such shifts in climate, Earth's deserts may expand, which will remove valuable crop and grazing lands.

With continued global warming, mean sea level will also rise as icecaps and glaciers melt and contribute their water to the world's oceans. It is predicted that by the 2050s, sea level will rise 21 cm, increasing the number of people at risk from flooding in coastal areas by approximately 20 million.

We would be remiss, however, if we did not point out that many other scientists are not convinced that the global warming trend is the direct result of increased human activity related to industrialization. They point out that although the level of greenhouse gases has increased, we are still uncertain about their rate of generation and rate of removal, and whether the rise in global temperature during the past century resulted from normal climatic variations through time or from human activity. Furthermore, these scientists point out that even if there is general global warming during the next hundred years, it is not certain that the dire predictions made by proponents of global warming will come true.

Earth, as we know, is a remarkably complex system, with many feedback mechanisms and interconnections throughout its various subsystems and cycles. It is very difficult to predict all of the consequences that global warming would have for atmospheric and oceanic circulation patterns.

What Would You Do?

An important environmental issue facing the world today is global warming. How can this problem be approached from a global systems perspective? What are the possible consequences of global warming, and can we really do anything about it? Are there ways to tell whether global warming occurred in the geologic past?

ORIGIN OF THE UNIVERSE AND SOLAR SYSTEM, AND EARTH'S PLACE IN THEM

How did the universe begin? What has been its history? What is its eventual fate, or is it infinite? These are just some of the basic questions people have asked and wondered about since they first looked into the nighttime sky and saw the vastness of the universe beyond Earth.

Origin of the Universe—Did It Begin with a Big Bang?

Most scientists think that the universe originated about 15 billion years ago in what is popularly called the **Big Bang.** The Big Bang is a model for the evolution of the universe in which a dense, hot state was followed by expansion, cooling, and a less dense state.

In a region infinitely smaller than an atom, both time and space were set at zero. Therefore, there is no "before the Big Bang," only what occurred after it. The reason is that space and time are unalterably linked to form a space–time continuum demonstrated by Einstein's theory of relativity. Without space, there can be no time.

How do we know the Big Bang took place approximately 15 billion years ago? Why couldn't the universe have always existed as we know it today? Two fundamental phenomena indicate that the Big Bang occurred. First, the universe is expanding. When astronomers look beyond our own solar system, they observe that everywhere in the universe galaxies are apparently moving away from each other at tremendous speeds. By measuring this expansion rate, astronomers can calculate how long ago the galaxies were all together at a single point. Second, everywhere in the universe there is a pervasive background radiation of 2.7 degrees above absolute zero (absolute zero equals −273°C). This background radiation is thought to be the faint afterglow of the Big Bang.

According to the currently accepted theory, matter as we know it did not exist at the moment of the Big Bang and the universe consisted of pure energy. During the first second following the Big Bang, the four basic forces—*gravity* (the attraction of one body toward another), *electromagnetic force* (combines electricity and magnetism into one force and binds atoms into molecules), *strong nuclear force* (binds protons and neutrons together), and *weak nuclear force* (responsible for the breakdown of an atom's nucleus, producing radioactive decay)—separated and the universe experienced enormous expansion. About 300,000 years later, the universe was cool enough for complete atoms of hydrogen and helium atoms to form, and photons (the energetic particles of light) separated from matter and light burst forth for the first time.

During the next 200 million years, as the universe continued expanding and cooling, stars and galaxies began to form and the chemical makeup of the universe changed. Initially, the universe was 100% hydrogen and helium, whereas today it is 98% hydrogen and helium and 2% all other elements by weight. How did such a change in the universe's composition occur? Throughout their life cycle, stars undergo many nuclear reactions in which lighter elements are converted into heavier elements by nuclear fusion. When a star dies, often explosively, the heavier elements that were formed in its core are returned to interstellar space and are available for inclusion in new stars. In this way, the composition of the universe is gradually enhanced in heavier elements.

Our Solar System—Its Origin and Evolution

Our solar system, which is part of the Milky Way Galaxy, consists of the Sun, nine planets, 101 known moons or satellites (although this number keeps changing with the discovery of new moons and satellites surrounding the Jovian planets), a tremendous number of asteroids—most of which orbit the Sun in a zone between Mars and Jupiter—and millions of comets and meteorites as well as interplanetary dust and gases (■ Figure 1.5). Any theory formulated to explain the origin and evolution of our solar system must therefore take into account its various features and characteristics.

Many scientific theories for the origin of the solar system have been proposed, modified, and discarded since the French scientist and philosopher René Descartes first proposed, in 1644, that the solar system formed from a gigantic whirlpool within a universal fluid. Today, the **solar nebula theory** for the origin of our solar system involves the condensation and collapse of interstellar material in a spiral arm of the Milky Way Galaxy.

The collapse of this cloud of gases and small grains into a counterclockwise-rotating disk concentrated about 90% of the material in the central part of the disk and formed an embryonic Sun, around which swirled a rotating cloud of material called a *solar nebula*. Within this solar nebula were

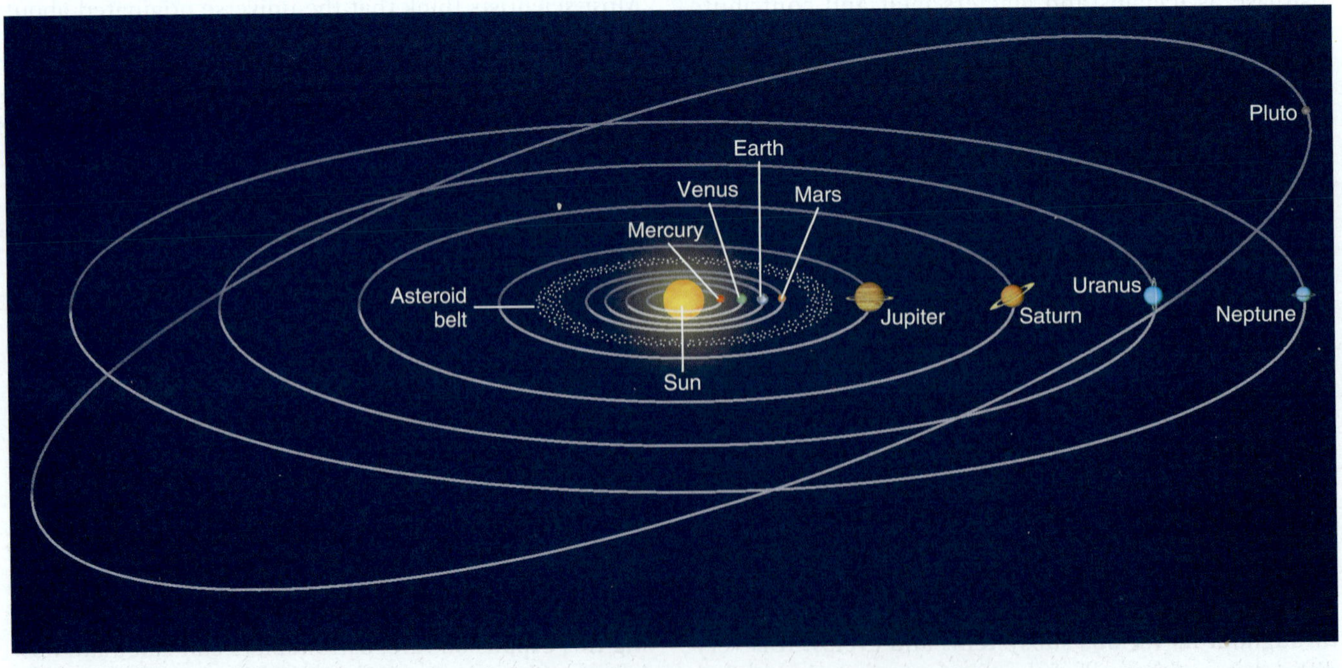

■ **Figure 1.5**

Diagrammatic representation of the solar system, showing the planets and their orbits around the Sun.

Dana Berry

■ **Figure 1.6**

At the stage of development shown here, planetesimals have formed in the inner solar system, and large eddies of gas and dust remain at great distances from the protosun.

localized eddies in which gases and solid particles condensed. During the condensation process, gaseous, liquid, and solid particles began to accrete into ever-larger masses called *planetesimals* (■ Figure 1.6) that collided and grew in size and mass until they eventually became planets.

The composition and evolutionary history of the planets are a consequence, in part, of their distance from the Sun (see "The Terrestrial and Jovian Planets" on pages 14 and 15). The **terrestrial planets**—Mercury, Venus, Earth, and Mars—so named because they are similar to *terra,* Latin for "earth," are all small and composed of rock and metallic elements that condensed at the high temperatures of the inner nebula. The **Jovian planets**—Jupiter, Saturn, Uranus, and Neptune—so named because they resemble Jupiter (the Roman god was also named Jove), all have small central rocky cores compared to their overall size and are composed mostly of hydrogen, helium, ammonia, and methane, which condense at low temperatures.

While the planets were accreting, material that had been pulled into the center of the nebula also condensed, collapsed, and was heated to several million degrees by gravitational compression. The result was the birth of a star, our Sun.

During the early accretionary phase of the solar system's history, collisions between various bodies were common, as indicated by the craters on many planets and moons. Asteroids probably formed as planetesimals in a localized eddy between what eventually became Mars and Jupiter in much the same way that other planetesimals formed the terrestrial planets. The tremendous gravitational field of Jupiter, however, prevented this material from ever accreting into a planet. Comets, which are interplanetary bodies composed of loosely bound rocky and icy material, are thought to have condensed near the orbits of Uranus and Neptune.

The solar nebula theory for the formation of the solar system thus accounts for most of the characteristics of the planets and their moons, the differences in composition between the terrestrial and Jovian planets, and the presence of the asteroid belt. Based on the available data, the solar nebula theory best explains the features of the solar system and provides a logical explanation for its evolutionary history.

Earth—Its Place in Our Solar System

Some 4.6 billion years ago, various planetesimals in our solar system gathered enough material together to form Earth and eight other planets. Scientists think that this early Earth was probably cool, of generally uniform composition and density throughout, and composed mostly of silicates, compounds consisting of silicon and oxygen, iron and magnesium oxides, and smaller amounts of all the other chemical elements (■ Figure 1.7a). Subsequently, when the combination of meteorite impacts, gravitational compression, and heat from radioactive decay increased the temperature of Earth enough to melt iron and nickel, this homogeneous composition disappeared (Figure 1.7b) and was replaced by a series of concentric layers of differing composition and density, resulting in a differentiated planet (Figure 1.7c).

This differentiation into a layered planet is probably the most significant event in Earth history. Not only did it lead to the formation of a crust and eventually to continents, but it also was probably responsible for the emission of gases from the interior that eventually led to the formation of the oceans and atmosphere.

The Terrestrial and Jovian Planets

The planets of our solar system can be divided into two major groups that are quite different, indicating that the two underwent very different evolutionary histories. The four inner planets—Mercury, Venus, Earth, and Mars—are the terrestrial planets; they are small and dense (composed of a metallic core and silicate mantle-crust), ranging from no atmosphere (Mercury) to an oppressively thick one (Venus). The outer four planets (excluding Pluto, which some astronomers don't regard as a planet at all)—Jupiter, Saturn, Uranus, and Neptune—are the Jovian planets; they are large, ringed, low-density planets with liquid interiors surrounded by thick atmospheres.

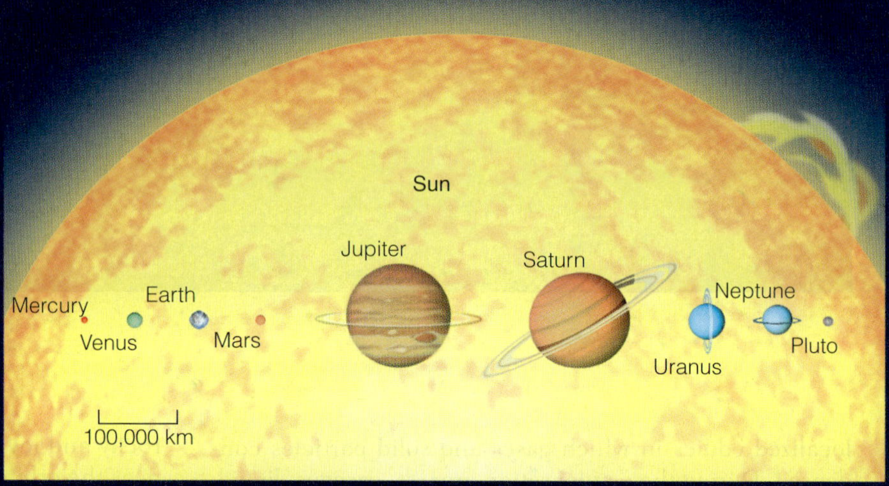

The relative sizes of the planets and the Sun. (Distances between planets are not to scale.)

Venus is surrounded by an oppressively thick atmosphere that completely obscures its surface. However, radar images from orbiting spacecraft reveal a wide variety of terrains, including volcanic features, folded mountain ranges, and a complex network of faults.

The **Moon** is one-fourth the diameter of Earth, has a low density relative to the terrestrial planets, and is extremely dry. Its surface is divided into low-lying dark-colored plains and light-colored highlands that are heavily cratered, attesting to a period of massive meteorite bombardment in our solar system more than 4 billion years ago. The hypothesis that best accounts for the origin of the Moon has a giant planetesimal, the size of Mars or larger, crashing into Earth 4.6 to 4.4 billion years ago, causing ejection of a large quantity of hot material that cooled and formed the Moon.

Mercury has a heavily cratered surface that has changed very little since its early history. Because Mercury is so small, its gravitational attraction is insufficient to retain atmospheric gases; any atmosphere that it may have held when it formed probably escaped into space quickly.

JPL/NASA

Earth is unique among our solar system's planets in that it has a hospitable atmosphere, oceans of water, and a variety of climates, and it supports life.

Mars has a thin atmosphere, little water, and distinct seasons. Its southern hemisphere is heavily cratered like the surfaces of Mercury and the Moon. The northern hemisphere has large smooth plains, fewer craters, and evidence of extensive volcanism. The largest volcano in the solar system is found in the northern hemisphere as are huge canyons, the largest of which, if present on Earth, would stretch from San Francisco to New York!

Jupiter is the largest of the Jovian planets. With its moons, rings, strong magnetic field, and intense radiation belts, Jupiter is the most complex and varied planet in our solar system. Jupiter's cloudy and violent atmosphere is divided into a series of different colored bands and a variety of spots (the Great Red Spot) that interact in incredibly complex motions.

Saturn's most conspicuous feature is its ring system, consisting of thousands of rippling, spiraling bands of countless particles. The width of Saturn's rings would just reach from Earth to the Moon.

Uranus is the only planet that lies on its side; that is, its axis of rotation nearly parallels the plane in which the planets revolve around the Sun. Some scientists think that a collision with an Earth-sized body early in its history may have knocked Uranus on its side. Like the other Jovian planets, Uranus has a ring system, albeit a faint one.

Neptune is a dynamic stormy planet with an atmosphere similar to those of the other Jovian planets. Winds up to 2000 km/hr blow over the planet, creating tremendous storms, the largest of which, the Great Dark Spot, seen in the center, is nearly as big as Earth and is similar to the Great Red Spot on Jupiter.

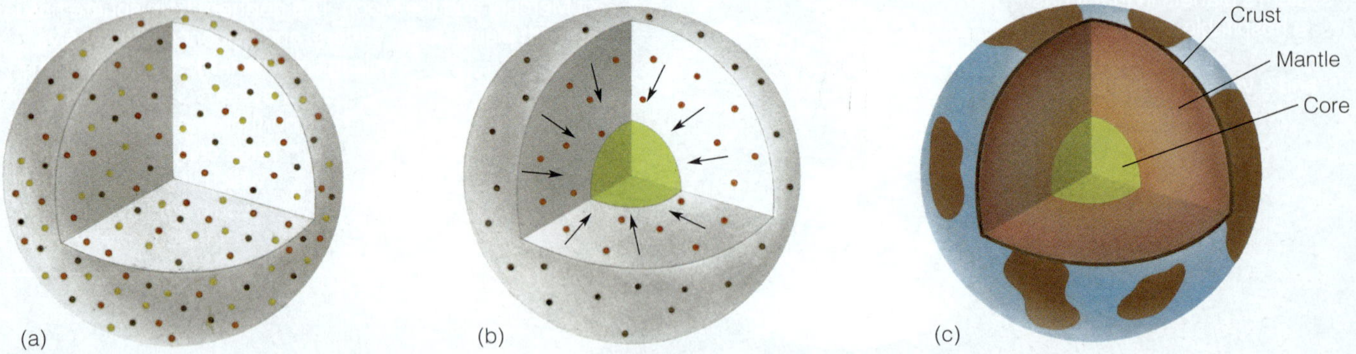

Figure 1.7

Homogeneous accretion theory for the formation of a differentiated Earth. (a) Early Earth was probably of uniform composition and density throughout. (b) Heating of early Earth reached the melting point of iron and nickel, which, being denser than silicate minerals, settled to Earth's center. At the same time, the lighter silicates flowed upward to form the mantle and the crust. (c) In this way, a differentiated Earth formed, consisting of a dense iron-nickel core, an iron-rich silicate mantle, and a silicate crust with continents and ocean basins.

WHY IS EARTH A DYNAMIC AND EVOLVING PLANET?

Earth is a dynamic planet that has changed continuously during its 4.6-billion-year existence. The size, shape, and geographic distribution of continents and ocean basins have changed through time, the composition of the atmosphere has evolved, and life-forms existing today differ from those that lived during the past. Mountains and hills have been worn away by erosion, and landscapes have been changed by the forces of wind, water, and ice. Volcanic eruptions and earthquakes reveal an active interior, and folded and fractured rocks indicate the tremendous power of Earth's internal forces.

Earth consists of three concentric layers: the core, the mantle, and the crust (■ Figure 1.8). This orderly division results from density differences between the layers as a function of variations in composition, temperature, and pressure.

The **core** has a calculated density of 10 to 13 grams per cubic centimeter (g/cm^3) and occupies about 16% of Earth's total volume. Seismic (earthquake) data indicate that the core consists of a small, solid, inner part and a larger, apparently liquid, outer portion. Both are thought to consist largely of iron and a small amount of nickel.

The **mantle** surrounds the core and comprises about 83% of Earth's volume. It is less dense than the core (3.3–5.7 g/cm^3) and is thought to be composed largely of *peridotite*, a dark, dense igneous rock containing abundant iron and magnesium. The mantle can be divided into three distinct zones based on physical characteristics. The lower mantle is solid and forms most of the volume of Earth's interior. The **asthenosphere** surrounds the lower mantle. It has the same composition as the lower mantle but behaves plastically and slowly flows. Partial melting within the asthenosphere generates *magma* (molten material), some of which rises to the surface because it is less dense than the rock from which it was derived. The upper mantle surrounds the asthenosphere. The solid upper mantle and the overlying crust constitute the **lithosphere,** which is broken into numerous individual pieces called **plates** that move over the asthenosphere as a result of underlying *convec-*

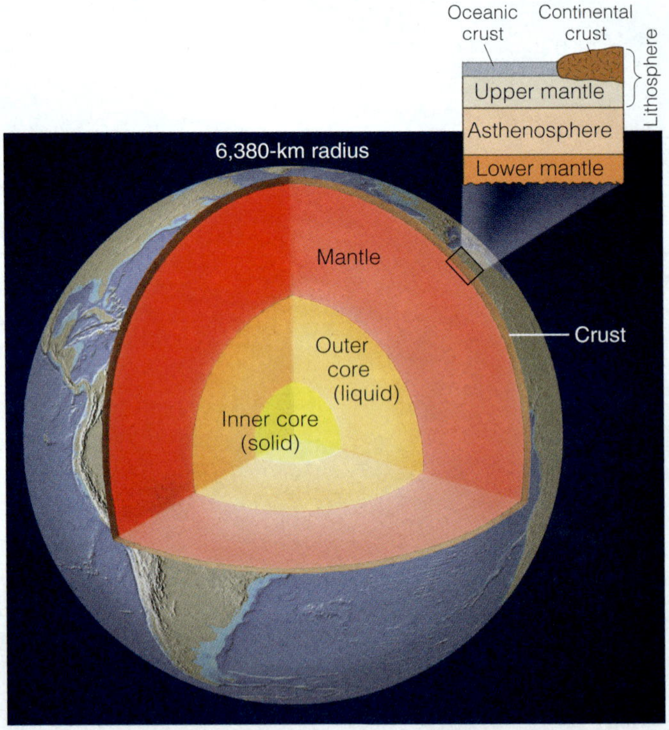

Figure 1.8

A cross section of Earth, illustrating the core, mantle, and crust. The enlarged portion shows the relationship between the lithosphere (composed of the continental crust, oceanic crust, and solid upper mantle) and the underlying asthenosphere and lower mantle.

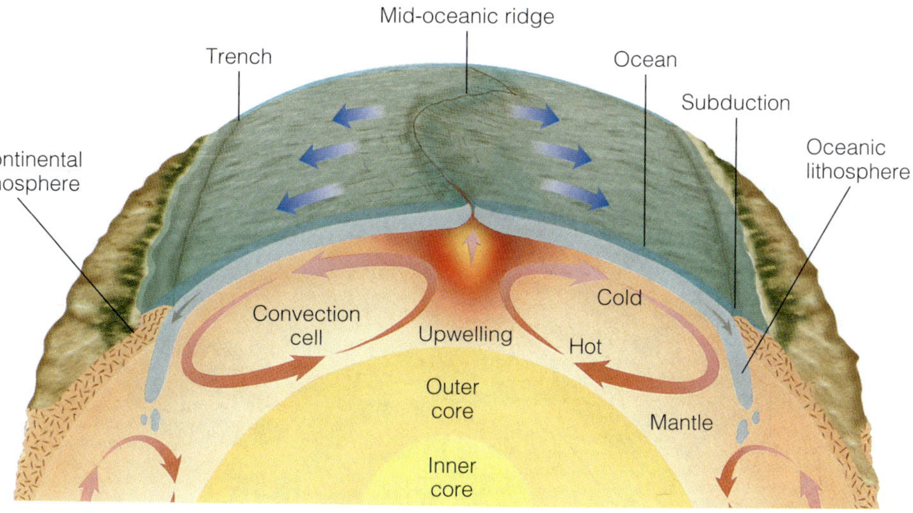

GeologyNow ■ **Active Figure 1.9**

Earth's plates are thought to move as a result of underlying mantle convection cells in which warm material from deep within Earth rises toward the surface, cools, and then, upon losing heat, descends back into the interior. The movement of these convection cells is thought to be the mechanism responsible for the movement of Earth's plates, as shown in this diagrammatic cross section.

tion cells (■ Figure 1.9). Interactions of these plates are responsible for such phenomena as earthquakes, volcanic eruptions, and the formation of mountain ranges and ocean basins.

The **crust**, Earth's outermost layer, consists of two types. *Continental crust* is thick (20–90 km), has an average density of 2.7 g/cm^3, and contains considerable silicon and aluminum. *Oceanic crust* is thin (5–10 km), denser than continental crust (3.0 g/cm^3), and is composed of the dark-colored igneous rock *basalt*.

GeologyNow Log into GeologyNow and select this chapter to work through a **Geology Interactive** activity on "Core Studies" (click Earth's Layers→Core Studies).

Plate Tectonic Theory

The recognition that the lithosphere is divided into rigid plates that move over the asthenosphere forms the foundation of **plate tectonic theory** (■ Figure 1.10). Zones of volcanic activity, earthquakes, or both mark most plate boundaries. Along these boundaries

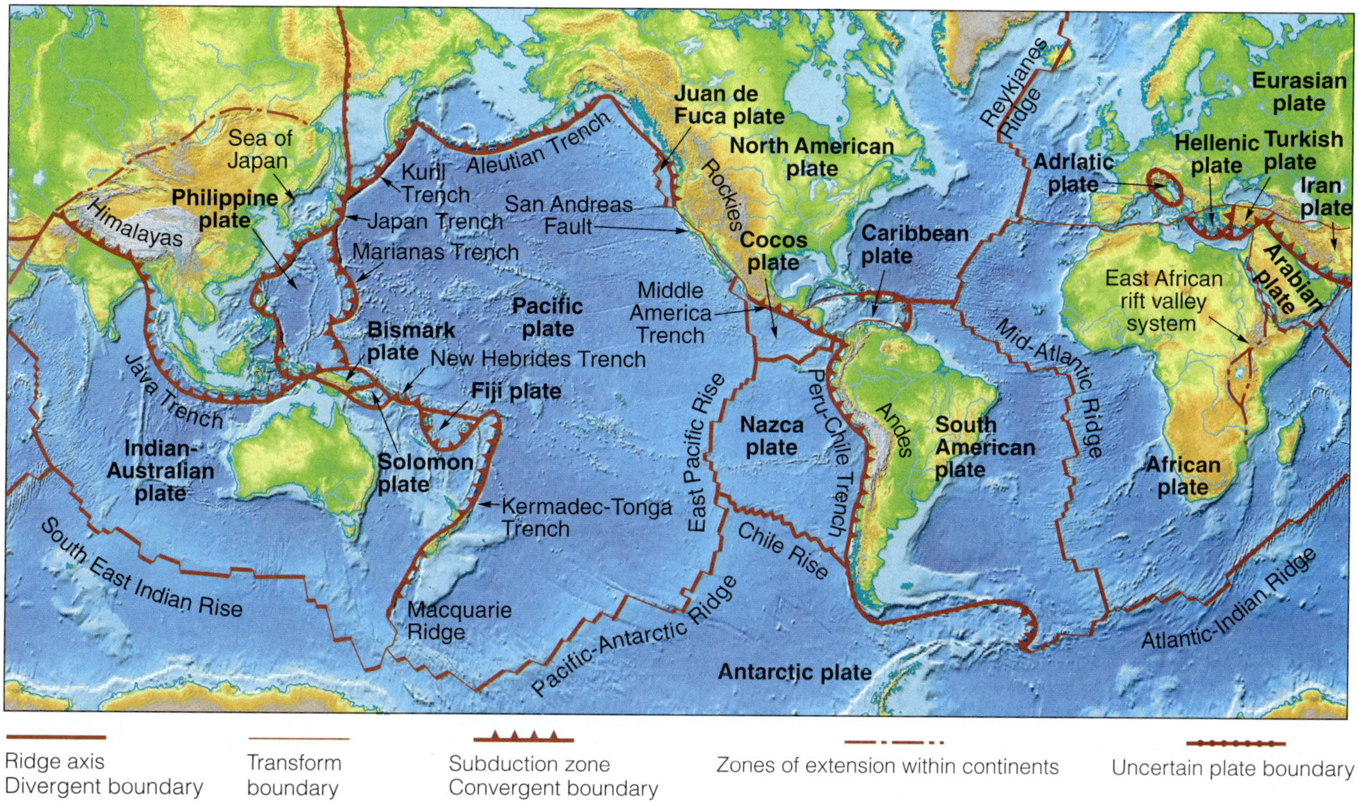

■ **Figure 1.10**

Earth's lithosphere is divided into rigid plates of various sizes that move over the asthenosphere.

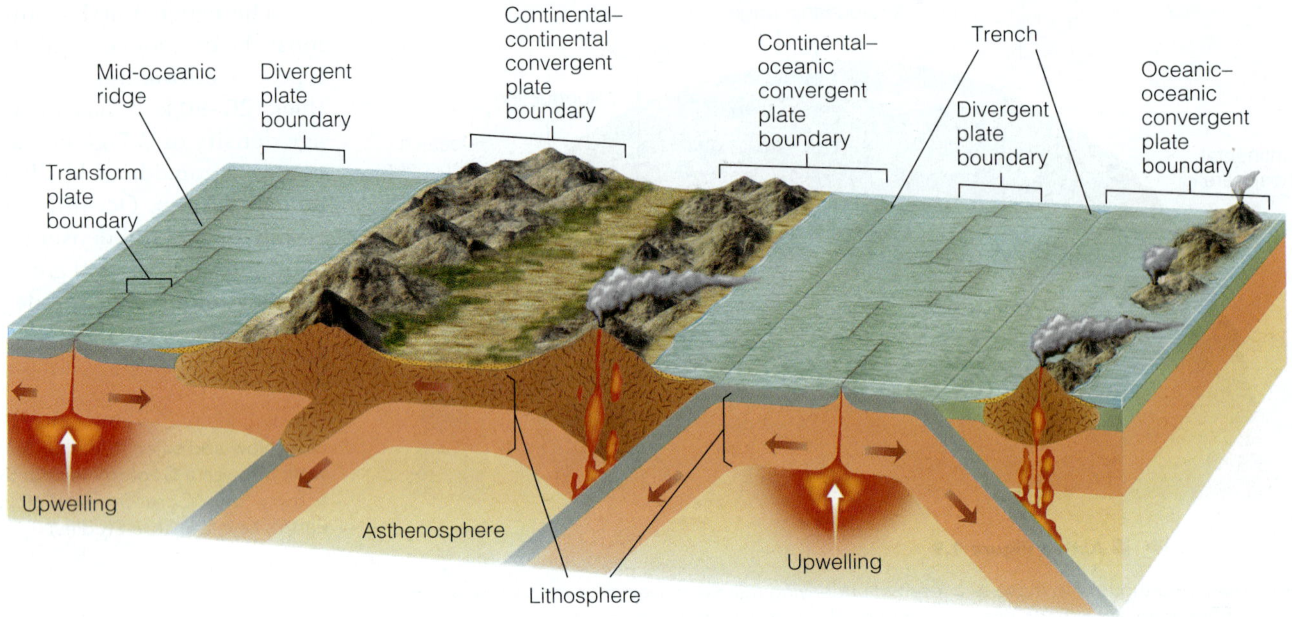

Figure 1.11

An idealized cross section illustrating the relationship between the lithosphere and the underlying asthenosphere and the three principal types of plate boundaries: divergent, convergent, and transform.

plates diverge, converge, or slide sideways past each other (Figure 1.11).

The acceptance of plate tectonic theory is recognized as a major milestone in the geologic sciences, comparable to the revolution Darwin's theory of evolution caused in biology. Plate tectonics has provided a framework for interpreting the composition, structure, and internal processes of Earth on a global scale. It has led to the realization that the continents and ocean basins are part of a lithosphere-asthenosphere-hydrosphere system that evolved together with Earth's interior (Table 1.3).

Table 1.3

Plate Tectonics and Earth Systems

Solid Earth	Plate tectonics is driven by convection in the mantle and in turn drives mountain-building and associated igneous and metamorphic activity.
Atmosphere	Arrangement of continents affects solar heating and cooling, and thus winds and weather systems. Rapid plate spreading and hot-spot activity may release volcanic carbon dioxide and affect global climate.
Hydrosphere	Continental arrangement affects ocean currents. Rate of spreading affects volume of mid-oceanic ridges and hence sea level. Placement of continents may contribute to onset of ice ages.
Biosphere	Movement of continents creates corridors or barriers to migration, the creation of ecological niches, and transport of habitats into more or less favorable climates.
Extraterrestrial	Arrangement of continents affects free circulation of ocean tides and influences tidal slowing of Earth's rotation.

Source: Adapted by permission from Stephen Dutch, James S. Monroe, and Joseph Moran, *Earth Science* (Minneapolis/St. Paul: West Publishing Co., 1997).

A revolutionary concept when it was proposed in the 1960s, plate tectonic theory has had far-reaching consequences in all fields of geology because it provides the basis for relating many seemingly unrelated phenomena. Besides being responsible for the major features of Earth's crust, plate movements also affect the formation and occurrence of Earth's natural resources as well as the distribution and evolution of the world's biota.

The impact of plate tectonic theory has been particularly notable in the interpretation of Earth history. For example, the Appalachian Mountains in eastern North America and the mountain ranges of Greenland, Scotland, Norway, and Sweden are not the result of unrelated mountain-building episodes but rather are part of a larger mountain-building event that involved closing an ancient "Atlantic Ocean" and the formation of the supercontinent Pangaea about 245 million years ago.

GeologyNow Log into GeologyNow and select this chapter to work through a **Geology Interactive** activity on "Plate Locations" (click Plate Tectonics→Plate Locations).

THE ROCK CYCLE

A **rock** is an aggregate of **minerals,** which are naturally occurring, inorganic, crystalline solids that have definite physical and chemical properties. Minerals are composed of elements such as oxygen, silicon, and aluminum, and elements are made up of atoms, the smallest particles of matter that retain the characteristics of an element. More than 3500 minerals have been identified and described, but only about a dozen make up the bulk of the rocks in Earth's crust (see Table 3.4).

Geologists recognize three major groups of rocks—*igneous, sedimentary,* and *metamorphic*—each of which is characterized by its mode of formation. Each group contains a variety of individual rock types that differ from one another on the basis of their composition or texture (the size, shape, and arrangement of mineral grains).

The **rock cycle** provides a way of viewing the interrelationships between Earth's internal and external processes (■ Figure 1.12). It relates the three rock groups to each other; to surficial processes such as weathering, transportation, and deposition; and to internal processes such as magma generation and metamorphism. Plate movement is the mechanism responsible for recycling rock materials and therefore drives the rock cycle.

Igneous rocks result when magma crystallizes or volcanic ejecta such as ash accumulate and consolidate. As magma cools, minerals crystallize, and the resulting rock is characterized by interlocking mineral grains. Magma that cools slowly beneath the surface produces *intrusive igneous rocks* (■ Figure 1.13a); magma that cools at the surface produces *extrusive igneous rocks* (Figure 1.13b).

Rocks exposed at Earth's surface are broken into particles and dissolved by weathering processes. The particles and dissolved materials may be transported by wind, water, or ice and eventually deposited as *sediment.* This sediment may then be compacted or cemented (lithified) into sedimentary rock.

Sedimentary rocks form in one of three ways: consolidation of mineral or rock fragments, precipitation of mineral matter from solution, or compaction of plant or animal remains (Figure 1.13c, d). Because sedimentary rocks form at or near Earth's surface, geologists can make inferences about the environment in which they were deposited, the transporting agent, and perhaps even something about the source from which the sediments were derived. Accordingly, sedimentary rocks are especially useful for interpreting Earth history.

Metamorphic rocks result from the alteration of other rocks, usually beneath the surface, by heat, pressure, and the chemical activity of fluids. For example, marble, a rock preferred by many sculptors and builders, is a metamorphic rock produced when the agents of metamorphism are applied to the sedimentary rocks limestone or dolostone. Metamorphic rocks are either *foliated* (Figure 1.13e) or *nonfoliated* (Figure 1.13f). Foliation, the parallel alignment of minerals due to pressure, gives the rock a layered or banded appearance.

How Are the Rock Cycle and Plate Tectonics Related?

Interactions between plates determine, to some extent, which of the three rock groups will form (■ Figure 1.14). For example, when plates converge, heat and pressure generated along the plate boundary may lead to igneous activity and metamorphism within the descending oceanic plate, thus producing various igneous and metamorphic rocks.

Some of the sediments and sedimentary rocks on the descending plate are melted, whereas other sediments and sedimentary rocks along the boundary of the nondescending plate are metamorphosed by the heat and pressure generated along the converging plate boundary. Later, the mountain range or chain of volcanic islands formed along the convergent plate boundary will be weathered and eroded, and the new sediments will be transported to the ocean to begin yet another cycle.

The interrelationship between the rock cycle and plate tectonics is just one example of how Earth's various subsystems and cycles are all interrelated. Heating within Earth's interior results in convection cells that power the movement of plates and also in magma, which forms intrusive and extrusive igneous rocks. Movement along plate boundaries may result in volcanic activity, earthquakes, and in some cases, mountain building. The interaction between the atmosphere, hydrosphere, and biosphere contributes to the weathering of rocks exposed on Earth's surface. Plates descending back into Earth's interior are subjected to increasing heat and pressure, which may lead to metamorphism as well as the generation of magma and yet another recycling of materials.

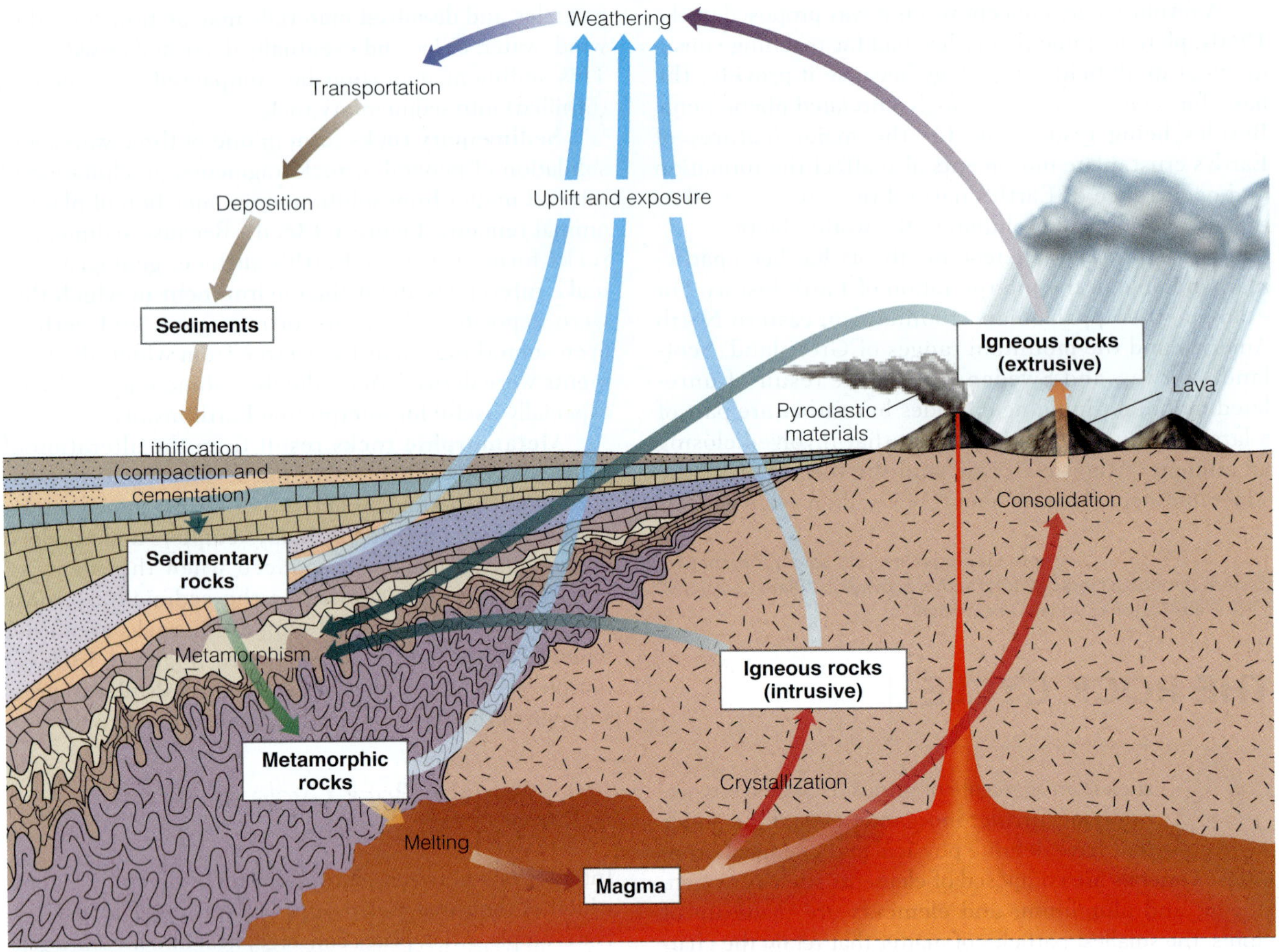

Figure 1.12

The rock cycle shows the interrelationships between Earth's internal and external processes and how the three major rock groups are related. Source: Modified from Figure 12, Dietrich, R. V., 1979, *Geology and Michigan: Fourty-nine Questions and Answers.*

Geology Now Log into GeologyNow and select this chapter to work through a **Geology Interactive** activity on "The Rock Cycle" (click Rocks and the Rock Cycle→Rock Cycle).

ORGANIC EVOLUTION AND THE HISTORY OF LIFE

Plate tectonic theory provides us with a model for understanding the internal workings of Earth and its effect on Earth's surface. The theory of **organic evolution** (all living things are related and have descended with modification from organisms living in the past) provides the conceptual framework for understanding the history of life. Together, the theories of plate tectonics and organic evolution have changed the way we view our planet, and we should not be surprised at the intimate association between them. Although the relationship between plate tectonic processes and the evolution of life is incredibly complex, paleontological data provide indisputable evidence of the influence of plate movement on the distribution of organisms.

The publication in 1859 of Darwin's *On the Origin of Species by Means of Natural Selection* revolutionized biology and marked the beginning of modern evolutionary biology. With its publication, most naturalists recognized that evolution provided a unifying theory that explained an otherwise encyclopedic collection of biologic facts.

The central thesis of organic evolution is that all present-day organisms are related, and that they have descended with modifications from organisms that lived during the past. When Darwin proposed his theory of organic evolution, he cited a wealth of supporting evidence, including the way organisms are classified, embryology, comparative anatomy, the geographic distribution of organisms, and, to a limited extent, the fossil record. Furthermore, Darwin proposed that *natural selection,* which results in the survival to reproductive age of those organisms best

Sue Monroe

Figure 1.13

Hand specimens of common igneous (a, b), sedimentary (c, d), and metamorphic (e, f) rocks. (a) Granite, an intrusive igneous rock. (b) Basalt, an extrusive igneous rock. (c) Conglomerate, a sedimentary rock formed by the consolidation of rock fragments. (d) Limestone, a sedimentary rock formed by the extraction of mineral matter from seawater by organisms or by the inorganic precipitation of the mineral calcite from seawater. (e) Gneiss, a foliated metamorphic rock. (f) Quartzite, a nonfoliated metamorphic rock.

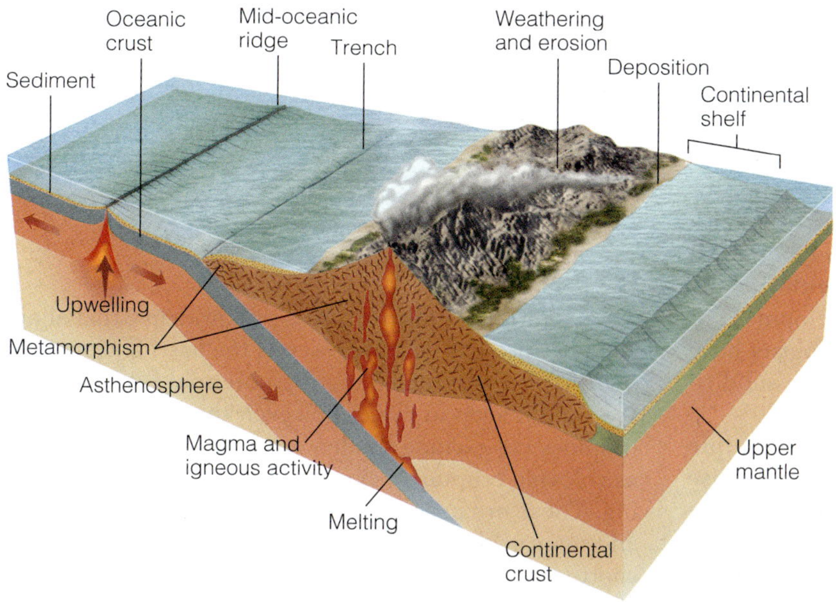

Figure 1.14

Plate tectonics and the rock cycle. The cross section shows how the three major rock groups—igneous, metamorphic, and sedimentary—are recycled through both the continental and oceanic regions.

adapted to their environment, is the mechanism that accounts for evolution.

Perhaps the most compelling evidence in favor of evolution can be found in the fossil record. Just as the geologic record allows geologists to interpret physical events and conditions in the geologic past, **fossils,** which are the remains or traces of once-living organisms, not only provide evidence that evolution has occurred, but also demonstrate that Earth has a history extending beyond that recorded by humans.

GEOLOGIC TIME AND UNIFORMITARIANISM

An appreciation of the immensity of geologic time is central to understanding the evolution of Earth and its biota. Indeed, time is one of the main aspects that sets geology apart from the other sciences, except astronomy. Most people have difficulty comprehending geologic time because they tend to think in terms of the human perspective—seconds, hours, days, and years. Ancient history is what occurred hundreds or even thousands of years ago. When geologists talk of ancient geologic history, however, they are referring to events that happened hundreds of millions or even billions of years ago. To a geologist, recent geologic events are those that occurred within the last million years or so.

It is also important to remember that Earth goes through cycles of much longer duration than the human perspective of time. Although they may have disastrous effects on the human species, global warming and cooling are part of a larger cycle that has resulted in numerous glacial advances and retreats during the past 1.6 million years. Because of their geologic perspective on time and how the various Earth subsystems and cycles are interrelated, geologists can make valuable contributions to many of the current environmental debates such as those involving global warming and sea level changes.

The **geologic time scale** resulted from the work of many 19th-century geologists who pieced together information from numerous rock exposures and constructed a sequential chronology based on changes in Earth's biota through time. Subsequently, with the discovery of radioactivity in 1895 and the development of various radiometric dating techniques, geologists have been able to assign absolute ages in years to the subdivisions of the geologic time scale (■ Figure 1.15).

One of the cornerstones of geology is the **principle of uniformitarianism,** which is based on the premise that present-day processes have operated throughout geologic time. Therefore, to understand and interpret geologic events from evidence preserved in rocks, we must first understand present-day processes and their results. In fact, uniformitarianism fits in completely with the system approach we are following for the study of Earth.

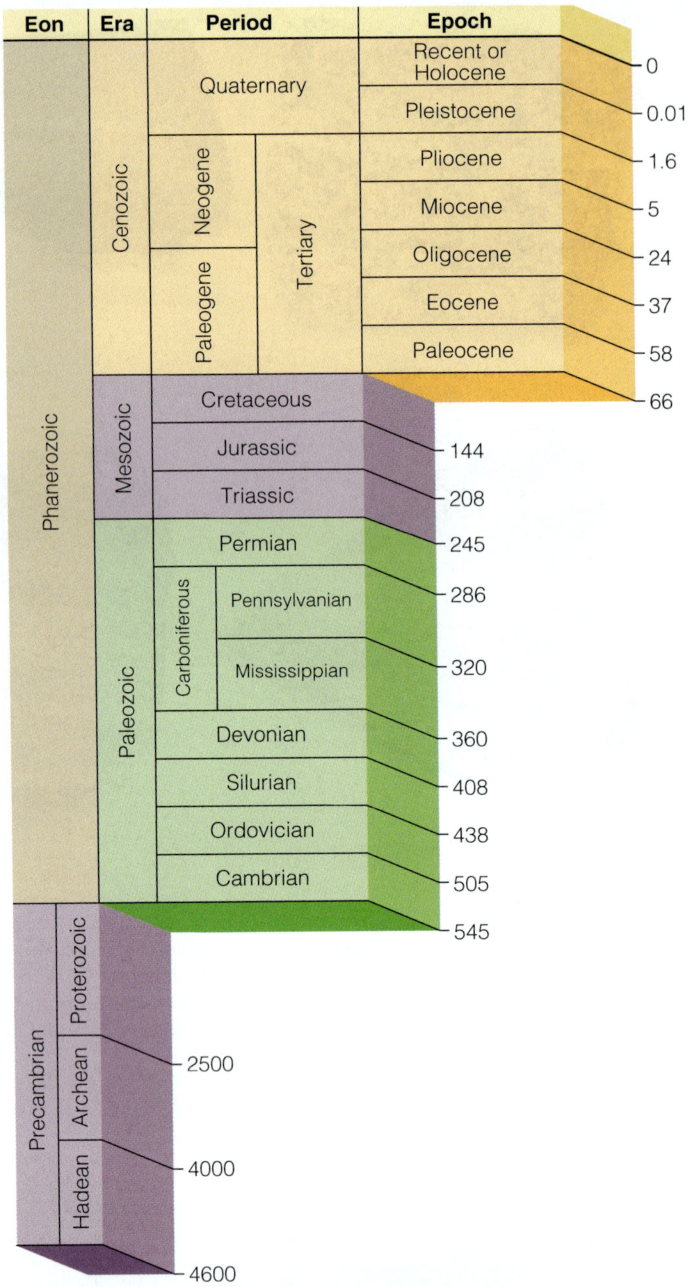

■ **Figure 1.15**

The geologic time scale. Numbers to the right of the columns are ages in millions of years before the present.

Uniformitarianism is a powerful principle that allows us to use present-day processes as the basis for interpreting the past and for predicting potential future events. We should keep in mind, however, that uniformitarianism does not exclude sudden or catastrophic events such as volcanic eruptions, earthquakes, landslides, or flooding. These are processes that shape our modern world, and some geologists view Earth history as a series of such short-term or punctuated events. This view is certainly in keeping with the modern principle of uniformitarianism.

Furthermore, uniformitarianism does not require that the rates and intensities of geologic processes be constant through time. We know that volcanic activity was more intense in North America 5 to 10 million years ago than it is today, and that glaciation has been more prevalent during the last several million years than in the previous 300 million years.

What uniformitarianism means is that even though the rates and intensities of geologic processes have varied during the past, the physical and chemical laws of nature have remained the same. Although Earth is in a dynamic state of change and has been ever since it formed, the processes that shaped it during the past are the same ones operating today.

HOW DOES THE STUDY OF GEOLOGY BENEFIT US?

The most meaningful lesson to learn from the study of geology is that Earth is an extremely complex planet in which interactions among its various subsystems are taking place and have been occurring for the past 4.6 billion years. If we want to ensure the survival of the human species, we must understand how the various subsystems work and interact with each other and, more important, how our actions affect the delicate balance between these systems.

The study of geology goes beyond learning numerous facts about Earth. In fact, we don't just study geology—we "live" it. Geology is an integral part of our lives. Our standard of living depends directly on our consumption of natural resources, resources that formed millions and billions of years ago. However, the ways we consume natural resources and interact with the environment, as individuals and as a society, also determine our ability to pass on this standard of living to the next generation.

As you study the various topics covered in this book, keep in mind the themes discussed in this chapter and how, like the parts of a system, they are interrelated. By relating each chapter's topic to its place in the entire Earth system, you will gain a greater appreciation of why geology is so integral to our lives.

1 GEO RECAP

Chapter Summary

- Earth can be viewed as a system of interconnected components that interact and affect one another. The principal subsystems of Earth are the atmosphere, hydrosphere, biosphere, lithosphere, mantle, and core. Earth is considered a dynamic planet that changes continuously because of the interactions among its various subsystems and cycles.
- Geology, the study of Earth, is divided into two broad areas: Physical geology is the study of Earth materials as well as the processes that operate within and on Earth's surface; historical geology examines the origin and evolution of Earth, its continents, oceans, atmosphere, and life.
- The scientific method is an orderly, logical approach that involves gathering and analyzing facts about a particular phenomenon, formulating hypotheses to explain the phenomenon, testing the hypotheses, and finally proposing a theory. A theory is a testable explanation for some natural phenomenon that has a large body of supporting evidence.
- Geology is part of the human experience. We can find examples of it in art, music, and literature. A basic understanding of geology is also important for dealing with the many environmental problems and issues facing society.
- Geologists engage in a variety of occupations, the main one being exploration for mineral and energy resources. They are also becoming increasingly involved in environmental issues and making short- and long-range predictions of the potential dangers from such natural disasters as volcanic eruptions and earthquakes.

- The universe began with a Big Bang approximately 15 billion years ago. Astronomers have deduced this age by measuring the rate at which celestial objects are moving away from each other in what appears to be an ever-expanding universe. Furthermore, the universe has a background radiation of 2.7 degrees above absolute zero, radiation that is thought to be the faint afterglow of the Big Bang.
- About 4.6 billion years ago, the solar system formed from a rotating cloud of interstellar matter. As this cloud condensed, it eventually collapsed under the influence of gravity and flattened into a counterclockwise-rotating disk. Within this rotating disk, the Sun, planets, and moons formed from the turbulent eddies of nebular gases and solids.
- Earth formed from a swirling eddy of nebular material 4.6 billion years ago. It probably accreted as a solid body and then soon underwent differentiation during a period of internal heating.
- Earth is differentiated into layers. The outermost layer is the crust, which is divided into continental and oceanic portions. The crust and the underlying solid part of the upper mantle, also known as the lithosphere, overlie the asthenosphere, a zone that slowly flows. The asthenosphere is underlain by the solid lower mantle. Earth's core consists of an outer liquid portion and an inner solid portion.
- The lithosphere is broken into a series of plates that diverge, converge, and slide sideways past one another.
- Plate tectonic theory provides a unifying explanation for many geologic features and events. The interaction between plates is responsible for volcanic eruptions, earthquakes, the formation of mountain ranges and ocean basins, and the recycling of rock materials.
- The three major rock groups are igneous, sedimentary, and metamorphic. Igneous rocks result from the crystallization of magma or the consolidation of volcanic ejecta. Sedimentary rocks are formed mostly by the consolidation of rock fragments, precipitation of mineral matter from solution, or compaction of plant or animal remains. Metamorphic rocks are produced from other rocks, generally beneath Earth's surface, by heat, pressure, and chemically active fluids.
- The rock cycle illustrates the interactions between internal and external Earth processes and shows how the three rock groups are interrelated.
- The central thesis of the theory of organic evolution is that all living organisms evolved (descended with modifications) from organisms that existed in the past.
- Time sets geology apart from the other sciences except astronomy, and an appreciation of the immensity of geologic time is central to understanding Earth's evolution. The geologic time scale is the calendar geologists use to date past events.
- The principle of uniformitarianism is basic to the interpretation of Earth history. This principle holds that the laws of nature have been constant through time and that the same processes operating today have operated in the past, though at different rates.
- Geology is an integral part of our lives. Our standard of living depends directly on our consumption of natural resources, resources that formed millions and billions of years ago.

Important Terms

asthenosphere (p. 14)
Big Bang (p. 11)
core (p. 14)
crust (p. 15)
fossil (p. 22)
geologic time scale (p. 22)
geology (p. 6)
hypothesis (p. 8)
igneous rock (p. 19)
Jovian planets (p. 13)
lithosphere (p. 14)
mantle (p. 14)
metamorphic rock (p. 19)
mineral (p. 19)
organic evolution (p. 20)
plate (p. 14)
plate tectonic theory (p. 15)
principle of uniformitarianism (p. 22)
rock (p. 19)
rock cycle (p. 19)
scientific method (p. 7)
sedimentary rock (p. 19)
solar nebula theory (p. 12)
system (p. 4)
terrestrial planets (p. 13)
theory (p. 7)

Review Questions

1. That all living organisms are the descendents of different life-forms that existed in the past is the central claim of
a._____the principle of fossil succession; b._____plate tectonics; c._____the principle of uniformitarianism; d._____organic evolution; e._____none of these.
2. Rocks that result from the alteration of other rocks, usually beneath the surface, by heat, pressure, and the chemical activity of fluids are

a._____igneous; b._____sedimentary; c._____metamorphic; d._____volcanic; e._____answers a and d.

3. A combination of related parts interacting in an organized fashion is

   a._____a cycle; b._____a theory; c._____uniformitarianism; d._____a hypothesis; e._____a system.

4. According to the currently accepted theory for the origin of the solar system,

   a._____a huge nebula collapsed under its own gravitational attraction; b._____the nebula formed a disc with the Sun in the center; c._____planetesimals accreted from gaseous, liquid, and solid particles; d._____all of these; e._____none of these.

5. The study of Earth materials is

   a._____economic geology; b._____physical geology; c._____historical geology; d._____structural geology; e._____environmental geology.

6. It is thought that plate movement results from

   a._____gravitational forces; b._____density differences between the mantle and core; c._____rotation of the mantle around the core; d._____convection cells; e._____the Coriolis effect.

7. Interaction between the atmosphere, hydrosphere, and biosphere is a major contributor to

   a._____mountain building; b._____the generation of magma; c._____weathering of Earth materials; d._____metamorphism; e._____plate movement.

8. Which of the following statements about a scientific theory is *not* true?

   a._____It is an explanation for some natural phenomenon; b._____Predictive statements can be derived from it; c._____It is a conjecture or guess; d._____It has a large body of supporting evidence; e._____It is testable.

9. The lithosphere consists of

   a._____the crust and the solid portion of the upper mantle; b._____the asthenosphere and the solid portion of the upper mantle; c._____the crust and asthenosphere; d._____continental and oceanic crust only; e._____the core and mantle.

10. The premise that present-day processes have operated throughout geologic time is the principle of

    a._____fossil succession; b._____uniformitarianism; c._____continental drift; d._____plate tectonics; e._____scientific deduction.

11. Which layer has the same composition as the mantle but behaves plastically?

    a._____continental crust; b._____oceanic crust; c._____outer core; d._____inner core; e._____asthenosphere.

12. Discuss what is meant by this statement: The health and well-being of the world's economy are completely dependent on geologic resources.

13. Why is an accurate geologic time scale particularly important for geologists in examining changes in global temperatures during the past?

14. Why is it important that everyone have a basic understanding of geology, even if they aren't going to become geologists?

15. Describe how you would use the scientific method to formulate a hypothesis explaining the similarity of mountain ranges on the east coast of North America and in England, Scotland, and the Scandinavian countries. How would you test your hypothesis?

16. Discuss how the three major layers of Earth differ from each other and why the differentiation into a layered planet is probably the most significant event in Earth history.

17. Explain how the principle of uniformitarianism allows for catastrophic events.

18. Discuss why plate tectonic theory is a unifying theory of geology.

19. Explain the advantage of using a system approach to the study of Earth.

20. Explain how a knowledge of geology would be useful in planning a military campaign against another country.

World Wide Web Activities

GeologyNow Assess your understanding of this chapter's topics with additional quizzing and comprehensive interactivities at

http://earthscience.brookscole.com/changingearth4e

as well as current and up-to-date weblinks, additional readings, and InfoTrac College Edition exercises.

Plate Tectonics: A Unifying Theory

CHAPTER 2

OUTLINE

- Introduction

  GEO-FOCUS 2.1: *Oil, Plate Tectonics, and Politics*

- What Were Some of the Early Ideas About Continental Drift?
- What Is the Evidence for Continental Drift?
- Paleomagnetism and Polar Wandering
- How Do Magnetic Reversals Relate to Seafloor Spreading?
- Why Is Plate Tectonics a Unifying Theory?
- What Are the Three Types of Plate Boundaries?
- What Are Hot Spots and Mantle Plumes?
- How Are Plate Movement and Motion Determined?
- What Is the Driving Mechanism of Plate Tectonics?
- How Does Plate Tectonics Affect the Distribution of Natural Resources?
- How Does Plate Tectonics Affect the Distribution of Life?
- Geo-Recap

GeologyNow *This icon, which appears throughout the book, indicates an opportunity to explore interactive tutorials, animations, or practice problems available on the GeologyNow website at http://earthscience.brookscole.com/changingearth4e.*

OBJECTIVES

At the end of this chapter, you will have learned that:

- Plate tectonics is the unifying theory of geology and has revolutionized geology.
- The hypothesis of continental drift was based on considerable geologic, paleontologic, and climatologic evidence.
- The hypothesis of seafloor spreading accounts for continental movement and the idea that thermal convection cells provide a mechanism for plate movement.
- The three types of plate boundaries are divergent, convergent, and transform. Along these boundaries new plates are formed, consumed, or slide past one another.
- Interaction along plate boundaries accounts for most of Earth's earthquake and volcanic activity.
- The rate of movement and motion of plates can be calculated in several ways.
- Some type of convective heat system is involved in plate movement.
- Plate movement affects the distribution of natural resources.
- Plate movement affects the distribution of the world's biota and has influenced evolution.

Image of South America generated from the Shuttle Radar Topography Mission aboard the Space Shuttle Endeavour, *launched on February 11, 2000. The Andes Mountains can be clearly seen along the Pacific Coast and are the result of subduction of the Nazca plate beneath the South American plate. The Amazon River can be seen to the east of the Andes Mountains draining much of northern South America.* Source: NASA

Introduction

At 8:46 A.M. on January 26, 2001, an earthquake of magnitude 7.7 rocked the Gujarat region of India as well as neighboring Pakistan. Flattening villages and toppling high-rise buildings in cities, this earthquake caused an estimated $1.3 billion in damages. It is estimated that more than 20,000 people died, 167,000 were injured, and 600,000 were left homeless. It was the most powerful earthquake to strike India since 1950, when an 8.5-magnitude earthquake killed more than 1500 people.

On June 15, 1991, Mount Pinatubo in the Philippines erupted violently, discharging huge quantities of ash and gases into the atmosphere. Fortunately, warnings of an impending eruption were heeded, and 200,000 people were evacuated from areas around the volcano, yet the eruption still caused 722 deaths.

What do these two tragic events and other equally destructive volcanic eruptions and earthquakes have in common? They are part of the dynamic interactions involving Earth's plates. When two plates come together, one plate is pushed or pulled under the other plate, triggering large earthquakes such as the one that shook India in 2001 and Iran in 2003. As the descending plate moves downward and is assimilated into Earth's interior, magma is generated. Being less dense than the surrounding material, the magma rises toward the surface, where it may erupt as a volcano such as Mount Pinatubo did in 1991 and others have since. It therefore should not be surprising that the distribution of volcanoes and earthquakes closely follows plate boundaries.

As we stated in Chapter 1, plate tectonic theory has had significant and far-reaching consequences in all fields of geology because it provides the basis for relating many seemingly unrelated phenomena. The interactions between moving plates determines the locations of continents, ocean basins, and mountain systems, which in turn affect atmospheric and oceanic circulation patterns that ultimately determine global climates (see Table 1.3). Plate movements have also profoundly influenced the geographic distribution, evolution, and extinction of plants and animals. Furthermore, the formation and distribution of many geologic resources, such as metal ores, are related to plate tectonic processes, so geologists incorporate plate tectonic theory into their prospecting efforts.

If you're like most people, you probably have no idea or only a vague notion of what plate tectonic theory is. Yet plate tectonics affects all of us, whether in terms of the destruction caused by volcanoes and earthquakes or politically and economically (see Geo-Focus 2.1). It is therefore important to understand this unifying theory, not only because it affects us as individuals and citizens of nation-states, but also because it ties together many aspects of the geology you will be studying.

WHAT WERE SOME OF THE EARLY IDEAS ABOUT CONTINENTAL DRIFT?

The idea that Earth's past geography was different from today is not new. The earliest maps showing the east coast of South America and the west coast of Africa probably provided people with the first evidence that continents may have once been joined together, then broken apart and moved to their present positions.

During the late 19th century, the Austrian geologist Edward Suess noted the similarities between the Late Paleozoic plant fossils of India, Australia, South Africa, and South America as well as evidence of glaciation in the rock sequences of these southern continents. The plant fossils comprise a unique flora that occurs in the coal layers just above the glacial deposits of these southern continents. This flora is very different from the contemporaneous coal swamp flora of the northern continents and is collectively known as the ***Glossopteris* flora** after its most conspicuous genus (■ Figure 2.1).

In his book, *The Face of the Earth*, published in 1885, Suess proposed the name *Gondwanaland* (or **Gondwana** as we will use here) for a supercontinent composed of the aforementioned southern continents. Abundant fossils of the *Glossopteris* flora are found in coal beds in Gondwana, a province in India. Suess thought these southern continents were connected by land bridges over which plants and animals migrated. Thus, in his view, the similarities of fossils on these continents were due to the appearance and disappearance of the connecting land bridges.

The American geologist Frank Taylor published a pamphlet in 1910 presenting his own theory of continental drift. He explained the formation of mountain ranges as a result of the lateral movement of continents. He also envisioned the present-day continents as parts of larger polar continents that eventually broke apart and migrated toward the equator after Earth's rotation was supposedly slowed by gigantic tidal forces. According to Taylor, these tidal forces were generated when Earth captured the Moon about 100 million years ago.

Although we now know that Taylor's mechanism is incorrect, one of his most significant contributions was his suggestion that the Mid-Atlantic Ridge, discovered by the 1872–1876 British HMS *Challenger* expeditions, might mark the site along which an ancient continent broke apart to form the present-day Atlantic Ocean.

GEOFOCUS 2.1

Oil, Plate Tectonics, and Politics

It is certainly not surprising that oil and politics are closely linked. The Iran–Iraq War of 1980–1989 and the Gulf War of 1990–1991 were both fought over oil (■ Figure 1). Indeed, many of the conflicts in the Middle East have had as their underlying cause, control of the vast deposits of petroleum in the region. Most people, however, are not aware of why there is so much oil in this part of the world.

Although large concentrations of petroleum occur in many areas of the world, more than 50% of all proven reserves are in the Persian Gulf region. Interestingly, however, this region did not become a significant petroleum-producing area until the economic recovery following World War II. After the war, Western Europe and Japan in particular became dependent on Persian Gulf oil, and they still rely heavily on this region for most of their supply. The United States is also dependent on imports from the Persian Gulf but receives significant quantities of petroleum from other sources such as Mexico and Venezuela.

Why is so much oil in the Persian Gulf region? The answer lies in the ancient geography and plate movements of this region during the Mesozoic and Cenozoic eras. During the Mesozoic Era, and particularly the Cretaceous Period when most of the petroleum formed, the Persian Gulf area was a broad marine shelf extending eastward from Africa. This continental margin lay near the equator where countless microorganisms lived in the surface waters. The remains of these organisms accumulated with the bottom sediments and were buried, beginning the complex process of petroleum generation and the formation of source beds.

As a consequence of rifting in the Red Sea and Gulf of Aden during the Cenozoic Era, the Arabian plate is moving northeast away from Africa and subducting beneath Iran. As the sediments of the passive continental margin were initially subducted, during the early stages of collision between Arabia and Iran, the heating broke down the organic molecules and led to the formation of petroleum. The tilting of the Arabian block to the northeast allowed the newly formed petroleum to migrate upward into the interior of the Arabian plate. The continued subduction and collision with Iran folded the rocks, creating traps for petroleum to accumulate, such that the vast area south of the collision zone (known as the Zagros suture) is a major oil-producing region.

Collection of the J. Paul Getty Museum

■ **Figure 1**

The Kuwaiti night skies were illuminated by 700 blazing oil wells set on fire by Iraqi troops during the 1991 Gulf War. The fires continued for nine months.

Patricia G. Gensel, University of North Carolina

■ **Figure 2.1**

Glossopteris leaves from the Upper Permian Dunedoo Formation, Australia. Fossils of the *Glossopteris* flora are found on all five of the Gondwana continents, thus providing evidence that these continents were at one time connected.

Alfred Wegener and the Continental Drift Hypothesis

Alfred Wegener, a German meteorologist (■ Figure 2.2), is generally credited with developing the hypothesis of **continental drift.** In his monumental book, *The Origin of Continents and Oceans* (first published in 1915), Wegener proposed that all landmasses were originally united in a single supercontinent that he named **Pangaea,** from the Greek meaning "all land." Wegener portrayed his grand concept of continental movement in a series of maps showing the breakup of Pangaea and the movement of the various continents to their present-day locations. Wegener amassed a tremendous amount of geologic, paleontologic, and climatologic evidence in support of continental drift, but the initial reaction of scientists to his then-heretical ideas can best be described as mixed.

Nevertheless, the eminent South African geologist Alexander du Toit further developed Wegener's arguments and gathered more geologic and paleontologic evidence in support of continental drift. In 1937 du Toit published *Our Wandering Continents,* in which he contrasted the glacial deposits of Gondwana with coal deposits of the same age found in the continents of the Northern Hemisphere. To resolve this apparent climatologic paradox, du Toit moved the Gondwana continents to the South Pole and brought the northern continents together such that the coal deposits were located at the equator. He named this northern landmass **Laurasia.** It consisted of present-day North America, Greenland, Europe, and Asia (except for India).

Bildarchiv Preussischer Kulterbestiz

■ **Figure 2.2**

Alfred Wegener, a German meteorologist, proposed the continental drift hypothesis in 1912 based on a tremendous amount of geologic, paleontologic, and climatologic evidence. He is shown here waiting out the Arctic winter in an expedition hut in Greenland.

WHAT IS THE EVIDENCE FOR CONTINENTAL DRIFT?

What then was the evidence Wegener, du Toit, and others used to support the hypothesis of continental drift? It includes the fit of the shorelines of continents, the appearance of the same rock sequences and mountain ranges of the same age on continents now widely separated, the matching of glacial deposits and paleoclimatic zones, and the similarities of many extinct plant and animal groups whose fossil remains are found today on widely separated continents.

Continental Fit

Wegener, like some before him, was impressed by the close resemblance of the coastlines of continents on opposite sides of the Atlantic Ocean, particularly South America and Africa. He cited these similarities as partial evidence that the continents were at one time joined together as a supercontinent that subsequently split apart. As his critics pointed out, though, the configuration of coastlines results from erosional and depositional processes and therefore is continually being modified. So even if the continents had separated during the Mesozoic Era, as Wegener proposed, it is not likely that the coastlines would fit exactly.

A more realistic approach is to fit the continents together along the continental slope where erosion would be minimal. In 1965 Sir Edward Bullard, an English geophysicist, and two associates showed that the best fit between the continents occurs at a depth of about 2000 m (■ Figure 2.3). Since then, other reconstructions using the latest ocean basin data have confirmed the close fit between continents when they are reassembled to form Pangaea.

Similarity of Rock Sequences and Mountain Ranges

If the continents were at one time joined, then the rocks and mountain ranges of the same age in adjoining locations on the opposite continents should closely match. Such is the case for the Gondwana continents (■ Figure 2.4). Marine, nonmarine, and glacial rock sequences of Pennsylvanian to Jurassic age are almost identical for all five Gondwana continents, strongly indicating that they were joined at one time.

The trends of several major mountain ranges also support the hypothesis of continental drift. These moun-

Geology Now ■ **Active Figure 2.3**

The best fit between continents occurs along the continental slope, where erosion would be minimal.

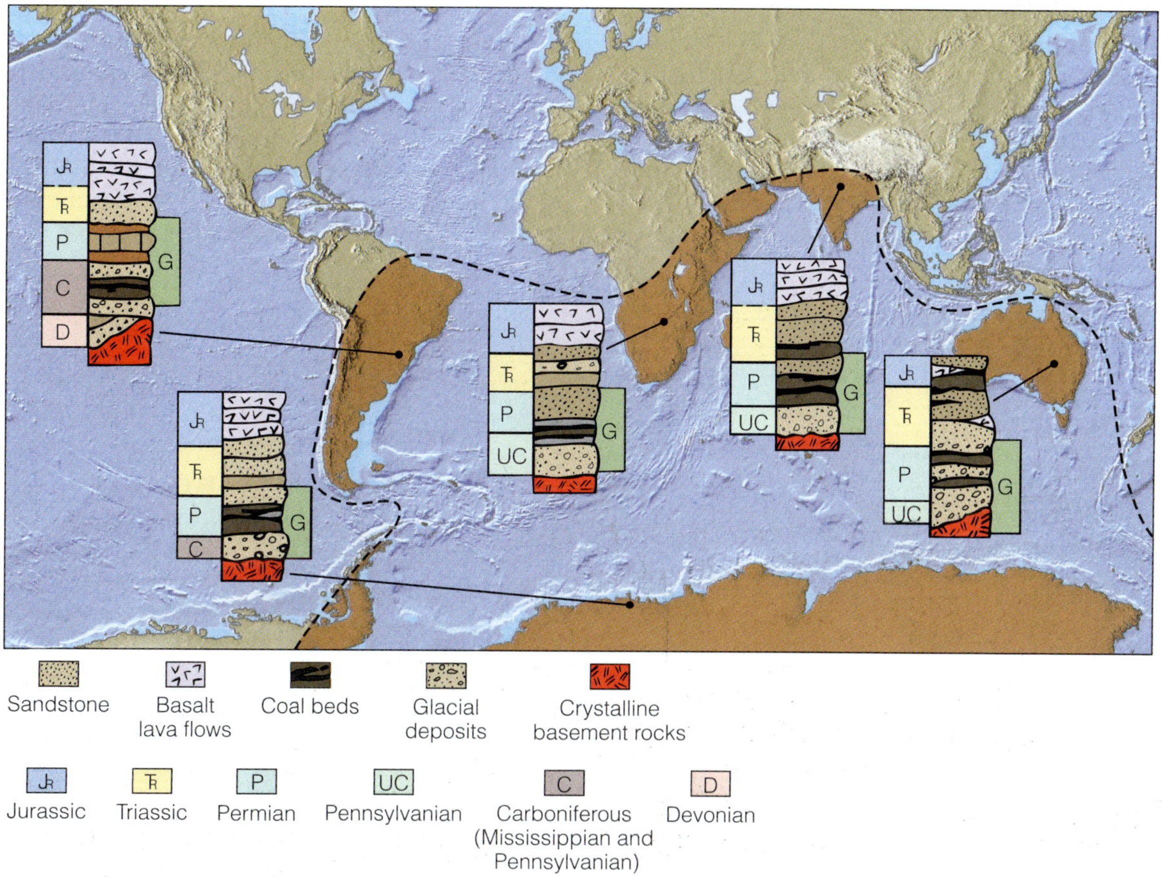

■ **Figure 2.4**

Marine, nonmarine, and glacial rock sequences of Pennsylvanian to Jurassic age are nearly the same for all Gondwana continents. Such close similarity strongly suggests that they were joined at one time. The range indicated by G is that of the *Glossopteris* flora. Source: From Robert J. Foster, *General Geology*, 4th Edition, © 1990. Reprinted by permission from Pearson Education, Inc., Upper Saddle River, NJ.

tain ranges seemingly end at the coastline of one continent only to apparently continue on another continent across the ocean. The folded Appalachian Mountains of North America, for example, trend northeastward through the eastern United States and Canada and terminate abruptly at the Newfoundland coastline. Mountain ranges of the same age and deformational style occur in eastern Greenland, Ireland, Great Britain, and Norway. Even though these mountain ranges are currently separated by the Atlantic Ocean, they form an essentially continuous mountain range when the continents are positioned next to each other (■ Figure 2.5).

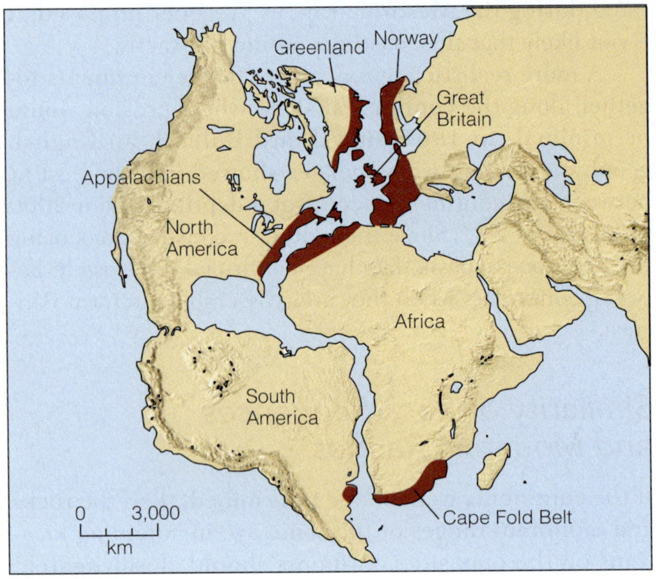

■ **Figure 2.5**

When continents are brought together, their mountain ranges form a single continuous range of the same age and style of deformation throughout. Such evidence indicates that the continents were at one time joined and were subsequently separated.

Glacial Evidence

During the Late Paleozoic Era, massive glaciers covered large continental areas of the Southern Hemisphere. Evidence for this glaciation includes layers of till (sediments deposited by glaciers) and striations (scratch marks) in the bedrock beneath the till. Fossils and sedimentary rocks of the same age from the Northern Hemisphere, however, give no indication of glaciation. Fossil plants found in coals indicate that the Northern Hemisphere had a tropical climate during the time that the Southern Hemisphere was glaciated.

All the Gondwana continents except Antarctica are currently located near the equator in subtropical to tropical climates. Mapping of glacial striations in bedrock in Australia, India, and South America indicates that the glaciers moved from the areas of the present-day oceans onto land. This would be highly unlikely because large continental glaciers (such as occurred on the Gondwana continents during the Late Paleozoic Era) flow outward from their central area of accumulation toward the sea.

If the continents did not move during the past, one would have to explain how glaciers moved from the oceans onto land and how large-scale continental glaciers formed near the equator. But if the continents are reassembled as a single landmass with South Africa located at the south pole, the direction of movement of Late Paleozoic continental glaciers makes sense (■ Figure 2.6).

■ **Figure 2.6**

(a) If the Gondwana continents are moved together so that South Africa is located at the South Pole, the glacial movements indicated by the striations (red arrows) makes sense. In this situation, the glacier (white area), located in a polar climate, moved radially outward from a thick central area toward its periphery.
(b) Permian-aged glacial striations in bedrock exposed at Hallet's Cove, Australia, indicate the direction of glacial movement more than 200 million years ago.

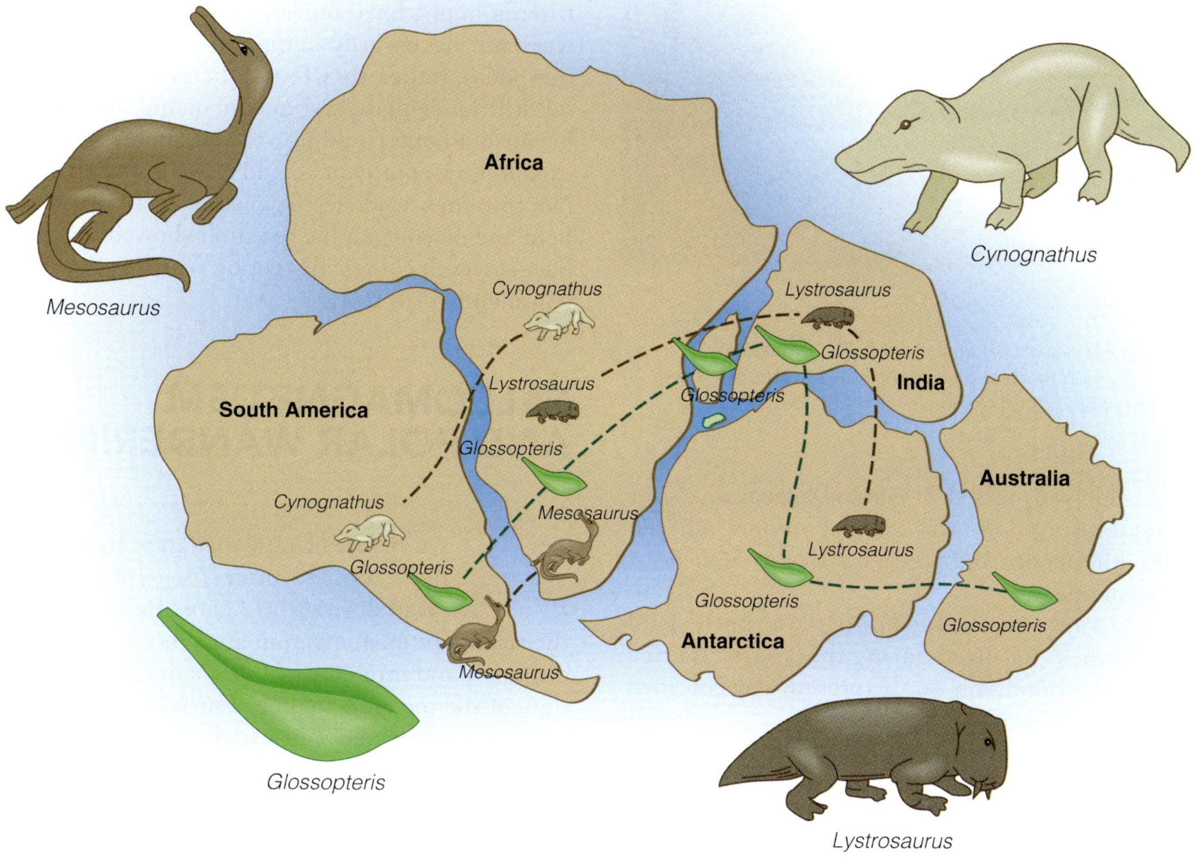

■ **Figure 2.7**

Some of the animals and plants whose fossils are found today on the widely separated continents of South America, Africa, India, Australia, and Antarctica. These continents were joined during the Late Paleozoic to form Gondwana, the southern landmass of Pangaea. *Glossopteris* and similar plants are found in Pennsylvanian- and Permian-aged deposits on all five continents. *Mesosaurus* is a freshwater reptile whose fossils are found in Permian-aged rocks in Brazil and South Africa. *Cynognathus* and *Lystrosaurus* are land reptiles that lived during the Early Triassic Period. Fossils of *Cynognathus* are found in South America and Africa, and fossils of *Lystrosaurus* have been recovered from Africa, India, and Antarctica. Source: Modified from E. H. Colbert, *Wandering Lands and Animals* (1973): 72, Figure 31.

Furthermore, this geographic arrangement places the northern continents nearer the tropics, which is consistent with the fossil and climatologic evidence from Laurasia.

Fossil Evidence

Some of the most compelling evidence for continental drift comes from the fossil record (■ Figure 2.7). Fossils of the *Glossopteris* flora are found in equivalent Pennsylvanian- and Permian-aged coal deposits on all five Gondwana continents. The *Glossopteris* flora is characterized by the seed fern *Glossopteris* (Figure 2.1) as well as by many other distinctive and easily identifiable plants. Pollen and spores of plants can be dispersed over great distances by wind, but *Glossopteris*-type plants produced seeds that are too large to have been carried by winds. Even if the seeds had floated across the ocean, they probably would not have remained viable for any length of time in saltwater.

The present-day climates of South America, Africa, India, Australia, and Antarctica range from tropical to polar and are much too diverse to support the type of plants in the *Glossopteris* flora. Wegener therefore reasoned that these continents must once have been joined so that these widely separated localities were all in the same latitudinal climatic belt (Figure 2.7).

The fossil remains of animals also provide strong evidence for continental drift. One of the best examples is *Mesosaurus*, a freshwater reptile whose fossils are found in Permian-aged rocks in certain regions of Brazil and South Africa and nowhere else in the world (Figures 2.7 and ■ 2.8). Because the physiologies of freshwater and marine animals are completely different, it is hard to imagine how a freshwater reptile could have swum across the Atlantic Ocean and found a freshwater environment nearly identical to its former habitat. Moreover, if *Mesosaurus* could have swum across the ocean, its fossil remains

■ **Figure 2.8**

Mesosaurus, a Permian-aged freshwater reptile whose fossil remains are found in Brazil and South Africa, indicates that these two continents were joined at the end of the Paleozoic Era.

should be widely dispersed. It is more logical to assume that *Mesosaurus* lived in lakes in what are now adjacent areas of South America and Africa, but that were then part of a single continent.

Lystrosaurus and *Cynognathus* are both land-dwelling reptiles that lived during the Triassic Period; their fossils are found only on the present-day continental fragments of Gondwana (Figure 2.7). Because they are both land animals, they certainly could not have swum across the oceans that currently separate the Gondwana continents. Therefore, the continents must once have been connected.

Notwithstanding all of the empirical evidence presented by Wegener and later by du Toit and others, most geologists simply refused to entertain the idea that continents might have moved in the past. The geologists were not necessarily being obstinate about accepting new ideas; rather, they found the evidence for continental drift inadequate and unconvincing. In part, this was because no one could provide a suitable mechanism to explain how continents could move over Earth's surface. Not until new evidence from studies of Earth's magnetic field and oceanographic research showed that the ocean basins were geologically young features did interest in continental drift revive.

PALEOMAGNETISM AND POLAR WANDERING

As a result of new evidence from paleomagnetic studies, interest in continental drift revived during the 1950s. **Paleomagnetism** is the remanent magnetism in ancient rocks recording the direction and intensity of Earth's magnetic poles at the time of the rock's formation. Earth can be thought of as a giant dipole magnet in which the magnetic poles essentially coincide with the geographic poles (■ Figure 2.9). Such an arrangement means that the strength of the magnetic field is not constant, but varies, being weakest at

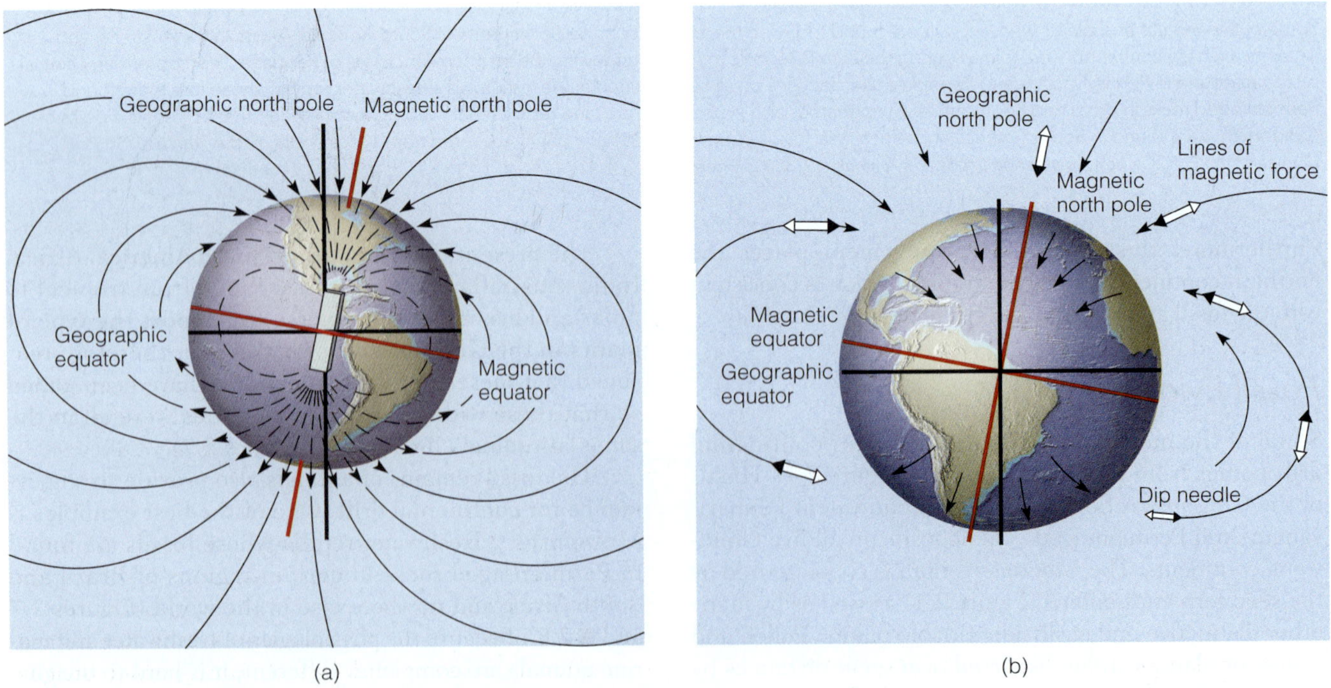

■ **Figure 2.9**

(a) Earth's magnetic field has lines of force just like those of a bar magnet. (b) The strength of the magnetic field changes uniformly from the magnetic equator to the magnetic poles. This change in strength causes a dip needle to parallel Earth's surface only at the magnetic equator, whereas its inclination with respect to the surface increases to 90 degrees at the magnetic poles.

the equator and strongest at the poles. Earth's magnetic field is thought to result from the different rotation speeds of the outer core and mantle.

When magma cools, the magnetic iron-bearing minerals align themselves with Earth's magnetic field, recording both its direction and strength. The temperature at which iron-bearing minerals gain their magnetization is called the **Curie point.** As long as the rock is not subsequently heated above the Curie point, it will preserve that remanent magnetism. Thus an ancient lava flow provides a record of the orientation and strength of Earth's magnetic field at the time the lava flow cooled.

As paleomagnetic research progressed during the 1950s, some unexpected results emerged. When geologists measured the paleomagnetism of geologically recent rocks, they found it was generally consistent with Earth's current magnetic field. The paleomagnetism of ancient rocks, though, showed different orientations. For example, paleomagnetic studies of Silurian lava flows in North America indicated that the north magnetic pole was located in the western Pacific Ocean at that time, whereas the paleomagnetic evidence from Permian lava flows pointed to yet another location in Asia. When plotted on a map, the paleomagnetic readings of numerous lava flows from all ages in North America trace the apparent movement of the magnetic pole through time (■ Figure 2.10). This paleomagnetic evidence from a single continent could be interpreted in three ways: The continent remained fixed and the north magnetic pole moved; the north magnetic pole stood still and the continent moved; or both the continent and the north magnetic pole moved.

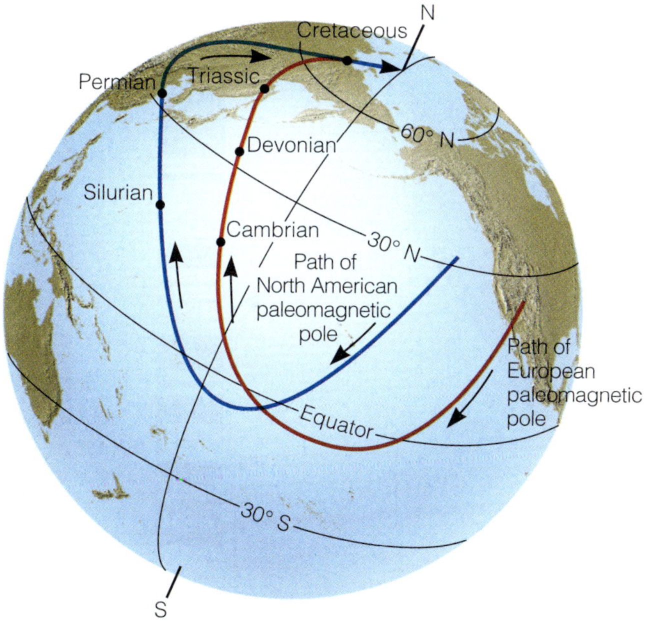

■ **Figure 2.10**

The apparent paths of polar wandering for North America and Europe. The apparent location of the north magnetic pole is shown for different periods on each continent's polar wandering path.

Source: From A. Cox and R. R. Doell, "Review of Paleomagnetism," *G. S. A. Bulletin*, vol. 71, figure 33, page 758, with permission of the publisher, the Geological Society of America, Boulder, Colorado. USA. Copyright © 1955 Geological Society of America.

Upon analysis, magnetic minerals from European Silurian and Permian lava flows pointed to a different magnetic pole location from those of the same age from North America (Figure 2.10). Furthermore, analysis of lava flows from all continents indicated that each continent had its own series of magnetic poles. Does this mean there were different north magnetic poles for each continent? That would be highly unlikely and difficult to reconcile with the theory accounting for Earth's magnetic field.

The best explanation for such data is that the magnetic poles have remained near their present locations at the geographic north and south poles and the continents have moved. When the continental margins are fitted together so that the paleomagnetic data point to only one magnetic pole, we find, just as Wegener did, that the rock sequences and glacial deposits match, and that the fossil evidence is consistent with the reconstructed paleogeography.

HOW DO MAGNETIC REVERSALS RELATE TO SEAFLOOR SPREADING?

Geologists refer to Earth's present magnetic field as being normal—that is, with the north and south magnetic poles located approximately at the north and south geographic poles. At various times in the geologic past, Earth's magnetic field has completely reversed. The existence of such **magnetic reversals** was discovered by dating and determining the orientation of the remanent magnetism in lava flows on land (■ Figure 2.11).

Once their existence was well established for continental lava flows, magnetic reversals were also discovered in igneous rocks of the oceanic crust as part of the extensive mapping of the ocean basins during the 1960s (■ Figure 2.12). Although the cause of magnetic reversals is still uncertain, their occurrence in the geologic record is well documented.

Besides the discovery of magnetic reversals, mapping of the ocean basins also revealed a ridge system 65,000 km long, constituting the most extensive mountain range in the world. Perhaps the best-known part of the ridge system is the Mid-Atlantic Ridge, which divides the Atlantic Ocean basin into two nearly equal parts (■ Figure 2.13).

As a result of oceanographic research conducted during the 1950s, Harry Hess of Princeton University

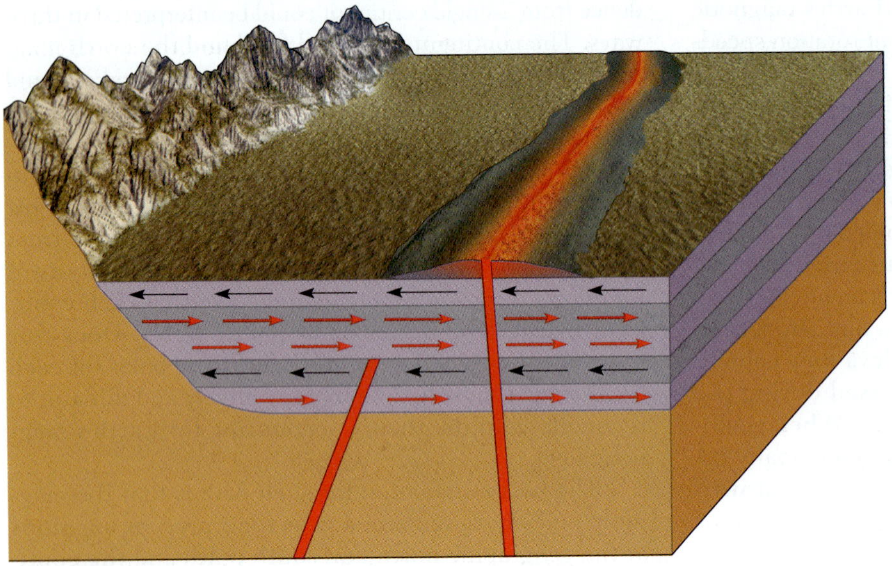

■ **Figure 2.11**

Magnetic reversals recorded in a succession of lava flows are shown diagrammatically by red arrows, and the record of normal polarity events is shown by black arrows.

proposed the theory of **seafloor spreading** in 1962 to account for continental movement. Hess suggested that continents do not move across oceanic crust, but rather that the continents and oceanic crust move together. He suggested that the seafloor separates at oceanic ridges where new crust is formed by upwelling magma. As the magma cools, the newly formed oceanic crust moves laterally away from the ridge.

As a mechanism to drive this system, Hess revived the idea of **thermal convection cells** in the mantle; that is, hot magma rises from the mantle, intrudes along fractures along oceanic ridges and thus forms new crust. Cold crust is subducted back into the mantle at oceanic trenches, where it is heated and recycled, thus completing a thermal convection cell (see Figure 1.9).

How could Hess's hypothesis be confirmed? Magnetic surveys of the oceanic crust revealed striped **magnetic anomalies** (deviations from the average strength of Earth's magnetic field) in the rocks that were both parallel to and symmetric with the oceanic ridges (Figure 2.12). Furthermore, the pattern of oceanic magnetic anomalies matched the pattern of magnetic reversals already known from studies of continental lava flows

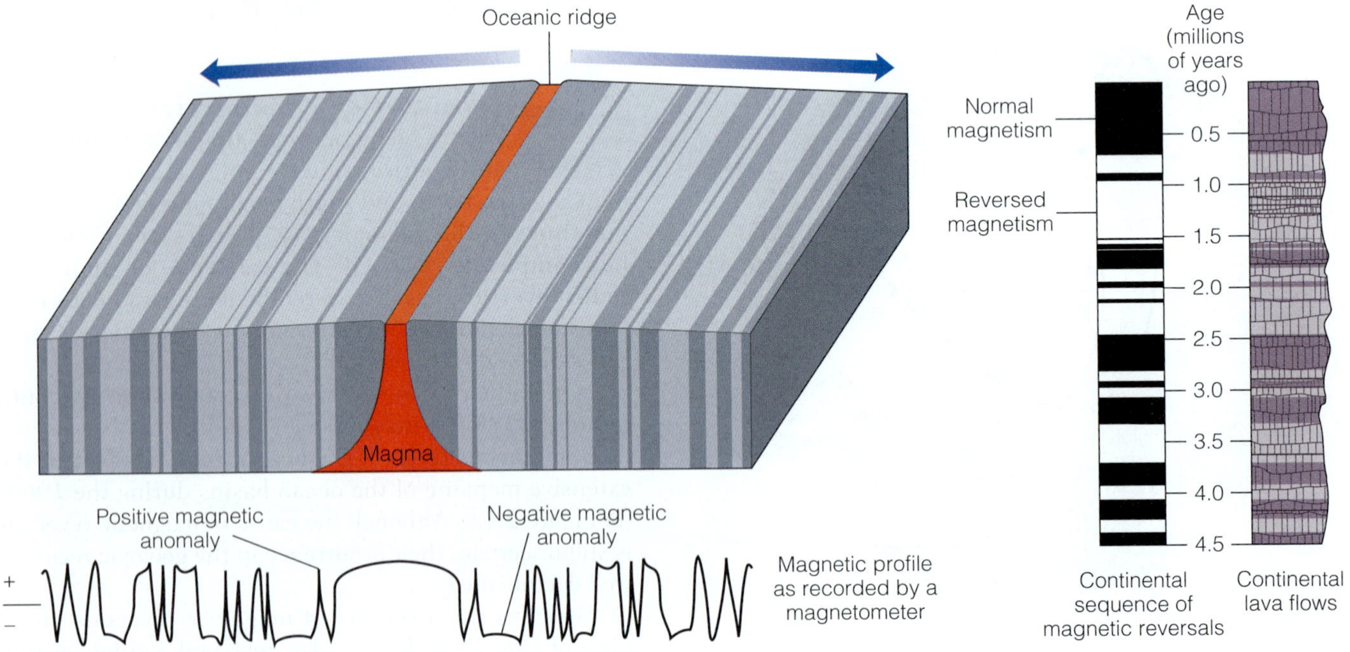

Geology Now ■ **Active Figure 2.12**

The sequence of magnetic anomalies preserved within the oceanic crust on both sides of an oceanic ridge is identical to the sequence of magnetic reversals already known from continental lava flows. Magnetic anomalies are formed when basaltic magma intrudes into oceanic ridges; when the magma cools below the Curie point, it records Earth's magnetic polarity at the time. Seafloor spreading splits the previously formed crust in half so that it moves laterally away from the oceanic ridge. Repeated intrusions record a symmetric series of magnetic anomalies that reflect periods of normal and reversed polarity. The magnetic anomalies are recorded by a magnetometer, which measures the strength of the magnetic field. Source: Reprinted with permission from A. Cox, "Geomagnetic Reversals," *Science 163*, January 17, 1969. Copyright © 1969 American Association for the Advancement of Science.

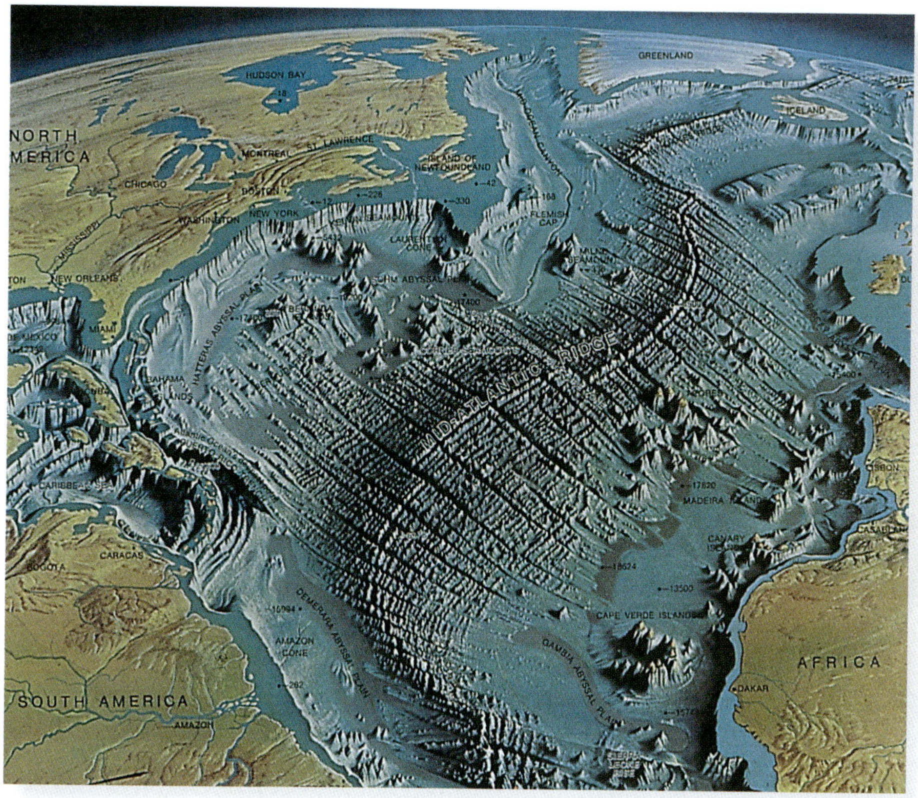

ALCOA

■ **Figure 2.13**

Artist's view of what the Atlantic Ocean basin would look like without water. The major feature is the Mid-Atlantic Ridge.

(Figure 2.11). When magma wells up and cools along a ridge summit, it records Earth's magnetic field at that time as either normal or reversed. As new crust forms at the summit, the previously formed crust moves laterally away from the ridge. These magnetic stripes, representing times of normal or reversed polarity, are parallel to and symmetrical around oceanic ridges (where upwelling magma forms new oceanic crust), conclusively confirming Hess's theory of seafloor spreading.

One of the consequences of the seafloor spreading theory is its confirmation that ocean basins are geologically young features whose openings and closings are partially responsible for continental movement (■ Figure 2.14). Radiometric dating reveals that the oldest oceanic crust is less than 180 million years old, whereas the oldest continental crust is 3.96 billion years old. Although geologists do not universally accept the idea of thermal convection cells as a driving mechanism for plate movement, most accept that plates are created at oceanic ridges and destroyed at deep-sea trenches, regardless of the driving mechanism involved.

Deep-Sea Drilling and the Confirmation of Seafloor Spreading

For many geologists, the paleomagnetic data amassed in support of continental drift and seafloor spreading were convincing. Results from the Deep Sea Drilling Project (see Chapter 9) have confirmed the interpretations made from earlier paleomagnetic studies.

According to the seafloor spreading hypothesis, oceanic crust is continuously forming at mid-oceanic ridges, moves away from these ridges by seafloor spreading, and is consumed at subduction zones. If this is the case, then oceanic crust should be youngest at the ridges and become progressively older with increasing distance away from them. Moreover, the age of the oceanic crust should be symmetrically distributed about the ridges. As we have just noted, paleomagnetic data confirm these statements. Furthermore, fossils from sediments overlying the oceanic crust and radiometric dating of rocks found on oceanic islands substantiate this predicted age distribution.

Sediments in the open ocean accumulate, on average, at a rate of less than 0.3 cm per 1000 years. If the ocean basins were as old as the continents, we would expect deep-sea sediments to be several kilometers thick. However, data from numerous drill holes indicate that deep-sea sediments are at most only a few hundred meters thick and are thin or absent at oceanic ridges. Their near-absence at the ridges should come as no surprise because these are the areas where new crust is continuously produced by volcanism and seafloor spreading. Accordingly, sediments have had little time to accumulate at or very close to spreading ridges where the oceanic crust is young, but their thickness increases with distance away from the ridges (■ Figure 2.15).

WHY IS PLATE TECTONICS A UNIFYING THEORY?

Plate tectonic theory is based on a simple model of Earth. The rigid lithosphere, consisting of both oceanic and continental crust, as well as the underlying upper mantle, consists of numerous variable-sized pieces called *plates* (■ Figure 2.16). The plates vary in thickness; those composed of upper mantle and continental crust are as much as 250 km thick, whereas those of upper mantle and oceanic crust are up to 100 km thick.

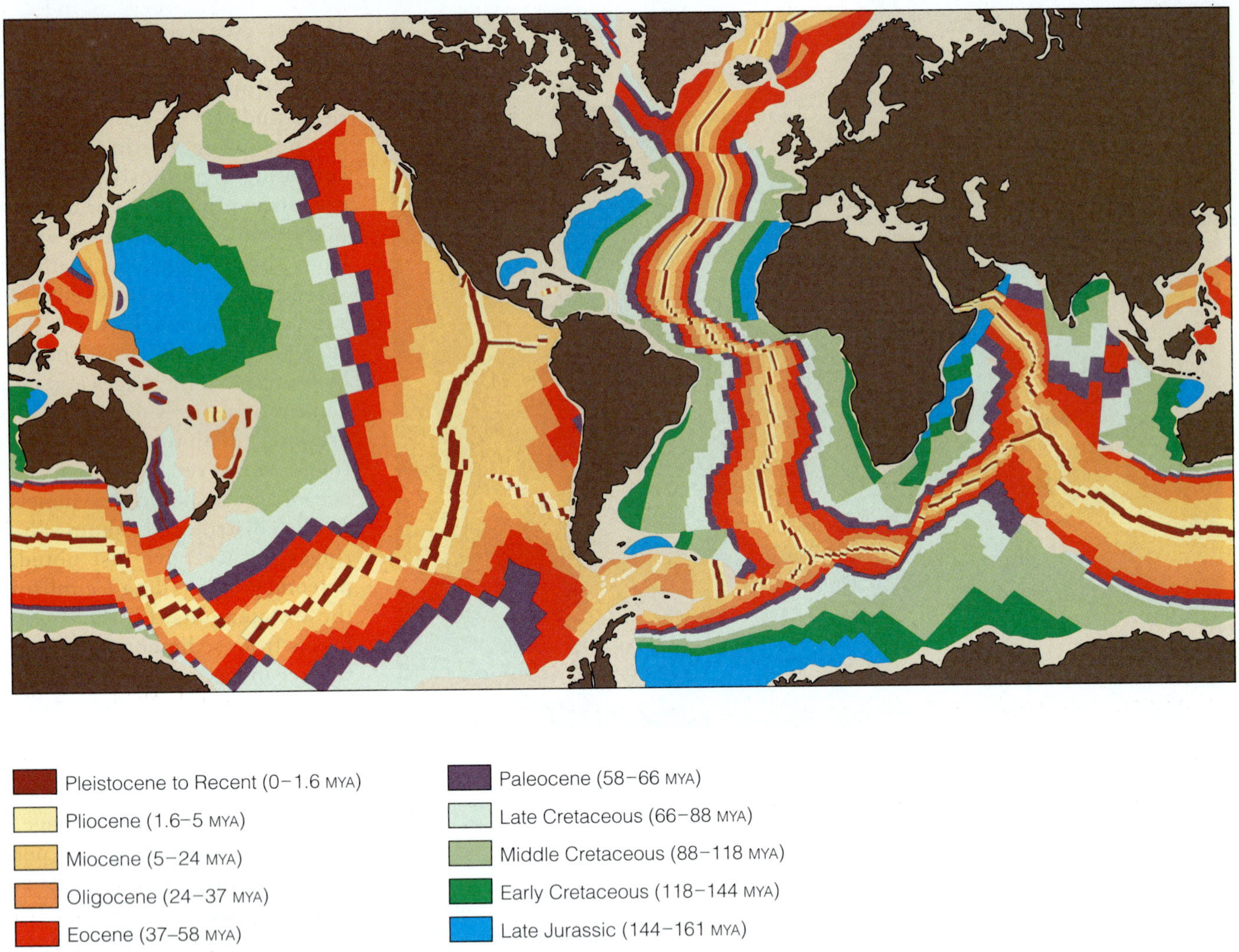

Figure 2.14

The age of the world's ocean basins established from magnetic anomalies demonstrates that the youngest oceanic crust is adjacent to the spreading ridges and that its age increases away from the ridge axis. Source: From Larson, R. L., et al. (1985). *The Bedrock Geology of the World*, W. H. Freeman and Co., New York, NY.

The lithosphere overlies the hotter and weaker semi-plastic asthenosphere. It is thought that movement resulting from some type of heat-transfer system within the asthenosphere causes the overlying plates to move. As plates move over the asthenosphere, they separate, mostly at oceanic ridges; in other areas such as at oceanic trenches, they collide and are subducted back into the mantle.

An easy way to visualize plate movement is to think of a conveyer belt moving luggage from an airplane's

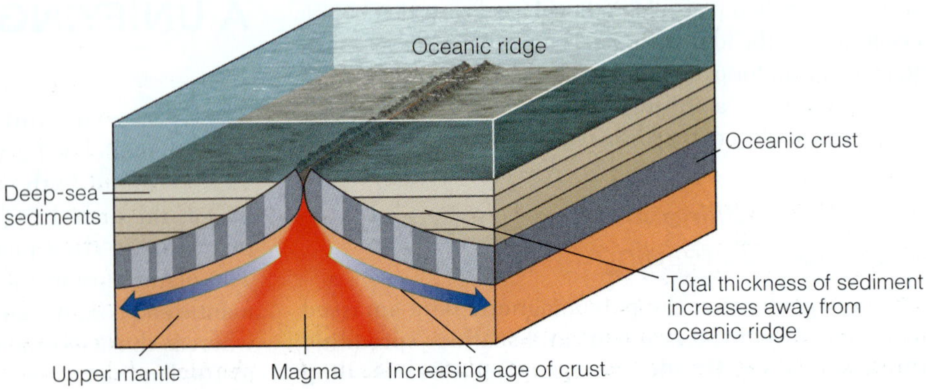

Figure 2.15

The total thickness of deep-sea sediments increases away from oceanic ridges. This is because oceanic crust becomes older away from oceanic ridges, and there has been more time for sediment to accumulate.

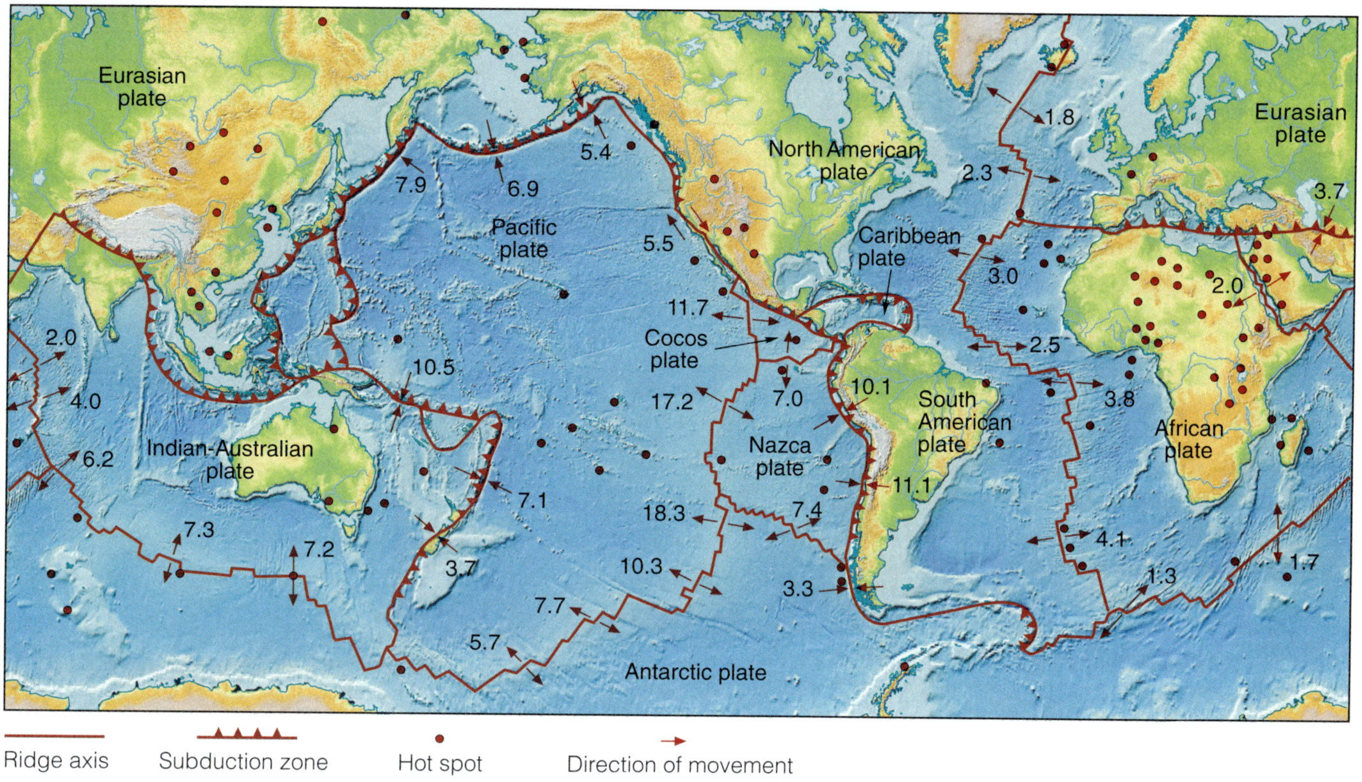

Figure 2.16

A map of the world showing the plates, their boundaries, relative motion and rates of movement in centimeters per year, and hot spots.

What Would You Do?

You've been selected to be part of the first astronaut team to go to Mars. While your two fellow crew members descend to the Martian surface, you'll be staying in the command module and circling the Red Planet. As part of the geologic investigation of Mars, one of the crew members will be mapping the geology around the landing site and deciphering the geologic history of the area. Your job will be to observe and photograph the planet's surface and try to determine whether Mars had an active plate tectonic regime in the past and whether there is current plate movement. What features would you look for, and what evidence might reveal current or previous plate activity?

cargo hold to a baggage cart. The conveyer belt represents convection currents within the mantle, and the luggage represents Earth's lithospheric plates. The luggage is moved along by the conveyer belt until it is dumped into the baggage cart in the same way plates are moved by convection cells until they are subducted into Earth's interior. Although this analogy allows you to visualize how the mechanism of plate movement takes place, remember that this analogy is limited. The major limitation is that, unlike the luggage, plates consist of continental and oceanic crust, which have different densities, and only oceanic crust is subducted into Earth's interior. Nonetheless, this analogy does provide an easy way to visualize plate movement.

Most geologists accept plate tectonic theory, in part because the evidence for it is overwhelming and it ties together many seemingly unrelated geologic features and events and shows how they are interrelated. Consequently, geologists now view such geologic processes as mountain building, seismicity, and volcanism from the perspective of plate tectonics. Furthermore, because all of the inner planets have had a similar origin and early history, geologists are interested in determining whether plate tectonics is unique to Earth or whether it operates in the same way on other planets (see "Tectonics of the Terrestrial Planets" on pages 40 and 41).

The Supercontinent Cycle

As a result of plate movement, all the continents came together to form the supercontinent Pangaea by the end of the Paleozoic Era. Pangaea began fragmenting during the Triassic Period and continues to do so, thus accounting for the present distribution of continents and ocean basins. It has been proposed that supercontinents consisting of all or most of Earth's landmasses form, break up, and re-form in a cycle spanning about 500 million years.

The *supercontinent cycle hypothesis* is an expansion on the ideas of the Canadian geologist J. Tuzo Wilson. During

Tectonics of the Terrestrial Planets

The four inner, or terrestrial, planets—Mercury, Venus, Earth, and Mars—all had a similar early history involving accretion, differentiation into a metallic core and silicate mantle and crust, and formation of an early atmosphere by outgassing. Their early history was also marked by widespread volcanism and meteorite impacts, both of which helped modify their surfaces.

Whereas the other three terrestrial planets as well as some of the Jovian moons display internal activity, Earth appears to be unique in that its surface is broken into a series of plates.

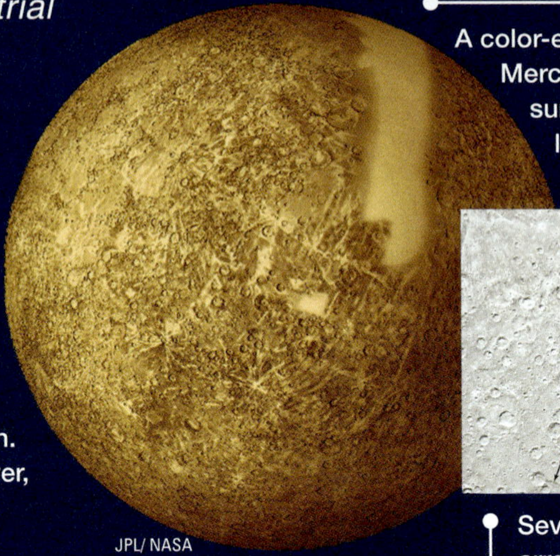

JPL/ NASA

A color-enhanced photomosaic of Mercury shows its heavily cratered surface, which has changed very little since its early history.

Courtesy of Victor Royer

Seven scarps (indicated by arrows) can clearly be seen in this image. These scarps might have formed when Mercury cooled and contracted early in its history.

Images of **Mercury** sent back by *Mariner 10* show a heavily cratered surface with the largest impact basins filled with what appear to be lava flows similar to the lava plains on Earth's Moon. The lava plains are not deformed, however, indicating that there has been little or no tectonic activity.

Another feature of Mercury's surface is a large number of scarps, a feature usually associated with earthquake activity. Yet, some scientists think that these scarps formed when Mercury cooled and contracted.

A color-enhanced photomosaic of Venus based on radar images beamed back to Earth by the *Magellan* spacecraft. This image shows impact craters and volcanic features characteristic of the planet.

Of all the planets, **Venus** is the most similar in size and mass to Earth, but it differs in most other respects. Whereas Earth is dominated by plate tectonics, volcanism seems to have been the dominant force in the evolution of the Venusian surface. Even though no active volcanism has been observed on Venus, the various-sized volcanic features and what appear to be folded mountains indicate a once-active planetary interior. All of these structures appear to be the products of rising convection currents of magma pushing up under the crust and then sinking back into the Venusian interior.

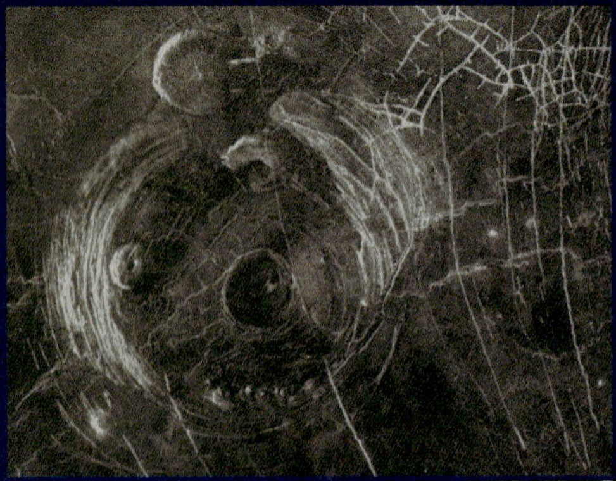

JPL/ NASA

Venus's Aine Corona, about 200 km in diameter, is ringed by concentric faults, suggesting that it was pushed up by rising magma. A network of fractures is visible in the upper right of this image as well as a recent lava flow at the center of the corona, several volcanic domes in the lower portion of the image, and a large volcanic pancake dome in the upper left of the image.

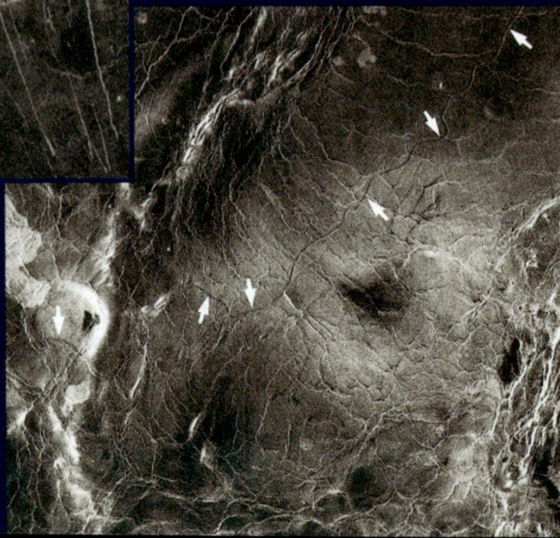

Arrows point to a 600-km segment of Venus's 6800-km long Baltis Vallis, the longest known lava flow channel in our solar system.

Volcano Sapas Mons contains two lava-filled calderas and is flanked by lava flows, attesting to the volcanic activity that was once common on Venus.

Mars, the Red Planet, has numerous features that indicate an extensive early period of volcanism. These include Olympus Mons, the solar system's largest volcano, lava flows, and uplifted regions thought to have resulted from mantle convection. In addition to volcanic features, Mars displays abundant evidence of tensional tectonics, including numerous faults and large fault-produced valley structures. Whereas Mars was tectonically active during the past, no evidence indicates that plate tectonics comparable to those on Earth has ever occurred there.

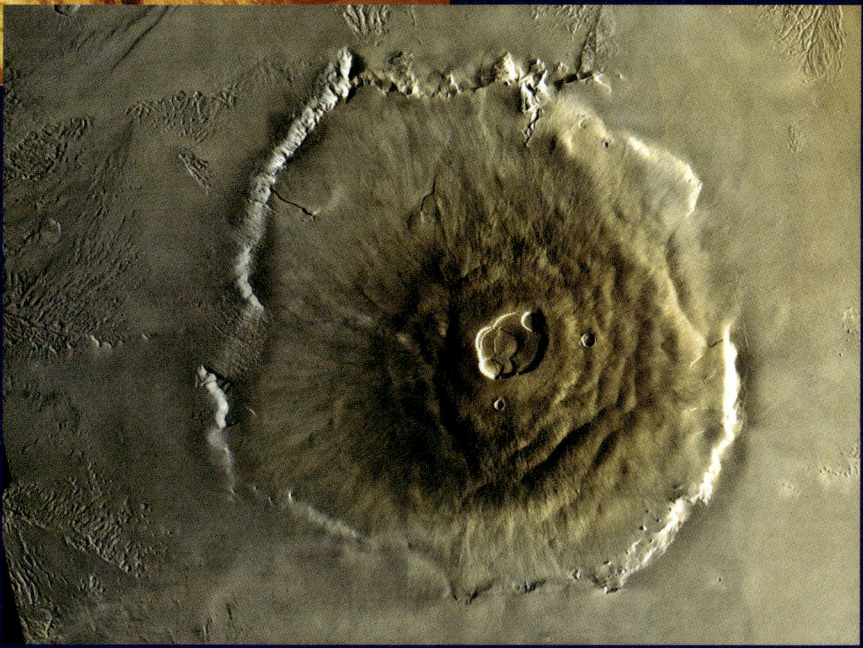

JPL/ NASA

A vertical view of Olympus Mons, a shield volcano and the largest volcano in our solar system. The edge of the Olympus Mons caldera is marked by a cliff several kilometers high rather than a moat as in Mauna Loa, Earth's largest shield volcano.

USGS

A photomosaic of Mars shows a variety of geologic structures, including the southern polar ice cap.

Although not a terrestrial planet, **Io**, the innermost of Jupiter's Galilean moons, must be mentioned. Images from the *Voyager* and *Galileo* spacecrafts show that Io has no impact craters. In fact, more than a hundred active volcanoes are visible on the moon's surface, and the sulfurous gas and ash erupted by these volcanoes bury any newly formed meteorite impact craters. Because of its proximity to Jupiter, the heat source of Io is probably tidal heating, in which the resulting friction is enough to at least partially melt Io's interior and drive its volcanoes.

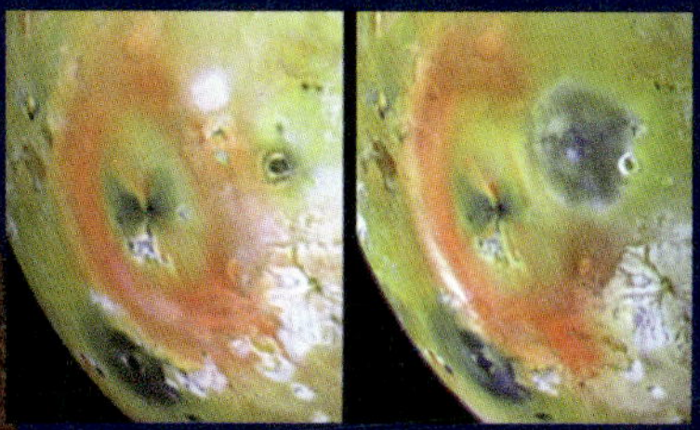

NASA

NASA

NASA

Volcanic features of Io, the innermost moon of Jupiter. As shown in these digitally enhanced color images, Io is a very volcanically active moon.

the early 1970s, Wilson proposed a cycle (now known as the Wilson cycle) that includes continental fragmentation, the opening and closing of an ocean basin, and reassembly of the continent. According to the supercontinent cycle hypothesis, heat accumulates beneath a supercontinent because rocks of continents are poor conductors of heat. As a result of the heat accumulation, the supercontinent domes upward and fractures. Basaltic magma rising from below fills the fractures. As a basalt-filled fracture widens, it begins subsiding and forms a long narrow ocean such as the present-day Red Sea. Continued rifting eventually forms an expansive ocean basin such as the Atlantic.

One of the most convincing arguments for proponents of the supercontinent cycle hypothesis is the "surprising regularity" of mountain building caused by compression during continental collisions. These mountain-building episodes occur about every 400 to 500 million years and are followed by an episode of rifting about 100 million years later. In other words, a supercontinent fragments and its individual plates disperse following a rifting episode, an interior ocean forms, and then the dispersed fragments reassemble to form another supercontinent.

The supercontinent cycle is yet another example of how interrelated the various systems and subsystems of Earth are and how they operate over vast periods of geologic time.

WHAT ARE THE THREE TYPES OF PLATE BOUNDARIES?

Because it appears that plate tectonics has operated since at least the Proterozoic Eon, it is important that we understand how plates move and interact with each other and how ancient plate boundaries are recognized. After all, the movement of plates has profoundly affected the geologic and biologic history of this planet.

Geologists recognize three major types of plate boundaries: *divergent, convergent,* and *transform* (Table 2.1). Along these boundaries new plates are formed, are consumed, or slide laterally past one another. Interaction of plates at their boundaries accounts for most of Earth's volcanic eruptions and earthquakes as well as the formation and evolution of its mountain systems.

Divergent Boundaries

Divergent plate boundaries or *spreading ridges* occur where plates are separating and new oceanic lithosphere is forming. Divergent boundaries are places where the crust is extended, thinned, and fractured as magma, derived from the partial melting of the mantle, rises to the surface. The magma is almost entirely basaltic and intrudes into vertical fractures to form dikes and pillow lava flows (see Figure 5.7). As successive injections of magma cool and solidify, they form new oceanic crust and record the intensity and orientation of Earth's magnetic field (Figure 2.12). Divergent boundaries most commonly occur along the crests of oceanic ridges—for example, the Mid-Atlantic Ridge. Oceanic ridges are thus characterized by rugged topography with high relief resulting from displacement of rocks along large fractures, shallow-focus earthquakes, high heat flow, and basaltic flows or pillow lavas.

Divergent plate boundaries are also present under continents during the early stages of continental breakup (■ Figure 2.17). When magma wells up beneath a continent, the crust is initially elevated, stretched, and thinned, producing fractures and rift valleys (Figure 2.17a).

Table 2.1

Types of Plate Boundaries

Type	Example	Landforms	Volcanism
Divergent			
Oceanic	Mid-Atlantic Ridge	Mid-oceanic ridge with axial rift valley	Basalt
Continental	East African Rift Valley	Rift valley	Basalt and rhyolite, no andesite
Convergent			
Oceanic–oceanic	Aleutian Islands	Volcanic island arc, offshore oceanic trench	Andesite
Oceanic–continental	Andes	Offshore oceanic trench, volcanic mountain chain, mountain belt	Andesite
Continental–continental	Himalayas	Mountain belt	Minor
Transform	San Andreas fault	Fault valley	Minor

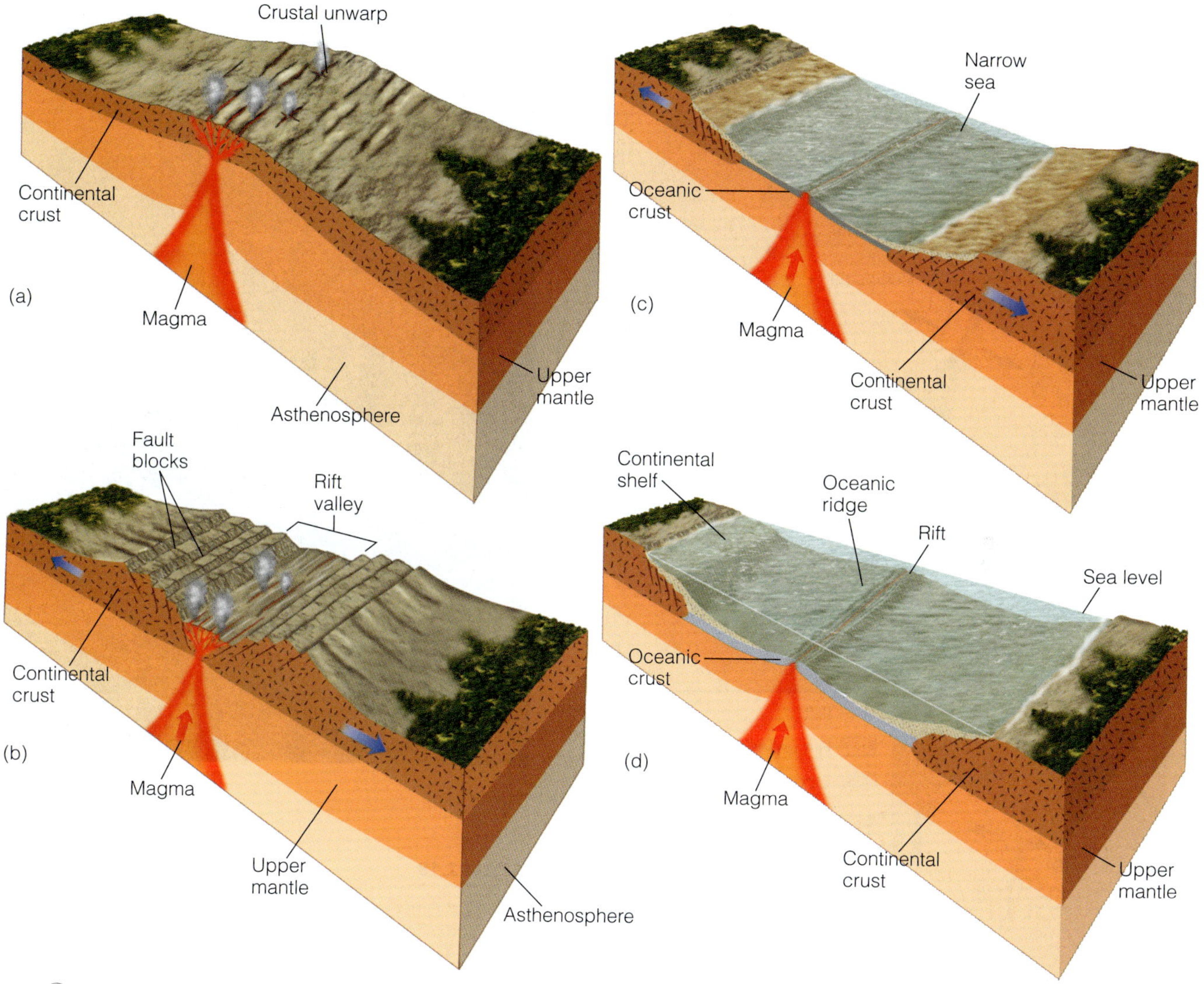

Geology Now ■ **Active Figure 2.17**

History of a divergent plate boundary. (a) Rising magma beneath a continent pushes the crust up, producing numerous cracks and fractures. (b) As the crust is stretched and thinned, rift valleys develop and lava flows onto the valley floors. (c) Continued spreading further separates the continent until a narrow seaway develops. (d) As spreading continues, an oceanic ridge system forms and an ocean basin develops and grows.

During this stage, magma typically intrudes into the faults and fractures, forming sills, dikes, and lava flows; the latter often cover the rift valley floor (Figure 2.17b). The East African Rift Valley is an excellent example of this stage of continental breakup (■ Figure 2.18).

As spreading proceeds, some rift valleys will continue to lengthen and deepen until the continental crust eventually breaks and a narrow linear sea is formed, separating two continental blocks (Figure 2.17c). The Red Sea separating the Arabian Peninsula from Africa is a good example of this stage of rifting (Figure 2.18a).

As a newly formed narrow sea continues to enlarge it may eventually become an expansive ocean basin such as the Atlantic Ocean basin is today, separating North and South America from Europe and Africa by thousands of kilometers (Figure 2.17d). The Mid-Atlantic Ridge is the boundary between these diverging plates; the American plates are moving westward, and the Eurasian and African plates are moving eastward.

An Example of Ancient Rifting What features in the geologic record can geologists use to recognize ancient rifting? Associated with regions of continental rifting are faults, dikes, sills, lava flows, and thick sedimentary sequences within rift valleys. The Triassic fault basins of the eastern United States are a good example of ancient continental rifting (see Figure 22.7). These fault basins mark the zone of rifting that occurred when North America split apart from Africa. The basins contain thousands of meters of continental sediment and are riddled with dikes and sills (see Chapter 22).

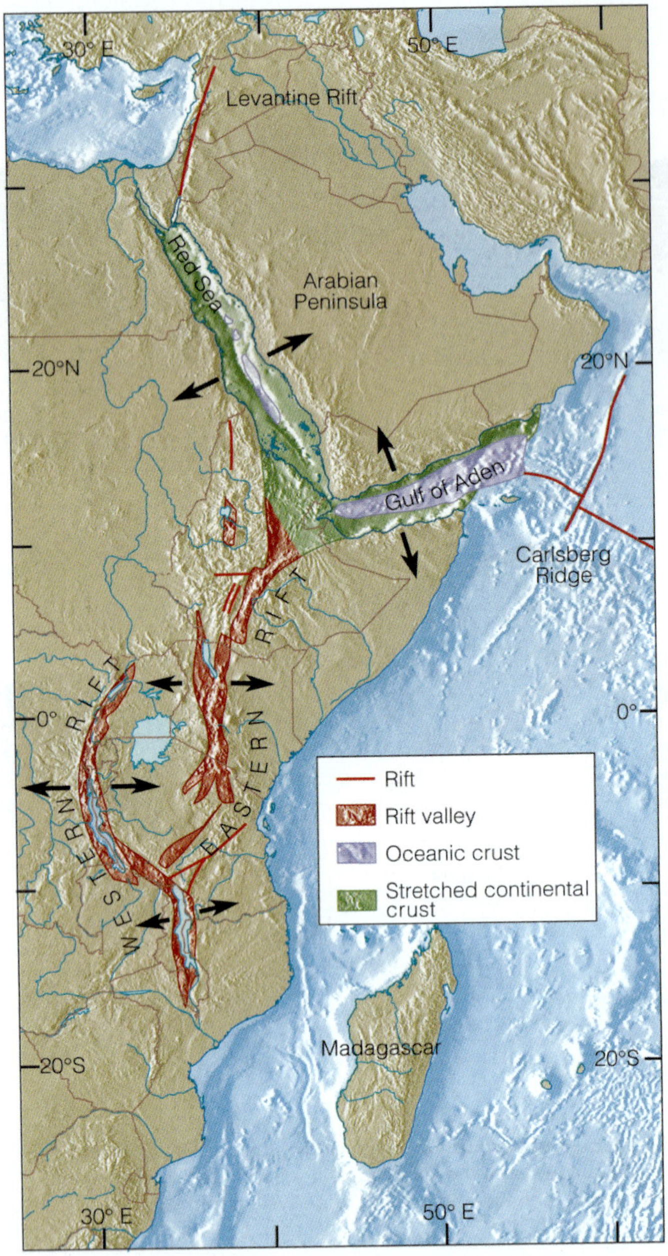

(a)

(b)

Figure 2.18

(a) The East African Rift Valley is being formed by the separation of eastern Africa from the rest of the continent along a divergent plate boundary. The Red Sea represents a more advanced stage of rifting, in which two continental blocks are separated by a narrow sea. (b) View looking down the Great Rift Valley of Africa. Little Magadi, seen in the background, is one of numerous soda lakes forming in the valley. Because of high evaporation rates and lack of any drainage outlets, these lakes are very saline. The Great Rift Valley is part of the system of rift valleys resulting from stretching of the crust as plates move away from each other in eastern Africa.

Convergent Boundaries

Whereas new crust forms at divergent plate boundaries, older crust must be destroyed and recycled in order for the entire surface area of Earth to remain the same. Otherwise, we would have an expanding Earth. Such plate destruction occurs at **convergent plate boundaries** where two plates collide and the leading edge of one plate is subducted beneath the margin of the other plate and eventually incorporated into the asthenosphere.

Convergent boundaries are characterized by deformation, volcanism, mountain building, metamorphism, earthquake activity, and important mineral deposits. Three types of convergent plate boundaries are recognized: *oceanic–oceanic*, *oceanic–continental*, and *continental–continental*.

Oceanic–Oceanic Boundaries

When two oceanic plates converge, one is subducted beneath the other along an **oceanic–oceanic plate boundary** (■ Figure 2.19). The subducting plate bends downward to form the outer wall of an oceanic trench. A *subduction complex*, composed of wedge-shaped slices of highly folded and faulted marine sediments and oceanic lithosphere scraped off the descending plate, forms along the inner wall of the oceanic trench. As the subducting plate descends into the mantle, it is heated and partially melted, generating magma commonly of andesitic composition. This magma is less dense than the surrounding mantle rocks and rises to the surface of the nonsubducted plate to form a curved chain of volcanic islands called a *volcanic island arc* (any plane intersecting a sphere makes an arc). This arc is nearly parallel to the oceanic trench and is separated from it by a distance of up

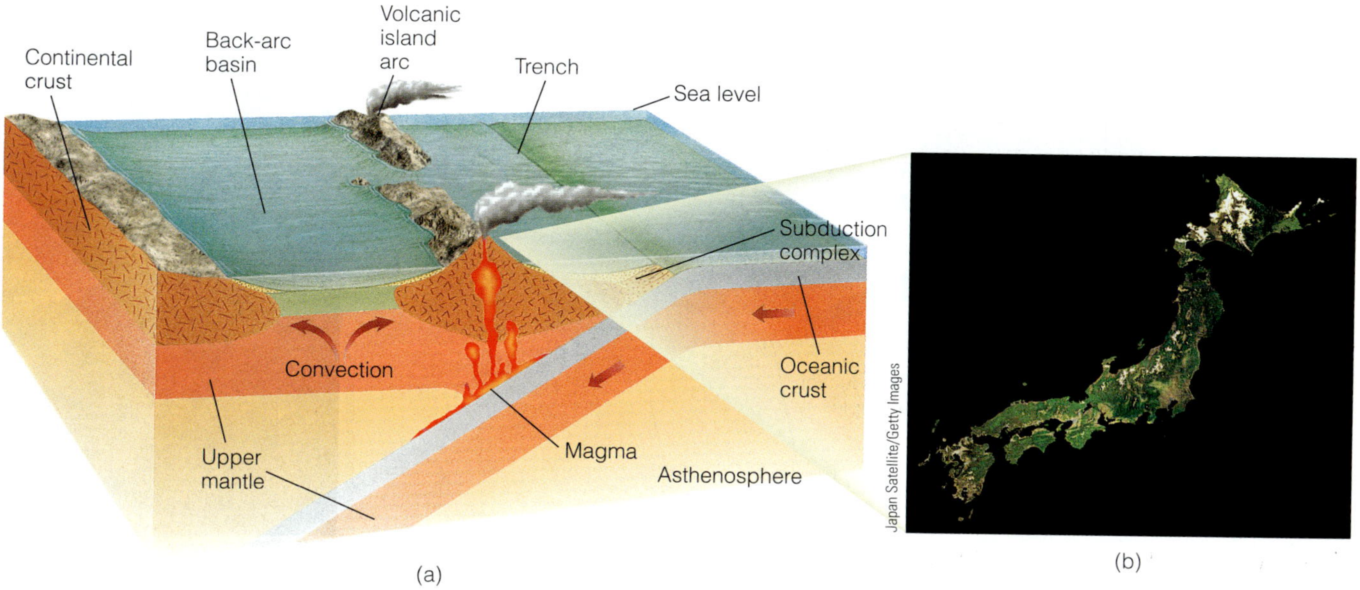

Geology Now ■ **Active Figure 2.19**

Oceanic–oceanic plate boundary. (a) An oceanic trench forms where one oceanic plate is subducted beneath another. On the nonsubducted plate, a volcanic island arc forms from the rising magma generated from the subducting plate. (b) Satellite image of Japan. The Japanese Islands are a volcanic island arc resulting from the subduction of one oceanic plate beneath another oceanic plate.

to several hundred kilometers—the distance depends on the angle of dip of the subducting plate (Figure 2.19).

In those areas where the rate of subduction is faster than the forward movement of the overriding plate, the lithosphere on the landward side of the volcanic island arc may be subjected to tensional stress and stretched and thinned, resulting in the formation of a *back-arc basin*. This back-arc basin may grow by spreading if magma breaks through the thin crust and forms new oceanic crust (Figure 2.19). A good example of a back-arc basin associated with an oceanic–oceanic plate boundary is the Sea of Japan between the Asian continent and the islands of Japan.

Most present-day active volcanic island arcs are in the Pacific Ocean basin and include the Aleutian Islands, the Kermadec-Tonga arc, and the Japanese (Figure 2.19b) and Philippine Islands. The Scotia and Antillean (Caribbean) island arcs are in the Atlantic Ocean basin.

Oceanic–Continental Boundaries When an oceanic and a continental plate converge, the denser oceanic plate is subducted under the continental plate along an **oceanic–continental plate boundary** (■ Figure 2.20). Just as at oceanic–oceanic plate boundaries, the descending oceanic plate forms the outer wall of an oceanic trench.

The magma generated by subduction rises beneath the continent and either crystallizes as large plutons before reaching the surface or erupts at the surface to produce a chain of andesitic volcanoes (also called a *volcanic arc*). An excellent example of an oceanic–continental plate boundary is the Pacific Coast of South America where the oceanic Nazca plate is currently being subducted under South America (Figure 2.16). The Peru–Chile Trench marks the site of subduction, and the Andes Mountains are the resulting volcanic mountain chain on the nonsubducting plate.

Continental–Continental Boundaries Two continents approaching each other are initially separated by an ocean floor that is being subducted under one continent. The edge of that continent displays the features characteristic of oceanic–continental convergence. As the ocean floor continues to be subducted, the two continents come closer together until they eventually collide. Because continental lithosphere, which consists of continental crust and the upper mantle, is less dense than oceanic lithosphere (oceanic crust and upper mantle), it cannot sink into the asthenosphere. Although one continent may partly slide under the other, it cannot be pulled or pushed down into a subduction zone (■ Figure 2.21).

When two continents collide, they are welded together along a zone marking the former site of subduction. At this **continental–continental plate boundary,** an interior mountain belt is formed consisting of deformed sediments and sedimentary rocks, igneous intrusions, metamorphic rocks, and fragments of oceanic crust. In addition, the entire region is subjected to numerous earthquakes. The Himalayas in central Asia, the world's highest mountain system, resulted from the collision between India and Asia that began 40 to 50 million years ago and is still continuing (see Chapter 10).

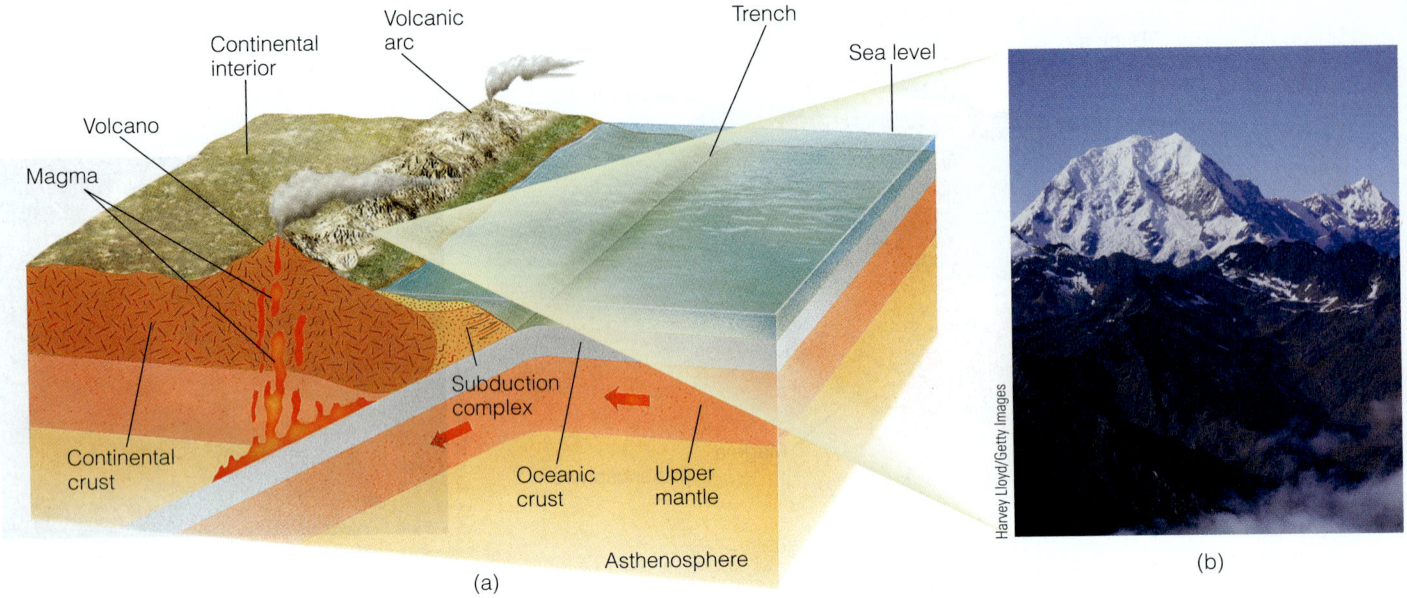

Geology Now ■ **Active Figure 2.20**

Oceanic–continental plate boundary. (a) When an oceanic plate is subducted beneath a continental plate, an andesitic volcanic mountain range is formed on the continental plate as a result of rising magma. (b) Aerial view of the Andes Mountains in Peru. The Andes are one of the best examples of continuing mountain building at an oceanic–continental plate boundary.

Recognizing Ancient Convergent Plate Boundaries

How can former subduction zones be recognized in the geologic record? Igneous rocks provide one clue. The magma erupted at the surface, forming island arc volcanoes and continental volcanoes, is of andesitic composition. Another clue can be found in the zone of intensely deformed rocks between the deep-sea trench where subduction is taking place and the area of igneous activity. Here, sediments and submarine rocks are folded, faulted, and metamorphosed into a chaotic mixture of rocks termed a *mélange*.

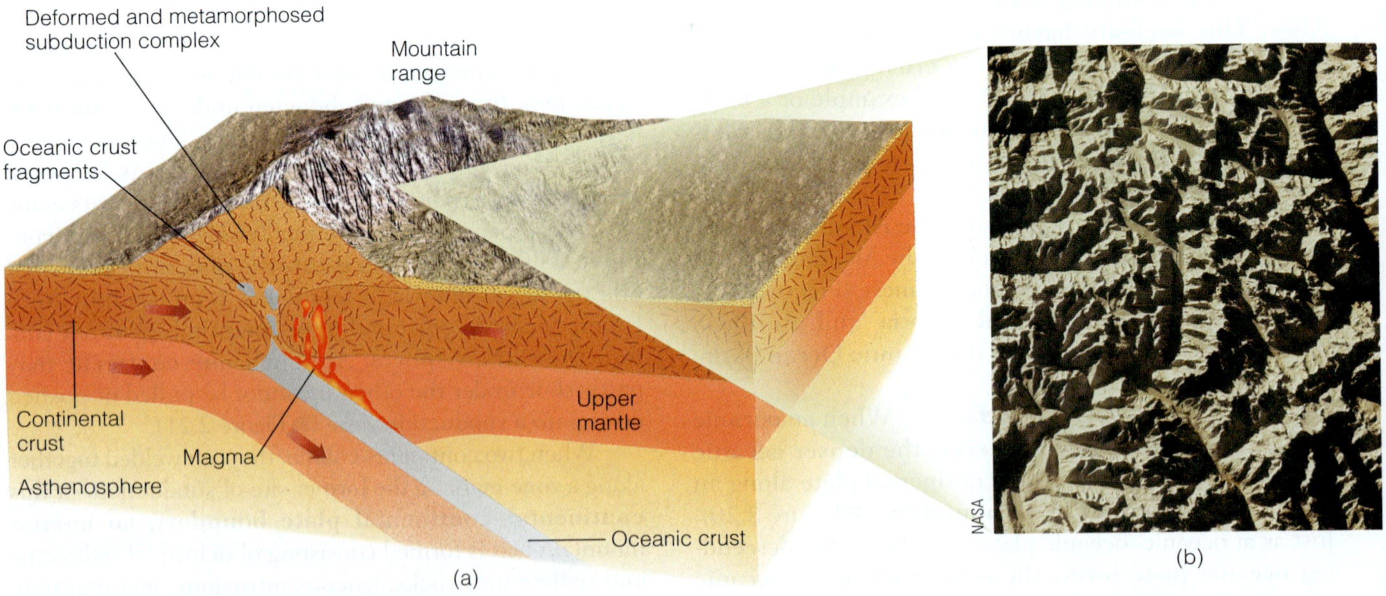

Geology Now ■ **Active Figure 2.21**

Continental–continental plate boundary. (a) When two continental plates converge, neither is subducted because of their great thickness and low and equal densities. As the two continental plates collide, a mountain range is formed in the interior of a new and larger continent. (b) Vertical view of the Himalayas, the highest mountain system in the world. The Himalayas began to form when India collided with Asia 40 to 50 million years ago.

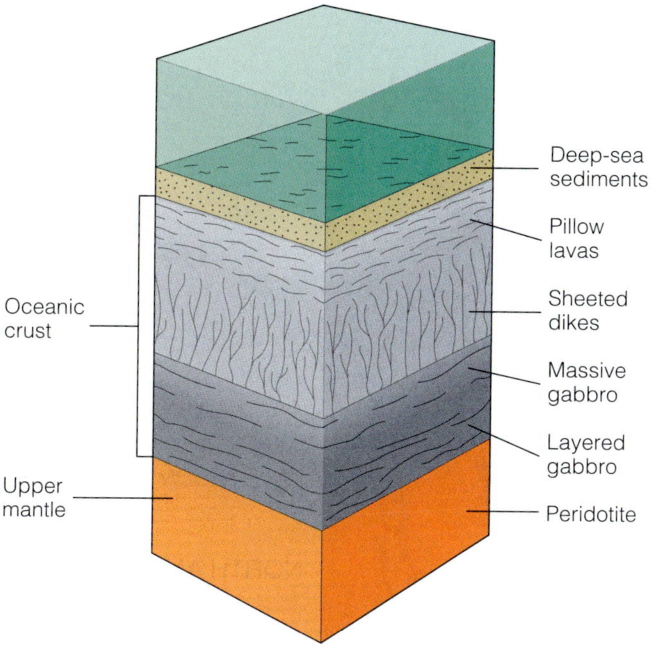

Figure 2.22

Ophiolites are sequences of rock on land consisting of deep-sea sediments, oceanic crust, and upper mantle.

During subduction, pieces of oceanic lithosphere are sometimes incorporated into the mélange and accreted onto the edge of the continent. Such slices of oceanic crust and upper mantle are called *ophiolites* (■ Figure 2.22). They consist of a layer of deep-sea sediments that include graywackes (poorly sorted sandstones containing abundant feldspars and rock fragments, usually in a clay-rich matrix), black shales, and cherts. These deep-sea sediments are underlain by pillow lavas, a sheeted dike complex, massive gabbro, and layered gabbro, all of which form the oceanic crust. Beneath the gabbro is peridotite, which probably represents the upper mantle. Ophiolites are key indicators of plate convergence along a subduction zone.

Elongate belts of folded and faulted marine sedimentary rocks, andesites, and ophiolites are found in the Appalachians, Alps, Himalayas, and Andes Mountains. The combination of such features is good evidence that these mountain ranges resulted from deformation along convergent plate boundaries.

Transform Boundaries

The third type of plate boundary is a **transform plate boundary.** These mostly occur along fractures in the seafloor, known as *transform faults,* where plates slide laterally past one another roughly parallel to the direction of plate movement. Although lithosphere is neither created nor destroyed along a transform boundary, the movement between plates results in a zone of intensely shattered rock and numerous shallow-focus earthquakes.

Transform faults "transform" or change one type of motion between plates into another type of motion. Most commonly, transform faults connect two oceanic ridge segments, but they can also connect ridges to trenches and trenches to trenches (■ Figure 2.23). Although the majority of transform faults are in oceanic crust and are marked by distinct fracture zones, they may also extend into continents.

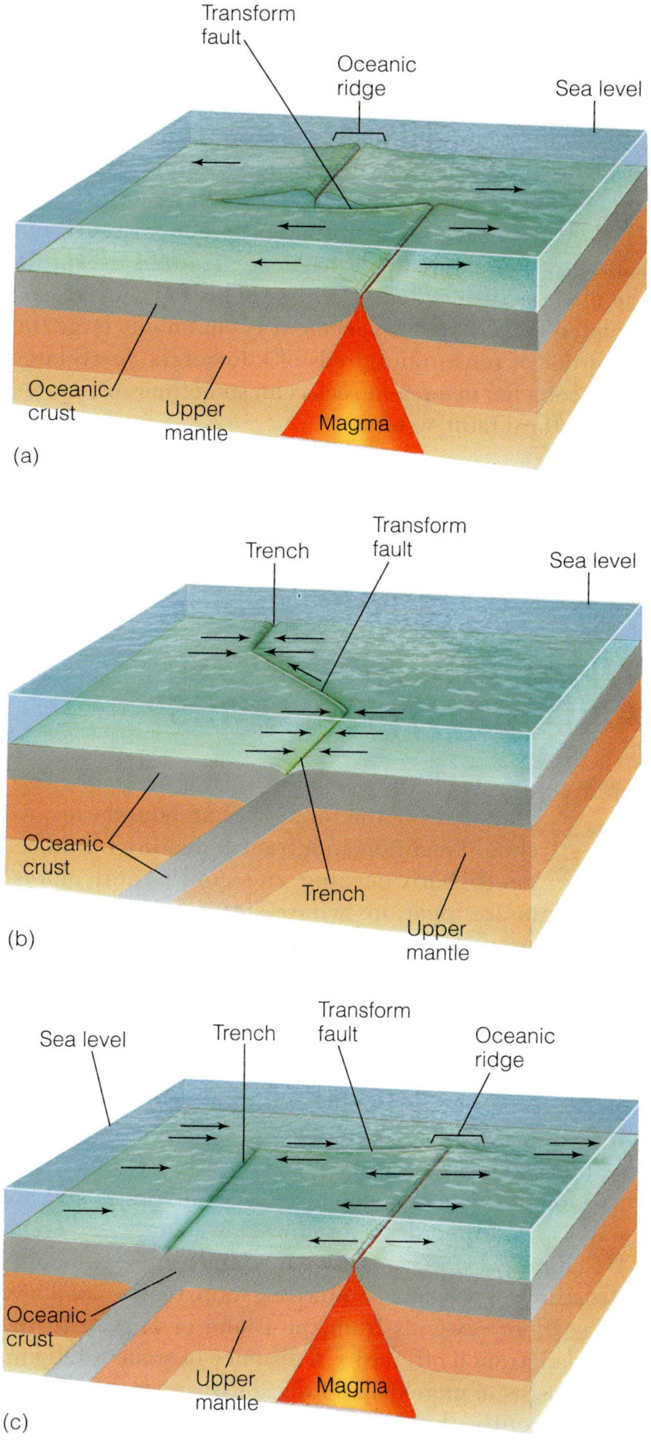

Figure 2.23

Horizontal movement between plates occurs along a transform fault. (a) The majority of transform faults connect two oceanic ridge segments. Note that relative motion between the plates occurs only between the two ridges. (b) A transform fault connecting two trenches. (c) A transform fault connecting a ridge and a trench.

One of the best-known transform faults is the San Andreas fault in California. It separates the Pacific plate from the North American plate and connects spreading ridges in the Gulf of California with the Juan de Fuca and Pacific plates off the coast of northern California (■ Figure 2.24). Many of the earthquakes that affect California are the result of movement along this fault.

Unfortunately, transform faults generally do not leave any characteristic or diagnostic features except for the obvious displacement of the rocks that they are associated with. This displacement is commonly large, on the order of tens to hundreds of kilometers. Such large displacements in ancient rocks can sometimes be related to transform fault systems.

GeologyNow Log into GeologyNow and select this chapter to work through **Geology Interactive** activities on "Plate Boundaries" (click Plate Tectonics→Plate Boundaries) and "Triple Junctions and Seafloor Studies" (click Plate Tectonics→Triple Junctions and Seafloor Studies).

■ **Figure 2.24**

Transform plate boundary. The San Andreas fault is a transform fault separating the Pacific plate from the North American plate. Movement along this fault has caused numerous earthquakes. The photograph shows a segment of the San Andreas fault as it cuts through the Carrizo Plain, California. Source: inset, U.S.G.S.

WHAT ARE HOT SPOTS AND MANTLE PLUMES?

Before leaving the topic of plate boundaries, we should mention an intraplate feature found beneath both oceanic and continental plates. **Hot spots** are locations where stationary columns of magma, originating deep within the mantle (*mantle plumes*), slowly rise to the surface and form volcanoes (Figure 2.16). Because the mantle plumes apparently remain stationary (although some evidence suggests that they might not) while the plates move over them, the resulting hot spots leave a trail of extinct, progressively older volcanoes called *aseismic ridges* that record the movement of the plate.

One of the best examples of aseismic ridges and hot spots is the Emperor Seamount–Hawaiian Island chain (■ Figure 2.25). This chain of islands and seamounts (structures of volcanic origin rising more than 1 km above the seafloor) extends from the island of Hawaii to the Aleutian Trench off Alaska, a distance of some 6000 km, and consists of more than 80 volcanic structures.

Currently, the only active volcanoes in this island chain are the island of Hawaii and the Loihi Seamount. The rest of the islands are extinct volcanic structures that become progressively older toward the north and northwest. This means that the Emperor Seamount–Hawaiian Island chain records the direction that the Pacific plate traveled as it moved over an apparently stationary mantle plume. In this case, the Pacific plate first moved in a north-northwesterly direction and then, as indicated by the sharp bend in the chain, changed to a west-northwesterly direction about 43 million years ago. The reason the Pacific plate changed directions is not known, but the shift might

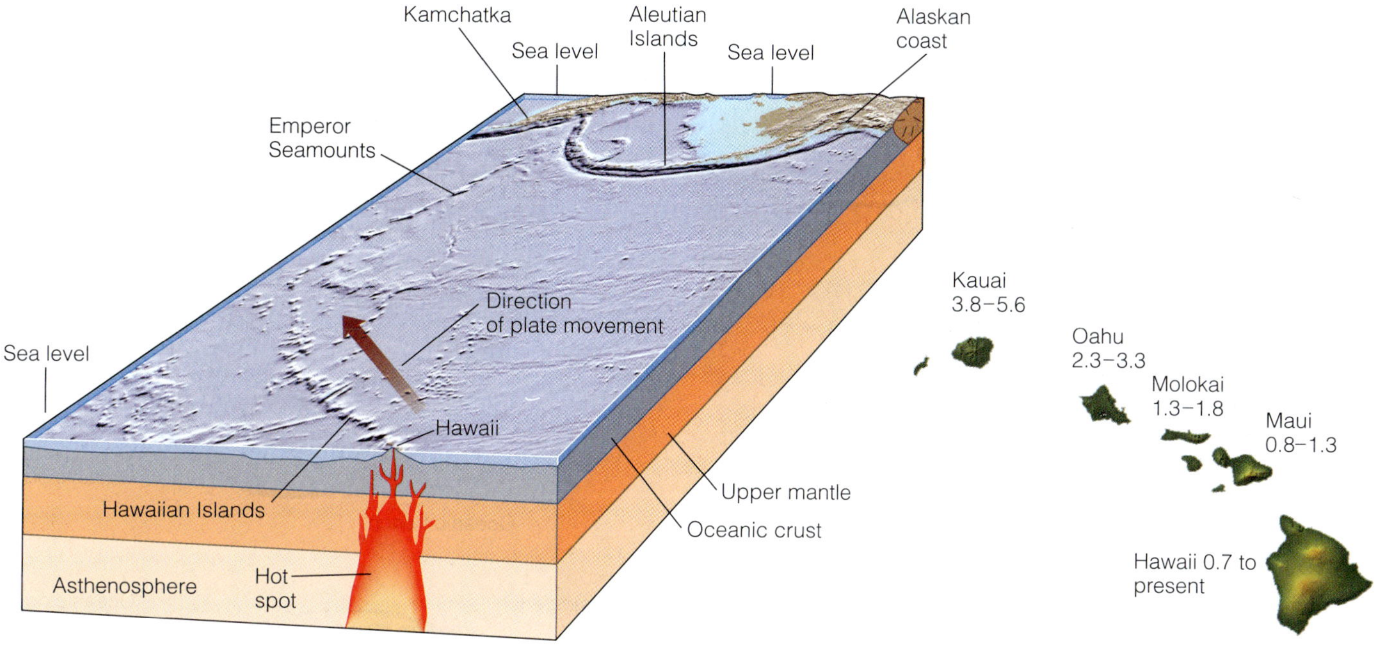

Geology Now ■ **Active Figure 2.25**

The Emperor Seamount–Hawaiian Island chain formed as a result of movement of the Pacific plate over a hot spot. The line of the volcanic islands traces the direction of plate movement. The numbers indicate the ages of the islands in millions of years.

be related to the collision of India with the Asian continent at around the same time (see Figure 10.23).

HOW ARE PLATE MOVEMENT AND MOTION DETERMINED?

How fast and in what direction are Earth's plates moving? Do they all move at the same rate? Rates of plate movement can be calculated in several ways. The least accurate method is to determine the age of the sediments immediately above any portion of the oceanic crust and divide that age by the distance from the spreading ridge. Such calculations give an average rate of movement.

A more accurate method of determining both the average rate of movement and relative motion is by dating the magnetic anomalies in the crust of the seafloor. The distance from an oceanic ridge axis to any magnetic anomaly indicates the width of new seafloor that formed during that time interval. Thus, for a given interval of time, the wider the strip of seafloor, the faster the plate has moved. In this way not only can the present average rate of movement and relative motion be determined (Figure 2.16), but the average rate of movement during the past can also be calculated by dividing the distance between anomalies by the amount of time elapsed between anomalies.

Geologists not only calculate the average rate of plate movement from magnetic anomalies but also use them to determine plate positions at various times in the past using magnetic anomalies. Because magnetic anomalies are parallel and symmetric with respect to spreading ridges, all one must do to determine the position of continents when particular anomalies formed is to move the anomalies back to the spreading ridge, which will also move the continents with them (■ Figure 2.26). Unfortunately, subduction destroys oceanic crust and the magnetic record it carries. Thus, we have an excellent record of plate movements since the breakup of Pangaea, but not as good an understanding of plate movement before that time.

The average rate of movement as well as the relative motion between any two plates can also be determined by satellite-laser ranging techniques. Laser beams from a station on one plate are bounced off a satellite (in geosynchronous orbit) and returned to a station on a different plate. As the plates move away from each other, the laser beam takes more time to go from the sending station to the stationary satellite and back to the receiving station. This difference in elapsed time is used to calculate the rate of movement and relative motion between plates.

Plate motions derived from magnetic reversals and satellite-laser ranging techniques give only the relative motion of one plate with respect to another. Hot spots allow geologists to determine absolute motion because they provide an apparently fixed reference from which the

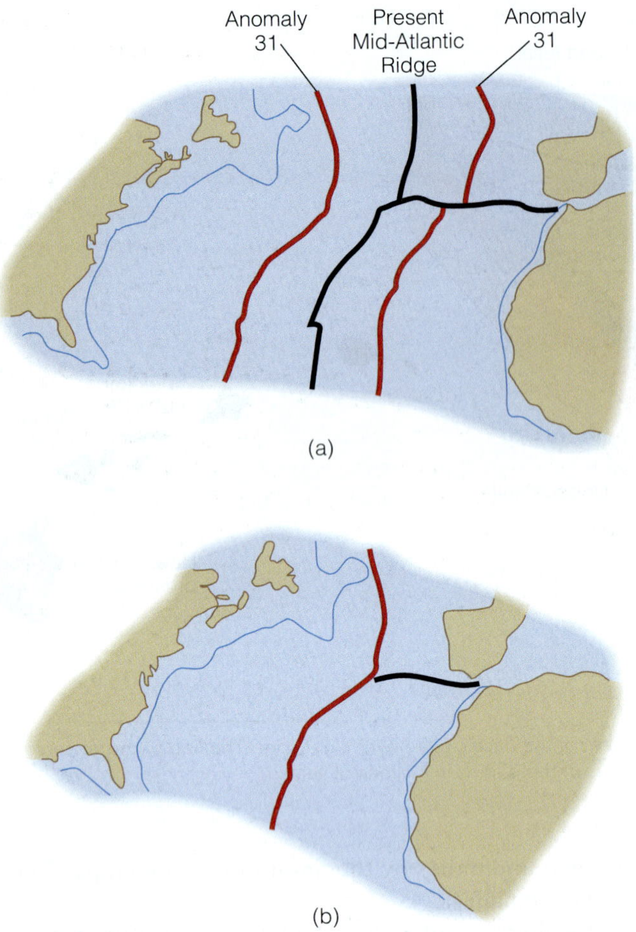

■ **Figure 2.26**

Reconstructing plate positions using magnetic anomalies. (a) The present North Atlantic, showing the Mid-Atlantic Ridge and magnetic anomaly 31, which formed 67 million years ago. (b) The Atlantic 67 million years ago. Anomaly 31 marks the plate boundary 67 million years ago. By moving the anomalies back together, along with the plates they are on, we reconstruct the former positions of the continents.

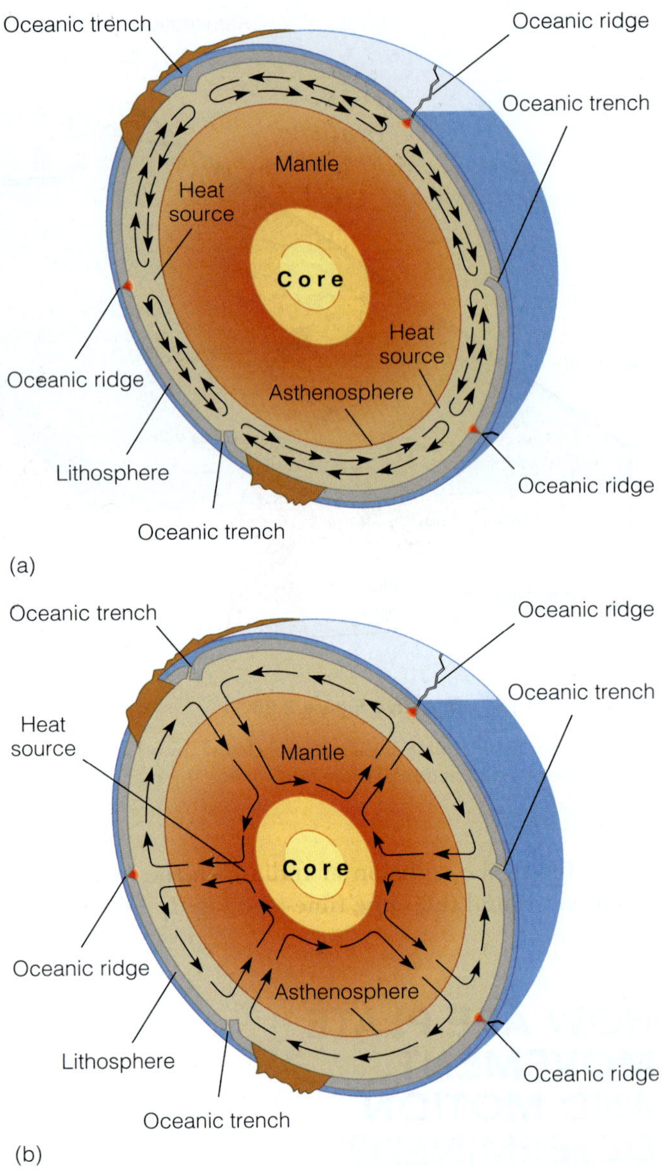

GeologyNow ■ **Active Figure 2.27**

Two models involving thermal convection cells have been proposed to explain plate movement. (a) In one model, thermal convection cells are restricted to the asthenosphere. (b) In the other model, thermal convection cells involve the entire mantle.

rate and direction of plate movement can be measured. The previously mentioned Emperor Seamount–Hawaiian Island chain formed as a result of movement over a hot spot. Thus, the line of the volcanic islands traces the direction of plate movement, and dating the volcanoes enables geologists to determine the rate of movement.

WHAT IS THE DRIVING MECHANISM OF PLATE TECTONICS?

A major obstacle to the acceptance of continental drift was the lack of a driving mechanism to explain continental movement. When it was shown that continents and ocean floors moved together, not separately, and that new crust formed at spreading ridges by rising magma, most geologists accepted some type of convective heat system as the basic process responsible for plate motion. The question still remains, however: What exactly drives the plates?

Two models involving thermal convection cells have been proposed to explain plate movement (■ Figure 2.27). In one model, thermal convection cells are restricted to the asthenosphere; in the second model, the entire mantle is involved. In both models, spreading ridges mark the ascending limbs of adjacent convection cells, and trenches are present where convection cells descend back into Earth's

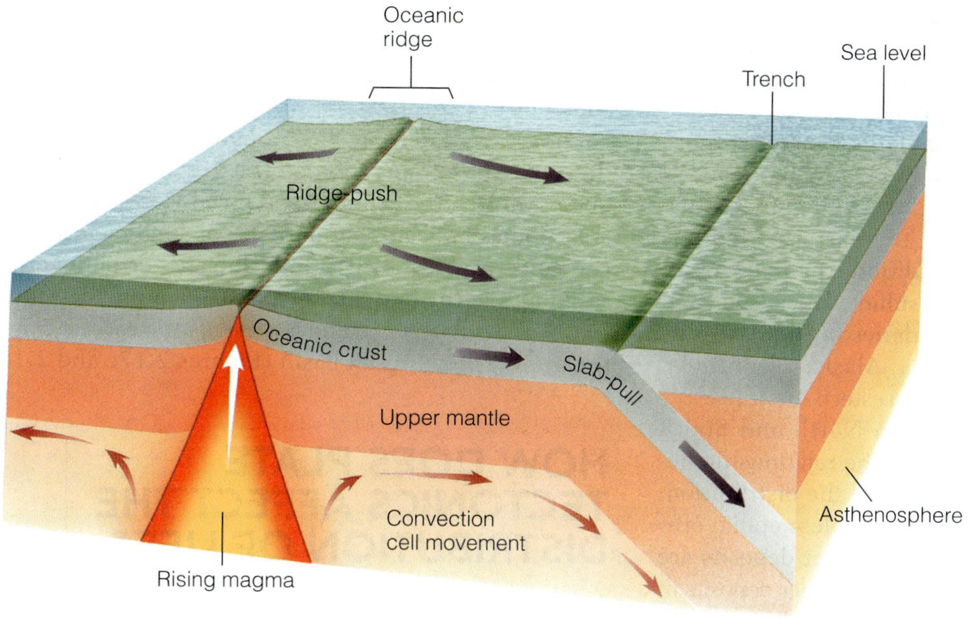

Figure 2.28

Plate movement is also thought to occur because of gravity-driven "slab-pull" or "ridge-push" mechanisms. In slab-pull, the edge of the subducting plate descends into the interior, and the rest of the plate is pulled downward. In ridge-push, rising magma pushes the oceanic ridges higher than the rest of the oceanic crust. Gravity thus pushes the oceanic lithosphere away from the ridges and toward the trenches.

interior. The locations of spreading ridges and trenches are therefore determined by the convection cells themselves, and the lithosphere lies above the thermal convection cell. Each plate thus corresponds to a single convection cell.

Although most geologists agree that Earth's internal heat plays an important role in plate movement, there are problems with both models. The major problem associated with the first model is the difficulty in explaining the source of heat for the convection cells and why they are restricted to the asthenosphere. In the second model, the source of heat comes from the outer core, but it is still not known how heat is transferred from the outer core to the mantle. Nor is it clear how convection can involve both the lower mantle and the asthenosphere.

In addition to some type of thermal convection system driving plate movement, some geologists think plate movement occurs because of a mechanism involving "slab-pull" or "ridge-push," both of which are gravity-driven but still depend on thermal differences within Earth (■ Figure 2. 28). In slab-pull, the subducting cold slab of lithosphere, being denser than the surrounding warmer asthenosphere, pulls the rest of the plate along as it descends into the asthenosphere. As the lithosphere moves downward, there is a corresponding upward flow back into the spreading ridge.

Operating in conjuction with slab-pull is the ridge-push mechanism. As a result of rising magma, the oceanic ridges are higher than the surrounding oceanic crust. It is thought that gravity pushes the oceanic lithosphere away from the higher spreading ridges and toward the trenches.

Currently, geologists are fairly certain that some type of convective system is involved in plate movement, but the extent to which other mechanisms, such as ridge-push and slab-pull, are involved is still unresolved. However, the fact that plates have moved in the past and are still moving today has been proven beyond a doubt. And although a comprehensive theory of plate movement has not yet been developed, more and more of the pieces are falling into place as geologists learn more about Earth's interior.

HOW DOES PLATE TECTONICS AFFECT THE DISTRIBUTION OF NATURAL RESOURCES?

Besides being responsible for the major features of Earth's crust and influencing the distribution and evolution of the world's biota, plate movements also affect the formation and distribution of some natural resources. Consequently, geologists are using plate tectonic theory in their search for petroleum (see Geo-Focus 2.1) and mineral deposits and in explaining the occurrence of these natural resources.

It is becoming increasingly clear that if we are to keep up with the continuing demands of a global industrialized society, the application of plate tectonic theory to the origin and distribution of natural resources is essential.

Mineral Deposits

Many metallic mineral deposits such as copper, gold, lead, silver, tin, and zinc are related to igneous and associated hydrothermal (hot water) activity, so it is not surprising that a close relationship exists between plate boundaries and the occurrence of these valuable deposits.

The magma generated by partial melting of a subducting plate rises toward the surface, and as it cools, it precipitates and concentrates various metallic ores. Many of the world's major metallic ore deposits are associated with convergent plate boundaries including those in the Andes of South America, the Coast Ranges and Rockies of North America, Japan, the Philippines, Russia, and a zone extending from the eastern Mediterranean region to Pakistan. In addition, the majority of the world's gold is associated with sulfide deposits located at ancient convergent plate boundaries in such areas as South Africa, Canada, California, Alaska, Venezuela, Brazil, southern India, Russia, and western Australia.

The copper deposits of western North and South America are an excellent example of the relationship between convergent plate boundaries and the distribution, concentration, and exploitation of valuable metallic ores (■ Figure 2.29). The world's largest copper deposits are found along this belt. The majority of the copper deposits in the Andes and the southwestern United States were formed less than 60 million years ago when oceanic plates were subducted under the North and South American plates. The rising magma and associated hydrothermal fluids carried minute amounts of copper, which was originally widely disseminated but eventually became concentrated in the cracks and fractures of the surrounding andesites. These low-grade copper deposits contain from 0.2% to 2% copper and are extracted from large open-pit mines (Figure 2.29b).

Divergent plate boundaries also yield valuable resources. The island of Cyprus in the Mediterranean is rich in copper and has been supplying all or part of the world's needs for the last 3000 years. The concentration of copper on Cyprus formed as a result of precipitation adjacent to hydrothermal vents along a divergent plate boundary. This deposit was brought to the surface when the copper-rich seafloor collided with the European plate, warping the seafloor and forming Cyprus.

Studies indicate that minerals of such metals as copper, gold, iron, lead, silver, and zinc are currently forming as sulfides in the Red Sea. The Red Sea is opening as a result of plate divergence and represents the earliest stage in the growth of an ocean basin (Figures 2.17c and 2.18a).

HOW DOES PLATE TECTONICS AFFECT THE DISTRIBUTION OF LIFE?

Plate tectonic theory is as revolutionary and far-reaching in its implications for geology as the theory of evolution was for biology when it was proposed. Interestingly, it was the fossil evidence that convinced Wegener, Suess, and du Toit, as well as many other geologists, of the correctness of continental drift. Together, the theories of plate tectonics and evolution have changed the way we view our planet, and we should not be surprised at the intimate association between them. Although the relationship between plate

(a)

R. V. Dietrich

(b)

■ **Figure 2.29**

(a) Important copper deposits are located along the west coasts of North and South America. (b) Bingham Mine in Utah is a huge open-pit copper mine with reserves estimated at 1.7 billion tons. More than 400,000 tons of rock are removed each day.

tectonic processes and the evolution of life is incredibly complex, paleontological data provide convincing evidence of the influence of plate movement on the distribution of organisms.

The present distribution of plants and animals is not random but is controlled largely by climate and geographic barriers. The world's biota occupy *biotic provinces,* regions characterized by distinctive assemblages of plants and animals. Organisms within a province have similar ecological requirements, and the boundaries that separate provinces are therefore natural ecological breaks. Climatic or geographic barriers are the most common province boundaries, and these are largely controlled by plate movements.

Because adjacent provinces usually have less than 20% of their species in common, global diversity is a direct reflection of the number of provinces; the more provinces there are, the greater the global diversity. When continents break up, for example, the opportunity for new provinces to form increases, with a resultant increase in diversity. Just the opposite occurs when continents come together. Plate tectonics thus plays an important role in the distribution of organisms and their evolutionary history.

Complex interactions of wind and ocean currents have a strong influence on the world's climates. These currents are influenced by the number, distribution, topography, and orientation of continents. For example, the southern Andes Mountains act as an effective barrier to moist, easterly blowing Pacific winds, resulting in a desert east of the southern Andes that is virtually uninhabitable. Temperature is one of the major limiting factors for organisms, and province boundaries often reflect temperature barriers. Because atmospheric and oceanic temperatures decrease from the equator to the poles, most species exhibit a strong climatic zonation. This biotic zonation parallels the world's latitudinal atmospheric and oceanic circulation patterns. Changes in climate thus have a profound effect on the distribution and evolution of organisms.

The distribution of continents and ocean basins not only influences wind and ocean currents, but also affects provinciality by creating physical barriers to, or pathways for, the migration of organisms. Intraplate volcanoes, island arcs, mid-oceanic ridges, mountain ranges, and subduction zones all result from the interaction of plates, and their orientation and distribution strongly influence the number of provinces and hence total global diversity. Thus, provinciality and diversity will be highest when there are numerous small continents spread across many zones of latitude.

When a geographic barrier separates a once-uniform fauna, species may undergo divergence. If conditions on opposite sides of the barrier are sufficiently different, then species must adapt to the new conditions, migrate, or become extinct. Adaptation to the new environment by various species may involve enough change that new species eventually evolve. The marine invertebrates found on opposite sides of the Isthmus of Panama provide an excellent example of divergence caused by the formation of a geographic barrier. Prior to the rise of this land connection between North and South America, a homogeneous population of bottom-dwelling invertebrates inhabited the shallow seas of the area. After the rise of the Isthmus of Panama by subduction of the Pacific plate about 5 million years ago, the original population was divided. In response to the changing environment, new species evolved on opposite sides of the isthmus (■ Figure 2.30).

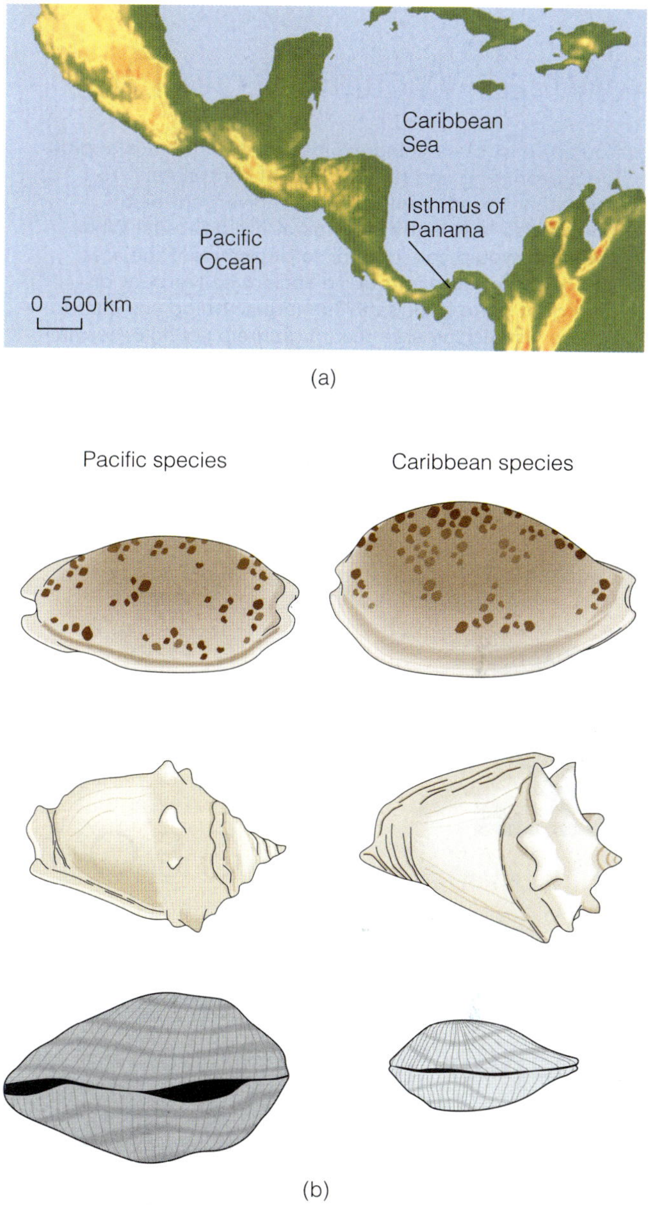

■ **Figure 2.30**

(a) The Isthmus of Panama forms a barrier that divided a once-uniform fauna. (b) Divergence of molluscan species after the formation of the Isthmus of Panama. Each pair belongs to the same genus but is a different species.

What Would You Do?

You are part of a mining exploration team that is exploring a promising and remote area of central Asia. You know that former convergent and divergent plate boundaries frequently are sites of ore deposits. What evidence would you look for to determine if the area you're exploring might be an ancient convergent or divergent plate boundary? Is there anything you can do before visiting the area that might help you in determining what the geology of the area is?

The formation of the Isthmus of Panama also influenced the evolution of the North and South American mammalian faunas. During most of the Cenozoic Era, South America was an island continent, and its mammalian fauna evolved in isolation from the rest of the world's faunas. When North and South America were connected by the Isthmus of Panama, most of the indigenous South American mammals were replaced by migrants from North America. Surprisingly, only a few South American mammal groups migrated northward.

2 GEO RECAP

Chapter Summary

- The concept of continental movement is not new. The earliest maps showing the similarity between the east coast of South America and the west coast of Africa provided the first evidence that continents may once have been united and subsequently separated from each other.
- Alfred Wegener is generally credited with developing the hypothesis of continental drift. He provided abundant geologic and paleontologic evidence to show that the continents were once united in one supercontinent he named Pangaea. Unfortunately, Wegener could not explain how the continents moved, and most geologists ignored his ideas.
- The hypothesis of continental drift was revived during the 1950s when paleomagnetic studies of rocks indicated the presence of multiple magnetic north poles instead of just one as there is today. This paradox was resolved by constructing a hypothetical map and moving the continents into different positions, making the paleomagnetic data consistent with a single magnetic north pole.
- Magnetic surveys of the oceanic crust revealed magnetic anomalies in the rocks, indicating that Earth's magnetic field has reversed itself in the past. Because the anomalies are parallel and form symmetric belts adjacent to oceanic ridges, new oceanic crust must have formed as the seafloor was spreading.
- Seafloor spreading has been confirmed by dating the sediments overlying the oceanic crust and by radiometric dating of rocks on oceanic islands. Such dating reveals that the oceanic crust becomes older with distance from spreading ridges.
- Plate tectonic theory became widely accepted by the 1970s because of the overwhelming evidence supporting it and because it provides geologists with a powerful theory for explaining such phenomena as volcanism, earthquake activity, mountain building,

global climatic changes, the distribution of the world's biota, and the distribution of mineral resources.

- The supercontinent cycle indicates that all or most of Earth's landmasses form, break up, and re-form in a cycle spanning about 500 million years.
- Three types of plate boundaries are recognized: divergent boundaries, where plates move away from each other; convergent boundaries, where two plates collide; and transform boundaries, where two plates slide past each other.
- Ancient plate boundaries can be recognized by their associated rock assemblages and geologic structures. For divergent boundaries, these may include rift valleys with thick sedimentary sequences and numerous dikes and sills. For convergent boundaries, ophiolites and andesitic rocks are two characteristic features. Transform faults generally do not leave any characteristic or diagnostic features in the geologic record.
- The average rate of movement and relative motion of plates can be calculated in several ways. The results of these different methods all agree and indicate that the plates move at different average velocities.
- Absolute motion of plates can be determined by the movement of plates over mantle plumes. A mantle plume is an apparently stationary column of magma that rises to the surface where it becomes a hot spot and forms a volcano.
- Although a comprehensive theory of plate movement has yet to be developed, geologists think that some type of convective heat system is the major driving force.
- A close relationship exists between the formation of some mineral deposits and petroleum, and plate boundaries. Furthermore, the formation and distribution of some natural resources are related to plate movements.
- The relationship between plate tectonic processes and the evolution of life is complex. The distribution of plants and animals is not random, but is controlled largely by climate and geographic barriers, which are controlled, to a large extent, by the movement of plates.

Important Terms

continental–continental plate boundary (p. 45)
continental drift (p. 30)
convergent plate boundary (p. 44)
Curie point (p. 35)
divergent plate boundary (p. 42)
***Glossopteris* flora** (p. 28)
Gondwana (p. 28)
hot spot (p. 48)
Laurasia (p. 30)
magnetic anomaly (p. 36)
magnetic reversal (p. 35)
oceanic–continental plate boundary (p. 45)
oceanic–oceanic plate boundary (p. 44)
paleomagnetism (p. 34)
Pangaea (p. 30)
plate tectonic theory (p. 37)
seafloor spreading (p. 36)
thermal convection cell (p. 36)
transform fault (p. 47)
transform plate boundary (p. 47)

Review Questions

1. The man credited with developing the continental drift hypothesis is
   a._____Wilson; b._____Wegener; c._____Hess; d._____du Toit; e._____Vine.
2. The name of the supercontinent that formed at the end of the Paleozoic Era is
   a._____Laurasia; b._____Gondwana; c._____Panthalassa; d._____Atlantis; e._____Pangaea.
3. Hot spots and aseismic ridges can be used to determine
   a _____the location of divergent plate boundaries; b._____the absolute motion of plates; c._____the location of magnetic anomalies in oceanic crust; d._____the relative motion of plates; e._____the location of convergent plate boundaries.

4. Subduction occurs along what type of boundary?
a._____divergent; b._____transform; c._____convergent; d._____answers a and b; e._____answers a and c.
5. The driving mechanism of plate movement is thought to be
a._____isostasy; b._____Earth's rotation; c._____thermal convection cells; d._____magnetism; e._____polar wandering.
6. The San Andreas fault is an example of what type of plate boundary?
a._____divergent; b._____convergent; c._____transform; d._____oceanic–continental; e._____continental–continental.
7. The most common biotic province boundaries are
a._____geographic barriers; b._____biologic barriers; c._____climatic barriers; d._____answers a and b; e._____answers a and c.
8. Magnetic surveys of the ocean basins indicate that
a._____the oceanic crust is youngest adjacent to mid-oceanic ridges; b._____the oceanic crust is oldest adjacent to mid-oceanic ridges; c._____the oceanic crust is youngest adjacent to the continents; d._____the oceanic crust is the same age everywhere; e._____answers b and c.
9. Convergent plate boundaries are zones where
a._____new continental lithosphere is forming; b._____new oceanic lithosphere is forming; c._____two plates come together; d._____two plates slide past each other; e._____two plates move away from each other.
10. Iron-bearing minerals in magma gain their magnetism and align themselves with the magnetic field when they cool through the
a._____Curie point; b._____magnetic anomaly point; c._____thermal convection point; d._____hot spot point; e._____isostatic point.
11. Using the age for each of the Hawaiian Islands in Figure 2.25 and an atlas in which you can measure the distance between islands, calculate the average rate of movement per year for the Pacific plate since each island formed. Is the average rate of movement the same for each island? Would you expect it to be? Explain why it may not be.

12. What evidence convinced Wegener that the continents were once joined together and subsequently broke apart?
13. Estimate the age of seafloor crust and the age and thickness of the oldest sediment off the East Coast of the United States (e.g., Virginia). In so doing, refer to Figure 2.14 for the ages and to the deep-sea sediment accumulation rate stated in this chapter.
14. How have plate tectonic processes affected the formation and distribution of natural resources?
15. If the movement along the San Andreas fault, which separates the Pacific plate from the North American plate, averages 5.5 cm per year, how long will it take before Los Angeles is opposite San Francisco?
16. Why is plate tectonics the unifying theory of geology?
17. Why is some type of thermal convection system thought to be the major force driving plate movement?
18. Explain how plate tectonics affects the evolution of life.
19. What is the supercontinent cycle? Who proposed this concept, and what kinds of geologic data were needed to support such a concept?
20. Explain why global diversity increases with an increase in the number of biotic provinces. How does plate movement affect the number of biotic provinces?

World Wide Web Activities

GeologyNow Assess your understanding of this chapter's topics with additional quizzing and comprehensive interactivities at

http://earthscience.brookscole.com/changingearth4e

as well as current and up-to-date weblinks, additional readings, and InfoTrac College Edition exercises.

Minerals—the Building Blocks of Rocks

CHAPTER 3

OUTLINE

Geology Now *This icon, which appears throughout the book, indicates an opportunity to explore interactive tutorials, animations, or practice problems available on the GeologyNow website at http://earthscience.brookscole.com/changingearth4e.*

OBJECTIVES

At the end of this chapter, you will have learned that:

- All matter, including minerals, is made up of atoms that bond to form elements and compounds.
- Geologists have a specific definition for the term *mineral*.
- You can distinguish minerals from other naturally occurring and manufactured substances.
- Minerals are incredibly varied, yet only a few are particularly common.
- Geologists use physical properties such as color, hardness, and density to identify minerals.
- Minerals originate in various ways and under varied conditions.
- Some minerals, designated rock-forming minerals, are particularly common in rocks, whereas others are found in small amounts.
- Some minerals and rocks are important natural resources that are essential to industrialized societies.

These black pearls valued at about $13,000 are on display at Maui Pearls in the town of Avarua on the island of Rarotonga, which is part of the Cook Islands in the South Pacific. Pearls are composed mostly of the mineral aragonite. Unlike other gemstones, pearls are essentially ready to use when found. Source: Sue Monroe

Introduction

We commonly use the term *mineral* for the dietary substances we need for good nutrition, such as calcium, iron, and magnesium, but these are actually chemical elements, not minerals, at least in the geologic sense. We also know that minerals are inorganic; however, not all inorganic substances are minerals. Water and water vapor are inorganic, but neither is a mineral—yet ice is a mineral. Thus minerals are solids as opposed to liquids or gases. So, *minerals* are naturally occurring inorganic solids that are further characterized as *crystalline,* meaning that their atoms are arranged in a specific way. Glass, in contrast, has no such ordered internal structure. Minerals also have a narrowly defined chemical composition and characteristic physical properties such as color, hardness, and density. We will examine all parts of this rather lengthy definition later in this chapter.

We cannot overstate the importance of minerals in many human endeavors. The ore deposits we rely on to sustain our industrialized societies are natural concentrations of minerals and rocks. Iron ore, industrial minerals for abrasives, glass, and cement, as well as minerals and rocks needed for animal feed supplements and fertilizers are essential to our economic well-being. The United States and Canada owe much of their economic success to the availability of abundant natural resources, although they must import some important commodities, thus accounting for political and economic ties with other nations.

One important reason to study minerals is that they are the building blocks of rocks, so rocks, with few exceptions, are combinations of one or more minerals. Granite, for instance, is made up of specified percentages of minerals known as quartz and feldspars along with other minerals in minor quantities. In several of the following chapters, we will have more to say about the importance of minerals in identifying and classifying rocks.

Some minerals are attractive and eagerly sought by private collectors and for museum displays (■ Figure 3.1a). Other

(a)

(b)

(c)

■ **Figure 3.1**

(a) Tourmaline and quartz (colorless) from the Himalaya Mine, San Diego County, California. (b) The diamond pendant in this necklace, housed in the Smithsonian Institution, is the 68-carat Victoria Transvaal diamond from South Africa. (c) Even though amber is an organic substance, it is nevertheless valued as a semiprecious gemstone.

minerals are known as *gemstones*—that is, precious or semiprecious minerals or rocks used for decorative purposes, especially jewelry. The precious gemstones such as diamond (Figure 3.1b), ruby, sapphire, and emerald are most desirable and most expensive. Many people have small precious gemstones and perhaps some semiprecious ones, such as garnet and turquoise. The lore associated with gemstones, such as relating them to one's birth month, makes them even more appealing to many people.

Amber and pearl are included among the semiprecious gemstones, but are they really minerals? Amber is hardened resin (sap) from coniferous trees and thus an organic substance and not a mineral, but it is nevertheless prized as a decorative "stone" (Figure 3.1c). It is best known from the Baltic Sea region of Europe where sun-worshiping cultures, noting its golden translucence resembling the Sun's rays, thought it possessed mystical powers. Pearls form when mollusks, such as clams or oysters, deposit successive layers of tiny mineral crystals around some irritant, perhaps a sand grain. Most pearls are lustrous white, but some are silver gray, green, or black (see the chapter opening photo).

From our discussion so far we have a formal definition of the term *mineral* and we know that minerals are the basic constituents of rocks. Now let's delve deeper into what minerals are made of by considering matter, atoms, elements, and bonding.

MATTER—WHAT IS IT?

Anything that has mass and occupies space is *matter.* Accordingly, water, plants, animals, the atmosphere, and minerals and rocks are composed of matter. Physicists recognize three states or phases of matter: *liquids, gases,* and *solids.** Liquids, such as surface water and groundwater, as well as atmospheric gases are important in our considerations of several surface processes, such as running water and wind, but here our main concern is with solids because by definition minerals are solids.

Atoms and Elements

Matter is made up of chemical **elements,** which in turn are composed of **atoms,** the smallest units of matter that retain the characteristics of a particular element (■ Figure 3.2). That is, elements cannot be changed into different substances except through radioactive decay (discussed in Chapter 17). Thus an element is made up of atoms, all of which have the same properties. Scientists have discovered 92 naturally occurring elements, some of which are listed in Table 3.1, and several others have been made in laboratories (see Appendix B). All naturally occurring elements and most artificial ones have a name and a symbol—for example, oxygen (O), aluminum (Al), and potassium (K).

At the center of an atom is a tiny **nucleus** made up of one or more particles known as **protons,** which have a positive electrical charge, and **neutrons,** which are electrically neutral (Figure 3.2). The nucleus is only about 1/100,000 of the diameter of an atom, yet it contains virtually all of the atom's mass. **Electrons,** particles with a negative electrical charge, orbit rapidly around the nucleus at specific distances in one or more **electron shells.** The electrons determine how an atom interacts with other atoms, but the nucleus determines how many electrons an atom has, because the positively charged protons attract and hold the negatively charged electrons in their orbits.

The number of protons in its nucleus determines an atom's identity and its **atomic number.** Hydrogen (H), for instance, has 1 proton in its nucleus and thus has an atomic number of 1. The nuclei of helium (He) atoms possess 2 protons, whereas those of carbon (C) have 6,

*Actually, scientists recognize a fourth state of matter known as *plasma,* an ionized gas as in fluorescent and neon lights and matter in the Sun and stars.

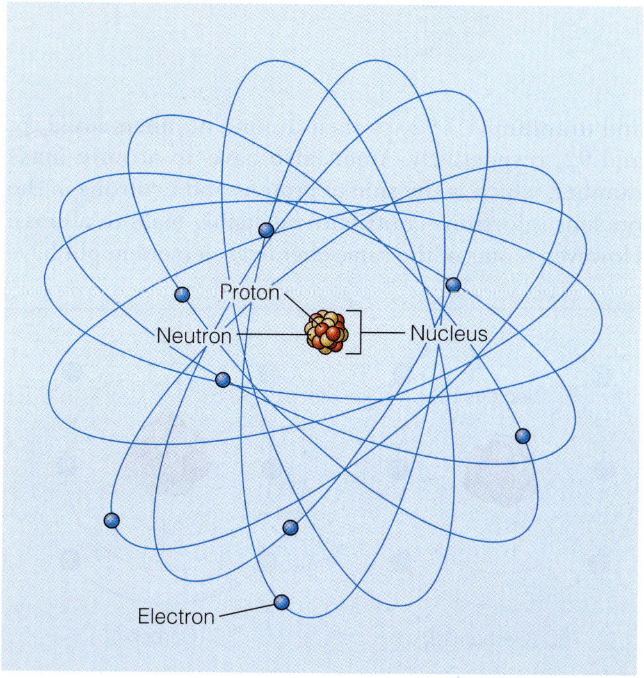

■ **Figure 3.2**

The structure of an atom. The dense nucleus consisting of protons and neutrons is surrounded by a cloud of orbiting electrons.

Table 3.1

Symbols, Atomic Numbers, and Electron Configurations for Some of the Naturally Occurring Elements

Element	Symbol	Atomic Number	Number of Electrons in Each Shell 1	2	3	4
Hydrogen	H	1	1			
Helium	He	2	2			
Lithium	Li	3	2	1		
Beryllium	Be	4	2	2		
Boron	B	5	2	3		
Carbon	C	6	2	4		
Nitrogen	N	7	2	5		
Oxygen	O	8	2	6		
Fluorine	F	9	2	7		
Neon	Ne	10	2	8		
Sodium	Na	11	2	8	1	
Magnesium	Mg	12	2	8	2	
Aluminum	Al	13	2	8	3	
Silicon	Si	14	2	8	4	
Phosphorus	P	15	2	8	5	
Sulfur	S	16	2	8	6	
Chlorine	Cl	17	2	8	7	
Argon	Ar	18	2	8	8	
Potassium	K	19	2	8	8	1
Calcium	Ca	20	2	8	8	2

and uranium (U) 92, so their atomic numbers are 2, 6, and 92, respectively. Atoms also have an **atomic mass number,** which is the sum of protons and neutrons in the nucleus (electrons contribute negligible mass to atoms). However, atoms of the same chemical element might have different atomic mass numbers because the number of neutrons can vary. All carbon (C) atoms have 6 protons—otherwise they would not be carbon—but the number of neutrons can be 12, 13, or 14. Thus we recognize three types of carbon, or what are known as *isotopes* (■ Figure 3.3), each with a different atomic mass number.

The isotopes of carbon, or those of any other element, behave the same chemically; carbon 12 and carbon 14 are both present in carbon dioxide (CO_2), for example. However, some isotopes are radioactive, meaning that they spontaneously decay or change to other elements. Carbon 14 is radioactive, whereas both carbon 12 and carbon 13 are not. Radioactive isotopes are

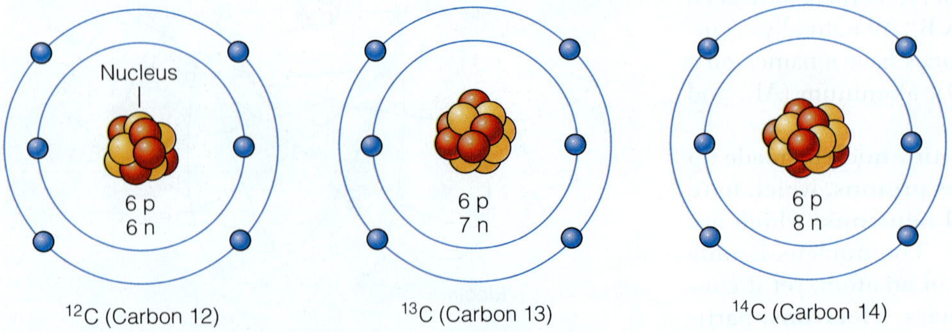

■ **Figure 3.3**

Schematic representation of the isotopes of carbon. Carbon has an atomic number of 6 and an atomic mass number of 12, 13, or 14, depending on the number of neutrons in its nucleus.

important for determining the absolute ages of rocks (see Chapter 17).

Bonding and Compounds

Interactions among electrons around atoms can result in two or more atoms joining together, a process known as **bonding.** If atoms of two or more elements bond, the resulting substance is a **compound.** Gaseous oxygen consists of only oxygen atoms and is thus an element, whereas the mineral quartz, consisting of silicon and oxygen atoms, is a compound. Most minerals are compounds, although gold, platinum, and several others are important exceptions.

To understand bonding, it is necessary to delve deeper into the structure of atoms. Recall that negatively charged electrons orbit the nuclei of atoms in electron shells. With the exception of hydrogen, which has only one proton and one electron, the innermost electron shell of an atom contains only two electrons. The other shells contain various numbers of electrons, but the outermost shell never has more than eight (Table 3.1). The electrons in the outermost shell are those that are usually involved in chemical bonding.

Two types of chemical bonds, *ionic* and *covalent,* are particularly important in minerals, and many minerals contain both types of bonds. Two other types of chemical bonds, *metallic* and *van der Waals,* are much less common but extremely important in determining the properties of some useful minerals.

Ionic Bonding

Notice in Table 3.1 that most atoms have fewer than eight electrons in their outermost electron shell. However, some elements, including neon and argon, have complete outer shells with eight electrons; because of this electron configuration, these elements, known as the *noble gases,* do not react readily with other elements to form compounds. Interactions among atoms tend to produce electron configurations similar to those of the noble gases. That is, atoms interact so that their outermost electron shell is filled with eight electrons, unless the first shell (with two electrons) is also the outermost electron shell, as in helium.

One way that the noble gas configuration is attained is by the transfer of one or more electrons from one atom to another. Common salt is composed of the elements sodium (Na) and chlorine (Cl); each element is poisonous, but when combined chemically they form the compound sodium chloride (NaCl), better known as the mineral halite. Notice in ■ Figure 3.4a that sodium has 11 protons and 11 electrons; thus the positive electrical charges of the protons are exactly balanced by the negative charges of the electrons, and the atom is electrically neutral. Likewise, chlorine with 17 protons and 17 electrons is electrically neutral (Figure 3.4a). But neither sodium nor chlorine has 8 electrons in its outermost electron shell; sodium has only 1, whereas chlorine has 7. To attain a stable configuration, sodium loses the electron in its outermost electron shell, leaving its next shell with 8 electrons as the outermost one (Figure 3.4a). Sodium now has one fewer electron (negative charge)

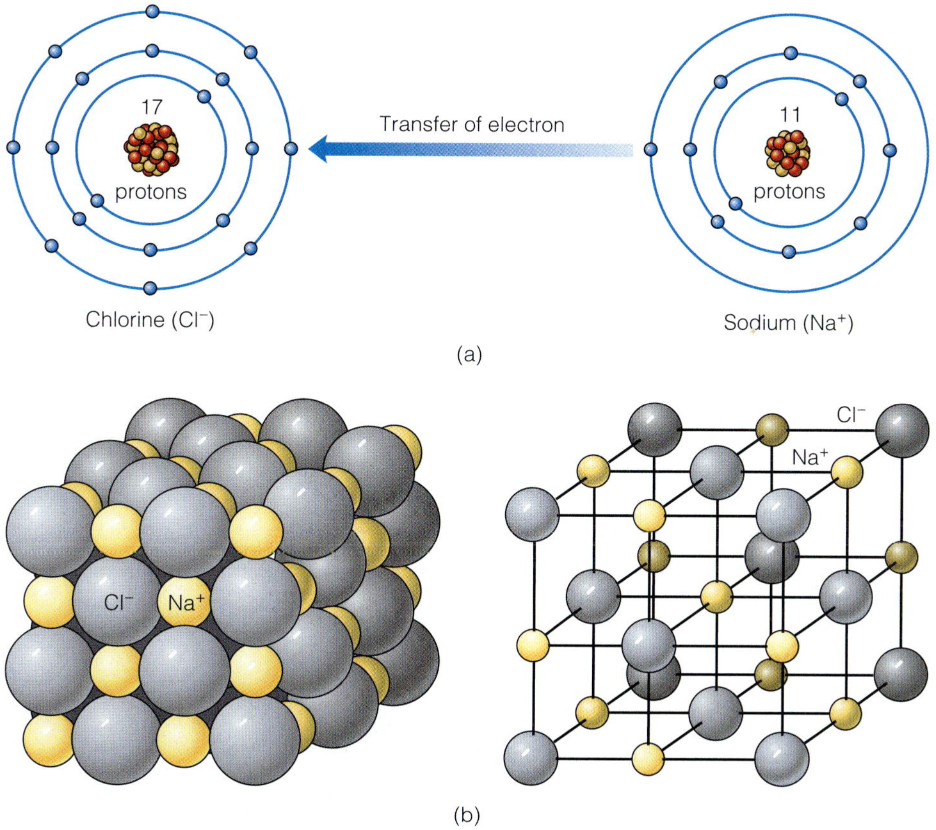

■ **Figure 3.4**

(a) Ionic bonding. The electron in the outermost shell of sodium is transferred to the outermost electron shell of chlorine. Once the transfer has occurred, sodium and chlorine are positively and negatively charged ions, respectively. (b) The crystal structure of sodium chloride, the mineral halite. The diagram on the left shows the relative sizes of sodium and chlorine ions, and the diagram on the right shows the locations of the ions in the crystal structure.

than it has protons (positive charge), so it is an electrically charged **ion** and is symbolized Na^+.

The electron lost by sodium is transferred to the outermost electron shell of chlorine, which had 7 electrons to begin with. The addition of one more electron gives chlorine an outermost electron shell of 8 electrons, the configuration of a noble gas. But its total number of electrons is now 18, which exceeds by 1 the number of protons. Accordingly, chlorine also becomes an ion, but it is negatively charged (Cl^-). An **ionic bond** forms between sodium and chlorine because of the attractive force between the positively charged sodium ion and the negatively charged chlorine ion (Figure 3.4a).

In ionic compounds, such as sodium chloride (the mineral halite), the ions are arranged in a three-dimensional framework that results in overall electrical neutrality. In halite, sodium ions are bonded to chlorine ions on all sides, and chlorine ions are surrounded by sodium ions (Figure 3.4b).

Covalent Bonding **Covalent bonds** form between atoms when their electron shells overlap and they share electrons. For example, atoms of the same element, such as carbon, cannot bond by transferring electrons from one atom to another. Carbon (C), which forms the minerals graphite and diamond, has four electrons in its outermost electron shell (■ Figure 3.5a). If these four electrons were transferred to another carbon atom, the atom receiving the electrons would have the noble gas configuration of eight electrons in its outermost electron shell, but the atom contributing the electrons would not.

In such situations, adjacent atoms share electrons by overlapping their electron shells. A carbon atom in diamond, for instance, shares all four of its outermost electrons with a neighbor to produce a stable noble gas configuration (Figure 3.5a).

Covalent bonds are not restricted to substances composed of atoms of a single kind. Among the most common minerals, the silicates (discussed later in this chapter), the element silicon forms partly covalent and partly ionic bonds with oxygen.

Metallic and van der Waals Bonds *Metallic bonding* results from an extreme type of electron sharing. The electrons of the outermost electron shell of metals such as gold, silver, and copper readily move about from one atom to another. This electron mobility accounts for the fact that metals have a metallic luster (their appearance in reflected light), provide good electrical and thermal conductivity, and can be easily reshaped. Only a few minerals possess metallic bonds, but those that do are very useful; copper, for example, is used for electrical wiring because of its high electrical conductivity.

Some electrically neutral atoms and molecules* have no electrons available for ionic, covalent, or metallic bonding. They nevertheless have a weak attractive force between them, called a *van der Waals* or *residual bond,* when in proximity. The carbon atoms in the mineral graphite are covalently bonded to form sheets, but the sheets are weakly held together by van der Waals bonds (Figure 3.5b). This type of bonding makes graphite useful for pencil leads; when a pencil is moved across a piece of paper, small pieces of graphite flake off along the planes held together by van der Waals bonds and adhere to the paper.

GeologyNow Log into GeologyNow and select this chapter to work through a **Geology Interactive** activity on "Atomic Behavior" (click Atoms and the Crystals→Atomic Behavior).

*A molecule is the smallest unit of a substance that has the properties of that substance. A water molecule (H_2O), for example, possesses two hydrogen atoms and one oxygen atom.

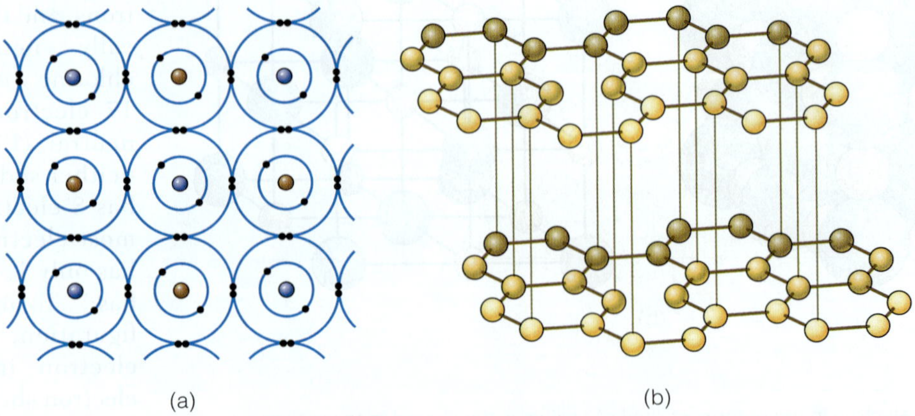

■ **Figure 3.5**

(a) Covalent bonds formed by adjacent atoms sharing electrons in diamond. (b) Covalent bonding also occurs in graphite, but here the carbon atoms are bonded together to form sheets that are held to one another by van der Waals bonds. The sheets themselves are strong, but the bonds between sheets are weak.

WHAT ARE MINERALS?

We defined a **mineral** as an inorganic, naturally occurring, crystalline solid with a narrowly defined chemical composition and characteristic physical properties. Furthermore, we know from the preceding section that most minerals are compounds of two or more chemically bonded elements as in quartz (SiO_2). In the following sections, we will examine each part of the formal definition of the term *mineral*.

Naturally Occurring Inorganic Substances

The criterion *naturally occurring* excludes from minerals all substances manufactured by humans, such as synthetic diamonds and rubies. This criterion is particularly important to those who buy and sell gemstones, most of which are minerals, because some human-made substances are very difficult to distinguish from natural gem minerals.

Some geologists think the term *inorganic* in the mineral definition is unnecessary. It does remind us that animal matter and vegetable matter are not minerals. Nevertheless, some organisms, including corals, clams, and a number of other animals and plants, construct their shells of the compound calcium carbonate ($CaCO_3$), which is either the mineral aragonite or calcite, or their shells are made of silicon dioxide (SiO_2) as in quartz.

Mineral Crystals

By definition minerals are **crystalline solids,** in which the constituent atoms are arranged in a regular, three-dimensional framework (Figure 3.4b). Under ideal conditions, such as in a cavity, mineral crystals can grow and form perfect crystals that possess planar surfaces (crystal faces), sharp corners, and straight edges (■ Figure 3.6). In other words, the regular geometric shape of a well-formed mineral crystal is the exterior manifestation of an ordered internal atomic arrangement. Not all rigid substances are crystalline solids; natural and manufactured glass lack the ordered arrangement of atoms and is said to be *amorphous*, meaning "without form."

Crystalline refers to a solid with a regular three-dimensional arrangement of atoms, whereas a **crystal** is a geometric shape with planar faces, sharp corners, and straight edges. Thus a crystal is the external manifestation of a crystalline structure. Not all crystalline solids yield well-formed crystals, however, because when many crystals form and grow adjacent to one another, they form an interlocking mosaic in which individual crystals are not apparent (■ Figure 3.7a, b). So how do we know that the mineral in Figure 3.7b is actually crystalline? X-ray beams and light transmitted through mineral crystals or crystalline solids behave in a predictable manner, which provides compelling evidence for an internal orderly structure.

Another way we can determine that minerals with no obvious crystals are actually crystalline is by their **cleavage,** the property of breaking or splitting repeatedly along smooth, closely spaced planes. Not all minerals have cleavage planes, but many do, and such regularity certainly indicates that splitting is controlled by internal structure.

As early as 1669, the Danish scientist Nicholas Steno determined that the angles of intersection of equivalent crystal faces on different specimens of quartz are identical. Since then, this *constancy of interfacial angles* has been demonstrated for many other minerals, regardless of their size, shape, age, or geographic occurrence (Figure 3.7c). Steno postulated that mineral crystals are made up of very small, identical building blocks, and that the arrangement of these building blocks determines the external form of mineral crystals, a proposal that has since been verified.

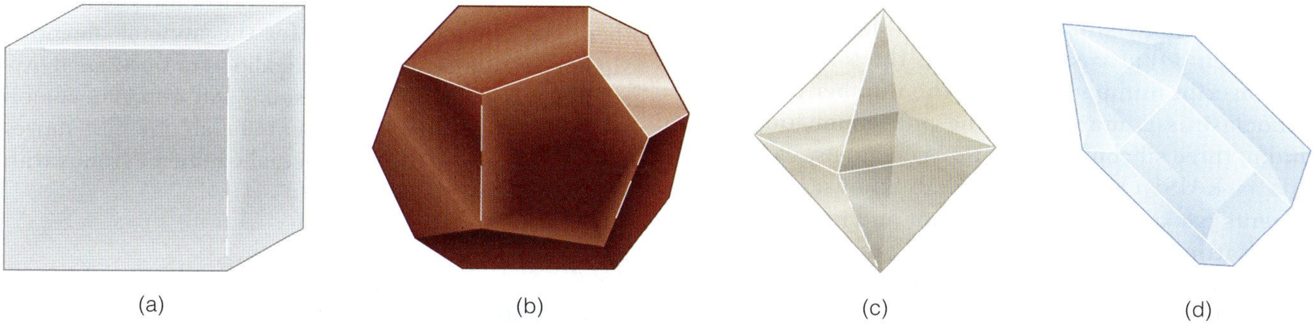

■ **Figure 3.6**

Mineral crystals occur in a variety of shapes. (a) Cubic crystals typically develop in the minerals halite and galena. (b) This crystal with 12 five-sided faces is a pyritohedron found in the mineral pyrite. (c) Diamond has octahedral or eight-sided crystals. (d) A prism terminated by pyramids is found in quartz.

■ **Figure 3.7**

(a) A well-shaped crystal of smoky quartz. (b) A specimen of rose quartz in which crystals cannot be seen. (c) Side views and cross sections of quartz crystals showing the constancy of interfacial angles. A well-shaped crystal (left), a larger well-shaped crystal (middle), and a poorly shaped crystal (right). The angles between equivalent crystal faces on different specimens of the same mineral are the same regardless of the size, age, shape, or geographic occurrence of the specimens. Source: (a) and (b), Sue Monroe.

Chemical Composition of Minerals

Mineral composition is shown by a chemical formula, which is a shorthand way of indicating the numbers of atoms of different elements that make up a mineral. The mineral quartz consists of one silicon (Si) atom for every two oxygen (O) atoms and thus has the formula SiO_2; the subscript number indicates the number of atoms. Orthoclase is composed of one potassium, one aluminum, three silicon, and eight oxygen atoms, so its formula is $KAlSi_3O_8$. Some minerals known as native elements consist of a single element and include silver (Ag), platinum (Pt), gold, (Au), and graphite and diamond, both of which are composed of carbon (C).

The definition of a mineral contains the phrase *a narrowly defined chemical composition* because some minerals actually have a range of compositions. For many minerals, the chemical composition does not vary. Quartz is composed of only silicon and oxygen (SiO_2), and halite contains only sodium and chlorine (NaCl). Other minerals have a range of compositions because one element can substitute for another if the atoms of two or more elements are nearly the same size and the same charge. Notice in ■ Figure 3.8 that iron and magnesium atoms are about the same size; therefore they can substitute for each other. The chemical formula for the mineral olivine is $(Mg,Fe)_2SiO_4$, meaning that, in addition to silicon and oxygen, it may contain only magnesium, only iron, or a combination of both. A number of other minerals also have ranges of compositions, so these are actually mineral groups with several members.

Physical Properties of Minerals

The last criterion in our definition of a mineral, *characteristic physical properties,* refers to such properties as hardness, color, and crystal form. These properties are controlled by composition and structure. We will have more to say about the physical properties of minerals later in this chapter.

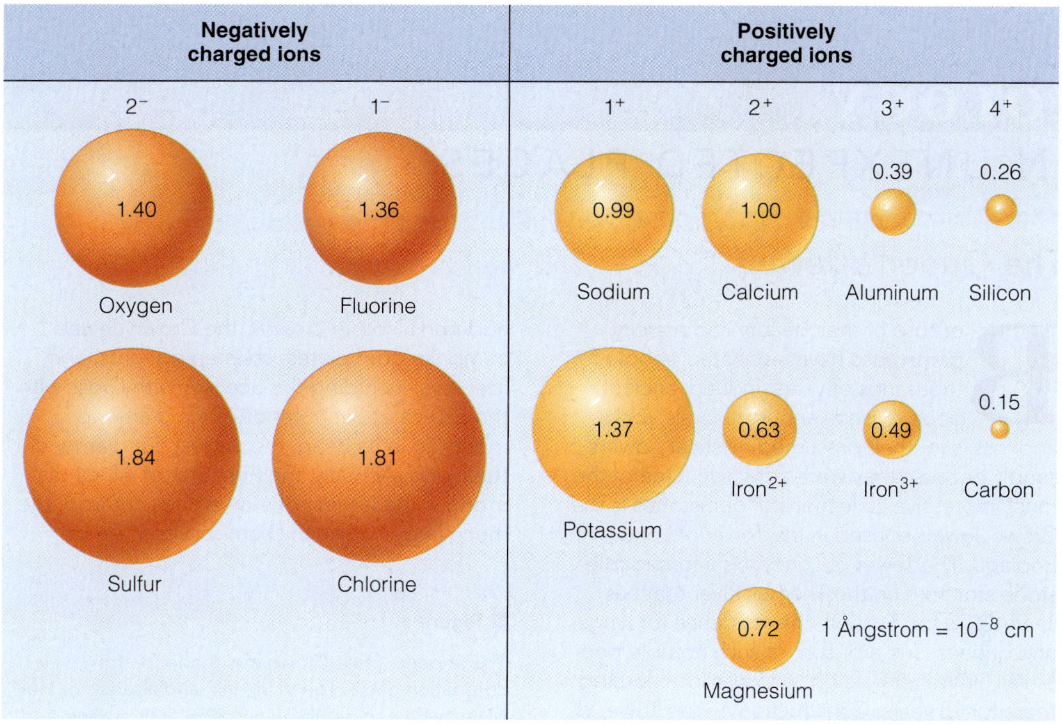

Figure 3.8

Electrical charges and relative sizes of ions common in minerals. The numbers within the ions are the radii shown in Angstrom units.

HOW MANY MINERALS ARE THERE?

Geologists have identified and described more than 3500 minerals, but only a few—perhaps two dozen—are common. One might think that an extremely large number of minerals could form from 92 naturally occurring elements, but several factors limit the number possible. For one thing, many combinations of elements simply do not occur; no compounds are composed of only potassium and sodium or of silicon and iron, for example. Another important factor is that the bulk of Earth's crust is made up of only eight chemical elements, and even among these eight, silicon and oxygen are by far the most common. In fact, most common minerals in the crust consist of silicon, oxygen, and one or more of the elements in ■ Figure 3.9.

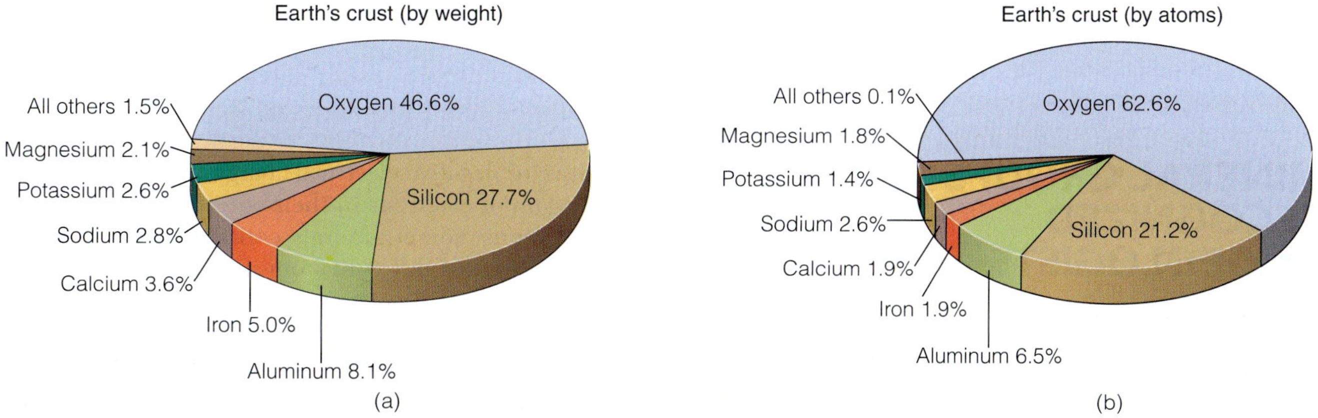

Figure 3.9

Common elements in Earth's crust. (a) Percentage of crust by weight, and (b) percentage of crust by atoms. Source: (a) From Miller, G. T., 1996. *Living in the Environment: Principles, Concepts, and Solutions.* Wadsworth Publishing. Figure 8.3.

GEOLOGY IN UNEXPECTED PLACES

The Queen's Jewels

Because of their beauty and scarcity, gemstones have fascinated people for thousands of years. Indeed, ancient people used various minerals, rocks, and fossils for their presumed mystical powers or simply because they were attractive. One of the most impressive collections of gemstones is the Crown Jewels housed in the Tower of London in England. The Tower of London is a formidable stone structure on the Thames River that has served as a fortification, the residence for kings and queens, and a prison for such notable people as Sir Walter Raleigh, who was incarcerated there for 13 years. Construction on the Tower of London began during the reign of William the Conqueror (1066–1087). It was enlarged and modified until about 1300, and since then it has remained much the same.

Within the Tower the Waterloo Barracks originally built for 1000 soldiers have housed the British Crown Jewels since the beginning of the 14th century. Only during World War II (1939–1945) were the Crown Jewels removed to a secret location for safekeeping, and later were returned to the Waterloo Barracks. Among the Crown Jewels is the crown made for the coronation of George VI in 1937 and later modified for Queen Elizabeth II in 1953. It is set with 2868 diamonds, 17 sapphires, 11 emeralds, 5 rubies, and 273 pearls (■ Figure 1). In addition to other crowns, the Crown Jewels comprise gold plates, christening fonts, and scepters, including the Scepter with Cross with the 530-carat First Star of Africa diamond mounted in its head, the largest cut diamond in the world. Actually, the First Star of Africa diamond is the largest of nine stones cut from the much larger Cullinan Diamond from Africa.

■ **Figure 1.**

The Imperial State Crown was made for the coronation of George VI in 1937 and altered for Her Majesty Queen Elizabeth II in 1953. Source: PhotoDisc Green/Getty Images.

MINERAL GROUPS RECOGNIZED BY GEOLOGISTS

Geologists recognize mineral classes or groups, each with members that share the same negatively charged ion or ion group (Table 3.2). We have mentioned that ions are atoms that have either a positive or negative electrical charge resulting from the loss or gain of electrons in their outermost shell. In addition to ions, some minerals contain tightly bonded, complex groups of different atoms known as *radicals* that act as single units. A good example is the carbonate radical, consisting of a carbon atom bonded to three oxygen atoms and thus having the formula CO_3 and a −2 electrical charge. Other common radicals and their charges are sulfate (SO_4, −2), hydroxyl (OH, −1), and silicate (SiO_4, −4) (■ Figure 3.10).

Table 3.2

Mineral Groups Recognized by Geologists

Mineral Group	Negatively Charged Ion or Radical	Examples	Composition
Carbonate	$(CO_3)^{-2}$	Calcite	$CaCO_3$
		Dolomite	$CaMg(CO_3)_2$
Halide	Cl^{-1}, F^{-1}	Halite	NaCl
		Fluorite	CaF_2
Hydroxide	$(OH)^{-1}$	Brucite	$Mg(OH)_2$
Native element	—	Gold	Au
		Silver	Ag*
		Diamond	C
Phosphate	$(PO_4)^{-3}$	Apatite	$Ca_5(PO_4)_3(F,Cl)$
Oxide	O^{-2}	Hematite	Fe_2O_3
		Magnetite	Fe_3O_4
Silicate	$(SiO_4)^{-4}$	Quartz	SiO_2
		Potassium feldspar	$KAlSi_3O_8$
		Olivine	$(Mg,Fe)_2SiO_4$
Sulfate	$(SO_4)^{-2}$	Anhydrite	$CaSO_4$
		Gypsum	$CaSO_4{\cdot}2H_2O$
Sulfide	S^{-2}	Galena	PbS
		Pyrite	FeS_2
		Argentite	Ag_2S*

*Note that silver is found as both a native element and a sulfide mineral.

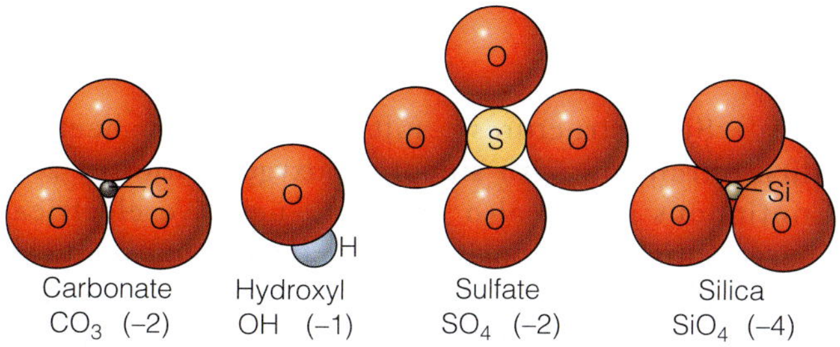

Figure 3.10

Many minerals contain radicals, which are complex groups of atoms tightly bonded together. The silica and carbonate radicals are particularly common in many minerals, such as quartz (SiO_2) and calcite ($CaCO_3$).

Silicate Minerals

Because silicon and oxygen are the two most abundant elements in Earth's crust, it is not surprising that many minerals contain these elements. A combination of silicon and oxygen is known as **silica,** and minerals that contain silica are **silicates.** Quartz (SiO_2) is pure silica because it is composed entirely of silicon and oxygen. But most silicates have one or more additional elements, as in orthoclase ($KAlSi_3O_8$) and olivine [$(Mg,Fe)_2SiO_4$]. Silicate minerals include about one-third of all known minerals, but their abundance is even more impressive when one considers that they make up perhaps 95% of Earth's crust.

The basic building block of all silicate minerals is the **silica tetrahedron,** consisting of one silicon atom and four oxygen atoms (Figure 3.11a). These atoms are arranged so that the four oxygen atoms surround a silicon atom, which occupies the space between the oxygen atoms, thus forming a four-faced pyramidal structure. The silicon atom has a positive charge of 4, and each of the four oxygen atoms has a negative charge of 2, resulting in a radical with a total negative charge of 4 $(SiO_4)^{-4}$.

Because the silica tetrahedron has a negative charge, it does not exist in nature as an isolated ion group; rather, it combines with positively charged ions or shares its oxygen atoms with other silica tetrahedra. In the simplest silicate minerals, the silica tetrahedra exist as single units bonded to positively charged ions. In minerals that contain isolated tetrahedra, the silicon-to-oxygen ratio is 1:4, and the negative charge of the silica ion is balanced by positive ions (Figure 3.11b). Olivine

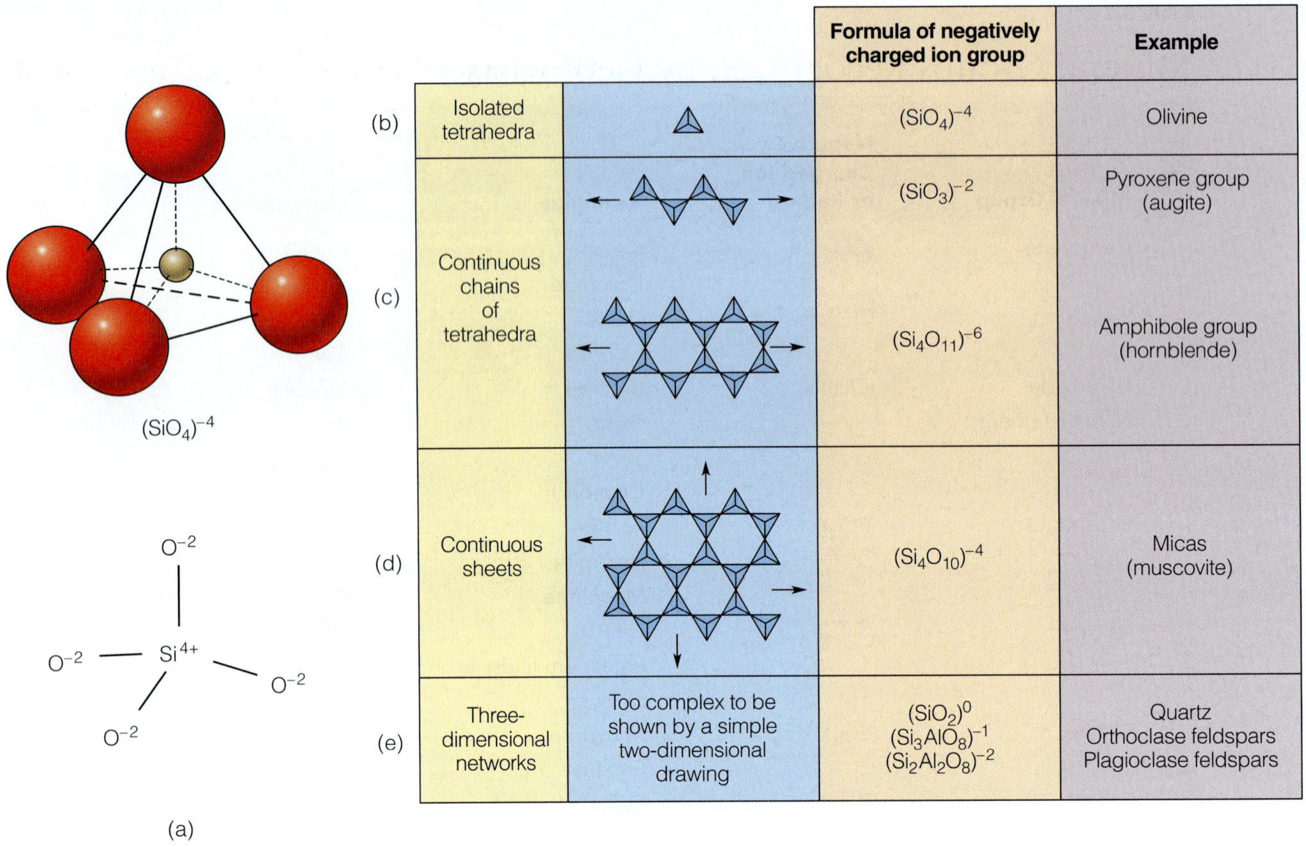

Geology Now ■ **Active Figure 3.11**

(a) Model of the silica tetrahedron, showing the unsatisfied negative charges at each oxygen. (b)–(e) Structures of the common silicate minerals shown by various arrangements of the silica tetrahedra. (b) Isolated tetrahedra. (c) Continuous chains. (d) Continuous sheets. (e) Networks. The arrows adjacent to single-chain, double-chain, and sheet silicates indicate that these structures continue indefinitely in the directions shown.

$[(Mg,Fe)_2SiO_4]$, for example, has either two magnesium (Mg^{+2}) ions, two iron (Fe^{+2}) ions, or one of each to offset the -4 charge of the silica ion.

Silica tetrahedra may also join together to form chains of indefinite length (Figure 3.11c). Single chains, as in the pyroxene minerals, form when each tetrahedron shares two of its oxygens with an adjacent tetrahedron, resulting in a silicon-to-oxygen ratio of 1:3. Enstatite, a pyroxene-group mineral, reflects this ratio in its chemical formula $MgSiO_3$. Individual chains, however, possess a net -2 electrical charge, so they are balanced by positive ions, such as Mg^{+2}, that link parallel chains together (Figure 3.11c).

The amphibole group of minerals is characterized by a double-chain structure in which alternate tetrahedra in two parallel rows are cross-linked (Figure 3.11c). The formation of double chains results in a silicon-to-oxygen ratio of 4:11, so each double chain possesses a -6 electrical charge. Mg^{+2}, Fe^{+2}, and Al^{+2} are usually involved in linking the double chains together.

In sheet structure silicates, three oxygens of each tetrahedron are shared by adjacent tetrahedra (Figure 3.11d). Such structures result in continuous sheets of silica tetrahedra with silicon-to-oxygen ratios of 2:5. Continuous sheets also possess a negative electrical charge satisfied by positive ions located between the sheets. This particular structure accounts for the characteristic sheet structure of the *micas,* such as biotite and muscovite, and the *clay minerals.*

Three-dimensional networks of silica tetrahedra form when all four oxygens of the silica tetrahedra are shared by adjacent tetrahedra (Figure 3.11e). Such sharing of oxygen atoms results in a silicon-to-oxygen ratio of 1:2, which is electrically neutral. Quartz is a common framework silicate.

Geologists recognize two subgroups of silicates: ferromagnesian and nonferromagnesian silicates. The **ferromagnesian silicates** are those that contain iron (Fe), magnesium (Mg), or both. These minerals are commonly dark and more dense than nonferromagnesian silicates. Some of the common ferromagnesian silicate minerals are olivine, the pyroxenes, the amphiboles, and biotite (■ Figure 3.12a).

The **nonferromagnesian silicates** lack iron and magnesium, are generally light colored, and are less

Figure 3.12

(a) Common ferromagnesian silicates and (b) nonferromagnesian silicates. Source: Sue Monroe

dense than ferromagnesian silicates (Figure 3.12b). The most common minerals in Earth's crust are nonferromagnesian silicates known as *feldspars*. Feldspar is a general name, however, and two distinct groups are recognized, each of which includes several species. The *potassium feldspars* are represented by microcline and orthoclase ($KAlSi_3O_8$). The second group of feldspars, the *plagioclase feldspars,* range from calcium-rich ($CaAl_2Si_2O_8$) to sodium-rich ($NaAlSi_3O_8$) varieties.

Quartz (SiO_2) is another common nonferromagnesian silicate. It is a framework silicate that can usually be recognized by its glassy appearance and hardness. Another fairly common nonferromagnesian silicate is muscovite, which is a mica (Figure 3.12b).

Carbonate Minerals

Carbonate minerals, those that contain the negatively charged carbonate radical $(CO_3)^{-2}$, include calcium carbonate ($CaCO_3$) as the minerals *aragonite* or *calcite* (■ Figure 3.13a). Aragonite is unstable and commonly changes to calcite, the main constituent of the sedimentary rock *limestone.* A number of other carbonate minerals are known, but only one of these need concern us: *Dolomite* $[CaMg(CO_3)_2]$ forms by the chemical alteration of calcite by the addition of magnesium. Sedimentary rock composed of the mineral dolomite is *dolostone* (see Chapter 7).

Other Mineral Groups

In addition to silicates and carbonates, geologists recognize several other mineral groups (Table 3.2). Even though minerals from these groups are less common than silicates and carbonates, many are found in rocks in small quantities and others are important resources. In the oxides, an element combines with oxygen as in hematite (Fe_2O_3) and magnetite (Fe_3O_4). Rocks with high concentrations of these minerals in the Lake Superior region of Canada and the United States are sources of iron ores for the manufacture of steel. The related hydroxides form mostly by the chemical alteration of other minerals.

We have noted that the *native elements* are minerals composed of a single element, such as diamond and graphite (C) and the precious metals gold (Au), silver (Ag), and platinum (Pt) (see "The Precious Metals" on pages 72 and 73). Some elements, such as silver and copper, are found both as native elements and as compounds and are thus also included in other mineral groups; argentite (Ag_2S), a silver sulfide, is an example.

Several minerals and rocks that contain the phosphate radical $(PO_4)^{-3}$ are important sources of phosphorus for fertilizers. The sulfides such as galena (PbS), the ore of lead, have a positively charged ion combined with sulfur (S^{-2}) (Figure 3.13b), whereas the sulfates have an element combined with the complex radical $(SO_4)^{-2}$, as in gypsum ($CaSO_4{\cdot}2H_2O$) (Figure 3.13c). The halides contain the halogen elements, fluorine (F^{-1}) and chlorine (Cl^{-1}); examples are halite (NaCl) (Figure 3.13d) and fluorite (CaF_2).

PHYSICAL PROPERTIES OF MINERALS

Internal structure and chemical composition determine the characteristic physical properties of all minerals. Many physical properties are remarkably constant for a given mineral species, but some, especially color, may vary. Although professional geologists

The Precious Metals

The discovery of gold by James Marshall at Sutter's Mill near Coloma in 1848 sparked the California gold rush (1849–1853) during which $200 million in gold was recovered.

Specimen of gold from Grass Valley, California. Gold is too heavy and too soft for tools and weapons, so it has been prized for jewelry and as a symbol of wealth, but it is also used in glass making, electrical circuitry, gold plating, the chemical industry, and dentistry.

Gold miners on the American River near Sacramento, California. Most of the gold came from placer deposits in which running water separated and concentrated minerals and rock fragments by their density.

A miner pans for gold (foreground) by swirling water, sand, and gravel in a broad, shallow pan. The heavier gold sinks to the bottom. At the far left a miner washes sediment in a cradle. As in panning, the cradle separates heavier gold from other materials.

Hydraulic mining in California in which strong jets of water washed gold-bearing sand and gravel into sluices. In this image taken in 1905 at Junction City, California, water is directed through a monitor onto a hillside. Hydraulic mining was efficient from the mining point of view but caused considerable environmental damage.

Reports in 1876 of gold in the Black Hills of South Dakota resulted in a flood of miners that led to hostilities with the Sioux Indians, and the annihilation of Lt. Col. George Armstrong Custer and 260 of his men at the Battle of the Little Big Horn in Montana. This view shows the headworks (upper right) of the Homestake Mine at Lead, South Dakota in 1900. The headworks is the cluster of buildings near the opening to a mine.

J.C.H. Grabill/Corbis

Like gold, silver is found as a native element as in this specimen, but it also occurs as a compound in the sulfide mineral argentite (Ag_2S). Silver is used in North America for silver halide film, jewelry, flatware, surgical instruments, and backing for mirrors.

James S. Monroe

This image shows the headworks of the Yellowjacket Mine at Gold Hill, Nevada, and the inset shows silver-bearing quartz (white) in volcanic rock. This largest silver discovery in North America, called the Comstock Lode, was responsible for bringing Nevada into the Union in 1864 during the Civil War, even though it had too few people to qualify for statehood. The Comstock Lode was mined for silver and gold from 1859 until 1898.

Sue. Monroe

(a) Calcite (b) Galena

(c) Gypsum (d) Halite

Sue Monroe

■ **Figure 3.13**

(a) Calcite ($CaCO_3$) is the most common carbonate mineral.
(b) The sulfide mineral galena (PbS) is the ore of lead.
(c) Gypsum ($CaSO_4 \cdot 2H_2O$) is a common sulfate mineral.
(d) Halite (NaCl) is a good example of a halide mineral.

use sophisticated techniques to study and identify minerals, most common minerals can be identified by using the physical properties described next (see Appendix C).

Luster and Color

Luster (not to be confused with *color*) is the quality and intensity of light reflected from a mineral's surface. Geologists define two basic types of luster: *metallic,* having the appearance of a metal, and *nonmetallic*. Notice that of the four minerals shown in Figure 3.13, only galena has a metallic luster. Among the several types of nonmetallic luster are glassy or vitreous (as in quartz), dull or earthy, waxy, greasy, and brilliant (as in diamond) (Figure 3.1b).

Beginning geology students are distressed by the fact that the color of some minerals varies considerably, making the most obvious physical property of little use for mineral identification. In any case, we can make some helpful generalizations about color. Ferromagnesian silicates are typically black, brown, or dark green, although olivine is olive green (Figure 3.12a). Nonferromagnesian silicates, on the other hand, vary considerably in color but are rarely very dark. White, cream, colorless, and shades of pink and pale green are more typical (Figure 3.12b).

Another helpful generalization is that the color of minerals with a metallic luster is more consistent than is the color of nonmetallic minerals. For example, galena is always lead-gray (Figure 3.13b) and pyrite is invariably brassy yellow. In contrast, quartz, a nonmetallic mineral, may be colorless, smoky brown to almost black, rose, yellow-brown, milky white, blue, or violet to purple (Figure 3.7a, b).

Crystal Form

As we noted, many mineral specimens do not show the perfect crystal form typical of that mineral species (Figures 3.6 and ■ 3.14). Keep in mind, however, that even though crystals may not be apparent, minerals nevertheless possess a crystalline structure.

Some minerals do typically occur as crystals. For example, 12-sided crystals of garnet are common, as are 6- and 12-sided crystals of pyrite. Minerals that grow in cavities or are precipitated from circulating hot water (hydrothermal solutions) in cracks and crevices in rocks also commonly occur as crystals (see Geo-Focus 3.1).

Crystal form can be a useful characteristic for mineral identification, but a number of minerals have the same crystal form. Pyrite (FeS_2), galena (PbS), and halite (NaCl) all occur as cubic crystals, but they can be easily identified by other properties such as color, luster, hardness, and density.

Cleavage and Fracture

Not all minerals possess cleavage, but those that do break, or split, along a smooth plane or planes of weak-

■ **Figure 3.14**

Mineral crystals. (a) Cubic crystals of fluorite (CaF_2). (b) Calcite ($CaCO_3$) crystal. (c) Garnet ($Fe_3Al_2(SiO_4)_3$) crystals. Source: (a & b) Sue Monroe, (c) Michael Penn/Juneau, Alaska, BLM.

ness determined by the strength of their chemical bonds. Cleavage is characterized in terms of quality (perfect, good, poor), direction, and angles of intersection of cleavage planes. Biotite, a common ferromagnesian silicate has perfect cleavage in one direction (■ Figure 3.15a). Biotite is a sheet silicate with the sheets of silica tetrahedra weakly bonded to one another by iron and magnesium ions (Figure 3.12a).

Feldspars possess two directions of cleavage that intersect at right angles (Figure 3.15b), and the mineral halite has three directions of cleavage, all of which intersect at right angles (Figure 3.15c). Calcite also possesses three directions of cleavage, but none of the intersection angles is a right angle, so cleavage fragments of calcite are rhombohedrons (Figure 3.15d). Minerals with four directions of cleavage include fluorite and diamond (Figure 3.15e). Ironically, diamond, the hardest mineral, can be cleaved easily. A few minerals such as sphalerite, an ore of zinc, have six directions of cleavage (Figure 3.15f).

Cleavage is an important diagnostic property of minerals, and recognizing it is essential in distinguishing between some minerals. The pyroxene mineral augite and the amphibole mineral hornblende, for example, look much alike: Both are dark green to black, have the same hardness, and possess two directions of cleavage. But the cleavage planes of augite intersect at about 90 degrees,

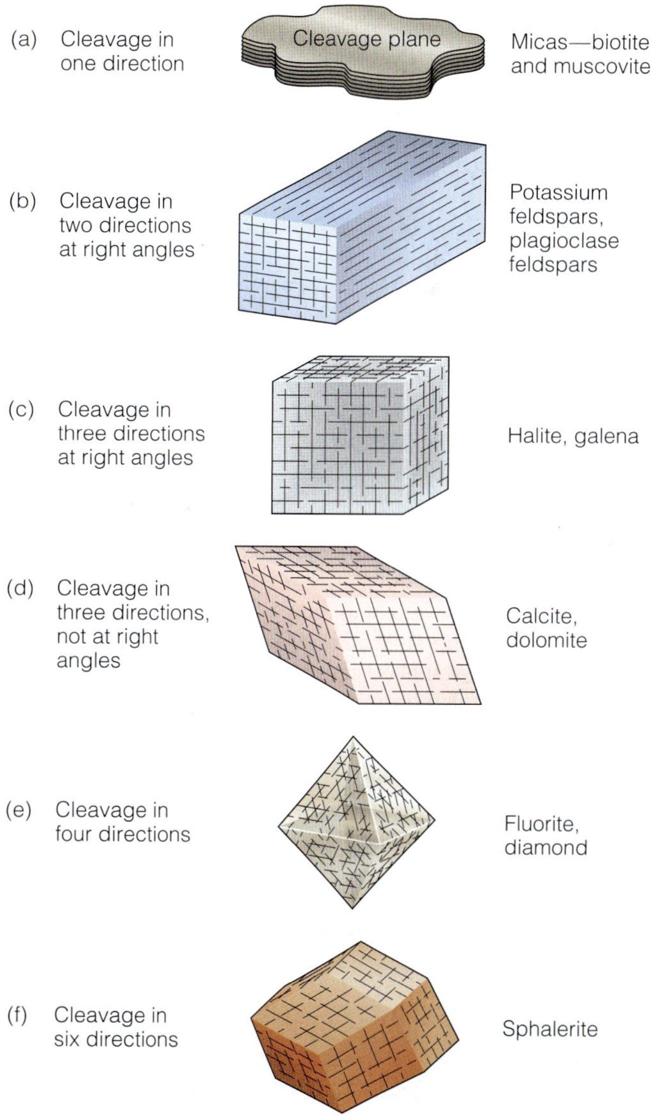

■ **Figure 3.15**

Several types of mineral cleavage. (a) One direction. (b) Two directions at right angles. (c) Three directions at right angles. (d) Three directions, not at right angles. (e) Four directions. (f) Six directions.

GEOFOCUS 3.1

Mineral Crystals

Certainly the most alluring features of minerals are crystals. Most crystals are rather small, measuring a few millimeters to centimeters long, but some reach gigantic proportions. Spodumene crystals up to 14 m long were mined in South Dakota for their lithium content, quartz crystals weighing several metric tons have been found in Russia, and sheets of muscovite measuring more than 2.4 m across come from mines in Ontario. Invariably, such large crystals grow in cavities where their growth is unrestricted, or they are found in *pegmatites*, a type of igneous rock with especially large minerals (see Chapter 4).

The most remarkable recent find of giant crystals occurred in April 2000 in a silver and lead mine in Chihuahua, Mexico. A cavern there is lined with hundreds of gypsum crystals more than 1 m long and what one author called "crystal moonbeams" —gypsum crystals 1.2 m in diameter and up to 15.2 m long (■ Figure 1). Perhaps these are the largest mineral crystals anywhere in the world. Fearing vandalism, the company that owned the mine kept the crystals a secret for some time, but the 65°C temperature and 100% humidity in the crystal-filled cavern would keep out all but the most determined vandals.

For many centuries, crystals and minerals were valued for their alleged healing powers and mystical properties. In fact, many minerals, especially mineral crystals, as well as some rocks and fossils have served as religious symbols and talismans, or have been carried, worn, applied externally, or ingested for their presumed mystical or curative powers. Diamond, according to one legend, wards off evil spirits, sickness, and floods, whereas topaz was thought to prevent mental disorders, and ruby was believed to preserve its owner's health and warn of imminent bad luck. Indeed, even today magazine and tabloid ads tout the healing qualities of various crystals and claim that they enhance emotional stability and clear thinking. Unfortunately for those who purchase crystals for these purposes, they provide no more benefit than artificial ones. Wishful thinking and the placebo effect are responsible for any perceived beneficial results.

One reason some people think crystals have favorable attributes is the curious property called the piezoelectric effect. When some crystals are compressed or an electrical current is applied, these minerals produce an electrical charge that enables them to be accurate timekeepers. For example, the electrical current from a watch's battery causes a quartz crystal to expand and contract very rapidly and regularly (about 100,000 times per second). Quartz crystal clocks were first developed in 1928, and now quartz clocks and watches are commonplace. Even inexpensive quartz timepieces are very accurate, and precision-manufactured quartz clocks used in astronomy do not gain or lose more than 1 second in 10 years.

An interesting historical note is that during World War II (1939–1945) the United States had difficulty obtaining Brazilian quartz crystals needed for making radios. This shortage prompted the development of artificially synthesized quartz, and now most quartz used in watches and clocks is synthetic. Even though the piezoelectric effect imparts no healing or protective powers to crystals, it is essential in applications in which precise measurements of time, pressure, or acceleration are needed. And, of course, many people are intrigued by crystals simply because they are so attractive.

Richard D. Fisher

■ **Figure 1**

Some of these gypsum crystals in a cavern in Chihuahua, Mexico, measure up to 15.2 m long and may be the world's largest crystals. They were discovered in April 2000.

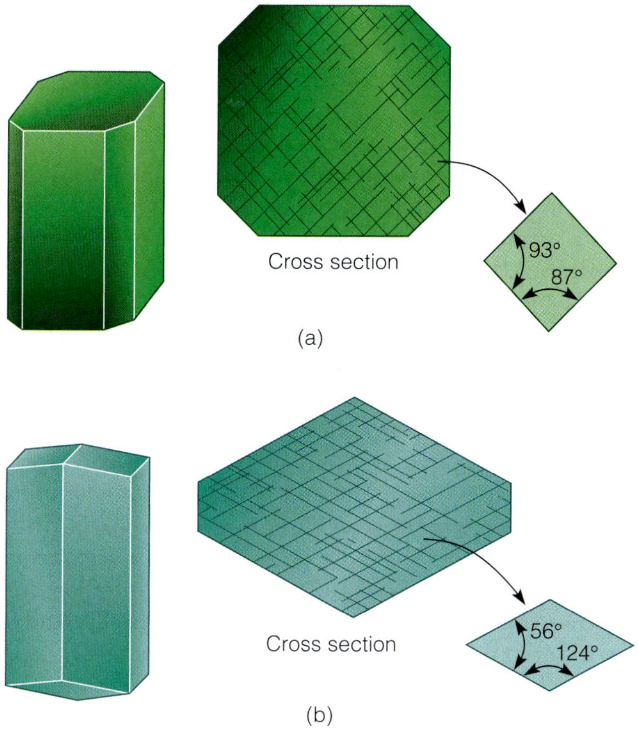

■ **Figure 3.16**

Cleavage in augite and hornblende. (a) Augite crystal and cross section of crystal showing cleavage. (b) Hornblende crystal and cross section of crystal showing cleavage.

whereas the cleavage planes of hornblende intersect at angles of 56 degrees and 124 degrees (■ Figure 3.16).

In contrast to cleavage, *fracture* is mineral breakage along irregular surfaces. Any mineral can be fractured if enough force is applied, but the fracture surfaces are commonly uneven or conchoidal (curved) rather than smooth.

Hardness

An Austrian geologist, Friedrich Mohs, devised a relative hardness scale for 10 minerals. He arbitrarily assigned a hardness value of 10 to diamond, the hardest mineral known, and lesser values to the other minerals. Relative hardness is easily determined by the use of Mohs hardness scale (Table 3.3). Quartz will scratch fluorite but cannot be scratched by fluorite, gypsum can be scratched by a fingernail, and so on. So **hardness** is defined as a mineral's resistance to abrasion and is controlled mostly by internal structure. For example, both graphite and diamond are composed of carbon, but the former has a hardness of 1 to 2, whereas the latter has a hardness of 10.

Specific Gravity (Density)

Specific gravity and density are two separate concepts, but here we will use them more or less as synonyms. A mineral's **specific gravity** is the ratio of its weight to the weight of an equal volume of pure water at 4°C. Thus a mineral with a specific gravity of 3.0 is three times as heavy as water. **Density,** in contrast, is a mineral's mass (weight) per unit of volume expressed in grams per cubic centimeters. So the specific gravity of galena (Figure 3.13b) is 7.58 and its density is 7.58 g/cm^3. In most instances we will refer to a mineral's density, and in some of the following chapters we will mention the density of various rocks.

Structure and composition control a mineral's specific gravity and density. Because ferromagnesian silicates contain iron, magnesium, or both, they tend to be denser than nonferromagnesian silicates. In general, the metallic minerals, such as galena and hematite, are denser than nonmetals. Pure gold with a density of 19.3 g/cm^3 is about two and one-half times as dense as lead. Diamond and graphite, both of which are composed of carbon (C), illustrate how structure controls specific gravity or density. The specific gravity of diamond is 3.5, whereas that of graphite varies from 2.09 to 2.33.

Table 3.3

Mohs Hardness Scale

Hardness	Mineral	Hardness of Some Common Objects
10	Diamond	
9	Corundum	
8	Topaz	
7	Quartz	
		Steel file (6½)
6	Orthoclase	
		Glass (5½–6)
5	Apatite	
4	Fluorite	
3	Calcite	Copper penny (3)
		Fingernail (2½)
2	Gypsum	
1	Talc	

Other Useful Mineral Properties

Other physical properties characterize some minerals. Talc has a distinctive soapy feel, graphite writes on paper, halite tastes salty, and magnetite is magnetic (■ Figure 3.17). Calcite possesses the property of *double refraction,* meaning that an object when viewed through a transparent piece of calcite will have a double image. Some sheet silicates are plastic and, when bent into a new shape, will retain that shape; others are flexible and, if bent, will return to their original position when the forces that bent them are removed.

Sue Monroe

(a)

Sue Monroe

(b)

Figure 3.17

Various mineral properties. (a) Graphite, the mineral used to make pencil "lead," writes on paper. (b) Magnetite, an important iron ore, is magnetic.

A simple chemical test to identify the minerals calcite and dolomite involves applying a drop of dilute hydrochloric acid to the mineral specimen. If the mineral is calcite, it will react vigorously with the acid and release carbon dioxide, which causes the acid to bubble or effervesce. Dolomite, in contrast, will not react with hydrochloric acid unless it is powdered.

HOW DO MINERALS FORM?

Thus far we have discussed the composition, structure, and physical properties of minerals but have not addressed how they originate. One phenomenon that accounts for the origin of minerals is the cooling of molten rock material known as *magma* (magma that reaches the surface is called *lava*). As magma or lava cools, minerals crystallize and grow, thereby determining the mineral composition of various igneous rocks such as basalt (dominated by ferromagnesian silicates) and granite (dominated by nonferromagnesian silicates) (see Chapter 4). Hot water solutions derived from magma commonly invade cracks and crevasses in adjacent rocks, and from these solutions a variety of minerals crystallize, some of economic importance. Minerals also originate when water in hot springs cools (see Chapter 13), and when hot, mineral-rich water discharges onto the seafloor at hot springs known as black smokers (see Chapter 9).

Dissolved materials in seawater, more rarely lake water, combine to form minerals such as halite (NaCl), gypsum ($CaSO_4 \cdot 2H_2O$), and several others when the water evaporates. Aragonite and/or calcite, both varieties of calcium carbonate ($CaCO_3$), might also form from evaporating water, but most originates when organisms such as clams, oysters, corals, and floating microorganisms use this compound to construct their shells. And a few plants and animals use silicon dioxide (SiO_2) for their skeletons, which accumulate as mineral matter on the seafloor when the organisms die (see Chapter 7).

Some clay minerals form when chemical processes compositionally and structurally alter other minerals, such as feldspars (see Chapter 6), and others originate when rocks are changed during metamorphism (see Chapter 7). In fact, the agents that cause metamorphism—heat, pressure, and chemically active fluids—are responsible for the origin of many minerals. A few minerals even originate when gasses such as hydrogen sulfide (H_2S) and sulfur dioxide (SO_2) react at volcanic vents to produce sulfur.

WHAT ARE ROCK-FORMING MINERALS?

Geologists use the term **rock** for a solid aggregate of one or more minerals, but the term also refers to masses of mineral-like matter as in the natural glass obsidian (see Chapter 4) and masses of solid organic matter as in coal (see Chapter 6). And even though some rocks may contain many minerals, only a few, designated **rock-forming minerals,** are sufficiently common for rock identification and classification (Table 3.4 and ■ Figure 3.18). Others, known as *accessory minerals,* are present in such small quantities that they can be disregarded.

Given that silicate minerals are by far the most common ones in Earth's crust, it follows that most rocks are composed of these minerals. Indeed, feldspar minerals (plagioclase feldspars and potassium feldspars) and quartz make up more than 60% of Earth's crust. So, even though there are hundreds of silicates, only a few are particularly common in rocks, although many others are present as accessory minerals.

The most common nonsilicate rock-forming minerals are the carbonates calcite ($CaCO_3$) and dolomite [$CaMg(CO_3)_2$], the main constituents of the sedimentary rocks limestone and dolostone, respectively (see Chapter 7). Among the sulfates and halides, gypsum ($CaSO_4 \cdot 2H_2O$) in rock gypsum and halite (NaCl) in rock salt (see Chapter 7) are common enough to qualify as rock-forming minerals. Even though these minerals and their corresponding rocks might be common in some areas, however, their overall abundance is limited compared to the silicate and carbonate rock-forming minerals.

Geology⇌Now Log into GeologyNow and select this chapter to work through a **Geology Interactive** activity on the "Mineral Lab" (click Atoms and Crystals→Mineral Lab).

Table 3.4

Important Rock-Forming Minerals

Mineral	Primary Occurrence
Ferromagnesian silicates	
Olivine	Igneous and metamorphic rocks
Pyroxene group	
Augite most common	Igneous and metamorphic rocks
Amphibole group	
Hornblende most common	Igneous and metamorphic rocks
Biotite	All rock types
Nonferromagnesian silicates	
Quartz	All rock types
Potassium feldspar group	
Orthoclase, microcline	All rock types
Plagioclase feldspar group	All rock types
Muscovite	All rock types
Clay mineral group	Soils, sedimentary rocks, and some metamorphic rocks
Carbonates	
Calcite	Sedimentary rocks
Dolomite	Sedimentary rocks
Sulfates	
Anhydrite	Sedimentary rocks
Gypsum	Sedimentary rocks
Halides	
Halite	Sedimentary rocks

NATURAL RESOURCES AND RESERVES

The United States and Canada have enjoyed considerable economic success because they have abundant natural resources. But what are resources, how and where do they form, and how are they found and exploited? Geologists at the U.S. Geological Survey use this definition: A **resource** is a concentration of naturally occurring solid, liquid, or gaseous material in or on Earth's crust in such form and amount that economic extraction of a commodity from the concentration is currently or potentially feasible.

Natural resources are mostly concentrations of minerals, rocks, or both, but liquid petroleum and natural gas are also included. In fact, some of the resources we refer to are *metallic resources* (copper, tin, iron ore, etc.), *nonmetallic resources* (sand and gravel, crushed stone, salt, sulfur, etc.), and *energy resources* (petroleum, natural gas, coal, and uranium). All of these are indeed resources, but we must make a distinction between a resource, the total amount of a commodity whether discovered or undiscovered, and a **reserve,** which is only that part of the resource base that is known and can be economically recovered. Aluminum can be extracted from aluminum-rich igneous and sedimentary rocks, but at present that cannot be done economically.

The distinction between a resource and a reserve is simple enough in principle, but in practice it depends on several factors, not all of which remain constant. Geographic location may be important. For instance, a resource in a remote region might not be mined because transportation costs are too high, and what might be deemed a resource rather than a reserve in the United States and Canada may be mined in a developing country where labor costs are low. The commodity in question is also important. Gold or diamonds in sufficient quantity can be mined profitably just about anywhere, whereas sand and gravel deposits must be close to their market areas.

Obviously the market price is important in evaluating any resource. From 1935 until 1968, the U.S. government

What Would You Do?

The distinction between minerals and rocks is not easy for beginning students to understand. As a teacher, you know that minerals are made up of chemical elements and that rocks consist of one or more minerals, but despite your best efforts to clearly define them, your students commonly mistake one for the other. Can you think of analogies that might help students understand the difference between minerals and rocks?

Figure 3.18

(a) Granite is made up of the minerals shown, so it is light-colored with a few black spots. (b) Basalt contains mostly dark minerals. Notice too that the minerals are clearly visible in granite, but in basalt they can be seen only when highly magnified. Source: Sue Monroe

maintained the price of gold at $35 per troy ounce (1 troy ounce = 31.1 g). When this restriction was removed, demand determined the market price and gold prices rose, reaching an all-time high of $843 per troy ounce in 1980. As a result, many marginal deposits became reserves and a number of abandoned mines were reopened.

The status of a resource is also affected by changes in technology. By the time of World War II (1939–1945), the richest iron ore deposits of the Great Lakes region in the United States and Canada had been mostly depleted. But the development of a method for separating the iron from unusable rock and shaping it into pellets ideal for use in blast furnaces made it profitable to mine rocks that contained less iron.

Most people know that industrialized societies depend on a variety of natural resources, but they know little about their occurrence, methods of recovery, and economics. Geologists are, of course, essential in finding and evaluating deposits, but extraction involves engineers, chemists, miners, and many people in support industries that supply mining equipment. Ultimately, though, the decision about whether a deposit should be mined or not is made by people trained in business and economics. In short, extraction must yield a profit. The extraction of resources other than oil, natural gas, and coal amounted to more than $40 billion during 2002 in the United States, and in Canada the extraction of nonfuel resources during the same year was nearly $18 billion (Canadian dollars).

In addition to resources such as petroleum, gold, and ores of iron and copper, some quite common minerals are also essential. For example, pure quartz sand is used to manufacture glass and optical instruments as well as sandpaper and steel alloys. Clay minerals are needed to make ceramics and paper; feldspars are used for porcelain, ceramics, enamel, and glass; and phosphate-bearing rock is used for fertilizers. Micas are used in a variety of products, including lipstick, glitter, and eye shadow as well as the lustrous paints on appliances and automobiles.

Access to resources is essential for industrialization and the high standard of living enjoyed in many countries. The United States and Canada are resource-rich nations, but many resources are *nonrenewable,* which means that there is a limited supply and they cannot be replenished by natural processes as fast as they are depleted. Accordingly, once a resource is depleted, suitable substitutes, if available, must be found. For some essential resources, the United States is totally dependent on imports; no cobalt was mined in this country during 2002. Yet the United States, the world's largest consumer of cobalt, uses this essential metal in gas-turbine aircraft engines and magnets and for corrosion- and wear-resistant alloys. Obviously all cobalt is imported, as is all manganese, an element essential for making steel.

The United States also imports all the aluminum ore it uses as well as all or some of many other resources (■ Figure 3.19). Canada, in contrast, is more self-reliant, meeting most of its domestic mineral and energy needs. Nevertheless, it must import phosphate, chromium, manganese, and aluminum ore. Canada also produces more crude oil and natural gas than it uses, and it is among the world leaders in producing and exporting uranium.

To ensure continued supplies of essential minerals and energy resources, geologists and other scientists, government agencies, and leaders in business and industry continually assess the status of resources in view of

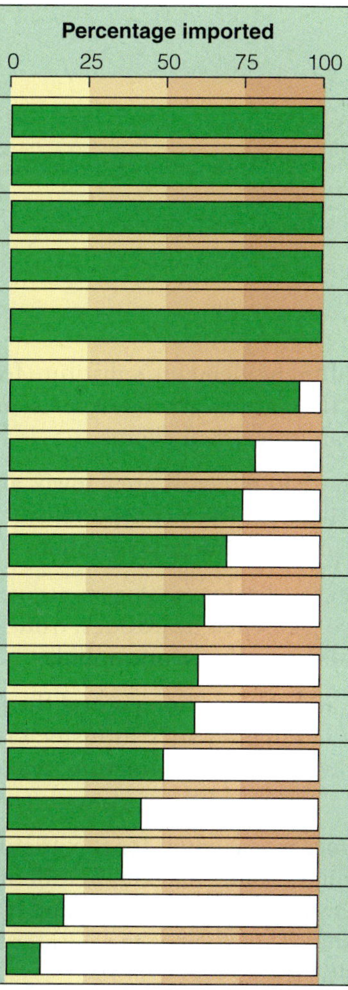

Commodity	Major Import Sources	Uses
Bauxite	Guinea, Jamaica, Brazil, Guyana	Ore of aluminum
Columbium	Brazil, Canada, Germany, Estonia	Carbon steel, superalloys
Graphite	China, Canada, Mexico, Brazil	Brake linings, lubricants
Manganese	Gabon, South Africa, Australia, Mexico	Steel production, dry cell batteries
Vanadium	South Africa, Canada, China, Czech Republic	Steel alloys
Platinum	South Africa, United Kingdom, Germany, Russia	Catalytic converters, jewelry
Tin	Peru, China, Indonesia, Brazil, Bolivia	Tin cans and containers
Cobalt	Finland, Norway, Russia, Canada	Superalloys
Tungsten	China, Russia	Carbide parts for cutting tools
Chromium	South Africa, Kazakhstan, Zimbabwe, Turkey, Russia	Stainless and heat-resistant steel
Silver	Canada, Mexico, Peru, United Kingdom	Silver halide film, jewelry
Zinc	Canada, Mexico, Kazakhstan	Galvanized metal, zinc alloys
Gold	Canada, Brazil, Peru, Australia	Jewelry and arts, electrical industry
Nickel	Canada, Norway, Russia, Australia	Stainless steel, electroplating
Copper	Canada, Chile, Peru, Mexico	Copper and copper alloys, wiring
Lead	Canada, Mexico, Australia, Peru	Lead for batteries, protective coatings
Iron ore	Canada, Brazil, Australia, Venezuela	Steel, cast iron

Sources: USGS Minerals Information: http://minerals.usgs.gov/minerals/
USGS Mineral Commodity Summaries 2003: http://usgs.gov/minerals/pubs/mcs/2003.pdf

■ **Figure 3.19**

The dependence of the United States on imports of various mineral commodities is apparent from this chart. The lengths of the bars correspond to the amounts of resources imported.

changing economic and political conditions and changes in science and technology. The U.S. Geological Survey, for instance, keeps detailed statistical records of mine production, imports, and exports, and regularly publishes reports on the status of numerous commodities. Similar reports appear regularly in the *Canadian Minerals Yearbook.* In several of the following chapters, we will discuss the geologic occurrence of resources.

What Would You Do?

Some reputable businesspeople tell you of opportunities to invest in natural resources. Two ventures look promising: a gold mine and a sand and gravel pit. Given that gold sells for about $420 per ounce, whereas sand and gravel are worth $4 or $5 per ton, would it be more prudent to invest in the gold mine? Explain not only how market price would influence your decision but also what other factors you might need to consider.

3 GEO RECAP

Chapter Summary

- Matter is composed of chemical elements, each of which consists of atoms. Protons and neutrons are present in an atom's nucleus, and electrons orbit around the nucleus in electron shells.
- The number of protons in an atom's nucleus determines its atomic number. The atomic mass number is the number of protons plus neutrons in the nucleus.
- Bonding results when atoms join with other atoms; different elements bond to form a compound. With few exceptions, minerals are compounds.
- Ionic and covalent bonds are most common in minerals, but metallic and van der Waals bonds are found in some.
- Minerals are crystalline solids, which means they possess an ordered internal arrangement of atoms.
- Mineral composition is indicated by a chemical formula, such as SiO_2 for quartz.
- Some minerals have a range of compositions because different elements substitute for one another if their atoms are about the same size and have the same electrical charge.
- More than 3500 minerals are known, and most of them are silicates. The two types of silicates are ferromagnesian and nonferromagnesian.
- In addition to silicates, geologists recognize carbonates, native elements, hydroxides, oxides, phosphates, halides, sulfates, and sulfides.
- Structure and composition control the physical properties of minerals such as luster, crystal form, hardness, color, cleavage, fracture, and specific gravity.
- Several processes account for the origin of minerals, including cooling magma, weathering, evaporation of seawater, metamorphism, and organisms using dissolved substances in seawater to build their shells.
- A few minerals, designated rock-forming minerals,

are common enough in rocks to be essential in their identification and classification. Most rock-forming minerals are silicates, but some carbonates are also common.

- Many resources are concentrations of minerals or rocks of economic importance. They are further characterized as metallic resources, nonmetallic resources, and energy resources.
- Reserves are that part of the resource base that can be extracted profitably. Distinguishing a resource from a reserve depends on market price, labor costs, geographic location, and developments in science and technology.
- The United States must import many resources to maintain its industrial capacity. Canada is more self-reliant, but it too must import some commodities.

Important Terms

atom (p. 61)
atomic mass number (p. 62)
atomic number (p. 61)
bonding (p. 63)
carbonate mineral (p. 71)
cleavage (p. 65)
compound (p. 63)
covalent bond (p. 64)
crystal (p. 65)
crystalline solid (p. 65)
density (p. 77)
electron (p. 61)
electron shell (p. 61)
element (p. 61)
ferromagnesian silicate (p. 70)
hardness (p. 77)
ion (p. 64)
ionic bond (p. 64)
luster (p. 74)
mineral (p. 65)
neutron (p. 61)
nonferromagnesian silicate (p. 70)
nucleus (p. 61)
proton (p. 61)
reserve (p. 79)
resource (p. 79)
rock (p. 78)
rock-forming mineral (p. 78)
silica (p. 69)
silica tetrahedron (p. 69)
silicate (p. 69)
specific gravity (p. 77)

Review Questions

1. A common rock-forming silicate mineral is _____, whereas the most common carbonate mineral is _____.
   a._____olivine/gypsum; b._____quartz/calcite; c._____hematite/galena; d._____halite/biotite; e._____muscovite/hornblende.
2. In what type of chemical bonding are electrons shared by adjacent atoms?
   a._____van der Waals; b._____silicate; c._____octahedral; d._____spherical; e._____covalent.
3. The two most abundant elements in Earth's crust are
   a._____oxygen and silicon; b._____iron and potassium; c._____aluminum and calcium; d._____granite and basalt; e._____magnesium and iridium.
4. An atom with 6 protons and 8 neutrons in its nucleus has an atomic mass number of
   a._____6; b._____8; c._____14; d._____48; e._____2.
5. Any mineral composed of an element combined with sulfur (S^{-2}) as in galena (PbS) is a(n)
   a._____oxide; b._____sulfide; c._____carbonate; d._____silicate; e._____hydroxide.
6. The atoms of the noble gases do not react to form compounds because they have
   a._____eight electrons in their outermost electron shell; b._____more positive charges than negative charges; c._____three directions of cleavage intersecting at right angles; d._____atomic mass numbers exceeding 92; e._____too much silica and not enough calcium.

7. A rock-forming mineral is any mineral
a._____found in rocks; b._____containing the $(CO_3)^{-2}$ radical; c._____in which oxygen combines with iron; d._____essential for the classification of rocks; e._____from the silicate group.

8. A mineral known as a native element is one in which
a._____one element can substitute for another; b._____composition is determined by reactions between oxygen and iron; c._____atoms bond to form continuous sheets; d._____at least silicon and oxygen are found; e._____only one chemical element is present.

9. Minerals that possess the property known as cleavage
a._____are denser than minerals that lack this property; b._____exhibit double refraction; c._____break along smooth internal planes of weakness; d._____include obsidian and coal; e._____are composed mostly of the noble gases.

10. The ferromagnesian silicate olivine has the chemical formula $(Mg,Fe)_2SiO_4$, which means that
a._____silicon and oxygen may or may not be present; b._____magnesium and iron can substitute for each another; c._____magnesium and iron are less abundant in Earth's crust than silicon and oxygen; d._____olivine contains either magnesium or iron but not both; e._____ferromagnesian silicates are darker than nonferromagnesian silicates.

11. Explain the distinction between rock-forming minerals and accessory minerals. Also, name some of the most common silicate rock-forming minerals and one carbonate rock-forming mineral.

12. Why must the United States, a resource-rich nation, import most or all of some of the resources it needs? What are some of the problems such a dependence on imports creates?

13. What is cleavage in minerals? How can it be used in "cutting" gemstones?
14. How do minerals characterized as silicates differ from carbonates and oxides?
15. Briefly discuss three ways in which minerals originate.
16. Compare ionic and covalent bonding.
17. Under what conditions do well-formed mineral crystals originate? Why are well-formed crystals not very common?
18. What accounts for the fact that some minerals, such as plagioclase feldspars, have a range of chemical compositions? Give an example from the ferromagnesian silicates.
19. How would the color and density of a rock composed mostly of ferromagnesian silicates differ from one made up primarily of nonferromagnesian silicates?
20. What is the basic distinction between minerals and rocks?

World Wide Web Activities

GeologyNow Assess your understanding of this chapter's topics with additional quizzing and comprehensive interactivities at

http://earthscience.brookscole.com/changingearth4e

as well as current and up-to-date weblinks, additional readings, and InfoTrac College Edition exercises.

Igneous Rocks and Intrusive Igneous Activity

CHAPTER 4

OUTLINE

Geology Now *This icon, which appears throughout the book, indicates an opportunity to explore interactive tutorials, animations, or practice problems available on the GeologyNow website at http://earthscience.brookscole.com/changingearth4e.*

his mass of granitic rock in Yosemite National Park in California is called l Capitan, meaning "The Chief." It is part of the Sierra Nevada atholith, a huge pluton mesuring 640 km long and 110 km wide. This ear-vertical cliff rises more than 900 m above the valley floor, making it ne highest unbroken cliff in the world.

OBJECTIVES

At the end of this chapter, you will have learned that:

- With few exceptions, magma is composed of silicon and oxygen with lesser amounts of several other chemical elements.
- Temperature and especially composition are the most important controls on the mobility of magma and lava.
- Most magma originates within Earth's upper mantle or lower crust at or near divergent and convergent plate boundaries.
- Several processes bring about chemical changes in magma, so magma may evolve from one kind into another.
- All igneous rocks form when magma or lava cools and crystallizes or by the consolidation of pyroclastic materials ejected during explosive eruptions.
- Geologists use texture and composition to classify igneous rocks.
- Intrusive igneous bodies called plutons form when magma cools below Earth's surface. The origin of the largest plutons is not fully understood.

Introduction

We mentioned that the term *rock* applies to a solid aggregate of one or more minerals as well as to mineral-like matter such as natural glass and solid masses of organic matter such as coal. Furthermore, in Chapter 1 we briefly discussed the three main families of rocks: igneous, sedimentary, and metamorphic. Recall that *igneous rocks* form when molten rock material known as *magma* or *lava* cools and crystallizes to form a variety of minerals, or when particulate matter called *pyroclastic materials* become consolidated. We are most familiar with igneous rocks formed from lava flows and pyroclastic materials because they are easily observed at Earth's surface, but you should be aware that most magma never reaches the surface. Indeed, much of it cools and crystallizes far underground and thus forms *plutons,* igneous bodies of various shapes and sizes.

Granite and several similar-appearing rocks are the most common rocks in the larger plutons such as those in the Sierra Nevada of California (see the chapter opening photo) and in Acadia National Park in Maine (■ Figure 4.1a). The images of Presidents Lincoln, Roosevelt, Jefferson, and Washington at Mount Rushmore National Memorial in South Dakota (Figure 4.1b) as well as the nearby Crazy Horse Memorial (under construction) are in the 1.7-billion-year-old Harney Peak Granite, which consists of a number of plutons. These huge plutons formed far below the surface, but subsequent uplift and deep erosion exposed them in their present form.

Some granite and related rocks are quite attractive, especially when sawed and polished. They are used for tombstones, mantlepieces, kitchen counters, facing stones on buildings, pedestals for statues, and statuary itself. More important, though, is the fact that fluids emanating from plutons account for many ore minerals of important metals, such as copper in adjacent rocks.

The origin of plutons, or intrusive igneous activity, and volcanism involving the eruption of lava flows, gases, and pyroclastic materials are closely related topics even though we discuss them in separate chapters. The same kinds of magmas are involved in both processes, but magma varies in its mobility, which explains why only some reaches the surface. Furthermore, plutons typically lie beneath areas of volcanism and, in fact, are the source of the overlying lava flows and pyroclastic materials. Plutons and most volcanoes are found at or near divergent and convergent plate boundaries, so the presence of igneous rocks is one criterion for recognizing ancient plate boundaries; igneous rocks also help us unravel the complexities of mountain-building episodes (see Chapter 10).

Richard Thom/Visuals Unlimited

(a)

National Park Service

(b)

■ **Figure 4.1**

(a) Light-colored granitic rocks exposed along the shoreline in Acadia National Park in Maine. The dark rock is basalt that formed when magma intruded along a fracture in the granitic rock. (b) The presidents' images at Mount Rushmore National Memorial in South Dakota are in the Harney Peak Granite.

One important reason to study igneous rocks and intrusive igneous activity is that igneous rocks are one of the three main families of rocks. In addition, igneous rocks make up large parts of all the continents and nearly all of the oceanic crust, which is formed continuously by igneous activity at divergent plate boundaries. And, as already mentioned, important mineral deposits are found adjacent to many plutons.

In this chapter, our main concerns are (1) the origin, composition, textures, and classification of igneous rocks, and (2) the origin, significance, and types of plutons. In Chapter 5, we will consider volcanism, volcanoes, and associated phenomena that result from magma reaching Earth's surface. Remember, though, that the origin of plutons and volcanism are related topics.

THE PROPERTIES AND BEHAVIOR OF MAGMA AND LAVA

In Chapter 3 we noted that one process that accounts for the origin of minerals, and thus rocks, is cooling and crystallization of magma and lava. **Magma** is molten rock below Earth's surface. Any magma is less dense than the rock from which it was derived, so it tends to move upward, but much of it cools and solidifies deep underground, thus accounting for intrusive igneous bodies known as *plutons*. However, some magma does rise to the surface where it issues forth as **lava flows,** or it is forcefully ejected into the atmosphere as particulate matter known as **pyroclastic materials** (from the Greek *pyro,* "fire," and *klastos,* "broken"). Certainly lava flows and eruptions of pyroclastic materials are the most impressive manifestations of all processes related to magma, but they result from only a small percentage of all magma that forms.

All **igneous rocks** derive from magma, but two separate processes account for them. They form when (1) magma or lava cools and crystallizes to form minerals, or (2) pyroclastic materials are consolidated to form a solid aggregate from the previously loose particles. Igneous rocks that result from the cooling of lava flows and the consolidation of pyroclastic materials are **volcanic rocks** or **extrusive igneous rocks**—that is, igneous rocks that form from materials erupted on the surface. In contrast, magma that cools below the surface forms **plutonic rocks** or **intrusive igneous rocks.**

Composition of Magma

In Chapter 3 we noted that by far the most abundant minerals in Earth's crust are silicates such as quartz, various feldspars, and several ferromagnesian silicates, all made up of silicon and oxygen, and other elements shown in Figure 3.9. As a result, melting of the crust yields mostly silica-rich magmas that also contain considerable aluminum, calcium, sodium, iron, magnesium, and potassium, and several other elements in lesser quantities. Another source of magma is Earth's upper mantle, which is composed of rocks that contain mostly ferromagnesian silicates. Thus magma from this source contains comparatively less silicon and oxygen (silica) and more iron and magnesium.

Although there are a few exceptions, the primary constituent of magma is silica, which varies enough to distinguish magmas classified as felsic, intermediate, and mafic.* **Felsic magma,** with more than 65% silica, is silica-rich and contains considerable sodium, potassium, and aluminum but little calcium, iron, and magnesium. In contrast, **mafic magma,** with less than 52% silica, is silica-poor and contains proportionately more calcium, iron, and magnesium. And as you would expect, **intermediate magma** has a composition between felsic and mafic magma (Table 4.1).

Table 4.1

The Most Common Types of Magmas and Their Characteristics

Type of Magma	Silica Content (%)	Sodium, Potassium, and Aluminum	Calcium, Iron, and Magnesium
Ultramafic	<45	↓	Increase ↑
Mafic	45–52	↓	↑
Intermediate	53–65	↓	↑
Felsic	>65	Increase	↑

How Hot Are Magma and Lava?

Everyone knows that lava is very hot, but how hot is hot? Erupting lava generally has a temperature in the range of 1000° to 1200°C, although a temperature of 1350°C was recorded above a lava lake in Hawaii where volcanic gases reacted with the atmosphere. Magma must be even hotter than lava, but no direct

*Lava from some volcanoes in Africa cools to form carbonitite, an igneous rock with at least 50% carbonate minerals, mostly calcite and dolomite.

■ **Figure 4.2**

A geologist using a thermocouple to determine the temperature of a lava flow in Hawaii.

measurements of magma temperatures have ever been made.

Most lava temperatures are taken at volcanoes that show little or no explosive activity, so our best information comes from mafic lava flows such as those in Hawaii (■ Figure 4.2). In contrast, eruptions of felsic lava are not as common, and the volcanoes that these flows issue from tend to be explosive and thus cannot be approached safely. Nevertheless, the temperatures of some bulbous masses of felsic lava in lava domes have been measured at a distance with an optical pyrometer. The surfaces of these lava domes are as hot as 900°C, but their interiors must surely be even hotter.

When Mount St. Helens erupted in 1980 in Washington State, it ejected felsic magma as particulate matter in pyroclastic flows. Two weeks later, these flows had temperatures between 300° and 420°C, and a steam explosion took place more than a year later when water encountered some of the still-hot pyroclastic materials. The reason that lava and magma retain heat so well is that rock conducts heat so poorly. Accordingly, the interiors of thick lava flows and pyroclastic flow deposits may remain hot for months or years, whereas plutons, depending on their size and depth, may not cool completely for thousands to millions of years.

Viscosity—Resistance to Flow

All liquids have the property of **viscosity,** or resistance to flow. For liquids such as water, viscosity is very low so they are highly fluid and flow readily. For other liquids, though, viscosity is so high that they flow much more slowly. Good examples are cold motor oil and syrup, both of which are quite viscous and thus flow only with difficulty. But when these liquids are heated, their viscosity is much lower and they flow more easily; that is, they become more fluid with increasing temperature. Accordingly, you might suspect that temperature controls the viscosity of magma and lava, and this inference is partly correct. We can generalize and say that hot magma or lava moves more readily than cooler magma or lava, but we must qualify this statement by noting that temperature is not the only control of viscosity.

Silica content strongly controls magma and lava viscosity. With increasing silica content, numerous networks of silica tetrahedra form and retard flow because for flow to take place, the strong bonds of the networks must be ruptured. Mafic magma and lava with 45–52% silica have fewer silica tetrahedra networks and as a result are more mobile than felsic magma and lava flows. One mafic flow in 1783 in Iceland flowed about 80 km, and geologists traced ancient flows in Washington State for more than 500 km. Felsic magma, in contrast, because of its higher viscosity, does not reach the surface as commonly as mafic magma. And when felsic lava flows do occur, they tend to be slow moving and thick and to move only short distances. A thick, pasty lava flow that erupted in 1915 from Lassen Peak in California flowed only about 300 m before it ceased moving.

HOW DOES MAGMA ORIGINATE AND CHANGE?

Most of us have not witnessed a volcanic eruption, but we have nevertheless seen news reports or documentaries showing magma issuing forth as lava flows or pyroclastic materials. In any case, we are familiar with some aspects of igneous activity, but most people are unaware of how and where magma originates, how it rises from its place of origin, and how it might change. Indeed, many believe the misconception that lava comes from a continuous layer of molten rock beneath the crust or that it comes from Earth's molten outer core.

First, let us address how and where magma originates. We know that the atoms in a solid are in constant motion and that, when a solid is heated, the energy of motion exceeds the binding forces and the solid melts. We are all familiar with this phenomenon, and we are also aware that not all solids melt at the same temperature. Once magma forms, it tends to rise because it is less dense than the rock that melted, and some actually makes it to the surface.

Magma may come from 100 to 300 km deep, but most forms at much shallower depths in the upper mantle or lower crust and accumulates in reservoirs known as **magma chambers.** Beneath spreading ridges, where the crust is thin, magma chambers exist at a depth of only a few kilometers, but along convergent plate boundaries, magma chambers are commonly a few tens of kilometers deep. The volume of a magma chamber ranges from a few to many hundreds of cubic kilometers of molten rock within the otherwise solid lithosphere. Some simply cools and crystallizes within Earth's crust, thus accounting for the origin of various plutons, whereas some rises to the surface and is erupted as lava flows or pyroclastic materials.

Bowen's Reaction Series

During the early part of the last century, N. L. Bowen hypothesized that mafic, intermediate, and felsic magmas could all derive from a parent mafic magma. He knew that minerals do not all crystallize simultaneously from cooling magma, but rather crystallize in a predictable sequence. Based on his observations and laboratory experiments, Bowen proposed a mechanism, now called **Bowen's reaction series,** to account for the derivation of intermediate and felsic magmas from mafic magma. Bowen's reaction series consists of two branches: a *discontinuous branch* and a *continuous branch* (■ Figure 4.3). As the temperature of magma decreases, minerals crystallize along both branches simultaneously, but for convenience we will discuss them separately.

In the discontinuous branch, which contains only ferromagnesian silicates, one mineral changes to another over specific temperature ranges (Figure 4.3). As the temperature decreases, a temperature range is reached in which a given mineral begins to crystallize. A previously formed mineral reacts with the remaining liquid magma (the melt) so that it forms the next mineral in the sequence. For instance, olivine $[(Mg,Fe)_2SiO_4]$ is the first ferromagnesian silicate to crystallize. As the magma continues to cool, it reaches the temperature range at which pyroxene is stable; a reaction occurs between the olivine and the remaining melt, and pyroxene forms.

With continued cooling, a similar reaction takes place between pyroxene and the melt, and the pyroxene structure is rearranged to form amphibole. Further cooling causes a reaction between the amphibole and the melt, and its structure is rearranged so that the sheet structure of biotite mica forms. Although the reactions just described tend to convert one mineral to the next in

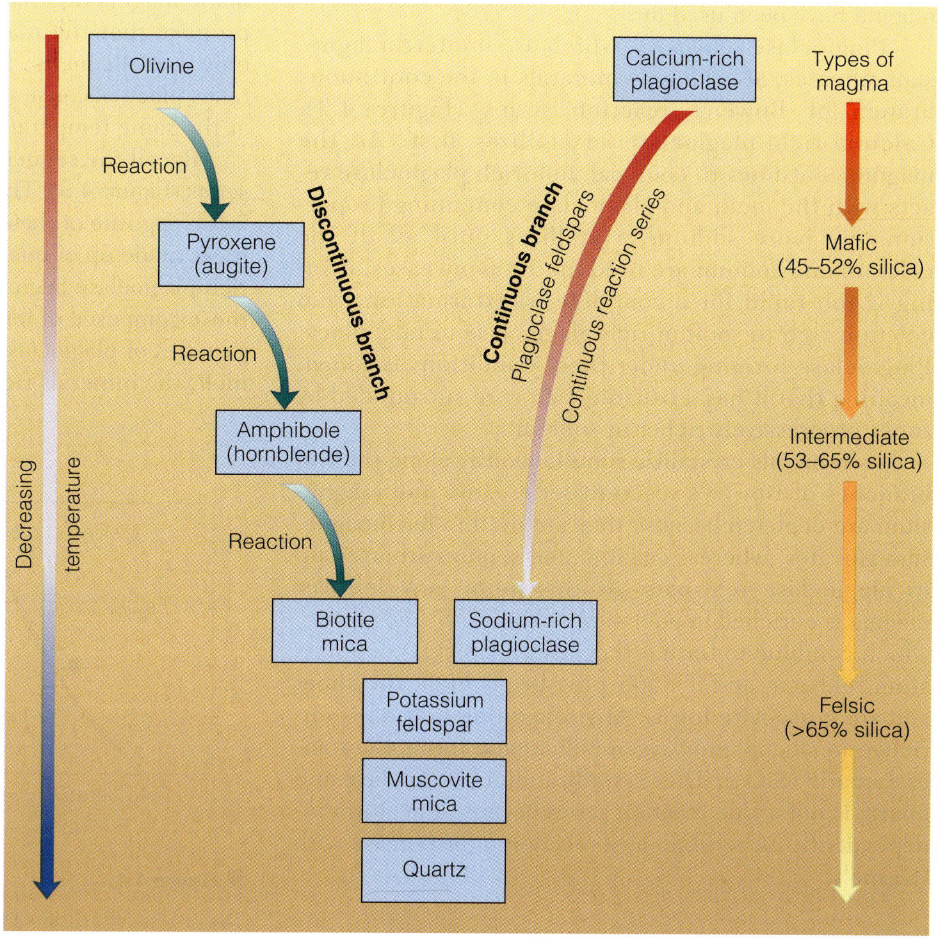

■ **Figure 4.3**

Bowen's reaction series consists of a discontinuous branch along which a succession of ferromagnesian silicates crystallize as the magma's temperature decreases, and a continuous branch along which plagioclase feldspars with increasing amounts of sodium crystallize. Notice also that the composition of the initial mafic magma changes as crystallization takes place along the two branches.

What Would You Do?

You are a high school science teacher interested in developing experiments to show your students that (1) composition and temperature affect the viscosity of a lava flow, and (2) when magma or lava cools, some minerals crystallize before others. Describe the experiments you might devise to illustrate these points.

the series, the reactions are not always complete. Olivine, for example, might have a rim of pyroxene, indicating an incomplete reaction. If magma cools rapidly enough, the early-formed minerals do not have time to react with the melt, and thus all the ferromagnesian silicates in the discontinuous branch can be in one rock. In any case, by the time biotite has crystallized, essentially all the magnesium and iron present in the original magma have been used up.

Plagioclase feldspars, which are nonferromagnesian silicates, are the only minerals in the continuous branch of Bowen's reaction series (Figure 4.3). Calcium-rich plagioclase crystallizes first. As the magma continues to cool, calcium-rich plagioclase reacts with the melt, and plagioclase containing proportionately more sodium crystallizes until all of the calcium and sodium are used up. In many cases, cooling is too rapid for a complete transformation from calcium-rich to sodium-rich plagioclase to take place. Plagioclase forming under these conditions is *zoned,* meaning that it has a calcium-rich core surrounded by zones progressively richer in sodium.

As minerals crystallize simultaneously along the two branches of Bowen's reaction series, iron and magnesium are depleted because they are used in ferromagnesian silicates, whereas calcium and sodium are used up in plagioclase feldspars. At this point, any leftover magma is enriched in potassium, aluminum, and silicon, which combine to form orthoclase ($KAlSi_3O_8$), a potassium feldspar, and if water pressure is high, the sheet silicate muscovite forms. Any remaining magma is enriched in silicon and oxygen (silica) and forms the mineral quartz (SiO_2). The crystallization of orthoclase and quartz is not a true reaction series because they form independently rather than by a reaction of orthoclase with the melt.

The Origin of Magma at Spreading Ridges

One fundamental observation regarding the origin of magma is that Earth's temperature increases with depth. Known as the *geothermal gradient,* this temperature increase averages about 25°C/km. Accordingly, rocks at depth are hot but remain solid because their melting temperature rises with increasing pressure (■ Figure 4.4a). However, beneath spreading ridges the temperature locally exceeds the melting temperature, at least in part because pressure decreases. That is, plate separation at ridges probably causes a decrease in pressure on the already hot rocks at depth, thus initiating melting (Figure 4.4a). In addition, the presence of water decreases the melting temperature beneath spreading ridges because water aids thermal energy in breaking the chemical bonds in minerals (Figure 4.4b).

Localized, cylindrical plumes of hot mantle material, called *mantle plumes,* rise beneath spreading ridges and elsewhere, and as they rise, pressure decreases and melting begins, thus yielding magma. Magma formed beneath spreading ridges is invariably mafic (45–52% silica). But the upper mantle rocks from which this magma is derived are characterized as ultramafic (<45% silica), consisting largely of ferromagnesian silicates and lesser amounts of nonferromagnesian silicates. To explain how mafic magma originates from ultramafic rock, geologists propose that the magma forms from source rock that only partially melts. This phenomenon of partial melting takes place because not all of the minerals in rocks melt at the same temperature.

Recall the sequence of minerals in Bowen's reaction series (Figure 4.3). The order in which these minerals melt is the opposite of their order of crystallization. Accordingly, rocks made up of quartz, potassium feldspar, and sodium-rich plagioclase begin melting at lower temperatures than those composed of ferromagnesian silicates and the calcic varieties of plagioclase. So when ultramafic rock starts to melt, the minerals richest in silica melt first, followed by

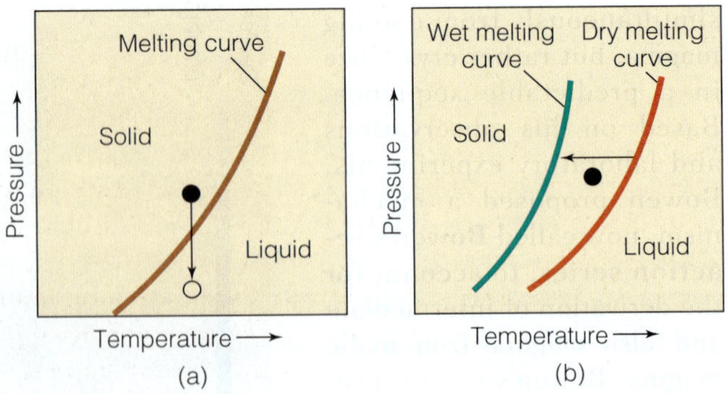

■ **Figure 4.4**

The effects of pressure and temperature on melting. (a) As pressure decreases, even when temperature remains constant, melting takes place. The black circle represents rock at high temperature. The same rock (open circle) melts at lower pressure. (b) If water is present, the melting curve shifts to the left because water provides an additional agent to break chemical bonds. Accordingly, rocks melt at a lower temperature (green melting curve) if water is present.

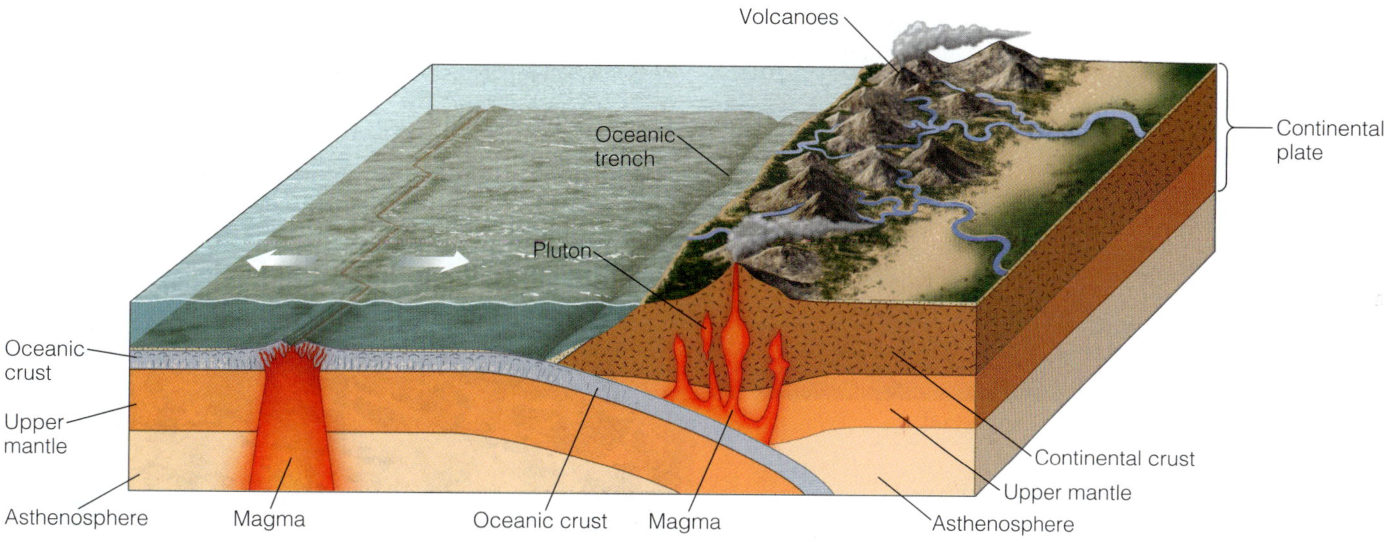

Figure 4.5

Both intrusive and extrusive igneous activity take place at divergent plate boundaries (spreading ridges) and where plates are subducted at convergent plate boundaries. Oceanic crust is composed largely of plutons and dark igneous rocks that cooled from submarine lava flows. Magma forms where an oceanic plate is subducted beneath another oceanic plate or beneath a continental plate as shown here. Much of the magma forms plutons, but some is erupted to form volcanoes (see Chapter 5).

those containing less silica. Therefore, if melting is not complete, mafic magma containing proportionately more silica than the source rock results.

Subduction Zones and the Origin of Magma

Another fundamental observation regarding magma is that, where an oceanic plate is subducted beneath either a continental plate or another oceanic plate, a belt of volcanoes and plutons is found near the leading edge of the overriding plate (■ Figure 4.5). It would seem, then, that subduction and the origin of magma must be related in some way, and indeed they are. Furthermore, magma at these convergent plate boundaries is mostly intermediate (53–65% silica) or felsic (>65% silica).

Once again, geologists invoke the phenomenon of partial melting to explain the origin and composition of magma at subduction zones. As a subducted plate descends toward the asthenosphere, it eventually reaches the depth where the temperature is high enough to initiate partial melting. In addition, the oceanic crust descends to a depth at which dewatering of hydrous minerals takes place, and as the water rises into the overlying mantle, it enhances melting and magma forms (Figure 4.4b).

Recall that partial melting of ultramafic rock at spreading ridges yields mafic magma. Similarly, partial melting of mafic rocks of the oceanic crust yields intermediate (53–65% silica) and felsic (>65% silica) magmas, both of which are richer in silica than the source rock. Moreover, some of the silica-rich sediments and sedimentary rocks of continental margins are probably carried downward with the subducted plate and contribute their silica to the magma. Also, mafic magma rising through the lower continental crust must be contaminated with silica-rich materials, which changes its composition.

Processes That Bring About Compositional Changes in Magma

Once magma forms, its composition may change by **crystal settling**, which involves the physical separation of minerals by crystallization and gravitational settling (■ Figure 4.6). Olivine, the first ferromagnesian silicate to form in the discontinuous branch of Bowen's reaction series, has a density greater than the remaining magma and tends to sink. Accordingly, the remaining magma becomes richer in silica, sodium, and potassium because much of the iron and magnesium were removed as olivine and perhaps pyroxene minerals crystallized.

Although crystal settling does take place, it does not do so on a scale that would yield very much felsic magma from mafic magma. In some thick, sheetlike plutons called *sills*, the first-formed ferromagnesian silicates are indeed concentrated in their lower parts, thus making

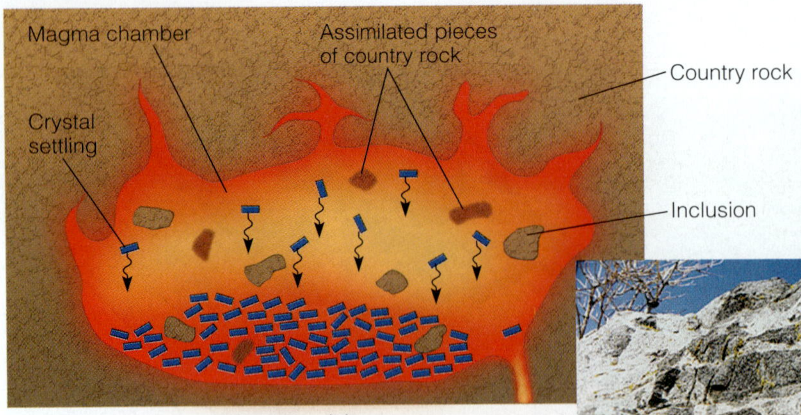

(a)

(b)

Figure 4.6

(a) Early-formed ferromagnesian silicates are denser than the magma and settle and accumulate in the magma chamber. Fragments of rock dislodged by upward-moving magma may melt and be incorporated into the magma, or they may remain as inclusions. (b) Dark inclusions in granitic rock.

their upper parts less mafic. But even in these plutons, crystal settling has yielded very little felsic magma.

If felsic magma could be derived on a large scale from mafic magma, there should be far more mafic magma than felsic magma. To yield a particular volume of granite (a felsic igneous rock), about 10 times as much mafic magma would have to be present initially for crystal settling to yield the volume of granite in question. If this were so, then mafic intrusive igneous rocks should be much more common than felsic ones. However, just the opposite is the case, so it appears that mechanisms other than crystal settling must account for the large volume of felsic magma. Partial melting of mafic oceanic crust and silica-rich sediments of continental margins during subduction yields magma richer in silica than the source rock. Furthermore, magma rising through the continental crust absorbs some felsic materials and becomes more enriched in silica.

The composition of magma also changes by **assimilation,** a process by which magma reacts with preexisting rock, called **country rock,** with which it comes in contact (Figure 4.6). The walls of a volcanic conduit or magma chamber are, of course, heated by the adjacent magma, which may reach temperatures of 1300°C. Some of these rocks partly or completely melt, provided their melting temperature is lower than that of the magma. Because the assimilated rocks seldom have the same composition as the magma, the composition of the magma changes.

The fact that assimilation occurs is indicated by *inclusions,* incompletely melted pieces of rock that are fairly common in igneous rocks. Many inclusions were simply wedged loose from the country rock as magma forced its way into preexisting fractures (Figure 4.6). No one doubts that assimilation takes place, but its effect on the bulk composition of magma must be slight. The reason is that the heat for melting comes from the magma itself, and this has the effect of cooling the magma. Only a limited amount of rock can be assimilated by magma, and that amount is insufficient to bring about a major compositional change.

Neither crystal settling nor assimilation can produce a significant amount of felsic magma from a mafic one. But both processes, if operating concurrently, can bring about greater changes than either process acting alone. Some geologists think that this is one way that intermediate magma forms where oceanic lithosphere is subducted beneath continental lithosphere.

A single volcano can erupt lavas of different composition, indicating that magmas of differing composition are present. It seems likely that some of these magmas would come into contact and mix with one another. If this is the case, we would expect that the composition of the magma resulting from **magma mixing** would be a modified version of the parent magmas. Suppose rising mafic magma mixes with felsic magma of about the same volume (■ Figure 4.7). The resulting "new" magma would have a more intermediate composition.

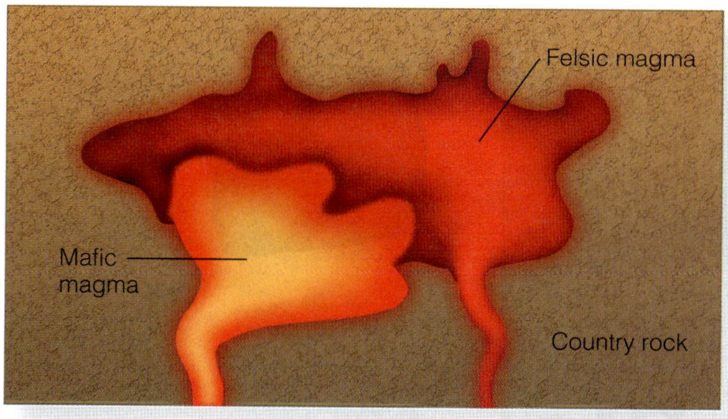

Figure 4.7

Magma mixing. Two magmas mix and produce magma with a composition different from either of the parent magmas. In this case, the resulting magma would have an intermediate composition.

IGNEOUS ROCKS—THEIR CHARACTERISTICS AND CLASSIFICATION

We have already defined *plutonic* or *intrusive igneous rocks* and *volcanic* or *extrusive igneous rocks*. Here we will have considerably more to say about the texture, composition, and classification of these rocks, which constitute one of the three major rock families depicted in the rock cycle (see Figure 1.12).

Igneous Rock Textures

The term *texture* refers to the size, shape, and arrangement of the minerals that make up igneous rocks. Size is the most important because mineral crystal size is related to the cooling history of magma or lava and generally indicates whether an igneous rock is volcanic or plutonic. The atoms in magma and lava are in constant motion, but when cooling begins, some atoms bond to form small nuclei. As other atoms in the liquid chemically bond to these nuclei, they do so in an orderly geometric arrangement and the nuclei grow into crystalline *mineral grains*, the individual particles that make up igneous rocks.

During rapid cooling, as takes place in lava flows, the rate at which mineral nuclei form exceeds the rate of growth and an aggregate of many small mineral grains is formed. The result is a fine-grained or **aphanitic texture,** in which individual minerals are too small to be seen without magnification (■ Figure 4.8a, b). With slow cooling, the rate of growth exceeds the rate of nuclei formation, and relatively large mineral grains form, thus yielding a coarse-grained or **phaneritic texture,** in which minerals are clearly visible (Figure 4.8c, d). Aphanitic textures generally indicate an extrusive origin, whereas rocks with phaneritic textures are usually intrusive. However, shallow plutons might have an aphanitic texture, and the rocks that form in the interiors of thick lava flows might be phaneritic.

Another common texture in igneous rocks is one termed **porphyritic,** in which minerals of markedly different size are present in the same rock. The larger minerals are *phenocrysts* and the smaller ones collectively make up the *groundmass,* which is simply the grains between phenocrysts (Figure 4.8e, f). The groundmass can be either aphanitic or phaneritic; the only requirement for a porphyritic texture is that the phenocrysts be considerably larger than the minerals in the groundmass. Igneous rocks with porphyritic textures are designated *porphyry,* as in basalt porphyry. These rocks have more complex cooling histories than those with aphanitic or phaneritic textures that might involve, for example, magma partly cooling beneath the surface followed by its eruption and rapid cooling at the surface.

Lava may cool so rapidly that its constituent atoms do not have time to become arranged in the ordered, three-dimensional frameworks of minerals. As a consequence *natural glass* such as *obsidian* forms (Figure 4.8g). Even though obsidian with its glassy texture is not composed of minerals, geologists nevertheless classify it as an igneous rock.

Some magmas contain large amounts of water vapor and other gases. These gases may be trapped in cooling lava where they form numerous small holes or cavities known as **vesicles;** rocks with many vesicles are termed *vesicular,* as in vesicular basalt (Figure 4.8h).

A **pyroclastic** or **fragmental texture** characterizes igneous rocks formed by explosive volcanic activity (Figure 4.8i). For example, ash discharged high into the atmosphere eventually settles to the surface where it accumulates; if consolidated, it forms pyroclastic igneous rock.

Composition of Igneous Rocks

Most igneous rocks, like the magma from which they originate, are characterized as mafic (45–52% silica), intermediate (53–65% silica), or felsic (>65% silica). A few are referred to as ultramafic (<45% silica), but these are probably derived from mafic magma by a process discussed later. The parent magma plays an important role in determining the mineral composition of igneous rocks, yet it is possible for the same magma to yield a variety of igneous rocks because its composition can change as a result of the sequence in which minerals crystallize, or by crystal settling, assimilation, and magma mixing (Figures 4.3, 4.6, and 4.7).

■ **Figure 4.8**

The various textures of igneous rocks. Texture is one criterion used to classify igneous rocks. (a, b) Rapid cooling as in lava flows results in many small minerals and an aphanitic (fine-grained) texture. (c, d) Slower cooling in plutons yields a phaneritic texture. (e, f) These porphyritic textures indicate a complex cooling history. (g) Obsidian has a glassy texture because magma cooled too quickly for mineral crystals to form. (h) Gases expand in lava and yield a vesicular texture. (i) Microscopic view of an igneous rock with a fragmental texture. The colorless, angular objects are pieces of volcanic glass measuring up to 2 mm.

Classifying Igneous Rocks

Geologists use texture and composition to classify most igneous rocks. Notice in ■ Figure 4.9 that all rocks except peridotite are in pairs; the members of a pair have the same composition but different textures. Basalt and gabbro, andesite and diorite, and rhyolite and granite are compositional (mineralogical) pairs, but basalt, andesite, and rhyolite are aphanitic and most commonly extrusive (volcanic), whereas gabbro, diorite, and granite are phaneritic and mostly intrusive (plutonic). The extrusive and intrusive members of each pair can usually be distinguished by texture, but remember that rocks in some shallow plutons may be aphanitic and rocks that formed in thick lava flows may be phaneritic. In other words, all of these rocks exist in a textural continuum.

The igneous rocks in Figure 4.9 are also differentiated by composition—that is, by their mineral content. Reading across the chart from rhyolite to andesite to basalt, for example, we see that the proportions of non-

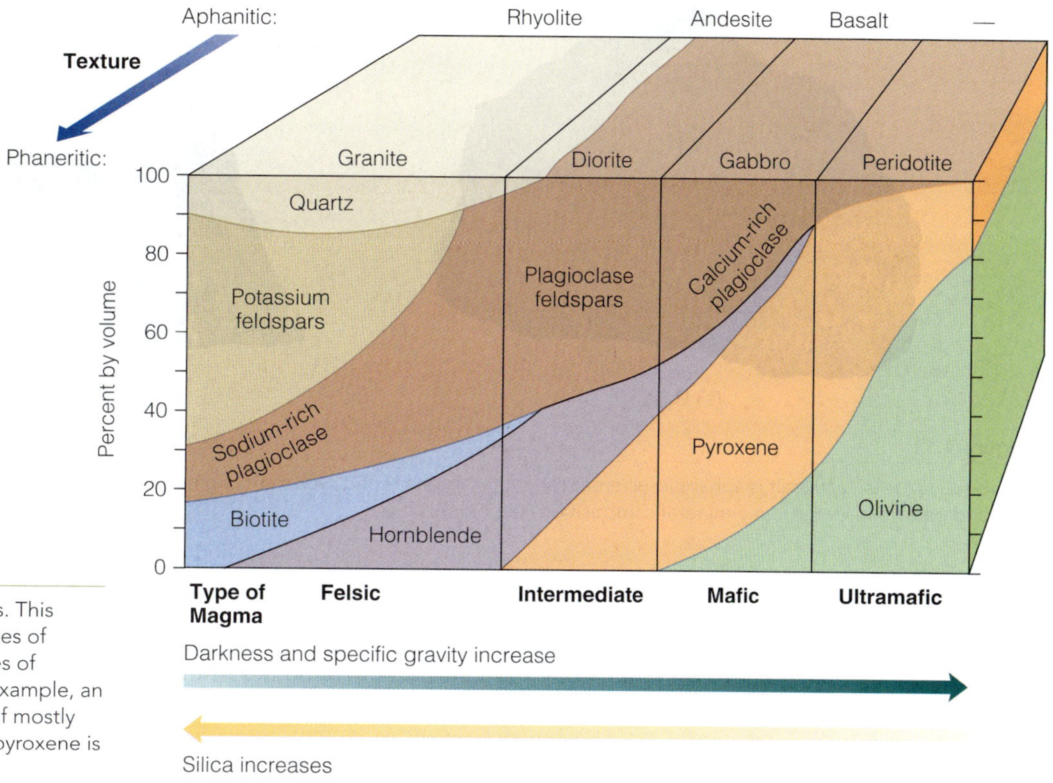

Figure 4.9

Classification of igneous rocks. This diagram shows the percentages of minerals as well as the textures of common igneous rocks. For example, an aphanitic (fine-grained) rock of mostly calcium-rich plagioclase and pyroxene is basalt.

ferromagnesian and ferromagnesian silicates change. The differences in composition, however, are gradual along a compositional continuum. In other words, there are rocks with compositions that correspond to the lines between granite and diorite, basalt and andesite, and so on.

Ultramafic Rocks Ultramafic rocks (<45% silica) are composed largely of ferromagnesian silicates. The ultramafic rock *peridotite* contains mostly olivine, lesser amounts of pyroxene, and usually a little plagioclase feldspar (Figures 4.9 and ■ 4.10). Pyroxenite, another ultramafic rock, is composed predominately of pyroxene. Because these minerals are dark, the rocks are generally black or dark green. Peridotite is probably the rock type that makes up the upper mantle (see Chapter 9). Ultramafic rocks in Earth's crust probably originate by concentration of the early-formed ferromagnesian minerals that separated from mafic magmas.

Ultramafic lava flows are known in rocks older than 2.5 billion years, but younger ones are rare or absent. The reason is that to erupt, ultramafic lava must have a near-surface temperature of about 1600°C; the surface temperatures of present-day mafic lava flows are between 1000° and 1200°C. During early Earth history, though, more radioactive decay heated the mantle to as much as 300°C hotter than now and ultramafic lavas could erupt onto the surface. Because the amount of heat has decreased over time, Earth has cooled, and eruptions of ultramafic lava flows ceased.

Basalt-Gabbro *Basalt* and *gabbro* are the aphanitic and phaneritic rocks, respectively, that crystallize from mafic magma (45–52% silica) (■ Figure 4.11). Thus, both have the same composition—mostly calcium-rich plagioclase and pyroxene, with smaller amounts of olivine and amphibole (Figure 4.9). Because they contain a large proportion of ferromagnesian silicates, basalt and gabbro are dark; those that are porphyritic typically contain calcium plagioclase or olivine phenocrysts.

Extensive basalt lava flows cover vast areas in Washington, Oregon, Idaho, and northern California (see Chapter 5). Oceanic islands such as Iceland, the Galápagos, the Azores, and the Hawaiian Islands are composed mostly of basalt, and basalt makes up the upper part of the oceanic crust.

Figure 4.10

This specimen of the ultramafic rock peridotite is made up mostly of olivine. Notice in Figure 4.9 that peridotite is the only phaneritic rock that does not have an aphanitic counterpart. Peridotite is rare at Earth's surface but is very likely the rock making up the mantle. Source: Sue Monroe

Figure 4.11

Mafic igneous rocks. (a) Basalt is aphanitic, whereas (b) gabbro is phaneritic. Notice the light reflected from the crystal faces in (b). Both basalt and gabbro have the same mineral composition (see Figure 4.9).

Gabbro is much less common than basalt, at least in the continental crust or where it can be easily observed. Small intrusive bodies of gabbro are present in the continental crust, but intermediate to felsic intrusive rocks are much more common. The lower part of the oceanic crust is composed of gabbro, however.

Andesite-Diorite Intermediate composition magma (53–65% silica) crystallizes to form *andesite* and *diorite,* which are compositionally equivalent fine- and coarse-grained igneous rocks (■ Figure 4.12). Andesite and diorite are composed predominately of plagioclase feldspar, with the typical ferromagnesian component being amphibole or biotite (Figure 4.9). Andesite is generally medium to dark gray, but diorite has a salt-and-pepper appearance because of its white to light gray plagioclase and dark ferromagnesian silicates (Figure 4.12).

Andesite is a common extrusive igneous rock formed from lava erupted in volcanic chains at convergent plate boundaries. The volcanoes of the Andes Mountains of South America and the Cascade Range in western North America are composed in part of andesite. Intrusive bodies of diorite are fairly common in the continental crust.

Figure 4.12

Intermediate igneous rocks. (a) Andesite has hornblende phenocrysts, so this is andesite porphyry. (b) Diorite has a salt-and-pepper appearance because it contains light-colored nonferromagnesian silicates and dark-colored ferromagnesian silicates.

Rhyolite-Granite *Rhyolite* and *granite* crystallize from felsic magma (>65% silica) and are therefore silica-rich rocks (■ Figure 4.13). They consist largely of potassium feldspar, sodium-rich plagioclase, and quartz, with perhaps some biotite and rarely amphibole (Figure 4.9). Because nonferromagnesian silicates predominate, rhyolite and granite are typically light colored. Rhyolite is fine grained, although most often it contains phenocrysts of potassium feldspar or quartz, and granite is coarse grained. Granite porphyry is also fairly common.

Rhyolite lava flows are much less common than andesite and basalt flows. Recall that the greatest control of magma viscosity is silica content. Thus, if felsic magma rises to the surface, it begins to cool, the pressure on it decreases, and gases are released explosively, usually yielding rhyolitic pyroclastic materials. The rhyolitic lava flows that do occur are thick and highly viscous and move only short distances.

Granite is a coarsely crystalline igneous rock with a composition corresponding to that of the field shown in Figure 4.9. Strictly speaking, not all rocks in this field are granites. For example, a rock with a composition close to the line separating granite and diorite is called *granodiorite*. To avoid the confusion that might result from introducing more rock names, we will follow the practice of referring to rocks to the left of the granite-diorite line in Figure 4.9 as *granitic*.

(a) Rhyolite

(b) Granite

Figure 4.13

Felsic igneous rocks. (a) Rhyolite and (b) granite are typically light colored because they contain mostly nonferromagnesian silicates. The dark spots in the granite are biotite mica. The white and pinkish minerals are feldspars, whereas the glassy minerals are quartz.

Granitic rocks are by far the most common intrusive igneous rocks, although they are restricted to the continents. Most granitic rocks were intruded at or near convergent plate margins during mountain-building episodes. When these mountainous regions are uplifted and eroded, the vast bodies of granitic rocks forming their cores are exposed. The granitic rocks of the Sierra Nevada of California form a composite body measuring about 640 km long and 110 km wide, and the granitic rocks of the Coast Ranges of British Columbia, Canada, are even more voluminous.

Pegmatite The term *pegmatite* refers to a particular texture rather than a specific composition, but most pegmatites are composed largely of quartz, potassium feldspar, and sodium-rich plagioclase, thus corresponding closely to granite. A few pegmatites are mafic or intermediate in composition and are appropriately called *gabbro* and *diorite pegmatites*. The most remarkable feature of pegmatites is the size of their minerals, which measure at least 1 cm across, and in some pegmatites they measure tens of centimeters or meters (■ Figure 4.14). Many pegmatites are associated with large granite intrusive bodies and are composed of minerals that formed from the water-rich magma that remained after most of the granite crystallized.

When magma cools and forms granite, the remaining water-rich magma has properties that differ from the magma from which it separated. It has a lower density and viscosity and commonly invades the adjacent rocks where minerals crystallize. This water-rich magma also contains a number of elements that rarely enter into the common minerals that form granite. Pegmatites that are essentially very coarsely crystalline granite are simple pegmatites, whereas those with minerals containing elements such as lithium, beryllium, cesium, boron, and several others are complex pegmatites. Some complex pegmatites contain 300 different mineral species, a few of which are important economically. In addition, several gem minerals such as emerald and aquamarine, both of which are varieties of the silicate mineral beryl, and tourmaline are found in some pegmatites. Many rare minerals of lesser value and well-formed crystals of common minerals, such as quartz, are also mined and sold to collectors and museums.

The formation and growth of mineral-crystal nuclei in pegmatites are similar to those processes in other magmas but with one critical difference: The water-rich magma from which pegmatites crystallize inhibits the formation of nuclei. However, some nuclei do form, and because the appropriate ions in the liquid can move easily and attach themselves to a growing crystal, individual minerals have the opportunity to grow very large.

Other Igneous Rocks A few igneous rocks, including tuff, volcanic breccia, obsidian, pumice, and scoria are identified primarily by their textures (■ Figure 4.15). Much of the fragmental material erupted by volcanoes is *ash*, a designation for pyroclastic materials measuring less than 2.0 mm, most of which consists of broken pieces or shards of volcanic glass (Figure 4.8i). The consolidation of ash forms the pyroclastic rock *tuff* (■ Figure 4.16a). Most tuff is silica-rich and light colored and is appropriately called *rhyolite tuff*. Some ash flows are so hot that as they come to rest, the ash particles fuse together and form a *welded tuff*. Consolidated deposits of larger pyroclastic materials, such as cinders, blocks, and bombs, are *volcanic breccia* (Figure 4.15).

Both *obsidian* and *pumice* are varieties of volcanic glass (Figure 4.16b, c). Obsidian may be black, dark gray, red, or brown, depending on the presence of iron.

Courtesy of Steve Stahl

(a)

Sue Monroel

(b)

Figure 4.14

(a) The light-colored rock is pegmatite exposed in the Black Hills of South Dakota. (b) A closeup view of a pegmatite specimen with minerals measuring 2 to 3 cm across. This is a simple pegmatite that has a composition much like that of granite.

Obsidian breaks with the conchoidal (smoothly curved) fracture typical of glass. Analyses of many samples indicate that most obsidian has a high silica content and is compositionally similar to rhyolite.

Pumice is a variety of volcanic glass containing numerous vesicles that develop when gas escapes through lava and forms a froth (Figure 4.16c). If pumice falls into water, it can be carried great distances because it is so porous and light that it floats. Another vesicular rock is *scoria*. It is more crystalline and denser than pumice, but it has more vesicles than solid rock (Figure 4.16d).

Geology Now Log into GeologyNow and select this chapter to work through a **Geology Interactive** activity on "Rock Laboratory" (click Rocks and the Rock Cycle→Rock Laboratory).

	Composition	Felsic ←→ Mafic
Texture	Vesicular	Pumice / Scoria
	Glassy	Obsidian
	Pyroclastic or fragmental	← Volcanic Breccia → Tuff/welded tuff

Figure 4.15

Classification of igneous rocks for which texture is the main consideration. Composition is shown, but it is not essential for naming these rocks.

What Would You Do ?

As the only member of your community with any geology background, you are considered the local expert on minerals and rocks. Suppose one of your friends brings you a rock specimen with the following features—composition: mostly potassium feldspar and plagioclase feldspar with about 10% quartz and minor amounts of biotite; texture: minerals average 3 mm across, but several potassium feldspars are up to 3 cm. Give the specimen a rock name and tell your friend as much as you can about the rock's history. Why are the minerals so large?

PLUTONS—THEIR CHARACTERISTICS AND ORIGINS

Unlike volcanism and the origin of volcanic rocks, we can study intrusive igneous activity only indirectly because **plutons,** intrusive igneous bodies, form when magma cools and crystallizes

Courtesy of David J. Matty

(a) Tuff

Sue Monroe

(b) Obsidian

Sue Monroe

(c) Pumice

Sue Monroe

(d) Scoria

Figure 4.16

Examples of igneous rocks classified primarily by their texture. (a) Tuff is made up of pyroclastic materials such as those shown in Figure 4.8i. (b) The natural glass obsidian. (c) Pumice is glassy and extremely vesicular. (d) Scoria is also vesicular, but it is darker, denser, and more crystalline than pumice.

within Earth's crust (see "Plutons" on pages 102 and 103). We can observe them only after erosion has exposed them at the surface. Furthermore, geolgists cannot duplicate the conditions under which plutons form except in small laboratory experiments. Accordingly, geologists face a greater challenge in interpreting the mechanisms whereby plutons form. Magma that cools to form plutons is emplaced in Earth's crust mostly at convergent and divergent plate boundaries, which are also areas of volcanism.

Geologists recognize several types of plutons based on their geometry (three-dimensional shape) and relationships to the country rock. In terms of their geometry, plutons are massive (irregular), tabular, cylindrical, or mushroom-shaped. Plutons are also either **concordant,** meaning they have boundaries that parallel the layering in the country rock, or **discordant,** with boundaries that cut across the country rock's layering (see "Plutons" on pages 102 and 103).

Dikes and Sills

Dikes and **sills** are tabular or sheetlike plutons, differing only in that dikes are discordant whereas sills are concordant (see "Plutons" on pages 102 and 103). Dikes are quite common; most of them are small bodies measuring 1 or 2 m across, but they range from a few centimeters to more than 100 m thick. Invariably they are emplaced within preexisting fractures or where fluid pressure is great enough for them to form their own fractures.

Erosion of the Hawaiian volcanoes exposes dikes in rift zones, the large fractures that cut across these volcanoes. The Columbia River basalts in Washington State (discussed in Chapter 5) issued from long fissures, and magma that cooled in the fissures formed dikes. Some of the large historic fissure eruptions are underlain by dikes; for example, dikes underlie both the Laki fissure eruption of 1783 in Iceland and the Eldgja fissure, also

Plutons

Intrusive bodies called plutons are common, but we see them at the surface only after deep erosion. Notice that they vary in geometry and their relationships to the country rock.

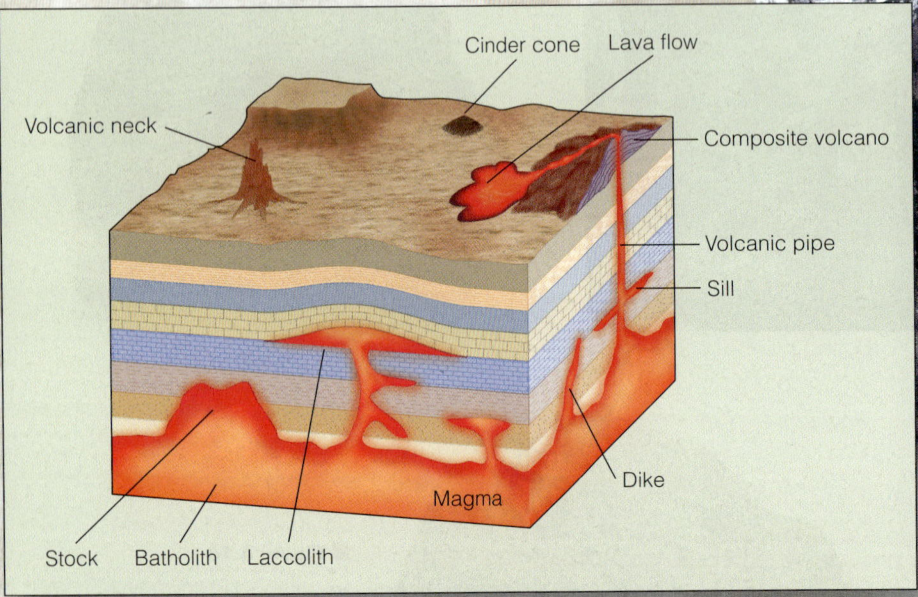

Block diagram showing various plutons. Some plutons cut across the layering in country rock and are discordant, whereas others parallel the layering and are concordant.

Part of the Sierra Nevada batholith in Yosemite National Park, California. The batholith, consisting of multiple intrusions of granitic rock, is more than 600 km long and up to 110 km wide. To appreciate the scale in this image, the waterfall has a descent of 435 m.

James S. Monroe

A volcanic neck in Monument Valley Tribal Park, Arizona. This landform is 457 m high. Most of the original volcano was eroded, leaving only this remnant.

Granitic rocks of a small stock at Castle Crags State Park, California.

The dark materials in this image are igneous rocks, whereas the light layers are sedimentary. Notice that the sill parallels the layering, so it is concordant. The dike, though, clearly cuts across the layering and is discordant. Sills and dikes have sheetlike geometry, but in this view we can see them in only two dimensions.

Martin G. Miller/ Visuals Unlimited

http://formontana.net/cb.html

Crown Butte in Montana is an eroded laccolith standing about 300 m above the surrounding plain. The magma making up this small pluton was intruded about 50 million years ago.

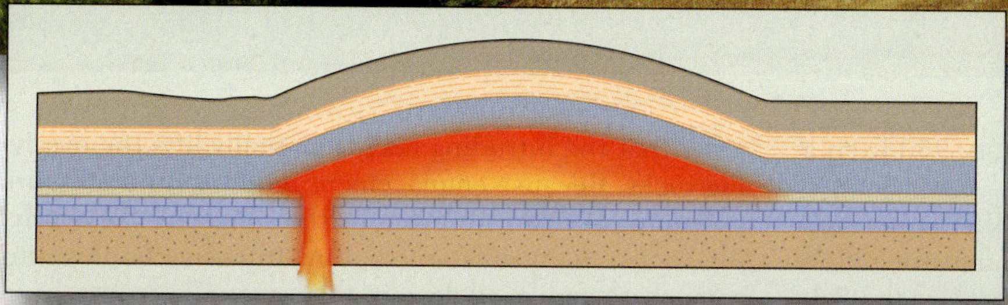

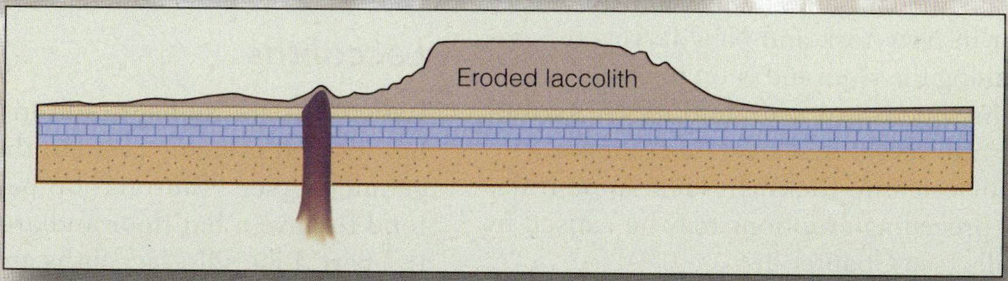

Diagrams showing the evolution of an eroded laccolith.

GEOFOCUS 4.1

Some Remarkable Volcanic Necks

We mentioned that as an extinct volcano weathers and erodes, a remnant of the original mountain may persist as a volcanic neck. The origin of volcanic necks is well known, but these isolated monoliths rising above otherwise rather flat land are scenic, awe-inspiring, and the subject of legends. They are found in many areas of recently active volcanism. A small volcanic neck rising only 79 m above the surface in the town of Le Puy, France, is the site of the 11th-century chapel of Saint Michel d'Aiguilhe (■ Figure 1). It is so steep that materials and tools used in its construction had to be hauled up in baskets.

Perhaps the most famous volcanic neck in the United States is Shiprock, New Mexico, which rises nearly 550 m above the surrounding plain and is visible from 160 km away. Radiating outward from this conical structure are three vertical dikes that stand like walls above the adjacent countryside (■ Figure 2a). According to one legend, Shiprock, or *Tsae-bidahi,* meaning "winged rock," represents a giant bird that brought the Navajo people from the north. The same legend holds that the dikes are snakes that turned to stone.

An absolute age determined for one of the dikes indicates that Shiprock is about 27 million years old. When the original volcano formed, apparently during explosive eruptions, rising magma penetrated various rocks including the Mancos Shale, the rock unit now exposed at the surface adjacent to Shiprock. The rock that makes up Shiprock itself is tuff-breccia, consisting of fragmented volcanic debris as well as pieces of metamorphic, sedimentary, and igneous rocks.

Geologists agree that Devil's Tower in northeastern Wyoming cooled from a small body of magma and that erosion has exposed it in its present form (■ Figure 2b). However, opinion is divided on whether it is a volcanic neck or an eroded laccolith. In either case, the rock that makes up Devil's Tower is 45 to 50 million years old, and President Theodore Roosevelt designated this impressive landform as our first national monument in 1906. At 260 m high, Devil's Tower is visible from 48 km away and served as a landmark for early travelers in this area. It achieved further distinction in 1977 when it was featured in the film *Close Encounters of the Third Kind.*

The Cheyenne and Sioux Indians call Devil's Tower *Mateo Tepee,* meaning "Grizzly Bear Lodge." It was also called the "Bad God's Tower," and reportedly "Devil's Tower" is a translation from this phrase. The tower's most conspicuous features are the near-vertical lines that, according to Cheyenne legends, are scratch marks made by a gigantic grizzly bear. One legend holds that the bear made the scratches while pursuing a group of children. An-

in Iceland, where eruptions occurred in A.D. 950 from a fissure nearly 30 km long.

Concordant sheetlike plutons are **sills;** many sills are a meter or less thick, although some are much thicker. A well-known sill in the United States is the Palisades sill that forms the Palisades along the west side of the Hudson River in New York and New Jersey. It is exposed for 60 km along the river and is up to 300 m thick. Most sills were intruded into sedimentary rocks, but eroded volcanoes also reveal that sills are commonly injected into piles of volcanic rocks. In fact, some inflation of volcanoes preceding eruptions may be caused by the injection of sills (see Chapter 5).

In contrast to dikes, which follow zones of weakness, sills are emplaced when the fluid pressure is so great that the intruding magma actually lifts the overlying rocks. Because emplacement requires fluid pressure exceeding the force exerted by the weight of the overlying rocks, many sills are shallow intrusive bodies, but some were emplaced deep in the crust.

Laccoliths

Laccoliths are similar to sills in that they are concordant, but instead of being tabular, they have a mushroomlike geometry (see "Plutons" on pages 102 and 103). They tend to have a flat floor and are domed up in their central part. Like sills, laccoliths are rather shallow intrusive bodies that actually lift up the overlying rocks when magma is intruded. In this case, however, the rock layers other tells of six brothers and a woman also pursued by a grizzly bear. One brother carried a rock, and when he sang a song it grew into Devil's Tower, safely carrying the brothers and woman out of the bear's reach.

Although not nearly as interesting as the Cheyenne legends, the origin for the "scratch marks" is well understood. These lines actually formed at the intersections of *columnar joints,* fractures that form in response to the cooling and contraction that occur in some plutons and lava flows (see Chapter 5). The columns outlined by these fractures are up to 2.5 m across, and the pile of rubble at the tower's base is simply an accumulation of collapsed columns.

Geology⇌Now Log into GeologyNow and select this chapter to work through a **Geology Interactive** activity on "Rock Laboratory" (click Rocks and the Rock Cycle→Rock Laboratory).

Richard List/Corbis

Figure 1

This volcanic neck in Le Puy, France, rises 79 m above the surrounding surface. Workers on the Chapel of Saint Michel d'Aiguilhe had to haul building materials and tools up in baskets.

Courtesy of Frank Hanna

(a)

James S. Monroe

(b)

Figure 2

(a) Shiprock, a volcanic neck in northwestern New Mexico, rises nearly 550 m above the surrounding plain. One of the dikes radiating from Shiprock is in the foreground. (b) Devil's Tower in northeastern Wyoming. The vertical lines result from intersections of fractures called columnar joints (see Chapter 5).

are arched up over the pluton. Most laccoliths are rather small bodies. Well-known laccoliths in the United States are in the Henry Mountains of southeastern Utah, and several buttes in Montana are eroded laccoliths.

Volcanic Pipes and Necks

A volcano has a cylindrical conduit known as a **volcanic pipe** that connects its crater with an underlying magma chamber. Through this structure magma rises to the surface. When a volcano ceases to erupt, its slopes are attacked by water, gases, and acids and it erodes, but the magma that solidified in the pipe is commonly more resistant to alteration and erosion. Consequently, much of the volcano is eroded but the pipe remains as a remnant called a **volcanic neck**. Several volcanic necks are found in the southwestern United States, especially in Arizona and New Mexico, and others are recognized elsewhere (see Geo-Focus 4.1 and "Plutons" on pages 102–103).

Batholiths and Stocks

By definition a **batholith**, the largest of all plutons, must have at least 100 km^2 of surface area, and most are far larger. A **stock**, in contrast, is similar but smaller. Some stocks are simply parts of large plutons that once exposed by erosion are batholiths (see "Plutons" on pages 102 and 103). Both batholiths and stocks are generally discordant, although locally they may be concordant, and batholiths, especially, consist of multiple intrusions. In other words, a batholith is a large composite body produced by repeated, voluminous intrusions of magma in the same region. The coastal batholith of Peru, for instance, was emplaced during a period of 60 to 70 million years and is made up of as many as 800 individual plutons.

The igneous rocks that make up batholiths are mostly granitic, although diorite may also be present. Batholiths and stocks are emplaced mostly near convergent plate boundaries during episodes of mountain building. One example is the Sierra Nevada batholith of California (see the chapter opening photo), which formed over millions of years during a mountain-building episode known as the Nevadan orogeny. Later uplift and erosion exposed this huge composite pluton at the surface. Other large batholiths in North America include the Idaho batholith, the Boulder batholith in Montana, and the Coast Range batholith in British Columbia, Canada.

Mineral resources are found in rocks of batholiths and stocks and in the adjacent country rocks. The copper deposits at Butte, Montana, are in rocks near the margins of the granitic rocks of the Boulder batholith. Near Salt Lake City, Utah, copper is mined from the mineralized rocks of the Bingham stock, a composite pluton composed of granite and granite porphyry. Granitic rocks also are the primary source of gold, which forms from mineral-rich solutions moving through cracks and fractures of the igneous body.

HOW ARE BATHOLITHS INTRUDED INTO EARTH'S CRUST?

Geologists realized long ago that the origin of batholiths posed a space problem. What happened to the rock that was once in the space now occupied by a batholith? One proposed answer was that no displacement had occurred, but rather that batholiths formed in place by alteration of the country rock through a process called *granitization*. According to this view, granite did not originate as magma but rather from hot, ion-rich solutions that simply altered the country rock and transformed it into granite. Granitization is a solid-state phenomenon, so it is essentially an extreme type of metamorphism (see Chapter 7).

Granitization is no doubt a real phenomenon, but most granitic rocks show clear evidence of an igneous origin. For one thing, if granitization had taken place, we would expect the change from country rock to granite to take place gradually over some distance. However, in almost all cases no such gradual change can be detected. In fact, most granitic rocks have what geologists refer to as sharp contacts with adjacent rocks. Another feature that indicates an igneous origin for granitic rocks is the alignment of elongate minerals parallel with their contacts, which must have occurred when magma was injected.

A few granitic rocks lack sharp contacts and gradually change in character until they resemble the adjacent country rock. These probably did originate by granitization. In the opinion of most geologists, only small quantities of granitic rock could form by this process, so it cannot account for the huge volume of granitic rocks of batholiths. Accordingly, geologists conclude that an igneous origin for almost all granite is clear, but they still must deal with the space problem.

One solution is that these large igneous bodies melted their way into the crust. In other words, they simply assimilated the country rock as they moved upward (Figure 4.6). The presence of inclusions, especially near the tops of some plutons, indicates that assimilation does occur. Nevertheless, as we noted, assimilation is a limited process because magma cools as country rock is assimilated. Calculations indicate that far too little heat is available in magma to assimilate the huge quantities of country rock necessary to make room for a batholith.

Geologists now generally agree that batholiths were emplaced by *forceful injection* as magma moved upward. Recall that granite is derived from viscous felsic magma and therefore rises slowly. It appears that the magma deforms and shoulders aside the country rock, and as it rises farther, some of the country rock fills the space beneath the magma (■ Figure 4.17a). A somewhat analogous situation was discovered in which large masses of sedimentary rock known as *rock salt* rise through the overlying rocks to form *salt domes* (Figure 4.17b–d).

Salt domes are recognized in several areas of the world, including the Gulf Coast of the United States. Layers of rock salt exist at some depth, but salt is less dense than most other types of rock materials. When under pressure, it rises toward the surface even though it remains solid, and as it moves up, it pushes aside and deforms the country rock. Natural examples of rock salt flowage are known, and it can easily be demonstrated experimentally. In the arid Middle East, for example, salt moving up in the manner described actually flows out at the surface.

Some batholiths do indeed show evidence of having been emplaced forcefully by shouldering aside and deforming the country rock. This mechanism probably occurs in the deeper parts of the crust where temperature and pressure are high and the country rocks are easily deformed in the manner described. At shallower depths, the crust is more rigid and tends to deform by fracturing. In this environment, batholiths may move upward by **stoping**, a process in which rising magma detaches and engulfs pieces of country rock (■ Figure 4.18).

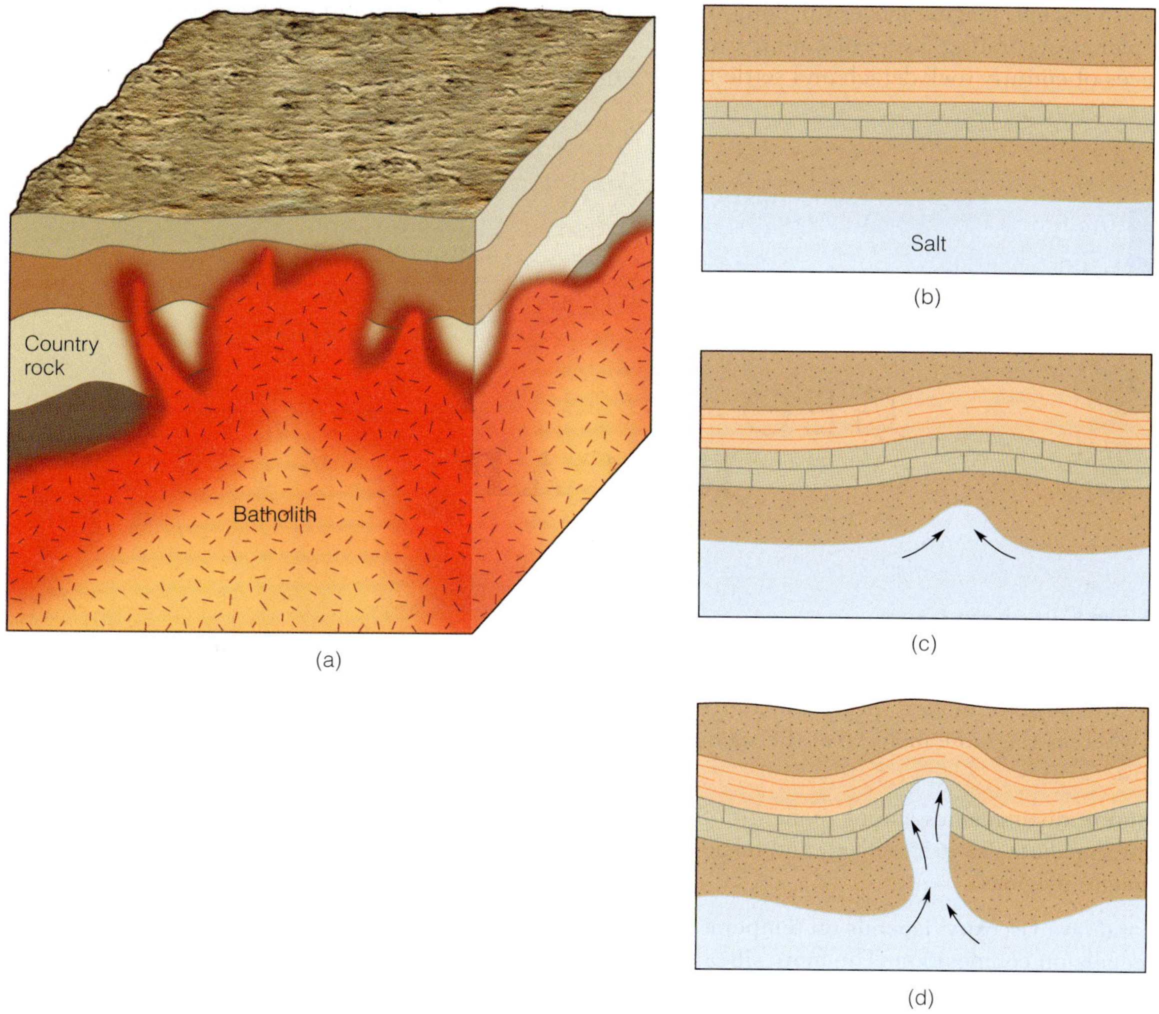

Figure 4.17

(a) Emplacement of a pluton by forceful injection. As the magma rises, it shoulders aside and deforms the country rock. (b–d) Three stages in a somewhat analogous situation when a salt dome forms by upward migration of the rock under pressure.

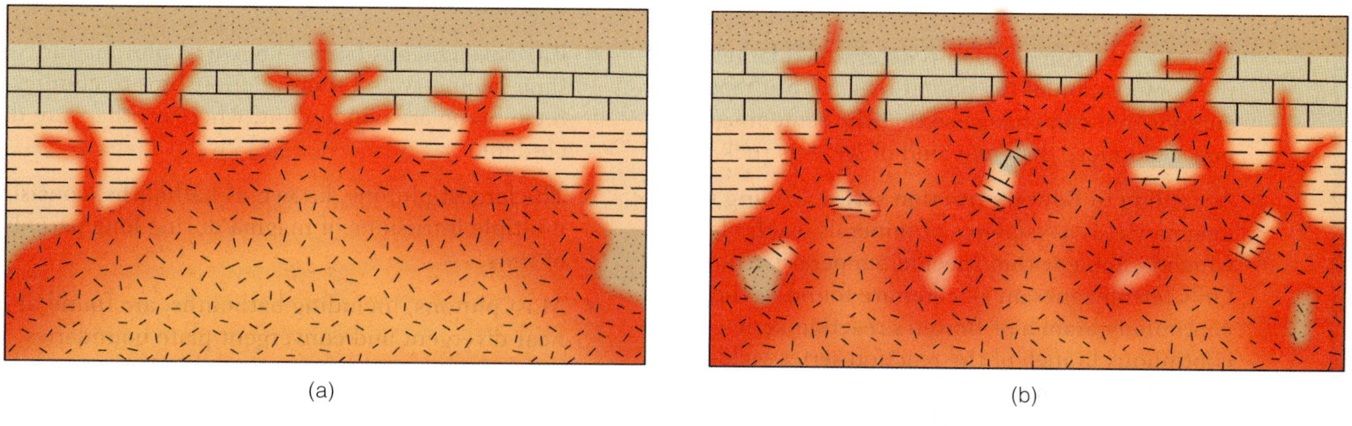

Figure 4.18

Emplacement of a batholith by stoping. (a) Magma is injected into fractures and planes between layers in the country rock. (b) Blocks of country rock are detached and engulfed in the magma, thereby making room for the magma to rise farther. Some of the engulfed blocks might be assimilated, and some may remain as inclusions (Figure 4.6).

According to this concept, magma moves up along fractures and the planes separating layers of country rock. Eventually, pieces of country rock detach and settle into the magma. No new room is created during stoping; the magma simply fills the space formerly occupied by country rock (Figure 4.18).

4 GEO RECAP

Chapter Summary

- *Magma* is the term for molten rock below Earth's surface, whereas the same material at the surface is called *lava*.
- Silica content distinguishes among mafic (45–52% silica), intermediate (53–65% silica), and felsic (>65% silica) magmas.
- Magma and lava viscosity depends on temperature and especially on composition: The more silica, the greater the viscosity.
- Minerals crystallize from magma and lava when small crystal nuclei form and grow.
- Rapid cooling accounts for the aphanitic textures of volcanic rocks, whereas comparatively slow cooling yields the phaneritic textures of plutonic rocks. Igneous rocks with markedly different sized minerals are porphyritic.
- Igneous rock composition is determined largely by the composition of the parent magma, but magma composition can change so that the same magma may yield more than one kind of igneous rock.
- According to Bowen's reaction series, cooling mafic magma yields a sequence of minerals, each of which is stable within specific temperature ranges. Only ferromagnesian silicates are found in the discontinuous branch of Bowen's reaction series. The continuous branch of the reaction series yields only plagioclase feldspars that become increasingly enriched in sodium as cooling occurs.
- A chemical change in magma may take place as early-formed ferromagnesian silicates form and, because of their density, settle in the magma.
- Compositional changes also take place in magma when it assimilates country rock or one magma mixes with another.
- Geologists recognize two broad categories of igneous rocks: volcanic or extrusive and plutonic or intrusive.
- Texture and composition are the criteria used to classify igneous rocks, although a few are defined only by texture.
- Crystallization from water-rich magma results in very large minerals in rocks known as pegmatite. Most pegmatite has an overall composition similar to granite.
- Intrusive igneous bodies known as plutons vary in their geometry and their relationships to country rock: Some are concordant, whereas others are discordant.
- The largest plutons, known as batholiths, consist of multiple intrusions of magma during long periods of time.
- Most plutons, including batholiths, are found at or near divergent and convergent plate boundaries.

Important Terms

aphanitic texture (p. 95)
assimilation (p. 94)
batholith (p. 105)
Bowen's reaction series (p. 91)
concordant pluton (p. 101)
country rock (p. 94)
crystal settling (p. 93)
dike (p. 101)
discordant pluton (p. 101)
felsic magma (p. 89)
igneous rock (p. 89)
intermediate magma (p. 89)
laccolith (p. 104)
lava flow (p. 89)
mafic magma (p. 89)
magma (p. 89)
magma chamber (p. 91)
magma mixing (p. 94)
phaneritic texture (p. 95)
pluton (p. 100)
plutonic (intrusive igneous) rock (p. 89)
porphyritic texture (p. 95)
pyroclastic (fragmental) texture (p. 95)
pyroclastic materials (p. 89)
sill (p. 104)
stock (p. 105)
stoping (p. 106)
vesicle (p. 95)
viscosity (p. 90)
volcanic neck (p. 105)
volcanic pipe (p. 105)
volcanic (extrusive igneous) rock (p. 89)

Review Questions

1. A dike is a discordant pluton, whereas a _____ is concordant.
a._____batholith; b._____volcanic neck; c._____laccolith; d._____stock; e._____ash fall.
2. An aphanitic igneous rock composed mostly of pyroxenes and calcium-rich plagioclase is
a._____granite; b._____obsidian; c._____rhyolite; d._____diorite; e._____basalt.
3. The size of the mineral grains that make up an igneous rock is a useful criterion for determining whether the rock is _____ or _____.
a._____volcanic/plutonic; b._____discordant/concordant; c._____vesicular/fragmental; d._____porphyritic/felsic; e._____ultramafic/igneous.
4. Magma characterized as intermediate
a._____flows more readily than mafic magma; b._____has between 53% and 65% silica; c._____crystallizes to form granite and rhyolite; d._____cools to form rocks that make up most of the oceanic crust; e._____is one from which ultramafic rocks are derived.
5. The phenomenon by which pieces of country rock are detached and engulfed by rising magma is known as
a._____stoping; b._____assimilation; c._____magma mixing; d._____Bowen's reaction series; e._____crystal settling.
6. An igneous rock that has minerals large enough to see without magnification is said to have a(n) _____ texture and is probably _____.
a._____laccolithic/pegmatite; b._____fragmental/felsic; c._____isometric/magmatic; d._____phaneritic/plutonic; e._____fragmental/obsidian.
7. Which one of the following statements about batholiths is false?
a._____They consist of multiple voluminous intrusions; b._____They form mostly at convergent plate boundaries during mountain building; c._____They consist of a variety of volcanic rocks, but especially basalt; d._____They must have at least 100 km^2 of surface area; e._____Though locally concordant, they are mostly discordant.

8. An igneous rock characterized as a porphyry is one
a._____that formed by crystal settling and assimilation; b._____possessing minerals of markedly different sizes; c._____made up largely of potassium feldspar and quartz; d._____that forms when pyroclastic materials are consolidated; e._____resulting from very rapid cooling.

9. Which pair of igneous rocks have the same texture?
a._____basalt-andesite; b._____granite-rhyolite; c._____pumice-obsidian; d._____tuff-diorite; e._____scoria-lapilli.

10. One process by which magma changes composition is
a._____crystal settling; b._____rapid cooling; c._____explosive volcanism; d._____fracturing; e._____plate convergence.

11. Two aphanitic igneous rocks have the following compositions: Specimen 1: 15% biotite; 15% sodium-rich plagioclase, 60% potassium feldspar, and 10% quartz. Specimen 2: 10% olivine, 55% pyroxene, 5% hornblende, and 30% calcium-rich plagioclase. Use Figure 4.9 to classify these rocks. Which would be the darkest and most dense?

12. How does a sill differ from a dike? (A diagram would be helpful.)

13. How do crystal settling and assimilation bring about compositional changes in magma? Also, give evidence that these processes actually take place.

14. Describe or diagram the sequence of events that lead to the origin of a volcanic neck.

15. Describe a porphyritic texture, and explain how it might originate.

16. Why are felsic lava flows so much more viscous than mafic ones?

17. How does a pegmatite form, and why are their mineral crystals so large?
18. Compare the continuous and discontinuous branches of Bowen's reaction series. Why are potassium feldspar and quartz not part of either branch?
19. You analyze the mineral composition of a thick sill and find that it has a mafic lower part but its upper part is more intermediate. Given that it all came from a single magma injected at one time, how can you account for its compositional differences?
20. What kind(s) of evidence would you look for to determine whether the granite in a batholith crystallized from magma or originated by granitization?

World Wide Web Activities

Geology Now Assess your understanding of this chapter's topics with additional quizzing and comprehensive interactivities at

http://earthscience.brookscole.com/changingearth4e

as well as current and up-to-date weblinks, additional readings, and InfoTrac College Edition exercises.

Volcanism and Volcanoes

CHAPTER 5
OUTLINE

Geology Now *This icon, which appears throughout the book, indicates an opportunity to explore interactive tutorials, animations, or practice problems available on the GeologyNow website at http://earthscience.brookscole.com/changingearth4e.*

unt Vesuvius has erupted 80 times since A.D. 79, most recently in 14. Naples, Italy, and other communities are either on the flanks of the cano or nearby. The Bay of Naples is on the right. Source: Stone/Getty ges

OBJECTIVES

At the end of this chapter, you will have learned that:

- In addition to lava flows, erupting volcanoes eject pyroclastic materials, especially ash, and various gases.
- Geologists identify the basic types of volcanoes by their eruptive style, composition, and shape.
- Although all volcanoes are unique, most are identified as shield volcanoes, cinder cones, or composite volcanoes.
- Volcanoes characterized as lava domes tend to erupt explosively and thus are dangerous.
- Active volcanoes in the United States are found in Hawaii, Alaska, and the Cascade Range of the Pacific Northwest.
- Eruptions in Hawaii and Alaska are commonplace, but only two eruptions have occurred in the continental United States during the 1900s and one in 2004. Canada has had no eruptions during historic time.
- Some eruptions yield vast sheets of lava or pyroclastic materials rather than volcanoes.
- Geologists have devised the volcanic explosivity index as a measure of an eruption's size.
- Some volcanoes are carefully monitored to help geologists anticipate eruptions.
- Most volcanoes are located in belts at or near divergent and convergent plate boundaries.

Introduction

No other geologic phenomenon has captured the public imagination more than volcanism. Eruptions are featured in television documentaries, and movies portray the destruction wrought by lava flows and explosive volcanism. Depictions of lava flows in movies notwithstanding, however, humans usually have little to fear from these incandescent streams of molten rock, although in 1977 and 2002 lava flows killed dozens of people in the Democratic Republic of the Congo. Lava flows may destroy buildings and roadways and cover otherwise productive agricultural land, and some eruptions, especially those at convergent plate boundaries, are explosive and pose considerable danger to nearby populated areas.

One of the best-known volcanic catastrophes ever recorded was the A.D. 79 eruption of Mount Vesuvius that destroyed the thriving Roman communities of Pompeii, Herculaneum, and Stabiae in what is now Italy (see the chapter opening photo and ■ Figure 5.1). Fortunately for us, Pliny the

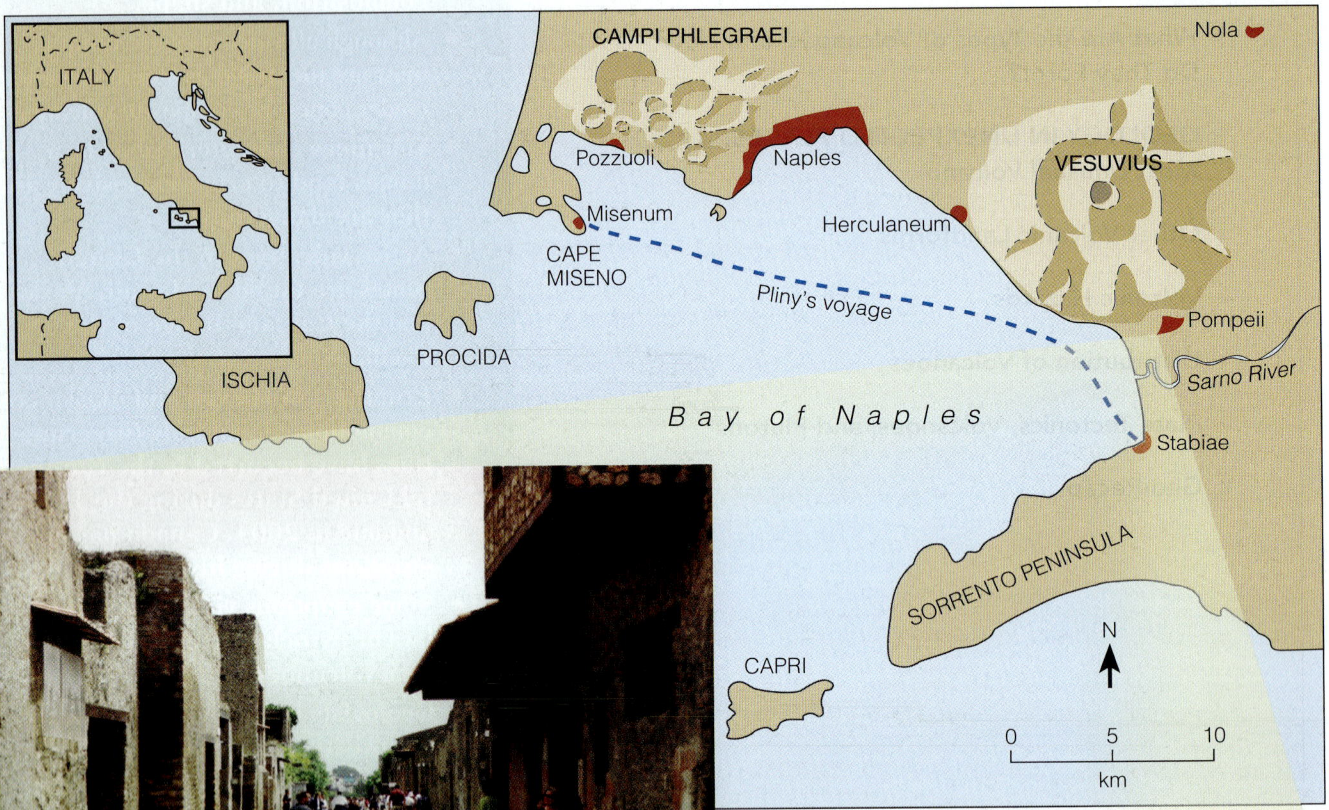

(a)

Sue Monroe

(b)

Sue Monroe

(c)

■ **Figure 5.1**

(a) The Mount Vesuvius region on the shore of the Bay of Naples in Italy. Vesuvius erupted in A.D. 79 and destroyed the cities of Pompeii, Herculaneum, and Stabiae. (b) The excavated ruins of Pompeii are a popular tourist attraction. (c) Body casts of some of the volcano's victims in Pompeii.

Table 5.1

Some Notable Volcanic Eruptions

Date	Volcano	Deaths
Apr. 10, 1815	Tambora, Indonesia	117,000 killed, including deaths from eruption, famine, and disease.
Oct. 8, 1822	Galunggung, Java	Pyroclastic flows and mudflows killed 4011.
Mar. 2, 1856	Awu, Indonesia	2806 died in pyroclastic flows.
Aug. 27, 1883	Krakatau, Indonesia	More than 36,000 died; most killed by tsunami.
June 7, 1892	Awu, Indonesia	1532 died in pyroclastic flows.
May 8, 1902	Mount Pelée, Martinique	Nuée ardente engulfed St. Pierre and killed 28,000.
Oct. 24, 1902	Santa Maria, Guatemala	5000 died during eruption.
May 19, 1919	Kelut, Java	Mudflows devastated 104 villages and killed 5110.
Jan. 21, 1951	Lamington, New Guinea	Pyroclastic flows killed 2942.
Mar. 17, 1963	Agung, Indonesia	1148 perished during eruption.
May 18, 1980	Mount St. Helens, Washington	63 killed; 600 km^2 of forest devastated.
Mar. 28, 1982	El Chichón, Mexico	Pyroclastic flows killed 1877.
Nov. 13, 1985	Nevado del Ruiz, Colombia	Minor eruption triggered mudflows that killed 23,000.
Aug. 21, 1986	Oku volcanic field, Cameroon	Cloud of CO_2 released from Lake Nyos killed 1746.
June 15, 1991	Mount Pinatubo, Philippines	About 281 killed during eruption; 83 died in later mudflows; 358 died of illness.
July 1999	Soufrière Hills, Montserrat	19 killed; 12,000 evacuated.
Jan. 17, 2002	Nyiragongo, Zaire	Lava flow killed 80–100 in Goma.

Younger recorded the event in detail; his uncle, Pliny the Elder, died while trying to investigate the eruption. In fact, Pliny the Younger's account is so vivid that Mount Vesuvius and similar eruptions during which huge quantities of pumice are blasted into the air are called *plinian.*

Pompeii, a city of about 20,000 people and only 9 km downwind from the volcano, was buried in nearly 3 m of pyroclastic materials that covered all but the tallest buildings (Figure 5.1). About 2000 victims have been discovered in the city, but certainly far more were killed. Pompeii was covered by volcanic debris rather gradually, but surges of incandescent volcanic materials in glowing avalanches swept through Herculaneum, quickly burying the town to a depth of about 20 m. Since A.D. 79, Mount Vesuvius has erupted 80 times, most violently in 1631 and 1906; it last erupted in 1944. Ongoing volcanic and seismic activity in this area poses a continuing threat to the many cities and towns along the shores of the Bay of Naples (Figure 5.1).

One very good reason to study volcanic eruptions is that they illustrate the complex interactions among Earth's systems. Volcanism, especially the emission of gases and pyroclastic materials, has an immediate and profound impact on the atmosphere, hydrosphere, and biosphere, at least in the vicinity of an eruption. And in some cases the effects are worldwide, as they were following the eruptions of Tambora in 1815, Krakatau in 1883, and Pinatubo in 1991. Furthermore, the fact that lava flows and explosive eruptions cause property damage, injuries, fatalities (Table 5.1), and at least short-term atmospheric changes indicates that volcanic eruptions are catastrophic events, at least from the human perspective.

Ironically, though, when considered in the context of Earth history, volcanism is actually a constructive process. The atmosphere and surface waters most likely resulted from the emission of gases during Earth's early history, and oceanic crust is continuously produced by volcanism at spreading ridges. Oceanic islands such as the Hawaiian Islands, Iceland, and the Azores owe their existence to volcanism, and weathering of lava flows, pyroclastic materials, and volcanic mudflows in tropical areas such as Indonesia converts them to productive soils.

People who live in Hawaii, southern Alaska, the Philippines, Japan, and Iceland are well aware of volcanic eruptions, but eruptions in the continental United States have occurred only three times since 1914, all in the **Cascade Range,** which stretches from northern California through Oregon and Washington and into southern British Columbia, Canada. Canada has had no eruptions during historic time. Ancient and ongoing volcanism in the western United States has yielded interesting features, several of which are featured in this chapter.

VOLCANISM

The term **volcanism** immediately brings to mind lava flows, and it does in fact encompass such activity, but the term refers specifically to those processes whereby lava and its contained gases as well as pyroclastic materials are extruded onto the surface or into the atmosphere. Volcanism yields distinctive landforms, particularly volcanoes, as well as volcanic (extrusive) igneous rocks. About 550 volcanoes are presently *active;* that is, they have erupted during historic time, but only 12 or so are erupting at any one time. Most of this activity is minor, although large eruptions are certainly not uncommon.

All of the terrestrial planets and Earth's Moon were volcanically active during their early histories, but now volcanoes are known only on Earth and on one or two other bodies in the solar system. Triton, one of Neptune's moons, probably has active volcanoes, and Jupiter's moon Io is by far the most volcanically active body in the solar system. Many of its hundred or so volcanoes are erupting at any given time.

In addition to active volcanoes, Earth has numerous *dormant* volcanoes that have not erupted during historic time but may do so in the future. Prior to its eruption in A.D. 79, Mount Vesuvius had not been active in human memory. The largest volcanic outburst in the last 50 years was when Mount Pinatubo in the Philippines erupted in 1991 after lying dormant for 600 years. Some volcanoes have not erupted in historic time and show no signs of doing so again; thousands of these *extinct* or *inactive* volcanoes are known.

Volcanic Gases

Samples from present-day volcanoes indicate that 50% to 80% of all volcanic gases are water vapor, with lesser amounts of carbon dioxide, nitrogen, sulfur gases, especially sulfur dioxide and hydrogen sulfide, and very small amounts of carbon monoxide, hydrogen, and chlorine. In areas of recent volcanism, such as Lassen Volcanic National Park in California, emission of gases continues, and one cannot help noticing the rotten-egg odor of hydrogen sulfide gas (■ Figure 5.2).

When magma rises toward the surface, the pressure is reduced and the contained gases begin to expand. In highly viscous, felsic magma, expansion is inhibited and gas pressure increases. Eventually, the pressure may become great enough to cause an explosion and produce pyroclastic materials such as ash. In contrast, low-viscosity mafic magma allows gases to expand and escape easily. Accordingly, mafic magma generally erupts rather quietly.

The amount of gases contained in magma varies but is rarely more than a few percent by weight. Even though volcanic gases constitute a small proportion of magma, they can be dangerous and in some cases have had far-reaching climatic effects.

Sue Monroe

■ **Figure 5.2**

Volcanic gases emitted at the Sulfur Works in Lassen Volcanic National Park, California.

Most volcanic gases quickly dissipate in the atmosphere and pose little danger to humans, but on several occasions they have caused fatalities. In 1783, toxic gases, probably sulfur dioxide, erupting from Laki fissure in Iceland had devastating effects. About 75% of the nation's livestock died, and the haze resulting from the gas caused lower temperatures and crop failures; about 24% of Iceland's population died as a result of the ensuing Blue Haze Famine. The country suffered its coldest winter in 225 years in 1783–1784, with temperatures 4.8°C below the long-term average. The eruption also produced what Benjamin Franklin called a "dry fog" that was responsible for dimming the intensity of sunlight in Europe. The severe winter of 1783–1784 in Europe and eastern North America is attributed to the presence of this "dry fog" in the upper atmosphere.

In 1986, in the African nation of Cameroon, 1746 people died when a cloud of carbon dioxide engulfed them. The gas accumulated in the waters of Lake Nyos, which occupies a volcanic caldera. Scientists disagree about what caused the gas to suddenly burst forth from the lake, but once it did, it flowed downhill along the surface because it was denser than air. In fact, the density and velocity of the gas cloud were great enough to flatten vegetation, including trees, a few kilometers from the lake. Unfortunately, thousands of animals and many people, some as far as 23 km from the lake, were asphyxiated.

Residents of the island of Hawaii have coined the term *vog* for volcanic smog. Kilauea volcano has been erupting

(a)

(b)

Figure 5.3

(a) This hollow beneath a now-solidified lava flow is a lava tube. (b) Part of this lava tube's roof has collapsed, forming a skylight through which the active flow can be seen.

continuously since 1983, releasing small amounts of lava, copious quantities of carbon dioxide, and about 1800 tons of sulfur dioxide per day. Carbon dioxide is no problem because it dissipates quickly in the atmosphere, but sulfur dioxide produces a haze and the unpleasant odor of sulfur. As long as Kilauea volcano erupts, Hawaii will have a vog problem. Vog probably poses little or no health risk for tourists, but a long-term threat exists for residents of the west side of the island where vog is most common.

Lava Flows

Lava flows are portrayed in movies and on television as fiery streams of incandescent rock that usually pose a great danger to humans (see Geo-Focus 5.1). Actually, lava flows are the least dangerous manifestation of volcanism, although they may destroy buildings and cover agricultural land. Most lava flows do not move particularly fast, and because they are fluid, they follow existing low areas. Thus, once a flow erupts from a volcano, determining the path it will take is fairly easy, and anyone in areas likely to be affected can be evacuated.

Even low-viscosity lava flows generally do not move very rapidly. Flows can move much faster, though, when their margins cool to form a channel, and especially when insulated on all sides as in a *lava tube,* where a speed of more than 50 km/hr has been recorded. A conduit known as a **lava tube** within a lava flow forms when the margins and upper surface of the flow solidify. Thus confined and insulated, the flow moves rapidly and over great distances. As an eruption ceases, the tube drains, leaving an empty tunnel-like structure (■ Figure 5.3a). Part of the roof of a lava tube may collapse to form a *skylight* through which an active flow can be observed (Figure 5.3b), or access can be gained to an inactive lava tube. In Hawaii, lava moves through lava tubes many kilometers long and in some cases discharges into the sea.

Geologists define two types of lava flows, both named for Hawaiian flows. A **pahoehoe** (pronounced *pah-hoy-hoy*) flow has a ropy surface much like taffy (■ Figure 5.4a). The surface of an **aa** (pronounced *ah-ah*) flow is characterized by rough, jagged, angular blocks and fragments (Figure 5.4b). Pahoehoe flows are less viscous than aa flows; indeed, the latter are viscous enough to break up into blocks and move forward as a wall of rubble.

Pressure on the partly solidified crust of a still-moving lava flow causes the surface to buckle into *pressure ridges* (■ Figure 5.5a). Gases escaping from a flow hurl globs of lava into the air, which fall back to the surface and adhere to one another, thus forming small, steep-sided *spatter cones,* or spatter ramparts if they are elongated (Figure 5.5b). Spatter cones a few meters high are common on lava flows in Hawaii, and you can see ancient ones in Craters of the Moon National Monument in Idaho.

Columnar jointing is common in many lava flows, especially mafic flows, but it is also found in other types of flows and in some intrusive igneous rocks. Once a lava flow stops moving, it contracts as it cools and produces forces that cause fractures called *joints* to open. On the surface of a lava flow, the joints are commonly polygonal (often six-sided) cracks that extend downward, thus forming parallel columns with their long axes perpendicular to the cooling surface (■ Figure 5.6). Excellent examples of columnar jointing are found in many areas.

Much of the igneous rock in the upper part of the oceanic crust is a distinctive type consisting of bulbous masses of basalt that resemble pillows, hence the name **pillow lava.** It was long recognized that pillow lava forms when lava is rapidly chilled beneath water, but its formation was not observed until 1971. Divers near

GEOFOCUS 5.1

Lava Flows Pose Little Danger to Humans—Usually

In the text we made the point that incandescent streams of molten rock are impressive and commonly portrayed in movies as a danger to humans. We also mentioned that lava flows are actually the least dangerous manifestation of volcanic activity, although they may destroy buildings, highways, and cropland. We should note, though, that on some occasions they have been directly or indirectly responsible for fatalities.

Some of the most recent fatalities caused by lava flows took place during January 2002 in the city of Goma in the Democratic Republic of the Congo (formerly Zaire). Nyiragongo, the volcano from which the flows issued, has erupted 19 times since 1884, and during two of these eruptions lava flows caused fatalities (■ Figure 1). One of several volcanoes in Africa's Virunga Volcanic Chain, Nyiragongo is a composite volcano along the East African Rift. A lava lake in its summit caldera was active for decades, and in 1977 the lake suddenly drained through fissures and covered several square kilometers with fluid lava flows. Unfortunately, about 70 people (300 in another estimate) and a herd of elephants perished in the flows that moved at speeds of 60 km/hr.

In the massive eruption at Nyiragongo on January 17, 2002, a huge plume of ash rose above the mountain and three fast-moving lava flows descended along its western and eastern flanks. Fourteen villages near the volcano were destroyed, and within one day one of the flows sliced through the city of Goma, 19 km south of Nyiragongo, destroying everything in a 60-m-wide path. The lava ignited many fires, and huge explosions took place where it came into contact with gasoline storage tanks.

The number of fatalities in Goma is uncertain, but most estimates are between 80 and 100. Explosions triggered by the lava flow rather than the flow itself were responsible for most of these deaths. About 400,000 people were evacuated from the city for three days. Though of little consolation to the survivors of those who died, the death toll was actually quite low considering that a lava flow moved quickly through a large, densely populated city.

Our statement that "lava flows are the least dangerous manifestation of volcanic activity" is correct, even though there are some exceptions. Far greater dangers are posed by explosive eruptions during which huge quantities of pyroclastic materials and gases are ejected into the atmosphere and form volcanic mudflows (lahars). Fatalities from these eruptions and associated activity might be in the thousands or tens of thousands (Table 5.1).

Images of the World

(a)

AFP/Corbis

(b)

■ **Figure 1**

(a) Nyiragongo is a composite volcano in central Africa that has erupted 19 times since 1884. It stands 3470 m high. (b) Part of one of the January 17, 2002, lava flows that killed dozens of people in Goma, Democratic Republic of the Congo. The lava flow is moving across the runway at Goma airport.

J. D. Griggs/USGS

(a)

Robert Tilling/USGS

(b)

■ **Figure 5.4**

(a) A pahoehoe lava flow in Hawaii. Notice the smooth lobes at the end of the flow and the smooth, folded texture of the flow's surface. (b) An aa flow advances over an older pahoehoe flow in Hawaii.

Hawaii saw pillows form when a blob of lava broke through the crust of an underwater lava flow and cooled almost instantly, forming a pillow-shaped structure with a glassy exterior. The remaining fluid inside then broke through the crust of the pillow, repeating the process and resulting in an accumulation of interconnected pillows (■ Figure 5.7).

Pyroclastic Materials

In addition to lava flows, erupting volcanoes eject pyroclastic materials, especially **ash,** a designation for pyroclastic particles that measure less than 2.0 mm (■ Figure 5.8). In some cases ash is ejected into the atmosphere and settles to the surface as an *ash fall.* In 1947, ash erupted from Mount

T. J. Takahashi/USGS

(a)

J. D. Griggs/USGS

(b)

■ **Figure 5.5**

(a) Pressure ridge on a 1982 lava flow in Hawaii. (b) Spatter rampart that formed in 1984 in Hawaii.

(a) (b)

■ **Figure 5.6**

(a) Columnar joints in a 60-million-year-old lava flow at the Giant's Causeway in Northern Ireland. As the lava cooled, fractures formed and intersected to form mostly five- and six-sided columns. (b) Surface view of columns on the island of Staffa in Scotland.

Hekla in Iceland fell 3800 km away on Helsinki, Finland. In contrast to an ash fall, an *ash flow* is a cloud of ash and gas that flows along or close to the land surface. Ash flows can move faster than 100 km/hr, and some cover vast areas.

In populated areas adjacent to volcanoes, ash falls and ash flows pose serious problems, and volcanic ash in the atmosphere is a hazard to aviation. Since 1980, about 80 aircraft have been damaged when they encoun-

(a) (b)

■ **Figure 5.7**

(a) These bulbous masses of pillow lava form when magma erupts under water. (b) Ancient pillow lava on land in Marin County, California. Two complete pillows and many broken ones are visible.

Sue Monroe

Figure 5.8

Pyroclastic materials. The large object on the left is a volcanic bomb; it is about 20 cm long. The streamlined shape of bombs indicates they were erupted as globs of magma that cooled and solidified as they descended. The granular objects in the upper right are pyroclastic materials known as lapilli. The pile of gray-white material on the lower right is ash.

tered clouds of volcanic ash. The most serious incident took place in 1989 when ash from Redoubt volcano in Alaska caused all four jet engines to fail on KLM Flight 867. The plane carrying 231 passengers nearly crashed when it fell more than 3 km before the crew could restart the engines. The plane landed safely in Anchorage, Alaska, but it required $80 million in repairs.

In addition to ash, volcanoes erupt *lapilli,* consisting of pyroclastic materials that measure from 2 to 64 mm, and *blocks* and *bombs,* both larger than 64 mm (Figure 5.8). Bombs have a twisted, streamlined shape, which indicates they were erupted as globs of magma that cooled and solidified during their flight through the air. Blocks, in contrast, are angular pieces of rock ripped from a volcanic conduit or pieces of a solidified crust of a lava flow. Because of their size, lapilli, bombs, and blocks are confined to the immediate area of an eruption.

WHAT ARE THE TYPES OF VOLCANOES, AND HOW DO THEY FORM?

Simply put, a **volcano** is a hill or mountain that forms around a vent where lava, pyroclastic materials, and gases erupt. Some volcanoes are conical, but others are bulbous, steep-sided masses of magma, and some resemble an inverted shield lying on the ground. In all cases, though, volcanoes have a conduit or conduits leading to a magma chamber beneath the surface. The Roman deity of fire, Vulcan, was the inspiration for calling these mountains volcanoes, and because of their danger and obvious connection to Earth's interior they have been held in awe by many cultures.

Probably no other geologic phenomenon, except perhaps earthquakes, has so much lore associated with it. In Hawaiian legends, the volcano goddess Pele resides in the crater of Kilauea on Hawaii. During one of her frequent rages, Pele causes earthquakes and lava flows, and she may hurl flaming boulders at those who offend her. Native Americans in the Pacific Northwest tell of a titanic battle between the volcano gods Skel and Llao to account for huge eruptions that took place about 6600 years ago in Oregon and California. Pliny the Elder (A.D. 23–79), mentioned in the Introduction, believed that before eruptions "the air is extremely calm and the sea quiet, because the winds have already plunged into the earth and are preparing to reemerge."*

Geologists recognize several major types of volcanoes, but one must realize that each volcano is unique in its history of eruptions and development. For instance, the frequency of eruptions varies considerably; the Hawaiian volcanoes and Mount Etna on Sicily have erupted repeatedly, whereas Pinatubo in the Philippines erupted in 1991 for the first time in 600 years. Some volcanoes are complex mountains that have the characteristics of more than one type of volcano.

Most volcanoes have a circular depression known as a **crater** at their summit, or on their flanks, that forms by explosions or collapse. Craters are generally less than 1 km across, whereas much larger rimmed depressions on volcanoes are **calderas.** In fact, some volcanoes have a summit crater within a caldera. Calderas are huge structures that form following voluminous eruptions during which part of a magma chamber drains and the mountain's summit collapses into the vacated space below. An excellent example is misnamed Crater Lake in Oregon (■ Figure 5.9). Crater Lake is actually a steep-rimmed caldera that formed about 6600 years ago in the manner just described; it is more than 1200 m deep and measures 9.7 × 6.5 km. As impressive as Crater Lake is, it is not nearly as large as some other calderas, such as the Toba caldera in Sumatra, which is 100 km long and 30 km wide.

Shield Volcanoes

Shield volcanoes resemble the outer surface of a shield lying on the ground with the convex side up (see "Types of Volcanoes" for pages 122 and 123). They have low, rounded profiles with gentle slopes ranging from about

*Quoted from page 40 in M. Krafft, *Volcanoes: Fire from the Earth* (New York: Harry N. Abrams, 1993).

Types of Volcanoes

All volcanoes are structures resulting from the eruption of lava and pyroclastic materials but they are all unique in their history of eruptions and development. Nevertheless, most are conveniently classified as one of the types shown here: shield, cinder cone, composite, and lava dome. There are also places where eruptions of very fluid lava take place along fissures and volcanoes do not develop.

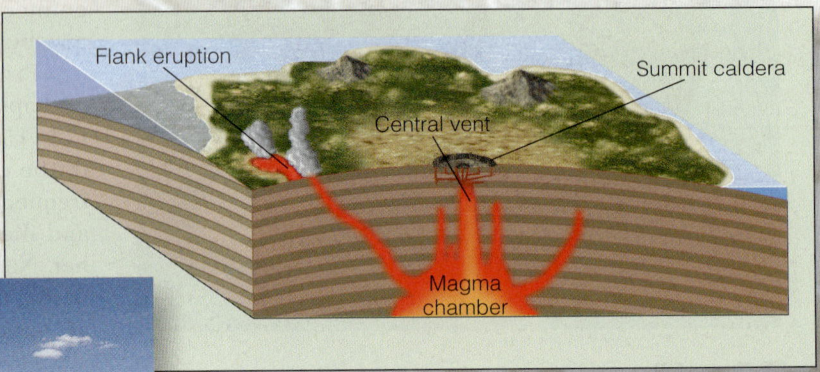

Shield volcanoes consist of numerous thin basalt lava flows that build up mountains with slopes rarely exceeding 10 degrees.

James S. Monroe

Crater Mountain in Lassen County, California is an extinct shield volcano. It is about 10 km across and stands 460 m high. The depression at its summit is a 2-km-wide crater.

Robert Tilling/USGS

View of Mauna Loa, an active shield volcano on Hawaii, with its upper 1.5 km covered by snow. Mauna Loa is the largest mountain on Earth; it measures about 100 km across its base, stands more than 9.5 km above the seafloor, and is made up of an estimated 50,000 km^3 of material.

James S. Monroe

A 230-m-high cinder cone in Lassen Volcanic National Park in California.

The image on the right shows the large, bowl-shaped crater at the summit of this cinder cone. It last erupted during the 1600's.

James S. Monroe

Composite volcanoes, or stratovolcanoes, are composed mostly of lava flows and pyroclastic materials of intermediate composition, although volcanic mudflow deposits are also common.

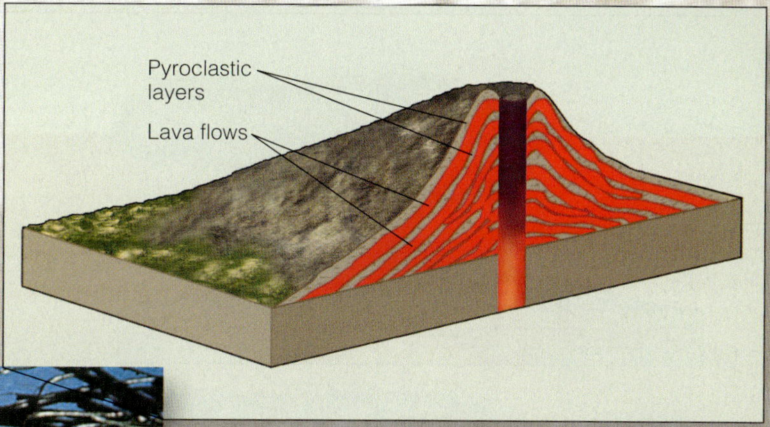

Mayon volcano in the Philippines a nearly symmetrical composite volcano that last erupted during 1999.

R. Solkowski/Consulting Geologist

Two views of Mount Shasta, a huge composite volcano in northern California. Mount Shasta is about 24 km across its base and rises more than 3400 m above its surroundings.

Sue Monroe

This view of Mount Shasta from the north shows a cone known as Shastina on the flank of the larger mountain.

Wayne E. Moore

T.P. Miller/USGS

This steep-sided lava dome lies atop Novarupta in Katmai National Park and Preserve in Alaska.

Chaos Crags in the distance are made up of at least four lava domes that formed less than 1200 years ago in Lassen Volcanic National Park in California. The debris in the foreground, called Chaos Jumbles, formed when parts of the domes collapsed.

James S. Monroe

Figure 5.9

Events leading to the origin of Crater Lake, Oregon, which is actually a caldera. (a, b) Ash clouds and ash flows partly drain the magma chamber. (c) The collapse of the summit and formation of the caldera. (c) Postcaldera eruptions partly cover the caldera floor, and a small cinder cone known as Wizard Island forms. (e) View from the rim of Crater Lake showing Wizard Island. Source: From Howell Williams, Crater Lake: The Story of Its Origin (Berkeley, Calif. University of California Press): Illustrations from p. 84 © 1941 Regents of the University of California, © renewed 1969, Howell Williams.

2 to 10 degrees; they are composed mostly of mafic flows that had low viscosity, so the flows spread out and formed thin, gently sloping layers. Eruptions from shield volcanoes, sometimes called *Hawaiian-type eruptions,* are quiet compared to those of volcanoes such as Mount St. Helens. Lava most commonly rises to the surface with little explosive activity, so it usually poses little danger to humans. Lava fountains, some up to 400 m high, contribute some pyroclastic materials to shield volcanoes, but otherwise they are composed largely of basalt lava flows; flows make up more than 99% of the Hawaiian volcanoes above sea level.

Although eruptions of shield volcanoes tend to be rather quiet, some of the Hawaiian volcanoes have, on occasion, produced sizable explosions when groundwater instantly vaporizes as it comes in contact with magma. One such explosion in 1790 killed about 80 warriors in a party headed by Chief Keoua, who was leading them across the summit of Kilauea volcano.

The current activity at Kilauea is impressive for another reason; it has been erupting continuously since January 3, 1983, making it the longest recorded eruption. During these 20 years, more than 2.3 km^3 of molten rock has flowed out at the surface, much of it reaching the sea and forming 2.2 km^2 of new property on the island of Hawaii. Unfortunately, lava flows from Kilauea have also destroyed about 200 homes and caused some $61 million in damages.

Shield volcanoes such as those of the Hawaiian Islands and Iceland are most common in the ocean basins, but some are also present on the continents—in East Africa, for instance. The island of Hawaii is made up of five huge shield volcanoes, two of which, Kilauea and Mauna Loa, are active much of the time. Mauna Loa is nearly 100 km across its base and stands more than 9.5 km above the surrounding seafloor; it has a volume estimated at 50,000 km^3, making it the world's largest volcano. By contrast, a very large volcano in the continental United States, Mount Shasta in California, has a volume of only about 350 km^3.

Solarfilm/GeoScience Features

■ **Figure 5.10**

Eldfell, a cinder cone in Iceland, began erupting in 1973 and in two days grew to 100 m high. The steam visible on the left side of the image resulted from aa lava flowing into the sea. Another cinder cone known as Helgafel is also visible.

Cinder Cones

Small, steep-sided **cinder cones** made up of particles resembling cinders form when pyroclastic materials accumulate around a vent from which they erupted (see "Types of Volcanoes" on pages 122 and 123). Cinder cones are small, rarely exceeding 400 m high, with slope angles up to 33 degrees, depending on the angle that can be maintained by the angular pyroclastic materials. Many of these small volcanoes have a large, bowl-shaped crater, and if they issue any lava flows, they usually break through the base or lower flanks of the mountains. Although all cinder cones are conical, their symmetry varies from those that are almost perfectly symmetrical to those that formed when prevailing winds caused pyroclastic materials to build up higher on the downwind side of the vent.

Many cinder cones form on the flanks or within the calderas of larger volcanoes and represent the final stages of activity, particularly in areas of basaltic volcanism. Wizard Island in Crater Lake, Oregon, is a small cinder cone that formed after the summit of Mount Mazama collapsed to form a caldera (Figure 5.9). Cinder cones are common in the southern Rocky Mountain states, particularly New Mexico and Arizona, and many others are in California, Oregon, and Washington.

In 1973, on the Icelandic island of Heimaey, the town of Vestmannaeyjar was threatened by a new cinder cone. The initial eruption began on January 23, and within two days a cinder cone, later named Eldfell, rose to about 100 m above the surrounding area (■ Figure 5.10). Pyroclastic materials from the volcano buried parts of the town, and by February a massive aa lava flow was advancing toward the town. The flow's leading edge ranged from 10 to 20 m thick, and its central part was as much as 100 m thick. The residents of Vestmannaeyjar sprayed the leading edge of the flow with seawater in an effort to divert it from the town. The flow was in fact diverted, but how effective the efforts of the townspeople were is not clear; they may have been simply lucky.

Composite Volcanoes (Stratovolcanoes)

Pyroclastic layers as well as lava flows, both of intermediate composition, are found in **composite volcanoes,** which are also called *stratovolcanoes* (see "Types of Volcanoes" on pages 122 and 123). As the lava flows cool, they typically form andesite; recall that intermediate lava flows are more viscous than mafic ones that yield basalt. Geologists use the term **lahar** for volcanic mudflows, which are common on composite volcanoes. A lahar may form when rain falls on unconsolidated pyroclastic materials and creates a muddy slurry that moves downslope (■ Figure 5.11). On November 13, 1985, a minor eruption of Nevado del Ruiz in Colombia melted snow and ice on the volcano, causing lahars that killed about 23,000 people (Table 5.1).

Composite volcanoes differ from shield volcanoes and cinder cones in composition as noted above, and their overall shape differs, too. Remember that shield volcanoes have very low slopes, whereas cinder cones are small, steep-sided, conical mountains. In marked contrast, composite volcanoes are steep-sided near their summits, perhaps as much as 30 degrees, but the slope decreases toward the base, where it may be no more than 5 degrees. Mayon volcano in the Philippines is one of the most nearly symmetrical composite volcanoes anywhere. It erupted in 1999 for the 13th time during the 1900s.

When most people think of volcanoes, they picture the graceful profiles of composite volcanoes, which are

GEOLOGY IN UNEXPECTED PLACES

A Most Unusual Volcano

Perhaps the East African Rift is not an unexpected place for volcanoes, but one known as Oldoinyo Lengai or Ol Doinyo Lengai in Tanzania is certainly among Earth's most peculiar volcanoes. Oldoinyo Lengai, which means "Mountain of God" in the Masai language, is an active composite volcano (it last erupted in August 2002), standing about 2890 m high. It is in an east–west belt of about 20 volcanoes near the southern part of the East African Rift. Its most notable feature, however, is its eruptions of magma that cools to form carbonitite, an igneous rock with at least 50% carbonate minerals, mostly calcite ($CaCO_3$) and dolomite [$CaMg(CO_3)_2$]. In fact, carbonitite looks much like the metamorphic rock marble (see Chapter 7).

Remember that based on silica content most magma varies from mafic to felsic, but only rarely does magma have a significant amount of carbonate minerals. At Oldoinyo Lengai, the carbonitite magma typically has very low viscosity and is fluid at temperatures of only 540° to 595°C, reflecting the low melting temperatures of carbonate minerals. As a result it flows rather quickly; it is not incandescent but rather looks like black mud. Its minerals are chemically unstable, however, so they react with water in the atmosphere and their color changes to pale gray very quickly. ■ Figure 1 shows a small cone within the volcano's crater spewing black lava; the older lava on the cone became white over several months.

■ Figure 1

This small cone lies in the crater of Oldoinyo Lengai in Tanzania. The black lava cools to form carbonitite, which consists of at least 50% carbonate minerals.

Courtesy of Frederick A. Belton

U.S. Dept. of Interior/USGS/David Johnson

(a)

Darrell G. Herd/USGS

(b)

■ Figure 5.11

(a) Homes partly buried by a volcanic mudflow, or lahar, on June 15, 1991, following the eruption of Mount Pinatubo in the Philippines. (b) Air view of Armero, Colombia, where at least 23,000 people died in lahars that inundated the area in 1985.

the typical large volcanoes found on the continents and island arcs. And some of these volcanoes are indeed large; Mount Shasta in northern California is made up of about 350 km^3 of material and measures about 20 km across its base. In fact, it dominates the skyline when approached from any direction. Other familiar composite volcanoes are several in the Cascade Range of the Pacific Northwest as well as Fujiyama in Japan and Mount Vesuvius in Italy (see the chapter opening photo).

Mount Pinatubo in the Philippines erupted violently on June 15, 1991. Huge quantities of gas and an estimated 3–5 km^3 of ash were discharged into the atmosphere, making this the world's largest eruption since 1912. Fortunately, warnings of an impending eruption were heeded, and 200,000 people were evacuated from around the volcano. The eruption was still responsible for 722 deaths (Table 5.1).

Lava Domes

Although some volcanoes show features of more than one of the types discussed so far, most can be classified as shield volcanoes, cinder cones, or composite volcanoes. One more type of volcano deserves our attention, however. **Lava domes,** also known as *volcanic domes* and *plug domes,* are steep-sided, bulbous mountains that form when viscous felsic magma, and occasionally intermediate magma, is forced toward the surface (see "Types of Volcanoes" on pages 122 and 123). Because felsic magma is so viscous, it moves upward very slowly and only when the pressure from below is great.

Beginning in 1980, a number of lava domes were emplaced in the crater of Mount St. Helens in Washington; most of these were destroyed during subsequent eruptions. Since 1983, Mount St. Helens has been characterized by sporadic dome growth, and renewed eruptions in 2004. In June 1991, a lava dome in Japan's Unzen volcano collapsed under its own weight, causing a flow of debris and hot ash that killed 43 people in a nearby town.

Lava dome eruptions are some of the most violent and destructive. In 1902, viscous magma accumulated beneath the summit of Mount Pelée on the island of Martinique. Eventually the pressure increased until the side of the mountain blew out in a tremendous explosion, ejecting a mobile, dense cloud of pyroclastic materials and a glowing cloud of gases and dust called a **nuée ardente** (French for "glowing cloud"). The pyroclastic flow followed a valley to the sea, but the nuée ardente jumped a ridge and engulfed the city of St. Pierre (■ Figure 5.12).

A tremendous blast hit St. Pierre leveling buildings; hurling boulders, trees, and pieces of masonry down the streets; and moving a 3-ton statue 16 m. Accompanying

Reuters/Corbis

(a)

American Museum of Natural History

(b)

■ **Figure 5.12**

(a) St. Pierre, Martinique, after it was destroyed by a nuée ardente erupted from Mount Pelée in 1902. Only 2 of the city's 28,000 inhabitants survived. (b) A large nuée ardente from Mount Pelée a few months after the one that destroyed St. Pierre.

the blast was a swirling cloud of incandescent ash and gases with an internal temperature of 700°C that incinerated everything in its path. The nuée ardente passed through St. Pierre in two or three minutes, only to be followed by a firestorm as combustible materials burned and casks of rum exploded. But by then most of the 28,000 residents of the city were already dead. In fact, in the area covered by the nuée ardente, only 2 survived!* One survivor was on the outer edge of the nuée ardente, but even there he was terribly burned and his family and neighbors were all killed. The other survivor, a stevedore incarcerated the night before for disorderly conduct, was in a windowless cell partly below ground level. He remained in his cell badly burned for four days after the eruption until rescue workers heard his cries for help. He later became an attraction in the Barnum and Bailey Circus where he was advertised as "The only living object that survived in the 'Silent City of Death' where 40,000 beings were suffocated, burned or buried by one belching blast of Mont Pelee's terrible volcanic eruption."**

Geology⇌Now Log into GeologyNow and select this chapter to work through a **Geology Interactive** activity on "Volcanic Landforms" (click Volcanism→Volcanic Landforms).

OTHER VOLCANIC LANDFORMS

The term *volcano* immediately brings to mind the magnificent composite volcanoes. However, as noted earlier, even though some volcanoes are the typical mountains we envision, numerous volcanoes with other shapes are found in many areas (see "Types of Volcanoes" on pages 122 and 123). In fact, in some areas of volcanism, volcanoes fail to develop at all. For instance, during *fissure eruptions,* fluid lava pours out and simply builds up rather flat-lying areas, whereas huge explosive eruptions might yield *pyroclastic sheet deposits,* which, as their name implies, have a sheetlike geometry.

Fissure Eruptions and Basalt Plateaus

Some 164,000 km^2 of eastern Washington and parts of Oregon and Idaho were covered by overlapping basalt lava flows between 17 and 5 million years ago. Now known as the Columbia River basalts, they are well exposed in canyons eroded by the Snake and Columbia Rivers (■ Figure 5.13a). Rather than being erupted from a central vent, these flows issued from long cracks or fissures and are thus known as **fissure eruptions.** Lava erupted from these fissures was so fluid (had such low viscosity) that it simply spread out, covering vast areas building up a **basalt plateau,** a broad, flat, elevated area underlain by lava flows (Figure 5.13b).

The Columbia River basalt flows have an aggregate thickness of about 1000 m, and some individual flows cover huge areas—for example, the Roza flow, which is 30 m thick, advanced along a front about 100 km wide and covered 40,000 km^2.

Fissure eruptions and basalt plateaus are not common, although several large areas of such features are known. Currently this kind of activity occurs only in Iceland. Iceland has a number of volcanic mountains, but the bulk of the island is composed of basalt flows erupted from fissures. Two major fissure eruptions, one in A.D. 930 and the other in 1783, account for about half of the magma erupted in Iceland during historic time. The 1783 eruption issued from the Laki fissure, which is 25 km long; lava flowed several tens of kilometers from the fissure and in one place filled a valley to a depth of about 200 m.

Pyroclastic Sheet Deposits

More than 100 years ago, geologists were aware of vast areas covered by felsic volcanic rocks a few meters to hundreds of meters thick. It seemed improbable that these could be vast lava flows, but it seemed equally unlikely that they were ash fall deposits. Based on observations of historic pyroclastic flows, such as the nuée ardente erupted by Mount Pelée in 1902, it seems that these ancient rocks originated as pyroclastic flows—hence the name **pyroclastic sheet deposits** (Figure 5.13c).

They cover far greater areas than any observed during historic time, however, and apparently erupted from long fissures rather than from a central vent. The pyroclastic materials of many of these flows were so hot that they fused together to form *welded tuff.*

Geologists now think that major pyroclastic flows issue from fissures formed during the origin of calderas. For instance, pyroclastic flows erupted during the formation of a large caldera now occupied by Crater Lake, Oregon (Figure 5.9).

Similarly, the Bishop Tuff of eastern California erupted shortly before the formation of the Long Valley caldera. Interestingly, earthquake activity in the Long Valley caldera and nearby areas beginning in 1978 may indicate that magma is moving upward beneath part of the caldera. Thus, the possibility of future eruptions in that area cannot be discounted.

*Although reports commonly claim that only two people survived the eruption, at least 69 and possibly as many as 111 people survived beyond the extreme margins of the nuée ardente and on ships in the harbor. Many, however, were badly injured.

**Quoted from A. Scarth, *Vulcan's Fury: Man Against the Volcano* (New Haven, CT: Yale University Press, 1999), p. 177.

Figure 5.13

(a) About 20 lava flows of the Columbia River basalts, exposed in the canyon of the Grand Ronde River in Washington. (b) Block diagram showing fissure eruptions and the origin of a basalt plateau. (c) Pyroclastic flow deposits that issued from Mount Pinatubo on June 16, 1991, in the Philippines. Some of the flows moved 16 km from the volcano and filled this valley to depths of 50 to 200 m.

VOLCANIC HAZARDS

We have discussed several volcanic hazards such as lava flows, nuée ardentes, gases, and lahars in this chapter. Lava flows and nuée ardentes obviously are threats only during an eruption, but lahars and landslides may take place even when no eruption has occurred for a long time (■ Figure 5.14). Certainly the most vulnerable areas in the United States are Alaska, Hawaii, California, Oregon, and Washington, but some other places in the western part of the country might also experience renewed volcanism.

What Would You Do?

No one doubts that some of the Cascade Range volcanoes will erupt again, but we don't know when or how large these eruptions will be. A job transfer takes you to a community in Oregon that has several nearby large volcanoes. You have some concerns about future eruptions. What kinds of information would you seek out before buying a home in this area? In addition, as a concerned citizen, can you make any suggestions about what should be done in the case of a large eruption?

How Large Is an Eruption, and How Long Do Eruptions Last?

Geologists have devised several ways of expressing the size of a volcanic eruption. One, called the destructiveness index, is based on the area covered by lava or pyroclastic materials during an eruption. Geologists also rank eruptions in terms of their intensity and magnitude, but these have both been incorporated into the more widely used **volcanic explosivity index (VEI)** (■ Figure 5.15). Unlike the Richter Magnitude Scale for earthquakes (see Chapter 8), the VEI is only semiquantitative, being based partly on subjective criteria.

The volcanic explosivity index (VEI) ranges from 0 (gentle) to 8 (cataclysmic) and is based on several aspects of an eruption, such as the volume of material explosively ejected and the height of the eruption plume. However, the volume of lava, fatalities, and property damage are not considered. For instance, the 1985 eruption of Nevado del Ruiz in Colombia killed 23,000 people, yet has a VEI of only 3. In contrast, the huge eruption (VEI = 6) of Novarupta in Alaska in 1912 caused no fatalities or injuries. Since A.D. 1500, only the 1815 eruption of Tambora had a value of 7; it was both large and deadly (Table 5.1). Nearly 5700 eruptions during the last 10,000 years have been assigned VEI numbers, but none has exceeded 7, and most (62%) were assigned a value of 2.

The duration of eruptions varies considerably. Fully 42% of about 3300 historic eruptions lasted less than one month. About 33% erupted for one to six months, but some 16 volcanoes have been active more or less continuously for more than 20 years. Stromboli and Mount Etna in Italy and Erta Ale in Ethiopia are good examples. For some explosive volcanoes, the time from the onset of their eruptions to the climactic event is weeks or months. A case in point is the colossal explosive eruption of Mount St. Helens on May 18, 1980, that occurred two months after eruptive activity began. Unfortunately, many volcanoes give little or no warning of such large-scale events; of 252 explosive eruptions, 42% erupted most violently during their first day of activity. As one might imagine, predicting eruptions is complicated by those volcanoes that give so little warning of impending activity.

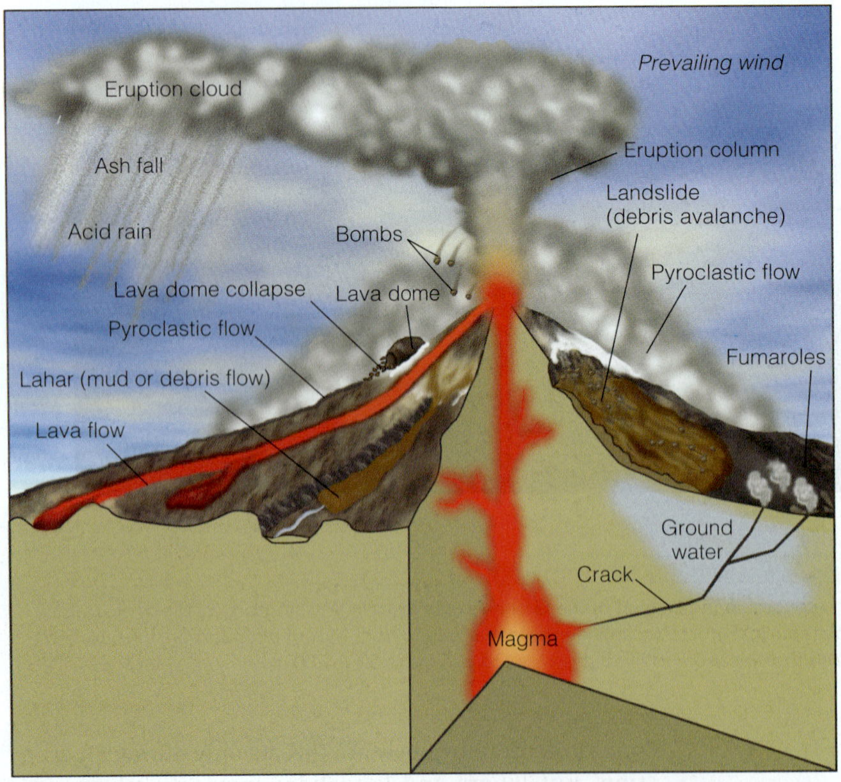

■ **Figure 5.14**

Some volcanic hazards, such as landslides and lahars, may occur even when a volcano is not erupting. This illustration shows a typical volcano in Alaska and the western United States, but volcanoes in Hawaii and elsewhere also pose hazards.

Is It Possible to Forecast Eruptions?

Most of the dangerous volcanoes are at or near the margins of tectonic plates, especially at convergent plate boundaries. At any one time about a dozen volcanoes are erupting, but most eruptions cause little or no property damage, injuries, or fatalities. Unfortunately, some do. The 1815 eruption of Tambora in Indonesia and a rather minor eruption of Nevado del Ruiz in Colombia in 1985 are good examples (Table 5.1).

Only a few of Earth's potentially dangerous volcanoes are monitored, including some in Japan, Italy, Russia, New Zealand, and the United States. Four facilities in the United States are devoted to volcano monitoring: the

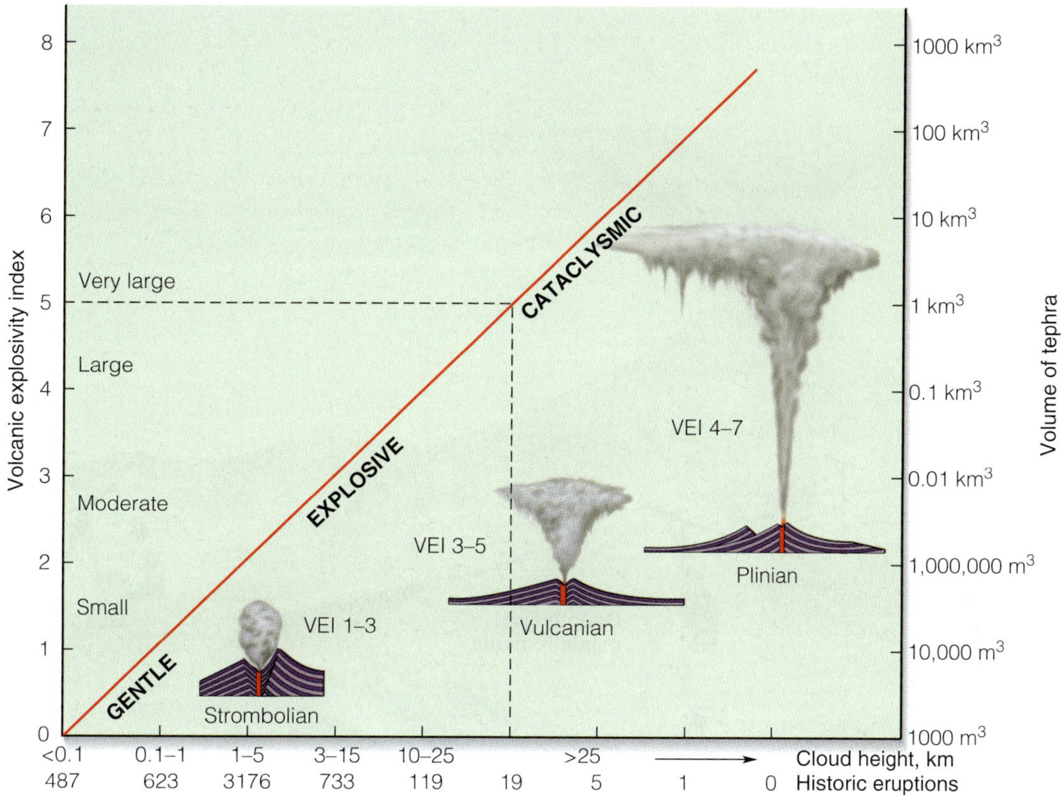

■ **Figure 5.15**

The volcanic explosivity index (VEI). In this example, an eruption with a VEI of 5 has an eruption cloud up to 25 km high and ejects at least 1 km^3 of tephra, a collective term for all pyroclastic materials. Geologists characterize eruptions as Hawaiian (nonexplosive), Strombolian, Vulcanian, and Plinian.

Hawaiian Volcano Observatory on Kilauea volcano; the David A. Johnston Cascades Volcano Observatory in Vancouver, Washington; the Alaska Volcano Observatory in Fairbanks, Alaska; and the Long Valley Observatory in Menlo Park, California. Many of the methods now used to monitor volcanoes were developed at the Hawaiian Volcano Observatory.

Volcano monitoring involves recording and analyzing physical and chemical changes at volcanoes (■ Figure 5.16a). Tiltmeters detect changes in the slopes of a volcano as it inflates when magma rises beneath it, and a geodimeter uses a laser beam to measure horizontal distances, which also change as a volcano inflates (Figure 5.16b, c). Geologists also monitor gas emissions, changes in groundwater level and temperature, hot springs activity, and changes in the local magnetic and electrical fields. Even the accumulating snow and ice, if any, are evaluated to anticipate hazards from floods should an eruption take place.

Of critical importance in volcano monitoring and warning of an imminent eruption is the detection of **volcanic tremor,** continuous ground motion that lasts for minutes to hours as opposed to the sudden, sharp jolts produced by most earthquakes. Volcanic tremor, also known as *harmonic tremor,* indicates that magma is moving beneath the surface.

To more fully anticipate the future activity of a volcano, its eruptive history must be known. Accordingly, geologists study the record of past eruptions preserved in rocks. Detailed studies before 1980 indicated that Mount St. Helens, Washington, had erupted explosively 14 or 15 times during the last 4500 years, so geologists concluded that it was one of the most likely Cascade Range volcanoes to erupt again. In fact, maps they prepared showing areas in which damage from an eruption could be expected were helpful in determining which areas should have restricted access and evacuations once an eruption did take place.

Geologists successfully gave timely warnings of impending eruptions of Mount St. Helens in Washington and Mount Pinatubo in the Philippines, but in both cases the climactic eruptions were preceded by eruptive activity of lesser intensity. In some cases, however, the warning signs are much more subtle and difficult to interpret. Numerous small earthquakes and other warning signs indicated to geologists of the U.S. Geological Survey (USGS) that magma was moving beneath the surface of the Long Valley caldera in eastern California, so in 1987 they issued a low-level warning, and then nothing happened.

Volcanic activity in the Long Valley caldera occurred as recently as 250 years ago, and there is every reason to

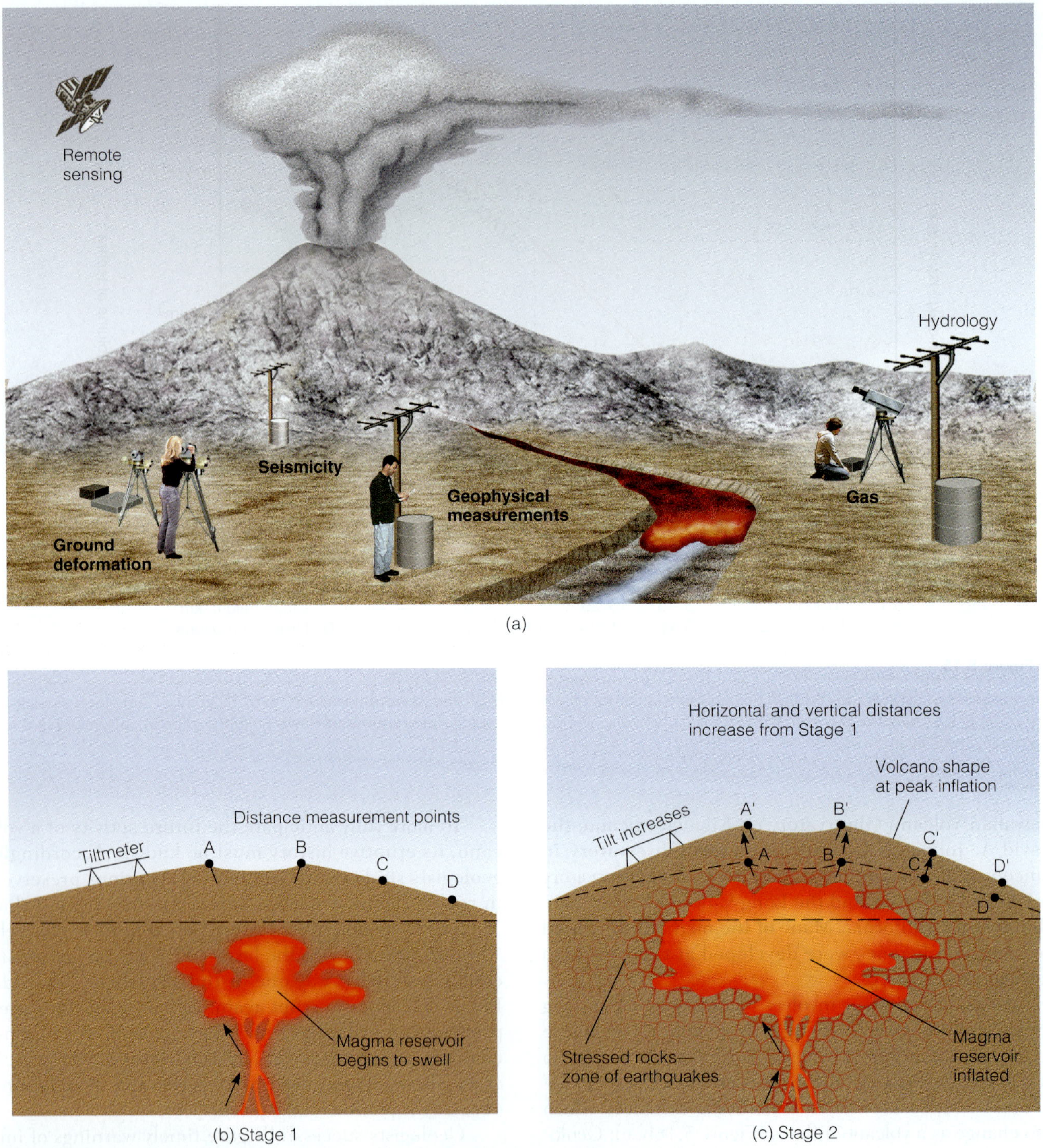

Figure 5.16

(a) Techniques used to monitor volcanoes. (b, c) Detection of ground deformation by tiltmeters and measurements of horizontal and vertical distances. As a volcano inflates when magma moves beneath it, volcanic tremor is also detected.

think it will occur again. Unfortunately, the local populace was largely unaware of the geologic history of the region, the USGS did a poor job in communicating its concerns, and premature news releases caused more concern than was justified. In any case, local residents were outraged because the warnings caused a decrease in tourism (Mammoth Mountain on the margins of the caldera is the second largest ski area in the country) and property values plummeted. Monitoring continues in the Long Valley caldera, and the signs of renewed volcanism, including earthquake swarms, trees being killed by carbon dioxide gas apparently emanating from magma, and hot spring activity, cannot be ignored.

GeologyNow Log into GeologyNow and select this chapter to work through **Geology Interactive** activities on "Magma Chemistry and Explosivity" (click Volcanism→Magma Chemistry and Explosivity) and on "Volcano Watch, USA" (click Volcanism→Volcano Watch, USA).

What Would You Do?

You are an enthusiast of natural history and would like to share your interests with your family. Accordingly, you plan a vacation to see some of the volcanic features in U.S. national parks and monuments. Let's assume your planned route will take you through Wyoming, Idaho, Washington, Oregon, and California. What specific areas might you visit, and what kinds of volcanic features would you see in these areas? What other parts of the United States might you visit in the future to see additional evidence for volcanism?

DISTRIBUTION OF VOLCANOES

Most of the world's active volcanoes are in well-defined zones or belts rather than randomly distributed. The **circum-Pacific belt,** with more than 60% of all active volcanoes, includes those in the Andes of South America; the volcanoes of Central America, Mexico, and the Cascade Range of North America; as well as the Alaskan volcanoes and those in Japan, the Philippines, Indonesia, and New Zealand (■ Figure 5.17). Also included in the circum-Pacific belt are the southernmost active volcanoes at Mount Erebus in Antarctica and a large caldera at Deception Island that erupted most recently during 1970. In fact, this belt nearly encircling the Pacific Ocean basin is popularly called the Ring of Fire.

The second area of active volcanism is the **Mediterranean belt** (Figure 5.17). About 20% of all active volcanism takes place in this belt, where the famous Italian volcanoes such as Mounts Etna and Vesuvius and the Greek volcano Santorini are found. Mount Etna has issued lava flows 190 times since 1500 B.C., when activity was first recorded. A particularly violent eruption of Santorini in 1390 B.C. might be the basis for the myth about the lost continent of Atlantis (see Chapter 9), and in A.D. 79 an eruption of Mount Vesuvius destroyed Pompeii and other nearby cities (see the Introduction).

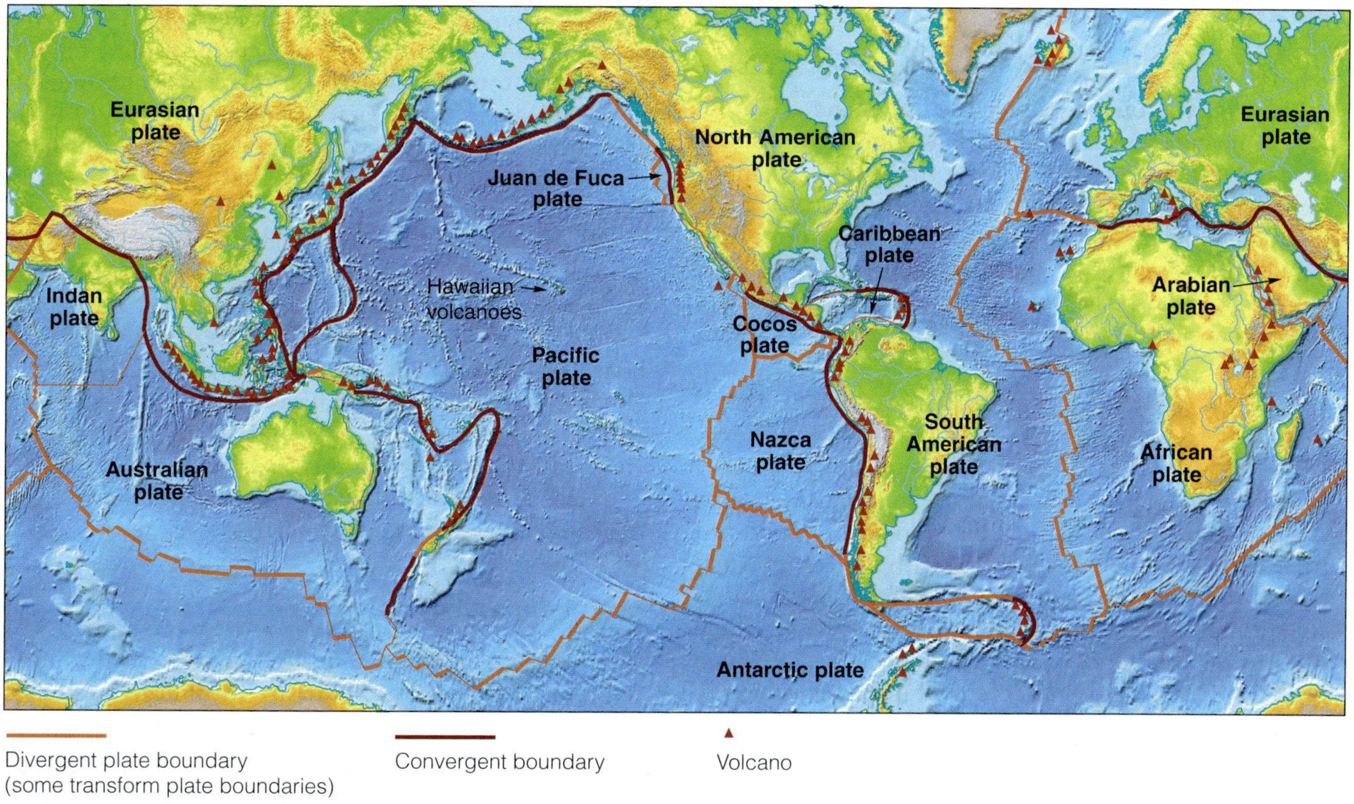

■ **Figure 5.17**

Most volcanoes are at or near convergent and divergent plate boundaries. The two major volcano belts are the circum-Pacific belt, commonly known as the Ring of Fire, with about 60% of all active volcanoes, and the Mediterranean belt, with 20% of active volcanoes. Most of the rest lie near the mid-oceanic ridges.

Nearly all the remaining active volcanoes are at or near mid-oceanic ridges or the extensions of these ridges onto land (Figure 5.17). These include the East Pacific Rise and the longest of all mid-oceanic ridges, the Mid-Atlantic Ridge. The latter is located near the center of the Atlantic Ocean basin, accounting for the volcanism in Iceland and elsewhere. It continues around the southern tip of Africa, where it connects with the Indian Ridge. Branches of the Indian Ridge extend into the Red Sea and East Africa, where such volcanoes as Kilamanjaro in Tanzania, Nyiragongo in Zaire (see Geo-Focus 5.1), and Erta Ale in Ethiopia with its continuously active lava lake are found.

Anyone familiar with volcanoes will have noticed that we have not mentioned the Hawaiian volcanoes. This is not an oversight as they are the notable exceptions to the distribution of volcanoes in well-defined belts. We will discuss their location and significance in the following section.

Geology Now Log into GeologyNow and select this chapter to work through a **Geology Interactive** activity on "Distribution of Volcanism" (click Volcanism→Distribution of Volcanism).

PLATE TECTONICS, VOLCANOES, AND PLUTONS

In Chapter 4 we discussed the origin and evolution of magma and concluded that (1) mafic magma is generated beneath spreading ridges, and (2) intermediate and felsic magma forms where an oceanic plate is subducted beneath another oceanic plate or a continental plate. Accordingly, most volcanism and emplacement of plutons take place at or near divergent and convergent plate boundaries.

Igneous Activity at Divergent Plate Boundaries

Much of the mafic magma that originates at spreading ridges is emplaced as vertical dikes and gabbro plutons, thus composing the lower part of the oceanic crust. However, some rises to the surface and issues forth as submarine lava flows and pillow lava (Figure 5.7), which constitutes the upper part of the oceanic crust. Much of this volcanism goes undetected, but researchers in submersibles have seen the results of eruptions shortly after they took place.

Mafic lava is very fluid, allowing gases to escape easily, and at great depth the water pressure is so great that explosive volcanism is prevented. In short, pyroclastic materials are rare to absent unless, of course, a volcanic center builds up above sea level. Even if this occurs, however, the mafic magma is so fluid that it forms the gently sloping layers found on shield volcanoes. Pyroclastic materials may be present on shield volcanoes but never in very great quantity.

Excellent examples of divergent plate boundary volcanism are found along the Mid-Atlantic Ridge, particularly where it rises above sea level as in Iceland (Figure 5.17). In November 1963, a new volcanic island, later named Surtsey, rose from the sea south of Iceland. The East Pacific Rise and the Indian Ridge are areas of similar volcanism. A divergent plate boundary is also present in Africa as the East African Rift system, which is well known for its volcanoes (Figure 5.17).

Igneous Activity at Convergent Plate Boundaries

Nearly all of the large active volcanoes in both the circum-Pacific and Mediterranean belts are composite volcanoes near the leading edges of overriding plates at convergent plate boundaries (Figure 5.17). The overriding plate, with its chain of volcanoes, may be oceanic as in the case of the Aleutian Islands, or it may be continental as is, for instance, the South American plate with its chain of volcanoes along its western edge.

As we noted, these volcanoes at convergent plate boundaries consist largely of lava flows and pyroclastic materials of intermediate to felsic composition. Remember that when mafic oceanic crust partially melts, some of the magma generated is emplaced near plate boundaries as plutons and some is erupted to build up composite volcanoes. More viscous magmas, usually of felsic composition, are emplaced as lava domes, thus accounting for the explosive eruptions that typically occur at convergent plate boundaries.

In previous sections, we alluded to several eruptions at convergent plate boundaries. Good examples are the explosive eruptions of Mount Pinatubo and Mayon volcano in the Philippines; both are near a plate boundary beneath which an oceanic plate is subducted. Mount St. Helens, Washington, is similarly situated, but it is on a continental rather than an oceanic plate. Mount Vesuvius in Italy, one of several active volcanoes in that region, lies on a plate that the northern margin of the African plate is subducted beneath.

Intraplate Volcanism

Mauna Loa and Kilauea on the island of Hawaii and Loihi just 32 km to the south are within the interior of a rigid plate far from any divergent or convergent plate boundary (Figure 5.17). The magma is derived from the upper mantle, as it is at spreading ridges, and accordingly is mafic, so it builds up shield volcanoes. Loihi is particularly interesting because it represents an early stage in the origin of a new Hawaiian island. It is a submarine volcano that rises more than 3000 m above the adjacent seafloor, but its summit is still about 940 m below sea level.

Even though the Hawaiian volcanoes are not at or near either a spreading ridge or a subduction zone, their evolution is nevertheless related to plate movements. Notice in Figure 2.25 that the ages of the rocks that make up the various Hawaiian islands increase toward the northwest. Kauai formed 3.8 to 5.6 million years ago, whereas Hawaii began forming less than 1 million years ago, and Loihi began to form even more recently. The islands have formed in succession as the Pacific plate moves continuously over a hot spot now beneath Hawaii and Loihi.

5

GEO RECAP

Chapter Summary

- Volcanism encompasses those processes by which magma rises to the surface as lava flows and pyroclastic materials and associated gases are released into the atmosphere.
- Gases make up only a few percent by weight of magma. Most is water vapor, but sulfur gases may have far-reaching climatic effects.
- Aa lava flows have surfaces of jagged, angular blocks, whereas the surfaces of pahoehoe flows are smoothly wrinkled.
- Several other features of lava flows are spatter cones, pressure ridges, lava tubes, and columnar joints. Lava erupted under water typically forms bulbous masses known as pillow lava.
- Volcanoes are found in various shapes and sizes, but all form where lava and pyroclastic materials are erupted from a vent.
- The summits of volcanoes have either a crater or a much larger caldera. Most calderas form following voluminous eruptions, and the volcanic peak collapses into a partially drained magma chamber.
- Shield volcanoes have low, rounded profiles and are composed mostly of mafic flows that cool and form basalt. Small, steep-sided cinder cones form around a vent where pyroclastic materials erupte and accumulate. Composite volcanoes are made up of lava flows and pyroclastic materials of intermediate composition and volcanic mudflows.
- Viscous bulbous masses of lava, generally of felsic composition, form lava domes, which are dangerous because they erupt explosively.
- Fluid mafic lava from fissure eruptions spreads over large areas to form a basalt plateau.
- Pyroclastic sheet deposits result from huge eruptions of ash and other pyroclastic materials, particularly when calderas form.
- Geologists have devised a volcanic explosivity index (VEI) to give a semiquantitative measure of the size of an eruption. Volume of material erupted and height of the eruption plume are criteria used to determine the VEI; fatalities and property damage are not considered.
- To effectively monitor volcanoes, geologists evaluate several physical and chemical aspects of volcanic regions. Of particular importance in monitoring volcanoes and forecasting eruptions is detecting volcanic tremor and determining the eruptive history of a volcano.
- About 80% of all volcanic eruptions take place in the circum-Pacific belt and the Mediterranean belt, mostly at convergent plate boundaries. Most of the rest of the eruptions occur along mid-oceanic ridges or their extensions onto land.
- The two active volcanoes on the island of Hawaii and one just to the south apparently lie above a hot spot over which the Pacific plate moves.

Important Terms

aa (p. 117)
ash (p. 119)
basalt plateau (p. 128)
caldera (p. 121)
Cascade Range (p. 115)
cinder cone (p. 125)
circum-Pacific belt (p. 133)
columnar jointing (p. 117)
composite volcano (stratovolcano) (p. 125)
crater (p. 121)
fissure eruption (p. 128)
lahar (p. 125)
lava dome (p. 127)
lava tube (p. 117)
Mediterranean belt (p. 133)
nuée ardente (p. 127)
pahoehoe (p. 117)
pillow lava (p. 117)
pyroclastic sheet deposit (p. 128)
shield volcano (p. 121)
volcanic explosivity index (VEI) (p. 130)
volcanic tremor (p. 131)
volcanism (p. 116)
volcano (p. 121)

Review Questions

1. One of the warning signs of an impending volcanic eruption is volcanic tremor, which is a._____inflation of a volcano as magma rises; b._____changes in groundwater temperature; c._____ground shaking that lasts for minutes or hours; d._____cooling and shrinkage in lava to form columnar joints; e._____ eruptions of fluid lava from long fissures.
2. Pillow lava forms when a._____pyroclastic materials accumulate in thick layers; b_____lava erupts under water; c._____globs of lava stick together on a lava flow's surface; d._____pressure within a flow causes buckling; e._____a volcano's summit collapses.
3. An incandescent cloud of gas and particles erupted from a volcano is a a._____nuée ardente; b._____pahoehoe; c._____spatter cone; d._____lapilli; e._____caldera.
4. _____ have slopes less than 10 degrees because they are composed of low-viscosity lava flows. a._____Lava tubes; b._____Pyroclastic sheet deposits; c._____Basalt plateaus; d._____Volcanic bombs; e._____Shield volcanoes.
5. Basalt plateaus form as a result of a._____repeated eruptions of felsic lava; b._____erosion of composite volcanoes; c._____inflation of a volcano as magma rises; d._____eruptions of fluid lava from fissures; e._____volcanic mudflows on cinder cones.
6. Most of the active volcanoes are found in the a._____mid-oceanic ridge volcanic zone; b._____Sierra Nevada–Cascade volcanic province; c._____circum-Pacific belt; d._____eastern Atlantic subduction zone; e._____Mediterranean divergent boundary.
7. The summits of some volcanoes have very wide, steep-sided depressions known as _________, most of which formed by ___________. a._____explosion pits/fissure eruptions; b._____calderas/summit collapse; c._____lava domes/forceful injection; d._____basalt plateaus/eruptions of cinders; e._____parasitic cones/lava flows.
8. Pahoehoe is a kind of lava flow with a a._____large component of pyroclastic materials; b._____lava tube; c._____smooth ropy surface; d._____mass of interconnected pillows; e._____fracture pattern outlining polygons.
9. Volcanoes emit several gases, but the most common one is a._____water vapor; b._____carbon dioxide; c._____hydrogen sulfide; d._____methane; e._____chlorine.
10. An area of active volcanism in the Pacific Northwest of the United States is the a._____Appalachian Mountains; b._____Cascade Range; c._____southern Rocky Mountains; d._____Marathon Mountains; e._____Teton Range.

11. Suppose you find rocks on land that consist of layers of pillow lava overlain by deep-sea sedimentary rocks. Where did the pillow lava originate, and what type of rock would you expect to lie beneath the pillow lava?
12. Why are most composite volcanoes at convergent plate boundaries, whereas most shield volcanoes are at or near divergent plate boundaries?
13. Explain how a caldera forms. Where would you go to see an example of a caldera?
14. What kinds of information do geologists evaluate when they monitor volcanoes and warn of imminent eruptions?
15. How do columnar joints and spatter cones form? Where are good places to see each?
16. What geologic events would have to occur for a chain of volcanoes to form along the East Coast of the United States and Canada?
17. What is a lava dome? Why are lava dome eruptions dangerous?
18. How do aa and pahoehoe lava flows differ? What accounts for their differences?
19. How does a crater differ from a caldera? How does each form?
20. Explain why eruptions of mafic lava are rather quiet, whereas eruptions of felsic lava are commonly explosive.

World Wide Web Activities

GeologyNow Assess your understanding of this chapter's topics with additional quizzing and comprehensive interactivities at

http://earthscience.brookscole.com/changingearth4e

as well as current and up-to-date weblinks, additional readings, and InfoTrac College Edition exercises.

Weathering, Soil, and Sedimentary Rocks

CHAPTER 6
OUTLINE

GeologyNow *This icon, which appears throughout the book, indicates an opportunity to explore interactive tutorials, animations, or practice problems available on the GeologyNow website at http://earthscience.brookscole.com/changingearth4e.*

Weathering and erosion along parallel fractures in sedimentary rocks have yielded the arches and other features such as isolated spires in Arches National Park in Utah. Delicate Arch is 9.7 m across and 14 m high. Source: James S. Monroe

OBJECTIVES

At the end of this chapter, you will have learned that:

- Weathering yields the raw materials for both soils and sedimentary rocks.
- Some weathering processes bring about physical changes in Earth materials with no change in composition, whereas others result in compositional changes.
- A variety of factors are important in the origin and evolution of soils.
- Soil degradation involves any loss of soil productivity that results from erosion, chemical pollution, or compaction.
- Sediments are deposited as aggregates of loose solids that may become sedimentary rocks if they are compacted and/or cemented.
- Geologists use texture and composition to classify sedimentary rocks.
- A variety of features preserved in sedimentary rocks are good indicators of how the original sediment was deposited.
- Most evidence of prehistoric life in the form of fossils is found in sedimentary rocks.
- Weathering is important in the origin and concentration of some resources, and sediments and sedimentary rocks are resources themselves or contain resources such as petroleum and natural gas.

Introduction

All rocks at or near Earth's surface as well as rocklike substances such as pavement and concrete in sidewalks, bridges, and foundations decay and crumble with age. In short, they experience **weathering,** defined as the physical breakdown and chemical alteration of Earth materials as they are exposed to the atmosphere, hydrosphere, and biosphere. Actually, weathering is a group of physical and chemical processes that alter Earth materials so that they are more nearly in equilibrium with a new set of environmental conditions. For instance, many rocks form within the crust where little or no oxygen or water is present, but at or near the surface they are exposed to both as well as to lower temperature and pressure and the activities of organisms.

During weathering, **parent material,** which is the rock acted on by weathering, is disaggregated to form smaller pieces (■ Figure 6.1), and some of its constituent minerals are altered or dissolved. Some of this weathered material simply accumulates in place and may be further modified to form *soil.* Much of it, however, is removed by **erosion,** which is the wearing away of soil and rock by geologic agents such as running water. This eroded material is carried elsewhere by running water, wind, glaciers, and marine currents and is eventually deposited as *sediment,* the raw material for *sedimentary rocks.*

Weathering, erosion, sediment deposition, and the origin of sedimentary rocks are essential parts of the *rock cycle* (see Figure 1.12). Earth's crust is composed mostly of *crystalline rock,* a term that refers loosely to metamorphic and igneous rocks, except those made up of pyroclastic materials. Nevertheless, sediment and sedimentary rocks, making up perhaps only 5% of the crust, are by far the most common materials in surface exposures and in the shallow subsurface. They cover about two-thirds of the continents and most of the seafloor, except spreading ridges. All rocks are important in deciphering Earth history, but sedimentary rocks have a special place in this endeavor because they preserve evidence of surface processes such as stream activity, wind, and glaciers responsible for them.

Several geologic processes such as volcanism, shoreline processes, and glaciation have

(a)

(b)

■ **Figure 6.1**

Most of the granite in this rock exposure has been so thoroughly weathered that only a few rounded masses of the original rock appear unaltered. The small cones in the foreground consist of individual minerals, mostly quartz and feldspars, and rock fragments—that is, small pieces of granite.
(b) Closeup view of the weathered material.

National Park Service

■ **Figure 6.2**

The spectacular scenery at Bryce Canyon National Park in Utah. The sedimentary rocks here belong to the 40- to 50-million-year-old Wasatch Formation that was deposited in an ancient lake. Weathering and erosion along closely spaced fractures have yielded spires, pillars, monuments, arches, fluted ridges, gullies, and ravines.

yielded many areas of exceptional scenery, but so have weathering and erosion. We marvel at the intricately sculpted landscape at Bryce Canyon National Park in Utah (■ Figure 6.2) and the rugged shoreline at Acadia National Park in Maine (see Figure 4.1a). Weathering and erosion of fractured rocks in Arches National Park in Utah have yielded a landscape of isolated spires and balanced rocks as well as the arches for which the park was named (see the chapter opening photo).

In addition to interesting landscapes, weathering is responsible for the origin of some natural resources such as aluminum ore, and it enriches others by removing soluble materials. Also, some sediments and sedimentary rocks are resources themselves or they are the host for resources such as petroleum and natural gas.

HOW ARE EARTH MATERIALS ALTERED?

Weathering is a surface or near-surface process, but the rocks it acts on are not structurally and compositionally homogeneous throughout, which accounts for **differential weathering.** That is, weathering takes place at different rates even in the same area, so it commonly results in uneven surfaces. Differential weathering and *differential erosion*—that is, variable rates of erosion—combine to yield some unusual and even bizarre features, such as hoodoos, spires, and arches (Figure 6.2).

The two recognized types of weathering, *mechanical* and *chemical,* both proceed simultaneously on parent material as well as on materials in transport and those deposited as sediment. In short, all surface or near-surface materials weather, although one type of weathering may predominate depending on such variables as climate and rock type. We will discuss mechanical and chemical weathering separately in the next sections only for convenience.

Mechanical Weathering

Mechanical weathering takes place when physical forces break Earth materials into smaller pieces that retain the composition of the parent material. Granite, for instance, might be mechanically weathered and yield smaller pieces of granite or individual grains of quartz, potassium feldspars, plagioclase feldspars, and biotite (Figure 6.1). Several physical processes are responsible for mechanical weathering.

Frost action involving water repeatedly freezing and thawing in cracks and pores in rocks is particularly effective where temperatures commonly fluctuate above and below freezing. In the high mountains of the western United States and Canada, frost action is effective even during summer months. But, as one would expect, it is of little or no importance in the tropics or where water is permanently frozen. The reason frost action is so effective is that water expands by about 9% when it freezes, thus exerting great force on the walls of a crack, widening and extending it by *frost wedging* (■ Figure 6.3a). Repeated freezing and thawing dislodge angular pieces of rock from the parent material that tumble downslope and accumulate as **talus** (Figure 6.3b).

Some rocks form at depth and are stable under tremendous pressure. Granite, for instance, crystallizes far below the surface, so when it is uplifted and eroded, its contained energy is released by outward expansion, a phenomenon known as **pressure release.** The outward expansion results in the origin of fractures called *sheet joints* that more or less parallel the exposed rock surface. Sheet-joint-bounded slabs of rock slip or slide off the parent rock, leaving large, rounded masses known as **exfoliation domes** (■ Figure 6.4b).

Figure 6.3

(a) Frost wedging takes place when water seeps into cracks and expands as it freezes. Angular pieces of rock are pried loose by repeated freezing and thawing. (b) Accumulation of talus (foreground) at the base of a slope. The parent material is highly fractured and quite susceptible to frost wedging, although other weathering processes also help break the rock into smaller pieces.

That solid rock expands and produces fractures might be counterintuitive but is nevertheless a well-known phenomenon. In deep mines, masses of rock detach from the sides of the excavation, often explosively. These *rock bursts* and less violent *popping* pose a danger to mine workers, and in South Africa they are responsible for about 20 deaths per year. In some quarries for building stone, excavations to only 7 or 8 m exposed rocks in which sheet joints formed (Figure 6.4c), in some cases with enough force to throw quarrying machines weighing more than a ton from their tracks.

During **thermal expansion and contraction** the volume of rocks changes as they heat up and then cool down. The temperature may vary as much as 30°C in a day in a desert, and rock, being a poor conductor of heat, heats and expands on its outside more than its inside. Even dark minerals absorb heat faster than light-colored ones, so differential expansion takes place between minerals. Surface expansion might generate enough stress to cause fracturing, but experiments in which rocks are heated and cooled repeatedly to simulate years of such activity indicate that thermal expansion and contraction are of minor importance in mechanical weathering.

The formation of salt crystals can exert enough force to widen cracks and dislodge particles in porous, granular rocks such as sandstone. And even in rocks with an interlocking mosaic of crystals, like granite, **salt crystal growth** pries loose individual minerals. It takes place mostly in hot, arid regions, but also probably affects rocks in some coastal areas.

Animals, plants, and bacteria all participate in the mechanical and chemical alteration of rocks (Figure 6.5). Burrowing animals, such as worms, reptiles, rodents, termites, and ants, constantly mix soil and sediment particles and bring material from depth to the surface where further weathering occurs. The roots of plants, especially large bushes and trees, wedge themselves into cracks in rocks and further widen them.

Chemical Weathering

Chemical weathering includes those processes by which rocks and minerals are decomposed by chemical alteration of the parent material. In contrast to mechanical weathering, chemical weathering changes the composition of weathered materials. For example, several clay minerals (sheet silicates) form by the chemical and structural alteration of other minerals such as potassium feldspars and plagioclase feldspars, both of which are framework silicates. Other minerals are completely decomposed during chemical weathering as their ions

Mark Gibson/Visuals Unlimited (b)

Sue Monroe (a) Courtesy of W. D. Lowry (c)

Figure 6.4

(a) Slabs of granitic rock bounded by sheet joints in the Sierra Nevada in California. Notice that these slabs are inclined down toward the roadway in the lower part of the image. (b) Stone Mountain is a large exfoliation dome in Georgia. (c) A sheet joint formed by expansion in the Mount Airy Granite in North Carolina. The hammer is about 30 cm long.

are taken into solution, but some chemically stable minerals are simply liberated from the parent material.

Important agents of chemical weathering include atmospheric gases, especially oxygen, water, and acids. Organisms also play an important role. Rocks with lichens (composite organisms made up of fungi and algae) on their surfaces undergo more rapid chemical alteration than lichen-free rocks (Figure 6.5b). In addition, plants remove ions from soil water and reduce the chemical stability of soil minerals, and plant roots release organic acids.

During **solution** the ions of a substance separate in a liquid, and the solid substance dissolves. Water is a remarkable solvent because its molecules have an asymmetric shape, consisting of one oxygen atom with two hydrogen atoms arranged so that the angle between the two hydrogen atoms is about 104 degrees (■ Figure 6.6). Because of this asymmetry, the oxygen end of the molecule retains a slight negative electrical charge, whereas the hydrogen end retains a slight positive charge. When a soluble substance such as the mineral halite (NaCl) comes in contact with a water molecule, the positively charged sodium ions are attracted to the negative end of the water molecule, and the negatively charged chloride ions are attracted to the positively charged end of the molecule (Figure 6.6). Thus, ions are liberated from the crystal structure, and the solid dissolves.

Most minerals are not very soluble in pure water because the attractive forces of water molecules are not sufficient to overcome the forces between particles in minerals. For instance, the mineral calcite ($CaCO_3$), the major constituent of the sedimentary rock limestone and the metamorphic rock marble, is practically insoluble in pure water, but it rapidly dissolves if a small amount of acid is present. One way to make water acidic is by dissociating the ions of carbonic acid as follows:

$$\begin{array}{ccccccccc} H_2O & + & CO_2 & \rightleftharpoons & H_2CO_3 & \rightleftharpoons & H^+ & + & HCO_3^- \\ \text{water} & & \text{carbon dioxide} & & \text{carbonic acid} & & \text{hydrogen ion} & & \text{bicarbonate ion} \end{array}$$

(a)

(b)

Figure 6.5

Organisms and weathering. (a) This tree near Anchorage, Alaska, is growing in a crack in the rocks and thus contributes to mechanical weathering. (b) The irregular orange masses on these rocks on a small island in the Irish Sea are lichens (composite organisms of fungi and algae). Lichens derive their nutrients from the rock and contribute to chemical weathering.

According to this chemical equation, water and carbon dioxide combine to form *carbonic acid,* a small amount of which dissociates to yield hydrogen and bicarbonate ions. The concentration of hydrogen ions determines the acidity of a solution; the more hydrogen ions present, the stronger the acid.

Carbon dioxide from several sources may combine with water and react to form acid solutions. The atmosphere is mostly nitrogen and oxygen, but about 0.03% is carbon dioxide, causing rain to be slightly acidic. Decaying organic matter and the respiration of organisms produce carbon dioxide in soils, so groundwater is also generally slightly acidic. Climate affects the acidity, however, with arid regions tending to have alkaline groundwater (that is, it has a low concentration of hydrogen ions).

Whatever the source of carbon dioxide, once an acidic solution is present, calcite rapidly dissolves according to the reaction

$$\underset{\text{calcite}}{CaCO_3} + \underset{\text{water}}{H_2O} + \underset{\text{carbon dioxide}}{CO_2} \rightleftharpoons \underset{\text{calcium ion}}{Ca^{++}} + \underset{\text{bicarbonate ion}}{2HCO_3^{-}}$$

The term **oxidation** has a variety of meanings for chemists, but in chemical weathering it refers to reactions with oxygen to form an oxide (one or more metallic elements combined with oxygen) or, if water is present, a hydroxide (a metallic element or radical combined with OH). For example, iron rusts when it combines with oxygen to form the iron oxide hematite:

$$\underset{\text{iron}}{4Fe} + \underset{\text{oxygen}}{3O_2} \rightarrow \underset{\text{iron oxide (hematite)}}{2Fe_2O_3}$$

Of course, atmospheric oxygen is abundantly available for oxidation reactions, but oxidation is generally a slow process unless water is present. Thus, most oxidation is carried out by oxygen dissolved in water.

Oxidation is important in the alteration of ferromagnesian silicates such as olivine, pyroxenes, amphiboles, and biotite. Iron in these minerals combines with oxygen to form the reddish iron oxide hematite (Fe_2O_3) or the yellowish or brown hydroxide limonite [$FeO(OH)\cdot nH_2O$]. The yellow, brown, and red colors of many soils and sedimentary rocks are caused by the presence of small amounts of hematite or limonite.

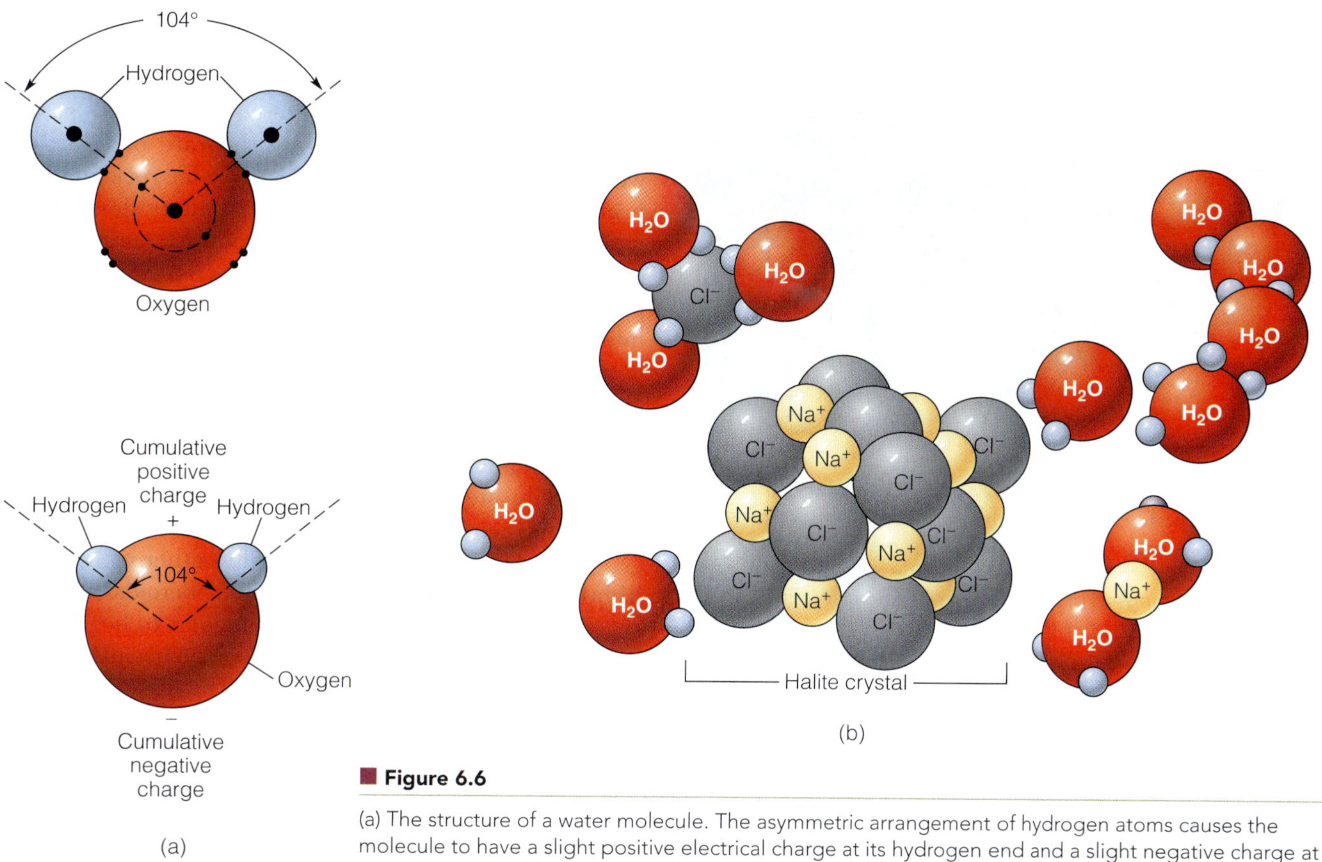

■ **Figure 6.6**

(a) The structure of a water molecule. The asymmetric arrangement of hydrogen atoms causes the molecule to have a slight positive electrical charge at its hydrogen end and a slight negative charge at its oxygen end. (b) Solution of sodium chloride (NaCl), the mineral halite, in water. Note that the sodium atoms are attracted to the oxygen end of a water molecule, whereas chloride ions are attracted to the hydrogen end of the molecule.

The chemical reaction between the hydrogen (H^+) ions and hydroxyl (OH^-) ions of water and a mineral's ions is known as **hydrolysis.** In hydrolysis, hydrogen ions actually replace positive ions in minerals. Such replacement changes the composition of minerals and liberates iron that then may be oxidized.

The chemical alteration of the potassium feldspar orthoclase provides a good example of hydrolysis. All feldspars are framework silicates, but when altered, they yield soluble salts and clay minerals, such as kaolinite, which are sheet silicates. The chemical weathering of orthoclase by hydrolysis occurs as follows:

$2KAlSi_3O_8$	+ $2H^+$	+ $2HCO_3^-$	+	$H_2O \rightarrow$
orthoclase	hydrogen ion	bicarbonate ion		water
$Al_2Si_2O_5(OH)_4$	+ $2K^+$	+ $2HCO_3^-$	+	$4SiO_2$
clay (kaolinite)	potassium ion	bicarbonate ion		silica

In this reaction hydrogen ions attack the ions in the orthoclase structure, and some liberated ions are incorporated in a developing clay mineral. The potassium and bicarbonate ions go into solution and combine to form a soluble salt. On the right side of the equation is excess silica that would not fit into the crystal structure of the clay mineral.

Factors That Control the Rate of Chemical Weathering

Chemical weathering processes operate on the surface of particles, so they alter rocks and minerals from the outside inward. In fact, if you break open a weathered stone, it is common to see a rind of weathering at and near the surface, but the stone is completely unaltered inside. The rate at which chemical weathering proceeds depends on several factors. One is simply the presence or absence of fractures because fluids seep along fractures, accounting for more intense chemical weathering along these surfaces (■ Figure 6.7). Of course, other factors also control chemical weathering, including particle size, climate, and parent material.

Because chemical weathering affects particle surfaces, the greater the surface area, the more effective is the weathering. It is important to realize that small particles have larger surface areas compared to their volume than do large particles. Notice in ■ Figure 6.8 that a block measuring 1 m on a side has a total surface area of 6 m^2, but when the block is broken into particles measuring 0.5 m on a side, the total surface area increases to 12 m^2. And if these particles are all reduced to 0.25 m on a

Sue Monroe

Figure 6.7

Fluids seep along fractures where chemical weathering is more intense than it is in unfractured parts of the same rock. Notice that a narrow white band stands out in relief near the left side of the image. This is composed of quartz, which is more resistant to chemical weathering than is the granitic host rock.

side, the total surface area increases to 24 m^2. Note that although the surface area in this example increases, the total volume remains the same at 1 m^3.

We can conclude that mechanical weathering contributes to chemical weathering by yielding smaller particles with greater surface area compared to their volume. Actually, your own experiences with particle size verify our contention about surface area and volume. Because of its very small particle size, powdered sugar gives an intense burst of sweetness as the tiny pieces dissolve rapidly, but otherwise it is the same as the granular sugar we use on our cereal or in our coffee. As an experiment, see how long it takes for crushed ice and an equal volume of block ice to melt, or determine the time it takes to boil an entire potato as opposed to one cut into small pieces.

It is not surprising that chemical weathering is more effective in the tropics than in arid and arctic regions because temperatures and rainfall are high and evaporation rates are low. In addition, vegetation and animal life are much more abundant. Consequently, the effects of weathering extend to depths of several tens of meters, but they commonly extend only centimeters to a few meters deep in arid and arctic regions.

Some rocks are more resistant to chemical alteration than others and thus are not altered as rapidly, so parent material is another control on the rate of chemical weathering. For example, the metamorphic rock quartzite is an extremely stable substance that alters slowly compared to most other rock types. In contrast, basalt, which contains large amounts of calcium-rich plagioclase and pyroxene minerals, decomposes rapidly because these minerals are chemically unstable. In fact, the stability of common minerals is just the opposite of their order of crystallization in Bowen's reaction series (Table 6.1, also see Figure 4.3): The minerals that form last in this series are chemically stable, whereas those that form early are more easily altered because they are most out of equilibrium with their conditions of formation.

One manifestation of chemical weathering is **spheroidal weathering** (■ Figure 6.9). In spheroidal weathering, a stone, even one that is rectangular to begin with, weathers to form a more spherical shape because that is the most stable shape it can assume. The reason is that on a rectangular stone the corners are attacked by weathering from three sides, and the edges are attacked from two sides, but the flat surfaces weather more or less uniformly (Figure 6.9). Consequently, the corners and edges are altered more rapidly, the material sloughs off, a more spherical shape develops (Figure 6.9), and all surfaces weather at the same rate.

HOW DOES SOIL FORM AND DETERIORATE?

A layer of **regolith,** a collective term for sediment as well as layers of pyroclastic materials and the residue formed in place by weathering, covers most of Earth's land surface. Some regolith consisting of weathered materials, air, water, and organic matter supports vegetation and is called **soil.** Almost all land-dwelling organisms depend directly or indirectly on soil for their existence. Plants grow in

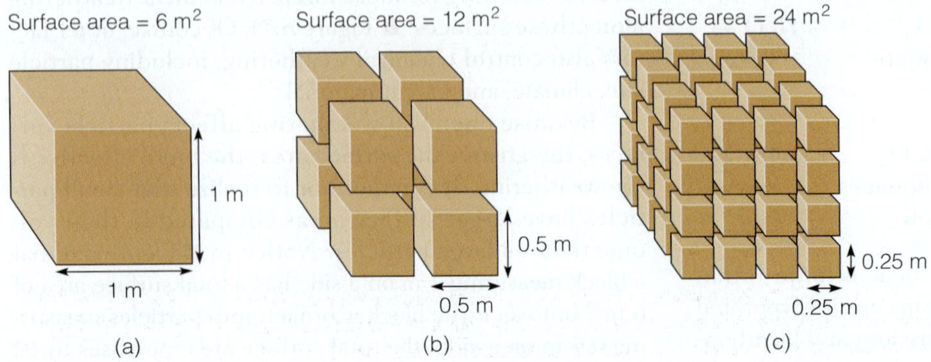

Figure 6.8

As a rock is divided into smaller and smaller particles, its surface area increases but its volume remains the same. In (a) the surface area is 6 m^2, and in (c) 24 m^2, but the volume remains the same at 1 m^3. Smaller particles have more surface area compared to their volume than do larger particles.

Table 6.1

Stability of Silicate Minerals

	Ferromagnesian Silicates	Nonferromagnesian Silicates
Increasing Stability ↓	Olivine	Calcium plagioclase
	Pyroxene	
	Amphibole	Sodium plagioclase
	Biotite	Potassium feldspar
		Muscovite
		Quartz

soil from which they derive their nutrients and most of their water, whereas many land-dwelling animals depend on plants for nutrients.

About 45% of good soil for farming and gardening is composed of weathered particles, with much of the remaining volume simply void spaces filled with air and/or water. In addition, a small but important amount of humus is usually present. *Humus* is carbon derived by bacterial decay of organic matter and is highly resistant to further decay. Even a fertile soil might have as little as 5% humus, but it is nevertheless important as a source of plant nutrients and it enhances a soil's capacity to retain moisture.

Some weathered materials in soils are simply sand- and silt-sized mineral grains, especially quartz, but other minerals may be present as well. These solids hold soil particles apart, allowing oxygen and water to circulate more freely. Clay minerals are also important in soils and aid in the retention of water as well as supplying nutrients to plants. Soils with excess clay minerals, however, drain poorly and are sticky when wet and hard when dry.

Residual soils form when parent material weathers in place. For example, if a body of granite weathers, and the weathering residue accumulates over the granite and is converted to soil, the soil thus formed is residual. In contrast, *transported soils* develop on weathered material that was eroded at the weathering site and carried to a new location, where it is altered to soil.

The Soil Profile

Observed in vertical cross section, soil consists of distinct layers or **soil horizons** that differ from one another in texture, structure, composition, and color (■ Figure 6.10). Starting from the top, the soil horizons are designated O, A, B, and C, but the boundaries between horizons are transitional. Because soil formation begins at the

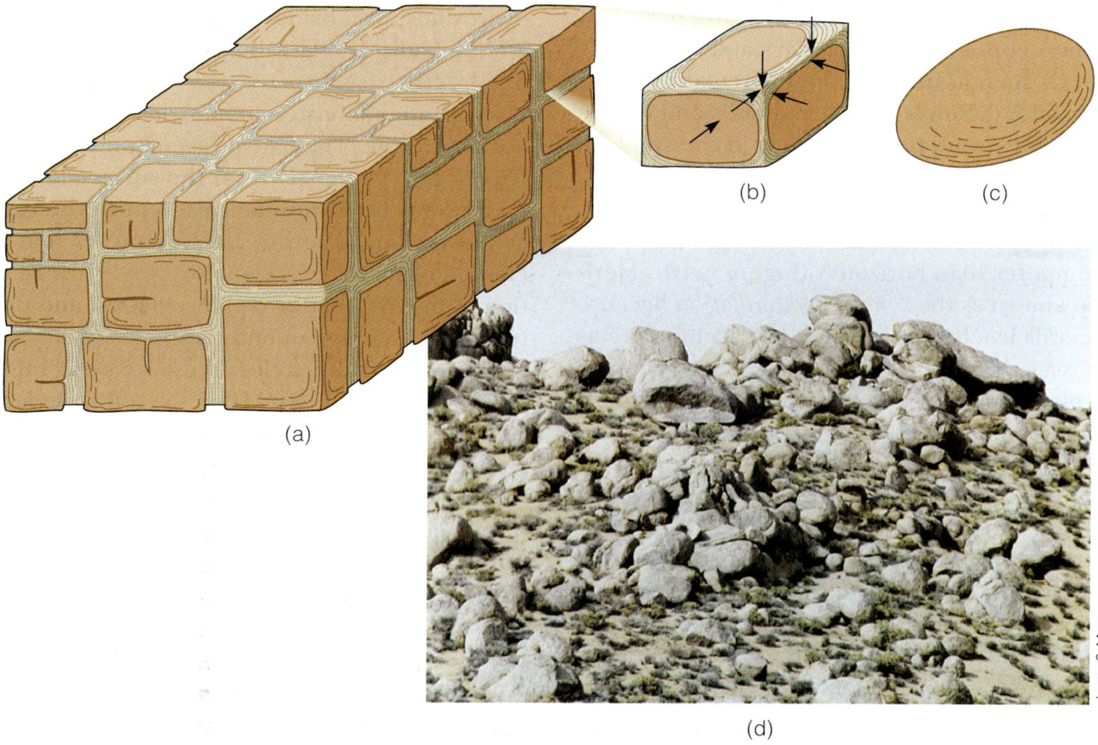

■ **Figure 6.9**

Spheroidal weathering. (a) The rectangular blocks outlined by fractures are attacked by chemical weathering processes, but (b) corners and edges weather most rapidly. (c) When the blocks are weathered so that their shape is more nearly spherical, their surface weathers evenly and no further change in shape takes place. (d) Exposure of granitic rocks reduced to spherical boulders.

■ **Figure 6.10**

The soil horizons in a fully developed soil.

surface and works downward, horizon A is more altered from the parent material than the layers below.

Horizon O, which is only a few centimeters thick, consists of organic matter. The remains of plant materials are clearly recognizable in the upper part of horizon O, but its lower part consists of humus.

Horizon A, called *topsoil,* contains more organic matter than horizons B and C below. It is also characterized by intense biological activity because plant roots, bacteria, fungi, and animals such as worms are abundant. Threadlike soil bacteria give freshly plowed soil its earthy aroma. In soils developed over a long period of time, horizon A consists mostly of clays and chemically stable minerals such as quartz. Water percolating down through horizon A dissolves soluble minerals and carries them away or downward to lower levels in the soil by a process called *leaching*. Accordingly, horizon A is also referred to as the *zone of leaching*.

Horizon B, or *subsoil,* contains fewer organisms and less organic matter than horizon A (Figure 6.10). Horizon B is also known as the *zone of accumulation* because soluble minerals leached from horizon A accumulate as irregular masses. If horizon A is eroded, leaving horizon B exposed, plants do not grow as well, and if it is clayey, it is harder when dry and stickier when wet than other soil horizons.

Horizon C has little organic matter and consists of partially altered parent material grading down into unaltered parent material (Figure 6.10). In horizons A and B, the composition and texture of the parent material have been so thoroughly altered that it is no longer recognizable. In contrast, rock fragments and mineral grains of the parent material retain their identity in horizon C.

Factors That Control Soil Formation

Soil scientists know that climate is the single most important factor in soil origins, but complex interactions among several factors account for soil type, thickness, and fertility (■ Figure 6.11). A very general classification recognizes three major soil types characteristic of different climatic settings. Soils that develop in humid regions such as the eastern United States and much of Canada are **pedalfers,** a name derived from the Greek word *pedon,* meaning "soil," and from the chemical symbols for aluminum (Al) and iron (Fe). Because these soils form where abundant moisture is present, most of the soluble minerals have been leached from horizon A. Although it may be gray, horizon A is commonly dark because of abundant organic matter, and aluminum-rich clays and iron oxides tend to accumulate in horizon B.

Soils found in much of the arid and semiarid western United States, especially the Southwest, are **pedocals.** Pedocal derives its name in part from the first three letters of "calcite." These soils contain less organic matter than pedalfers, so horizon A is lighter colored and contains more unstable minerals because of less intense chemical weathering. As soil water evaporates, calcium carbonate leached from above precipitates in horizon B where it forms irregular masses of *caliche.* Precipitation of sodium salts in some desert areas where soil water evaporation is intense yields *alkali soils* that are so alkaline they cannot support plants.

Laterite forms in the tropics where chemical weathering is intense and leaching of soluble minerals is complete. These soils are red, extend to depths of several tens of meters, and are composed largely of aluminum hydroxides, iron oxides, and clay minerals; even quartz, a chemically stable mineral, is leached out (■ Figure 6.12).

Although laterites support lush vegetation, they are not very fertile. The native vegetation is sustained by nutrients derived mostly from the surface layer of organic matter, but little humus is present in the soil itself because bacterial action destroys it. When laterite is cleared of its native vegetation, the surface accumulation of organic matter rapidly oxidizes, and there is little to replace it. Consequently, societies that practice slash-and-burn agriculture clear these soils and raise crops for only a few years at best. Then the soil is depleted of plant nutrients, the clay-rich laterite bakes brick hard in the tropical sun, and the farmers move on to another area where the process is repeated.

The same rock type can yield different soils in different climatic regimes, and in the same climatic regime the same soils can develop on different rock types. Thus, it seems that climate is more important than parent material in determining the type of soil. Nevertheless, rock type does exert some control. For example, the metamorphic rock quartzite will have a thin soil over it because it is chemically stable, whereas an adjacent body of granite will have a much deeper soil.

Soils depend on organisms for their fertility, and in return they provide a suitable habitat for many organisms. Earthworms—as many as a million per acre—ants,

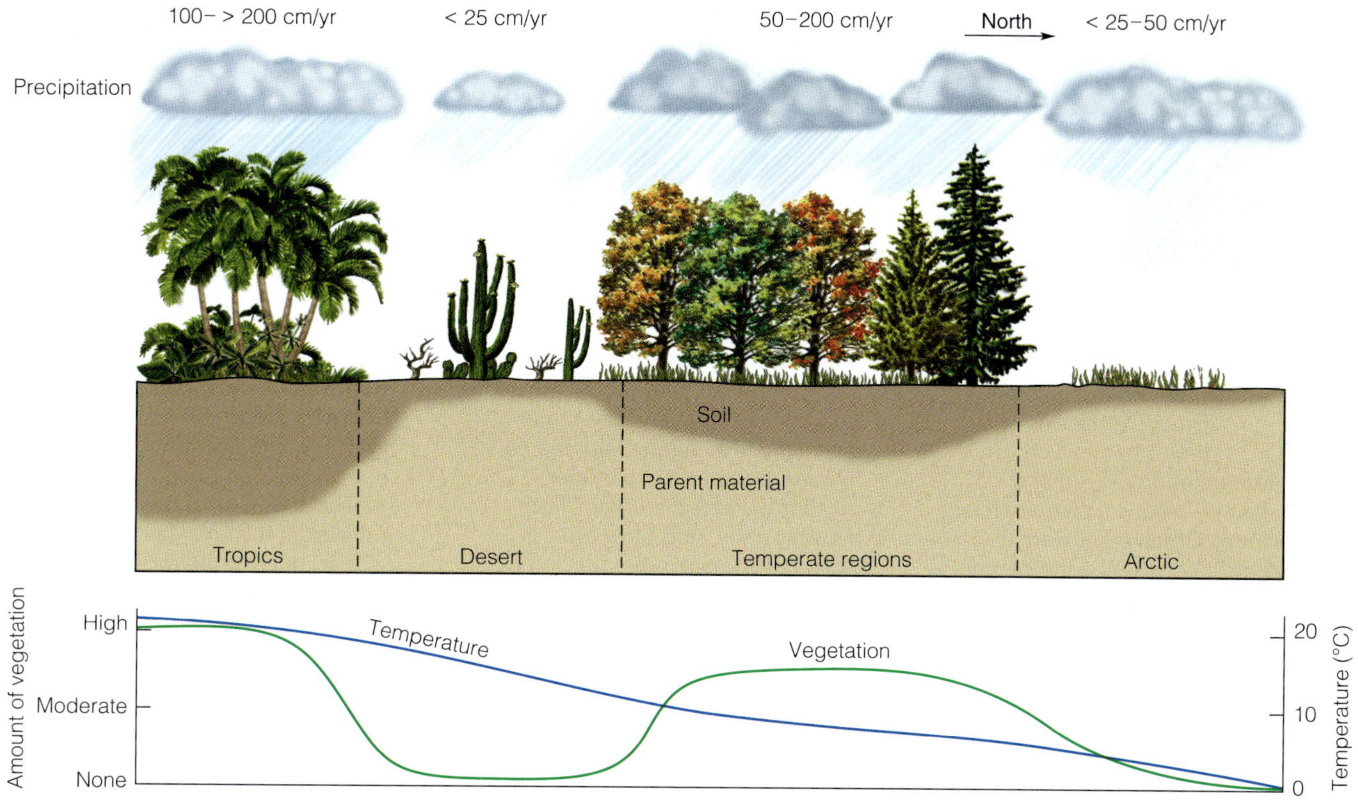

Figure 6.11

Generalized diagram showing soil formation as a function of climate and vegetation altering parent material over time. Soil-forming processes operate most vigorously where precipitation and temperatures are high, as in the tropics.

sowbugs, termites, centipedes, millipedes, and nematodes, along with various types of fungi, algae, and single-celled organisms, make their homes in soil. All contribute to the formation of soils and provide humus when they die and are decomposed by bacterial action.

Much of the humus in soils is provided by grasses or leaf litter that microorganisms decompose to obtain food. In so doing, they break down organic compounds in plants and release nutrients back into the soil. In addition, organic acids produced by decaying soil organisms are important in further weathering of parent materials and soil particles.

Burrowing animals constantly churn and mix soils, and their burrows provide avenues for gases and water. Soil organisms, especially some types of bacteria, are extremely important in changing atmospheric nitrogen into a form of soil nitrogen suitable for use by plants.

The difference in elevation between high and low points in a region is called *relief*. And because climate is such an important factor in soil formation and climate changes with elevation, areas with considerable relief have different soils in mountains and adjacent lowlands. *Slope* also is an important control, but it actually influences soil formation in two ways. One is simply *slope angle;* steep slopes have little or no soil because weathered materials are eroded faster than soil-forming processes can operate. The other factor is *slope direction*. In the Northern Hemisphere, north-facing slopes receive less sunlight than south-facing slopes and have cooler internal temperatures, support different vegetation, and if in a cold climate, remain snow covered or frozen longer.

How much time is needed to develop a centimeter of soil or a fully developed soil a meter or so deep? We

Walt Anderson/Visuals Unlimited

Figure 6.12

Laterite, shown here in Madagascar, is a deep red soil that forms in the tropics.

GEOFOCUS 6.1

The Dust Bowl—An American Tragedy

The stock market crash of 1929 ushered in the Great Depression, a time when millions of people were unemployed and many had no means to acquire food and shelter. Urban areas were affected most severely by the depression, but rural areas suffered as well, especially during the great drought of the 1930s. Prior to the 1930s, farmers had enjoyed a degree of success unparalleled in U.S. history. During World War I (1914–1918), the price of wheat soared, and after the war when Europe was recovering, the government subsidized wheat prices. High prices and mechanized farming resulted in more and more land being tilled. Even the weather cooperated, and land in the western United States that would otherwise have been marginally productive was plowed. Deep-rooted prairie grasses that held the soil in place were replaced by shallow-rooted wheat.

Beginning in about 1930, drought prevailed throughout the country; only two states—Maine and Vermont—were not drought-stricken. Drought conditions varied from moderate to severe, but the consequences were particularly severe in the southern Great Plains. Some rain fell, but not enough to maintain agricultural production. And because the land, even marginal land, had been tilled, the native vegetation was no longer present to keep the topsoil from blowing away. And blow away it did—in huge quantities.

A large region in the southern Great Plains that was particularly hard hit by drought, dust storms, and soil erosion came to be known as the Dust Bowl. Although its boundaries were not well defined, it included parts of Kansas, Colorado, and New Mexico as well as the panhandles of Oklahoma and Texas (■ Figure 1a); together the Dust Bowl and its less affected fringe area covered more than 400,000 km^2!

Dust storms were common during the 1930s, and some reached phenomenal sizes (Figure 1b). One of the largest storms occurred in 1934 and covered more than 3.5 million km^2. It lifted dust nearly 5 km into the air, obscured the sky over large parts of six states, and blew hundreds of millions of tons of soil eastward where it settled on New York City, Washington, DC, and other eastern cities, as well as on ships as far as 480 km out in the Atlantic Ocean. The Soil Conservation Service reported dust storms of regional extent on 140 occasions during 1936 and 1937. Dust was everywhere. It seeped into houses, suffocated wild animals and livestock, and adversely affected human health.

The dust was, of course, the material from the tilled lands; in other words, much of the topsoil simply blew away. Blowing dust was not the only problem; sand piled up along fences, drifted against houses and farm machinery, and covered what otherwise might have been productive soils. Agricultural production fell precipitously in the Dust Bowl, farmers could not meet their mortgage payments, and by 1935 tens of thousands were homeless, on relief, or leaving (Figure 1c). Many of these people went west to California and became the migrant farm workers immortalized in John Steinbeck's novel *The Grapes of Wrath*.

The Dust Bowl was an economic disaster of great magnitude. Droughts had stricken the southern Great Plains before and have done so since—from August 1995 well into the summer of 1996, for instance—but the drought of the 1930s was especially severe. Political and economic factors contributed to the disaster. Due in part to the artificially inflated wheat prices, many farmers were deeply in

can give no definite answer because weathering proceeds at vastly different rates depending on climate and parent material, but an overall average might be about 2.5 cm per century. Nevertheless, a lava flow a few centuries old in Hawaii may have a well-developed soil on it, whereas a flow the same age in Iceland will have considerably less soil. Given the same climatic conditions, soil develops faster on unconsolidated sediment than it does on bedrock.

Under optimum conditions, soil-forming processes operate rapidly in the context of geologic time. From the human perspective, though, soil formation is a slow process; consequently, soil is a nonrenewable resource.

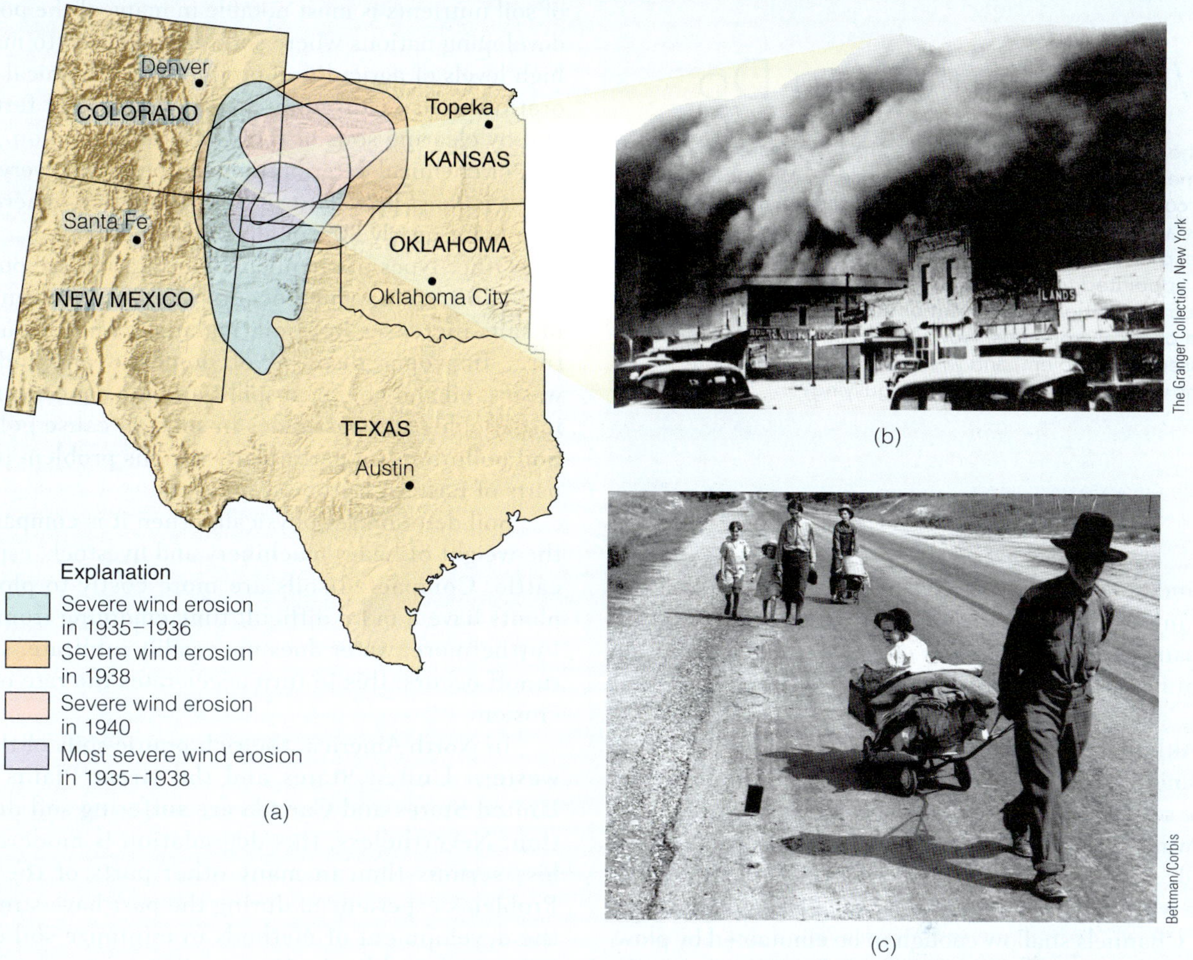

Figure 1

(a) Map of the Dust Bowl of the 1930s. (b) This huge dust storm was photographed at Lamar, Colorado, in 1934. (c) By the mid-1930s, tens of thousands of people were on relief, homeless, or leaving the Dust Bowl. In 1939, Dorthea Lange photographed this homeless family of seven in Pittsburgh County, Oklahoma. Source: (a) "Map: Extent of Area Subject to Severe Wind Erosion," p. 30, from *Dust Bowl: The Southern Plains in the 1930s*, by Donald Worster. Copyright © 1979, 1982 by Oxford University Press, Inc. Used by permission of Oxford University Press, Inc.

debt—mostly because they had purchased farm machinery in order to produce more and benefit from the high prices. Feeling economic pressure because of their huge debts, they tilled marginal land and employed few, if any, soil conservation measures.

If the Dust Bowl has a bright side, it is that the government, farmers, and the public in general no longer take soil for granted or regard it as a substance that needs no nurturing. In addition, a number of soil conservation methods developed then have now become standard practices.

Soil Degradation

Soils are nonrenewable, so soil losses that exceed the rate of formation are viewed with alarm. Likewise any reduction in soil fertility and productivity is cause for concern, especially in areas where soils already provide only a marginal existence. Erosion and chemical and physical deterioration are all forms of **soil degradation** and are serious problems in many parts of the world.

Erosion, an ongoing natural process, is usually slow enough for soil formation to keep pace, but unfortunately, some human practices add to the problem. Removing natural vegetation by plowing, overgrazing, overexploitation for fire wood, and deforestation all contribute to erosion

What Would You Do?

In the past few years many gullies have appeared in farmer's fields in your area, and residents of your area are concerned because agriculture is the main source of jobs and tax revenue. Obviously a decrease in agricultural production would be an economic disaster. You are appointed to a county board charged with making recommendations to prevent or at least minimize erosion on local croplands. How would you determine what caused the problem, and what specific recommendations would you make to reduce gullying?

by wind and running water. The Dust Bowl that developed in several Great Plains states during the 1930s is a poignant example of just how effective wind erosion is on soil pulverized and exposed by plowing (see Geo-Focus 6.1).

Although wind has caused considerable soil erosion in some areas, running water is much more powerful. Some soil is removed by *sheet erosion,* which involves the removal of thin layers of soil more or less evenly over a broad, sloping surface. *Rill erosion,* in contrast, takes place when running water scours small, troughlike channels. Channels shallow enough to be eliminated by plowing are *rills,* but those too deep (about 30 cm) to be plowed over are *gullies* (■ Figure 6.13). Where gullying is extensive, croplands can no longer be tilled and must be abandoned.

Soil undergoes chemical deterioration when its nutrients are depleted and its productivity decreases. Loss of soil nutrients is most notable in many of the populous developing nations where soils are overused to maintain high levels of agricultural productivity. Chemical deterioration is also caused by insufficient use of fertilizers and by clearing soils of their natural vegetation. Examples of chemical deterioration can be found everywhere, but it is most prevalent in South America, where it accounts for nearly 30% of all soil degradation.

Other types of chemical deterioration are pollution and *salinization,* which occurs when the concentration of salts increases in a soil, making it unfit for agriculture. Improper disposal of domestic and industrial wastes, oil and chemical spills, and the concentration of insecticides and pesticides in soils all cause pollution. Soil pollution is a particularly serious problem in some parts of Eastern Europe.

Soil deteriorates physically when it is compacted by the weight of heavy machinery and livestock, especially cattle. Compacted soils are more costly to plow, and plants have a more difficult time emerging from them. Furthermore, water does not readily infiltrate, so more runoff occurs; this in turn accelerates the rate of water erosion.

In North America, the rich prairie soils of the midwestern United States and the Great Plains of the United States and Canada are suffering soil degradation. Nevertheless, this degradation is moderate and less serious than in many other parts of the world. Problems experienced during the past have stimulated the development of methods to minimize soil erosion on agricultural lands. Crop rotation, contour plowing, and the construction of terraces have all proved helpful (■ Figure 6.14). So has no-till planting, in which the residue from the harvested crop is left on the ground to protect the surface from the ravages of wind and water.

James S. Monroe

(a)

H. H. Walson/USGS

(b)

■ **Figure 6.13**

(a) Rill erosion in a field in Michigan during a rainstorm. The rill was later plowed over. (b) A large gully in the upper basin of the Rio Reventado in Costa Rica.

Science VU/Visuals Unlimited

Figure 6.14

Contour plowing and strip cropping are two soil conservation practices used on this farm. Contour plowing involves plowing parallel to the contours of the land to inhibit runoff and soil erosion. In strip cropping, row crops such as corn alternate with other crops such as grass.

WEATHERING AND RESOURCES

We have discussed various aspects of soils, which are certainly one of our most precious natural resources. Indeed, if it were not for soils, food production on Earth would be vastly different and capable of supporting far fewer people. In addition, other aspects of soils are important economically. We discussed the origin of laterite in response to intense chemical weathering in the tropics, and we noted further that they are not very productive. If the parent material is rich in aluminum, however, the ore of aluminum called *bauxite* accumulates in horizon B. Some bauxite is found in Arkansas, Alabama, and Georgia, but at present it is cheaper to import rather than mine these deposits, so both the United States and Canada depend on foreign sources of aluminum ore.

Bauxite and other accumulations of valuable minerals by the selective removal of soluble substances during chemical weathering are known as *residual concentrations*. Certainly bauxite is a good example of a residual concentration, but other deposits that formed in a similar fashion are those rich in iron, manganese, clays, nickel, phosphate, tin, diamonds, and gold. Some of the sedimentary iron deposits in the Lake Superior region of the United States and Canada were enriched by chemical weathering when soluble parts of the deposits were carried away. Some kaolinite deposits in the southern United States formed when chemical weathering altered feldspars in pegmatites or as residual concentrations of clay-rich limestones and dolostones. Kaolinite is a clay mineral used in the manufacture of paper and ceramics.

Chemical weathering is also responsible for gossans and ore deposits that lie beneath them. A *gossan* is a yellow to red deposit made up mostly of hydrated iron oxides that formed by oxidation and leaching of sulfide minerals such as pyrite (FeS_2). The dissolution of pyrite and other sulfides forms sulfuric acid, which causes other metallic minerals to dissolve, and these tend to be carried down toward the groundwater table, where the descending solutions form minerals containing copper, lead, and zinc. Gossans have been mined for iron, but they are far more important as indicators of underlying ore deposits.

SEDIMENT AND SEDIMENTARY ROCKS

Weathering, erosion, transport, and deposition are essential parts of the rock cycle (see Figure 1.12) because they are responsible for the origin and deposition of *sediment* that may become *sedimentary rock*. The term **sediment** refers to (1) all solid particles of preexisting rocks yielded by weathering, (2) minerals derived from solutions that contain materials dissolved during chemical weathering, and (3) minerals extracted from water, mostly seawater, by organisms to build their shells. **Sedimentary rock** is any rock made up of consolidated sediment.

One important criterion for classifying sedimentary particles is their size, particularly for solid particles, or *detrital sediment,* as opposed to *chemical sediment,* which consists of minerals extracted from solution by inorganic chemical processes or the activities of organisms. Particles described as *gravel* measure more than 2 mm, whereas sand measures $\frac{1}{16}$–2 mm, and silt is any particle between $\frac{1}{256}$ and $\frac{1}{16}$ mm. None of these designations implies anything about composition; most gravel is made up of rock fragments—that is, small pieces of granite, basalt, or any other rock type—but sand and silt grains are usually single minerals, especially quartz. Particles smaller than $\frac{1}{256}$ mm are termed clay, but clay has two meanings. One is simply a size designation, but the term also refers to certain types of sheet silicates known as *clay minerals.* However, most clay minerals are also clay sized.

Sediment Transport and Deposition

Weathering is fundamental to the origin of sediment and sedimentary rocks, and so are erosion and *deposition*—that is, the movement of sediment by natural processes and its accumulation in some area. Because glaciers are moving solids, they can carry sediment of any size, whereas wind transports only sand and smaller sediment. Waves and marine currents transport sediment along shorelines, but running water is by far the most common way to transport sediment from its source to other locations.

During transport, *abrasion* reduces the size of particles, and the sharp corners and edges are worn smooth, a process known as *rounding,* as pieces of sand and gravel collide with one another (■ Figure 6.15a, b). Transport and processes that operate where sediment accumulates also result in *sorting,* which refers to the particle-size distribution in a sedimentary deposit. Sediment is characterized as well sorted if all particles are about the same size, and poorly sorted if a wide range of particle sizes is present (Figure 6.15c). Both rounding and sorting have important implications for other aspects of sediment and sedimentary rocks, such as how readily fluids move through them, and they also help geologists decipher the history of a deposit.

Regardless of how sediment is transported, it is eventually deposited in some geographic area known as a **depositional environment.** Deposition might take place on a floodplain, in a stream channel, on a beach, on the seafloor, or in a variety of other depositional environments where physical, chemical, and biological processes impart various characteristics to the accumulating sediment. Geologists recognize three major depositional settings: continental (on the land), transitional (on or near seashores), and marine, each with several specific depositional environments (■ Figure 6.16).

How Does Sediment Become Sedimentary Rock?

A deposit of detrital sediment consists of a loose aggregate of particles. Mud accumulating in lakes and sand and gravel in stream channels or on beaches are good examples (Figure 6.15). To convert these aggregates of particles into sedimentary rocks requires **lithification** by compaction, cementation, or both (■ Figure 6.17).

To illustrate the relative importance of compaction and cementation, consider a detrital sedimentary deposit made up of mud and another composed of sand. In both cases, the sediment consists of solid particles and *pore spaces,* the voids between particles. These deposits are subjected to **compaction** from their own weight and the weight of any additional sediment deposited on top of them, thereby reducing the amount of pore space and the volume of the deposit. Our hypothetical mud deposit may

■ **Figure 6.15**

Rounding and sorting. (a) Beginning students often mistake rounding to mean ball-shaped or spherical. All three of these stones are well rounded, meaning that their sharp corners and edges have been worn smooth. (b) Deposit of well-rounded, well-sorted gravel. The particles average about 5 cm across. (c) Angular, poorly sorted gravel. Note the quarter to provide scale.

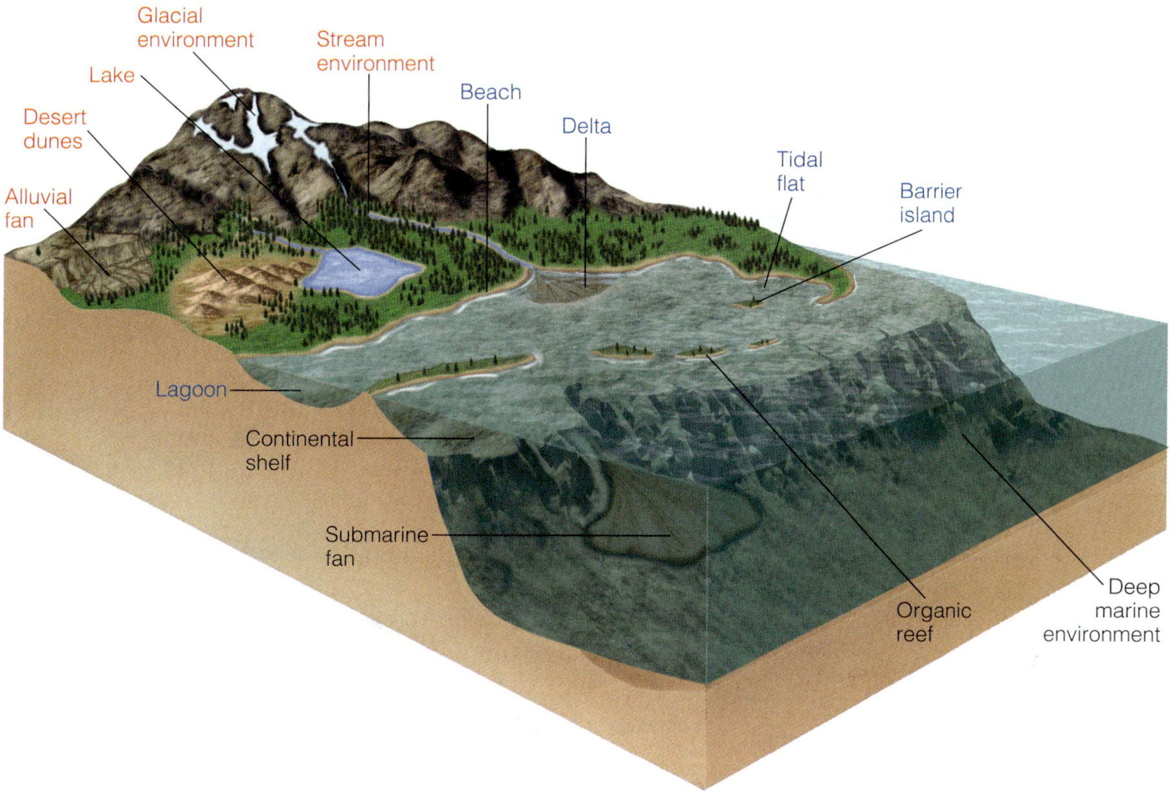

Figure 6.16

Depositional environments. Continental environments are shown in red type. The environments along the seashore, shown in blue type, are transitional from continental to marine. The others, shown in black type, are marine environments.

have 80% water-filled pore space, but after compaction its volume is reduced by as much as 40% (Figure 6.17). The sand deposit with as much as 50% pore space is also compacted, but far less than the mud deposit, so that the grains fit more tightly together (Figure 6.17).

Compaction alone is sufficient for lithification of mud, but for sand and gravel **cementation** involving the precipitation of minerals in pore spaces is also necessary. The two most common chemical cements are calcium carbonate ($CaCO_3$) and silicon dioxide (SiO_2), but iron oxide and hydroxide cement, such as hematite (Fe_2O_3) and limonite [$FeO(OH){\cdot}nH_2O$], are found in some sedimentary rocks. Recall that calcium carbonate readily dissolves in water that contains a small amount of carbonic acid, and chemical weathering of feldspars and other minerals yields silica in solution. Cementation takes place when minerals precipitate in the pore spaces of sediment from circulating water, thereby binding the loose particles together. Iron oxide and hydroxide cements account for the red, yellow, and brown sedimentary rocks found in many areas (see the chapter opening photo).

We have explained lithification of detrital sediments, but we have not yet considered this process in chemical sediments. By far the most common chemical sediments are calcium carbonate mud and sand- and gravel-sized accumulations of calcium carbonate grains, such as shells and shell fragments. Compaction and cementation also take place in these sediments, converting them into various types of limestone, but compaction is generally less effective because cementation takes place soon after deposition. In any case, the cement is calcium carbonate derived by partial solution of some of the particles in the deposit.

GeologyNow Log into GeologyNow and select this chapter to work through a **Geology Interactive** activity on "The Rock Cycle" (click Rocks and the Rock Cycle→Rock Cycle).

TYPES OF SEDIMENTARY ROCKS

Thus far, we have considered the origin of sediment, its transport, deposition, and lithification. We now turn to the types of sedimentary rocks and how they are classified. The two broad classes or types of sedimentary rocks are *detrital* and *chemical,* although the latter has a subcategory known as *biochemical* (Table 6.2).

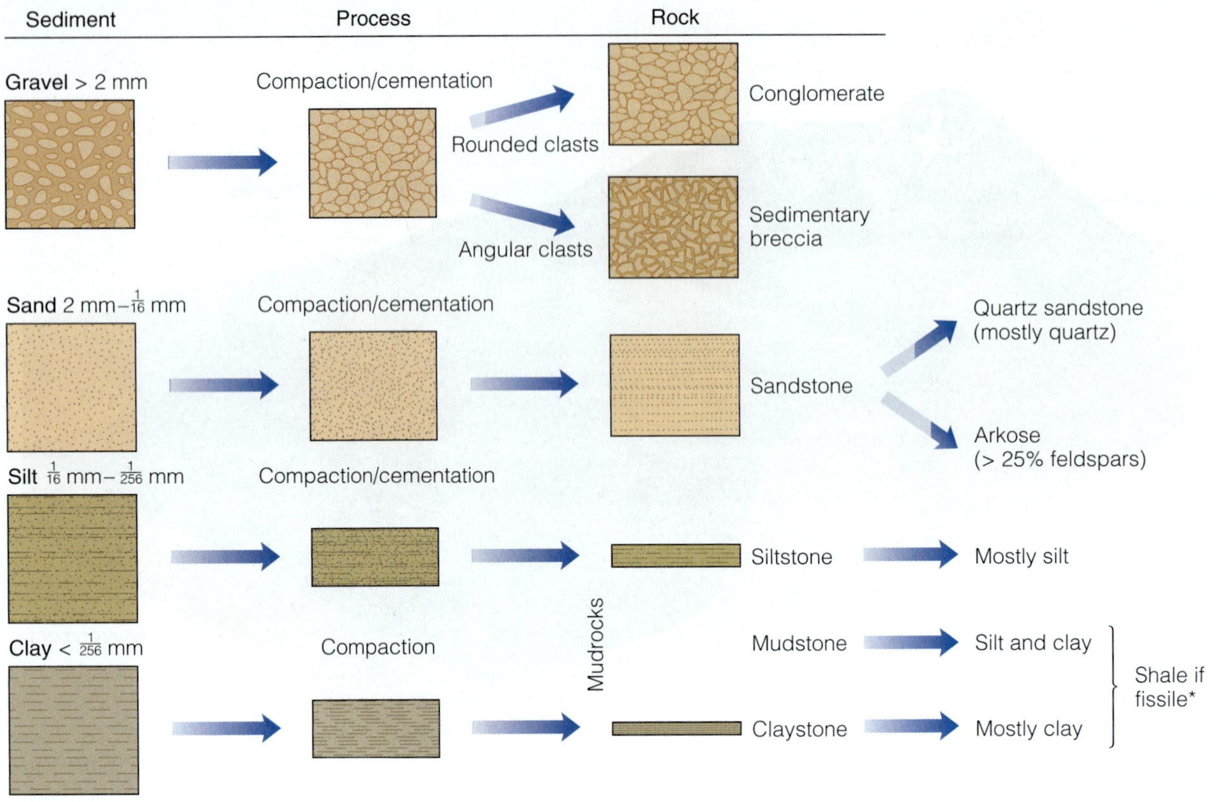

*Fissile refers to rocks capable of splitting along closely spaced planes.

Figure 6.17

Lithification of detrital sediments and the classification of detrital sedimentary rocks. Notice that little compaction takes place in sand and gravel.

Detrital Sedimentary Rocks

Detrital sedimentary rocks are made up of *detritus,* the solid particles such as sand and gravel derived from parent material. All detrital sedimentary rocks have a *clastic texture,* meaning they are composed of particles or fragments known as *clasts.* The several varieties in this broad category are classified by the size of their constituent particles, although composition is used to modify some rock names.

Both *conglomerate* and *sedimentary breccia* are composed of gravel-sized particles (Figure 6.17 and Figure

Table 6.2

Classification of Chemical and Biochemical Sedimentary Rocks

Texture	Composition	Rock Name	
CHEMICAL SEDIMENTARY ROCKS			
Varies	Calcite ($CaCO_3$)	Limestone	Carbonate rocks
Varies	Dolomite [$CaMg(CO_3)_2$]	Dolostone	Carbonate rocks
Crystalline	Gypsum ($CaSO_4 \cdot 2H_2O$)	Rock gypsum	Evaporites
Crystalline	Halite (NaCl)	Rock salt	Evaporites
BIOCHEMICAL SEDIMENTARY ROCKS			
Clastic	Calcite ($CaCO_3$) shells	Limestone (various types such as chalk and coquina)	
Usually crystalline	Altered microscopic shells of SiO_2	Chert (various color varieties)	
—	Carbon from altered land plants	Coal (lignite, bituminous, anthracite)	

(a) Conglomerate

(b) Sedimentary breccia

(c) Quartz sandstone

(d)

Figure 6.18

Detrital sedimentary rocks. (a) Conglomerate with rounded gravel particles measuring 4 to 5 cm on average. (b) Sedimentary breccia is made up of angular gravel. (c) Quartz sandstone. (d) Exposure of shale in Tennessee. Source: Sue Monroe

6.18a, b), but conglomerate has rounded gravel, whereas sedimentary breccia has angular gravel. Conglomerate is common, but sedimentary breccia is rare because gravel-sized particles become rounded very quickly during transport. Thus, if you encounter sedimentary breccia, you can conclude that its angular gravel has experienced little transport, probably less than a kilometer. Considerable energy is needed to transport gravel, so conglomerate is usually found in environments such as stream channels and beaches.

Sand is simply a size designation for particles between $\frac{1}{16}$ and 2 mm, so any mineral or rock fragment can be in *sandstone*. Geologists recognize varieties of sandstone based on mineral content (Figures 6.17 and 6.18c). *Quartz sandstone* is the most common and, as the name implies, is made up mostly of quartz grains. Another variety of sandstone called *arkose* contains at least 25% feldspar minerals. Sandstone is found in a number of depositional environments, including stream channels, sand dunes, beaches, barrier islands, deltas, and the continental shelf.

Mudrock is a general term that encompasses all detrital sedimentary rocks composed of silt- and clay-size particles (Figure 6.17). Varieties include *siltstone,* composed mostly of silt-sized particles, *mudstone,* a mixture of silt and clay, and *claystone,* composed primarily of clay-sized particles. Some mudstones and claystones are designated *shale* if they are fissile, meaning that they break along closely spaced parallel planes (Figure 6.18d). Even weak currents can transport silt- and clay-sized particles, and deposition takes place only where currents and fluid turbulence are minimal, as in the quiet offshore waters of lakes or in lagoons.

Chemical and Biochemical Sedimentary Rocks

Several compounds and ions taken into solution during chemical weathering are the raw materials for **chemical sedimentary rocks.** Some of these rocks have a *crystalline texture,* meaning they are composed of a mosaic of interlocking mineral crystals. Others, though, have a clastic texture; some limestones, for instance, are composed of fragmented seashells. Organisms play an important role in the origin of chemical sedimentary rocks designated **biochemical sedimentary rocks.**

Limestone and dolostone, the most abundant chemical sedimentary rocks, are known as **carbonate rocks** because each is made up of minerals that contain the carbonate radical (CO_3). Limestone consists of calcite ($CaCO_3$), and dolostone is made up of dolomite [$CaMg(CO_3)_2$] (see Chapter 3). Recall that calcite rapidly dissolves in acidic water, but the chemical reaction leading to dissolution is reversible, so calcite can precipitate from solution under some circumstances. Thus, some limestone, though probably not very much, forms by inorganic chemical precipitation. Most limestone is biochemical because organisms are so important in its origin—the rock in coral reefs and limestone composed of seashells, for instance (■ Figure 6.19a).

A type of limestone composed almost entirely of fragmented seashells is known as *coquina* (Figure 6.19b), and chalk is a soft variety of limestone made up largely of microscopic shells (Figure 6.19c). One distinctive variety of limestone contains small spherical grains called *ooids* that have a small nucleus around which concentric layers of calcite precipitated (Figure 6.19d). Lithified deposits of ooids form *oolitic limestones.*

Dolostone is similar to limestone, but most or all of it formed secondarily by the alteration of limestone. The consensus among geologists is that dolostone originates when magnesium replaces some of the calcium in calcite, thereby converting calcite to dolomite.

Some of the dissolved substances derived by chemical weathering precipitate from evaporating water and thus form chemical sedimentary rocks known as **evaporites** (Table 6.2). *Rock salt,* composed of halite (NaCl), and *rock gypsum* ($CaSO_4 \cdot 2H_2O$) are the most common (■ Figure 6.20a, b), although several others are known and some are important resources. Compared with mudrocks, sandstone, and limestone, evaporites are not very common but nevertheless are significant deposits in areas such as Michigan, Ohio, New York, the Gulf Coast region, and Saskatchewan, Canada.

Sue Monroe

(a) Limestone with fossils

Sue Monroe

(b) Coquina

Richard V. Dietrich

(c) Chalk

Rex Elliot

(d) Ooids

■ **Figure 6.19**

(a) Limestone with numerous fossil shells. (b) Coquina is composed of broken shells. (c) Chalk cliffs in Denmark. Chalk is made up of microscopic shells. (d) Present-day ooids up to 2 mm across from the Bahamas.

Sue Monroe

(a) Rock salt

Sue Monroe

(b) Rock gypsum

Sue Monroe

(c) Chert

Sue Monroe

(d) Bituminous coal

Figure 6.20

Chemical and biochemical sedimentary rocks. (a) Core of rock salt from an oil well in Michigan. (b) Rock gypsum. (c) Chert, a dense, hard rock made up of microscopic crystals of quartz. (d) Bituminous coal.

Chert is a hard rock composed of microscopic crystals of quartz (Table 6.2 and Figure 6.20c). Some of the color varieties of chert are *flint,* which is black because of inclusions of organic matter, and *jasper,* which is colored red or brown by iron oxides. Because chert is hard and lacks cleavage, it can be shaped to form sharp cutting edges, so it has been used to manufacture tools, spear points, and arrowheads. Chert is found as irregular masses or *nodules* in other rocks, especially limestone, and as distinct layers of *bedded chert* made up of tiny shells of silica-secreting organisms.

Coal consists of compressed, altered remains of land plants, but it is nevertheless a biochemical sedimentary rock (Figure 6.20d). It forms in swamps and bogs where the water is oxygen deficient or where organic matter accumulates faster than it decomposes. In oxygen-deficient swamps and bogs, the bacteria that decompose vegetation can live without oxygen, but their wastes must be oxidized, and because little or no oxygen is present, wastes accumulate and kill the bacteria. Bacterial decay ceases, and the vegetation is not completely decomposed and forms organic muck. When buried and compressed, the muck becomes *peat,* which looks rather like coarse pipe tobacco. Where peat is abundant, as in Ireland and Scotland, it is used for fuel.

Peat represents the first step in forming coal. If peat is more deeply buried and compressed, and especially if it is heated too, it is converted to dull black coal called *lignite.* During this change, the easily vaporized or volatile elements are driven off, enriching the residue in carbon; lignite has about 70% carbon, whereas only about 50% is present in peat. *Bituminous coal,* with about 80% carbon, is dense, black, and so thoroughly altered that plant remains are rarely seen. It burns more efficiently than lignite, but the highest-grade coal is *anthracite,* a metamorphic type of coal (see Chapter 7), with up to 98% carbon.

GeologyNow Log into GeologyNow and select this chapter to work through a **Geology Interactive** activity on "Rock Laboratory" (click Rocks and the Rock Cycle→Rock Laboratory).

SEDIMENTARY FACIES

If a layer of sediment or sedimentary rock is traced laterally, it generally changes in composition, texture, or both. It changes by lateral gradation resulting from the simultaneous operation of different

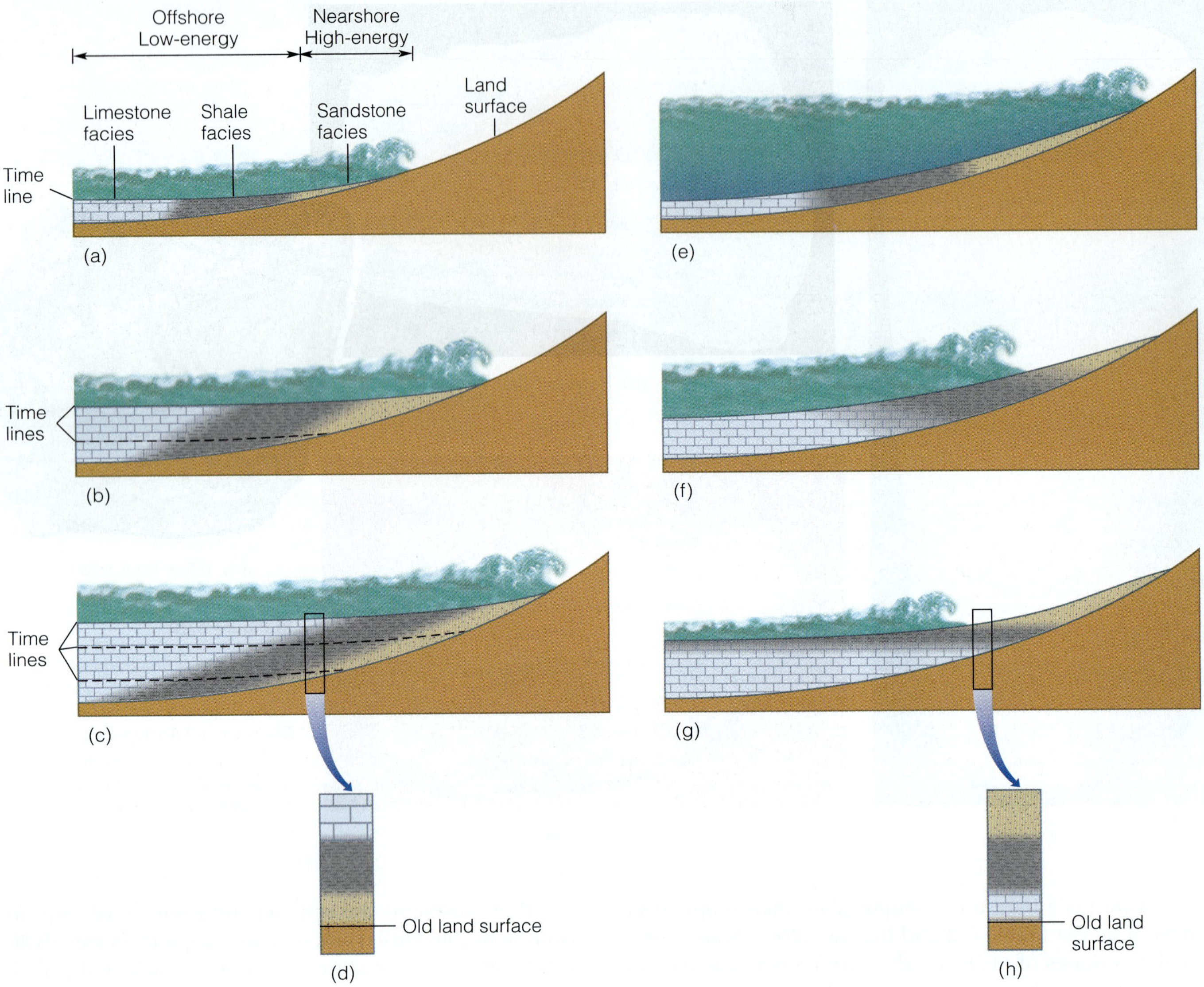

Figure 6.21

(a–c) Three stages of a marine transgression. (d) Diagrammatic view of the vertical sequence of facies resulting from a transgression. (e–g) Three stages of a marine regression. (h) Vertical sequence of facies resulting from a regression.

processes in adjacent depositional environments. For example, sand may be deposited in a high-energy nearshore marine environment whereas mud and carbonate sediments accumulate simultaneously in the laterally adjacent low-energy offshore environments (■ Figure 6.21). Deposition in each of these environments produces **sedimentary facies,** bodies of sediment each possessing distinctive physical, chemical, and biological attributes. Figure 6.21 illustrates three sedimentary facies: a sand facies, a mud facies, and a carbonate facies. If these sediments become lithified, they are sandstone, mudstone (or shale), and limestone facies, respectively.

Many sedimentary rocks in the interiors of continents show clear evidence of deposition in marine environments. The rock layers in Figure 6.21d, for example, consist of a sandstone facies that was deposited in a nearshore marine environment overlain by shale and limestone facies deposited in offshore environments. Geologists explain this vertical sequence of facies by deposition occurring during a time when sea level rose with respect to the continents. As sea level rises, the shoreline moves inland, giving rise to a **marine transgression** (Figure 6.21), and the depositional environments parallel to the shoreline migrate landward. As a result of a marine transgression, offshore facies are superimposed over nearshore facies, thus accounting for the vertical succession of sedimentary facies. Even though the nearshore environment is long and narrow at any particular time, deposition takes place continuously as the environment migrates landward. The sand deposit may be tens to hundreds of meters thick but have horizontal dimensions of length and width measured in hundreds of kilometers.

The opposite of a marine transgression is a **marine regression** (Figure 6.21e–h). If sea level falls with respect to a continent, the shoreline and environments that parallel the shoreline move seaward. The vertical sequence produced by a marine regression has facies of the nearshore environment superposed over facies of offshore environments. Marine regressions also account for the deposition of a facies over a large geographic area.

What Would You Do?

You live in the continental interior where flat-lying sedimentary rock layers are well exposed. Some local residents tell you of a location nearby where sandstone and mudstone with dinosaur fossils are overlain first by a seashell-bearing sandstone, followed upward by shale and finally limestone containing the remains of clams, oysters, and corals. How would you explain the presence of fossils, especially marine fossils so far from the sea, and how this vertical sequence of rocks came to be deposited?

READING THE STORY IN SEDIMENTARY ROCKS

We mentioned in the Introduction that sedimentary rocks preserve a record of the conditions under which they formed. However, no one was present when ancient sediments were deposited, so geologists must evaluate those aspects of sedimentary rocks that allow them to make inferences about the original depositional environment. And making such determinations is of more than academic interest. For instance, barrier island sand deposits make good reservoirs for hydrocarbons, so knowing the environment of deposition and the geometry of these deposits is helpful in exploration for resources.

Sedimentary textures such as sorting and rounding can give clues to depositional processes. Windblown dune sands tend to be well sorted and well rounded, but poor sorting is typical of glacial deposits. The geometry or three-dimensional shape is another important aspect of sedimentary rock bodies. Marine transgressions and regressions yield sediment bodies with a blanket or sheetlike geometry, but sand deposits in stream channels are long and narrow and described as having a shoestring geometry. Sedimentary textures and geometry alone are usually insufficient to determine depositional environment, but when considered with other sedimentary rock properties, especially *sedimentary structures* and *fossils,* they enable geologists to reliably determine the history of a deposit.

Sedimentary Structures

Physical and biological processes operating in depositional environments are responsible for a variety of features known as **sedimentary structures.** One of the most common is distinct layers known as **strata** or **beds** (■ Figure 6.22a), with individual layers less than a millimeter up to many meters thick. These strata or beds are separated from one another by surfaces above and below in which the rocks differ in composition, texture, color, or a combination of features. Layering of some kind is present in almost all sedimentary rocks, but a few, such as limestone that formed in coral reefs, lack this feature.

Many sedimentary rocks are characterized by **cross-bedding,** in which layers are arranged at an angle to the

Sue Monroe

(a)

Sue Monroe

(b)

■ **Figure 6.22**

(a) Bedding or stratification is obvious in these alternating layers of mudrock (shale in this case) and sandstone. (b) Cross-beds in ancient sandstone in Montana. The hammer is about 30 cm long.

surface on which they are deposited (Figure 6.22b). Cross-beds are found in many depositional environments such as sand dunes in deserts and along shorelines, as well as in stream-channel deposits and shallow marine sediments. Invariably, cross-beds result from transport and deposition by wind or water currents, and the cross-beds are inclined downward in the same direction the current flowed. So, ancient deposits with cross-beds inclined down toward the south, for example, indicate that the currents responsible for them flowed from north to south.

Some individual sedimentary rock layers show an upward decrease in grain size, termed **graded bedding,** mostly formed by turbidity current deposition. A *turbidity current* is an underwater flow of sediment and water with a greater density than sediment-free water. Because of its greater density, a turbidity current flows downslope until it reaches the relatively flat seafloor, or lakefloor, where it slows and begins depositing large particles followed by progressively smaller ones (■ Figure 6.23). Some graded bedding also forms in stream channels during the waning stages of floods.

The surfaces that separate layers in sand deposits commonly have **ripple marks,** small ridges with intervening troughs, giving them a somewhat corrugated appearance. Some ripple marks are asymmetrical in cross section, with a gentle slope on one side and a steep slope on the other. Currents that flow in one direction as in stream channels generate these so-called *current ripple marks* (■ Figure 6.24a, b). And because the steep slope of these ripples is on the downstream side, they are good indications of ancient current directions. In contrast, *wave-formed ripple marks* tend to be symmetrical in cross section and, as their name implies, are generated by the to-and-fro motion of waves (Figure 6.24c, d).

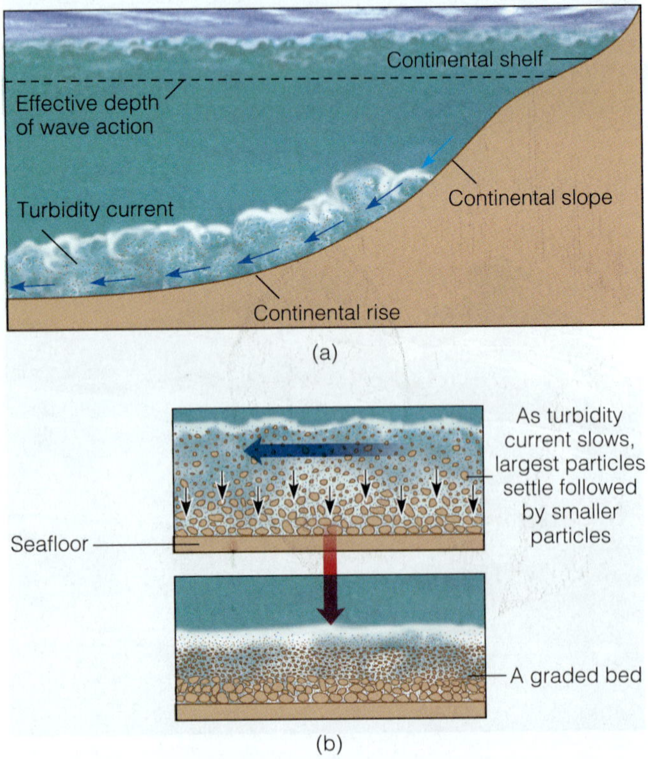

■ Figure 6.23

Graded bedding. (a) Turbidity current flows downslope along the seafloor (or lake bottom) because it is denser than sediment-free water. (b) Deposition of a graded bed takes place as the flow slows and deposits progressively smaller particles.

When clay-rich sediment dries, it shrinks and develops intersecting fractures called **mud cracks** (■ Figure 6.25). Mud cracks in ancient sedimentary rocks indicate that the sediment was deposited in an environment where periodic drying took place, such as on a river floodplain, near a lakeshore, or where muddy deposits are exposed along seacoasts at low tide.

Fossils—Remains and Traces of Ancient Life

Fossils, the remains or traces of ancient organisms, are interesting as evidence of prehistoric life (■ Figure 6.26), and are also important in determining depositional environments. Most people are familiar with fossil dinosaurs and some other land-dwelling animals but are unaware that fossils of invertebrates, animals lacking a segmented vertebral column, such as corals, clams, oysters, and a variety of microorganisms, are much more useful because they are so common. It is true that the remains of land-dwelling creatures and plants can be washed into marine environments, but most are preserved in rocks deposited on land or perhaps transitional environments such as deltas. In contrast, fossils of corals tell us that the rocks in which they are preserved were deposited in the ocean.

Clams with heavily constructed shells typically live in shallow turbulent seawater, whereas organisms living in low-energy environments commonly have thin, fragile shells. Marine organisms that carry on photosynthesis are restricted to the zone of sunlight penetration, which is usually less than 200 m. The amount of sediment is also a limiting factor on the distribution of organisms. Many corals live in shallow, clear seawater because suspended sediment clogs their respiratory and food-gathering organs, and some have photosynthesizing algae living in their tissues.

Microfossils are particularly useful for environmental studies because hundreds or even thousands can be recovered from small rock samples. In oil-drilling operations, small rock chips known as *well cuttings* are brought to the surface. These samples may contain numerous microfossils, but rarely have entire fossils of larger organisms. These fossils are routinely used to determine depositional environments and to match up rocks of the same relative age (see Chapter 17).

Determining the Environment of Deposition

Geologists rely on textures, sedimentary structures, and fossils to interpret how a particular sedimentary rock

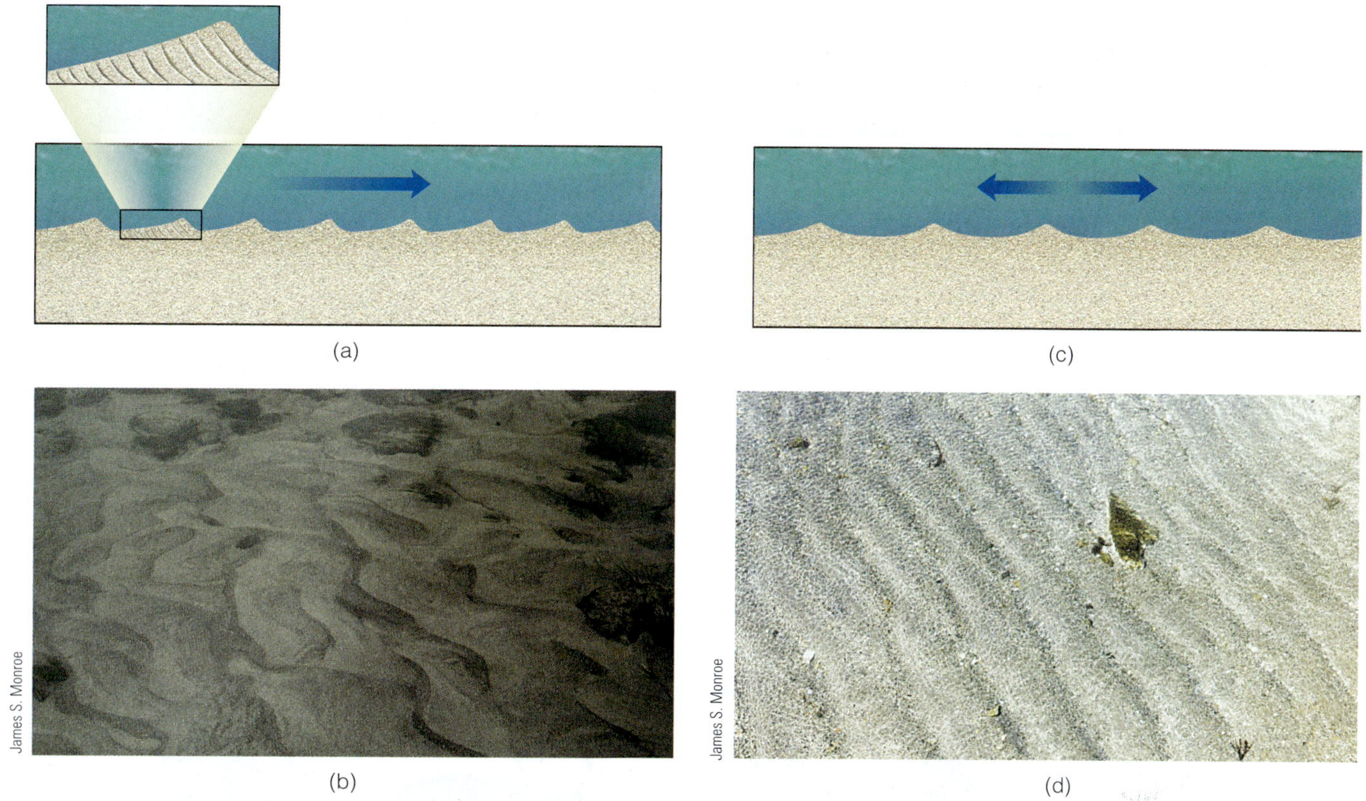

Figure 6.24

Ripple marks. (a) Current ripple marks form in response to flow in one direction, as in a stream channel. The enlargement of one ripple shows its internal structure. Note that individual layers within the ripple are inclined, showing an example of cross-bedding. (b) Current ripples that formed in a small stream channel; flow was from right to left. (c) The to-and-fro currents of waves in shallow water deform the surface of the sand layer into wave-formed ripple marks. (d) Wave-formed ripple marks in sand in shallow seawater.

body was deposited. Furthermore, they compare the features seen in ancient rocks with those in deposits forming today. In short, sedimentary rocks provide a record of many events that took place during the past.

But are we justified in using present-day processes and environments to make inferences about what happened when no human observers were present? Perhaps some examples will help answer this question.

Figure 6.25

(a) Mud cracks form in clay-rich sediments when they dry and shrink. (b) Mud cracks in ancient rocks in Glacier National Park, Montana. Notice that the cracks have been filled by sediment.

GEOLOGY IN UNEXPECTED PLACES

Sandstone Lion

The 9-m-long Lion Monument in Lucerne, Switzerland, was chiseled into sandstone in 1821 as a memorial to the approximately 850 Swiss soldiers who died during the French Revolution of 1792 in Paris (■ Figure 1a). Lukas Ahorn chiseled the monument into the sandstone wall of a quarry; the inscription above the lion pays honor to the "loyalty and courage of the Swiss." An officer on leave from the army at the time of the battle in Paris took the first steps to set up the monument.

Notice that the sandstone layers are inclined downward, or dip, to the left at about 50 degrees. We could postulate that (1) the original layers were horizontal and simply tilted 50 degrees into this position, or (2) perhaps they were rotated 140 degrees from their original position so that the layers are now upside down, or *overturned* in geologic parlance. To resolve this problem, you must determine which of the layers shown was at the top of the original sequence of beds and thus the youngest. In Figure 1b, notice that the cross-beds have a sharp angular contact with the younger rocks above them, whereas they are nearly parallel with the older rocks below. Accordingly, we conclude that the youngest rock layer is the one toward the upper left and the rock layers have not been overturned.

Having determined which layer is oldest and which is youngest, we now know that any rocks exposed to the right of the image are older than the ones shown and, of course, any to the left are younger. However, it is important to note that we have determined *relative ages* only—that is, which layers are older versus younger. Nothing in this image tells us the absolute age in number of years before the present. We consider relative and absolute ages more fully in Chapter 17.

■ **Figure 1**

(a) Lion Monument in Lucerne, Switzerland.
(b) Cross-bedding shows sharp angular contact with younger rocks above and nearly parallel contact with older rocks below. Source: Sue Monroe

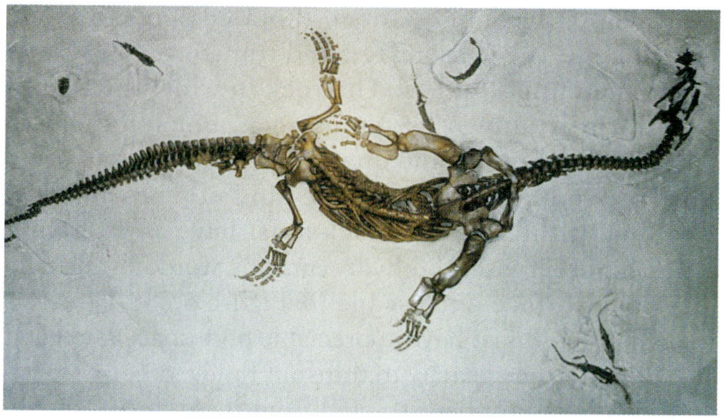
Sue Monroe

(a)

Sue Monroe

(b)

Figure 6.26

Fossils. (a) Bones of a 2.3-m-long, Mesozoic-aged marine reptile in the museum at the Glacier Garden in Lucerne, Switzerland. (b) Shells of extinct ocean-dwelling animals known as horn corals.

The Navajo Sandstone of the southwestern United States is an ancient desert dune deposit that formed when the prevailing winds blew from the northeast. What evidence justifies this conclusion? This 300-m-thick sandstone is made up of well-sorted, well-rounded sand grains measuring 0.2–0.5 mm in diameter. Furthermore, it has cross-beds up to 30 m high (■ Figure 6.27a) and current ripple marks, both typical of desert dunes. Some of the sand layers have preserved dinosaur tracks and tracks of other land-dwelling animals, ruling out the possibility of a marine origin. In short, the Navajo Sandstone possesses several features that point to a desert dune depositional environment. Finally, the cross-beds are inclined downward toward the southwest, indicating that the prevailing winds were from the northeast.

In the Grand Canyon of Arizona several formations are well exposed; a *formation* is a widespread unit of rock, especially sedimentary rock, that is recognizably different from the rocks above and below. A vertical sequence consisting of the Tapeats Sandstone, Bright Angel Shale, and Muav Limestone is present in the lower part of the canyon (Figure 6.27b), all of which contain features, including fossils, clearly indicating that they were deposited in transitional and marine environments. As a matter of fact, all three were forming simultaneously in different adjacent environments, and during a marine transgression they were deposited in the vertical se-

James S. Monroe

(a)

Alan Mayo/GeoPhoto

(b)

Figure 6.27

Ancient sedimentary rocks and their interpretation. (a) The Jurassic-aged Navajo Sandstone in Zion National Park, Utah, is a wind-blown dune deposit. Vertical fractures intersect cross-beds, giving this cliff its checkerboard appearance—hence, the name Checkerboard Mesa. (b) View of three formations in the Grand Canyon in Arizona. These rocks were deposited during a marine transgression. Compare with the vertical sequence of rocks in Figure 6.21d.

quence now seen. They conform closely to the sequence shown in Figure 6.21d.

IMPORTANT RESOURCES IN SEDIMENTARY ROCKS

The uses of sediments and sedimentary rocks or the materials they contain vary considerably. Sand and gravel are essential to the construction industry, pure clay deposits are used for ceramics, and limestone is used in the manufacture of cement and in blast furnaces where iron ore is refined to make steel. Evaporites are the source of table salt as well as a number of chemical compounds, and rock gypsum is used to manufacture wallboard. Phosphate-bearing sedimentary rock is used in fertilizers and animal feed supplements.

Some valuable sedimentary deposits are found in streams and on beaches where minerals were concentrated during transport and deposition. These *placer deposits,* as they are called, are surface accumulations resulting from the separation and concentration of materials of greater density from those of lesser density. Much of the gold recovered during the initial stages of the California gold rush (1849–1853) was mined from placer deposits, and placers of a variety of other minerals such as diamonds and tin are important.

Historically, most coal mined in the United States has been bituminous coal from the Appalachian region that formed in coastal swamps during the Pennsylvanian Period (286 and 320 million years ago). Huge lignite and subbituminous coal deposits in the western United States are becoming increasingly important. During 2002, more than a billion tons of coal were mined in this country, more than half of it from mines in Wyoming, West Virginia, and Kentucky.

Anthracite coal (see Chapter 7) is especially desirable because it burns more efficiently than other types of coal. Unfortunately, it is the least common variety, so most coal used for heating buildings and generating electricity is bituminous (Figure 6.20d). *Coke,* a hard, gray substance consisting of the fused ash of bituminous coal, is used in blast furnaces where steel is produced. Synthetic oil and gas and a number of other products are also made from bituminous coal and lignite.

Petroleum and Natural Gas

Petroleum and natural gas are both *hydrocarbons,* meaning that they are composed of hydrogen and carbon. The remains of microscopic organisms settle to the seafloor, or lakefloor in some cases, where little oxygen is present to decompose them. If buried beneath layers of sediment, they are heated and transformed into petroleum and natural gas. The rock in which hydrocarbons form is known as *source rock,* but for them to accumulate in economic quantities, they must migrate from the source rock into some kind of *reservoir rock.* And finally, the reservoir rock must have an overlying *cap rock;* otherwise, the hydrocarbons would eventually reach the surface and escape (■ Figure 6.28). Effective reservoir rocks must have appreciable pore space and good permeability, the capacity to transmit fluids; otherwise, hydrocarbons cannot be extracted from them in reasonable quantities.

Many hydrocarbon reservoirs consist of nearshore marine sandstones with nearby fine-grained, organic-rich source rocks. Such oil and gas traps are called *stratigraphic traps* because they owe their existence to variations in the strata (Figure 6.28a). Ancient coral reefs are also good stratigraphic traps. Indeed, some of the oil in the Persian Gulf region and Michigan is trapped in ancient reefs. *Structural traps* result when rocks are deformed by folding, fracturing, or both. In sedimentary rocks that have been deformed into a series of folds, hydrocarbons migrate to the high parts of these structures (Figure 6.28b). Displacement of rocks along faults (fractures along which movement has occurred) also yields traps for hydrocarbons (Figure 6.28b).

Other sources of petroleum that will probably become increasingly important in the future include *oil shales* and *tar sands.* The United States has about two-thirds of all known oil shales, although large deposits are known in South America, and all continents have some oil shale. The richest deposits in the United States are in the Green River Formation of Colorado, Utah, and Wyoming. When the appropriate extraction processes are used, liquid oil and combustible gases can be produced from an organic substance called *kerogen* of oil shale. Oil shales in the Green River Formation yield between 10 and 140 gallons of oil per ton of rock processed, and the total amount of oil recoverable with present processes is estimated at 80 billion barrels. Currently, no oil is produced from oil shale in the United States because conventional drilling and pumping are less expensive.

Tar sand is a type of sandstone in which viscous, asphaltlike hydrocarbons fill the pore spaces. This substance is the sticky residue of once-liquid petroleum from which the volatile constituents have been lost. Liquid petroleum can be recovered from tar sand, but for this to happen, large quantities of rock must be mined and processed. Because the United States has few tar sand deposits, it cannot look to this source as a significant future energy resource. The Athabaska tar sands in Alberta, Canada, however, are one of the largest deposits of this type. These deposits are currently being mined, and it is estimated that they contain several hundred billion barrels of recoverable petroleum.

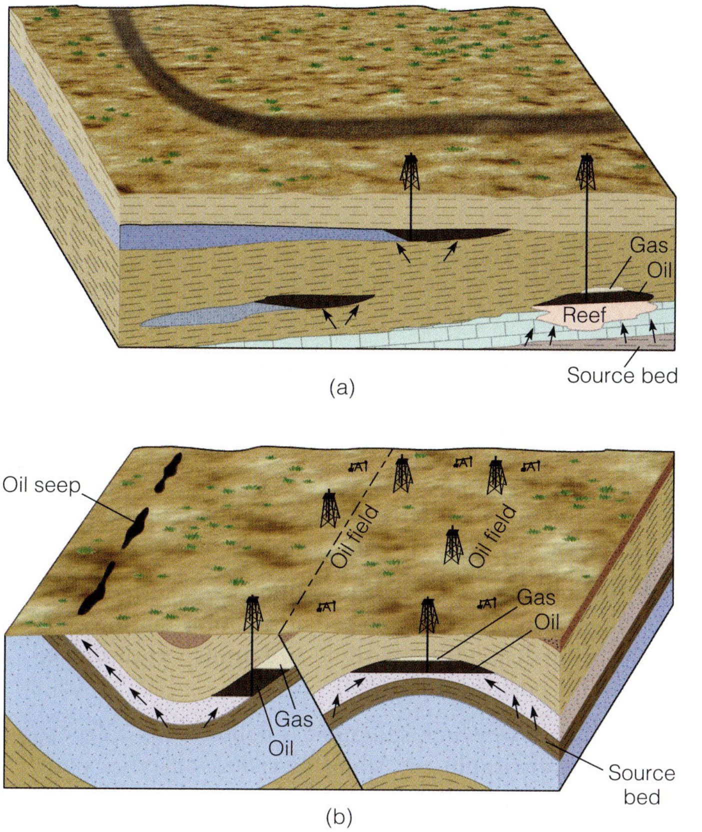

Figure 6.28

Oil and natural gas traps. The arrows indicate the migration of hydrocarbons. (a) Two examples of stratigraphic traps. (b) Two examples of structural traps, one formed by folding, the other by faulting.

Uranium

Most of the uranium used in nuclear reactors in North America comes from the complex potassium-, uranium-, vanadium-bearing mineral *carnotite* found in some sedimentary rocks. Some uranium is also derived from *uraninite* (UO_2), a uranium oxide in granitic rocks and hydrothermal veins. Uraninite is easily oxidized and dissolved in groundwater, transported elsewhere, and chemically reduced and precipitated in the presence of organic matter.

The richest uranium ores in the United States are widespread in the Colorado Plateau area of Colorado and adjoining parts of Wyoming, Utah, Arizona, and New Mexico. These ores, consisting of fairly pure masses and encrustations of carnotite, are associated with plant remains in sandstones that formed in ancient stream channels. Although most of these ores are associated with fragmentary plant remains, some petrified trees also contain large quantities of uranium.

Large reserves of low-grade uranium ore also are found in the Chattanooga Shale. The uranium is finely disseminated in this black, organic-rich mudrock that underlies large parts of several states including Illinois, Indiana, Ohio, Kentucky, and Tennessee. Canada is the world's largest producer and exporter of uranium.

Banded Iron Formation

The chemical sedimentary rock known as *banded iron formation* consists of alternating thin layers of chert and iron minerals, mostly the iron oxides hematite and magnetite. Banded iron formations are present on all the continents and account for most of the iron ore mined in the world today. Vast banded iron formations are present in the Lake Superior region of the United States and Canada and in the Labrador Trough of eastern Canada. We will consider the origin of banded iron formations in Chapter 19.

6 GEO RECAP

Chapter Summary

- Mechanical and chemical weathering disintegrate and decompose parent material so that it is more nearly in equilibrium with new physical and chemical conditions. The products of weathering include solid particles and substances in solution.
- Mechanical weathering includes such processes as frost action, pressure release, salt crystal growth, thermal expansion and contraction, and the activities of organisms. Particles liberated by mechanical weathering retain the chemical composition of the parent material.
- The chemical weathering processes of solution, oxidation, and hydrolysis bring about chemical changes of the parent material. Clay minerals and substances in solution form during chemical weathering.
- Mechanical weathering aids chemical weathering by breaking parent material into smaller pieces, thereby exposing more surface area.
- Mechanical and chemical weathering produce regolith, some of which is soil if it consists of solids, air, water, and humus and supports plant growth.
- Soils are characterized by horizons that are designated, in descending order, as O, A, B, and C. Soil horizons differ from one another in texture, structure, composition, and color.
- Soils called pedalfers develop in humid regions, whereas arid and semiarid region soils are pedocals. Laterite is a soil that results from intense chemical weathering in the tropics. Laterites are deep and red and are sources of aluminum ores if derived from aluminum-rich parent material.
- Soil erosion, caused mostly by sheet and rill erosion, is a problem in some areas. Human practices such as construction, agriculture, and deforestation can accelerate losses of soil to erosion.
- Sedimentary particles are designated in order of decreasing size as gravel, sand, silt, and clay.
- Sedimentary particles are rounded and sorted during transport, although the degree of rounding and sorting depends on particle size, transport distance, and depositional process.
- Any area in which sediment is deposited is a depositional environment. Major depositional settings are continental, transitional, and marine, each of which includes several specific depositional environments.
- Lithification involves compaction and cementation, which convert sediment into sedimentary rock. Silica and calcium carbonate are the most common chemical cements, but iron oxide and iron hydroxide cements are important in some rocks.
- Detrital sedimentary rocks consist of solid particles derived from preexisting rocks. Chemical sedimentary rocks are derived from substances in solution by inorganic chemical processes or the biochemical activities or organisms. Geologists also recognize a subcategory called biochemical sedimentary rocks.
- Sedimentary facies are bodies of sediment or sedimentary rock that are recognizably different from adjacent sediments or rocks.
- Some sedimentary facies are geographically widespread because they were deposited during marine transgressions or marine regressions.
- Sedimentary structures such as bedding, cross-bedding, and ripple marks commonly form in sediments when, or shortly after, they are deposited.
- Geologists determine the depositional environments of ancient sedimentary rocks by studying sedimentary textures and structures, examining fossils, and making comparisons with present-day depositional processes.

- Intense chemical weathering is responsible for the origin of residual concentrations, many of which contain valuable minerals such as iron, lead, copper, and clay.
- Many sediments and sedimentary rocks, including sand, gravel, evaporites, coal, and banded iron formations, are important resources. Most oil and natural gas are found in sedimentary rocks.

Important Terms

bed (p. 161)
biochemical sedimentary rock (p. 158)
carbonate rock (p. 158)
cementation (p. 155)
chemical sedimentary rock (p. 158)
chemical weathering (p. 142)
compaction (p. 154)
cross-bedding (p. 161)
depositional environment (p. 154)
detrital sedimentary rock (p. 156)
differential weathering (p. 141)
erosion (p. 140)
evaporite (p. 158)
exfoliation dome (p. 141)
fossil (p. 162)
frost action (p. 141)
graded bedding (p. 162)
hydrolysis (p. 145)
laterite (p. 148)
lithification (p. 154)
marine regression (p. 161)
marine transgression (p. 160)
mechanical weathering (p. 141)
mud crack (p. 162)
oxidation (p. 144)
parent material (p. 140)
pedalfer (p. 148)
pedocal (p. 148)
pressure release (p. 141)
regolith (p. 146)
ripple mark (p. 162)
salt crystal growth (p. 142)
sediment (p. 153)
sedimentary facies (p. 160)
sedimentary rock (p. 153)
sedimentary structure (p. 161)
soil (p. 146)
soil degradation (p. 151)
soil horizon (p. 147)
solution (p. 143)
spheroidal weathering (p. 146)
strata (p. 161)
talus (p. 141)
thermal expansion and contraction (p. 142)
weathering (p. 140)

Review Questions

1. A vertical sequence of sedimentary rocks in which nearshore facies overlie offshore facies resulted from
a._____deposition by turbidity currents; b._____a marine regression; c._____meandering stream deposition; d._____compaction and cementation of evaporites; e._____granitization.
2. An essential component of soils is partly decomposed organic matter known as
a._____humus; b._____regolith; c._____talus; d._____gossan; e._____carbonic acid.
3. If a small amount of carbonic acid is present in groundwater, _____ dissolves rapidly.
a._____pedocal; b._____exfoliation domes; c._____limestone; d._____manganese; e._____laterite.
4. Cross-bedding preserved in sedimentary rocks is a good indication of
a._____the intensity of organic activity; b._____ancient current directions; c._____the amount of silica cement; d._____how old the containing rocks are; e._____whether or not the rocks contain important resources.
5. Dolostone forms from limestone when
a._____limestone loses some of its water; b._____evaporite deposition takes place in a lagoon; c._____organic matter accumulates in a swamp; d._____sand is deposed over a layer of mud; e._____some of the calcium in limestone is replaced by magnesium.
6. Which one of the following is *not* a chemical weathering process?
a._____salt crystal growth; b._____frost wedging; c._____oxidation; d._____pressure release; e._____thermal expansion and contraction.
7. Horizon C differs from other soil horizons in that it
a._____is the most fertile; b._____has weathered the longest; c._____is made up of sodium sulfate; d._____contains the most humus; e._____grades down into parent material.

8. A deposit of detrital sediment characterized as poorly sorted has
a.______a large amount of calcium carbonate cement; b.______cross-bedding and current ripple marks; c.______particles of markedly different sizes; d.______more ferromagnesian silicates than nonferromagnesian silicates; e.______iron oxide cement.

9. Pressure release is the primary process responsible for
a.______spheroidal weathering; b.______exfoliation domes; c.______residual ores; d.______frost heaving; e.______soil degradation.

10. Spheroidal weathering takes place because
a.______corners and edges of stones weather faster than flat surfaces; b.______aluminum oxides are nearly insoluble; c.______oxidation changes limestone to dolostone; d.______naturally occurring rocks are spherical to begin with; e.______thermal expansion and contraction are so effective.

11. Lithification involves cementation and ______
a.______replacement; b.______compaction; c.______inversion; d.______granitization; e.______stoping.

12. Traps for petroleum and natural gas formed by the folding and fracturing of rocks are known as ______ traps.
a.______lithologic; b.______compaction; c.______stratigraphic; d.______compositional; e.______structural.

13. In one of our national parks you observe a vertical sequence of rocks with sandstone at the base followed upward by shale and limestone, each of which contains fossil clams and corals. Give an account of the history of these rocks. That is, how were they deposited, and how did they come to be superposed in the sequence observed?

14. How does mechanical weathering differ from and contribute to chemical weathering?

15. Draw soil profiles for semiarid and humid regions, and list the characteristics of each.

16. In what fundamental way(s) do detrital sedimentary rocks differ from chemical sedimentary rocks?

17. Explain how exfoliation domes form. In what kinds of rocks do they devleop, and where would you go to see examples?

18. Describe the processes leading to the lithification of deposits of sand and mud.
19. Illustrate and describe two sedimentary structures that can be used to determine ancient current directions.
20. How do the factors of climate, parent material, and time determine the depth and fertility of soil?
21. How does coal form, and what varieties of coal do geologists recognize? Which of these varieties makes the best fuel?
22. Describe the types of soil degradation. What practices are used to prevent or at least minimize soil erosion?
23. Explain what structural and stratigraphic traps are and how they differ from each other.
24. Explain how particle size, climate, and parent material control the rate of chemical weathering.

World Wide Web Activities

GeologyNow Assess your understanding of this chapter's topics with additional quizzing and comprehensive interactivities at

http://earthscience.brookscole.com/changingearth4e

as well as current and up-to-date weblinks, additional readings, and InfoTrac College Edition exercises.

Sedimentary Environments

Sedimentary Structures

Sedimentary rocks contain a variety of features called **sedimentary structures** that formed during deposition or shortly thereafter.* All are manifestations of physical and biological processes that operated in the depositional environment. Analysis of sedimentary structures is the single most useful aspect of sedimentary rocks for determining how they were deposited in the first place. The origin of most sedimentary structures is well known because geologists have seen them forming in present-day environments, and many have been produced experimentally (Table 6.1).

Most sedimentary rocks have a layered aspect known as **stratification** or **bedding** (see the chapter opening photo and • Figure 6.3). Layers less than 1 cm thick are called *laminations*, whereas *beds* are thicker. Laminations are common in mudrocks, but coarser-grained sedimentary rocks are more commonly bedded. Individual beds are separated from one

TABLE 6.1 Summary of Sedimentary Structures

Physical Sedimentary Structures (produced by processes such as currents)	Characteristics	Origin
Laminations (or laminae)	Layers less than 1 cm thick	Form mostly as particles settle from suspension
Beds	Layers more than 1 cm thick	Form as particles settle from suspension and from moving sediment as sand in a stream channel
Graded bedding	Individual layers with an upward decrease in grain size	Deposition by turbidity currents or during the waning stages of floods
Cross-bedding	Layers deposited at an angle to the surface on which they accumulated	Deposition on a sloping surface as the downwind side of a sand dune
Ripple marks	Small (<3 cm high) ridges and troughs on bedding planes	
Current ripple marks	Asymmetric ripple marks	Result from deposition by water or air currents flowing in one direction
Wave-formed ripple marks	Symmetric ripple marks; generally with a sharp crest and broad trough	Formed by oscillating currents (waves)
Mud cracks	Intersecting cracks in clay-rich sediments	Drying and shrinkage of mud along a lakeshore or on tidal flats
Biogenic Sedimentary Structures (produced by organisms)		
Trace fossils	Tracks, trails, tubes, and burrows	Indications of organic activity. Intense activity results in *bioturbation* involving disruption of sediment

James S. Monroe

(a)

James S. Monroe

(b)

• **Figure 6.3 Stratification (a)** Alternating beds of sandstone and shale. The sandstone beds are 10 to 15 cm thick. **(b)** Laminations, layers less than 1 cm thick, in the chemical sedimentary rock called banded iron formation. The pencil measures about 14 cm long.

*There are also structures that form long after deposition, mostly by chemical processes, which are not useful for determining depositional environments.

(a) (b) (c)

Sue Monroe

• **Figure 6.9 Bioturbation** Bioturbation results from the activities of organisms, especially those that burrow through sediment. **(a)** U-shaped burrows and **(b)** vertical burrows. **(c)** The vertical, dark-colored areas in this rock are sediment-filled burrows. Could you use burrows such as these to determine which layers are youngest in deformed sedimentary rocks? How?

hundreds or thousands of square kilometers but are perhaps only a few tens to hundreds of meters thick. That is, they are not very thick compared to their dimensions of length and width. Thus they are characterized as having a *blanket* or *sheet geometry.*

Some sand deposits have an *elongate* or *shoestring geometry,* especially those deposited in stream channels or barrier islands. Delta deposits tend to be lens shaped when viewed in cross profile or long profile, but lobate when observed from above. Buried reefs are irregular, but many are long and narrow, although rather circular ones are known, too.

Fossils—The Biologic Content of Sedimentary Rocks

We already defined *fossils* as the remains or traces of prehistoric organisms, and we discussed how geologists use fossils in some aspects of stratigraphy—to establish biostratigraphic units, for example (see Chapter 5). However, fossils are also important constituents of some rocks, especially limestones that may be composed largely of shells of marine-dwelling animals such as brachiopods, clams, and corals (see Figure 2.15a), or even the droppings (pellets) of these organisms. Furthermore, much of the sediment on the deep seafloor is made up of the microscopic shells of organisms. Fossils are not present in all sedimentary rocks, but if they are, they are important for determining depositional environments.

We must consider two factors when using fossils in environmental analyses. First, did the organisms in question live where they were buried, or were their remains transported there? Fossil dinosaurs, for example, usually indicate deposition in some land environment such as a river floodplain, but if their bones are found in rocks with clams, corals, and sea lilies, we must assume a carcass was washed out to sea. Second, what kind of habitat did the organisms originally occupy? Studies of a fossil's structure and its living relatives, if any, are helpful. Clams with heavy, thick shells typically live in shallow, turbulent water, whereas those with thin shells are found in low-energy environments. Most corals live in warm, clear, shallow marine environments where symbiotic bacteria can carry out photosynthesis.

Microfossils are particularly useful because geologists can recover many individuals from small rock samples. In oil-drilling operations, small rock chips called *well cuttings* are brought to the surface. These cuttings rarely contain complete fossils of large organisms, but they may have thousands of microfossils that aid in relative age determinations and environmental analyses. Even some trace fossils, which of course are not transported from their place of origin, are also characteristic of particular environments.

Depositional Environments

We defined a **depositional environment** as any area where sediment accumulates, but more specifically it entails a particular area where physical, chemical, and biological processes operate to yield a distinctive kind of deposit. Geologists recognize three broad areas of deposition—*continental, transitional,* and *marine*—each of which has several specific environments (• Figure 6.10, Table 6.2).

Continental Environments

Deposition on the continents—that is, on land—may take place in fluvial systems (rivers and streams), lakes, deserts, and areas covered by or adjacent to glaciers. The sedimentary deposits in each of these environments show combinations of features that allow us to determine that they were in fact deposited on land and to distinguish one from the other.

The term **fluvial** refers to river and stream activity and to their deposits. These fluvial deposits accumulate in either **braided stream** systems or **meandering stream** systems. Braided streams have multiple broad, shallow channels in which mostly sheets of gravel and cross-bedded sand are

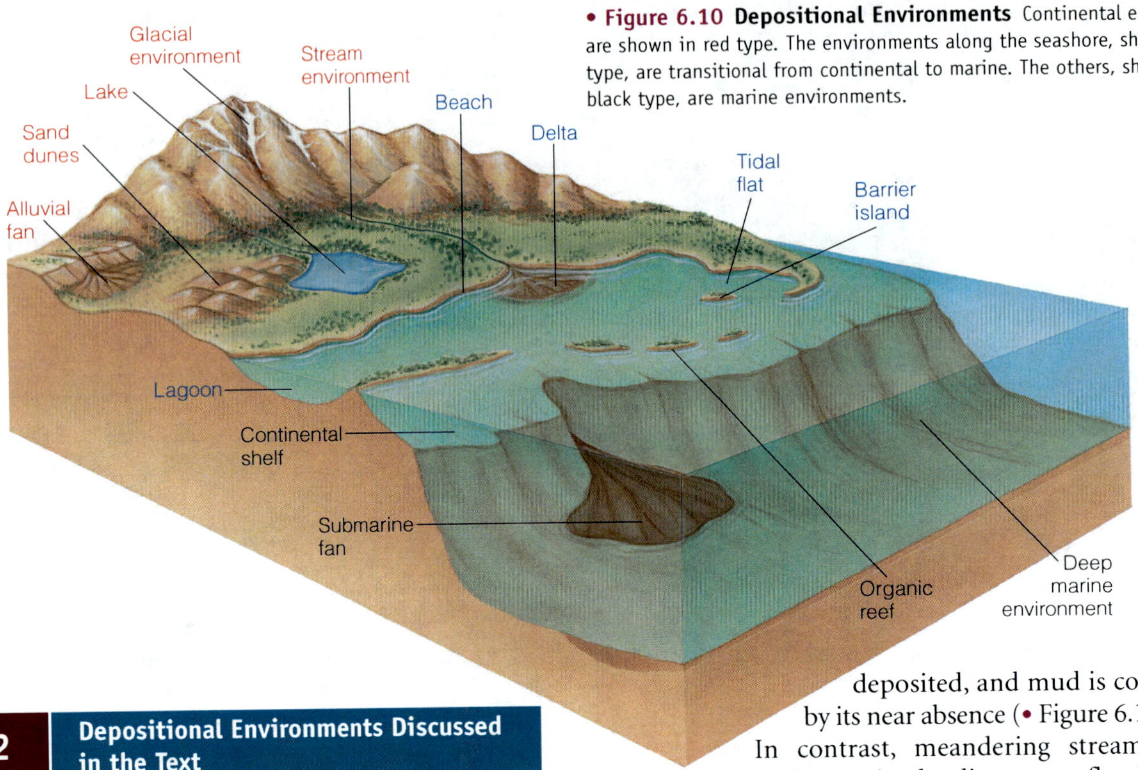

• **Figure 6.10 Depositional Environments** Continental environments are shown in red type. The environments along the seashore, shown in blue type, are transitional from continental to marine. The others, shown in black type, are marine environments.

deposited, and mud is conspicuous by its near absence (• Figure 6.11a and b). In contrast, meandering streams deposit mostly fine-grained sediments on floodplains but do have cross-bedded sand bodies, each with a shoestring geometry (Figure 6.11c and d). One of the most distinctive features of meandering streams is *point bar* deposits consisting of a sand body overlying an erosion surface that developed on the convex side of a meander loop (Figure 6.11c).

Desert environments are commonly inferred from an association of features found in **sand dune, alluvial fan,** and **playa lake** deposits (• Figure 6.12).Wind-blown dunes are typically composed of well-sorted, well-rounded sand with cross-beds meters to tens of meters high (Figure 6.12b and c). And, of course, any fossils are those of land-dwelling plants and animals. Alluvial fans form best along the margins of desert basins where streams and debris flows discharge from mountains onto a valley floor and form a triangular deposit of sand and gravel (Figure 6.12a). The more central part of a desert basin might be the site of a temporary lake (playa lake), in which laminated mud and evaporites accumulate.

TABLE 6.2 Depositional Environments Discussed in the Text

Continental Environments	Fluvial	Braided stream
		Meandering stream
	Desert	Sand dunes
		Alluvial fans
		Playa lakes
	Glacial	Ice deposition (moraines)
		Fluvial deposition (outwash)
		Glacial lakes
Transitional Environments	Delta	Stream dominated
		Wave dominated
		Tide dominated
	Beach	
	Barrier island	Beach
		Sand dune
		Lagoon
Marine Environments	Continental shelf	Tidal flats
		Detrital deposition
	Carbonate platform	Carbonate deposition
	Continental slope and rise	Submarine fans (turbidities)
	Deep-ocean basin	Oozes
		Pelagic clay
	Evaporite environments*	

*Evaporites may be deposited in a variety of environments including playa lakes, saline lakes, lagoons, and tidal flats in arid regions, and in marine environments.

What Would You Do?

No one was present millions of years ago to record data about the climate, geography, and geologic processes. So how is it possible to decipher unobserved past events? In other words, what features in rocks, and especially in sedimentary rocks, would you look for to determine what happened in the far distant past? Can you think of any economic reasons to decipher Earth history from the record preserved in rocks?

(a)

(c)

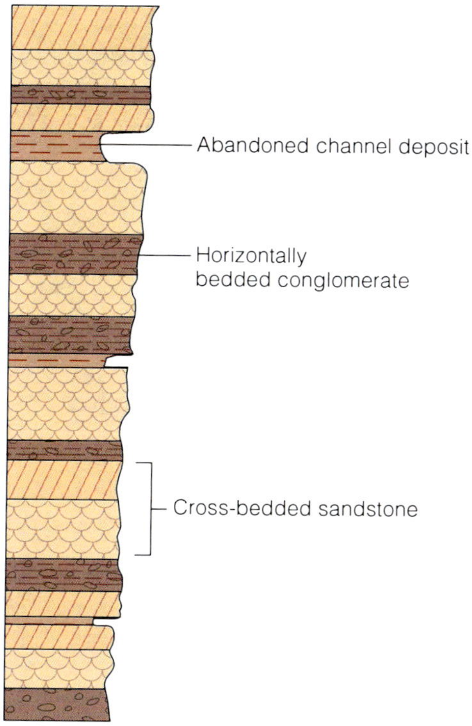

(b)

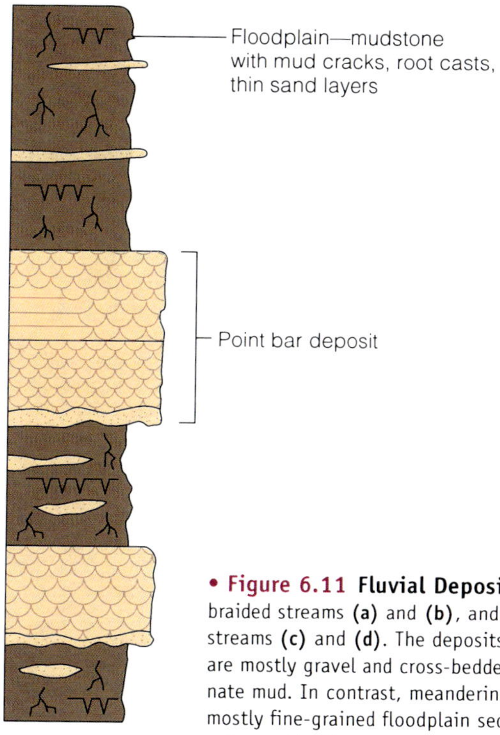

(d)

• **Figure 6.11 Fluvial Deposits** Fluvial deposits in braided streams **(a)** and **(b)**, and in meandering streams **(c)** and **(d)**. The deposits of braided streams are mostly gravel and cross-bedded sand with subordinate mud. In contrast, meandering stream deposits are mostly fine-grained floodplain sediment with subordinate sand bodies.

All sediments deposited in glacial environments are collectively called **drift,** but we must distinguish between two kinds of drift. **Till** is poorly sorted, nonstratified drift deposited directly by glacial ice, mostly in ridgelike deposits called *moraines* (• Figure 6.13). A second type of drift, called **outwash,** is sand and gravel deposited by braided streams issuing from melting glaciers. The association of these deposits along with scratched (striated) and polished bedrock generally indicates that glaciers were responsible for deposition. Another feature that may be helpful is glacial lake deposits showing alternating dark and light laminations; each dark–light couplet is a **varve** (Figure 6.13d). Varves represent yearly accumulations of sediment—the light layers form in spring and summer and consist of silt and clay, whereas the dark layers consist of clay and organic matter that settled when the lake froze over. Dropstones liberated from icebergs might also be present in glacial lake deposits (Figure 6.13d).

Transitional Environments

Transitional environments include those in which both marine processes and processes typical of continental environments operate (Figure 6.10). For instance, deposition where a river or stream (fluvial system) enters the sea yields a body of sediment called a **delta,** but the deposit is modified by marine processes, especially waves and tides. Small deltas in

Marli Bryant Miller

(a)

Royalty-Free/Corbis

(b)

Sue Monroe

(c)

• **Figure 6.12 Deposits in Deserts** **(a)** This view in Death Valley, California, shows huge alluvial fans along the base of the Panamint Mountains. **(b)** Sand dunes also in Death Valley. An association of the deposits in **(a)** and **(b)** is found in desert depositional environments. **(c)** Large-scale cross-beds in a Permian-aged wind-blown dune deposit in Arizona.

lakes are also common, but the ones along seashores are much larger and far more complex. Furthermore, many deltas along seashores are important economically because they form the reservoir rocks for petroleum and natural gas.

The simplest deltas, those in lakes, have a threefold subdivision of bottomset, foreset, and topset beds much like the one shown in • Figure 6.14. As delta sedimentation takes place, bottomset beds are deposited some distance offshore, over which foreset beds come to rest. And finally deposition also takes place in stream channels forming the topset beds. As a result, the delta builds seaward (or lakeward), or **progrades,** and forms a vertical sequence of rocks that become coarser-grained from bottom to top. In addition, the bottomset beds may contain marine fossils (or fossils of lake-dwelling animals), whereas the topset beds usually have fossils of land plants or animals.

Marine deltas rarely conform precisely to this simple threefold division, because they are strongly influenced by fluvial processes, as well as by waves and tides. Thus a stream/river–dominated delta has long distributary channels extending far seaward, as in the Mississippi River delta (• Figure 6.15a). Wave-dominated deltas also have distributary channels, but as their entire seaward margin progrades, it is simultaneously modified by wave action (Figure 6.15b). Tidal sand bodies that parallel the direction of tidal flow are characteristic of tide-dominated deltas (Figure 6.15c).

On broad continental margins with abundant sand, long **barrier islands** lie offshore separated from the mainland by a lagoon. These islands are common along the Gulf Coast, and much of the Atlantic Coast of the United States and many ancient deposits formed in this environment. Notice from • Figure 6.16a that a barrier island complex has subenvironments in which beach sands grade offshore into finer deposits of the shoreface, dune sands that are differentiated from desert dune sands by the presence of shell fragments, and fine-grained lagoon deposits with marine fossils and bioturbation.

Tidal flats are present along many coastlines where part of the shoreline environment is periodically covered by seawater at high tide and then exposed at low tide. Many

• **Figure 6.13 Glacial Deposits** **(a)** and **(b)** The origin of glacial drift—that is, till and outwash. **(c)** The deposit at the terminus of a glacier is an end moraine that is made up of till that is not stratified or sorted. **(d)** Dropstone in glacial lake deposits showing varves in Canada.

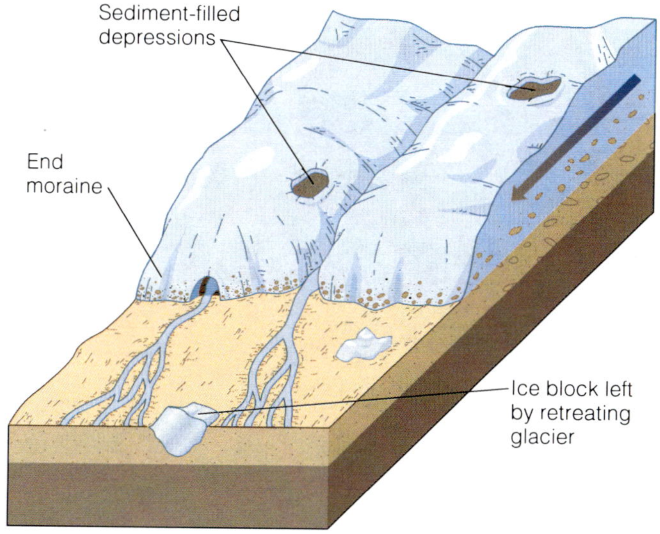

(a)

(c)

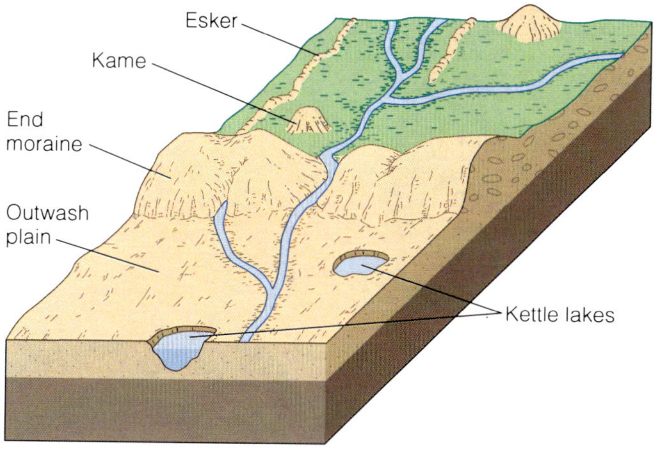

(b)

(d)

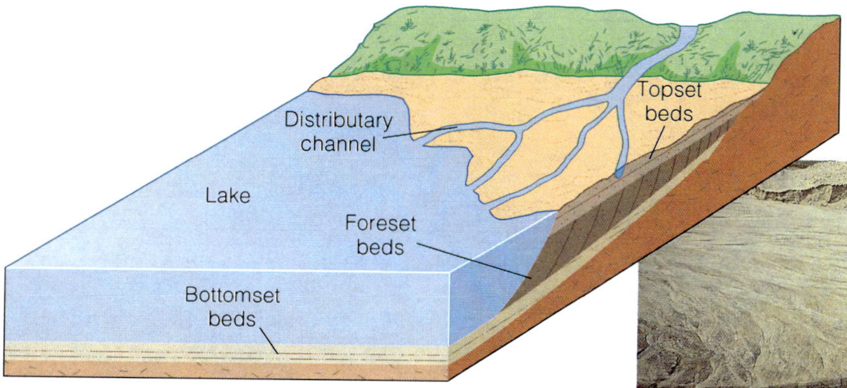

(a)

(b)

• **Figure 6.14 Origin of a Delta** **(a)** Surface view and internal structure of the simplest type of prograding delta. **(b)** This small delta that measures about 20 m across is prograding into a small lake.

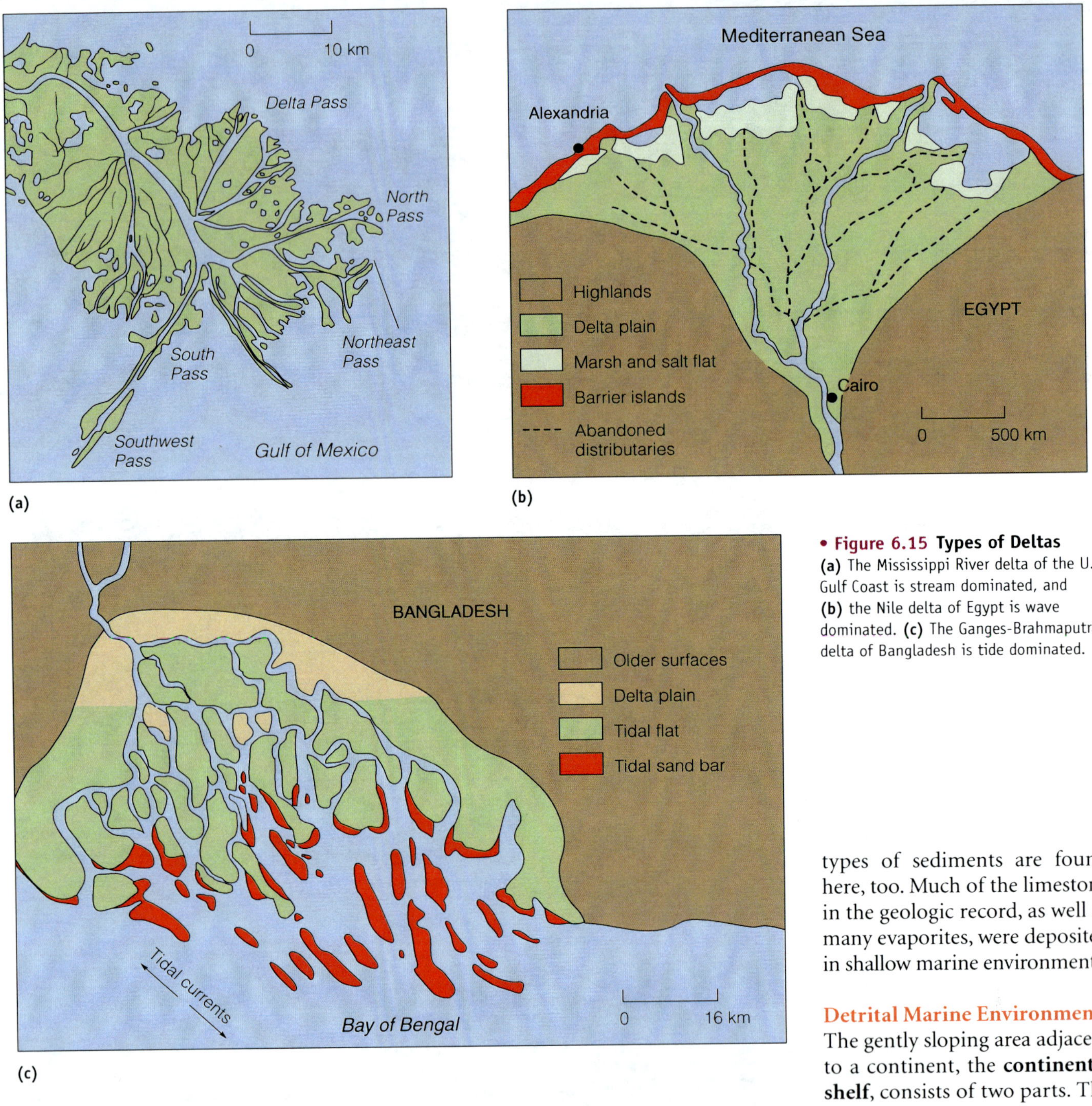

• Figure 6.15 Types of Deltas
(a) The Mississippi River delta of the U.S. Gulf Coast is stream dominated, and **(b)** the Nile delta of Egypt is wave dominated. **(c)** The Ganges-Brahmaputra delta of Bangladesh is tide dominated.

tidal flats build or prograde seaward and yield a sequence of rocks grading upward from sand to mud. One of their most distinctive features is herringbone cross-bedding—sets of cross-beds that dip in opposite directions (Figure 6.16b and c).

Marine Environments

Marine environments include depositional settings on the continental shelf, slope, and rise, and the deep seafloor (• Figure 6.17). Much of the detritus eroded from continents is eventually deposited in marine environments, but other types of sediments are found here, too. Much of the limestone in the geologic record, as well as many evaporites, were deposited in shallow marine environments.

Detrital Marine Environments

The gently sloping area adjacent to a continent, the **continental shelf**, consists of two parts. The high-energy inner part is periodically stirred up by waves and tidal currents. Its sediment is mostly sand shaped into large cross-bedded dunes, and bedding planes are commonly marked by wave-formed ripple marks. Of course, marine fossils are typical, and so is bioturbation. The low-energy outer part of the shelf has mostly mud with marine fossils, but at the transition between the inner and outer shelf, layers of sand and mud intertongue.

Much of the sediment derived from continents crosses the continental shelf and is funneled into deeper water through submarine canyons. It eventually comes to rest on the **continental slope** and **continental rise** as a series of overlapping submarine fans (Figure 6.17). Once sediment passes the outer margin of the shelf, turbidity currents transport it, so sands with graded bedding are common—but so is mud that settled from seawater (Figure 6.4).

Beyond the continental shelf, the seafloor is nearly covered by fine-grained deposits known as *pelagic clay* and *ooze* (• Figure 6.18). The notable exceptions are the mid-oceanic ridges, which are areas where new oceanic crust forms. In fact, one prediction of plate tectonic theory is that the thickness of seafloor sediments should increase with increasing distance from spreading ridges (see Chapter 3). Sand and gravel are notable by their absence on the deep seafloor, because there are few mechanisms that can effectively transport coarse-grained sediment far from land. However, ice rafting accounts for a band of sand and gravel adjacent to Greenland and Antarctica.

The main sources of deep-sea sediments are dust blown from continents or oceanic islands, volcanic ash, and shells of microorganisms that dwelled in the ocean's surface waters. One type of sediment called pelagic clay consists of clay covering most of the deeper parts of the seafloor. Calcareous ($CaCO_3$) and siliceous (SiO_2) oozes, in contrast, are made up of microscopic shells.

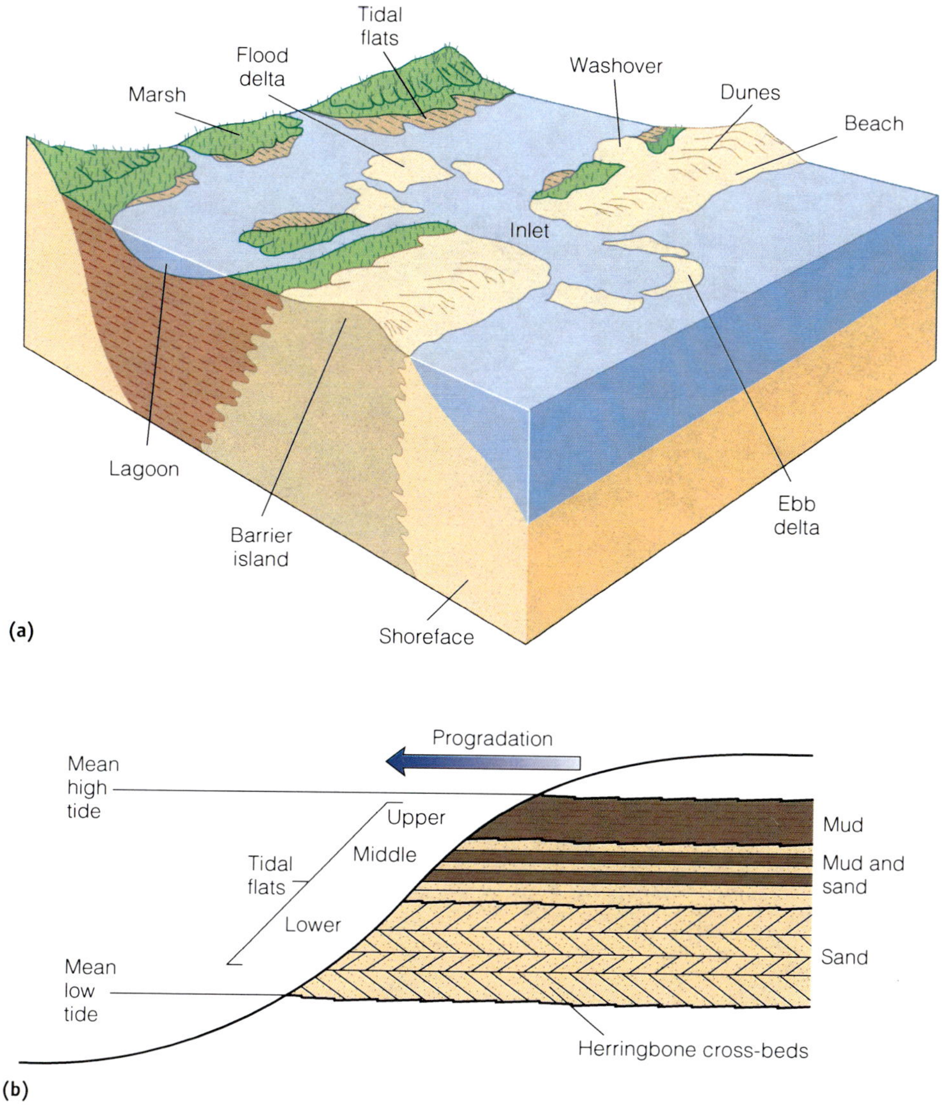

Gary M. Fleeger/Pennsylvania Geological Survey

(c)

• **Figure 6.16 Barrier Island Complex** **(a)** A barrier island is made up mostly of sand, but it has subenvironments where mud or mixtures of mud and sand are deposited. **(b)** These tidal flat deposits formed where a shoreline prograded into the sea. Notice the distinctive herringbone cross-beds that dip in opposite directions because of reversing tidal currents. **(c)** Herringbone cross-beds in the Loyalhanna Formation in Pennsylvania.

Carbonate Depositional Environments Various types of limestone (composed of calcite) and dolostone (composed of dolomite) are the only widespread carbonate rocks, and we already know that most dolostone is altered limestone. Thus our discussion focuses on the deposition of sediment that when lithified becomes limestone.

In some respects limestone is similar to detrital sedimentary rocks—many limestones are made up of gravel- and sand-sized grains, and microcrystalline carbonate mud called **micrite.** However, these materials form in the environment of deposition rather than being transported there. Nevertheless, local currents and waves sort them and account for various sedimentary structures such as cross-bedding and ripple marks, and micrite along shorelines commonly shows mud cracks.

Some limestone forms in lakes, but by far most of it was deposited in warm, shallow seas on carbonate shelves

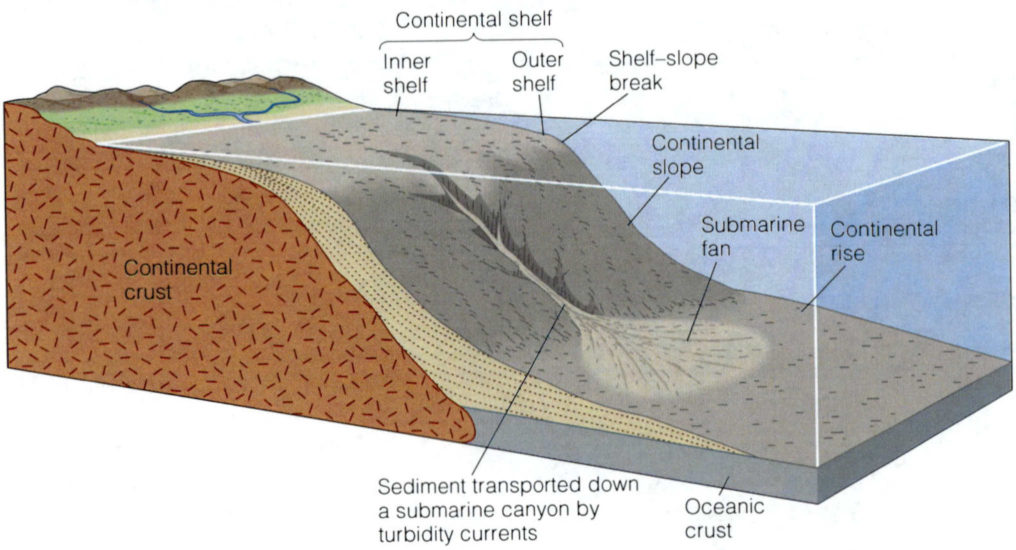

• **Figure 6.17 Depositional Environments on the Continental Shelf, Slope, and Rise** Submarine canyons are the main avenues of sediment transport across the shelf. Turbidity currents carry sediment beyond the shelf, where they accumulate as submarine fans.

or on carbonate platforms rising from oceanic depths (• Figure 6.19). In either case, deposition occurs where little detrital sediment is present, especially mud. Carbonate barriers form in high-energy areas and may be reefs, banks of skeletal particles, or accumulations of spherical carbonate grains known as *ooids*—limestone composed mostly of ooids is called *oolitic limestone.* Reef rock tends to be structureless, composed of skeletons of corals, mollusks, sponges, and many other organisms, whereas carbonate banks are made up of layers with horizontal beds, cross-beds, and wave-formed ripple marks. The lagoons, however, tend to have micrite with marine fossils and bioturbation.

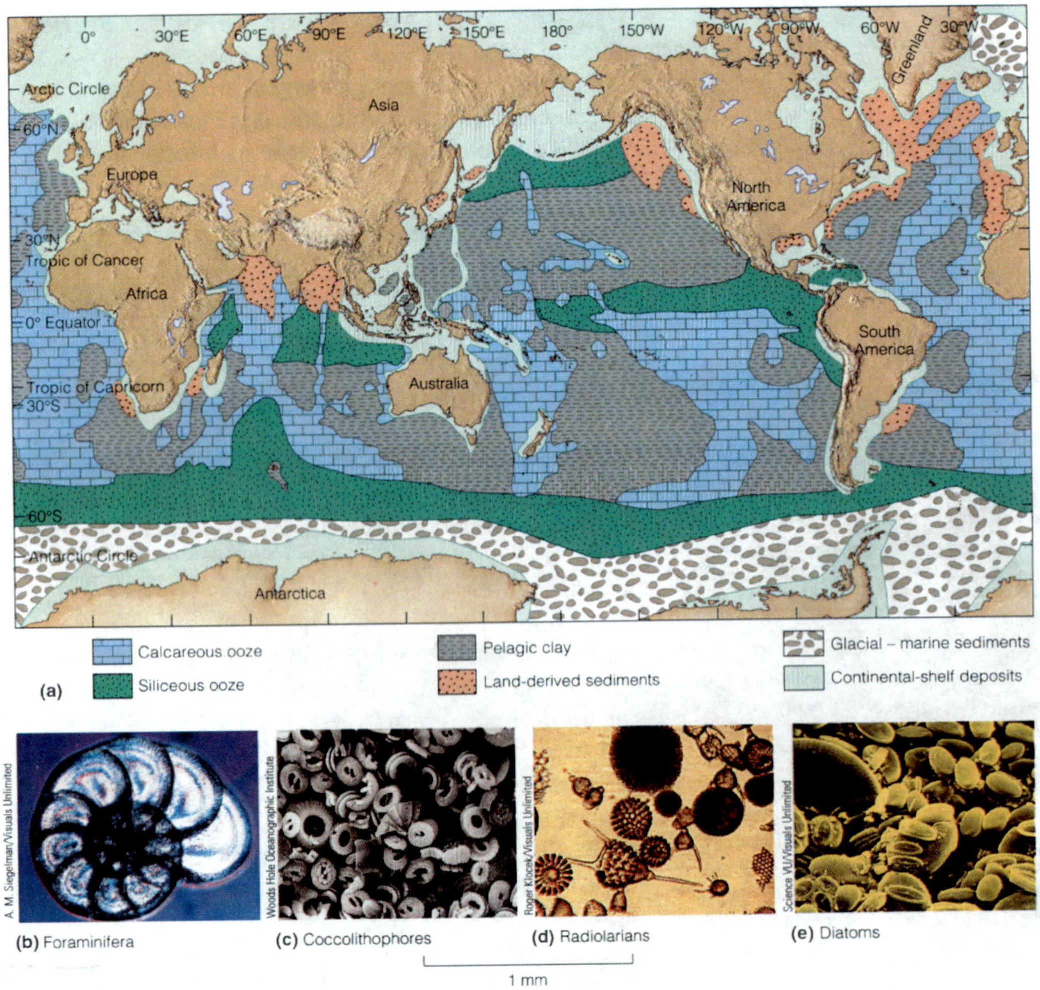

• **Figure 6.18 Deep-Seafloor Sediments** **(a)** Most of the seafloor is covered by pelagic clay and calcareous and siliceous ooze. Calcareous ooze is made up of skeletons of **(b)** foraminifera (floating single-celled animals) and **(c)** coccolithophores (floating single-celled plants), whereas siliceous ooze is composed of **(d)** radiolarians (single-celled floating animals) and (e) diatoms (single-celled floating plants).

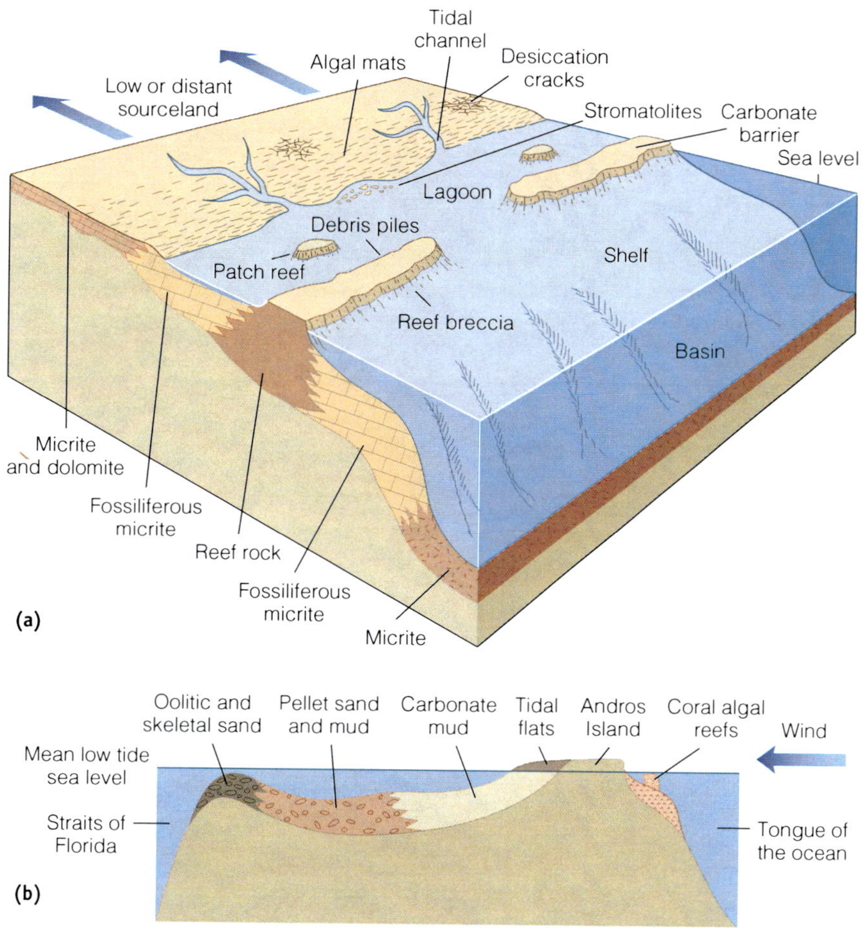

• **Figure 6.19 Carbonate Depositional Environments (a)** Deposition on a shelf attached to a continent is now taking place in southern Florida and the Persian Gulf. **(b)** The Great Bahama Bank in the Atlantic Ocean is an isolated carbonate shelf that rises from oceanic depths.

Evaporite Environments Evaporites consisting mostly of rock salt and rock gypsum* are found in a variety of environments, including playa lakes and saline lakes, but most of the extensive deposits formed in the seas. Although locally abundant, evaporites are not nearly as common as sandstone, mudrocks, and limestone. Deposits more than 2 km thick lie beneath the floor of the Mediterranean Sea, and large deposits are also known in Michigan, Ohio, and New York as well as in several Gulf Coast states and western Canada.

Exactly how some of these deposits originated is controversial, but all agree they formed where evaporation rates were high enough for seawater to become sufficiently concentrated for minerals to precipitate from solution. These conditions are best met in coastal environments in arid regions such as the present-day Persian Gulf. If the area shown in Figure 6.16a were an arid region, inflow of normal seawater into the lagoon would be restricted, and evaporation might lead to increased salinity and perhaps evaporite deposition.

Interpreting Depositional Environments

We began this chapter by stating that the Navajo Sandstone of the southwestern United States was deposited as a vast blanket of sand dunes (Figure 6.1),

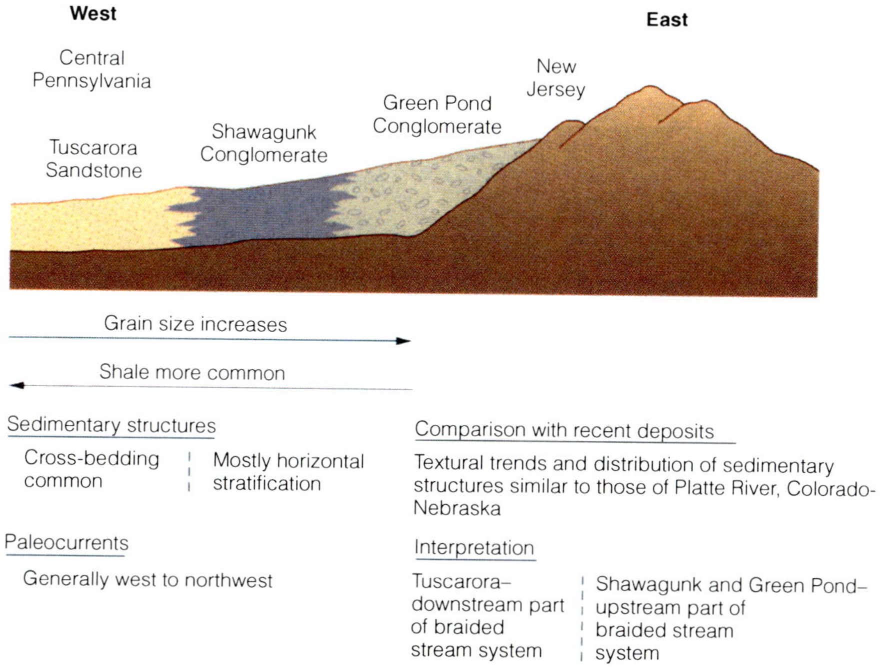

• **Figure 6.20 Lower Silurian Strata in the Eastern United States** Simplified cross section showing the probable lateral relationships for the Green Pond Conglomerate, Shawangunk Conglomerate, and Tuscarora Sandstone and the criteria used to interpret these rocks as braided stream deposits.

*Gypsum ($CaSO_4 \cdot 2H_2O$) is the common sulfate mineral precipitated from seawater, but when deeply buried, gypsum loses its water and is converted to anhydrite ($CaSO_4$).

but couldn't it just as well have been deposited along the seashore or in river channels? Studies show that the sandstone is made up mostly of well-sorted, well-rounded quartz grains measuring 0.2 to 0.5 mm in diameter. In addition, the rocks have tracks of land-dwelling animals, including dinosaurs, excluding the possibility of a marine origin, and the sandstone has cross-beds up to 30 m high and current ripple marks like those produced by wind. All the evidence taken together justifies our interpretation of a wind-blown dune deposit. In fact, the cross-beds dip generally southwest, so we can confidently say that the wind blew mostly from the northeast.

Another good example of using the combined features of sedimentary rocks and comparisons with present-day deposits is Lower Silurian strata exposed in New Jersey and Pennsylvania. The features of these rocks are summarized in • Figure 6.20, all of which lead to the conclusion that they were deposited in braided streams that flowed generally from east to west. The preceding examples were taken from continental depositional environments, but the same reasoning is used for marine deposits. For example, geologists have evaluated vertical facies relationships, rock types, sedimentary structures, and fossils in Ordovician rocks in Arkansas and conclude that they formed as transgressive shelf carbonate deposits (• Figure 6.21).

In several of the following chapters we refer to fluvial deposits, ancient deltas, carbonate shelf deposits, and others as we examine the geologic record, especially for North America. We cannot include the supporting evidence for all of these interpretations, but we can say that they are based on the kinds of criteria discussed in this chapter.

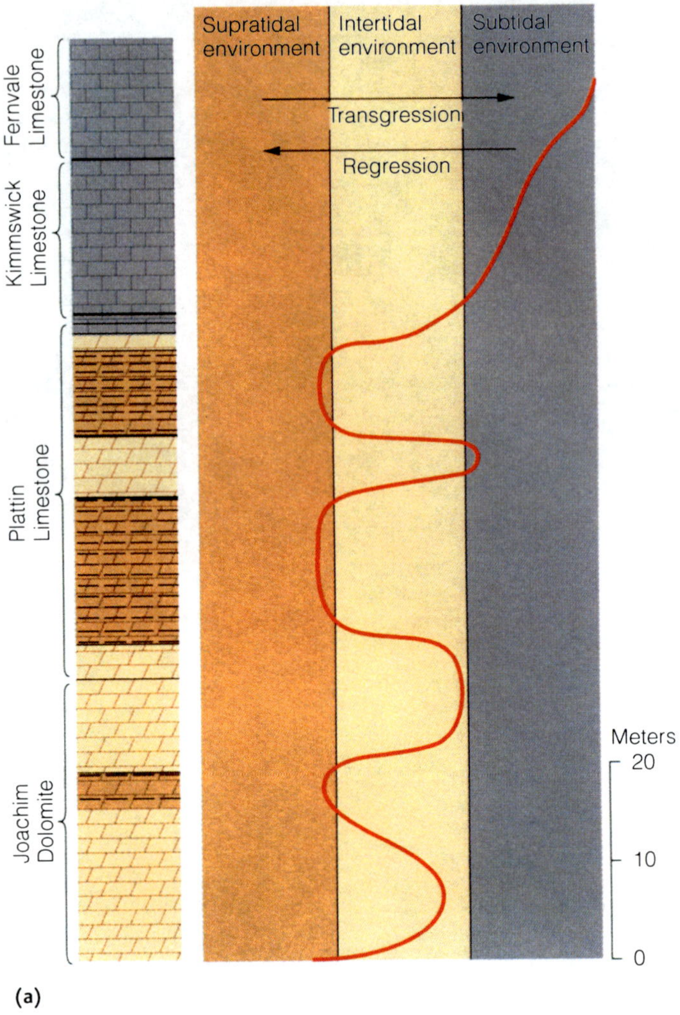

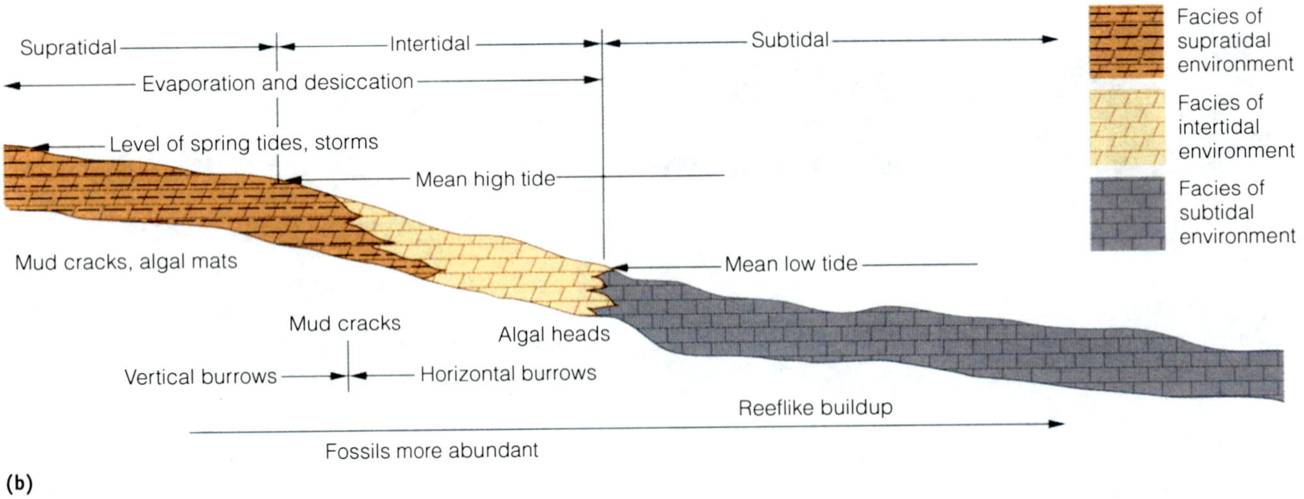

• **Figure 6.21 Middle and Upper Ordovician Strata in Northern Arkansas** **(a)** Vertical stratigraphic relationships and inferred environments of deposition. Notice that the trend was transgressive even though several minor regressions also occurred. **(b)** Simplified cross section showing lateral facies relationships.

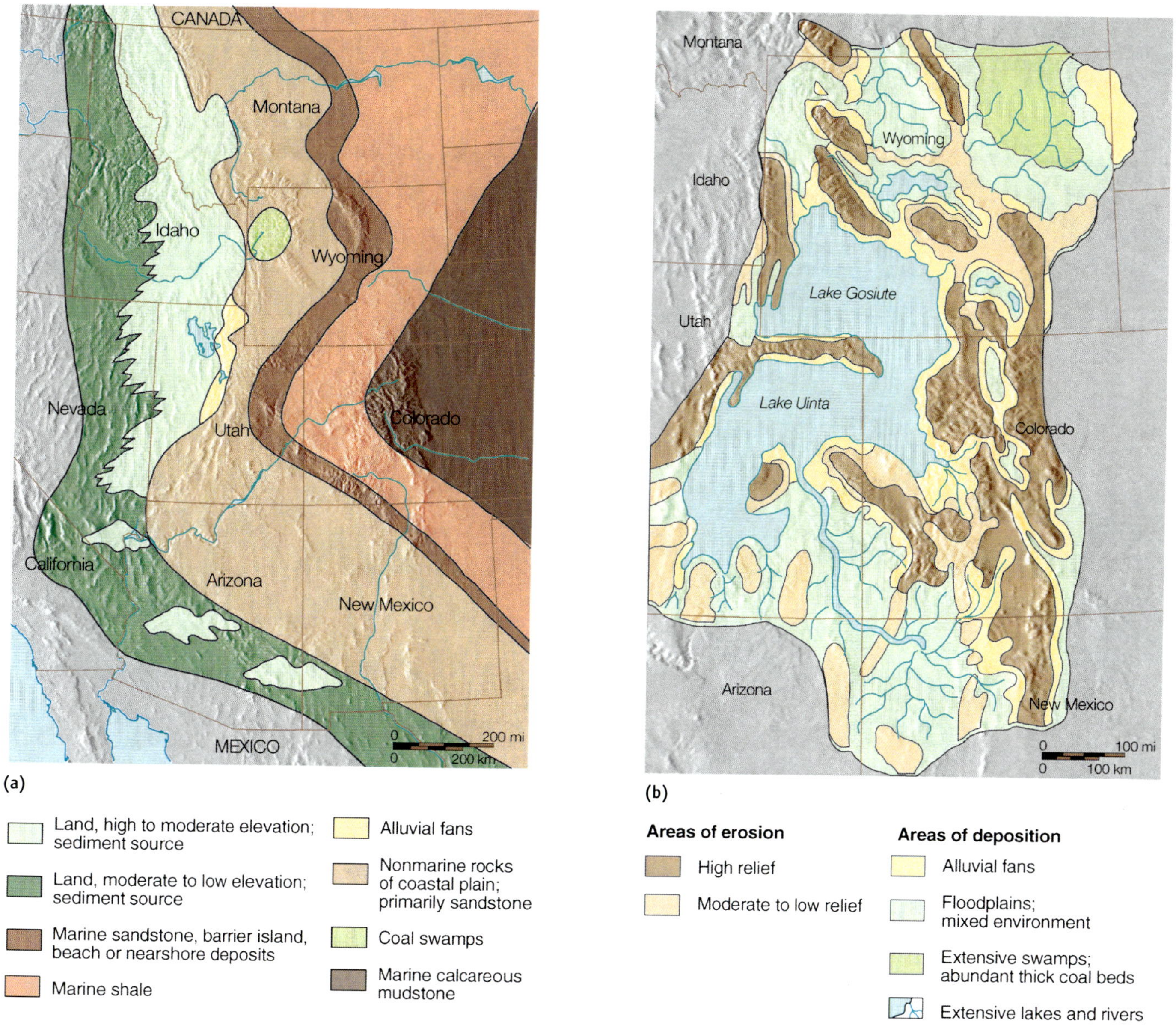

• **Figure 6.22 Paleogeographic Maps** Maps showing the **(a)** Late Cretaceous and **(b)** Eocene paleogeography of several western states. Notice that marine deposition took place over much of the area during the Late Cretaceous, but by Eocene time all deposition was in continental environments.

Paleogeography

The Navajo Sandstone is a wind-blown desert dune deposit, so we know that during the Jurassic Period a vast desert was present in what is now the southwestern United States (Figure 6.1). We also know that a marine transgression took place from the Late Precambrian to Middle Cambrian in North America as the shoreline migrated inland from east and west (see Figure 5.8). In short, Earth's geography—that is, the distribution of its various surface features—has varied through time.

Paleogeography deals with Earth's geography of the past. In the chapters on geologic history, paleogeographic maps show the distribution of continents at various times; they have not remained fixed because of plate movements. The distribution of land and sea constitutes Earth's first-order features, but paleogeography also applies on a more local scale. For example, detailed studies of rocks in several western states allow us to determine with some accuracy how the area appeared during the Late Cretaceous (• Figure 6.22a). According to this paleogeographic map, a broad coastal plain sloped gently eastward from a mountainous region to the sea. The rocks also tell us that many animals lived here, including dinosaurs and mammals. Later, however, vast lakes, river floodplains, and alluvial fans covered much of this area, and the sea had withdrawn from the continent (Figure 6.22b).

Metamorphism and Metamorphic Rocks

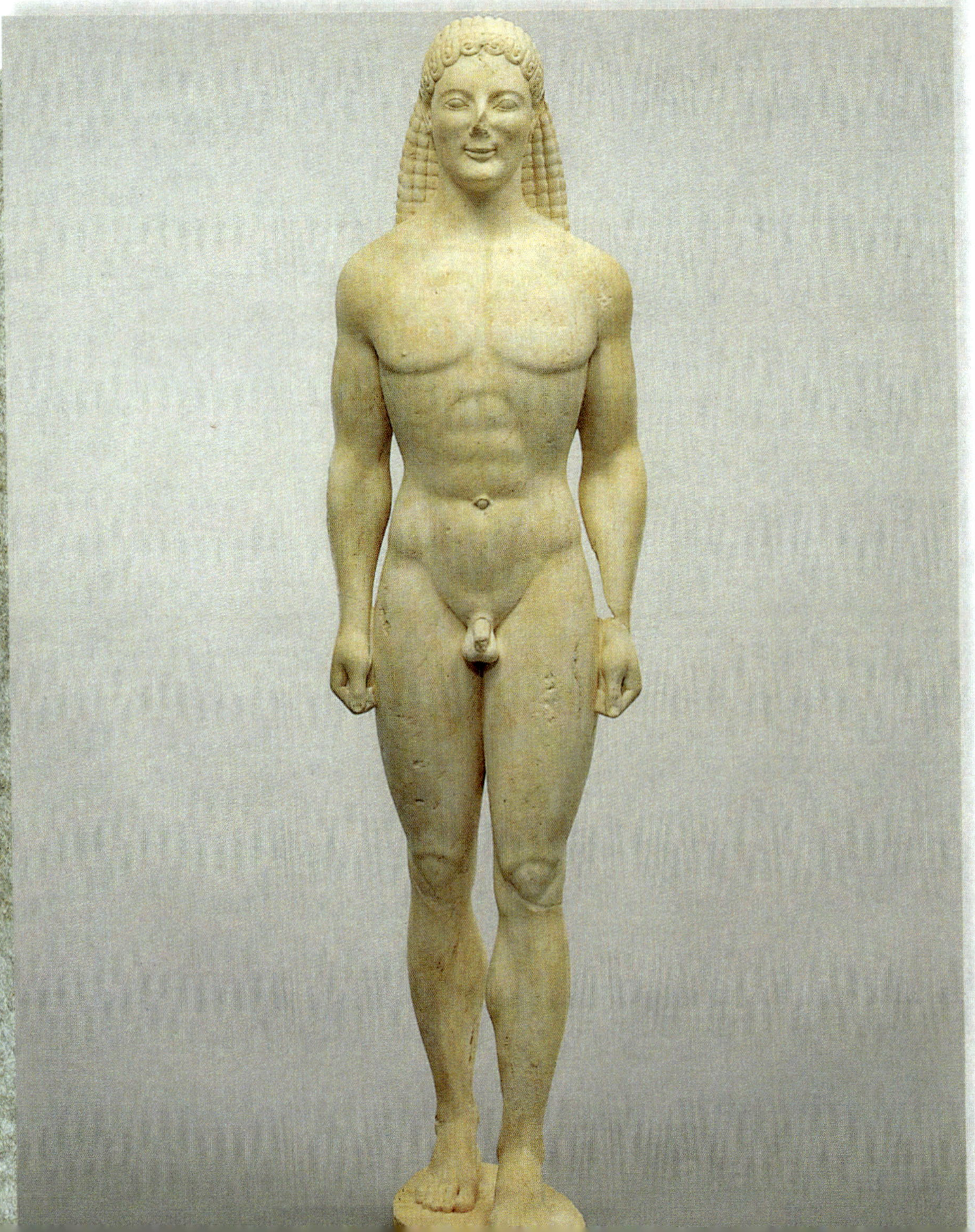

CHAPTER 7

OUTLINE

Geology⇌Now *This icon, which appears throughout the book, indicates an opportunity to explore interactive tutorials, animations, or practice problems available on the GeologyNow website at http://earthscience.brookscole.com/changingearth4e.*

OBJECTIVES

At the end of this chapter, you will have learned that:

- Metamorphic rocks result from the transformation of other rocks by various processes occurring beneath Earth's surface.
- Heat, pressure, and fluid activity are the three agents of metamorphism.
- Contact, dynamic, and regional metamorphism are the three types of metamorphism.
- Metamorphic rocks are typically divided into two groups, foliated and nonfoliated, primarily on the basis of texture.
- Metamorphic rocks with a foliated texture include slate, phyllite, schist, gneiss, and amphibolite.
- Metamorphic rocks with a nonfoliated texture include marble, quartzite, greenstone, and hornfels.
- Metamorphic rocks can be grouped into metamorphic zones based on the presence of index minerals that form under specific temperature and pressure conditions.
- The successive appearance of particular metamorphic minerals indicates increasing or decreasing metamorphic intensity.
- Metamorphism is associated with all three types of plate boundaries but is most widespread along convergent plate boundaries.
- Many metamorphic minerals and rocks are valuable metallic ores, building materials, and gemstones.

This Greek kouros, which stands 206 cm tall, has been the object of an extensive authentication study by the Getty Museum. Using a variety of geologic tests, scientists have determined that the kouros was carved from dolomitic marble that probably came from the Cape Vathy quarries on the island of Thasos. Source: Garry Hobart/GeoImagery

Introduction

Its homogeneity, softness, and varying textures have made marble, a metamorphic rock formed from limestone or dolostone, a favorite rock of sculptors throughout history. As the value of authentic marble sculptures has increased over the years, the number of forgeries has also increased. With the price of some marble sculptures in the millions of dollars, private collectors and museums need some means of ensuring the authenticity of the work they are buying. Aside from the monetary considerations, it is important that forgeries not become part of the historical and artistic legacy of human endeavor.

Experts have traditionally relied on artistic style and weathering characteristics to determine whether a marble sculpture is authentic or a forgery. Because marble is not very resistant to weathering, however, forgers have been able to produce the weathered appearance of an authentic work. Using newly developed techniques, geologists can now distinguish a naturally weathered marble surface from one that has been artificially altered. Yet, there are examples in which expert opinion is still divided on whether a sculpture is authentic.

One of the best examples is the Greek kouros (a sculptured figure of a Greek youth) the J. Paul Getty Museum in Malibu, California, purchased for a reputed price of $7 million in 1984 (see the chapter opening photo). Because certain stylistic features caused some experts to question its authenticity, the museum had a variety of geochemical and mineralogical tests performed in an effort to authenticate the kouros.

Although numerous scientific tests have not unequivocally proved authenticity, they have shown that the weathered surface layer of the kouros bears more similarities to naturally occurring weathered surfaces of dolomitic marble than to known artificially produced surfaces. Furthermore, no evidence indicates that the surface alteration of the kouros is of modern origin.

Unfortunately, despite intensive study by scientists, archaeologists, and art historians, opinion is still divided as to the authenticity of the Getty kouros. Most scientists accept that the kouros was carved sometime around 530 B.C. Pointing to inconsistencies in its style of sculpture for that period, other art historians think that it is a modern forgery.

Regardless of whether the Getty kouros is proven to be authentic or a forgery, geologic testing to authenticate marble sculptures is now an important part of many museums' curatorial functions. To help geologists authenticate marble sculptures, a large body of data about the characteristics and origin of marble is being amassed as more sculptures and marble quarries are analyzed.

Metamorphic rocks (from the Greek *meta,* "change," and *morpho,* "shape") are the third major group of rocks. They result from the transformation of other rocks by metamorphic processes that usually occur beneath Earth's surface (see Figure 1.12). During metamorphism, rocks are subjected to sufficient heat, pressure, and fluid activity to change their mineral composition, texture, or both, thus forming new rocks. These transformations take place below the melting temperature of the rock; otherwise, an igneous rock would result.

A useful analogy for metamorphism is baking a cake. Just like a metamorphic rock, the cake depends on the ingredients, their proportions, how they are mixed together, how much water or milk is added, and the temperature and length of time used for baking the cake.

Except for marble and slate, most people are not familiar with metamorphic rocks. Students frequently ask us why it is important to study metamorphic rocks and processes. Our answer is: look around you.

A large portion of Earth's continental crust is composed of metamorphic and igneous rocks. Together they form the crystalline basement rocks underlying the sedimentary rocks of a continent's surface. These basement rocks are widely exposed in regions of the continents known as *shields,* which have been very stable during the past 600 million years (■ Figure 7.1).

What Would You Do?

As the director of a major museum, you have the opportunity to purchase, for a considerable sum of money, a newly discovered marble bust by a famous ancient sculptor. You want to be sure it is not a forgery. What would you do to ensure that the bust is authentic and not a clever forgery? After all, you are spending a large sum of the museum's money. As a nonscientist, how would you go about making sure the proper tests are performed to ensure the bust's authenticity?

Metamorphic rocks also make up a sizable portion of the crystalline core of large mountain ranges. Some of the oldest known rocks, dated at 3.96 billion years from the Canadian Shield, are metamorphic, so they formed from even older rocks!

Metamorphic rocks such as marble and slate are used as building materials, and certain metamorphic minerals are economically important. Garnets, for example, are used as gemstones or abrasives; talc is used in cosmetics, in manufacturing paint, and as a lubricant; and kyanite is used to produce heat-resistant materials in sparkplugs. Therefore, knowledge of metamorphic rocks and processes is of economic value.

Asbestos, a metamorphic mineral, is used for insulation and fireproofing and is widespread in buildings and building materials. Asbestos has different forms, however, and they do not all pose the same health hazards. Recognizing this fact would have been useful during the debates over the dangers asbestos posed to the public's health (see Geo-Focus 7.1).

What Would You Do?

The problem of removing asbestos from public buildings is an important national health and political issue. The current policy of the Environmental Protection Agency (EPA) mandates that all forms of asbestos are treated as identical hazards. Yet studies indicate only one form of asbestos is a known health hazard. Because the cost of asbestos removal has been estimated to be as high as $100 billion, many people are questioning whether it is cost effective to remove asbestos from all public buildings where it has been installed.

As a leading researcher on the health hazards of asbestos, you have been asked to testify before a congressional committee on whether it is worthwhile to spend so much money for asbestos removal. How would you address this issue to formulate a policy that balances the risks and benefits of removing asbestos from public buildings? What role would geologists play in formulating this policy?

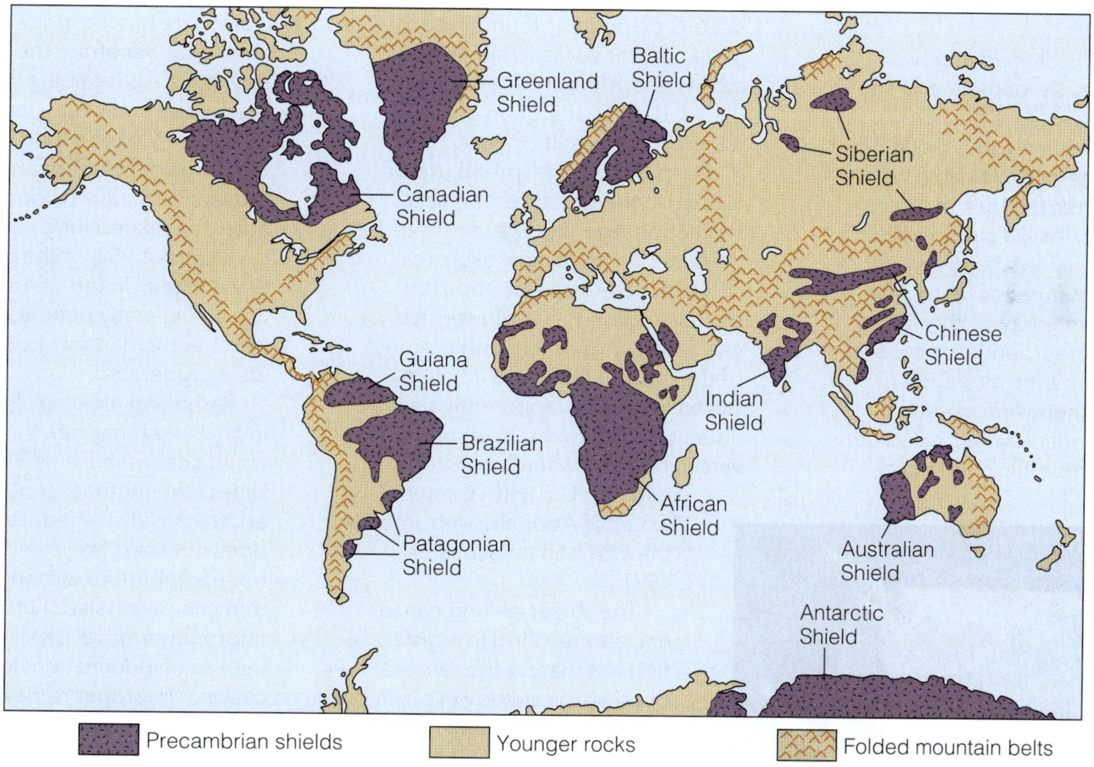

Figure 7.1

Metamorphic rock occurrences. Shields are the exposed portions of the crystalline basement rocks underlying each continent; these areas have been very stable during the past 600 million years. Metamorphic rocks also constitute the crystalline core of major mountain belts.

GEOFOCUS 7.1

Asbestos: Good or Bad?

Asbestos (from the Latin, "unquenchable") is a general term applied to any silicate mineral that easily separates into flexible fibers. The combination of such features as noncombustibility and flexibility makes asbestos an important industrial material of considerable value. In fact, asbestos has more than 3000 known uses, including brake linings, fireproof fabrics, and heat insulators.

Asbestos is divided into two broad groups: serpentine and amphibole. *Chrysotile* is the fibrous form of serpentine asbestos (■ Figure 1); it is the most valuable type and constitutes the bulk of all commercial asbestos. Its strong, silky fibers are easily spun and can withstand temperatures of up to 2750°C.

The vast majority of chrysotile asbestos is in serpentine, a type of rock formed by the alteration of ultramafic igneous rocks such as peridotite under low- and medium-grade metamorphic conditions. Other chrysotile results when the metamorphism of magnesium limestone or dolostone produces discontinuous serpentine bands within the carbonate beds.

Among the varieties of amphibole asbestos, *crocidolite* is the most common. Also known as blue asbestos, crocidolite is a long, coarse, spinning fiber that is stronger but more brittle than chrysotile and also less resistant to heat. Crocidolite is found in such metamorphic rocks as slates and schists and is thought to form by the solid-state alteration of other minerals as a result of deep burial.

Despite the widespread use of asbestos, the U.S. Environmental Protection Agency (EPA) instituted a gradual ban on all new asbestos products. The ban was imposed because some forms of asbestos can cause lung cancer and scarring of the lungs if fibers are inhaled. Because the EPA apparently paid little attention to the issue of risks versus benefits when it enacted this rule, the U.S. Fifth Circuit Court of Appeals overturned the EPA ban on asbestos in 1991.

The threat of lung cancer has also resulted in legislation mandating the removal of asbestos already in place in all public buildings, including all public and private schools. However, important questions have been raised concerning the threat posed by asbestos and the additional potential hazards that may arise from its improper removal.

Current EPA policy mandates that all forms of asbestos are to be treated as identical hazards. Yet studies indicate that only the amphibole forms constitute a known health hazard. Chrysotile, whose fibers tend to be curly, does not become lodged in the lungs. Furthermore, its fibers are generally soluble and disappear in tissue. In contrast, crocidolite has long, straight, thin fibers that penetrate the lungs and stay there. These fibers irritate the lung tissue and over a long period of time can lead to lung cancer. Thus crocidolite, and not chrysotile, is overwhelmingly responsible for asbestos-related lung cancer. Because about 95% of the asbestos in place in the United States is chrysotile, many people question whether the dangers from asbestos are exaggerated.

Removing asbestos from buildings where it has been installed could cost as much as $100 billion. Unless the material containing the asbestos is disturbed, asbestos does not shed fibers and thus does not contribute to airborne asbestos that can be inhaled. Furthermore, improper removal of asbestos can lead to contamination. In most cases of improper removal, the concentration of airborne asbestos fibers is far higher than if the asbestos had been left in place.

The problem of asbestos contamination is a good example of how geology affects our lives and why a basic knowledge of science is important.

Charles D. Winters/Photo Researchers, Inc.

■ **Figure 1**

Specimen of chrysotile. Chrysotile is the fibrous form of serpentine asbestos and the most commonly used in buildings and other structures.

WHAT ARE THE AGENTS OF METAMORPHISM?

The three agents of metamorphism are heat, pressure, and fluid activity. During metamorphism, the original rock undergoes change to achieve equilibrium with its new environment. The changes may result in the formation of new minerals and/or a change in the texture of the rock brought about by the reorientation of the original minerals. In some instances, the change is minor, and features of the original rock can still be recognized. In other cases, the rock changes so much that the identity of the original rock can be determined only with great difficulty, if at all.

Besides heat, pressure, and fluid activity, time is also important to the metamorphic process. Chemical reactions proceed at different rates and thus require different amounts of time to complete. Reactions involving silicate compounds are particularly slow, and because most metamorphic rocks are composed of silicate minerals, it is thought that metamorphism is a slow geologic process.

Heat

Heat is an important agent of metamorphism because it increases the rate of chemical reactions that may produce minerals different from those in the original rock. Heat may come from extrusive lavas or from intrusive magmas or deep burial in the crust such as occurs during subduction along a convergent plate boundary.

When rocks are intruded by bodies of magma, they are subjected to intense heat that affects the surrounding rock; the most intense heating usually occurs adjacent to the magma body and gradually decreases with distance from the intrusion. The zone of metamorphosed rocks that forms in the country rock adjacent to an intrusive igneous body is usually distinct and easy to recognize.

Recall that temperature increases with depth and that Earth's geothermal gradient averages about 25°C/km. Rocks that form at the surface may be transported to great depths by subduction along a convergent plate boundary and thus subjected to increasing temperature and pressure. During subduction, some minerals may be transformed into other minerals that are more stable under the higher temperature and pressure conditions.

Pressure

When rocks are buried, they are subjected to increasingly greater **lithostatic pressure;** this pressure, which results from the weight of the overlying rocks, is applied equally in all directions (■ Figure 7.2a). A similar situation occurs when an object is immersed in water. For example, the deeper a Styrofoam cup is submerged in the ocean, the smaller it gets because pressure increases with depth and is exerted on the cup equally in all directions, thereby compressing the Styrofoam (Figure 7.2b).

Just as in the Styrofoam cup example, rocks are subjected to increasing lithostatic pressure with depth such that the mineral grains within a rock may become more closely packed. Under these conditions, the minerals may *recrystallize;* that is, they become smaller and denser minerals.

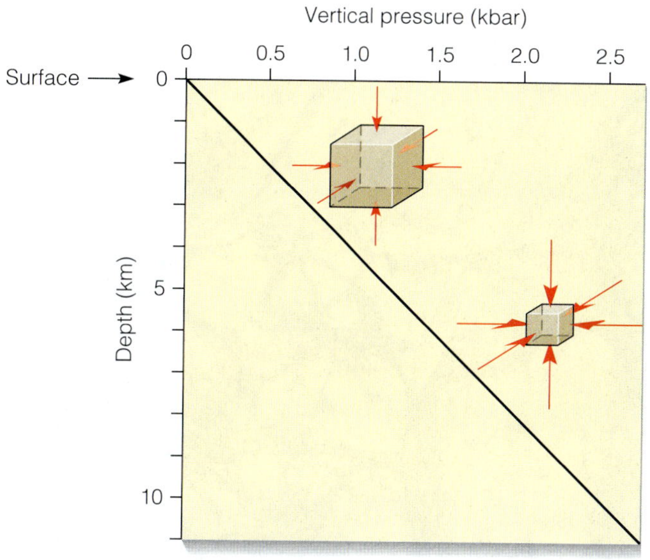

(a)

Courtesy of David J. and Jane M. Matty

(b)

■ **Figure 7.2**

(a) Lithostatic pressure is applied equally in all directions in Earth's crust due to the weight of overlying rocks. Thus pressure increases with depth, as indicated by the sloping black line. (b) A similar situation occurs when 200-ml Styrofoam cups are lowered to ocean depths of approximately 750 m and 1500 m. Increased water pressure is exerted equally in all directions on the cups, and they consequently decrease in volume while still maintaining their general shape. Source: (a): From C. Gillen, *Metamorphic Geology,* Figure 4.4, p. 73. Copyright © 1982 Kluwer Academic Publishers. Reprinted by permission of the author.

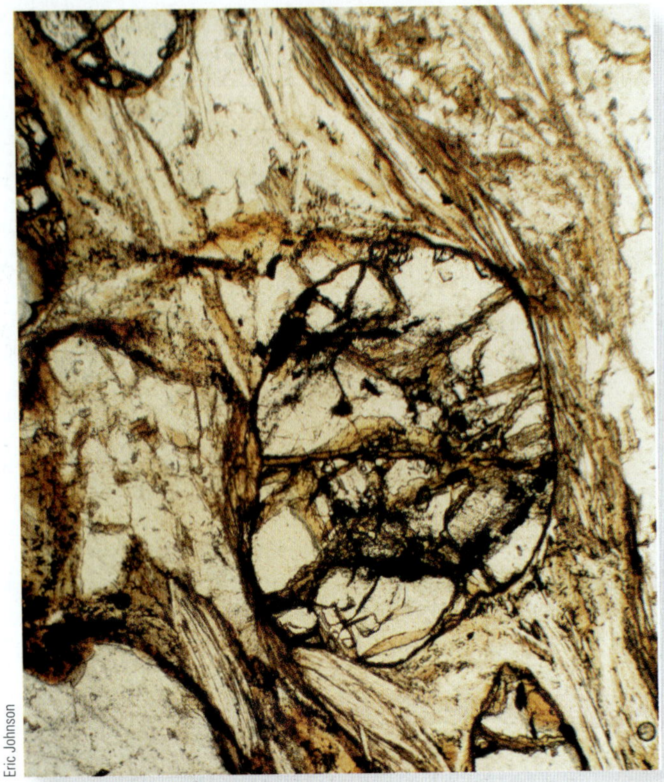

Eric Johnson

■ **Figure 7.3**

Differential pressure is pressure that is unequally applied to an object. Rotated garnets are a good example of the effects of differential pressure applied to a rock during metamorphism. This rotated garnet (center) came from a schist in northeast Sardinia.

Along with lithostatic pressure resulting from burial, rocks may also experience **differential pressures** (■ Figure 7.3). In this case, the pressures are not equal on all sides, and the rock is consequently distorted. Differential pressures typically occur during deformation associated with mountain building and can produce distinctive metamorphic textures and features.

Fluid Activity

In almost every region of metamorphism, water and carbon dioxide (CO_2) are present in varying amounts along mineral grain boundaries or in the pore spaces of rocks. These fluids, which may contain ions in solution, enhance metamorphism by increasing the rate of chemical reactions. Under dry conditions, most minerals react very slowly, but when even small amounts of fluid are introduced, reaction rates increase, mainly because ions can move readily through the fluid and thus enhance chemical reactions and the formation of new minerals.

The following reaction provides a good example of how new minerals can be formed by **fluid activity.** Seawater moving through hot basaltic rock of the oceanic crust transforms olivine into the metamorphic mineral serpentine:

$$\underset{\text{olivine}}{2Mg_2SiO_4} + \underset{\text{water}}{2H_2O} \rightarrow \underset{\text{serpentine}}{Mg_3Si_2O_5(OH)_4} + \underset{\substack{\text{carried}\\ \text{away in}\\ \text{solution}}}{MgO}$$

The chemically active fluids important in the metamorphic process come primarily from three sources. The first is water trapped in the pore spaces of sedimentary rocks as they form. The second is the volatile fluid within magma. The third source is the dehydration of water-bearing minerals such as gypsum ($CaSO_4 \cdot 2H_2O$) and some clays.

WHAT ARE THE THREE TYPES OF METAMORPHISM?

Geologists recognize three major types of metamorphism: *contact metamorphism,* in which magmatic heat and fluids act to produce change; *dynamic metamorphism,* which is principally the result of high differential pressures associated with intense deformation; and *regional metamorphism,* which occurs within a large area and is caused primarily by mountain-building forces. Even though we will discuss each type of metamorphism separately, the boundary between them is not always distinct and depends largely on which of the three metamorphic agents was dominant.

Contact Metamorphism

Contact metamorphism takes place when a body of magma alters the surrounding country rock. At shallow depths, intruding magma raises the temperature of the surrounding rock, causing thermal alteration. Furthermore, the release of hot fluids into the country rock by the cooling intrusion can aid in the formation of new minerals.

Important factors in contact metamorphism are the initial temperature and size of the intrusion as well as the fluid content of the magma and/or country rock. The initial temperature of an intrusion is controlled, in part, by its composition: mafic magmas are hotter than felsic magmas and hence have a greater thermal effect on the rocks surrounding them. The size of the intrusion is also important. In the case of small intrusions, such as dikes and sills, usually only those rocks in immediate contact with the intrusion are affected. Because large intrusions, such as batholiths, take a long time to cool, the increased temperature in the surrounding rock may last long enough for a larger area to be affected.

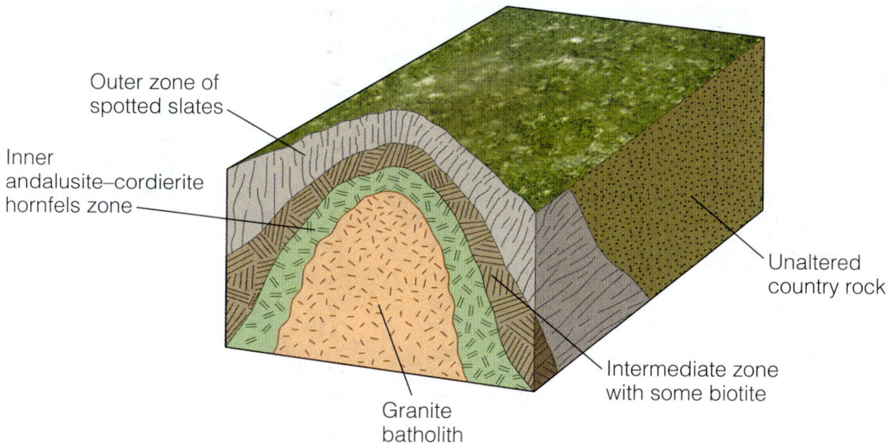

■ Figure 7.4

A metamorphic aureole typically surrounds many igneous intrusions. The metamorphic aureole associated with this idealized granite batholith contains three zones of mineral assemblages reflecting the decreases in temperature with distance from the intrusion. An andalusite–cordierite hornfels forms adjacent to the batholith. This is followed by an intermediate zone of extensive recrystallization in which some biotite develops, and farthest from the intrusion is the outer zone, which is characterized by spotted slates.

Temperatures can reach nearly 900°C adjacent to an intrusion, but they gradually decrease with distance. The effects of such heat and the resulting chemical reactions usually occur in concentric zones known as **aureoles** (■ Figure 7.4). The boundary between an intrusion and its aureole may be either sharp or transitional.

Metamorphic aureoles vary in width depending on the size, temperature, and composition of the intrusion as well as the mineralogy of the surrounding country rock. Typically, large intrusive bodies have several metamorphic zones, each characterized by distinctive mineral assemblages indicating the decrease in temperature with distance from the intrusion (Figure 7.4). The zone closest to the intrusion, and hence subject to the highest temperatures, may contain high-temperature metamorphic minerals (that is, minerals in equilibrium with the higher temperature environment) such as sillimanite. The outer zones may be characterized by lower temperature metamorphic minerals such as chlorite, talc, and epidote.

Contact metamorphism can result not only from igneous intrusions, but also from lava flows (■ Figure 7.5). Lava flowing over land may thermally alter the underlying rocks. Whereas recognizing a recent lava flow and the resulting contact metamorphism of the rocks below it is easy, less obvious is whether an igneous body is intrusive or extrusive in a rock outcrop where sedimentary rocks occur above and below the igneous body. Recognizing which sedimentary rock units have been metamorphosed enables geologists to determine whether the igneous body is intrusive (such as a sill or dike) or extrusive (lava flow). Such a determination is critical in reconstructing the geologic history of an area (see Chapter 17) and may have important economic implications as well.

Fluids also play an important role in contact metamorphism. Many magmas are wet and contain hot, chemically active fluids that may emanate into the surrounding rock. These fluids can react with the rock and aid in the formation of new minerals. In addition, the country rock may contain pore fluids that, when heated by magma, also increase reaction rates.

The formation of new minerals by contact metamorphism depends not only on proximity to the intrusion but also on the composition of the country rock. Shales, mudstones, impure limestones, and impure dolostones are particularly susceptible to the formation of new minerals by contact metamorphism, whereas pure sandstones or pure limestones typically are not.

Because heat and fluids are the primary agents of contact metamorphism, two types of contact metamorphic rocks are generally recognized: those resulting from baking of country rock and those altered by hot solutions. Many of the rocks that result from contact metamorphism have the texture of porcelain; that is, they are hard and fine grained. This is particularly true for rocks with a high clay content, such as shale. Such texture results because the clay minerals in the rock are baked, just as a clay pot is baked when fired in a kiln.

During the final stages of cooling when an intruding magma begins to crystallize, large amounts of hot, watery solutions are often released. These solutions may react with the country rock and produce new metamorphic minerals. This process, which usually occurs near Earth's surface, is called *hydrothermal alteration* (from the Greek *hydro*, "water," and *therme*, "heat") and may result in valuable mineral deposits. Geologists think that many of the world's ore deposits result from the migration of metallic ions in hydrothermal solutions. Examples are copper, gold, iron ores, tin, and zinc in various localities including Australia, Canada, China, Cyprus, Finland, Russia, and the western United States.

Dynamic Metamorphism

Most **dynamic metamorphism** is associated with fault (fractures along which movement has occurred) zones where rocks are subjected to high differential pressures.

■ **Figure 7.5**

A highly weathered basaltic lava flow near Susanville, California, has altered an underlying rhyolitic volcanic ash by contact metamorphism. The red zone below the lava flow has been baked by the heat of the lava when it flowed over the ash layer. The lava flow displays spheroidal weathering, a type of weathering common in fractured rocks (see Chapter 6).

The metamorphic rocks that result from pure dynamic metamorphism are called *mylonites* and are typically restricted to narrow zones adjacent to faults. Mylonites are hard, dense, fine-grained rocks, many characterized by thin laminations (■ Figure 7.6). Tectonic settings where mylonites occur include the Moine Thrust Zone in northwest Scotland and portions of the San Andreas fault in California (see Chapter 2).

■ **Figure 7.6**

Mylonite from the Adirondack Highlands, New York. Note the thin laminations.

Regional Metamorphism

Most metamorphic rocks result from **regional metamorphism,** which occurs over a large area and is usually caused by tremendous temperatures, pressures, and deformation within the deeper portions of the crust. Regional metamorphism is most obvious along convergent plate boundaries where rocks are intensely deformed and recrystallized during convergence and subduction. Within these metamorphic rocks, there is usually a gradation of metamorphic intensity from areas that were subjected to the most intense pressures and/or highest temperatures to areas of lower pressures and temperatures. Such a gradation in metamorphism can be recognized by the metamorphic minerals that are present.

Regional metamorphism is not confined to only convergent margins. It also occurs in areas where plates diverge, although usually at much shallower depths because of the high geothermal gradient associated with these areas.

From field studies and laboratory experiments, certain minerals are known to form only within specific temperature and pressure ranges. Such minerals are known as **index minerals** because their presence allows geologists to recognize low-, intermediate-, and high-grade metamorphic zones (■ Figure 7.7).

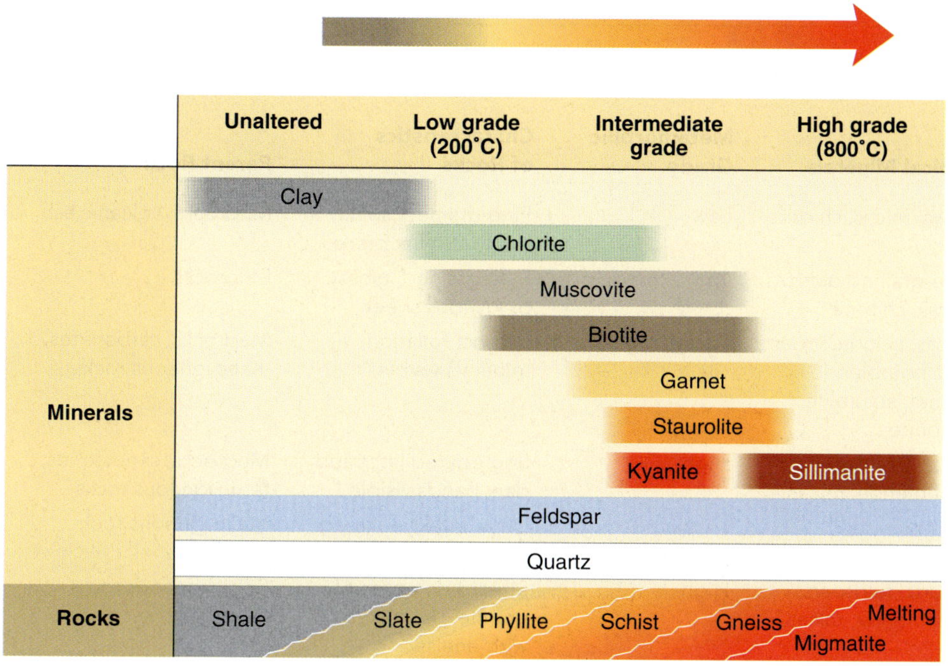

■ **Figure 7.7**

Change in mineral assemblage and rock type with increasing metamorphism of shale. When a clay-rich rock such as shale is subjected to increasing metamorphism, new minerals form, as shown by the colored bars. The progressive appearance of particular minerals allows geologists to recognize low-, intermediate-, and high-grade metamorphic zones.

When a clay-rich rock such as shale is metamorphosed, new minerals form as a result of metamorphic processes. The mineral chlorite, for example, is produced under relatively low temperatures of about 200°C, so its presence indicates low-grade metamorphism. As temperatures and pressures continue to increase, new minerals form that are stable under those conditions. Thus, there is a progression in the appearance of new minerals from chlorite, whose presence indicates low-grade metamorphism, to sillimanite, whose presence indicates high-grade metamorphism and temperatures exceeding 500°C.

Different rock compositions develop different index minerals. When sandy dolomites are metamorphosed, for example, they produce an entirely different set of index minerals. Thus, a specific set of index minerals commonly forms in specific rock types as metamorphism progresses.

Although such common minerals as mica, quartz, and feldspar can occur in both igneous and metamorphic rocks, other minerals such as andalusite, sillimanite, and kyanite generally occur only in metamorphic rocks derived from clay-rich sediments. Although these three minerals all have the same chemical formula (Al_2SiO_5), they differ in crystal structure and other physical properties because each forms under a different range of pressures and temperatures. Thus, they are sometimes used as index minerals for metamorphic rocks formed from clay-rich sediments.

Geology⇌Now Log into GeologyNow and select this chapter to work through a **Geology Interactive** activity on "The Rock Cycle" (click Rocks and the Rock Cycle→Rock Cycle).

HOW ARE METAMORPHIC ROCKS CLASSIFIED?

For purposes of classification, metamorphic rocks are commonly divided into two groups: those exhibiting a *foliated texture* (from the Latin *folium*, "leaf") and those with a *nonfoliated texture* (Table 7.1).

Foliated Metamorphic Rocks

Rocks subjected to heat and differential pressure during metamorphism typically have minerals arranged in a parallel fashion, giving them a **foliated texture** (■ Figure 7.8). The size and shape of the mineral grains determine whether the foliation is fine or coarse. If the foliation is such that the individual grains cannot be recognized without magnification, the rock is slate (■ Figure 7.9a). A coarse foliation results when granular minerals such as quartz and feldspar are segregated into roughly parallel and streaky zones that differ in composition and color, as in gneiss. Foliated metamorphic rocks can be arranged in order of increasingly coarse grain size and perfection of foliation.

Slate is a very fine-grained metamorphic rock that commonly exhibits *slaty cleavage* (Figure 7.9b). Slate is the result of low-grade regional metamorphism of shale or, more rarely, volcanic ash. Because it can easily be split along cleavage planes into flat pieces, slate is an excellent rock for roofing and floor tiles, billiard and pool

Table 7.1

Classification of Common Metamorphic Rocks

Texture	Metamorphic Rock	Typical Minerals	Metamorphic Grade	Characteristics of Rocks	Parent Rock
Foliated	Slate	Clays, micas, chlorite	Low	Fine-grained, splits easily into flat pieces	Mudrocks, volcanic ash
	Phyllite	Fine-grained quartz, micas, chlorite	Low to medium	Fine-grained, glossy or lustrous sheen	Mudrocks
	Schist	Micas, chlorite, quartz, talc, hornblende, garnet, staurolite, graphite	Low to high	Distinct foliation, minerals visible	Mudrocks, carbonates, mafic igneous rocks
	Gneiss	Quartz, feldspars, hornblende, micas	High	Segregated light and dark bands visible	Mudrocks, sandstones, felsic igneous rocks
	Amphibolite	Hornblende, plagioclase	Medium to high	Dark, weakly foliated	Mafic igneous rocks
	Migmatite	Quartz, feldspars, hornblende, micas	High	Streaks or lenses of granite intermixed with gneiss	Felsic igneous rocks mixed with sedimentary rocks
Nonfoliated	Marble	Calcite, dolomite	Low to high	Interlocking grains of calcite or dolomite, reacts with HCl	Limestone or dolostone
	Quartzite	Quartz	Medium to high	Interlocking quartz grains, hard, dense	Quartz sandstone
	Greenstone	Chlorite, epidote, hornblende	Low to high	Fine-grained, green	Mafic igneous rocks
	Hornfels	Micas, garnets, andalusite, cordierite, quartz	Low to medium	Fine-grained, equidimensional grains, hard, dense	Mudrocks
	Anthracite	Carbon	High	Black, lustrous, subconcoidal fracture	Coal

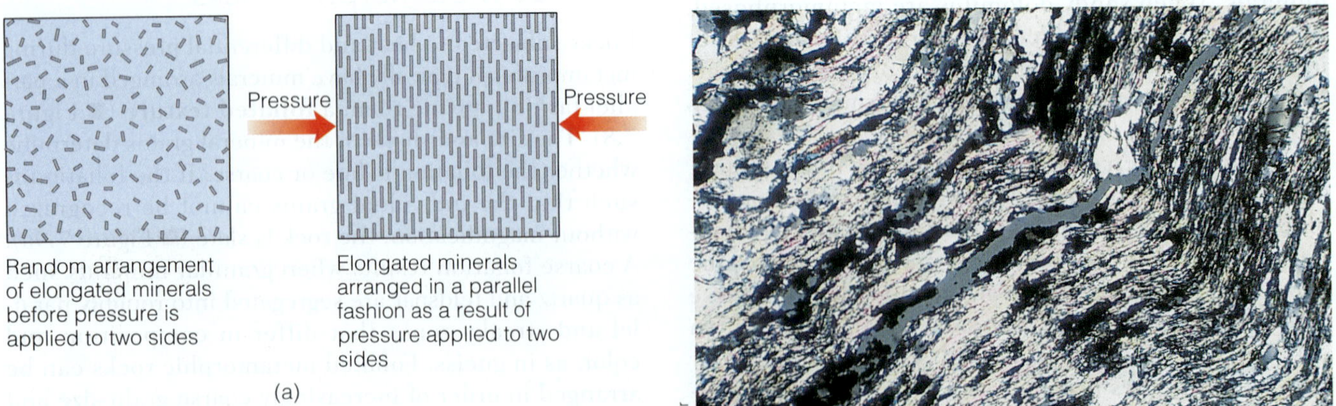

Figure 7.8

(a) When rocks are subjected to differential pressure, the mineral grains are typically arranged in a parallel fashion, producing a foliated texture. (b) Photomicrograph of a metamorphic rock with a foliated texture showing the parallel arrangement of mineral grains.

table tops, and blackboards (Figure 7.9c). The different colors of most slates are caused by minute amounts of graphite (black), iron oxide (red and purple), and chlorite (green).

Phyllite is similar in composition to slate but coarser grained. The minerals, however, are still too small to be identified without magnification. Phyllite can be distinguished from slate by its glossy or lustrous sheen (■ Figure 7.10). It represents an intermediate grain size between slate and schist.

Schist is most commonly produced by regional metamorphism. The type of schist formed depends on the intensity of metamorphism and the character of the original rock (■ Figure 7.11). Metamorphism of many rock types can yield schist, but most schist appears to have formed from clay-rich sedimentary rocks (Table 7.1).

■ **Figure 7.9**

(a) Hand specimen of slate. (b) This panel of Arvonia Slate from Albemarne Slate Quarry, Virginia, shows bedding (upper right to lower left) at an angle to the slaty cleavage. (c) Slate roof of Chalet Enzian, Switzerland.

■ **Figure 7.10**

Specimen of phyllite. Note the lustrous sheen as well as the bedding (upper left to lower right) at an angle to the cleavage of the specimen.

■ **Figure 7.11**

Schist. (a) Garnet–mica schist. (b) Hornblende–mica–garnet schist.

GEOLOGY IN UNEXPECTED PLACES

Starting Off with a Clean Slate

Slate is a common metamorphic rock that has many uses. Two familiar uses are in the playing surface of billiard tables and roofing shingles.

Although slate is abundant throughout the world, most of it is unsuitable for billiard tables. For billiard tables, the slate must have a very fine grain so it can be honed to a smooth surface, somewhat elastic so it will expand and contract with the table's wood frame, and essentially nonabsorbent. Presently Brazil, China, India, and Italy are the major exporters of billiard table–quality slate, with the best coming from the Liguarian region of northern Italy. Most quality tables use at least 1-inch-thick slate that is split into three pieces. Although using three slabs requires extra work to ensure a tight fit and smooth surface, a table with three pieces is preferred over a single piece because it is less likely to fracture. Furthermore, the slate is usually slightly larger than the playing surface so that it extends below the rails of the table, thus giving additional strength to the rails and stability to the table. In addition, a quality table will have a wood backing glued to the underside of the slate so the felt cloth that is stretched tightly over the slate's surface can be stapled to the wood to provide a smooth playing surface.

Slate has been used as a roofing material for centuries. When properly installed and maintained, slate normally lasts for 60 to 125 years; many slate roofs have been around for more than 200 years. In the United States, slate roofing tiles typically come in shades of gray, green, purple, black, and red (■ Figure 1). There are 36 common sizes of tiles, ranging from 12 to 24 inches long with the width about half the length. The typical slate tile is usually 1/4 inch thick. Thicker tiles may be used, but they are harder to work with and greatly increase the weight of the roof.

The years between 1897 and 1914 witnessed the height of the U.S. roofing slate industry in both quantity and value of output. By the end of the 19th century, more than 200 slate quarries were operating in 13 states. With the introduction of asphalt shingles, which can be mass produced, easily transported, and installed at a much lower cost than slate shingles, the slate shingle industry in the United States began to decline around 1915. The renewed popularity of historic preservation and the recognition of slate's durability, however, have brought a resurgence in the slate roofing industry. It's not that unusual these days for geology to be overhead as well as underfoot.

■ Figure 1

Different colored slates make up the roof of this elementary school in Mount Pleasant, Michigan.
Source: Reed Wicander

All schists contain more than 50% platy and elongated minerals, all of which are large enough to be clearly visible. Their mineral composition imparts a *schistosity* or *schistose foliation* to the rock that usually produces a wavy type of parting when split. Schistosity is common in low- to high-grade metamorphic environments, and each type of schist is known by its most conspicuous mineral or minerals, such as mica schist, chlorite schist, and talc schist.

Gneiss is a metamorphic rock that is streaked or has segregated bands of light and dark minerals. Gneisses are composed mostly of granular minerals such as quartz

Reed Wicander

■ **Figure 7.12**

Gneiss is characterized by segregated bands of light and dark minerals. This folded gneiss is exposed at Wawa, Ontario, Canada.

and/or feldspar, with lesser percentages of platy or elongated minerals such as micas or amphiboles (■ Figure 7.12). Quartz and feldspar are the principal light-colored minerals, whereas biotite and hornblende are the typical dark minerals. Gneiss typically breaks in an irregular manner, much like coarsely crystalline nonfoliated rocks.

Most gneiss probably results from recrystallization of clay-rich sedimentary rocks during regional metamorphism (Table 7.1). Gneiss also can form from igneous rocks such as granite or older metamorphic rocks.

Another fairly common foliated metamorphic rock is *amphibolite*. A dark rock, it is composed mainly of hornblende and plagioclase. The alignment of the hornblende crystals produces a slightly foliated texture. Many amphibolites result from medium- to high-grade metamorphism of such ferromagnesian silicate-rich igneous rocks as basalt.

Some areas of regional metamorphism have exposures of "mixed rocks" with both igneous and high-grade metamorphic characteristics. In these rocks, called *migmatites*, streaks or lenses of granite are usually intermixed with high-grade ferromagnesian-rich metamorphic rocks, imparting a wavy appearance to the rock (■ Figure 7.13).

Most migmatites are thought to be the product of extremely high-grade metamorphism, and several models for their origin have been proposed. Part of the problem in determining the origin of migmatites is explaining how the granitic component formed. According to one model, the granitic magma formed in place by the partial melting of rock during intense metamorphism. Such

Ed Bartram

■ **Figure 7.13**

Migmatites consist of high-grade metamorphic rock intermixed with streaks or lenses of granite. This migmatite is exposed at Thirty Thousand Islands of Georgian Bay, Lake Huron, Ontario, Canada.

■ **Figure 7.14**

Nonfoliated textures are characterized by a mosaic of roughly equidimensional minerals, as in this photomicrograph of marble.

an origin is possible provided that the host rocks contained quartz and feldspars and that water was present. Another possibility is that the granitic components formed by the redistribution of minerals by recrystallization in the solid state—that is, by pure metamorphism.

Nonfoliated Metamorphic Rocks

In some metamorphic rocks, the mineral grains do not show a discernable preferred orientation. Instead, these rocks consist of a mosaic of roughly equidimensional minerals and are characterized as having a **nonfoliated texture** (■ Figure 7.14). Most nonfoliated metamorphic rocks result from contact or regional metamorphism of rocks with no platy or elongate minerals. Frequently, the only indication that a granular rock has been metamorphosed is the large grain size resulting from recrystallization. Nonfoliated metamorphic rocks are generally of two types: those composed mainly of only one mineral—for example, marble or quartzite, and those in which the different mineral grains are too small to be seen without magnification—such as greenstone and hornfels.

Marble is a well-known metamorphic rock composed predominantly of calcite or dolomite; its grain size ranges from fine to coarsely granular (see the chapter opening photo and ■ Figure 7.15). Marble results from either contact or regional metamorphism of limestones or dolostones (Table 7.1). Pure marble is snowy white or bluish, but many color varieties exist because of the presence of mineral impurities in the original sedimentary rock. The softness of marble, its uniform texture, and its varying colors have made it the favorite rock of builders and sculptors throughout history (see the Introduction and "The Many Uses of Marble" on pages 188 and 189).

Quartzite is a hard, compact rock typically formed from quartz sandstone under medium- to high-grade metamorphic conditions during contact or regional metamorphism (■ Figure 7.16). Because recrystallization is so complete, metamorphic quartzite is of uniform strength and therefore usually breaks across the component quartz grains rather than around them when it is struck. Pure quartzite is white, but iron and other impurities commonly impart a reddish or other color to it. Quartzite is commonly used as foundation material for road and railway beds.

The name *greenstone* is applied to any compact, dark-green, altered, mafic igneous rock that formed under low- to high-grade metamorphic conditions. The green color results from the presence of chlorite, epidote, and hornblende.

Hornfels, a fine-grained, nonfoliated metamorphic rock resulting from contact metamorphism, is composed of various equidimensional mineral grains. The composition of hornfels directly depends on the composition of the original rock, and many compositional varieties are known. The majority of hornfels, however, are ap-

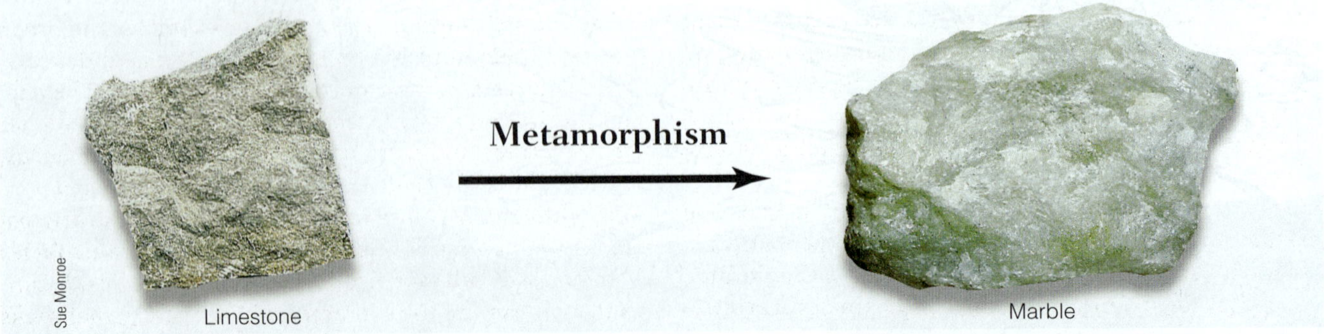

■ **Figure 7.15**

Marble results from the metamorphism of the sedimentary rock limestone or dolostone.

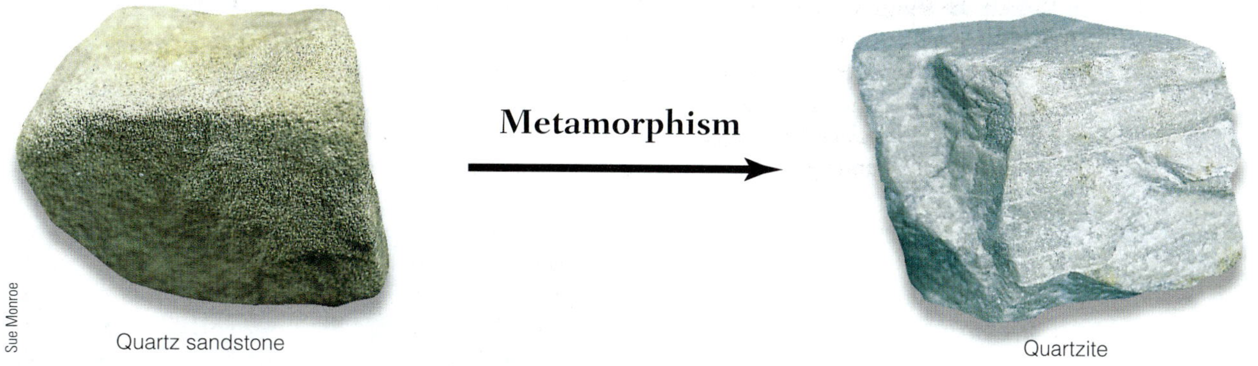

■ **Figure 7.16**

Quartzite results from the metamorphism of quartz sandstone.

parently derived from contact metamorphism of clay-rich sedimentary rocks or impure dolostones.

Anthracite is a black, lustrous, hard coal that contains a high percentage of fixed carbon and a low percentage of volatile matter. It usually forms from the metamorphism of lower-grade coals by heat and pressure and is thus considered by many geologists to be a metamorphic rock.

Geology⇌Now Log into GeologyNow and select this chapter to work through a **Geology Interactive** activity on "Rock Laboratory" (click Rocks and the Rock Cycle→Rock Laboratory).

WHAT ARE METAMORPHIC ZONES AND FACIES?

The first systematic study of metamorphic zones was conducted in the late 1800s by George Barrow and other British geologists working in the Dalradian schists of the southwestern Scottish Highlands. Here, clay-rich sedimentary rocks have been subjected to regional metamorphism, and the resulting metamorphic rocks can be divided into different zones based on the presence of distinctive silicate mineral assemblages. These mineral assemblages, each recognized by the presence of one or more index minerals, indicate different degrees of metamorphism. The index minerals Barrow and his associates chose to represent increasing metamorphic intensity were chlorite, biotite, garnet, staurolite, kyanite, and sillimanite (Figure 7.7). Note that these are the metamorphic minerals produced from clay-rich sedimentary rocks. Other mineral assemblages and index minerals are produced from rocks with different original compositions.

The successive appearance of metamorphic index minerals indicates gradually increasing or decreasing intensity of metamorphism. Going from lower- to higher-grade zones, the first appearance of a particular index mineral indicates the location of the minimum temperature and pressure conditions needed for the formation of that mineral. When the locations of the first appearances of that index mineral are connected on a map, the result is a line of equal metamorphic intensity or an *isograd*. The region between isograds is known as a **metamorphic zone.** By noting the occurrence of metamorphic index minerals, geologists can construct a map showing the metamorphic zones of an entire area (■ Figure 7.17).

Numerous studies of different metamorphic rocks have demonstrated that although the texture and composition of any rock may be altered by metamorphism, the overall chemical composition may be little changed. Thus, the different mineral assemblages found in increasingly higher grade metamorphic rocks derived from the same original rock result from changes in temperature and pressure.

A **metamorphic facies** is a group of metamorphic rocks characterized by particular mineral assemblages formed under the same broad temperature–pressure conditions (■ Figure 7.18). Each facies is named after its most characteristic rock or mineral. For example, the green metamorphic mineral chlorite, which forms under relatively low temperatures and pressures, yields rocks belonging to the *greenschist facies.* Under increasingly higher temperatures and pressures, other metamorphic facies, such as the *amphibolite* and *granulite facies,* develop.

Although usually applied to areas where the original rocks were clay-rich, the concept of metamorphic facies can be used with modification in other situations. It cannot, however, be used in areas where the original rocks were pure quartz sandstones or pure limestones or dolostones. Such rocks would yield only quartzites and marbles, respectively.

The Many Uses of Marble

Marble is a remarkable stone that has a variety of uses. Formed from limestone or dolostone by the metamorphic processes of heat and pressure, marble comes in a variety of colors and textures.

Marble has been used by sculptors and architects for many centuries in statuary, monuments, as a facing and main stone in buildings and structures, as well as for floor tiling and other ornamental and structural uses. Marble can also be found in toothpaste and as a source of lime in agricultural fertilizers.

Reed Wicander

Aphrodite of Melos, also known as *Venus de Milo*, is one of the most recognizable works of art in the world. Dated at around 150 B.C., *Venus de Milo* was created by an unknown artist during the Hellenistic period and carved from the world-famous Parian marble from Paros in the Cyclades. Today *Venus de Milo* attracts thousands of visitors a year to the Louvre Museum in Paris, where she can be viewed and appreciated.

Photodisc/ Getty Images

Marble has been used extensively as a building stone through the ages and throughout the world. For example, the Greek Parthenon was constructed of white Pentelic marble from Mt. Pentelicus in Attica.

Photodisc/Getty Images

Galen Rowell/ Corbis

The Taj Mahal in India is largely constructed of Makrana marble quarried from hills just southwest of Jaipur in Rajasthan. In addition to its main use as a building material, marble was used throughout the structure in art works and intricately carved marble flowers (right). All in all, it took more than 20,000 workers 17 years to build the Taj Mahal from A.D 1631 to 1648.

In the United States, marble is used as a building stone in many structures and is quarried from many locations. Marble is used in a variety of buildings and monuments in Washington, D.C. The Washington Monument is built from three different kinds of marble. The first 152 feet of the monument, built between 1848 and 1854, is faced with marble from the Texas, Maryland, quarry. Following 25 years of virtual inactivity, construction resumed with four rows of white marble from Lee, Massachusetts, added above the Texas marble. This marble proved to be too costly, so the upper part of the monument was finished with Cockeysville marble from quarries at Cockeysville, Maryland. The three different marbles can be distinguished by the slight differences in color.

The Peace Monument at Pennsylvania Avenue on the west side of the Capitol is constructed from white marble from Carrara, Italy, a locality famous for its marble.

Photodisc/ Getty Images

A marble quarry in northcentral Vermont. Vermont is known for producing some of the finest marble in the United States.

Richard V. Dietrich

Photodisc/ Getty Images

Another example of a marble building in Washington, DC, is the Lincoln Monument, built from the Colorado Yule Marble, which is quarried at Marble, Colorado. This very pure white marble has been used not only for the Lincoln Monument but also for many other prominent buildings throughout the United States.

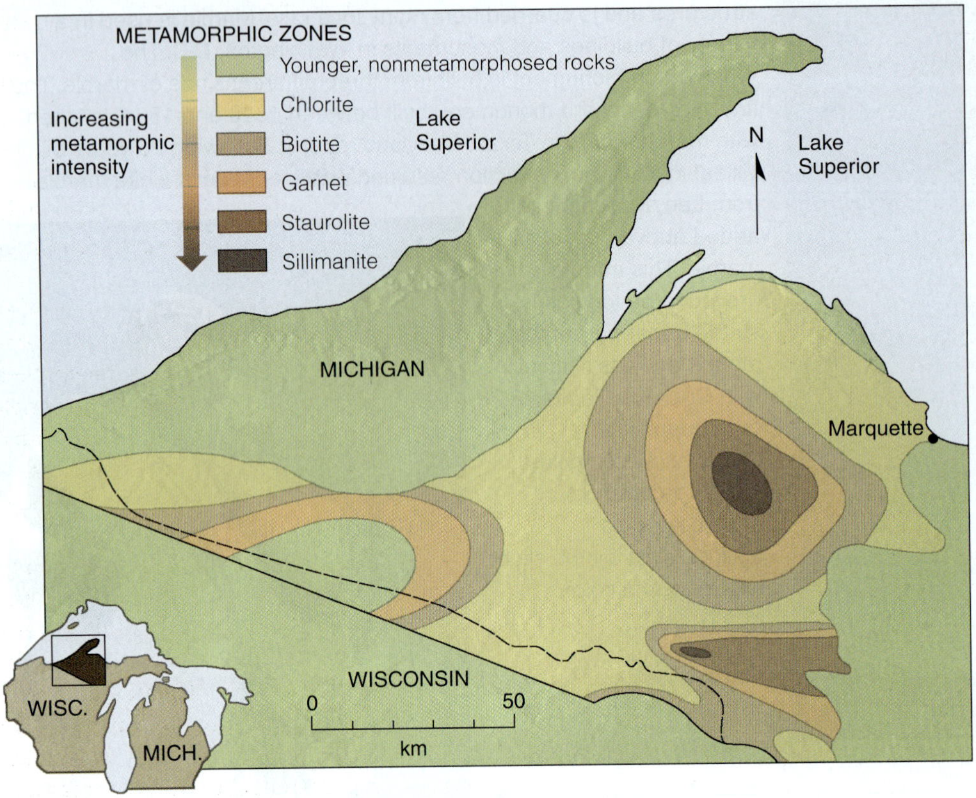

■ **Figure 7.17**

Metamorphic zones in the Upper Peninsula of Michigan. The zones in this region are based on the presence of distinctive silicate mineral assemblages resulting from the metamorphism of sedimentary rocks during an interval of mountain building and minor granitic intrusion during the Proterozoic Eon, about 1.5 billion years ago. The lines separating the different metamorphic zones are isograds. Source: From H. L. James, *G. S. A. Bulletin*, vol. 66 (1955), plate 1, page 1454, with permission of the publisher, the Geological Society of America, Boulder, Colorado. USA. Copyright © 1955 Geological Society of America.

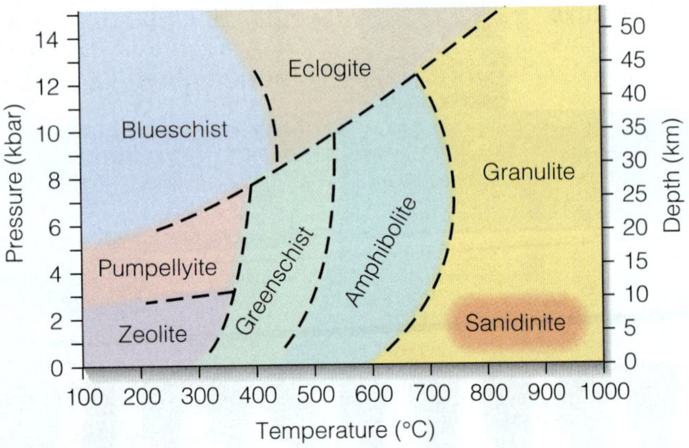

■ **Figure 7.18**

A pressure–temperature diagram showing where various metamorphic facies occur. A facies is characterized by a particular mineral assemblage that formed under the same broad temperature–pressure conditions. Each facies is named after its most characteristic rock or mineral. Source: From *AGI Data Sheet 35.4*, AGI Data Sheets, 3rd edition (1989) with the kind permission of the American Geological Institute. For more information, log on to www.agiweb.org

HOW DOES PLATE TECTONICS AFFECT METAMORPHISM?

Although metamorphism is associated with all three types of plate boundaries (see Figure 1.11), it is most common along convergent plate margins. Metamorphic rocks form at convergent plate boundaries because temperature and pressure increase as a result of plate collisions.

■ Figure 7.19 illustrates the various temperature–pressure regimes produced along an oceanic–continental convergent plate boundary and the type of metamorphic facies and rocks that can result. When an oceanic plate collides with a continental plate, tremendous pressure is generated as the oceanic plate is subducted. Because rock is a poor heat conductor, the cold descending oceanic plate heats slowly, and metamorphism is caused mostly by increasing pressure with depth. Metamorphism in such an environment produces rocks typical of the *blueschist facies* (low temperature, high pressure), which is characterized by the blue amphibole mineral glaucophane (Figure 7.18). Geologists use the occurrence of blueschist facies rocks as evidence of ancient subduction zones. An excellent example of blueschist metamorphism can be found in the California Coast

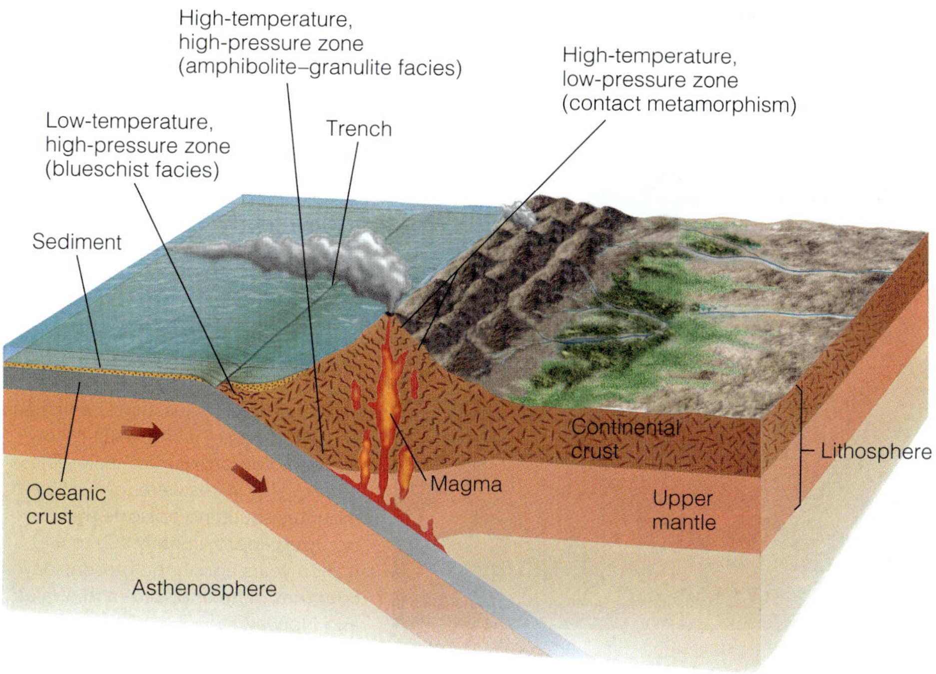

■ **Figure 7.19**

Metamorphic facies resulting from various temperature–pressure conditions produced along an oceanic–continental convergent plate boundary.

As subduction along the oceanic–continental convergent plate boundary continues, both temperature and pressure increase with depth and yield high-grade metamorphic rocks. Eventually, the descending plate begins to melt and generates magma that moves upward. This rising magma may alter the surrounding rock by contact metamorphism, producing migmatites in the deeper portions of the crust and hornfels at shallower depths. Such an environment is characterized by high temperatures and low to medium pressures.

Although metamorphism is most common along convergent plate margins, many divergent plate boundaries are characterized by contact metamorphism. Rising magma at mid-oceanic ridges heats the adjacent rocks, producing contact metamorphic minerals and textures. Besides contact metamorphism, fluids emanating from the rising magma—and its reaction with seawater—very commonly produce metal-bearing hydrothermal solutions that may precipitate minerals of economic value. These deposits may eventually be

Ranges. Here rocks of the Franciscan Complex were metamorphosed under low-temperature, high-pressure conditions that clearly indicate the presence of a former subduction zone (■ Figure 7.20).

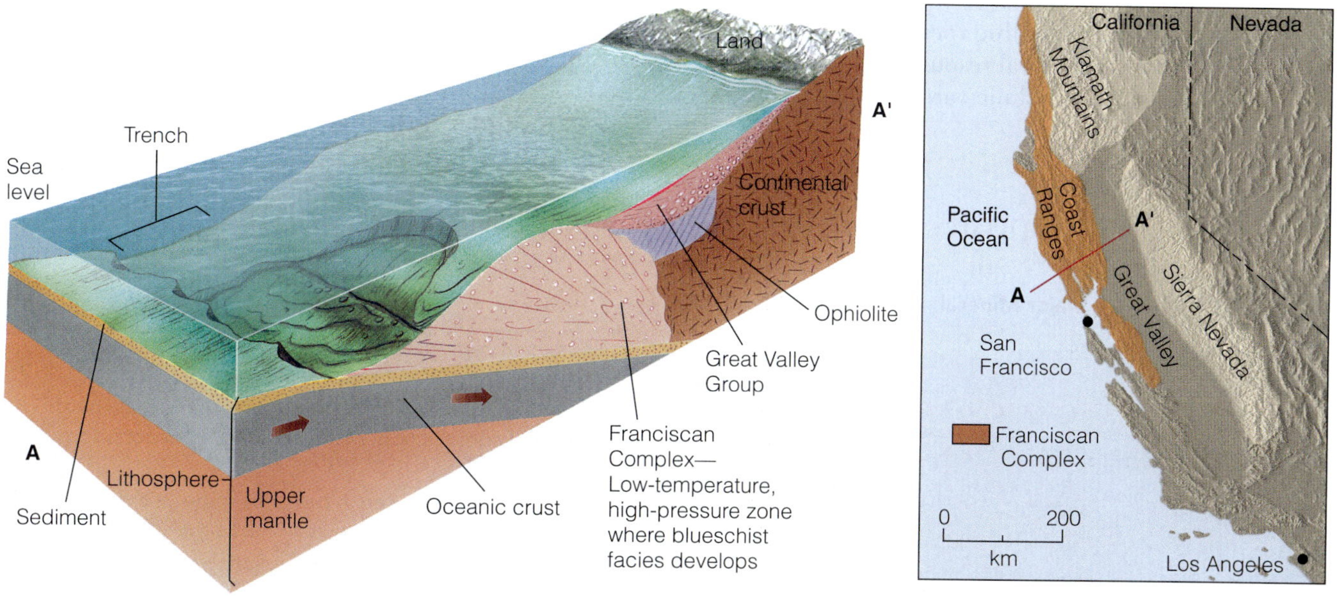

■ **Figure 7.20**

Index map of California showing the location of the Franciscan Complex and a diagrammatic reconstruction of the environment in which it was regionally metamorphosed under low-temperature, high-pressure subduction conditions approximately 150 million years ago. The red line on the index map shows the orientation of the reconstruction to the current geography. Source: From "Effects of Late Jurassic-Early Tertiary Subduction in California," *San Joaquin Geological Society Short Course*, 1977, 66, Figure 5-9. Reprinted with permission.

Peter Hulme/Ecoscene/Corbis

■ **Figure 7.21**

Slate quarry in Wales. Slate, which has a variety of uses, is the result of low-grade metamorphism of shale. These high-quality slates were formed by a mountain-building episode that occurred approximately 400 to 440 million years ago in the present-day countries of Ireland, Scotland, Wales, and Norway.

brought to Earth's surface by later tectonic activity. The copper ores of Cyprus are a good example of such hydrothermal activity (see Chapter 2).

METAMORPHISM AND NATURAL RESOURCES

Many metamorphic rocks and minerals are valuable natural resources. Although these resources include various types of ore deposits, the two most familiar and widely used metamorphic rocks, as such, are marble and slate, which, as we discussed, have been used for centuries in a variety of ways (■ Figure 7.21).

Many ore deposits result from contact metamorphism during which hot, ion-rich fluids migrate from igneous intrusions into the surrounding rock, thereby producing rich ore deposits. The most common sulfide ore minerals associated with contact metamorphism are bornite, chalcopyrite, galena, pyrite, and sphalerite; two common oxide ore minerals are hematite and magnetite. Tin and tungsten are also important ores associated with contact metamorphism (Table 7.2).

Table 7.2

The Main Ore Deposits Resulting from Contact Metamorphism

Ore Deposit	Major Mineral	Formula	Use
Copper	Bornite Chalcopyrite	Cu_5FeS_4 $CuFeS_2$	Important sources of copper, which is used in manufacturing, transportation, communications, and construction
Iron	Hematite Magnetite	Fe_2O_3 Fe_3O_4	Major sources of iron for manufacture of steel, which is used in nearly every form of construction, manufacturing, transportation, and communications
Lead	Galena	PbS	Chief source of lead, which is used in batteries, pipes, solder, and elsewhere where resistance to corrosion is required
Tin	Cassiterite	SnO_2	Principal source of tin, which is used for tin plating, solder, alloys, and chemicals
Tungsten	Scheelite Wolframite	$CaWO_4$ $(Fe,Mn)WO_4$	Chief sources of tungsten, which is used in hardening metals and manufacturing carbides
Zinc	Sphalerite	$(Zn, Fe)S$	Major source of zinc, which is used in batteries and in galvanizing iron and making brass

Other economically important metamorphic minerals include talc for talcum powder; graphite for pencils and dry lubricants; garnets and corundum, which are used as abrasives or gemstones, depending on their quality; and andalusite, kyanite, and sillimanite, all of which are used in manufacturing high-temperature porcelains and temperature-resistant minerals for products such as sparkplugs and the linings of furnaces.

7 GEO RECAP

Chapter Summary

- Metamorphic rocks result from the transformation of other rocks, usually beneath Earth's surface, as a consequence of one or a combination of three agents: heat, pressure, and fluid activity.
- Heat for metamorphism comes from intrusive magmas, extrusive lava flows, or deep burial. Pressure is either lithostatic or differential. Fluids trapped in sedimentary rocks or emanating from intruding magmas can enhance chemical changes and the formation of new minerals.
- The three major types of metamorphism are contact, dynamic, and regional.
- Index minerals—minerals that form only within specific temperature and pressure ranges—allow geologists to recognize low-, intermediate-, and high-grade metamorphic zones.
- Metamorphic rocks are primarily classified according to their texture. In a foliated texture, platy and elongate minerals have a preferred orientation. A nonfoliated texture does not exhibit any discernable preferred orientation of the mineral grains.
- Foliated metamorphic rocks can be arranged in order of grain size and/or perfection of their foliation. Slate is fine grained, followed by (in coarser-grained order) phyllite and schist; gneiss displays segregated bands of minerals. Amphibolite is another fairly common foliated metamorphic rock.
- Marble, quartzite, greenstone, and hornfels are common nonfoliated metamorphic rocks.
- Metamorphic zones are based on index minerals and are areas of equal metamorphic intensity. Metamorphic facies are characterized by particular assemblages of minerals that formed under specific metamorphic conditions. These facies are named for a characteristic constituent mineral or rock type.
- Metamorphism occurs along all three kinds of plate boundaries but is most widespread at convergent plate margins.
- Metamorphic rocks formed near Earth's surface along an oceanic–continental convergent plate boundary result from low-temperature, high-pressure conditions. As a subducted oceanic plate descends, it is subjected to increasingly higher temperatures and pressures that result in higher-grade metamorphism.
- Many metamorphic rocks and minerals, such as marble, slate, graphite, talc, and asbestos, are valuable natural resources.

Important Terms

aureole (p. 179)
contact metamorphism (p. 178)
differential pressure (p. 178)
dynamic metamorphism (p. 179)
fluid activity (p. 178)
foliated texture (p. 181)
heat (p. 177)
index mineral (p. 180)
lithostatic pressure (p. 177)
metamorphic facies (p. 187)
metamorphic rock (p. 174)
metamorphic zone (p. 187)
nonfoliated texture (p. 186)
regional metamorphism (p. 180)

Review Questions

1. Magmatic heat and fluid activity are the primary agents involved in what type of metamorphism?
   a._____ dynamic; b._____ lithostatic; c._____ contact; d._____ regional; e._____ thermodynamic.
2. Which of the following is *not* an agent or process of metamorphism?
   a._____ pressure; b._____ heat; c._____ fluid activity; d._____ time; e._____ gravity.
3. Which is the order of increasingly coarser grain size and perfection of foliation?
   a._____ gneiss → schist → phyllite → slate; b._____ phyllite → slate → schist → gneiss; c._____ schist → slate → gneiss → phyllite; d._____ slate → phyllite → schist → gneiss; e._____ slate → schist → phyllite → gneiss.
4. Along what type of plate boundary is metamorphism most common?
   a._____ divergent; b._____ transform; c._____ aseismic; d._____ convergent; e._____ lithospheric.
5. Metamorphic zones
   a._____ reflect a metamorphic grade; b._____ are characterized by distinctive mineral assemblages; c._____ are separated from each other by isograds; d._____ all of these; e._____ none of these.
6. The nonfoliated metamorphic rock formed from limestone or dolostone is called
   a._____ quartzite; b._____ marble; c._____ hornfels; d._____ greenstone; e._____ schist.
7. Concentric zones surrounding an igneous intrusion and characterized by distinctive mineral assemblages are
   a._____ thermodynamic rings; b._____ hydrothermal regions; c._____ metamorphic layers; d._____ regional facies; e._____ aureoles.
8. Metamorphic rocks can form from what type of original rock?
   a._____ igneous; b._____ sedimentary; c._____ metamorphic; d._____ volcanic; e._____ all of these.
9. Pressure resulting from deep burial and applied equally in all directions on a rock is
   a._____ directional; b._____ differential; c._____ lithostatic; d._____ shear; e._____ unilateral.
10. The majority of metamorphic rocks result from which type of metamorphism?
    a._____ lithostatic; b._____ contact; c._____ regional; d._____ local; e._____ dynamic.
11. What specific features about foliated metamorphic rocks make them unsuitable as foundations for dams? Are there any metamorphic rocks that would make good foundations? Why?
12. Discuss the role each of the three agents of metamorphism plays in transforming any rock into a metamorphic rock.
13. What is regional metamorphism, and under what conditions does it occur?
14. Describe the two types of metamorphic texture, and discuss how they are produced.
15. Why is metamorphism more widespread along convergent plate boundaries than along any other type of plate boundary?

16. Name several economically valuable metamorphic minerals or rocks and discuss why they are valuable.
17. How can aureoles be used to determine the effects of metamorphism?
18. Why should the average citizen know about metamorphic rocks and how they form?
19. Using Figure 7.18, go to a point that is represented by 450°C and 6 kbar of pressure. What metamorphic facies is represented by those conditions? If the pressure is raised to 10 kbar, what facies is represented by the new conditions? What change in depth of burial is required to effect a pressure change of 6 to 10 kbar?
20. If plate tectonic movement did not exist, could there be metamorphism? Do you think metamorphic rocks exist on other planets in our solar system? Why?

World Wide Web Activities

GeologyNow Assess your understanding of this chapter's topics with additional quizzing and comprehensive interactivities at

http://earthscience.brookscole.com/changingearth4e

as well as current and up-to-date weblinks, additional readings, and InfoTrac College Edition exercises.

Earthquakes and Earth's Interior

CHAPTER 8

OUTLINE

Geology Now *This icon, which appears throughout the book, indicates an opportunity to explore interactive tutorials, animations, or practice problems available on the GeologyNow website at http://earthscience.brookscole.com/changingearth4e.*

OBJECTIVES

At the end of this chapter, you will have learned that:

- Energy is stored in rocks and is released when they fracture, thus producing various types of waves that travel outward in all directions from their source.
- Most earthquakes take place in well-defined zones at transform, divergent, and convergent plate boundaries.
- An earthquake's epicenter is found by analyzing earthquake waves at no fewer than three seismic stations.
- Intensity is a qualitative assessment of the damage done by an earthquake. The Richter Magnitude Scale and Moment Magnitude Scale are used to express the amount of energy released during an earthquake.
- Great hazards are associated with earthquakes, such as ground shaking, fire, tsunami, and ground failure.
- Efforts by scientists to make accurate, short-term earthquake predictions have thus far met with only limited success.
- Geologists use seismic waves to determine Earth's internal structure.
- Earth has a central core overlain by a thick mantle and a thin outer layer of crust.
- Earth possesses considerable internal heat that continuously escapes at the surface.

woman and her child stand amid the rubble of their home in Bam, Iran, n January 5, 2004. An estimated 43,000 people died in this, the most ecent major earthquake to hit Iran. Source: Morteza Nikoubaz/Reuters/Corbis

Introduction

At 5:27 A.M., on December 26, 2003, violent shaking from an earthquake awakened hundreds of thousands of people in the Bam area of southeastern Iran. When the earthquake was over, an estimated 43,000 people were dead, at least 30,000 were injured, and approximately 75,000 survivors were left homeless. The amount of destruction this 6.6-magnitude earthquake caused is staggering. At least 85% of the structures in the Bam area were destroyed or damaged. Collapsed buildings were everywhere, streets were strewn with rubble, and all communications were knocked out. All in all, this was a disaster of epic proportions. Yet it was not the first, nor will it be the last, major devastating earthquake in this region or other parts of the world.

As one of nature's most frightening and destructive phenomena, earthquakes have always aroused a sense of fear and have been the subject of myths and legends. What makes an earthquake so frightening is that when it begins, there is no way to tell how long it will last or how violent it will be. About 13 million people have died in earthquakes during the past 4000 years, with about 2.7 million of these deaths occurring during the last century (Table 8.1).

Geologists define an **earthquake** as the shaking or trembling caused by the sudden release of energy, usually as a result of faulting, which involves the displacement of rocks along fractures (we discuss the different types of faults in Chapter 10). After an earthquake, continuing adjustments along a fault may generate a series of earthquakes known as *aftershocks*. Most aftershocks are smaller than the main shock, but they can still cause considerable damage to already weakened structures, as happened in the 2003 Iran earthquake.

Although the geologic definition of an earthquake is accurate, it is not nearly as imaginative or colorful as the explanations held in the past. Many cultures attributed the cause of earthquakes to movements of some kind of animal on which Earth rested. In Japan it was a giant catfish, in Mongolia a giant frog, in China an ox, in South America a whale, and to the Algonquin of North America an immense tortoise. A legend from Mexico holds that earthquakes occur when the devil, El Diablo, rips open the crust so he and his friends can reach the surface.

If earthquakes are not the result of animal movement or the devil ripping open the crust, what does cause earthquakes? Geologists know that most earthquakes result from energy released along plate boundaries, and as such, earthquakes are a manifestation of Earth's dynamic nature and the fact that Earth is an internally active planet.

Why should you study earthquakes? The obvious answer is because they are destructive and cause many deaths and injuries to people living in earthquake-prone areas. Earthquakes also affect the economies of many countries in terms of cleanup costs, lost jobs, and lost business revenues. From a purely personal standpoint, you someday may be caught in an earthquake. Even if you don't plan to live in an area subject to earthquakes, you probably will someday travel where there is the threat of earthquakes, and you should know what to do if you experience one.

In this chapter, you will learn what you can do to make your home more earthquake resistant, precautions to take if you live where earthquakes are common, and what to do during and after an earthquake to minimize the chances of serious injury to yourself or even death!

Table 8.1

Some Significant Earthquakes

Year	Location	Magnitude (estimated before 1935)	Deaths (estimated)
1556	China (Shanxi Province)	8.0	1,000,000
1755	Portugal (Lisbon)	8.6	70,000
1906	USA (San Francisco, California)	8.3	3000
1923	Japan (Tokyo)	8.3	143,000
1976	China (Tangshan)	8.0	242,000
1985	Mexico (Mexico City)	8.1	9500
1988	Armenia	7.0	25,000
1990	Iran	7.3	40,000
1993	India	6.4	30,000
1995	Japan (Kobe)	7.2	5000+
1998	Afghanistan	6.1	5000+
1999	Turkey	7.4	17,000
2001	India	7.9	14,000+
2003	Iran	6.6	43,000
2004	Indonesia	9.0	>156,000

WHAT IS THE ELASTIC REBOUND THEORY?

Based on studies conducted after the 1906 San Francisco earthquake, H. F. Reid of The Johns Hopkins University proposed the **elastic rebound theory** to explain how energy is released during earthquakes. Reid studied three sets of measurements taken across a portion of the San Andreas fault that had broken during the 1906 earthquake. The measurements revealed that points on opposite sides of the fault had moved 3.2 m during the 50-year period prior to breakage in 1906, with the west side moving northward (■ Figure 8.1).

According to Reid, rocks on opposite sides of the San Andreas fault had been storing energy and bending slightly for at least 50 years before the 1906 earthquake. Any straight line such as a fence or road that crossed the San Andreas fault was gradually bent because rocks on one side of the fault moved relative to rocks on the other side (Figure 8.1). Eventually, the strength of the rocks was exceeded, the rocks on opposite sides of the fault rebounded or "snapped back" to their former undeformed shape, and the energy stored was released as earthquake waves radiating outward from the break (Figure 8.1). Additional field and laboratory studies conducted by Reid and others have confirmed that elastic rebound is the mechanism by which energy is released during earthquakes.

The energy stored in rocks undergoing elastic deformation is analogous to the energy stored in a tightly wound watch spring. The tighter the spring is wound, the more energy is stored, so more energy is available for release. If the spring is wound so tightly that it breaks, then the stored energy is released as the spring rapidly unwinds and partially regains its original shape. Perhaps an even more meaningful analogy is simply bending a long, straight stick over one's knee. As the stick bends, it deforms and eventually reaches the point at which it breaks. When this happens, the two pieces of the original stick snap back into their original straight position. Likewise, rocks subjected to intense forces bend until they break and then return to their original position, releasing energy in the process.

WHAT IS SEISMOLOGY?

Seismology, the study of earthquakes, emerged as a true science during the 1880s with the development of **seismographs**, instruments that detect, record, and measure the vibrations produced by an earthquake (■ Figure 8.2). The record made by a seismograph is called a *seismogram*. Although most seismographs

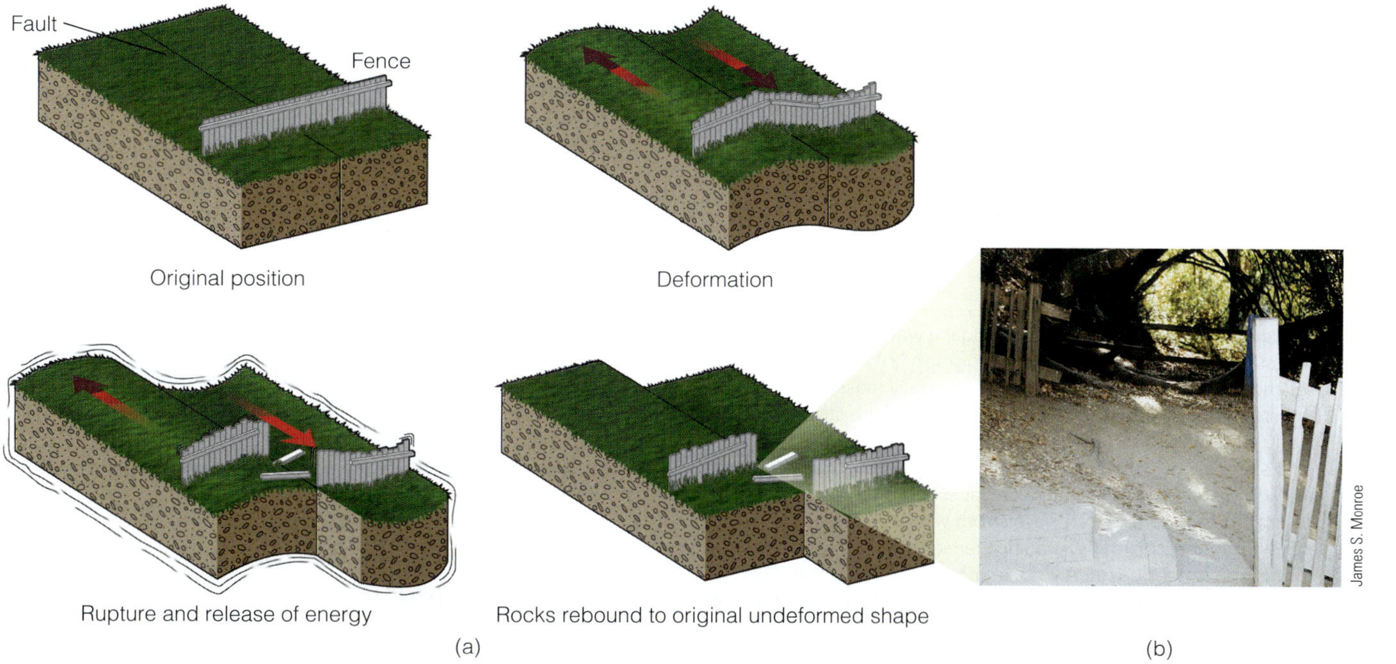

■ **Figure 8.1**

(a) According to the elastic rebound theory, when rocks are deformed, they store energy and bend. When the internal strength of the rocks is exceeded, they fracture, releasing the energy as they rebound to their former undeformed shape. This sudden release of energy causes an earthquake. (b) During the 1906 San Francisco earthquake, this fence in Marin County was displaced nearly 5 m.

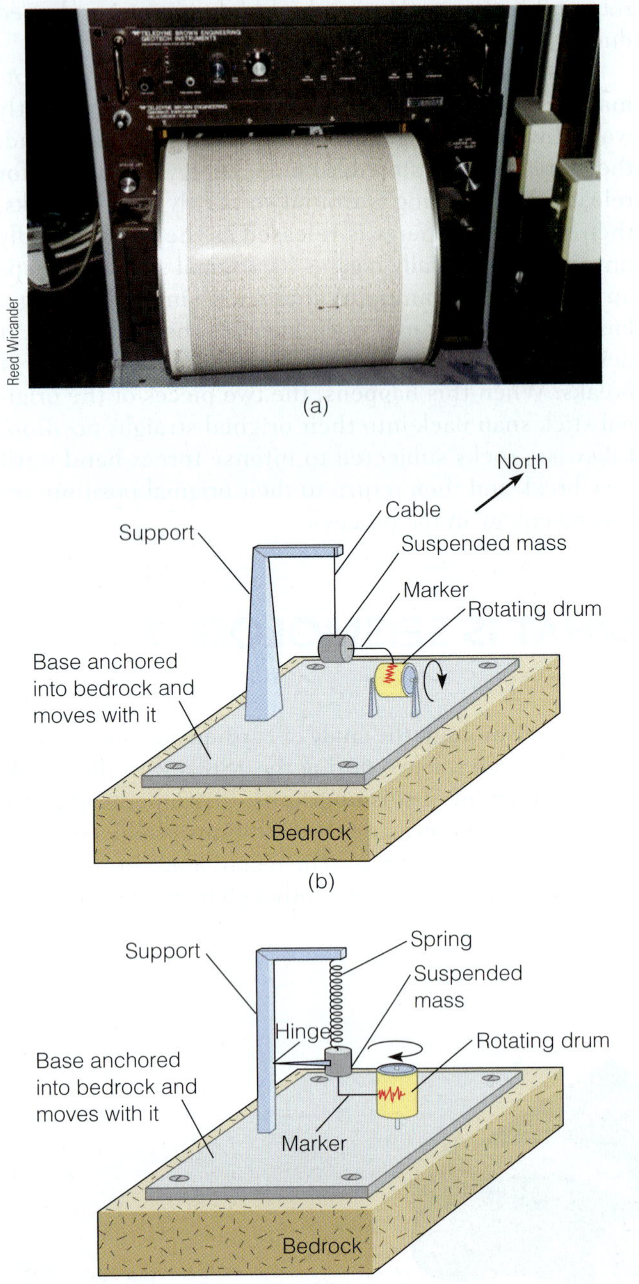

■ **Figure 8.2**

Modern seismographs record earthquake waves electronically. (a) Earthquakes recorded by a seismograph. (b) A horizontal-motion seismograph. Because of its inertia, the heavy mass that contains the marker remains stationary while the rest of the structure moves along with the ground during an earthquake. As long as the length of the arm is not parallel to the direction of ground movement, the marker will record the earthquake waves on the rotating drum. This seismograph would record waves from west or east, but to record waves from the north or south another seismograph at right angles to this one is needed. (c) A vertical-motion seismograph. This seismograph operates on the same principle as a horizontal-motion instrument and records vertical ground movement.

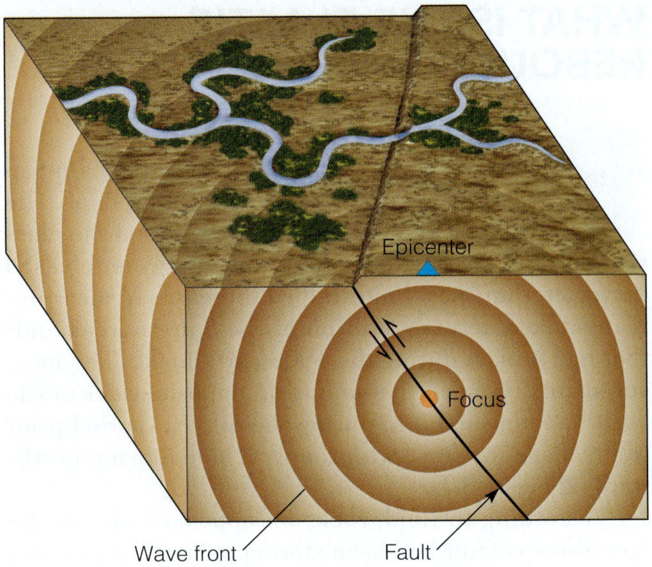

GeologyNow ■ **Active Figure 8.3**

The focus of an earthquake is the location where rupture begins and energy is released. The place on the surface vertically above the focus is the epicenter. Seismic wave fronts move out in all directions from their source, the focus of an earthquake.

today have electronic sensors, computer printouts have largely replaced the strip-chart seismograms of earlier seismographs.

When an earthquake occurs, energy in the form of *seismic waves* radiates out from the point of release (■ Figure 8.3). These waves are somewhat analogous to the ripples that move out concentrically from the point where a stone is thrown into a pond. Unlike waves on a pond, however, seismic waves move outward in all directions from their source.

Earthquakes take place because rocks are capable of storing energy but their strength is limited, so if enough force is present, they rupture and thus release their stored energy. In other words, most earthquakes result when movement occurs along fractures (faults), most of which are related to plate movements. Once a fracture begins, it moves along the fault at several kilometers per second for as long as conditions for failure exist. The longer the fracture along which movement occurs, the more time it takes for the stored energy to be released, and thus the longer the ground will shake. During some very large earthquakes, the ground might shake for 3 minutes, a seemingly brief time but interminable if you are experiencing the earthquake firsthand!

The Focus and Epicenter of an Earthquake

The point within Earth where fracturing begins—that is, the point at which energy is first released—is an earthquake's **focus,** or *hypocenter.* What we usually hear in news reports, however, is the location of the **epicenter,**

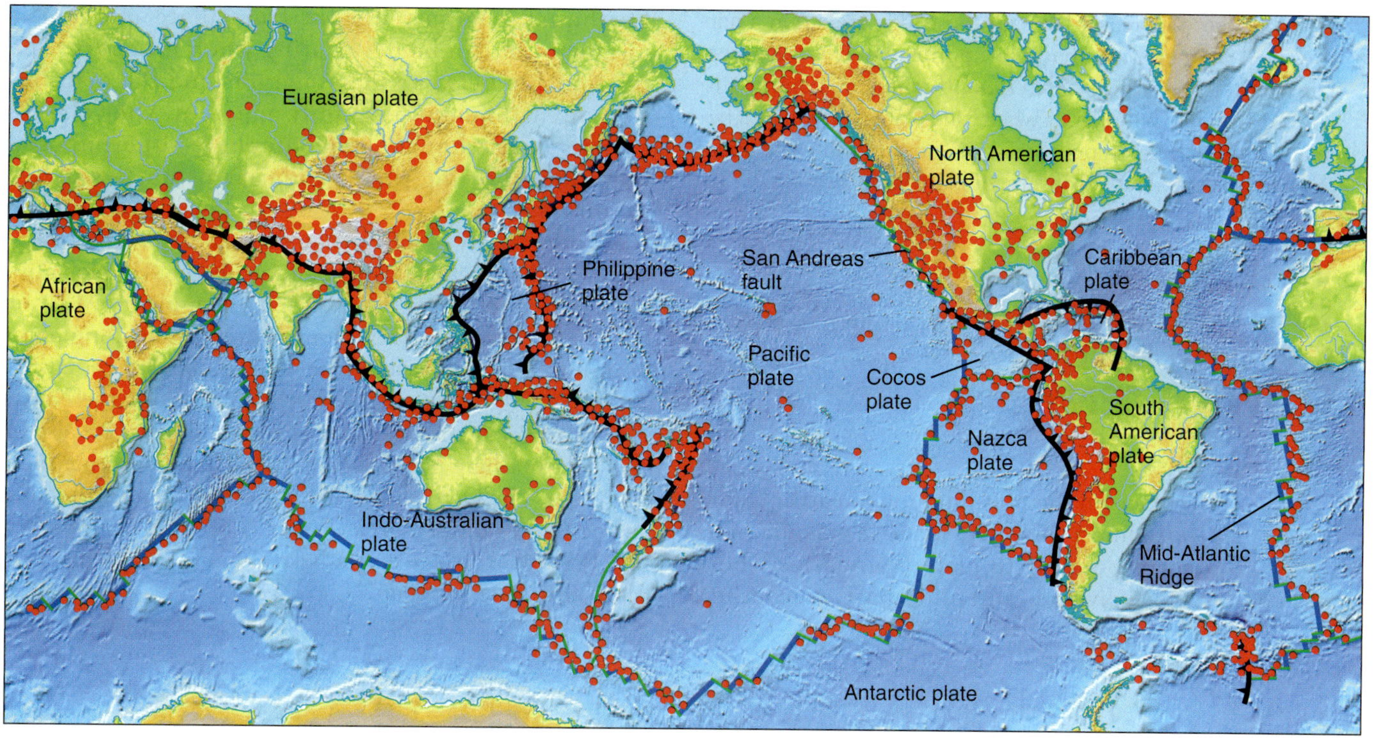

Figure 8.4

The relationship between earthquake epicenters and plate boundaries. Approximately 80% of earthquakes occur within the circum-Pacific belt, 15% within the Mediterranean–Asiatic belt, and the remaining 5% within plate interiors or along oceanic spreading ridges. Each dot represents a single earthquake epicenter. Source: Data from National Oceanic and Atmospheric Administration.

the point on Earth's surface directly above the focus (Figure 8.3). For instance, according to a report by the U.S. Geological Survey, the August 1999 earthquake in Turkey had an epicenter about 11 km southeast of the city of Izmit, and its focal depth (the depth below Earth's surface of an earthquake's focus) was about 17 km.

Seismologists recognize three categories of earthquakes based on focal depth. *Shallow-focus* earthquakes have focal depths of less than 70 km from the surface, whereas those with foci between 70 and 300 km are *intermediate-focus*, and the foci of those characterized as *deep-focus* are more than 300 km deep. However, earthquakes are not evenly distributed among these three categories. Approximately 90% of all earthquake foci are at depths of less than 100 km, whereas only about 3% of all earthquakes are deep. Shallow-focus earthquakes are, with few exceptions, the most destructive.

An interesting relationship exists between earthquake foci and plate boundaries. Earthquakes generated along divergent or transform plate boundaries are invariably shallow focus, while many shallow- and nearly all intermediate- and deep-focus earthquakes occur along convergent margins (Figure 8.4). Furthermore, a pattern emerges when the focal depths of earthquakes near island arcs and their adjacent ocean trenches are plotted. Notice in Figure 8.5 that

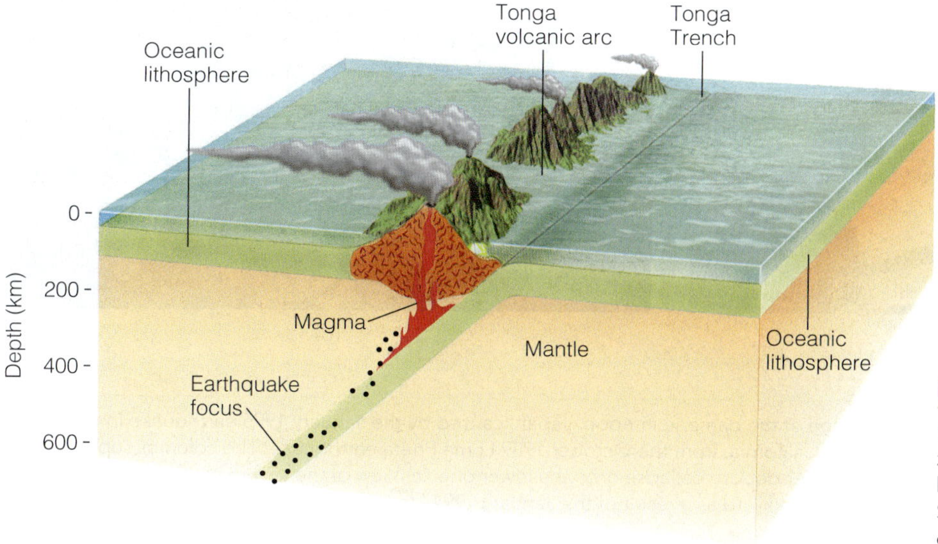

Figure 8.5

Focal depth increases in a well-defined zone that dips approximately 45 degrees beneath the Tonga volcanic arc in the South Pacific. Dipping seismic zones are called Benioff zones.

the focal depth increases beneath the Tonga Trench in a narrow, well-defined zone that dips approximately 45 degrees. Dipping seismic zones, called *Benioff zones,* are common to convergent plate boundaries where one plate is subducted beneath another. Such dipping seismic zones indicate the angle of plate descent along a convergent plate boundary.

WHERE DO EARTHQUAKES OCCUR, AND HOW OFTEN?

No place on Earth is immune to earthquakes, but almost 95% take place in seismic belts corresponding to plate boundaries where plates converge, diverge, and slide past each other. Earthquake activity distant from plate margins is minimal but can be devastating when it occurs. The relationship between plate margins and the distribution of earthquakes is readily apparent when the locations of earthquake epicenters are superimposed on a map showing the boundaries of Earth's plates (Figure 8.4).

The majority of all earthquakes (approximately 80%) occur in the *circum-Pacific belt,* a zone of seismic activity nearly encircling the Pacific Ocean basin. Most of these earthquakes result from convergence along plate margins, as in the case of the 1995 Kobe, Japan, earthquake (■ Figure 8.6a). The earthquakes along the North American Pacific Coast, especially in California, are also in this belt, but here plates slide past one another rather than converge. The October 17, 1989, Loma Prieta earthquake in the San Francisco area (Figure 8.6b) and the January 17, 1994, Northridge earthquake (Figure 8.6c) happened along this plate boundary.

Bunyo Ishikawa/Corbis Sygma

Dennis Fox

Roger Ressmeyer/Corbis

(a) (b) (c)

■ **Figure 8.6**

Earthquake damage in the circum-Pacific belt. (a) Some of the damage in Kobe, Japan, caused by the January 1995 earthquake in which more than 5000 people died. (b) Damage in Oakland, California, from the October 1989 Loma Prieta earthquake. The columns supporting the upper deck of Interstate 880 failed, causing the upper deck to collapse onto the lower one. (c) View of the severe exterior damage to the Northridge Meadows apartments in which 16 people were killed as a result of the January 1994 Northridge, California, earthquake.

South Carolina Library, University of South Carolina

■ **Figure 8.7**

Damage done to Charleston, South Carolina, by the earthquake of August 31, 1886. This earthquake is the largest reported in the eastern United States.

ing something so large without its internal parts shifting slightly is impossible. Similarly, plates are not likely to move without some internal stresses that occasionally cause earthquakes. Interestingly, many intraplate earthquakes are associated with very ancient and presumed inactive faults that are reactivated at various intervals.

More than 150,000 earthquakes strong enough to be felt are recorded every year by the worldwide network of seismograph stations. In addition, an estimated 900,000 earthquakes are recorded annually by seismographs but are too small to be individually cataloged. These small earthquakes result from the energy released as continual adjustments take place between the various plates.

The second major seismic belt, accounting for 15% of all earthquakes, is the *Mediterranean–Asiatic belt*. This belt extends westward from Indonesia through the Himalayas, across Iran and Turkey, and westward through the Mediterranean region of Europe. The devastating 1990 and 2003 earthquakes in Iran that killed 40,000 and 43,000 people, respectively, the 1999 Turkey earthquake that killed about 17,000, and the 2001 India earthquake that killed more than 14,000 people are recent examples of the destructive earthquakes that strike this region (Table 8.1).

The remaining 5% of earthquakes occur mostly in the interiors of plates and along oceanic spreading-ridge systems. Most of these earthquakes are not strong, although several major intraplate earthquakes are worthy of mention. For example, the 1811 and 1812 earthquakes near New Madrid, Missouri, killed approximately 20 people and nearly destroyed the town. So strong were these earthquakes that they were felt from the Rocky Mountains to the Atlantic Ocean and from the Canadian border to the Gulf of Mexico. Another major intraplate earthquake struck Charleston, South Carolina, on August 31, 1886, killing 60 people and causing $23 million in property damage (■ Figure 8.7).

The cause of intraplate earthquakes is not well understood, but geologists think they arise from localized stresses caused by the compression that most plates experience along their margins. A useful analogy is moving a house. Regardless of how careful the movers are, mov-

GeologyNow Log into GeologyNow and select this chapter to work through a **Geology Interactive** activity on "Earthquakes in Space and Time" (click Earthquakes and Tsunami→Earthquakes in Space and Time).

WHAT ARE SEISMIC WAVES?

Many people have experienced an earthquake but are probably unaware that the shaking they feel and the damage to structures are caused by the arrival of *seismic waves*, a general term encompassing all waves generated by an earthquake. When movement on a fault takes place, energy is released in the form of two kinds of waves that radiate outward in all directions from an earthquake's focus. *Body waves*, so called because they travel through the solid body of Earth, are somewhat like sound waves, and *surface waves*, which travel along the ground surface, are analogous to undulations or waves on water surfaces.

Body Waves

An earthquake generates two types of body waves: P-waves and S-waves (■ Figure 8.8). **P-waves** or *primary waves* are the fastest seismic waves and can travel through solids, liquids, and gases. P-waves are compressional, or push-pull, waves and are similar to sound waves in that they move

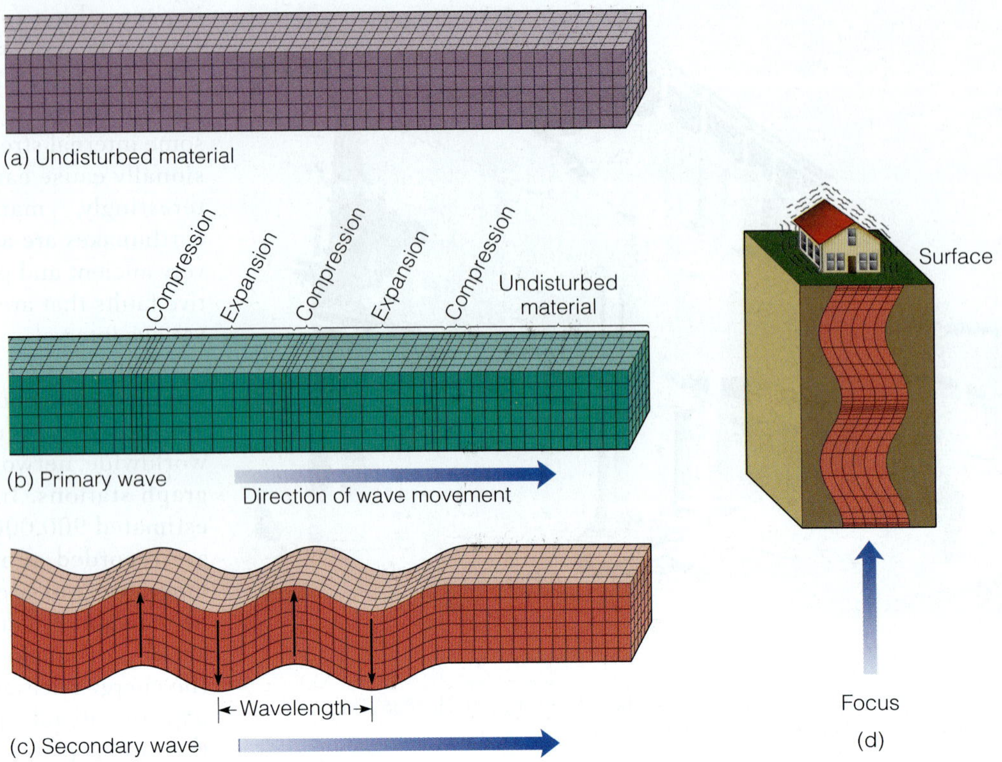

Geology Now ■ **Active Figure 8.8**

(a) Undisturbed material for reference. (b) and (c) show how body waves travel through Earth. (b) Primary waves (P-waves) compress and expand material in the same direction they travel. (c) Secondary waves (S-waves) move material perpendicular to the direction of wave movement. (d) P- and S-waves and their effect on a surface structure. Source: (a, b, c): From *Nuclear Explosions and Earthquakes: The Parted Veil,* by Bruce A. Bolt. Copyright © 1976 by W. H. Freeman and Company. Used with permission.

material forward and backward along a line in the same direction that the waves themselves are moving (Figure 8.8b). Thus, the material through which P-waves travel is expanded and compressed as the waves move through it and returns to its original size and shape after the wave passes by.

S-waves or *secondary waves* are somewhat slower than P-waves and can travel only through solids. S-waves are *shear waves* because they move the material perpendicular to the direction of travel, thereby producing shear stresses in the material they move through (Figure 8.8c). Because liquids (as well as gases) are not rigid, they have no shear strength and S-waves cannot be transmitted through them.

The velocities of P- and S-waves are determined by the density and elasticity of the materials through which they travel. For example, seismic waves travel more slowly through rocks of greater density but more rapidly through rocks with greater elasticity. *Elasticity* is a property of solids, such as rocks, and means that once they have been deformed by an applied force, they return to their original shape when the force is no longer present. Because P-wave velocity is greater than S-wave velocity in all materials, P-waves always arrive at seismic stations first.

Surface Waves

Surface waves travel along the surface of the ground, or just below it, and are slower than body waves. Unlike the sharp jolting and shaking that body waves cause, surface waves generally produce a rolling or swaying motion, much like the experience of being on a boat.

Several types of surface waves are recognized. The two most important are **Rayleigh waves (R-waves)** and **Love waves (L-waves)**, named after the British scientists who discovered them, Lord Rayleigh and A. E. H. Love. Rayleigh waves are generally the slower of the two and behave like water waves in that they move forward while the individual particles of material move in an elliptical path within a vertical plane oriented in the direction of wave movement (■ Figure 8.9b).

The motion of a Love wave is similar to that of an S-wave, but the individual particles of the material move only back and forth in a horizontal plane perpendicular to the direction of wave travel (Figure 8.9c). This type of lateral motion can be particularly damaging to building foundations.

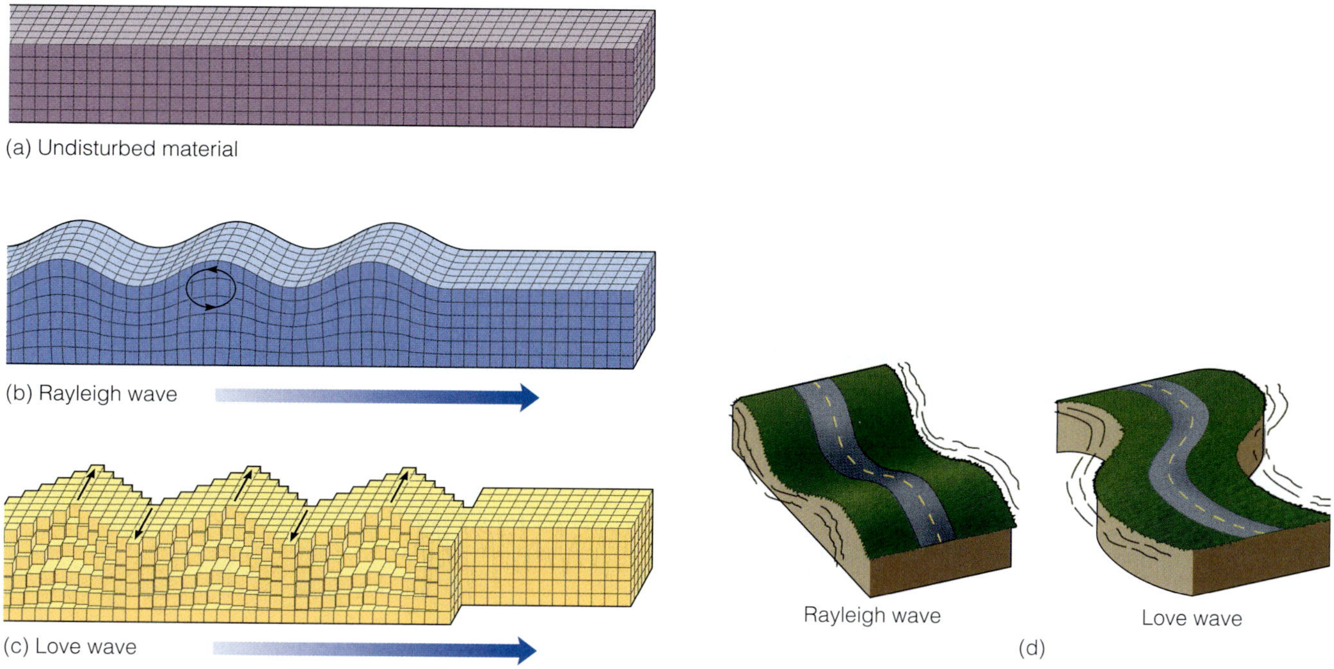

GeologyNow ■ **Active Figure 8.9**

Surface waves. (a) Undisturbed material for reference. (b) and (c) show how surface waves travel along Earth's surface or just below it. (b) Rayleigh waves (R-waves) move material in an elliptical path in a plane oriented parallel to the direction of wave movement. (c) Love waves (L-waves) move material back and forth in a horizontal plane perpendicular to the direction of wave movement. (d) The arrival of R- and L-waves causes the surface to undulate and shake from side to side. Source: (a, b, c) From *Nuclear Explosions and Earthquakes: The Parted Veil,* by Bruce A. Bolt. Copyright © 1976 by W. H. Freeman and Company. Used with permission.

HOW IS AN EARTHQUAKE'S EPICENTER LOCATED?

We mentioned that news articles commonly report an earthquake's epicenter, but just how is the location of an epicenter determined? Once again, geologists rely on the study of seismic waves. We know that P-waves travel faster than S-waves, nearly twice as fast in all substances, so P-waves arrive at a seismograph station first followed some time later by S-waves. Both P- and S-waves travel directly from the focus to the seismograph station through Earth's interior, but L- and R-waves arrive last because they are the slowest, and they also travel the longest route along the surface (■ Figure 8.10a, b). L- and R-waves cause much of the damage during earthquakes, but only P- and S-waves need concern us here because they are the ones important in finding an epicenter.

Seismologists, geologists who study seismology, have accumulated a tremendous amount of data over the years and now know the average speeds of P- and S-waves for any specific distance from their source. These P- and S-wave travel times are published in *time–distance graphs,* which illustrate that the difference between the arrival times of the two waves is a function of the distance between a seismograph and an earthquake's focus (Figure 8.10c). That is, the farther the waves travel, the greater the *P–S time interval* or simply the time difference between the arrivals of P- and S-waves (Figure 8.10a, c).

If the P–S time intervals are known from at least three seismograph stations, then the epicenter of any earthquake can be determined (■ Figure 8.11). Here is how it works. Subtracting the arrival time of the first P-wave from the arrival time of the first S-wave gives the P–S time interval for each seismic station. Each of these time intervals is plotted on a time–distance graph, and a line is drawn straight down to the distance axis of the graph, thus giving the distance from the focus to each seismic station (Figure 8.10c). Next, a circle whose radius equals the distance shown on the time–distance graph from each of the seismic stations is drawn on a map (Figure 8.11). The intersection of the three circles is the location of the earthquake's epicenter. It should be obvious from Figure 8.11 that P–S time intervals from at least three seismic stations are needed. If only one were used, the epicenter could be at any location on the circle drawn around that station, and two stations would give two possible locations for the epicenter.

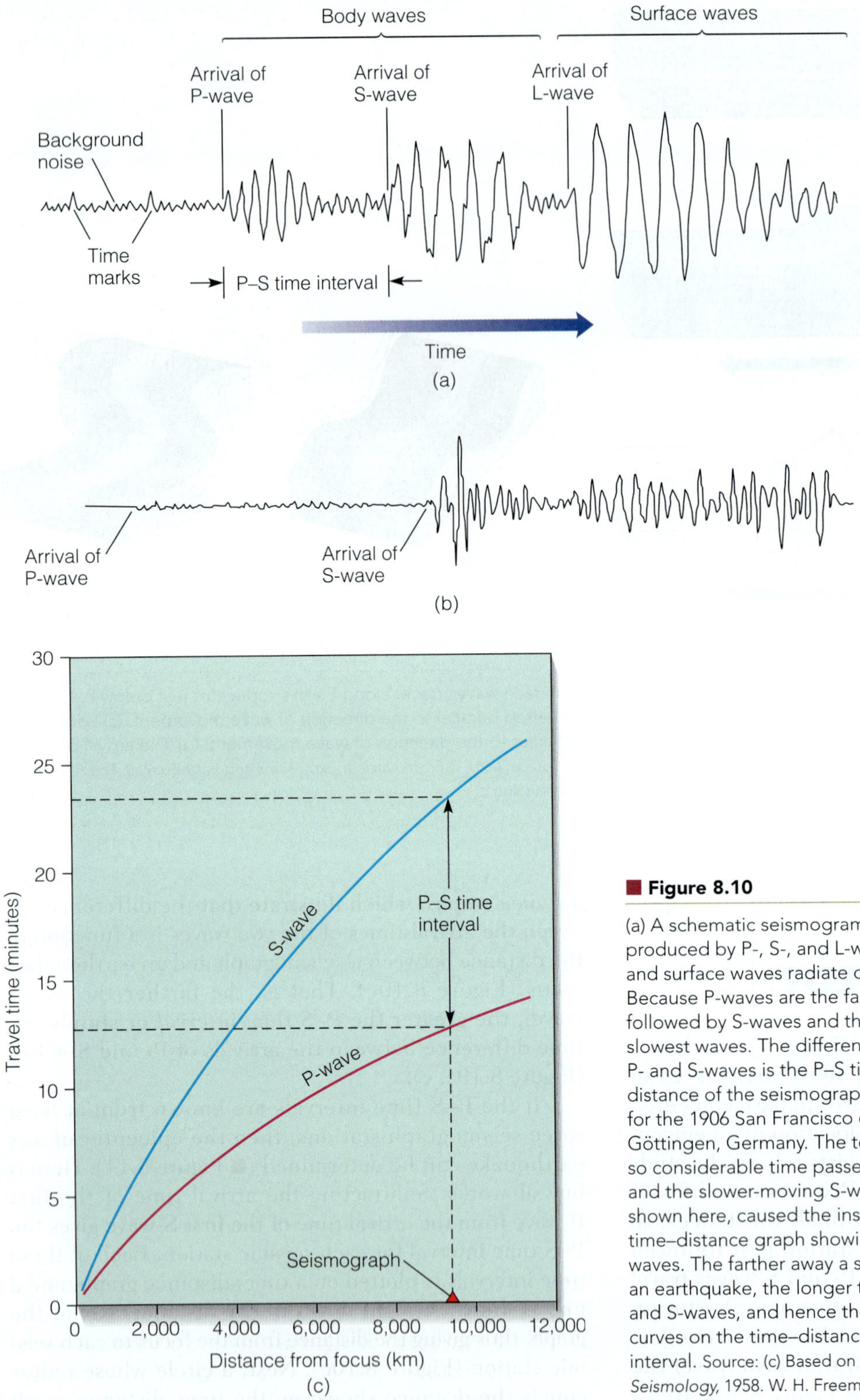

■ **Figure 8.10**

(a) A schematic seismogram showing the arrival order and patterns produced by P-, S-, and L-waves. When an earthquake occurs, body and surface waves radiate out from the focus at the same time. Because P-waves are the fastest, they arrive at a seismograph first, followed by S-waves and then by surface waves, which are the slowest waves. The difference between the arrival times of the P- and S-waves is the P–S time interval; it is a function of the distance of the seismograph station from the focus. (b) Seismogram for the 1906 San Francisco earthquake, recorded 14,668 km away in Göttingen, Germany. The total record represents about 26 minutes, so considerable time passed between the arrival of the P-waves and the slower-moving S-waves. The arrival of surface waves, not shown here, caused the instrument to go off the scale. (c) A time–distance graph showing the average travel times for P- and S-waves. The farther away a seismograph station is from the focus of an earthquake, the longer the interval between the arrivals of the P- and S-waves, and hence the greater the distance between the curves on the time–distance graph as indicated by the P–S time interval. Source: (c) Based on data from C. F. Richter, *Elementary Seismology*, 1958. W. H. Freeman and Company.

Determining the focal depth of an earthquake is much more difficult and considerably less precise than finding its epicenter. The focal depth is usually found by making computations based on several assumptions, comparing the results with those obtained at other seismic stations, and then recalculating and approximating the depth as closely as possible. Even so, the results are not highly accurate, but they do tell us that most earthquakes, probably 75%, have foci no deeper than 10 to 15 km and that a few are as deep as 680 km.

GeologyNow Log into GeologyNow and select this chapter to work through a **Geology Interactive** activity on "P and S Waves" (click Earth's Layers→P and S Waves).

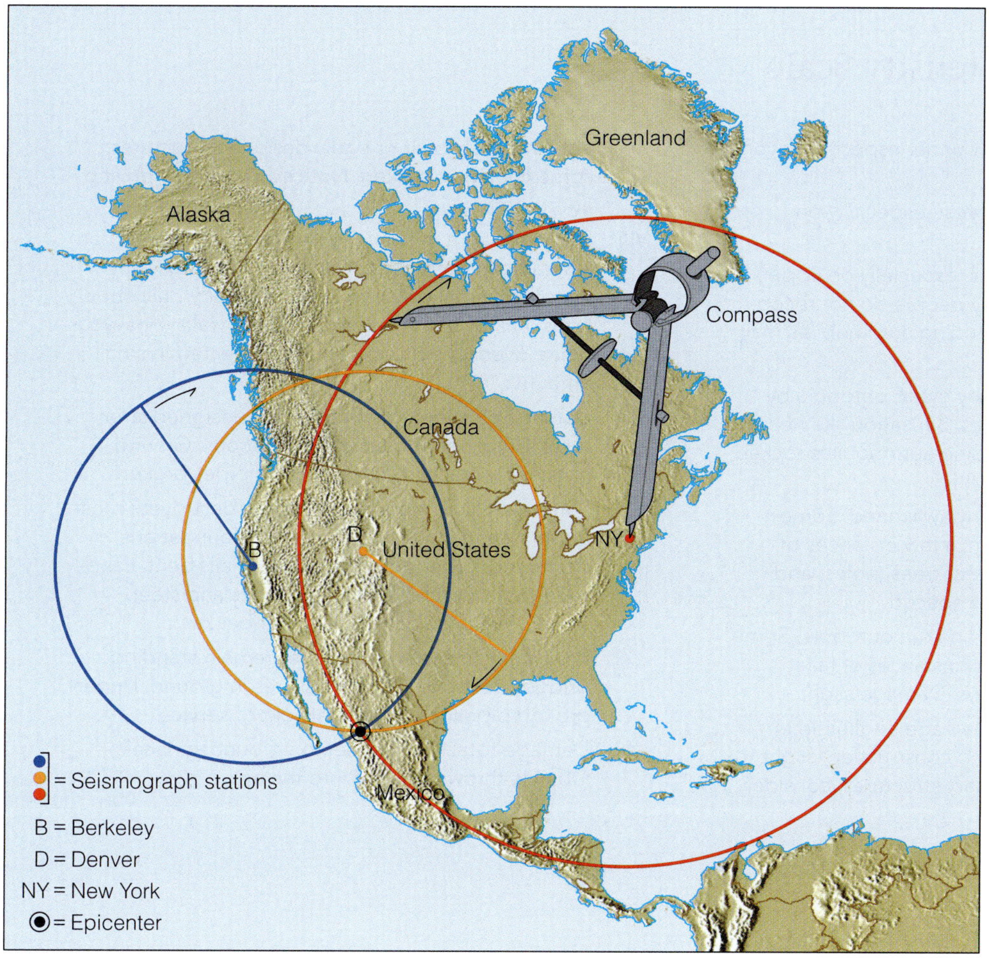

Geology Now ■ **Active Figure 8.11**

Three seismograph stations are needed to locate the epicenter of an earthquake. The P–S time interval is plotted on a time–distance graph for each seismograph station to determine the distance that station is from the epicenter. A circle with that radius is drawn from each station, and the intersection of the three circles is the epicenter of the earthquake.

HOW ARE THE SIZE AND STRENGTH OF AN EARTHQUAKE MEASURED?

Following any earthquake that causes extensive damage, fatalities, and injuries, graphic reports of the quake's violence and human suffering are common. Headlines tell us that thousands died, many more were injured or left homeless, and property damage is in the millions and possibly billions of dollars. Few other natural processes account for such tragic consequences. Although descriptions of fatalities and damage give some indication of the size of an earthquake, geologists are interested in more reliable methods of determining an earthquake's size and strength.

Two measures of an earthquake's strength are commonly used. One is *intensity*, a qualitative assessment of the kinds of damage done by an earthquake. The other, *magnitude*, is a quantitative measure of the amount of energy released by an earthquake. Each method provides important information that can be used to prepare for future earthquakes.

Intensity

Intensity is a subjective measure of the kind of damage done by an earthquake as well as people's reaction to it. Since the mid-19th century, geologists have used intensity as a rough approximation of the size and strength of an earthquake. The most common intensity scale used in the United States is the **Modified Mercalli Intensity Scale,** which has values ranging from I to XII (Table 8.2).

Intensity maps can be constructed for regions hit by earthquakes by dividing the affected region into various intensity zones. The intensity value given for each zone is the maximum intensity that the earthquake produced for that zone. Even though intensity maps are not precise because of the subjective nature of the measurements, they do provide geologists with a rough approximation of the location of the earthquake, the kind and extent of the damage done, and the effects of local geology on different types of building construction (■ Figure 8.12). Because intensity is a measure of the kind of damage done by an earthquake, insurance companies still classify earthquakes on the basis of intensity.

Generally, a large earthquake will produce higher intensity values than a small earthquake, but many other factors besides the amount of energy released by an earthquake also affect its intensity. These include distance from the epicenter, focal depth of the earthquake, population density and local geology of the area, type of building construction employed, and duration of shaking.

Table 8.2

Modified Mercalli Intensity Scale

I	Not felt except by a very few under especially favorable circumstances.
II	Felt by only a few people at rest, especially on upper floors of buildings.
III	Felt quite noticeably indoors, especially on upper floors of buildings, but many people do not recognize it as an earthquake. Standing automobiles may rock slightly.
IV	During the day felt indoors by many, outdoors by few. At night some awakened. Sensation like heavy truck striking building, standing automobiles rocked noticeably.
V	Felt by nearly everyone, many awakened. Some dishes, windows, etc. broken, a few instances of cracked plaster. Disturbance of trees, poles, and other tall objects sometimes noticed.
VI	Felt by all, many frightened and run outdoors. Some heavy furniture moved, a few instances of fallen plaster or damaged chimneys. Damage slight.
VII	Everybody runs outdoors. Damage negligible in buildings of good design and construction; slight to moderate in well-built ordinary structures; considerable in poorly built or badly designed structures; some chimneys broken. Noticed by people driving automobiles.
VIII	Damage slight in specially designed structures; considerable in normally constructed buildings with possible partial collapse; great in poorly built structures. Fall of chimneys, monuments, walls. Heavy furniture overturned. Sand and mud ejected in small amounts.
IX	Damage considerable in specially designed structures. Buildings shifted off foundations. Ground noticeably cracked. Underground pipes broken.
X	Some well-built wooden structures destroyed; most masonry and frame structures with foundations destroyed; ground badly cracked. Rails bent. Landslides considerable from river banks and steep slopes. Water splashed over river banks.
XI	Few, if any (masonry) structures remain standing. Bridges destroyed. Broad fissures in ground. Underground pipelines completely out of service.
XII	Damage total. Waves seen on ground surfaces. Objects thrown upward into the air.

Source: U.S. Geological Survey.

Magnitude

If earthquakes are to be compared quantitatively, we must use a scale that measures the amount of energy released and is independent of intensity. Such a scale was developed in 1935 by Charles F. Richter, a seismologist at the California Institute of Technology. The **Richter Magnitude Scale** measures earthquake **magnitude,** which is the total amount of energy released by an earthquake at its source. It is an open-ended scale with values beginning at 1. The largest magnitude recorded has been 8.6, and though values greater than 9 are theoretically possible, they are highly improbable because rocks are not able to store the energy necessary to generate earthquakes of this magnitude.

Scientists determine the magnitude of an earthquake by measuring the amplitude of the largest seismic wave as recorded on a seismogram (■ Figure 8.13). To avoid large numbers, Richter used a conventional base-10 logarithmic scale to convert the amplitude of the largest recorded seismic wave to a numerical magnitude value (Figure 8.13). Therefore, each whole-number increase in magnitude represents a 10-fold increase in wave amplitude. For example, the amplitude of the largest seismic wave for an earthquake of magnitude 6 is 10 times the amplitude produced by an earthquake of magnitude 5, 100 times as large as a magnitude-4 earthquake, and 1000 times that of an earthquake of magnitude 3 ($10 \times 10 \times 10 = 1000$).

A common misconception about the size of earthquakes is that an increase of one unit on the Richter Magnitude Scale—a 7 versus a 6, for instance—means a 10-fold increase in size. It is true that each whole-number increase in magnitude represents a 10-fold increase in the wave amplitude, but each magnitude increase corresponds to a roughly 30-fold increase in the amount of energy released (actually it is 31.5, but 30 is close enough for our purposes). This means that it would take about 30 earthquakes of magnitude 6 to equal the energy released in one earthquake with a magnitude of 7. The 1964 Alaska earthquake with a magnitude of 8.6 released almost 900 times more energy than the 1994 Northridge, California, earthquake of magnitude 6.7! And the Alaska earthquake released more than 27,000 times as much energy as one with a magnitude of 5.6 would have.

We mentioned that more than 900,000 earthquakes are recorded around the world each year. Table 8.3 shows that the vast majority of earthquakes have a Richter magnitude of less than 2.5, and that great earthquakes (those with a magnitude greater than 8.0) occur, on average, only once every five years.

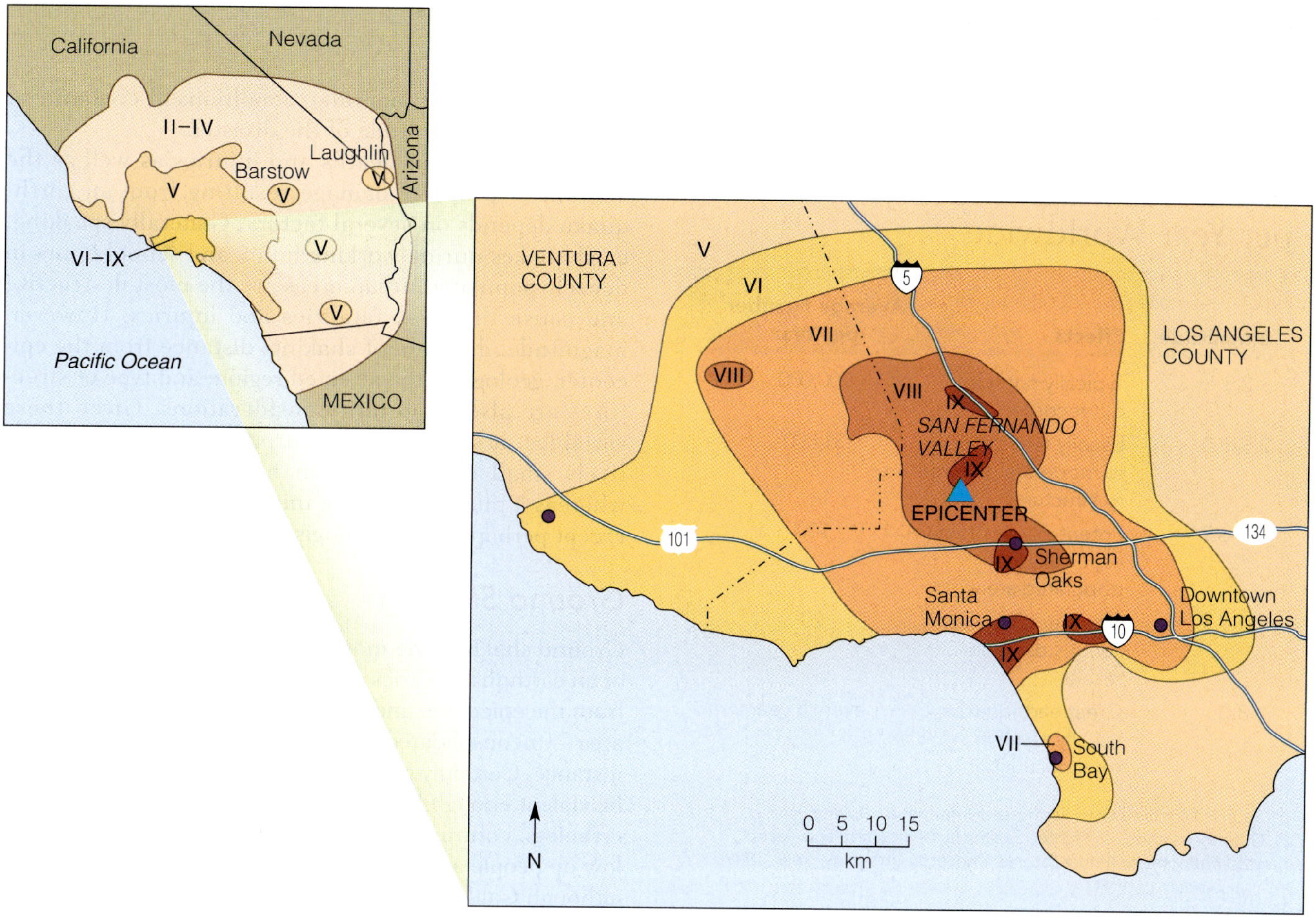

■ **Figure 8.12**

Preliminary Modified Mercalli Intensity map for the 1994 Northridge, California, earthquake, showing the region divided into intensity zones based on the kind of damage done. This earthquake had a magnitude of 6.7.

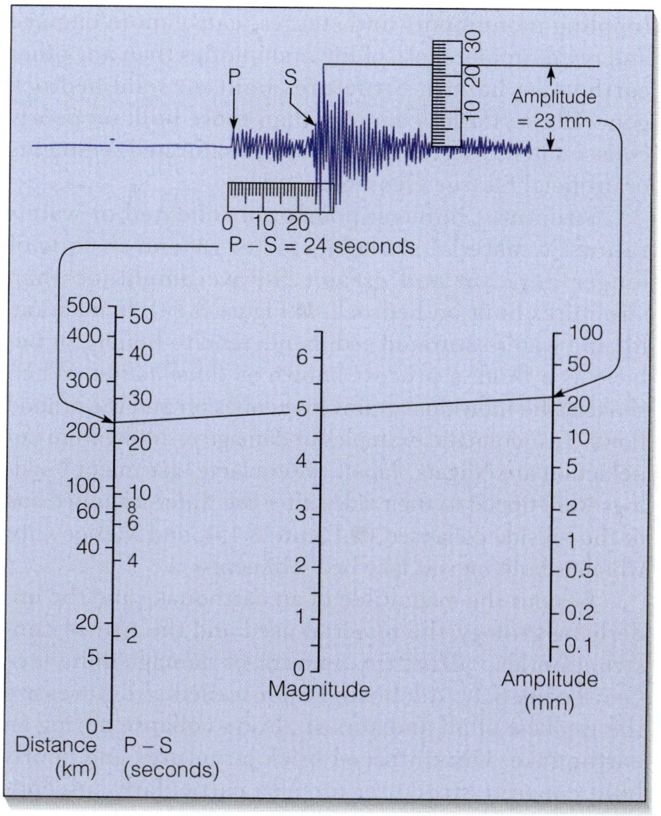

The Richter Magnitude Scale was devised to measure earthquake waves on a particular seismograph and at a specific distance from an earthquake. One of its limitations is that it underestimates the energy of very large earthquakes because it measures the highest peak on a seismogram, which represents only an instant during an earthquake. For large earthquakes, the energy might be released over several minutes and along hundreds of kilometers of a fault. For example, during the 1857 Fort Tejon, California, earthquake, the ground shook for more than 2 minutes and energy was released for 360 km along the fault. Despite the shortcomings, Richter magnitudes still commonly appear in news releases. More recently, seismologists developed a

■ **Figure 8.13**

The Richter Magnitude Scale measures the total amount of energy released by an earthquake at its source. The magnitude is determined by measuring the maximum amplitude of the largest seismic wave and marking it on the right-hand scale. The difference between the arrival times of the P- and S-waves (recorded in seconds) is marked on the left-hand scale. When a line is drawn between the two points, the magnitude of the earthquake is the point at which the line crosses the center scale. Source: From *Earthquakes*, by Bruce A. Bolt. Copyright © 1988 by W. H. Freeman and Company, Used with permission.

Table 8.3

Average Number of Earthquakes of Various Magnitudes per Year Worldwide

Magnitude	Effects	Average Number per Year
<2.5	Typically not felt but recorded	900,000
2.5–6.0	Usually felt; minor to moderate damage to structures	31,000
6.1–6.9	Potentially destructive, especially in populated areas	100
7.0–7.9	Major earthquakes; serious damage results	20
>8.0	Great earthquakes; usually result in total destruction	1 every 5 years

Source: Modified from *Earthquake Information Bulletin,* and B. Gutenberg and C. F. Richter, *Seismicity of the Earth and Associated Phenomena* (Princeton, NJ: Princeton University Press, 1949).

Seismic-Moment Magnitude Scale that considers the area of a fault along which rupture occurred and the amount of movement of rocks adjacent to the fault. Seismologists are confident that they now have a scale with which they can not only compare different size earthquakes but also evaluate the size of earthquakes that occurred before instruments were available to record them.

GeologyNow Log into GeologyNow and select this chapter to work through a **Geology Interactive** activity on "Seismic Case Study Alaska, 1964" (click Earthquakes and Tsunami→Seismic Case Study Alaska, 1964).

WHAT ARE THE DESTRUCTIVE EFFECTS OF EARTHQUAKES?

Certainly, earthquakes are one of nature's most destructive phenomena. Little or no warning precedes earthquakes, and once they begin, little or nothing can be done to minimize their effects, although planning before an earthquake can help. However, earthquake prediction may become a reality in the future (discussed in a later section). The destructive effects of earthquakes include ground shaking, fire, seismic sea waves, and landslides, as well as panic, disruption of vital services, and psychological shock. In some cases, rescue attempts are hampered by inadequate resources or planning, conditions of civil unrest, or simply the magnitude of the disaster.

The number of deaths and injuries as well as the amount of property damage resulting from an earthquake depends on several factors. Generally speaking, earthquakes during working hours and school hours in densely populated urban areas are the most destructive and cause the most fatalities and injuries. However, magnitude, duration of shaking, distance from the epicenter, geology of the affected region, and type of structures are also important considerations. Given these variables, it should not be surprising that a comparatively small earthquake can have disastrous effects, whereas a much larger one might go largely unnoticed, except perhaps by seismologists.

Ground Shaking

Ground shaking, the most obvious and immediate effect of an earthquake, varies depending on magnitude, distance from the epicenter, and type of underlying materials in the area—unconsolidated sediment or fill versus bedrock, for instance. Certainly ground shaking is terrifying, and it may be violent enough for fissures to open in the ground. Nevertheless, contrary to popular myth, fissures do not swallow up people and buildings and then close on them. And although California will no doubt have big earthquakes in the future, rocks cannot store enough energy to displace a landmass as large as California into the Pacific Ocean, as the tabloids sometimes suggest will happen.

The effects of ground shaking, such as collapsing buildings, falling building facades and window glass, and toppling monuments and statues, cause more damage and result in more loss of life and injuries than any other earthquake hazard. Structures built on solid bedrock generally suffer less damage than those built on poorly consolidated material such as water-saturated sediments or artificial fill (see Geo-Focus 8.1).

Structures built on poorly consolidated or water-saturated material are subjected to ground shaking of longer duration and greater S-wave amplitude than structures built on bedrock (■ Figure 8.14). In addition, fill and water-saturated sediments tend to liquefy, or behave as a fluid, a process known as *liquefaction.* When shaken, the individual grains lose cohesion and the ground flows. Two dramatic examples of damage resulting from liquefaction are Niigata, Japan, where large apartment buildings were tipped to their sides after the water-saturated soil of the hillside collapsed (■ Figure 8.15), and Mexico City, which is built on soft lake bed sediments.

Besides the magnitude of an earthquake and the underlying geology, the material used and the type of construction also affect the amount of damage done (see Geo-Focus 8.1). Adobe and mud-walled structures are the weakest of all and almost always collapse during an earthquake. Unreinforced brick structures and poorly built concrete structures are also particularly susceptible to collapse. For example, the 1976 earthquake in

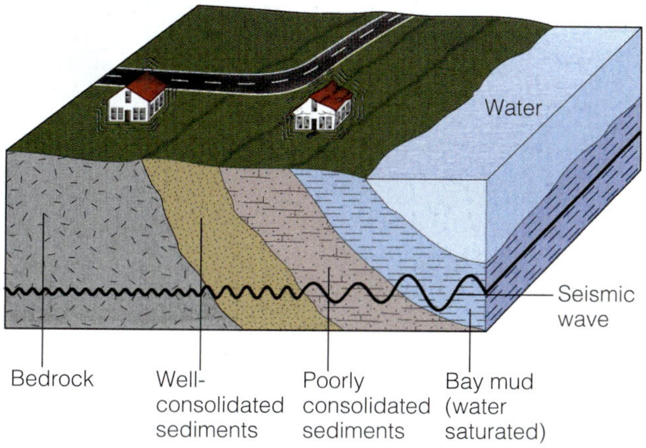

■ **Figure 8.14**

The amplitude and duration of seismic waves generally increase as they pass from bedrock into poorly consolidated or water-saturated material. Thus structures built on weaker material typically suffer greater damage than similar structures built on bedrock.

Tangshan, China, completely leveled the city because hardly any structures were built to resist seismic forces. In fact, most had unreinforced brick walls, which have no flexibility, and consequently they collapsed during the shaking (■ Figure 8.16a).

The 6.4-magnitude earthquake that struck India in 1993 killed about 30,000 people, whereas the 6.7-magnitude Northridge, California, earthquake one year later resulted in only 61 deaths. Why such a difference in the death toll? Both earthquakes occurred in densely populated regions, but in India the brick and stone buildings could not withstand ground shaking; most collapsed and entombed their occupants (Figure 8.16b).

Fire

In many earthquakes, particularly in urban areas, fire is a major hazard. Almost 90% of the damage done in the 1906 San Francisco earthquake was caused by fire. The shaking severed many electrical and gas lines, which touched off flames and started fires all over the city. Because water mains were ruptured by the earthquake, there was no effective way to fight the fires that raged out of control for three days, destroying much of the city.

Eighty-three years later during the 1989 Loma Prieta earthquake, a fire broke out in the Marina district of San Francisco (■ Figure 8.17). This time, however, the fire was contained within a small area because San Francisco had a system of valves throughout its water and gas pipeline system so that lines could be isolated from breaks (see "The San Andreas Fault" on pages 216 and 217).

During the September 1, 1923, earthquake in Japan, fires destroyed 71% of the houses in Tokyo and practically all the houses in Yokohama. In all, 576,262 houses were destroyed by fire, and 143,000 people died, many as a result of fire.

Tsunami: Killer Waves

On December 26, 2004, a magnitude 9.0 earthquake struck 160 km off the west coast of northern Sumatra, Indonesia, generating the deadliest tsunami in history. Within hours, walls of water up to 10.5 m high pounded the coasts of Indonesia, Sri Lanka, India, Thailand, Somalia, Myanmar, Malaysia, and the Maldives, killing more than 156,000 people and causing billions of dollars in damage (■ Figure 8.18).

This earthquake-generated wave is popularly called a "tidal wave" but is more correctly termed a *seismic sea wave* or **tsunami,** a Japanese term meaning "harbor wave." The term *tidal wave* nevertheless persists in popular literature and some news accounts, but these waves are not caused by or related to tides. Indeed, tsunami are destructive sea waves generated when large amounts of energy are rapidly released into a body of water. Many result from submarine earthquakes, but volcanoes at sea or submarine landslides can also cause them. For example, the 1883 eruption of Krakatau between Java and Sumatra generated a large sea wave that killed 36,000 on nearby islands.

National Geophysical Data Center

■ **Figure 8.15**

The effects of ground shaking on water-saturated soil are dramatically illustrated by the collapse of these buildings in Niigata, Japan, during a 1964 earthquake. The buildings were designed to be earthquake resistant and fell over on their sides intact.

GEOFOCUS 8.1

Designing Earthquake-Resistant Structures

One way to reduce property damage, injuries, and loss of life is to design and build structures that are as earthquake resistant as possible. Many things can be done to improve the safety of current structures and of new buildings.

To design earthquake-resistant structures, engineers must understand the dynamics and mechanics of earthquakes, including the type and duration of the ground motion and how rapidly the ground accelerates during an earthquake. An understanding of the area's geology is also important because certain ground materials such as water-saturated sediments or landfill can lose their strength and cohesiveness during an earthquake (Figure 8.14). Finally, engineers must be aware of how different structures behave under different earthquake conditions.

With the level of technology currently available, a well-designed, properly constructed building should be able to withstand small, short-duration earthquakes of less than 5.5 magnitude with little or no damage. In moderate earthquakes (5.5–7.0 magnitude), the damage suffered should not be serious and should be repairable. In a major earthquake of greater than 7.0 magnitude, the building should not collapse, although it may later have to be demolished.

Many factors enter into the design of an earthquake-resistant structure, but the most important is that the building be tied together; that is, the foundation, walls, floors, and roof should all be joined together to create a structure that can withstand both horizontal and vertical shaking (■ Figure 1). Almost all the structural failures resulting from earthquake ground movement occur at weak connections, where the various parts of a structure are not securely tied together. Buildings with open or unsupported first stories are particularly susceptible to damage. Some reinforcement must be done, or collapse is a distinct possibility (Figure 1).

Tall buildings, such as skyscrapers, must be designed so that a certain amount of swaying or flexing can occur, but not so much that they touch neighboring buildings during swaying. If a building is brittle and does not give, it will crack and fail. Besides designed flexibility, engineers must ensure that a building does not vibrate at the same frequency as the ground does during an earthquake. When that happens, the force applied by the seismic waves at ground level is multiplied several times by the time they reach the top of the building.

Damage to high-rise structures can also be minimized or prevented by using diagonal steel beams to help prevent swaying. In addition, tall buildings in earthquake-prone areas are now commonly placed on layered steel and rubber structures

Once a tsunami is generated, it can travel across an entire ocean and cause devastation far from its source. In the open sea, tsunami travel at several hundred kilometers per hour and commonly go unnoticed as they pass beneath ships because they are usually less than 1 m high and the distance between wave crests is typically hundreds of kilometers. When they enter shallow water, however, the wave slows down and water piles up to heights from a meter or two to many meters high. The 1946 tsunami that struck Hilo, Hawaii, was 16.5 m high! In any case, the tremendous energy possessed by a tsunami is concentrated on a shoreline when it hits either as a large breaking wave or, in some cases, as what appears to be a very rapidly rising tide.

A common popular belief is that a tsunami is a single large wave that crashes onto a shoreline. Any tsunami consists of a series of waves that pour onshore for as long as 30 minutes followed by an equal time during which water rushes back to sea. Furthermore, after the first wave hits, more waves follow at 20- to 60-minute intervals. About 80 minutes after the 1755 earthquake in Lisbon, Portugal, the first of three tsunami, the largest more than 12 m high, destroyed the waterfront area and killed numerous people. Following the arrival of a 2-m-high tsunami in Crescent City, California, in 1964, curious people went to the waterfront to inspect the damage. Unfortunately, 10 were killed by a following 4-m-high wave!

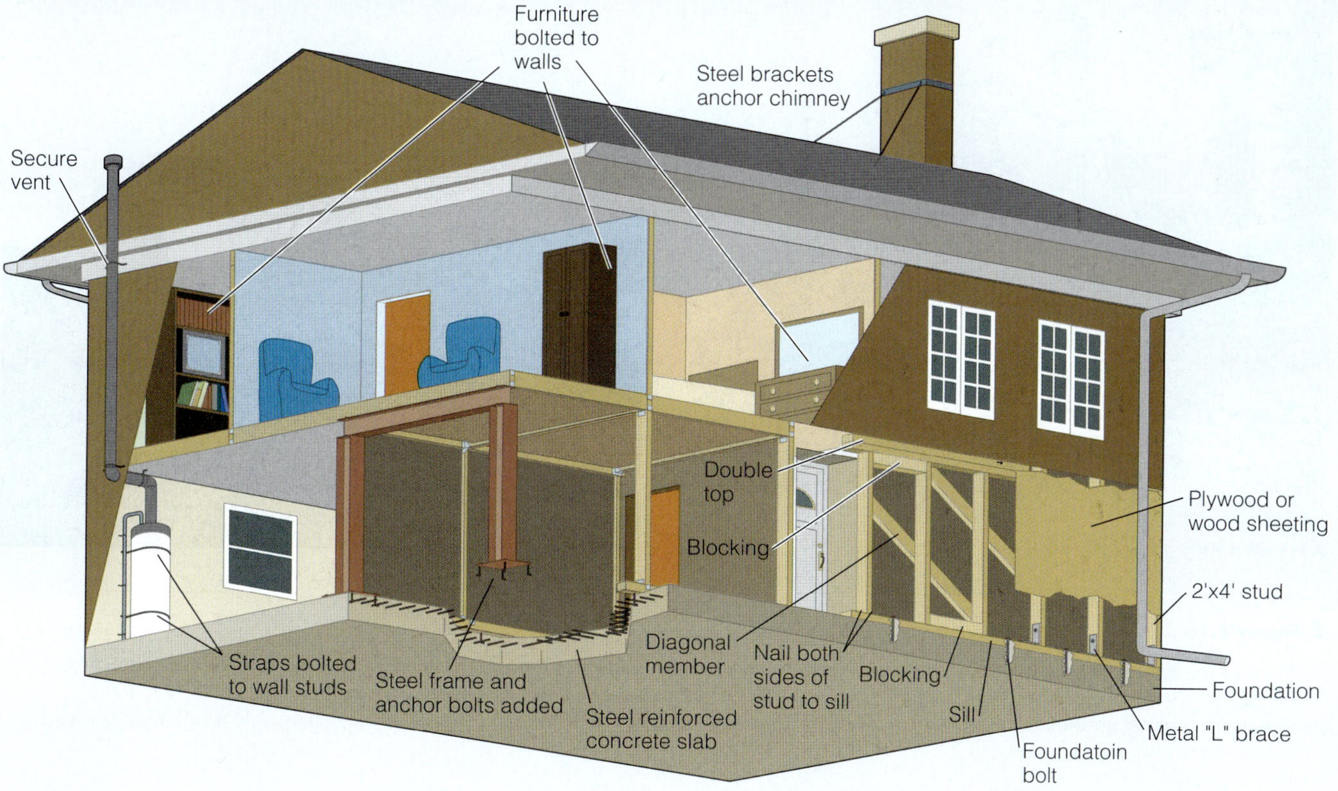

Figure 1

This illustration shows some of the things a homeowner can do to reduce damage to a building because of ground shaking during an earthquake. Notice that the structure must be solidly attached to its foundation, and bracing the walls helps prevent damage from horizontal motion.

and devices similar to shock absorbers that help decrease the amount of sway.

What about structures built many years ago? Just as in new buildings, the most important thing that can be done to increase the stability and safety of older structures is to tie together the different components of each building. This can be done by adding a steel frame to unreinforced parts of a building such as a garage, bolting the walls to the foundation, adding reinforced beams to the exterior, and using beam and joist connectors whenever possible. Although such modifications are expensive, they are usually cheaper than having to replace a building that was destroyed by an earthquake.

One of nature's warning signs of some approaching tsunami is a sudden withdrawal of the sea from a coastal region. In fact, the sea might withdraw so far that it cannot even be seen and the seafloor is laid bare over a huge area. On more than one occasion, people have rushed out to inspect exposed reefs or collect fish and shells only to be swept away when the tsunami arrived.

Following the tragic 1946 tsunami that hit Hilo, Hawaii, the U.S. Coast and Geodetic Survey established a Tsunami Early Warning System in Honolulu, Hawaii, in an attempt to minimize tsunami devastation. This system combines seismographs and instruments that detect earthquake-generated sea waves. Whenever a strong earthquake takes place anywhere within the Pacific basin, its location is determined, and instruments are checked to see whether a tsunami has been generated. If it has, a warning is sent out to evacuate people from low-lying areas that may be affected. Nevertheless, tsunami remain a threat to people in coastal areas, especially around the Pacific Ocean (Table 8.4). Unfortunately, no such warning system exists for the Indian Ocean. If one had been in place, it is possible that the death toll from the December 26, 2004 tsunami might not have been as high.

GeologyNow Log into GeologyNow and select this chapter to work through a **Geology Interactive** activity on "Tsunami" (click Earthquakes and Tsunami→Tsunami).

China Stock

(a)

Cindy Andrews/Liaison Agency, Inc.

(b)

Figure 8.16

(a) Many of the approximately 242,000 people who died in the 1976 earthquake in Tangshan, China, were killed by collapsing structures. Many buildings were constructed from unreinforced brick, which has no flexibility and quickly fell during the quake. (b) In 1993 India experienced its worst earthquake in more than 50 years. Thousands of brick and stone houses collapsed, killing at least 30,000 people.

Figure 8.17

San Francisco Marina district fire caused by broken gas lines during the 1989 Loma Prieta earthquake.

Martin E. Klimek, *Marin Independent Journal*

Figure 8.18

Aerial view of the coastal outskirts of Banda Aceh, capital city of Indonesia's Aceh province, taken on Monday, December 27, 2004, the day after a magnitude 9.0 earthquake struck off the west coast of northern Sumatra, causing a tsunami that devastated most of the coastal areas of the eastern Indian Ocean region. As can be seen, the giant wall of water that struck this area caused tremendous damage and flooding, leaving thousands homeless, and many more thousands dead.

Ground Failure

Earthquake-triggered landslides are particularly dangerous in mountainous regions and have been responsible for tremendous amounts of damage and many deaths. The 1959 earthquake in Madison Canyon, Montana, for example, caused a huge rock slide (■ Figure 8.19), whereas the 1970 Peru earthquake caused an avalanche that destroyed the town of Yungay and killed an estimated 66,000 people. Most of the 100,000 deaths from the 1920 earthquake in Gansu, China, resulted when cliffs composed of loess (wind-deposited silt) collapsed. More than 20,000 people were killed when two-thirds of the town of Port Royal, Jamaica, slid into the sea following an earthquake on June 7, 1692.

CAN EARTHQUAKES BE PREDICTED?

A successful prediction must include a time frame for the occurrence of an earthquake, its location, and its strength. Despite the tremendous amount of information geologists have

Table 8.4

Tsunami Fatalities Since 1990

Date	Location	Maximum Wave Height	Fatalities
September 2, 1992	Nicaragua	10 m	170
December 12, 1992	Flores Island	26 m	>1000
July 12, 1993	Okushiri, Japan	31 m	239
June 2, 1994	East Java	14 m	238
November 14, 1994	Mindoro Island	7 m	49
October 9, 1995	Jalisco, Mexico	11 m	1
January 1, 1996	Sulawesi Island	3.4 m	9
February 17, 1996	Irian Jaya	7.7 m	161
February 21, 1996	North coast of Peru	5 m	12
July 17, 1998	Papua New Guinea	15 m	>2200
December 26, 2004	Sumatra, Indonesia	10.5 m	>156,000

Source: F. I. Gonzales, Tsunami! *Scientific American* 280, no. 5 (1999): 59.

The San Andreas Fault

The circum-Pacific belt is well known for its volcanic activity and earthquakes. Indeed, about 60% of all volcanic eruptions and 80% of all earthquakes take place in this belt that nearly encircles the Pacific Ocean basin (Figure 8.4).

One well-known and well-studied segment of the circum-Pacific belt is the 1300-km-long San Andreas fault extending from the Gulf of California north through coastal California until it terminates at the Mendocino fracture zone off California's north coast. In plate tectonic terminology, it marks a transform plate boundary between the North American and Pacific plates (see Chapter 2).

Earthquakes along the San Andreas and related faults will continue to occur. But other segments of the circum-Pacific belt as well as the Mediterranean–Asiatic belt are also active and will continue to experience earthquakes.

USGS

Aerial view of the San Andreas fault. Noti the gulleys have been displaced by the fa

James S. Monroe

View across the San Andreas fault at Tomales Bay, north of San Francisco. The low area occupied by the bay is underlain by shattered rocks of the San Andreas fault zone. Rocks underlying the hills in the distance are on the North American plate, whereas those at the point where this photograph was taken are on the Pacific plate.

This shop in Olema, California, is rather whimsically called The Epicenter, alluding to the fact that it is in the San Andreas fault zone.

James S. Monroe

Steinbrugge Collection/ UC Berkeley

San Francisco following the 1906 earthquake. This view along Sacramento Street shows damaged buildings and the approaching fire.

Rocks on opposite sides of the San Andreas fault periodically lurch past one another, generating large earthquakes. The most famous one destroyed San Francisco on April 18, 1906. It resulted when 465 km of the fault ruptured, causing about 6 m of horizontal displacment in some areas (see Figure 8.1b). It is estimated that 3000 people died. The shaking lasted nearly 1 minute and caused property damage estimated at $400 million in 1906 dollars! About 28,000 buildings were destroyed, many of them by the three-day fire that raged out of control and devastated about 12 km^2 of the city.

Since 1906 the San Andreas fault and its subsidiary faults have spawned many more earthquakes. One of the most tragic was centered at Northridge, California, a small community north of Los Angeles. During the early morning hours of January 17, 1994, Northridge and surrounding areas were shaken for 40 seconds. When it was over, 61 people were dead and thousands injured; an oil main and at least 250 gas lines had ruptured, igniting numerous fires; nine freeways were destroyed; and thousands of homes and other buildings were damaged or destroyed.

Don Bloomer/ Time

Severe damage caused by ground shaking during the 1994 Northridge earthquake. Sixteen died in this building.

Don Bloomer/ Time

A portion of Interstate 5 Golden State Freeway collapsed during the 1994 Northridge earthquake. Fortunately, no one was killed at this location.

Ted Soqui/ Sygma/ Corbis

A spectacular fire on Balboa Boulevard, Northridge, was caused by a gas-main explosion during the earthquake.

Reed Wicander

(a)

Reed Wicander

(b)

■ Figure 8.19

On August 17, 1959, an earthquake with a Richter magnitude of 7.3 shook southwestern Montana and a large area in adjacent states. (a) The fault scarp in this image was produced when the block in the background moved up several meters compared to the one in the foreground. (b) The earthquake triggered a landslide (visible in the distance) that blocked the Madison River in Montana and created Earthquake Lake (foreground). The slide entombed about 26 people in a campground at the valley bottom.

gathered about the cause of earthquakes, successful predictions are still rare. Nevertheless, if reliable predictions can be made, they can greatly reduce the number of deaths and injuries.

From an analysis of historic records and the distribution of known faults, geologists construct *seismic risk maps* that indicate the likelihood and potential severity of future earthquakes based on the intensity of past earthquakes.

An international effort by scientists from several countries resulted in the publication of the first Global Seismic Hazard Assessment Map in December 1999 (■ Figure 8.20). Although such maps cannot be used to predict when an earthquake will take place in any particular area, they are useful in anticipating future earthquakes and helping people plan and prepare for them.

Earthquake Precursors

Studies conducted during the past several decades indicate that most earthquakes are preceded by both short-term and long-term changes within Earth. Such changes are called *precursors*.

Changes in elevation and tilting of the land surface have frequently preceded earthquakes and may be warnings of impending quakes. Extremely slight changes in the angle of the ground surface can be measured by tiltmeters. Tiltmeters have been placed on both sides of the San Andreas fault to measure tilting of the ground surface that is thought to result from increasing pressure in the rocks. Data from measurements in central California indicate significant tilting immediately preceding small earthquakes. Furthermore, extensive tiltmeter

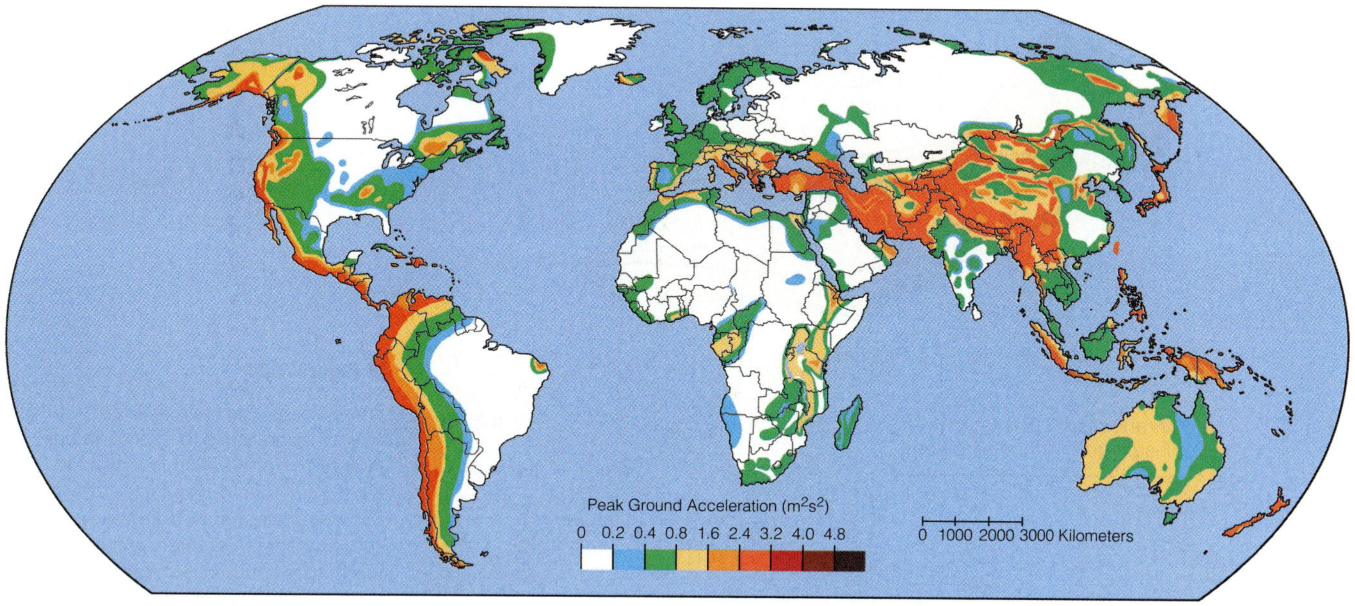

Figure 8.20

The Global Seismic Hazard Assessment Program published this seismic hazard map showing peak ground accelerations. The values are based on a 90% probability that the indicated horizontal ground acceleration during an earthquake is not likely to be exceeded in 50 years. The higher the number, the greater the hazard. As expected, the greatest seismic risks are in the circum-Pacific belt and the Mediterranean–Asiatic belt.

work performed in Japan prior to the 1964 Niigata earthquake clearly showed a relationship between increased tilting and the main shock. Although more research is needed, such precursors appear to be useful in making short-term earthquake predictions.

Other earthquake precursors include fluctuations in the water level of wells and changes in Earth's magnetic field and the electrical resistance of the ground. These fluctuations are thought to result from changes in the amount of pore space in rocks because of increasing pressure.

Besides the various precursors just discussed, one long-range prediction technique used in seismically active areas involves plotting the location of major earthquakes and their aftershocks to detect areas that have had major earthquakes in the past but are currently inactive. Such regions are locked and not releasing energy. Nevertheless, pressure is continuing to accumulate in these regions because of plate motions, making these *seismic gaps* prime locations for future earthquakes. Several seismic gaps along the San Andreas fault have the potential for future major earthquakes (■ Figure 8.21).

Earthquake Prediction Programs

Currently, only four nations—the United States, Japan, Russia, and China—have government-sponsored earthquake prediction programs. These programs include laboratory and field studies of rock behavior before, during, and after large earthquakes as well as monitoring activity along major active faults. Most earthquake prediction work in the United States is done by the U.S. Geological Survey (USGS) and involves research into all aspects of earthquake-related phenomena.

The Chinese have perhaps the most ambitious earthquake prediction program in the world, which is understandable considering their long history of destructive earthquakes. Their earthquake prediction program was initiated soon after two large earthquakes occurred at Xingtai (300 km southwest of Beijing) in 1966. The program includes extensive study and monitoring of all possible earthquake precursors. In addition, the Chinese emphasize changes in phenomena that can be observed and heard without the use of sophisticated instruments. They successfully predicted the 1975 Haicheng earthquake but failed to predict the devastating 1976 Tangshan earthquake that killed at least 242,000 people (Figure 8.16a).

What Would You Do?

Your city has experienced moderate to large earthquakes in the past, and as a result, the local planning committee, of which you are a member, has been charged with making recommendations as to how your city can best reduce damage as well as potential injuries and fatalities resulting from future earthquakes. You are told to consider zoning regulations; building codes for private dwellings, hospitals, public buildings, and high-rise structures; and emergency contingency plans. What kinds of recommendations would you make, and what and who would you ask for professional guidance?

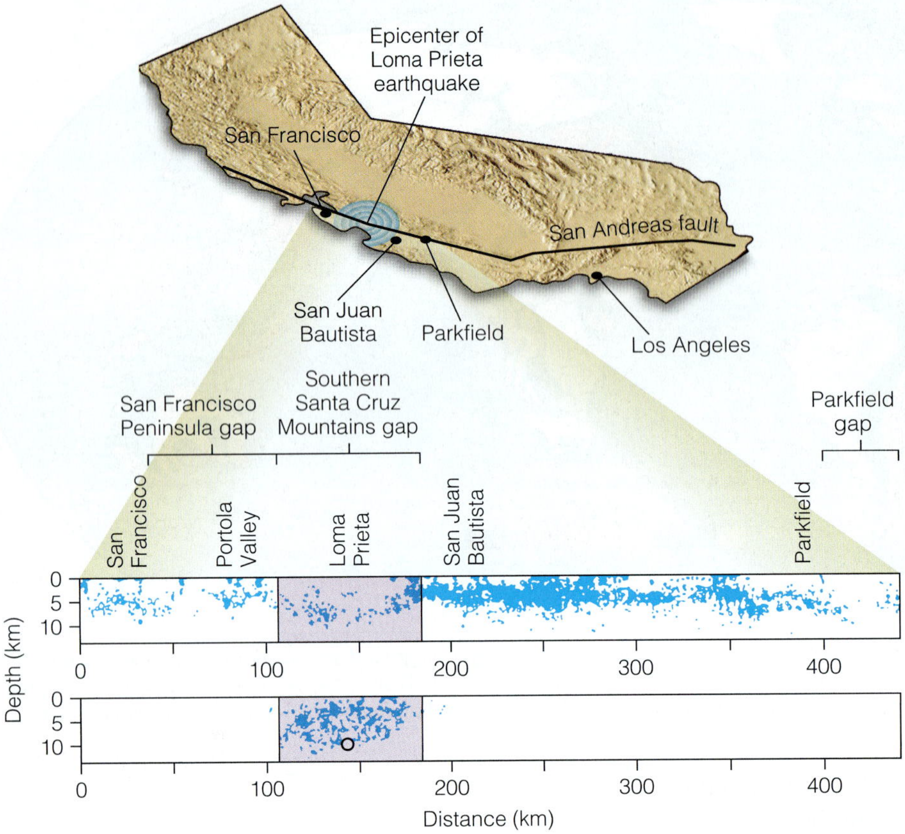

Figure 8.21

Three seismic gaps are evident in this cross section along the San Andreas fault from north of San Francisco to south of Parkfield. The first is between San Francisco and Portola Valley, the second near Loma Prieta Mountain, and the third is southeast of Parkfield. The top section shows the epicenters of earthquakes between January 1969 and July 1989. The bottom section shows the southern Santa Cruz Mountains gap after it was filled by the October 17, 1989, Loma Prieta earthquake (open circle) and its aftershocks. Source: Data from *The Loma Prieta Earthquake of October 17, 1989.* U.S. Geological Survey.

Progress is being made toward dependable, accurate earthquake predictions, and studies are under way to assess public reactions to long-, medium-, and short-term earthquake warnings. However, unless short-term warnings are actually followed by an earthquake, most people will probably ignore the warnings as they frequently do now for hurricanes, tornadoes, and tsunami. Perhaps the best we can hope for is that people in seismically active areas will take measures to minimize their risk from the next major earthquake (Table 8.5).

GeologyNow Log into GeologyNow and select this chapter to work through a **Geology Interactive** activity on "Seismic Risk USA" (click Earthquakes and Tsunami→Seismic Risk USA).

CAN EARTHQUAKES BE CONTROLLED?

Reliable earthquake prediction is still in the future, but can anything be done to control or at least partly control earthquakes? Because of the tremendous energy involved, it seems unlikely that humans will ever be able to prevent earthquakes. However, it may be possible to gradually release the energy stored in rocks, thus decreasing the probability of a large earthquake and extensive damage.

During the early to mid-1960s, Denver, Colorado, experienced numerous small earthquakes. This was surprising because Denver had not been prone to earthquakes in the past. In 1962, geologist David M. Evans suggested that Denver's earthquakes were directly related to the injection of contaminated wastewater into a disposal well 3674 m deep at the Rocky Mountain Arsenal, northeast of Denver (■ Figure 8.22a). The U.S. Army initially denied that a connection existed, but a USGS study concluded that the pumping of waste fluids into the disposal well was the cause of the earthquakes.

Figure 8.22b shows the relationship between the average number of earthquakes in Denver per month and the average amount of contaminated fluids injected into the disposal well per month. Obviously, a high degree of correlation between the two exists, and the correlation is particularly convincing considering that during the time when no waste fluids were injected, earthquake activity decreased dramatically.

The area beneath the Rocky Mountain Arsenal consists of highly fractured gneiss overlain by sedimentary rocks. When water was pumped into these fractures, it decreased the friction on opposite sides of the fractures and, in essence, lubricated them so that movement occurred, causing the earthquakes that Denver experienced.

Experiments conducted in 1969 at an abandoned oil field near Rangely, Colorado, confirmed the arsenal hypothesis. Water was pumped in and out of

Table 8.5

What You Can Do to Prepare for an Earthquake

Anyone who lives in an area that is subject to earthquakes or who will be visiting or moving to such an area can take certain precautions to reduce the risks and losses resulting from an earthquake.

Before an earthquake:

1. Become familiar with the geologic hazards of the area where you live and work.
2. Make sure your house is securely attached to the foundation by anchor bolts and that the walls, floors, and roof are all firmly connected together.
3. Heavy furniture such as bookcases should be bolted to the walls; semiflexible natural gas lines should be used so that they can give without breaking; water heaters and furnaces should be strapped and the straps bolted to wall studs to prevent gas-line rupture and fire. Brick chimneys should have a bracket or brace that can be anchored to the roof.
4. Maintain a several-day supply of fresh water and canned foods, and keep a fresh supply of flashlight and radio batteries as well as a fire extinguisher.
5. Maintain a basic first-aid kit, and have a working knowledge of first-aid procedures.
6. Learn how to turn off the utilities at your house.
7. Above all, have a planned course of action for when an earthquake strikes.

During an earthquake:

1. Remain calm and avoid panic.
2. If you are indoors, get under a desk or table if possible, or stand in an interior doorway or room corners as these are the structurally strongest parts of a room; avoid windows and falling debris.
3. In a tall building, do not rush for the stairwells or elevators.
4. In an unreinforced or other hazardous building, it may be better to get out of the building rather than to stay in it. Be on the alert for fallen power lines and the possibility of falling debris.
5. If you are outside, get to an open area away from buildings if possible.
6. If you are in an automobile, stay in the car, and avoid tall buildings, overpasses, and bridges if possible.

After an earthquake:

1. If you are uninjured, remain calm and assess the situation.
2. Help anyone who is injured.
3. Make sure there are no fires or fire hazards.
4. Check for damage to utilities and turn off gas valves if you smell gas.
5. Use your telephone only for emergencies.
6. Do not go sightseeing or move around the streets unnecessarily.
7. Avoid landslide and beach areas.
8. Be prepared for aftershocks.

abandoned oil wells, the pore-water pressure in these wells was measured, and seismographs were installed in the area to measure any seismic activity. Monitoring showed that small earthquakes were occurring in the area when fluids were injected and that earthquake activity declined when fluids were pumped out. What the geologists were doing was starting and stopping earthquakes at will, and the relationship between pore-water pressures and earthquakes was established.

Based on these results, some geologists have proposed that fluids be pumped into the locked segments or seismic gaps of active faults to cause small- to moderate-sized earthquakes. They think that this would relieve the pressure on the fault and prevent a major earthquake from occurring. Although this plan is intriguing, it also has many potential problems. For instance, there is no guarantee that only a small earthquake might result. Instead a major earthquake might occur, causing tremendous property damage and loss of life. Who would be responsible? Certainly, a great deal more research is needed before such an experiment is performed, even in an area of low population density.

It appears that until such time as earthquakes can be accurately predicted or controlled, the best means of defense is careful planning and preparation (Table 8.5).

WHAT IS EARTH'S INTERIOR LIKE?

During most of historic time, Earth's interior was perceived as an underground world of vast caverns, heat, and sulfur gases, populated by demons. By the 1860s, scientists knew what the average density of Earth was and that pressure and temperature increase with depth. And even though Earth's interior is hidden from direct observation, scientists now have a reasonably good idea of its internal structure and composition.

Earth is generally depicted as consisting of concentric layers that differ in composition and density separated from adjacent layers by rather distinct boundaries (■ Figure 8.23). Recall that the outermost layer, or the *crust,* is Earth's thin skin. Below the crust and extending about halfway to Earth's center is the *mantle,* which comprises more than 80% of Earth's volume. The central part of Earth consists of a *core,*

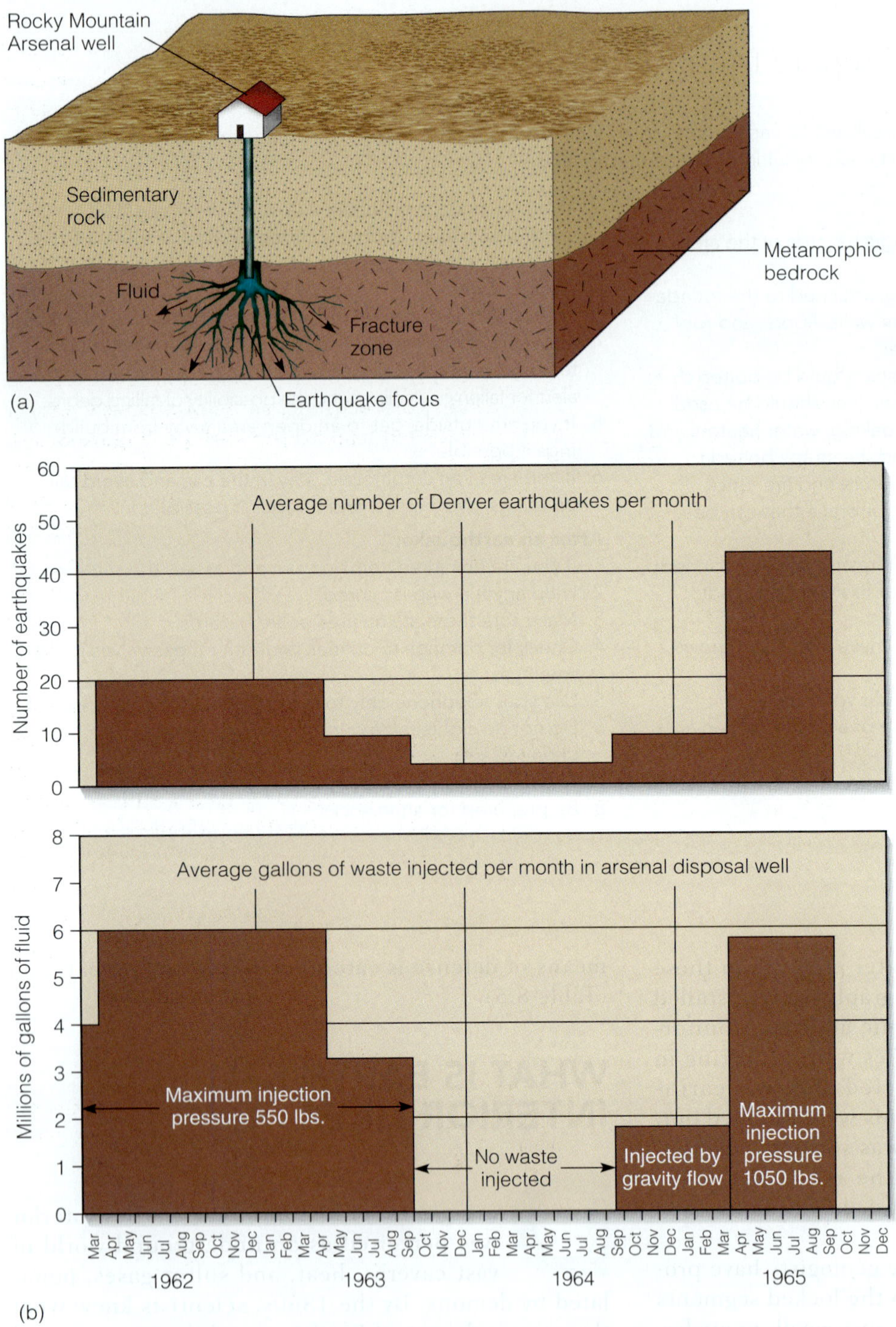

■ **Figure 8.22**

(a) A block diagram of the Rocky Mountain Arsenal well and the underlying geology. (b) A graph showing the relationship between the amount of wastewater injected into the well per month and the average number of Denver earthquakes per month. There have been no significant earthquakes in Denver since injection of wastewater into the disposal well ceased in 1965. Source: From Figure 6, page 17, *Geotimes* Vol. 10, No. 9 (1966) with the kind permission of the American Geological Institute. For more information, log onto www.agiweb.org

which is divided into a solid inner portion and a liquid outer part (Figure 8.23).

The behavior and travel times of P- and S-waves provide geologists with much information about Earth's internal structure. Seismic waves travel outward as wave fronts from their source areas, although it is most convenient to depict them as *wave rays,* which are lines showing the direction of movement of small parts of wave fronts (Figure 8.3). Any disturbance, such as a passing train or construction equipment, can cause seismic waves, but only those generated by large earthquakes, explosive volcanism, asteroid impacts, and nuclear explosions can travel completely through Earth.

As we noted earlier, P- and S-wave velocity is determined by the density and elasticity of the materials they travel through, both of which increase with depth. Wave velocity is slowed by increasing density but increases in materials with greater elasticity. Because elasticity increases with depth faster than density, a general increase in seismic wave velocity takes place as the waves penetrate to greater depths. P-waves travel faster than S-waves under all circumstances, but unlike P-waves, S-waves are not transmitted through a liquid because liquids have no shear strength (rigidity); liquids simply flow in response to shear stress.

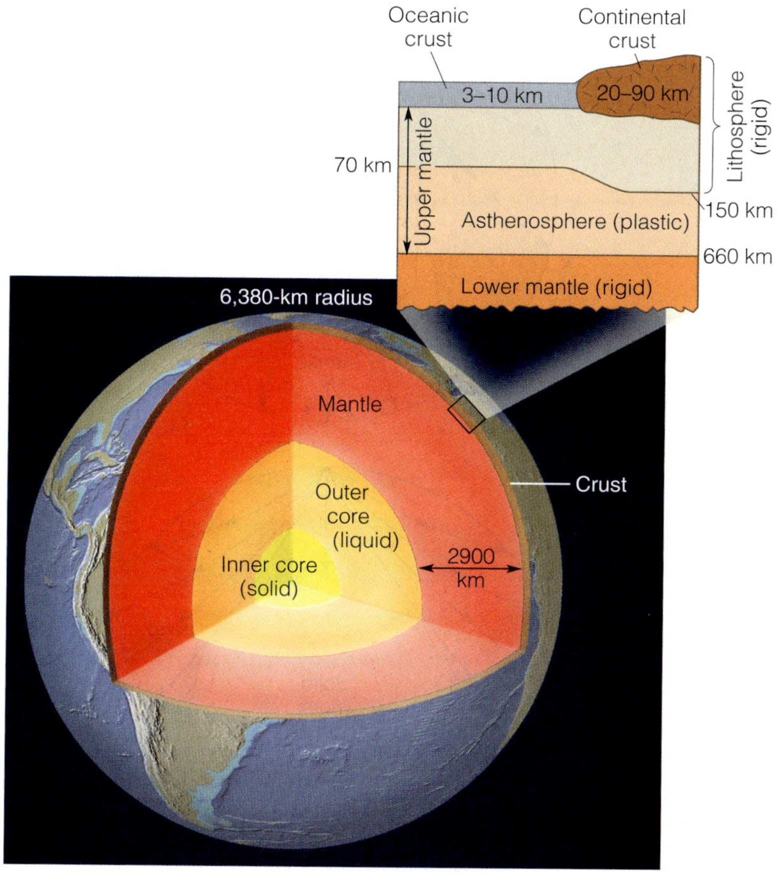

■ **Figure 8.23**

Earth's internal structure. The inset shows Earth's outer part in more detail. The asthenosphere is solid but behaves plastically and flows.

As a seismic wave travels from one material into another of different density and elasticity, its velocity and direction of travel change. That is, the wave is bent, a phenomenon known as **refraction**, in much the same way as light waves are refracted as they pass from air into more dense water (■ Figure 8.24). Because seismic waves pass through materials of differing density and elasticity, they are continually refracted so that their paths are curved; wave rays travel only in a straight line and are not refracted when their direction of travel is perpendicular to a boundary (Figure 8.24).

In addition to refraction, seismic rays are **reflected**, much as light is reflected from a mirror. When seismic rays encounter a boundary separating materials of different density or elasticity, some of a wave's energy is reflected back to the surface (Figure 8.24). If we know the wave velocity and the time required for the wave to travel from its source to the boundary and back to the surface, we can calculate the depth of the reflecting boundary. Such information is useful in determining not only Earth's internal structure but also the depths of sedimentary rocks that may contain petroleum.

Although changes in seismic wave velocity occur continuously with depth, P-wave velocity increases suddenly at the base of the crust and decreases abruptly at a depth of about 2900 km (■ Figure 8.25). Such marked changes in seismic wave velocity indicate a boundary called a **discontinuity** across which a significant change in Earth materials or their properties occurs. These discontinuities are the basis for subdividing Earth's interior into concentric layers.

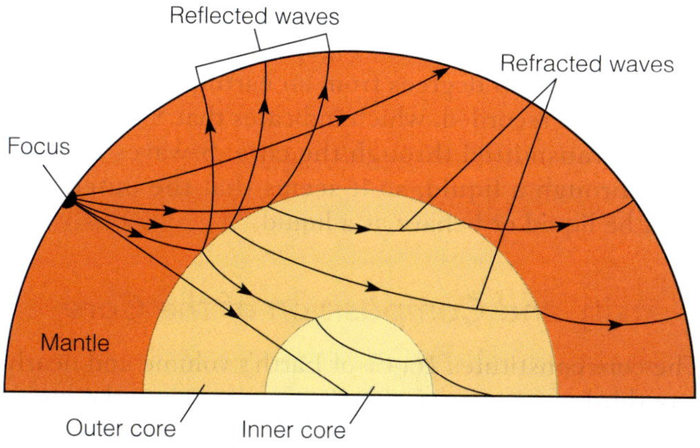

■ **Figure 8.24**

Refraction and reflection of P-waves. When seismic waves pass through a boundary separating Earth materials of different density or elasticity, they are refracted, and some of their energy is reflected back to the surface. Notice that the only wave ray not refracted is the one perpendicular to boundaries.

GeologyNow Log into GeologyNow and select this chapter to work through a **Geology Interactive** activity on "Reflection and Refraction" (click Earth's Layers→Reflection and Refraction).

EARTH'S CORE

In 1906, R. D. Oldham of the Geological Survey of India realized that seismic waves arrived later than expected at seismic stations more than 130 degrees from an earthquake focus. He postulated that Earth has a core that transmits seismic waves more slowly than shallower Earth materials. We now know that P-wave velocity decreases markedly at a depth of 2900 km, which indicates an important discontinuity now recognized as the core–mantle boundary (Figure 8.25).

Because of the sudden decrease in P-wave velocity at the core–mantle boundary, P-waves are refracted in the core so that little P-wave energy reaches the surface in the area between 103 degrees and 143 degrees from

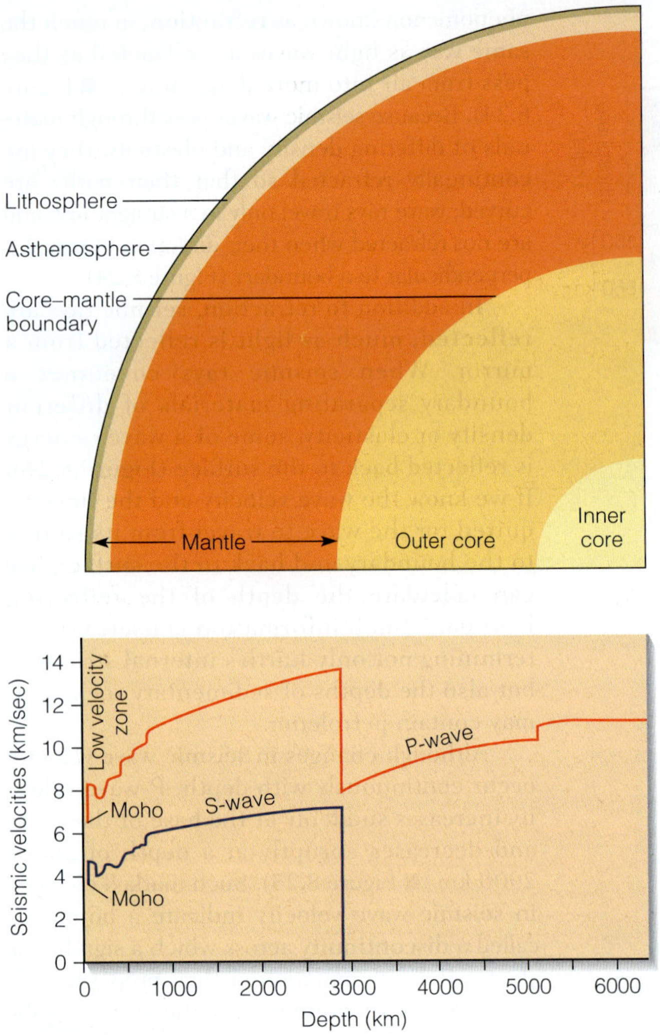

■ **Figure 8.25**

Profiles showing seismic wave velocities versus depth. Several discontinuities are shown across which seismic wave velocities change rapidly. Source: From G. C. Brown and A. E. Musset, *The Inaccessible Earth* (Kluwer Academic Publishers, 1981), Figure 12.7a. Reprinted with permission of the author.

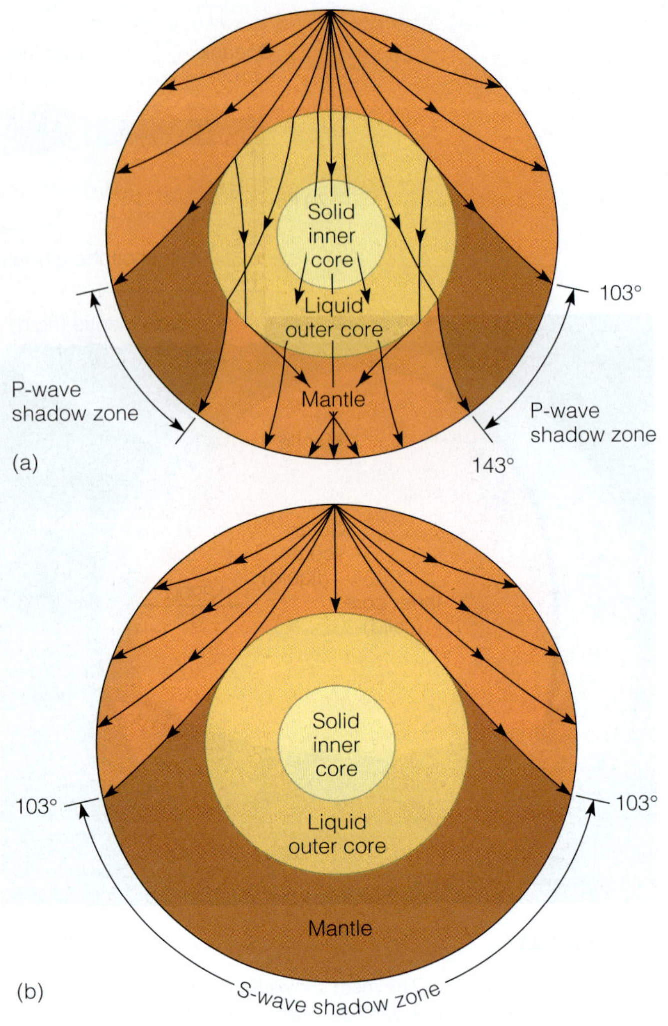

GeologyNow ■ **Active Figure 8.26**

(a) P-waves are refracted so that little P-wave energy reaches the surface in the P-wave shadow zone. (b) The presence of an S-wave shadow zone indicates that S-waves are being blocked within Earth.

an earthquake focus (■ Figure 8.26). This area in which little P-wave energy is recorded by seismographs is a **P-wave shadow zone.**

The P-wave shadow zone is not a perfect shadow zone because some weak P-wave energy reaches the surface within the zone. Scientists proposed several hypotheses to account for this observation, but all were rejected by the Danish seismologist Inge Lehmann, who in 1936 postulated that the core is not entirely liquid as previously thought. She proposed that seismic wave reflection from a solid inner core accounted for the arrival of weak P-wave energy in the P-wave shadow zone, a proposal that was quickly accepted by seismologists.

In 1926, the British physicist Harold Jeffreys realized that S-waves were not simply slowed by the core but were completely blocked by it. So, besides a P-wave shadow zone, a much larger and more complete **S-wave shadow zone** also exists (Figure 8.26b). At locations greater than 103 degrees from an earthquake focus, no S-waves are recorded, which indicates that S-waves cannot be transmitted through the core. S-waves will not pass through a liquid, so it seems that the outer core must be liquid or behave as a liquid.

Density and Composition of the Core

The core constitutes 16.4% of Earth's volume and nearly one-third of its mass. Geologists can estimate the core's density and composition by using seismic evidence and laboratory experiments. Furthermore, meteorites, which are thought to represent remnants of the material from which the solar system formed, are used to make estimates of density and composition. For example, meteorites composed of iron and nickel alloys may represent the differentiated interiors of large asteroids and approximate the density and composition of Earth's core. The density of the outer core varies from 9.9 to 12.2 g/cm^3. At Earth's center, the pressure is equivalent to about 3.5 million times normal atmospheric pressure.

The core cannot be composed of the minerals that are most common at the surface because even under the tremendous pressures at great depth they would still not be dense enough to yield an average density of 5.5 g/cm^3 for Earth. Both the outer and inner cores are thought to be composed largely of iron, but pure iron is too dense to be the sole constituent of the outer core. Thus, it must be "diluted" with elements of lesser density. Laboratory experiments and comparisons with iron meteorites indicate that about 12% of the outer core may consist of sulfur and perhaps some silicon and small amounts of nickel and potassium.

In contrast, pure iron is not dense enough to account for the estimated density of the inner core, so perhaps 10–20% of the inner core consists of nickel. These metals form an iron–nickel alloy thought to be sufficiently dense under the pressure at that depth to account for the density of the inner core.

When the core formed during early Earth history, it was probably molten and has since cooled so that its interior has crystallized. Indeed, the inner core continues to grow as Earth slowly cools, and liquid of the outer core crystallizes as iron. Recent evidence also indicates that the inner core rotates faster than the outer core, moving about 20 km/yr relative to the outer core.

GeologyNow Log into GeologyNow and select this chapter to work through a **Geology Interactive** activity on "Core Studies" (click Earth's Layers→Core Studies).

EARTH'S MANTLE

Another significant discovery about Earth's interior was made in 1909 when the Yugoslavian seismologist Andrija Mohorovičić detected a discontinuity at a depth of about 30 km. While studying the arrival times of seismic waves from Balkan earthquakes, Mohorovičić noticed that seismic stations a few hundred kilometers from an earthquake's epicenter were recording two distinct sets of P- and S-waves.

From his observations, Mohorovičić concluded that a sharp boundary separates rocks with different properties at a depth of about 30 km. He postulated that P-waves below this boundary travel at 8 km/sec, whereas those above the boundary travel at 6.75 km/sec. When an earthquake occurs, some waves travel directly from the focus to a seismic station, whereas others travel through the deeper layer and some of their energy is refracted back to the surface (■ Figure 8.27). Waves traveling through the deeper layer travel farther to a seismic station but they do so more rapidly and arrive before those in the shallower layer. The boundary identified by Mohorovičić separates the crust from the mantle and is now called the **Mohorovičić discontinuity,** or simply the **Moho.** It is present everywhere except beneath spreading ridges, but its depth varies. Beneath the continents it ranges from 20 to 90 km with an average of 35 km; beneath the sea floor it is 5 to 10 km deep.

The Mantle's Structure, Density, and Composition

Although seismic wave velocity in the mantle increases with depth, several discontinuities exist. Between depths of 100 and 250 km, both P- and S-wave velocities decrease markedly (■ Figure 8.28). This 100- to 250-km-deep layer is the *low-velocity zone,* which corresponds closely to the *asthenosphere,* a layer in which the rocks are close to their melting point and are less elastic, accounting for the observed decrease in seismic wave velocity. The asthenosphere is an important zone because it may be where some magma is generated. Furthermore, it lacks strength, flows plastically, and is thought to be the layer over which the plates of the outer, rigid *lithosphere* move.

Other discontinuities are also present at deeper levels within the mantle. But unlike those between the crust and mantle or between the mantle and core, these probably represent structural changes in minerals rather than compositional changes. In other words, geologists think the mantle is composed of the same material throughout, but the structural states of minerals such as

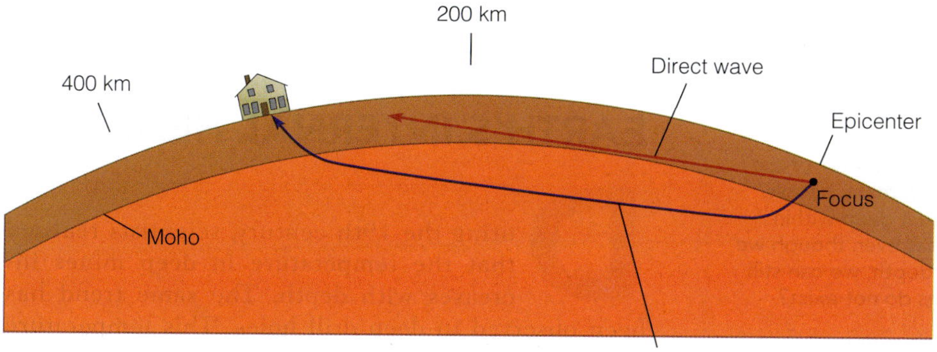

■ **Figure 8.27**

Andrija Mohorovičić studied seismic waves and detected a seismic discontinuity at a depth of about 30 km. The deeper, faster seismic waves arrive at seismic stations first, even though they travel farther. This discontinuity, now known as the Moho, is between the crust and mantle.

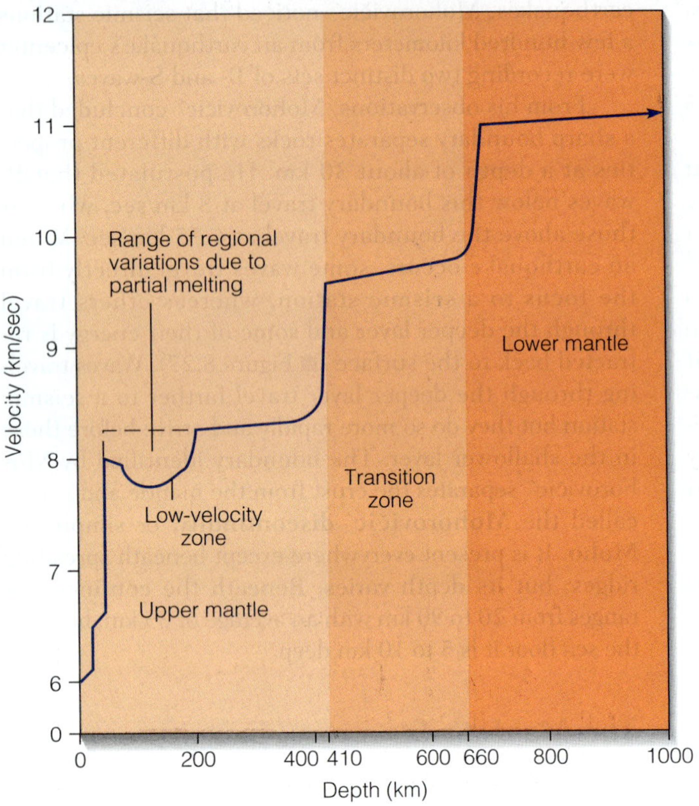

Figure 8.28

Variations in P-wave velocity in the upper mantle and transition zone. Source: From G. C. Brown and A. E. Musset, *The Inaccessible Earth* (Kluwer Academic Publishers, 1981), Figure 7.11. Reprinted with permission of the author.

olivine change with depth. At a depth of 410 km, seismic wave velocity increases slightly as a consequence of such changes in mineral structure (Figure 8.28). Another velocity increase occurs at about 660 km, where the minerals break down into metal oxides, such as FeO (iron oxide) and MgO (magnesium oxide), and silicon dioxide (SiO_2). These two discontinuities define the top and base of a *transition zone* separating the upper mantle from the lower mantle (Figure 8.28).

What Would You Do?

Of course, novels such as *Journey to the Center of the Earth* are fiction, but it is surprising how many people think that vast caverns and cavities exist deep within the planet. How would you explain that even though we have no direct observations at great depth, we can still be sure that these proposed openings do not exist?

Although the mantle's density, which varies from 3.3 to 5.7 g/cm^3, can be inferred rather accurately from seismic waves, its composition is less certain. The igneous rock *peridotite,* containing mostly ferromagnesian silicates, is considered the most likely component. Laboratory experiments indicate that it possesses physical properties that account for the mantle's density and observed rates of seismic wave transmissions. Peridotite also forms the lower parts of igneous rock sequences thought to be fragments of the oceanic crust and upper mantle emplaced on land. In addition, peridotite occurs as inclusions in volcanic rock bodies such as *kimberlite pipes* that are known to have come from depths of 100 to 300 km. These inclusions appear to be pieces of the mantle.

SEISMIC TOMOGRAPHY

The model of Earth's interior consisting of an iron-rich core and a rocky mantle is probably accurate, but not very precise. In recent years, geophysicists have developed a technique called *seismic tomography* that allows them to develop more accurate models of Earth's interior. In seismic tomography, numerous crossing seismic waves are analyzed in much the same way radiologists analyze CAT (computerized axial tomography) scans. In CAT scans, X rays penetrate the body and a two-dimensional image of its interior is formed. Repeated CAT scans from slightly different angles are computer analyzed and stacked to produce a three-dimensional image.

In a similar manner, geophysicists use seismic waves to probe Earth's interior. In seismic tomography, the average velocities of numerous crossing seismic waves are analyzed so that "slow" and "fast" areas of wave travel are detected (■ Figure 8.29). Remember that seismic wave velocity depends partly on elasticity; cold rocks have greater elasticity and therefore transmit seismic waves faster than hotter rocks.

As a result of studies in seismic tomography, a much clearer picture of Earth's interior is emerging. It has already given us a better understanding of complex convection within the mantle and a clearer picture of the nature of the core–mantle boundary.

EARTH'S INTERNAL HEAT

During the 19th century, scientists realized that the temperature in deep mines increases with depth. The same trend has been observed in deep drill holes. This temperature

GEOLOGY IN UNEXPECTED PLACES

Diamonds and Earth's Interior

Diamond—a jewel, a shape, a tool. Chances are you or someone you know owns a diamond ring, necklace, or bracelet, because diamond, which symbolizes strength and purity, is the most popular and sought-after gemstone. The value of gem-quality diamonds is determined by their color (colorless ones are most desirable), clarity (lack of flaws), and carat (1 carat = 200 milligrams). A diamond's value also depends on the cut, the way it is cleaved, cut, and polished to yield small plane surfaces known as facets, which enhance the quality of reflected light (■ Figure 1 and see Figure 3.1b).

Most of the world's diamonds come from stream and beach placer deposits, but the ultimate source of most gem-quality diamonds and industrial diamonds is kimberlite pipes composed of dark gray or blue igneous rock that originated at great depths. In fact, diamond is composed of carbon that forms at pressures found at least 100 km deep—that is, in Earth's mantle. Diamond establishes a minimum depth for the magma that cools to form kimberlite, and a form of silica also in these rocks that indicates that the magma originated between 100 and 300 km below the surface. In addition to diamonds and silica, kimberlite commonly contains inclusions of peridotite (see Figure 4.10) that are most likely pieces of the mantle.

Because diamond is so hard (see Table 3.3) it is used for abrasives and for tools that cut other hard substances, such as other gemstones, eyeglasses, and even computer chips. In road construction, diamonds are used to grind down old pavement before a new layer of blacktop is poured. Even petroleum companies use diamond-studded drill bits.

■ **Figure 1**

A diamond's value is determined by color, clarity, carat, and cut, the four C's.

increase with depth, or **geothermal gradient,** near the surface is about 25°C/km. In areas of active or recently active volcanism, the geothermal gradient is greater than in adjacent nonvolcanic areas, and temperature rises faster beneath spreading ridges than elsewhere beneath the seafloor.

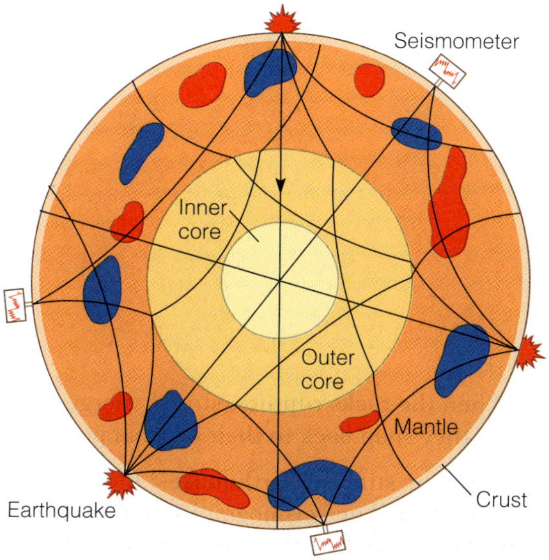

■ **Figure 8.29**

Numerous earthquake waves are analyzed to detect areas within Earth that transmit seismic waves faster or slower than adjacent areas. Areas of fast wave travel correspond to "cold" regions (blue), whereas "hot" regions (red) transmit seismic waves more slowly.

Source: From "Journey to the Center of the Earth," by T. A. Heppenheimer, *Discover,* v. 8, no. 11, Nov. 1987. Illustration by Andrew Christie, copyright © 1987.

Most of Earth's internal heat is generated by radioactive decay, especially the decay of isotopes of uranium and thorium and to a lesser degree of potassium 40. When these isotopes decay, they emit energetic particles and gamma rays that heat surrounding rocks. And because rock is such a poor conductor of heat, it takes little radioactive decay to build up considerable heat, given enough time.

Unfortunately, the geothermal gradient is not useful for estimating temperatures at great depth. If we were simply to extrapolate from the surface downward, the temperature at 100 km would be so high that, despite the great pressure, all known rocks would melt. Yet except for pockets of magma, it appears that the mantle is solid rather than liquid because it transmits S-waves. Accordingly, the geothermal gradient must decrease markedly.

Current estimates of the temperature at the base of the crust are 800° to 1200°C. The latter figure seems to be an upper limit; if it were any higher, melting would be expected. Furthermore, fragments of mantle rock in kimberlite pipes, thought to have come from depths of 100–300 km, appear to have reached equilibrium at these depths at a temperature of about 1200°C. At the core–mantle boundary, the temperature is probably between 2500° and 5000°C; the wide spread of values indicates the uncertainties of such estimates. If these figures are reasonably accurate, the geothermal gradient in the mantle is only about 1°C/km.

Because the core is so remote and its composition uncertain, only very general estimates of its temperature are possible. Based on various experiments, the maximum temperature at the center of the core is thought to be 6500°C, very close to the estimated temperature for the surface of the Sun!

EARTH'S CRUST

Our main concern in the latter part of this chapter is Earth's interior, but to be complete we must briefly discuss the crust, which along with the upper mantle constitutes the lithosphere.

Continental crust is complex, consisting of all rock types, but it is usually described as "granitic," meaning that its overall composition is similar to that of granitic rocks. With the exception of metal-rich rocks such as iron ore deposits, most rocks of the continental crust have densities between 2.0 and 3.0 g/cm^3, with the average density of the crust being about 2.70 g/cm^3. P-wave velocity in continental crust is about 6.75 km/sec. Continental crust averages 35 km thick, but it is much thicker beneath mountain ranges and considerably thinner in such areas as the Rift Valleys of East Africa and a large area called the Basin and Range Province in the western United States where it is being stretched and thinned.

In contrast to continental crust, oceanic crust is simpler, consisting of gabbro in its lower part and overlain by basalt. It is thinnest, about 5 km, at spreading ridges, and nowhere is it thicker than 10 km. Its density of 3.0 g/cm^3 accounts for the fact that it transmits P-waves at about 7 km/sec. In fact, this P-wave velocity is what one would expect if oceanic crust is composed of basalt and gabbro. We present a more detailed description of the oceanic crust's composition and structure in Chapter 9.

8 GEO RECAP

Chapter Summary

- Earthquakes are vibrations caused by the sudden release of energy, usually along a fault.
- The elastic rebound theory holds that pressure builds in rocks on opposite sides of a fault until the strength of the rocks is exceeded and rupture occurs. When the rocks rupture, stored energy is released as they snap back to their original position.
- Seismology is the study of earthquakes. Earthquakes are recorded on seismographs, and the record of an earthquake is a seismogram.

- The point where energy is released is an earthquake's focus, and its epicenter is directly above the focus on the surface.
- Approximately 80% of all earthquakes occur in the circum-Pacific belt, 15% within the Mediterranean–Asiatic belt, and the remaining 5% mostly in the interior of the plates or along oceanic spreading-ridge systems.
- The two types of body waves are P-waves and S-waves. P-waves travel through all materials, whereas S-waves do not travel through liquids. P-waves are the fastest waves and are compressional, whereas S-waves are shear.
- The two types of surface waves are Rayleigh and Love waves. They travel along or just below the surface.
- Scientists locate the epicenter of an earthquake by using a time–distance graph of the P- and S-waves from any given distance. Three seismographs are needed to locate the epicenter.
- Intensity is a measure of the kinds of damage done by an earthquake and is expressed by values from I to XII in the Modified Mercalli Intensity Scale.
- Magnitude measures the amount of energy released by an earthquake and is expressed in the Richter Magnitude Scale. Each increase in the magnitude number represents about a 30-fold increase in energy released.
- The Seismic-Moment Magnitude Scale more accurately estimates the energy released during very large earthquakes.
- Ground shaking is the most destructive of all earthquake hazards. The amount of damage done by an earthquake depends on the geology of the area, type of building construction, magnitude of the earthquake, and duration of shaking.
- Tsunami are seismic sea waves that are produced by earthquakes, submarine landslides, and eruptions of volcanoes at sea.
- Seismic risk maps help geologists make long-term predictions about the severity of earthquakes based on past occurrences.
- Earthquake precursors are changes preceding an earthquake that can be used to predict earthquakes. Precursors include seismic gaps, changes in surface elevations, and fluctuations of water levels in wells.
- A variety of earthquake research programs are under way in the United States, Japan, Russia, and China. Studies indicate that most people would probably not heed a short-term earthquake warning.
- Injecting fluids into locked segments of an active fault holds some promise as a possible means of earthquake control.
- Earth has an outer layer of oceanic and continental crust below which lies a rocky mantle and an iron-rich core with a solid inner part and a liquid outer part.
- Studies of P- and S-waves, laboratory experiments, comparisons with meteorites, and studies of inclusions in volcanic rocks provide evidence about the composition and structure of Earth's interior.
- Density and elasticity of Earth materials determine the velocity of seismic waves. Seismic waves are refracted when their direction of travel changes. Wave reflection occurs at boundaries across which the properties of rocks change.
- Geologists use the behavior of P- and S-waves and the presence of P- and S-wave shadow zones to estimate the density and composition of Earth's interior and to estimate the size and depth of the core and mantle.
- Earth's inner core is probably made up of iron and nickel, whereas the outer core is mostly iron with 10–20% other substances.
- Peridotite is the most likely rock making up Earth's mantle.
- Oceanic crust is composed of basalt and gabbro, whereas continental crust has an overall composition similar to granite. The Moho is the boundary between the crust and the mantle.
- The geothermal gradient of 25°C/km cannot continue to great depths; within the mantle and core it is probably about 1°C/km. The temperature at Earth's center is estimated to be 6500°C.

Important Terms

discontinuity (p. 223)
earthquake (p. 198)
elastic rebound theory (p. 199)
epicenter (p. 200)
focus (p. 200)
geothermal gradient (p. 227)
intensity (p. 207)
Love wave (L-wave) (p. 204)
magnitude (p. 208)
Modified Mercalli Intensity Scale (p. 207)

Mohorovičić' discontinuity (Moho) (p. 225)
P-wave (p. 203)
P-wave shadow zone (p. 224)
Rayleigh wave (R-wave) (p. 204)
reflection (p. 223)
refraction (p. 223)
Richter Magnitude Scale (p. 208)
seismograph (p. 199)
seismology (p. 199)
S-wave (p. 204)
S-wave shadow zone (p. 224)
tsunami (p. 211)

Review Questions

1. Most earthquakes take place in the
   a. _____ spreading-ridge zone; b. _____ Mediterranean–Asiatic belt; c. _____ rifts in continental interiors; d. _____ circum-Pacific belt; e. _____ Appalachian fault zone.
2. A P-wave is one in which
   a. _____ movement is perpendicular to the direction of wave travel; b. _____ Earth's surface moves as a series of waves; c. _____ materials move forward and back along a line in the same direction that the wave moves; d. _____ large waves crash onto a shoreline following a submarine earthquake; e. _____ movement at the surface is similar to movement in water waves.
3. With few exceptions, the most damaging earthquakes are
   a. _____ deep focus; b. _____ caused by volcanic eruptions; c. _____ those with Richter magnitudes of about 2; d. _____ shallow focus; e. _____ those that occur along spreading ridges.
4. A tsunami is a(n)
   a. _____ part of a fault with a seismic gap; b. _____ precursor to an earthquake; c. _____ seismic sea wave; d. _____ particularly large and destructive earthquake; e. _____ earthquake with a focal depth exceeding 300 km.
5. A qualitative assessment of the damage done by an earthquake is expressed by
   a. _____ intensity; b. _____ dilatancy; c. _____ seismicity; d. _____ magnitude; e. _____ liquefaction.
6. It would take about _____ earthquakes with a Richter magnitude of 3 to equal the energy released in one earthquake with a magnitude of 6.
   a. _____ 9; b. _____ 2,000,000; c. _____ 27,000; d. _____ 30; e. _____ 250.
7. An earthquake's epicenter is
   a. _____ usually in the lower part of the mantle; b. _____ a point on the surface directly above the focus; c. _____ determined by analyzing surface wave arrival times at seismic stations; d. _____ a measure of the energy released during an earthquake; e. _____ the damage corresponding to a value of IV on the Modified Mercalli Intensity Scale.
8. The seismic discontinuity at the base of the crust is known as the
   a. _____ transition zone; b. _____ magnetic reflection point; c. _____ low-velocity zone; d. _____ Moho; e. _____ high-velocity zone.
9. The geothermal gradient is Earth's
   a. _____ capacity to reflect and refract seismic waves; b _____ most destructive aspect of earthquakes; c. _____ temperature increase with depth; d. _____ average rate of seismic wave velocity in the mantle; e. _____ elastic rebound potential.
10. Oceanic crust is composed of
    a. _____ granite and gabbro; b. _____ peridotite and gabbro; c. _____ granite and peridotite; d. _____ peridotite and basalt; e. _____ basalt and gabbro.
11. Why are structures built on solid bedrock usually damaged less during an earthquake than those built on unconsolidated material?
12. What are precursors, and how can they be used to predict earthquakes?
13. What are the differences between intensity and magnitude?
14. Describe the composition, density, and depth of Earth's core, mantle, and crust.
15. What accounts for the various seismic discontinuities found with the mantle?
16. How does the elastic rebound theory account for energy released during an earthquake?
17. How would P- and S-waves behave if Earth were completely solid and had the same composition and density throughout?
18. Discuss why insurance companies use the qualitative Modified Mercalli Intensity Scale instead of the quantitative Richter Magnitude Scale in classifying earthquakes.

19. From the arrival times of P- and S-waves shown in the accompanying chart and the graph in Figure 8.10c, calculate how far away from each seismograph station the earthquake occurred. How would you determine the epicenter of this earthquake?

	Arrival Time of P-Wave	Arrival Time of S-Wave
Station A:	2:59:03 P.M.	3:04:03 P.M.
Station B:	2:51:16 P.M.	3:01:16 P.M.
Station C:	2:48:25 P.M.	2:55:55 P.M.

20. Use the graph in Figure 8.13 to answer this question. A seismograph in Berkeley, California, records the arrival time of an earthquake's P-waves as 6:59:54 P.M. and the S-waves as 7:00:02 P.M. The maximum amplitude of the S-waves as recorded on the seismogram was 75 mm. What was the magnitude of the earthquake, and how far away from Berkeley did it occur?

World Wide Web Activities

Geology Now Assess your understanding of this chapter's topics with additional quizzing and comprehensive interactivities at

http://earthscience.brookscole.com/changingearth4e

as well as current and up-to-date weblinks, additional readings, and InfoTrac College Edition exercises.

The Seafloor

CHAPTER 9

OUTLINE

- Introduction
- Exploring the Oceans
- Oceanic Crust—Its Structure and Composition
- What Are Continental Margins?
- What Features Are Found in the Deep-Ocean Basins?
- Sedimentation and Sediments on the Deep Seafloor
- Reefs
- Seawater and Seafloor Resources

  GEO-FOCUS 9.1: *Oceanic Circulation and Resources from the Sea*
- Geo-Recap

Geology Now *This icon, which appears throughout the book, indicates an opportunity to explore interactive tutorials, animations, or practice problems available on the GeologyNow website at http://earthscience.brookscole.com/changingearth4e.*

Pillow lava on the Mid-Atlantic Ridge. Much of the upper part of the oceanic crust is made up of pillow lava and sheet flow. Source: Ralph White/Corbis

OBJECTIVES

At the end of this chapter, you will have learned that:

- Scientists use echo sounding, seismic profiling, sampling, and observations from submersibles to study the largely hidden seafloor.
- Oceanic crust is thinner and compositionally less complex than continental crust.
- The margins of continents consist of a continental shelf and slope and in some cases a continental rise with adjacent abyssal plains. The elements that make up a continental margin depend on the geologic activity that takes place in these marginal areas.
- Although the seafloor is flat and featureless in some places, it also has ridges, trenches, seamounts, and other features.
- Geologic activities at or near divergent and convergent plate boundaries account for distinctive seafloor features such as submarine volcanoes and deep-sea trenches.
- Most seafloor sediment comes from the weathering and erosion of continents and oceanic islands and from the shells of tiny marine organisms.
- Organisms in warm, shallow seas build wave-resistant structures known as reefs.
- Several important resources such as common salt come from seawater, and hydrocarbons are found in some seafloor sediments.

Introduction

According to two dialogues written in about 350 B.C. by the Greek philosopher Plato, there was a huge continent called Atlantis in the Atlantic Ocean west of the Pillars of Hercules, or what we now call the Strait of Gibraltar (■ Figure 9.1). According to Plato's account, Atlantis controlled a large area extending as far east as Egypt. Yet despite its vast wealth, advanced technology, and large army and navy, Atlantis was defeated in war by Athens. Following the conquest of Atlantis,

> there were violent earthquakes and floods and one terrible day and night came when . . . Atlantis . . . disappeared beneath the sea. And for this reason even now the sea there has become unnavigable and unsearchable, blocked as it is by the mud shallows which the island produced as it sank.*

No "mud shallows" exist in the Atlantic, as Plato asserted. In fact, no geologic evidence indicates that Atlantis ever existed, so why has the legend persisted for so long?

One reason is that sensational stories of lost civilizations are popular, but another is that until recently no one had much knowledge of what lies beneath the oceans. Much of the seafloor is a hidden domain, so myths and legends were widely accepted. The most basic observation we can make about Earth is that it has vast water-covered areas and continents, which at first glance might seem to be nothing more than parts of the planet not covered by water. Nevertheless, the continents and the ocean basins are very different.

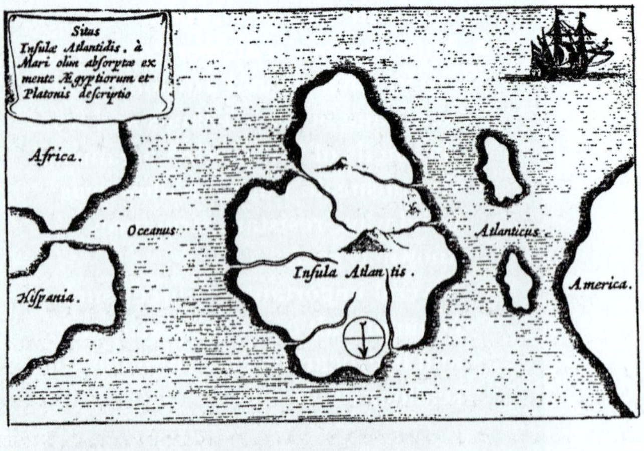

■ **Figure 9.1**

According to Plato, Atlantis was a continent west of the Pillars of Hercules, now called the Strait of Gibraltar. In this map from Anthanasium Kircher's *Mundus Subterraneus* (1664), north is toward the bottom of the map. The Strait of Gibraltar is the narrow area between Hispania (Spain) and Africa.

Continental crust is thicker and less dense than oceanic crust. Oceanic crust is composed of basalt and gabbro, whereas continental crust is made up of all rock types, although its overall composition compares closely to granite. Oceanic crust is produced continuously at spreading ridges and consumed at subduction zones, so none of it is very old, geologically speaking (see Chapter 2). The oldest oceanic crust is about 180 million years old, but the age of rocks on continents varies from recent to 3.96 billion years.

One important reason to study the seafloor is that it makes up the largest part of Earth's surface (■ Figure 9.2). Despite the commonly held misconception that the seafloor is flat and featureless, its topography is as varied as that of the continents. Furthermore, many seafloor features as well as several aspects of the oceanic crust provide important evidence for plate tectonic theory (see Chapter 2). And finally, natural resources are found on the marginal parts of continents, in seawater, and on the seafloor.

As we begin our investigation of the seafloor, you should be aware that our discussion focuses on (1) the physical attributes and composition of the oceanic crust, (2) the composition and distribution of seafloor sediments, (3) seafloor topography, and (4) the origin and evolution of the continental margins. *Oceanographers* study these topics too, but they also study the chemistry and physics of seawater as well as oceanic circulation patterns and marine biology. We should also point out that whereas the oceans and their marginal seas (Figure 9.2) are largely underlain by oceanic crust, the same is not true of the Dead Sea, Salton Sea, and Caspian Sea; these are actually saline lakes on the continents.

EXPLORING THE OCEANS

An interconnected body of saltwater that we call oceans and seas covers 71% of Earth's surface. Nevertheless, this world ocean has areas distinct enough for us to recognize the Pacific, Atlantic, Indian, and Arctic Oceans. The term *ocean* refers to these large areas of saltwater, whereas *sea* designates a smaller body of water, usually a marginal part of an ocean (Figure 9.2).

During most of historic time, people knew little of the oceans, and until recently they thought the seafloor was a vast, featureless plain. In fact, through most of this time the seafloor, in one sense, was more remote than the Moon's surface because it could not be observed.

*From the *Timaeus*, quoted in E. W. Ramage, Ed., *Atlantis: Fact or Fiction?* (Bloomington: Indiana University Press, 1978), p. 13.

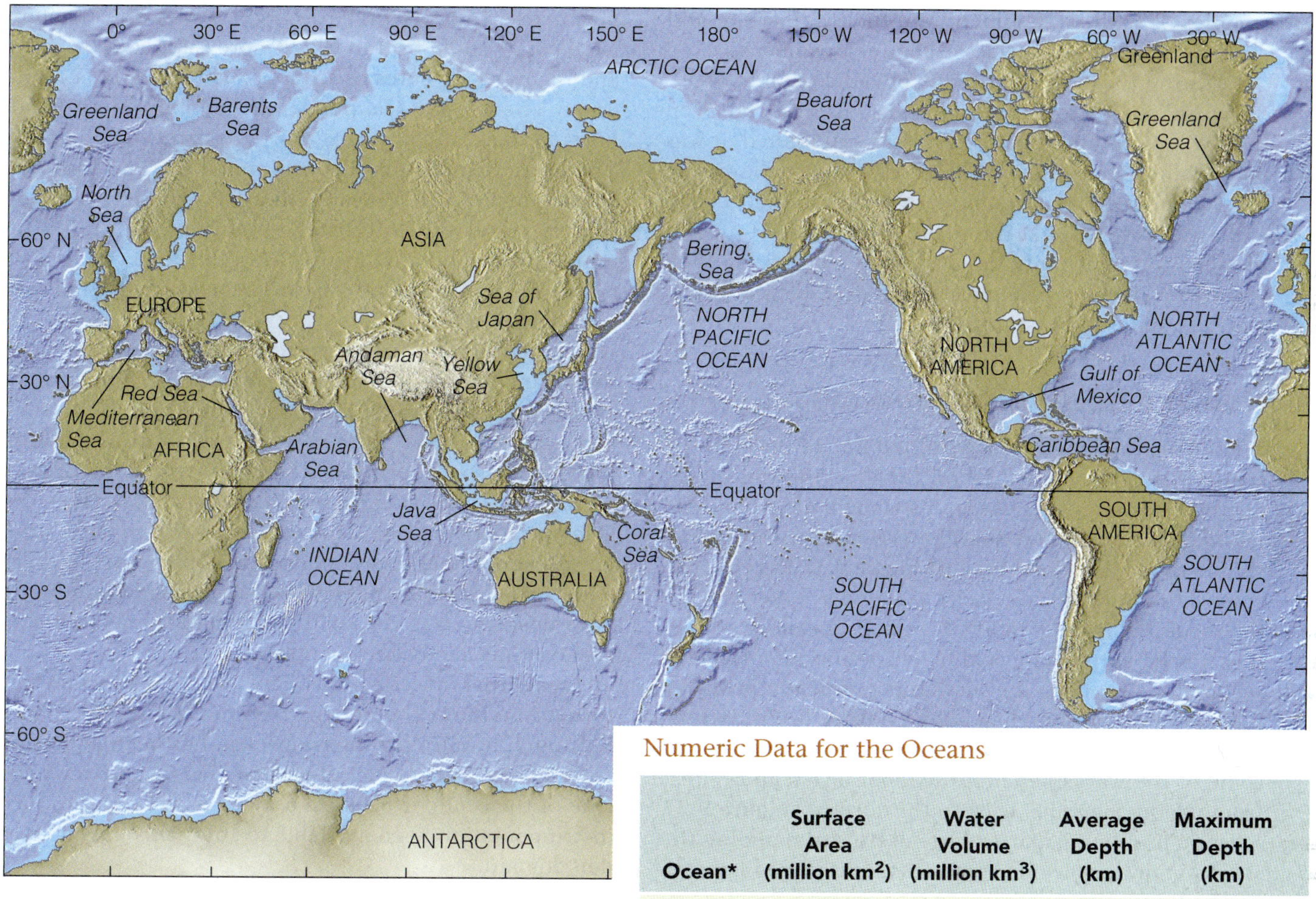

Numeric Data for the Oceans

Ocean*	Surface Area (million km^2)	Water Volume (million km^3)	Average Depth (km)	Maximum Depth (km)
Pacific	180	700	4.0	11.0
Atlantic	93	335	3.6	9.2
Indian	77	285	3.7	7.5
Arctic	15	17	1.1	5.2

Source: P. R. Pinet, 1992. *Oceanography* (St. Paul, MN: West, 1992).

*Excludes adjacent seas, such as the Caribbean Sea and Sea of Japan, which are marginal parts of oceans.

Figure 9.2

A map of the four major oceans and many of the world's seas, which are marginal parts of oceans.

Early Exploration

The ancient Greeks had determined Earth's size and shape rather accurately, but western Europeans were not aware of the vastness of the oceans until the 1400s and 1500s, when explorers sought trade routes to the Indies. Even when Christopher Columbus set sail on August 3, 1492, in an effort to find a route to the Indies, he greatly underestimated the width of the Atlantic Ocean. Contrary to popular belief, he was not attempting to demonstrate Earth's spherical shape; its shape was well accepted by then. The controversy was over Earth's circumference and the shortest route to China; on these points, Columbus's critics were correct.

These and similar voyages added considerably to the growing body of knowledge about the oceans, but truly scientific investigations did not begin until the late 1700s. At that time Great Britain was the dominant maritime power, and to maintain that dominance the British sought to increase their knowledge of the oceans. So, scientific voyages led by Captain James Cook were launched in 1768, 1772, and 1777. From 1831 until 1836, the HMS *Beagle* sailed the seas. Aboard was Charles Darwin, who is well known for his views of organic evolution but who also proposed a theory on the evolution of coral reefs. In 1872, the converted British warship HMS *Challenger* began a four-year voyage to sample seawater, determine oceanic depths, collect samples of seafloor sediment and rock, and name and classify thousands of species of marine organisms.

During these voyages many oceanic islands previously unknown to Europeans were visited. And even though exploration of the oceans was still limited, it was becoming increasingly apparent that the seafloor was not flat and featureless as formerly believed. Indeed, scientists discovered that the seafloor has varied topography just as continents do, and they recognized such features as oceanic trenches, submarine ridges, broad plateaus, hills, and vast plains.

How Are Oceans Explored Today?

Measuring the length of a weighted line lowered to the seafloor was the first method for determining ocean depths. Now scientists use an instrument called an *echo sounder,* which detects sound waves that travel from a ship to the seafloor and back (■ Figure 9.3). Depth is calculated by knowing the velocity of sound in water and the time required for the waves to reach the seafloor and return to the ship, thus yielding a continuous profile of seafloor depths along the ship's route. **Seismic profiling** is similar to echo sounding but even more useful. Strong waves from an energy source reflect from the seafloor, and some of the waves penetrate seafloor layers and reflect from various horizons back to the surface (Figure 9.3). Seismic profiling is particularly useful for mapping the structure of the oceanic crust where it is buried beneath seafloor sediments.

The Deep Sea Drilling Project, an international program sponsored by several oceanographic institutions, began in 1968. Its first research vessel, the *Glomar Challenger,* could drill in water more than 6000 m deep and recover long cores of seafloor sediment and oceanic crust. The *Glomar Challenger* drilled more than 1000 holes in the seafloor during the 15 years of the program. The Deep Sea Drilling Project ended in 1983, but beginning in 1985 the Ocean Drilling Program with its research vessel the JOIDES* *Resolution* continued to explore the seafloor (■ Figure 9.4a). Research vessels also sample the seafloor using *clamshell samplers* and *piston corers* (■ Figure 9.5).

In addition to surface vessels, submersibles are now important vehicles for seafloor exploration. Some, such as the *Argo,* are remotely controlled and towed by a surface vessel. In 1985 the *Argo,* equipped with sonar and television systems, provided the first views of the British ocean liner HMS *Titanic* since it sank in 1912. The U.S. Geological Survey uses a towed device with sonar to produce seafloor images resembling aerial photographs. Scientists aboard submersibles such as *Alvin* (Figure 9.4b) have descended to the seafloor in many areas to make observations and collect samples.

Scientific investigations have yielded important information about the oceans for more than 200 years, but much of our current knowledge has been acquired since World War II (1939–1945). This is particularly true of the seafloor because only in recent decades has instrumentation been available to study this largely hidden domain.

*JOIDES is an acronym for Joint Oceanographic Institutions for Deep Earth Sampling.

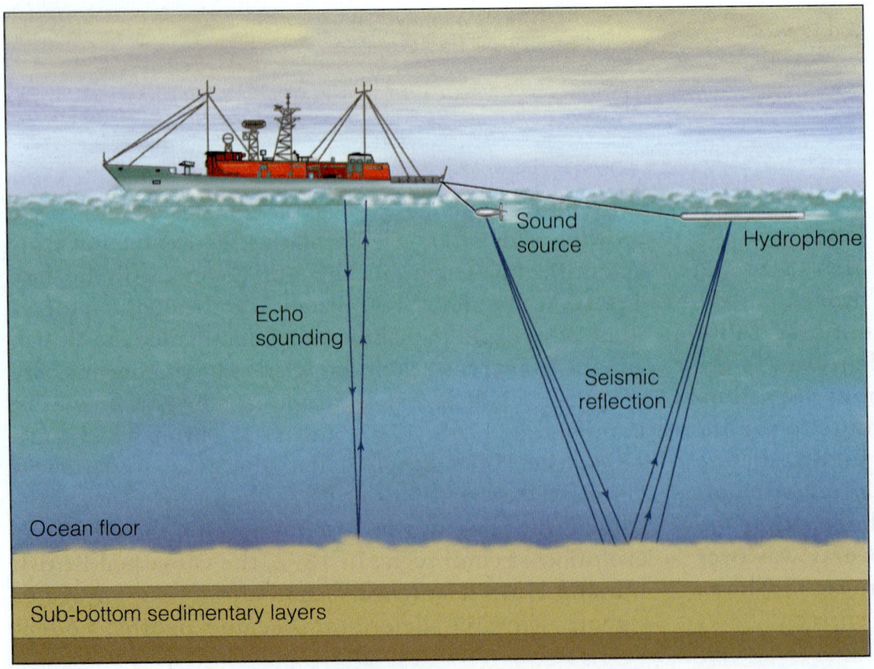

■ **Figure 9.3**

Diagram showing how echo sounding and seismic profiling are used to study the seafloor. Some of the energy generated at the energy source is reflected from various horizons back to the surface, where it is detected by hydrophones.

(a)

(b)

Figure 9.4

Oceanographic research vessels. (a) The JOIDES *Resolution* is capable of drilling the deep seafloor. (b) The submersible *Alvin* is used for observing and sampling the deep seafloor.

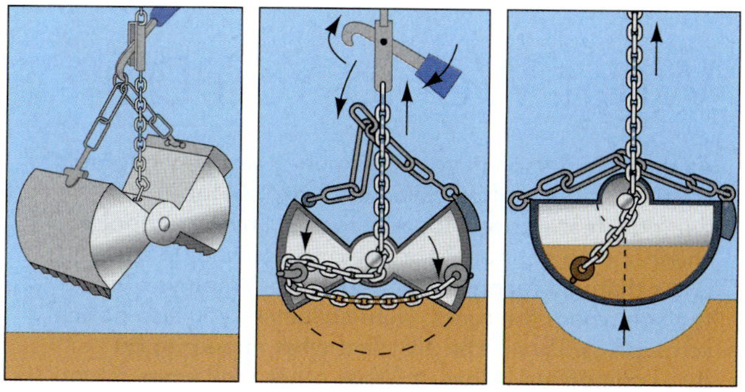

(a)

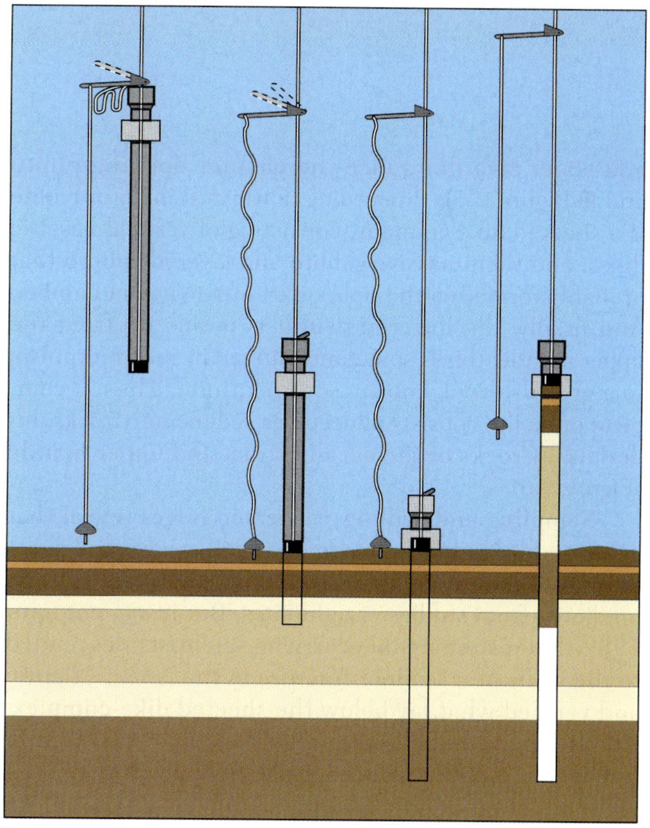

(b)

Figure 9.5

Sampling the seafloor. (a) A clamshell sampler taking a seafloor sample. (b) A piston corer falls to the seafloor, penetrates the sediment, and then is retrieved.

OCEANIC CRUST— ITS STRUCTURE AND COMPOSITION

We have mentioned that oceanic crust is composed of basalt and gabbro and is generated continuously at spreading ridges. Of course, drilling into the oceanic crust provides some details about its composition and structure, but it has never been completely penetrated and sampled. So how do we know what it is composed of and how it varies with depth? Actually, even before it was sampled and observed, these details were known.

Remember that oceanic crust is consumed at subduction zones and thus most of it is recycled, but a small amount is found in mountain ranges on land where it was emplaced by moving along large fractures called thrust faults (faults are discussed more fully in Chapter 10). These preserved slivers of oceanic crust along with part of the underlying upper mantle are known as **ophiolites.** Detailed studies reveal that an ideal ophiolite consists of deep-sea sedimentary rocks underlain by rocks of the upper oceanic crust, especially pillow lava

What Would You Do?

As the only person in your community with any geologic training, you are often called onto explain local geologic features and identify fossils. Several school children on a natural history field trip picked up some rocks that you recognize as peridotite. When you visit the site where the rocks, were collected, you also notice some pillow lava in the area and what appear to be dikes composed of basalt. What other rock types might you expect to find here? How would you explain (1) the association of these rocks with one another, and (2) how they came to be on land?

and sheet lava flows (see the chapter opening photo and ■ Figure 9.6). Proceeding downward in an ophiolite is a sheeted dike complex, consisting of vertical basaltic dikes, and then massive gabbro and layered gabbro that probably formed in the upper part of a magma chamber. And finally, the lowermost unit is peridotite from the upper mantle; this is sometimes altered by metamorphism to a greenish rock known as serpentinite. Thus, a complete ophiolite consists of deep-sea sedimentary rocks underlain by rocks of the oceanic crust and upper mantle (Figure 9.6).

Sampling and drilling at oceanic ridges reveal that oceanic crust is indeed made up of pillow lava and sheet lava flows underlain by a sheeted dike complex, just as predicted from studies of ophiolites. But it was not until 1989 that a submersible carrying scientists descended to the walls of a seafloor fracture in the North Atlantic and verified what lay below the sheeted dike complex. Just as expected, the lower oceanic crust consists of gabbro and the upper mantle is made up of peridotite.

WHAT ARE CONTINENTAL MARGINS?

In the Introduction we made the point that continents are not simply areas above sea level, although most people perceive of continents as land areas outlined by the oceans. The true geologic margin of a continent—that is, where granitic continental crust changes to oceanic crust composed of basalt and gabbro—is below sea level. Accordingly, the margins of continents are submerged, and we recognize **continental margins** as separating the part of a continent above sea level from the deep seafloor.

A continental margin is made up of a gently sloping continental shelf, a more steeply inclined continental slope, and in some cases, a deeper, gently sloping continental rise (■ Figure 9.7). Seaward of the continental margin lies the deep ocean basin. Thus, the continental margins extend to increasingly greater depths until they merge with the deep seafloor. Continental crust changes to oceanic crust somewhere beneath the continental rise, so part of the continental slope and the continental rise actually rest on oceanic crust.

The Continental Shelf

As one proceeds seaward from the shoreline across the continental margin, the first area encountered is a gently sloping **continental shelf** lying between the shore and the more steeply dipping continental slope (Figure 9.7). The width of the continental shelf varies considerably, ranging from a few tens of meters to more than 1000 km; the shelf terminates where the inclination of the seafloor increases abruptly from 1 degree or less to several degrees. The outer margin of the continental shelf, or simply the *shelf–slope break,* is at an av-

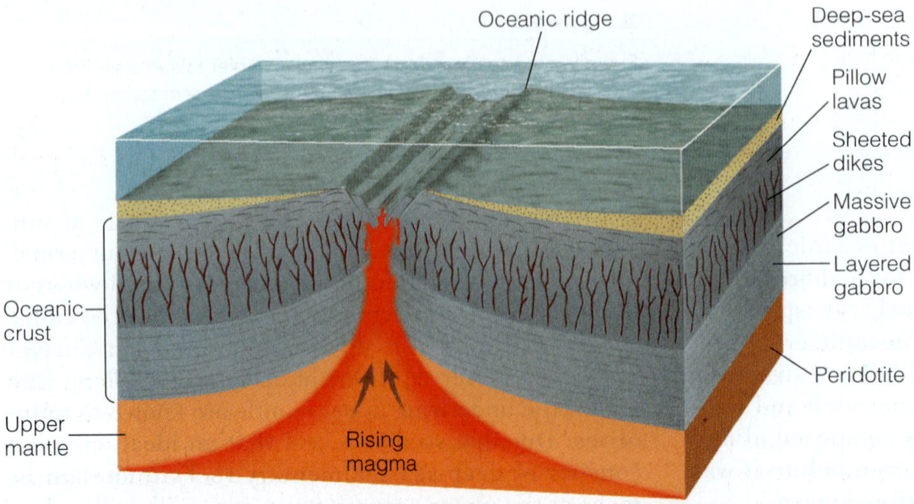

■ **Figure 9.6**

Oceanic crust consisting of the layers shown here forms as magma rises beneath oceanic ridges. Fragments of oceanic crust and upper mantle on land are known as ophiolites.

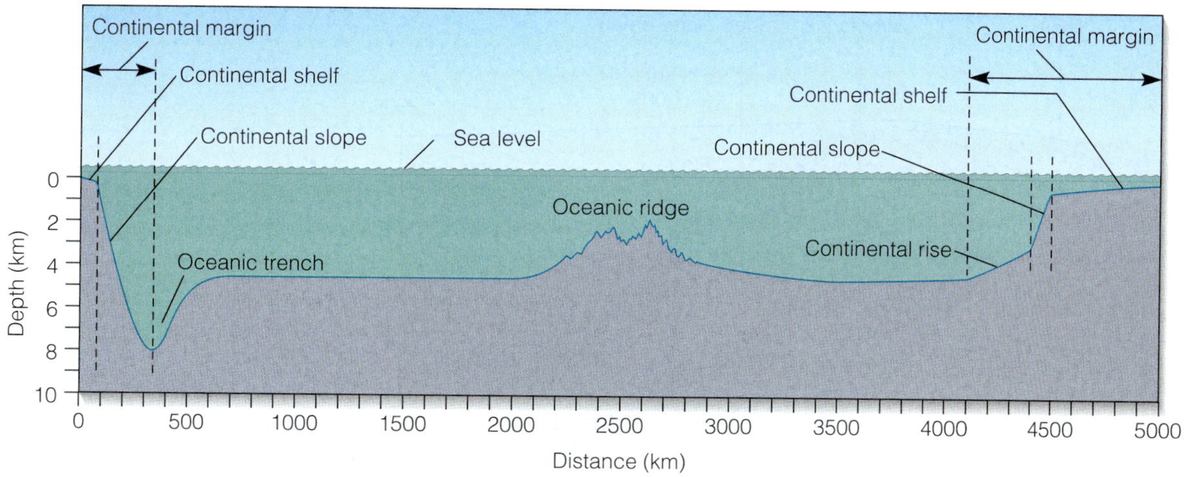

■ **Figure 9.7**

A generalized profile of the seafloor showing features of the continental margins. The vertical dimensions of the features in this profile are greatly exaggerated because the vertical and horizontal scales differ.

erage depth of 135 m, so by oceanic standards the continental shelves are covered by shallow water.

At times during the Pleistocene Epoch (1.6 million to 10,000 years ago), sea level was as much as 130 m lower than it is now. As a result, the continental shelves were above sea level and were areas of stream channel and floodplain deposition. In addition, in many parts of northern Europe and North America, glaciers extended well out onto the continental shelves and deposited gravel, sand, and mud. Since the Pleistocene ended, sea level has risen, submerging these deposits, which are now being reworked by marine processes. Evidence that these sediments were in fact deposited on land includes remains of human settlements and fossils of a variety of land-dwelling animals (see Chapter 23).

The Continental Slope and Rise

The seaward margin of the continental shelf is marked by the *shelf–slope break* (at an average depth of 135 m), where the more steeply inclined **continental slope** begins (Figure 9.7). In most areas around the margins of the Atlantic, the continental slope merges with a more gently sloping **continental rise.** This rise is absent around the margins of the Pacific where continental slopes descend directly into an oceanic trench (Figure 9.7).

The shelf–slope break, marking the boundary between the shelf and slope, is an important feature in terms of sediment transport and deposition. Landward of the break—that is, on the shelf—sediments are affected by waves and tidal currents, but these processes have no effect on sediments seaward of the break, where gravity is responsible for their transport and deposition on the slope and rise. In fact, much of the land-derived sediment crosses the shelves and is eventually deposited on the continental slopes and rises, where more than 70% of all sediments in the oceans are found. Much of this sediment is transported through submarine canyons by turbidity currents.

Submarine Canyons, Turbidity Currents, and Submarine Fans

In Chapter 6, we discussed the origin of graded bedding, most of which results from **turbidity currents,** underwater flows of sediment–water mixtures with densities greater than sediment-free water. As a turbidity current flows onto the relatively flat seafloor, it slows and begins depositing sediment, the largest particles first, followed by progressively smaller particles, thus forming a layer with graded bedding (see Figure 6.23). Deposition by turbidity currents yields a series of overlapping **submarine fans,** which constitute a large part of the continental rise (■ Figure 9.8). Submarine fans are distinctive features, but their outer margins are difficult to discern because they grade into deposits of the deep-ocean basin.

No one has ever observed a turbidity current in progress in the oceans, so for many years some doubted their existence; however, in 1971 abnormally turbid water was sampled just above the seafloor in the North Atlantic, indicating that a turbidity current had recently occurred. In addition, seafloor samples from many areas show a succession of layers with graded bedding and the remains of shallow-water organisms that were displaced into deeper water by turbidity currents (Figure 9.8).

Perhaps the most compelling evidence for turbidity currents is the pattern of trans-Atlantic cable breaks that took place in the North Atlantic near Newfoundland on November 18, 1929. Initially, an earthquake was assumed to have ruptured telephone and telegraph cables. However, whereas the breaks on the continental shelf

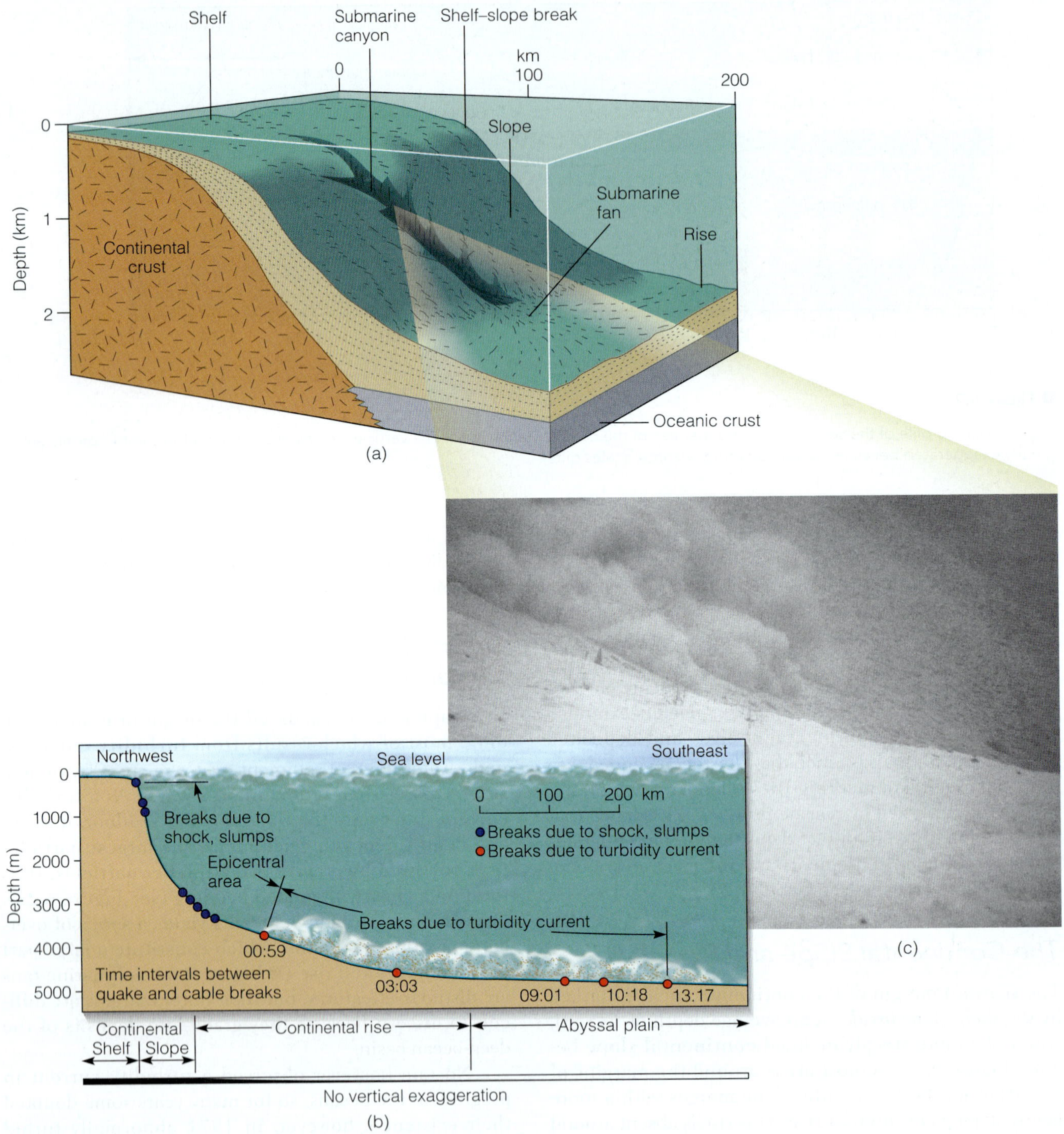

Figure 9.8

(a) Much of the continental rise is made up of submarine fans deposited by turbidity currents that moved through submarine canyons. (b) Submarine cable breaks caused by a turbidity current south of Newfoundland in 1929. The profile labeled "No vertical exaggeration" shows what the seafloor actually looks like in this area. (c) The propeller of a submarine caused this turbidity current that flowed down a slope near Jamaica.

near the epicenter occurred when the earthquake struck, cables farther seaward were broken later and in succession (Figure 9.8b). The last cable to break was 720 km from the source of the earthquake, and it did not snap until 13 hours after the first break. In 1949, geologists realized that an earthquake-generated turbidity current had moved downslope, breaking the cables in succession. The precise time at which each cable broke was known, so calculating the velocity of the turbidity current was simple. It moved at about 80 km/hr on the continental slope, but slowed to about 27 km/hr when it reached the continental rise.

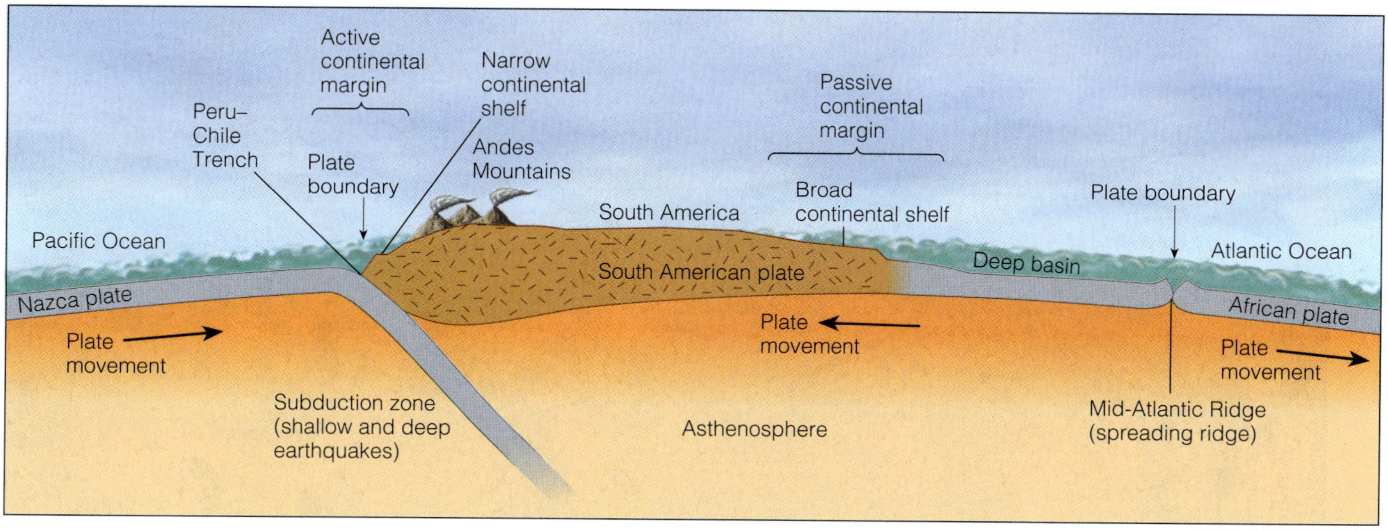

Figure 9.9

Active and passive continental margins along the west and east coasts of South America. Notice that passive continental margins are much wider than active ones. Seafloor sediment is not shown.

Deep, steep-sided **submarine canyons** are present on continental shelves, but they are best developed on continental slopes (Figure 9.8a). Some submarine canyons extend across the shelf to rivers on land; they apparently formed as river valleys when sea level was lower during the Pleistocene. However, many have no such association, and some extend far deeper than can be accounted for by river erosion during times of lower sea level. Scientists know that strong currents move through submarine canyons. Turbidity currents periodically move through these canyons and are now thought to be the primary agent responsible for their erosion.

Types of Continental Margins

Continental margins are *active* or *passive,* depending on their relationship to plate boundaries. An **active continental margin** develops at the leading edge of a continental plate where oceanic lithosphere is subducted. The western margin of South America is a good example (■ Figure 9.9). Here, an oceanic plate is subducted beneath the continent, resulting in seismic activity, a geologically young mountain range, and active volcanism. In addition, the continental shelf is narrow, and the continental slope descends directly into the Peru–Chile Trench, so sediment is dumped into the trench and no continental rise develops. The western margin of North America is also considered an active continental margin, although much of it is now bounded by transform faults rather than a subduction zone. However, plate convergence and subduction still take place in the Pacific Northwest along the continental margins of northern California, Oregon, and Washington.

The continental margins of eastern North America and South America differ considerably from their western margins. For one thing, they possess broad continental shelves as well as a continental slope and rise; also, vast, flat *abyssal plains* are present adjacent to the rises (■ Figure 9.9). Furthermore, these **passive continental margins** are within a plate rather than at a plate boundary, and they lack the typical volcanic and seismic activity found at active continental margins. Nevertheless, earthquakes do take place occasionally at these margins.

Active and passive continental margins share some features, but they are notably different in the widths of their continental shelves, and active margins have an oceanic trench but no continental rise. Why the differences? At both types of continental margins, turbidity currents transport sediment into deeper water. At passive margins, the sediment forms a series of overlapping submarine fans and thus develops a continental rise, whereas at an active margin, sediment is simply dumped into the trench and no rise forms. The proximity of a trench to a continent also explains why the continental shelves of active margins are so narrow. In contrast, land-derived sedimentary deposits at passive margins have built a broad platform extending far out into the ocean.

WHAT FEATURES ARE FOUND IN THE DEEP-OCEAN BASINS?

Most of the seafloor, with an average depth of 3.8 km, lies far below the depth of sunlight penetration, which is generally less than 100 m. Accordingly, most of the seafloor is completely dark, no plant life exists, the temperature is just above freezing, and the pressure varies from 200 to

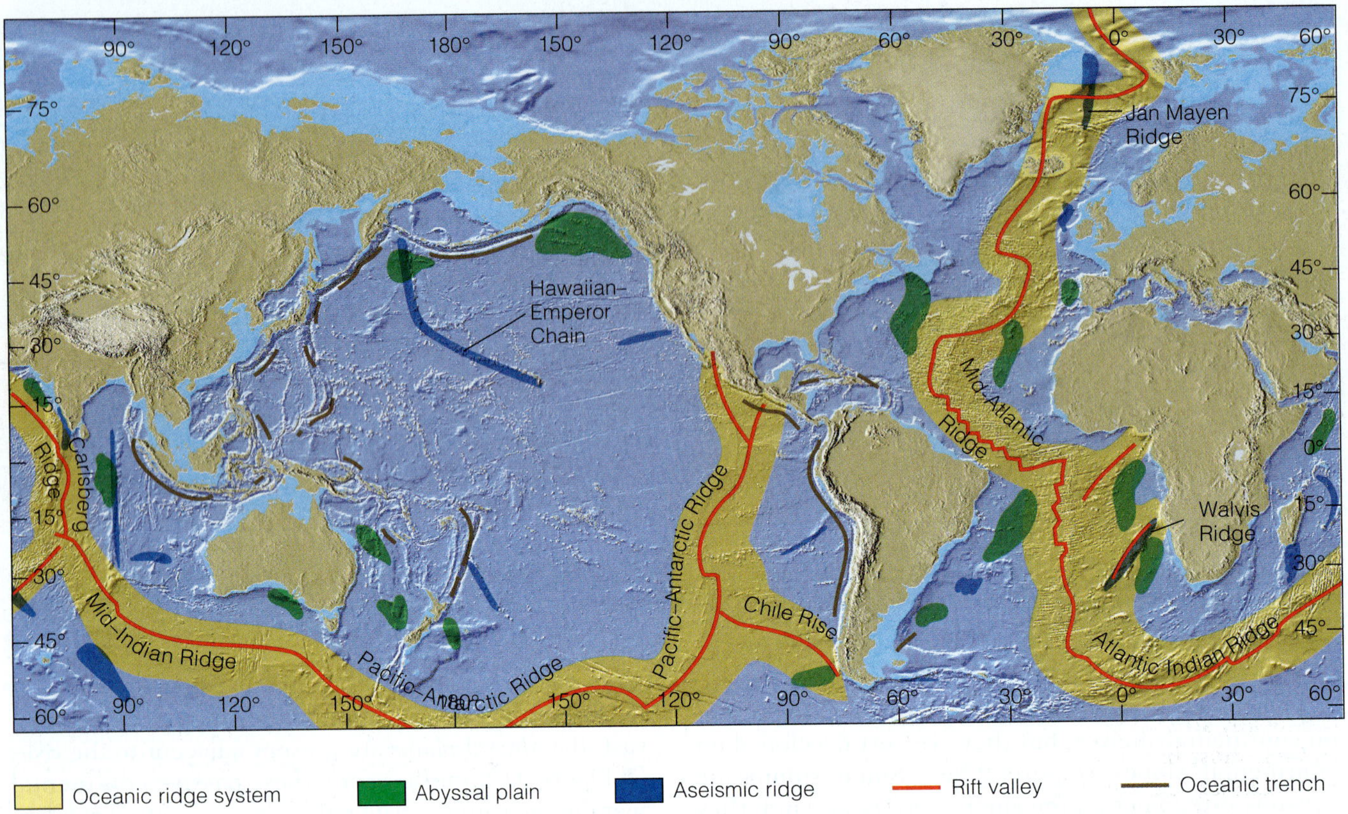

Figure 9.10

The distribution of oceanic trenches (brown), abyssal plains (green), the oceanic ridge system (yellow) and rift valleys (red), and some of the aseismic ridges (blue). Source: From Alyn and Alison Duxbury, *An Introduction to the World's Oceans*. Copyright © 1984 McGraw-Hill Companies, Inc. Reprinted with permission.

more than 1000 atmospheres depending on depth. In fact, biologic productivity is low on the deep seafloor with the exception of hydrothermal vent communities (discussed later).

Scientists have descended to the greatest oceanic depths, submarine ridges, and elsewhere in submersibles, so they have observed some of the seafloor. Nevertheless, much of the seafloor has been studied only by echo sounding, seismic profiling, sampling of seafloor sediments and oceanic crust, and remotely controlled submersibles. Oceanographers are developing a more thorough understanding of the oceans and the deep seafloor, and they now know of many deep-ocean features such as vast plains, trenches, and ridges.

Abyssal Plains

Beyond the continental rises of passive continental margins are **abyssal plains,** flat surfaces covering vast areas of the seafloor. In some areas they are interrupted by peaks rising more than 1 km, but abyssal plains are nevertheless the flattest, most featureless areas on Earth (Figure 9.10). Their flatness is a result of sediment deposition covering the rugged topography of the seafloor.

Abyssal plains are invariably found adjacent to continental rises, which are composed mostly of overlapping submarine fans. Along active continental margins, sediments derived from the shelf and slope are trapped in an oceanic trench, and abyssal plains fail to develop. Accordingly, abyssal plains are common in the Atlantic Ocean basin but rare in the Pacific Ocean basin (Figure 9.10).

Oceanic Trenches

Long, steep-sided depressions on the seafloor near convergent plate boundaries, or simply **oceanic trenches,** constitute no more than 2% of the seafloor, but they are important features because it is here that lithospheric plates are consumed by subduction (see Chapter 2). Because oceanic trenches are found along active continental margins, they are common in the Pacific Ocean basin but largely lacking in the Atlantic, those in the Caribbean being notable exceptions (Figure 9.10). On the landward sides of oceanic trenches, the continental slope descends into them at up to 25 degrees, and many have thick accumulations of sediments. The greatest oceanic depths are found in these trenches; the Challenger Deep of the Marianas Trench in the Pacific is more than 11,000 m deep!

Sensitive instruments can detect the amount of heat energy escaping from Earth's interior by the phenomenon of *heat flow*. As one might expect, heat flow is greatest in areas of active or recently active volcanism. For instance, higher-than-average heat flow takes place at spreading ridges, but at subduction zones heat flow values are less than the average for Earth as a whole. It seems that oceanic crust at oceanic trenches is cooler and slightly denser than elsewhere.

Seismic activity also takes place at or near oceanic trenches along planes dipping at about 45 degrees. In Chapter 8 we discussed these inclined seismic zones called Benioff zones (see Figure 8.5), where most of Earth's intermediate and deep earthquakes occur. Volcanism does not take place in trenches, but because these are zones where oceanic lithosphere is subducted beneath either oceanic or continental lithosphere, an arcuate chain of volcanoes is found on the overriding plate (Figure 9.9). The Aleutian Islands and the volcanoes along the western margin of South America are good examples of such chains.

Oceanic Ridges

When the first submarine cable was laid between North America and Europe during the late 1800s, a feature called the Telegraph Plateau was discovered in the North Atlantic. Using data from the 1925–27 voyage of the German research vessel *Meteor,* scientists proposed that the plateau was actually a continuous ridge extending the length of the Atlantic Ocean basin. Subsequent investigations revealed that this conjecture was correct, and we now call this feature the Mid-Atlantic Ridge (Figure 9.10).

The Mid-Atlantic Ridge is more than 2000 km wide and rises 2 to 2.5 km above the adjacent seafloor. Furthermore, it is part of a much larger **oceanic ridge** system of mostly submarine mountainous topography. This system runs from the Arctic Ocean through the middle of the Atlantic and curves around South Africa, where the Indian Ridge continues into the Indian Ocean; the Atlantic–Pacific Ridge extends eastward and a branch of this, the East Pacific Rise, trends northeast until it reaches the Gulf of California (Figure 9.10). The entire system is at least 65,000 km long, far exceeding the length of any mountain system on land. Oceanic ridges are composed almost entirely of basalt and gabbro and possess features produced by tensional forces. Mountain ranges on land, in contrast, consist of igneous, metamorphic, and sedimentary rocks, and they formed when rocks were folded and fractured by compressive forces (see Chapter 10).

Oceanic ridges are mostly below sea level, but they rise above the sea in Iceland, the Azores, Easter Island, and several other places. Of course, oceanic ridges are the sites where new oceanic crust is generated and plates diverge (see Chapter 2). The rate of plate divergence is important because it determines the cross-sectional profile of a ridge. For example, the Mid-Atlantic Ridge has a comparatively steep profile because divergence is slow, allowing the new oceanic crust to cool, shrink, and subside closer to the ridge crest than it does in areas of faster divergence such as the East Pacific Rise. A ridge may also have a rift along its crest that opens in response to tension (■ Figure 9.11a). A rift is particularly obvious along the Mid-Atlantic Ridge, but it is absent along parts of the East Pacific Rise. These rifts are commonly 1 to 2 km deep and several kilometers wide. They open as seafloor spreading takes place (discussed in Chapter 2) and are characterized by shallow-focus earthquakes, basaltic volcanism, and high heat flow.

Even though most oceanographic research is still done by echo sounding, seismic profiling, and seafloor sampling, scientists have been making direct observations of oceanic ridges and their rifts since 1974. As part of Project FAMOUS (French-American Mid-Ocean Undersea Study), submersibles have descended to the ridges and into their rifts in several areas. Researchers have not seen any active volcanism, but they did see pillow lavas (see Figure 5.7), lava tubes, and sheet lava flows, some of which formed very recently. In fact, on return visits to a site, they have seen the effects of volcanism that occurred since their previous visit. And on January 25, 1998, a submarine volcano began erupting along the Juan de Fuca Ridge west of Oregon. Researchers aboard submersibles have also observed hot water being discharged from the seafloor at or near ridges in submarine hydrothermal vents.

Submarine Hydrothermal Vents

Scientists first saw **submarine hydrothermal vents** on the seafloor in 1979 when they descended about 2500 m to the Galapagos Rift in the eastern Pacific Ocean. Since 1979, they have seen similar vents in several other areas in the Pacific (Figure 9.11b), Atlantic, and Sea of Japan. The vents are at or near spreading ridges where cold seawater seeps through oceanic crust, is heated by the hot rocks at depth, and then rises and discharges into the seawater as plumes of hot water with temperatures as high as 400°C. Many of the plumes are black because dissolved minerals giving them the appearance of black smoke—hence the name **black smoker.**

Submarine hydrothermal vents are interesting from the biologic, geologic, and economic points of view. Near the vents live communities of organisms, such as bacteria, crabs, mussels, starfish, and tube worms, many of which had never been seen before (Figure 9.11c). No sunlight is available, so the organisms in these communities depend on bacteria that oxidize sulfur compounds for their ultimate source of nutrients. The vents are also interesting because of their economic potential. The heated seawater reacts with oceanic crust, transforming

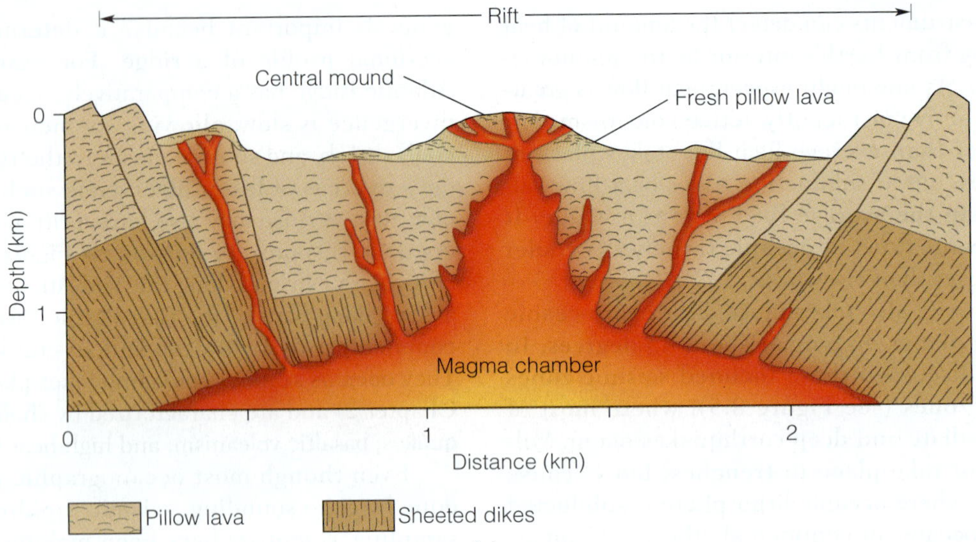

(a)

(b)

(c)

Figure 9.11

(a) Cross section of the Mid-Atlantic Ridge showing its central rift with moundlike accumulations of volcanic rocks, mostly pillow lava. (b) A hydrothermal vent known as a black smoker at 2800 m on the East Pacific Rise. The plume of "black smoke" is hot water saturated with dissolved minerals. (c) Several types of organisms, including these 1.5-m-long tube worms, live near black smokers.

it into a metal-rich solution that discharges into seawater, and cools, precipitating iron, copper, and zinc sulfides and other minerals. A chimneylike vent forms that eventually collapses and forms a mound of sediments rich in the elements mentioned above.

Apparently the chimneys through which black smokers discharge grow rapidly. A 10-m-high chimney accidentally knocked over in 1991 by the submersible *Alvin* grew to 6 m in just three months. Also in 1991 scientists aboard *Alvin* saw the results of a submarine eruption on the East Pacific Rise, which they missed by less than two weeks. Fresh lava and ash covered the area as well as the remains of tube worms killed during the eruption. And in a nearby area, a new fissure opened in the seafloor; by December 1993, a new hydrothermal vent community had become well established.

In 2001 scientists announced another kind of seafloor vent in the North Atlantic responsible for massive pillars and spires as tall as 60 m. Unlike the black smokers, though, these vents are 14–15 km from spreading ridges, and they consist of light-colored minerals that were derived by chemical reaction between seawater and minerals in the oceanic crust.

What Would You Do?

Hydrothermal vents on the seafloor are known sites of several metals of great importance to industrialized societies. Furthermore, it appears that these metals are being deposited even now, so if we mine one area, more of the same resources form elsewhere. Given these conditions, it would seem that our problems of diminishing resources are solved. So why not simply mine the seafloor? Also, many chemical elements are present in seawater. The technology exists to extract elements such as gold, uranium, and others, so why not do so?

Seafloor Fractures

Oceanic ridges are not continuous features winding without interruption around the globe. They abruptly terminate where they are offset along fractures oriented more or less at right angles to ridge axes (■ Figure 9.12). These large-scale fractures are hundreds of kilometers long, although they are difficult to trace where they are buried beneath seafloor sediments. Many geologists are convinced that some geologic features on the continents are best explained by the extension of these fractures into continents.

Shallow-focus earthquakes take place along these fractures, but only between the displaced ridge segments. Furthermore, because ridges are higher than the adjacent seafloor, the offset segments yield nearly vertical escarpments 2 or 3 km high (Figure 9.12). The reason oceanic ridges have so many fractures is that plate divergence takes place irregularly on a sphere, resulting in stresses that cause fracturing. We discussed these fractures between offset ridge segments more fully in Chapter 2, where they are termed *transform faults*.

Seamounts, Guyots, and Aseismic Ridges

As noted, the seafloor is not a flat, featureless plain except for the abyssal plains, and even these are underlain by rugged topography. In fact, a large number of volcanic hills, seamounts, and guyots rise above the seafloor in all ocean basins, but they are particularly abundant in the Pacific. All are of volcanic origin and differ mostly in size. **Seamounts** rise more than 1 km above the seafloor, and if flat topped, they are called

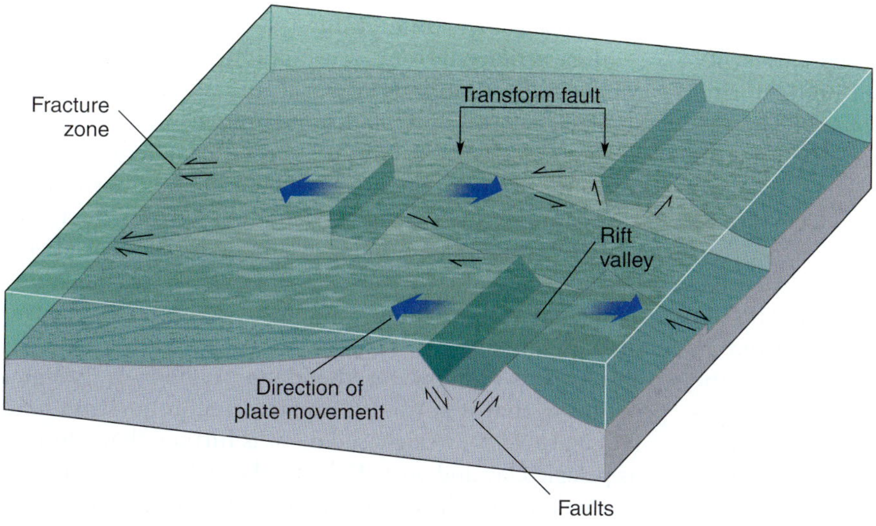

■ **Figure 9.12**

Diagrammatic view of an oceanic ridge offset along fractures. That part of a fracture between displaced segments of the ridge crest is known as a transform fault (see Chapter 2).

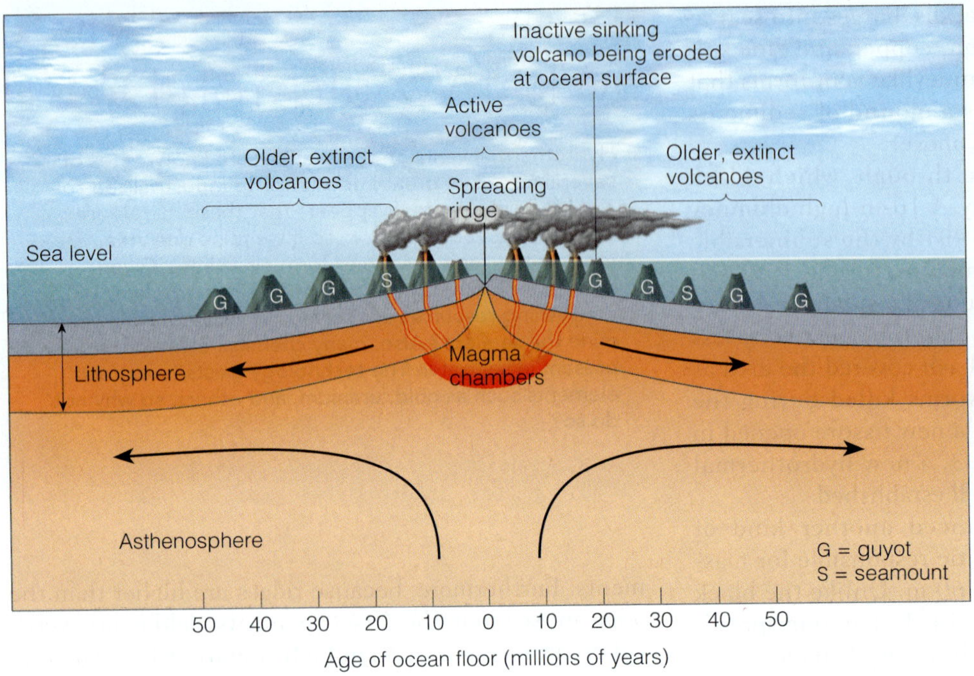

Figure 9.13

The origin of seamounts (S) and guyots (G). As a plate on which a volcano rests moves into greater depths, a volcanic island may be eroded and become a flat-topped guyot.

guyots (■ Figure 9.13). Guyots are volcanoes that originally extended above sea level. However, as the plate upon which they were situated continued to move, they were carried away from a spreading ridge, and as the oceanic crust cooled it descended to greater depths. Thus, what was once an island slowly sank beneath the sea, and as it did, wave erosion produced the typical flat-topped appearance (Figure 9.13). Many other volcanic features smaller than seamounts exist on the seafloor, but they probably originated in the same way. These so-called *abyssal hills* average only about 250 m high.

Other common features in the ocean basins are long, narrow ridges and broad plateaulike features rising as much as 2 to 3 km above the surrounding seafloor. These are **aseismic ridges,** so called because they lack seismic activity. A few of these ridges are probably small fragments separated from continents during rifting and are referred to as *microcontinents*. The Jan Mayen Ridge in the North Atlantic is probably a microcontinent (Figure 9.10).

Most aseismic ridges form as a linear succession of hot-spot volcanoes. These may develop at or near an oceanic ridge, but each volcano so formed is carried laterally with the plate upon which it originated. The net result is a line of seamounts/guyots extending from an oceanic ridge (Figure 9.13); the Walvis Ridge in the South Atlantic is a good example (Figure 9.10). Aseismic ridges also form over hot spots unrelated to ridges—the Hawaiian–Emperor chain in the Pacific, for instance (Figure 9.10).

SEDIMENTATION AND SEDIMENTS ON THE DEEP SEAFLOOR

Sediments on the deep seafloor are mostly fine grained, consisting of silt- and clay-sized particles, because few processes transport sand and gravel very far from land. Certainly icebergs carry sand and gravel, and, in fact, a broad band of glacial-marine sediment is adjacent to Antarctica and Greenland. Floating vegetation might also carry large particles far out to sea, but it contributes very little sediment to the deep seafloor.

Most of the fine-grained sediment on the deep seafloor is derived from (1) windblown dust and volcanic ash from the continents and volcanic islands, and (2) the shells of microscopic plants and animals that live in the near-surface waters. Minor sources are chemical reactions in seawater that yield manganese nodules found in all ocean basins (■ Figure 9.14) and cosmic dust. Researchers think that as many as 40,000 metric tons of cosmic dust fall to Earth each year, but this is a trivial quantity compared to the volume of sediment derived from the two primary sources.

Most sediment on the deep seafloor is *pelagic*, meaning that it settled from suspension far from land (■ Figure 9.15). Pelagic sediment is further characterized as pelagic clay and ooze. **Pelagic clay** is brown or red and, as its name implies, is composed of clay-sized particles

Institute of Oceanographic Sciences/Nerc/SPL/Photo Researchers, Inc.

(a)

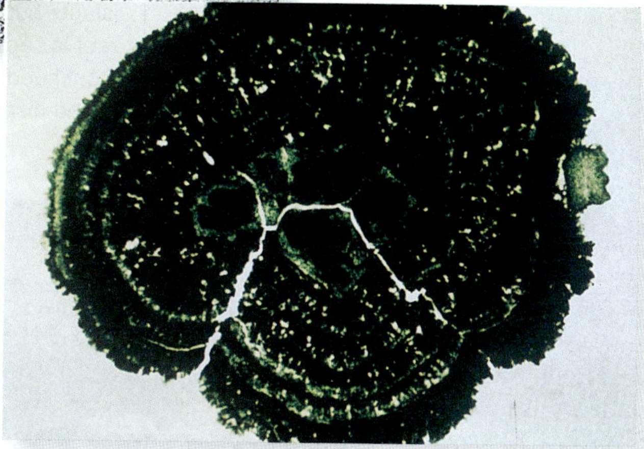

Dr. James Andrews

(b)

Figure 9.14

(a) Manganese nodules on the seafloor. (b) Sectioned manganese nodule showing its internal structure consisting of concentric layers. This nodule measures about 6 cm across.

from the continents or oceanic islands. **Ooze,** in contrast, is made up mostly of tiny shells of marine organisms. *Calcareous ooze* consists primarily of calcium carbonate ($CaCO_3$) skeletons of marine organisms such as foraminifera, and *siliceous ooze* is composed of the silica (SiO_2) skeletons of such single-celled organisms such as radiolarians (animals) and diatoms (plants).

REEFS

The term **reef** has a variety of meanings such as shallowly submerged rocks that pose a hazard to navigation, but here we restrict it to mean a moundlike, wave-resistant structure composed of the skeletons of marine organisms (see "Reefs: Rocks Made by Organisms" on pages 248 and 249). Although commonly called coral reefs, they actually have a solid framework composed of skeletons of corals and various mollusks, such as clams, and encrusting organisms, including sponges and algae. Reefs are restricted to shallow, tropical seas where the water is clear and its temperature does not fall below about 20°C. The depth to which reefs grow, rarely more than 50 m, depends on sunlight penetration because many of the corals rely on symbiotic algae that must have sunlight for energy.

Reefs of many shapes are known, but most are one of three basic varieties: fringing, barrier, and atoll. *Fringing reefs* are solidly attached to the margins of an island or continent. They have a rough, tablelike surface, are as much as 1 km wide, and on their seaward side slope steeply down to the seafloor. *Barrier reefs* are similar to fringing reefs, except a lagoon separates them from the mainland. The best-known barrier reef in the world is the 2000-km-long Great Barrier Reef of Australia.

Circular to oval reefs surrounding a lagoon are known as *atolls*. Atolls form around volcanic islands that subside below sea level as the plate on which they rest is carried progressively farther from an oceanic ridge. As subsidence proceeds, the reef organisms construct the

Reefs: Rocks Made by Organisms

Reefs are wave-resistant structures composed of the skeletons of corals, mollusks, sponges, and encrusting algae. Most reefs are characterized as fringing, barrier, or atolls, all of which actively grow in shallow, warm seawater where there is little or no influx of detrital sediment, especially mud. Reef rock is a type of limestone that forms directly as a solid, rather than from sediment that is later lithified. Ancient reefs are important reservoirs for hydrocarbons in some areas.

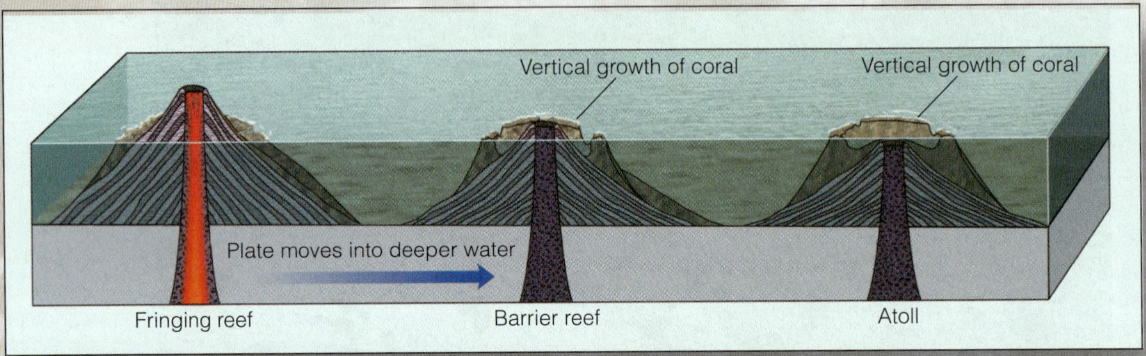

Three stages in the evolution of a reef. A fringing reef forms around a volcanic island, but as the island is carried into deeper water on a moving plate, the reef is separated from the island by a lagoon and becomes a barrier reef. Continued plate movement carries the island into even deeper water. The island disappears below sea level but the reef grows upward, forming an atoll.

Courtesy of Carl Roessler

Courtesy of Jeff and Dianna Monroe

Underwater views of reefs in the Red Sea (left) and in Hawaii.

Patrick Ward/Corbis

Sue Monroe

The white line of breaking waves marks the site of a barrier reef around Rarotonga in the Cook Islands in the Pacific Ocean. The island is only about 12 km long.

This oval reef with a central lagoon in the Pacific Ocean is an atoll. How does it differ from the reef shown at the bottom of the previous page?

An ancient reef in Australia. You can see the reef talus on the left side of the image sloping away from the reef core, which has no layering. To the right of the reef core are back-reef deposits, which show horizontal bedding.

Block diagram showing the various environments in a reef complex.

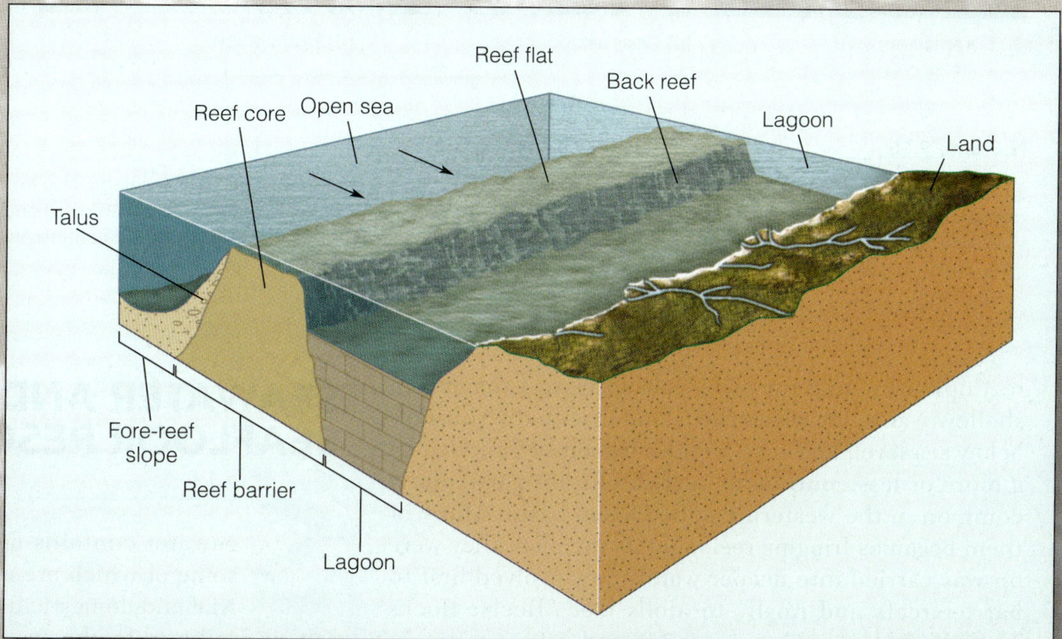

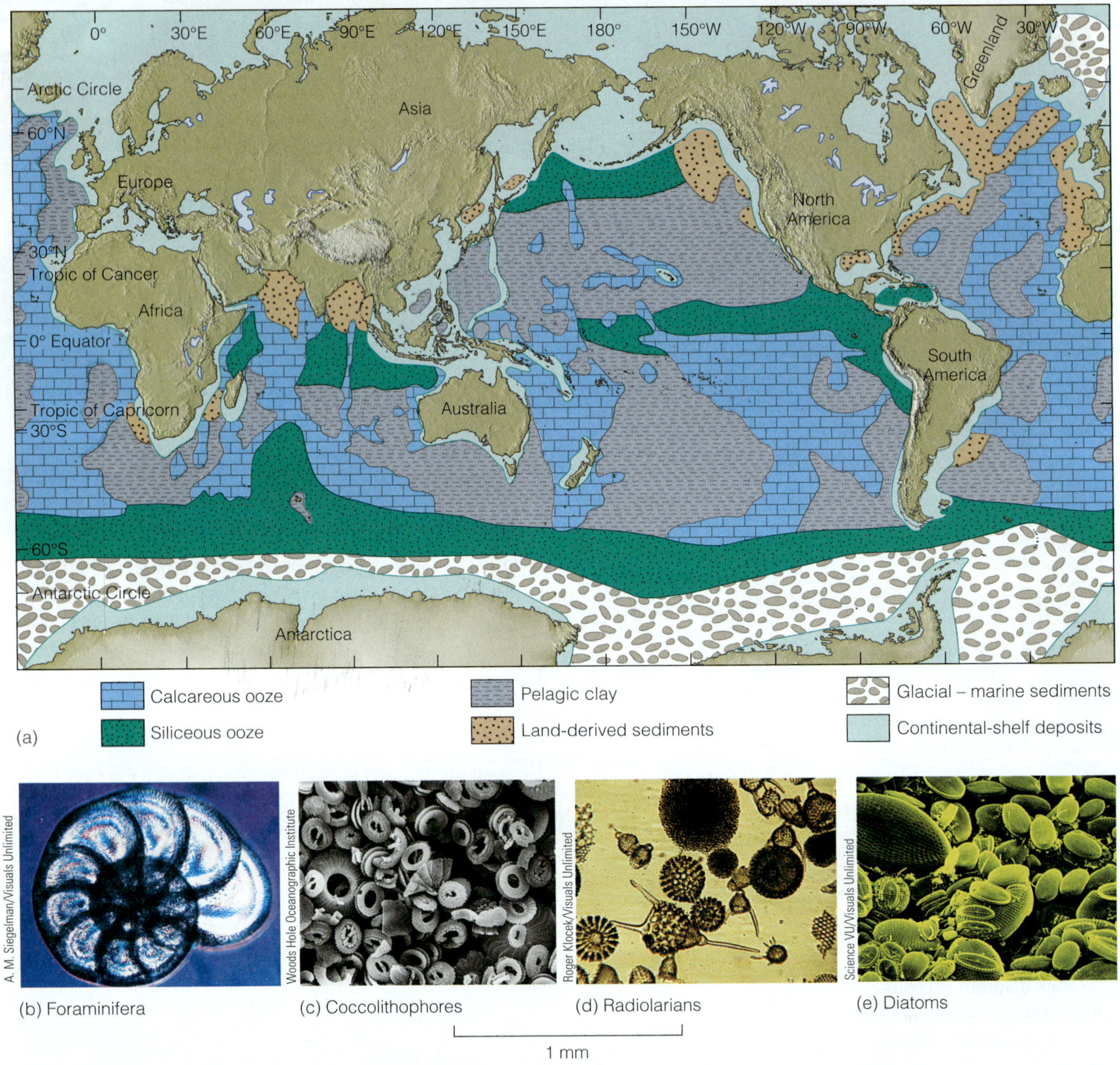

Figure 9.15

(a) A variety of sediments are present in the ocean basins, but most on the deep seafloor are pelagic clay and calcareous and siliceous ooze. Common constituents of calcareous ooze are skeletons of (b) foraminifera (floating single-celled animals) and (c) coccolithophores (floating single-celled plants), whereas siliceous ooze consists of skeletons of (d) radiolarians (single-celled floating animals) and (e) diatoms (single-celled floating plants).

reef upward so that the living part of the reef remains in shallow water. However, the island eventually subsides below sea level, leaving a circular lagoon surrounded by a more or less continuous reef. Atolls are particularly common in the western Pacific Ocean basin. Many of them began as fringing reefs, but as the plate they were on was carried into deeper water, they evolved first to barrier reefs and finally to atolls (see "Reefs: Rocks Made by Organisms" on pages 248 and 249).

SEAWATER AND SEAFLOOR RESOURCES

Seawater contains many elements in solution, some of which are extracted for various industrial and domestic uses. Sodium chloride (table salt) is produced by the evaporation of seawater, and a

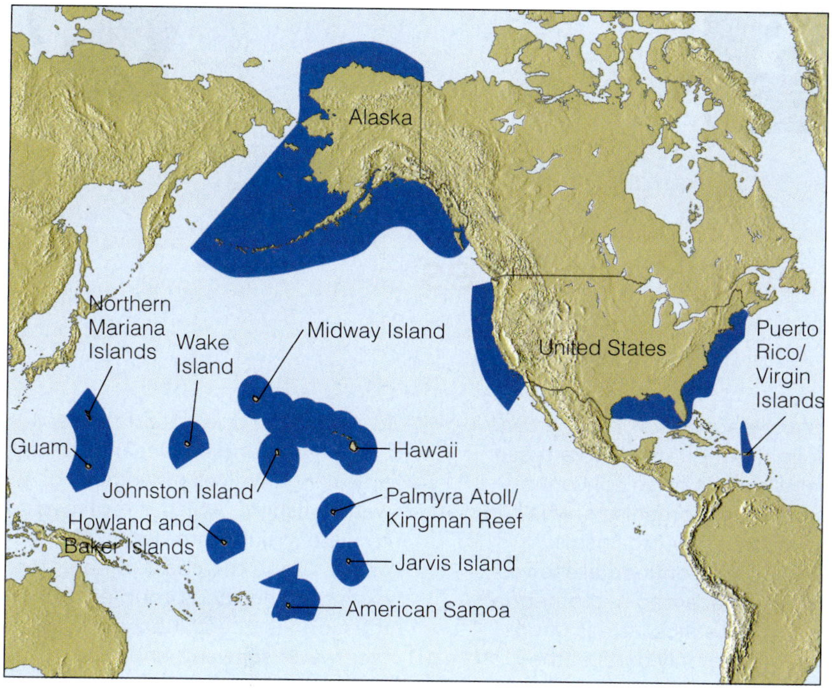

Figure 9.16

The Exclusive Economic Zone (EEZ), shown in dark blue, includes a vast area adjacent to the United States and its possessions.

large proportion of the world's magnesium comes from seawater. Numerous other elements and compounds can be extracted from seawater, but for many, such as gold, the cost is prohibitive.

In addition to substances in seawater, deposits on the seafloor or within seafloor sediments are becoming increasingly important. Many of these potential resources lie well beyond continental margins, so their ownership is a political and legal problem that has not yet been resolved.

Most nations bordering the ocean claim those resources within their adjacent continental margin. The United States by a presidential proclamation issued on March 10, 1983, claims sovereign rights over an area designated the **Exclusive Economic Zone (EEZ)** (Figure 9.16). The EEZ extends seaward 200 nautical miles (371 km) from the coast and includes areas adjacent to U.S. territories such as Guam, American Samoa, Wake Island, and Puerto Rico. In short, the United States claims rights to all resources within an area about 1.7 times larger than its land area. A number of other nations make similar claims.

Numerous resources are found within the EEZ, some of which have been exploited for many years. Sand and gravel for construction are mined from the continental shelf in several areas, and about 17% of U.S. oil and natural gas production comes from wells on the continental shelf (Figure 9.17). Ancient shelf deposits in the Persian Gulf region contain the world's largest reserves of oil.

A potential resource within the EEZ is methane hydrate, consisting of single methane molecules bound up in networks formed by frozen water. These methane hydrates are stable at water depths of more than 500 m and near-freezing temperatures. According to one estimate, the carbon in these deposits is double that in all coal, oil, and natural gas reserves. However, no one knows yet whether methane hydrates can be effectively recovered and used as an energy source. Additionally, their contribution to global warming must be assessed because a volume of methane 3000 times greater than in the atmosphere is present in seafloor

Bruce Hall

Figure 9.17

An oil-drilling platform off the coast of southern California. Although this one is in fairly shallow water, a similar platform off the Louisiana coast is anchored in water 872 m deep. About 17% of all U.S. oil and natural gas production comes from wells on the continental shelves.

GEOFOCUS 9.1

Oceanic Circulation and Resources from the Sea

Earth's oceans are in constant motion. Huge quantities of water circulate in surface and deep currents as water is transferred from one part of an ocean basin to another. The Gulf Stream and South Equatorial Current carry great quantities of water toward the poles and have an important modifying effect on climate. In addition to surface and deep currents that carry water horizontally, vertical circulation takes place when *upwelling* slowly transfers cold water from depth to the surface and *downwelling* transfers warm surface water to depth.

Upwelling is of more than academic interest. It not only transfers water from depth to the surface but also carries nutrients, especially nitrates and phosphate, into the zone of sunlight penetration. Here these nutrients sustain huge concentrations of floating organisms, which in turn support other organisms. Other than the continental shelves and areas adjacent to hydrothermal vents on the seafloor, areas of upwelling are the only parts of the oceans where biological productivity is very high. In fact, they are so productive that even though constituting less than 1% of the ocean surface, they support more than 50% (by weight) of all fishes.

Scientists recognize three types of upwelling, but only *coastal upwelling* need concern us here. Most coastal upwelling takes place along the west coasts of Africa, North America, and South America, although one notable exception is in the Indian Ocean. Coastal upwelling involves movement of water offshore, which is replaced by water rising from depth (■ Figure 1). Along the coast of Peru, for example, the winds coupled with the Coriolis effect* transport surface water seaward, and cold, nutrient-rich water rises to replace it. This area is a major fishery, and

*The Coriolis effect is the apparent deflection of a moving object from its anticipated course resulting from Earth's rotation. Oceanic currents are deflected clockwise in the Northern Hemisphere and counterclockwise in the Southern Hemisphere.

deposits—and methane is 10 times more effective than carbon dioxide as a greenhouse gas.

The manganese nodules previously discussed are another potential seafloor resource (Figure 9.14). These spherical objects are composed mostly of manganese and iron oxides but also contain copper, nickel, and cobalt. The United States, which must import most of the manganese and cobalt it uses, is particularly interested in these nodules as a potential resource.

Other seafloor resources of interest include massive sulfide deposits that form by submarine hydrothermal activity at spreading ridges. These deposits containing iron, copper, zinc, and other metals have been identified within the EEZ at the Gorda Ridge off the coasts of Cal-

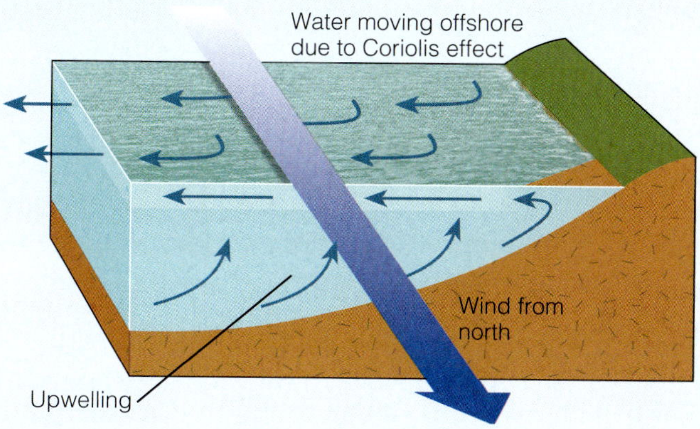

Figure 1

Wind from the north along the west coast of a continent coupled with the Coriolis effect causes surface water to move offshore, resulting in upwelling of cold, nutrient-rich deep water.

changes in the surface-water circulation every three to seven years adjacent to South America are associated with the onset of El Niño, a weather phenomenon with far-reaching consequences.

Among the nutrients in upwelling oceanic waters is considerable phosphorus, an essential element for animal and plant nutrition. Although present in minute quantities in many sedimentary rocks, most commercial phosphorus is derived from *phosphorite,* a sedimentary rock with such phosphate-rich minerals as fluorapatite [$Ca_5(PO_4)_3F$]. Areas of upwelling along the outer margins of continental shelves are the depositional sites of most of the so-called bedded phosphorites, which are interlayered with carbonate rocks, chert, shale, and sandstone. Vast deposits in the Permian-aged Phosphoria Formation of Montana, Wyoming, and Idaho formed in this manner.

Upwelling accounts for most of Earth's phosphate-rich sedimentary rocks, but some forms by other processes. In *phosphatization,* carbonate grains such as animal skeletons and ooids are replaced by phosphate, and *guano* is made up of calcium phosphate from bird or bat excrement. Another type of phosphate deposit is essentially a placer deposit where the skeletons of vertebrate animals are found in large numbers [vertebrate skeletons are made up mostly of hydroxyapatite, [$Ca_5(PO_4)_3OH$]. The 3- to 15-million-year-old Bone Valley Formation in Florida is a good example.

The United States is the world leader in production and consumption of phosphate rock, most of it coming from deposits in Florida and North Carolina, but some is also mined in Idaho and Utah. More than 90% of all phosphate rock mined in this country is used to make chemical fertilizers and animal feed supplements. It also has several other uses in metallurgy, preserved foods, ceramics, and matches.

Geology Now Log into GeologyNow and select this chapter to work through **Geology Interactive** activities on "World Currents," (click Waves, Tides, and Current→World Currents) and "El Niño," (click Weather and Climate→El Niño).

ifornia and Oregon; similar deposits occur at the Juan de Fuca Ridge within the Canadian EEZ.

Within the EEZ, manganese nodules are found near Johnston Island in the Pacific Ocean and on the Blake Plateau off the east coast of South Carolina and Georgia. In addition, seamounts and seamount chains within the EEZ in the Pacific are known to have metalliferous oxide crusts several centimeters thick from which cobalt and manganese could be mined.

Another important resource found in shallow marine deposits is phosphate-rich sedimentary rock known as *phosphorite* (see Geo-Focus 9.1).

9 GEO RECAP

Chapter Summary

- Scientific investigations of the oceans began more than 200 years ago, but much of our knowledge comes from studies done during the last few decades.
- Present-day research vessels are equipped to investigate the seafloor by sampling and drilling, echo sounding, and seismic profiling. Scientists also use submersibles in their studies.
- Deep-sea drilling and observations on land and on the seafloor confirm that oceanic crust is made up, in descending order, of pillow lava/sheet lava flows, sheeted dikes, and gabbro.
- Continental margins consist of a gently sloping continental shelf, a more steeply inclined continental slope, and in some cases a continental rise.
- The width of continental shelves varies considerably. They slope seaward to the shelf–slope break at a depth averaging 135 m, where the seafloor slope increases abruptly.
- Submarine canyons, mostly on continental slopes, carry huge quantities of sediment by turbidity currents into deeper water, where it is deposited as overlapping submarine fans that make up a large part of the continental rise.
- Active continental margins at the leading edge of a tectonic plate have a narrow shelf and a slope that descends directly into an oceanic trench. Volcanism and seismic activity also characterize these margins.
- Passive continental margins lie within a tectonic plate and have wide continental shelves, and the slope merges with a continental rise that grades into an abyssal plain. These margins show little seismic activity and no volcanism.
- Long, narrow oceanic trenches are found where oceanic lithosphere is subducted beneath either oceanic lithosphere or continental lithosphere. The trenches are the sites of the greatest oceanic depths and low heat flow.
- Oceanic ridges are composed of volcanic rocks, and many have a central rift caused by tensional forces. Basaltic volcanism, hydrothermal vents, and shallow-focus earthquakes occur at ridges, which are offset by fractures that cut across them.
- Seamounts, guyots, and abyssal hills rising from the seafloor are common features that differ mostly in scale and shape. Many aseismic ridges on the seafloor consist of chains of seamounts, guyots, or both.
- Submarine hydrothermal vents known as black smokers found at or near spreading ridges support biologic communities and are potential sources of several resources.
- Moundlike, wave-resistant structures called reefs, consisting of animal skeletons, are found in a variety of shapes, but most are classified as fringing reefs, barrier reefs, or atolls.
- Sediments called pelagic clay and ooze cover vast areas of the seafloor.
- The United States claims rights to all resources within 200 nautical miles of its shorelines. Resources including sand and gravel as well as metals are found within this Exclusive Economic Zone.

Important Terms

abyssal plain (p. 242)
active continental margin (p. 241)
aseismic ridge (p. 246)
black smoker (p. 243)
continental margin (p. 238)
continental rise (p. 239)
continental shelf (p. 238)
continental slope (p. 239)
Exclusive Economic Zone (EEZ) (p. 251)
guyot (p. 246)
oceanic ridge (p. 243)
oceanic trench (p. 242)
ooze (p. 247)
ophiolite (p. 237)
passive continental margin (p. 241)
pelagic clay (p. 246)
reef (p. 247)
seamount (p. 245)
seismic profiling (p. 236)
submarine canyon (p. 241)
submarine fan (p. 239)
submarine hydrothermal vent (p. 243)
turbidity current (p. 239)

Review Questions

1. An atoll is
a._____a circular to oval reef enclosing a lagoon; b._____a type of deep-sea sediment made up of clay; c._____found on the deep seafloor adjacent to a black smoker; d._____a deposit composed of copper and zinc sulfides; e._____made up of sediment that settles from suspension far from land.
2. The greatest oceanic depths are found at
a._____aseismic ridges; b._____seamounts; c._____oceanic trenches; d._____passive continental margins; e._____submarine canyons.
3. A large part of a continental rise consists of overlapping
a._____guyots; b._____hydrothermal vents; c._____continental margins; d._____abyssal plains; e._____submarine fans.
4. The two types of sediment most common on the deep seafloor are
a._____gravel and limestone; b._____pelagic clay and ooze; c._____manganese nodules and cosmic dust; d._____reefs and sand; e._____seamounts and guyots.
5. The gently sloping part of the continental margin adjacent to the continent is the continental
a._____slope; b._____plain; c._____profile; d._____shelf; e._____ridge.
6. A flat-topped seamount rising more than 1 km above the seafloor is a(n)
a._____guyot; b._____reef; c._____black smoker; d._____oceanic ridge; e._____aseismic plateau.
7. Which one of the following is characteristic of an active continental margin?
a._____wide continental shelf; b._____continental shelf merging with a continental rise; c._____seismic activity; d._____vast abyssal plains; e._____submarine hydrothermal vents.
8. Turbidity current deposits typically show
a._____graded bedding; b._____a large component of siliceous ooze; c._____many skeletons of corals; d._____sulfide minerals; e._____ophiolites.
9. Which one of the following statements is *incorrect*?
a._____Most intermediate- and deep-focus earthquakes occur at active continental margins; b._____Submarine hydrothermal vents are found near spreading ridges; c._____Oceanic crust is made up of granite and sandstone; d._____Most continental margins around the Pacific are active; e._____Pelagic clay covers much of the deep seafloor.
10. A broad, flat area adjacent to the continental rise is called
a._____oceanic crust; b._____submarine canyon; c._____abyssal plain; d._____seamount; e._____passive rise.
11. Identify the types of continental margins in Figure 9.7. What are the characteristics of each?

12. Describe submarine canyons, and explain how scientists think they form. Is there any evidence to support their ideas?
13. Explain how a reef evolves from fringing to barrier to an atoll.
14. How do mid-oceanic ridges form, and how do they differ from mountain ranges on land?
15. Why are abyssal plains common around the margins of the Atlantic but rare in the Pacific Ocean basin?
16. How did geologists figure out the nature of the upper mantle and oceanic crust even before they observed mantle and crust rocks in the ocean basins?
17. What is calcareous ooze and pelagic clay, and where are they found?
18. How are seismic profiling and echo sounding used to study the seafloor?
19. The most distant part of a 30-million-year-old aseismic ridge is 1000 km from an oceanic ridge. How fast, on average, did the plate with this ridge move in centimeters per year?
20. During the Pleistocene Epoch (Ice Age), sea level was as much as 130 m lower than it is today. What effect did this lower sea level have on rivers? Is there any evidence from continental shelves that might bear on this question? If so, what?

World Wide Web Activities

GeologyNow Assess your understanding of this chapter's topics with additional quizzing and comprehensive interactivities at

http://earthscience.brookscole.com/changingearth4e

as well as current and up-to-date weblinks, additional readings, and InfoTrac College Edition exercises.

Deformation, Mountain Building, and the Continents

:HAPTER 10

OUTLINE

- Introduction
- Rock Deformation—How Does It Occur?
  GEOLOGY IN UNEXPECTED PLACES: *Ancient Ruins and Geology*
- Strike and Dip—The Orientation of Deformed Rock Layers
- Deformation and Geologic Structures
- Deformation and the Origin of Mountains
  GEO-FOCUS 10.1: *Geologic Maps—Their Construction and Uses*
- Earth's Continental Crust
- Geo-Recap

Geology⇌Now *This icon, which appears throughout the book, indicates an opportunity to explore interactive tutorials, animations, or practice problems available on the GeologyNow website at http://earthscience.brookscole.com/changingearth4e.*

OBJECTIVES

At the end of this chapter, you will have learned that:

- Rock deformation involves changes in the shape or volume or both of rocks in response to applied forces.
- Geologists use several criteria to differentiate among geologic structures such as folds, joints, and faults.
- Correctly interpreting geologic structures is important in human endeavors such as constructing highways and dams, choosing sites for power plants, and finding and extracting some resources.
- Deformation and the origin of geologic structures are important in the origin and evolution of mountains.
- Most of Earth's large mountain systems formed, and in some cases continue to form, at or near the three types of convergent plate boundaries.
- Terranes have special significance in mountain building.
- Earth's continental crust, and especially mountains, stands higher than adjacent crust because of its composition and thickness.

his sawed and polished specimen of the Moine Schist on display in the oyal Museum in Edinburgh, Scotland, shows intense deformation. Jotice that many layers in the rock are intricately folded, and some layrs have been displaced along small fractures. Source: Sue Monroe

Introduction

"Solid as a rock" implies permanence and durability, but you know from earlier chapters that physical and chemical processes disaggregate and decompose rocks, and rocks behave very differently at great depth than they do at or near Earth's surface. Indeed, under the tremendous pressures and high temperature at several kilometers below the surface, rock layers actually crumple or fold yet remain solid, and at shallower depths they yield by fracturing or a combination of folding and fracturing. In either case, dynamic forces within Earth cause **deformation,** a general term encompassing all changes in the shape or volume (or both) of rocks (see the chapter opening photo and ■ Figure 10.1).

The action of dynamic forces within Earth is obvious from ongoing seismic activity, volcanism, plate movements, and the continuing evolution of mountains in South America, Asia, and elsewhere. In short, Earth is an active planet with a variety of processes driven by internal heat, particularly plate movements; most of Earth's seismic activity, volcanism, and rock deformation take place at divergent, convergent, and transform plate boundaries.

The origin of Earth's truly large mountain ranges on the continents involves tremendous deformation, usually accompanied by emplacement of plutons, volcanism, and metamorphism, at convergent plate boundaries. The Appalachians of North America, the Alps in Europe, the Himalayas of Asia, and the Andes in South America all owe their existence to deformation at convergent plate boundaries. And in some cases this activity continues even now. Thus, deformation and mountain building are closely related topics and accordingly we consider both in this chapter.

The past and continuing evolution of continents involves not only deformation at continental margins but also additions of new material to existing continents, a phenomenon known as *continental accretion* (see Chapter 19). North America, for instance, has not always had its present shape and area. Indeed, it began evolving during the Archean Eon (4.0–2.5 billion years ago) as new material was added to the continent at deformation belts along its margins.

Much of this chapter is devoted to a review of *geologic structures,* such as folded and fractured rock layers resulting from deformation, their descriptive terminology, and the forces responsible for them. Even so, there are several practical reasons to study deformation and mountain building. For one thing, crumpled and fractured rock layers provide a record of the kinds and intensities of forces that operated during the past. Thus, interpretations of these structures allow us to satisfy our curiosity about Earth history, and in addition such studies are essential in engineering endeavors such as choosing sites for dams, bridges, and nuclear power plants, especially if they are in areas of ongoing deformation. Also, many aspects of mining and exploration for petroleum and natural gas rely on correctly identifying geologic structures.

David J. Matty

(a)

Reed Wicander

(b)

■ **Figure 10.1**

Many rocks show the effect of deformation. (a) Small-scale folds in sedimentary rocks. The pen is 13.5 cm long. (b) These rocks have been deformed by folding and fracturing. Notice the light pole for scale. The nearly vertical fracture where light-colored rocks were displaced is a fault, a fracture along which rocks on opposite sides of the fracture have moved parallel with the fracture surface.

ROCK DEFORMATION—HOW DOES IT OCCUR?

We defined *deformation* as a general term referring to changes in the shape or volume (or both) of rocks; that is, rocks may be crumpled into folds or fractured as a result of **stress,** which results from force applied to a given area of rock. If the intensity of the stress is greater than the rock's internal strength, the rock undergoes **strain,** which is simply deformation caused by stress. The terminology is a little confusing at first, but keep in mind that *deformation* and *strain* are synonyms, and stress is the force that causes deformation or strain. The following discussion

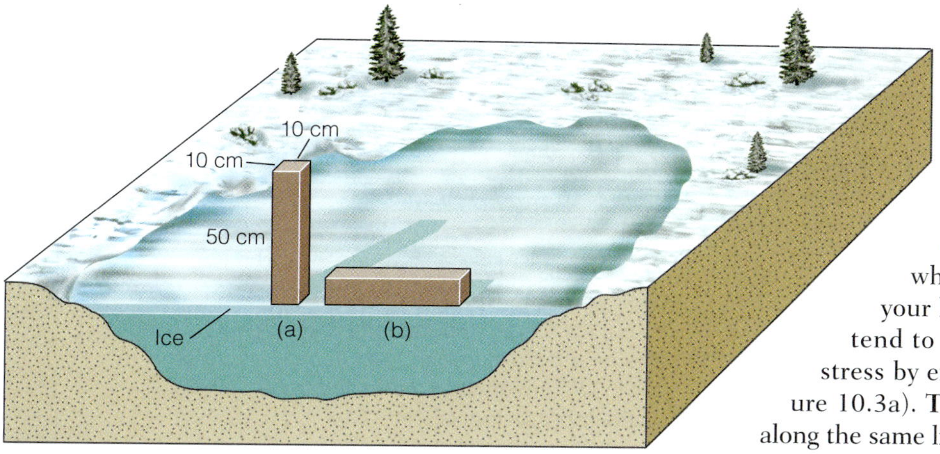

Figure 10.2

Stress and strain exerted on an ice-covered pond. The vertical object (a) has a density of 1 g/cm^3 and a volume of 5000 cm^3. The area of the object on the ice is 100 cm^2, so the stress exerted on the ice is 50 g/cm^2. The object on its side (b) has an area of 500 cm^2 in contact with the ice. Accordingly, the stress exerted on the ice is only 10 g/cm^2.

and Figure 10.2 will help clarify the meaning of stress and the distinction between stress and strain.

Stress and Strain

Remember that stress is the force applied to a given area of rock, usually expressed in kilograms per square centimeter (kg/cm^2). For example, the stress, or force, exerted by a person walking on an ice-covered pond is a function of the person's weight and the area beneath her or his feet. The ice's internal strength resists the stress unless the stress is too great, in which case the ice may bend or crack as it is strained (deformed). In Figure 10.2, we use a rectangular object rather than a person to simplify the calculations. To avoid breaking through the ice, the person may lie down; this does not reduce the total stress but it does distribute it over a larger area, thus reducing the stress per unit area.

Although stress is force per unit area, it comes in three varieties: *compression*, *tension*, and *shear*, depending on the direction of the applied forces. In **compression,** rocks or any other object are squeezed or compressed by forces directed toward one another along the same line, as when you squeeze a rubber ball in your hand. Rock layers in compression tend to be shortened in the direction of stress by either folding or fracturing (Figure 10.3a). **Tension** results from forces acting along the same line but in opposite directions. Tension tends to lengthen rocks or pull them apart (Figure 10.3b). Incidentally, rocks are much stronger in compression than they are in tension. In **shear stress,** forces act parallel to one another but in opposite directions, resulting in deformation by displacement along closely spaced planes (Figure 10.3c).

Types of Strain

Geologists characterize strain as **elastic strain** if deformed rocks return to their original shape when the deforming forces are relaxed. In Figure 10.2, the ice on the pond may bend under a person's weight but return to its original shape once the person leaves. As you might expect, rocks are not very elastic, but Earth's

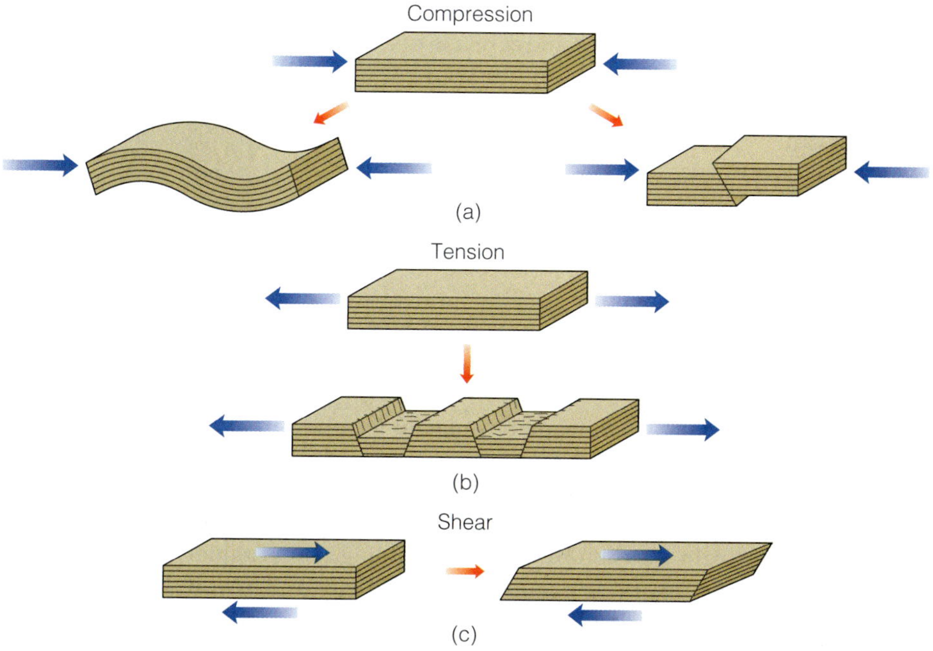

Figure 10.3

Stress and possible types of resulting deformation. (a) Compression causes shortening of rock layers by folding or faulting. (b) Tension lengthens rock layers and causes faulting. (c) Shear stress causes deformation by displacement along closely spaced planes.

GEOLOGY IN UNEXPECTED PLACES

Ancient Ruins and Geology

Ancient Roman and Greek ruins might not seem a good place to study geology, but they do tell us something about the use of stone in construction and about stress and strain. Stone is incredibly strong, but its strength varies depending on how it is used. Remember that it is stronger in compression than it is in tension. Thus, when stones are used for a chimney, for example, all stresses act vertically with the lower stones supporting the weight of those above. In other words, the stones are in compression.

Given that stone is so strong, why did ancient Greek and Roman builders erect buildings in which horizontal beams are supported by closely spaced vertical columns (■ Figure 1)? Wouldn't it have been more efficient to eliminate every other column and save on both materials and labor? The problem is the strength of stone in compression versus tension. If the horizontal beams spanned great distances, they would simply collapse under their own weight because a stone beam is compressed on its upper side but subjected to tension on its underside. The stress on its top is about the same as on the bottom, but the top contributes little to the beam's overall strength. So if used this way, tension fractures would form on the bottom and travel upward to the top, causing failure.

Ancient builders were aware of stone's strength, probably from trial and error, and accordingly used closely spaced columns to support horizontal beams. Builders today are similarly restricted in the use of stone in comparable applications. Can you figure out why simply doubling our hypothetical beam's dimensions, other than its length, would not solve the problem of spanning great distances between columns?

■ **Figure 1**

Ancient Greek ruins with closely spaced vertical columns supporting horizontal beams.

Jim Winkley/Econscene/Corbis

crust behaves elastically when loaded by glacial ice and is depressed into the mantle.

As stress is applied, rocks respond first by elastic strain, but when strained beyond their elastic limit, they undergo **plastic strain** as when they yield by folding, or they behave like brittle solids and **fracture** (■ Figure 10.4). In either folding or fracturing, the strain is permanent; that is, the rocks do not recover their original shape or volume even if the stress is removed.

Whether strain is elastic, plastic, or fracture depends on the kind of stress applied, pressure and temperature, rock type, and the length of time rocks are subjected to stress. A small stress applied over a long period, as on a mantelpiece supported only at its ends, will cause the rock to sag; that is, the rock deforms plastically (Figure 10.4). By contrast, a large stress applied rapidly to the same object, as when struck by a hammer, results in fracture. Rock type is important because not all rocks have the same internal strength and thus respond to stress differently. Some rocks are *ductile* whereas others are *brittle*, depending on the amount of plastic strain they exhibit. Brittle rocks show little or no plastic strain before they fracture, but ductile rocks exhibit a great deal (Figure 10.4).

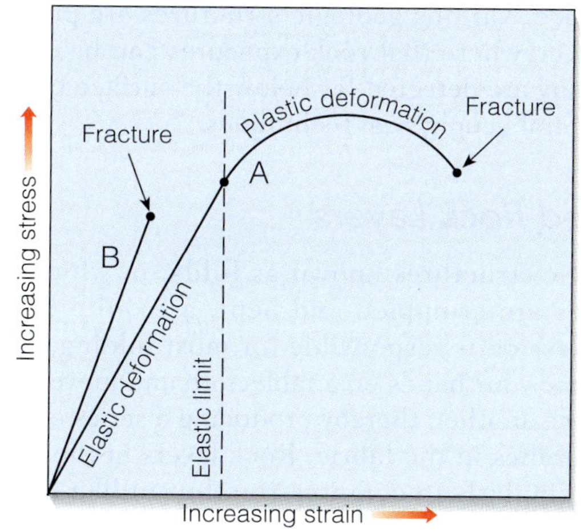

■ **Figure 10.4**

Rocks initially respond to stress by elastic deformation and return to their original shape when the stress is released. If the elastic limit is exceeded as in curve A, rocks deform plastically, which is permanent deformation. The amount of plastic deformation rocks exhibit before fracturing depends on their ductility: if they are ductile, they show considerable plastic deformation (curve A), but if they are brittle, they show little or no plastic deformation before failing by fracture (curve B).

Many rocks show the effects of plastic deformation that must have taken place deep within the crust. At or near the surface, rocks commonly behave like brittle solids and fracture, but at depth they more often yield by plastic deformation; they become more ductile with increasing pressure and temperature. Most earthquake foci are at depths of less than 30 km, indicating that deformation by fracturing becomes increasingly difficult with depth and no fracturing is known deeper than about 700 km.

What Would You Do?

The types of stresses as well as elastic versus plastic strain might seem rather esoteric, but perhaps understanding these concepts has some practical applications. What relevance do you think knowing about stress and strain has to some professions, other than geology, and what professions might these be? Can you think of stresses and strain that we contend with in our daily lives? As an example, what happens when a car smashes into a tree?

STRIKE AND DIP—THE ORIENTATION OF DEFORMED ROCK LAYERS

During the 1660s, Nicholas Steno, a Danish anatomist, proposed several principles essential for deciphering Earth history from the record preserved in rocks. One is the *principle of original horizontality,* meaning that sediments accumulate in horizontal or nearly horizontal layers. Thus, if we observe steeply inclined sedimentary rocks, we are justified in inferring that they were deposited nearly horizontally, lithified, and then tilted into their present position (■ Figure 10.5). Rock layers deformed by folding, faulting, or both are no longer in their original position, so geologists use *strike* and *dip* to describe their orientation with respect to a horizontal plane.

Sue Monroe

(a)

James S. Monroe

(b)

■ **Figure 10.5**

(a) These rock layers in Valley of the Gods, Utah, are horizontal as when they were deposited and lithified. (b) We can infer that these sandstone beds in Colorado were deposited horizontally, lithified, and then tilted into their present position.

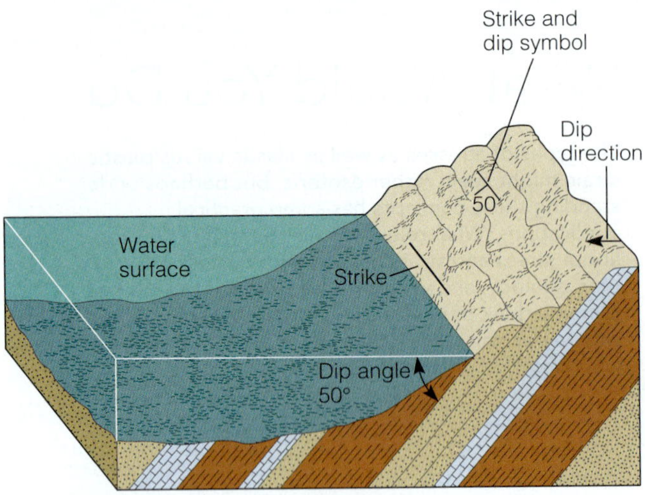

■ **Figure 10.6**

Strike and dip. The intersection of a horizontal plane (the water surface) and an inclined plane (the surface of any of the rock layers) forms a line known as strike. The dip of these layers is their maximum angular deviation from horizontal. Notice the strike and dip symbol with 50 adjacent to it indicating the angle of dip.

By definition **strike** is the direction of a line formed by the intersection of a horizontal plane and an inclined plane. The surfaces of the rock layers in ■ Figure 10.6 are good examples of inclined planes, whereas the water surface is a horizontal plane. The direction of the line formed at the intersection of these planes is the strike of the rock layers. The strike line's orientation is determined by using a compass to measure its angle with respect to north. **Dip** is a measure of an inclined plane's deviation from horizontal, so it must be measured at right angles to strike direction (Figure 10.6).

Geologic maps showing the age, aerial distribution, and geologic structures of rocks in an area use a special symbol to indicate strike and dip. A long line oriented in the appropriate direction indicates strike, and a short line perpendicular to the strike line shows the direction of dip (Figure 10.6). Adjacent to the strike and dip symbol is a number corresponding to the dip angle. The usefulness of strike and dip symbols will become apparent in the following sections on folds and faults.

DEFORMATION AND GEOLOGIC STRUCTURES

Remember that deformation and its synonym strain refer to changes in the shape or volume of rocks. During deformation rocks might be crumpled into folds, or they might be fractured, or perhaps folded and fractured. Any of these features resulting from deformation is referred to as a **geologic structure.** Various geologic structures are present almost everywhere that rock exposures can be observed, and many are detected far below the surface by drilling and several geophysical techniques.

Folded Rock Layers

Geologic structures known as **folds,** in which planar features are crumpled and bent, are quite common. Compression is responsible for most folding, as when you place your hands on a tablecloth and move them toward one another, thereby producing a series of up- and down-arches in the fabric. Rock layers in the crust respond similarly to compression, but unlike the tablecloth, folding in rock layers is permanent. That is, plastic strain has taken place: so once folded, the rocks stay folded. Most folding probably takes place deep in the crust where rocks are more ductile than they are at or near the surface. The configuration of folds and the intensity of folding vary considerably, but only three basic types of folds are recognized: *monoclines, anticlines,* and *synclines.*

Monoclines

A simple bend or flexure in otherwise horizontal or uniformly dipping rock layers is a **monocline** (■ Figure 10.7a). The large monocline in Figure 10.7b formed when the Bighorn Mountains in Wyoming rose vertically along a fracture. The fracture did not penetrate to the surface, so as uplift of the mountains proceeded, the near-surface rocks were bent so that they now appear to be draped over the margin of the uplifted block. In a manner of speaking, a monocline is simply one half of an anticline or syncline.

Anticlines and Synclines

Monoclines are not rare, but they are not nearly as common as anticlines and synclines. An **anticline** is an uparched or convex upward fold with the oldest rock layers in its core, whereas a **syncline** is a down-arched or concave downward fold in which the youngest rock layers are in its core (■ Figure 10.8). Anticlines and synclines have an axial plane connecting the points of maximum curvature of each folded layer (■ Figure 10.9); the axial plane divides folds into halves, each half being a *limb.* Because folds are most often found in a series of anticlines alternating with synclines, an anticline and adjacent syncline share a limb. It is important to remember that anticlines and synclines are simply folded rock layers and do not necessarily correspond to high and low areas at the surface (■ Figure 10.10).

Folds are commonly exposed to view in areas of deep erosion, but even where eroded, strike and dip and the relative ages of the folded rock layers easily distinguish anticlines from synclines. Notice in ■ Figure 10.11 that in the surface view of the anticline, each limb dips outward or away from the center of the fold, and the oldest exposed

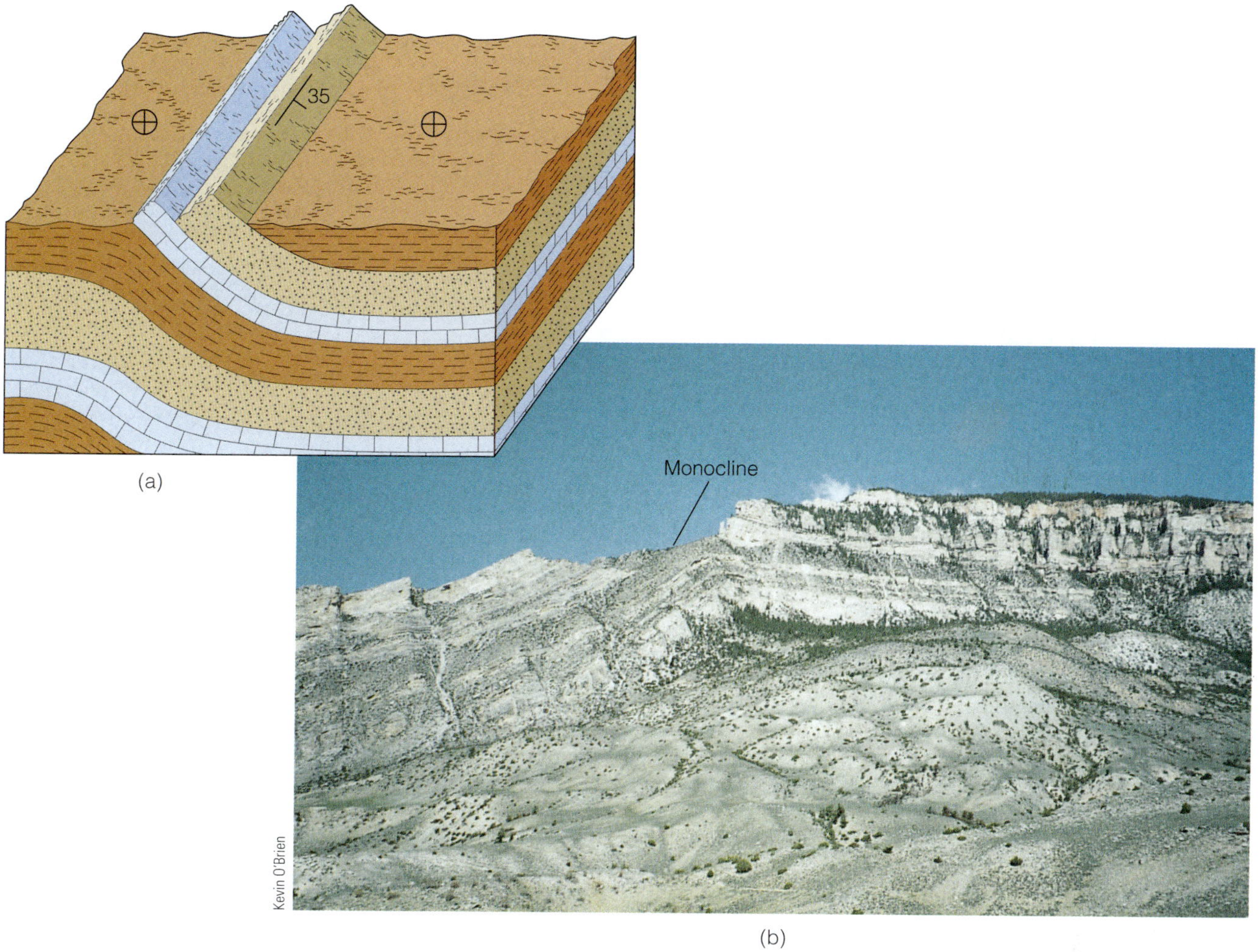

Figure 10.7

(a) A monocline. Notice the strike and dip symbol and the circled cross, which is the symbol for horizontal layers. (b) A monocline in the Bighorn Mountains in Wyoming.

rocks are in the fold's core. In an eroded syncline, though, each limb dips inward toward the fold's center, where the youngest exposed rocks are found.

The folds described so far are *upright,* meaning that their axial planes are vertical and both fold limbs dip at the same angle (Figure 10.11). In many folds, though, the axial plane is not vertical, the limbs dip at different angles, and the folds are characterized as *inclined* (■ Figure 10.12a). If both limbs dip in the same direction, the fold is *overturned.* That is, one limb has been rotated more than 90 degrees from its original position so that it is now upside down (Figure 10.12b). In some areas, deformation has been so intense that axial planes of folds are now horizontal, giving rise to what geologists call *recumbent folds* (Figure 10.12c). Overturned and recumbent folds are particularly common in mountains resulting from compression at convergent plate boundaries (discussed later in this chapter).

For upright folds, the distinction between anticlines and synclines is straightforward, but interpreting complex folds that have been tipped on their sides or turned completely upside down is more difficult. Can you determine which of the two folds shown in Figure 10.12c is an anticline? Even if strike and dip symbols were shown, you could still not resolve this question, but knowing the relative ages of the folded rock layers provides a solution. Remember that an anticline has the oldest rock layers in its core, so the fold nearest the surface is an anticline and the lower fold is a syncline.

Plunging Folds As if upright and inclined folds were not enough, geologists further characterize folds as *nonplunging* or *plunging.* In some folds, the fold axis, a line formed by the intersection of the axial plane with the folded layers, is horizontal and the folds are nonplunging (Figure 10.11). Much more commonly, though, fold axes are inclined so that they appear to plunge beneath adjacent rocks, and the folds are said to be plunging (■ Figure 10.13).

It might seem that with this additional complication, differentiating plunging anticlines from plunging synclines

Reed Wicander

■ **Figure 10.8**

Folded rocks in the Calico Mountains of southeastern California. Three folds are visible from left to right: a syncline, an anticline, and another syncline. We can infer that compression was responsible for these folds.

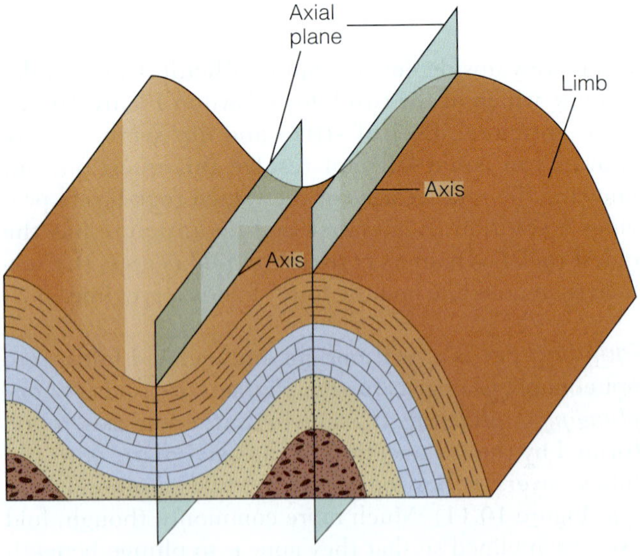

■ **Figure 10.9**

Syncline and anticline showing the axial plane, axis, and fold limbs.

would be much more difficult, but geologists use exactly the same criteria they use for nonplunging folds. Therefore, all rock layers dip away from the fold axis in plunging anticlines and toward the axis in plunging synclines. The oldest exposed rocks are in the core of an eroded plunging anticline, whereas the youngest exposed rock layers are found in the core of an eroded plunging syncline (Figure 10.13b).

In Chapter 6 we noted that anticlines form one type of structural trap in which petroleum and natural gas might accumulate (see Figure 6.28b). As a matter of fact, most of the world's petroleum production comes from anticlines, although other geologic structures and stratigraphic traps are important, too. Accordingly, geologists and their employers are particularly interested in correctly identifying geologic structures in areas of potential hydrocarbon production.

Domes and Basins **Domes** and **basins** are circular to oval folds (■ Figure 10.14). We can think of them as the circular to oval equivalents of anticlines and synclines, respectively. In an eroded dome, all rock layers dip outward from a central point and the oldest exposed rocks are

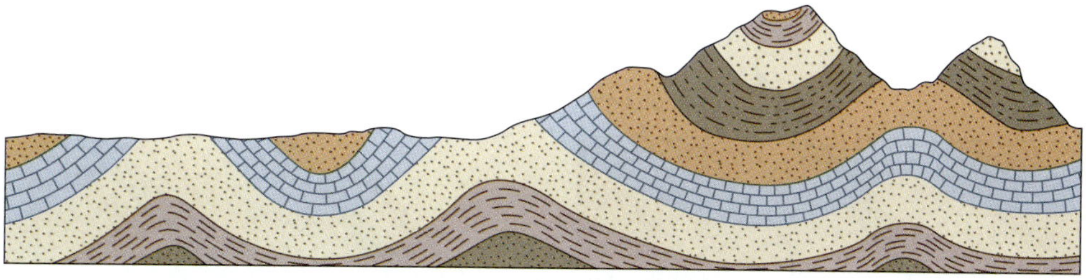

(a)

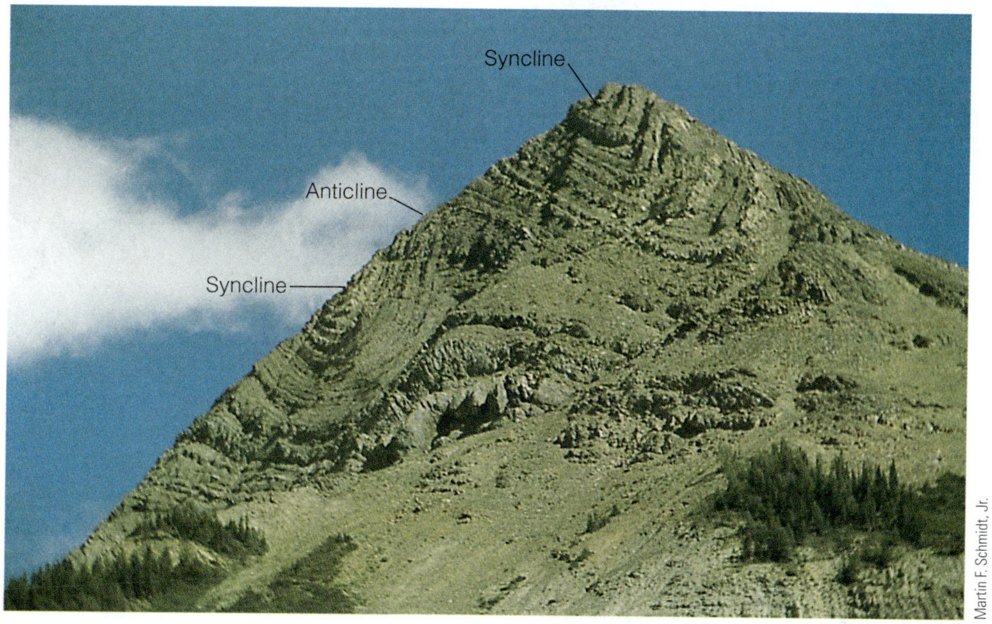

(b)

Figure 10.10

Folds and their relationship to topography. (a) Cross section illustrating that anticlines and synclines do not correspond to high and low areas of the surface. Notice that folds even underlie the rather flat area. (b) A syncline is at the peak of this mountain in Kootenay National Park, British Columbia, Canada. Lower on the left flank of the mountain, an anticline and another syncline are also visible.

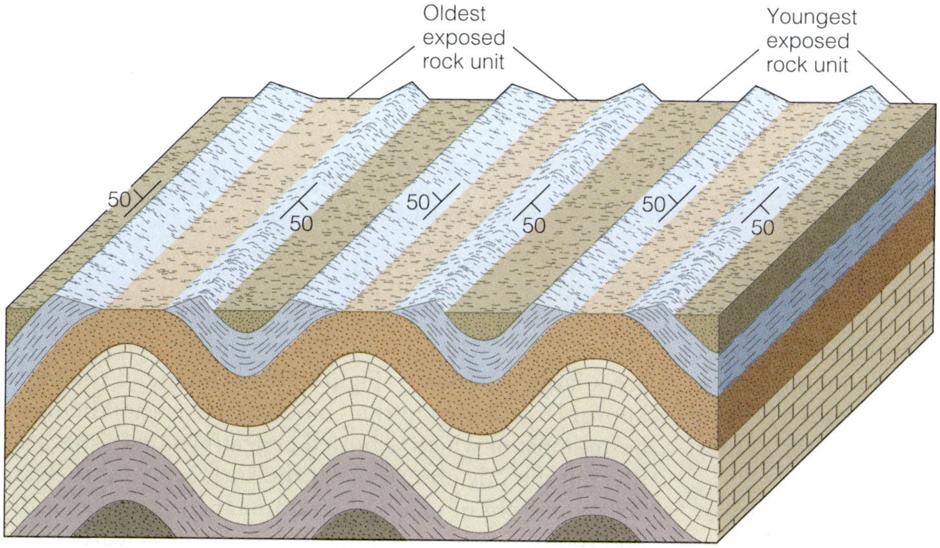

Figure 10.11

Identifying eroded anticlines and synclines by strike and dip and the relative ages of the folded rock layers.

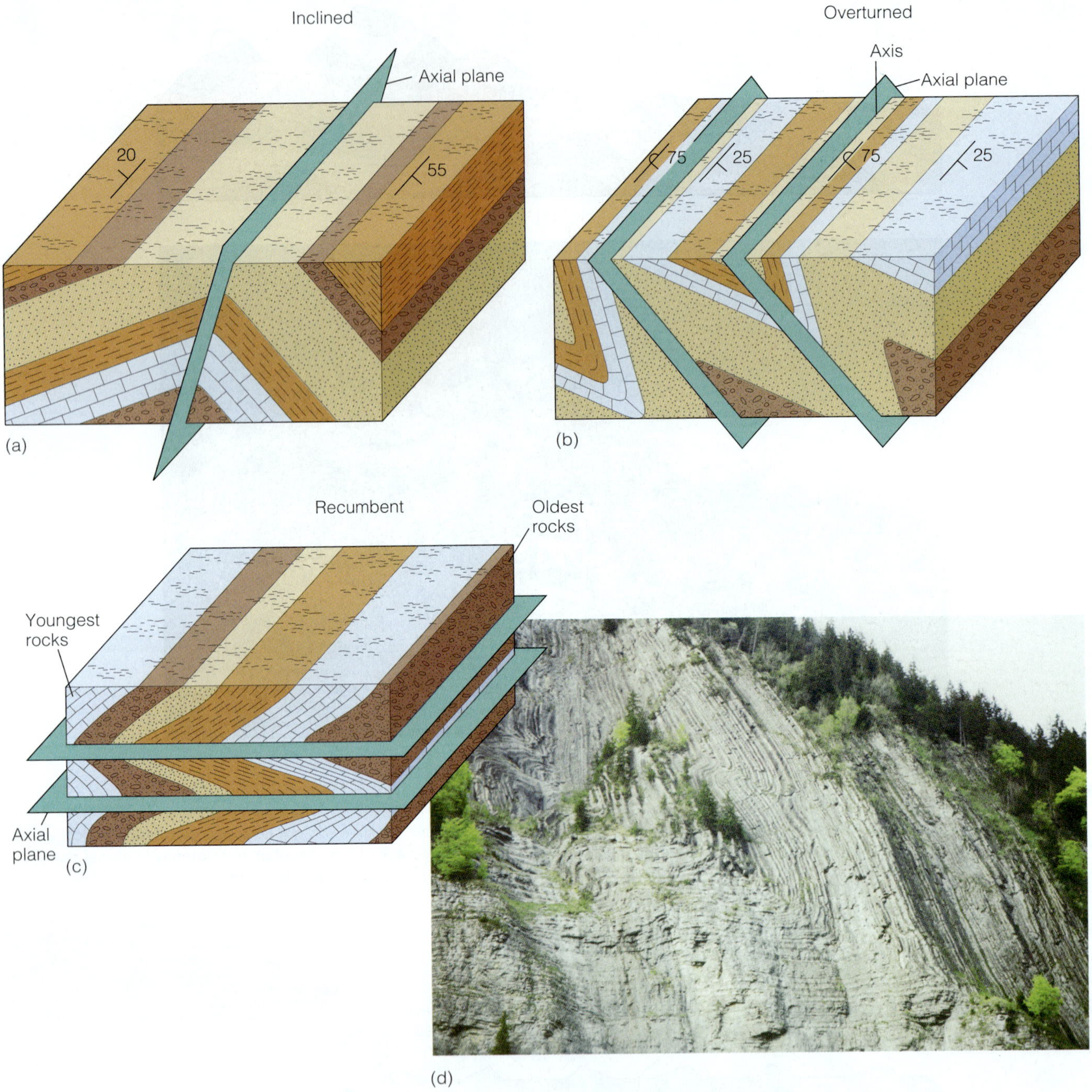

■ **Figure 10.12**

(a) An inclined fold. The axial plane is not vertical, and the fold limbs dip at different angles. (b) Overturned folds. Both fold limbs dip in the same direction, but one limb is inverted. Notice the special strike and dip symbol to indicate overturned beds. (c) Recumbent folds. (d) Recumbent fold in Switzerland.

at the center of the structure. Just the opposite is true in a basin—that is, all rock layers dip inward toward a central point and the youngest exposed rocks are at the center (Figure 10.14).

Many domes and basins are so large that they can be visualized only on geologic maps or aerial photographs. The Black Hills of South Dakota, for example, are a large oval dome (Figure 10.14c). One of the best-known large basins in the United States is the Michigan basin, most of which is buried beneath younger strata so it is not directly observable at the surface. Nevertheless, strike and dip of exposed strata near the basin margin and thousands of drill holes for oil and gas clearly show that the strata are deformed into a large basin.

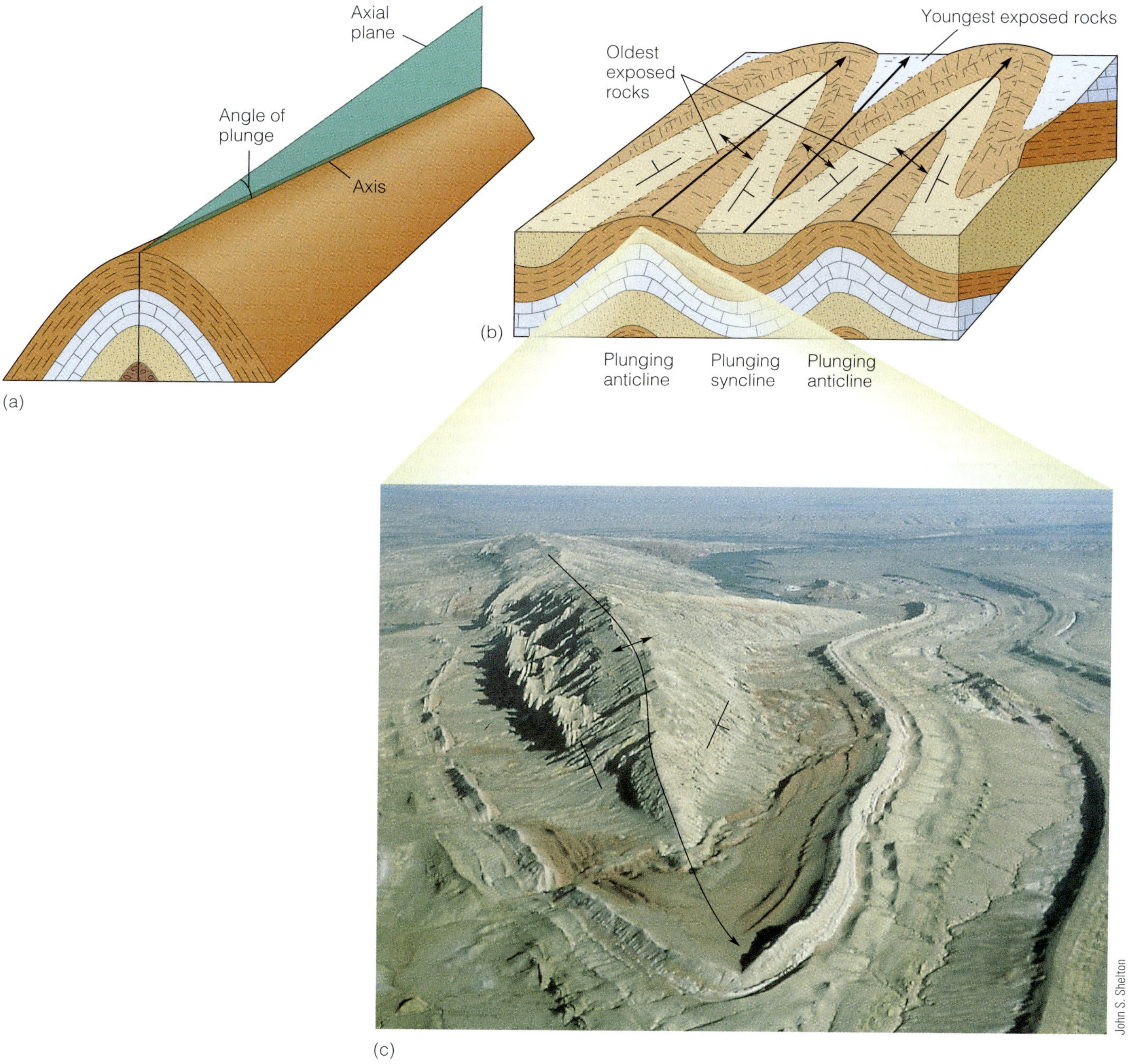

Figure 10.13

(a) A plunging fold. (b) Surface and cross-sectional views of plunging folds. The long arrow at the center of each fold is the standard geologic symbol used to depict plunging anticlines and synclines. The arrow at the end of the line shows the direction of plunge. (c) View of the eroded plunging Sheep Mountain anticline in Wyoming. Notice the smaller fold on the right flank of the larger one. Can you tell from the strike and dip symbols that this is a plunging anticline?

Unfortunately, the terms *dome* and *basin* are also used to distinguish high and low areas of Earth's surface, but as with anticlines and synclines, domes and basins resulting from deformation do not necessarily correspond with mountains or valleys. In some of the following discussions, we will have occasion to use these terms in other contexts, but we will try to be clear when we refer to surface elevations as opposed to geologic structures.

Joints

Besides folding, rocks are also permanently deformed by fracturing. **Joints** are fractures along which no movement has taken place parallel with the fracture surface (■ Figure 10.15), although they may open up; that is, they show movement perpendicular to the fracture. Coal miners used the term *joint* long ago for cracks in rocks that they thought were surfaces where the adjacent blocks were "joined" together.

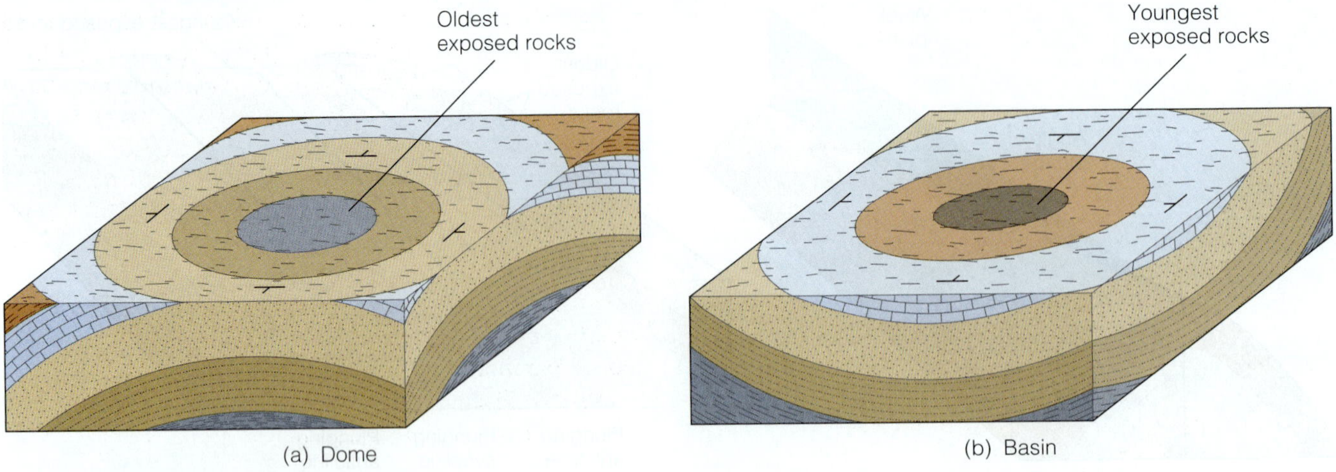

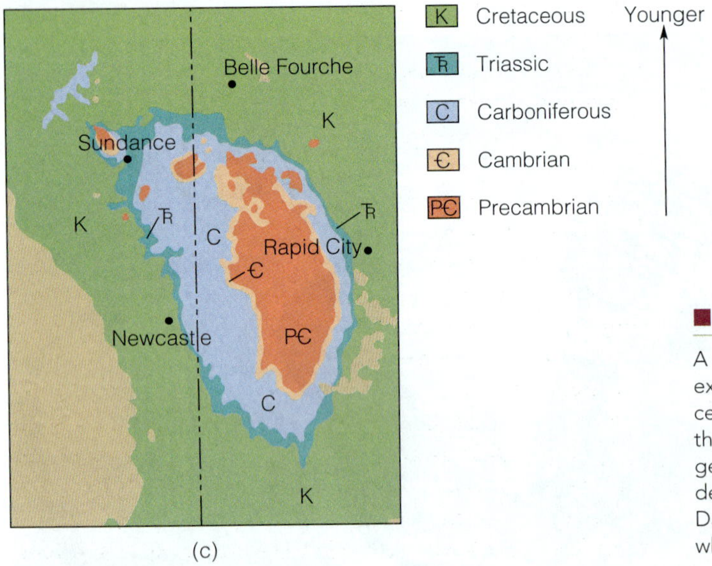

■ **Figure 10.14**

A dome (a) and a basin (b) Notice that in a dome the oldest exposed rocks are in the center and all rocks dip outward from a central point, whereas in a basin the youngest exposed rocks are in the center and all rocks dip inward toward a central point. (c) This geologic map, a surface view only, uses colors and symbols to depict rocks and geologic structures in the Black Hills of South Dakota. From the information provided, can you determine whether this map depicts a dome or a basin?

Remember that rocks near the surface are brittle and therefore commonly fail by fracturing when subjected to stress. Hence, almost all near-surface rocks have joints that form in response to compression, tension, and shearing. They vary from minute fractures to those extending for many kilometers and are often arranged in two or perhaps three prominent sets. Regional mapping reveals that joints and sets of joints are usually related to other geologic structures such as large folds and faults.

We have discussed columnar joints that form when lava or magma in some shallow plutons cools and contracts (see Figure 5.6). A different type of jointing previously discussed is sheet jointing that forms in response to pressure release (see Figure 6.4).

Faults

Another type of fracture known as a **fault** is one along which blocks of rock on opposite sides of the fracture have moved parallel with the fracture surface, and the surface along which movement takes place is a **fault plane** (■ Figure 10.16a). Not all faults penetrate to the surface, but those that do might show a *fault scarp,* a bluff or cliff formed by vertical movement (Figure 10.16b). Fault scarps are usually quickly eroded and obscured. When movement takes place on a fault plane, the rocks on opposite sides may be scratched and polished (Figure 10.16b) or crushed and shattered into angular blocks, forming *fault breccia* (Figure 10.16c).

Refer to Figure 10.16a and notice the designations *hanging wall block* and *footwall block.* The **hanging wall block** consists of the rock overlying the fault, whereas the **footwall block** lies beneath the fault plane. You can recognize these two blocks on any fault except a vertical one—that is, one that dips at 90 degrees. To identify some kinds of faults you must not only correctly identify these two blocks but also determine which one moved relatively up or down. We use the phrase *relative move-*

(a)

(b)

James S. Monroe

■ **Figure 10.15**

(a) Erosion along parallel joints in Arches National Park, Utah. (b) Joints intersecting at right angles yield this rectangular pattern in Wales.

ment because you cannot usually tell which block moved or if both moved. In Figure 10.16a, the footwall block may have moved up, the hanging wall block may have moved down, or both could have moved. Nevertheless, the hanging wall block appears to have moved down relative to the footwall block.

Remember our discussion of strike and dip of rock layers. Fault planes are also inclined planes and they too are characterized by strike and dip (Figure 10.16a). In fact, the two basic varieties of faults are defined by whether the blocks on opposite sides of the fault plane moved parallel to the direction of dip (dip-slip faults) or along the direction of strike (strike-slip faults) (see "Types of Faults" on 272 and 273).

Dip-Slip Faults All movement on **dip-slip faults** takes place parallel with the fault's dip; that is, movement is vertical, either up or down the fault plane. In ■ Figure 10.17a on page 275, for instance, the hanging wall block moved down relative to the footwall block, giving rise to a **normal fault.** In contrast, in a **reverse fault,** the hanging wall block moves up relative to the footwall block (Figure 10.17b). In Figure 10.17c, the hanging wall block also moved up relative to the footwall block, but the fault has a dip of less than 45 degrees and is a special variety of reverse fault known as a **thrust fault.**

Reference Figure 10.3b, which shows that normal faults result from tension. Numerous normal faults are present along one or both sides of mountain ranges in the Basin and Range Province of the western United States where the crust is being stretched and thinned. The Sierra Nevada at the western margin of the Basin and Range is bounded by normal faults, and the range has risen along these faults so that it now stands more than 3000 m above the lowlands to the east (see Chapter 23). Also, an active normal fault is found along the eastern margin of the Teton Range in Wyoming, accounting for the 2100-m elevation difference between the valley floor and the highest peaks in the mountains (■ Figure 10.18).

Reverse and thrust faults both result from compression (Figure 10.17b, c). Large-scale examples of both are found in mountain ranges that formed at convergent plate margins, where one would expect compression (discussed later in this chapter). A well-known thrust fault is the Lewis overthrust of Montana. (An overthrust is a low-angle thrust fault with movement measured in kilometers.) On this fault a huge slab of Precambrian-aged rocks moved at least 75 km eastward and now rests upon much younger Cretaceous-aged rocks (see "Types of Faults" on pages 272 and 273).

Strike-Slip Faults **Strike-slip faults,** resulting from shear stresses, show horizontal movement with blocks on opposite sides of the fault sliding past one another (Figure 10.17d). In other words, all movement is in the direction of the fault plane's strike—hence the name

Types of Faults

Faults are very common geologic structures. They are fractures along which movement takes place parallel to the fracture surface. A block of rock adjacent to a fault may move up or down a fault plane—that is, up or down the dip of the fault. These are thus called dip-slip faults. On the other hand, movement may take place along a fault's strike, giving rise to strike-slip faults. Movement on faults and the release of stored energy are responsible for earthquakes (see Chapter 8). Most faults are found at the three major types of plate boundaries: convergent, divergent, and transform.

James S. Monroe

Two small normal faults cutting through layers of volcanic ash in Oregon (left). Notice that the sandstone layers to the right of the hammer are cut by a reverse fault (below). Compare the sense of movement of the hanging wall blocks in these two images.

Martin Miller/ Visuals Unlimited

David J. Matty

Sue Monroe

Can you identify the type of faults shown in these two images?

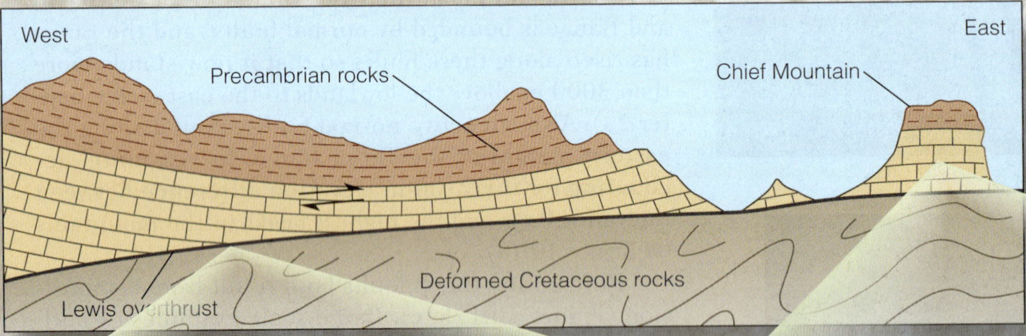

Diagrammatic view of the Lewis overthrust fault (a low-angle thrust fault) in Glacier National Park, Montana. Ancient Precambrian-aged rocks now rest on deformed Cretaceous-aged sedimentary rocks.

James S. Monroe

View from Marias Pass reveals the fault as a light-colored line on the mountainside.

James S. Monroe

Erosion has isolated Chief Mountain from the rest of the slab of overthrust rock.

Map showing the location of the Great Glen fault, a left-lateral strike-slip fault that cuts across Scotland.

View to the southwest along Loch Ness, Scotland, lying in the Great Glen fault zone, which at this point is more than 1.5 km wide.

Right-lateral offset of a gully by the San Andreas fault in central California. The gully is offset about 21 m.

Oblique slip took place on this fault in central Nevada during a 1915 earthquake. Notice the fence that shows right-lateral displacement and dip-slip displacement.

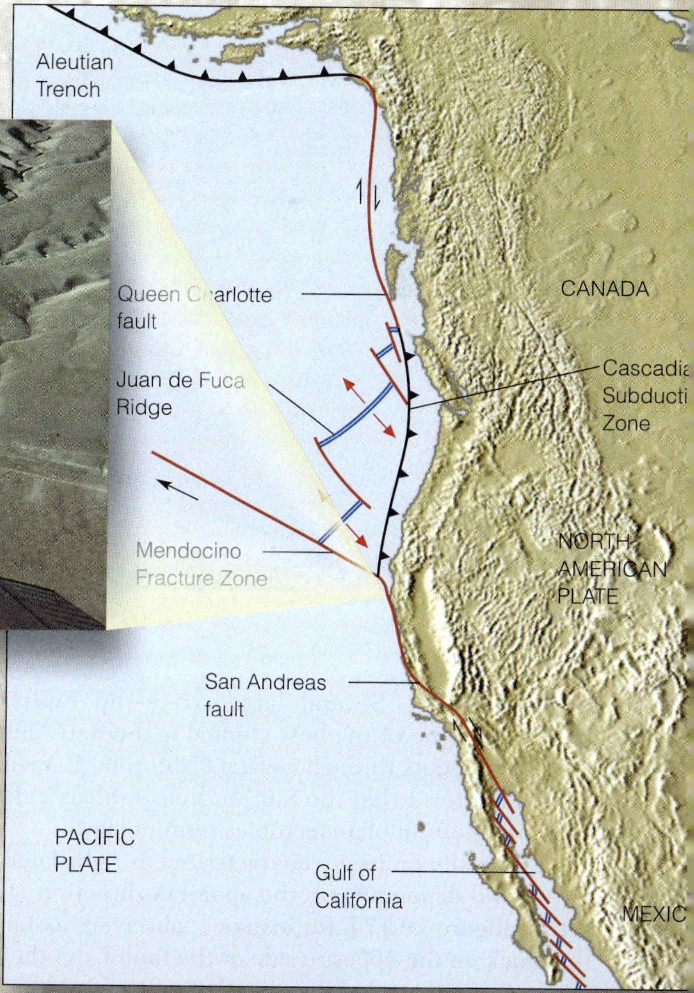

Plate tectonic setting for the San Andreas fault, a strike-slip fault. Remember that in plate tectonics terminology this is a transform fault.

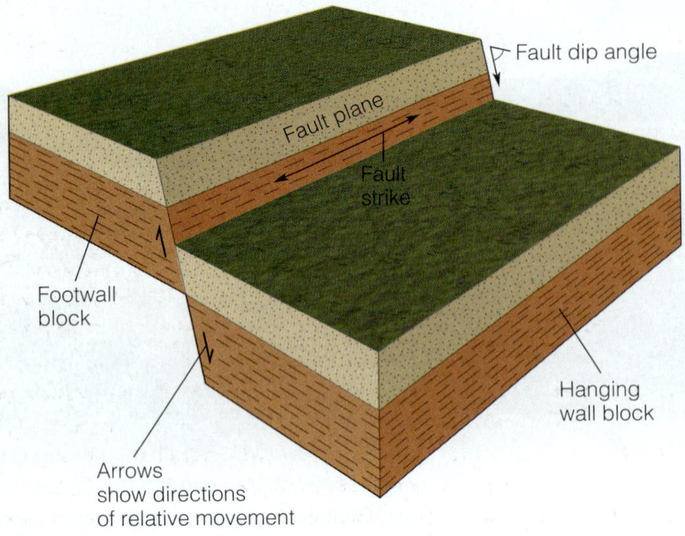

■ **Figure 10.16**

(a) Fault terminology. (b) Polished, scratched fault plane and fault scarp near Klamath Falls, Oregon. (c) Fault breccia, the zone of rubble along a fault in the Bighorn Mountains, Wyoming. If arrows were not shown, could you determine which side of the fault moved relatively up?

strike-slip fault. Several large strike-slip faults are known, but one of the best studied is the San Andreas fault, which cuts through coastal California. Remember from Chapter 2 that the San Andreas fault is called a *transform fault* in plate tectonics terminology.

Strike-slip faults are characterized as right-lateral or left-lateral depending on the apparent direction of offset. In Figure 10.17d, for instance, observers looking at the block on the opposite side of the fault from their location notice that it appears to have moved to the left. Accordingly, this is a *left-lateral strike-slip fault.* If it had been a *right-lateral strike-slip fault,* the block across the fault from the observers would appear to have moved to the right. The San Andreas fault in California is a right-lateral strike-slip fault, whereas the Great Glen fault in Scotland is a left-lateral strike-slip fault (see "Types of Faults" on pages 272 and 273).

Oblique-Slip Faults The movement on most faults is primarily dip-slip or strike-slip, but on **oblique-slip faults** both types of movement take place. Strike-slip movement might be accompanied by a component of dip-slip, giving rise to a combined movement that includes left-lateral and reverse, or right-lateral and normal (Figure 10.17e and see "Types of Faults" on pages 272 and 273).

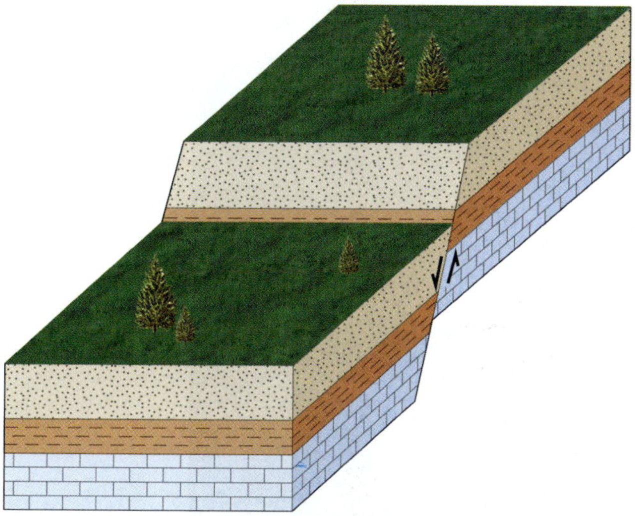

(a) Normal fault

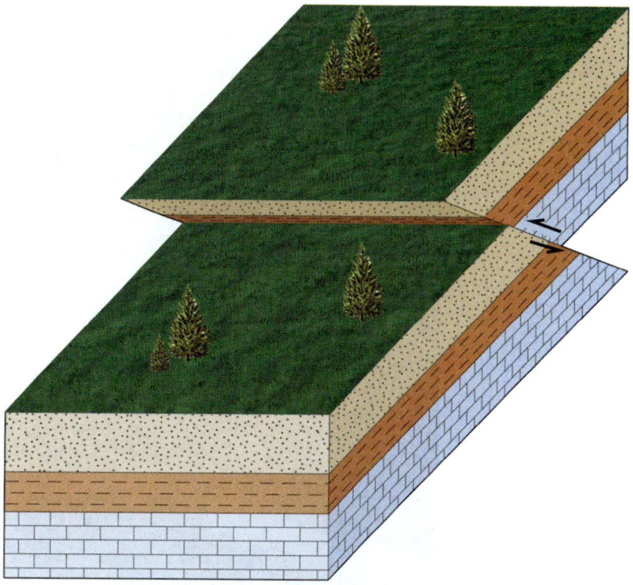

(b) Reverse fault

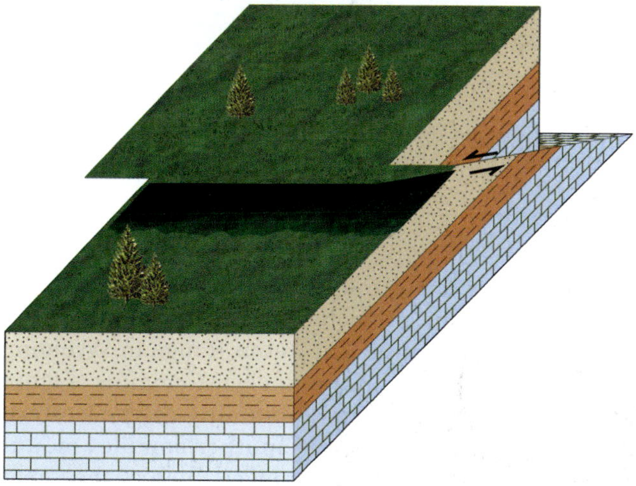

(c) Thrust fault

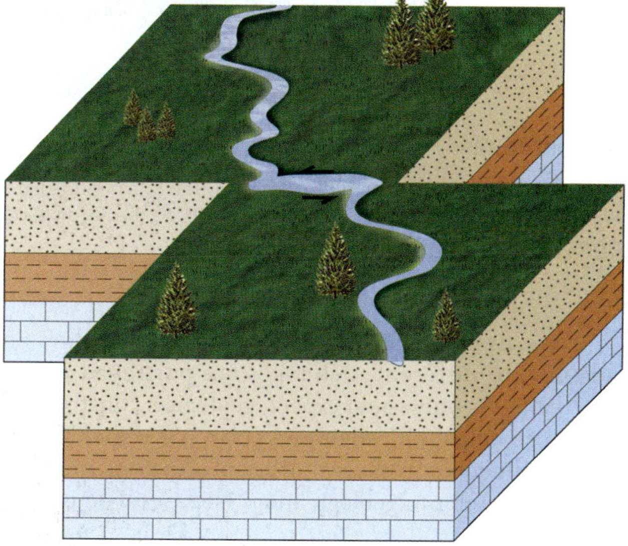

(d) Strike-slip fault

(e) Oblique-slip fault

Figure 10.17

Types of faults. (a), (b), and (c) are dip-slip faults. (a) Normal fault—hanging wall block moves down relative to footwall block. (b, c) Reverse and thrust faults—hanging wall block moves up relative to footwall block. (d) Strike-slip fault—all movement is parallel to the strike of the fault. (e) Oblique-slip fault—combination of dip-slip and strike-slip movements.

James S. Monroe

■ **Figure 10.18**

The Teton Range in Wyoming is one of many mountain ranges in the Rocky Mountain system. The range began forming about 10 million years ago as uplift took place on a normal fault. As the Teton block was uplifted, it was also eroded by running water, glaciers, and gravity-driven processes.

Various geologic structures along with colors and symbols for different rock types are portrayed on *geologic maps*. As one would expect, geologists construct and use these maps, but engineers, city and regional planners, and people in several other professions may have occasion to refer to geologic maps (see Geo-Focus 10.1).

DEFORMATION AND THE ORIGIN OF MOUNTAINS

Mountains originate in several ways, but the truly large mountains on continents result mostly from compression-induced deformation at convergent plate boundaries. Before discussing mountain building, though, we should define what we mean by the term *mountain* and briefly discuss the types of mountains. *Mountain* is a designation for any area of land that stands significantly higher, at least 300 m, than the surrounding country and has a restricted summit area. Some mountains are single, isolated peaks, but more commonly they are parts of linear associations of peaks and ridges known as *mountain ranges* that are related in age and origin. A *mountain system*, a complex linear zone of deformation and crustal thickening, on the other hand, consists of several or many mountain ranges. The Teton Range in Wyoming is one of many ranges in the Rocky Mountains (Figure 10.18). The Appalachian Mountains of the eastern United States and Canada is another complex mountain system made up of many ranges, such as the Great Smoky Mountains of North Carolina and Tennessee, the Adirondack Mountains of New York, and the Green Mountains of Vermont.

Mountain Building

Mountains develop in a variety of ways, some involving little or no deformation. For instance, differential weathering and erosion have yielded high areas with adjacent lowlands in the southwestern United States, but these erosional remnants are rather flat topped or pinnacle-shaped and go by the names *mesa* and *butte*, respectively (see Chapter 15). A single volcanic mountain might develop over a hot spot, although more commonly a series of volcanoes forms as a plate moves over a hot spot, as in the Hawaiian Islands (see Figure 2.25). And keep in mind that the oceanic ridge system consists of mountains that exceed the size of any mountains on land. But the oceanic ridges form by volcanism at divergent plate boundaries and show features produced by tensional stresses. Large mountains on land, however, are composed of all rock types and show clear indications of compression.

Block-faulting is yet another way to form mountains, but it involves considerable deformation (■ Figure 10.19). Block-faulting entails movement on normal faults so that one or more blocks are elevated relative to adjacent areas. A classic example is the large-scale active block-faulting in the Basin and Range Province of the western United States, a large area centered on Nevada but extending into several adjacent states and northern Mexico. Here the crust is stretched in an east-west direction; thus, tensional stresses produce north-south–oriented, range-bounding faults. Differential movement on these faults has produced uplifted blocks called *horsts* and down-dropped blocks called *grabens* (Figure 10.19) bounded on both sides by parallel normal faults. Erosion of the horsts has yielded the mountainous topography, and the grabens have filled with sediments eroded from the horsts (Figure 10.19).

Plate Tectonics and Mountain Building

Geologists define the term **orogeny** as an episode of mountain building during which intense deformation takes place, generally accompanied by metamorphism, the emplacement of plutons, especially batholiths, and thickening of Earth's crust. The processes responsible for an orogeny are still not fully understood, but it is known that mountain building is related to plate movements. In fact, the advent of plate tectonic theory completely changed the way geologists view the origin of mountain systems.

Any theory that accounts for mountain building must adequately explain the characteristics of mountains, such as their geometry and location; they tend to be long and narrow and at or near plate margins. Mountains also show intense deformation, especially compression-induced overturned and recumbent folds as well as reverse and thrust faults. Furthermore, granitic plutons and regional metamorphism characterize the interiors or

■ **Figure 10.19**

(a) Block-faulting and the origin of horsts and grabens. Many of the mountain ranges in the Basin and Range Province of the western United States and northern Mexico formed in this manner. (b) View of the Humboldt Range in Nevada, which is a horst bounded by normal faults.

GEOFOCUS 10.1

Geologic Maps—Their Construction and Uses

Geologic maps use lines, symbols, and colors to depict the distribution of rock types, show age relationships among rocks, and delineate geologic structures such as folds and faults (■ Figure 1). Most geologic maps are printed on some kind of base map—that is, a map that shows geographic locations and perhaps elevations. In short, geologic maps provide considerable information about Earth materials in a region, and like all maps they have a scale and a legend to explain the colors and symbols used.

The area depicted on a geologic map depends on the purpose of the original study. Some maps show entire continents or countries, but because they display such large areas, they cannot show small-scale features. These maps are useful for depicting the regional distribution of rocks of a given age and for delineating large geologic structures, but geologic maps of much smaller areas are of more immediate use (Figures 1 and 2).

Constructing a geologic map is simple in concept but not always easy in reality. Geologists visit surface rock exposures, or outcrops, where they record pertinent information such as strike and dip, rock composition, and relative age relationships. Notice in ■ Figure 2 that outcrops are discontinuous, the most common situation encountered by geologists. Nevertheless, the outcrops allow us to infer what lies beneath the covered areas between outcrops. Thus, the finished geologic map shows the area as if there were no soil cover (Figure 2c).

Once a geologic map is completed, geologists determine the geologic history of the area shown. And because strike and dip symbols show the orientation of rock layers and faults, geologists can construct cross sections and thereby illustrate three-dimensional relationships among the various rocks. This step is particularly important in many economic ventures because cross sections may indicate where an oil well should be drilled or where other resources may be present below the surface.

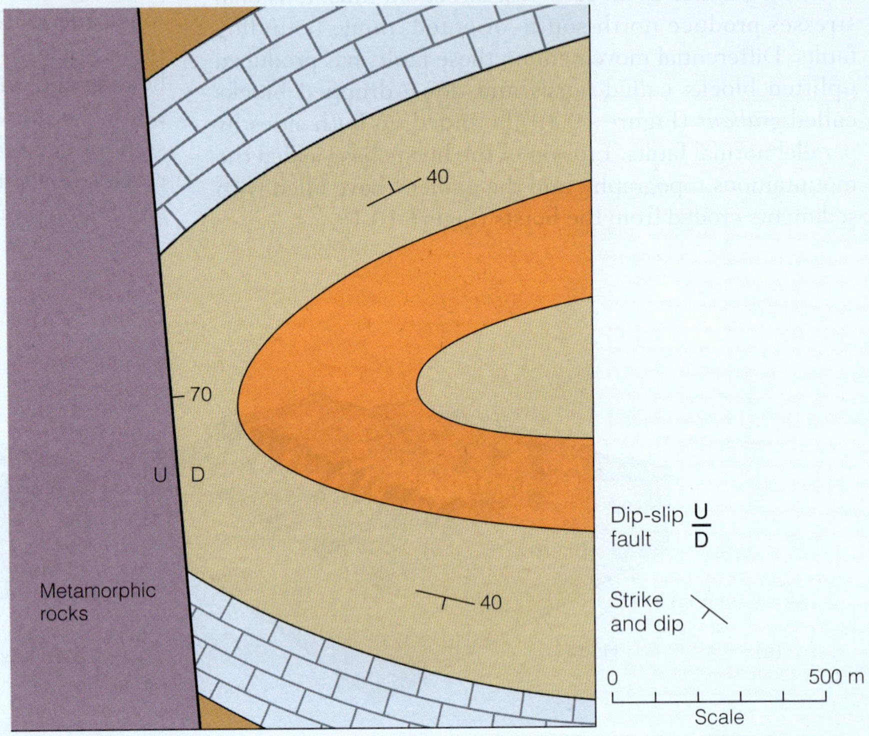

■ **Figure 1**

Geologic map showing the distribution of rocks and structures of an area.

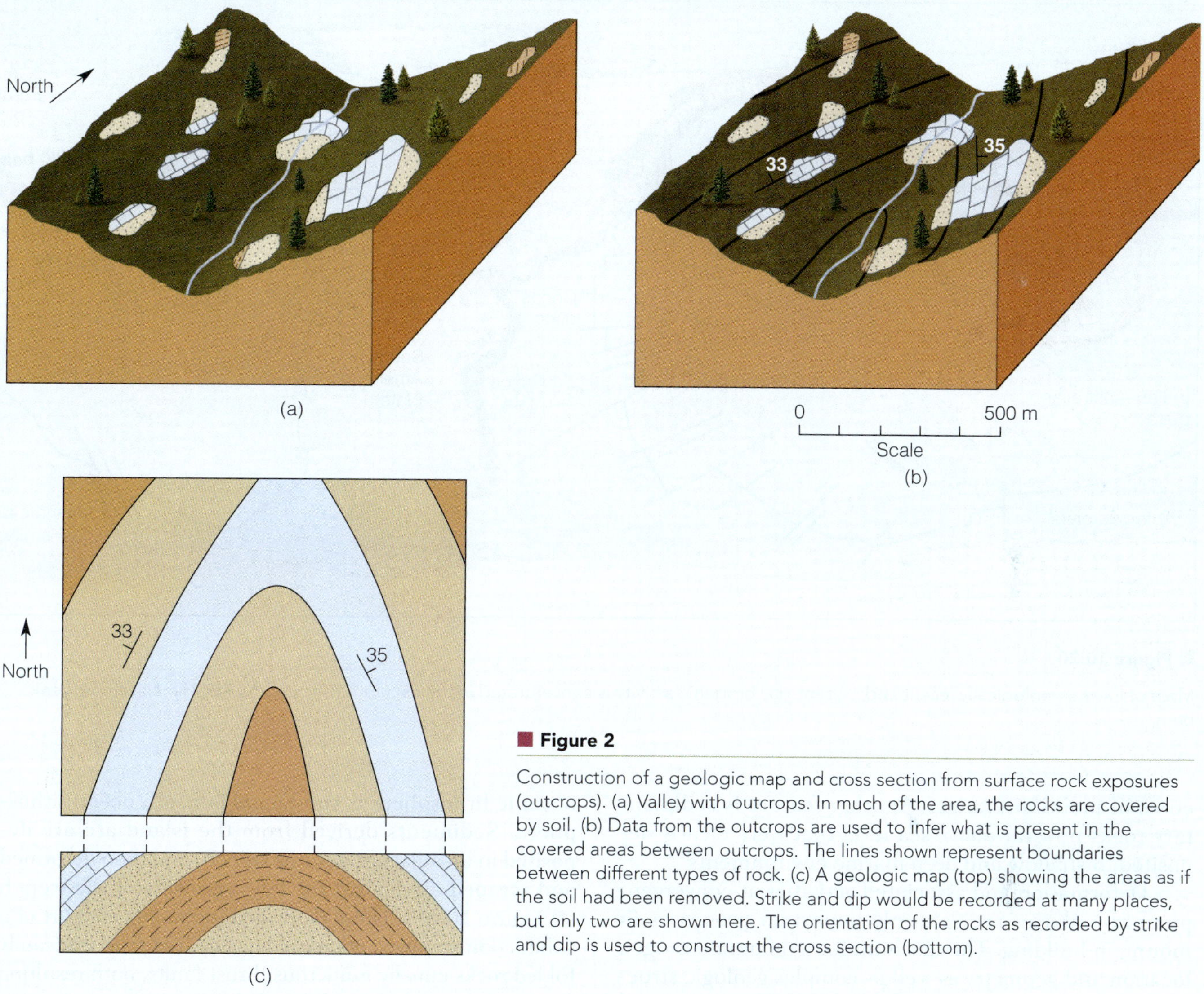

Figure 2

Construction of a geologic map and cross section from surface rock exposures (outcrops). (a) Valley with outcrops. In much of the area, the rocks are covered by soil. (b) Data from the outcrops are used to infer what is present in the covered areas between outcrops. The lines shown represent boundaries between different types of rock. (c) A geologic map (top) showing the areas as if the soil had been removed. Strike and dip would be recorded at many places, but only two are shown here. The orientation of the rocks as recorded by strike and dip is used to construct the cross section (bottom).

Besides geologists, people in other professions use geologic maps. You, for example, may one day be a member of a safety planning board of a city with known active faults. You may use geologic maps to develop zoning regulations and construction codes. Given the position of the active fault in Figure 1, zoning this area for a housing development would not be prudent, but it might be perfectly satisfactory for agriculture or some other land use. Perhaps you will become a land-use planner or member of a county commission charged with selecting a site for a sanitary landfill or securing an adequate supply of groundwater for your community. You would almost certainly use geologic maps in these endeavors.

Geologic maps are used extensively in planning and constructing dams, choosing the best route for a highway through a mountainous region, and selecting the sites for nuclear reactors. It would certainly be unwise to build a dam in a valley with a known active fault or with rocks too weak to support such a large structure. And in areas of continuing volcanic activity, geologic maps provide essential information about the kinds of volcanic processes that might occur in the future, such as lava flows, ashfalls, or mudflows. Recall from Chapter 5 that one way to access volcanic hazards is by mapping and determining the geologic history of an area.

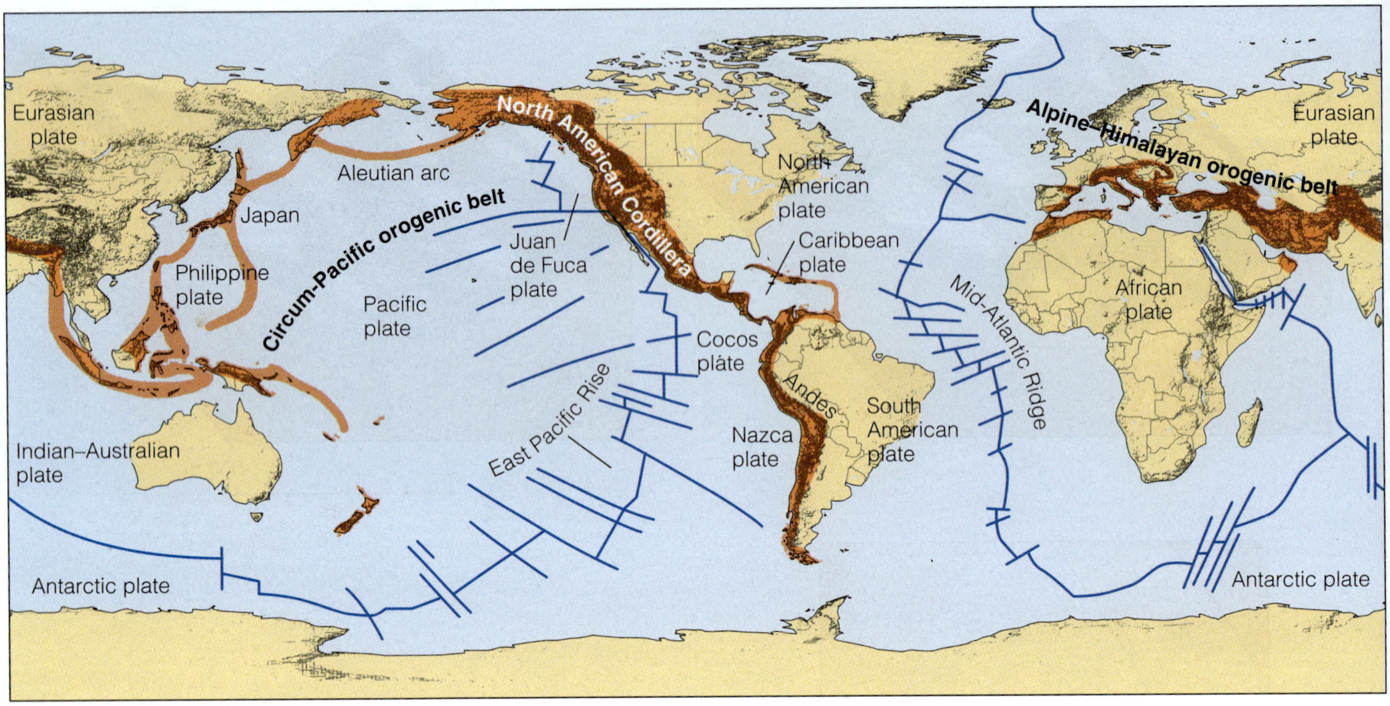

■ **Figure 10.20**

Most of Earth's geologically recent and present-day orogenic activity is concentrated in the circum-Pacific and Alpine–Himalayan orogenic belts.

cores of mountain ranges. Another feature is sedimentary rocks now far above sea level that were clearly deposited in shallow and deep marine environments.

Deformation and associated activities at convergent plate boundaries are certainly important processes in mountain building. They account for a mountain range's location and geometry as well as complex geologic structures, plutons, and metamophism. Yet the present-day topographic expression of mountains is also related to several surface processes such as mass wasting (gravity-driven processes including landslides), glaciers, and running water. In other words, erosion also plays an important role in the evolution of mountains.

Most of Earth's geologically recent and present-day orogenies are found in two major zones or belts: the *Alpine–Himalayan orogenic belt* and the *circum-Pacific orogenic belt* (■ Figure 10.20). Both belts are composed of a number of smaller segments knows as *orogens,* each a zone of deformed rocks and many of which have been metamorphosed and intruded by plutons. In fact, we can explain most of Earth's past and present orogenies in terms of the geologic activity at convergent plate boundaries. Recall from Chapter 2 that convergent plate boundaries might be oceanic–oceanic, oceanic–continental, or continental–continental.

Orogenies at Oceanic–Oceanic Plate Boundaries

Deformation, igneous activity, and the origin of a volcanic island arc characterize orogenies that take place where oceanic lithosphere is subducted beneath ocean lithosphere. Sediments derived from the island arc are deposited in an adjacent oceanic trench, and then deformed and scraped off against the landward side of the trench (■ Figure 10.21). These deformed sediments are part of a subduction complex, or an *accretionary wedge,* of intricately folded rocks cuts by numerous thrust faults, both resulting from compression. In addition, orogenies in this setting are characterized by low-temperature, high-pressure metamorphism of the blueschist facies (see Figure 7.18).

Deformation caused largely by the emplacement of plutons also takes place in the island arc system where many rocks show evidence of high-temperature, low-pressure metamorphism. The overall effect of island arc orogenesis is the origin of two more or less parallel orogenic belts consisting of a landward volcanic island arc underlain by batholiths and a seaward belt of deformed trench rocks (Figure 10.21). The Japanese Islands are a good example of this type of deformation.

In the area between an island arc and its nearby continent, the back-arc basin, volcanic rocks and sediments derived from the island arc and the adjacent continent are also deformed as the plates continue to converge. The sediments are intensely folded and displaced toward the continent along low-angle thrust faults. Eventually, the entire island arc complex is fused to the edge of the continent, and the back-arc basin sediments are thrust onto the continent, forming a thick stack of thrust sheets (Figure 10.21).

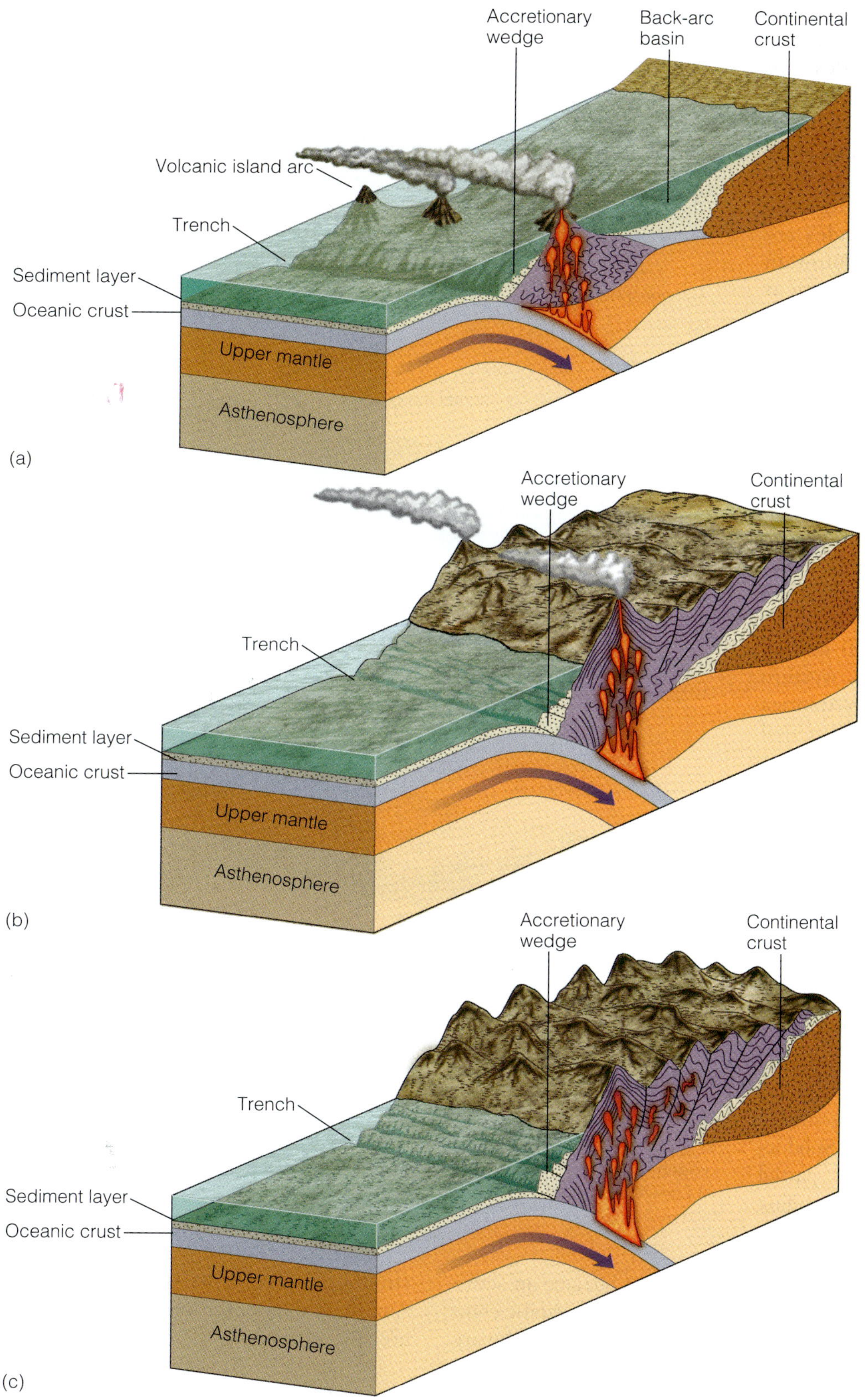

Figure 10.21

Orogeny and the origin of a volcanic island arc at an oceanic–oceanic plate boundary. (a) Subduction of an oceanic plate beneath an island arc. (b) Continued subduction, emplacement of plutons, and beginning of deformation by thrusting and folding of back-arc basin sediments. (c) Thrusting of back-arc basin sediments onto the adjacent continent and suturing of the island arc to the continent.

Orogenies at Oceanic–Continental Plate Boundaries The Andes of South America are perhaps the best example of continuing orogeny at an oceanic–continental plate boundary. Among the ranges of the Andes are the highest mountain peaks in the Americas and many active volcanoes. Furthermore, the west coast of South America is an extremely active segment of the circum-Pacific earthquake belt, and one of Earth's great oceanic trench systems, the Peru–Chile Trench, which lies just off the west coast of South America.

Prior to 200 million years ago, the western margin of South America was a passive continental margin where sediments accumulated much as they do now along the East Coast of North America. However, when Pangaea split apart along what is now the Mid-Atlantic Ridge, the South American plate moved westward. As a consequence, the oceanic lithosphere west of South America began subducting beneath the continent (■ Figure 10.22). Subduction resulted in partial melting of the descending plate, which produced the andesitic volcanic arc of composite volcanoes, and the west coast became an active continental margin. Felsic magmas, mostly of granitic composition, were emplaced as large plutons beneath the arc (Figure 10.22).

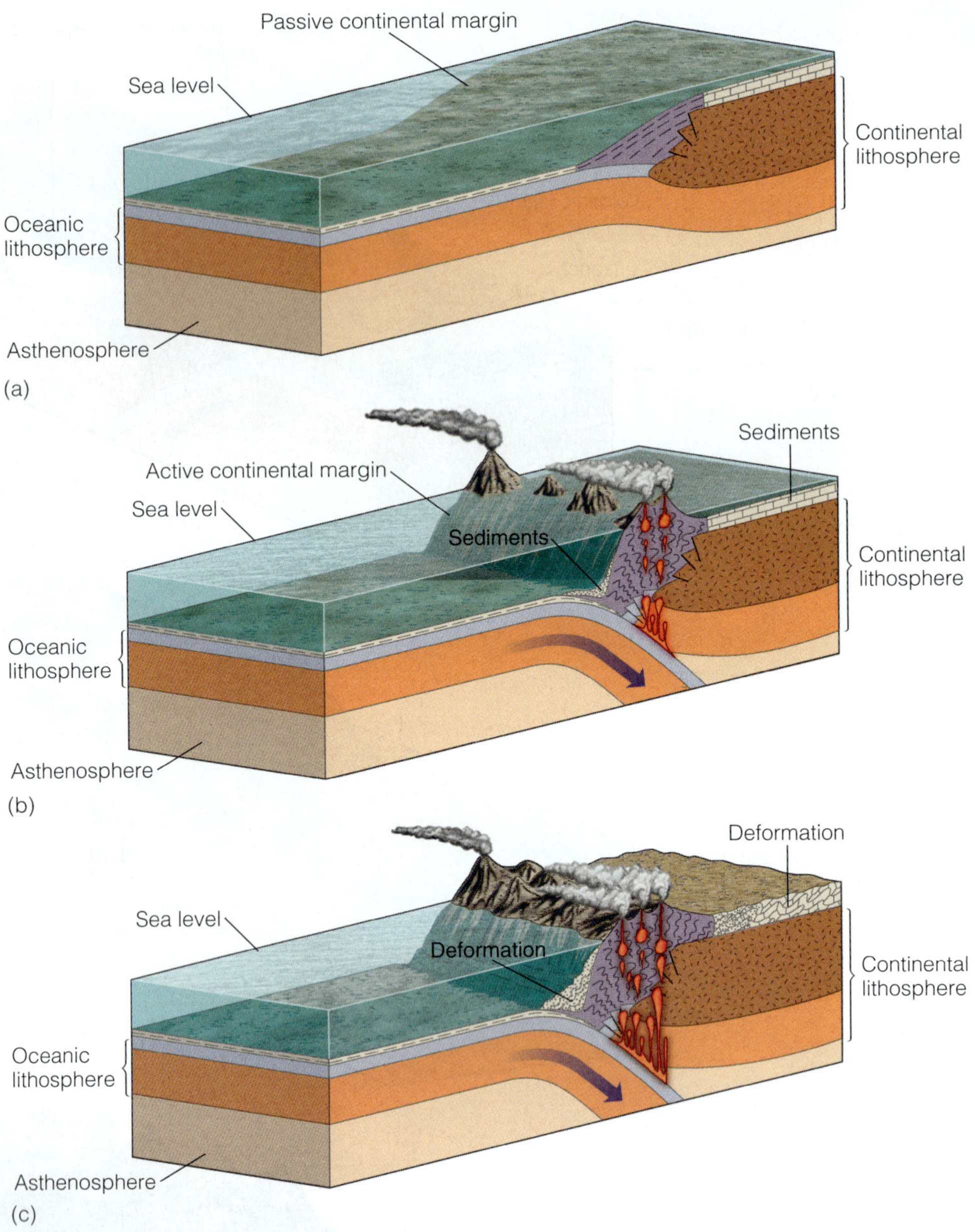

■ **Figure 10.22**

Generalized diagrams showing three stages in the development of the Andes of South America. (a) Prior to 200 million years ago, the west coast of South America was a passive continental margin. (b) An orogeny began when the west coast of South America became an active continental margin. (c) Continued deformation, volcanism, and plutonism.

As a result of the events just described, the Andes Mountains consist of a central core of granitic rocks capped by andesitic volcanoes. To the west of this central core along the coast are the deformed rocks of the accretionary wedge. And to the east of the central core are intensely folded sedimentary rocks that were thrust eastward onto the continent (Figure 10.22). Present-day subduction, volcanism, and seismicity along South America's west coast indicate that the Andes Mountains are still forming.

Orogenies at Continental–Continental Plate Boundaries The best example of an orogeny along a continental–continental plate boundary is the Himalayas of Asia. The Himalayas began forming when India collided with Asia about 40 to 50 million years ago. Prior to that time, India was far south of Asia and separated from it by an ocean basin (■ Figure 10.23a).

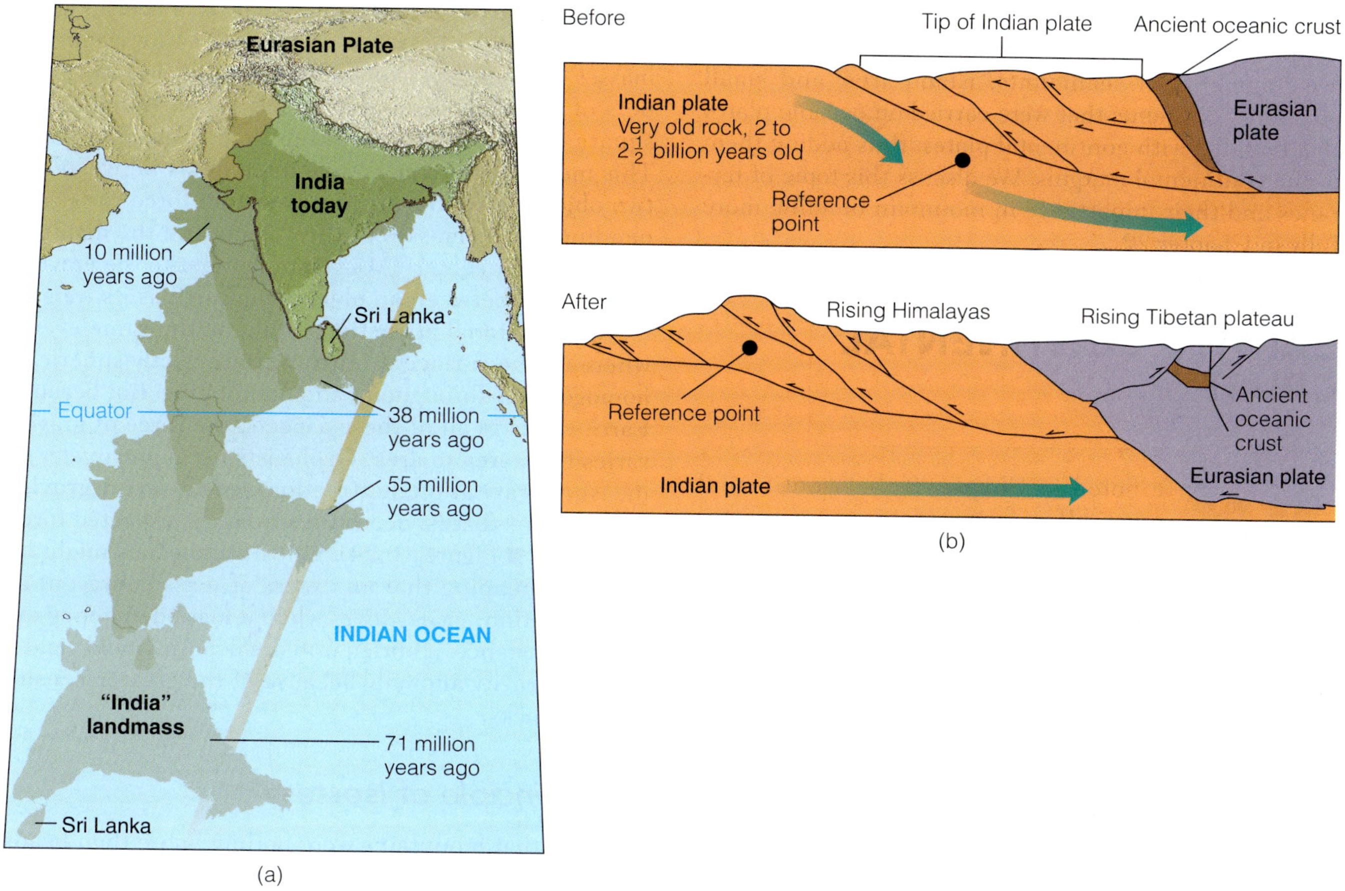

Figure 10.23

(a) During its long journey northward, India moved at about 10 cm per year, but beginning between 40 and 50 million years ago, the rate of movement decreased by about half as India collided with the Eurasian plate. (b) These cross sections show the Indian and Eurasian plates before and after their collision, which resulted in uplift of the Himalayas and the Tibetan Plateau.

As the Indian plate moved northward, a subduction zone formed along the southern margin of Asia where oceanic lithosphere was consumed. Partial melting generated magma, which rose to form a volcanic arc, and large granite plutons were emplaced into what is now Tibet. At this stage, the activity along Asia's southern margin was similar to what is now occurring along the west coast of South America.

The ocean separating India from Asia continued to close and India eventually collided with Asia (Figure 10.23b). As a result, two continental plates became welded, or sutured, together. Thus the Himalayas are now within a continent rather than along a continental margin. The exact time of India's collision with Asia is uncertain, but between 40 and 50 million years ago, India's rate of northward drift decreased abruptly from about 10 cm per year to about 5 cm per year. Because continental lithosphere is not dense enough to be subducted, this decrease seems to mark the time of collision and India's resistance to subduction. Consequently, the leading margin of India was thrust beneath Asia, causing crustal thickening, thrusting, and uplift. Sedimentary rocks that had been deposited in the sea south of Asia were thrust northward, and two major thrust faults carried rocks of Asian origin onto the Indian plate (Figure 10.23b). Rocks deposited in the shallow seas along India's northern margin now form the higher parts of the Himalayas. Since its collision with Asia, India has been thrust horizontally about 2000 km beneath Asia and now moves north at several cm per year.

Other mountain systems also formed as a result of collisions between two continental plates. The Urals in Russia and the Appalachians of North America formed by such collisions. In addition, the Arabian plate is now colliding with Asia along the Zagros Mountains of Iran.

Terranes and the Origin of Mountains

In the preceding section, we discussed orogenies along convergent plate boundaries that result in adding material to a continent, a process termed **continental accretion.** Much of the material added to continental margins is eroded older continental crust, but some plutonic and volcanic rocks are new additions. During the 1970s and 1980s, however, geologists discovered that parts of many mountain systems are also made up of small, accreted lithospheric blocks that clearly originated elsewhere. These **terranes,*** as they are called, are fragments of seamounts, island arcs, and small pieces of continents that were carried on oceanic plates that collided with continental plates, thus adding them to the continental margins. We discuss this topic of terranes and their importance in mountain building more fully in Chapter 22.

EARTH'S CONTINENTAL CRUST

We noted in Chapter 9 that continental crust differs from oceanic crust in composition, density, and topographic expression. Obviously oceanic crust is lower than continental crust, but why is this so? Furthermore, why do mountains stand higher than surrounding continental crust? To answer these questions we must examine continental crust in more detail. As you already know, oceanic crust is composed of basalt and gabbro, whereas continental crust is characterized as granitic, meaning it has a composition similar to granite. Nevertheless, it contains a wide variety of igneous, sedimentary, and metamorphic rocks, has an average density of 2.7 g/cm^3, and varies from 20 to 90 km thick. In short, it differs from oceanic crust in several important aspects, and it is also considerably more complex.

In most places continental crust is about 35 km thick, but it is much thicker beneath the Rocky Mountains, the Appalachians, the Alps in Europe, and the Himalayas of Asia. (Oceanic crust is only 5 to 10 km thick.) As a matter of fact, continental crust is thicker beneath all of Earth's mountain systems. And it is this difference in thickness coupled with the fact that continental crust is less dense than oceanic crust that explains why mountains stand high.

Floating Continents?

The term *floating* immediately brings to mind a ship at sea or some other buoyant object in a fluid, but it certainly does not evoke an image of a continent buoyed up by some kind of fluid below. Nevertheless, continental crust, and oceanic crust for that matter, is floating in a manner of speaking on a more dense substance. To understand why this is so, you must be familiar with the concepts of gravity and the principle of isostasy.

Isaac Newton formulated the law of universal gravitation in which the force of gravity (F) between two masses (m_1 and m_2) is directly proportional to the products of their masses and inversely proportional to the square of the distance between their centers of mass. This means that an attractive force exists between any two objects, and the magnitude of that force varies depending on the masses of the objects and the distance between their centers. We generally refer to the gravitational force between an object and Earth as its *weight.*

Gravitational attraction would be the same everywhere on the surface if Earth were perfectly spherical, homogeneous throughout, and not rotating. But because Earth varies in all of these aspects, the force of gravity varies from area to area. Geologists use a *gravimeter* to measure gravitational attraction and to detect **gravity anomalies**—that is, departures from the expected force of gravity (■ Figure 10.24). Gravity anomalies might be *positive,* meaning that an excess of mass is present at some location, or *negative,* when a mass deficiency exists. For instance, a buried iron ore deposit would yield a positive gravity anomaly because of the greater density of these rocks.

The Principle of Isostasy

Suppose that mountains were nothing more than heaps of material piled on the continental crust, as shown in Figure 10.24a. If this were so, we would expect a gravity survey across this mountainous area to reveal a huge positive gravity anomaly—that is, an excess of mass between the surface and Earth's center. The fact that no such anomaly exists implies that some of the dense mantle material at depth must be displaced by less dense crustal rocks (Figure 10.24b).

According to the **principle of isostasy,** Earth's crust is in floating equilibrium with the denser mantle below. This phenomenon is easy to understand by analogy to an iceberg. Ice is slightly less dense than water, so it floats. But according to Archimedes' *principle of buoyancy,* an iceberg sinks in the water until it displaces a volume of water equal to the weight of the iceberg. When the iceberg has sunk to its equilibrium position, only about 10% of its volume projects above water level. And if some of the ice above water level melts, the iceberg rises in order to maintain the same proportion of ice above and below water.

Earth's crust is similar to the iceberg in our analogy in that it sinks into the mantle to its equilibrium level. Where the crust is thickest, as beneath mountain ranges, it sinks farther down into the mantle but also rises higher above the equilibrium surface (Figure 10.24). Both continental and oceanic crust are less dense than the upper mantle (its density is 3.3 g/cm^3), but continental crust, being thicker and less dense than oceanic crust, stands higher.

*Some geologists prefer the terms *suspect terrane, exotic terrane,* or *displaced terrane.* Notice also the spelling of *terrane* as opposed to the more familiar *terrain,* the latter a geographic term indicating a particular area of land.

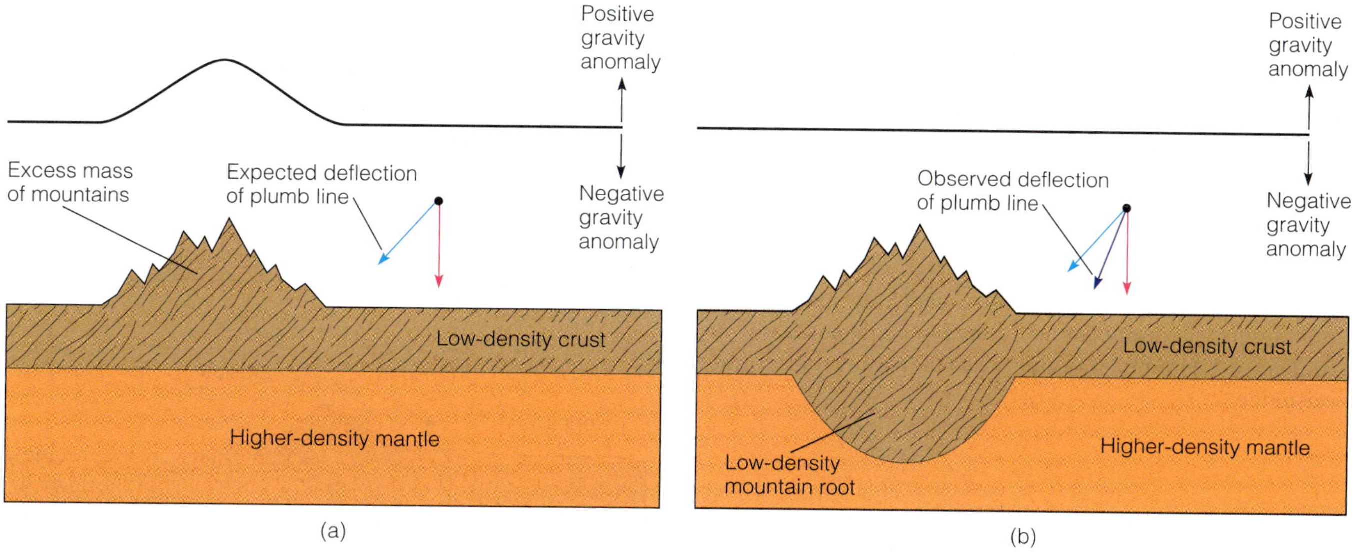

Figure 10.24

(a) A plumb line (a cord with a suspended weight) is normally vertical, pointing to Earth's center of gravity. Near a mountain range, the plumb line should be deflected as shown if the mountains are simply thicker, low-density material resting on denser material, and a gravity survey across the mountains would indicate a positive gravity anomaly. (b) The actual deflection of the plumb line during the survey in India was less than expected. It was explained by postulating that the Himalayas have a low-density root. A gravity survey in this case would show no anomaly because the mass of the mountains above the surface is compensated for at depth by low-density material displacing denser material.

Some of you might realize that crust floating on the mantle raises an apparent contradiction. Remember from Chapter 8 we said that the mantle is a solid because it transmits S-waves, which do not move through a fluid. But according to the principle of isostasy, the mantle behaves as a fluid. How can this apparent paradox be resolved? When considered in terms of the brief time required for S-waves to pass through it, the mantle is indeed solid. But when subjected to stress over long periods, it yields by flowage and thus at this time scale it can be regarded as a viscous fluid. A familiar substance that has the properties of a fluid or a solid depending on how rapidly deforming stress is applied is Silly Putty. Given sufficient time, it will flow under its own weight, but it shatters as a brittle solid if struck a sharp blow.

What Would You Do?

As a member of a planning commission, you are charged with developing zoning regulations and building codes for an area with known active faults, steep hills, and deep soils. A number of contractors as well as developers and citizens in your community are demanding action because they want to begin several badly needed housing developments. How might geologic maps and an appreciation of geologic structures influence you in this endeavor? Do you think, considering the probable economic gains for your community, that the regulations you draft should be rather lenient or very strict? If the latter, how would you explain why you favored regulations that would involve additional cost for houses?

Isostatic Rebound

What happens when a ship is loaded with cargo and then later unloaded? Of course it first sinks lower in the water and then rises, but it always finds its equilibrium position. Earth's crust responds similarly to loading and unloading but much more slowly. For example, if the crust is loaded, as when widespread glaciers accumulate, the crust sinks farther into the mantle to maintain equilibrium. The crust behaves similarly in areas where huge quantities of sediment accumulate.

If loading by glacial ice or sediment depresses Earth's crust farther into the mantle, it follows that when vast glaciers melt or where deep erosion takes place, the crust should rise back up to its equilibrium level. And in fact it does. This phenomenon, known as **isostatic rebound,** is taking place in Scandinavia, which was covered by a thick ice sheet until about 10,000 years ago; it is now rebounding at about 1 m per century. In fact, coastal cities in Scandinavia have rebounded rapidly enough that docks constructed several centuries ago are now far from shore. Isostatic rebound has also occurred in eastern Canada where the crust has risen as much as 100 m in the last 6000 years.

■ Figure 10.25 shows the response of Earth's continental crust to loading and unloading as mountains form and evolve. Recall that during an orogeny, emplacement of plutons, metamorphism, and general thickening of the crust accompany deformation. Consequently, the crust rises higher above and projects farther below the equilibrium surface than adjacent thinner crust does. However, as the mountains erode, isostatic rebound takes place and the mountains rise whereas adjacent areas of sedimentation subside (Figure 10.25). If continued long enough, the mountains will disappear and then can be detected only by the plutons and metamorphic rocks that show their former existence.

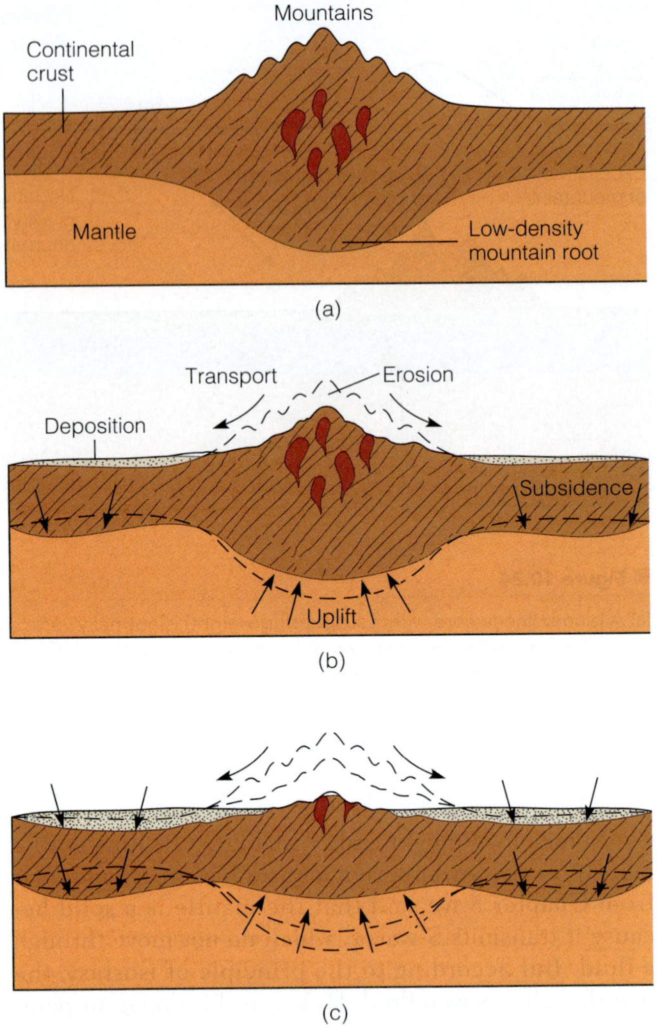

Geology Now ■ **Active Figure 10.25**

A diagrammatic representation showing the isostatic response of the crust to erosion (unloading) and widespread deposition (loading).

10 GEO RECAP

Chapter Summary

- Folded and fractured rocks have been deformed or strained by applied stresses.
- Stress is compression, tension, or shear. Elastic strain is not permanent, but plastic strain and fracture are, meaning that rocks do not return to their original shape or volume when the deforming forces are removed.
- Strike and dip are used to define the orientation of deformed rock layers. This same concept applies to other planar features such as fault planes.
- Anticlines and synclines are up- and down-arched folds, respectively. They are identified by strike and dip of the folded rocks and the relative ages of rocks in these folds.
- Domes and basins are the circular to oval equivalents of anticlines and synclines, but they are commonly much larger structures.
- The two structures that result from fracture are joints and faults. Joints may open up but they show no movement parallel with the fracture surface, whereas faults do show movement parallel with the fracture surface.
- Joints are very common and form in response to compression, tension, and shear.
- On dip-slip faults, all movement is up or down the dip of the fault. If the hanging wall moves relatively down it is a normal fault, but if the hanging wall moves up it is a reverse fault. Normal faults result from tension; reverse faults from compression.
- In strike-slip faults, all movement is along the strike of the fault. These faults are either right-lateral or left-lateral, depending on the apparent direction of offset of one block relative to the other.
- Oblique-slip faults show components of both dip-slip and strike-slip movement.
- A variety of processes account for the origin of mountains. Some involve little or no deformation, but the large mountain systems on the continents resulted from deformation at convergent plate boundaries.
- A volcanic island arc, deformation, igneous activity, and metamorphism characterize orogenies at oceanic–oceanic plate boundaries, whereas subduction at an oceanic–continental plate boundary also results in orogeny.
- Some mountain systems are within continents far from a present-day plate boundary. These mountains formed when two continental plates collided and became sutured.
- Geologists now realize that orogenies also involve collisions of terranes with continents.
- Continental crust is characterized as granitic, and it is much thicker and less dense than oceanic crust that is composed of basalt and gabbro.
- According to the principle of isostasy, Earth's crust floats in equilibrium in the denser mantle below. Continental crust stands higher than oceanic crust because it is thicker and less dense.

Important Terms

anticline (p. 264)
basin (p. 266)
compression (p. 261)
continental accretion (p. 283)
deformation (p. 260)
dip (p. 264)
dip-slip fault (p. 271)
dome (p. 266)
elastic strain (p. 261)
fault (p. 270)
fault plane (p. 270)
fold (p. 264)
footwall block (p. 270)
fracture (p. 262)
geologic structure (p. 264)
gravity anomaly (p. 284)
hanging wall block (p. 270)
isostatic rebound (p. 285)
joint (p. 269)
monocline (p. 264)
normal fault (p. 271)
oblique-slip fault (p. 274)
orogeny (p. 277)
plastic strain (p. 262)
principle of isostasy (p. 284)
reverse fault (p. 271)
shear stress (p. 261)
strain (p. 260)
stress (p. 260)
strike (p. 264)
strike-slip fault (p. 271)
syncline (p. 264)
tension (p. 261)
terrane (p. 284)
thrust fault (p. 271)

Review Questions

1. Rocks characterized as ductile
a._____fracture easily when in compression; b._____show a great amount of plastic strain; c._____are found along the crests of anticlines; d._____are common along dip-slip faults; e._____are the main rocks in terranes.

2. A basin is a(n)
a._____fault on which the hanging wall block moved down; b._____elongate fold with all strata dipping away from the fold axis; c._____kind of fracturing that occurs in lava as it cools; d._____oval fold with the youngest exposed rocks at its center; e._____type of deformation found adjacent to strike-slip faults.

3. The process by which new material is added to the margins of plates is
a._____continental accretion; b._____shear stress; c._____thrust faulting; d._____elastic strain; e._____platform evolution.

4. The line formed by the intersection of a horizontal plane and an inclined plane is the definition of
a._____stress; b._____brittle behavior; c._____jointing; d._____uplift; e._____strike.

5. The fault illustrated in Figure 10.17e shows both_____and_____faulting.
a._____hanging wall block uplift/footwall block subsidence; b._____low-angle thrust/normal; c. _____thrust/reverse; d._____reverse dip-slip/left-lateral strike-slip; e._____right-lateral strike-slip/normal.

6. *Orogeny* is the geologic term used for
a._____deformation with little or no plastic strain; b._____the origin of large circular folds; c._____an episode of deformation and the origin of mountains; d._____rocks that show both elastic and plastic deformation; e._____a type of fold with its axis inclined.

7. According to the principle of isostasy,
a._____oceanic crust is less dense than continental crust; b._____continents are buoyed up by the more dense mantle; c._____loading by glaciers causes Earth's crust to rise; d._____overturned folds and thrust faults result from compression; e._____the San Andreas fault is a right-lateral strike-slip fault.

8. An oceanic plate colliding with a continental plate accounts for the ongoing mountain building
a._____in the Rocky Mountains; b._____along the west coast of South America; c._____where the Pacific plate collides with Japan; d._____along North America's eastern margin; e._____in a large region in Africa.

9. Most folding of rock layers results from
a._____tension; b._____shear stress; c._____convection; d._____compression; e._____fracturing.

10. A fault along which the hanging wall block moves down relative to the footwall block is a _____ fault.
    a._____normal; b._____strike-slip; c._____thrust; d._____oblique; e._____reverse.
11. What kinds of evidence indicate that mountain building took place in an area where mountains are no longer present?
12. Rocks are displaced 200 km along a strike-slip fault during a period of 5 million years. What was the average rate of movement per year? Is the average likely to represent the actual rate of displacement on this fault? Explain.
13. How would you explain stress and strain to someone unfamiliar with the concepts?
14. Discuss the features of mountains formed at an oceanic–continental plate boundary, and give an example of where such activity is presently taking place.
15. Explain what is meant by the term *terrane* and how terranes are incorporated into continents.
16. Describe how time, rock type, pressure, and temperature influence rock deformation.
17. What are the similarities and differences between a syncline and a basin?
18. Illustrate a recumbent anticline, and explain what criteria are necessary to distinguish it from a recumbent syncline.
19. Notice that the monocline in Figure 10.7b is deeply eroded where flexure of the rock layers is greatest. Why is this so?
20. What is meant by the *elastic limit* of rocks, and what happens when rocks are strained beyond their elastic limit?

World Wide Web Activities

Geology⇌Now Assess your understanding of this chapter's topics with additional quizzing and comprehensive interactivities at

http://earthscience.brookscole.com/changingearth4e

as well as current and up-to-date weblinks, additional readings, and InfoTrac College Edition exercises.

Mass Wasting

CHAPTER 11

OUTLINE

GeologyNow *This icon, which appears throughout the book, indicates an opportunity to explore interactive tutorials, animations, or practice problems available on the GeologyNow website at http://earthscience.brookscole.com/changingearth4e.*

OBJECTIVES

At the end of this chapter, you will have learned that:

- It is important to understand the different types of mass wasting because mass wasting affects us all and causes significant destruction.
- Factors such as slope angle, weathering and climate, water content, vegetation, and overloading are interrelated, and all contribute to mass wasting.
- Mass movements can be triggered by such factors as overloading, soil saturation, and ground shaking.
- Mass wasting is categorized as either rapid mass movements or slow mass movements.
- The different types of rapid mass movements are rockfalls, slumps, rock slides, mudflows, debris flows, and quick clays; each type has recognizable characteristics.
- The different types of slow mass movements are earthflows, solifluction, and creep; each type has recognizable characteristics.
- People can minimize the effects of mass wasting by conducting geologic investigations of an area and stabilizing slopes to prevent and ameliorate movement.

Residents of Caracas, Venezuela, clean up the debris from massive flooding and mudslides that devastated large areas of the country during December 1999. Source: AP/Wide World Photos

Introduction

Triggered by relentless torrential rains that began on December 15, 1999, the floods and mudslides that devastated Venezuela were some of the worst ever to strike that country. Although an accurate death toll is impossible to determine, it is estimated that between 10,000 and 30,000 people were killed; 100,000 to 150,000 were left homeless; 35,000 to 40,000 homes were destroyed or buried by mudslides; and $10 to $20 billion in damage was done before the rains and slides abated (see the chapter opening photo). It is easy to cite numbers of dead and homeless, but the human side of the disaster was most vividly brought home by a mother who described standing helplessly by and watching her four small children buried alive in the family car as a raging mudslide carried it away.

Mudslides engulfed and buried not only homes, buildings, and roads, but even entire communities. Some areas were covered with as much as 7 m of mud. In addition, flooding and the accompanying mudslides swept away large parts of many of Venezuela's northern coastal communities, leaving huge areas uninhabitable.

This terrible tragedy illustrates the close link between geology and individuals, governments, and society in general, a theme we stressed in Chapter 1. The underlying causes of the mudslides are not unique to Venezuela; they can be found anywhere in the world. By being able to recognize and understand these causes and what the result may be, we can find ways to reduce hazards and minimize damage in terms of both human suffering and property damage. This tragedy shows how geology affects all our lives and how interconnected the various systems and subsystems of Earth are.

The topography of land areas is the result of the interaction among Earth's internal processes, type of rocks exposed at the surface, effects of weathering, and the erosional agents of water, ice, and wind. The specific type of landscape developed depends, in part, on which agent of erosion is dominant. *Landslides* (a general term for mass movements), which can be very destructive, are part of the normal adjustments of slopes to changing surface conditions.

Mass wasting (also called *mass movement*) is defined as the downslope movement of material under the direct influence of gravity. Most types of mass wasting are aided by weathering and involve surficial material. The material moves at rates ranging from almost imperceptible, as in the case of creep, to extremely fast, as in a rockfall or slide. Although water can play an important role, the relentless pull of gravity is the major force behind mass wasting.

Mass wasting is an important geologic process that can occur at any time and almost any place. It is thus important to study this phenomenon because it affects all of us, no matter

Table 11.1

Selected Landslides, Their Cause, and the Number of People Killed

Date	Location	Type	Deaths
218 B.C.	Alps (European)	Avalanche—destroyed Hannibal's army	18,000
1556	China (Hsian)	Landslides—earthquake triggered	1,000,000
1806	Switzerland (Goldau)	Rock slide	457
1903	Canada (Frank, Alberta)	Rock slide	70
1920	China (Kansu)	Landslides—earthquake triggered	~200,000
1941	Peru (Huaraz)	Avalanche and mudflow	7000
1962	Peru (Mt. Hauscarán)	Ice avalanche and mudflow	~4000
1963	Italy (Vaiont Dam)	Landslide—subsequent flood	~2000
1966	United Kingdom (Aberfan, South Wales)	Debris flow—collapse of mining-waste tip	144
1970	Peru (Mt. Hauscarán)	Rockfall and debris avalanche—earthquake triggered	25,000
1981	Indonesia (West Irian)	Landslide—earthquake triggered	261
1987	El Salvador (San Salvador)	Landslide	1000
1989	Tadzhikistan	Mudflow—earthquake triggered	274
1994	Colombia (Paez River Valley)	Avalanche—earthquake triggered	>300
1999	Venezuela	Mudflow	>10,000
2003	USA (Southern California)	Mudflow	10
2005	USA (Southern California)	Mudflow	10

Source: Data from J. Whittow, *Disasters: The Anatomy of Environmental Hazards* (Athens: University of Georgia Press, 1979); *Geotimes;* and *Earth.*

where we live (Table 11.1). In the United States, mass wasting occurs in all 50 states and causes economically significant destruction in more than 25 states. Furthermore, between 25 and 50 people, on average, are killed each year by landslides in the United States, and the annual cost in damages exceeds $1.5 billion! Although all major landslides have natural causes, many smaller ones are the result of human activity and could have been prevented or their damage minimized. In this chapter, we examine the factors that lead to mass wasting and discuss ways to prevent or minimize the damage it causes.

What Would You Do?

You are a member of a planning board for your seaside community. A developer wants to rezone some coastal property to build 20 condominiums. This would be a boon to the local economy because it would provide jobs and increase the tax base. However, because the area is somewhat hilly and fronts the ocean, you are concerned about how safe the buildings would be. What types of studies would need to be done before any rezoning could take place? Is it possible to build safe structures along a hilly coastline? What specifically would you ask the environmental consulting firm the planning board has hired to look for in terms of actual or potential geologic hazards, if the condominiums are built?

WHAT FACTORS INFLUENCE MASS WASTING?

When the gravitational force acting on a slope exceeds its resisting force, slope failure (mass wasting) occurs. The resisting forces that help to maintain slope stability include the slope material's strength and cohesion, the amount of internal friction between grains (individual particles of material), and any external support of the slope (■ Figure 11.1). These factors collectively define a slope's **shear strength.**

Opposing a slope's shear strength is the force of gravity. Gravity operates vertically but has a component acting parallel to the slope, thereby causing instability (Figure 11.1). The steeper a slope's angle, the greater the component of force acting parallel to the slope, and the greater the chance for mass wasting. The steepest angle that a slope can maintain without collapsing is its *angle of repose*. At this angle, the shear strength of the slope's material exactly counterbalances the force of gravity. For unconsolidated material, the angle of repose normally ranges from 25 to 40 degrees. Slopes steeper than 40 degrees usually consist of unweathered solid rock.

All slopes are in a state of *dynamic equilibrium,* which means that they are constantly adjusting to new conditions. Although we tend to view mass wasting as a disruptive and usually destructive event, it is one of the ways that a slope adjusts to new conditions. Whenever a building or road is constructed on a hillside, the equilibrium of that slope is affected. The slope must then adjust, perhaps by mass wasting, to this new set of conditions.

Many factors can cause mass wasting: a change in slope angle, weakening of material by weathering, increased water content, changes in the vegetation cover, and overloading. Although most of these are interrelated, we will examine them separately for ease of discussion, but we will also show how they individually and collectively affect a slope's equilibrium.

Slope Angle

Slope angle is probably the major cause of mass wasting. Generally speaking, the steeper the slope, the less stable it is. Therefore, steep slopes are more likely to experience mass wasting than gentle ones.

A number of processes can oversteepen a slope. One of the most common is undercutting by stream or wave action (■ Figure 11.2). This removes the slope's base, increases

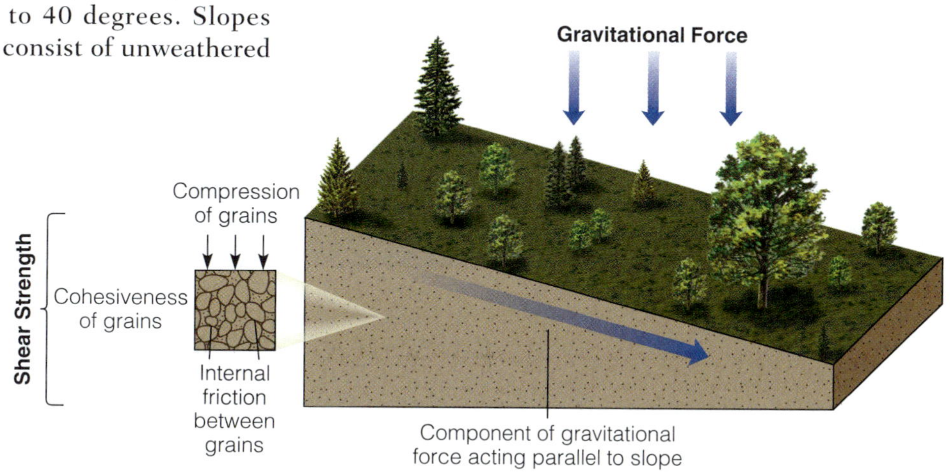

■ **Figure 11.1**

A slope's shear strength depends on the slope material's strength and cohesion, the amount of internal friction between grains, and any external support of the slope. These factors promote slope stability. The force of gravity operates vertically but has a component acting parallel to the slope. When this force, which promotes instability, exceeds a slope's shear strength, slope failure occurs.

■ **Figure 11.2**

Undercutting by stream erosion (a) removes a slope's base, which increases the slope angle, and (b) can lead to slope failure. (c) Undercutting by stream erosion caused slumping along this stream near Weidman, Michigan.

the slope angle, and thereby increases the gravitational force acting parallel to the slope. Wave action, especially during storms, often results in mass movements along the shores of oceans or large lakes (■ Figure 11.3).

Excavations for road cuts and hillside building sites are another major cause of slope failure (■ Figure 11.4). Grading the slope too steeply or cutting into its side increases the stress in the rock or soil until it is no longer strong enough to remain at the steeper angle, and mass movement ensues. Such action is analogous to undercutting by streams or waves and has the same result, thus explaining why so many mountain roads are plagued by frequent mass movements.

Weathering and Climate

Mass wasting is more likely to occur in loose or poorly consolidated slope material than in bedrock. As soon as rock is exposed at Earth's surface, weathering begins to disintegrate and decompose it, reducing its shear strength and increasing its susceptibility to mass wasting. The deeper the weathering zone extends, the greater the likelihood of some type of mass movement.

Recall that some rocks are more susceptible to weathering than others and that climate plays an important role in the rate and type of weathering. In the tropics, where temperatures are high and considerable rain falls, the effects of weathering extend to depths of several tens of meters, and mass movements most commonly occur in the deep weathering zone. In arid and semiarid regions, the weathering zone is usually considerably shallower. Nevertheless, intense, localized cloudbursts can drop large quantities of water on an area in a short time. With little vegetation to absorb this water, runoff is rapid and frequently results in mudflows.

Water Content

The amount of water in rock or soil influences slope stability. Large quantities of water from melting snow or heavy rainfall greatly increase the likelihood of slope failure. The additional weight that water adds to a slope can be enough to cause mass movement. Furthermore, water percolating through a slope's material helps to decrease friction between grains, contributing to a loss of cohesion. For example, slopes composed of dry clay are usually quite stable, but when wetted, they quickly lose cohesiveness and internal friction and become an unstable slurry. This occurs because clay, which can hold large quantities of water, consists of platy particles that easily slide over

James S. Monroe

■ **Figure 11.3**

This sea cliff north of Bodega Bay, California, was undercut by waves during the winter of 1997–1998. As a result, part of the land slid into the ocean, damaging several houses.

■ **Figure 11.4**

(a) Highway excavations disturb the equilibrium of a slope by (b) removing a portion of its support as well as oversteepening it at the point of excavation. (c) Such action can result in landslides. (d) Cutting into the hillside to construct this portion of the Pan-American Highway in Mexico resulted in a rockfall that completely blocked the road.

(a)

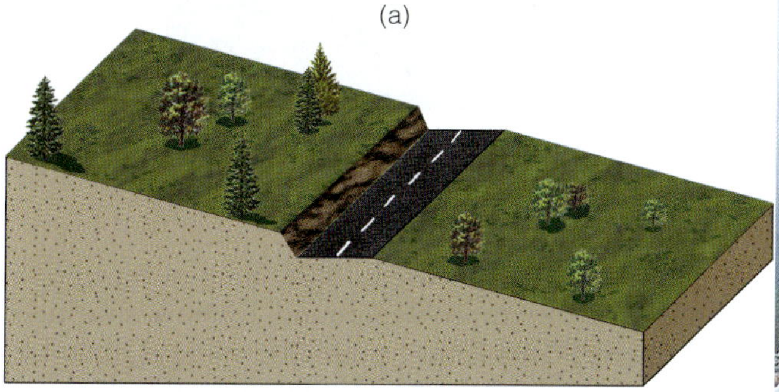

(b)

(c)

R. V. Dietrich

(d)

GEOLOGY IN UNEXPECTED PLACES

New Hampshire Says Good-Bye to the "Old Man"

Imagine waking up one morning and discovering that your state symbol had suddenly disappeared during the night. That's exactly what happened to the residents of New Hampshire when they discovered that "The Old Man of the Mountain," a landmark symbolizing their state's independence and stubbornness, had collapsed from natural causes sometime during the night of May 5, 2003.

The Old Man of the Mountain, located in Franconia Notch State Park, was composed of a series of five horizontal granite ledges that had weathered over millions of years to form a man's profile. Overlooking Profile Lake 400 m below, the Old Man measured about 13 m from chin to forehead and jutted out about 8 m from the main part of the mountain (■ Figure 1).

The natural forces that shaped the Old Man also brought about its demise. Freezing temperatures, high winds, and heavy rains all contributed to finally overcoming the forces that held the rock together. Despite nearly 100 years of effort to protect the landmark from destruction, the forces of nature eventually won. All that's left of the Old Man now is a pile of rubble at the mountain's base and some stabilizing cables and epoxy where the Old Man once looked out on the state it symbolized for so long.

■ Figure 1

The Old Man of the Mountain before it collapsed.

White Mountains Attractions Association

each other when wet. For this reason, clay beds are frequently the slippery layer along which overlying rock units slide downslope.

Vegetation

Vegetation affects slope stability in several ways. By absorbing water from a rainstorm, vegetation decreases the water saturation of a slope's material that would otherwise lead to a loss of shear strength. Vegetation's root system also helps stabilize a slope by binding soil particles together and holding the soil to bedrock.

The removal of vegetation by either natural or human activity is a major cause of many mass movements. Summer brush and forest fires in southern California frequently leave the hillsides bare of vegetation. Fall rainstorms saturate the ground, causing mudslides that do tremendous damage and cost millions of dollars to clean up (■ Figure 11.5). The soils of many hillsides in New Zealand are sliding because deep-rooted native bushes have been replaced by shallow-rooted grasses used for sheep grazing. When heavy rains saturate the soil, the shallow-rooted grasses cannot hold the slope in place, and parts of it slide downhill.

Overloading

Overloading is almost always the result of human activity and typically results from dumping, filling, or piling up of material. Under natural conditions, a ma-

Boris Yaro/Los Angeles Times

■ **Figure 11.5**

A California Highway Patrol officer stands on top of a 2-m-high wall of mud that rolled over a patrol car near the Golden State Freeway on October 23, 1987. Flooding and mudslides also trapped other vehicles and closed the freeway.

Triggering Mechanisms

The factors discussed thus far all contribute to slope instability. Most—though not all—rapid mass movements are triggered by a force that temporarily disturbs slope equilibrium. The most common triggering mechanisms are strong vibrations from earthquakes and excessive amounts of water from a winter snow melt or a heavy rainstorm (■ Figure 11.7).

Volcanic eruptions, explosions, and even loud claps of thunder may be enough to trigger a landslide if the slope is sufficiently unstable. Many *avalanches,* which are rapid movements of snow and ice down steep mountain slopes, are triggered by a loud gunshot or, in rare cases, even a person's shout.

terial's load is carried by its grain-to-grain contacts, with the friction between the grains maintaining a slope. The additional weight created by overloading increases the water pressure within the material, which in turn decreases its shear strength, thereby weakening the slope material. If enough material is added, the slope will eventually fail, sometimes with tragic consequences.

Geology and Slope Stability

The relationship between the topography and the geology of an area is important in determining slope stability (■Figure 11.6). If the rocks underlying a slope dip in the same direction as the slope, mass wasting is more likely to occur than if the rocks are horizontal or dip in the opposite direction. When the rocks dip in the same direction as the slope, water can percolate along the various bedding planes and decrease the cohesiveness and friction between adjacent rock units (Figure 11.6a). This is particularly true when clay layers are present because clay becomes slippery when wet.

Even if the rocks are horizontal or dip in a direction opposite to that of the slope, joints may dip in the same direction as the slope. Water migrating through them weathers the rock and expands these openings until the weight of the overlying rock causes it to fall (Figure 11.6b).

WHAT ARE THE DIFFERENT TYPES OF MASS WASTING?

Geologists recognize a variety of mass movements (Table 11.2). Some are of one distinct type, whereas others are a combination of different types. It is not uncommon for one type of mass movement to change into another along its course. Even though many slope failures are combinations of different materials and movements, it is still convenient to classify them according to their dominant behavior.

Mass movements are generally classified on the basis of three major criteria (Table 11.2): (1) rate of movement (rapid or slow); (2) type of movement (primarily falling, sliding, or flowing); and (3) type of material involved (rock, soil, or debris).

Rapid mass movements involve a visible movement of material. Such movements usually occur quite suddenly, and the material moves quickly downslope. Rapid mass movements are potentially dangerous and frequently result in loss of life and property damage. Most rapid mass movements occur on relatively steep slopes and can involve rock, soil, or debris.

Slow mass movements advance at an imperceptible rate and are usually detectable only by the effects of their movement, such as tilted trees and power poles or

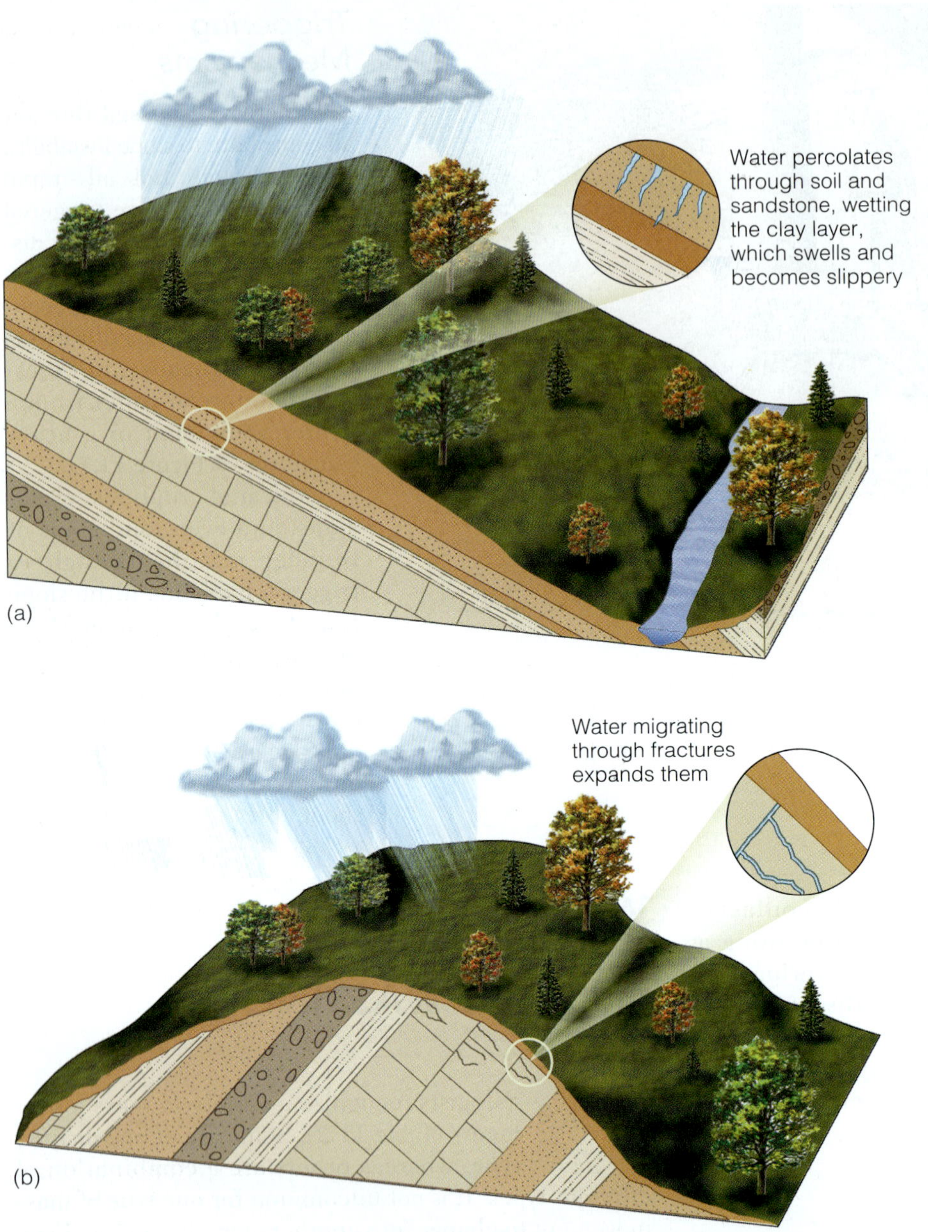

■ **Figure 11.6**

(a) Rocks dipping in the same direction as a hill's slope are particularly susceptible to mass wasting. Undercutting of the base of the slope by a stream removes support and steepens the slope at the base. Water percolating through the soil and into the underlying rocks increases the weight and, if clay layers are present, wets the clay, making the layers slippery. (b) Fractures dipping in the same direction as a slope are enlarged by chemical weathering, which can weaken the rocks and cause mass wasting.

cracked foundations. Although rapid mass movements are more dramatic, slow mass movements are responsible for the downslope transport of a much greater volume of weathered material.

Falls

Rockfalls are a common type of extremely rapid mass movement in which rocks of any size fall through the air (■Figure 11.8a). Rockfalls occur along steep canyons, cliffs, and road cuts and build up accumulations of loose rocks and rock fragments at their base called *talus* (see Figure 6.3b).

Rockfalls result from failure along joints or bedding planes in the bedrock and are commonly triggered by natural or human undercutting of slopes, or by earthquakes. Many rockfalls in cold climates are the result of frost wedging. Chemical weathering caused by water percolating through the fissures in carbonate rocks (limestone, dolostone, and marble) is also responsible for many rockfalls.

Rockfalls range in size from small rocks falling from a cliff to massive falls involving millions of cubic meters of debris that destroy buildings, bury towns, and block highways (Figure 11.8b). Rockfalls are a particularly common hazard in mountainous areas where roads have been built by blasting and grading through steep hillsides of bedrock. Anyone who has ever driven through the Appalachians, the Rocky Mountains, or the Sierra Nevada is familiar with the "Watch for Falling Rocks" signs posted to warn drivers of the danger. Slopes that are particularly prone to rockfalls are sometimes covered with wire mesh in an effort to prevent dislodged rocks from falling to the road below (■ Figure 11.9a). Another tactic is to put up wire mesh fences along the base of the slope to catch or slow down bouncing or rolling rocks (Figure 11.9b).

Slides

A slide involves movement of material along one or more surfaces of failure. The type of material may be soil, rock, or a combination of the two, and it may break apart during movement or remain intact. A slide's rate of movement can vary from extremely slow to very rapid (Table 11.2).

Two types of slides are generally recognized: (1) slumps or rotational slides, in which movement occurs along a curved surface, and (2) rock or block slides, which move along a more or less planar surface.

Rod Rolle/Liaison/Getty Images

A **slump** involves the downward movement of material along a curved surface of a rupture and is characterized by the backward rotation of the slump block (■ Figure 11.10). Slumps usually occur in unconsolidated or weakly consolidated material and range in size from small individual sets, such as occur along stream banks, to massive, multiple sets that affect large areas and cause considerable damage.

Slumps can be caused by a variety of factors, but the most common is erosion along the base of a slope, which removes support for the overlying material. This local steepening may be caused naturally by stream erosion along its banks (Figure 11.2c) or by wave action at the base of a coastal cliff (■ Figure 11.11). Slope oversteepening can also be caused by human activity, such as the construction of highways and housing developments. Slumps are particularly prevalent along highway cuts, where they are generally the most frequent type of slope failure observed.

Although many slumps are merely a nuisance, large-scale slumps involving populated areas and highways can cause extensive damage. Such is the case in coastal southern California where slumping and sliding have

■ Figure 11.7

Heavy winter rains caused this 200,000-yd^3 landslide in March 1995 at La Conchita, California, 75 miles northeast of Los Angeles. Although no casualties occurred, nine homes were destroyed or badly damaged.

Table 11.2

Classification of Mass Movements and Their Characteristics

Type of Movement	Subdivision	Characteristics	Rate of Movement
Falls	Rockfall	Rocks of any size fall through the air from steep cliffs, canyons, and road cuts	Extremely rapid
Slides	Slump	Movement occurs along a curved surface of rupture; most commonly involves unconsolidated or weakly consolidated material	Extremely slow to moderate
	Rock slide	Movement occurs along a generally planar surface	Rapid to very rapid
Flows	Mudflow	Consists of at least 50% silt- and clay-sized particles and up to 30% water	Very rapid
	Debris flow	Contains larger-sized particles and less water than mudflows	Rapid to very rapid
	Earthflow	Thick, viscous, tongue-shaped mass of wet regolith	Slow to moderate
	Quick clays	Composed of fine silt and clay particles saturated with water; when disturbed by a sudden shock, lose their cohesiveness and flow like a liquid	Rapid to very rapid
	Solifluction	Water-saturated surface sediment	Slow
	Creep	Downslope movement of soil and rock	Extremely slow
Complex movements		Combination of different movement types	Slow to extremely rapid

■ **Figure 11.8**

(a) Rockfalls result from failure along cracks, fractures, or bedding planes in the bedrock and are common features in areas of steep cliffs. (b) A huge rockfall closed both lanes of traffic on Highway 70 near Rogers Flat, California, on July 25, 2003. Rocks the size of large dump trucks had to be blasted into smaller pieces to clear the highway. Despite the pavement cracking caused by the falling boulders, geologists determined that the roadbase was undamaged and the road would be safe following cleanup operations.

■ **Figure 11.9**

Minimizing damage from rock falls.
(a) Wire mesh is used to cover this steep slope in Hawaii. This is a common practice in mountainous areas to prevent rocks from falling on the road. (b) A wire mesh fence along the base of this hillside of Highway 44 in California has caught many boulders and prevented them from rolling onto the highway.

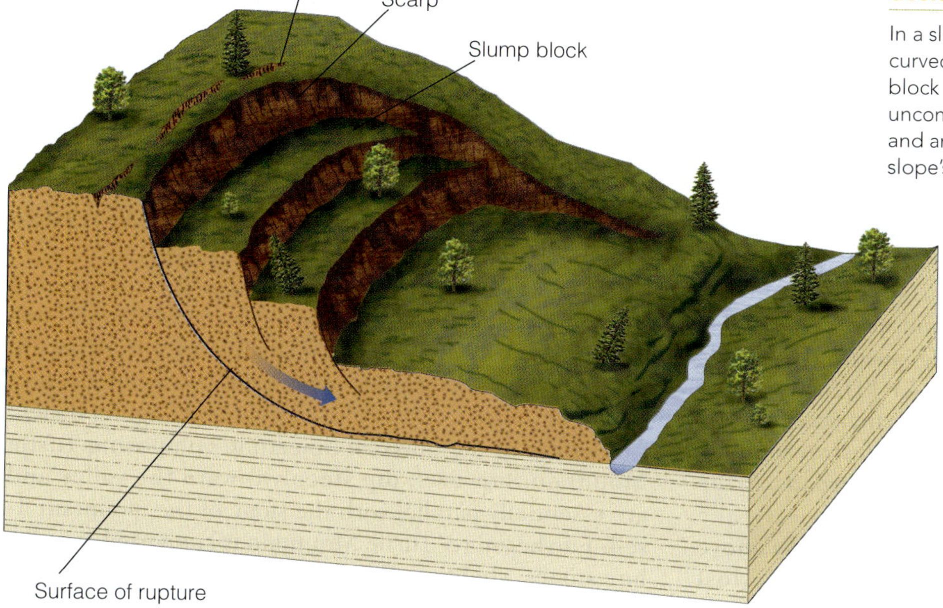

Geology Now ■ **Active Figure 11.10**

In a slump, material moves downward along the curved surface of a rupture, causing the slump block to rotate backward. Most slumps involve unconsolidated or weakly consolidated material and are typically caused by erosion along the slope's base.

John S. Shelton

■ **Figure 11.11**

Undercutting of steep sea cliffs by wave action resulted in massive slumping in the Pacific Palisades area of southern California on March 31 and April 3, 1958. Highway 1 was completely blocked. Note the heavy earth-moving equipment for scale.

been a constant problem. Many areas along the coast are underlain by poorly to weakly consolidated silts, sands, and gravels interbedded with clay layers, some of which are weathered ash falls. In addition, southern California is tectonically active so that many of these deposits are cut by faults and joints, which allow the infrequent rains to percolate downward rapidly, wetting and lubricating the clay layers.

Southern California has a semiarid climate and is dry most of the year. When it does rain, typically between November and March, large amounts of rain can fall in a short time. Thus, the ground quickly becomes saturated, leading to landslides along steep canyon walls as well as along coastal cliffs (Figure 11.11). Most of the slope failures along the southern California coast are the result of slumping. These slumps have destroyed many expensive homes and forced numerous roads to be closed and relocated.

A **rock** or ***block*** **slide** occurs when rocks move downslope along a more or less planar surface. Most rock slides take place because the local slopes and rock layers dip in the same direction (■ Figure 11.12), although they can also occur along fractures parallel to a slope. In addition to slumping, rock slides are common occurrences along the southern California coast. At Point Fermin, seaward-dipping rocks with interbedded slippery clay layers are undercut by waves, causing numerous slides (see "Point Fermin—Slip Sliding Away" on pages 306 and 307).

Farther south in the town of Laguna Beach, startled residents watched as a rock slide destroyed or damaged 50 homes on October 2, 1978 (■ Figure 11.13). Just as at Point Fermin, the rocks at Laguna Beach dip about 25 degrees in the same direction as the slope of the canyon walls and contain clay beds that "lubricate" the overlying rock layers, causing the rocks and the houses built on them to slide. In addition, percolating water from the previous winter's heavy rains wet a subsurface clayey siltstone, thus reducing its shear strength and helping to activate the slide. Although the 1978 slide covered only about 5 acres, it was part of a larger ancient slide complex.

Not all rock slides are the result of rocks dipping in the same direction as a hill's slope. The rock slide at Frank, Alberta, Canada, on April 29, 1903, illustrates how nature and human activity can combine to create a situation with tragic results (■ Figure 11.14).

It would appear at first glance that the coal-mining town of Frank, lying at the base of Turtle Mountain, was in no danger from a landslide (Figure 11.14). After all, many of the rocks dipped away from the mining valley, unlike the situations at Point Fermin and Laguna Beach. The joints in the massive limestone composing Turtle Mountain, however, dip steeply toward the valley and are essentially parallel with the slope of the mountain itself. Furthermore, Turtle Mountain is supported by weak limestones, shales, and coal layers that underwent slow plastic deformation from the weight of the overlying massive limestone. Coal mining along the base of the valley also contributed to the stress on the rocks by removing some of the underlying support. All these factors, as well as frost action and chemical weathering that widened the joints, finally resulted in a massive rock slide. Almost 40 million m^3 of rock slid down Turtle Mountain along joint planes, killing 70 people and partially burying the town of Frank.

Flows

Mass movements in which material flows as a viscous fluid or displays plastic movement are termed *flows.* Their rate of movement ranges from extremely slow to extremely rapid (Table 11.2). In many cases, mass

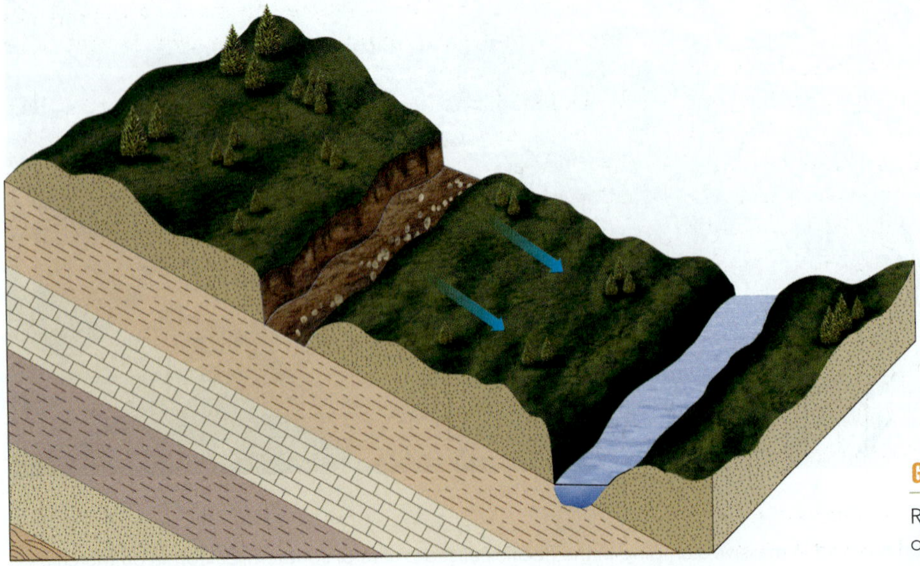

Geology Now ■ **Active Figure 11.12**

Rock slides occur when material moves downslope along a generally planar surface.

Figure 11.13

A combination of interbedded clay beds that become slippery when wet, rocks dipping in the same direction as the slope of the sea cliffs, and undercutting of the sea cliffs by wave action activated a rock slide at Laguna Beach, California, that destroyed numerous homes and cars on October 2, 1978.

movements begin as falls, slumps, or slides and change into flows farther downslope.

Of the major mass movement types, **mudflows** are the most fluid and move most rapidly (at speeds up to 80 km per hour). They consist of at least 50% silt- and clay-sized material combined with a significant amount of water (up to 30%). Mudflows are common in arid and semiarid environments where they are triggered by heavy rainstorms that quickly saturate the regolith, turning it into a raging flow of mud that engulfs everything in its path. Mudflows can also occur in mountain regions (■ Figure 11.15) and in areas covered by volcanic ash where they can be particularly destructive (see Chapter 5). Because mudflows are so fluid, they generally follow preexisting channels until the slope decreases or the channel widens, at which point they fan out.

As urban areas in arid and semiarid climates continue to expand, mudflows and the damage they create are becoming problems. Mudflows are common, for example, in the steep hillsides around Los Angeles where they have damaged or destroyed many homes.

Debris flows are composed of larger particles than mudflows and do not contain as much water. Consequently, they are usually more viscous than mudflows, typically do not move as rapidly, and rarely are confined to preexisting channels. Debris flows can be just as damaging, though, because they can transport large objects (■ Figure 11.16).

Earthflows move more slowly than either mudflows or debris flows. An earthflow slumps from the upper part of a hillside, leaving a scarp, and flows slowly downslope as a thick, viscous, tongue-shaped mass of wet regolith (■ Figure 11.17). Like mudflows and debris flows, earthflows can be of any size and are frequently destructive. They occur most commonly in humid climates on grassy, soil-covered slopes following heavy rains.

B. Bradley, Geology Dept. of the University of Colorado

Figure 11.14

(a) The tragic Turtle Mountain rock slide that killed 70 people and partially buried the town of Frank, Alberta, Canada, on April 29, 1903, was caused by a combination of factors. These included joints that dipped in the same direction as the slope of Turtle Mountain, a fault partway down the mountain, weak shale and siltstone beds underlying the base of the mountain, and mined-out coal seams. (b) Results of the 1903 rock slide at Frank.

■ **Figure 11.15**

A mudflow near Estes Park, Colorado.

■ **Figure 11.16**

A debris flow and damaged house in lower Ophir Creek, western Nevada. Note the many large boulders that are part of the debris flow.

Some clays spontaneously liquefy and flow like water when they are disturbed. Such **quick clays** have caused serious damage and loss of lives in Sweden, Norway, eastern Canada (■ Figure 11.18), and Alaska (Table 11.1). Quick clays are composed of fine silt and clay particles made by the grinding action of glaciers. Geologists think these fine sediments were originally deposited in a marine environment where their pore space was filled with saltwater. The ions in saltwater helped establish strong

■ **Figure 11.17**

(a) Earthflows form tongue-shaped masses of wet regolith that move slowly downslope. They occur most commonly in humid climates on grassy, soil-covered slopes. (b) An earthflow near Baraga, Michigan.

Point Fermin—Slip Sliding Away

Dubbed the "sunken city" by residents of the area, Point Fermin in southern California is famous for its numerous examples of mass wasting. The area is underlain by fine-grained sedimentary rocks interbedded with diatomite layers and volcanic ash. When these layers get wet, they get slippery and tend to slide easily. The rocks also dip slightly toward the ocean and form steep coastal bluffs that are being undercut by constant wave action at their base. This wave action results in oversteepening of the cliffs, which causes slumping.

Mass wasting began in 1929 with minor slumping in the area. In the early 1940s, water mains in the region were broken and several individual blocks began slumping. Movement largely ceased following this main phase of slumping, but it has continued intermittently until the present and residents have paid the price for living in an unstable coastal area.

Eleanora Robbins/ USGS

A map of southern California showing the location of Point Fermin and an aerial view at low tide of the sliding that has taken place. Note the numerous slump blocks and oversteepened cliffs. The continuous pounding of waves and surf along the base of the cliffs further erodes and undercuts them, leading to even more slumping and sliding.

Reed Wicander

A view of one portion of the Point Fermin slide area showing the fine-grained sedimentary rocks dipping slightly toward the ocean and the oversteepened cliffs resulting from slumping and sliding in the foreground.

Reed Wicander

A view of one slump block shows remnants of a former road and a palm tree still growing as if nothing has happened.

Reed Wicander

An abandoned house on the edge of an oversteepened coastal bluff. Note the concrete blocks placed along the beach to absorb the erosive energy of the incoming waves and slow down the erosion to the cliffs.

Slumping along the oversteepened cliffs at Point Fermin. The abandoned house in the photo above is just to the right of this view at the top of the cliff.

Reed Wicander

Creep and minor slumping are evident in this photo. Note the two small slump scarps. The smaller one in the background is mostly grass-covered, whereas the one in the foreground has bare spots and is apparently moving at a slightly faster rate. Notice the effect of creep on the right-hand wall of the house. The bottom part of the wall is moving toward the right of the photo as a result of creep, producing a bend of the wall that can be seen clearly near its base.

Canadian Air Force

■ **Figure 11.18**

Quick-clay slide at Nicolet, Quebec, Canada. The house on the slide (to the right of the bridge) traveled several hundred feet with relatively little damage.

Alaska Earthquake Collection, #43 Ct, USGS

■ **Figure 11.19**

(a) Groundshaking by the 1964 Alaska earthquake turned parts of the Bootlegger Cove Clay into a quick clay, causing numerous slides. (b) Low-altitude photograph of the Turnagain Heights subdivision of Anchorage shows some of the landslide fissures that developed as well as the extensive damage to buildings in the area. The remains of the Four Seasons apartment building can be seen in the background.

bonds between the clay particles, thus stabilizing and strengthening the clay. When the clays were subsequently uplifted above sea level, the saltwater was flushed out by fresh groundwater, reducing the effectiveness of the ionic bonds between the clay particles and thereby reducing the overall strength and cohesiveness of the clay. Consequently, when the clay is disturbed by a sudden shock or shaking, it essentially turns to a liquid and flows.

An example of the damage that can be done by quick clays occurred in the Turnagain Heights area of Anchorage, Alaska, in 1964 (■ Figure 11.19). Underlying most of the Anchorage area is the Bootlegger Cove Clay, a massive clay unit of poor permeability. Because the Bootlegger Cove Clay forms a barrier preventing groundwater from flowing through the adjacent glacial deposits to the sea, considerable hydraulic pressure builds up behind the clay. Some of this water has flushed out the saltwater in the clay and has saturated the lenses of sand and silt associated with the clay beds. When the magnitude 8.6 Good Friday earthquake struck on March 27, 1964, the shaking turned parts of the Bootlegger Cove Clay into a quick clay and precipitated a series of massive slides in the coastal bluffs that destroyed most of the homes in the Turnagain Heights subdivision (Figure 11.19b).

Solifluction is the slow downslope movement of water-saturated surface sediment. Solifluction can occur in any climate where the ground becomes saturated with water, but is most common in areas of permafrost.

Permafrost, ground that remains permanently frozen, covers nearly 20% of the world's land surface (■ Figure 11.20a). During the warmer season when the upper portion of the permafrost thaws, water and surface sediment form a soggy mass that flows by solifluction and produces a characteristic lobate topography (Figure 11.20b).

As might be expected, many problems are associated with construction in a permafrost environment. A good example is what happens when an uninsulated building is constructed directly on permafrost. Heat escapes through the floor, thaws the ground below, and turns it into a soggy, unstable mush. Because the ground is no longer solid, the building settles unevenly into the ground and numerous structural problems result (■ Figure 11.21).

Construction of the Alaska pipeline from the oil fields in Prudhoe Bay to the ice-free port of Valdez raised numerous concerns about the effect it might have on the permafrost and the potential for solifluction. Some thought that oil flowing through the pipeline would be warm enough to melt the permafrost, causing the pipeline to sink farther into the ground and possibly rupture. After numerous studies were conducted, scientists concluded that the pipeline, completed in 1977, could safely be buried for more than half of its 1280-km length; where melting of the permafrost might cause structural problems to the pipe, it was insulated and installed above ground.

Creep, the slowest type of flow, is the most widespread and significant mass wasting process in terms of the total amount of material moved downslope and the monetary damage it does annually. Creep involves extremely slow downhill movement of soil or rock. Although it can occur anywhere and in any climate, it is most effective and significant as a geologic agent in humid regions. In fact, it is the most common form of mass wasting in the southeastern United States and the southern Appalachian Mountains.

Because the rate of movement is essentially imperceptible, we are frequently unaware of creep's existence until we notice its effects: tilted trees and power poles, broken streets and sidewalks, or cracked retaining walls or foundations (■ Figure 11.22). Creep usually involves the whole hillside and probably occurs, to some extent, on any weathered or soil-covered, sloping surface.

What Would You Do?

You've found your dream parcel of land in the hills of northern Baja California, where you plan to retire someday. Because you want to make sure the area is safe to build a house, you decide to do your own geologic investigation of the area to make sure there aren't any obvious geologic hazards. What specific things would you look for that might indicate mass wasting in the past? Even if there is no obvious evidence of rapid mass wasting, what features would you look for that might indicate a problem with slow types of mass wasting such as creep?

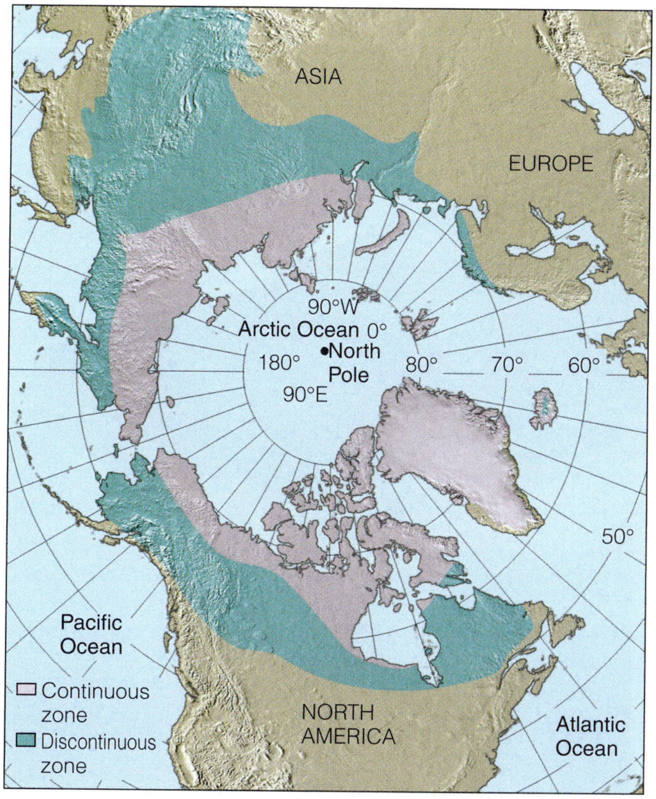

(a)

(b)

■ **Figure 11.20**

(a) Distribution of permafrost areas in the Northern Hemisphere. (b) Solifluction flows near Suslositna Creek, Alaska, show the typical lobate topography that is characteristic of solifluction conditions.

■ **Figure 11.21**

This house, south of Fairbanks, Alaska, has settled unevenly because the underlying permafrost in fine-grained silts and sands has thawed.

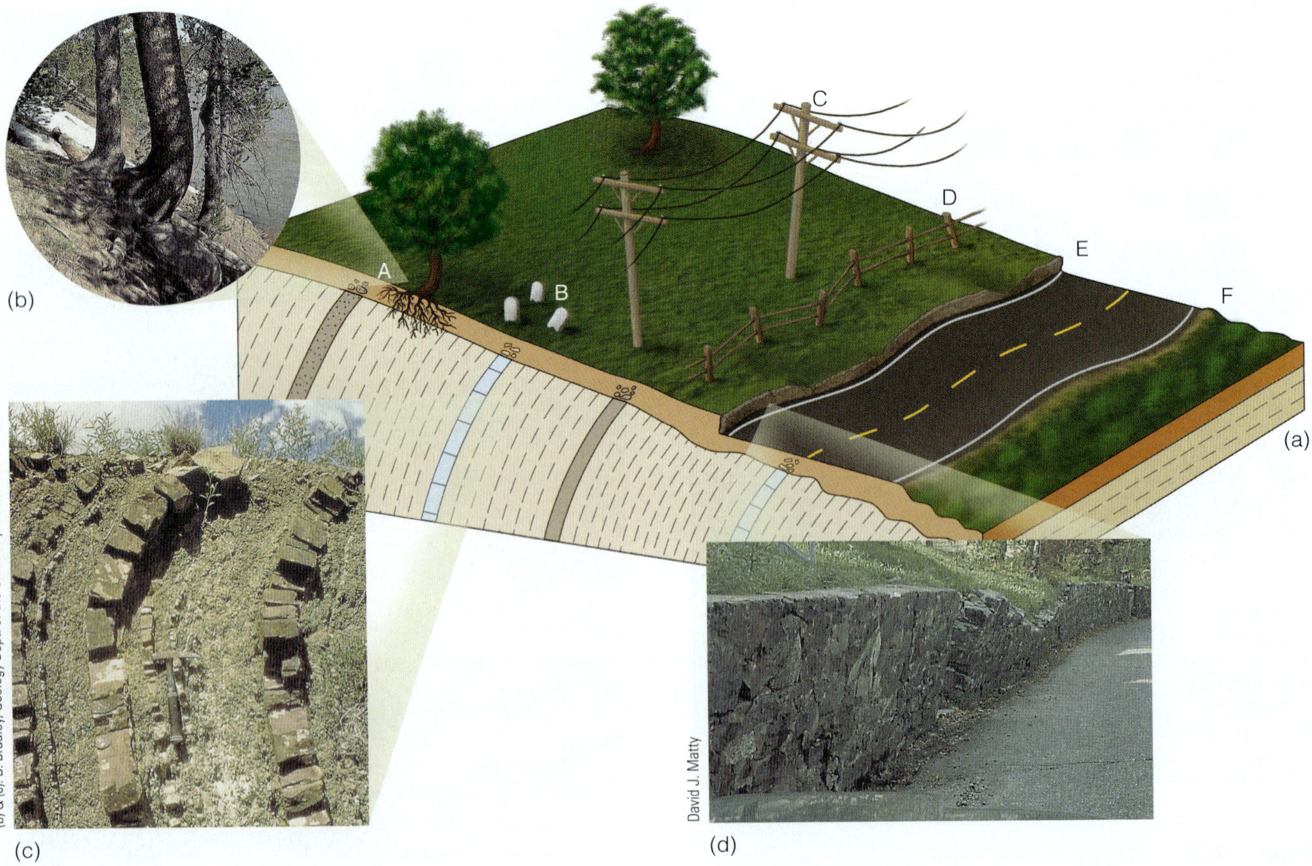

■ **Figure 11.22**

(a) Some evidence of creep: (A) curved tree trunks; (B) displaced monuments; (C) tilted power poles; (D) displaced and tilted fences; (E) roadways moved out of alignment; (F) hummocky surface. (b) Trees bent by creep, Wyoming. (c) Creep has bent these sandstone and shale beds of the Haymond Formation near Marathon, Texas. (d) Stone wall tilted by creep, Champion, Michigan.

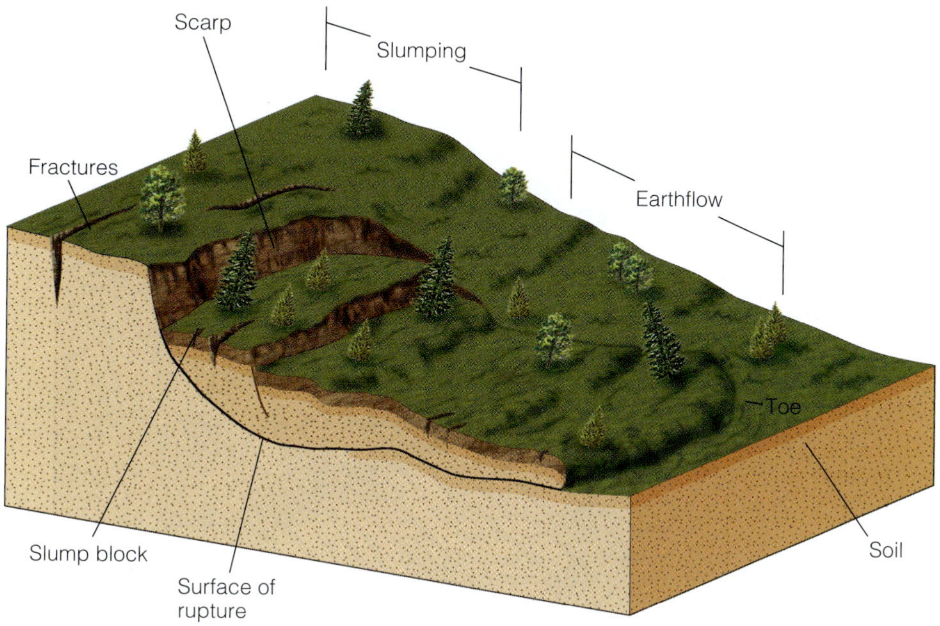

■ **Figure 11.23**

A complex movement in which slumping occurs at the head, followed by an earthflow.

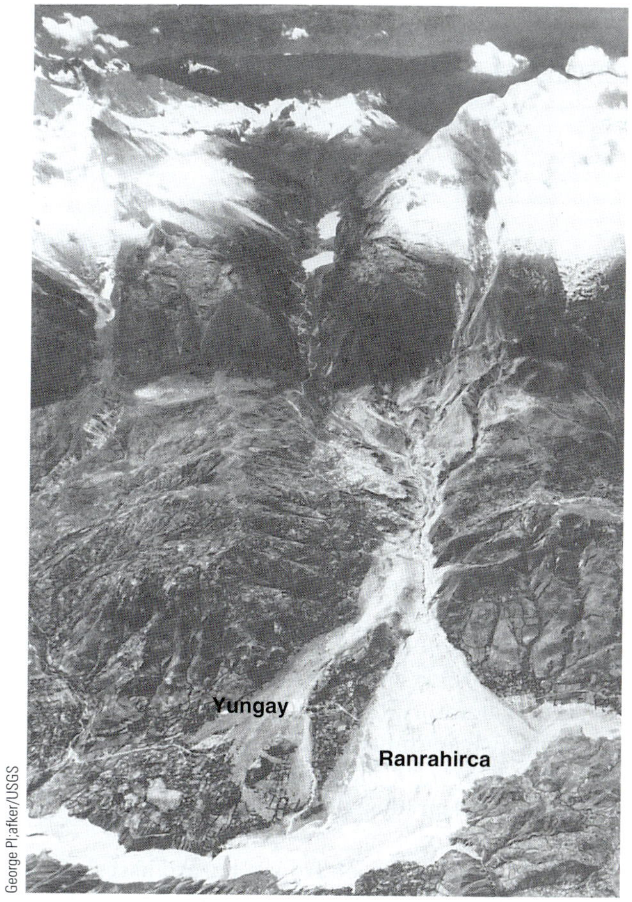

■ **Figure 11.24**

An earthquake 65 km away triggered a landslide on Nevado Huascarán, Peru, that destroyed the towns of Yungay and Ranrahirca and killed more than 25,000 people.

Creep is difficult not only to recognize but also to control. Although engineers can sometimes slow or stabilize creep, many times the only course of action is to simply avoid the area if at all possible or, if the zone of creep is relatively thin, design structures that can be anchored into the bedrock.

Complex Movements

Recall that many mass movements are combinations of different movement types. When one type is dominant, the movement can be classified as one of those described thus far. If several types are more or less equally involved, however, it is called a **complex movement.**

The most common type of complex movement is the slide-flow, in which there is sliding at the head and then some type of flowage farther along its course. Most slide-flow landslides involve well-defined slumping at the head, followed by a debris flow or earthflow (■ Figure 11.23). Any combination of different mass movement types can, however, be classified as a complex movement.

A *debris avalanche* is a complex movement that often occurs in very steep mountain ranges. Debris avalanches typically start out as rockfalls when large quantities of rock, ice, and snow are dislodged from a mountainside, frequently as a result of an earthquake. The material then slides or flows down the mountainside, picking up additional surface material and increasing in speed. The 1970 Peru earthquake (Table 11.1) set in motion the debris avalanche that destroyed the towns of Yungay and Ranrahirca, Peru, and killed more than 25,000 people (■ Figure 11.24).

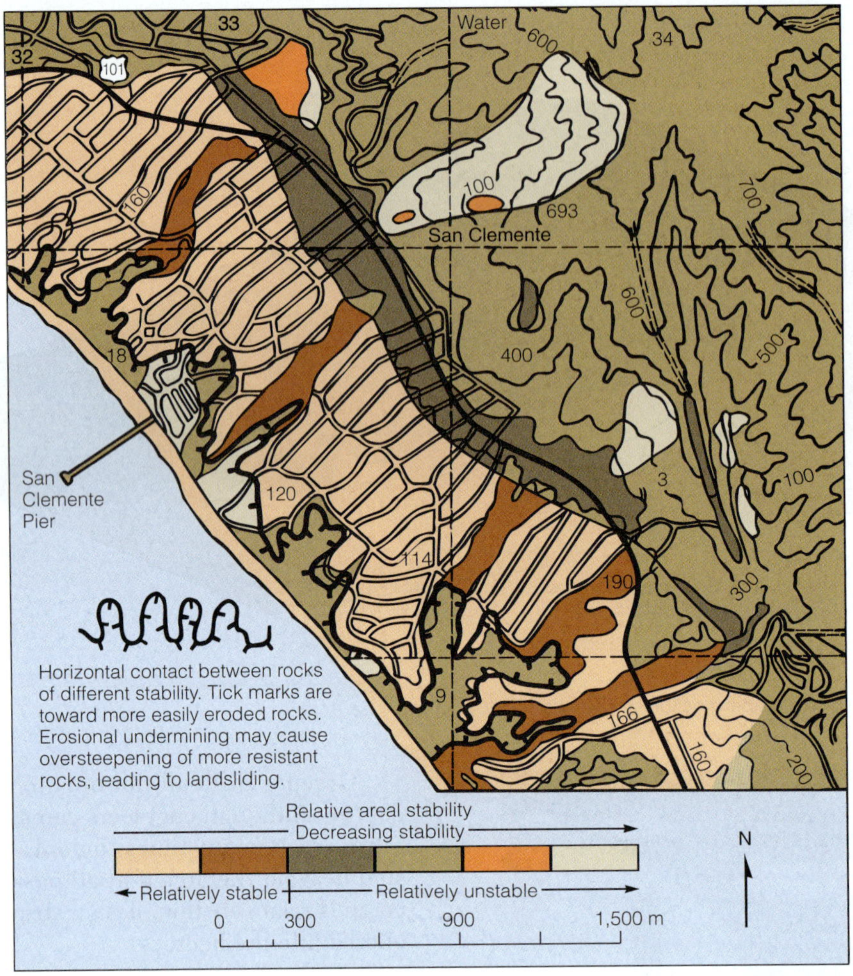

Figure 11.25

Relative slope-stability map of part of San Clemente, California, showing areas delineated according to relative stability.

HOW CAN WE RECOGNIZE AND MINIMIZE THE EFFECTS OF MASS MOVEMENTS?

The most important factor in eliminating or minimizing the damaging effects of mass wasting is a thorough geologic investigation of the region in question. In this way, former landslides and areas susceptible to mass movements can be identified and perhaps avoided. By assessing the risks of possible mass wasting before construction begins, engineers can take steps to eliminate or minimize the effects of such events.

Identifying areas with a high potential for slope failure is important in any hazard assessment study; these studies include identifying former landslides as well as sites of potential mass movement. Scarps, open fissures, displaced or tilted objects, a hummocky surface, and sudden changes in vegetation are some of the features that indicate former landslides or an area susceptible to slope failure. The effects of weathering, erosion, and vegetation may, however, obscure the evidence of previous mass wasting.

Soil and bedrock samples are also studied, in both the field and laboratory, to assess such characteristics as composition, susceptibility to weathering, cohesiveness, and ability to transmit fluids. These studies help geologists and engineers predict slope stability under a variety of conditions.

The information derived from a hazard assessment study can be used to produce *slope-stability maps* of the area (■ Figure 11.25). These maps allow planners and developers to make decisions about where to site roads, utility lines, and housing or industrial developments based on the

■ **Figure 11.26**

(a) Driving drainpipes that are perforated on one side into a hillside, with the perforated side up, can remove some subsurface water and help stabilize the hillside. (b) A drainpipe driven into the hillside at Point Fermin, California.

relative stability or instability of a particular location. In addition, the maps indicate the extent of an area's landslide problem and the type of mass movement that may occur. This information is important for grading slopes or building structures to prevent or minimize slope-failure damage.

Although most large mass movements usually cannot be prevented, geologists and engineers can use various methods to minimize the danger and damage resulting from them. Because water plays such an important role in many landslides, one of the most effective and inexpensive ways to reduce the potential for slope failure or to increase existing slope stability is surface and subsurface drainage of a hillside. Drainage serves two purposes. It reduces the weight of the material likely to slide and increases the shear strength of the slope material by lowering pore pressure.

Surface waters can be drained and diverted by ditches, gutters, or culverts designed to direct water away from slopes. Drainpipes perforated along one surface and driven into a hillside can help remove subsurface water (■ Figure 11.26). Finally, planting vegetation on hillsides helps stabilize slopes by holding the soil together and reducing the amount of water in the soil.

Another way to help stabilize a hillside is to reduce its slope. Recall that overloading and oversteepening by grading are common causes of slope failure. Reducing the

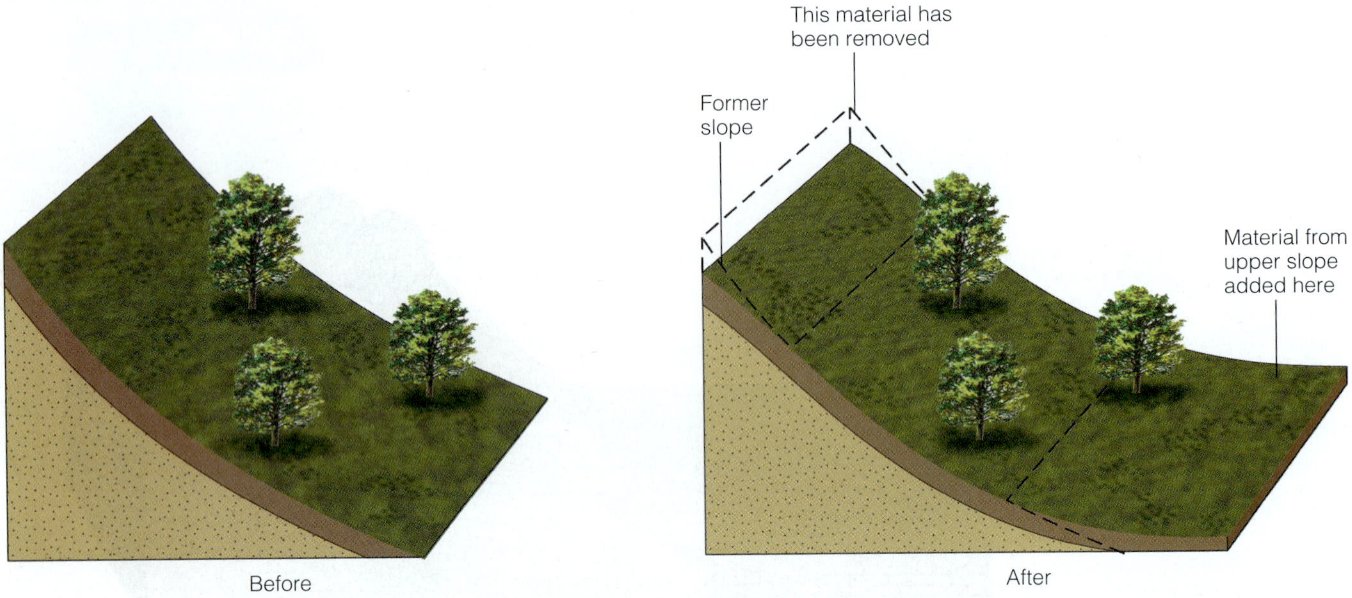

Figure 11.27

One common method used to help stabilize a hillside and reduce its slope is the cut-and-fill method. Material from the steeper upper part of the hillside is removed, thereby reducing the slope angle, and is used to fill in the base. This provides some additional support at the base of the slope.

Figure 11.28

(a) Another common method used to stabilize a hillside and reduce its slope is benching. This process involves making several cuts along a hillside to reduce the overall slope. Furthermore, individual slope failures are now limited in size, and the material collects on the benches. (b) Benching is used in nearly all road cuts.

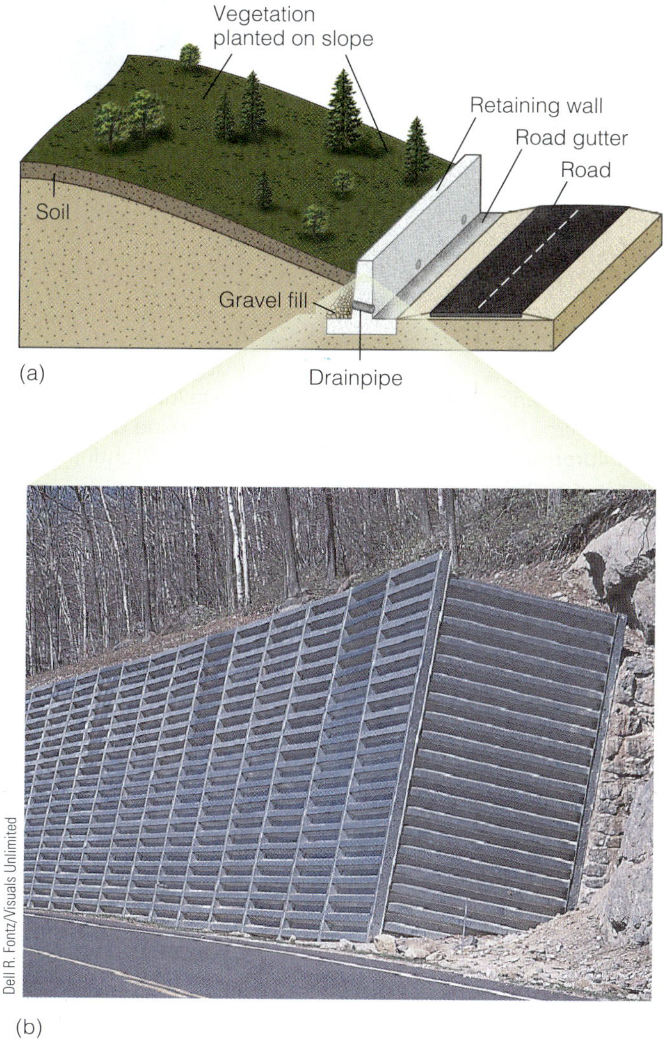

Dell R. Fontz/Visuals Unlimited

Figure 11.29

(a) Retaining walls anchored into bedrock, backfilled with gravel, and provided with drainpipes can support a slope's base and reduce landslides. (b) Steel retaining wall built to stabilize the slope and keep falling and sliding rocks off the highway.

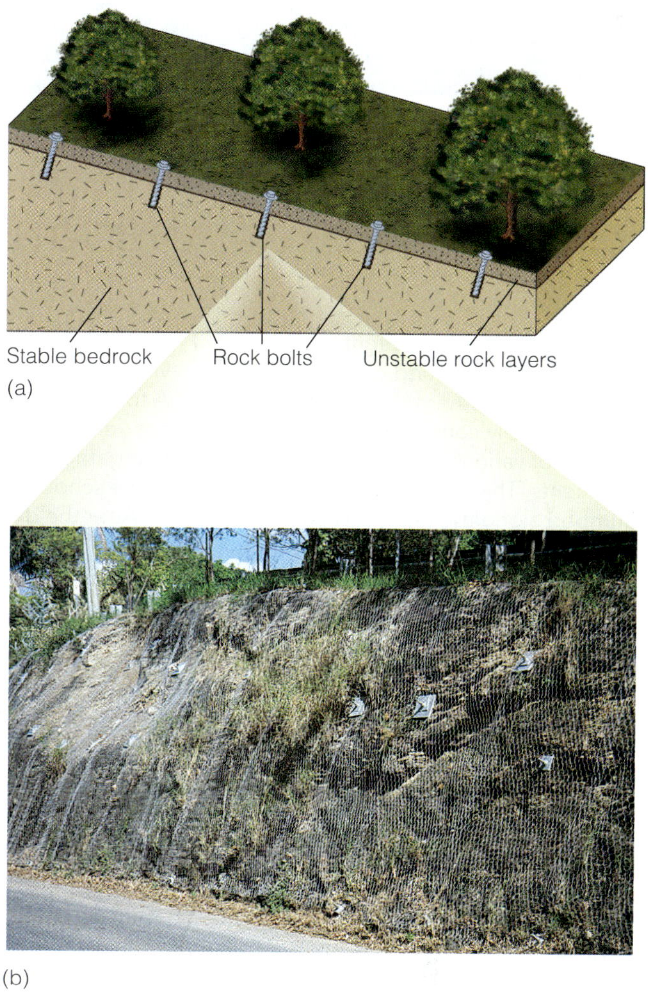

Reed Wicander

Figure 11.30

(a) Rock bolts secured in bedrock can help stabilize a slope and reduce landslides. (b) Rock bolts and wire mesh are used to help secure rock on a steep hillside in Brisbane, Australia.

angle of a hillside decreases the potential for slope failure. Two methods are usually employed to reduce a slope's angle. In the *cut-and-fill* method, material is removed from the upper part of the slope and used as fill at the base, thus providing a flat surface for construction and reducing the slope (■ Figure 11.27). The second method, which is called *benching,* involves cutting a series of benches or steps into a hillside (■ Figure 11.28). This process reduces the overall average slope, and the benches serve as collecting sites for small landslides or rockfalls that might occur. Benching is most commonly used on steep hillsides in conjunction with a system of surface drains to divert runoff.

In some situations, retaining walls are constructed to provide support for the base of the slope (■ Figure 11.29). These are usually anchored well into bedrock, backfilled with crushed rock, and provided with drain holes to prevent the buildup of water pressure in the hillside.

Rock bolts, similar to those employed in tunneling and mining, can sometimes be used to fasten potentially unstable rock masses into the underlying stable bedrock (■ Figure 11.30). This technique has been used successfully on the hillsides of Rio de Janeiro, Brazil, and to help secure the slopes at the Glen Canyon Dam on the Colorado River.

Recognition, prevention, and control of landslide-prone areas are expensive, but not nearly as expensive as the damage can be when such warning signs are ignored or not recognized. The collapse of Tip No. 7 at Aberfan, Wales (see Geo-Focus 11.1), is just one of many tragic examples in which the warning signs of impending disaster were ignored.

GEOFOCUS 11.1

The Tragedy at Aberfan, Wales

The debris brought out of underground coal mines in southern Wales typically consists of a wet mixture of various sedimentary rock fragments. This material is usually dumped along the nearest valley slope, where it builds up into large waste piles called *tips*. A tip is fairly stable as long as the material composing it is relatively dry and its sides are not too steep.

Between 1918 and 1966, seven large tips composed of mine debris were built at various elevations on the valley slopes above the small coal-mining village of Aberfan. Shortly after 9:00 A.M. on October 21, 1966, the 250-m-high, rain-soaked Tip No. 7 collapsed, and a black sludge flowed down the valley with the roar of a loud train (■ Figure 1). Before it came to a halt 800 m from its starting place, the flow had destroyed two farm cottages, crossed a canal, and buried Pantglas Junior School, suffocating virtually all the children of Aberfan. A total of 144 people died in the flow, among them 116 children who had gathered for morning assembly in the school.

After the disaster, everyone asked, "Why did this tragedy occur, and could it have been prevented?" The subsequent investigation revealed that no stability studies had ever been done on the tips and that repeated warnings about potential failure of the tips, as well as previous slides, had all been ignored.

In 1939, 8 km to the south, a tip constructed under conditions almost identical to those of Tip No. 7 collapsed. Luckily, no one was injured, but unfortunately, the failure was soon forgotten and the Aberfan tips continued to grow. In 1944 Tip No. 4 failed, and again no one was injured.

In 1958 Tip No. 7 was sited solely on the basis of available space, with no regard to the area's geology. Despite previous tip failures and warnings of slope failure by tip workers and others, mine debris was piled onto Tip No. 7 until the day of the disaster.

What exactly caused Tip No. 7 and the others to fail? The official investigation revealed that the foundation of the tips had become saturated with water from the springs over which they were built. In the case of the collapsed tips, the pore pressure from the water exceeded the friction between grains, and the entire mass liquefied like a "quicksand." Behaving like a liquid, the mass quickly moved downhill, spreading out laterally. As it flowed, water escaped from the mass, and the sedimentary particles regained their cohesion.

Following the inquiry, it was recommended that a National Tip Safety Committee be established to assess the dangers of existing tips and advise on the construction of new tip sites.

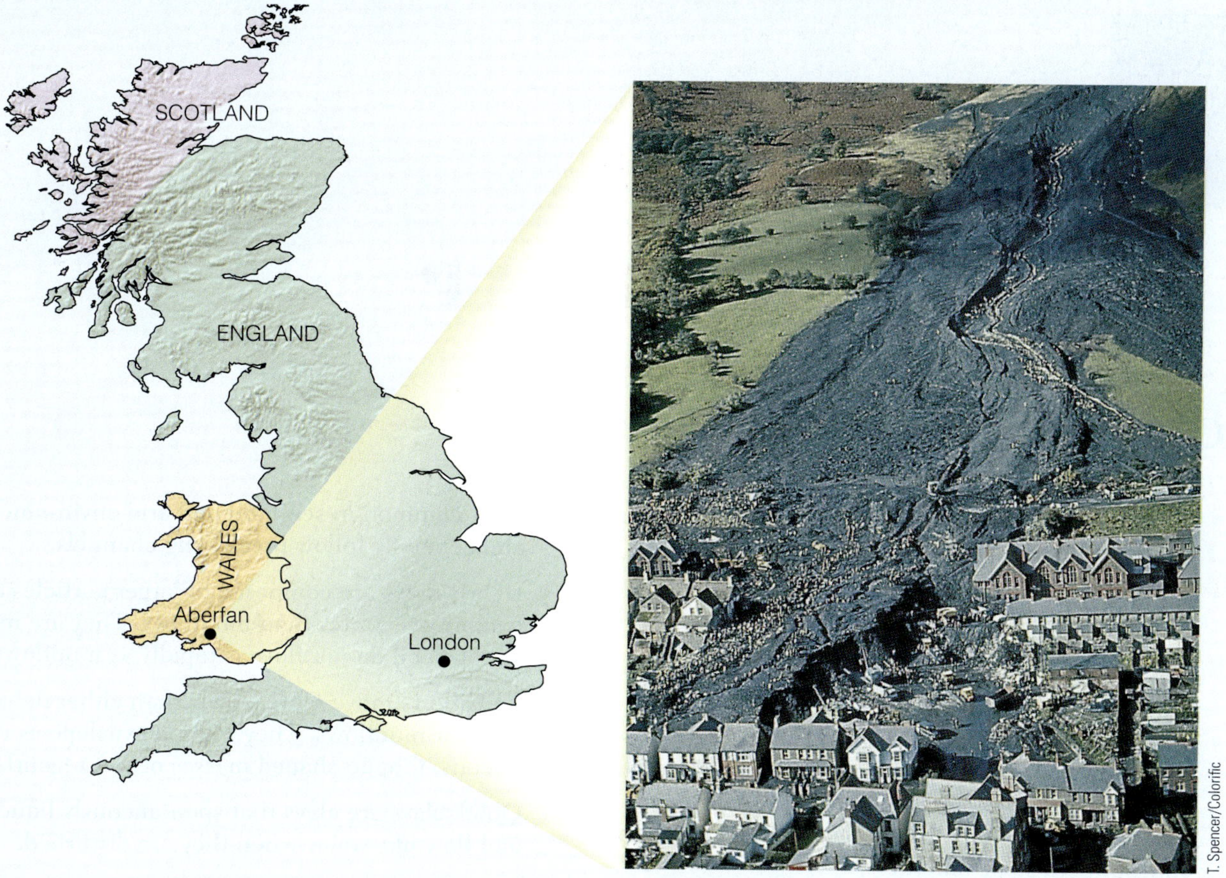

■ **Figure 1**

Location map and aerial view of the Aberfan tip disaster in which 144 people died.

11 GEO RECAP

Chapter Summary

- Mass wasting is the downslope movement of material under the influence of gravity. It occurs when the gravitational force acting parallel to a slope exceeds the slope's strength.
- Mass wasting frequently results in loss of life as well as millions of dollars in damage annually.
- Mass wasting can be caused by many factors, including slope angle, weathering of slope material, water content, overloading, and removal of vegetation. Usually several of these factors in combination contribute to slope failure.
- Mass movements are generally classified on the basis of their rate of movement (rapid versus slow), type of movement (falling, sliding, or flowing), and type of material (rock, soil, or debris).
- Rockfalls are a common mass movement in which rocks free-fall.
- Two types of slides are recognized. Slumps are rotational slides that involve movement along a curved surface; they are most common in poorly consolidated or unconsolidated material. Rock slides occur when movement takes place along a more or less planar surface; they usually involve solid pieces of rock.
- Several types of flows are recognized on the basis of their rate of movement (rapid versus slow), type of material (rock, sediment, or soil), and amount of water.
- Mudflows consist of mostly clay- and silt-sized particles and contain more than 30% water. They are most common in semiarid and arid environments and generally follow preexisting channels.
- Debris flows are composed of larger particles and contain less water than mudflows. They are more viscous and do not flow as rapidly as mudflows.
- Earthflows move more slowly than either debris flows or mudflows. They move downslope as thick, viscous, tongue-shaped masses of wet regolith.
- Quick clays are clays that spontaneously liquefy and flow like water when they are disturbed.
- Solifluction is the slow downslope movement of water-saturated surface material and is most common in areas of permafrost.
- Creep, the slowest type of flow, is the imperceptible downslope movement of soil or rock. It is the most widespread of all types of mass wasting.
- Complex movements are combinations of different types of mass movements in which no single type is dominant. Most complex movements involve sliding and flowing.
- The most important factor in reducing or eliminating the damaging effects of mass wasting is a thorough geologic investigation to outline areas susceptible to mass movements.
- Slopes can be stabilized by building retaining walls, draining excess water, regrading slopes, and planting vegetation.

Important Terms

complex movement (p. 312)
creep (p. 310)
debris flow (p. 308)
earthflow (p. 308)
mass wasting (p. 292)
mudflow (p. 308)
permafrost (p. 310)
quick clay (p. 309)
rapid mass movement (p. 297)
rock slide (p. 300)
rockfall (p. 298)
shear strength (p. 293)
slide (p. 299)
slow mass movement (p. 297)
slump (p. 299)
solifluction (p. 310)

Review Questions

1. The force opposing a slope's shear strength is
   a. _____external support; b. _____gravity;
   c. _____internal support; d. _____cohesion;
   e. _____internal friction.
2. The downslope movement of material along a curved surface of rupture is a(n)
   a. _____rockfall; b. _____slump;
   c. _____mudflow; d. _____earthflow;
   e. _____rock slide.
3. The most widespread and costly of all mass wasting processes is
   a. _____slumps; b. _____creep; c. _____mudflows; d. _____rockfalls; e. _____quick clays.
4. Where can mass wasting occur?
   a. _____only on steep slopes; b. _____only in temperate climates; c. _____anywhere;
   d. _____only on gentle slopes; e. _____only where bedrock is exposed.
5. Which of the following are the most fluid of mass movements?
   a. _____earthflows; b. _____mudflows;
   c. _____debris flows; d. _____slumps;
   e. _____solifluction.
6. Movement of material along a surface or surfaces of failure is a
   a. _____flow; b. _____fall; c. _____slide;
   d. _____all of these; e. _____none of these.
7. Which of the following is a factor influencing mass wasting?
   a. _____slope angle; b. _____vegetation;
   c. _____weathering; d. _____water content;
   e. _____all of these.
8. Solifluction occurs most commonly in which areas?
   a. _____beaches; b. _____deserts;
   c. _____permafrost; d. _____tropical forests;
   e. _____none of these.
9. Which of the following helps reduce the slope angle, or provides support at the base, of a hillside?
   a. _____cut and fill; b. _____retaining walls;
   c. _____benching; d. _____all of these;
   e. _____none of these.
10. Former landslides and areas currently susceptible to slope failure can be identified by which of the following features?
    a. _____tilted objects; b. _____open fissures;
    c. _____scarps; d. _____hummocky surfaces;
    e. _____all of these.
11. Discuss some of the ways slope stability can be maintained to reduce the likelihood of mass movements.
12. What potential value would a slope stability map be to a person seeking to purchase new home property? Using the slope stability map in Figure 11.25, locate the line that shows a horizontal contact between rocks of different stability. What is the potential for mass wasting along this line, and why?
13. Discuss how topography and the underlying geology contribute to slope failure.
14. Why are slumps such a problem along highways and railroad tracks in areas with relief?
15. What is the relationship between geologic planes of weakness, such as bedding planes, and water in causing mass wasting? Using Figure 11.12 as an example, explain how the geologic planes of weakness in this slope plus water from rainfall influenced the development of the depicted slide.
16. Discuss how the different factors that influence mass wasting are interrelated.
17. How could mass wasting be recognized on other planets or moons, and what would that tell us about the geology and perhaps atmosphere of the planet or moon on which it occurred?
18. If an area has a documented history of mass wasting that has endangered or taken human life, how should people and governments prevent such events from happening again? Are most large mass wasting events preventable or predictable?
19. Why is it important to know about the different types of mass wasting?
20. Why is creep so prevalent, and why does it do so much damage? What are some of the ways that creep might be controlled?

World Wide Web Activities

GeologyNow Assess your understanding of this chapter's topics with additional quizzing and comprehensive interactivities at

http://earthscience.brookscole.com/changingearth4e

as well as current and up-to-date weblinks, additional readings, and InfoTrac College Edition exercises.

Running Water

CHAPTER 12

OUTLINE

- Introduction
- Water on Earth

  GEO-FOCUS 12.1: *Dams, Reservoirs, and Hydroelectric Power Plants*
- Running Water
- How Does Running Water Erode and Transport Sediment?
- Deposition by Running Water

  GEOLOGY IN UNEXPECTED PLACES: *Floating Burial Chambers*
- Can Floods Be Predicted and Controlled?
- Drainage Basins and Drainage Patterns
- How Do Valleys Form and Evolve?
- Geo-Recap

GeologyNow *This icon, which appears throughout the book, indicates an opportunity to explore interactive tutorials, animations, or practice problems available on the GeologyNow website at http://earthscience.brookscole.com/changingearth4e.*

OBJECTIVES

At the end of this chapter, you will have learned that:

- Running water, one part of the hydrologic cycle, does considerable geologic work.
- Water is continuously cycled from the oceans to land and back to the oceans.
- Running water transports large quantities of sediment and deposits sediment in or adjacent to braided and meandering rivers and streams.
- Alluvial fans (on land) and deltas (in a standing body of water) are deposited when a stream's capacity to transport sediment decreases.
- Flooding is a natural part of stream activity that takes place when a channel receives more water than it can handle.
- The several types of structures to control floods are only partly effective.
- Rivers and streams continuously adjust to changes.
- The concept of a graded stream is an ideal, although many rivers and streams approach the graded condition.
- Most valleys form and change in response to erosion by running water coupled with other geologic processes such as mass wasting.

Cumberland Falls on the Big South Fork River in Cumberland Falls State Resort Park in Kentucky plunges 18 m, making it the second highest waterfall east of the Rocky Mountains. Source: Kentucky Department of Parks

Introduction

Among the terrestrial planets, Earth is unique in having abundant liquid water. Both Mercury and Earth's Moon are too small to retain water, and Venus, because of its high temperature, is too hot for surface water to exist. At present, Mars has some frozen water and trace amounts of water in its atmosphere, but spacecraft images and other data reveal areas with winding valleys that were probably carved by running water during Mars's early history. In contrast, oceans and seas cover 71% of Earth's surface (see Figure 9.2), and small but important quantities of water are present in the atmosphere and on land.

Certainly the hydrosphere has a tremendous impact on the surface, but the hydrosphere consists of water vapor in the atmosphere, groundwater (see Chapter 13), water frozen in glaciers (see Chapter 14), water in the oceans (see Chapters 9 and 16), and a small amount of water on land (see the chapter opening photo). Water on land is also found in lakes, swamps, and bogs, but our main concern in this chapter is the small percentage of running water confined to channels.

To appreciate the power of running water, you need only read some of the vivid accounts of floods. For example, at 4:07 P.M. on May 31, 1889, the residents of Johnstown, Pennsylvania, heard "a roar like thunder," and within 10 minutes a catastrophic flood destroyed the town, leaving at least 2200 people dead. An 18-m-high wall of water roared through the town at more than 60 km/hr, sweeping up houses, debris, and hundreds of people (■ Figure 12.1). According to one account, "Thousands of people desperately tried to escape the wave. Those caught by the wave found themselves swept up in a torrent of oily, muddy water, surrounded by tons of grinding debris. . . . Many became hopelessly entangled in miles of barbed wire from the destroyed wire works."*

The fact that Johnstown was built on a floodplain in a narrow valley contributed to the tragedy, but the failure of the South Fork Dam about 22 km upstream on the Conemaugh River was the main cause of the disaster. The dam was poorly maintained and unable to withstand the added stress from the 20–25 cm of rain that fell in 24 hours. This was not the last dam failure in the United States, but it certainly was the most tragic one.

Floods that result from natural causes as well as from human carelessness are a continuing threat. Indeed, they are so common that reports of damage, injuries, and fatalities from floods appear in the news regularly. Several large floods occur in North America every year, but the last flood of truly vast proportions was the Flood of '93 that inundated large parts of the central United States, especially Iowa, Wisconsin, Illinois, Missouri, Minnesota, North and South Dakota, Kansas, and parts of some adjacent states. At last 50 people died, 70,000 were homeless, and property damage was estimated at $15 to $20 billion.

Nevertheless, many benefits come from running water and even from some floods. Running water in channels—that is, streams and rivers—is one important source of freshwater for industry, agriculture, domestic use, and recreation, and about 8% of all electricity used in North America is generated by falling water at hydroelectric power plants (see GeoFocus 12.1). Large waterways throughout the world are major avenues of commerce, and when Europeans explored the interior of North America, they followed the St. Lawrence, Mississippi, Missouri, Columbia, and Ohio Rivers.

Schenectaday Museum; Hall of Electrical History/Corbis

(a)

Bettmann/Corbis

(b)

■ **Figure 12.1**

Johnstown, Pennsylvania, before (a) and after (b) the May 31, 1889, flood.

*National Park Service—U.S. Department of Interior, Johnstown Information Service Online.

WATER ON EARTH

According to one estimate, 1.36 billion km^3 of water is present on Earth, most of it (97.2%) in the oceans (Table 12.1). About 2.15% is frozen in glaciers on land, especially in Antarctica and Greenland, with most of the remaining 0.65% in the atmosphere, groundwater, lakes, swamps, and bogs. Only about 0.0001% of all water in the hydrosphere is in stream and river channels at any one time. Nevertheless, running water, with its resultant erosion, transport, and deposition, is with few exceptions the most important geologic agent bringing about changes to Earth's land surface. Only in areas covered by vast glaciers or parts of some deserts are other geologic agents more important than running water. Even in most deserts, though, the effects of running water are conspicuous, although channels are dry most of the time.

Much of our discussion of running water is descriptive, but always be aware that streams and rivers are dynamic systems that must continuously respond to change. For example, paving in urban areas increases surface runoff to waterways, and other human activities such as building dams and impounding reservoirs also alter the dynamics of stream and river systems. Natural changes, too, affect the complex interacting parts of stream and river systems.

Table 12.1

Water on Earth

Location	Volume (km^3)	Percent of Total
Oceans	1,327,500,000	97.20
Ice caps and glaciers	29,315,000	2.15
Groundwater	8,442,580	0.625
Freshwater and saline lakes and inland seas	230,325	0.017
Atmosphere at sea level	12,982	0.001
Stream channels	1,255	0.0001

The Hydrologic Cycle

The connection between precipitation and clouds is obvious, but where does the moisture for rain and snow come from in the first place? In the previous section we noted that 97.2% of all water on Earth is in the oceans, so one might immediately suspect that the oceans are the ultimate source of precipitation. In fact, water is continuously recycled from the oceans, through the atmosphere, to the continents, and back to the oceans. This **hydrologic cycle,** as it is called, is powered by solar radiation and is possible because water changes easily from liquid to gas (water vapor) under surface conditions. About 85% of all water entering the atmosphere comes from water, which corresponds to a layer about 1 m thick evaporating from the oceans each year. The remaining 15% comes from water on land, but this water originally came from the oceans as well.

Regardless of its source, water vapor rises into the atmosphere where the complex processes of cloud formation and condensation take place. Much of the world's precipitation, about 80%, falls directly back into the oceans, in which case the hydrologic cycle is a three-step process of evaporation, condensation, and precipitation. For the 20% of all precipitation that falls on land, the hydrologic cycle is more complex, involving evaporation, condensation, movement of water vapor from the oceans to land, precipitation, and runoff. Although some precipitation evaporates as it falls and reenters the cycle, about 36,000 km^3 of the precipitation that falls on land returns to the oceans by **runoff,** the surface flow in streams and rivers.

Not all precipitation returns directly to the oceans, though. Some is temporarily stored in lakes and swamps, snowfields and glaciers, or seeps below the surface where it enters the groundwater system (see Chapter 13). Water might remain in some of these reservoirs for thousands of years, but eventually glaciers melt, lakes and groundwater feed streams and rivers, and this water returns to the oceans. Even the water used by plants evaporates, a process known as *transpiration,* and returns to the atmosphere. In short, all water derived from the oceans eventually makes it back to the oceans and can thus begin the hydrologic cycle again (■ Figure 12.2).

Fluid Flow

Solids are rigid substances that retain their shapes unless deformed by a force, but fluids—that is, liquids and gases—have no strength so they flow in response to any force no matter how slight. Liquid water, which is our concern here, flows downslope in response to gravity; its flow may be *laminar* or *turbulent.* In laminar flow, lines of flow called streamlines parallel one another with little or no mixing between adjacent layers (■ Figure 12.3a). All flow is in one direction only, and it remains unchanged through time. In turbulent flow, streamlines intertwine, causing complex mixing within the moving fluid (Figure 12.3b). If we could trace a single water molecule in turbulent flow, it may move in any direction at a particular time although its overall movement is in the direction of flow.

Runoff during a rainstorm depends on the **infiltration capacity,** the maximum rate at which surface

GEOFOCUS 12.1

Dams, Reservoirs, and Hydroelectric Power Plants

Flip a switch and we illuminate our homes and workplaces; turn a dial and we heat our homes or perhaps cook our food or wash our clothes; and some of our public transportation relies on an unseen but important energy source—electricity. In fact, about 40% of the total energy used in industrialized nations is converted to electricity. Most electricity is generated at power plants that burn fossil fuels (oil, natural gas, and especially coal), but nuclear power plants are common in some areas. Geothermal energy (see Chapter 13), wind, and tidal power (see Chapter 16) account for only a small percentage of all electricity.

Electricity generated at *hydroelectric power plants*—that is, power plants that use moving water—accounts for only 8% of the total electricity generated in North America, but its importance varies from area to area. For instance, less than 2% of the electricity generated in Florida, Ohio, and Texas comes from hydroelectric plants, whereas Washington, Oregon, and Idaho derive more than 80% of their electricity from this source.

At all power plants, a spinning turbine connected to a generator containing a coil of wire produces electricity. However, the energy used to turn turbines differs. In plants that burn fossil fuel, for example, coal, oil, or natural gas is burned to heat water and generate steam, which in turn serves as the energy source to turn turbines. Steam is also generated in nuclear power plants.

In hydroelectric power plants, moving water rather than steam is used to turn turbines. To provide the necessary water, a dam is built to impound a reservoir where the water is higher than the power plant. Water moves from the reservoir through a large pipe called a *penstock* and spins a turbine (■ Figure 1). In other words, the potential energy of the water in the reservoir is converted to electrical energy at the power plant. Regardless of the power source—fossil fuels, moving water, or uranium—once electricity is generated, it is transmitted to areas of use by power lines.

Falling water was first used to generate electricity in 1892 at Appleton, Wisconsin, and since then hydroelectric power plants have become common features on many of the world's waterways. One of the largest in terms of generating capacity is on the Parana River on the Paraguay–Brazil border, and an even larger one was recently completed in the People's Republic of China; the reservoir behind the Three Gorges Dam on the Yangtze River began filling in April 2003. There are many other similar facilities in North America. A large region of the Pacific Northwest depends on electricity generated at the Grand Coulee Dam in Washington, and hydroelectric plants on the Niagara River on the U.S.–Canadian border provide electricity to a large area.

One might be curious about why agencies and governments do not simply increase their hydroelectric output. After all, hydroelectric power generation has several appealing aspects, not the least of which is that it is a renewable resource. However, not all areas have this potential; suitable sites for dams and reservoirs might not be present, for instance. In addition, dams are very expensive to build, reservoirs fill with sediment, and during droughts enough water might not be available to keep reservoirs sufficiently full. And, of course, people must be relocated from areas where reservoirs are impounded and from the discharge areas downstream from dams. So, although hydroelectric dams remain important, we cannot realistically look forward to very much additional use of this energy source.

materials can absorb water. Several factors control the infiltration capacity, including intensity and duration of rainfall. If rain is absorbed as fast as it falls, no surface runoff takes place. For instance, loosely packed dry soil absorbs water faster than tightly packed wet soil, and thus more rain must fall on loose dry soil before runoff begins. Regardless of the initial condition of surface materials, once they are saturated, excess water collects on the surface and, if on a slope, it moves downhill.

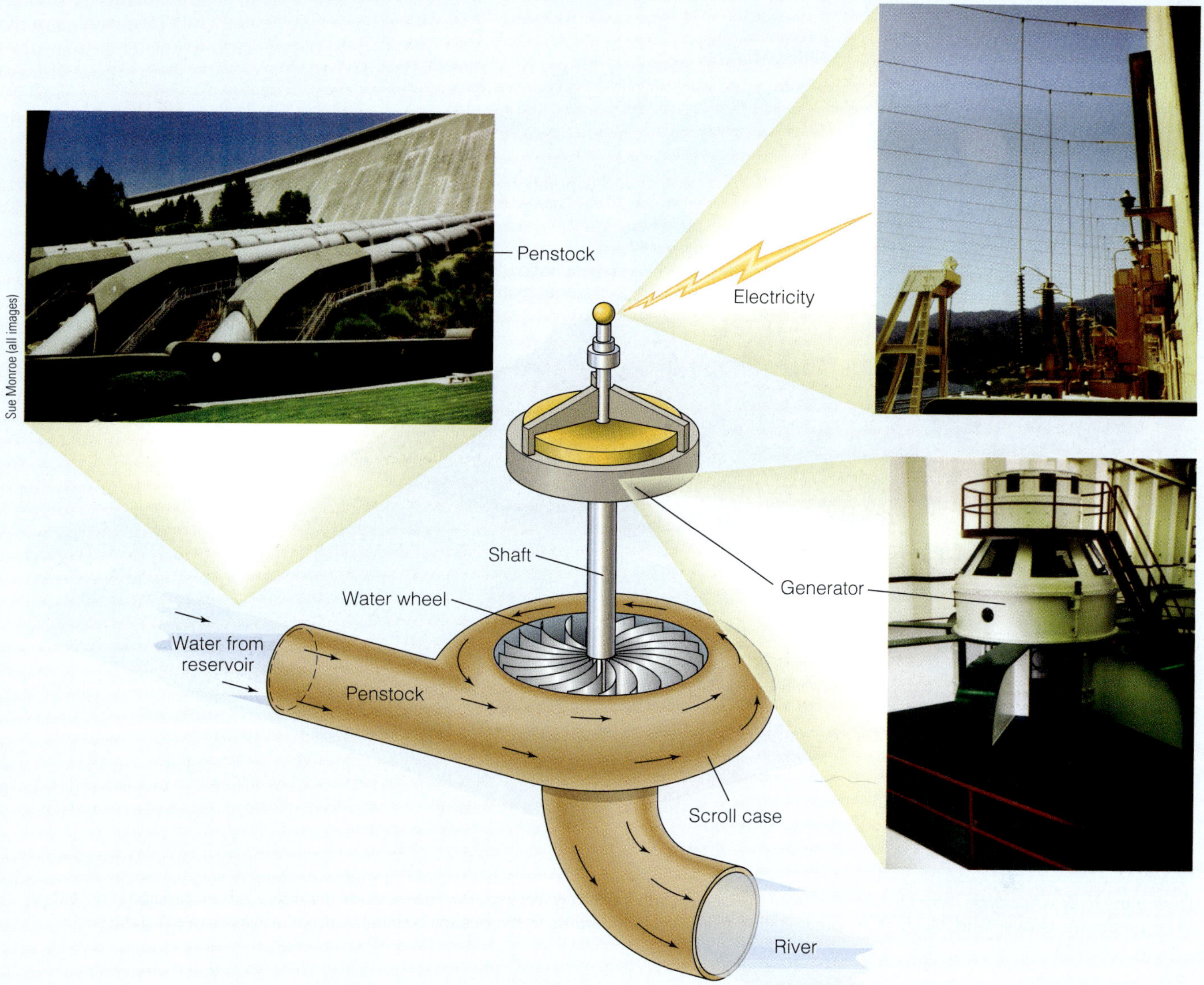

Figure 1

Electricity is produced at a generator when water rushes through the penstock, where it spins a water wheel connected by a shaft to an electromagnet within a coil of wire. Each penstock shown here measures about 4 m in diameter and carries water at 16 to 22 km/hr. The dam from which water enters the penstocks is visible in the background. Once electricity is generated, it is transmitted to areas of use by power lines.

RUNNING WATER

We have mentioned that moving water is the most important geologic agent modifying Earth's land surface. The only exceptions are areas covered by vast glaciers and in parts of some deserts; otherwise, the effects of running water are ubiquitous.

Sheet Flow and Channel Flow

Even on steep slopes, flow is initially slow and hence causes little or no erosion. As water moves downslope,

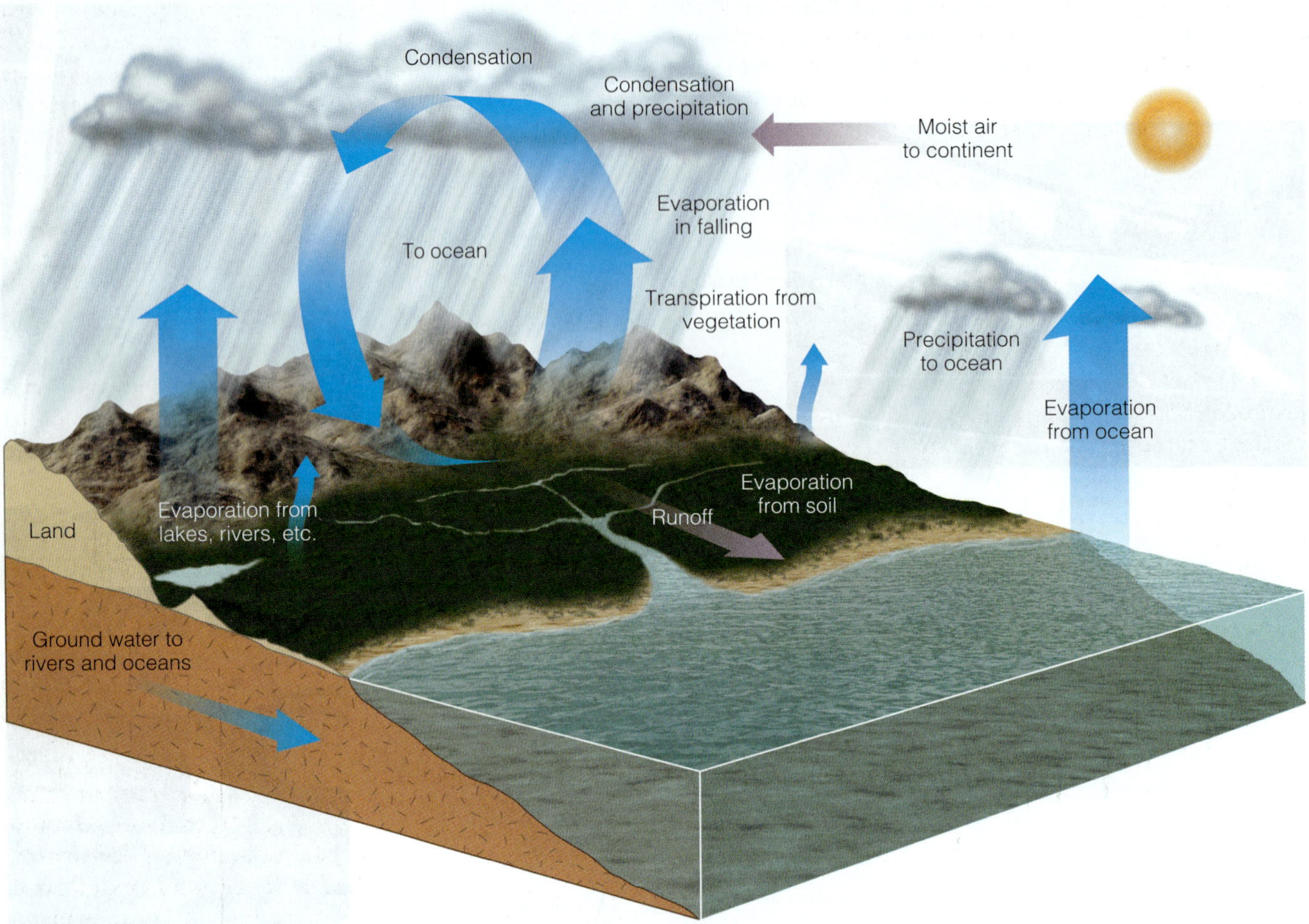

Figure 12.2

During the hydrologic cycle, water evaporates from the oceans and rises as water vapor to form clouds that release their precipitation either over the oceans or over land. Much of the precipitation falling on land returns to the oceans by surface runoff, thus completing the cycle.

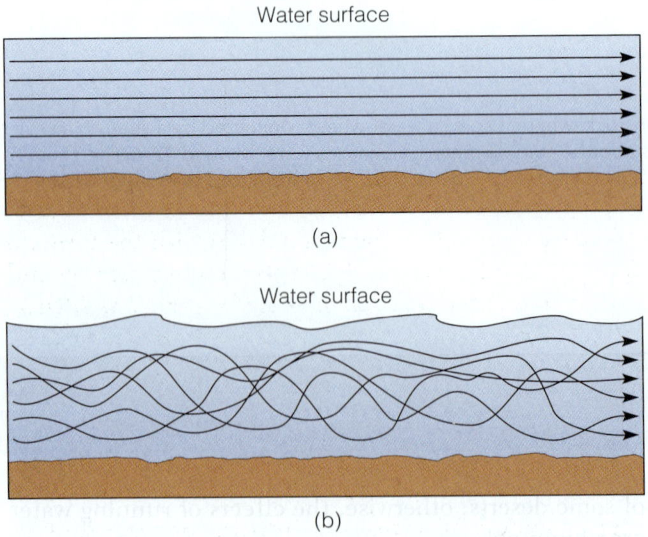

Figure 12.3

(a) In laminar flow, streamlines are parallel to one another, and little or no mixing takes place between adjacent layers in the fluid. (b) In turbulent flow, streamlines are complexly intertwined, indicating mixing between layers. Most flow in streams is turbulent.

though, it accelerates and may move by *sheet flow,* a more or less continuous film of water flowing over the surface. Sheet flow is not confined to depressions, and it accounts for *sheet erosion,* a particular problem on some agricultural lands (see Chapter 6).

In *channel flow,* surface runoff is confined to troughlike depressions that vary in size from tiny rills with a trickling stream of water to the Amazon River in South America, which is 6450 km long and at one place 2.4 km wide and 90 m deep. We describe flow in channels with terms such as *rill, brook, creek, stream,* and *river,* most of which are distinguished by size and volume. Here we use the terms *stream* and *river* more or less interchangeably, although the latter usually refers to a larger body of running water.

Streams and rivers receive water from several sources, including sheet flow and rain that falls directly into their channels. Far more important, though, is water supplied by soil moisture and groundwater, both of which flow downslope and discharge into waterways. In areas where groundwater is plentiful, streams and rivers maintain a fairly stable flow year-round because their water supply is continuous. In contrast, the amount of water in streams and rivers of arid and semi-arid regions fluctuates widely because they depend more on infrequent rainstorms and surface runoff for their water.

Gradient, Velocity, and Discharge

Water in any channel flows downhill over a slope known as its **gradient.** Suppose a river has its headwaters (source) 1000 m above sea level and its flows 500 km to the sea, so it drops vertically 1000 m over a horizontal distance of 500 km. Its gradient is found by dividing the vertical drop by the horizontal distance, which in this example is 1000 m/500 km = 2 m/km (■ Figure 12.4). On the average this river drops vertically 2 m for every kilometer along its course.

In the preceding example, we calculated the average gradient for a hypothetical river, but gradients vary not only among channels but even along the course of a single channel. Rivers and streams are steeper in their upper reaches (near their headwaters) where they may have gradients of several tens of meters per kilometer, but they have gradients of only a few centimeters per kilometer where they discharge into the sea.

The **velocity** of running water is a measure of the downstream distance water travels in a given time. It is usually expressed in meters per second (m/sec) or feet per second (ft/sec), and it varies across a channel's width as well as along its length. Water moves more slowly and with greater turbulence near a channel's bed and banks because friction is greater there than it is some distance from these boundaries (■ Figure 12.5a). Channel shape and roughness also influence flow velocity. Broad, shallow channels and narrow, deep channels have proportionately more water in contact with their perimeters than channels with semicircular cross sections (Figure 12.5b). So, if other variables are the same, water flows faster in a semicircular channel because there is less frictional resistance. As one would expect, rough channels, such as those strewn with boulders, offer more frictional resistance to flow than do channels with a bed and banks composed of sand or mud.

Intuitively you might suspect that the gradient is the most important control on velocity—the steeper the gradient, the greater the velocity. In fact, a channel's average velocity actually increases downstream even though its gradient decreases! Keep in mind that we are talking about average velocity for a long segment of a channel, not velocity at a single point. Three factors account for this downstream increase in velocity. First, velocity increases even with decreasing gradient in response to the acceleration of gravity. Second, the upstream reaches of channels tend to be boulder-strewn, broad, and shallow, so frictional resistance to flow is high, whereas downstream segments of the same channels are more semicircular and have banks composed of finer materials. And finally, the number of smaller tributaries joining a larger channel increases downstream. Thus the total volume of water (discharge) increases, and increasing discharge results in greater velocity.

We mentioned discharge in the preceding paragraph but noted only that it refers to the volume of water. More specifically, **discharge** is the volume of water that passes a particular point in a given period of time. Discharge is found from the dimensions of a water-filled channel—that is, its cross-sectional area (A) and its flow velocity (V). Discharge (Q) is then calculated with the formula $Q = VA$ and is expressed in cubic meters per second (m^3/sec) or cubic feet per second (ft^3/sec). The Mississippi River has an average discharge of 18,000 m^3/sec, and the average discharge for the Amazon River in South America is 200,000 m^3/sec.

In most rivers and streams, discharge increases downstream as more and more water enters a channel. However, there are a few exceptions. Because of high evaporation rates and infiltration, the flow in some desert waterways actually decreases downstream until the water disappears. And even in perennial rivers and streams, discharge is obviously highest during times of heavy rainfall and at a minimum during the dry season.

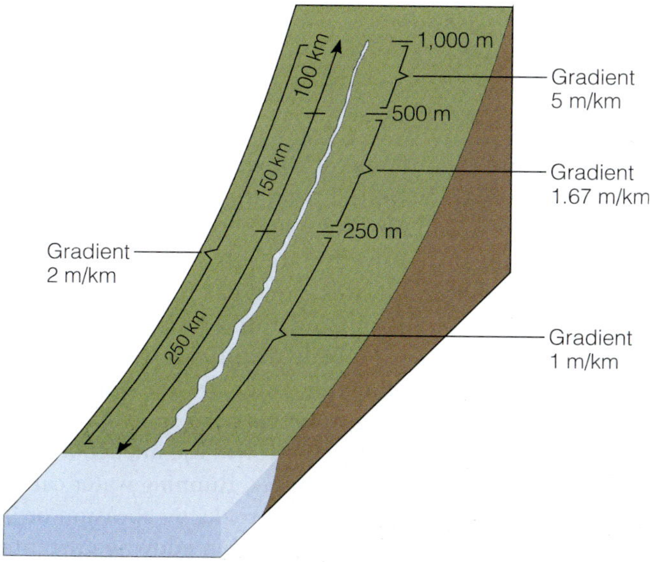

■ **Figure 12.4**

The average gradient of this stream is 2 m/km, but gradient can be calculated for any segment of a stream, as shown in this example. Notice that the gradient is steepest in the headwaters area and decreases in a downstream direction.

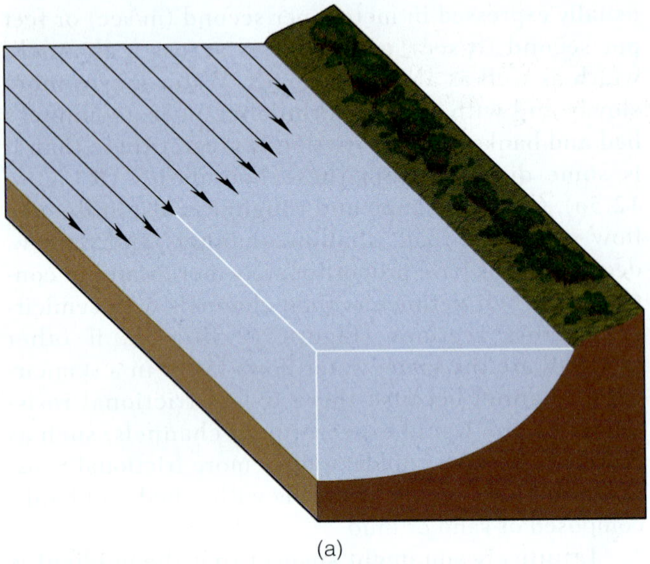

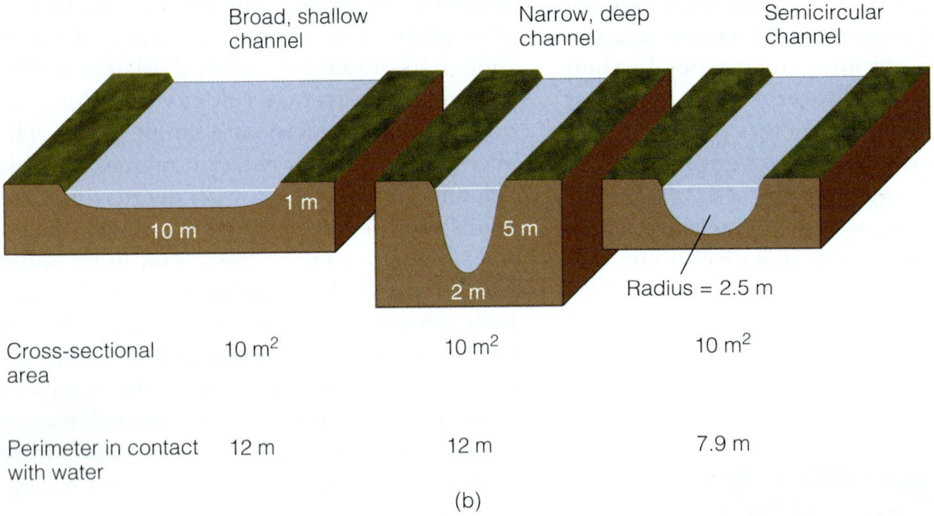

Figure 12.5

Flow velocity in rivers and streams varies as a result of friction with their banks and beds. (a) The maximum flow velocity is near the center and top of a straight channel where friction is least. The arrows are proportional to velocity. (b) These three differently shaped channels have the same cross-sectional area. The semicircular channel has less water in contact with its perimeter and thus less frictional resistance to flow.

HOW DOES RUNNING WATER ERODE AND TRANSPORT SEDIMENT?

Running water's role in erosion cannot be underestimated. Water possesses two kinds of energy: potential and kinetic. *Potential energy* is the energy of position, such as the energy possessed by water behind a dam or at high elevation. In running water, the energy of position is converted to *kinetic energy,* the energy of motion. Most of this kinetic energy is used up by fluid turbulence, but a small amount is available to erode and transport sediment.

Erosion involves the removal from a source area of dissolved substances as well as loose particles of soil, minerals, and rock. Some of the dissolved materials in running water are acquired from the beds and banks of channels where soluble rocks such as limestone are exposed, but most dissolved materials are derived from sheet flow and groundwater. Solid particles from channel perimeters or introduced into channels by mass wasting are set in motion by **hydraulic action**—that is, the direct impact of moving water on loose materials (Figure 12.6). Running water carrying sand and gravel also erodes by **abrasion,** which involves the impact of solid particles with exposed rock surfaces. Running water carrying sediment may polish exposed rocks by abrasion or it may scour out potholes when gravel in eddying currents abrades stream beds, yielding circular to oval depressions (Figure 12.7).

Once materials are eroded, they are transported by running water for some distance from their source and

(a)

(b)

■ **Figure 12.6**

(a) This stream acquires some of its sediment load by undercutting its banks. (b) Some of the sediment in the Snake River of Idaho comes from these talus cones that accumulated as a result of mass wasting.

eventually deposited. Transport involves moving both a **dissolved load,** consisting of materials taken into solution during chemical weathering, and a *solid load* ranging from clay-sized particles to large boulders (■ Figure 12.8). This solid load is further divided into a suspended load and a bed load. In the **suspended load,** the smallest particles in transport, such as silt and clay, are kept suspended above the channel bed by fluid turbulence (Figure 12.8). Suspended load in rivers and streams is what gives the water its murky appearance. Running water also transports a **bed load** of larger particles, especially sand and gravel, that cannot be kept suspended by fluid turbulence. Some of the sand, however, might be temporarily suspended when an eddying current swirls across a channel's bed and lifts grains into the water. These grains move forward with the water but also settle and come to rest on the bed, where they may be moved again by the same processes of intermittent bouncing and skipping, a phenomenon known as *saltation* (Figure 12.8). Particles too large to be suspended even temporarily move by rolling and sliding.

DEPOSITION BY RUNNING WATER

Some of the sediment now being deposited in the Gulf of Mexico by the Mississippi River came from such distant sources as Pennsylvania, Minnesota, and Alberta, Canada. Transport might be lengthy, but deposition eventually takes place. Some deposits accumulate along the way in channels, on adjacent floodplains, or where rivers and streams discharge from mountains onto nearby low-lands or where they flow into lakes or seas.

Rivers and streams constantly erode, transport, and deposit sediment, but they do most of their geologic work when they flood. Consequently, their deposits, collectively called **alluvium,** do not represent the day-to-day activities of running water, but rather the periodic, large-scale events of sedimentation that

■ **Figure 12.7**

(a) Abrasion has imparted a lustrous sheen to this otherwise ordinary basalt. (b) These potholes in the bed of the Chippewa River in Ontario, Canada, measure about 1 m across. Two potholes at the top center have merged to form a larger, composite pothole. (c) These stones measuring 7 to 8 cm from a pothole are remarkably spherical because of abrasion.

take place during floods. Remember from Chapter 6 that sediments accumulate in *depositional environments* characterized as continental, transitional, and marine. Deposits of rivers and streams are found mostly in the first two of these settings; however, much of the detrital sediment found on continental margins is derived from the land and transported to the oceans by running water.

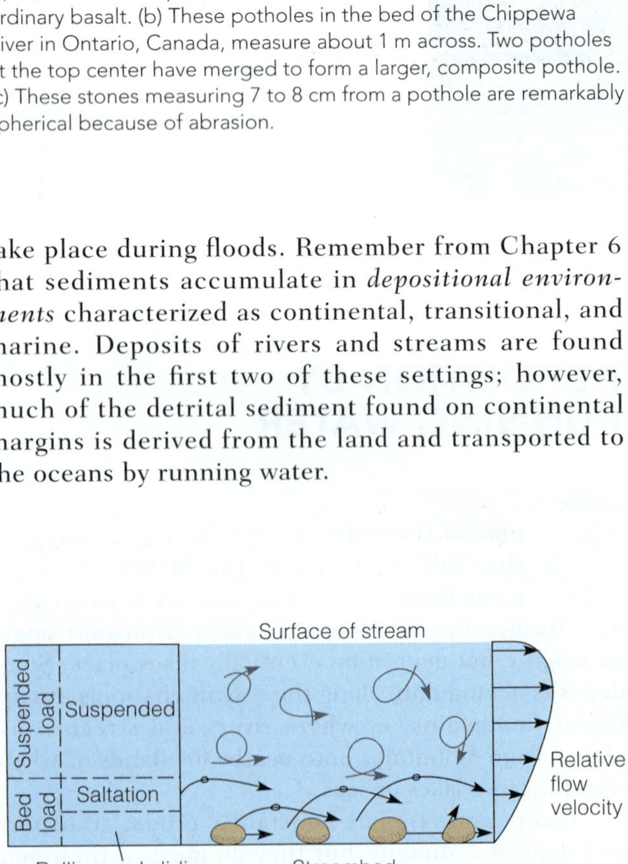

■ **Figure 12.8**

Sediment transport by running water. The velocity profile at the right indicates that the water flows fastest near the surface and slowest along the streambed.

The Deposits of Braided and Meandering Channels

Most rivers and streams possess channels characterized as *braided* or *meandering*. A **braided stream** has an intricate network of dividing and rejoining channels separated from one another by sand and gravel bars (■ Figure 12.9). Seen from above, the channels resemble the complex strands of a braid. Braided channels develop when the sediment supply exceeds the transport capacity of running water, resulting in the deposition of sand and gravel bars. During high-water stages, the bars are submerged, but when the water is low, they are exposed and divide a single channel into multiple channels. Braided streams have broad, shallow channels and are characterized as bed-load transport streams because they transport and deposit mostly sand and gravel (Figure 12.9).

Braided channels are common in arid and semiarid regions with sparse vegetation and surface materials that are unprotected and easily eroded. So much sediment is released from melting glaciers that rivers and streams discharging from them are also commonly braided (see Chapter 14).

Meandering streams have a single sinuous channel with broadly looping curves known as *meanders* (■ Figure 12.10). Channels of meandering streams are semicir-

(a)

(b)

■ **Figure 12.9**

Braided streams and their deposits. (a) A braided stream in Alaska. The deposits of this stream are composed mostly of sand. (b) A braided stream with gravel bars near Chester, California.

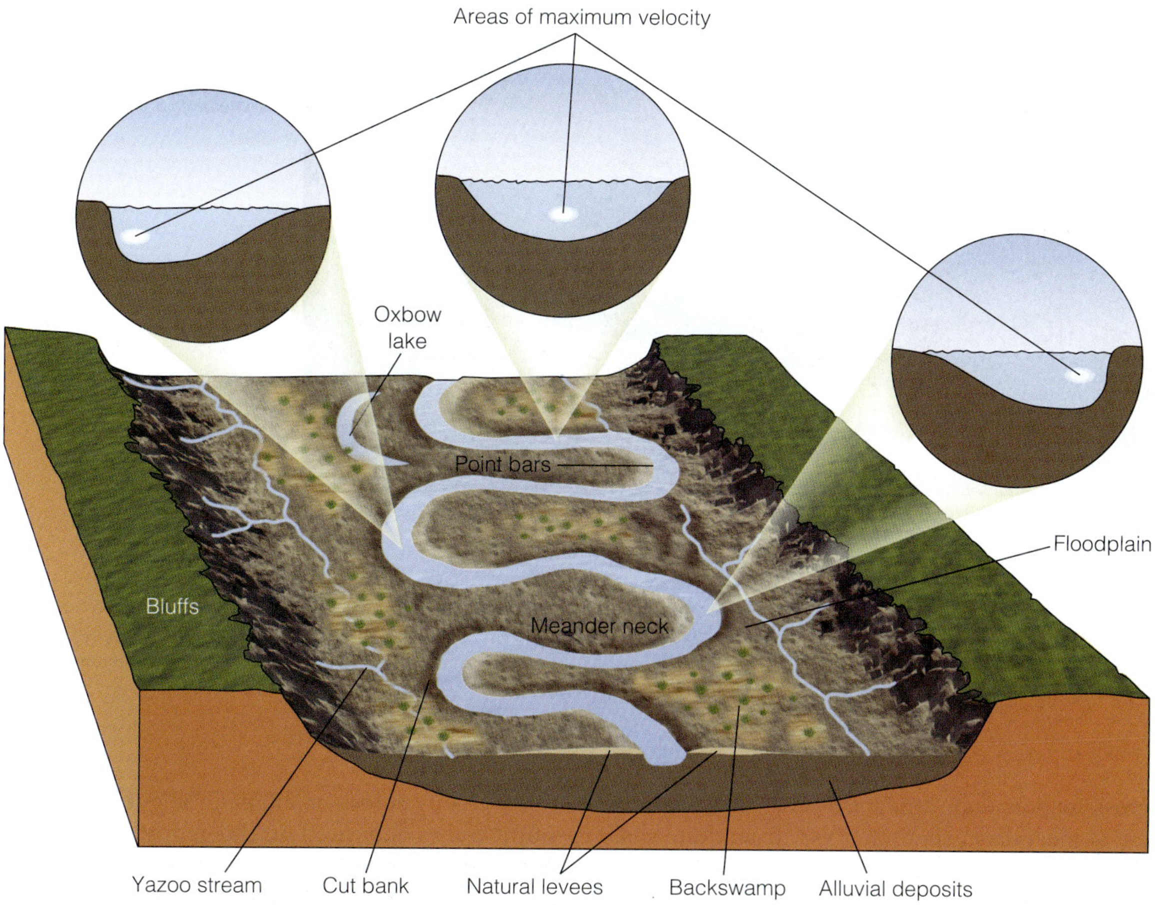

■ **Figure 12.10**

Diagrammatic view of a meandering river.

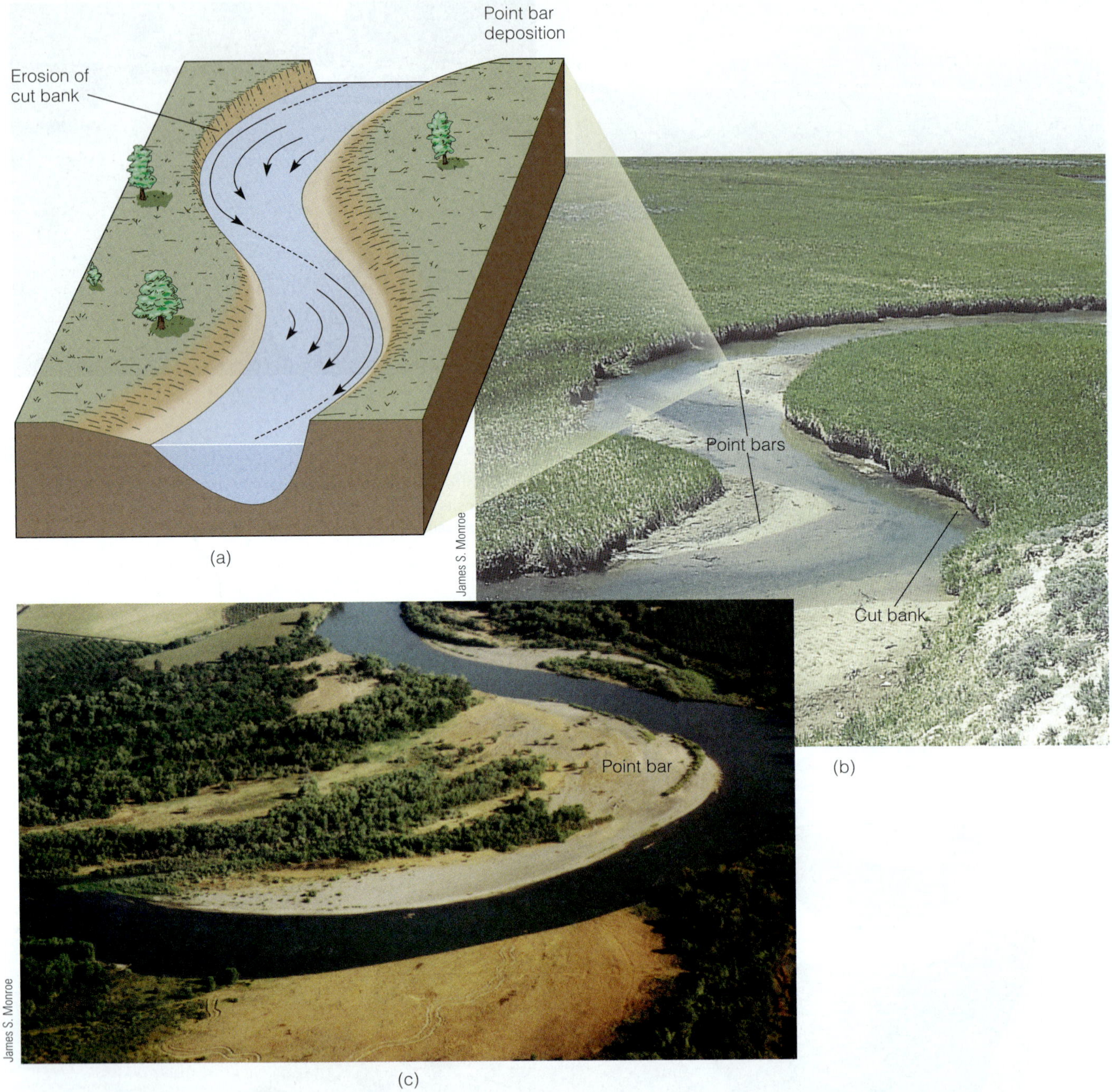

Figure 12.11

(a) In a meandering channel, flow velocity is greatest near the outer bank. The dashed line follows the path of maximum velocity, and the solid arrows are proportional to velocity. Because velocity varies across the channel, the outer bank or cut bank is eroded but a point bar is deposited on the opposite side of the channel. (b) Two small point bars of sand in a meandering stream. Notice how they are inclined into the deeper part of the channel. Also note the cut bank. (c) This point bar measures several hundred meters across.

cular in cross section along straight reaches but markedly asymmetric at meanders, where they vary from quite shallow to deep across the meander. The deeper side of the channel is known as the *cut bank* because greater velocity and fluid turbulence erode it. In contrast, flow velocity is at a minimum on the opposite bank, which slopes gently into the channel. As a result of this unequal distribution of flow velocity across meanders, the cut bank erodes and a **point bar** is deposited on the gently sloping inner bank. Most point bars are composed of cross-bedded sand, but some consist of gravel (■ Figure 12.11).

Meanders commonly become so sinuous that the thin neck of land between adjacent ones is cut off during a flood. Many of the floors of valleys with meandering channels are marked by crescent-shaped **oxbow lakes,** which are simply cutoff meanders (■ Figures 12.10 and 12.12). Oxbow lakes may persist for some time, but they eventually fill with organic matter and fine-grained sediments carried by floods.

Floodplain Deposits

Rivers and streams periodically receive more water than their channels can accommodate, so they overflow their banks and spread across adjacent low-lying, relatively flat **floodplains** (Figure 12.10). Floodplain sediments might be sand and gravel that accumulated when meandering streams deposited a succession of point bars

What Would You Do?

Given what is known about the dynamics of running water in channels, it is remarkable that houses are still built on the cut banks of meandering rivers. No doubt property owners think these locations provide good views because they sit high above the adjacent channel. Explain why you would or would not build a house in such a location? What recommendations would you make to a planning commission on land use for areas as described here? Are there any specific zoning regulations or building codes you might favor?

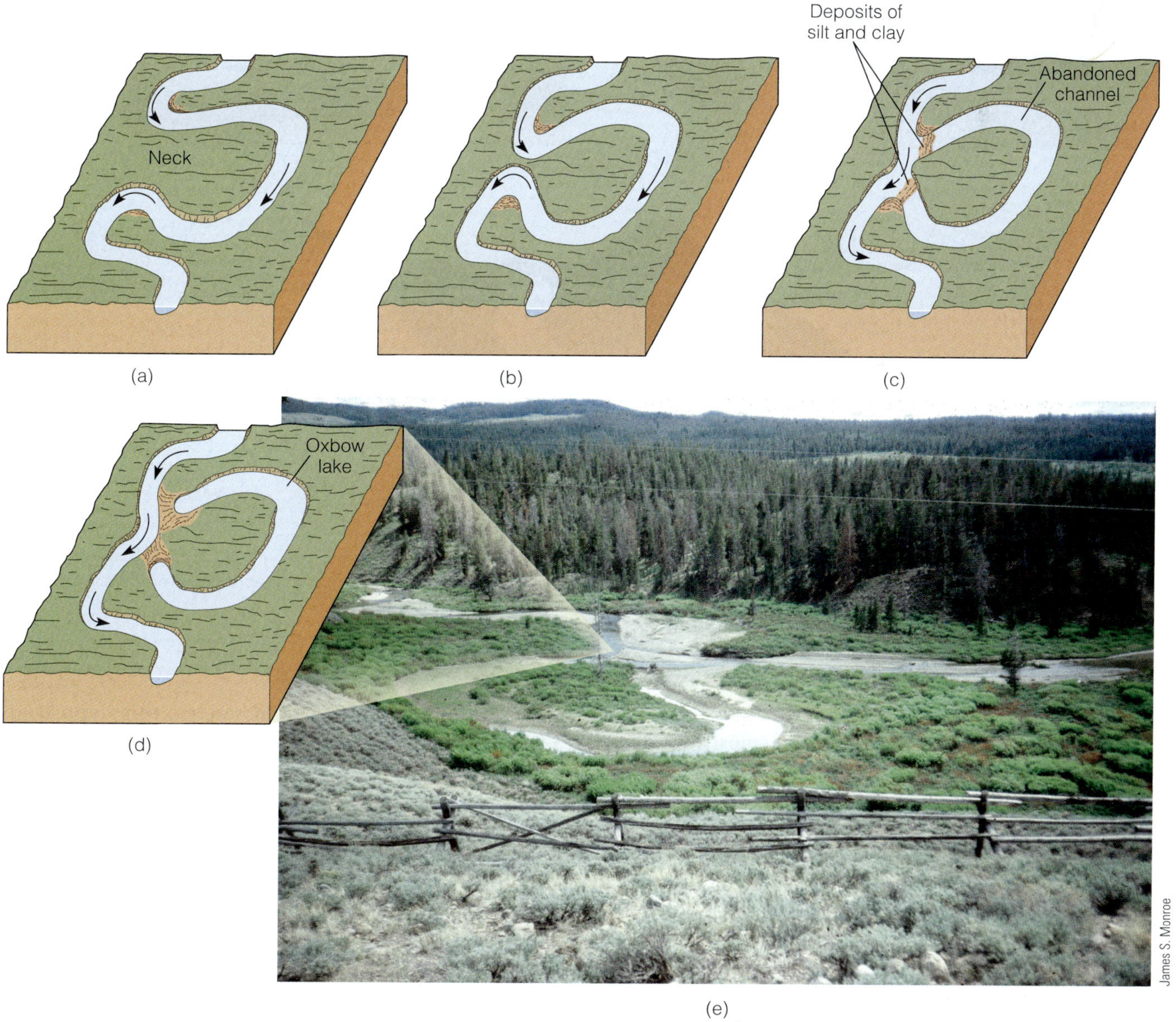

Geology Now ■ Active Figure 12.12

Four stages in the origin of an oxbow lake. In (a) and (b) the meander neck becomes narrower, and in (c) it cuts off part of the original channel to form an oxbow lake (d). (e) This oxbow lake in Wyoming formed recently.

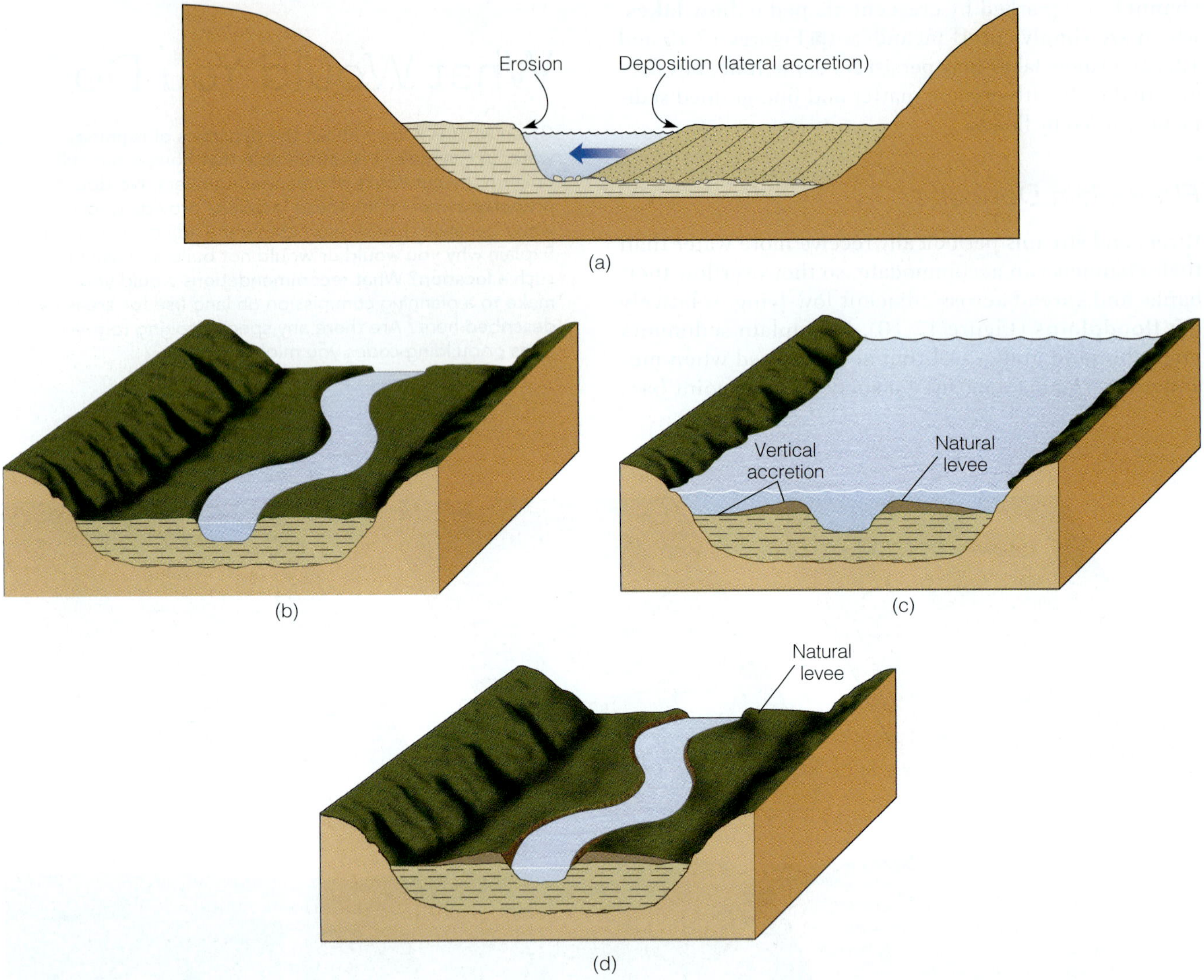

Figure 12.13

(a) Floodplain deposits form by lateral accretion of point bars. (b–d) Three stages in the origin of floodplain deposits by vertical accretion.

as they migrated laterally (■ Figure 12.13a). More commonly, however, fine-grained sediments, mostly mud, are dominant on floodplains. During a flood, a stream overtops its banks and water pours onto the floodplain, but as it does so its velocity and depth rapidly decrease. As a result, ridges of sandy alluvium known as **natural levees** are deposited along the channel margins, and mud is carried beyond the natural levees into the floodplain where it settles from suspension (Figure 12.13b–d).

Deltas

Where a river or stream flows into a standing body of water, such as a lake or the ocean, its flow velocity rapidly diminishes and any sediment in transport is deposited. Under some circumstances, this deposition creates a **delta,** an alluvial deposit that causes the shoreline to build outward into the lake or sea, a process called *progradation.* The simplest prograding deltas have a characteristic vertical sequence of *bottomset beds* overlain successively by *foreset beds* and *topset beds* (■ Figure 12.14). This vertical sequence develops when a river or stream enters another body of water where the finest sediment (silt and clay) is carried some distance out into the lake or sea; there it settles to form bottomset beds. Nearer shore, foreset beds are deposited as gently inclined layers, and topset beds, consisting of the coarsest sediments, are deposited in a network of *distributary channels* traversing the top of the delta (Figure 12.14).

■ **Figure 12.14**

(a) Internal structure of the simplest type of prograding delta. (b) A small delta, measuring about 20 m across.

Many small deltas in lakes have the three-part sequence described above, but deltas deposited along seacoasts are much larger, far more complex, and considerably more important as potential areas of natural resources. In fact, depending on the relative importance of stream (or river), wave, and tide processes, geologists identify three main types of marine deltas (■ Figure 12.15). *Stream-dominated deltas* have long fingerlike sand bodies, each deposited in a distributary channel that progrades far seaward. The Mississippi delta is a good example. In contrast, the Nile delta in Egypt is *wave-dominated.* It also has distributary channels, but the seaward margin of the delta consists of islands reworked by waves, and the entire margin of the delta progrades. *Tide-dominated deltas* are continuously modified into tidal sand bodies that parallel the direction of tidal flow.

Alluvial Fans

Lobate deposits of alluvium on land known as **alluvial fans** form best on lowlands with adjacent highlands in arid and semiarid regions where little vegetation exists to stabilize surface materials (■ Figure 12.16). During periodic rainstorms, surface materials are quickly saturated and surface runoff is funneled into a mountain canyon leading to adjacent lowlands. In the mountain

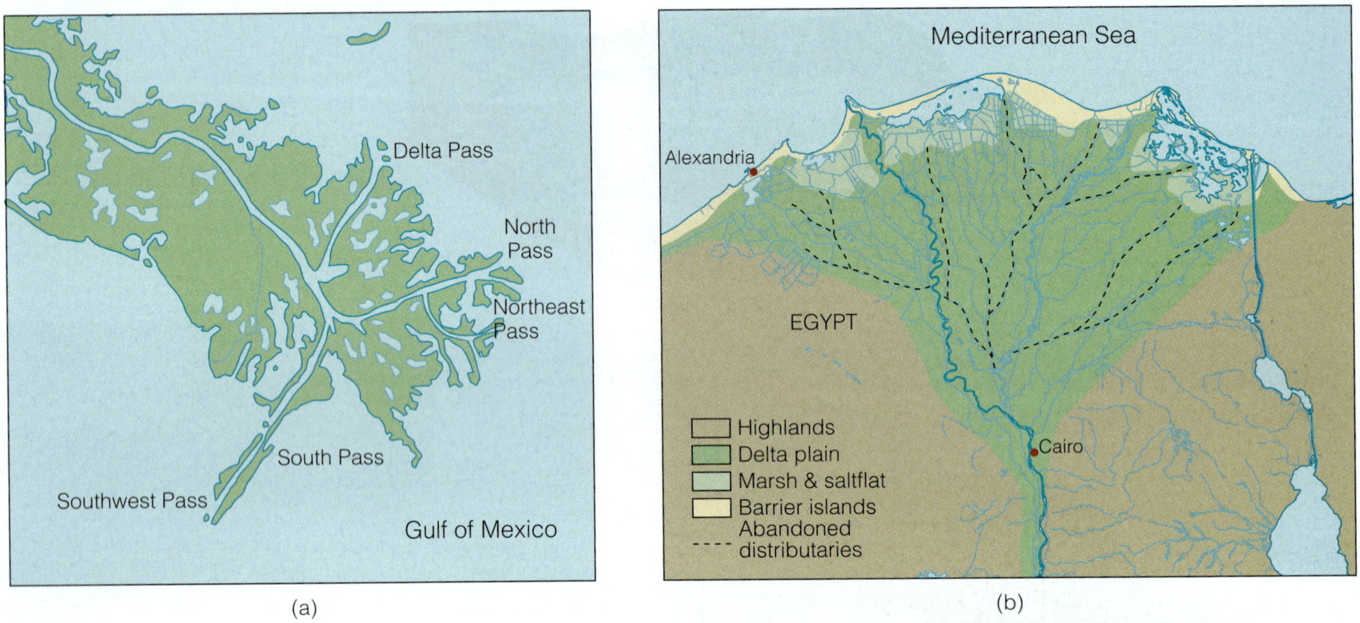

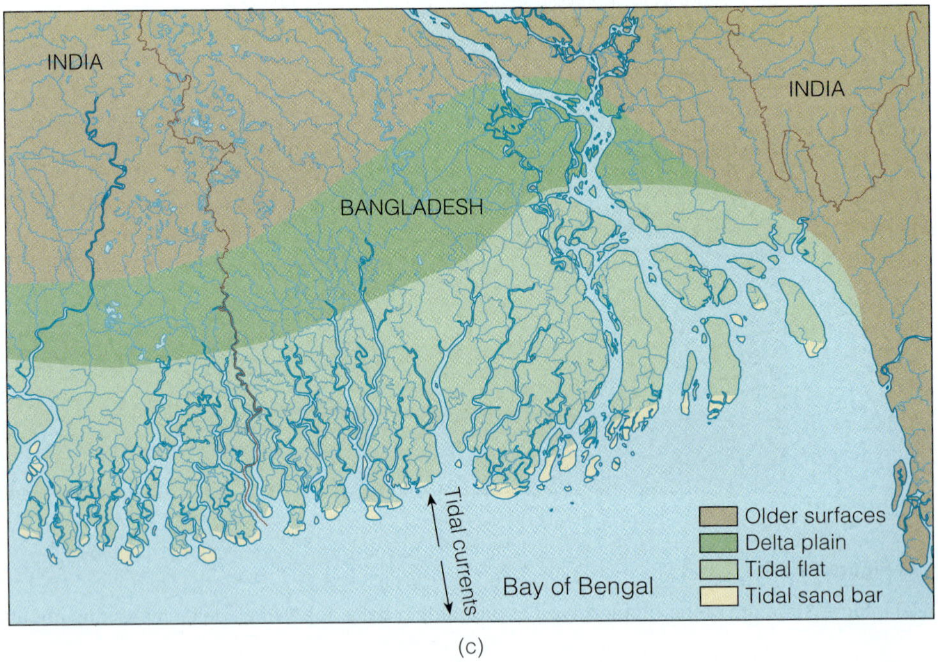

Figure 12.15

(a) The Mississippi River delta of the U.S. Gulf Coast is stream-dominated. (b) The Nile delta of Egypt is wave-dominated. (c) The Ganges-Brahmaputra delta of Bangladesh is tide-dominated.

What Would You Do?

The largest part of many states' mineral revenues comes from sand and gravel, most of which is used in construction. You become aware of a sand and gravel deposit that you can acquire for a small investment. How would the proximity of the deposit to potential markets influence your decision? Assuming the deposit was stream deposited, would it be important to know whether deposition took place in braided or meandering channels? How could you tell one from the other?

canyon, the runoff is confined so it cannot spread laterally, but when it discharges onto the lowlands, it quickly spreads out, its velocity diminishes, and deposition ensues. Repeated episodes of sedimentation result in the accumulation of a fan-shaped body of alluvium.

Deposition by running water in the manner just described is responsible for many alluvial fans. In this case they are composed mostly of sand and gravel, both of which contain a variety of sedimentary structures. In some cases, though, the water flowing through a canyon picks up so much sediment that it becomes a viscous debris flow. Consequently, some alluvial fans consist mostly of debris flow deposits that show little or no lay-

GEOLOGY IN UNEXPECTED PLACES

Floating Burial Chambers

Have you ever thought about how geologic conditions affect where and how cemeteries are constructed? Did you know that architects design cemeteries and must consider the geology of an area when planning a site? Consider the cemetery in Leesville, Louisiana, where stone burial chambers rest on the ground surface because of the area's unusually high water table. Many of the burial chambers are now nearly afloat as the land subsides, resulting in a concurrent rise in sea level, and of course the water table rises as well.

The subsidence-rising water table problem is not confined to Leesville but affects much of southern Louisiana, especially the southern part of the Mississippi River delta (■ Figure 1). Flood-control projects on the Mississippi River prevent most sediment from reaching the seaward margin of the delta, which contributes to loss of land due to wave erosion. Even more important is the natural propensity for delta deposits, especially mud, to compact under their own weight and subside. As a result, many areas on the delta, including New Orleans, are now below sea level and must be protected by systems of levees.

Each year about 65 km^2 of Louisiana's coastline is lost to these processes and, ironically, human efforts to prevent floods have only worsened the situation. And what about global warming, which causes thermal expansion of the ocean's surface waters and melting of glaciers? This will definitely contribute to the region's problems as sea level rises and much of the delta continues to subside. What do you think the future holds for the cemetery at Leesville, Louisiana?

■ **Figure 1**

The Mississippi River delta from space.

JPL/NASA

ering. Of course, the dominant type of deposition can change through time, so a particular fan might have both types of deposits.

CAN FLOODS BE PREDICTED AND CONTROLLED?

When a river or stream receives more water than its channel can handle, it floods, occupying part or all of its floodplain. Indeed, floods are so common that unless they cause considerable property damage or fatalities, they rarely rate more than a passing notice in the news. A major catastrophic flooding in the United States took place in 1993 (see "The Flood of '93" on pages 338 and 339), but since then several other areas in North America and elsewhere have experienced serious flooding. One of the most disastrous was during December 1999 when floods and mudslides killed tens of thousands in Venezuela (see Chapter 11).

People have tried to control floods for thousands of years. Common practices are to construct dams that impound reservoirs and to build levees along stream banks (■ Figure 12.17a, b). Levees raise the banks of a stream, thereby restricting flow during floods. Unfortunately, deposition within the channel raises the streambed, making the

The Flood of '93

Although several floods take place each year that cause damage, injuries, and fatalities, the last truly vast flooding in North America occurred during June and July of 1993. Now called the Flood of '93, it was responsible for 50 deaths and 70,000 were left homeless. Extensive property damage occured in several states, but particularly hard hit were Missouri and Iowa (see chart). Unusual behavior of the jet stream and the convergence of air masses over the Midwest were responsible for numerous thunderstorms that caused the flooding.

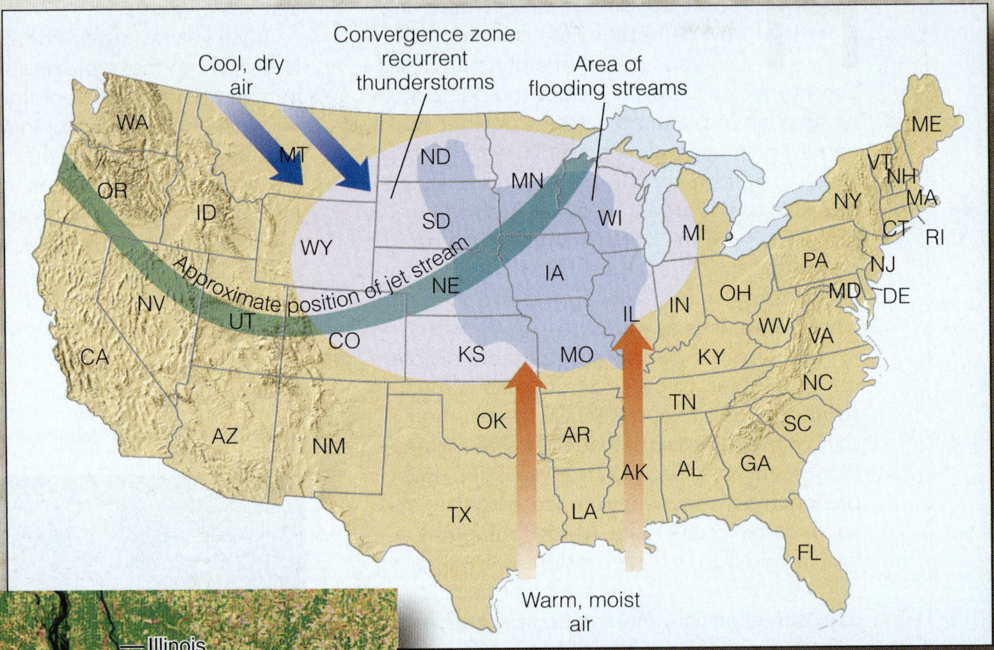

The dominant weather pattern for June and July 1993. The jet stream remained over the Midwest during the summer rather than shifting north over Canada as it usually does. Thunderstorms developed in the convergence zone where warm, moist air and cool, dry air met.

Satellite images of the Mississippi, Missouri, and Illinois rivers near the juncture of three rivers during the drought of 1988 (left), and during the flood of 1993 (below). The "x" marks the site of Portage des Sioux, Missouri (see facing page).

Landsat Imagery courtesy of Earth Observation Satellite Co.

Portage des Sioux, St. Charles County, Missouri, on July 16, 1993. The channel of the Mississippi River is on the far right.

Sue Monroe

Floodwaters in Portage des Sioux covered the 5.5-m-high pedestal of this statue on the bank of the Mississippi River.

Breached levee on the Mississippi River near Davenport, Iowa. This is one of 800 levees that failed or was overtopped during the flood.

Damage caused by the Flood of '93. Compiled by the U. S. Army Corps of Engineers, figures are rounded to the nearest $1000.

State	Residential	Agricultural	Other	Total
Illinois	$176,833,000	$166,502,000	$409,020,000	$752,355,000
Iowa	57,827,000	1,030,030,000	334,835,000	1,422,692,000
Kansas	35,829,000	855,849,000	176,162,000	1,067,840,000
Minnesota	16,940,000	694,041,000	286,540,000	997,521,000
Missouri	405,175,000	540,666,000	1,255,191,000	2,201,032,000
Nebraska	18,584,000	120,521,000	67,899,000	207,004,000
North Dakota	10,138,000	35,039,000	25,806,000	70,983,000
South Dakota	21,919,000	276,218,000	195,408,000	493,545,000
Wisconsin	17,747,000	133,835,000	135,917,000	287,499,000
Total—All States	760,992,000	3,852,701,000	2,886,778,000	7,500,471,000

■ **Figure 12.16**

(a) Alluvial fans form where streams discharge from mountain canyons onto adjacent lowlands. (b) Alluvial fans adjacent to the Panamint Range on the margin of Death Valley, California.

levees useless unless they too are raised. Levees along the banks of the Huang He in China caused the streambed to rise more than 20 m above its surrounding floodplain in 4000 years. When the Huang He breached its levees in 1887, more than 1 million people were killed. Sacramento, California, lying at the junction of two rivers, is among the most flood-prone cities in the United States. Some of the levees that protect the city are 150 years old and in poor condition; the cost of repairing them has risen to as much as $250,000 per 100 m.

Dams and levees alone are insufficient to control large floods, so in many areas floodways are also used. A floodway is a channel constructed to divert part of the excess water in a stream around populated areas or areas of economic importance (Figure 12.17c). Reforestation of cleared land also reduces the potential for flooding because vegetated soil helps prevent runoff by absorbing more water.

When flood-control projects are well planned and constructed, they are functional. What many people fail to realize is that these projects are designed to contain floods of a given size; should larger floods occur, rivers spill onto floodplains anyway. Furthermore, dams occasionally collapse, and reservoirs eventually fill with sediment unless dredged. In short, flood-control projects not only are initially expensive but also require constant, costly maintenance.

(a)

(b)

(c)

■ **Figure 12.17**

Flood control. (a) The 235-m-high Oroville Dam on the Feather River in California is the highest in the United States. It helps control flooding, supplies water for irrigation, and is a popular recreation area. (b) This levee, an artifical embankment along a waterway, helps protect nearby areas from floods. (c) In addition to dams and levees for flood control, floodways carry excess water from a river (not visible) around communities.

DRAINAGE BASINS AND DRAINAGE PATTERNS

Thousands of waterways, which are parts of larger drainage systems, flow directly or indirectly into the oceans. The only exceptions are some rivers and streams that flow into desert basins surrounded by higher areas. But even these are parts of larger systems consisting of a main channel with all its tributaries—that is, streams that contribute water to another stream. The Mississippi River and its tributaries such as the Ohio, Missouri, Arkansas, and Red Rivers and thousands of smaller ones, or any other drainage system for that matter, carry runoff from an area known as a **drainage basin.** A topographically high area called a **divide** separates a drainage basin from adjoining ones (■ Figure 12.18). The continental divide along the crest of the Rocky Mountains in North America, for instance, separates drainage in opposite directions; drainage to the west goes to the Pacific, whereas drainage to the east eventually reaches the Gulf of Mexico.

Various arrangements of channels within an area are classified as types of **drainage patterns.** *Dendritic drainage,* consisting of a network of channels resembling tree branching, is the most common (■ Figure 12.19a). It develops on gently sloping surfaces composed of materials that respond more or less homogeneously to erosion, such as areas underlain by nearly horizontal sedimentary rocks.

In dendritic drainage, tributaries join larger channels at various angles, but *rectangular drainage* is characterized by right-angle bends and tributaries joining

(a)

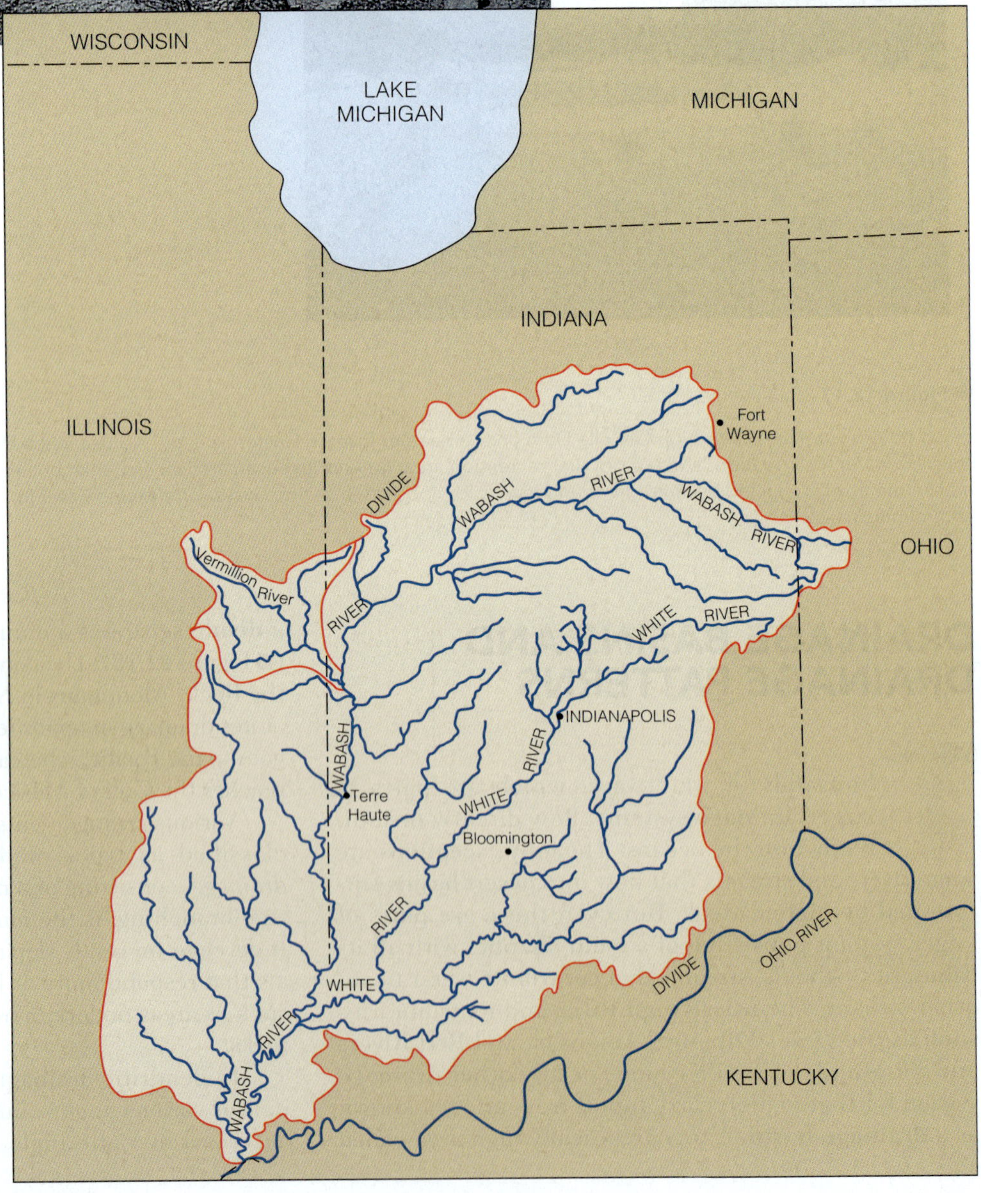

(b)

Figure 12.18

(a) Small drainage basins separated from one another by divides (dashed lines), which are along the crests of the ridges between channels, (solid lines). (b) The drainage basin of the Wabash River, one of the tributaries of the Ohio River. All tributary streams within the drainage basin, such as the Vermillion River, have their own smaller drainage basins. Divides are shown by red lines.

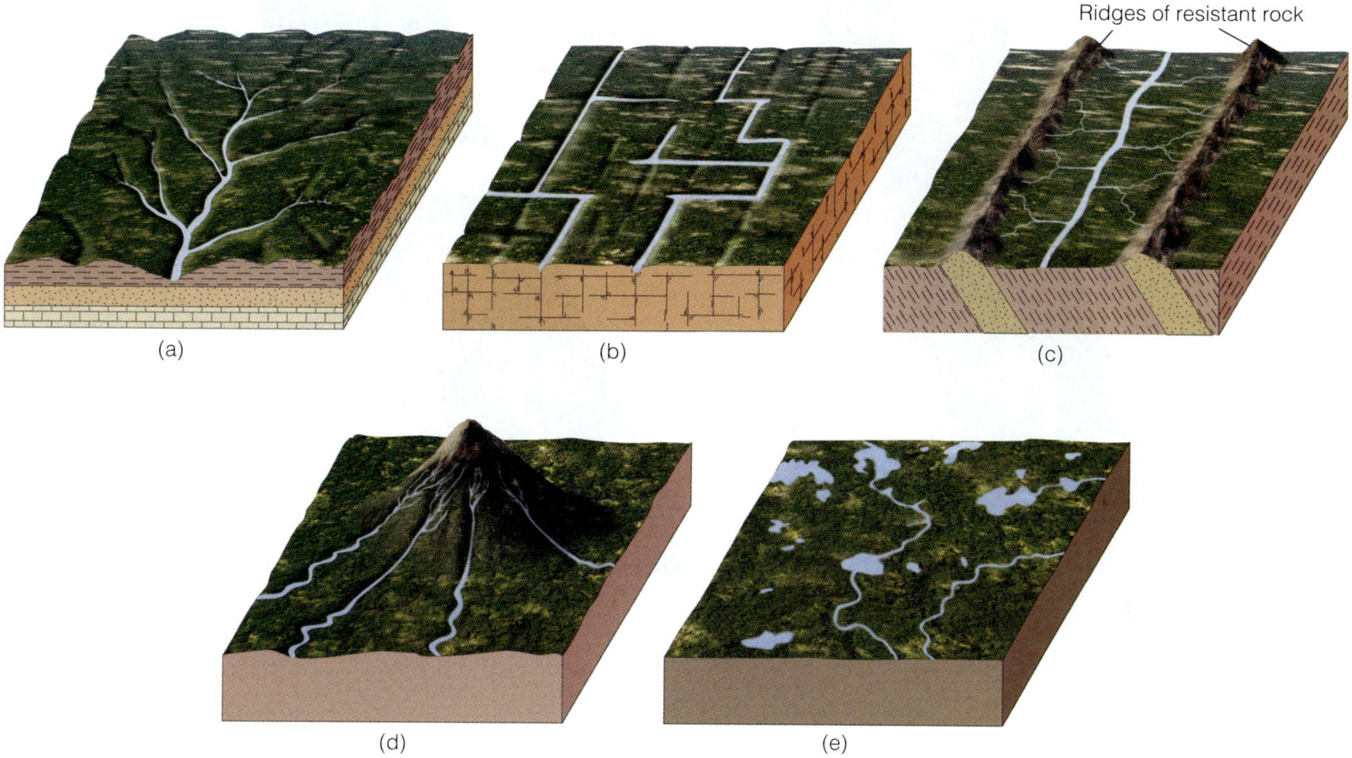

Figure 12.19

Examples of drainage patterns: (a) dendritic drainage, (b) rectangular drainage, (c) trellis drainage, (d) radial drainage, and (e) deranged drainage.

larger channels at right angles (Figure 12.19b). Such regularity in channels is strongly controlled by geologic structures, particularly regional joint systems that intersect at right angles.

Trellis drainage consisting of a network of nearly parallel main streams with tributaries joining them at right angles is common in some parts of the eastern United States. In Virginia and Pennsylvania, erosion of folded sedimentary rocks developed a landscape of alternating ridges on resistant rocks and valleys underlain by easily eroded rocks. Main waterways follow the valleys, and short tributaries flowing from the nearby ridges join the main channels at nearly right angles (Figure 12.19c).

In *radial drainage,* streams flow outward in all directions from a central high point, such as a large volcano (Figure 12.19d). Many of the volcanoes in the Cascade Range of western North America have radial drainage patterns.

In all the types of drainage mentioned so far, some kind of pattern is easily recognized. *Deranged drainage,* in contrast, is characterized by irregularity, with streams flowing into and out of swamps and lakes, streams with only a few short tributaries, and vast swampy areas between channels (Figure 12.19e). This kind of drainage developed recently and has not yet formed a fully organized drainage system. In parts of Minnesota, Wisconsin, and Michigan, where glaciers obliterated the previous drainage, only 10,000 years have elapsed since the glaciers melted. As a result, drainage systems have not fully developed and large areas remain undrained.

The Significance of Base Level

Just how deeply can a stream or river erode? The Grand Canyon in Arizona is more than 1.6 km deep, but the bottom of the canyon is still far above sea level. Obviously, channels must maintain some gradient, so they are restricted by **base level,** the lowest limit to which running water can erode. Sea level is called *ultimate base level,* and theoretically a channel could erode deeply enough so that its gradient rose ever so gently inland from the sea (■ Figure 12.20). Ultimate base level applies to an entire stream or river system, but channels may also have *local* or *temporary base levels.* For instance, a local base level may be a lake or another stream, or where a stream or river flows across particularly resistant rocks and a waterfall develops (Figure 12.20 and the chapter opening photo).

Ultimate base level is sea level, but suppose sea level dropped or rose with respect to the land, or suppose the land rose or subsided? In these cases, base level would change and bring about changes in stream and river systems. For example, during the Pleistocene Epoch (Ice Age), sea level was about 130 m lower than it is now, and

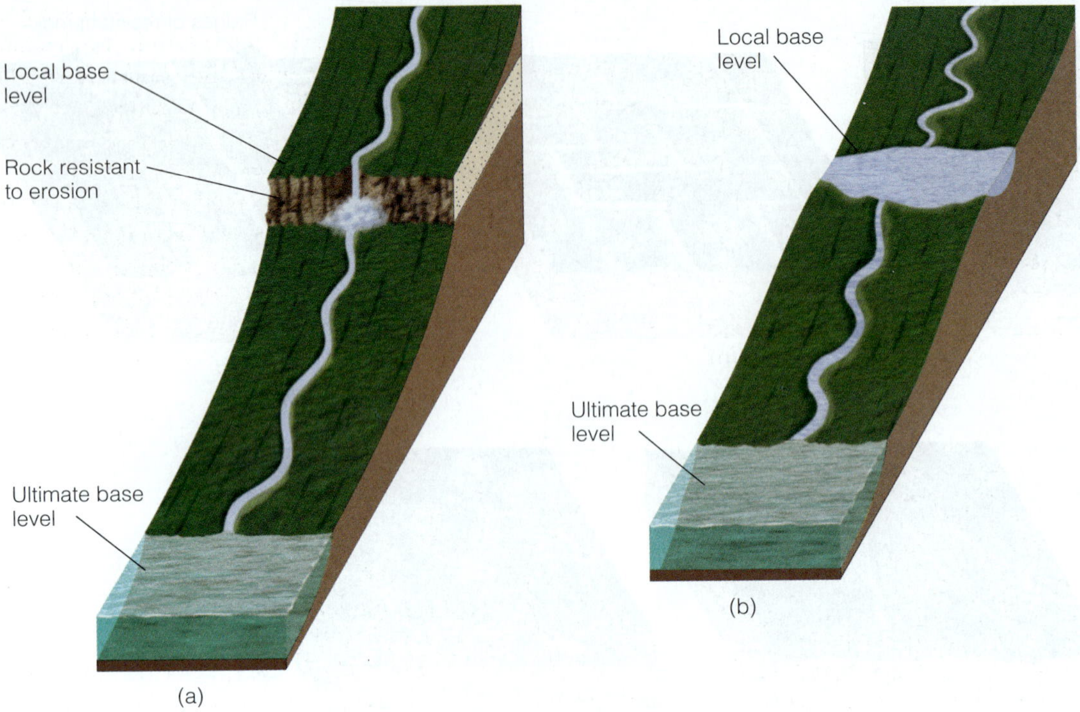

■ **Figure 12.20**

(a) Sea level is ultimate base level, but a resistant rock layer over which a waterfall plunges forms a local base level. (b) Local base level where a stream flows into a lake.

streams adjusted by eroding deeper valleys (they had steeper gradients) and extending well out onto the continental shelves. Rising sea level at the end of the Ice Age accounted for rising base level, decreased stream gradients, and deposition within channels.

Natural changes, such as fluctuations in sea level during the Pleistocene, alter the dynamics of rivers and streams, but so does human intervention. Geologists and engineers are well aware that building a dam to impound a reservoir creates a local base level (■ Figure 12.21a). A stream entering a reservoir slows down and deposits sediment, so unless dredged, reservoirs eventually fill with sediment. In addition, the water discharged at a dam is largely sediment-free but still possesses energy to carry a sediment load. As a result it is not uncommon for streams to erode vigorously downstream from a dam to acquire a sediment load.

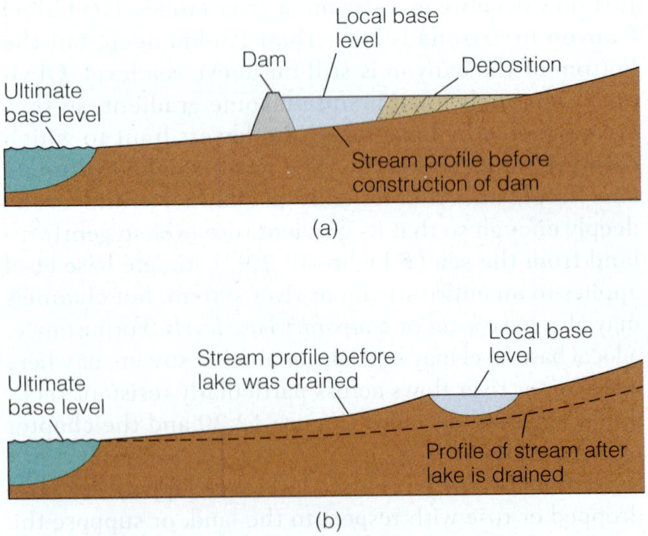

■ **Figure 12.21**

(a) Constructing a dam and impounding a reservoir create a local base level. A stream deposits much of its sediment load where it flows into a reservoir. (b) A stream adjusts to a lower base level when a lake is drained.

Draining a lake may seem like a small change and well worth the time and expense to expose dry land for agriculture or commercial development. But draining a lake eliminates a local base level, and a stream that originally flowed into the lake responds by rapidly eroding a deeper valley as it adjusts to a new base level (Figure 12.21b).

What Is a Graded Stream?

The *longitudinal profile* of any waterway shows the elevations of a channel along its length as viewed in cross section (■ Figure 12.22). For some rivers and streams, the longitudinal profile is smooth, but others show irregularities such as lakes and waterfalls, all of which are local base levels. Over time these irregularities tend to be eliminated because deposition takes place where the gradient is insufficient to maintain sediment transport, and erosion decreases the gradient where it is steep. So, given enough time, rivers and streams develop a smooth, concave longitudinal profile of equilibrium, meaning that all parts of the system dynamically adjust to one another.

A **graded stream** is one with an equilibrium profile in which a delicate balance exists among gradient, dis-

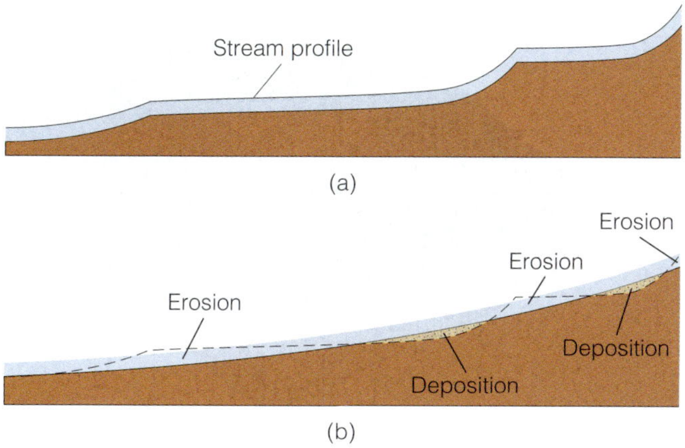

■ **Figure 12.22**

(a) An ungraded stream had irregularities in its longitudinal profile. (b) Erosion and deposition along the course of a stream eliminate irregularities and cause it to develop the smooth, concave profile typical of a graded stream.

charge, flow velocity, channel shape, and sediment load so that neither significant erosion nor deposition takes place within its channel. Such a delicate balance is rarely attained, so the concept of a graded stream is an ideal. Nevertheless, the graded condition is closely approached in many streams, although only temporarily and not necessarily along their entire lengths.

Even though the concept of a graded stream is an ideal, we can anticipate the response of a graded stream to changes that alter its equilibrium. For instance, a change in base level would cause a stream to adjust as previously discussed. Increased rainfall in a stream's drainage basin would result in greater discharge and flow velocity. In short, the stream would now possess greater energy—energy that must be dissipated within the stream system by, for example, a change from a semicircular to a broad, shallow channel that would dissipate more energy by friction. On the other hand, the stream may respond by eroding a deeper valley and effectively reduce its gradient until it is once again graded.

Vegetation inhibits erosion by stabilizing soil and other loose surface materials. So a decrease in vegetation in a drainage basin might lead to higher erosion rates, causing more sediment to be washed into a stream than it can effectively carry. Accordingly, the stream may respond by deposition within its channel, which increases its gradient until it is sufficiently steep to transport the greater sediment load.

HOW DO VALLEYS FORM AND EVOLVE?

Low areas on land known as **valleys** are bounded by higher land, and most of them have a river or stream running their length, with tributaries draining the nearby high areas. Valleys are common landforms, and with few exceptions they form and evolve in response to erosion by running water, although other processes, especially mass wasting, contribute. The shapes and sizes of valleys vary from small, steep-sided *gullies* to those that are broad with gently sloping valley walls (■ Figure 12.23). Steep-walled, deep valleys of vast size are *canyons*, and particularly narrow and deep ones are *gorges*.

A valley might start to erode where runoff has sufficient energy to dislodge surface materials and excavate a small rill. Once formed, a rill collects more runoff and becomes deeper and wider and continues to do so until a full-fledged valley develops. Processes related to running water that contribute to valley formation include

■ **Figure 12.23**

Valleys of various sizes and shapes form mostly by running water and mass wasting, but they may be modified by other processes such as glaciers (see Chapter 14). (a) Many valleys have gently sloping walls much like this one. The river at the bottom of this valley is not visible. (b) The valley of the Colorado River near Moab, Utah, has steep to nearly vertical walls.

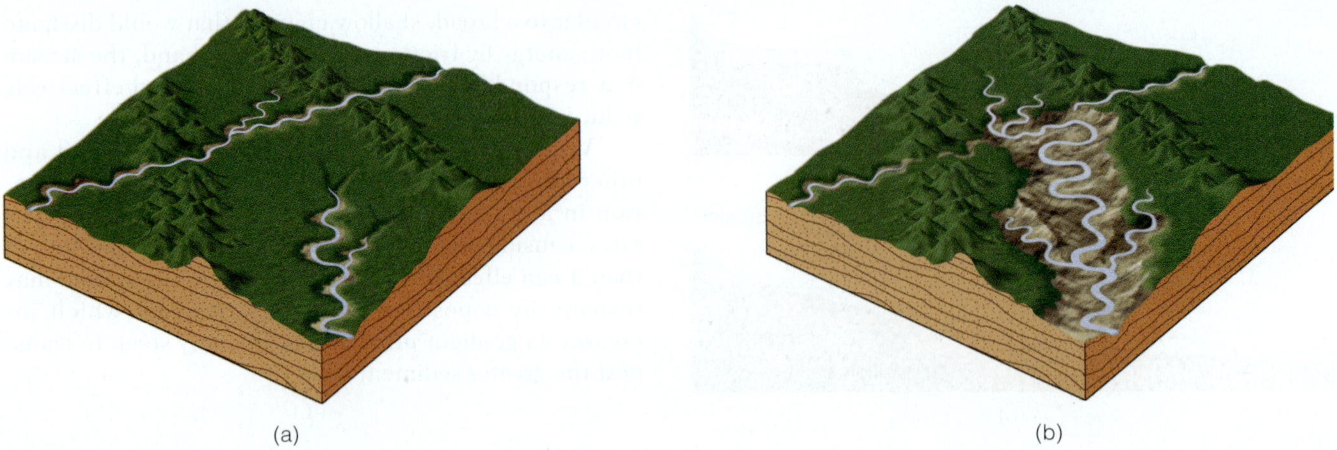

Figure 12.24

Two stages in the evolution of a valley. (a) The stream widens its valley by lateral erosion and mass wasting while simultaneously extending its valley by headward erosion. (b) As the larger stream continues to erode headward, stream piracy takes place when it captures some of the drainage of the smaller stream. Notice also that the larger valley is wider in (b) than it was in (a).

downcutting, lateral erosion, headward erosion, and sheet wash. A variety of mass wasting processes are also important.

Downcutting takes place when a river or stream has more energy than it needs to transport sediment, so some of its excess energy is used to deepen its valley. If downcutting were the only process operating, valleys would be narrow and steep sided. In most cases, though, the valley walls are undercut by stream action, a process called *lateral erosion*, creating unstable slopes that may fail by one or more mass wasting processes. Furthermore, erosion by sheet wash and erosion by tributary streams carry materials from the valley walls into the main stream in the valley.

Valleys not only become deeper and wider but also become longer by *headward erosion*, a phenomenon involving erosion by entering runoff at the upstream end of a valley (Figure 12.24). Continued headward erosion commonly results in *stream piracy*, the breaching of a drainage divide and diversion of part of the drainage of another stream (Figure 12.24). Once stream piracy takes place, both drainage systems must adjust to these new conditions; one system now has greater discharge and the potential to do more erosion and sediment transport, whereas the other is diminished in its ability to accomplish these tasks.

According to one concept, stream erosion of an area uplifted above sea level yields a distinctive series of landscapes. When erosion begins, streams erode downward; their valleys are deep, narrow, and V-shaped, and their profiles have a number of irregularities (Figure 12.25a).

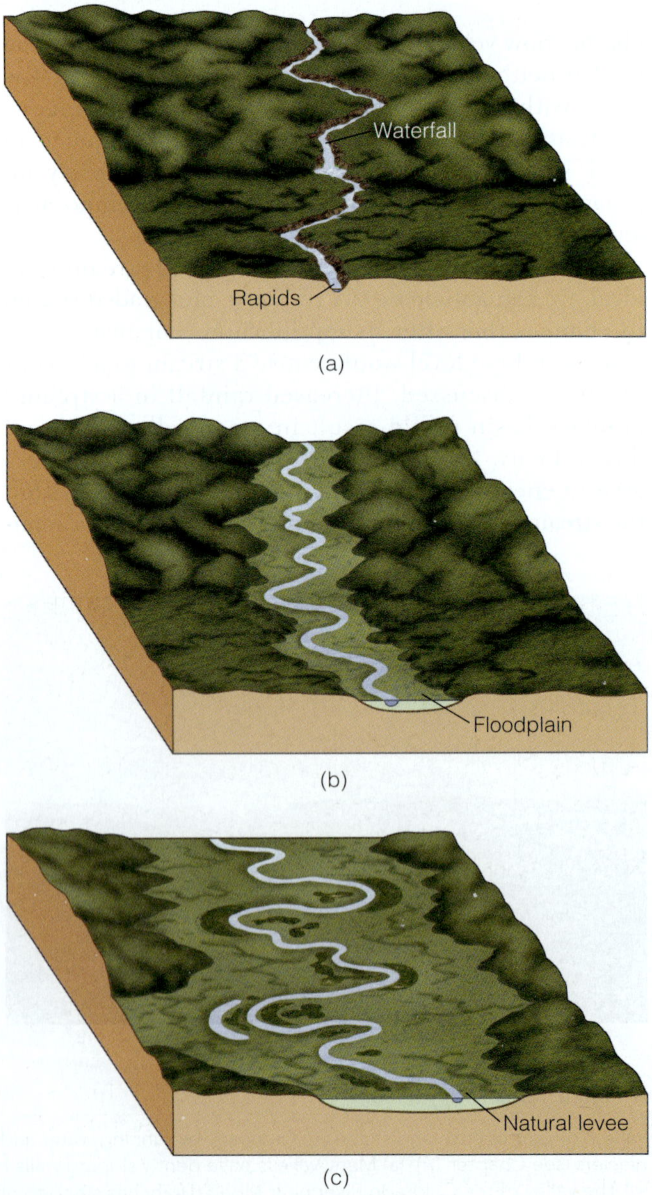

Figure 12.25

Idealized stages in the development of a stream and its associated landforms. According to this idea, an uplifted area begins eroding as in (a), and with time a landscape evolves as illustrated in (b) and finally in (c).

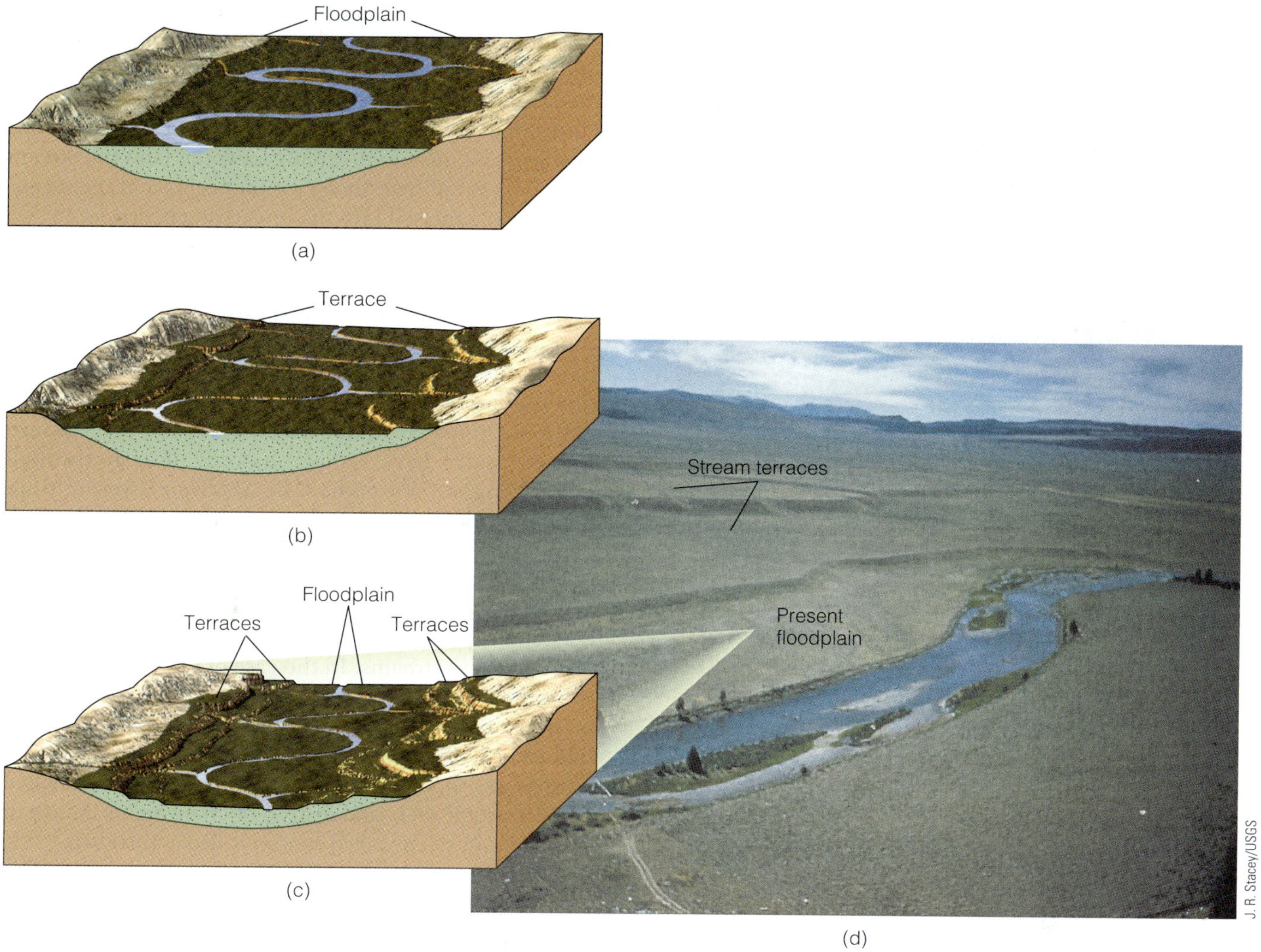

■ **Figure 12.26**

Origin of stream terraces. (a) A stream has a broad floodplain. (b) The stream erodes downward and establishes a new floodplain at a lower level. Remnants of its old, higher floodplain are stream terraces. (c) Another level of stream terraces forms as the stream erodes downward again. (d) Stream terraces along the Madison River in Montana.

As streams cease eroding downward, they start eroding laterally, thereby establishing a meandering pattern and a broad floodplain (Figure 12.25b). Finally, with continued erosion, a vast, rather featureless plain develops (Figure 12.25c).

Many streams do indeed show the features typical of these stages. For instance, the Colorado River flows through the Grand Canyon and closely matches the features in the initial stage shown in Figure 12.25a. Streams in many areas approximate the second stage of development, and certainly the lower Mississippi closely resembles the last stage. Nevertheless, the idea of a sequential development of stream-eroded landscapes has been largely abandoned because there is no reason to think that streams necessarily follow this idealized progression. Indeed, a stream on a gently sloping surface near sea level could develop features of the last stage very early in its history. In addition, as long as the rate of uplift exceeds the rate of downcutting, a stream will continue to erode downward and be confined to a narrow canyon.

Stream Terraces

Adjacent to many channels are erosional remnants of floodplains that formed when the streams were flowing at a higher level. These **stream terraces** consist of a fairly flat upper surface and a steep slope descending to the level of the lower, present-day floodplain (■ Figure 12.26). In some cases, a stream has several steplike surfaces above its present-day floodplain, indicating that stream terraces formed several times.

Although all stream terraces result from erosion, they are preceded by an episode of floodplain formation and sediment deposition. Subsequent erosion causes the stream to cut downward until it is once again graded (Figure 12.26). Then it begins to erode laterally and establishes a new floodplain at a lower level. Several such episodes account for the multiple terrace levels adjacent to some channels.

Renewed erosion and the formation of stream terraces are usually attributed to a change in base level. Either

Figure 12.27

The Colorado River at Dead Horse State Park in Utah is incised to a depth of 600 m.

uplift of the land a stream flows over or lowering of sea level yields a steeper gradient and increased flow velocity, thus initiating an episode of downcutting. When the stream reaches a level at which it is once again graded, downcutting ceases. Although changes in base level no doubt account for many stream terraces, greater runoff in a stream's drainage basin can also result in the formation of terraces.

Incised Meanders

Some streams are restricted to deep, meandering canyons cut into bedrock, where they form features called **incised meanders.** For example, the Colorado River in Utah occupies a meandering canyon more than 600 m deep (■ Figure 12.27). Streams restricted by rock walls usually cannot erode laterally; thus, they lack a floodplain and occupy the entire width of the canyon floor.

It is not difficult to understand how a stream can cut downward into rock, but how a stream forms a meandering pattern in bedrock is another matter. Because lateral erosion is inhibited once downcutting begins, one must infer that the meandering course was established when the stream flowed across an area covered by alluvium. For example, suppose that a stream near base level has established a meandering pattern. If the land the stream flows over is uplifted, then erosion begins and the meanders become incised into the underlying bedrock.

Superposed Streams

Streams flow downhill in response to gravity, so their courses are determined by preexisting topography. Yet a number of streams seem, at first glance, to have defied this fundamental control. For example, the Delaware, Potomac, and Susquehanna Rivers in the eastern United States have valleys that cut directly through ridges lying in their paths. The Madison River in Montana meanders northward through a broad valley and then enters a narrow canyon cut into bedrock that leads to the next valley where the river resumes meandering.

These are examples of **superposed streams.** To understand superposition, one must know the geologic histories of these streams. In the case of the Madison River, the valleys it now occupies were once filled with sedimentary rocks, so the river flowed on a surface at a higher level (■ Figure 12.28). As the river eroded downward, it was superposed directly upon a preexisting knob of more resistant rock, and instead of changing its course, it cut a narrow, steep-walled canyon called a *water gap.*

Superposition also accounts for the fact that the Delaware, Potomac, and Susquehanna Rivers flow through water gaps. During the Mesozoic Era, the Appalachian Mountain region was eroded to a sediment-covered plain across which numerous streams flowed generally eastward. During the Cenozoic Era, regional uplift commenced, and as a result of the uplift, the streams began to erode downward and were superposed on resistant strata, thus forming water gaps (Figure 12.28).

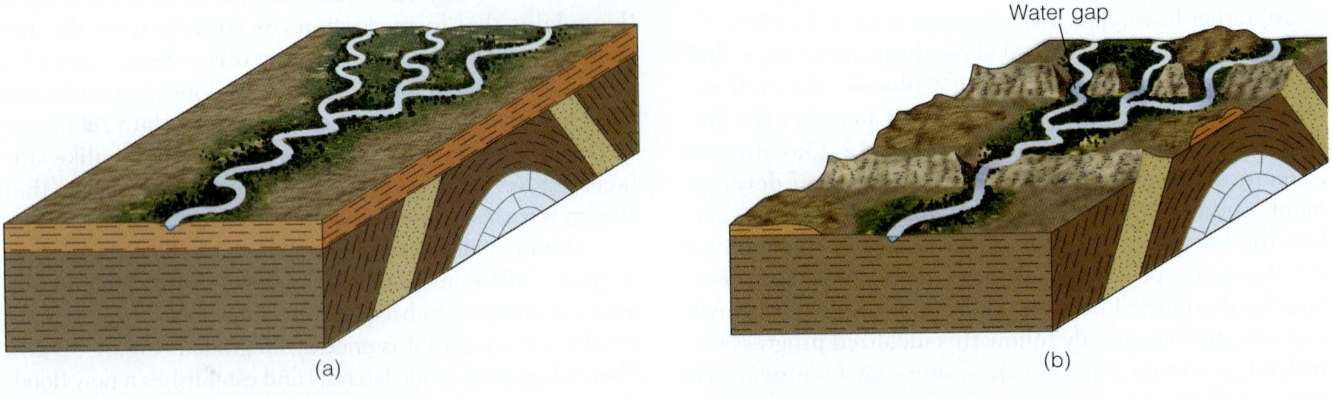

Figure 12.28

The origin of a superposed stream. (a) A stream begins cutting down into horizontal strata. (b) A horizontal layer is removed by erosion, exposing the underlying structure. The stream flows across resistant beds that form the ridges.

12 GEO RECAP

Chapter Summary

- Water continuously evaporates from the oceans, rises as water vapor, condenses, and falls as precipitation, about 20% of which falls on land and eventually returns to the oceans, mostly by surface runoff.
- Running water moves by laminar flow, in which streamlines parallel one another, and by turbulent flow, in which streamlines complexly intertwine. Almost all flow in channels is turbulent.
- Runoff takes place by sheet flow, a thin, more or less continuous sheet of water, and by channel flow, confined to long troughlike stream and river channels.
- The vertical drop in a given distance, or the gradient, for a channel varies from steep in its upper reaches to more gentle in its lower reaches.
- Flow velocity and discharge are related, so that if either changes, the other changes as well.
- Erosion by running water takes place by hydraulic action, abrasion, and dissolution of soluble substances.
- The bed load in channels is made up of sand and gravel, whereas suspended load consists of silt- and clay-sized particles. Running water also transports a dissolved load.
- Braided waterways have a complex of dividing and rejoining channels, and their deposits are mostly sheets of sand and gravel.
- A single sinuous channel is typical of meandering streams that deposit mostly mud with subordinate point bar deposits of sand or more rarely gravel.
- The broad, flat floodplains adjacent to channels are the sites of oxbow lakes, which are simply abandoned meanders.
- An alluvial deposit at a river's mouth is a delta. Some deltas conform to the three-part division of bottomset, foreset, and topset beds, but large marine deltas are much more complex and are characterized as stream/river-, wave-, or tide-dominated.
- Alluvial fans are lobate deposits of sand and gravel on land that form best in semiarid regions. They form mostly by deposition from running water, but debris flows are also important.
- Rivers and streams carry runoff from their drainage basins, which are separated from one another by divides.
- Sea level is ultimate base level, the lowest level to which streams or rivers can erode. Local base levels may be lakes or where streams or rivers flow across resistant rocks.
- Graded streams tend to eliminate irregularities in their channels, so they develop a smooth, concave profile of equilibrium.
- A combination of processes, including downcutting, lateral erosion, sheet wash, mass wasting, and headward erosion, are responsible for the origin and evolution of valleys.
- Stream terraces and incised meanders usually form when a stream or river that was formerly in equilibrium begins a new episode of downcutting.

Important Terms

abrasion (p. 328)
alluvial fan (p. 335)
alluvium (p. 329)
base level (p. 343)
bed load (p. 329)
braided stream (p. 330)
delta (p. 334)
discharge (p. 327)
dissolved load (p. 329)
divide (p. 339)
drainage basin (p. 339)
drainage pattern (p. 339)
floodplain (p. 333)
graded stream (p. 344)
gradient (p. 327)
hydraulic action (p. 328)
hydrologic cycle (p. 323)
incised meanders (p. 348)
infiltration capacity (p. 323)
meandering stream (p. 330)
natural levee (p. 334)
oxbow lake (p. 333)
point bar (p. 332)
runoff (p. 323)
stream terrace (p. 347)
superposed stream (p. 348)
suspended load (p. 329)
valley (p. 345)
velocity (p. 327)

Review Questions

1. The discharge of a stream or river is
a._____the quantity of water moving past a specific place in a given amount of time; b._____how fast water moves around the outside of a meander; c._____the distance water flows from its source to the ocean; d._____the amount of water lost to evaporation and infiltration; e._____a measure of its total load of sand, gravel, and dissolved materials.

2. Running water carrying sand and gravel effectively erodes by
a._____solution; b._____piracy; c._____abrasion; d._____deposition; e._____sheetwash.

3. A stream characterized as braided has
a._____a single sinuous channel; b._____a deep, narrow valley; c._____numerous point bar deposits; d._____a delta dominated by wave action; e._____multiple interconnected channels.

4. In which one of the following areas would radial drainage likely develop?
a._____composite volcano; b._____alluvial fan; c._____point bar; d._____oxbow lake; e._____stream terrace.

5. The bed load of a stream is made up of
a._____dissolved materials and organic matter; b._____clay, silt, and runoff; c._____alluvial fans and deltas; d._____sand and gravel; e._____all materials in solution.

6. The lowest level to which a stream or river can erode is known as
a._____base level; b._____incised piracy; c._____dendritic drainage; d._____natural levee; e._____gradient.

7. Erosion by the direct impact of water is
a._____point bar collapse; b._____capacity; c._____hydraulic action; d._____superposition; e._____alluvial disruption.

8. The topographically high area between adjacent drainage basins is a
a._____profile of equilibrium; b._____divide; c._____floodplain; d._____delta; e._____discharge.

9. An erosional remnant of a floodplain higher than a stream's current floodplain is a(n)
a._____lateral accretion deposit; b._____terrace; c._____incised meander; d._____point bar; e._____divide.

10. A narrow, deep valley is known as a(n)
a._____canyon; b._____gully; c._____gorge; d._____alluvium; e._____graded stream.

11. Why is Earth the only planet in the solar system with abundant liquid water?

12. Calculate the *daily discharge* for a river 148 m wide and 2.6 m deep, with a flow velocity of 0.3 m/sec.

13. Describe how a point bar and a natural levee form.

14. How is it possible for a meandering stream to erode laterally and yet maintain a more or less constant channel width? (A diagram would be helpful.)

15. A river 2000 m above sea level flows 1500 km to the ocean. What is its gradient? Do you think your calculated gradient is valid for all segments of this river? Explain.

16. How do alluvial fans and deltas compare and differ?

17. Explain how a stream can lengthen its channel at both its upper and lower ends.
18. About 10.75 km^3 of sediment erodes from the continents annually, and the volume of the continents above sea level is 93,000,000 km^3. Thus the continents should erode to sea level in just a little more than 8,600,000 years. Our erosion rate and volume of continents are reasonably accurate, but something is seriously wrong with the assumption that the continents will be leveled in so short a time. Explain.
19. Explain the concept of a graded stream, and describe conditions that might upset the graded condition.
20. The steeper a stream or river's gradient, the greater the flow velocity. Right? Explain.

World Wide Web Activities

GeologyNow Assess your understanding of this chapter's topics with additional quizzing and comprehensive interactivities at

http://earthscience.brookscole.com/changingearth4e

as well as current and up-to-date weblinks, additional readings, and InfoTrac College Edition exercises.

Groundwater

CHAPTER 13

OUTLINE

GeologyNow *This icon, which appears throughout the book, indicates an opportunity to explore interactive tutorials, animations, or practice problems available on the GeologyNow website at http://earthscience.brookscole.com/changingearth4e.*

OBJECTIVES

At the end of this chapter, you will have learned that:

- Groundwater is one reservoir of the hydrologic cycle and accounts for approximately 22% of the world's supply of freshwater.
- Porosity and permeability are largely responsible for the amount, availability, and movement of groundwater.
- The water table separates the zone of aeration from the underlying zone of saturation and is a subdued replica of the overlying land surface.
- Groundwater moves downward because of the force of gravity.
- In an artesian system, groundwater is confined and builds up high hydrostatic pressure.
- Groundwater is an important agent of both erosion and deposition and is responsible for karst topography and a variety of cave features.
- Modifications of the groundwater system may result in lowering of the water table, saltwater incursion, subsidence, and contamination.
- Hot springs and geysers result when groundwater is heated, typically in regions of recent volcanic activity.
- Geothermal energy is a desirable and relatively nonpolluting alternative form of energy.

. variety of cave deposits such as stalactites, stalagmites, columns, and urtains are present in Mammoth Cave, Kentucky. Source: David Muench/Corbis

Introduction

Within the limestone region of western Kentucky lies the largest cave system in the world. In 1941 approximately 51,000 acres were set aside and designated as Mammoth Cave National Park. In 1981 it became a World Heritage Site.

From ground level, the topography of the area is unimposing with gently rolling hills. Beneath the surface, however, are more than 540 km of interconnected passageways whose spectacular geologic features have been enjoyed by both cave explorers and tourists.

During the War of 1812, approximately 180 metric tons of saltpeter, used in the manufacture of gunpowder, were mined from Mammoth Cave. At the end of the war, the saltpeter market collapsed, and Mammoth Cave was developed as a tourist attraction, easily overshadowing the other caves in the area. During the next 150 years, the discovery of new passageways and caverns helped establish Mammoth Cave as the world's premier cave and the standard against which all others are measured.

The formation of the caves themselves began about 3 million years ago when groundwater began dissolving the region's underlying St. Genevieve Limestone to produce a complex network of openings, passageways, and huge chambers that constitute present-day Mammoth Cave. Flowing through the various caverns is the Echo River, a system of streams that eventually joins the Green River at the surface.

The colorful cave deposits are the primary reason millions of tourists have visited Mammoth Cave over the years. Hanging down from the ceiling and growing up from the floor are spectacular icicle-like structures as well as columns and curtains in a variety of colors (see the chapter opening photo). Moreover, intricate passageways connect various-sized rooms. The cave is also home to more than 200 species of insects and other animals, including about 45 blind species.

In addition to the beautiful caves, caverns, and cave deposits produced by groundwater movement, groundwater also is an important source of freshwater for agriculture, industry, and domestic users. More than 65% of the groundwater used in the United States each year goes for irrigation, with industrial use second, followed by domestic needs. These demands have severely depleted the groundwater supply in many areas and led to such problems as ground subsidence and saltwater contamination. In other areas, pollution from landfills, toxic waste, and agriculture has rendered the groundwater supply unsafe.

As the world's population and industrial development expand, the demand for water, particularly groundwater, will increase. Not only must new groundwater sources be located, but once found, these sources must be protected from pollution and managed properly to ensure that users do not withdraw more water than can be replenished. It is therefore important that people become aware of what a valuable resource groundwater is, so that they can assure future generations of a clean and adequate supply of this water source.

GROUNDWATER AND THE HYDROLOGIC CYCLE

Groundwater, water that fills open spaces in rocks, sediment, and soil beneath the surface, is one reservoir in the hydrologic cycle, accounting for approximately 22% (8.4 million km^3) of the world's supply of freshwater (see Table 12.1). Like all other water in the hydrologic cycle, the ultimate source of groundwater is the oceans, but its more immediate source is the precipitation that infiltrates the ground and seeps down through the voids in soil, sediment, and rocks. Groundwater may also come from water infiltrating from streams, lakes, swamps, artificial recharge ponds, and water-treatment systems.

Regardless of its source, groundwater moving through the tiny openings between soil and sediment particles and the spaces in rocks filters out many impurities such as disease-causing microorganisms and many pollutants. Not all soils and rocks are good filters, though, and sometimes so much undesirable material

Table 13.1

Porosity Values for Different Materials

Material	Percentage Porosity
Unconsolidated sediment	
Soil	55
Gravel	20–40
Sand	25–50
Silt	35–50
Clay	50–70
Rocks	
Sandstone	5–30
Shale	0–10
Solution activity in limestone, dolostone	10–30
Fractured basalt	5–40
Fractured granite	10

Source: U.S. Geological Survey, Water Supply Paper 2220 (1983) and others.

may be present that it contaminates the groundwater. Groundwater movement and its recovery at wells depend on two critical aspects of the materials it moves through: *porosity and permeability.*

HOW DO EARTH MATERIALS ABSORB WATER?

Porosity and permeability are important physical properties of Earth materials and are largely responsible for the amount, availability, and movement of groundwater. Water soaks into the ground because the soil, sediment, and rock have open spaces or pores. **Porosity** is the percentage of a material's total volume that is pore space. Porosity most often consists of the spaces between particles in soil, sediments, and sedimentary rocks, but other types of porosity include cracks, fractures, faults, and vesicles in volcanic rocks (■ Figure 13.1).

Porosity varies among different rock types and is dependent on the size, shape, and arrangement of the material composing the rock (Table 13.1). Most igneous and metamorphic rocks as well as many limestones and dolostones have very low porosity because they consist of tightly interlocking crystals. Their porosity can be increased, however, if they have been fractured or weathered by groundwater. This is particularly true for massive limestone and dolostone, whose fractures can be enlarged by acidic groundwater.

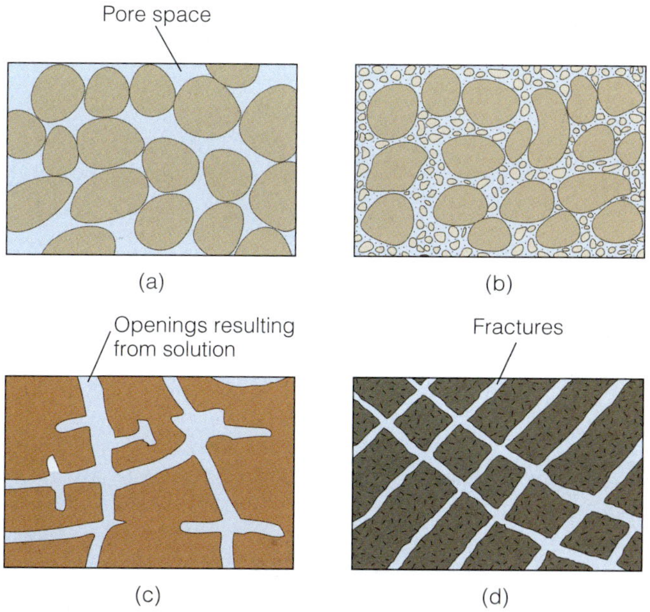

■ **Figure 13.1**

A rock's porosity depends on the size, shape, and arrangement of the material composing the rock. (a) A well-sorted sedimentary rock has high porosity, whereas (b) a poorly sorted one has lower porosity. (c) In soluble rocks such as limestone, porosity can be increased by solution, whereas (d) crystalline metamorphic and igneous rocks can be rendered porous by fracturing.
Source: Modified from *U.S. News & World Report* (8 March 1991): 72–73.

By contrast, detrital sedimentary rocks composed of well-sorted and well-rounded grains can have high porosity because any two grains touch at only a single point, leaving relatively large open spaces between the grains (Figure 13.1a). Poorly sorted sedimentary rocks, in contrast, typically have low porosity because smaller grains fill in the spaces between the larger grains, further reducing porosity (Figure 13.1b). In addition, the amount of cement between grains can decrease porosity.

Porosity determines the amount of groundwater Earth materials can hold, but it does not guarantee that the water can be easily extracted as well. So in addition to being porous, Earth materials must have the capacity to transmit fluids, a property known as **permeability.** Thus, both porosity and permeability play important roles in groundwater movement and recovery. Permeability is dependent not only on porosity but also on the size of the pores or fractures and their interconnections. For example, deposits of silt or clay are typically more porous than sand or gravel, but they have low permeability because pores between the particles are very small, and molecular attraction between the particles and water is great, thereby preventing movement of the water. In contrast, pore spaces between grains in sandstone and conglomerate are much larger, and molecular attraction on the water is therefore low. Chemical and biochemical sedimentary rocks, such as limestone and dolostone, and many igneous and metamorphic rocks that are highly fractured can also be very permeable provided the fractures are interconnected.

The contrasting porosity and permeability of familiar substances are well demonstrated by sand versus clay. Pour some water on sand and it rapidly sinks in, whereas water poured on clay simply remains on the surface. Furthermore, wet sand dries quickly, but once clay absorbs water, it may take days to dry out because of its low permeability. Neither sand nor clay makes a good substance in which to grow crops or gardens, but a mixture of the two plus some organic matter in the form of humus makes an excellent soil for farming and gardening (see Chapter 6).

A permeable layer transporting groundwater is an *aquifer,* from the Latin *aqua,* "water." The most effective aquifers are deposits of well-sorted and well-rounded sand and gravel. Limestones in which fractures and bedding planes have been enlarged by solution are also good aquifers. Shales and many igneous and metamorphic rocks make poor aquifers because they are typically impermeable, unless fractured. Rocks such as these and any other materials that prevent the movement of groundwater are *aquicludes.*

WHAT IS THE WATER TABLE?

Some of the precipitation on land evaporates, and some enters streams and returns to the oceans by surface runoff; the remainder seeps into the ground. As this water moves down from the surface, a small amount adheres to the material it moves through and halts its downward progress. With the exception of this *suspended water*, however, the rest seeps further downward and collects until it fills all the available pore spaces. Thus, two zones are defined by whether their pore spaces contain mostly air, the **zone of aeration**, or mostly water, the underlying **zone of saturation**. The surface that separates these two zones is the **water table** (■ Figure 13.2).

The base of the zone of saturation varies from place to place but usually extends to a depth where an impermeable layer is encountered or to a depth where confining pressure closes all open space. Extending irregularly upward a few centimeters to several meters from the zone of saturation is the *capillary fringe*. Water moves upward in this region because of surface tension, much as water moves upward through a paper towel.

In general, the configuration of the water table is a subdued replica of the overlying land surface; that is, it rises beneath hills and has its lowest elevations beneath valleys. Several factors contribute to the surface configuration of a region's water table, including regional differences in amount of rainfall, permeability, and rate of groundwater movement. During periods of high rainfall, groundwater tends to rise beneath hills because it cannot flow fast enough into adjacent valleys to maintain a level surface. During droughts, the water table falls and tends to flatten out because it is not being replenished. In arid and semiarid regions, the water table is usually quite flat regardless of the overlying land surface.

HOW DOES GROUNDWATER MOVE?

Gravity provides the energy for the downward movement of groundwater. Water entering the ground moves through the zone of aeration to the zone of saturation (■ Figure 13.3). When water reaches the water table, it continues to move through the zone of saturation from areas where the water table is high toward areas where it is lower, such as streams, lakes, or swamps. Only some of the water follows the direct route along the slope of the water table. Most of it takes longer curving paths down and then enters a stream, lake, or swamp from below, because it moves from areas of high pressure toward areas of lower pressure within the saturated zone.

Groundwater velocity varies greatly and depends on many factors. Velocities range from 250 m per day in some extremely permeable material to less than a few centimeters per year in nearly impermeable material. In

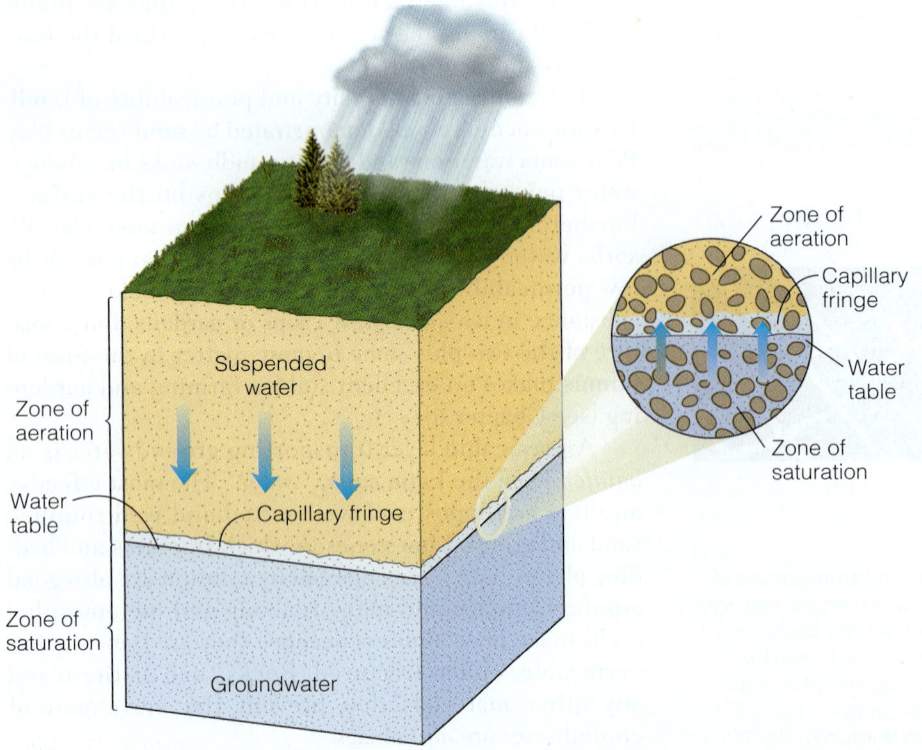

■ **Figure 13.2**

The zone of aeration contains both air and water within its open spaces, whereas all open spaces in the zone of saturation are filled with groundwater. The water table is the surface that separates the zones of aeration and saturation. Within the capillary fringe, water rises by surface tension from the zone of saturation into the zone of aeration.

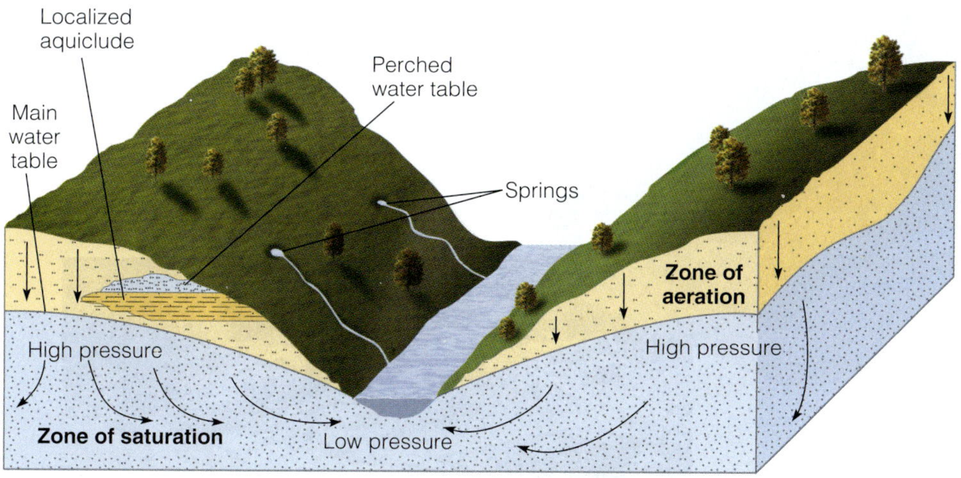

Figure 13.3

Groundwater moves down through the zone of aeration to the zone of saturation. Then some of it moves along the slope of the water table, and the rest moves through the zone of saturation from areas of high pressure toward areas of low pressure. Some water might collect over a local aquiclude, such as a shale layer, thus forming a perched water table.

most ordinary aquifers, the average velocity of groundwater is a few centimeters per day.

Geology⊜Now Log into GeologyNow and select this chapter to work through a **Geology Interactive** activity on "Groundwater Basics" (click Groundwater→Groundwater Basics).

WHAT ARE SPRINGS, WATER WELLS, AND ARTESIAN SYSTEMS?

You can think of the water in the zone of saturation much like a reservoir whose surface rises or falls depending on additions as opposed to natural and artificial withdrawals. *Recharge*—that is, additions to the zone of saturation—comes from rainfall or melting snow, or water might be added artificially at wastewater-treatment plants or recharge ponds constructed for just this purpose. But if groundwater is discharged naturally or withdrawn at wells without sufficient recharge, the water table drops just as a savings account diminishes if withdrawals exceed deposits. Withdrawals from the groundwater system take place where groundwater flows laterally into streams, lakes, or swamps, where it discharges at the surface as *springs,* and where it is withdrawn from the system at water wells.

Springs

Places where groundwater flows or seeps out of the ground as **springs** have always fascinated people. The water flows out of the ground for no apparent reason and from no readily identifiable source. So it is not surprising that springs have long been regarded with superstition and revered for their supposed medicinal value and healing powers. Nevertheless, there is nothing mystical or mysterious about springs.

Although springs can occur under a wide variety of geologic conditions, they all form in basically the same way (■ Figure 13.4a). When percolating water reaches the water table or an impermeable layer, it flows laterally, and if this flow intersects the surface, the water discharges as a spring (Figure 13.4b). The Mammoth Cave area in Kentucky is underlain by fractured limestones whose fractures have been enlarged into caves by solution activity (see the chapter opening photo). In this geologic environment, springs occur where the fractures and caves intersect the ground surface, allowing groundwater to exit onto the surface. Most springs are along valley walls where streams have cut valleys below the regional water table.

Springs can also develop wherever a perched water table intersects the surface (Figure 13.3). A *perched water table* may occur wherever a local aquiclude is present within a larger aquifer, such as a lens of shale within sandstone. As water migrates through the zone of aeration, it is stopped by the local aquiclude, and a localized zone of saturation "perched" above the main water table forms. Water moving laterally along the perched water table may intersect the surface to produce a spring.

Water Wells

Water wells are openings made by digging or drilling down into the zone of saturation. Once the zone of saturation has been penetrated, water percolates into the well, filling it to the level of the water table. A few wells are free flowing (see the next section), but for most the water must be brought to the surface by pumping. In some parts of the world, water is raised to the surface with nothing more than a bucket on a rope or a hand-operated pump. In many parts of the United States and Canada, one can see windmills from times past that used wind power to pump water (■ Figure 13.5a). Most of these are no longer in use, having been replaced by more efficient electric pumps (Figure 13.5b).

When groundwater is pumped from a well, the water table in the area around the well is lowered, forming a **cone of depression** (■ Figure 13.6). A cone of depression

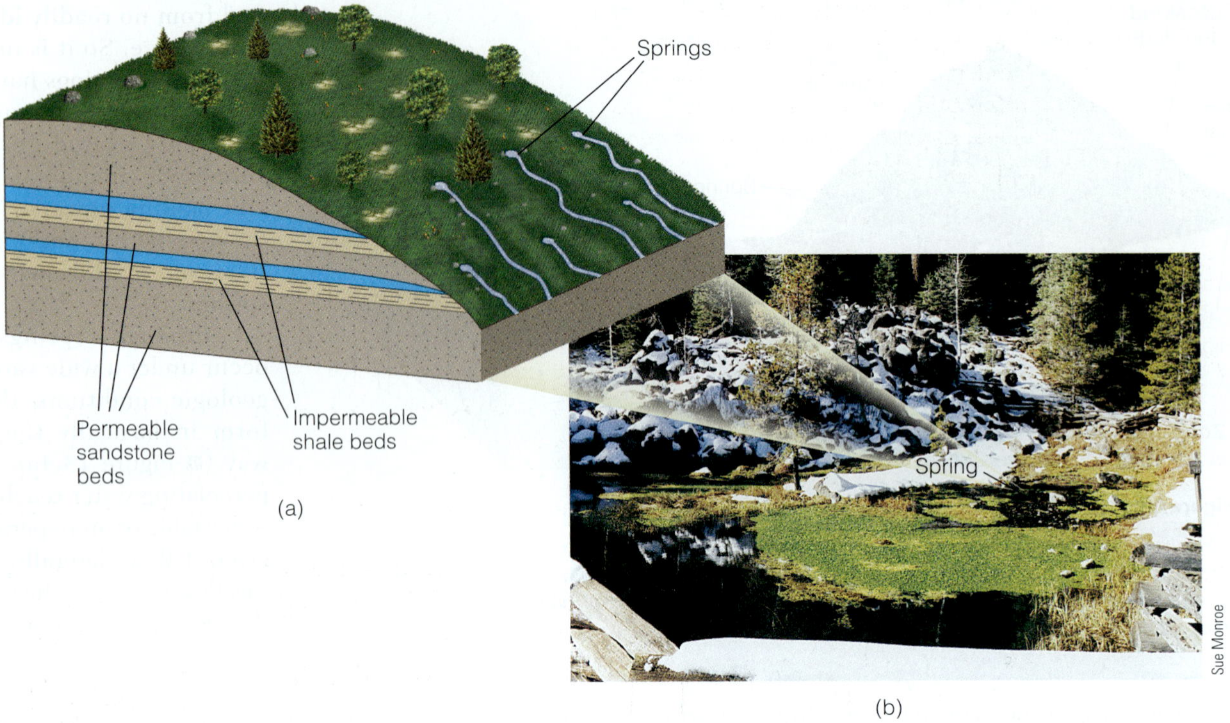

■ **Figure 13.4**

Springs form wherever laterally moving groundwater intersects Earth's surface. (a) Most commonly, springs form when percolating water reaches an impermeable layer and migrates laterally until it seeps out at the surface. (b) This spring issues from rocks at the base of the mountain in the background.

forms because the rate of water withdrawal from the well exceeds the rate of water inflow to the well, thus lowering the water table around the well. This lowering of the water table normally does not pose a problem for the average domestic well, provided the well is drilled deep enough into the zone of saturation. The tremendous amounts of water used by industry and for irrigation, however, may create a large cone of depression that lowers the water table sufficiently to cause shallow wells in the immediate area to go dry (Figure 13.6). This situation is not uncommon and fre-

■ **Figure 13.5**

(a) During the past, windmills like this one were used extensively to pump water to the surface. This windmill still functions and supplies the water for the small cemetery in the foreground. (b) Electric pumps such as this one in a farmer's field have largely replaced windmills.

quently results in lawsuits by the owners of the shallow dry wells. Furthermore, lowering of the regional water table is becoming a serious problem in many areas, particularly in the southwestern United States where rapid growth has placed tremendous demands on the groundwater system. Unrestricted withdrawal of groundwater cannot continue indefinitely, and the rising costs and decreasing supply of groundwater should soon limit the growth of this region of the United States.

Geology Now Log into GeologyNow and select this chapter to work through a **Geology Interactive** activity on "Tapping the Ground" (click Groundwater→Tapping the Ground).

What Would You Do?

Your well periodically goes dry and never produces the amount of water you would like or need. Is your well too shallow, is it drilled into Earth materials with low permeability, or are nearby irrigation pumps causing your problems? What can you do to solve this problem?

Artesian Systems

The word *artesian* comes from the French town and province of Artois (called Artesium during Roman times) near Calais, where the first European artesian well was drilled in A.D. 1126 and is still flowing today. The term **artesian system** can be applied to any system in which groundwater is confined and builds up high hydrostatic (fluid) pressure. Water in such a system is able to rise above the level of the aquifer if a well is drilled through the confining layer, thereby reducing the pressure and forcing the water upward. For an artesian system to develop, three geologic conditions must be present (■ Figure 13.7): (1) The aquifer must be confined above and below by aquicludes to prevent water from escaping; (2) the rock sequence is usually tilted and exposed at the surface, enabling the aquifer to be recharged; and (3) precipitation in the recharge area is sufficient to keep the aquifer filled.

The elevation of the water table in the recharge area and the distance of the well from the recharge area determine the height to which artesian water rises in a well. The surface defined by the water table in the recharge area, called the *artesian-pressure surface,* is indicated by the sloping dashed line in Figure 13.7. If there were no friction in the aquifer, well water from an artesian aquifer would rise exactly to the elevation of the artesian-pressure surface. Friction, however, slightly reduces the pressure of the aquifer water and consequently the level to which artesian water rises. This is why the pressure surface slopes.

An artesian well will flow freely at the ground surface only if the wellhead is at an elevation below the artesian-pressure surface. In this situation, the water flows out of the well because it rises toward the artesian-pressure surface, which is at a higher elevation than the wellhead. In a nonflowing artesian well, the wellhead is above the artesian-pressure surface, and the water will rise in the well only as high as the artesian-pressure surface.

In addition to artesian wells, many artesian springs exist. Such springs form if a fault or fracture intersects the confined aquifer, allowing water to rise above the aquifer. Oases in deserts are commonly artesian springs.

Because the geologic conditions necessary for artesian water can occur in a variety of ways, artesian systems are common in many areas of the world underlain by sedimentary rocks. One of the best-known artesian systems in the United States underlies South Dakota and extends southward to central Texas. The majority of the artesian water from this system is used for irrigation. The aquifer of this artesian system, the Dakota Sandstone, is recharged where it is exposed along the margins of the Black Hills of South Dakota. The hydrostatic

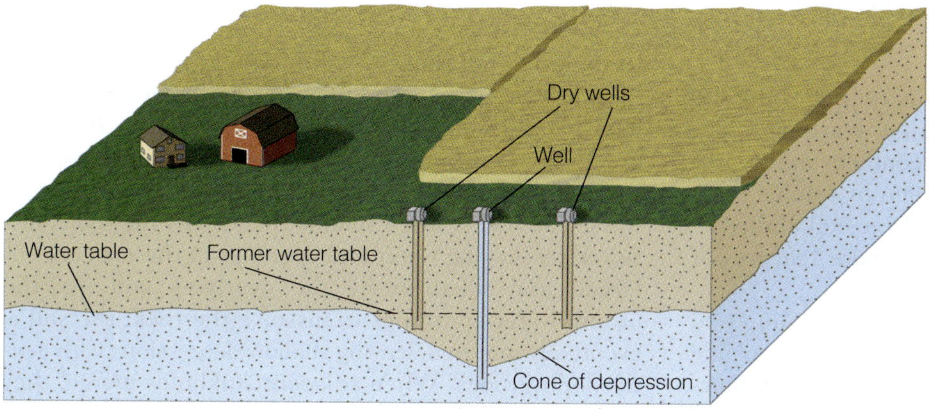

■ **Figure 13.6**

A cone of depression forms whenever water is withdrawn from a well. If water is withdrawn faster than it can be replenished, the cone of depression will grow in depth and circumference, lowering the water table in the area and causing nearby shallow wells to go dry.

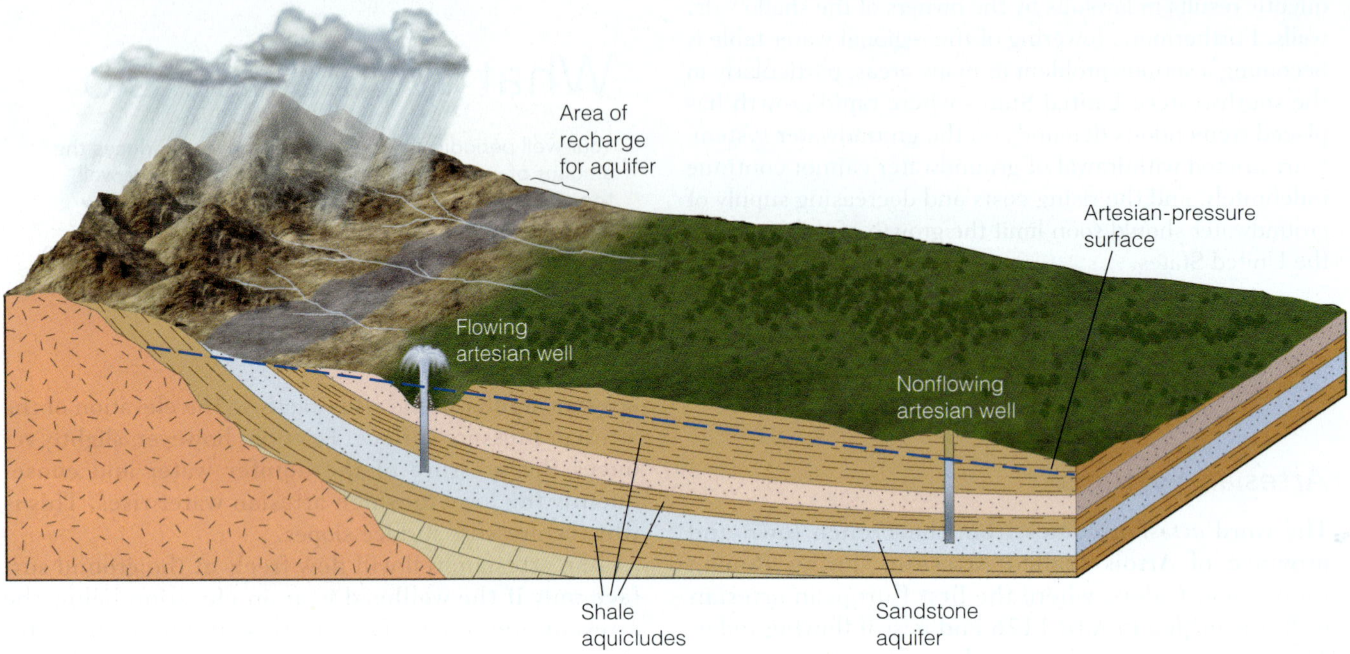

GeologyNow ■ **Active Figure 13.7**

An artesian system must have an aquifer confined above and below by aquicludes, the aquifer must be exposed at the surface, and precipitation in the recharge area must be sufficient to keep the aquifer filled. The elevation of the water table in the recharge area, which is indicated by a sloping dashed line (the artesian-pressure surface), defines the highest level to which well water can rise. If the elevation of a wellhead is below the elevation of the artesian-pressure surface, the well will be free-flowing because the water will rise toward the artesian-pressure surface, which is at a higher elevation than the wellhead. If the elevation of a wellhead is at or above that of the artesian-pressure surface, the well will be nonflowing.

pressure in this system was originally great enough to produce free-flowing wells and to operate waterwheels. The extensive use of water for irrigation over the years has reduced the pressure in many of the wells so that they are no longer free-flowing and the water must be pumped.

GeologyNow Log into GeologyNow and select this chapter to work through a **Geology Interactive** activity on "Potentiometric Surface" (click Groundwater→Potentiometric Surface).

HOW DOES GROUNDWATER ERODE AND DEPOSIT MATERIAL?

When rainwater begins to seep into the ground, it immediately starts to react with the minerals it contacts and weathers them chemically. In areas underlain by soluble rock, groundwater is the principal agent of erosion and is responsible for the formation of many major features of the landscape.

Limestone, a common sedimentary rock composed primarily of the mineral calcite ($CaCO_3$), underlies large areas of Earth's surface (■ Figure 13.8). Although limestone is practically insoluble in pure water, it readily dissolves if a small amount of acid is present. Carbonic acid (H_2CO_3) is a weak acid that forms when carbon dioxide combines with water ($H_2O + CO_2 \rightarrow H_2CO_3$) (see Chapter 6). Because the atmosphere contains a small amount of carbon dioxide (0.03%) and carbon dioxide is also produced in soil by the decay of organic matter, most groundwater is slightly acidic. When groundwater percolates through the various openings in limestone, the slightly acidic water readily reacts with the calcite to dissolve the rock by forming soluble calcium bicarbonate, which is carried away in solution (see Chapter 6).

Sinkholes and Karst Topography

In regions underlain by soluble rock, the ground surface may be pitted with numerous depressions that vary in size and shape. These depressions, called **sinkholes** or merely *sinks,* mark areas with underlying soluble rock (■ Figure 13.9). Most sinkholes form in one of two ways. The first is when soluble rock below the soil is dissolved by seeping water and openings in the rock are enlarged and filled in by the overlying soil. As the groundwater continues to dissolve the rock, the soil is eventually removed, leaving shallow depressions with gently sloping sides. When adjacent sinkholes merge, they

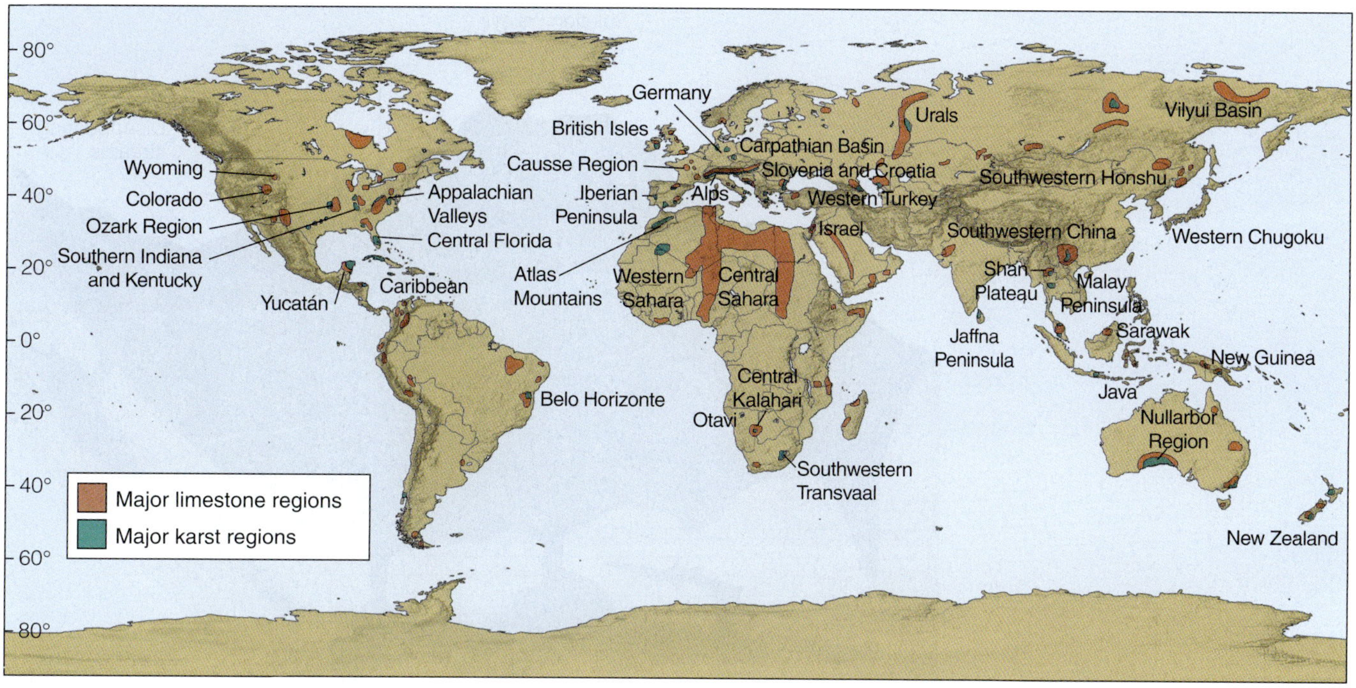

Figure 13.8

The distribution of the major limestone and karst areas of the world.

Frank Kujawa/University of Central Florida. GeoPhoto

(a)

James S. Monroe

(b)

Figure 13.9

(a) This sinkhole formed on May 8 and 9, 1981, in Winter Park, Florida. It formed in previously dissolved limestone following a drop in the water table. The 100-m-wide, 35-m-deep sinkhole destroyed a house, numerous cars, and a municipal swimming pool. (b) A small sinkhole in Montana now occupied by a lake. The water enters the lake from a hot spring so it remains warm year round. In fact, tropical fish have been introduced into the lake and thus live in an otherwise inhospitable climate.

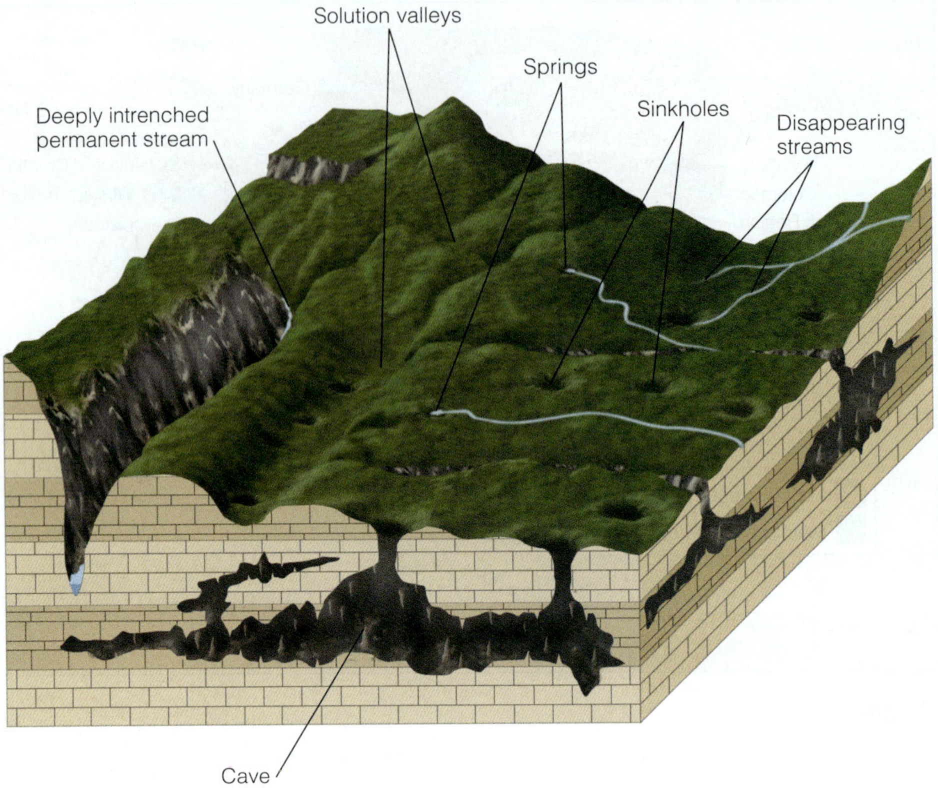

■ **Figure 13.10**

Some of the features of karst topography.

form a network of larger, irregular, closed depressions called *solution valleys.*

Sinkholes also form when a cave's roof collapses, usually producing a steep-sided crater. Sinkholes formed in this way are a serious hazard, particularly in populated areas. In regions prone to sinkhole formation, the depth and extent of underlying cave systems must be mapped before any development to ensure that the underlying rocks are thick enough to support planned structures.

Karst topography, or simply *karst,* develops largely by groundwater erosion in many areas underlain by soluble rocks (■ Figure 13.10). The name *karst* is derived from the plateau region of the border area of Slovenia, Croatia, and northeastern Italy where this type of topography is well developed. In the United States, regions of karst topography include large areas of southwestern Illinois, southern Indiana, Kentucky, Tennessee, northern Missouri, Alabama, and central and northern Florida (Figure 13.8).

Karst topography is characterized by numerous caves, springs, sinkholes, solution valleys, and disappearing streams (Figure 13.10). *Disappearing streams* are so named because they typically flow only a short distance at the surface and then disappear into a sinkhole. The water continues flowing underground through fractures or caves until it surfaces again at a spring or other stream.

Karst topography varies from the spectacular high relief landscapes of China to the subdued and pockmarked landforms of Kentucky (■ Figure 13.11). Common to all karst topography, though, is the presence of thick-bedded, readily soluble rock at the surface or just below the soil, and enough water for solution activity to occur (see "The Burren Area of Ireland" on pages 364 and 365). Karst topography is therefore typically restricted to humid and temperate climates.

Caves and Cave Deposits

Caves are perhaps the most spectacular examples of the combined effects of weathering and erosion by groundwater. As groundwater percolates through carbonate rocks, it dissolves and enlarges fractures and openings to form a complex interconnecting system of crevices, caves, caverns, and underground streams. A **cave** is usually defined as a naturally formed subsurface opening that is generally connected to the surface and is large enough for a person to enter. A *cavern* is a very large cave or a system of interconnected caves.

More than 17,000 caves are known in the United States. Most of them are small, but some are quite large and spectacular. Some of the more famous ones are Mammoth Cave, Kentucky (see the Introduction); Carlsbad Caverns, New Mexico; Lewis and Clark Caverns, Montana; Wind Cave and Jewel Cave, South

Reed Wicander

(a)

John S. Shelton

(b)

■ **Figure 13.11**

(a) The Stone Forest, 125 km southeast of Kunming, China, is a high-relief karst landscape formed by the dissolution of carbonate rocks.
(b) Solution valleys, sinkholes, and sinkhole lakes dominate the subdued karst topography east of Bowling Green, Kentucky.

Dakota; Lehman Cave, Nevada; and Meramec Caverns, Missouri, which Jesse James and his outlaw band often used as a hideout. The United States has many famous caves, but so has Canada, including 536-m-deep Arctomys Cave in Mount Robson Provincial Park, British Columbia, the deepest known cave in North America.

Caves and caverns form as a result of the dissolution of carbonate rocks by weakly acidic groundwater (■ Figure 13.12). Groundwater percolating through the zone of aeration slowly dissolves the carbonate rock and enlarges its fractures and bedding planes. On reaching the water table, the groundwater migrates toward the region's surface streams. As the groundwater moves through the zone of saturation, it continues to dissolve the rock and gradually forms a system of horizontal passageways through which the dissolved rock is carried to the streams. As the surface streams erode deeper valleys, the water table drops in response to the lower elevation of the streams. The water

The Burren Area of Ireland

The Burren region in northwest county Clare, Ireland, covers more than 100 square miles and is one of the finest examples of karst topography in Europe. Although the Burren landscape is frequently referred to as lunarlike because it seems barren and lifeless, it is actually teeming with life and is world-famous for its variety of vegetation. In fact, because of the heat-retention capacity of the massive limestones and the soil trapped in the vertical joints of bare limestone pavement, an extremely diverse community of plants abounds.

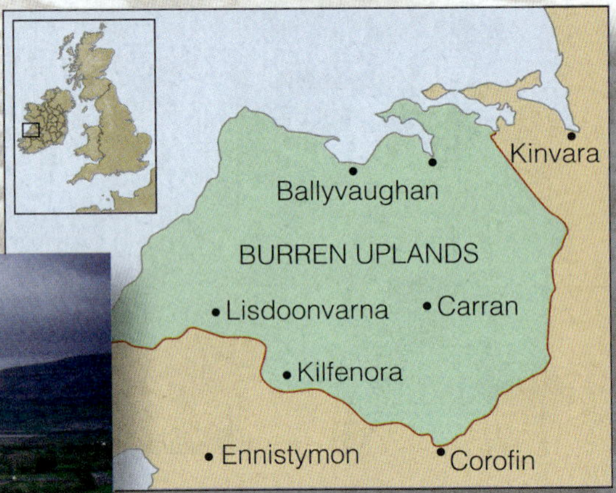

A map of Ireland showing the Burren region of county Clare.

The present landscape is best described as glaciated karst. Like most of Ireland, the Burren was covered by a warm, shallow sea some 340 million years ago. As much as 780 m of interbedded marine limestones and shales were deposited at this time. These marine limestones and shales were then covered by nearly 330 m of sandstones, siltstones, and shales. During the Pleistocene Epoch, glaciers stripped off most of the detrital rocks, thus exposing the underlying limestones to weathering and erosion. The main erosional agents in the Burren are rainwater, which is slightly acidic, and the humic acids produced by localized vegetation. Together they have been the driving force producing the distinctive karst topography we find today.

Bare limestone pavement with a small wedge tomb from the Neolithic period—approximately 6000 years ago—in the background.

Bare limestone pavements typically display a blocky appearance. The network of vertical cracks is the result of weathering of the joint pattern produced in the limestone during uplift of the region.

A closeup shows the characteristic karren weathering pattern produced by solution of the limestone. *Karren* is used to describe the various microsolutional features of limestone pavement.

Reed Wicander

Despite a lack of significant soil cover, a profusion of plants live in the soil trapped in the cracks, joints, and solution cavities of the limestone beds in the Burren.

Perhaps the most famous Irish Neolithic dolmen is the Poulnabrone portal tomb, which dates from about 5800 years ago. The name *Poulnabrone* literally means "the hole of the sorrows." The capstone of the Poulnabrone dolmen rests on two 1.8-m-high portal stones to create a chamber in a low circular cairn, 9 m in diameter.

Reed Wicander

Reed Wicander

A typical small wedge tomb. The Burren is famous for its monuments, characteristic of every period from the Neolithic to the present. The legacy of early settlers is the many Neolithic tombs as well as stone structures and walls.

Excavation of the Poulnabrone dolmen revealed that the tomb contained the remains of as many as 22 people, buried over a period of six centuries, as well as a polished stone axe, two stone disc beads, flint and chert arrowheads and scrapers, and more than 60 shards of pottery.

Reed Wicander

A Burren limestone wall. Burren limestone has been used for centuries in the construction of tombs, forts, houses, and walls. Today it is in demand for limestone-finished houses, walls, and garden features.

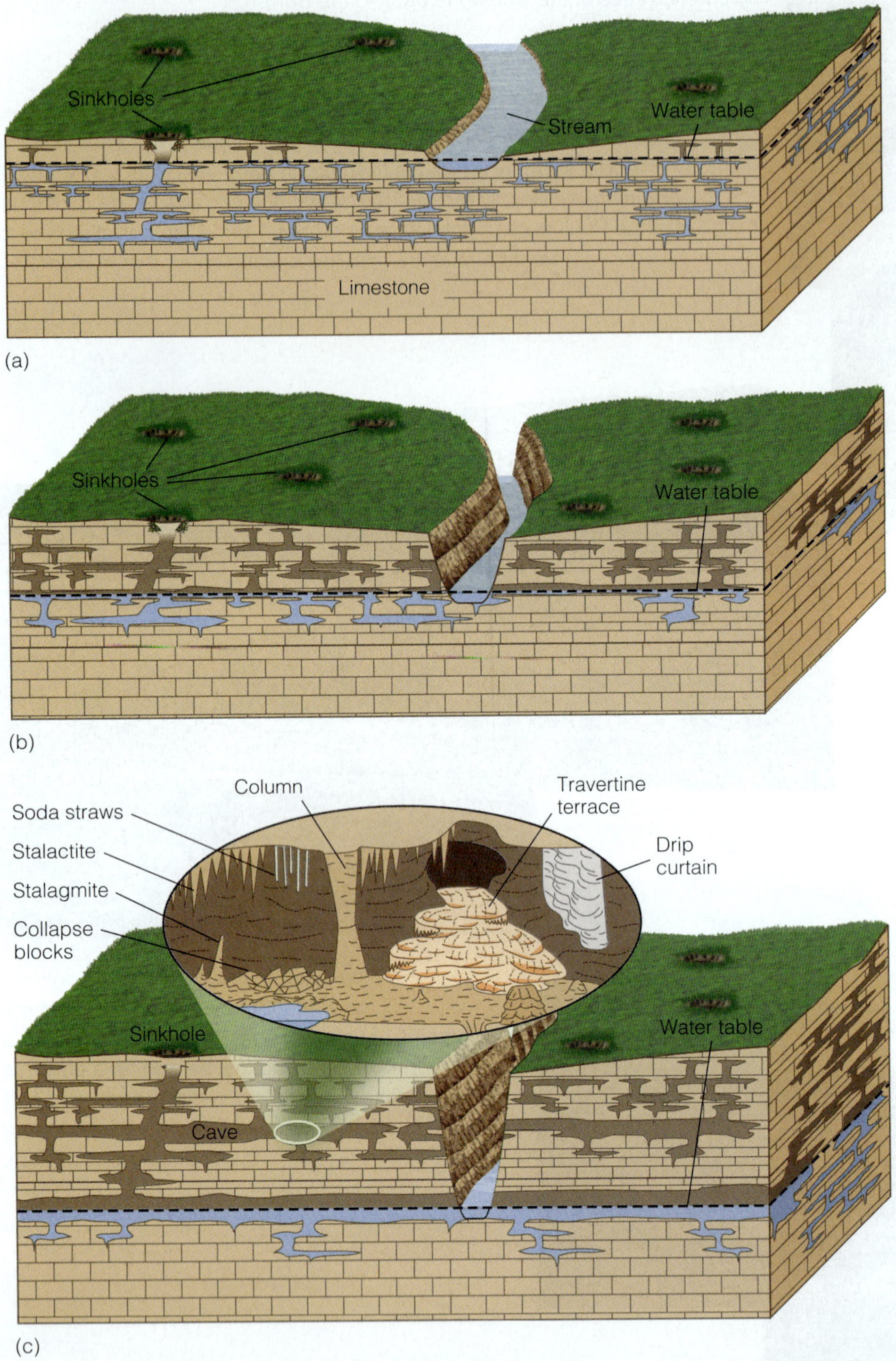

■ **Figure 13.12**

The formation of caves. (a) As groundwater percolates through the zone of aeration and flows through the zone of saturation, it dissolves the carbonate rocks and gradually forms a system of passageways. (b) Groundwater moves along the surface of the water table, forming a system of horizontal passageways through which dissolved rock is carried to the surface streams, thus enlarging the passageways. (c) As the surface streams erode deeper valleys, the water table drops, and the abandoned channelways form an interconnecting system of caves and caverns.

that flowed through the system of horizontal passageways now percolates to the lower water table where a new system of passageways begins to form. The abandoned channelways form an interconnecting system of caves and caverns. Caves eventually become unstable and collapse, littering the floor with fallen debris.

When most people think of caves, they think of the seemingly endless variety of colorful and bizarre-shaped deposits found in them. Although a great many different types of cave deposits exist, most form in essentially the same manner and are collectively known as *dripstone*. As water seeps into a cave, some of the dissolved carbon dioxide in the water escapes, and a small amount of calcite is precipitated. In this manner, the various dripstone deposits are formed.

Stalactites are icicle-shaped structures hanging from cave ceilings that form as a result of precipitation from dripping water (■ Figure 13.13). With each drop of water, a thin layer of calcite is deposited over the previous layer, forming a cone-shaped projection that grows down from the ceiling.

The water that drips from a cave's ceiling also precipitates a small amount of calcite when it hits the floor. As additional calcite is deposited, an upward-growing projection called a *stalagmite* forms (see Geo-Focus 17.1 and Figure 13.13). If a stalactite and stalagmite meet, they form a *column*. Groundwater seeping from a crack in a cave's ceiling may form a vertical sheet of rock called a *drip curtain*, whereas water flowing across a cave's floor may produce *travertine terraces* (Figure 13.12).

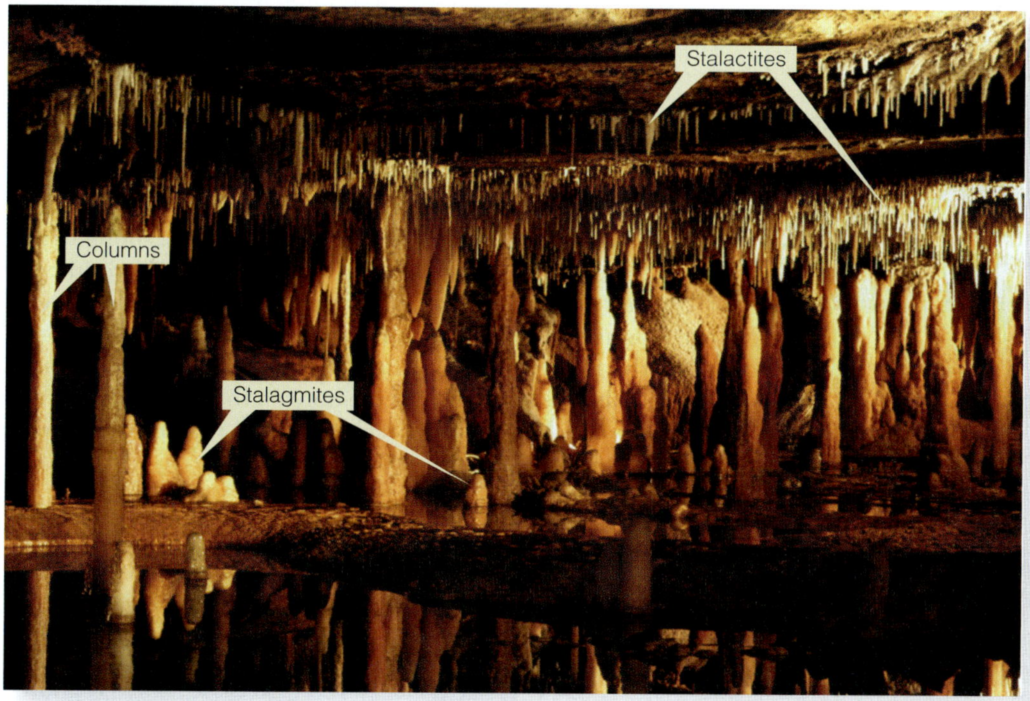

■ **Figure 13.13**

Stalactites are the icicle-shaped structures hanging from the cave's ceiling, whereas the upward-pointing structures on the floor are stalagmites. Columns result when stalactites and stalagmites meet. All three structures are visible in Buchan Cave, Victoria, Australia.

HOW DO HUMANS AFFECT THE GROUNDWATER SYSTEM?

Groundwater is a valuable natural resource that is rapidly being exploited with little regard to the effects of overuse and misuse. Currently, about 20% of all water used in the United States is groundwater. This percentage is rapidly increasing, and unless this resource is used more wisely, sufficient amounts of clean groundwater will not be available in the future. Modifications of the groundwater system may have many consequences, including (1) lowering of the water table, causing wells to dry up; (2) loss of hydrostatic pressure, causing once free-flowing wells to require pumping; (3) saltwater incursion; (4) subsidence; and (5) contamination.

Lowering the Water Table

Withdrawing groundwater at a significantly greater rate than it is replaced by either natural or artificial recharge can have serious effects. For example, the High Plains aquifer is one of the most important aquifers in the United States. It underlies more than 450,000 km^2, including most of Nebraska, large parts of Colorado and Kansas, portions of South Dakota, Wyoming, and New Mexico, as well as the panhandle regions of Oklahoma and Texas, and accounts for approximately 30% of the groundwater used for irrigation in the United States (■ Figure 13.14). Irrigation from the High Plains aquifer is largely responsible for the region's agricultural productivity, which includes a significant percentage of the nation's corn, cotton, and wheat, and half of U.S. beef cattle. Large areas of land (about 18 million acres) are currently irrigated with water pumped from the High Plains aquifer. Irrigation is popular because yields from irrigated lands can be triple what they would be without irrigation.

Although the High Plains aquifer has contributed to the high productivity of the region, it cannot continue to provide the quantities of water that it has in the past. In some parts of the High Plains, from 2 to 100 times more water is being pumped annually than is being recharged. Consequently, water is being removed from the aquifer faster than it is being replenished, causing the water table to drop significantly in many areas (Figure 13.14).

What will happen to this region's economy if long-term withdrawal of water from the High Plains aquifer greatly exceeds its recharge rate so that it can no longer supply the quantities of water necessary for irrigation? Solutions range from going back to farming without irrigation to diverting water from other regions such as the Great Lakes. Farming without irrigation would result in greatly decreased yields and higher costs and prices for agricultural products. The diversion of water from elsewhere would cost billions of dollars and the price of agricultural products would still rise.

Another excellent example of what we might call deficit spending with regard to groundwater took place

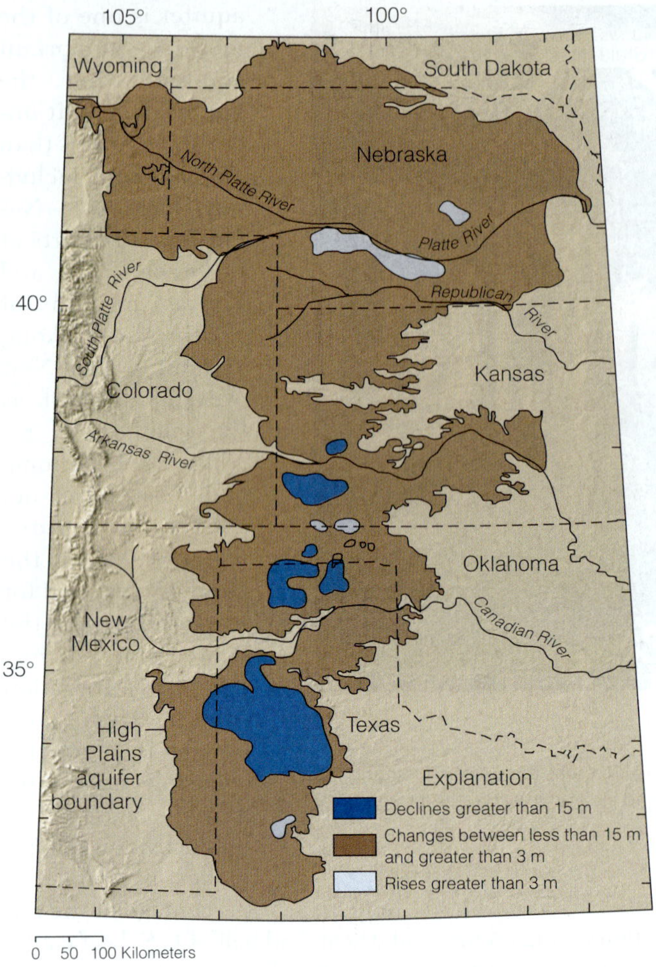

■ **Figure 13.14**

Geographic extent of the High Plains aquifer and changes in water level from predevelopment through 1993.
Source: From J. B. Weeks et al., *U.S. Geological Survey Professional Paper* 1400-A, 1988.

in California during the drought of 1987–1992. During that time, the state's aquifers were overdrawn at the rate of 10 million acre/feet per year (an acre/foot is the amount of water that covers 1 acre 1 foot deep). In short, each year of the drought California was withdrawing over 12 km^3 of groundwater more than was being replaced. Unfortunately, excessive depletion of the groundwater reservoir has other consequences, such as subsidence involving sinking or settling of the ground surface (discussed in a later section).

Water supply problems certainly exist in many areas, but on the positive side, water use in the United States actually declined during the five years following 1980 and has remained nearly constant since then, even though the population has increased. This downturn in demand resulted largely from improved techniques in irrigation, more efficient industrial water use, and a general public awareness of water problems coupled with conservation practices. Nevertheless, the rates of withdrawal of ground-water from some aquifers still exceeds their rate of recharge, and population growth in the arid to semiarid Southwest is putting larger demands on an already limited water supply.

Saltwater Incursion

The excessive pumping of groundwater in coastal areas has resulted in *saltwater incursion* such as occurred on Long Island, New York, during the 1960s. Along coastlines where permeable rocks or sediments are in contact with the ocean, the fresh groundwater, being less dense than seawater, forms a lens-shaped body above the underlying saltwater (■ Figure 13.15a). The weight of the freshwater exerts pressure on the underlying saltwater. As long as rates of recharge equal rates of withdrawal,

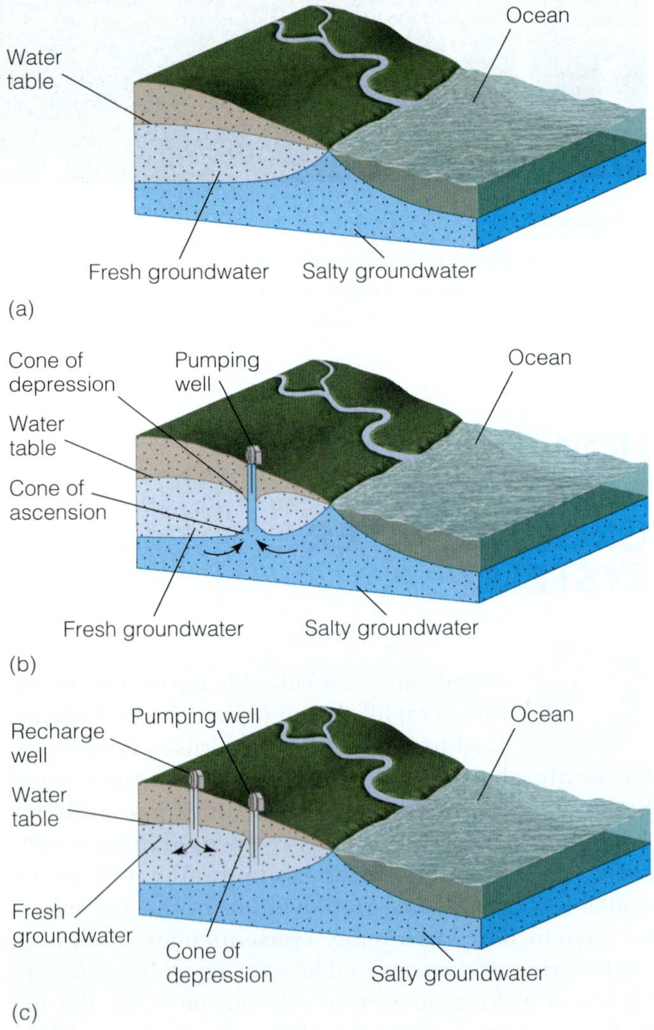

■ **Figure 13.15**

Saltwater incursion. (a) Because freshwater is not as dense as saltwater, it forms a lens-shaped body above the underlying saltwater. (b) If excessive pumping occurs, a cone of depression develops in the fresh groundwater, and a cone of ascension forms in the underlying salty groundwater that may result in saltwater contamination of the well. (c) Pumping water back into the groundwater system through recharge wells can help lower the interface between the fresh groundwater and the salty groundwater and reduce saltwater incursion.

the contact between the fresh groundwater and the seawater remains the same. If excessive pumping occurs, however, a deep cone of depression forms in the fresh groundwater (Figure 13.15b). Because some of the pressure from the overlying freshwater has been removed, saltwater forms a *cone of ascension* as it rises to fill the pore space that formerly contained freshwater. When this occurs, wells become contaminated with saltwater and remain contaminated until recharge by freshwater restores the former level of the fresh-groundwater water table.

Saltwater incursion is a major problem in many rapidly growing coastal communities. As the population in these areas grows, greater demand for groundwater creates an even greater imbalance between recharge and withdrawal.

To counteract the effects of saltwater incursion, recharge wells are often drilled to pump water back into the groundwater system (Figure 13.15c). Recharge ponds that allow large quantities of fresh surface water to infiltrate the groundwater supply may also be constructed. Both of these methods are used successfully on Long Island, New York, which has had a saltwater incursion problem for several decades.

Subsidence

As excessive amounts of groundwater are withdrawn from poorly consolidated sediments and sedimentary rocks, the water pressure between grains is reduced, and the weight of the overlying materials causes the grains to pack closer together, resulting in *subsidence* of the ground. As more and more groundwater is pumped to meet the increasing needs of agriculture, industry, and population growth, subsidence is becoming more prevalent.

The San Joaquin Valley of California is a major agricultural region that relies largely on groundwater for irrigation. Between 1925 and 1977, groundwater withdrawals in parts of the valley caused subsidence of nearly 9 m (■ Figure 13.16). Other areas in the United States that have experienced subsidence are New Orleans, Louisiana, and Houston, Texas, both of which have subsided more than 2 m, and Las Vegas, Nevada, where 8.5 m of subsidence has taken place (Table 13.2).

Elsewhere in the world, the tilt of the Leaning Tower of Pisa in Italy is partly due to groundwater withdrawal (■ Figure 13.17). The tower started tilting soon after construction began in 1173 because of differential compaction of the foundation. During the 1960s, the city of Pisa withdrew ever-larger amounts of groundwater, causing the ground to subside further; as a result, the tilt of the tower increased until it was considered in danger of falling over. Strict control of groundwater withdrawal and stabilization of the foundation have reduced the amount of tilting to about 1 mm per year, thus ensuring that the tower should stand for several more centuries.

A spectacular example of continuing subsidence is taking place in Mexico City, which is built on a former lake bed. As groundwater is removed for the increasing needs of the city's 15.6 million people, the water table has been lowered up to 10 m. As a result, the fine-grained lake sediments are compacting, and Mexico City is slowly and unevenly subsiding. Its opera house has settled more than 3 m, and half of the first floor is now below ground level. Other parts of the city have subsided as much as 7.5 m, creating similar problems for other structures (■ Figure 13.18).

The extraction of oil can also cause subsidence. Long Beach, California, has subsided 9 m as a result of 34 years of oil production. More than $100 million of damage was done to the pumping, transportation, and harbor facilities in this area because of subsidence and encroachment of the sea (■ Figure 13.19). Once water was pumped back into the oil reservoir, thus stabilizing it, subsidence virtually stopped.

USGS

■ **Figure 13.16**

The dates on this power pole dramatically illustrate the amount of subsidence in the San Joaquin Valley, California. Because of groundwater withdrawals and subsequent sediment compaction, the ground subsided nearly 9 m between 1925 and 1977. For a time, surface water use reduced subsidence, but during the drought of 1987 to 1992 it started again as more groundwater was withdrawn.

Table 13.2

Subsidence of Cities and Regions Due Primarily to Groundwater Removal

Location	Maximum Subsidence (m)	Area Affected (km^2)
Mexico City, Mexico	8.0	25
Long Beach and Los Angeles, California	9.0	50
Taipei Basin, Taiwan	1.0	100
Shanghai, China	2.6	121
Venice, Italy	0.2	150
New Orleans, Louisiana	2.0	175
London, England	0.3	295
Las Vegas, Nevada	8.5	500
Santa Clara Valley, California	4.0	600
Bangkok, Thailand	1.0	800
Osaka and Tokyo, Japan	4.0	3000
San Joaquin Valley, California	9.0	9000
Houston, Texas	2.7	12,100

Source: Data from R. Dolan and H. G. Goodell, "Sinking Cities," *American Scientist 74* (1986): 38–47; and J. Whittow, *Disasters: The Anatomy of Environmental Hazards* (Athens: University of Georgia Press, 1979).

Royalty-Free/Corbis

■ **Figure 13.17**

The Leaning Tower of Pisa, Italy. The tilting is partly the result of subsidence due to the removal of groundwater.

Groundwater Contamination

A major problem facing our society is the safe disposal of the pollutant by-products of an industrialized economy. We are becoming increasingly aware that streams, lakes, and oceans are not unlimited reservoirs for waste and that we must find new safe ways to dispose of pollutants.

The most common sources of groundwater contamination are sewage, landfills, toxic-waste-disposal sites, and agriculture. Once pollutants get into the groundwater system, they spread wherever groundwater travels, which can make their containment difficult (see Geo-Focus 13.1). Furthermore, because groundwater moves so slowly, it takes a long time to cleanse a groundwater reservoir once it has become contaminated.

In many areas, septic tanks are the most common way of disposing of sewage. A septic tank slowly releases sewage into the ground, where it is decomposed by oxidation and microorganisms and filtered by the sediment as it percolates through the zone of aeration. In most situations, by the time the water from the sewage reaches the zone of saturation, it has been cleansed of any im-

R. V. Dietrich

Figure 13.18

Excessive withdrawal of groundwater from beneath Mexico City has resulted in subsidence and uneven settling of buildings. The right side of this church (Our Lady of Guadalupe) has settled slightly more than a meter.

What Would You Do?

Americans generate tremendous amounts of waste. Some of this waste, such as battery acid, paint, cleaning agents, insecticides, and pesticides, can easily contaminate the groundwater system. Your community is planning to construct a city dump to contain waste products, but it simply wants to dig a hole, dump waste in, and then bury it. What do you think of this plan? Are you skeptical of this plan's merits, and if so, what would you suggest to remedy any potential problems?

purities and is safe to use (Figure 13.20a). If the water table is close to the surface or if the rocks are very permeable, however, water entering the zone of saturation may still be contaminated and unfit to use.

Landfills are also potential sources of groundwater contamination (Figure 13.20b). Not only does liquid waste seep into the ground, but rainwater also carries dissolved chemicals and other pollutants down into the groundwater reservoir. Unless the landfill is carefully designed and lined with an impermeable layer such as clay, many toxic compounds such as paints, solvents, cleansers, pesticides, and battery acid will find their way into the groundwater system.

Toxic-waste sites where dangerous chemicals are either buried or pumped underground are an increasing source of groundwater contamination. The United States alone must dispose of several thousand metric tons of hazardous chemical waste per year. Unfortunately, much of this waste has been, and still is, being

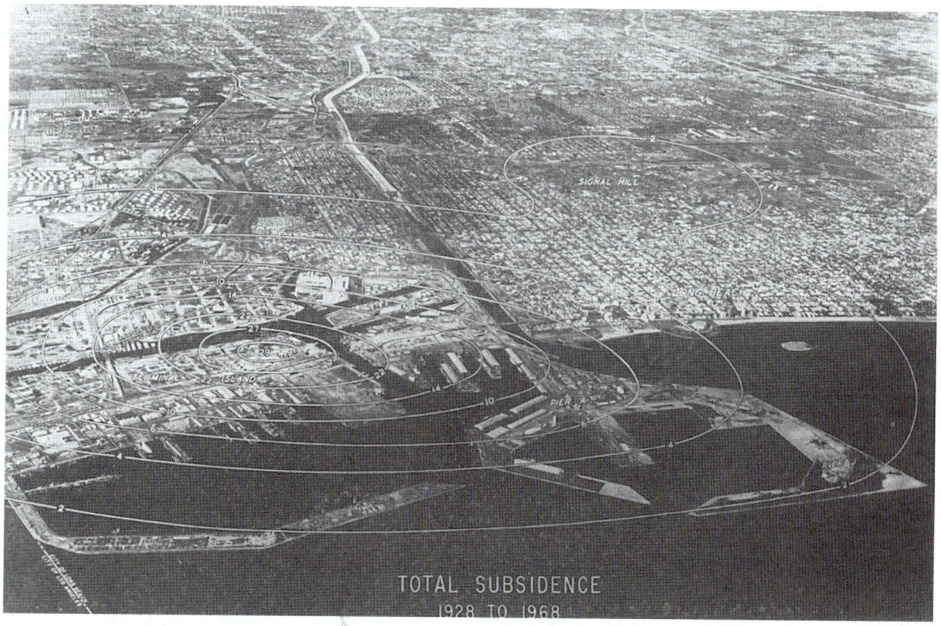

City of Long Beach, Dept. of Oil Properties

Figure 13.19

The withdrawal of petroleum from the oil field in Long Beach, California, resulted in up to 9 m of ground subsidence because of sediment compaction. It was not until water was pumped back into the reservoir to replace the petroleum that ground subsidence essentially ceased.

GEOFOCUS 13.1

Arsenic and Old Lace

Many people probably learned that arsenic is a poison from either reading or seeing the play *Arsenic and Old Lace*, written by Joseph Kesselring. In the play, the elderly Brewster sisters poison lonely old men by adding a small amount of arsenic to their homemade elderberry wine.

Arsenic is a naturally occurring toxic element found in the environment, and several types of cancer have been linked to arsenic in water. Arsenic also harms the central and peripheral nervous systems and may cause birth defects and reproductive problems. In fact, because of arsenic's prevalence in the environment and its adverse health effects, Congress included it in the amendments to the Safe Drinking Water Act in 1996. Arsenic gets into the groundwater system mainly as arsenic-bearing minerals dissolve in the natural weathering process of rocks and soils.

A map published in 2001 by the U.S. Geological Survey (USGS) shows the extent and concentration

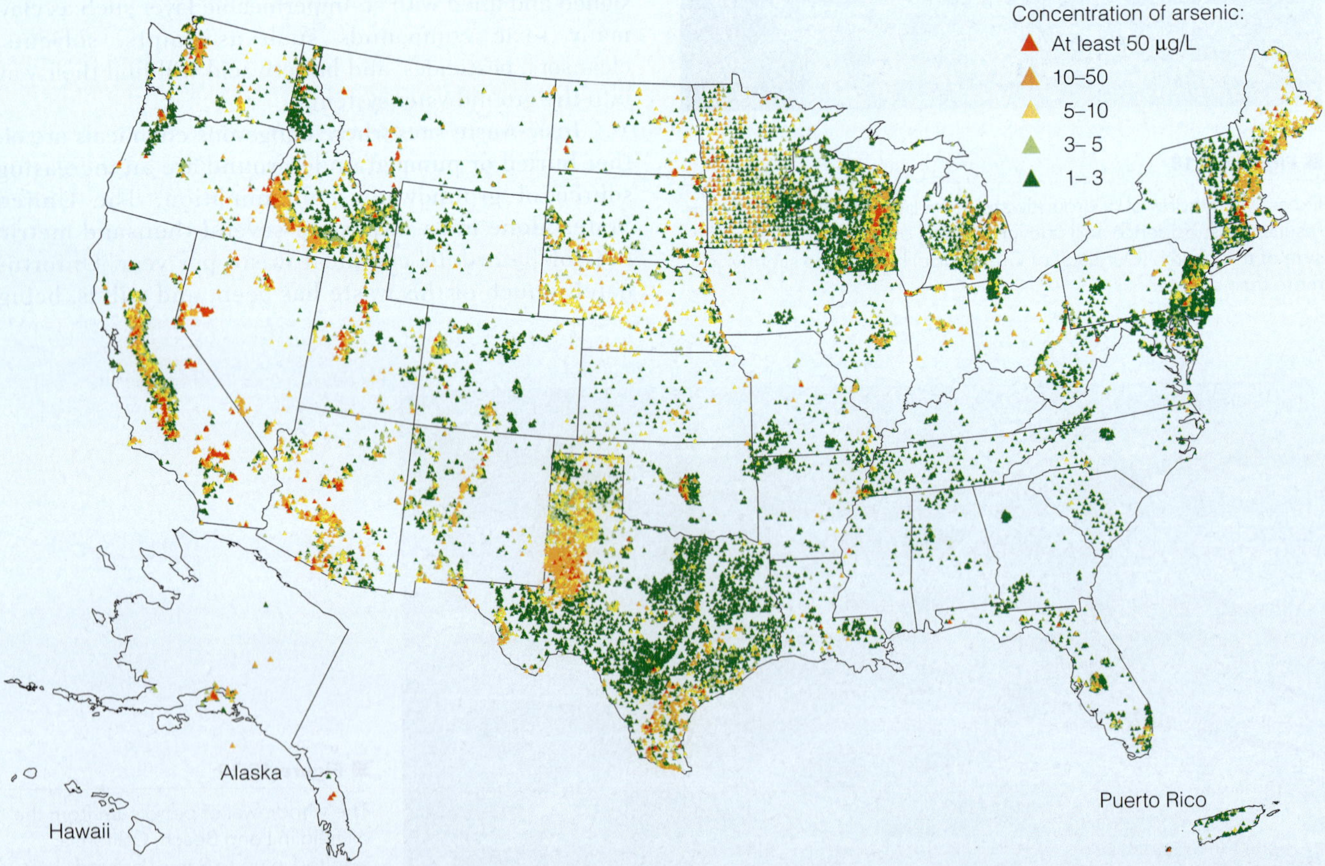

Figure 1

Arsenic concentrations for 31,350 groundwater samples collected from 1973 to 2001.

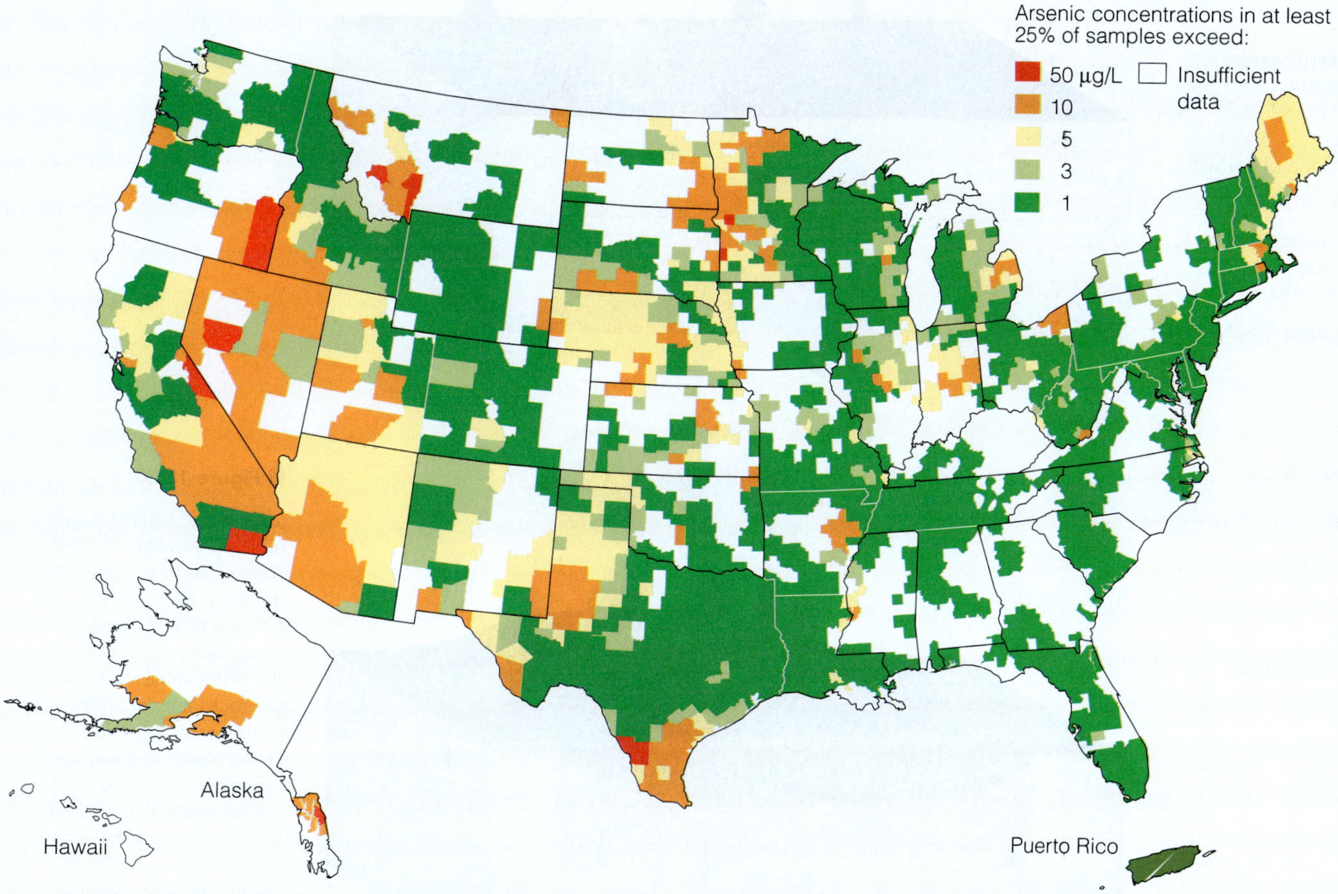

■ **Figure 2**

Arsenic concentrations found in at least 25% of groundwater samples per county. The map is based on 31,350 groundwater samples collected between 1973 and 2001.

of arsenic in the nation's groundwater supply (■ Figure 1). The highest concentrations of groundwater arsenic were found throughout the West and in parts of the Midwest and Northeast. Although the map is not intended to provide specific information for individual wells or even a locality within a county, it helps researchers and policymakers identify areas of high concentration so that they can make informed decisions about water use. We should point out, however, that a high degree of local variability in the amount of arsenic in the groundwater can be caused by local geology, type of aquifer, depth of well, and other factors. The only way to learn the arsenic concentration in any well is to have it tested.

What is considered a safe level of arsenic in drinking water? In 2001 the U.S. Environmental Protection Agency (USEPA, or EPA) lowered the maximum level of arsenic permitted in drinking water from 50 to 10 µg of arsenic per liter. This is the same figure used by the World Health Organization. From the data in Figure 1, additional maps were created. ■ Figure 2 shows arsenic concentrations found in at least 25% of groundwater samples per county. Based on these data, approximately 10% of the samples in the USGS study exceed the new standard of 10 µg of arsenic per liter of drinking water.

Public water supply systems that exceed the existing EPA arsenic standard are required to either treat the water to remove the arsenic or find an alternative supply. Although reducing the acceptable level of arsenic in drinking water will surely increase the cost of water to consumers, it will also decrease their exposure to arsenic and the possible adverse health effects associated with this toxic element.

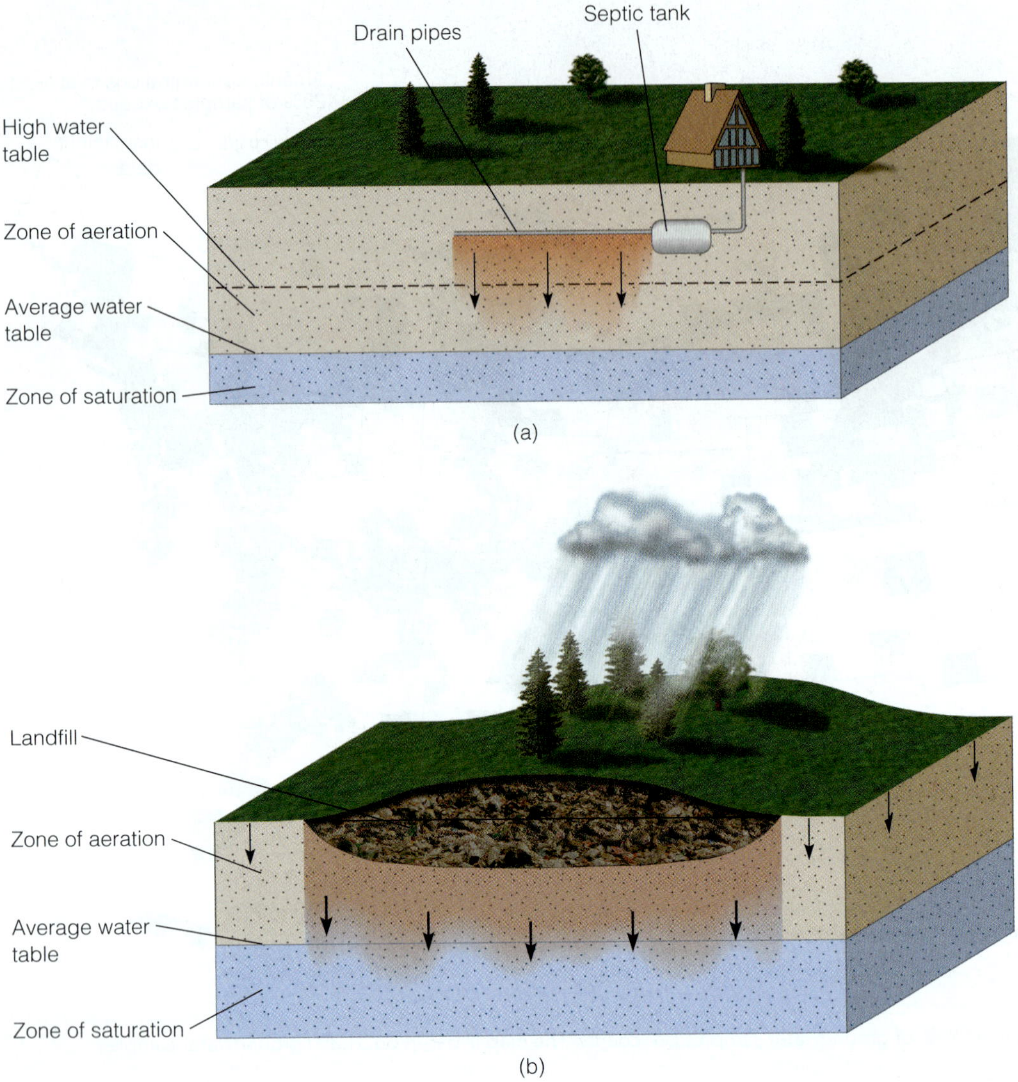

Figure 13.20

(a) A septic system slowly releases sewage into the zone of aeration. Oxidation, bacterial degradation, and filtering usually remove impurities before they reach the water table. However, if the rocks are very permeable or the water table is too close to the septic system, the groundwater may be contaminated. (b) Unless there is an impermeable barrier between a landfill and the water table, pollutants can be carried into the zone of saturation and contaminate the groundwater supply.

improperly dumped and is contaminating the surface water, soil, and groundwater.

Examples of indiscriminate dumping of dangerous and toxic chemicals can be found in every state. Perhaps the most famous is the Love Canal, near Niagara Falls, New York. During the 1940s, the Hooker Chemical Company dumped approximately 19,000 tons of chemical waste into Love Canal. In 1953 it covered one of the dump sites with dirt and sold it for one dollar to the Niagara Falls Board of Education, which built an elementary school and playground on the site. Heavy rains and snow during the winter of 1976–1977 raised the water table and turned the area into a muddy swamp in the spring of 1977. Mixed with the mud were thousands of toxic, noxious chemicals that formed puddles in the playground, oozed into people's basements, and covered gardens and lawns. Trees, lawns, and gardens began to die, and many of the residents of the area suffered from serious illnesses. The cost of cleaning up the Love Canal site and relocating its residents exceeded $100 million, and the site and neighborhood are now vacant.

Groundwater Quality

Finding groundwater is rather easy because it is present beneath the land surface nearly everywhere, although the depth to the water table varies considerably. But just finding water is not enough. Sufficient amounts must be located in porous and permeable materials if groundwater is to be withdrawn for agricultural, industrial, or domestic use. The availability of groundwater was important in the westward expansion in both Canada and the United States, and now more than one-third of all water for irrigation comes from the groundwater system. More than 90% of the water used for domestic purposes in rural America and the water for a number of large cities come from groundwater, and, as one would expect, quality is more important here than it is for most other purposes.

Discounting contamination by humans from landfills, septic systems, toxic-waste sites, and industrial effluents, groundwater quality is mostly a function of (1) the kinds

GEOLOGY IN UNEXPECTED PLACES

Water-Treatment Plants

Your local or regional water-treatment plant may not be the first place you think of when it comes to geology, but surprisingly a fair amount of geology is associated with it. To start off, there is the source of water, which frequently comes from wells drilled below the local water table. The quality of this water depends on the local geology. The water frequently contains high concentrations of iron, calcium, and magnesium; the latter two contribute to the hardness of the water. These elements are usually removed to soften the water.

The first step in softening the water is to aerate it with oxygen to help remove iron, carbon dioxide, and other gases from the water. Removing the carbon dioxide in the water reduces the amount of chemicals needed in subsequent stages of the softening process.

The next step in softening the water is adding lime and sodium hydroxide to precipitate the dissolved calcium and magnesium. This precipitate settles out to form a layer of sludge that is pumped to holding ponds; it can be used later as an agricultural soil conditioner. Following the removal of calcium and magnesium, carbon dioxide is again added to the water to reduce its pH to about 9, which means it is slightly alkaline. This step stops the softening process and stabilizes the water's chemical composition.

The final process is to pass the treated water through filters to remove final particles and then add chemicals to ensure that the water is safe for drinking. The water is now ready to be pumped to the plant's storage reservoirs, from which it is distributed to the final customers.

Figure 1

One of the sludge ponds of the Mt. Pleasant, Michigan Water-Treatment Plant. The sludge produced by the softening process is diverted into a sludge pond. The sludge will be reused as an agricultural soil conditioner.

Reed Wicander

of materials that make up an aquifer, (2) the residence time of water in an aquifer, and (3) the solubility of rocks and minerals. These factors account for the amount of dissolved materials in groundwater, such as calcium, iron, fluoride, and several others. Most pose no health problems, but some have undesirable effects such as deposition of minerals in water pipes and water heaters and offensive taste or smell, or they may stain clothing and fixtures or inhibit the effectiveness of detergents.

GeologyNow Log into GeologyNow and select this chapter to work through a **Geology Interactive** activity on "Groundwater Contamination" (click Groundwater→Contamination).

HYDROTHERMAL ACTIVITY—WHAT IS IT, AND WHERE DOES IT OCCUR?

Hydrothermal is a term referring to hot water. Some geologists restrict the meaning to include only water heated by magma, but here we use it to refer to any hot subsurface water and the surface activity that results from its discharge. One manifestation of hydrothermal activity in areas of active or

Figure 13.21

Hot springs. (a) Hot spring in the West Thumb Geyser Basin, Yellowstone National Park, Wyoming. (b) The water in this hot spring at Bumpass Hell in Lassen Volcanic National Park, California, is boiling. (c) Mud pot at the Sulfur Works, also in Lassen Volcanic National Park. (d) The Park Service warns of the dangers in hydrothermal areas, but some people ignore the warnings and are injured or killed.

recently active volcanism is the discharge of gases, such as steam, at vents known as *fumeroles* (see Figure 5.2). Of more immediate concern here, however, is the groundwater that rises to the surface as *hot springs* or *geysers*. It may be heated by its proximity to magma or by Earth's geothermal gradient because it circulates deeply.

Hot Springs

A **hot spring** (also called a *thermal spring* or *warm spring*) is any spring in which the water temperature is higher than 37°C, the temperature of the human body (Figure 13.21a). Some hot springs are much hotter, with

British Tourist Authority

■ **Figure 13.22**

One of the many bathhouses in Bath, England, that were built around hot springs shortly after the Roman conquest in A.D. 43.

temperatures up to the boiling point in many instances (Figure 13.21b). Another type of hot spring, called a *mud pot,* results when chemically altered rocks yield clays that bubble as hot water and steam rise through them (Figure 13.21c). Of the approximately 1100 known hot springs in the United States, more than 1000 are in the far West, with the others in the Black Hills of South Dakota, Georgia, the Ouachita region of Arkansas, and the Appalachian region.

Hot springs are also common in other parts of the world. One of the most famous is at Bath, England, where shortly after the Roman conquest of Britain in A.D. 43, numerous bathhouses and a temple were built around the hot springs (■ Figure 13.22).

The heat for most hot springs comes from magma or cooling igneous rocks. The geologically recent igneous activity in the western United States accounts for the large number of hot springs in that region. The water in some hot springs, however, circulates deep into Earth, where it is warmed by the normal increase in temperature, the geothermal gradient. For example, the spring water of Warm Springs, Georgia, is heated in this manner. This hot spring was a health and bathing resort long before the Civil War (1861–1865); later, with the establishment of the Georgia Warm Springs Foundation, it was used to help treat polio victims.

Geysers

Hot springs that intermittently eject hot water and steam with tremendous force are known as **geysers.** The word comes from the Icelandic *geysir,* "to gush" or "to rush forth." One of the most famous geysers in the world is Old Faithful in Yellowstone National Park in Wyoming (■ Figure 13.23a). With a thunderous roar, it erupts a column of hot water and steam every 30 to 90 minutes. Other well-known geyser areas are found in Iceland and New Zealand.

Geysers are the surface expression of an extensive underground system of interconnected fractures within hot igneous rocks (■ Figure 13.24). Groundwater percolating down into the network of fractures is heated as it comes into contact with the hot rocks. Because the water near the bottom of the fracture system is under greater pressure than that near the top, it must be heated to a higher temperature before it will boil. Thus, when the deeper water is heated to near the boiling point, a slight rise in temperature or a drop in pressure, such as from escaping gas, will instantly change it to steam. The expanding steam quickly pushes the water above it out of the ground and into the air, producing a geyser eruption. After the eruption, relatively cool groundwater starts to seep back into the fracture system where it is heated to near its boiling temperature and the eruption cycle begins again. Such a process explains how geysers can erupt with some regularity.

Hot spring and geyser water typically contains large quantities of dissolved minerals because most minerals dissolve more rapidly in warm water than in cold water. Because of this high mineral content, some believe the waters of many hot springs have medicinal properties. Numerous spas and bathhouses have been built at hot springs throughout the world to take advantage of these supposed healing properties.

When the highly mineralized water of hot springs or geysers cools at the surface, some of the material in solution is precipitated, forming various types of deposits. The amount and type of precipitated minerals depend on the solubility and composition of the material that the groundwater flows through. If the groundwater contains dissolved calcium carbonate ($CaCO_3$), then *travertine* or *calcareous tufa* (both of which are varieties of limestone) are precipitated. Spectacular examples of hot spring travertine deposits are found at Pamukhale in Turkey and at Mammoth Hot Springs in Yellowstone National Park (■ Figure 13.25a). Groundwater containing dissolved silica will, upon reaching the surface, precipitate a soft, white, hydrated mineral called *siliceous sinter* or *geyserite,* which can accumulate around a geyser's opening (Figure 13.25b).

■ **Figure 13.23**

Geysers in Yellowstone National Park, Wyoming. (a) Old Faithful Geyser erupts every 30 to 90 minutes, spewing water 32 to 56 m high. (b) A small geyser erupting in Norris Geyser Basin.

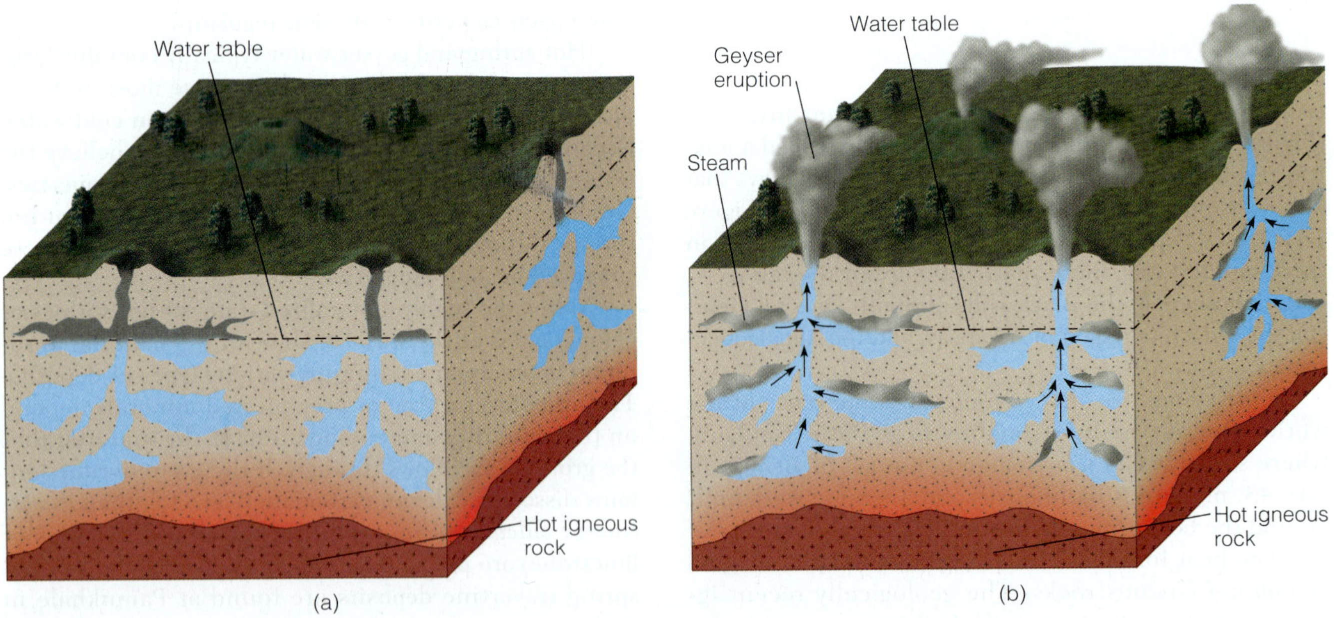

■ **Figure 13.24**

The eruption of a geyser. (a) Groundwater percolates down into a network of interconnected openings and is heated by the hot igneous rocks. The water near the bottom of the fracture system is under greater pressure than that near the top and consequently must be heated to a higher temperature before it will boil. (b) Any rise in temperature of the water above its boiling point or a drop in pressure will cause the water to change to steam, which quickly pushes the water above it up and out of the ground, producing a geyser eruption.

Reed Wicander

(a)

Reed Wicander

(b)

■ **Figure 13.25**

Hot-spring deposits in Yellowstone National Park, Wyoming. (a) Minerva Terrace formed when calcium-carbonate–rich hot-spring water cooled, precipitating travertine. (b) Liberty Cap is a geyserite mound formed by numerous geyser eruptions of silicon-dioxide–rich hot-spring water.

Julie Donnelly-Nolan/USGS

■ **Figure 13.26**

Steam rising from geothermal power plants at The Geysers in Sonoma County, California.

Geothermal Energy

Geothermal energy is defined as any energy produced from Earth's internal heat. In fact, the term *geothermal* comes from *geo,* "Earth," and *thermal,* "heat." Several forms of internal heat are known, such as hot dry rocks and magma, but so far only hot water and steam are used.

Approximately 1 to 2% of the world's current energy needs could be met by geothermal energy. In those areas where it is plentiful, geothermal energy can supply most, if not all, of the energy needs, sometimes at a fraction of the cost of other types of energy. Some of the countries currently using geothermal energy in one form or another are Iceland, the United States, Mexico, Italy, New Zealand, Japan, the Philippines, and Indonesia.

In the United States, the first commercial geothermal electricity-generating plant was built in 1960 at The Geysers, about 120 km north of San Francisco, California. Wells were drilled into the numerous near-vertical fractures underlying the region. As pressure on the rising groundwater decreases, the water changes to steam, which is piped directly to electricity-generating turbines and generators (■ Figure 13.26).

As oil reserves decline, geothermal energy is becoming an attractive alternative, particularly in parts of the western United States, such as the Salton Sea area of southern California, where geothermal exploration and development have begun.

13 GEO RECAP

Chapter Summary

- Groundwater consists of all subsurface water trapped in the pores and other open spaces in rocks, sediment, and soil.
- About 22% of the world's supply of freshwater is groundwater, which constitutes one reservoir in the hydrologic cycle.
- For groundwater to move through materials, they must be porous and permeable. Any material that transmits groundwater is an aquifer, whereas materials that prevent groundwater movement are aquicludes.
- The zone of saturation (in which pores are filled with water) is separated from the zone of aeration (in which pores are filled with air and water) by the water table. The water table is a subdued replica of the land surface in most places.
- Groundwater moves slowly through the pore spaces in the zone of aeration and moves through the zone of saturation to outlets such as streams, lakes, and swamps.
- Springs are found wherever the water table intersects the surface. Some springs are the result of a perched water table—that is, a localized aquiclude within an aquifer and above the regional water table.
- Water wells are made by digging or drilling into the zone of saturation. When water is pumped out of a well, a cone of depression forms.
- In an artesian system, confined groundwater builds up high hydrostatic pressure. Three conditions must generally be met for an artesian system to form: The aquifer must be confined above and below by aquicludes, the aquifer is usually tilted and exposed at the surface so it can be recharged, and precipitation must be sufficient to keep the aquifer filled.
- Karst topography results from groundwater weathering and erosion and is characterized by sinkholes, caves, solution valleys, and disappearing streams.
- Caves form when groundwater in the zone of saturation weathers and erodes soluble rock such as limestone. Cave deposits, called dripstone, result from the precipitation of calcite.
- Modifications of the groundwater system can cause serious problems. Excessive withdrawal of groundwater may result in dry wells, loss of hydrostatic pressure, saltwater incursion, and ground subsidence.
- Groundwater contamination is becoming a serious problem and can result from sewage, landfills, and toxic waste.
- Groundwater may be heated by magma or by the geothermal gradient as it circulates deeply. In either case, the water commonly rises to the surface, thus accounting for hydrothermal activity in the form of hot springs, geysers, and several other features.
- Geothermal energy comes from the steam and hot water trapped within Earth's crust. It is a relatively nonpolluting form of energy that is used as a source of heat and to generate electricity.

Important Terms

artesian system (p. 359)
cave (p. 364)
cone of depression (p. 357)
geothermal energy (p. 380)
geyser (p. 377)
groundwater (p. 354)
hot spring (p. 376)
hydrothermal (p. 375)
karst topography (p. 364)
permeability (p. 355)
porosity (p. 355)
sinkhole (p. 360)
spring (p. 357)
water table (p. 356)
water well (p. 357)
zone of aeration (p. 356)
zone of saturation (p. 356)

Review Questions

1. Two features typical of areas of karst topography are
   a._____ geysers and hot springs;
   b._____ hydrothermal activity and springs;
   c._____ saltwater incursion and pollution;
   d._____ sinkholes and disappearing streams;
   e._____ dripstone and a cone of depression.
2. What is the correct order, from highest to lowest, of groundwater usage in the United States?
   a._____ industrial, agricultural, domestic;
   b._____ agricultural, industrial, domestic;
   c._____ agricultural, domestic, industrial;
   d._____ domestic, agricultural, industrial;
   e._____ industrial, domestic, agricultural.
3. The porosity of Earth materials is defined as
   a._____ their ability to transmit fluids;
   b._____ the depth of the zone of saturation;
   c._____ the percentage of void spaces;
   d._____ their solubility in the presence of weak acids; e._____ the temperature of groundwater.
4. A cone of depression forms when
   a._____ a stream flows into a sinkhole;
   b._____ water in the zone of aeration is replaced by water from the zone of saturation;
   c._____ a spring forms where a perched water table intersects the surface; d._____ water is withdrawn faster than it can be replaced;
   e._____ the ceiling of a cave collapses, forming a steep-sided crater.
5. A hot spring that periodically erupts is known as a
   a._____ mud pot; b._____ geyser;
   c._____ cone of ascension; d._____ travertine terrace; e._____ stalactite.
6. Which of the following conditions must exist for an artesian system to form?
   a._____ an aquifer must be confined above and below by aquicludes; b._____ groundwater must circulate near magma; c._____ water must rise very high in the capillary fringe;
   d._____ the rocks below the surface must be especially resistant to solution; e._____ the water table must be at or very near the surface.
7. When water is pumped from wells in some coastal areas, a problem arises known as
   a._____ artesian recharge; b._____ dripstone deposition; c._____ geothermal depression;
   d._____ saltwater incursion; e._____ permeability decrease.
8. *Hydrothermal* is a term referring to
   a._____ calcareous tufa; b._____ groundwater contamination; c._____ hot water;
   d._____ sinkhole formation; e._____ artesian wells.
9. Which of the following is a cave deposit?
   a._____ aquiclude; b._____ chamber;
   c._____ artesian spring; d._____ stalagmite;
   e._____ sinkhole.
10. Why isn't geothermal energy a virtually unlimited source of energy?
11. Describe three features you might see in an active hydrothermal area. Where in the United States would you go to see such activity?
12. Describe the configuration of the water table beneath a humid area and beneath an arid region. Why are the configurations different?
13. Explain how some Earth materials can be porous yet not permeable. Give an example.

14. Explain how groundwater weathers and erodes Earth materials.
15. Why does groundwater move so much slower than surface water?
16. Explain how saltwater incursion takes place and why it is a problem in coastal areas.
17. Discuss the role of groundwater in the hydrologic cycle.
18. Why should we be concerned about how fast the groundwater supply is being depleted in some areas?
19. Describe some ways to quantitatively measure the rate of groundwater movement.
20. One concern geologists have about burying nuclear waste in present-day arid regions such as Nevada is that the climate may change during the next several thousand years and become more humid, thus allowing more water to percolate through the zone of aeration. Why is this a concern? What would the average rate of groundwater movement have to be during the next 5000 years to reach buried canisters containing radioactive waste buried at a depth of 400 m?

World Wide Web Activities

GeologyNow Assess your understanding of this chapter's topics with additional quizzing and comprehensive interactivities at

http://earthscience.brookscole.com/changingearth4e

as well as current and up-to-date weblinks, additional readings, and InfoTrac College Edition exercises.

Glaciers and Glaciation

CHAPTER 14

OUTLINE

GeologyNow *This icon, which appears throughout the book, indicates an opportunity to explore interactive tutorials, animations, or practice problems available on the GeologyNow website at http://earthscience.brookscole.com/changingearth4e.*

OBJECTIVES

At the end of this chapter, you will have learned that:

- Moving bodies of ice on land known as glaciers cover about 10% of Earth's land surface.
- During the Pleistocene Epoch (Ice Age), glaciers were much more widespread than they are now.
- Water frozen in glaciers constitutes one reservoir in the hydrologic cycle.
- In any area with a yearly net accumulation of snow, the snow is first converted to granular ice known as firn and eventually into glacial ice.
- The concept of the glacial budget is important to understanding the dynamics of any glacier.
- Glaciers move by a combination of basal slip and plastic flow, but several factors determine their rates of movement, and under some conditions they may move rapidly.
- Glaciers effectively erode, transport, and deposit sediment, thus accounting for the origin of several distinctive landforms.
- As a result of glaciation during the Ice Age, Earth's crust was depressed into the mantle and has since risen, sea level fluctuated widely, and lakes were present in areas that are now quite arid.
- A current widely accepted theory explaining the onset of ice ages points to irregularities in Earth's rotation and orbit.

Harvard Glacier is one of several glaciers that flow into College Fiord in Alaska. The glacier is about 2.4 km wide at this point. During the Ice Age a much larger glacier occupied the 40-km-long fiord.

Introduction

Scientists know that Earth's surface temperatures have increased during the last few decades, although they disagree about how much humans have contributed to climatic change. Is the temperature increase simply part of a normal climatic fluctuation, or does the introduction of greenhouse gases into the atmosphere actually have an adverse effect on climate? These questions are not fully answered, but we do know from the geologic record that an Ice Age took place between 1.6 million and 10,000 years ago, and since the end of the Ice Age, Earth has experienced several climatic fluctuations. About 6000 years ago, during the Holocene Maximum, temperatures on average were slightly warmer than they are now, and some of today's arid regions such as the Sahara Desert of north Africa had enough precipitation to support lush vegetation, swamps, and lakes.

A time of cooler temperatures followed the Holocene Maximum, but from about A.D. 1000 to 1300, Europe experienced what is known as the Medieval Warm Period during which wine grapes grew 480 km farther north than they do now. Then a cooling trend beginning in about A.D. 1300 led to the *Little Ice Age,* from about 1500 to the middle or late 1800s. During this time, glaciers expanded (■ Figure 14.1), winters were colder, summers were cooler and wetter, sea ice at high latitudes persisted for long periods, and famines were widespread.

Conditions varied considerably during the Little Ice Age. Some winters were mild and the growing seasons long enough to support the largely agrarian society of Europe. But at other times, cool, wet summers and colder winters contributed to poor harvests and famine. During the coldest part of the Little Ice Age, from about 1680 to 1730, the growing season in England was about five weeks shorter than it was during the 1900s, and in 1695 Iceland was surrounded by sea ice for much of the year. To the surprise of people living in the Orkney Islands, off Scotland's north coast, Eskimos were sighted offshore on several occasions during the late 1600s and early 1700s.

Most of you have some idea of what a glacier is and have probably heard of the Ice Age. Geologists define a **glacier** as a mass of ice on land consisting of compacted, recrystallized snow that flows either downslope or outward from a central area (see the chapter opening photo). The def-

Reprinted with permission from T. H. Van Andel, *New Views on an Old Planet.*

(a)

Sue Monroe

(b)

■ **Figure 14.1**

(a) During the Little Ice Age, many glaciers in Europe extended much farther down their valleys than they do now. Samuel Birmann (1793–1847) painted this view, titled *The Unterer Grindlewald,* in 1826. (b) The same glacier today, its terminus now hidden behind a rock projection at the lower end of its valley.

inition excludes sea ice as in the North Polar region and frozen seawater adjacent to Antarctica. Drifting icebergs are not glaciers either, although they may have come from glaciers that flowed into the sea.

Among the various surface processes that modify Earth's land surface, glaciers are particularly effective at erosion, transport, and deposition. They deeply scour the land they move over, producing a number of easily recognized landforms, and they deposit huge quantities of sediment. Indeed, in many northern states and Canada, glacial deposits are important sources of sand and gravel for construction and reservoirs for groundwater. Glaciers covered far more area during the Pleistocene Epoch (Ice Age), but they still cover about 10% of Earth's land surface.

Unfortunately, our period of recordkeeping is too short to resolve the question of whether the last Ice Age and the Little Ice Age are truly events of the past or simply phases of long-term climatic events and likely to occur again. Nevertheless, studying glaciers and their possible causes may help clarify some aspects of long-term climatic changes and possibly tell us something about the debate on global warming. Glaciers are very sensitive to even short-term climatic changes, so geologists closely monitor them to see if they advance, remain stationary, or retreat.

GLACIERS

Presently, glaciers cover nearly 15 million km^2, or about 10% of Earth's land surface (Table 14.1). As a matter of fact, if all the glacial ice on Earth were in the United States and Canada, these countries would be covered by ice about 1.5 km thick! Small glaciers are common in the high mountains of the western United States, especially Alaska, and western Canada as well as the Andes in South America, the Alps of Europe, and the Himalayas in Asia. Even some of the highest peaks in Africa, though near the equator, have glaciers. Australia is the only continent with no glaciers. The small glaciers in mountains are picturesque, but the truly vast glaciers are in Antarctica and Greenland, which contain most of Earth's glacial ice (Table 14.1).

At first glance, glaciers appear static. Even briefly visiting a glacier may not dispel this impression because, although glaciers move, they usually do so slowly. Nevertheless, they do move, and just like other geologic agents such as running water, glaciers are dynamic systems that continuously adjust to changes. For example, a stream's capacity to erode and transport varies depending

Table 14.1

Present-Day Ice-Covered Areas

	Area (km^2)	Volume (km^3)	Percent of Total
Antarctica	12,588,000	30,109,800	91.49
Greenland	1,802,600	2,620,000	7.96
Canadian Arctic islands	153,169		
Iceland	12,173		
Svalbard	58,016		
Russian Arctic islands	55,541		
Alaska	51,476		
United States (other than Alaska)	513		
Western Canada	24,880		
South America	26,500		
Europe	7,410		
Asia	116,854		
Africa	12		
New Zealand	1,000		
Others	176	180,000[a]	0.55
	14,898,276	32,909,800	100.00

[a]Total volume of glacial ice outside Antarctica and Greenland.

Source: U.S. Geological Survey Professional Paper 1386-A.

on its velocity and discharge. Likewise, the amount of ice in a glacier determines its rate of movement and thus its ability to erode and transport sediment. So glaciers also respond to changes; they just do so more slowly.

During the Ice Age, glaciers covered far more area than they do now, and as a result sea level was as much as 130 m lower. Consequently, the continental shelves were largely exposed and quickly covered by vegetation, and streams responded to a lower base level and eroded deep canyons across the shelves. Also, land connections existed between the British Isles and mainland Europe and between Siberia and Alaska, and various animals migrated across them.

When the Ice Age glaciers formed, their tremendous weight caused the crust to subside as much as 300 m below preglacial levels. When the glaciers melted, though, isostatic rebound began and continues even now in some areas. More than 100 m of rebound has taken place in northeastern Canada during the last 6000 years, and parts of Scandinavia are rebounding at about 1 m per century. Indeed, some coastal areas have rebounded enough so that docks built a few centuries ago are now far from the shore.

Glaciers—Part of the Hydrologic Cycle

Recall from Chapter 1 that one of Earth's systems, the hydrosphere, consists of all the surface water in the oceans and on land, including water frozen in glaciers. Glaciers contain only 2.15% of all the water on Earth, but fully 75% of all freshwater is in glaciers (see Table 12.1). This quantity of frozen water in glaciers, the dynamics of glaciers, and the geologic work done by glaciers are our main interests here.

Glaciers constitute one reservoir in the hydrologic cycle where water is stored for long periods, but even this water eventually returns to its original source, the oceans (see Figure 12.2). Many glaciers at high latitudes, as in Antarctica, Greenland, Alaska, and northern Canada, flow directly into the seas where they melt, or icebergs break off from them (a process known as *calving*) and drift out to sea where they eventually melt. At lower latitudes or areas more remote from the sea, glaciers flow from higher to lower elevations where melting takes place and the resulting water enters the groundwater system (another reservoir in the hydrologic cycle) or returns to the seas by surface runoff. In some parts of the western United States and Canada, glaciers are important freshwater reservoirs that release water to streams during the dry season.

In addition to melting, glaciers lose water by *sublimation,* a process in which ice changes to water vapor without an intermediate liquid phase. Sublimation is easy to understand if you think of ice cubes stored in a container in a freezer. Because of sublimation the older ice cubes at the bottom of the container are much smaller than the more recently formed ones. The water vapor derived by sublimation from glaciers enters the atmosphere where it may condense and fall again as rain or snow, but in the long run all water in glaciers returns to the oceans.

How Do Glaciers Form and Move?

In Chapter 3 we mentioned that ice is crystalline and possesses characteristic physical and chemical properties, and thus is a mineral. Accordingly, glacial ice is a type of rock, but one that is easily deformed. Glacial ice forms in a very straightforward manner (■ Figure 14.2). In any area where more snow falls than melts during the warmer seasons, a net accumulation occurs. Freshly fallen snow has about 80% air-filled pore space and 20% solids, but it compacts as it accumulates, partly thaws, and refreezes, converting to a granular type of snow known as **firn.** As more snow accumulates, firn is buried and further compacted and recrystallized until it becomes **glacial ice,** consisting of about 90% solids (Figure 14.2). As we mentioned in the Introduction, *glaciers* are moving masses of compacted, recrystallized snow on land, but just how do they move?

At this time it is useful to recall some terms about deformation from Chapter 10. Remember that stress is force per unit area and strain is a change in the shape or volume or both of solids. When snow and ice reach a critical thickness of about 40 m, the stress on the ice at depth is great enough to induce **plastic flow,** a type of permanent deformation involving no fracturing. Glaciers move primarily by plastic flow, but they may also slide over their underlying surface by **basal slip** (■ Figure 14.3). Basal slip is facilitated by the presence of water, which reduces friction between the underlying surface and a glacier. The total movement of a glacier, then, results from a combination of plastic flow and basal slip, although plastic flow occurs continuously whereas basal slip varies depending on the season. Indeed, if a glacier is solidly frozen to the surface below, it moves only by plastic flow.

You now have some idea of how glaciers form in the first place, but what controls their distribution? As you probably suspect, temperature and the amount of snowfall are important factors. Of course temperature varies with elevation and latitude, so we expect to find glaciers in high mountains and at high latitudes, if the areas receive enough snow. Many small glaciers are in the Sierra Nevada of California but only at elevations exceeding 3900 m. In fact, the high mountains in California, Oregon, and Washington all possess glaciers because in addition to their elevations they receive huge amounts of snow. Indeed, Mount Baker in Washington had almost 29 m of snow during the winter of 1998–1999, and average snowfalls of 10 m or more are common in many parts of these mountains. Glaciers are also present in the mountains along the Pacific Coast of Canada, which

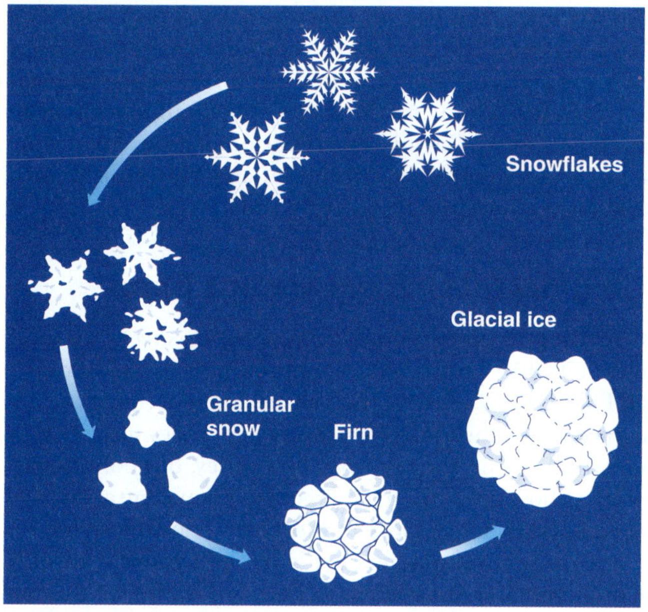

■ **Figure 14.2**

(a) Conversion of freshly fallen snow to firn and glacial ice. (b) This iceberg, derived from Portage Glacier in Alaska, shows the typical blue color of glacial ice. The ice absorbs the longer wavelengths of white light, but the blue (short wavelength) is transmitted into the ice and scattered, accounting for the blue color.

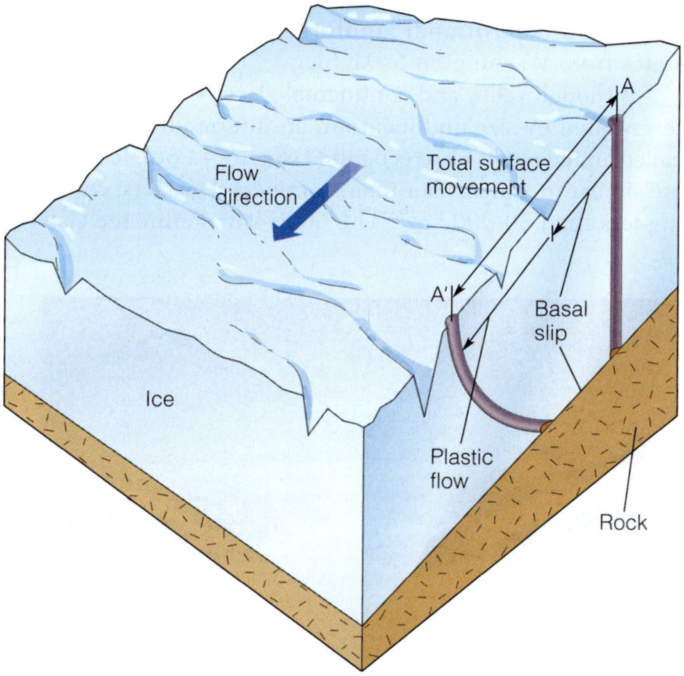

■ **Figure 14.3**

Part of a glacier showing movement by a combination of plastic flow and basal slip. Plastic flow involves internal deformation within the ice, whereas basal slip is sliding over the underlying surface. If a glacier is solidly frozen to its bed, it moves only by plastic flow. Notice that the top of the glacier moves farther in a given time than the bottom does.

What Would You Do?

Suppose you are a high school Earth science teacher trying to explain to students that ice is a mineral and a rock, and how a solid such as ice can flow like a fluid. Furthermore, you explain that the upper 40 m or so of a glacier is brittle and fractures, and yet the ice below that depth simply flows when subjected to pressure. Now that your students are thoroughly confused, how will you explain and demonstrate that ice can behave like a solid yet show properties of fluid flow? (*Hint:* Refer to some of the discussion in Chapter 10.)

also receive considerable snow and in addition are farther north. Some of the higher peaks in the Rocky Mountains of both the United States and Canada support glaciers, too.

WHAT KINDS OF GLACIERS ARE THERE?

All glaciers share some characteristics, but they also vary in several ways. Some are confined to mountain valleys or bowl-shaped depressions on mountainsides and flow from higher to lower elevations. Others are much thicker and extend farther; they flow outward from centers of accumulation and are unconfined by topography. Thus, we recognize two basic types of glaciers: *valley* and *continental,* and some variations on these basic types.

Valley Glaciers

A **valley glacier,** as its name implies, is confined to a mountain valley through which it flows from higher to lower elevations (■ Figure 14.4). We use the term *valley glacier* here, but *alpine glacier* and *mountain glacier* are synonyms. Many valley glaciers have smaller tributary glaciers entering them, just as rivers and streams have tributaries, thus forming a network of glaciers in a system of interconnected valleys. A valley glacier's shape is obviously controlled by the shape of the valley it occupies, so these glaciers are long, narrow tongues of moving ice. Where a valley glacier flows from a valley onto a wider plain and spreads out, or where two or more valley glaciers coalesce at the base of a mountain range, they form a more extensive ice cover called a *piedmont glacier.*

Valley glaciers are small compared to continental glaciers, but even so they may be several kilometers across, up to 200 km long, and hundreds of meters thick. For instance, the Bering Glacier in Alaska is about 200 km long, and the Saskatchewan Glacier in Canada is 555 m thick. Erosion and deposition by valley glaciers are responsible for much of the spectacular scenery in such places as Grand Teton National Park, Wyoming; Glacier National Park, Montana; and Waterton, Banff, and Jasper National Parks in Canada.

Continental Glaciers and Ice Caps

Continental glaciers, also known as *ice sheets,* cover at least 50,000 km^2 and are unconfined by topography (■ Figure 14.5). That is, their shape and movement are not controlled by the underlying landscape. Valley glaciers flow downhill within the confines of a valley, but continental glaciers flow outward in all directions from central areas of accumulation in response to variations in ice thickness.

Continental glaciers are currently present in only Greenland and Antarctica. In both areas, the ice is more than 3000 m thick in their central areas, becomes thinner toward the margins, and covers all but the highest mountains (Figure 14.5b). The aerial extent of the continental glacier in Greenland is about 1,800,000 km^2, and in Antarctica the East and West Antarctic Glaciers merge to form a continuous ice sheet covering more than 12,650,000 km^2. During the Pleistocene Epoch, continental glaciers covered large parts of the Northern Hemisphere continents. They account for many erosional and depositional landforms in Canada and the states from Washington to Maine.

Although valley and continental glaciers are easily differentiated by size and location, an intermediate variety called an **ice cap** is also recognized (Figure 14.5c). Ice caps are similar to, but smaller than, continental glaciers, covering less than 50,000 km^2. The 6000-km^2 Penny Ice Cap

Engineering Mechanics, Virginia Polytechnic Institute and State University

(a)

Sue Monroe

(b)

■ **Figure 14.4**

(a) View of a valley glacier in Alaska. Notice the tributaries to the large glacier. (b) This view is from the upper end of a valley glacier in Switzerland. This glacier's terminus is about 22 km in the distance.

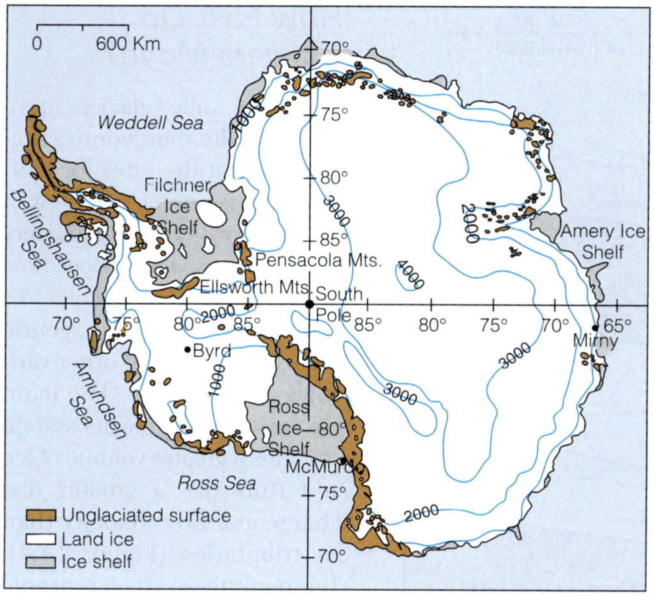

(a)

James S. Monroe

(b)

James S. Monroe

(c)

Figure 14.5

(a) The West Antarctic and much larger East Antarctic ice sheets merge to form a nearly continuous ice cover averaging 2160 m thick and reaching a maximum thickness of about 4000 m. (b) View of part of the ice sheet that covers most of Greenland. (c) An ice cap on Baffin Island, Canada.

on Baffin Island, Canada, and the 3900-km^2 Juneau Icefield in Alaska and Canada are good examples. Some ice caps form when valley glaciers grow, overtop the divides and passes between adjacent valleys, and coalesce to form a continuous ice cap. They also form on fairly flat terrain in Iceland and some of the islands of the Canadian Arctic.

ACCUMULATION AND WASTAGE—THE GLACIAL BUDGET

Just as a savings account grows and shrinks as funds are deposited and withdrawn, glaciers expand and contract in response to accumulation and wastage. We describe their behavior in terms of a **glacial budget,** which is essentially a balance sheet of accumulation and wastage. The upper part of a valley glacier is a **zone of accumulation,** where additions exceed losses and the glacier's surface is perennially covered by snow. In contrast, the same glacier at a lower elevation is in a **zone of wastage,** where losses from melting, sublimation, and calving of icebergs exceed the rate of accumulation (■ Figure 14.6).

At the end of winter, a glacier's surface is completely covered with the accumulated seasonal snowfall. During spring and summer, the snow begins to melt, first at lower elevations and then progressively higher up the glacier. The elevation to which snow recedes during a wastage season is called the *firn limit* (Figure 14.6). One can easily identify the zones of accumulation and wastage by noting the position of the firn limit.

Observations of a single glacier reveal that the position of the firn limit usually changes from year to year. If it does not change or shows only minor fluctuations, then the glacier has a balanced budget; that is, additions in the zone of accumulation are exactly balanced by losses in the zone of wastage, and the distal end or *terminus* of the glacier remains stationary (Figure 14.6a). When the firn limit moves down the glacier, the glacier has a positive budget; its additions exceed its losses, and its terminus advances (Figure 14.6b). If the budget is negative, the glacier recedes and its terminus retreats up the glacial valley (Figure 14.6c). Even though a glacier's terminus may be receding, the glacial ice continues to move toward the terminus by plastic flow and basal slip. If a negative budget persists long enough, a glacier recedes and thins until it no longer flows; it then becomes a *stagnant glacier.*

Although we used a valley glacier as an example, the same budget considerations control the behavior of

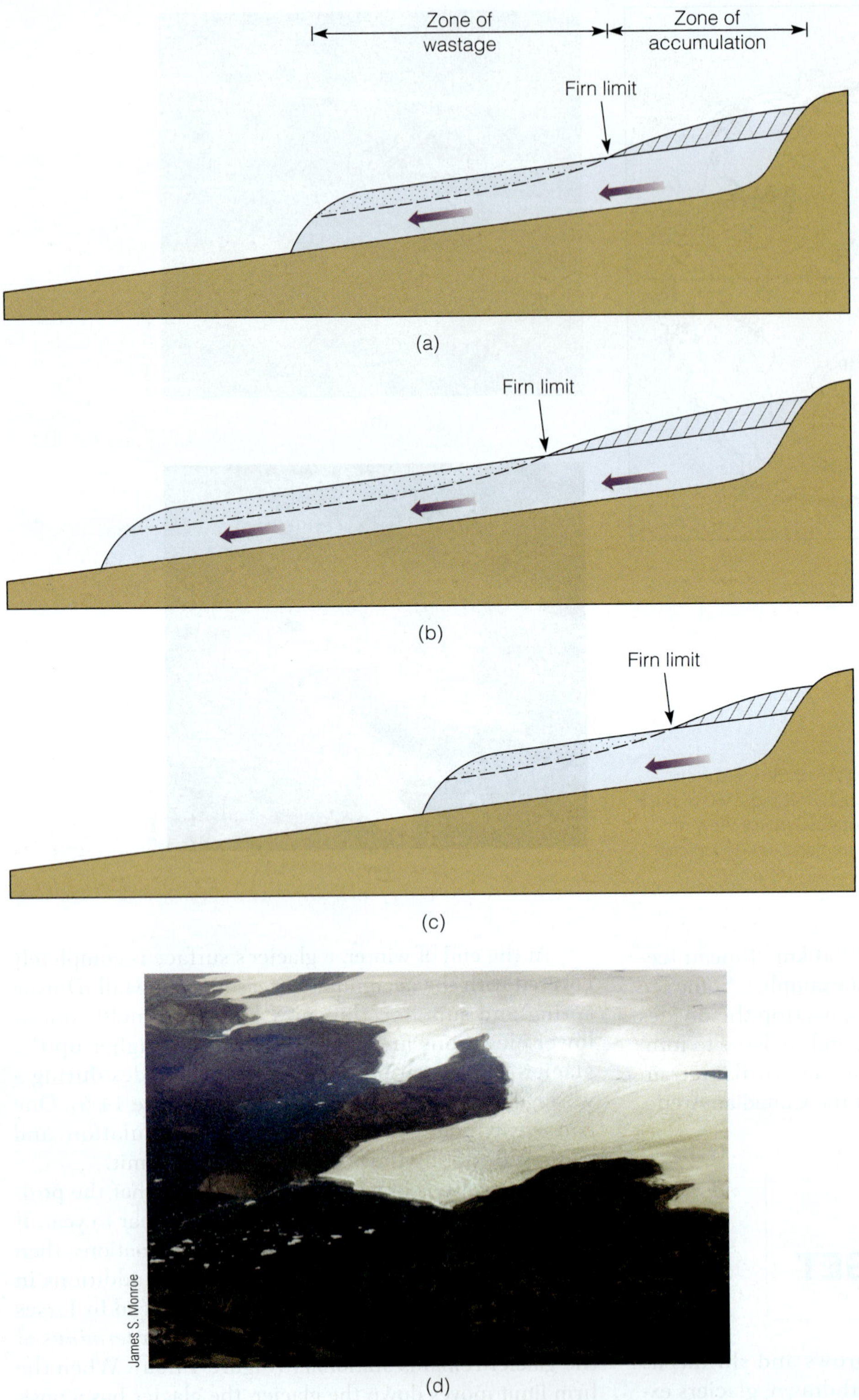

Figure 14.6

Response of a glacier to budget changes. (a) If losses in the zone of wastage (stippled area) equal additions in the zone of accumulation (crosshatched area), the terminus of the glacier remains stationary. (b) If gains exceed losses, the glacier's terminus advances. (c) If losses exceed gains, the glacier's terminus retreats. (d) The entire Greenland ice sheet is in the zone of accumulation, but as shown here it flows into the ocean where icebergs break off.

continental glaciers. For example, the entire Greenland ice sheet is in the zone of accumulation, but it flows into the ocean where wastage occurs (Figure 14.6d).

How Fast Do Glaciers Move?

In general, valley glaciers move more rapidly than continental glaciers, but the rates for both vary from centimeters to tens of meters per day. Valley glaciers moving down steep slopes flow more rapidly than glaciers of comparable size on gentle slopes, assuming all other variables are the same. The main glacier in a valley glacier system contains a greater volume of ice and thus has a greater discharge and flow velocity than its tributaries (Figure 14.4). Temperature exerts a seasonal control on valley glaciers because, although plastic flow remains rather constant year-round, basal slip is more important during warmer months when meltwater is more abundant.

Flow rates also vary within the ice itself. For example, flow velocity increases in the zone of accumulation until the firn limit is reached; from that point, the velocity becomes progressively slower toward the glacier's terminus. Valley glaciers are similar to rivers in that the valley walls and floor cause frictional resistance to flow, so the ice in contact with the walls and floor moves more slowly than the ice some distance away (■ Figure 14.7).

Notice in Figure 14.7 that the flow velocity increases upward until the top few tens of meters of ice are reached, but little or no additional increase occurs after that point. This upper ice constitutes the rigid part of the glacier that is moving as a consequence of basal slip and plastic flow below. The fact that this upper 40 m or so of ice behaves like a brittle solid is clearly demonstrated by large fractures called *crevasses* that develop when a valley glacier flows over a step in its valley floor where the slope increases or where it flows around a corner (■ Figure 14.8a). In ei-

■ **Figure 14.7**

Flow velocity in a valley glacier varies both horizontally and vertically. Velocity is greatest at the top center of the glacier because friction with the walls and floor of the trough slows the flow adjacent to these boundaries. The lengths of the arrows in the figure are proportional to velocity.

ther case, the glacial ice is stretched (subjected to tension), and large crevasses develop, but they extend downward only to the zone of plastic flow. In some cases, a valley glacier descends over such a steep precipice that crevasses break up the ice into a jumble of blocks and spires, and an *ice fall* develops (Figure 14.8b).

One reason continental glaciers move comparatively slowly is that they are at higher latitudes and are frozen to the underlying surface most of the time, which limits the amount of basal slip. Some basal slip does take place even beneath the Antarctic ice sheet, but most glacial movement is by plastic flow. Nevertheless, some parts of continental glaciers manage to achieve extremely high flow rates. Near the margins of the Greenland ice sheet, the ice is forced between mountains in what are called *outlet glaciers*. In some of these outlets, flow velocities exceed 100 m per day.

Some areas of rapid flow known as ice streams in West Antarctica have flow rates considerably greater than in adjacent glacial ice. Drilling revealed a 5-m-thick layer of water-saturated sediment beneath these ice streams, which apparently facilitates movement of the ice above. Some geologists think that geothermal heat from active volcanism melts the underside of the ice.

Glacial Surges

A **glacial surge** is a short-lived episode of accelerated flow in a glacier during which the glacier's surface breaks into a maze of cravasses and its terminus advances markedly. Although surges are best documented in valley glaciers, they also take place in ice caps and perhaps continental glaciers. During a surge, a glacier may advance several tens of meters per day for weeks or months and then return to its normal flow rate. Surging glaciers constitute only a tiny proportion of all glaciers, and there are none in the United States outside Alaska. Even in Canada they are found only in the Yukon Territory and the Queen Elizabeth Islands.

The fastest glacial surge ever recorded was in 1953 in the Kutiah Glacier in Pakistan; the glacier advanced 12 km in three months. In 1986 the terminus of Hubbard Glacier in Alaska began advancing at about 10 m per day, and in 1993 Alaska's Bering Glacier advanced more than 1.5 km in just three weeks.

The onset of a glacial surge is heralded by a thickened bulge in the upper part of a glacier that begins to move at several times the glacier's normal velocity toward the terminus. When the bulge reaches the terminus, it causes rapid movement and displacement of the terminus by as much as 20 km. Surges are probably related to accelerated rates of basal slip rather than more

R. V. Dietrich

(a)

Sue Monroe

(b)

■ **Figure 14.8**

(a) Crevasses on glaciers in Alaska. (b) The area of chaotic ice in the middle of this image is an ice fall on a small glacier in Switzerland.

rapid plastic flow. One theory holds that thickening in the zone of accumulation with concurrent thinning in the zone of wastage increases the glacier's slope and accounts for accelerated flow. But another theory holds that pressure on soft sediment beneath a glacier forces fluids through the sediment, thereby allowing the overlying glacier to slide more effectively.

GLACIAL EROSION AND TRANSPORT

Glaciers are moving solids that erode and transport huge quantities of materials, especially unconsolidated sediment and soil. Important erosional processes include bulldozing, plucking, and abrasion. Bulldozing, although not a formal geologic term, is fairly self-explanatory: a glacier simply shoves or pushes unconsolidated materials in its path. This effective process was aptly described in 1744 during the Little Ice Age by an observer in Norway:

> *When at times [the glacier] pushes forward a great sound is heard, like that of an organ and it pushes in front of it unmeasurable masses of soil, grit and rocks bigger than any house could be, which it then crushes small like sand.**

Plucking, also called *quarrying*, results when glacial ice freezes in the cracks and crevices of a bedrock projection and eventually pulls it loose. One manifestation of plucking is a landform known as a *roche moutonnée*, French for "rock sheep." As shown in ■ Figure 14.9, a glacier smooths the "upstream" side of an obstacle, such as a small hill, and plucks pieces of rock from the "downstream" side by repeatedly freezing and pulling away from the obstacle.

*Quoted in C. Officer and J. Page, *Tales of the Earth* (New York: Oxford University Press, 1993), p. 99.

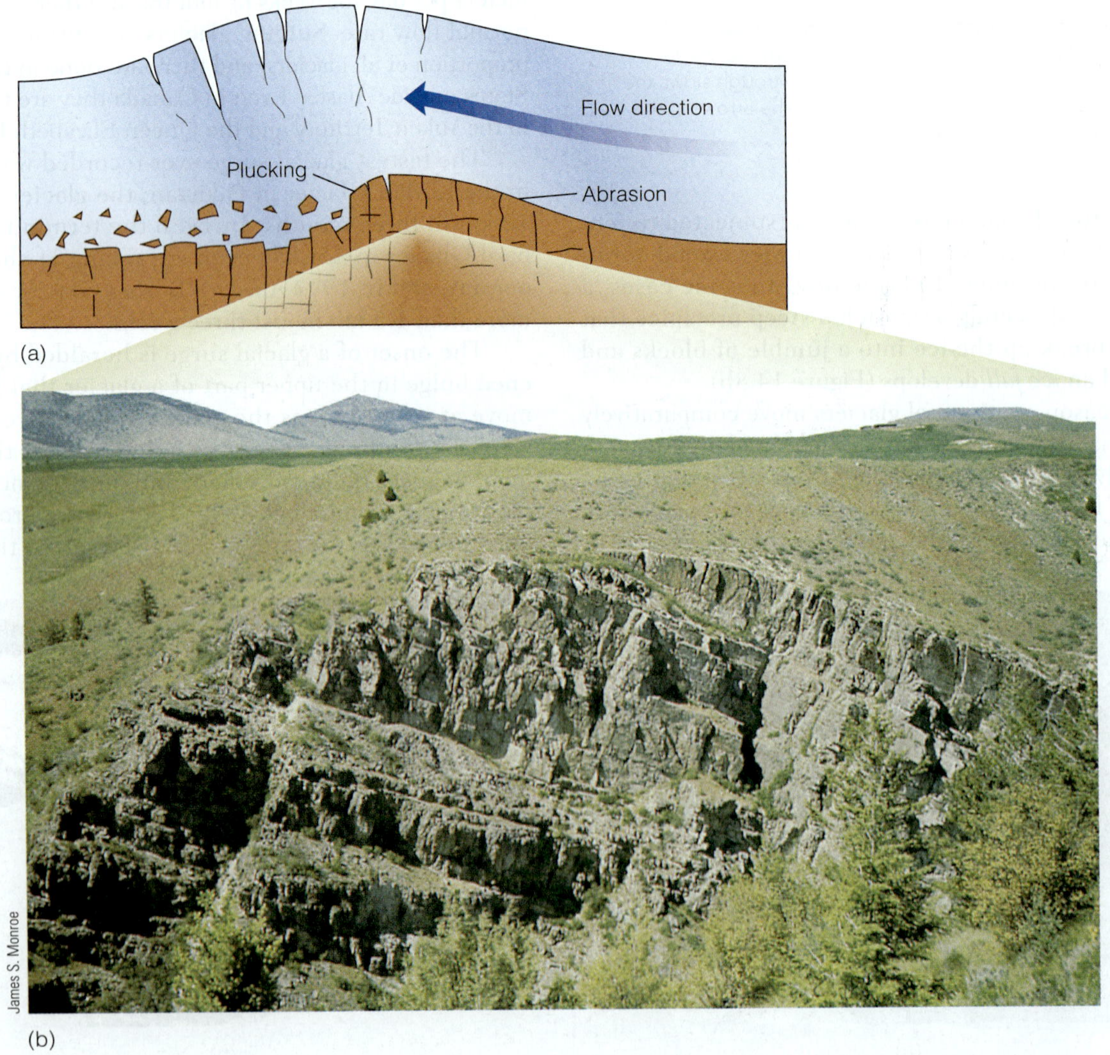

■ **Figure 14.9**

(a) Origin of a roche moutonnée. As the ice moves over a hill, it smooths the "upstream" side by abrasion and shapes the "downstream" side by plucking. (b) A roche moutonnée in Montana.

James S. Monroe

(a)

James S. Monroe

(b)

James S. Monroe

(c)

■ **Figure 14.10**

When sediment-laden ice moves over rocks, it abrades them and imparts a sheen known as glacial polish (a), as on this gneiss in Michigan. Glacial polish is also visible in (b) and so are straight scratches called glacial striations. The rock is basalt at Devil's Postpile National Monument, California. (c) The water in this stream in Switzerland is discolored by rock flour, small particles formed by glacial abrasion.

Bedrock over which sediment-laden glacial ice moves is effectively eroded by **abrasion** and commonly develops a **glacial polish,** a smooth surface that glistens in reflected light (■ Figure 14.10a). Abrasion also yields *glacial striations,* rather straight scratches on rock surfaces (Figure 14.10b). Glacial striations are rarely more than a few millimeters deep, whereas *glacial grooves* are similar but much larger and deeper. Abrasion also thoroughly pulverizes rocks so that they yield an aggregate of clay- and silt-sized particles that have the consistency of flour—hence the name *rock flour* (Figure 14.10c). Rock flour is so common in streams discharging from glaciers that the water has a milky appearance.

Continental glaciers derive sediment from mountains projecting through them, and windblown dust settles on their surfaces. Otherwise, they obtain most of their sediment from the surface they move over, which is transported in the lower part of the ice sheet. In contrast, valley glaciers carry sediment in all parts of the ice, but it is concentrated at the base and along the margins (■ Figure 14.11). Some of the marginal sediment is derived

Sue Monroe

■ **Figure 14.11**

Debris on the surface of the Mendenhall Glacier in Alaska. The largest boulder measures about 2 m across. Notice the ice fall in the background. To get some idea of scale, notice the person at the left center of the image.

by abrasion and plucking, but much of it is supplied by mass wasting processes, as when soil, sediment, or rock falls or slides onto a glacier's surface.

Erosion by Valley Glaciers

Erosion by valley glaciers has yielded some of the world's most inspiring scenery. Many mountain ranges are scenic to begin with, but when modified by valley glaciers, they take on a unique appearance of angular ridges and peaks in the midst of broad valleys. Several national parks and monuments in the western United States and in Canada owe their scenic appeal to erosion by valley glaciers (see Geo-Focus 14.1). The erosional landforms that resulted from valley glaciation are easily recognized and enable us to appreciate the tremendous erosive power of moving ice (■ Figure 14.12).

U-Shaped Glacial Troughs One of the most distinctive features of valley glaciation is a **U-shaped glacial trough** (Figure 14.12c). Mountain valleys eroded by running water are typically V-shaped in cross section; that is, they have valley walls that descend to a narrow valley bottom (see "Valley Glaciers and Erosion" on pages 400 and 401). In contrast, valleys scoured by glaciers are deepened, widened, and straightened so that they have very steep or vertical walls but broad, rather flat valley floors; thus, they exhibit a U-shaped profile. Many glacial troughs contain triangular *truncated spurs,* which are cutoff or truncated ridges that extend into the preglacial valley (Figure 14.12c).

During the Pleistocene, when glaciers were more extensive, sea level was about 130 m lower than at present, so glaciers flowing into the sea eroded their valleys to much greater depths than they do now. When the glaciers melted at the end of the Pleistocene, sea level rose and the ocean filled the lower ends of the glacial troughs, so that now they are long, steep-walled embayments called **fiords.**

Lower sea level during the Pleistocene was not responsible for the formation of all fiords. Unlike running water, glaciers can erode a considerable distance below sea level. In fact, a glacier 500 m thick can stay in contact with the sea floor and effectively erode it to a depth of about 450 m before the buoyant effects of water cause the glacial ice to float! The depth of some fiords is impressive; some in Norway and southern Chile are about 1300 m deep.

Hanging Valleys Waterfalls form in several ways, but some of the world's highest and most spectacular are found in recently glaciated areas. For example, several waterfall in Yosemite National Park, California, plunge from a **hanging valley,** which is a tributary valley whose floor is at a higher level than that of the main valley (see "Valley Glaciers and Erosion" on pages 400 and 401).

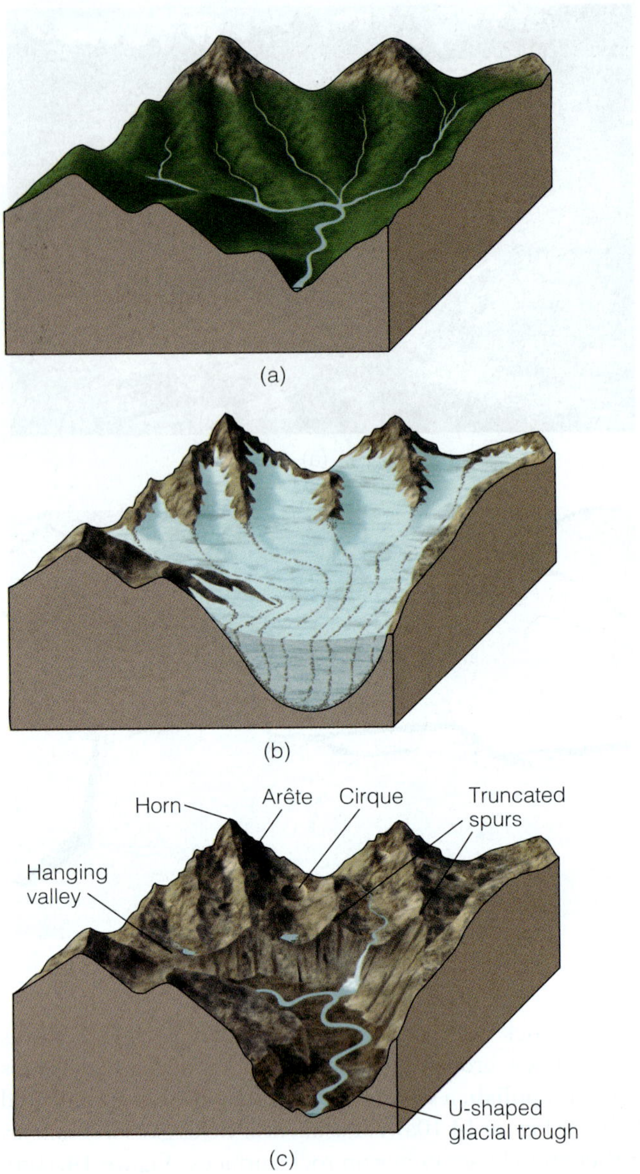

■ **Figure 14.12**

Erosional landforms produced by valley glaciers. (a) A mountain area before glaciation. (b) The same area during the maximum extent of valley glaciers. (c) After glaciation.

As Figure 14.12 shows, the large glacier in the main valley vigorously erodes, whereas the smaller glaciers in tributary valleys are less capable of erosion. When the glaciers disappear, the smaller tributary valleys remain as hanging valleys. Accordingly, streams flowing through hanging valleys plunge over vertical or steep precipices.

Cirques, Arêtes, and Horns Perhaps the most spectacular erosional landforms in areas of valley glaciation are at the upper ends of glacial troughs and along the divides that separate adjacent glacial troughs. Valley glaciers form and move out from steep-walled, bowl-shaped depressions called **cirques** at the upper end of their troughs (Figure 14.12c). Cirques are typically steep-walled on three sides, but one side opens into a glacial trough.

The origin of cirques is not fully understood, but these depressions apparently form by erosion of a preexisting depression on a mountain side. As snow and ice accumulate in the depression, frost wedging and plucking enlarge it until it takes on the typical cirque shape. Abrasion, plucking, and several mass wasting processes cut deeper into mountain sides by headward erosion so cirques become wider and deeper. Thus, a combination of processes can erode a small mountainside depression into a large cirque; the largest one known is the Walcott Cirque in Antarctica, which is 16 km wide and 3 km deep. Many cirques have a lip or threshold, indicating that the glacial ice not only moves outward but also rotates, scouring out a depression rimmed by rock. These depressions commonly contain a small lake known as a *tarn* (see "Valley Glaciers and Erosion" on pages 400 and 401).

The fact that cirques expand laterally and by headward erosion accounts for the origin of two other distinctive erosional features, *arêtes* and *horns*. **Arêtes**—narrow, serrated ridges—form in two ways. In many cases, cirques form on opposite sides of a ridge, and headward erosion reduces the ridge until only a thin partition of rock remains (Figure 14.12c). The same effect results when erosion in two parallel glacial troughs reduces the intervening ridge to a thin spine of rock.

The most majestic of all mountain peaks are **horns;** these steep-walled, pyramidal peaks are formed by headward erosion of cirques. In order for a horn to form, a mountain peak must have at least three cirques on its flanks, all eroding headward (Figure 14.12c). Excellent examples of horns are Mount Assiniboine in the Canadian Rockies, the Grant Teton in Wyoming, and the most famous of all, the Matterhorn in Switzerland.

Continental Glaciers and Erosional Landforms

Areas eroded by continental glaciers tend to be smooth and rounded because these glaciers bevel and abrade high areas that projected into the ice. Rather than yielding the sharp, angular landforms typical of valley glaciation, continental glaciers produce a landscape of rather flat topography interrupted by rounded hills. These areas have deranged drainage (see Figure 12.19e), numerous lakes and swamps, low relief, extensive bedrock exposures, and little or no soil. They are referred to as *ice-scoured plains* (■ Figure 14.13).

In a large part of Canada, particularly the vast Canadian Shield region, continental glaciation has stripped off the soil and unconsolidated surface sediment to reveal extensive exposures of striated and polished bedrock. Similar though smaller bedrock exposures are also widespread in the northern United States from Maine through Minnesota.

GLACIAL DEPOSITS

Both valley and continental glaciers effectively erode and transport, but eventually they deposit their sediment load as **glacial drift,** a general term for all deposits that result from glacial activity. A vast sheet of Pleistocene glacial drift is in the northern tier of the United States and adjacent parts of Canada. Smaller but similar deposits are found where valley glaciers existed or remain active. The appearance of these deposits may not be as inspiring as are some landforms that result from glacial erosion, but they are important as reservoirs for groundwater and in many areas they are exploited for their sand and gravel. As a matter of fact, glacial sand and gravel constitute a large part of the mineral extraction economies of several states and provinces.

All glacial drift has been carried from its source area and subsequently deposited elsewhere.

Alan Kelleheim/Mary Pat Ziter, JLM Visuals

■ **Figure 14.13**

An ice-scoured plain in the Northwest Territories of Canada.

GEOFOCUS 14.1

Glaciers in U.S. and Canadian National Parks

Glaciers are in several national parks in the United States and Canada, but here we concentrate on only two: Waterton Lakes National Park in Alberta and Glacier National Park in Montana. In 1932 both parks were designated an international peace park, the first of its kind.

Both parks have spectacular scenery; interesting wildlife such as mountain goats, bighorn sheep, and grizzly bears; and an impressive geologic history. The present-day landscapes resulted from deformation and uplift from Cretaceous to Eocene times, followed by deep erosion by glaciers and running water. Park visitors can see the results of the phenomenal forces at work during deformation by visiting sites where a large fault is visible, and glacial landforms such as U-shaped glacial troughs, arêtes, cirques, and horns are some of the finest in North America (■ Figure 1).

Most of the rocks exposed in Glacier National Park belong to the Late Proterozoic-aged Belt Supergroup,* whereas those in Waterton Lakes National Park are assigned to the Purcell Supergroup. The names differ north and south of the border, but the rocks are the same. These Belt–Purcell rocks are nearly 4000 m thick and were deposited between 1.45 billion and 850 million years ago. The rocks themselves are interesting and some are attractive, especially red and green rocks consisting mostly of mud and thick limestone formations. In addition, many of the rocks contain a variety of sedimentary structures such as mud cracks, ripple marks, and cross-bedding that help geologists interpret how they were deposited in the first place.

Dark mudstones and sandstones were deposited during the Cretaceous Period, when a marine transgression covered a large part of North America.

The most impressive geologic structure in the parks is the Lewis overthrust,** a large fault along which Belt–Purcell rocks have moved at least 75 km eastward so that they now rest on much younger

James S. Monroe

(a)

Ron Watts/Corbis

(b)

■ **Figure 1**

The sharp angular peaks and ridges and rounded valleys are typical of areas eroded by valley glaciers such as (a) Glacier National Park, Montana, and (b) Waterton National Park, Alberta, Canada.

**Supergroup* is a geologic term for two or more groups that in turn are composed of two or more formations.

**An overthrust fault is a very low-angle thrust fault along which movement is usually measured in kilometers.

Cretaceous-aged rocks. (See "Types of Faults" on pages 272 and 273.)

During the Pleistocene Epoch, glaciers formed and grew, overtopping the divides between valleys and thus forming an ice cap that nearly buried the entire area. In fact, several episodes of Pleistocene glaciation took place, but the evidence for the most recent one is most obvious. These glaciers flowed outward in all directions, and in the east they merged with the continental glacier covering most of Canada and the northern states. Much of the parks' landscapes developed during these glacial episodes as valleys were gouged deeper and widened, and cirques, arêtes, and horns developed (Figure 1).

Today only a few dozen small glaciers remain active in the parks. But just like earlier glaciers, they continue to erode, transport, and deposit sediment, only at a considerably reduced rate. In fact, many of the 150 or so glaciers in Glacier National Park in 1850 are now gone or remain as only patches of stagnant ice. And even among the others it is difficult to determine exactly how many are active because they are so small and move so slowly, only a few meters per year. They did, however, expand markedly during the Little Ice Age but have since retreated. For example, Grinnell Glacier covered only 0.88 km^2 in 1993 as opposed to 2.33 km^2 in 1850 (■ Figure 2), and during the same time Sperry Glacier was down to 0.87 km^2 from 3.76 km^2.

It seems that these small glaciers are shrinking as a result of the 1°C increase in average summer temperatures in this region since 1900. According to one U.S. Geological Survey report, expected increased warming will eliminate the glaciers by 2030 and certainly by 2100 even if no additional warming takes place.

■ **Figure 2**

Grinnell Glacier in Glacier National Park, Montana. In 1850, at the end of the Little Ice Age, the glacier extended much farther and covered about 2.33 km^2. By 1981, its terminus had retreated to the position shown, and in 1993, it covered only about 0.88 km^2.

None of the active glaciers in either park can be reached by road, but several are visible from a distance. Nevertheless, Pleistocene glaciers and the remaining active ones were responsible for much of the striking scenery. Now weathering, mass wasting, and streams are modifying the glacial landscape.

Valley Glaciers and Erosion

Valley glaciers effectively erode and produce several easily recognized landforms. Where glaciers move through mountain valleys, the valleys are deepened and widened, giving them a distinctive U-shaped profile. The peaks and ridges rising above valley glaciers are also eroded, and they become jagged and angular. Much of the spectacular scenery in Grand Teton National Park, Wyoming, Yosemite National Park, California, and Glacier National Park, Montana, resulted from erosion by valley glaciers. In fact, valley glaciers remain active in some of the mountains of western North America, especially in Alaska and Canada.

Part of southwestern Greenland (right) where valley glaciers merged to form an ice cap. Should these glaciers melt, several landforms like those in Figure 14.12c would be present. The Teton Range in Wyoming (above) acquired its angular peaks and ridges and broadly rounded valleys as a result of erosion by valley glaciers.

U-shaped glacial troughs. The glacial trough above is in northern Montana, whereas the one above right is in southern Germany. The lake is impounded behind a glacial deposit known as an end moraine. The steep-walled glacial trough in Norway (right) extends below sea level, so it is a fiord.

The bowl-shaped depression on Mount Wheeler in Great Basin National Park, Nevada, is a cirque. It has steep walls on three sides and opens out into a glacial trough.

Lake Helen on Lassen Peak in Lassen Volcanic National Park, California, is a tarn—that is, a lake in a cirque.

Nevada Falls in Yosemite National Park, California, plunges 181 m from a hanging valley. The valley in the foreground is a huge U-shaped glacial trough.

The Matterhorn (above) in Switzerland is a well-known horn. This view on the Jungfrau (left) in Switzerland shows two small Glaciers, a cirque headwall, and an arête.

R. V. Dietrich

(a)

James S. Monroe

(b)

■ Figure 14.14

Glacial erratics at (a) Hammond, New York, and (b) Beaver Island in Lake Michigan. Both boulders have been transported far from their sources and are unlike the rocks at their present locations.

The many glacially deposited rock fragments and boulders scattered about the surface that were obviously not derived from the area in which they now rest are known as **erratics** (■ Figure 14.14). A good example is the popular decorative stone in Michigan known as puddingstone, consisting of quartzite with conspicuous pieces of red jasper that came from rock exposures in Ontario, Canada.

Geologists define two types of glacial drift: till and stratified drift. **Till** consists of sediments deposited directly by glacial ice. These deposits are not sorted by particle size or density, and they exhibit no layering or stratification. The till of both valley and continental glaciers is similar, but that of continental glaciers is much more extensive and usually has been transported farther.

In contrast to till, **stratified drift** is layered (stratified) and invariably exhibits some degree of sorting. As a matter of fact, stratified drift is actually layers of sand and gravel that accumulated in braided stream channels. In Chapter 12 we mentioned that streams issuing from melting glaciers are commonly braided because they receive more sediment than they can effectively transport.

Landforms Composed of Till

Landforms composed of till include several types of *moraines* and elongated hills known as *drumlins*.

End Moraines

The terminus of any glacier may become stabilized in one position for some period of time, perhaps a few years or even decades. Stabilization of the ice front does not mean that the glacier has ceased to flow, only that it has a balanced budget. When an ice front is stationary, the glacier continues to flow, and the sediment transported within or upon the ice is dumped as a pile of rubble at the glacier's terminus. These deposits are **end moraines,** which continue to grow as long as the ice front remains stabilized (■ Figure 14.15). End moraines of valley glaciers are commonly crescent-shaped ridges of till spanning the valley occupied by the glacier. Those of continental glaciers similarly parallel the ice front but are much more extensive.

Following a period of stabilization, a glacier may advance or retreat, depending on changes in its budget. If it advances, the ice front overrides and modifies its former moraine. If it has a negative budget, though, the ice front retreats toward the zone of accumulation. As the ice front recedes, till is deposited as it is liberated from the melting ice and forms a layer of **ground moraine** (Figure 14.15b). Ground moraine has an irregular, rolling topography, whereas end moraine consists of long ridgelike accumulations of sediment.

After a glacier has retreated for some time, its terminus may once again stabilize, and it deposits another end moraine. Because the ice front has receded, such moraines are called **recessional moraines** (Figure 14.15b). During the Pleistocene, continental glaciers in the mid-continent region extended as far south as southern Ohio, Indiana, and Illinois. Their outermost end moraines, marking the greatest extent of the glaciers, go by the special name **terminal moraine** (valley glaciers also deposit terminal moraines) (■ Figure 14.16). As the glaciers retreated from the positions where their terminal moraines were deposited, they temporarily stopped retreating numerous times and deposited dozens of recessional moraines.

Lateral and Medial Moraines

As we discussed, valley glaciers transport considerable sediment along their margins. Much of this sediment is abraded and plucked from the valley walls, but a significant amount falls or slides onto the glacier's surface by mass wasting processes. In any case, when a glacier melts, this sediment is deposited as long ridges of till called **lateral moraines** along the margin of the glacier (■ Figure 14.17).

Where two lateral moraines merge, as when a tributary glacier flows into a larger glacier, a **medial moraine**

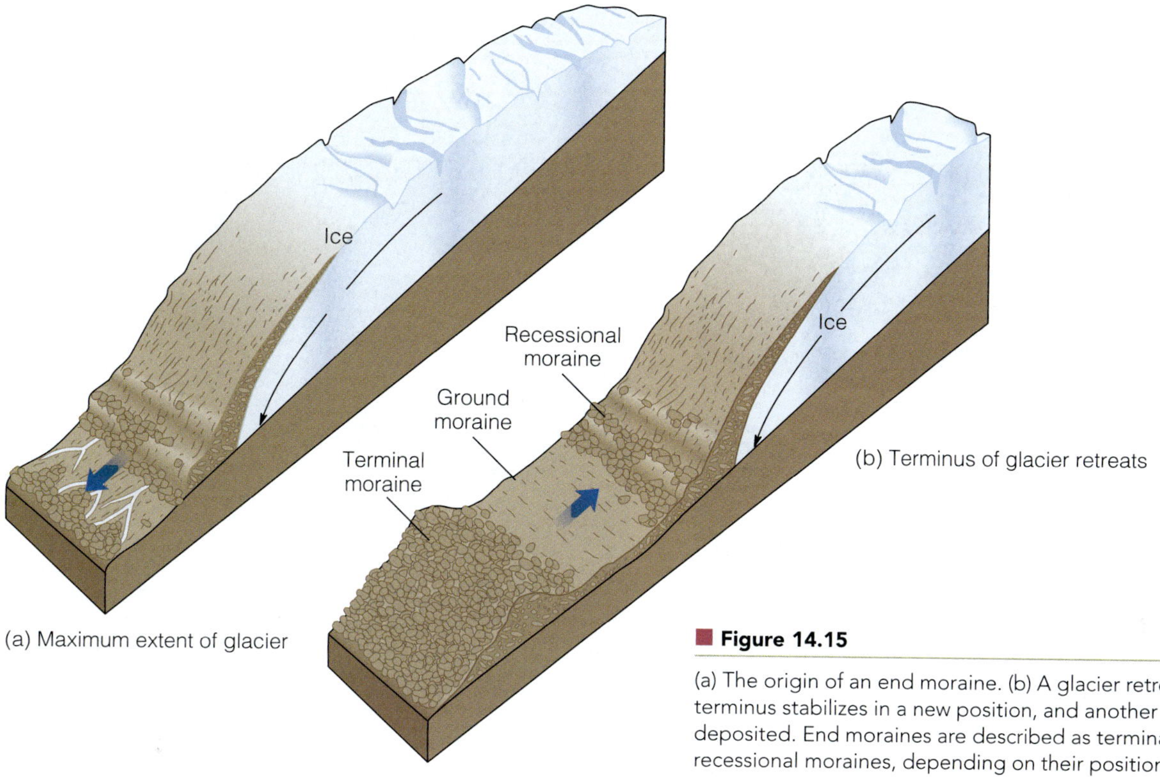

■ **Figure 14.15**

(a) The origin of an end moraine. (b) A glacier retreats and its terminus stabilizes in a new position, and another end moraine is deposited. End moraines are described as terminal moraines or recessional moraines, depending on their positions.

(a)

(b)

■ **Figure 14.16**

(a) An end moraine deposited by a valley glacier. This particular end moraine is also a terminal moraine because it is the one most distant from the glacier's source. (b) Closeup of an end moraine. Notice that the deposit is not sorted by particle size, and it shows no layering or stratification.

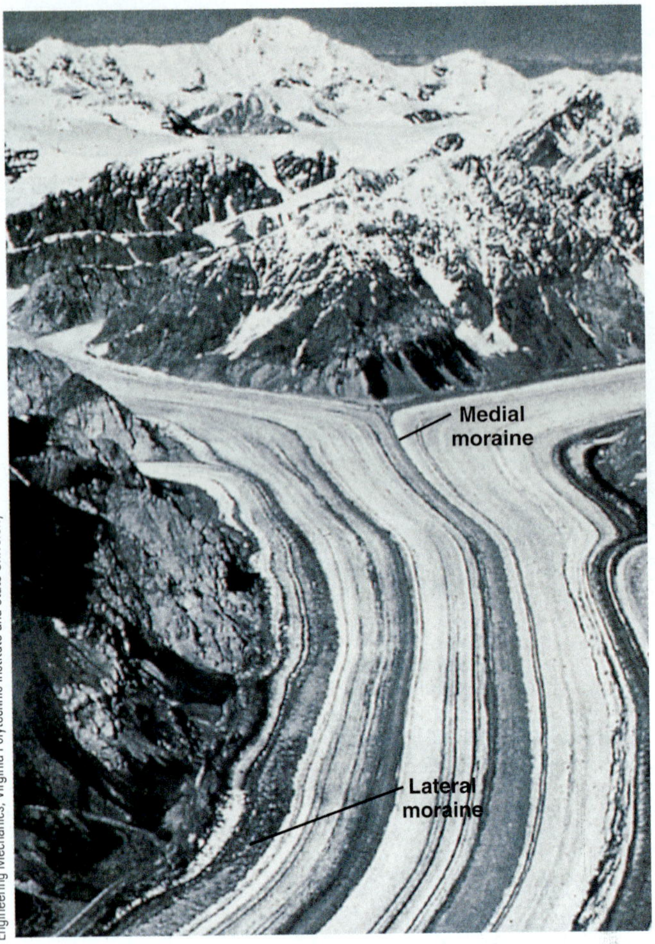

(a)

(b)

Figure 14.17

(a) Lateral and medial moraines on a glacier in Alaska. Notice that where the two tributary glaciers converge, two lateral moraines merge to form a medial moraine. (b) The two parallel ridges extending from this mountain valley are lateral moraines.

forms (Figure 14.17). Although medial moraines are identified by their position on a valley glacier, they are, in fact, formed from the coalescence of two lateral moraines. One can generally determine how many tributaries a valley glacier has by the number of its medial moraines.

Drumlins In many areas where continental glaciers deposited till, the till has been reshaped into elongated hills known as **drumlins.** Some drumlins are as large as 50 m high and 1 km long, but most are much smaller. From the side, a drumlin looks like an inverted spoon with the steep end on the side from which the glacial ice advanced and the gently sloping end pointing in the direction of ice movement (■ Figure 14.18).

One hypothesis for the origin of drumlins holds that they form in the zone of plastic flow as glacial ice modifies till into streamlined hills. According to another hypothesis, drumlins form when huge floods of glacial meltwater modify deposits of till.

Drumlins rarely occur as single, isolated hills; instead they are usually found in *drumlin fields* that contain hundreds or thousands of drumlins. Drumlin fields are found in several states and Ontario, Canada, but perhaps the finest example is near Palmyra, New York.

Landforms Composed of Stratified Drift

Stratified drift is found in areas of both valley and continental glaciation, but as one would expect, it is more extensive in association with continental glaciation.

Outwash Plains and Valley Trains Glaciers discharge meltwater laden with sediment most of the time, except perhaps during the coldest months. This meltwater forms a series of braided streams that radiate out from the front of continental glaciers over a wide region. So much sediment is supplied to these streams that much of it is deposited within the channels as sand and gravel bars. The vast blanket of sediments so formed is an **outwash plain** (■ Figure 14.19).

Valley glaciers also discharge huge amounts of meltwater and, like continental glaciers, have braided streams extending from them. However, these streams are confined to the lower parts of glacial troughs, and their long, narrow deposits of stratified drift are known as **valley trains** (■ Figures 14.19 and 14.20a).

Outwash plains and valley trains commonly contain numerous circular to oval depressions, many of which contain small lakes. These depressions are *kettles* that

■ **Figure 14.18**

These streamlined hills are drumlins. Can you tell the direction of ice movement from their shape?

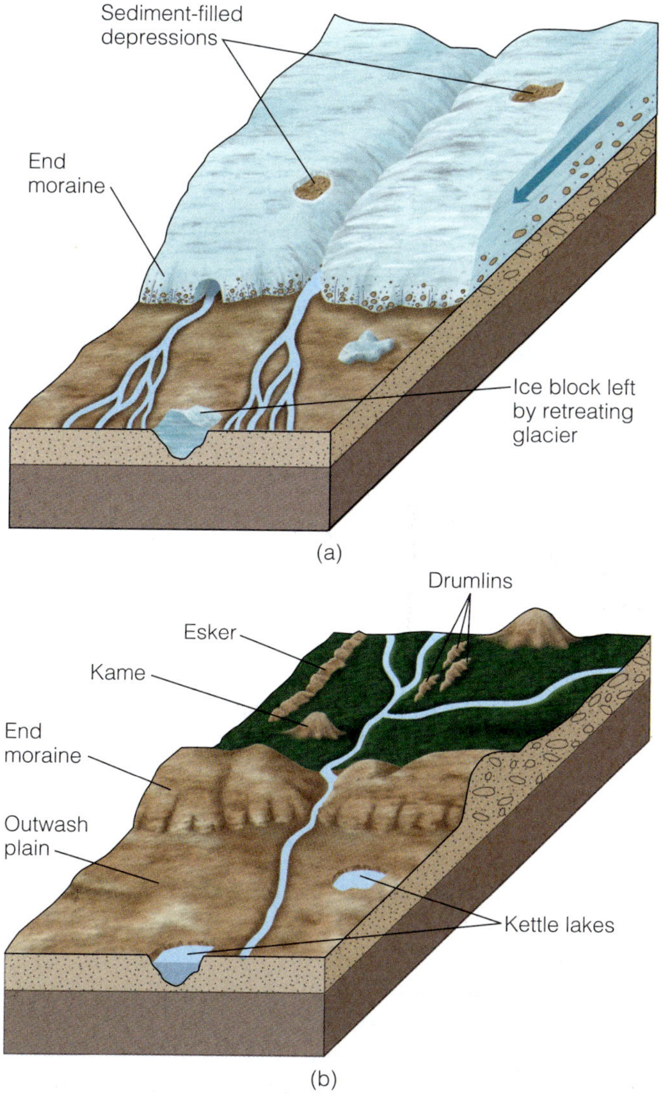

form when a retreating ice sheet or valley glacier leaves a block of ice that is subsequently partly or wholly buried (Figure 14.19 and 14.20b). When the ice block eventually melts, it leaves a depression; if the depression extends below the water table, it becomes the site of a small lake. Some outwash plains have so many kettles that they are called *pitted outwash plains.*

Kames and Eskers **Kames** are conical hills as high as 50 m composed of stratified drift (Figures 14.19 and 14.20c). Many kames form when a stream deposits sediment in a depression on a glacier's surface; as the ice melts, the deposit is lowered to the surface. Kames also form in cavities within or beneath stagnant ice.

Long, sinuous ridges of stratified drift, many of which meander and have tributaries, are **eskers** (Figures 14.19 and 14.20d). Some eskers are as high as 100 m and can be traced for more than 100 km. Most eskers are in areas once covered by continental glaciers, but they are also found beneath large valley glaciers. The sorting and stratification of the sediments in eskers clearly indicate deposition by running water. The properties of ancient eskers and observations of present-day glaciers indicate that they form in tunnels beneath stagnant ice (Figure 14.19).

■ **Figure 14.19**

Two stages in the origin of kettles, kames, eskers, drumlins, and outwash plains: (a) during glaciation and (b) after glaciation.

Figure 14.20

(a) A valley train in Alaska made up of stratified drift. (b) A kettle in a moraine in Alaska. (c) This small hill in Wisconsin is a kame. (d) This sinuous ridge near Dahlen, North Dakota, is an esker.

James S. Monroe

(a)

James S. Monroe

(b)

B.M.C. Pape

(c)

Tom Bean/Corbis

(d)

Glacial Lake Deposits

Numerous lakes exist in areas of glaciation. Some formed as a result of glaciers scouring out depressions, others are found where a stream's drainage was blocked, and others are the result of water accumulating behind moraines or in kettles. Regardless of how they formed, glacial lakes, like all lakes, are areas of deposition. Sediment may be carried into them and deposited as small deltas, but of special interest are the fine-grained deposits. Mud deposits in glacial lakes are commonly finely laminated (having layers less than 1 cm thick) and consist of alternating light and dark layers. Each light-dark couplet is a *varve* (■ Figure 14.21), which represents an annual episode of deposition. The light layer forms during the spring and summer and consists of silt and clay; the dark layer forms during the winter when the smallest particles of clay and organic matter settle from suspension as the lake freezes over. The number of varves indicates how many years a glacial lake existed.

Another distinctive feature of glacial lakes containing varves is *dropstones* (Figure 14.21). These are pieces of gravel, some of boulder size, in otherwise very fine-grained deposits. Most of them were probably carried into the lakes by icebergs that eventually melted and released sediment contained in the ice.

WHAT CAUSES ICE AGES?

How an individual glacier forms is well understood: if more snow falls than melts during the warm season, a net accumulation takes place; the snow gets deeper and deeper and at depth is

Canadian Geological Survey

■ **Figure 14.21**

Glacial varves with a dropstone.

What Would You Do?

During a visit to the Pacific Northwest, you notice that a stream has cut a gorge. In the lower part of the gorge is a bedrock projection similar to the one in Figure 14.9b. Overlying this bedrock is a sequence of thin (2–4 mm thick) alternating layers of dark and light, very fine-grained materials (silt and clay) with a few boulders measuring 10–15 cm across. In the uppermost part of the gorge is a deposit of mud, sand, and gravel that shows no layering or sorting. How would you decipher the geologic history of these deposits?

converted to glacial ice. And, as previously discussed, flow begins when the critical thickness of about 40 m is reached. So, we know how glaciers form, something of their dynamics, and how they affect the land surface, but we have not addressed two questions: (1) What causes large-scale episodes of glaciation? and (2) Why have there been so few episodes of widespread glaciation? Glaciers not only were much more widespread during the Pleistocene Epoch but also expanded and contracted several times. (See Chapter 23, "Cenozoic Earth and Life History," for more on the effects of Pleistocene glaciers.)

Only a few periods of glaciation are recognized in the geologic record, each separated from the others by long intervals of mild climate. Such long-term climatic changes probably result from slow geographic changes related to plate tectonic activity. Moving plates carry continents to high latitudes where glaciers can exist, provided they receive enough precipitation as snow. Plate collisions, the subsequent uplift of vast areas far above sea level, and the changing atmospheric and oceanic circulation patterns caused by the changing shapes and positions of plates also contribute to long-term climatic change.

A theory explaining ice ages must address the fact that during the Pleistocene Ice Age (1.6 million to 10,000 years ago) several intervals of glacial expansion were separated by warmer interglacial periods. At least four major episodes of glaciation have been recognized in North America, and six or seven glacial advances and retreats occurred in Europe. These intermediate-term climatic events take place on time scales of tens to hundreds of thousands of years. The cyclic nature of this most recent episode of glaciation has long been a problem in formulating a comprehensive theory of climatic change.

The Milankovitch Theory

A particularly interesting hypothesis for intermediate-term climatic events was put forth by the Yugoslavian astronomer Milutin Milankovitch during the 1920s. He proposed that minor irregularities in Earth's rotation and orbit are sufficient to alter the amount of solar radiation that Earth receives at any given latitude and hence can affect climatic changes. Now called the **Milankovitch theory,** it was initially ignored but has received renewed interest during the last 25 years.

Milankovitch attributed the onset of the Pleistocene Ice Age to variations in three parameters of Earth's orbit (■ Figure 14.22). The first of these is orbital eccentricity, which is the degree to which the orbit departs from a perfect circle. Calculations indicate a roughly 100,000-year cycle between times of maximum eccentricity. This corresponds closely to 20 warm-cold climatic cycles that occurred during the Pleistocene. The second parameter is the angle between Earth's axis and a line perpendicular to the plane of its orbit around the Sun. This angle shifts about 1.5 degrees from its current value of 23.5 degrees during a 41,000-year cycle. The third parameter is the precession of the equinoxes, which causes the position of the equinoxes and solstices to shift slowly around Earth's elliptical orbit in a 23,000-year cycle.

Continuous changes in these three parameters cause the amount of solar heat received at any latitude to vary slightly over time. The total heat received by the planet, however, remains little changed. Milankovitch proposed, and now many scientists agree, that the interaction of these three parameters provides the triggering mechanism for the glacial-interglacial episodes of the Pleistocene.

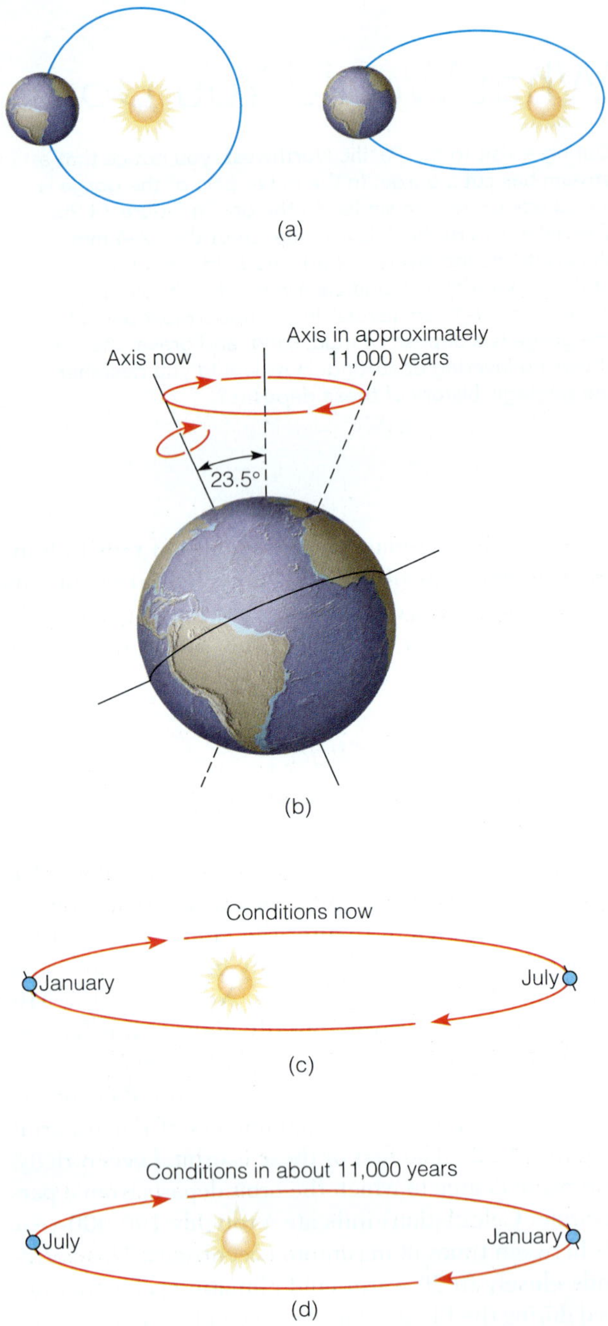

Short-Term Climatic Events

Climatic events with durations of several centuries, such as the Little Ice Age, are too short to be accounted for by plate tectonics or Milankovitch cycles. Several hypotheses have been proposed, including variations in solar energy and volcanism.

Variations in solar energy could result from changes within the Sun itself or from anything that would reduce the amount of energy Earth receives from the Sun. The latter could result from the solar system passing through clouds of interstellar dust and gas or from substances in the atmosphere reflecting solar radiation back into space. Records kept during the past 80 years, however, indicate that during this time the amount of solar radiation has varied only slightly. Thus, although variations in solar energy may influence short-term climatic events, such a correlation has not been demonstrated.

During large volcanic eruptions, tremendous amounts of ash and gases are spewed into the atmosphere where they reflect incoming solar radiation and thus reduce atmospheric temperatures. Recall from Chapter 5 that small droplets of sulfur gases remain in the atmosphere for years and can have a significant effect on climate. Several large-scale volcanic events have occurred, such as the 1815 eruption of Tambora, and are known to have had climatic effects. However, no relationship between periods of volcanic activity and periods of glaciation has yet been established.

Figure 14.22

Minor irregularities in Earth's rotation and orbit may affect climatic changes. (a) Earth's orbit varies from nearly a circle (left) to an ellipse (right) and back again in about 100,000 years. (b) Earth moves around its orbit while spinning about its axis, which is tilted to the plane of its orbit around the Sun at 23.5 degrees and points toward the North Star. Earth's axis of rotation slowly moves and traces out the path of a cone in space. (c) At present, Earth is closest to the Sun in January when the Northern Hemisphere experiences winter. (d) In about 11,000 years, as a result of precession, Earth will be closer to the Sun in July, when summer occurs in the Northern Hemisphere

14 GEO RECAP

Chapter Summary

- Glaciers currently cover about 10% of the land surface and contain about 2.15% of all water on Earth.
- A glacier forms when winter snowfall exceeds summer melt and therefore accumulates year after year. Snow is compacted and converted to glacial ice, and when the ice is about 40 m thick, pressure causes it to flow.
- Glaciers move by plastic flow and basal slip.
- Valley glaciers are confined to mountain valleys and flow from higher to lower elevations, whereas continental glaciers cover vast areas and flow outward in all directions from a zone of accumulation.
- The behavior of a glacier depends on its budget, which is the relationship between accumulation and wastage. If a glacier has a balanced budget, its terminus remains stationary; a positive or negative budget results in the advance or retreat of the terminus, respectively.
- Glaciers move at varying rates depending on slope, discharge, and season. Valley glaciers tend to flow more rapidly than continental glaciers.
- Glaciers effectively erode and transport because they are solids in motion. They are particularly effective at eroding soil and unconsolidated sediment, and they can transport any size sediment supplied to them.
- Continental glaciers transport most of their sediment in the lower part of the ice, whereas valley glaciers may carry sediment in all parts of the ice.
- Erosion of mountains by valley glaciers yields several sharp, angular landforms including cirques, arêtes, and horns. U-shaped glacial troughs, fiords, and hanging valleys are also products of valley glaciation.
- Continental glaciers abrade and bevel high areas, producing a smooth, rounded landscape known as an ice-scoured plain.
- Depositional landforms include moraines, which are ridge-like accumulations of till. The several types of moraines are terminal, recessional, lateral, and medial.
- Drumlins are composed of till that was apparently reshaped into streamlined hills by continental glaciers or floods.
- Stratified drift in outwash and valley trains consists of sand and gravel deposited by meltwater streams issuing from glaciers. Ridges known as eskers, and conical hills called kames are also composed of stratified drift.
- Major glacial intervals separated by tens or hundreds of millions of years probably occur as a result of the changing positions of tectonic plates, which in turn cause changes in oceanic and atmospheric circulation patterns.
- Currently, the Milankovitch theory is widely accepted as the explanation for glacial-interglacial intervals.
- The reasons for short-term climatic changes, such as the Little Ice Age, are not understood. Two proposed causes are changes in the amount of solar energy received by Earth and volcanism.

Important Terms

abrasion (p. 395)
arête (p. 397)
basal slip (p. 388)
cirque (p. 397)
continental glacier (p. 390)
drumlin (p. 404)
end moraine (p. 402)
erratic (p. 402)
esker (p. 405)
fiord (p. 396)
firn (p. 388)
glacial budget (p. 391)
glacial drift (p. 397)
glacial ice (p. 388)
glacial polish (p. 395)
glacial surge (p. 393)
glacier (p. 386)
ground moraine (p. 402)
hanging valley (p. 396)
horn (p. 397)
ice cap (p. 390)
kame (p. 405)
lateral moraine (p. 402)
medial moraine (p. 402)
Milankovitch theory (p. 407)
outwash plain (p. 404)
plastic flow (p. 388)
recessional moraine (p. 402)
stratified drift (p. 402)
terminal moraine (p. 402)
till (p. 402)
U-shaped glacial trough (p. 396)
valley glacier (p. 390)
valley train (p. 404)
zone of accumulation (p. 391)
zone of wastage (p. 391)

Review Questions

1. Which one of the following statements is correct?
a._____A cirque forms where two valley glaciers merge; b._____Ice Age glacial deposits are found as far south as Alabama; c._____Most of northern Europe was covered by glacial ice during the Little Ice Age; d._____Many of the valley glaciers in the United States are in the Appalachian Mountains; e._____Less than 1% of all glacial ice is found outside Antarctica and Greenland.

2. If a glacier deposits a terminal moraine and then retreats and deposits another moraine, the latter is known as a(n)_____ moraine.
a._____lateral; b._____medial; c._____recessional; d._____optimal; e._____outwash.

3. An ârete is a(n)
a._____knifelike ridge between glacial troughs; b._____deposit of unsorted sand and gravel; c._____pyramid-shaped peak eroded by continental glaciers; d._____type of glacier found in mountain valleys; e._____outwash plain with numerous kettles.

4. When freshly fallen snow compacts and partly melts and refreezes, it forms granular ice known as
a._____kame; b._____till; c._____firn; d._____drift; e._____cirque.

5. Glaciers move mostly by
a._____surging; b._____basal slip; c._____lateral compression; d._____plastic flow; e._____abrasion.

6. A bowl-shaped depression on a mountainside at the upper end of a glacial trough is a(n)
a._____valley train; b._____cirque; c._____esker; d._____drumlin; e._____erratic.

7. During the Ice Age or _____ Epoch, glaciers covered about _____ of the land surface.
a._____Cretaceous/75%; b._____Proterozoic/10%; c._____Paleozoic/50%; d._____Pleistocene/30%; e._____Mesozoic/15%.

8. If a glacier has a balanced budget
a._____it stops moving; b._____its terminus remains stationary; c._____its rate of wastage exceeds its rate of accumulation; d._____the glacier's length decreases; e._____crevasses no longer form.

9. An ice-scoured plain is a(n)
a._____subdued landscape resulting from erosion by a continental glacier; b._____area with many cirques, horns, and âretes; c._____vast area covered by outwash deposits; d._____region in which drumlins and eskers are commonly found; e._____type of deposit consisting of alternating dark and light layers of clay.

10. The two present-day areas that have continental glaciers are
a._____Glacier National Park, Montana, and Waterton National Park, Canada; b._____Mount Baker, Washington, and the Sierra Nevada of California; c._____Antarctica and Greenland; d._____Scandinavia and Canada; e._____Iceland and Baffin Island.

11. When at least three glaciers erode a single mountain peak, it yields a pyramid-shaped peak known as a(n)
a._____horn; b._____drift; c._____esker;
d._____striation; e._____fiord.
12. A glacially transported boulder now lying far from its source is a(n)
a._____firn; b._____erratic; c._____moraine;
d._____varve; e._____dropstone.
13. How is it possible for glaciers to erode below sea level whereas steams cannot?
14. A valley glacier has a cross-sectional area of 400,000 m^2 and a flow velocity of 2 m/day. How long will it take for 1 km^3 of ice to move past a given point?
15. In your travels you encounter a roadside rock exposure that consists of alternating layers of dark and light laminated mud containing a few boulders measuring 20 to 40 cm across. Explain the sequence of events responsible for deposition.
16. Several of the Cascade Range volcanoes in California, Oregon, and Washington have glaciers on them. What two factors account for glaciers on these mountains?
17. Explain in terms of the glacial budget how a once-active glacier becomes stagnant.
18. What kinds of evidence would indicate that a now ice-free area was once covered by a continental glacier?
19. What are basal slip and plastic flow, what causes each, and how do they vary seasonally?
20. How does the Milankovitch theory explain the onset of Pleistocene episodes of glaciation?
21. What is the firn limit on a glacier, and how does its position relate to a glacier's budget?
22. What are hanging valleys, and how do they originate?
23. How do terminal and recessional moraines form?
24. How does glacial ice originate, and why is it considered a rock?

World Wide Web Activities

Geology Now Assess your understanding of this chapter's topics with additional quizzing and comprehensive interactivities at

http://earthscience.brookscole.com/changingearth4e

as well as current and up-to-date weblinks, additional readings, and InfoTrac College Edition exercises.

The Work of Wind and Deserts

CHAPTER 15
OUTLINE

- Introduction
- How Does Wind Transport Sediment?
- How Does Wind Erode Landforms?

  GEO-FOCUS 15.1: *Radioactive Waste Disposal—Safe or Sorry?*

  GEOLOGY IN UNEXPECTED PLACES: *Blowing in the Wind*
- What Are the Different Types of Wind Deposits?
- How Are Air-Pressure Belts and Global Wind Patterns Distributed?
- Where Do Deserts Occur?
- What Are the Characteristics of Deserts?
- What Types of Landforms Are Found in Deserts?
- Geo-Recap

Geology⇌Now *This icon, which appears throughout the book, indicates an opportunity to explore interactive tutorials, animations, or practice problems available on the GeologyNow website at http://earthscience.brookscole.com/changingearth4e.*

The Saharan community of El Gedida in western Egypt is slowly being overwhelmed by advancing sand. Source: George Gerster/The National Audubon Society/Photo Researchers

OBJECTIVES

At the end of this chapter, you will have learned that:

- Wind transports sediment and modifies the landscape through the processes of abrasion and deflation.
- Dunes and loess are the result of wind depositing material.
- Dunes form when wind flows over and around an obstruction.
- The four major dune types are barchan, longitudinal, transverse, and parabolic.
- Loess is formed from wind-blown silt and clay and is derived from three main sources: deserts, Pleistocene glacial outwash deposits, and river floodplains in semiarid regions.
- The global pattern of air-pressure belts and winds is responsible for Earth's atmospheric circulation patterns.
- Deserts are dry and receive less than 25 cm of rain per year, have high evaporation rates, typically have poorly developed soils, and are mostly or completely devoid of vegetation.
- The majority of deserts are found in the dry climates of the low and middle latitudes.
- Deserts have many distinctive landforms produced by both wind and running water.

Introduction

During the past several decades, deserts have been advancing across millions of acres of productive land, destroying rangeland, croplands, and even villages (see the chapter opening photo). Such expansion, estimated at 70,000 km^2 per year, has exacted a terrible toll in human suffering. Because of the relentless advance of deserts, hundreds of thousands of people have died of starvation or been forced to migrate as "environmental refugees" from their homelands to camps, where the majority are severely malnourished. This expansion of deserts into formerly productive lands is called **desertification** and is a major problem in many countries.

Most regions undergoing desertification lie along the margins of existing deserts where a delicately balanced ecosystem serves as a buffer between the desert on one side and a more humid environment on the other. Their potential to adjust to increasing environmental pressures from natural causes as well as human activity is limited. Ordinarily, desert regions expand and contract gradually in response to natural processes such as climatic change, but much recent desertification has been greatly accelerated by human activities.

In many areas, the natural vegetation has been cleared as crop cultivation has expanded into increasingly drier fringes to support growing populations. Because grasses are the dominant natural vegetation in most fringe areas, raising livestock is a common economic activity. However, increasing numbers of livestock in many areas have greatly exceeded the land's ability to support them. Consequently, the vegetation cover that protects the soil has diminished, causing the soil to crumble and be stripped away by wind and water, which results in increased desertification.

One particularly hard-hit area of desertification is the Sahel of Africa (a belt 300–1100 km wide, lying south of the Sahara). Because drought is common in the Sahel, the region can support only a limited population of livestock and humans. Unfortunately, growing human and animal populations and more intensive agriculture have increased the demands on the lands. Also experiencing periodic droughts, this region has suffered tremendously as crops have failed and livestock has overgrazed the natural vegetation, resulting in thousands of deaths, displaced people, and the encroachment of the Sahara.

The tragedy of the Sahel and prolonged droughts in other desert fringe areas remind us of the delicate equilibrium of ecosystems in such regions. Once the fragile soil cover has been removed by erosion, it takes centuries for new soil to form (see Chapter 6).

There are many important reasons to study deserts and the processes that are responsible for their formation. First, deserts cover large regions of Earth's surface. More than 40% of Australia is desert, and the Sahara occupies a vast part of northern Africa. Although deserts are generally sparsely populated, some desert regions are experiencing an influx of people, such as Las Vegas, Nevada, the high desert area of southern California, and various locations in Arizona. Many of these places already have problems with population growth.

Furthermore, with the current debate about global warming, it is important to understand how desert processes operate and how global climate changes affect the various Earth systems and subsystems. By understanding how desertification operates, people can take steps to eliminate or reduce the destruction done, particularly in terms of human suffering. Learning about the underlying causes of climate change by examining ancient desert regions may provide insight into the possible duration and severity of present and future climatic changes. This can have important ramifications in decisions about whether burying nuclear waste in a desert, such as Yucca Mountain, Nevada, is as safe as some claim and in our best interests as a society (see GeoFocus 15.1).

More than 6000 years ago, the Sahara was a fertile savannah supporting a diverse fauna and flora, including humans. Then the climate changed, and the area became a desert. How did this happen? Will this region change back again in the future? These are some of the questions geoscientists hope to answer by studying deserts.

And last, many agents and processes that have shaped deserts do not appear to be limited to our planet. Many features found on Mars, especially as seen in images transmitted by the *Spirit* and *Opportunity* rovers, are apparently the result of the same wind-driven processes that operate on Earth.

What Would You Do?

You have been asked to testify before a congressional committee charged with determining whether the National Science Foundation should continue to fund research devoted to the study of climate changes during the Cenozoic Era. Your specialty is desert landforms and the formation of deserts. What arguments would you make to convince the committee to continue funding research on ancient climates?

HOW DOES WIND TRANSPORT SEDIMENT?

Wind is a turbulent fluid and therefore transports sediment in much the same way as running water. Although wind typically flows at a greater velocity than water, it has a lower density and thus can carry only clay- and silt-size particles as *suspended load*. Sand and larger particles are moved along the ground as *bed load*.

Bed Load

Sediments too large or heavy to be carried in suspension by water or wind are moved as bed load either by *saltation* or by rolling and sliding. As we discussed in Chapter 12, saltation is the process by which a portion of the bed load moves by intermittent bouncing along a streambed. Saltation also occurs on land. Wind starts sand grains rolling and lifts and carries some grains short distances before they fall back to the surface. As the descending sand grains hit the surface, they strike other grains, causing them to bounce along by saltation (■ Figure 15.1). Wind-tunnel experiments show that once sand grains begin moving, they continue to move, even if the wind drops below the speed necessary to start them moving! This happens because once saltation begins, it sets off a chain reaction of collisions between sand grains that keeps the grains in constant motion.

Saltating sand usually moves near the surface, and even when winds are strong, grains are rarely lifted higher than about a meter. If the winds are very strong, these wind-whipped grains can cause extensive abrasion. A car's paint can be removed by sandblasting in a short time, and its windshield will become completely frosted and translucent from pitting.

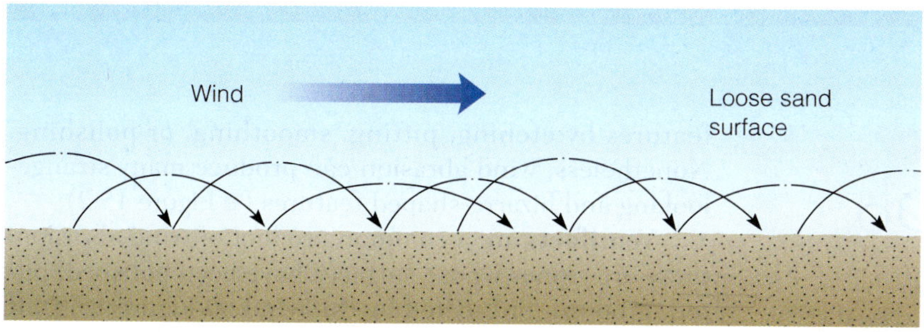

■ **Figure 15.1**

Most sand is moved near the ground surface by saltation. Sand grains are picked up by the wind and carried a short distance before falling back to the ground, where they usually hit other grains, causing them to bounce and move in the direction of the wind.

Suspended Load

Silt- and clay-sized particles constitute most of a wind's suspended load. Even though these particles are much smaller and lighter than sand-sized particles, wind usually starts the latter moving first. The reason for this phenomenon is that a thin layer of motionless air lies next to the ground where the small silt and clay particles remain undisturbed. The larger sand grains, however, stick up into the turbulent air zone where they can be moved. Unless the stationary air layer is disrupted, the silt and clay particles remain on the ground, providing a smooth surface. This phenomenon can be observed on a dirt road on a windy day. Unless a vehicle travels over the road, little dust is raised even though it is windy. When a vehicle moves over the road, it breaks the calm boundary layer of air and disturbs the smooth layer of dust, which is picked up by the wind and forms a dust cloud in the vehicle's wake.

In a similar manner, when a sediment layer is disturbed, silt- and clay-sized particles are easily picked up and carried in suspension by the wind, creating clouds of dust or even dust storms. Once these fine particles are lifted into the atmosphere, they may be carried thousands of kilometers from their source. For example, large quantities of fine dust from the southwestern United States were blown eastward and fell on New England during the Dust Bowl of the 1930s (see Geo-Focus 6.1, Figure 1b).

HOW DOES WIND ERODE LANDFORMS?

Although wind action produces many distinctive erosional features and is an extremely efficient sorting agent, running water is responsible for most erosional landforms in arid regions, even though stream channels are typically dry. Wind erodes material in two ways: abrasion and deflation.

Abrasion

Abrasion involves the impact of saltating sand grains on an object and is analogous to sandblasting. The effects of abrasion are usually minor because sand, the most common agent of abrasion, is rarely carried more than a meter above the surface. Rather than creating major erosional features, wind abrasion typically modifies existing

GEOFOCUS 15.1

Radioactive Waste Disposal—Safe or Sorry?

One problem of the nuclear age is finding safe storage sites for radioactive waste from nuclear power plants, the manufacture of nuclear weapons, and the radioactive by-products of nuclear medicine. Radioactive waste can be grouped into two categories: low-level and high-level waste. Low-level wastes are low enough in radioactivity that, when properly handled, they do not pose a significant environmental threat. Most low-level wastes can be safely buried in controlled dump sites where the geology and groundwater system are well known and careful monitoring is provided.

High-level radioactive waste, such as the spent uranium-fuel assemblies used in nuclear reactors and the material used in nuclear weapons, is extremely dangerous because of large amounts of radioactivity; it therefore presents a major environmental problem. Currently some 77,000 tons of spent nuclear fuel are temporarily stored at 131 sites in 39 states, while awaiting shipment to a permanent site in Yucca Mountain, Nevada (■ Figure 1).

In 1987 Congress amended the Nuclear Waste Policy Act and directed the Department of Energy (DOE) to study only Yucca Mountain as a possible nuclear waste repository. With the passage and signing by President Bush of House Joint Resolution 87 in 2002, the way was cleared for the DOE to prepare an application for a Nuclear Regulatory Commission license to begin constructing a nuclear waste repository at Yucca Mountain. When completed, the repository could begin receiving shipments of spent nuclear fuel by 2010.

Why Yucca Mountain? After more than 20 years of study and $8 billion spent on researching the region, many scientists think Yucca Mountain has the features necessary to isolate high-level nuclear waste from the environment for at least 10,000 years, which is the minimum time the waste will remain dangerous. What makes Yucca Mountain so appealing is its remote location and long distance from a large population center—in this case, Las Vegas, which is about 166 km southeast of Yucca Mountain. It has a very dry climate, with less than 15 cm of rain per year, and a deep water table that is 500 to 720 m below the repository. In fact, the radioactive waste will be buried in volcanic tuff at a depth of about 300 m in canisters designed to remain leakproof for at least 300 years.

What then are the concerns and why the opposition to Yucca Mountain? One of the main concerns is whether the climate will change during the next 10,000 years. If the region should become more humid,

What Would You Do?

As an expert in desert processes, you have been assigned the job of teaching the first astronaut crew that will explore Mars all about deserts and their landforms. The reason is that many Martian features display evidence of having formed as a result of wind processes, and many landforms are the same as those found in deserts on Earth. Describe how you would teach the astronauts to recognize wind-formed features and where on Earth you would take the astronauts to show them the types of landforms they may find on Mars.

features by etching, pitting, smoothing, or polishing. Nonetheless, wind abrasion can produce many strange-looking and bizarre-shaped features (■ Figure 15.2).

Ventifacts are a common product of wind abrasion; these are stones whose surfaces have been polished, pitted, grooved, or faceted by the wind (■ Figure 15.3). If the wind blows from different directions, or if the stone is moved, the ventifact will have multiple facets. Ventifacts are most common in deserts, yet they can form wherever stones are exposed to saltating sand grains, as on beaches in humid regions and some outwash plains in New England.

Yardangs are larger features than ventifacts and also result from wind erosion (■ Figure 15.4). They are elongated, streamlined ridges that look like an overturned ship's

U.S. Department of Energy

■ **Figure 1**

The location and aerial view of Nevada's Yucca Mountain.

more water will percolate through the zone of aeration. This will increase the corrosion rate of the canisters and could cause the water table to rise, thereby decreasing the travel time between the repository and the zone of saturation. This area of the country was much more humid during the Ice Age, 1.6 million to 10,000 years ago (see Chapter 14).

Another concern is the seismic activity in the area. It is in fact riddled with faults and has experienced numerous earthquakes during historic time. Nevertheless, based on underground inspections at Yucca Mountain as well as the tunnels at the Nevada Test site nearby, the DOE is convinced that earthquakes pose little danger to the underground repository itself because the disruptive effects of an earthquake are usually confined to the surface. Furthermore, it is required that the facilities, both above and below ground, be designed to withstand any severe earthquake likely to strike the area.

Finally, some people worry about the possibility of sabotage to the facility as well as the problems of transporting high-level nuclear waste to the facility. Being buried 300 m below ground renders it virtually impenetrable to acts of terrorism or sabotage, but the possibility of an accident or terrorist attack on the way to the repository is still a concern to many.

Although it appears that Yucca Mountain meets all the requirements for a safe high-level radioactive waste repository, the site is still controversial and at the time of this writing still embroiled in lawsuits seeking to block its construction and funding battles in Congress.

O. Alamany & W. Vicens/Corbis

■ **Figure 15.2**

Wind abrasion has formed these structures by eroding the exposed limestone in Desierto Libico, Egypt.

■ **Figure 15.3**

(a) A ventifact forms when wind-borne particles (1) abrade the surface of a rock (2) forming a flat surface. If the rock is moved, (3) additional flat surfaces are formed, (b) Large ventifacts lying on desert pavement in Death Valley National Monument, California.

■ **Figure 15.4**

Profile view of a streamlined yardang in the Roman playa deposits of the Kharga Depression, Egypt.

hull. Yardangs are typically found in clusters aligned parallel to the prevailing winds. They probably form by differential erosion in which depressions, parallel to the direction of wind, are carved out of a rock body, leaving sharp, elongated ridges. These ridges may then be further modified by wind abrasion into their characteristic shape. Although yardangs are fairly common desert features, interest in them was renewed when images radioed back from Mars showed that they are also widespread features on the Martian surface.

Deflation

Another important mechanism of wind erosion is **deflation,** which is the removal of loose surface sediment by the wind. Among the characteristic features of deflation in many arid and semiarid regions are *deflation hollows* or *blowouts* (■ Figure 15.5). These shallow depressions of variable dimensions result from differential erosion of surface materials. Ranging in size from several kilometers in diameter and tens of meters deep to small depressions only a few meters wide and less than a meter deep, deflation hollows are common in the southern Great Plains region of the United States.

In many dry regions, the removal of sand-sized and smaller particles by wind leaves a surface of pebbles, cobbles, and boulders. As the wind removes the fine-grained material from the surface, the effects of gravity and occasional floodwaters rearrange the remaining coarse particles into a mosaic of close fitting rocks called **desert pavement** (Figures 15.3b and ■ 15.6). Once desert pavement forms, it protects the underlying material from further deflation.

■ **Figure 15.5**

A deflation hollow in Death Valley, California.

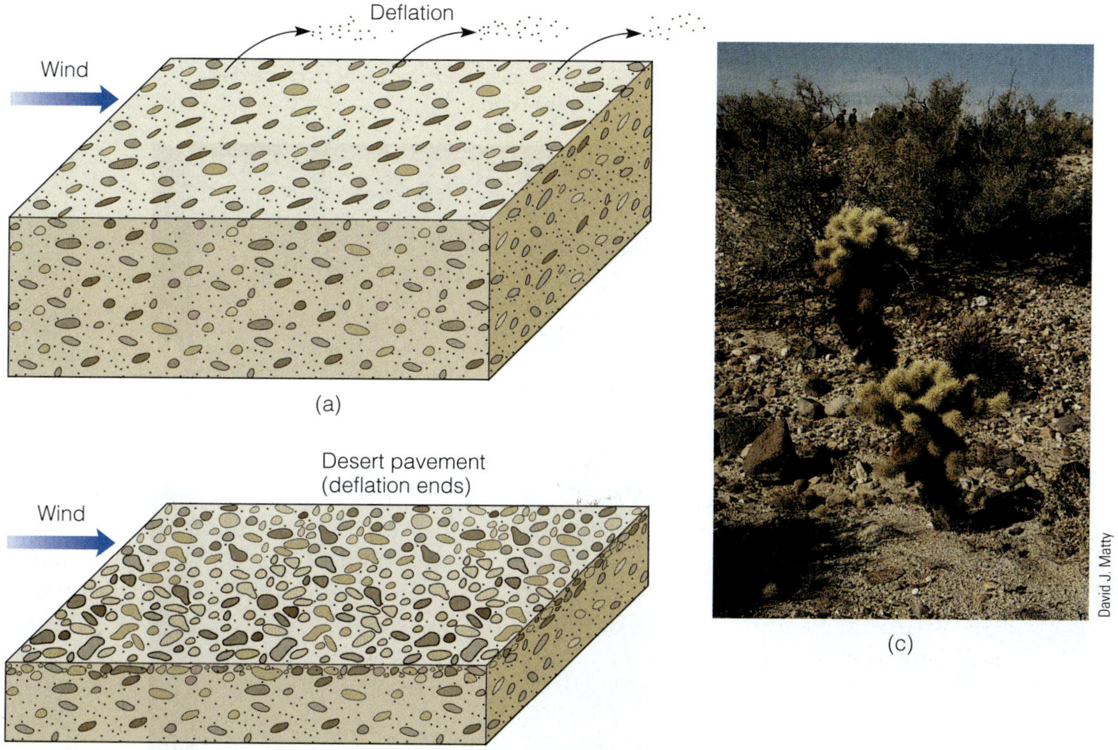

Geology Now ■ **Active Figure 15.6**

Deflation and the origin of desert pavement. (a) Fine-grained material is removed by wind, (b) leaving a concentration of larger particles that form desert pavement. (c) Desert pavement in the Mojave Desert, California. Several ventifacts can be seen in the lower left of the photo.

GEOLOGY IN UNEXPECTED PLACES

Blowing in the Wind

On a hot summer day, a light breeze can make us feel cooler and more comfortable. That same breeze on a cold day in winter can make us feel even colder and more miserable. Usually we don't pay much attention to the wind and most people don't associate it with geology. Yet wind is an important geologic agent, not only in deserts but almost everywhere.

Wind sculpts unusual shapes in rocks (Figure 15.2), and ventifacts are found not only in the desert (Figure 15.3) but also on beaches and even out of this world (■ Figure 1)! Anyone who has ever been caught in a strong windstorm can attest to the erosive power of wind. Many a car's paint and windows have been ruined by sandblasting when caught in a sudden sandstorm. In areas with lots of sand, wooden telephone and telegraph poles need a metal skirting from their base up to about 6 to 8 feet to prevent them from being "cut down" by the erosive effects of wind-blown sand.

Look at the base of many cemetery headstones and you'll see that they typically are more weathered than the rest of the headstone (■ Figure 2). This is because wind can move larger particles along the ground by saltation, resulting in more erosion near the base of headstones than farther up, where only very small particles can be carried. The next time you hear or feel the wind blowing, remember that it is also a geologic agent at work.

■ Figure 1

Wind has not only sandblasted the surface of these boulders strewn on the Martian surface but also probably helped shape them.

NASA

■ Figure 2

An 1864 headstone in a cemetery in Mount Pleasant, Michigan, shows the differential effects of blowing wind. Notice how the lower portion of the headstone is more weathered than the rest of it. This is partly because wind can move larger particles by saltation, and these particles in turn have greater erosive power than smaller particles that can be carried higher by the wind.

Reed Wicander

WHAT ARE THE DIFFERENT TYPES OF WIND DEPOSITS?

Although wind is of minor importance as an erosional agent, it is responsible for impressive deposits, which are primarily of two types. The first, *dunes,* occur in several distinctive types, all of which consist of sand-sized particles that are usually deposited near their source. The second is *loess,* which consists of layers of wind-blown silt and clay deposited over large areas downwind and commonly far from their source.

The Formation and Migration of Dunes

The most characteristic features associated with sand-covered regions are **dunes,** which are mounds or ridges of wind-deposited sand (■ Figure 15.7). Dunes form when wind flows over and around an obstruction, resulting in the deposition of sand grains, which accumulate and build up a deposit of sand. As they grow, these sand deposits become self-generating in that they form ever-larger wind barriers that further reduce the wind's velocity, resulting in more sand deposition and growth of the dune.

Most dunes have an asymmetrical profile, with a gentle windward slope and a steeper downwind or leeward slope that is inclined in the direction of the prevailing wind (■ Figure 15.8a). Sand grains move up the gentle windward slope by saltation and accumulate on the leeward side, forming an angle between 30 and 34 degrees from the horizontal, which is the angle of repose of dry sand. When this angle is exceeded by accumulating sand, the slope collapses, and the sand slides down the leeward slope, coming to rest at its base. As sand moves from a dune's windward side and periodically slides down its leeward slope, the dune slowly migrates in the direction of the prevailing wind (Figure 15.8b). When preserved in the geologic record, dunes help geologists determine the prevailing direction of ancient winds (■ Figure 15.9).

PhotoDisc

■ **Figure 15.7**

Large sand dunes in Death Valley, California. Well-developed ripple marks can be seen on the surface of the dunes. The prevailing wind direction is from left to right.

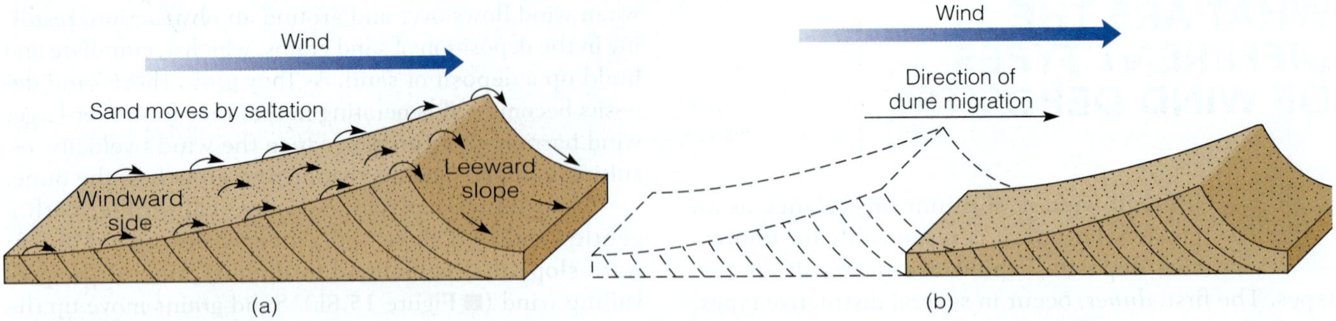

Geology Now ■ **Active Figure 15.8**

(a) Profile view of a sand dune. (b) Dunes migrate when sand moves up the windward side and slides down the leeward slope. Such movement of the sand grains produces a series of cross-beds that slope in the direction of wind movement.

■ **Figure 15.9**

Cross-bedding in this sandstone in Zion National Park, Utah, helps geologists determine the prevailing direction of the wind that formed these ancient sand dunes.

Dune Types

Geologists recognize four major dune types (barchan, longitudinal, transverse, and parabolic), although intermediate forms also exist. The size, shape, and arrangement of dunes result from the interaction of such factors as sand supply, the direction and velocity of the prevailing wind, and the amount of vegetation. Although dunes are usually found in deserts, they can also develop wherever sand is abundant, such as along the upper parts of many beaches.

Barchan dunes are crescent-shaped dunes whose tips point downwind (■ Figure 15.10). They form in areas that have a generally flat, dry surface with little vegetation, a limited supply of sand, and a nearly constant wind direction. Most barchans are small, with the largest reaching about 30 m high. Barchans are the most mobile of the major dune types, moving at rates that can exceed 10 m per year.

Longitudinal dunes (also called *seif dunes*) are long, parallel ridges of sand aligned generally parallel to the direction of the prevailing winds; they form where the sand supply is somewhat limited (■ Figure 15.11). Longitudinal dunes result when winds converge from slightly different directions to produce the prevailing wind. They range in height from about 3 m to more than 100 m, and some stretch for more than 100 km. These dunes are especially well developed in central Australia, where they cover nearly one-fourth of the continent. They also cover extensive areas in Saudi Arabia, Egypt, and Iran.

Transverse dunes form long ridges perpendicular to the prevailing wind direction in areas that have abun-

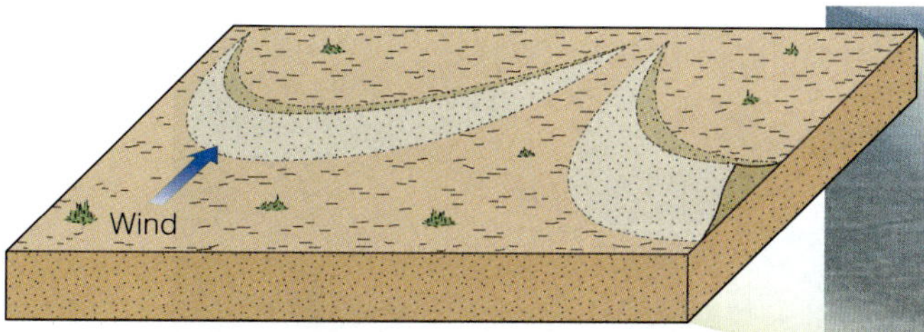

(a)

John S. Shelton

(b)

■ **Figure 15.10**

(a) Barchan dunes form in areas that have a limited amount of sand, a nearly constant wind direction, and a generally flat, dry surface with little vegetation. The tips of barchan dunes point downwind. (b) Several barchan dunes west of the Salton Sea, California. Can you tell the direction of the prevailing wind?

dant sand and little or no vegetation (■ Figure 15.12). When viewed from the air, transverse dunes have a wave-like appearance and are therefore sometimes called *sand seas*. The crests of transverse dunes can be as high as 200 m, and the dunes may be as wide as 3 km. Some transverse dunes develop a clearly distinguishable barchan form and may separate into individual barchan dunes along the edges of the dune field where there is less sand. Such intermediate-form dunes are known as *barchanoid dunes*.

Parabolic dunes are most common in coastal areas that have abundant sand, strong onshore winds, and a partial cover of vegetation (■ Figure 15.13). Although parabolic dunes have a crescent shape like barchan dunes, their tips point upwind. Parabolic dunes form when the vegetation cover is broken and deflation produces a deflation hollow or blowout. As the wind transports the sand out of the depression, it builds up on the convex downwind dune crest. The central part of the dune is excavated by the wind, while vegetation holds the ends and sides fairly well in place.

Another type of dune commonly found in the deserts of North Africa and Saudi Arabia is the *star dune*, so named because of its resemblance to a multipointed star (■ Figure 15.14). Star dunes are among the tallest in the world, rising, in some cases, more than 100 m above the surrounding desert plain. They consist of pyramidal hills of sand, from which radiate several ridges of sand, and they develop where the wind direction is variable. Star dunes can remain stationary for centuries at a time and have served as desert landmarks for nomadic peoples.

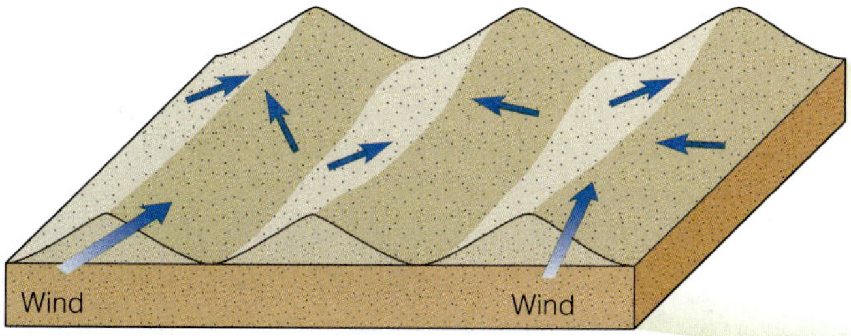

(a)

©1994 CNES. Provided by SPOT Image Corporation

(b)

■ **Figure 15.11**

(a) Longitudinal dunes form long, parallel ridges of sand aligned roughly parallel to the prevailing wind direction. They typically form where sand supplies are limited. (b) Longitudinal dunes, 15 m high, in the Gibson Desert, west central Australia. The bright blue areas between the dunes are shallow pools of rainwater, and the darkest patches are areas where the Aborigines have set fires to encourage the growth of spring grasses.

Figure 15.12

(a) Transverse dunes form long ridges of sand that are perpendicular to the prevailing wind direction in areas that have little or no vegetation and abundant sand. (b) Transverse dunes, Great Sand Dunes National Monument, Colorado. The prevailing wind direction is from lower left to upper right.

Figure 15.13

(a) Parabolic dunes typically form in coastal areas that have a partial cover of vegetation, a strong onshore wind, and abundant sand.
(b) Parabolic dune developed along the Lake Michigan shoreline west of St. Ignace, Michigan.

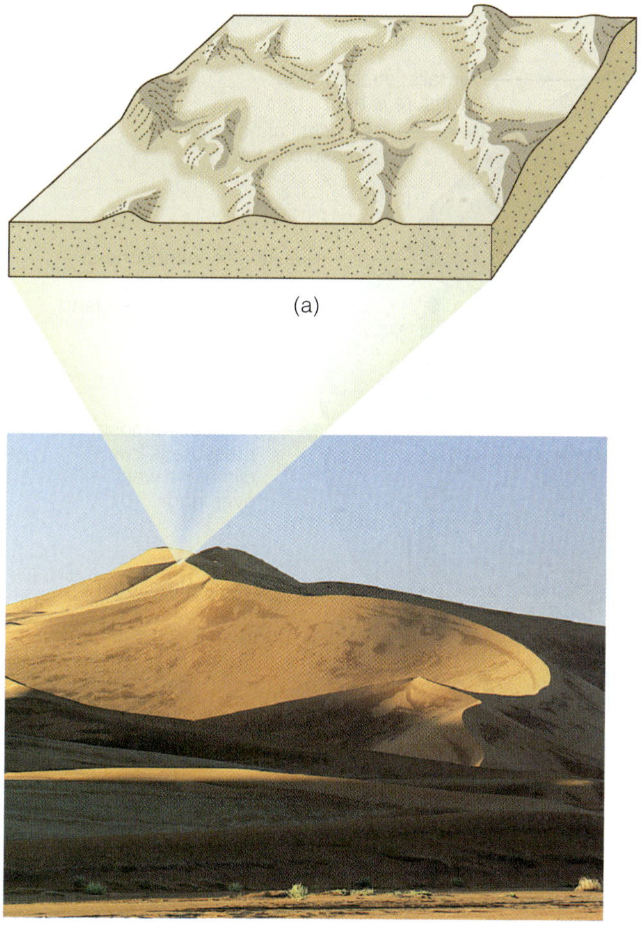

Nigel J. Dennis/Photo Researchers, Inc.

Figure 15.14

(a) Star dunes are pyramidal hills of sand that develop where the wind direction is variable. (b) Ground-level view of star dunes in Namib-Naukluft Park, Namibia.

Lowell Georgia/Corbis

Figure 15.15

Terraced wheat fields in the loess soil at Tangwa Village, China. Because of its unconsolidated nature, many farmers live in hillside caves they carved from the loess.

Loess

Wind-blown silt and clay deposits composed of angular quartz grains, feldspar, micas, and calcite are known as **loess.** The distribution of loess shows that it is derived from three main sources: deserts, Pleistocene glacial outwash deposits, and the floodplains of rivers in semiarid regions. Loess must be stabilized by moisture and vegetation to accumulate. Consequently, loess is not found in deserts, even though deserts provide much of its material. Because of its unconsolidated nature, loess is easily eroded, and as a result, eroded loess areas are characterized by steep cliffs and rapid lateral and headward stream erosion (■ Figure 15.15).

At present, loess deposits cover approximately 10% of Earth's land surface and 30% of the United States. The most extensive and thickest loess deposits are in northeast China, where accumulations more than 30 m thick are common. The extensive deserts in central Asia are the source for this loess. Other important loess deposits are on the North European Plain from Belgium eastward to Ukraine, in Central Asia, and the Pampas of Argentina. The United States has loess deposits in the Great Plains, the Midwest, the Mississippi River Valley, and eastern Washington.

Loess-derived soils are some of the world's most fertile (Figure 15.15). It is therefore not surprising that the world's major grain-producing regions correspond to the distribution of large loess deposits, such as the North European Plain, Ukraine, and the Great Plains of North America.

HOW ARE AIR-PRESSURE BELTS AND GLOBAL WIND PATTERNS DISTRIBUTED?

To understand the work of wind and the distribution of deserts, we need to consider the global pattern of air-pressure belts and winds, which are responsible for Earth's atmospheric circulation patterns. Air pressure is the density of air exerted on its surroundings (that is, its weight). When air is heated, it expands and rises, reducing its mass for a given volume and causing a decrease in air pressure. Conversely, when air is cooled, it contracts and air pressure increases. Therefore, those areas of Earth's surface that receive the most solar radiation, such as the equatorial regions, have low air pressure, whereas the colder areas, such as the polar regions, have high air pressure.

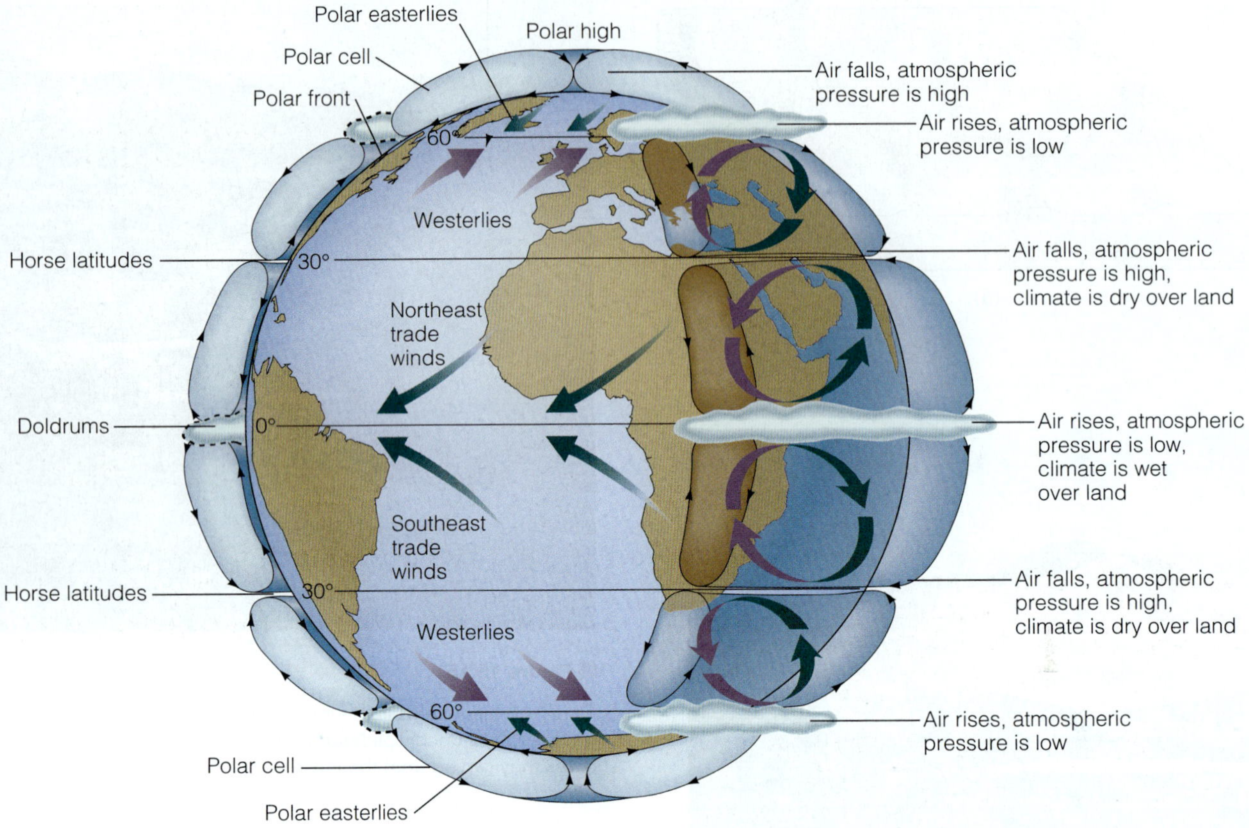

Figure 15.16

The general circulation of Earth's atmosphere.

Air flows from high-pressure zones to low-pressure zones. If Earth did not rotate, winds would move in a straight line from one zone to another. Because Earth rotates, however, winds are deflected to the right of their direction of motion (clockwise) in the Northern Hemisphere and to the left of their direction of motion (counterclockwise) in the Southern Hemisphere. This deflection of air between latitudinal zones resulting from Earth's rotation is known as the **Coriolis effect.** The combination of latitudinal pressure differences and the Coriolis effect produces a worldwide pattern of east-west–oriented wind belts (■ Figure 15.16).

Earth's equatorial zone receives the most solar energy, which heats the surface air and causes it to rise. As the air rises, it cools and releases moisture that falls as rain in the equatorial region (Figure 15.16). The rising air is now much drier as it moves northward and southward toward each pole. By the time it reaches 20 to 30 degrees north and south latitudes, the air has become cooler and denser and begins to descend. Compression of the atmosphere warms the descending air mass and produces a warm, dry, high-pressure area, the perfect conditions for the formation of the low-latitude deserts of the Northern and Southern Hemispheres (■ Figure 15.17).

WHERE DO DESERTS OCCUR?

Dry climates occur in the low and middle latitudes where the potential loss of water by evaporation exceeds the yearly precipitation (Figure 15.17). Dry climates cover 30% of Earth's land surface and are subdivided into semiarid and arid regions. *Semiarid regions* receive more precipitation than arid regions, yet are moderately dry. Their soils are usually well developed and fertile and support a natural grass cover. *Arid regions,* generally described as **deserts,** are dry; they receive less than 25 cm of rain per year, have high evaporation rates, typically have poorly developed soils, and are mostly or completely devoid of vegetation.

The majority of the world's deserts are in the dry climates of the low and middle latitudes (Figure 15.17). In North America, most of the southwestern United States and northern Mexico are characterized by this hot, dry climate, whereas in South America, this climate is primarily restricted to the Atacama Desert of coastal Chile and Peru. The Sahara in northern Africa, the Arabian

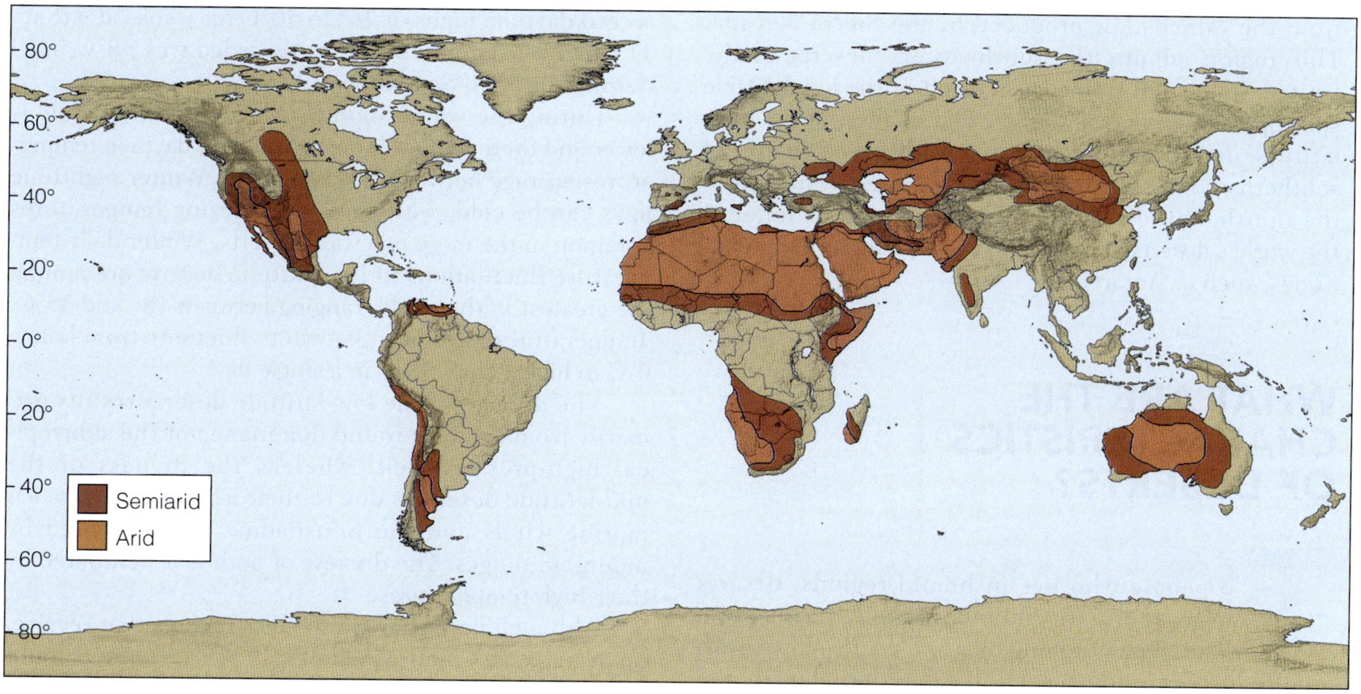

Figure 15.17

The distribution of Earth's arid and semiarid regions.

Desert in the Middle East, and the majority of Pakistan and western India form the largest essentially unbroken desert environment in the Northern Hemisphere. More than 40% of Australia is desert, and most of the rest of it is semiarid.

The remaining dry climates of the world are found in the middle and high latitudes, mostly within continental interiors in the Northern Hemisphere (Figure 15.17). Many of these areas are dry because of their remoteness from moist maritime air and the presence of mountain ranges that produce a **rainshadow desert** (Figure 15.18). When moist marine air moves inland and meets a mountain range, it is forced upward. As it rises, it cools, forming clouds and producing precipitation that falls on the windward side of the mountains. The air that descends on the leeward side of the mountain range is much warmer and drier, producing a rainshadow desert.

Three widely separated areas are included within the mid-latitude dry climate zone (Figure 15.17). The largest is the central part of Eurasia extending from just north of the Black Sea eastward to north-central China. The Gobi Desert in China is the largest desert in this region. The Great Basin area of North America is the second largest mid-latitude dry-climate zone and results

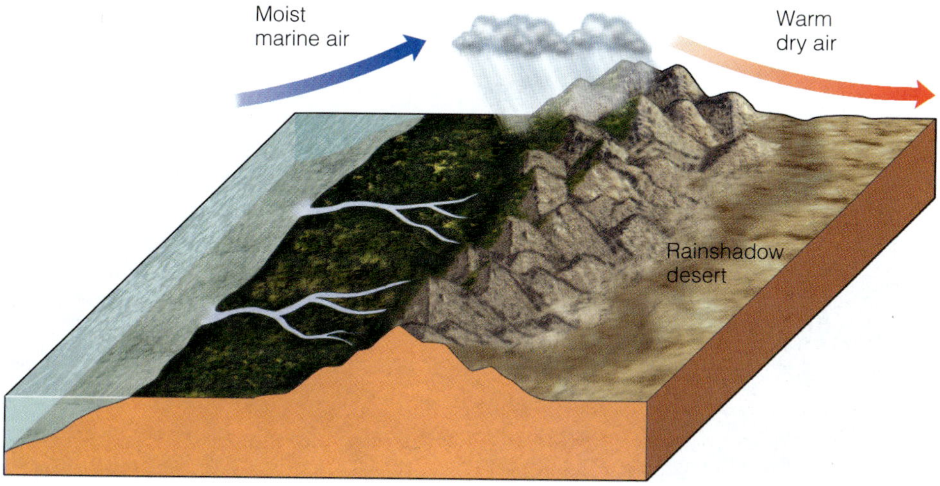

Figure 15.18

Many deserts in the middle and high latitudes are rainshadow deserts, so named because they form on the leeward side of mountain ranges. When moist marine air moving inland meets a mountain range, it is forced upward where it cools and forms clouds that produce rain. This rain falls on the windward side of the mountains. The air descending on the leeward side is much warmer and drier, producing a rainshadow desert.

from the rainshadow produced by the Sierra Nevada. This region adjoins the southwestern deserts of the United States that formed as a result of the low-latitude subtropical high-pressure zone. The smallest of the mid-latitude dry-climate areas is the Patagonian region of southern and western Argentina. Its dryness results from the rainshadow effect of the Andes. The remainder of the world's deserts are found in the cold but dry high latitudes, such as Antarctica.

WHAT ARE THE CHARACTERISTICS OF DESERTS?

To people who live in humid regions, deserts may seem stark and inhospitable. Instead of a landscape of rolling hills and gentle slopes with an almost continuous cover of vegetation, deserts are dry, have little vegetation, and consist of nearly continuous rock exposures, desert pavement, or sand dunes. Yet despite the great contrast between deserts and more humid areas, the same geologic processes are at work, only operating under different climatic conditions.

Temperature, Precipitation, and Vegetation

The heat and dryness of deserts are well known. Many of the deserts of the low latitudes have average summer temperatures that range between 32° and 38°C. It is not uncommon for some low-elevation inland deserts to record daytime highs of 46° to 50°C for weeks at a time. The highest temperature ever recorded was 58°C in El Azizia, Libya, on September 13, 1922.

During the winter months when the Sun's angle is lower and there are fewer daylight hours, daytime temperatures average between 10° and 18°C. Winter nighttime lows can be cold, with frost and freezing temperatures common in the more poleward deserts. Winter daily temperature fluctuations in low-latitude deserts are among the greatest in the world, ranging between 18° and 35°C. Temperatures have been known to fluctuate from below 0°C to higher than 38°C in a single day!

The dryness of the low-latitude deserts results primarily from the year-round dominance of the subtropical high-pressure belt, whereas the dryness of the mid-latitude deserts is due to their isolation from moist marine winds and the rainshadow effect created by mountain ranges. The dryness of both is accentuated by their high temperatures.

Although deserts are defined as regions that receive, on average, less than 25 cm of rain per year, the amount of rain that falls each year is unpredictable and unreliable. It is not uncommon for an area to receive more than an entire year's average rainfall in one cloudburst and then to receive little rain for several years. Thus, yearly rainfall averages can be misleading.

Deserts display a wide variety of vegetation (■ Figure 15.19). Although the driest deserts, or those with large areas of shifting sand, are almost devoid of vegetation, most deserts support at least a sparse plant cover. Compared to humid areas, desert vegetation may appear monotonous. A closer examination, however, reveals an amazing diversity of plants that have evolved the ability to live in the near-absence of water.

Desert plants are widely spaced, typically small, and grow slowly. Their stems and leaves are usually hard and

Charlie Ott, The National Audubon Society Collection/Photo Researchers, Inc.

■ **Figure 15.19**

Desert vegetation is typically sparse, widely spaced, and characterized by slow growth rates. The vegetation shown here in Organ Pipe National Monument, Arizona, includes saguaro and cholla cacti, paloverde trees, and jojoba bushes and is characteristic of the vegetation found in the Sonoran Desert of North America.

waxy to minimize water loss by evaporation and protect the plant from sand erosion. Most plants have a widespread shallow root system to absorb the dew that forms each morning in all but the driest deserts and to help anchor the plant in what little soil there may be. In extreme cases, many plants lie dormant during particularly dry years and spring to life after the first rain shower with a beautiful profusion of flowers.

Weathering and Soils

Mechanical weathering is dominant in desert regions. Daily temperature fluctuations and frost wedging are the primary forms of mechanical weathering (see Chapter 6). The breakdown of rocks by roots and from salt crystal growth is of minor importance. Some chemical weathering does occur, but its rate is greatly reduced by aridity and the scarcity of organic acids produced by the sparse vegetation. Most chemical weathering takes place during the winter months when there is more precipitation, particularly in the midlatitude deserts.

An interesting feature seen in many deserts is a thin, red, brown, or black shiny coating on the surface of many rocks (see "Rock Art for the Ages" on pages 430 and 431). This coating, called *rock varnish,* is composed of iron and manganese oxides (■ Figure 15.20). Because many of the varnished rocks contain little or no iron and manganese oxides, the varnish is thought to result either from windblown iron and manganese dust that settles on the ground or from the precipitated waste of microorganisms.

James S. Monroe

■ Figure 15.20

The shiny black coating on this rock exposed at Castle Valley, Utah, is rock varnish. It is composed of iron and manganese oxides.

Desert soils, if developed, are usually thin and patchy because the limited rainfall and the resultant scarcity of vegetation reduce the efficiency of chemical weathering and hence soil formation. Furthermore, the sparseness of the vegetative cover enhances wind and water erosion of what little soil actually forms.

Mass Wasting, Streams, and Groundwater

When traveling through a desert, most people are impressed by such wind-formed features as moving sand, sand dunes, and sand and dust storms. They may also notice the dry washes and dry streambeds. Because of the lack of running water, most people would conclude that wind is the most important erosional geologic agent in deserts. They would be wrong. Running water, even though it occurs infrequently, causes most of the erosion in deserts. The dry conditions and sparse vegetation characteristic of deserts enhance water erosion. If you look closely, you can see the evidence of erosion and transportation by running water nearly everywhere except in areas covered by sand dunes.

Most of a desert's average annual rainfall of 25 cm or less comes in brief, heavy, localized cloudbursts. During these times, considerable erosion takes place because the ground cannot absorb all of the rainwater. With so little vegetation to hinder its flow, runoff is rapid, especially on moderately to steeply sloping surfaces, resulting in flash floods and sheetflows. Dry stream channels quickly fill with raging torrents of muddy water and mudflows, which carve out steep-sided gullies and overflow their banks. During these times, a tremendous amount of sediment is rapidly transported and deposited far downstream.

Although water is the major erosive agent in deserts today, it was even more important during the Pleistocene Epoch when these regions were more humid (see Chapter 23). During that time, many of the major topographic features of deserts were forming. Today that topography is modified by wind and infrequently flowing streams.

Most desert streams are poorly integrated and flow only intermittently. Many of them never reach the sea because the water table is usually far deeper than the channels of most streams, so they cannot draw upon groundwater to replace water lost to evaporation and absorption into the ground. This type of drainage in which a stream's load is deposited within the desert is called *internal drainage* and is common in most arid regions.

Although most deserts have internal drainage, some deserts have permanent through-flowing streams such as the Nile and Niger Rivers in Africa, the Rio Grande and Colorado River in the southwestern United States, and the Indus River in Asia. These streams can flow through desert regions because their headwaters are well outside the desert and water is plentiful enough to offset losses resulting from evaporation and infiltration. Demands for greater amounts of water for agriculture and domestic use from the Colorado River, however, are leading to increased salt concentrations in its lower reaches and causing political problems between the United States and Mexico.

Rock Art for the Ages

Rock art includes rock paintings (where paints made from natural pigments are applied to a rock surface) and petroglyphs (from the Greek petro, *meaning "rock," and* glyph, *meaning "carving or engraving"), which are the abraded, pecked, incised, or scratched marks made by humans on boulders, cliffs, and cave walls.*

Rock art has been found on every continent except Antarctica and is a valuable archaeological resource that provides graphic evidence of the cultural, social, and religious relationships and practices of ancient peoples. The oldest known rock art was made by hunters in western Europe and dates back to the Pleistocene Epoch. Africa has more rock art sites than any other continent. The oldest known African rock art, found in the southern part of the continent, is estimated to be 27,000 years old.

Petroglyphs are a fragile and nonrenewable cultural resource that cannot be replaced if they are damaged or destroyed. A commitment to their preservation is essential so that future generations can study them as well as enjoy their beauty and mystery.

In the arid Southwest and Great Basin area of North America where rock art is plentiful, rock paintings and petroglyphs extend back to about 2000 B.C. Here, rock art can be divided into two categories. *Representational art* deals with life-forms such as humans, birds, snakes, and humanlike supernatural beings. Rarely exact replicas, they are more or less stylized versions of the beings depicted. *Abstract art,* in contrast, bears no resemblance to any real-life images.

Reed Wicander

Various petroglyphs exposed at an outcrop along Cub Creek Road in Dinosaur National Monument, Utah.

Reed Wicander

A humanlike petroglyph, an example of representational art, exposed at an outcrop along Cub Creek Road in Dinosaur National Monument, Utah. Note the contrast between the fresh exposure of the rock where the upper part of the petroglyph's head has been removed, the weathered brown surface of the rest of the petroglyph, and the black rock varnish coating the rock surface.

Petroglyphs are the most common form of rock art in North America and are made by pecking, incising, or scratching the rock surface with a tool harder than the rock itself. In arid regions, many rock surfaces display a patination, or thin brown or black coating, known as rock varnish (see Figure 15.20). When this coating is broken by pecking, incising, or scratching, the underlying lighter colored natural-rock surface provides an excellent contrast for the petroglyphs.

Petroglyphs are especially abundant in the Southwest and Great Basin area, where they occur by the thousands, having been made by Native Americans from many cultures during the past several thousand years. Petroglyphs can be seen in many of the U.S. national parks and monuments, such as Petrified Forest National Park, Arizona; Dinosaur National Monument, Colorado and Utah; Canyonlands National Park, Utah; and Petroglyph National Monument, New Mexico—to name a few.

Reed Wicander

Examples of both representational and abstract art are displayed in these petroglyphs from Arizona.

Tom Bean/Corbis

Several petroglyphs on basalt boulders in Rinconada Canyon, Petroglyph National Monument, Albuquerque, New Mexico.

Thomas Abercrombie/National Geographic/Getty Images

Examples of rock art from Tassili-n-Ajjer in southern Africa.

Rock art found in a rock shelter open to visitors in Uluṟu, Australia. These rock paintings are the work of A̲nagu artists, local Aboriginal people, and are created for religious and ceremonial expression as well as for teaching and storytelling. Each abstract symbol can have many levels of meaning and can help illustrate a story.

Reed Wicander

The paints used in this rock art are made from natural mineral substances mixed with water and sometimes with animal fat. Red, yellow, and orange pigments come from iron-stained clays, whereas the white pigments come from either ash or the mineral calcite. The black colors come from charcoal.

Wind

Although running water does most of the erosional work in deserts, wind can also be an effective geologic agent capable of producing a variety of distinctive erosional and depositional features (Figure 15.2). It is effective in transporting and depositing unconsolidated sand-, silt-, and dust-sized particles. Contrary to popular belief, most deserts are not sand-covered wastelands, but rather vast areas of rock exposures and desert pavement. Sand-covered regions, or sandy deserts, constitute less than 25% of the world's deserts. The sand in these areas has accumulated primarily by the action of wind.

WHAT TYPES OF LANDFORMS ARE FOUND IN DESERTS?

Because of differences in temperature, precipitation, and wind, as well as the underlying rocks and recent tectonic events, landforms in arid regions vary considerably. Although wind is an important geologic agent, many distinctive landforms are produced and modified by running water.

After an infrequent and particularly intense rainstorm, excess water not absorbed by the ground may accumulate in low areas and form *playa lakes* (■ Figure 15.21a). These lakes are temporary, lasting from a few hours to several months. Most of them are shallow and have rapidly shifting boundaries as water flows in or leaves by evaporation and seepage into the ground. The water is often very saline.

When a playa lake evaporates, the dry lake bed is called a **playa** or *salt pan* and is characterized by mud cracks and precipitated salt crystals (Figure 15.21b). Salts in some playas are thick enough to be mined commercially. For example, borates have been mined in Death Valley, California, for more than one hundred years.

Other common features of deserts, particularly in the Basin and Range region of the United States, are alluvial fans and bajadas. **Alluvial fans** form when sediment-laden streams flowing out from the generally straight, steep mountain fronts deposit their load on the relatively flat desert floor. Once beyond the mountain front where no valley walls contain streams, the sediment spreads out laterally, forming a gently sloping and poorly sorted fan-

(a)

(b)

■ **Figure 15.21**

(a) Playa lake formed after a rainstorm filled Croneis Dry Lake, Mojave Desert, California. (b) Racetrack Playa, Death Valley, California. The Inyo Mountains can be seen in the background.

Martin G. Miller/Visuals Unlimited

■ **Figure 15.22**

Aerial view of an alluvial fan, Death Valley, California.

shaped sedimentary deposit (■ Figure 15.22). Alluvial fans are similar in origin and shape to deltas (see Chapter 12) but are formed entirely on land. Alluvial fans may coalesce to form a *bajada*, a broad alluvial apron that typically has an undulating surface resulting from the overlap of adjacent fans (■ Figure 15.23).

Large alluvial fans and bajadas are frequently important sources of groundwater for domestic and agricultural use. Their outer portions are typically composed of fine-grained sediments suitable for cultivation, and their gentle slopes allow good drainage of water. Many alluvial fans and bajadas are also the sites of large towns and cities, such as San Bernardino, California; Salt Lake City, Utah; and Teheran, Iran.

Most mountains in desert regions, including those of the Basin and Range region, rise abruptly from gently sloping surfaces called pediments. **Pediments** are erosional bedrock surfaces of low relief that slope gently away from mountain bases (■ Figure 15.24). Most pediments are covered by a thin layer of debris, alluvial fans, or bajadas.

The origin of pediments has been the subject of much controversy. Most geologists agree that they are erosional features developed on bedrock in association with the erosion and retreat of a mountain front (Figure 15.24a). The disagreement concerns how the erosion has occurred. Although not all geologists would agree, it appears that pediments are produced by the combined activities of lateral erosion by streams, sheet flooding,

Alan L. and Linda D. Mayo, GeoPhoto Publishing Co.

■ **Figure 15.23**

Coalescing alluvial fans forming a bajada at the base of the Black Mountains, Death Valley, California.

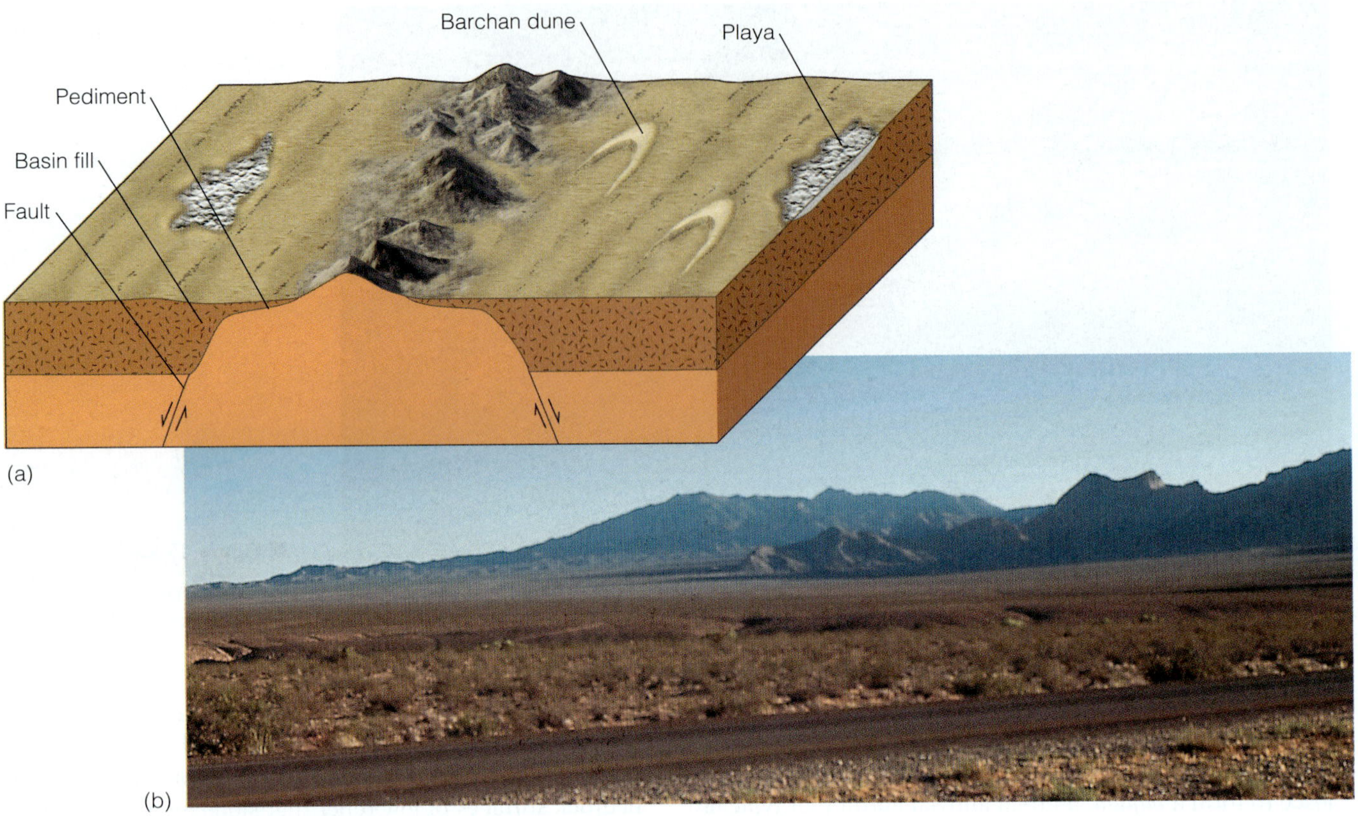

Reed Wicander

■ **Figure 15.24**

(a) Pediments are erosional bedrock surfaces formed by erosion along a mountain front. (b) Pediment north of Mesquite, Nevada.

Royalty Free/Corbis

Tom Bean/Corbis

■ **Figure 15.25**

(a) Several mesas as well as buttes can be seen in this aerial view of Monument Valley Navajo Tribal Park, Navajo Indian Reservation. The mesas are the broad, flat-topped structures, one of which is prominent in the foreground, whereas the buttes are more pillarlike structures. (b) Left Mitten Butte and Right Mitten Butte in Monument Valley on the border of Arizona and Utah.

and various weathering processes along the retreating mountain front. Thus, pediments grow at the expense of the mountains, and they will continue to expand as the mountains are eroded away or partially buried.

Rising conspicuously above the flat plains of many deserts are isolated steep-sided erosional remnants called **inselbergs,** a German word meaning "island mountain." Inselbergs have survived for a longer period of time than other mountains because of their greater resistance to weathering.

Other easily recognized erosional remnants common to arid and semiarid regions are mesas and buttes (■ Figure 15.25). A **mesa** is a broad, flat-topped erosional remnant bounded on all sides by steep slopes. Continued weathering and stream erosion form isolated pillarlike structures known as **buttes.** Buttes and mesas consist of relatively easily weathered sedimentary rocks capped by nearly horizontal, resistant rocks such as sandstone, limestone, or basalt. They form when the resistant rock layer is breached, allowing rapid erosion of the less resistant underlying sediment. One of the best-known areas of mesas and buttes in the United States is Monument Valley on the Arizona–Utah border (Figure 15.25).

15 GEO RECAP

Chapter Summary

- Wind transports sediment in suspension or as bed load, which involves saltation and surface creep.
- Wind erodes material by either abrasion or deflation. Abrasion is a near-surface effect caused by the impact of saltating sand grains. Ventifacts are common wind-abraded features.
- Deflation is the removal of loose surface material by wind. Deflation hollows resulting from differential erosion of surface material are common features of many deserts, as is desert pavement, which effectively protects the underlying surface from additional deflation.
- The two major wind deposits are dunes and loess. Dunes are mounds or ridges of wind-deposited sand, whereas loess is wind-deposited silt and clay.
- The four major dune types are barchan, longitudinal, transverse, and parabolic. The amount of sand available, the prevailing wind direction, the wind velocity, and the amount of vegetation determine which type will form.
- Loess is derived from deserts, Pleistocene glacial outwash deposits, and river floodplains in semiarid regions. Loess covers approximately 10% of Earth's land surface and weathers to a rich and productive soil.
- The winds of the major air-pressure belts, oriented east-west, resulting from rising and cooling air, are deflected by the Coriolis effect. These belts help control the world's climate.
- Deserts are very dry (averaging less than 25 cm of rain per year), have poorly developed soils, and are mostly or completely devoid of vegetation.
- Dry climates are located in the low and middle latitudes where the potential loss of water by evaporation exceeds the yearly precipitation. Dry climates cover 30% of Earth's surface and are subdivided into semiarid and arid regions.
- The majority of the world's deserts are in the low-latitude dry-climate zone between 20 and 30 degrees north and south latitudes. Their dry climate results from a high-pressure belt of descending dry air. The remaining deserts are in the middle latitudes, where their distribution is related to the rain-shadow effect, and in the dry polar regions.

- Deserts are characterized by little precipitation and high evaporation rates. Furthermore, rainfall is unpredictable and, when it does occur, tends to be intense and of short duration. As a consequence of such aridity, desert vegetation and animals are scarce.
- Mechanical weathering is the dominant form of weathering in deserts. The sparse precipitation and slow rates of chemical weathering result in poorly developed soils.
- Running water is the dominant agent of erosion in deserts and was even more important during the Pleistocene, when wetter climates resulted in humid conditions.
- Wind is an erosional agent in deserts and is very effective in transporting and depositing unconsolidated fine-grained sediments.
- Important desert landforms include playas, which are dry lakebeds; when temporarily filled with water, they form playa lakes. Alluvial fans are fan-shaped sedimentary deposits that may coalesce to form bajadas.
- Pediments are erosional bedrock surfaces of low relief that slope gently away from mountain bases.
- Inselbergs are isolated steep-sided erosional remnants that rise above the surrounding desert plains. Buttes and mesas are, respectively, pinnacle-like and flat-topped erosional remnants with steep sides.

Important Terms

abrasion (p. 415)
alluvial fan (p. 432)
barchan dune (p. 422)
butte (p. 435)
Coriolis effect (p. 426)
deflation (p. 418)
desert (p. 426)
desert pavement (p. 418)
desertification (p. 414)
dune (p. 421)
inselberg (p. 435)
loess (p. 425)
longitudinal dune (p. 422)
mesa (p. 435)
parabolic dune (p. 423)
pediment (p. 433)
playa (p. 432)
rainshadow desert (p. 427)
transverse dune (p. 422)
ventifact (p. 416)

Review Questions

1. Bed load is primarily transported by
a._____precipitation; b._____deflation;
c._____saltation; d._____abrasion;
e._____suspension.

2. The Coriolis effect causes wind to be deflected
a._____only to the right in both hemispheres;
b._____to the right in the Northern Hemisphere and to the left in the Southern Hemisphere; c._____only to the left in both hemispheres; d._____to the left in the Northern Hemisphere and to the right in the Southern Hemisphere; e._____not at all.

3. Which of the following dunes are among the tallest in the world, consist of pyramidal hills of sand, and develop where the wind direction is variable?
a._____barchan; b._____longitudinal;
c._____transverse; d._____star;
e._____parabolic.

4. The major agent of erosion in deserts today is
a._____glaciers; b._____wind;
c._____abrasion; d._____running water;
e._____none of these.

5. Deserts
a._____are found in the low, middle, and high latitudes; b._____receive more than 25 cm of rain per year; c._____are mostly or completely devoid of vegetation; d._____answers a and c;
e._____answers b and c.

6. Coalescing alluvial fans form
a._____buttes; b._____mesas;
c._____playas; d._____inselbergs;
e._____bajadas.

7. A crescent-shaped dune whose tips point downwind is a_____dune.
a._____star; b._____longitudinal;
c._____parabolic; d._____barchan;
e._____transverse.

8. Airborne particles abrading the surface of a rock produce

   a._____inselbergs; b._____dunes; c._____ventifacts; d._____playas; e._____loess.

9. In what area of the world are the thickest and most extensive loess deposits found?

   a._____United States; b._____Ukraine; c._____Argentina; d._____China; e._____Belgium.

10. The primary cause of dryness in low-latitude deserts is

    a._____isolation from moist marine winds; b._____dominance of the subtropical high-pressure belt; c._____Coriolis effect; d._____rainshadow effect; e._____all of these.

11. If deserts are dry regions in which mechanical weathering predominates, why are so many of their distinctive landforms the result of running water and not of wind?
12. Considering what you now know about deserts, their location, how they form, and the various landforms found in them, how can you use this information to determine where deserts may have existed in the past?
13. Much of the recent desertification has been greatly accelerated by human activity. Can anything be done to slow the process?
14. Why are so many desert rock formations red?
15. Why is desert pavement important in a desert environment?
16. Is it possible to get the same type of sand dunes on Mars as on Earth? What does that tell us about the climate and geology of Mars?
17. How do dunes form and migrate? Why is dune migration a problem in some areas?
18. Why are low-latitude deserts so common?
19. What is loess, and why is it important?
20. As noted in the text, some large cities are built on alluvial fans. What are the advantages and disadvantages of such an arrangement?

World Wide Web Activities

GeologyNow Assess your understanding of this chapter's topics with additional quizzing and comprehensive interactivities at

http://earthscience.brookscole.com/changingearth4e

as well as current and up-to-date weblinks, additional readings, and InfoTrac College Edition exercises.

Shorelines and Shoreline Processes

CHAPTER 16

OUTLINE

- Introduction
- Tides, Waves, and Nearshore Currents

  GEO-FOCUS 16.1: *Energy from the Oceans*
- Deposition Along Shorelines
- How Are Shorelines Eroded?
- The Nearshore Sediment Budget
- How Are Coastal Areas Managed As Sea Level Rises?

  GEOLOGY IN UNEXPECTED PLACES: *Erosion and the Cape Hatteras Lighthouse*
- Storm Waves and Coastal Flooding
- Types of Coasts
- Geo-Recap

Geology Now *This icon, which appears throughout the book, indicates an opportunity to explore interactive tutorials, animations, or practice problems available on the GeologyNow website at http://earthscience.brookscole.com/changingearth4e.*

OBJECTIVES

At the end of this chapter, you will have learned that:

- Wind-generated waves and their associated nearshore currents effectively modify shorelines by erosion and deposition.
- The gravitational attraction by the Moon and Sun and Earth's rotation are responsible for the rhythmic daily rise and fall of sea level known as tides.
- Seacoasts and lakeshores are both modified by waves and nearshore currents, but seacoasts also experience tides, which are insignificant in even the largest lakes.
- The concept of a nearshore sediment budget considers equilibrium, losses, and gains in the amount of sediment in a coastal area.
- Distinctive erosional and depositional landforms such as wave-cut platforms, spits, and barrier islands are found along shorelines.
- Shoreline management is complicated by rising sea level, and structures originally far from the shoreline are now threatened or have already been destroyed.
- Several types of coasts are recognized based on criteria such as deposition and erosion and fluctuations in sea level.

Waves pounding the shoreline near Bodega Bay, California. Most of the geologic work on shorelines is done by waves, especially large waves generated during storms. Source: Sue Monroe

Introduction

Intuitively, we know that a **shoreline** is the area of land in contact with the sea, but we can expand this definition by noting that shorelines include the land between low tide and the highest level on land affected by storm waves. Thus, a shoreline is a long, narrow zone where marine processes are active, particularly waves (see the chapter opening photo) and nearshore currents. In short, shorelines are dynamic systems where energy is expended, erosion takes place, and sediment is transported and deposited. As a matter of fact, shorelines continuously adjust to changes such as increased or diminished sediment supply and increased wave activity.

Our main concern in this chapter is with ocean shorelines, or seashores, but waves and nearshore currents are also active in large lakes. Along the shores of the Great Lakes, many depositional and erosional features typical of seacoasts are well represented (■ Figure 16.1). The most notable differences between lakeshores and seashores are that waves and nearshore currents are much more energetic along seashores, and even the largest lakes have insignificant tides.

You already know that the hydrosphere is a vast system consisting of all water on Earth, the bulk of which is in the oceans. Through this enormous body of water, energy is transferred to shorelines, so understanding the geologic processes operating on shorelines is important to many people. Indeed, many of the world's centers of commerce and much of Earth's population are concentrated in a narrow band at or near shorelines. Furthermore, a number of coastal communities such as Myrtle Beach, South Carolina, Fort Lauderdale, Florida, and Padre Island, Texas, depend heavily on tourists visiting and enjoying their beaches.

Geologists, oceanographers, marine biologists, and coastal engineers, among others, are interested in the dynamic nature of shorelines. Elected officials and city planners of coastal communities must be familiar with shoreline processes so they can develop policies and zoning regula-

(a)

(b)

■ **Figure 16.1**

Seashores (a) and lakeshores (b) have many similar features. Both are modified by waves and nearshore currents, although these processes are more vigorous along seacoasts, and even the largest lakes have no tides of any significance. (a) The Pacific Coast of the United States. (b) This erosional feature called Miner's Castle in Lake Superior in Michigan forms just as similar features do along seashores.

tions that serve the public while protecting the fragile shoreline environment. Such understanding is especially important now because in many parts of the world sea level is rising, so buildings that were far inland are now in peril or have been destroyed. Furthermore, hurricanes expend much of their energy on shorelines, resulting in extensive coastal flooding, numerous fatalities, and considerable property damage.

The geologic processes operating on shorelines provide another excellent example of systems interactions—in this case, between part of the hydrosphere and the solid Earth. But the atmosphere is also involved in transferring energy from wind to water, causing waves, which in turn generate nearshore currents. And, of course, the gravitational attraction of the Moon and Sun on oceanic waters is responsible for the rhythmic rise and fall of tides.

The continents have more than 400,000 km of shoreline, some of it rocky and steep as along North America's West Coast and its northeast coast in Maine and the Maritime Provinces of Canada. Other areas, including much of the East Coast of the United States as well as the Gulf Coast, have broad, sandy beaches or long, narrow islands of sand just offshore. Whatever the type of shoreline, waves, nearshore currents, and tides continuously bring about changes, some of which are not desirable at least from the human perspective.

TIDES, WAVES, AND NEARSHORE CURRENTS

In contrast to other geologic agents such as running water, wind, and glaciers, which operate over vast areas, shoreline processes are restricted to a narrow zone at any particular time. But shorelines might migrate landward or seaward depending on changing sea level or uplift or subsidence of coastal regions. During a rise in sea level, for instance, the shoreline migrates landward, and then wave, tide, and nearshore-current activity shifts landward as well; during times when sea level falls, just the opposite takes place. Recall from Chapter 6 that during marine transgressions and regressions, beach and nearshore sediments are deposited over vast regions. (See Figure 6.21)

In the marine realm, several biological, chemical, and physical processes are operating continuously. Organisms change the local chemistry of seawater and contribute their skeletons to nearshore sediments, and temperature and salinity changes and internal waves occur in the oceans. However, the processes most important for modifying shorelines are purely physical ones, especially waves, tides, and nearshore currents. We cannot totally discount some of these other processes, though; offshore reefs composed of the skeletons of organisms, for instance, may protect a shoreline area from most of the energy of waves.

Tides

The surface of the oceans rises and falls twice daily in response to the gravitational attraction of the Moon and Sun. These regular fluctuations in the ocean's surface, or **tides,** result in most seashores having two daily high tides and two low tides as sea level rises and falls anywhere from a few centimeters to more than 15 m (■ Figure 16.2). A complete tidal cycle includes a *flood tide* that progressively covers more and more of a nearshore area until high tide is reached, followed by *ebb tide* during which the nearshore area is once again exposed (Figure 16.2). These regular fluctuations in sea level constitute one largely untapped source of energy as do waves, ocean currents, and temperature differences in seawater (see Geo-Focus 16.1).

James S. Monroe

(a)

James S. Monroe

(b)

■ **Figure 16.2**

Low tide (a) and high tide (b) in Turnagain Arm, part of Cook Inlet in Alaska. The tidal range here is about 10 m. Turnagain Arm is a huge fiord now being filled in by sediment carried in by rivers. Notice the mudflats in (a).

GEOFOCUS 16.1

Energy from the Oceans

Complex interactions among systems have been a recurring theme throughout this book, and here too we emphasize these interactions as we discuss the potential to extract energy from the oceans. Solar energy is responsible for differential heating of the atmosphere and thus air circulation, which is responsible for waves and oceanic currents, although Earth's rotation also plays an important role in currents. And gravitational attraction between Earth and the Sun and Moon coupled with Earth's rotation accounts for the twice-daily rise and fall of tides seen in most coastal areas.

If we could harness the energy of waves, oceanic currents, temperature differences in oceanic waters, and tides, an almost limitless, largely nonpolluting energy supply would be ensured. Unfortunately, the ocean's energy is diffuse, meaning that the energy for a given volume of water is small and thus difficult to concentrate and use. Of the several sources of ocean energy, only tides show much promise for the near future.

Ocean thermal energy conversion (OTEC) exploits the temperature difference between surface waters and those at depth to run turbines and generate electricity. The amount of energy from this source is enormous, but a number of practical problems must be solved. For one thing, enormous quantities of water must be circulated through a power plant, which requires large surface areas devoted to this purpose. Despite several decades of research, no

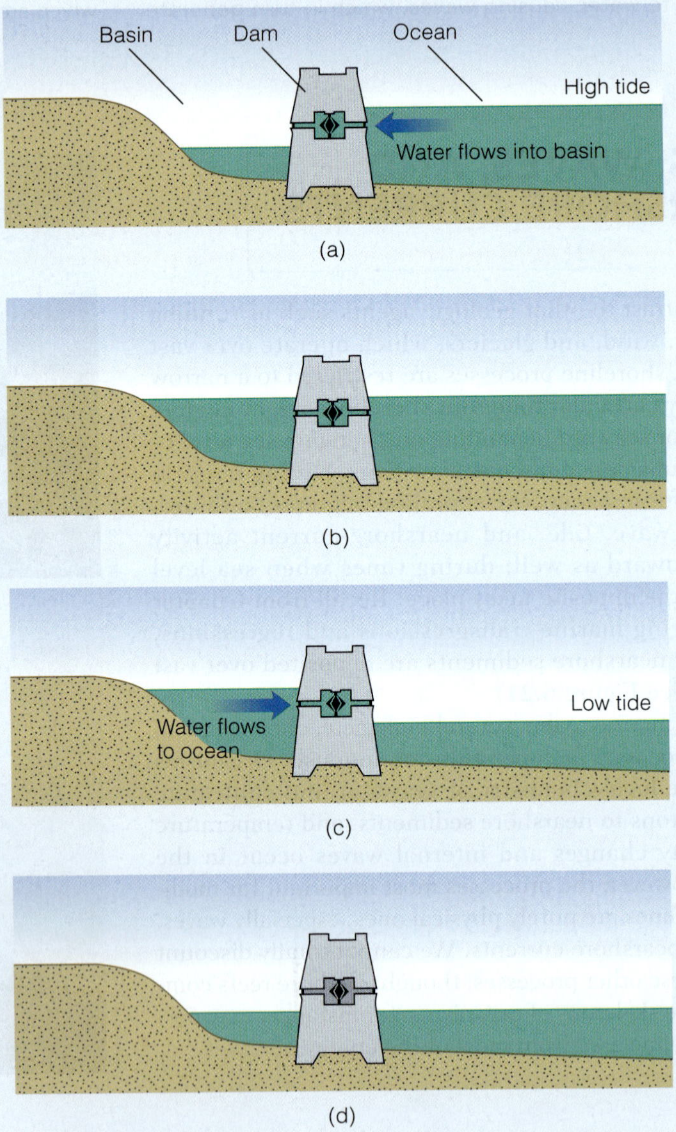

Figure 1

Rising and falling tides produce electricity by spinning turbines connected to generators, just as at hydroelectric plants. (a) Water flows from ocean to basin during flood tide. (b) Basin full. (c) Water flows from basin to ocean during ebb tide. (d) Basin water is depleted, and the cycle begins once more.

Figure 2

The La Rance River estuary tidal power–generating plant in western France. This 850-m-long dam was built where the maximum tidal range is 13.4 m. Electricity is generated during both flood and ebb tides.

commercial OTEC plants are operating or under construction, although small experimental ones have been tested in Hawaii and Japan.

Oceanic currents such as the Gulf Stream also possess energy that might be tapped to generate electricity. Unfortunately, these currents flow at rates of only a few kilometers per hour at most, whereas hydroelectric power plants on land rely on water moving rapidly from higher to lower elevations. And unlike streams on land, oceanic currents cannot be dammed, their energy is diffuse, and any power plant would have to contend with unpredictable changes in flow direction.

Harnessing wave energy to generate electricity is not a new idea, and in fact it is used on an extremely limited scale. Any facility using this form of energy would obviously have to be designed to withstand the effects of storms and saltwater corrosion. No large-scale wave-energy plants are operating, but the Japanese have developed devices to power lighthouses and buoys, and a facility with a capacity to provide power to about 300 homes began operating in Scotland during September 2000.

Tidal power has been used for centuries in some coastal areas to run mills, but its use at present for generating electricity is limited. Most coastal areas experience a twice-daily rise and fall of tides, but only a few areas are suitable for exploiting this energy source. One limitation is that the tidal range must be at least 5 m, and there must also be a coastal region where water can be stored following high tide.

Many areas along the U.S. Gulf Coast would certainly benefit from a tidal power plant, but the tidal range of generally less than 1 m precludes the possibility of development. An area with an appropriate tidal range in some remote location such as southern Chile or the Arctic Islands of Canada offers little potential because of the great distance from population centers. Accordingly, in North America only a few areas show much potential for developing tidal energy: Cook Inlet, Alaska, the Bay of Fundy on the U.S.–Canadian border, and some areas along the New England coast, for instance.

The idea behind tidal power is simple, although putting it into practice is not easy. First, a dam with sluice gates to regulate water flow must be built across the entrance to a bay or estuary. When the water level has risen sufficiently high during flood tide, the sluice gates are closed. Water held on the landward side of the dam is then released and electricity is generated just as it is at a hydroelectric dam. Actually, a tidal power plant can operate during both flood and ebb tides (Figure 1).

The first tidal power-generating facility was constructed at the La Rance River estuary in France (Figure 2). This 240-megawatt (MW) plant began operations in 1966 and has been quite successful. In North America, a much smaller 20-MW tidal power plant has been operating in the Bay of Fundy, Nova Scotia, where the tidal range, the greatest in the world, exceeds 16 m. The Bay of Fundy is part of the more extensive Gulf of Maine, where the United States and Canada have been considering a much larger tidal power–generating project for several decades.

Although tidal power shows some promise, it will not solve our energy needs even if developed to its fullest potential. Most analysts think that only 100 to 150 sites worldwide have sufficiently high tidal ranges and the appropriate coastal configuration to exploit this energy resource. This, coupled with the facts that construction costs are high and tidal energy systems can have disastrous effects on the ecology (biosphere) of estuaries, makes it unlikely that tidal energy will ever contribute more than a small percentage of all energy production.

Both the Moon and the Sun have sufficient gravitational attraction to exert tide-generating forces strong enough to deform the solid body of Earth, but they have a much greater influence on the oceans. The Sun is 27 million times more massive than the Moon, but it is 390 times as far from Earth, and its tide-generating force is only 46% as strong as that of the Moon. Accordingly, the tides are dominated by the Moon, but the Sun plays an important role as well.

If we consider only the Moon acting on a spherical, water-covered Earth, its tide-generating forces produce two bulges on the ocean surface (■ Figure 16.3a). One bulge points toward the Moon because it is on the side of Earth where the Moon's gravitational attraction is greatest. The other bulge is on the opposite side of Earth, where the Moon's gravitational attraction is least. These two bulges always point toward and away from the Moon (Figure 16.3a), so as Earth rotates and the Moon's position changes, an observer at a particular shoreline location experiences the rhythmic rise and fall of tides twice daily, but the heights of two successive high tides may vary depending on the Moon's inclination with respect to the equator.

The Moon revolves around Earth every 28 days, so its position with respect to any latitude changes slightly each day. That is, as the Moon moves in its orbit and Earth rotates on its axis, it takes the Moon 50 minutes longer each day to return to the same position it was in the previous day. Thus, an observer would experience a high tide at 1:00 P.M. on one day, for example, and at 1:50 P.M. on the following day.

Tides are also complicated by the combined effects of the Moon and the Sun. Even though the Sun's tide-generating force is weaker than the Moon's, when the two are aligned every two weeks, their forces added together generate *spring tides* about 20% higher than average tides (Figure 16.3b). When the Moon and Sun are at right angles to each other, also at two-week intervals, the Sun's tide-generating force cancels some of the Moon's, and *neap tides* about 20% lower than average occur (Figure 16.3c).

Tidal ranges are also affected by shoreline configuration. Broad, gently sloping continental shelves as in the Gulf of Mexico have low tidal ranges, whereas steep, irregular shorelines experience much greater rise and fall of tides. Tidal ranges are greatest in some narrow, funnel-shaped bays and inlets. The Bay of Fundy in Nova Scotia has a tidal range of 16.5 m, and ranges greater than 10 m occur in several other areas.

Tides have an important impact on shorelines because the area of wave attack constantly shifts onshore and offshore as the tides rise and fall. Tidal currents themselves, however, have little modifying effect on shorelines, except in narrow passages where tidal current velocity is great enough to erode and transport sediment. Indeed, if it were not for strong tidal currents, some passageways would be blocked by sediments deposited by nearshore currents.

GeologyNow Log into GeologyNow and select this chapter to work through Geology Interactive activities on "High and Low Tides" (click Waves, Tides and Current → High/Low Tides), "Tidal

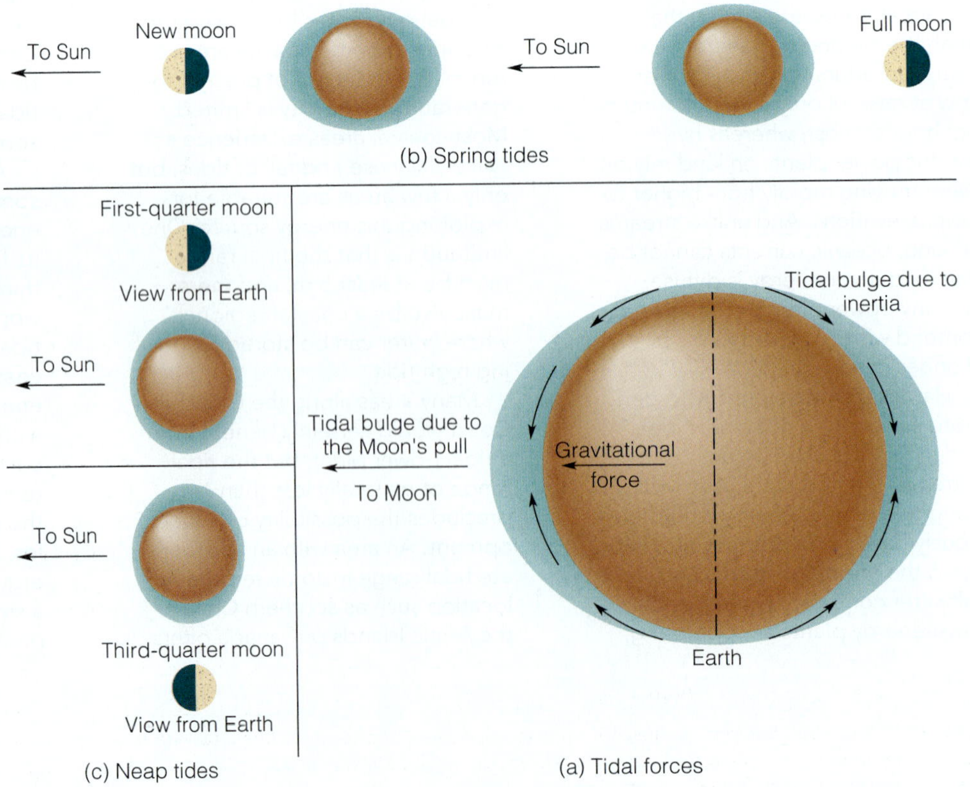

GeologyNow

■ **Active Figure 16.3**

(a) Tides are caused by the gravitational pull of the Moon and, to a lesser degree, the Sun. The Earth–Moon–Sun alignments at the times of the (b) spring and (c) neap tides are shown.

Forces" (click Waves, Tides and Current → Tidal Forces), and on "Tidal Shift" (click Waves, Tides and Current → Tidal Shift).

Waves

Oscillations of a water surface, or simply **waves,** are seen on all bodies of water, but they are best developed in the oceans and they have their greatest impact on seashores. In fact, waves are directly or indirectly responsible for most erosion, sediment transport, and deposition in coastal areas. A typical series of waves with the terms applied to them is illustrated in ■ Figure 16.4a. A **crest,** as you would expect, is the highest part of a wave, whereas the low area between crests is a **trough.** The distance from crest to crest (or trough to trough) is the *wavelength,* and the vertical distance from trough to crest is *wave height.* You can calculate the speed at which a wave advances, called celerity (C), by the formula

$$C = L/T$$

where L is wavelength and T is wave period—that is, the time it takes for two successive wave crests, or troughs, to pass a given point.

The speed of wave advance (C) is actually a measure of the velocity of the wave form rather than the speed of the molecules of water in a wave. In fact, water waves are somewhat similar to waves moving across a grass-covered

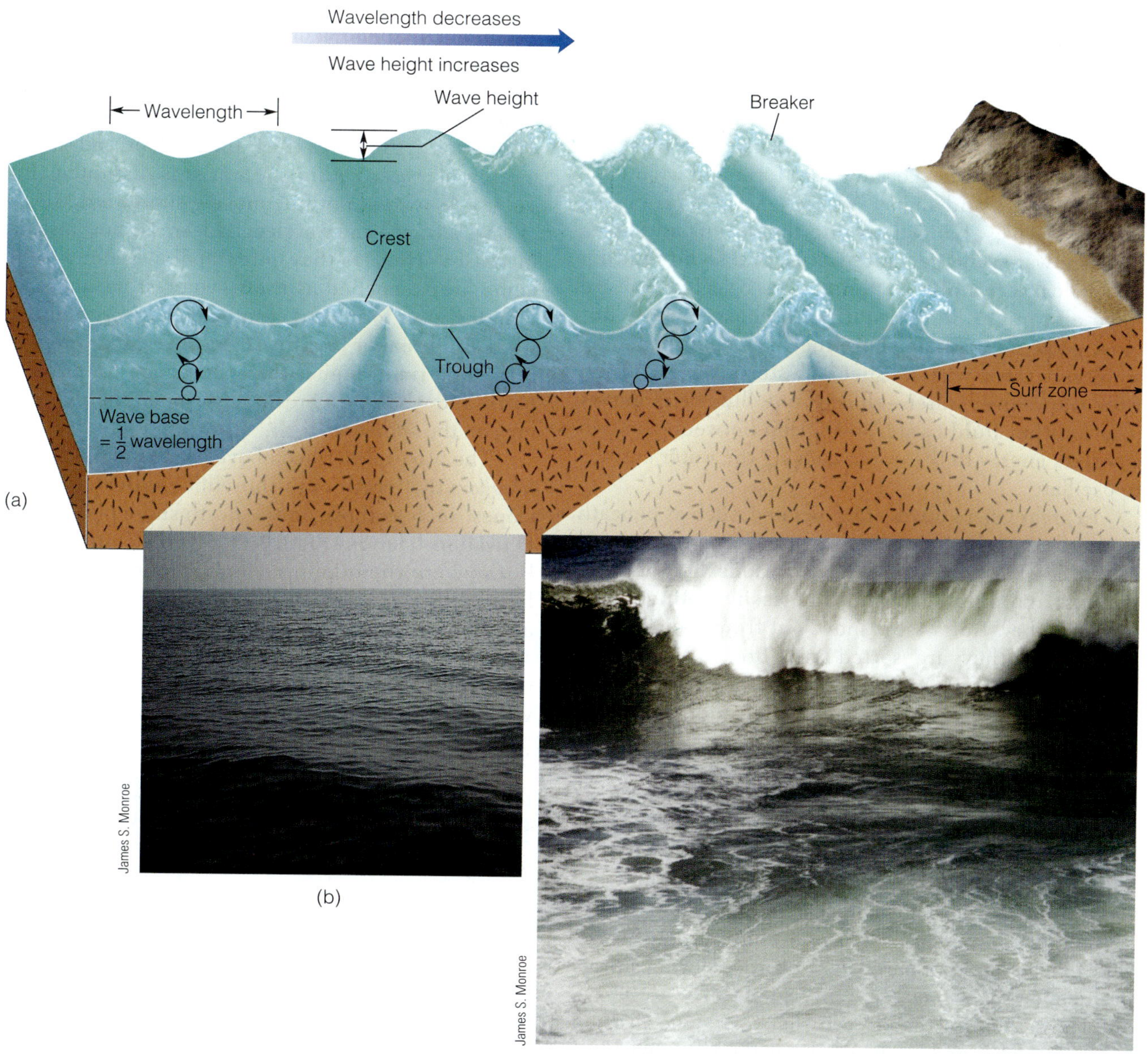

Geology Now ■ **Active Figure 16.4**

(a) Waves and the terminology applied to them. (b) As swells in deep water move toward shore, the orbital motion of water within them is disrupted when they enter water shallower than wave base. (c) Wavelength decreases while wave height increases, causing the waves to oversteepen and plunge forward as breakers.

field; the grass moves forward and back as the wave passes, but the individual blades of grass remain in their original position. When waves move across a water surface, the water moves in circular orbits but shows little or no net forward movement (Figure 16.4a). Only the wave form moves forward, and as it does it transfers energy in the direction of wave movement.

The diameters of the orbits that water follows in waves diminish rapidly with depth, and at a depth of about one-half wavelength ($L/2$), called **wave base,** they are essentially zero. Thus, at a depth exceeding wave base, the water and seafloor or lake floor, are unaffected by surface waves (Figure 16.4a). Wave base is an important consideration in some aspects of shoreline modification; we will explore it more fully in later sections of this chapter.

Geology Now Log into GeologyNow and select this chapter to work through a **Geology Interactive** activity on "Wave Properties" (click Waves, Tides and Current → Wave Properties).

Wave Generation

Everyone has seen waves but probably has little idea of how they form and what controls their size and shape. Actually, several processes generate waves. Landslides into the oceans and lakes displace water and generate waves, and so do faulting and volcanic eruptions. Some waves so formed are huge and might be devastating to coastal areas, but most geologic work on shorelines is accomplished by wind-generated waves, especially storm waves. When wind blows over water—that is, one fluid (air) moves over another fluid (water)—friction between the two transfers energy to the water, causing the water surface to oscillate.

In areas where waves are generated, such as beneath a storm center at sea, sharp-crested, irregular waves called *seas* develop. Seas have various heights and lengths, and one wave cannot be easily distinguished from another. But as seas move out from their area of generation, they are sorted into broad *swells* with rounded, long crests and all are about the same size.

As one would expect, the harder and longer the wind blows, the larger are the waves. Wind velocity and duration, however, are not the only factors that control wave size. High-velocity wind blowing over a small pond will never generate large waves regardless of how long it blows. In fact, waves on ponds and most lakes appear only while the wind is blowing; once the wind stops, the water quickly smooths out. In contrast, the surface of the ocean is always in motion, and waves with heights of 34 m have been recorded during storms in the open sea.

The reason for the disparity between wave sizes on ponds and lakes and on the oceans is the **fetch,** which is the distance the wind blows over a continuous water surface. Fetch is limited by the available water surface, so on ponds and lakes it corresponds to their length or width, depending on wind direction. To produce waves of greater length and height, more energy must be transferred from wind to water; hence large waves form beneath large storms at sea.

Shallow-Water Waves and Breakers

Swells moving out from an area of wave generation lose little energy as they travel long distances across the ocean. In these deep-water swells, the water surface oscillates and water moves in circular orbits, but little net displacement of water takes place in the direction of wave travel (Figure 16.4a). Of course, wind blows some water from waves, thus forming white caps with foamy white crests, and surface currents transport water great distances, but deep-water waves themselves accomplish little actual water movement. When these waves enter progressively shallow water, however, the wave shape changes and water is displaced in the direction of wave advance.

Broad, undulating deep-water waves are transformed into sharp-crested waves as they enter shallow water. This transformation begins at a water depth corresponding to wave base—that is, one-half wavelength (■ Figure 16.5a). At this point, the waves "feel" the seafloor, and the orbital motion of water within the waves is disrupted (Figure 16.4a). As waves continue moving shoreward, the speed of wave advance and wavelength decrease but wave height increases. Thus, as they enter shallow water, waves become oversteepened as the wave crest advances faster than the wave form, and eventually the crest plunges forward as a **breaker** (Figure 16.4c). Breaking waves might be several times higher than their deep-water counterparts, and when they break they expend their kinetic energy on the shoreline.

The waves described above are the classic *plunging breakers* that crash onto shorelines with steep offshore slopes, such as those on the north shore of Oahu in the Hawaiian Islands (Figure 16.5b). In contrast, shorelines where the offshore slope is more gentle usually have *spilling breakers,* where the waves build up slowly and the wave's crest spills down the wave front (Figure 16.5c). Whether the breakers are plunging or spilling, the water rushes onto the shore and then returns seaward to become part of another breaking wave.

Geology Now Log into GeologyNow and select this chapter to work through **Geology Interactive** activities on "Wave Progression" (click Waves, Tides and Current → Wave Progression) and "Wind-Wave Relationships" (click Waves, Tides and Current → Wind-Wave Relationships).

Nearshore Currents and Sediment Transport

The area extending seaward from the upper limit of the shoreline to just beyond the area of breaking waves is conveniently designated as the *nearshore zone.* Within the nearshore zone are the areas where waves break, the breaker zone, and a surf zone, where water from breaking waves rushes forward and then flows seaward as backwash. The nearshore zone's width varies depending on the length of approaching waves because long waves break at a greater depth, and thus farther offshore, than do short waves. Incoming waves are responsible for two types of currents in the nearshore zone: *longshore currents* and *rip currents*.

Wave Refraction and Longshore Currents

Deep-water waves have long, continuous crests, but rarely are their crests parallel with the shoreline (■ Figure 16.6). In other words, they seldom approach a shoreline head-on, but rather at some angle. Thus, one part of a wave enters shallow water where it encounters wave base and begins breaking before other parts of the same wave. As a wave begins to break, its velocity diminishes, but the part of the wave still in deep water races ahead until it too encounters wave base. The net effect of this oblique approach is that waves bend so that they more nearly parallel the shoreline, a phenomenon known as **wave refraction** (Figure 16.6).

Even though waves are refracted, they still usually strike the shoreline at some angle, causing the water between the breaker zone and the beach to flow parallel to the shoreline. These **longshore currents**, as they are called, are long and narrow and flow in the same general direction as the approaching waves. These currents are particularly important in transporting and depositing sediment in the nearshore zone.

What Would You Do?

While swimming at your favorite beach, you suddenly notice that you are far down the beach from the point where you entered the water. Furthermore, the size of the waves you were swimming in has diminished considerably. You decide to swim back to shore and then walk back to your original staring place, but no matter how hard you swim, you are carried farther and farther away from the shore. Assuming that you survive this incident, explain what happened and what you did to remedy the situation.

Rip Currents

Waves carry water into the nearshore zone, so there must be a mechanism for mass transfer of

James S. Monroe

(a)

Tom Servais/Surfer Magazine

(b)

Tom Servais/Surfer Magazine

(c)

■ **Figure 16.5**

Wave base and breakers. (a) The waves in this lake have a wavelength of about 2 m, so wave base is 1 m. Where the waves encounter wave base, they stir up the bottom sediment, accounting for the nearshore turbid water. (b) A plunging breaker on the north shore of Oahu, Hawaii. (c) A spilling breaker.

Figure 16.6

Wave refraction (wave crests are indicated by dashed lines). These waves are refracted as they enter shallow water and more nearly parallel the shoreline. The waves generate a longshore current that flows in the direction of wave approach, from upper left to lower right (arrow) in this example.

water back out to sea. One way in which water moves seaward from the nearshore zone is in **rip currents,** narrow surface currents that flow out to sea through the breaker zone (■ Figure 16.7). Surfers commonly take advantage of rip currents for an easy ride out beyond the breaker zone, but these currents pose a danger to inexperienced swimmers. Some rip currents flow at several kilometers per hour, so if a swimmer is caught in one, it is useless to try to swim directly back to shore. Instead, because rip currents are narrow and usually nearly perpendicular to the shore, one can swim parallel to the shoreline for a short distance and then turn shoreward with no difficulty.

Rip currents are circulating cells fed by longshore currents. When waves approach a shoreline, the amount of water builds up until the excess moves out to sea through the breaker zone. The rip currents are fed by nearshore currents that increase in velocity from midway between each rip current (Figure 16.7a).

Relief on the seafloor plays an important role in determining the location of rip currents. They commonly develop where wave heights are lower than in adjacent areas, and differences in wave height are controlled by variations in water depth. For instance, if waves move over a depression, the height of the waves over the depression tends to be less than in adjacent areas.

DEPOSITION ALONG SHORELINES

Depositional features of shorelines include *beaches, spits, baymouth bars, tombolos,* and *barrier islands.* The characteristics of beaches are determined by wave energy and shoreline materials, and they are continuously modified by waves and longshore currents. Spits, baymouth bars, and tombolos result from deposition by longshore currents, but the origin of barrier islands is not fully resolved. Rip currents play only a minor role in the configuration of shorelines, but they do transport fine-grained sediment seaward through the breaker zone.

Beaches

Beaches are the most familiar coastal landforms, attracting millions of visitors each year and providing the economic base for many communities. Depending on shoreline materials and wave intensity, beaches may be discontinuous, existing as only *pocket beaches* in protected areas such as embayments, or they may be continuous for long distances.

Figure 16.7

(a) Rip currents are fed on each side by currents moving parallel to the shoreline. (b) Suspended sediment, indicated by discolored water, is being carried seaward in this rip current. (c) Sign along the California coast warning swimmers of dangerous rip currents, higher than normal (sleeper) waves, and backwash.

By definition a **beach** is a deposit of unconsolidated sediment extending landward from low tide to a change in topography such as a line of sand dunes, a sea cliff, or the point where permanent vegetation begins. Typically, a beach has several component parts (see "Shoreline Processes and Beaches" on pages 450 and 451), including a **backshore** that is usually dry, being covered by water only during storms or exceptionally high tides. The backshore consists of one or more **berms,** platforms composed of sediment deposited by waves; the berms are nearly horizontal or slope gently landward. The sloping area below a berm exposed to wave swash is the **beach face.** The beach face is part of the **foreshore,** an area covered by water during high tide but exposed during low tide.

Some of the sediment on beaches is derived from weathering and wave erosion of the shoreline, but most of it is transported to the coast by streams and redistributed along the shoreline by longshore currents. As we noted, waves usually strike beaches at some angle, causing the sand grains to move up the beach face at a similar angle; as the sand grains are carried seaward in the backwash, however, they move perpendicular to the long axis of the beach. Thus, individual sand grains move in a zigzag pattern in the direction of longshore currents. This movement is not restricted to the beach; it extends seaward to the outer edge of the breaker zone.

In an attempt to widen a beach or prevent erosion, shoreline residents often build *groins,* structures that project seaward at right angles from the shoreline. Groins interrupt the flow of longshore currents, causing sand deposition on their upcurrent sides and widening the beach at that location. However, erosion inevitably occurs on the downcurrent side of a groin.

Shoreline Processes and Beaches

Beaches are the most familiar depositional landforms along shorelines. They are found in a variety of sizes and shapes, with long sandy beaches typical of the East and Gulf Coasts, and smaller, mostly protected beaches along the West Coast. The sand on most beaches is primarily quartz, but there are some notable exceptions: shell sand beaches in Florida and black sand beaches in Hawaii. Beaches are dynamic systems where waves, tides, and marine currents constantly bring about change.

The Grand Strand of South Carolina (left), shown here at Myrtle Beach, is 100 km of nearly continuous beach. A small pocket beach (above) at Julia Pfeiffer Burns State Park, California.

Diagram of a beach showing its component parts.

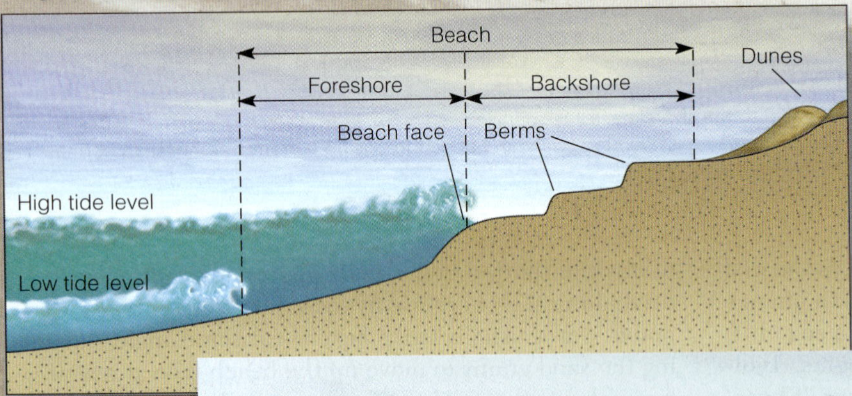

The backshore area of a pocket beach along the Pacific coast. Note that the berm ends at the rocks on the right and the beach face slopes steeply seaward.

Sand dunes are present at the upper part of this beach at Oregon Dunes National Recreation Area.

James S. Monroe

Longshore currents transport sediment along the shoreline, a phenomenon called longshore drift, between the breaker zone and the upper limit of wave action (below). These groins (right) at Cape May, New Jersey, interrupt the flow of longshore currents. Sand is trapped on their upcurrent sides, whereas erosion takes place on their downcurrent sides. Can you tell the direction of longshore drift in this image?

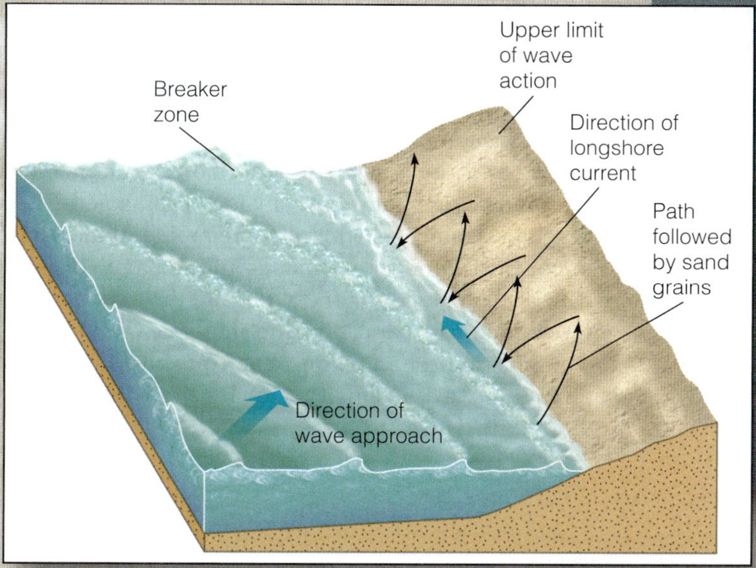

John S. Shelton

Sue Monroe

Although most beaches are made up predominantly of quartz sand, there are exceptions. This black sand beach on Maui in Hawaii is made up of small basalt rock fragments and particles of obsidian that formed when lava flowed into the sea.

Sue Monroe

The California beach is made up of quartz and sand- and gravel-sized rock fragments.

Sue Monroe

This beach shown on Oahu in Hawaii is composed mostly of fragmented shells of marine organisms, although some minerals from basalt are also included.

Quartz is the most common mineral in most beach sands, but there are some notable exceptions. For example, the black sand beaches of Hawaii are composed of sand- and gravel-sized basalt rock fragments or small grains of volcanic glass, and some Florida beaches are composed of the fragmented calcium carbonate shells of marine organisms. In short, beaches are composed of whatever material is available; quartz is most abundant simply because it is available in most areas and is the most durable and stable of the common rock-forming minerals.

Seasonal Changes in Beaches

The loose grains that make up beaches are constantly moved by waves, but the overall configuration of a beach remains unchanged as long as equilibrium conditions persist. We can think of the beach profile consisting of a berm or berms and a beach face shown on page 450 as a profile of equilibrium; that is, all parts of the beach are adjusted to the prevailing conditions of wave intensity, nearshore currents, and materials composing the beach.

Tides and longshore currents affect the configuration of beaches to some degree, but storm waves are by far the most important agent modifying their equilibrium profile. In many areas, beach profiles change with the seasons; so we recognize *summer beaches* and *winter beaches,* each of which is adjusted to the conditions prevailing at those times (■ Figure 16.8). Summer beaches are sand covered and have a wide berm, a gently sloping beach face, and a smooth offshore profile. Winter beaches, in contrast, tend to be coarser grained and steeper; they have a small berm or none at all, and their offshore profiles reveal sand bars paralleling the shoreline (Figure 16.8a).

Seasonal changes in beach profiles are related to changing wave intensity. During the winter, energetic

■ **Figure 16.8**

(a) Seasonal changes in a beach profile. (b) and (c) San Gregorio State Beach, California. These photos were taken from nearly the same place but (c) was taken two years after (b). Much of the change from (b) to (c) is accounted for by beach erosion during 1997–1998 winter storms.

■ **Figure 16.9**

(a) Spits and baymouth bars form where longshore currents deposit sand in deeper water, as at the entrance to a bay. (b) A spit across a river's mouth. (c) A baymouth bar.

storm waves erode the sand from beaches and transport it offshore where it is stored in sand bars (Figure 16.8a). The same sand that was eroded from a beach during the winter returns the next summer when it is driven onshore by more gentle swells. The volume of sand in the system remains more or less constant; it simply moves farther offshore or onshore depending on wave energy.

The terms *winter beach* and *summer beach,* though widely used, are somewhat misleading. A winter beach profile can develop at any time if there is a large storm, and likewise a summer beach profile can develop anytime during a prolonged winter calm period.

Spits, Baymouth Bars, and Tombolos

Beaches are the most familiar depositional features of coasts, but spits, baymouth bars, and tombolos are common, too. In fact, these features are simply continuations of a beach. A **spit,** for instance, is a fingerlike projection of a beach into a body of water such as a bay, and a **baymouth bar** is a spit that has grown until it completely closes off a bay from the open sea (■ Figure 16.9). Both are composed of sand, more rarely gravel, that was transported and deposited by longshore currents where they weakened as they entered the deeper water of a bay's opening. Some spits are modified by waves so that their free ends are curved; they go by the name *hook* or *recurved spit* (Figure 16.9)

A rarer type of spit is a **tombolo** that extends out from the shoreline to an island (■ Figure 16.10). A tombolo forms on the shoreward side of an island as wave refraction around the island creates converging currents that turn seaward and deposit a sand bar. So, in contrast to spits and baymouth bars, which are usually nearly parallel with the shoreline, the long axes of tombolos are nearly at right angles to the shoreline.

Spits, baymouth bars, and tombolos are most common along irregular seashores, but they can also be found in large lakes. Regardless of their setting, spits and baymouth bars especially constitute a continuing problem where bays must be kept open for pleasure boating, commercial shipping, or both. Obviously a bay closed off by a sand bar is of little use for either endeavor, so a bay must be regularly dredged or protected from deposition by longshore currents. In some areas *jetties* are constructed that extend seaward (or lakeward) to interrupt the flow of longshore currents and thus protect the opening to a bay.

Figure 16.10

(a) Wave refraction around an island and the origin of a tombolo. (b) A small tombolo along the Pacific Coast of the United States. (c) Soon after this breakwater, an offshore barrier, was constructed at Santa Monica, California, a bulge similar to a tombolo appeared in the beach.

Barrier Islands

Long, narrow islands of sand lying a short distance offshore from the mainland are **barrier islands** (■ Figure 16.11). On their seaward sides they are smoothed by waves, but their landward margins are irregular because storm waves carry sediment over the island and deposit it in a lagoon where it is little modified by further wave activity. The component parts of a barrier island include a beach, wind-blown sand dunes, and a marshy area on their landward sides.

Everyone agrees that barrier islands form on gently sloping continental shelves where abundant sand is available and where both wave energy and the tidal range are low. In fact, these are the reasons that so many are along the United States' Atlantic and Gulf Coasts. But even though it is well known where barrier islands form, the details of their origin are still unresolved. According to one model, they formed as spits that became detached from land, whereas another model holds that they formed as beach ridges that subsequently subsided.

Most barrier islands are migrating landward as a result of erosion on their seaward sides and deposition on their landward sides. This is a natural part of barrier island evolution, and it takes place rather slowly. However, it takes place fast enough to cause many problems for island residents and communities.

(a)

(b)

■ **Figure 16.11**

(a) View from space of the barrier islands along the Gulf Coast of Texas. Notice that a lagoon up to 20 km wide separates the long, narrow barrier islands from the mainland. (b) Aerial view of Padre Island on the Texas Gulf Coast. Laguna Madre is on the left, and the Gulf of Mexico is on the right.

HOW ARE SHORELINES ERODED?

Beaches are absent, poorly developed, or restricted to protected areas on seacoasts where erosion rather than deposition predominates (see "Shoreline Processes and Beaches" on pages 450 and 451). Erosion creates steep or vertical slopes known as *sea cliffs*. During storms, these cliffs are pounded by waves (hydraulic action), worn by the impact of sand and gravel (abrasion) (■ Figure 16.12), and more or less continuously eroded by dissolution involving the chemical breakdown of rocks by the solvent action of seawater. Tremendous energy from waves is concentrated on the bases of sea cliffs and is most effective on those composed of sediments or highly fractured rocks. In any case, the net effect of these processes is erosion of the sea cliff and retreat of the cliff face landward.

Wave-Cut Platforms

Wave intensity and the resistance of shoreline materials to erosion determine the rate at which a sea cliff retreats landward. A sea cliff of glacial drift on Cape Cod, Massachusetts, erodes as much as 30 m per century, and some parts of the White Cliffs of Dover in England retreat landward at more than 100 m per century. By comparison, sea cliffs of dense igneous or metamorphic rocks erode and retreat much more slowly.

Sea cliffs erode mostly as a result of hydraulic action and abrasion at their bases. As a sea cliff is undercut by erosion, the upper part is left unsupported and susceptible to mass wasting processes. Thus, sea cliffs retreat little by little, and as they do, they leave a beveled surface called a **wave-cut platform** that slopes gently seaward (■ Figure 16.13). Broad wave-cut platforms exist in many areas, but invariably the water over them is shallow because the abrasive planing action of waves is effective to a depth of only about 10 m. The sediment eroded from sea cliffs is transported seaward until it reaches deeper water at the edge of the wave-cut platform. There it is deposited and forms a *wave-built platform,* which is a

What Would You Do?

While visiting a barrier island, you notice some fine-grained, organic-rich deposits along the beach. In fact, there is so much organic matter that the sediments are black and have a foul odor. Their presence on the beach does not seem to make sense because you know that sediments like these were almost certainly deposited in a marsh on the landward (opposite) side of the island. How can you explain your observations?

James S. Monroe

(a)

James S. Monroe

(b)

■ **Figure 16.12**

(a) Rocks in the lower part of this image of a small island in the Irish Sea have been smoothed by abrasion, but the rocks higher up are out of the reach of waves. (b) A combination of processes accounts for erosion of these sea cliffs near Bodega Bay, California. Although erosion was already in progress, it was particularly intense during storms of February 1998. Attempts to stabilize the shoreline have failed, and some of the houses have been abandoned.

seaward extension of the wave-cut platform (Figure 16.13). Wave-cut platforms now above sea level are known as **marine terraces** (Figure 16.13c).

Sea Caves, Arches, and Stacks

Sea cliffs do not retreat uniformly because some of the materials of which they are composed are more resistant to erosion than others. **Headlands** are seaward-projecting parts of the shoreline that are eroded on both sides by wave refraction (■ Figure 16.14). *Sea caves* form on opposite sides of a headland, and if these caves join, they form a *sea arch* (Figure 16.14a, b). Continued erosion causes the span of an arch to collapse, creating isolated *sea stacks* on wave-cut platforms (Figure 16.14c).

In the long run, shoreline processes tend to straighten an initially irregular shoreline. Wave refraction causes more wave energy to be expended on headlands and less on embayments. Thus, headlands erode, and some of the sediment yielded by erosion is deposited in the embayments.

THE NEARSHORE SEDIMENT BUDGET

We can think of the gains and losses of sediment in the nearshore zone in terms of a **nearshore sediment budget** (■ Figure 16.15). If a nearshore system has a balanced budget, sediment is supplied to it as fast as it is removed, and the volume of sediment remains more or less constant, although sand may shift offshore and onshore with the changing seasons (Figure 16.8). A positive budget means gains exceed losses, whereas a negative budget means losses exceed gains. If a negative budget prevails long enough, a nearshore system is depleted and beaches may disappear.

Erosion of sea cliffs provides some sediment to beaches, but in most areas probably no more than 5–10% of the total sediment comes from this source. There are exceptions, though; almost all the sediment on the beaches of Maine is derived from the erosion of shoreline

Figure 16.13

(a) Wave erosion causes a sea cliff to migrate landward, leaving a gently sloping surface known as a wave-cut platform. A wave-built platform originates by deposition at the seaward margin of the wave-cut platform. (b) Sea cliffs and a wave-cut platform. (c) This gently sloping surface along the Pacific Coast of California is a marine terrace. Notice the sea stacks rising above the terrace and others formed by erosion of the sea cliff.

rocks. Most sediment on typical beaches is transported to the shoreline by streams and then redistributed along the shoreline by longshore currents. Thus, longshore currents also play a role in the nearshore sediment budget because they continuously move sediment into and away from beach systems.

The primary ways that a nearshore system loses sediment are offshore transport, wind, and deposition in submarine canyons. Offshore transport mostly involves fine-grained sediment carried seaward where it eventually settles in deeper water. Wind is an important process because it removes sand from beaches and blows it inland where it piles up as sand dunes.

If the heads of submarine canyons are nearshore, huge quantities of sand are funneled into them and deposited in deeper water. La Jolla and Scripps submarine canyons off the coast of southern California funnel off an estimated 2 million m^3 of sand each year. In most areas, however, submarine canyons are too far offshore to interrupt the flow of sand in the nearshore zone.

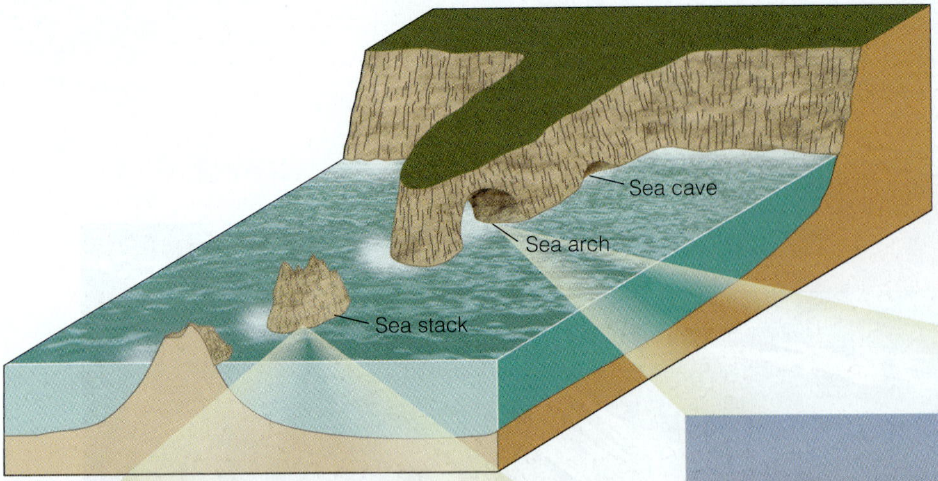

(a)

(b)

(c)

Figure 16.14

(a) Erosion of a headland and the origin of sea caves, sea arches, and sea stacks. (b) An arch has developed in this sea stack in Australia. (c) Sea stacks on California's central coast.

It should be apparent from this discussion that if a nearshore system is in equilibrium, its incoming supply of sediment exactly offsets its losses. Such a delicate balance tends to continue unless the system is somehow disrupted. One common change that affects this balance is the construction of dams across the streams that supply sand. Once dams have been built, all sediment from the upper reaches of the drainage systems is trapped in reservoirs and thus cannot reach the shoreline.

HOW ARE COASTAL AREAS MANAGED AS SEA LEVEL RISES?

During the last century, sea level rose about 12 cm worldwide, and all indications are that it will continue to rise. The absolute rate of sea-level rise in a shoreline region depends on two factors. The first is the volume of water in the ocean basins, which is increasing as a result of glacial ice melting and the thermal expansion of near-surface seawater. Many scientists think that sea level will continue to rise because of global warming caused by increasing concentrations of greenhouse gases in the atmosphere.

The second factor that controls sea level is the rate of uplift or subsidence of a coastal area. In some areas, uplift is occurring fast enough that sea level is actually falling with respect to the land. In other areas, sea level is rising while the coastal region is simultaneously subsiding, resulting in a net change in sea level of as much as 30 cm per century. Perhaps such a "slow" rate of sea level change seems insignificant; after all, it amounts to only a few millimeters per year. But in gently sloping coastal areas, as in the eastern United States from New Jersey southward, even a slight rise in sea level will eventually have widespread effects.

Many of the nearly 300 barrier islands along the East and Gulf Coasts of the United States are migrating

Inputs	+	Outputs	=	Balance
Longshore transport into beach River supply Cliff erosion Onshore transport		Longshore transport out of beach Offshore transport Wind transportation into dunes		Accretion Erosion Steady state

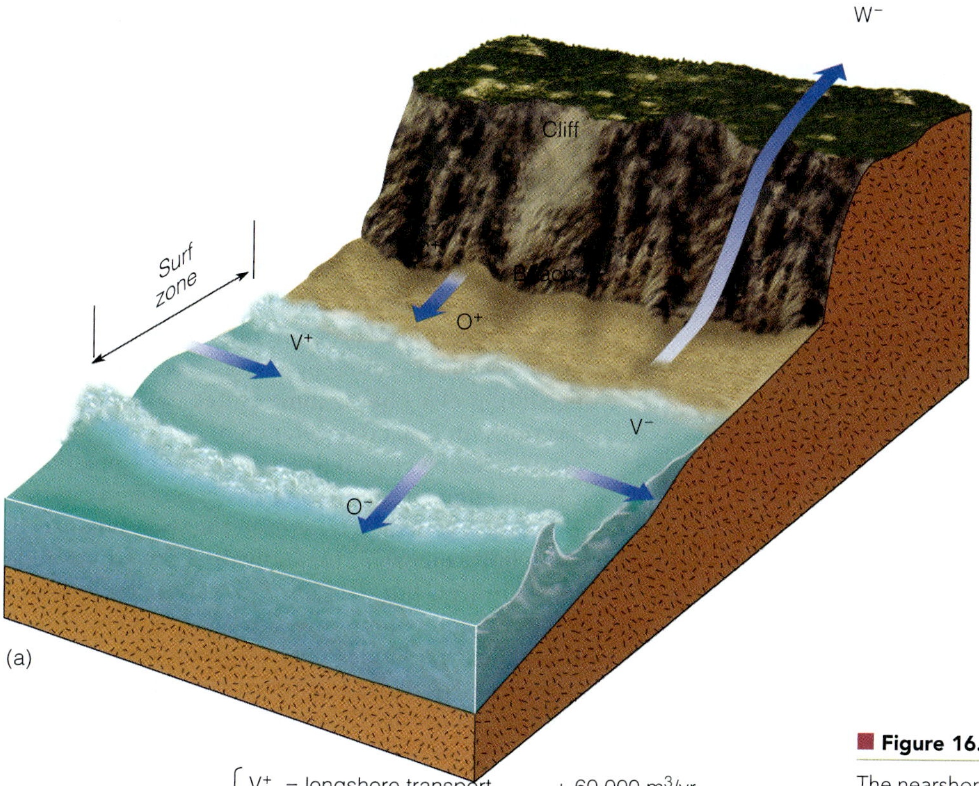

(a)

Inputs	V$^+$ = longshore transport into beach	+ 60,000 m^3/yr
	C$^+$ = cliff erosion	+ 5,000 m^3/yr
	O$^+$ = onshore transport	+ 5,000 m^3/yr
Outputs	W$^-$ = wind	−1,000 m^3/yr
	V$^-$ = longshore transport out of beach	− 54,000 m^3/yr
	O$^-$ = offshore transport (includes transport to submarine canyons)	− 20,000 m^3/yr
	Balance	− 5,000 m^3/yr (net erosion)

(b)

Figure 16.15

The nearshore sediment budget. (a) The long-term sediment budget can be assessed by considering inputs versus outputs. If inputs and outputs are equal, a beach is in a steady state or equilibrium. If outputs exceed inputs, however, the beach has a negative budget and erosion occurs. Accretion occurs when the beach has a positive budget with inputs exceeding outputs. (b) A hypothetical example of a negative nearshore sediment budget. The beach is losing 5,000 m^3 a year to erosion.

landward as sea level rises (Figure 16.16). Landward migration of barrier islands would pose few problems if it were not for the numerous communities, resorts, and vacation homes located on them. Moreover, barrier islands are not the only threatened areas. For example, Louisiana's coastal wetlands, an important wildlife habitat and seafood-producing area, are currently being lost at a rate of about 90 km^2 per year. Much of this loss results from sediment compaction, but rising sea level exacerbates the problem.

Rising sea level also directly threatens many beaches upon which communities depend for revenue. The beach at Miami Beach, Florida, for instance, was disappearing at an alarming rate until the Army Corps of Engineers began replacing the eroded beach sand (Figure 16.17). The problem is even more serious in other countries. A rise in sea level of only 2 m would inundate large areas of the U.S. East and Gulf Coasts, but would cover 20% of the entire country of Bangladesh. Other problems associated with rising sea level include increased coastal flooding during storms and saltwater incursions that may threaten groundwater supplies (see Chapter 13).

Armoring shorelines with *seawalls* (embankments of reinforced concrete or stone) (Figure 16.18) and using *riprap* (piles of stones) protect beachfront structures, but both are initially expensive and during large storms are commonly damaged or destroyed. Seawalls do afford some

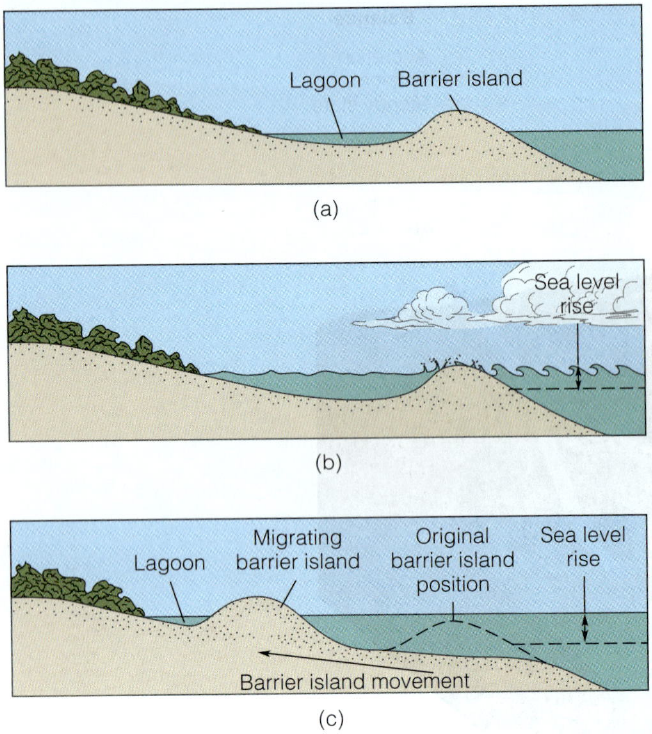

■ **Figure 16.16**

Rising sea level and the landward migration of barrier islands. (a) Barrier island before landward migration in response to rising sea level. (b) Sea level rises and the barrier island migrates landward when storm waves wash sand from its seaward side and deposit it in the lagoon. (c) Over time, the entire complex migrates landward.

protection and are seen in many coastal areas along the oceans and large lakes, but some states, including North and South Carolina, Rhode Island, Oregon, and Maine, no longer allow their construction. The futility of artificially maintaining beaches is aptly shown by efforts to protect homes on a South Carolina barrier island. After each spring tide, heavy equipment builds a sand berm to protect homes from the next spring tide (■ Figure 16.19b), only to see the berm disappear and then have to rebuild it in a never-ending cycle of erosion and expensive artificial replacement.

Because we can do nothing to prevent sea level from rising, engineers, scientists, planners, and political leaders must examine what they can do to prevent or minimize the effects of shoreline erosion. At present, only a few viable options exist. One is to put strict controls on coastal development. North Carolina, for example, permits large structures no closer to the shoreline than 60 times the annual erosion rate. Although a growing awareness of shoreline processes has resulted in similar legislation elsewhere, some states have virtually no restrictions on coastal development.

Regulating coastal development is commendable, but it has no impact on existing structures and coastal communities. A general retreat from the shoreline may be possible, but expensive, for individual dwellings and small communities, but it is impractical for large population centers. Such communities as Atlantic City, New Jersey; Miami Beach, Florida; and Galveston, Texas, have adopted one of two strategies to combat coastal erosion. One is to build protective barriers such as seawalls. Seawalls, such

U.S. Army Corps of Engineers

(a)

U.S. Army Corps of Engineers

(b)

■ **Figure 16.17**

The beach at Miami Beach, Florida, (a) before and (b) after the U.S. Army Corps of Engineers' beach nourishment project.

Rosenberg Library, Galveston, Texas

Figure 16.18

Construction of this seawall to protect Galveston, Texas, from storm waves began in 1902. Notice that the wall is curved to deflect waves upward.

(b)

(a)

Figure 16.19

(a) These large blocks of basalt were piled along this beach to prevent erosion and protect a luxury hotel just to the left out of the image. (b) Heavy equipment builds a berm on the seaward side of beach homes on the Isle of Palms, South Carolina, to protect them from waves. During the next spring tide, the berms disappear and must be rebuilt.

GEOLOGY IN UNEXPECTED PLACES

Erosion and the Cape Hatteras Lighthouse

The Outer Banks of North Carolina are a series of barrier islands separated from the mainland by a lagoon up to 48 km wide. Actually, the Outer Banks are made up of several barrier islands and spits extending about 240 km from the Virginia state line to Shackleford Banks near Beaufort, North Carolina (■ Figure 1a). On September 18, 2003, Hurricane Isabel roared across the Outer Banks, causing wind damage and flooding there and in several adjacent states, particularly Virginia.

Since 1526, more than 600 ships have been lost on the dangerous sand bars adjacent to the Outer Banks, such as Diamond Shoals, an area of very shallow seawater. To reduce ship losses, several lighthouses were built on the islands; the most famous is the Cape Hatteras Lighthouse. It was constructed in 1802, although the current lighthouse was built in 1870. In 1999, the National Park Service, which administers Cape Hatteras National Seashore, had the lighthouse moved 500 m inland and 760 m to the southwest (Figure 1b). The reason? The Outer Banks, just like other barrier islands, are migrating toward the mainland as storm waves carry sand from the seaward margins into the lagoon. Indeed, Hurricane Isabel cut a new channel through Hatteras Island, thereby isolating communities to the north and south.

When the Cape Hatteras Lighthouse was built in 1870, it was 457 m from the shoreline, but when it was moved in 1999, it stood only 36.5 m inland (Figure 1b). Given that the annual erosion rate is about 3 m per year and all previous attempts to stabilize the shoreline had failed, the Park Service found it prudent to act when it did. The Cape Hatteras Lighthouse, the tallest brick lighthouse in the world, should now be safe for several centuries.

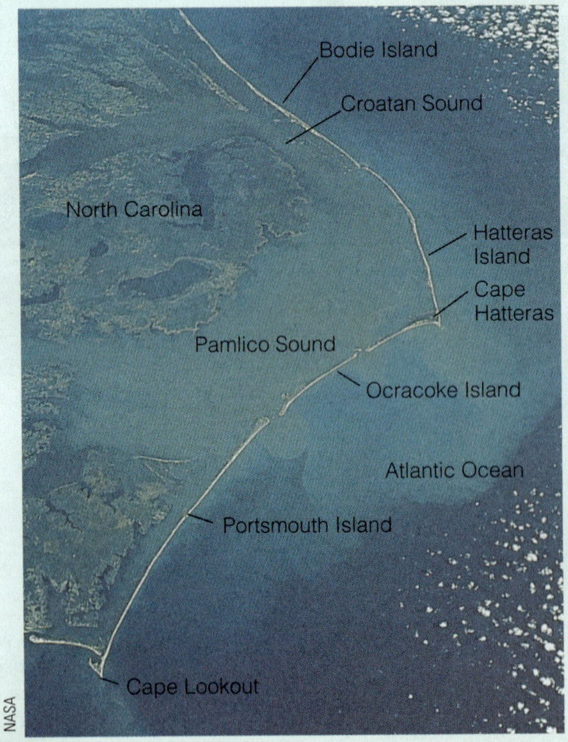

(a)

(b)

■ **Figure 1**

(a) These barrier islands make up the Outer Banks of North Carolina. (b) Cape Hatteras Lighthouse during the 1999 move of the 60-m-high, 4381-metric-ton structure. Its previous position was at the top of the image.

as the one at Galveston, Texas, are effective, but they are tremendously expensive to construct and maintain. More than $50 million was spent in just five years to replenish the beach and build a seawall at Ocean City, Maryland. Furthermore, barriers retard erosion only in the area directly behind them; Galveston Island west of its seawall has been eroded back about 45 m.

Another option, adopted by both Atlantic City, New Jersey, and Miami Beach, Florida, is to pump sand onto the beaches to replace that lost to erosion (Figure 16.17). This, too, is expensive as the sand must be replenished periodically because erosion is a continuing process. In many areas, groins are constructed to preserve beaches, but unless additional sand is artificially supplied to the beaches, longshore currents invariably erode sand from the downcurrent sides of the groins.

STORM WAVES AND COASTAL FLOODING

Coastal flooding caused by storm-generated waves and intense rainfall is of considerable concern to shoreline residents. In fact, coastal flooding during large storms causes more damage, injuries, and fatalities than high winds do. Flooding during hurricanes results when huge waves are driven onshore and as much as 60 cm of rain falls in as little as 24 hours. In addition, as a hurricane moves over the ocean, low atmospheric pressure beneath the eye of the storm causes the ocean surface to bulge upward as much as 0.5 m. When the eye reaches the shoreline, the bulge coupled with wind-driven waves piles up in a *storm surge* that may rise several meters above normal high tide and inundate areas far inland (■ Figure 16.20).

In 1900 Galveston, Texas, was hit by a hurricane and storm waves surged inland, eventually covering the entire barrier island the city was built on. Buildings and other structures near the shoreline were battered to pieces, and "great beams and railway ties were lifted by the [waves] and driven like battering rams into dwellings and business houses"* farther inland. Between 6000 and 8000 people died. In an effort to protect the city from future storms, a huge seawall was constructed (Figure 16.18) and the entire city was elevated to the level of the top of the seawall.

More recently, in 1989 Charleston, South Carolina, and nearby areas were flooded by the 2.5- to 6-m-high storm surge generated by Hurricane Hugo, which caused 21 deaths and more than $7 billion in property damage. Bangladesh is even more susceptible to storm surges; in 1970, 300,000 people drowned, and in 1991 another 139,000 were lost. Large-scale coastal flooding took place when Hurricane Isabel hit the East Coast along the Outer Banks of North Carolina on September 19, 2003. Of course, the four hurricanes that hit Florida and parts of the Gulf Coast during the 2004 hurricane season caused widespread wind damage as well as coastal flooding.

The problem of coastal flooding is, of course, exacerbated by rising sea level. To gain some perspective on the magnitude of the problem, consider this: Of the nearly $2 billion paid out by the federal government's National Flood Insurance Program since 1974, exclusive of Hurricane Isabel, most has gone to owners of beachfront homes.

*L. W. Bates, Jr., "Galveston—A City Built upon Sand," *Scientific American,* 95 (1906): 64.

Geoff Mackley

■ **Figure 16.20**

Storm surge at Nags Head, North Carolina, when Hurricane Isabel came ashore in September 2003.

TYPES OF COASTS

Coasts are difficult to classify because of variations in the factors that control their development and variations in their composition and configuration. Rather than attempt to categorize all coasts, we shall simply note that two types of coasts have already been discussed: those dominated by deposition and those dominated by erosion. Now we look further at the changing relationships between coasts and sea level.

Depositional and Erosional Coasts

Depositional coasts, such as the U.S. Gulf Coast, are characterized by an abundance of detrital sediment and such depositional landforms as wide sandy beaches, deltas, and barrier islands. In contrast, erosional coasts are steep and irregular and typically lack well-developed beaches except in protected areas (see "Shorelines Processes and Beaches" on pages 450 and 451). They are further characterized by sea cliffs, wave-cut platforms, and sea stacks. Many of the coasts along the West Coast of North America fall into this category.

The following section will examine coasts in terms of their changing relationships to sea level. But note that although some coasts, such as those in southern California, are described as emergent (uplifted), these same coasts may be erosional as well. In other words, coasts commonly possess features that allow them to be classified in more than one way.

Submergent and Emergent Coasts

If sea level rises with respect to the land or the land subsides, coastal regions are flooded and said to be **submergent** or *drowned* **coasts** (■ Figure 16.21). Much of the East Coast of North America from Maine southward

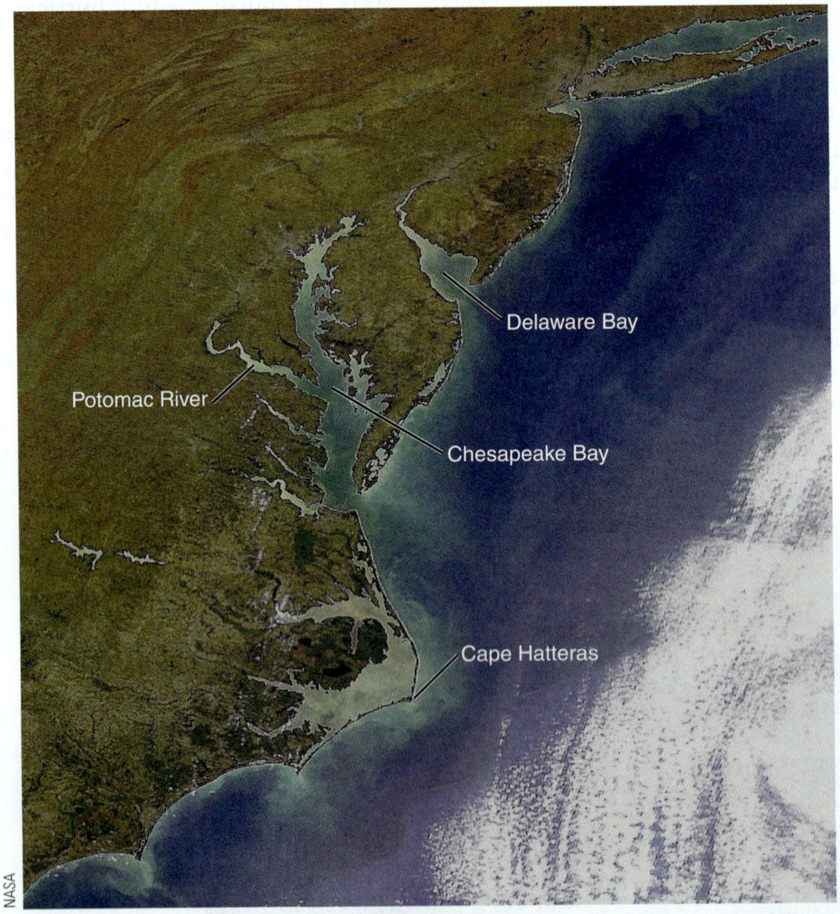

■ **Figure 16.21**

Submergent coasts tend to be extremely irregular with estuaries such as Chesapeake Bay. It formed when the East Coast of the United States was flooded as sea level rose following the Pleistocene Epoch.

through South Carolina was flooded during the rise in sea level following the Pleistocene Epoch, so it is extremely irregular. Recall that during the expansion of glaciers during the Pleistocene, sea level was as much as 130 m lower than at present, and that streams eroded their valleys more deeply and extended across continental shelves. When sea level rose, the lower ends of these valleys were drowned, forming *estuaries* such as Delaware and Chesapeake Bays (Figure 16.21). Estuaries are simply the seaward ends of river valleys where seawater and freshwater mix. The divides between adjacent drainage systems on submergent coasts project seaward as broad headlands or a line of islands.

Submerged coasts are also present at higher latitudes where Pleistocene glaciers flowed into the sea. When sea level rose, the lower ends of the glacial troughs were drowned, forming fiords (see "Valley Glaciers and Erosion" on pages 400 and 401).

Emergent coasts are found where the land has risen with respect to sea level (■ Figure 16.22). Emergence takes place when water is withdrawn from the oceans, as occurred during the Pleistocene expansion of glaciers. Presently coasts are emerging as a result of isostasy or tectonism. In northeastern Canada and the Scandinavian countries, for instance, the coasts are irregular because isostatic rebound is elevating formerly glaciated terrain from beneath the sea.

Coasts that form in response to tectonism, on the other hand, tend to be straighter because the seafloor topography being exposed as uplift is smooth. The west coasts of North and South America are rising as a consequence of plate tectonics. Distinctive features of these coasts are marine terraces (Figures 16.13c and 16.22), which are old wave-cut platforms now elevated above sea level. Uplift in such areas appears to be episodic rather than continuous, as indicated by the multiple levels of terraces in some places. In southern California, several terrace levels are present, each of which probably represents a period of tectonic stability followed by uplift. The highest terrace is now about 425 m above sea level.

■ **Figure 16.22**

Emergent coasts tend to be steep and straighter than submergent coasts. Notice several sea stacks and the sea arch. Also, a marine terrace is visible in the distance.

16 GEO RECAP

Chapter Summary

- Shorelines are continuously modified by the energy of waves and longshore currents and, to a limited degree, by tidal currents.
- The gravitational attraction of the Moon and Sun causes the ocean surface to rise and fall as tides twice daily in most shoreline areas.
- Waves are oscillations on water surfaces that transmit energy in the direction of wave movement. Surface waves affect the water and seafloor only to wave base, which is equal to half the wavelength.
- Little or no net forward motion of water occurs in waves in the open sea. When waves enter shallow water, they are transformed into waves in which water moves in the direction of wave advance.
- Wind-generated waves, especially storm waves, are responsible for most geologic work on shorelines, but waves can also be generated by faulting, volcanic explosions, and rockfalls.
- Breakers form where waves enter shallow water and disrupt the orbital motion of water particles. The waves become oversteepened and plunge forward or spill onto the shoreline, thus expending their kinetic energy.
- Waves approaching a shoreline at an angle generate a longshore current. These currents are capable of considerable erosion, transport, and deposition.
- Narrow surface currents called rip currents carry water from the nearshore zone seaward through the breaker zone.
- Beaches, the most common shoreline depositional features, are continuously modified by nearshore processes, and their profiles generally exhibit seasonal changes.
- Spits, baymouth bars, and tombolos all form and grow as a result of longshore current transport and deposition.
- Barrier islands are nearshore sediment deposits of uncertain origin. They parallel the mainland but are separated from it by a lagoon.
- The volume of sediment, or nearshore sediment budget, in a nearshore system remains rather constant unless the system is somehow disrupted, as when dams are built across streams that supply sand to the system.
- Coastal flooding during storms by waves and storm surge is an ongoing problem in many areas.
- Many shorelines are characterized by erosion rather than deposition. Such shorelines have sea cliffs and wave-cut platforms. Other features commonly present include sea caves, sea arches, and sea stacks. Depositional coasts are characterized by long sandy beaches, deltas, and barrier islands.
- Submergent and emergent coasts are defined on the basis of their relationships to changes in sea level.

Important Terms

backshore (p. 449)
barrier island (p. 454)
baymouth bar (p. 453)
beach (p. 449)
beach face (p. 449)
berm (p. 449)
breaker (p. 446)
crest (wave) (p. 445)
emergent coast (p. 465)
fetch (p. 446)
foreshore (p. 449)
headland (p. 456)
longshore current (p. 447)
marine terrace (p. 456)
nearshore sediment budget (p. 456)
rip current (p. 448)
shoreline (p. 440)
spit (p. 453)
submergent coast (p. 464)
tide (p. 441)
tombolo (p. 453)
trough (wave) (p. 445)
wave (p. 445)
wave base (p. 446)
wave-cut platform (p. 455)
wave refraction (p. 447)

Review Questions

1. A berm is
   a._____a gently landward-sloping part of the backshore zone; b._____the highest level reached during flood tide; c._____the lowest part of a wave between crests; d._____the distance that wind blows over a continuous water surface; e._____the angle at which a wave hits the shoreline.
2. A wave-cut platform now above sea level is a
   a._____wave trough; b._____submergent coast; c._____longshore current; d._____marine terrace; e._____baymouth bar.
3. A sand deposit that extends into the mouth of a bay is a
   a._____tombolo; b._____spit; c._____berm; d._____crest; e._____beach face.
4. A longshore current is generated by
   a._____a rapidly rising tide followed by ebb tide; b._____erosion of headlands and deposition in embayments; c._____a diminished amount of sand reaching a beach; d._____erosion of shoreline rocks; e._____waves approaching a shoreline at an angle.
5. A long, narrow sand deposit separated from the mainland by a lagoon is a
   a._____baymouth bar; b._____tidal delta; c._____barrier island; d._____beach cusp; e._____wave-built platform.
6. The time it takes for two successive wave crests (or troughs) to pass a given point is known as the
   a._____wave period; b._____wave celerity; c._____wave form; d._____wave height; e._____wave fetch.
7. Fetch is the
   a._____excess water in the nearshore zone that moves seaward in rip currents; b._____distance wind blows over a continuous water surface; c._____time it takes for two waves to pass a given point; d._____part of a wave that encounters wave base first; e._____effective depth to which waves can erode.
8. A spit that connects an island with the shoreline is a(n)
   a._____tombolo; b._____breaker; c._____emergent coast; d._____drift; e._____sea stack.
9. Although there are some exceptions, most beaches receive most of their sediment from
   a._____coastal submergence; b._____erosion of shoreline rocks; c._____erosion of reefs; d._____streams and rivers; e._____breakers.
10. Erosion by water carrying sand and gravel is known as
    a._____collusion; b._____hydraulic action; c._____longshore drift; d._____abrasion; e._____wave base.

11. A hypothetical shoreline has a balanced budget, but a dam is built on the river that supplies sediment and a wall is built to protect sea cliffs from erosion. What will likely happen to beaches in this area? Explain.
12. Why are long sandy beaches common along North America's East Coast but found mostly in protected areas on the West Coast?
13. How do barrier islands migrate landward? Is there any evidence indicating that U.S. barrier islands are migrating? Give some specific examples.
14. Why does an observer at a shoreline location notice two high and two low tides daily?
15. What is wave base, and how does it affect waves as they enter progressively shallower water?
16. While driving along North America's West Coast, you notice a rather flat surface near the seashore with several isolated masses or rock rising above it. What is this surface, and how did it form?
17. What are rip currents, and how would you recognize one? If caught in a rip current, how would you get out?
18. Name and explain how three deposition landforms originate along coasts.
19. How are waves responsible for transporting sediment great distances along shorelines?
20. How and why do summer and winter beaches differ?

World Wide Web Activities

Geology Now Assess your understanding of this chapter's topics with additional quizzing and comprehensive interactivities at

http://earthscience.brookscole.com/changingearth4e

as well as current and up-to-date weblinks, additional readings, and InfoTrac College Edition exercises.

Geologic Time: Concepts and Principles

CHAPTER 17

OUTLINE

- Introduction
- How Has the Concept of Geologic Time and Earth's Age Changed Throughout Human History?

  GEO-FOCUS 17.1: *Geologic Time and Climate Change*
- Why Are James Hutton's Contributions to Geology Important?

  GEOLOGY IN UNEXPECTED PLACES: *Time Marches On—The Great Wall of China*
- What Are Relative Dating Methods?
- How Do Geologists Correlate Rock Units?
- What Are Absolute Dating Methods?
- How Was the Geologic Time Scale Developed?
- Stratigraphy and Stratigraphic Terminology
- Geo-Recap

Geology Now *This icon, which appears throughout the book, indicates an opportunity to explore interactive tutorials, animations, or practice problems available on the GeologyNow website at http://earthscience.brookscole.com/changingearth4e.*

The Grand Canyon, Arizona. Major John Wesley Powell led two expeditions down the Colorado River and through the canyon in 1869 and 1871. He was struck by the seemingly limitless time represented by the rocks exposed in the canyon walls and by the recognition that these rock layers, like the pages in a book, contain the geologic history of this region.
Source: Royalty-Free/Corbis

OBJECTIVES

At the end of this chapter, you will have learned that:

- The concept of geologic time and its measurements have changed throughout human history.
- The principle of uniformitarianism is fundamental to geology.
- Relative dating, placing geologic events in a sequential order, provides a means to interpret geologic history.
- The three types of unconformities—disconformities, angular unconformities, and nonconformities—are erosional surfaces separating younger from older rocks and represent significant intervals of geologic time for which we have no record at a particular location.
- Time equivalency of rock units can be demonstrated by various correlation techniques.
- Absolute dating methods are used to date geologic events in terms of years before present.
- The most accurate radiometric dates are obtained from igneous rocks.
- The geologic time scale evolved primarily during the 19th century through the efforts of many people.
- Stratigraphic terminology includes units based on content and units related to geologic time.

Introduction

In 1869, Major John Wesley Powell, a Civil War veteran who lost his right arm in the battle of Shiloh, led a group of hardy explorers down the uncharted Colorado River through the Grand Canyon. With no maps or other information, Powell and his group ran the many rapids of the Colorado River in fragile wooden boats, hastily recording what they saw. Powell wrote in his diary, "All about me are interesting geologic records. The book is open and I read as I run."

From this initial reconnaissance, Powell led a second expedition down the Colorado River in 1871. This second trip included a photographer, a surveyor, and three topographers. The expedition made detailed topographic and geologic maps of the Grand Canyon area as well as the first photographic record of the region.

Probably no one has contributed as much to the understanding of the Grand Canyon as Major Powell. In recognition of his contributions, the Powell Memorial was erected on the South Rim of the Grand Canyon in 1969 to commemorate the 100th anniversary of this history-making first expedition.

Most tourists today, like Powell and his fellow explorers in 1869, are astonished by the seemingly limitless time represented by the rocks exposed in the walls of the Grand Canyon. For most people, staring down 1.5 km at the rocks in the canyon is their only exposure to the concept of geologic time. When standing on the rim and looking down into the Grand Canyon, we are really looking far back in time, all the way back to the early history of our planet. In fact, more than a billion years of history are preserved in the rocks of the Grand Canyon.

Just as we read the pages of a book, we can read the rock layers of the Grand Canyon and learn that this region underwent episodes of mountain building as well as periods of advancing and retreating shallow seas. How do we know this? The answer lies in applying the principles of relative dating to the rocks we see exposed in the canyon, as well as recognizing that the geologic processes we see operating today have operated throughout Earth history.

We begin this chapter by asking the question, What is time? We seem obsessed with time and we organize our lives around it with the help of clocks, calendars, and appointment books. Yet most of us feel we don't have enough of it; we are always running "behind" or "out of time." Whereas physicists deal with extremely short intervals of time and geologists deal with incredibly long periods of time, most of us tend to view time from the perspective of our own existence; that is, we partition our lives into seconds, hours, days, weeks, months, and years. Ancient history is what occurred hundreds or even thousands of years ago. Yet when geologists talk of ancient geologic history, they are referring to events that happened millions or even billions of years ago!

Vast time periods set geology apart from most of the other sciences, and an appreciation of the immensity of geologic time is fundamental to understanding the physical and biological history of our planet. In fact, understanding and accepting the magnitude of geologic time are major contributions geology has made to the sciences.

In some respects, time is defined by the methods used to measure it. Geologists use two different frames of reference when discussing geologic time. **Relative dating** is placing geologic events in a sequential order as determined from their position in the geologic record. Relative dating does not tell us how long ago a particular event took place, only that one event preceded another. A useful analogy for relative dating is a television guide that does not list the times programs are shown. You cannot tell what time a particular program will be shown, but by watching a few shows and checking the guide, you can determine whether you have missed the show or how many shows are scheduled before the one you want to see.

The various principles used to determine relative dating were discovered hundreds of years ago, and since then they have been used to construct the *relative geologic time scale* (■ Figure 17.1). Furthermore, these principles are still widely used by geologists today.

Absolute dating provides specific dates for rock units or events expressed in years before the present. In our analogy of the television guide, the time when the programs are actually shown are the absolute dates. In this way, you not only can determine whether you have missed a show (relative dating), but also know how long it will be until a show you want to see will be shown (absolute dating).

Radiometric dating is the most common method of obtaining absolute ages. Dates are calculated from the natural decay rates of various radioactive elements present in trace amounts in some rocks. It was not until the discovery of radioactivity near the end of the 19th century that absolute ages could be accurately applied to the relative geologic time scale. Today, the geologic time scale is really a dual scale: a relative scale based on rock sequences with radiometric dates expressed as years before the present (Figure 17.1).

Besides providing an appreciation for the immensity of geologic time, why is the study of geologic time important? One of the most important lessons to be learned in this chapter is how to reason and apply the fundamental geologic principles to solve geologic problems. The logic used in applying the principles of relative dating to interpret the geologic history of an area involves basic reasoning skills that can be transferred to and used in almost any profession.

Advances and refinements in absolute dating techniques during the 20th century have changed the way we view Earth in terms of when events occurred in the past and the rates of geologic change through time. The ability to accurately determine past climatic changes and their causes has important implications for the current debate on global warming and its effects on humans (see Geo-Focus 17.1).

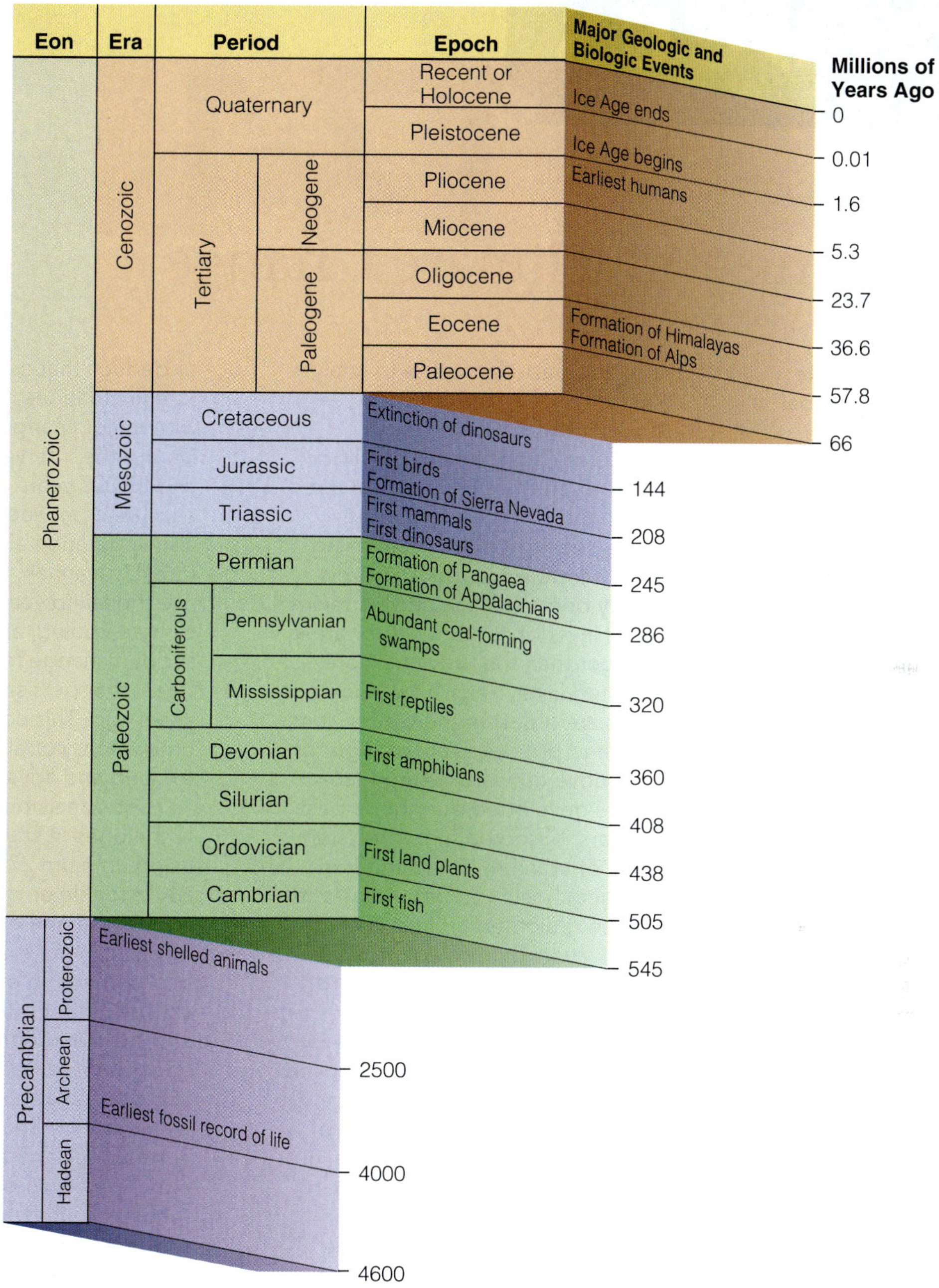

Figure 17.1

The geologic time scale. Some of the major geologic and biologic events are indicated along the right-hand margin.
Source: Modified from A. R. Palmer, "The Decade of North American Geology, 1983 Geologic Time Scale," *Geology* (Geological Society of America, 1983), p. 504.

HOW HAS THE CONCEPT OF GEOLOGIC TIME AND EARTH'S AGE CHANGED THROUGHOUT HUMAN HISTORY?

The concept of geologic time and its measurement have changed throughout human history. Many early Christian scholars and clerics tried to establish the date of creation by analyzing historical records and the genealogies found in Scripture. Based on their analyses, they generally believed that Earth and all of its features were no more than about 6000 years old. The idea of a very young Earth provided the basis for most Western chronologies of Earth history prior to the 18th century.

During the 18th and 19th centuries, several attempts were made to determine Earth's age on the basis of scientific evidence rather than revelation. The French zoologist Georges Louis de Buffon (1707–1788) assumed Earth gradually cooled to its present condition

GEOFOCUS 17.1

Geologic Time and Climate Change

With all the debate concerning global warming and its possible implications, it is extremely important to be able to reconstruct past climatic regimes as accurately as possible. To model how Earth's climate system has responded to changes in the past and to use that information for simulations of future climatic scenarios, we must have as precise a geologic calendar as possible.

One way to study climatic changes is to examine lake sediment or ice cores that contain organic matter. By taking closely spaced samples and dating the organic matter in the cores using the carbon 14 dating technique, geologists can construct a detailed chronology for each core examined. Changes in isotope ratios, pollen, and plant and invertebrate fossil assemblages can then be accurately dated, and the time and duration of climate changes correlated over increasingly larger areas. Without a means of precise dating, we would have no way to accurately model past climatic changes with the precision needed to predict possible future climate changes on a human time scale.

An interesting method that is becoming more common in reconstructing past climates is to analyze stalagmites from caves. Stalagmites are icicle-shaped structures rising from a cave floor and formed of calcium carbonate precipitated from evaporating water (see Figure 13.13). A stalagmite therefore records a layered history because each newly precipitated layer of calcium carbonate is younger than the previously precipitated layer. Thus, a stalagmite's layers are oldest in the center at its base and progressively younger as they move outward. Using techniques developed during the past 10 years, based on high-precision ratios of uranium 234 to thorium 230, geologists can determine very precise radiometric dates on individual layers of a stalagmite. This technique enables geologists to determine the age of materials much older than they can date by carbon 14, and it is reliable back to about 500,000 years.

An interesting recent study of stalagmites from Crevice Cave in Missouri revealed a history of climatic and vegetation change in the mid-continent region of the United States during the interval between 75,000 and 25,000 years ago. By analyzing carbon 13 and oxygen 18 isotope profiles during this interval, geologists were able to deduce that average temperature fluctuations of 4°C correlated with major changes in vegetation. During the interval between 75,000 and 55,000 years ago, the climate oscillated between warm and cold, and vegetation alternated among forest, savannah, and prairie. Fifty-five thousand years ago the climate cooled, and there was a sudden change from grasslands to forest that persisted until 25,000 years ago. This corresponds to the time when global ice sheets began to build and advance.

High-precision dating techniques in stalagmite studies, using uranium 234 and thorium 230, provide an accurate chronology that allows geologists to model climate systems of the past and perhaps to determine what causes global climatic changes and their duration. Without these sophisticated dating techniques and others like them, geologists would not be able to make precise correlations and accurately reconstruct past environments and climates. By analyzing past environmental and climate changes and their duration, geologists hope they can use these data, sometime in the near future, to predict and possibly modify regional climatic changes.

from a molten beginning. To simulate this history, he melted iron balls of various diameters and allowed them to cool to the surrounding temperature. By extrapolating their cooling rate to a ball the size of Earth, he determined that Earth was at least 75,000 years old. Although this age was much older than that derived from Scripture, it was vastly younger than we now know the planet to be.

Other scholars were equally ingenious in attempting to calculate Earth's age. For example, if deposition rates could be determined for various sediments, geologists reasoned that they could calculate how long it would take to deposit any rock layer. They could then extrapolate how old Earth was from the total thickness of sedimentary rock in its crust. Rates of deposition vary, however, even for the same type of rock. Furthermore, it is impossible to estimate how much a rock has been removed by erosion, or how much a rock sequence has been reduced by compaction. As a result of these variables, estimates ranged from less than 1 million years to more than 2 billion years.

Another attempt to determine Earth's age involved ocean salinity. Scholars assumed that Earth's ocean waters were originally fresh and that their present salinity was the result of dissolved salt being carried into the ocean basins by streams. Knowing the volume of ocean water and its salinity, John Joly, a 19th-century Irish geologist, measured the amount of salt currently in the world's streams. He then calculated that it would have taken at least 90 million years for the oceans to reach their present salinity level. This was still much younger than the now-accepted age of 4.6 billion years for Earth, mainly because Joly had no way to calculate how much salt had been recycled or the amount of salt stored in continental salt deposits and seafloor clay deposits.

Besides trying to determine Earth's age, the naturalists of the 18th and 19th centuries were formulating some of the fundamental geologic principles that are used in deciphering Earth history. From the evidence preserved in the geologic record, it was clear to them that Earth is very old and that geologic processes have operated over long periods of time. A good example of geologic processes operating to produce a spectacular landscape is the evolution of Uluru and Kata Tjuta (see "Uluru and Kata Tjuta" on pages 476 and 477).

WHY ARE JAMES HUTTON'S CONTRIBUTIONS TO GEOLOGY IMPORTANT?

Many consider the Scottish geologist James Hutton (1726–1797) to be the father of modern geology. His detailed studies and observations of rock exposures and present-day geologic processes were instrumental in establishing the **principle of uniformitarianism** (see Chapter 1), the concept that the same processes have operated over vast amounts of time. Because Hutton relied on known processes to account for Earth history, he concluded that Earth must be very old and wrote that "we find no vestige of a beginning, and no prospect of an end."

Unfortunately, Hutton was not a particularly good writer, so his ideas were not widely disseminated or accepted. In 1830, Charles Lyell published a landmark book, *Principles of Geology,* in which he championed Hutton's concept of uniformitarianism. Instead of relying on catastrophic events to explain various Earth features, Lyell recognized that imperceptible changes brought about by present-day processes could, over long periods of time, have tremendous cumulative effects. Through his writings, Lyell firmly established uniformitarianism as the guiding principle of geology. Furthermore, the recognition of virtually limitless amounts of time was also necessary for, and instrumental in, the acceptance of Darwin's 1859 theory of evolution (see Chapter 18).

After finally establishing that present-day processes have operated over vast periods of time, geologists were nevertheless nearly forced to accept a very young age for Earth when a highly respected English physicist, Lord Kelvin (1824–1907), claimed, in a paper written in 1866, to have overturned the uniformitarian foundation of geology. Starting with the generally accepted belief that Earth was originally molten, Kelvin assumed that it has gradually been losing heat and that, by measuring this heat loss, he could determine its age.

Kelvin knew from deep mines in Europe that Earth's temperature increases with depth, and he reasoned that Earth is losing heat from its interior. By knowing the size of Earth, the melting temperatures of rocks, and the rate of heat loss, Kelvin calculated the age at which Earth was entirely molten. From these calculations, he concluded that Earth could not be older than 400 million years or younger than 20 million years. This wide discrepancy in age reflected uncertainties in average temperature increases with depth and the various melting points of Earth's constituent materials.

After finally establishing that Earth was very old and that present-day processes operating over long periods of time account for geologic features, geologists were in a quandary. If they accepted Kelvin's dates, they would have to abandon the concept of seemingly limitless time that was the underpinning of uniformitarian geology and one of the foundations of Darwinian evolution and squeeze events into a shorter time frame.

Kelvin's reasoning and calculations were sound, but his basic premises were false, thereby invalidating his conclusions. Kelvin was unaware that Earth has an internal heat source, radioactivity, that has allowed it to maintain a fairly constant temperature through time.* His 40-year campaign for a young Earth ended with the discovery of radioactivity near the end of the 19th century. His calculations were no longer valid and his proof for a geologically young Earth collapsed.

Although the discovery of radioactivity destroyed Kelvin's arguments, it provided geologists with a clock that could measure Earth's age and validate what geologists

*Actually, Earth's temperature has decreased through time because the original amount of radioactive materials has been decreasing and thus is not supplying as much heat. However, the temperature is decreasing at a rate considerably slower than would be required to lend any credence to Kelvin's calculations.

Uluṟu and Kata Tjuṯa

Rising majestically above the surrounding flat desert of central Australia are Uluṟu and Kata Tjuṯa. Uluṟu and Kata Tjuṯa are the aboriginal names for what most people know as Ayers Rock and The Olgas. The history of Uluṟu and Kata Tjuṯa began about 550 million years ago when a huge mountain range formed in what is now central Australia. It subsequently eroded, and vast quantities of gravel were transported by streams and deposited along its base to form large alluvial fans. Marine sediments then covered the alluvial fans and the entire region was uplifted by tectonic forces between 400 and 300 million years ago and then subjected to weathering.

The spectacular and varied rock shapes of Uluṟu and Kata Tjuṯa are the result of millions of years of weathering and erosion by water and, to a lesser extent, wind acting on the fractures formed during uplift. Differences in the composition and texture of the rocks also played a role in sculpting these colorful and magnificent structures.

Reed Wicander

Aerial view of Uluṟu with Kata Tjuṯa in the background. Contrary to popular belief, Uluṟu is not a giant boulder. Rather it is the exposed portion of the nearly vertically tilted Uluṟu Arkose. The caves, caverns, and depressions visible on the northeastern side are the result of weathering.

A closeup view of the brain- and honeycomb-like small caves seen on the northeast side of Uluṟu.

Uluṟu at sunset. The near-vertical tilting of the sedimentary beds of the Uluṟu Arkose that make up Uluṟu can be seen clearly. Differential weathering of the sedimentary layers has produced the distinct parallel ridges and other features characteristic of Uluṟu.

Reed Wicander

Aerial view of Kata Tjuṯa with Uluṟu in the background. Kata Tjuṯa is composed of the Mount Currie Conglomerate, a coarse-grained and poorly sorted conglomerate. The sediments that were lithified into the Mount Currie Conglomerate were deposited, like the Uluṟu Arkose, as an alluvial fan beginning approximately 550 million years ago.

Reed Wicander

A view of the rounded domes of Kata Tjuṯa and typical vegetation as seen from within one of its canyons. The red color of the rocks is the result of oxidation of iron in the sediments.

Reed Wicander

An aerial closeup view of Kata Tjuṯa. The distinctive dome shape of the rocks is the result of weathering and erosion of the Mount Currie Conglomerate. In addition to weathering, the release of pressure on the buried rocks as they were exposed at the surface by tectonic forces contributed to the characteristic rounded shapes of Kata Tjuṯa.

GEOLOGY IN UNEXPECTED PLACES

Time Marches On—The Great Wall of China

The Great Wall of China, built over many centuries as a military fortification against invasion by enemies from the north, has largely succumbed to the ravages of nature and human activity. Originally begun as a series of short walls during the Zhou Dynasty (770–476 B.C.), the wall grew as successive dynasties connected different parts, with final improvements made during the Ming Dynasty (A.D. 1368–1644). Even though the Great Wall is not continuous, it stretches more than 5000 km across northern China from the east coast to the central part of the country (■ Figure 1). Contrary to popular belief, the Great Wall of China is not the only man-made structure visible from space or the Moon. From low Earth orbit, many artificial objects are visible, such as highways, cities, and railroads. When viewed from a distance of a few thousand kilometers, no man-made objects are visible and the Great Wall can barely be seen with the naked eye from the shuttle, according to NASA. In fact, China's first astronaut, Yang Liwei, told state television on his return from space, "I did not see the Great Wall from space."

So with that short history and debunking of an urban legend, what is the Great Wall made of? Basically, the Great Wall was constructed with whatever material was available in the area, including sedimentary, metamorphic, and igneous rocks, bricks, sand, gravel, and even dirt and straw. Regardless of the material, the wall was built by hand by thousands of Chinese over many centuries.

In the Badaling area of Beijing, which is the part of the Great Wall most tourists visit and has been restored, it was built using the igneous and sedimentary rocks from the mountains around Badaling. The sedimentary rocks are mudstones, sandstones, and limestones, whereas the igneous rocks are granite. The sides of the walls in this area are constructed of rectangular slabs of granite, and the top or roof of the Great Wall here is paved with large gray bricks (■ Figure 2).

The average height of the Great Wall is 8.5 m, and it is 6.5 m wide along its base. The wall along the top averages 5.7 m, wide enough for 5 horses or 10 warriors to walk side by side.

Most visitors to the Great Wall are so impressed by its size and history that they don't even notice what it is made from. Now, however, you know and should you visit this impressive structure you can tell your fellow travelers all about it.

■ **Figure 1**

The Great Wall of China winding across the top of the hills at Badaling, just outside Beijing.

Reed Wicander

■ **Figure 2**

The top of the Great Wall at Badaling. Note the original rocks in the lower portion of the side of the wall and the paving bricks in the foreground that make up the top of the wall.

Reed Wicander

had been saying all along—namely, that Earth was indeed very old!

WHAT ARE RELATIVE DATING METHODS?

Before the development of radiometric dating techniques, geologists had no reliable means of absolute dating and therefore depended solely on relative dating methods. Relative dating puts events in sequential order but does not tell us how long ago an event took place. Although the principles of relative dating may now seem self-evident, their discovery was an important scientific achievement because they provided geologists with a means to interpret geologic history and develop a relative geologic time scale.

Fundamental Principles of Relative Dating

The 17th century was an important time in the development of geology as a science because of the widely circulated writings of the Danish anatomist Nicolas Steno (1638–1686). Steno observed that when streams flood, they spread out across their floodplains and deposit layers of sediment that bury organisms dwelling on the floodplain. Subsequent floods produce new layers of sediments that are deposited or superposed over previous deposits. When lithified, these layers of sediment become sedimentary rock. Thus, in an undisturbed succession of sedimentary rock layers, the oldest layer is at the bottom and the youngest layer is at the top. This **principle of superposition** is the basis for relative-age determinations of strata and their contained fossils (■ Figure 17.2).

Steno also observed that because sedimentary particles settle from water under the influence of gravity, sediment is deposited in essentially horizontal layers, thus illustrating the **principle of original horizontality** (Figure 17.2). Therefore, a sequence of sedimentary rock layers that is steeply inclined from the horizontal must have been tilted after deposition and lithification.

Steno's third principle, the **principle of lateral continuity,** states that a layer of sediment extends laterally in all directions until it thins and pinches out or terminates against the edge of the depositional basin (Figure 17.2).

James Hutton is credited with discovering the **principle of cross-cutting relationships.** Based on his detailed studies and observations of rock exposures in Scotland, Hutton recognized that an igneous intrusion or fault must be younger than the rocks it intrudes or displaces (■ Figure 17.3).

Although this principle states that an intrusive igneous structure is younger than the rocks it intrudes, the association of sedimentary and igneous rocks may cause problems in relative dating. Buried lava flows and intrusive igneous bodies such as sills look very similar in a sequence of strata (■ Figure 17.4). A buried lava flow, however, is older than the rocks above it (principle of superposition), whereas a sill is younger than all the beds below it and younger than the bed immediately above it.

To resolve such relative-age problems as these, geologists look to see whether the sedimentary rocks in contact with the igneous rocks show signs of baking or alteration by heat (see Chapter 7). A sedimentary rock that shows such effects must be older than the igneous rock with which it is in contact. In Figure 17.4, for example, a sill produces a zone of baking immediately above and below it because it intruded into previously existing sedimentary rocks. A lava flow, in contrast, bakes only those rocks below it.

Another way to determine relative ages is by using the **principle of inclusions.** This principle holds that inclusions, or fragments of one rock contained within a layer of another, are older than the rock layer itself. The batholith shown in ■ Figure 17.5a contains sandstone inclusions, and the sandstone unit shows the effects of baking. Accordingly, we conclude that the sandstone is older than the batholith. In Figure 17.5b, however, the sandstone contains granite rock fragments, indicating that the batholith was the source rock for the inclusions and is therefore older than the sandstone.

Fossils have been known for centuries (see Chapter 18), yet their utility in relative dating and geologic mapping was not fully appreciated until the early 19th century.

Reed Wicander

■ **Figure 17.2**

Badlands National Monument in South Dakota illustrates three of the six fundamental principles of relative dating. The sedimentary rocks of the Badlands were originally deposited horizontally by sluggish streams in a variety of continental environments (principle of original horizontality). The oldest rocks are at the bottom of this highly dissected landscape, and the youngest rocks are at the top, forming the rims (principle of superposition). The exposed rock layers extend laterally in all directions for some distance (principle of lateral continuity).

■ **Figure 17.3**

The principle of cross-cutting relationships. (a) A dark dike has been intruded into older light-colored granite along the north shore of Lake Superior, Ontario, Canada. (b) A small fault (arrows show the direction of movement) displacing tilted beds along Templin Highway, Castaic, California.

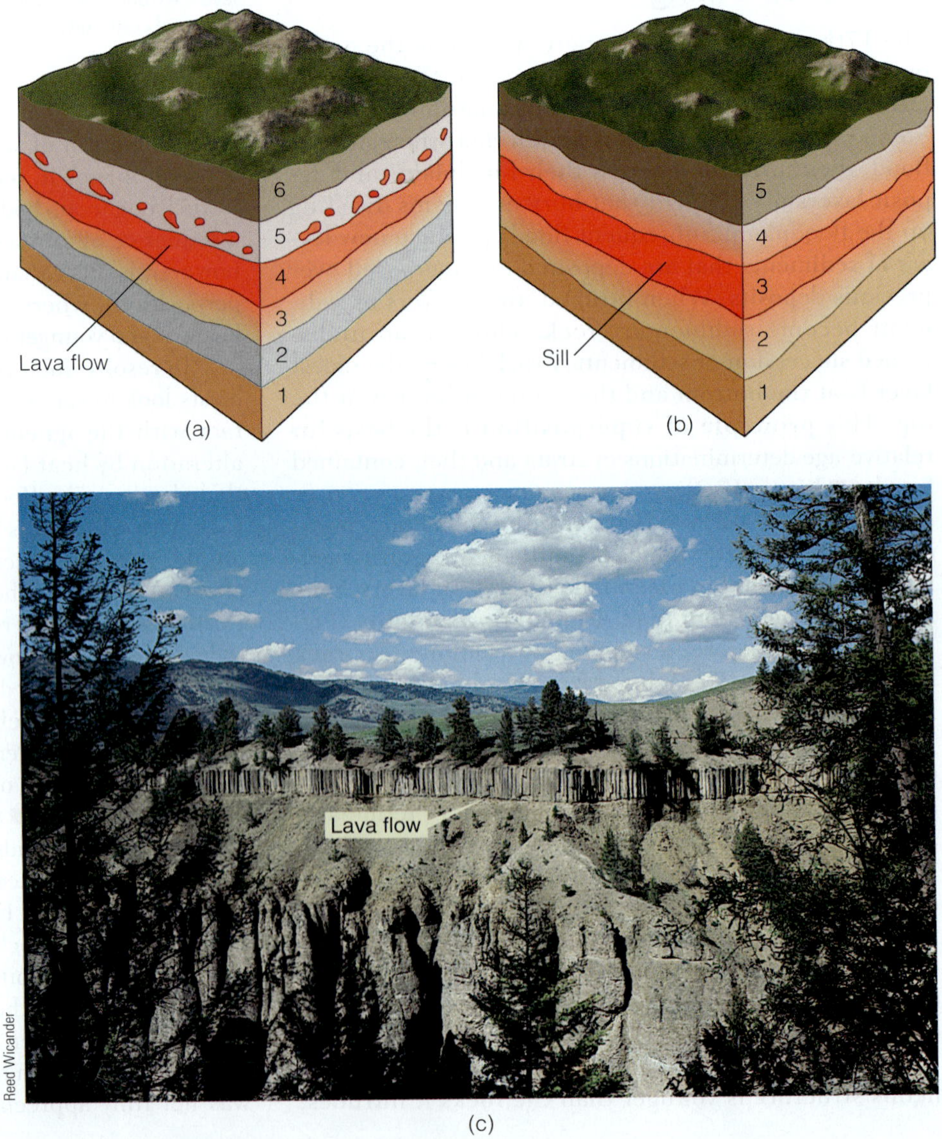

■ **Figure 17.4**

Relative ages of lava flows, sills, and associated sedimentary rocks may be difficult to determine. (a) A buried lava flow in bed 4 baked the underlying bed, and bed 5 contains inclusions of the lava flow. The lava flow is younger than bed 3 and older than beds 5 and 6. (b) The rock units above and below the sill in bed 3 have been baked, indicating that the sill is younger than beds 2 and 4, but its age relative to bed 5 cannot be determined. (c) This buried lava flow in Yellowstone National Park, Wyoming, displays columnar jointing.

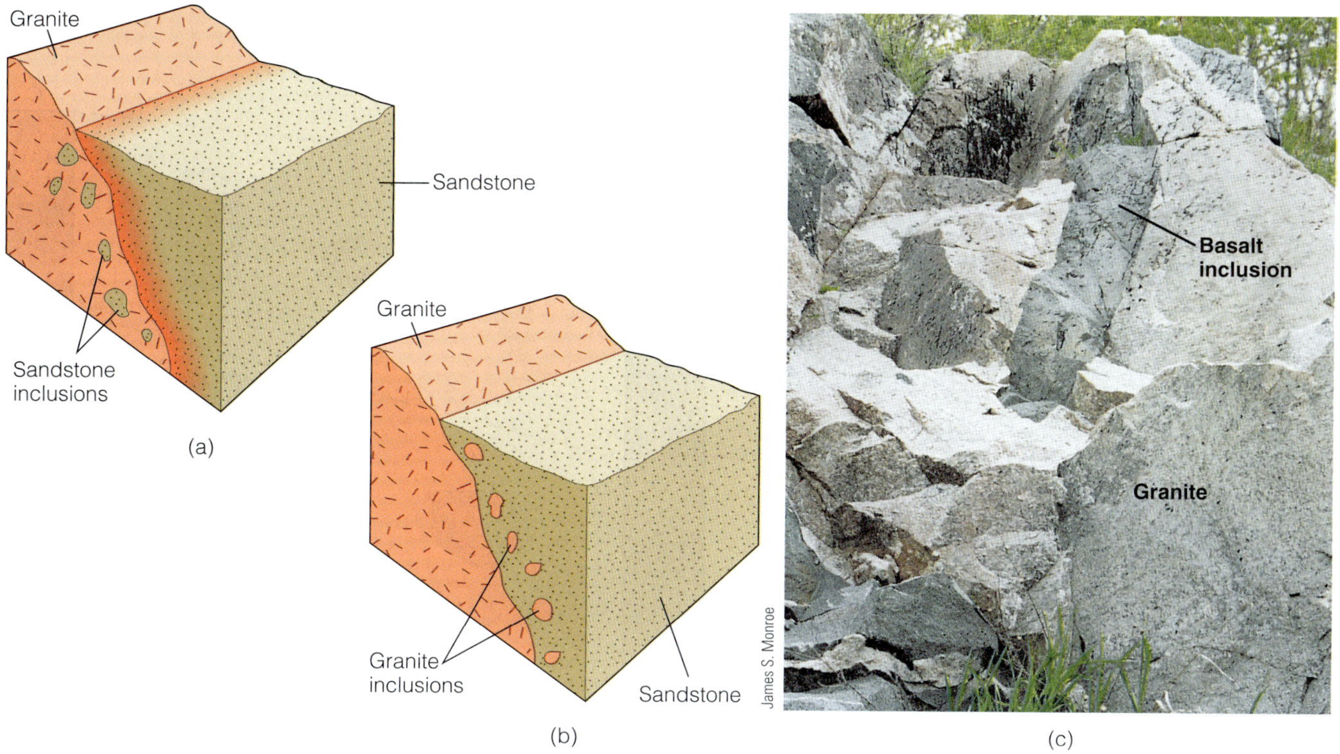

Figure 17.5

The principle of inclusions. (a) The batholith is younger than the sandstone because the sandstone has been baked at its contact with the granite and the granite contains sandstone inclusions. (b) Granite inclusions in the sandstone indicate that the batholith was the source of the sandstone and therefore is older. (c) Outcrop in northern Wisconsin showing basalt inclusions (black) in granite (white). Accordingly, the basalt inclusions are older than the granite.

William Smith (1769–1839), an English civil engineer involved in surveying and building canals in southern England, independently recognized the principle of superposition by reasoning that the fossils at the bottom of a sequence of strata are older than those at the top of the sequence. This recognition served as the basis for the **principle of fossil succession** or the *principle of faunal and floral succession,* as it is sometimes called (Figure 17.6).

According to this principle, fossil assemblages succeed one another through time in a regular and predictable order. The validity and successful use of this principle depend on three points: (1) Life has varied through time, (2) fossil assemblages are recognizably different from one another, and (3) the relative ages of the fossil assemblages can be determined. Observations of fossils in older versus younger strata clearly demonstrate that life-forms have changed. Because this is true, fossil assemblages (point 2) are recognizably different. Furthermore, superposition can be used to demonstrate the relative ages of the fossil assemblages.

Unconformities

Our discussion so far has been concerned with vertical relationships among conformable strata—that is, sequences of rock in which deposition was more or less continuous. A bedding plane between strata may represent a depositional break of anywhere from minutes to tens of years, but it is inconsequential in the context of geologic time. However, in some sequences of strata, surfaces known as **unconformities** may be present, representing times of nondeposition, erosion, or both. These unconformities encompass long periods of geologic time, perhaps millions or tens of millions of years. Accordingly, the geologic record is incomplete at that particular location, just as a book with missing pages is incomplete, and the interval of geologic time not represented by strata is called a *hiatus* (Figure 17.7).

The general term *unconformity* encompasses three specific types of surfaces. First, a **disconformity** is a surface of erosion or nondeposition separating younger from older rocks, both of which are parallel with one another (Figure 17.8). Unless the erosional surface separating the older from the younger parallel beds is well defined or distinct, the disconformity frequently resembles an ordinary bedding plane. Accordingly, many disconformities are difficult to recognize and must be identified on the basis of fossil assemblages.

Second, an **angular unconformity** is an erosional surface on tilted or folded strata over which younger strata were deposited (Figure 17.9). The strata below the unconformable surface generally dip more steeply than those above, producing an angular relationship.

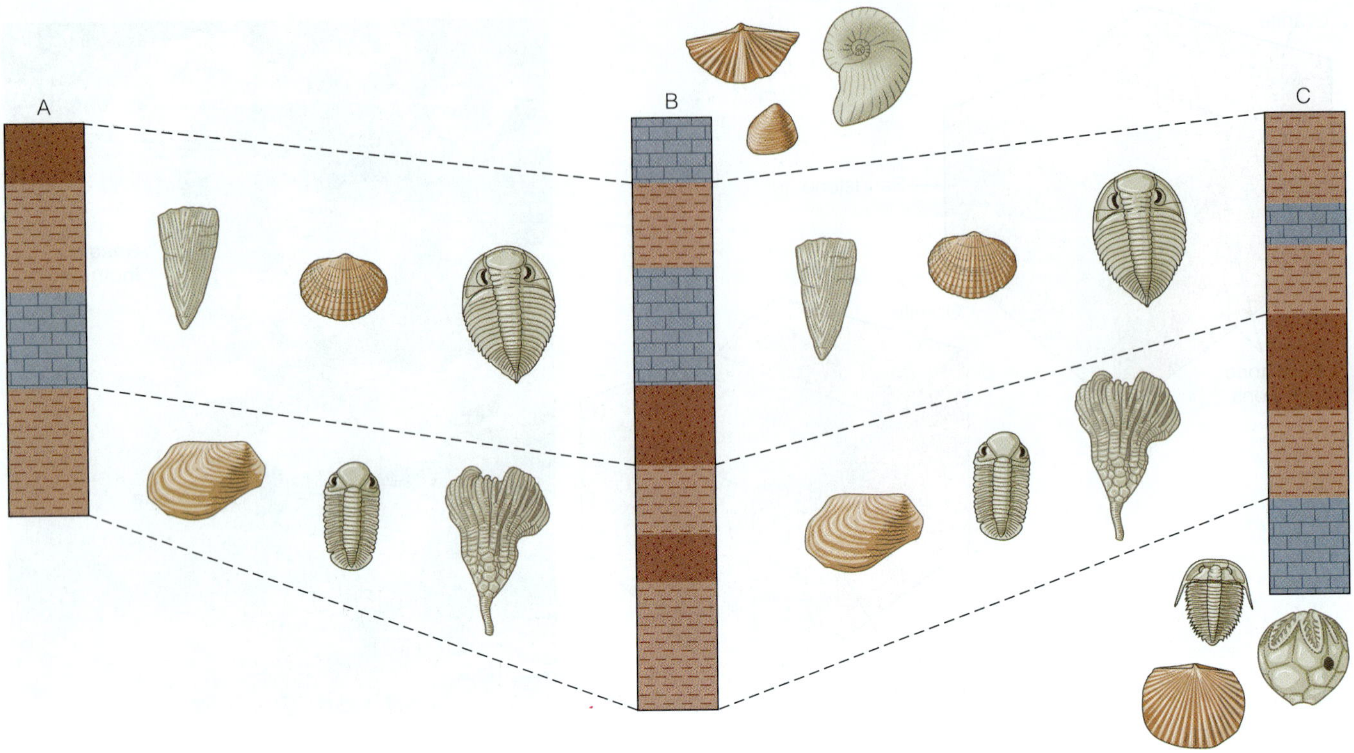

■ **Figure 17.6**

This generalized diagram shows how geologists use the principle of fossil succession to identify strata of the same age in different areas. The rocks in the three sections filled in with dashed lines contain similar fossils and are thus the same age. Note that the youngest rocks in this region are in section B, whereas the oldest rocks are in section C.

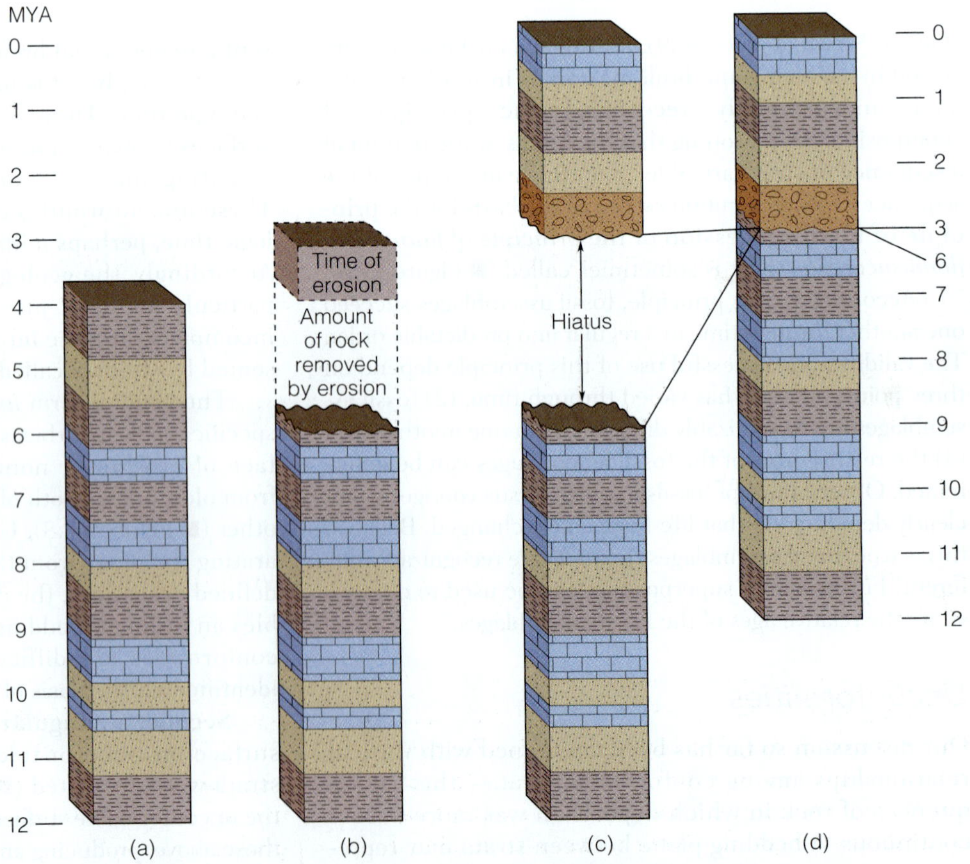

■ **Figure 17.7**

A simplified diagram showing the development of an unconformity and a hiatus. (a) Deposition began 12 million years ago (MYA) and continued more or less uninterrupted until 4 MYA. (b) A 1-million-year episode of erosion occurred, and during that time strata representing 2 million years of geologic time were eroded. (c) A hiatus of 3 million years exists between the older strata and the strata that formed during a renewed episode of deposition that began 3 MYA. (d) The actual stratigraphic record. The unconformity is the surface separating the strata and represents a major break in our record of geologic time.

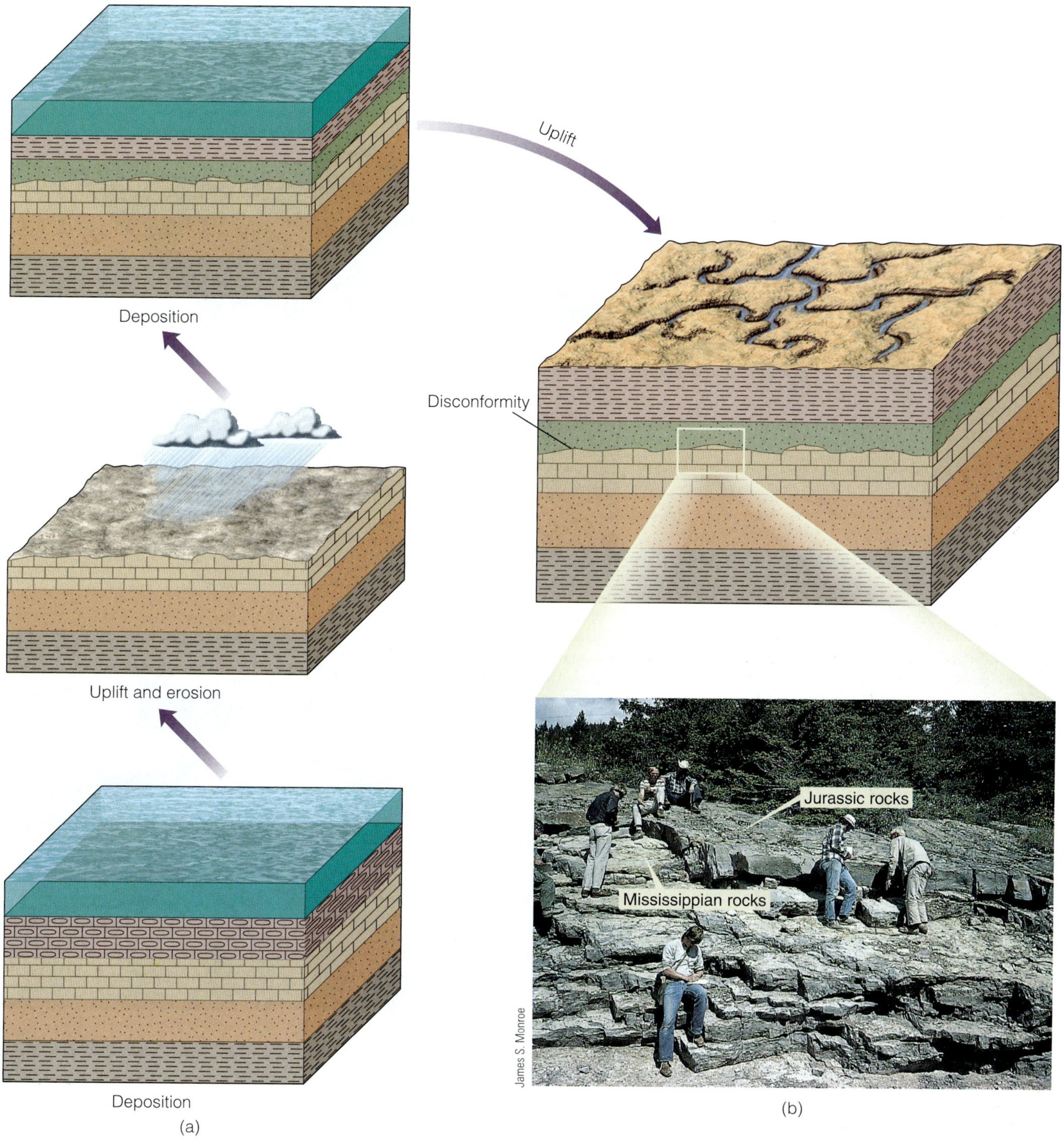

Figure 17.8

(a) Formation of a disconformity. (b) Disconformity between Mississippian and Jurassic strata in Montana. The geologist at the upper left is sitting on Jurassic strata, and his right foot is resting on Mississippian rocks.

The angular unconformity illustrated in Figure 17.9b is probably the most famous in the world. It was here at Siccar Point, Scotland, that James Hutton realized that severe upheavals had tilted the lower rocks and formed mountains that were then worn away and covered by younger, flat-lying rocks. The erosional surface between the older tilted rocks and the younger flat-lying strata meant that a significant gap existed in the geologic record. Although Hutton did not use the term *unconformity,* he was the first to understand and explain the significance of such discontinuities in the geologic record.

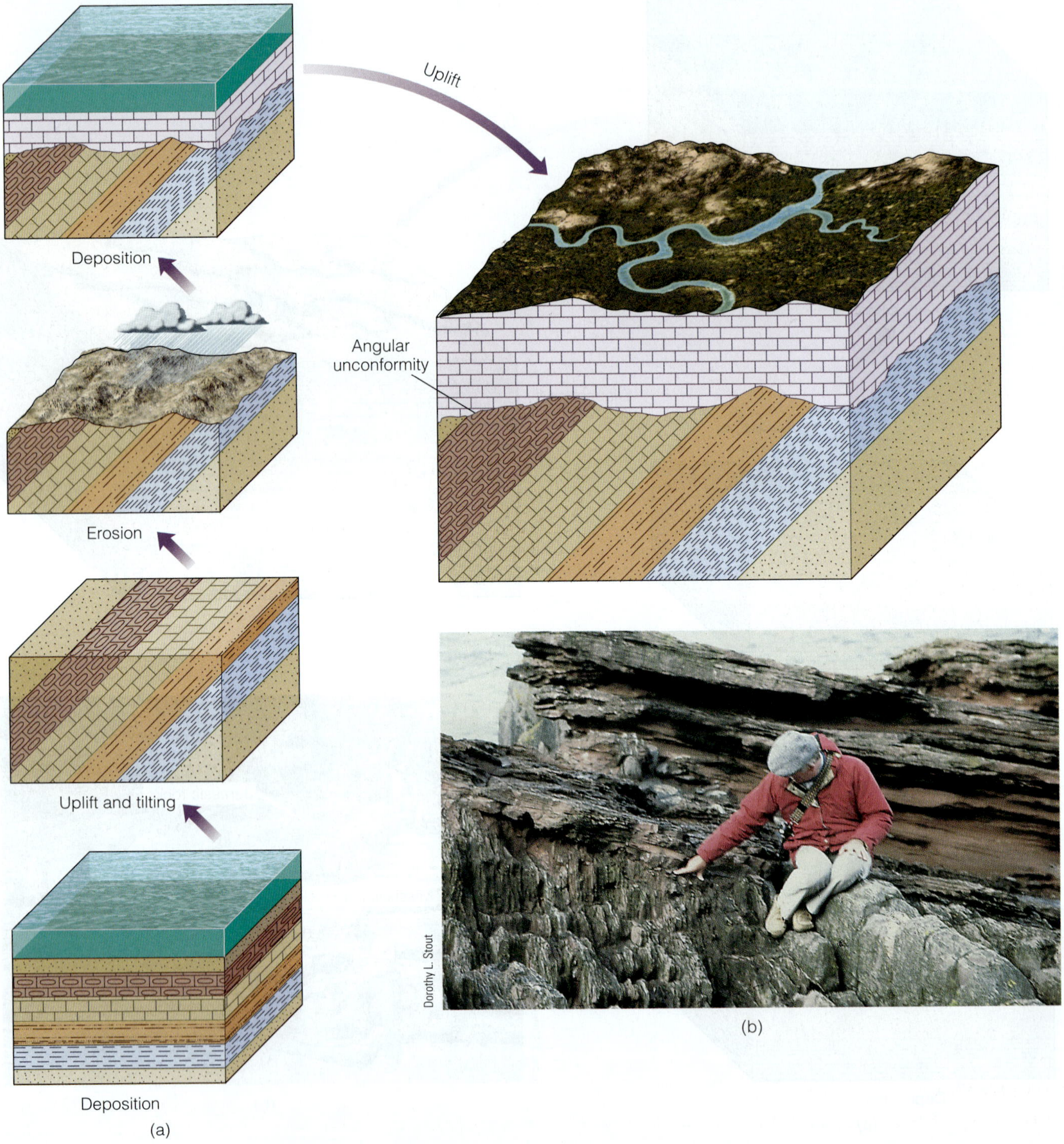

■ **Figure 17.9**

(a) Formation of an angular unconformity. (b) Angular unconformity at Siccar Point, Scotland. James Hutton first realized the significance of unconformities at this site in 1788.

A **nonconformity** is the third type of unconformity. Here an erosional surface cut into metamorphic or igneous rocks is covered by sedimentary rocks (■ Figure 17.10). This type of unconformity closely resembles an intrusive igneous contact with sedimentary rocks. The principle of inclusions is helpful in determining whether the relationship between the underlying igneous rocks and the overlying sedimentary rocks is the result of an intrusion or erosion (Figure 17.5). In the case of an intrusion, the igneous rocks are younger, whereas in the case of erosion, the sedimentary rocks are younger. Being able to distinguish between a nonconformity and an intrusive contact is important because they represent different sequences of events.

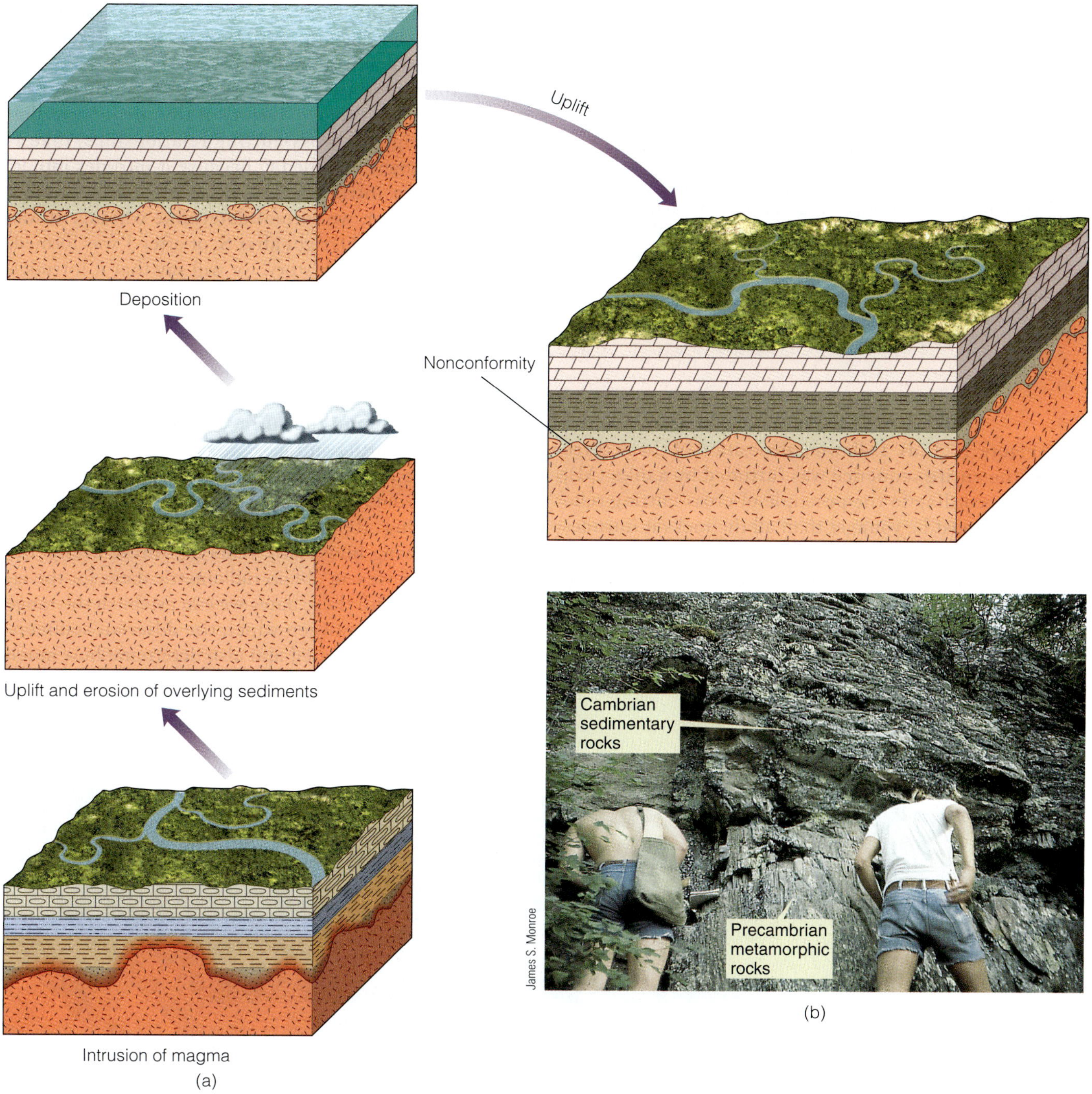

■ **Figure 17.10**

(a) Formation of a nonconformity. (b) Nonconformity between Precambrian metamorphic rocks and the overlying Cambrian-age Deadwood Formation, South Dakota.

Applying the Principles of Relative Dating

We can decipher the geologic history of the area represented by the block diagram in ■ Figure 17.11 by applying the various relative-dating principles just discussed. The methods and logic used in this example are the same as those applied by 19th-century geologists in constructing the geologic time scale.

According to the principles of superposition and original horizontality, beds A–G were deposited horizontally; then either they were tilted, faulted (H), and eroded, or after deposition, they were faulted (H), tilted, and then eroded (■ Figure 17.12a–c). Because the fault cuts beds A–G, it must be younger than the beds according to the principle of cross-cutting relationships.

Beds J–L were then deposited horizontally over this erosional surface, producing an angular unconformity

What Would You Do?

You have been chosen to be part of the first astronaut crew to land on Mars. You were selected because you are a geologist, and therefore your primary responsibility is to map the geology of the landing site area. An important goal of the mission is to work out the geologic history of the area. How will you go about this? Will you be able to use the principles of relative dating? How will you go about correlating the various rock units? Will you be able to determine absolute ages? How would you do this?

(I) (Figure 17.12d). Following deposition of these three beds, the entire sequence was intruded by a dike (M), which, according to the principle of cross-cutting relationships, must be younger than all the rocks it intrudes (Figure 17.12e).

The entire area was then uplifted and eroded; next beds P and Q were deposited, producing a disconformity (N) between beds L and P and a nonconformity (O) between the igneous intrusion M and the sedimentary bed P (Figure 17.12f, g). We know that the relationship between igneous intrusion M and the overlying sedimentary bed P is a nonconformity because of the inclusions of M in P (principle of inclusions).

At this point, there are several possibilities for reconstructing the geologic history of this area. According to the principle of cross-cutting relationships, dike R must be younger than bed Q because it intrudes into it. It could have intruded anytime *after* bed Q was deposited; however, we cannot determine whether R was formed right after Q, right after S, or after T was formed. For purposes of this history, we will say that it intruded after the deposition of bed Q (Figure 17.12g, h).

Following the intrusion of dike R, lava S flowed over bed Q, followed by the deposition of bed T (Figure 17.12i, j). Although the lava flow (S) is not a sedimentary unit, the principle of superposition still applies because it flowed onto the surface, just as sediments are deposited on Earth's surface.

We have established a relative chronology for the rocks and events of this area by using the principles of relative dating. Remember, however, that we have no way of knowing how many years ago these events occurred unless we can obtain radiometric dates for the igneous rocks. With these dates we can establish the range of absolute ages between which the different sedimentary units were deposited and also determine how much time is represented by the unconformities.

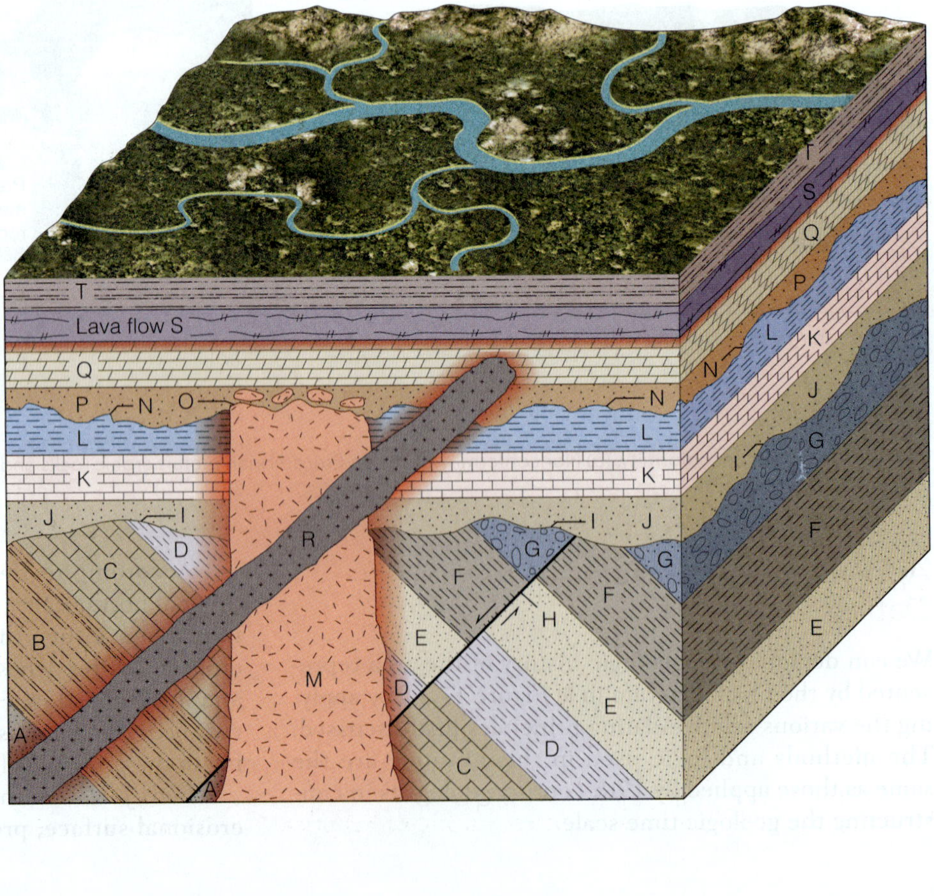

Figure 17.11

A block diagram of a hypothetical area in which the various relative-dating principles can be applied to determine its geologic history.

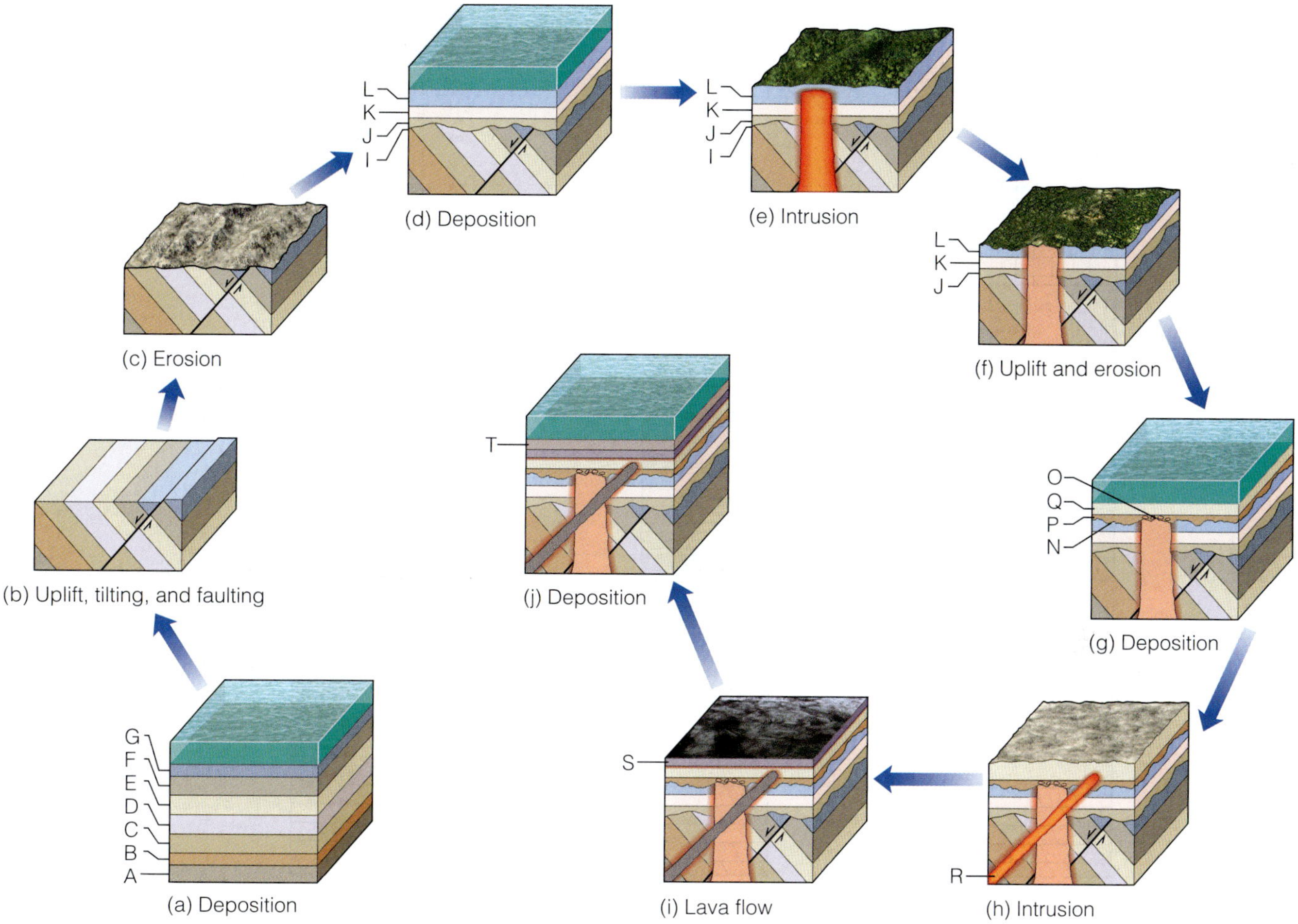

Figure 17.12

(a) Beds A–G are deposited. (b) The preceding beds are tilted and faulted. (c) Erosion. (d) Beds J–L are deposited, producing an angular unconformity I. (e) The entire sequence is intruded by a dike. (f) The entire sequence is uplifted and eroded. (g) Beds P and Q are deposited, producing a disconformity (N) and a nonconformity (O). (h) Dike R intrudes. (i) Lava (S) flows over bed Q, baking it. (j) Bed T is deposited.

HOW DO GEOLOGISTS CORRELATE ROCK UNITS?

To decipher Earth history, geologists must demonstrate the time equivalency of rock units in different areas. This process is known as **correlation.**

If surface exposures are adequate, units may simply be traced laterally (principle of lateral continuity), even if occasional gaps exist (Figure 17.13). Other criteria used to correlate units are similarity of rock type, position in a sequence, and key beds. *Key beds* are units, such as coal beds or volcanic ash layers, that are sufficiently distinctive to allow identification of the same unit in different areas (Figure 17.13).

Generally, no single location in a region has a geologic record of all events that occurred during its history; therefore geologists must correlate from one area to another to determine the complete geologic history of the region. An excellent example is the history of the Colorado Plateau (Figure 17.14). This region provides a record of events that occurred over approximately 2 billion years. Because of the forces of erosion, the entire record is not preserved at any single location. Within the walls of the Grand Canyon are rocks of the Precambrian and Paleozoic eras, whereas Paleozoic and Mesozoic Era rocks are found in Zion National Park, and Mesozoic and Cenozoic Era rocks are exposed in Bryce Canyon (Figure 17.14). By correlating the uppermost rocks at one location with the lowermost equivalent rocks of another area, geologists can decipher the history of the entire region.

Although geologists can match up rocks on the basis of similar rock type and superposition, correlation of this type can be done only in a limited area where beds can be traced from one site to another. To correlate rock units over a large area or to correlate age-equivalent

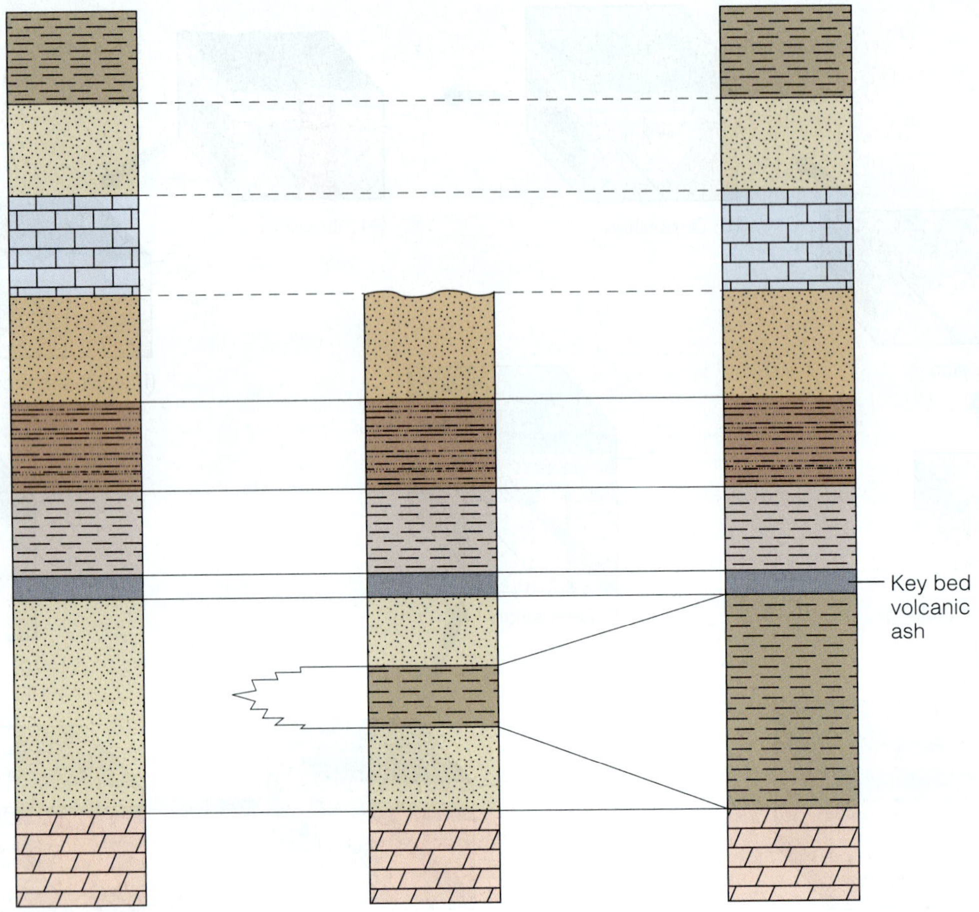

■ **Figure 17.13**

In areas of adequate exposures, rock units can be traced laterally, even if occasional gaps exist, and correlated on the basis of similarity in rock type and position in a sequence. Rocks can also be correlated by a key bed—in this case, volcanic ash.

Source: From *History of the Earth; An Introduction to Historical Geology,* 2nd ed., by Bernhard Kummel. © 1961, 1970 by W. H. Freeman Co. Used with permission.

units of different composition, fossils and the principle of fossil succession must be used.

Fossils are useful as relative time indicators because they are the remains of organisms that lived for a certain length of time during the geologic past. Fossils that are easily identified, are geographically widespread, and existed for a rather short interval of geologic time are particularly useful. Such fossils are **guide fossils** or *index fossils* (■ Figure 17.15). The trilobite *Paradoxides* and the brachiopod *Atrypa* meet these criteria and are therefore good guide fossils. In contrast, the brachiopod *Lingula* is easily identified and widespread, but its long geologic range of Ordovician to Recent makes it of little use in correlation.

Because most fossils have fairly long geologic ranges, geologists construct *concurrent range zones* to determine the age of the sedimentary rocks that contain the fossils. Concurrent range zones are established by plotting the overlapping geologic ranges of two or more fossils that have different geologic ranges (■ Figure 17.16). The first and last occurrences of fossils are used to determine zone boundaries. Correlating concurrent range zones is probably the most accurate method of determining time equivalence.

Subsurface Correlation

In addition to surface geology, geologists are interested in subsurface geology because it provides additional information about geologic features beneath Earth's surface. A variety of techniques and methods are used to acquire and interpret data about the subsurface geology of an area.

When drilling is done for oil or natural gas, cores or rock chips called *well cuttings* are commonly recovered from the drill hole. These samples are studied under the microscope and reveal such important information as rock type, porosity (the amount of pore space), permeability (the ability to transmit fluids), and the presence of oil stains. In addition, the samples can be processed for a variety of microfossils that aid in determining the geologic age of the rock and the environment of deposition.

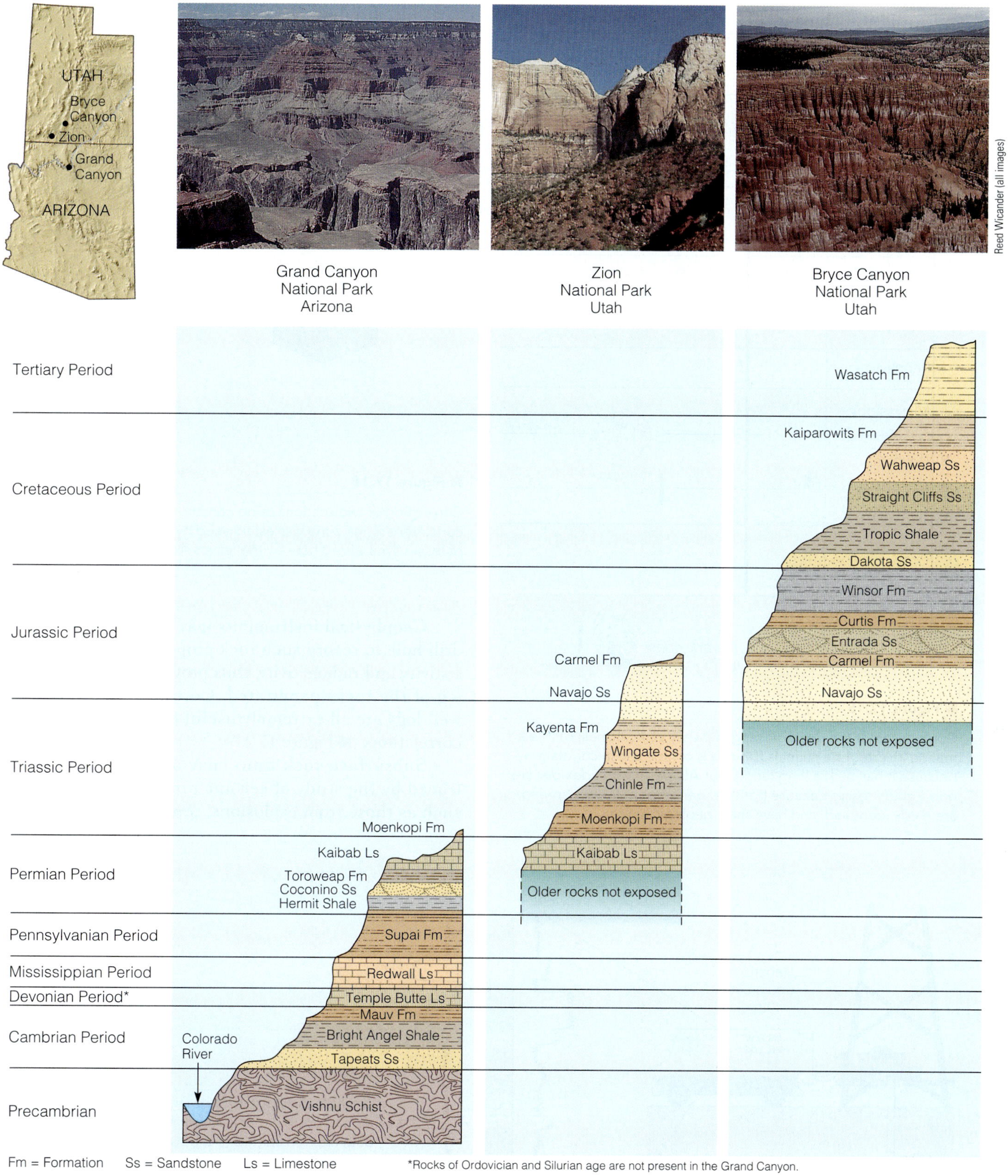

■ **Figure 17.14**

Correlation of rocks within the Colorado Plateau. By correlating rocks from various locations, the history of the entire region can be deciphered.

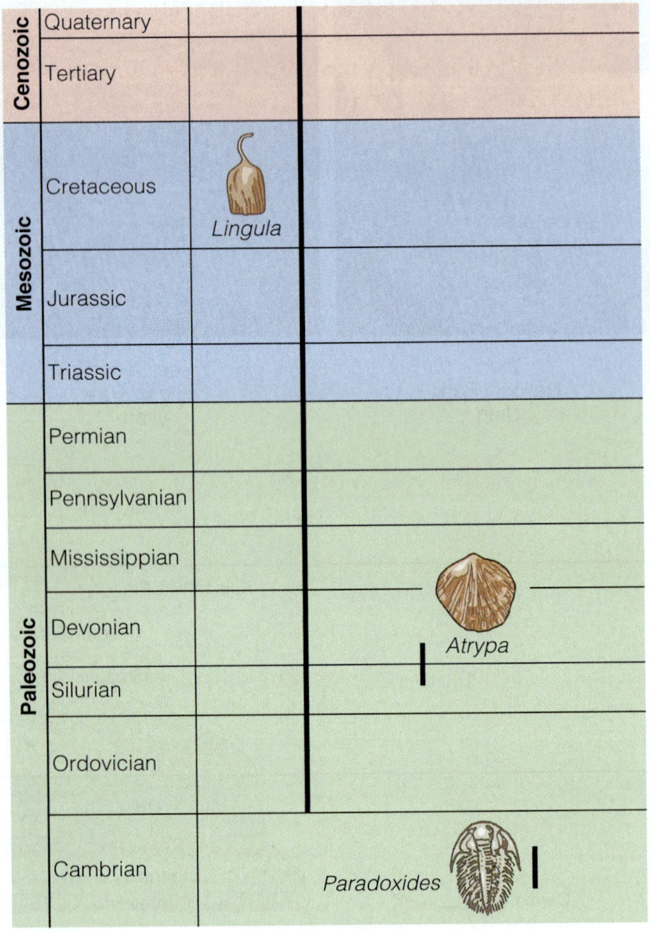

■ **Figure 17.15**

Comparison of the geologic ranges (heavy vertical lines) of three marine invertebrate animals. *Lingula* is of little use in correlation because it has such a long range. But *Atrypa* and *Paradoxides* are good guide fossils because both are geographically widespread, are easily identified, and have short geologic ranges.

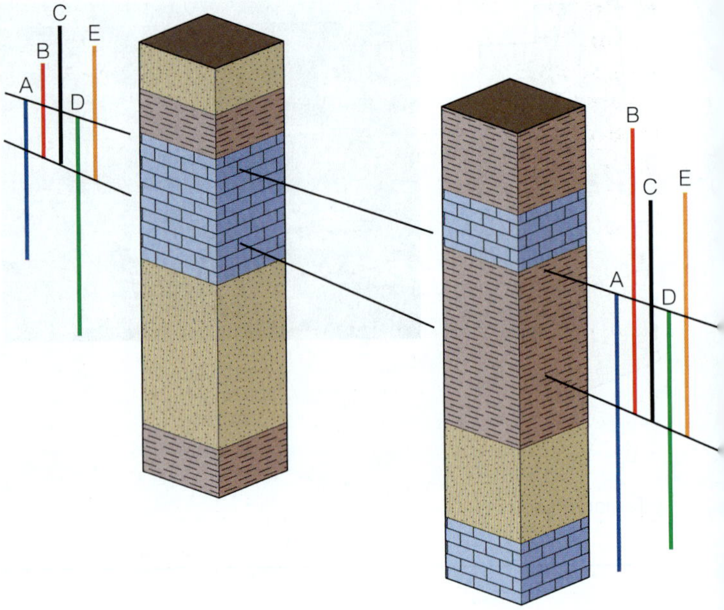

■ **Figure 17.16**

Correlation of two sections using concurrent range zones. This concurrent range zone was established by the overlapping ranges of fossils symbolized here by the letters A through E.

Geophysical instruments may be lowered down the drill hole to record such rock properties as electrical resistivity and radioactivity, thus providing a record or *well log* of the rocks penetrated. Cores, well cuttings, and well logs are all extremely useful in making subsurface correlations (■ Figure 17.17).

Subsurface rock units may also be detected and traced by the study of seismic profiles. Energy pulses, such as those from explosions, travel through rocks at a

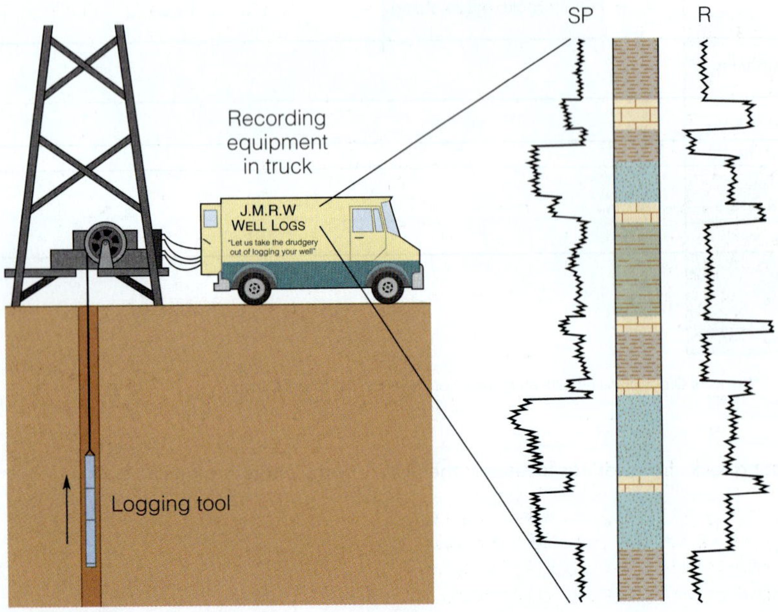

■ **Figure 17.17**

A schematic diagram showing how well logs are made. As the logging tool is withdrawn from the drill hole, data are transmitted to the surface, where they are recorded and printed as a well log. The curve labeled SP in this diagrammatic electric log is a plot of self-potential (electrical potential caused by different conductors in a solution that conducts electricity) with depth. The curve labeled R is a plot of electrical resistivity with depth. Electric logs yield information about the rock type and fluid content of subsurface formations. Electric logs are also used to correlate from well to well.

velocity determined by rock density, and some of this energy is reflected from various horizons (contacts between contrasting layers) back to the surface, where it is recorded. Seismic stratigraphy is particularly useful in tracing units in areas such as the continental shelves where it is very expensive to drill holes and other techniques have limited use.

WHAT ARE ABSOLUTE DATING METHODS?

Although most of the isotopes of the 92 naturally occurring elements are stable, some are radioactive and spontaneously decay to other more stable isotopes of elements, releasing energy in the process. The discovery, in 1903 by Pierre and Marie Curie, that radioactive decay produces heat meant that geologists finally had a mechanism for explaining Earth's internal heat that did not rely on residual cooling from a molten origin. Furthermore, geologists now had a powerful tool to date geologic events accurately and to verify the long time periods postulated by Hutton, Lyell, and Darwin.

Atoms, Elements, and Isotopes

As discussed in Chapter 3, all matter is made up of chemical elements, each composed of extremely small particles called *atoms*. The nucleus of an atom is composed of *protons* and *neutrons* with *electrons* encircling it (see Figure 3.2). The number of protons is an element's *atomic number* and helps determine its properties and characteristics. The combined number of protons and neutrons in an atom is its *atomic mass number*. However, not all atoms of the same element have the same number of neutrons in their nuclei. These variable forms of the same element are called *isotopes*. Most isotopes are stable, but some are unstable and spontaneously decay to a more stable form. It is the decay rate of unstable isotopes that geologists measure to determine the absolute ages of rocks.

Radioactive Decay and Half-Lives

Radioactive decay is the process by which an unstable atomic nucleus is spontaneously transformed into an atomic nucleus of a different element. Scientists recognize three types of radioactive decay, all of which result in a change of atomic structure (■ Figure 17.18). In *alpha decay* 2 protons and 2 neutrons are emitted from the nucleus, resulting in the loss of 2 atomic numbers and 4 atomic mass numbers. In *beta decay*, a fast-moving electron is emitted from a neutron in the nucleus, changing that neutron to a proton and consequently increasing the atomic number by 1 with no resultant atomic mass number change. *Electron capture* is when a proton captures an electron from an electron shell and thereby converts to a neutron, resulting in the loss of 1 atomic number but not changing the atomic mass number.

Some elements undergo only 1 decay step in the conversion from an unstable form to a stable form. For example, rubidium 87 decays to strontium 87 by a single beta emission, and potassium 40 decays to argon 40 by a single electron capture. Other radioactive elements undergo several decay steps. Uranium 235 decays to lead 207 by 7 alpha and 6 beta steps, whereas uranium 238 decays to lead 206 by 8 alpha and 6 beta steps (■ Figure 17.19).

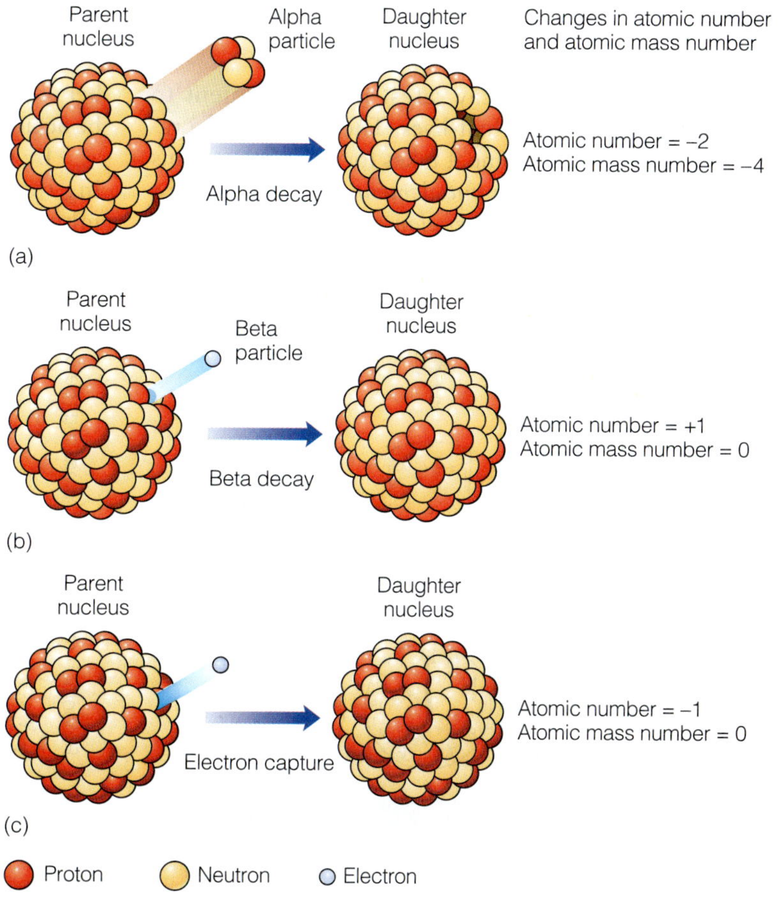

■ **Figure 17.18**

Three types of radioactive decay. (a) Alpha decay, in which an unstable parent nucleus emits 2 protons and 2 neutrons. (b) Beta decay, in which an electron is emitted from the nucleus. (c) Electron capture, in which a proton captures an electron and is thereby converted to a neutron.

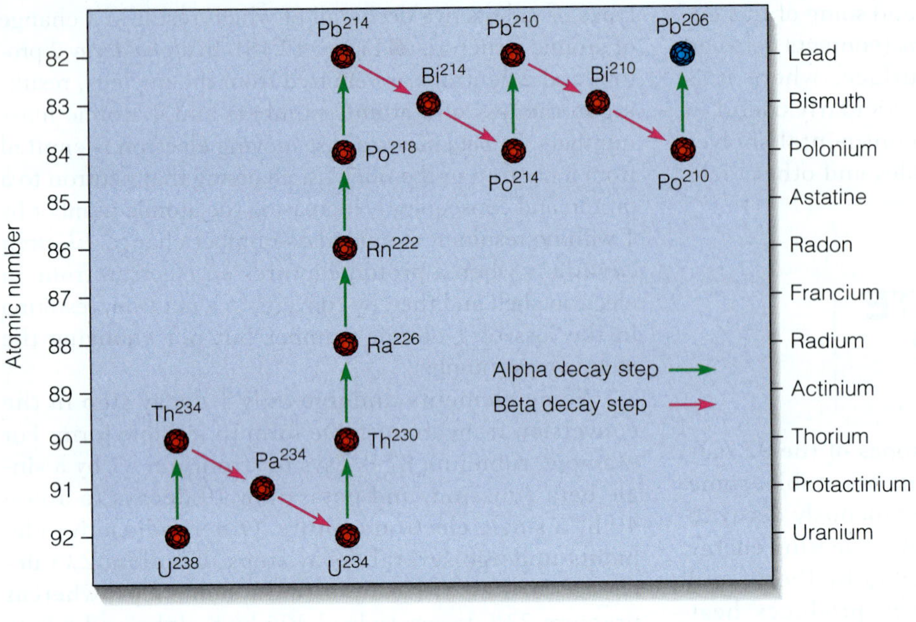

■ Figure 17.19

Radioactive decay series for uranium 238 to lead 206. Radioactive uranium 238 decays to its stable daughter product, lead 206, by 8 alpha and 6 beta decay steps. Many different isotopes are produced as intermediate steps in the decay series. Source: Based on data from S. M. Richardson and H. Y. McSween, Jr., *Geochemistry—Pathways and Processes*, Prentice-Hall.

When we discuss decay rates, it is convenient to refer to them in terms of half-lives. The **half-life** of a radioactive element is the time it takes for half of the atoms of the original unstable *parent element* to decay to atoms of a new, more stable *daughter element*. The half-life of a given radioactive element is constant and can be precisely measured. Half-lives of various radioactive elements range from less than a billionth of a second to 49 billion years.

Radioactive decay occurs at a geometric rate rather than a linear rate. Therefore, a graph of the decay rate produces a curve rather than a straight line (■ Figure 17.20). For example, an element with *1,000,000* parent atoms will have *500,000* parent atoms and 500,000 daughter atoms after one half-life. After two half-lives, it will have *250,000* parent atoms (one-half of the previous parent atoms, which is equivalent to one-fourth of the original parent atoms) and 750,000 daughter atoms. After three half-lives, it will have *125,000* parent atoms (one-half of the previous parent atoms, or one-eighth of the original parent atoms) and 875,000 daughter atoms, and so on until the number of parent atoms that remains is so few that they cannot be accurately measured by present-day instruments.

By measuring the parent–daughter ratio and knowing the half-life of the parent (which has been determined in the laboratory), geologists can calculate the age of a sample that contains the radioactive element. The

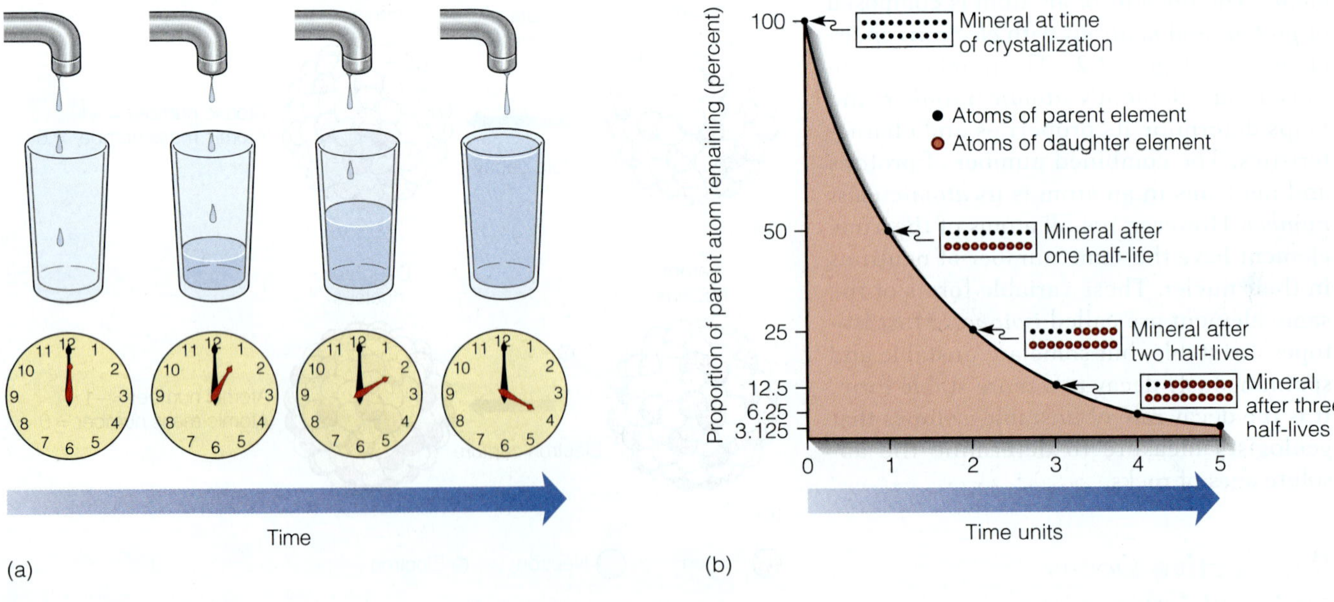

Geology⇌Now ■ **Active Figure 17.20**

(a) Uniform linear change is characteristic of many familiar processes. In this example, water is being added to a glass at a constant rate.
(b) Geometric radioactive decay curve, in which each time unit represents one half-life, and each half-life is the time it takes for half of the parent element to decay to the daughter element.

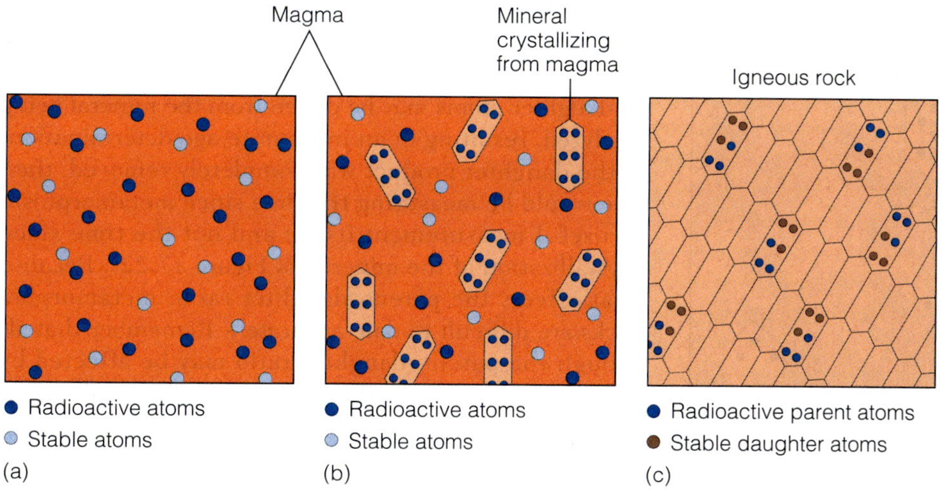

Figure 17.21

(a) Magma contains both radioactive and stable atoms. (b) As magma cools and begins to crystallize, some radioactive atoms are incorporated into certain minerals because they are the right size and can fit into the crystal structure. Therefore, at the time of crystallization, the mineral will contain 100% radioactive parent atoms and 0% stable daughter atoms. (c) After one half-life, 50% of the radioactive parent atoms will have decayed to stable daughter atoms.

parent–daughter ratio is usually determined by a *mass spectrometer,* an instrument that measures the proportions of atoms of different masses.

GeologyNow Click Geology Interactive to work through an activity on Absolute Dating through Geologic Time.

Sources of Uncertainty

The most accurate radiometric dates are obtained from igneous rocks. As a magma cools and begins to crystallize, radioactive parent atoms are separated from previously formed daughter atoms. Because they are the right size, some radioactive parent atoms are incorporated into the crystal structure of certain minerals. The stable daughter atoms, though, are a different size from the radioactive parent atoms and consequently cannot fit into the crystal structure of the same mineral as the parent atoms. Therefore, a mineral that crystallizes in a cooling magma will contain radioactive parent atoms but no stable daughter atoms (■ Figure 17.21). Thus, the time that is being measured is the time of crystallization of the mineral that contains the radioactive atoms, and not the time of formation of the radioactive atoms.

Except in unusual circumstances, sedimentary rocks cannot be radiometrically dated because one would be measuring the age of a particular mineral rather than the time that it was deposited as a sedimentary particle. One of the few instances in which radiometric dates can be obtained on sedimentary rocks is when the mineral glauconite is present. Glauconite is a greenish mineral that contains radioactive potassium 40, which decays to argon 40 (Table 17.1). It forms in

Table 17.1

Five of the Principal Long-Lived Radioactive Isotope Pairs Used in Radiometric Dating

ISOTOPES		Half-Life of Parent (years)	Effective Dating Range (years)	Minerals and Rocks That Can Be Dated
Parent	**Daughter**			
Uranium 238	Lead 206	4.5 billion	10 million to 4.6 billion	Zircon Uraninite
Uranium 235	Lead 207	704 million		
Thorium 232	Lead 208	14 billion		
Rubidium 87	Strontium 87	48.8 billion	10 million to 4.6 billion	Muscovite Biotite Potassium feldspar Whole metamorphic or igneous rock
Potassium 40	Argon 40	1.3 billion	100,000 to 4.6 billion	Glauconite, Hornblende Muscovite, Whole volcanic rock Biotite

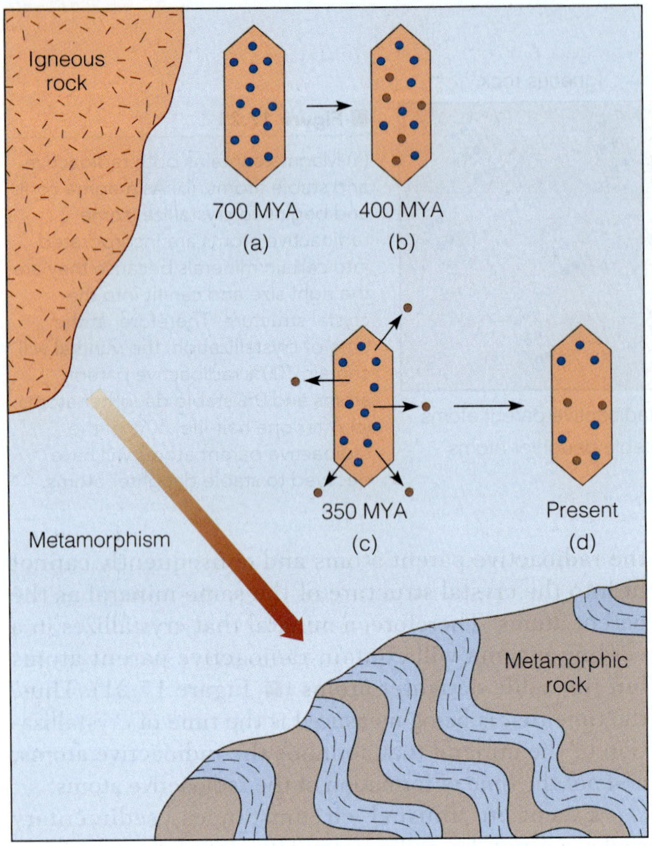

■ **Figure 17.22**

The effect of metamorphism in driving out daughter atoms from a mineral that crystallized 700 million years ago (MYA). The mineral is shown immediately after crystallization (a), then at 400 million years (b), when some of the parent atoms had decayed to daughter atoms. Metamorphism at 350 MYA (c) drives the daughter atoms out of the mineral into the surrounding rock. (d) Assuming the rock has remained a closed chemical system throughout its history, dating the mineral today yields the time of metamorphism, whereas dating the whole rock provides the time of its crystallization, 700 MYA.

certain marine environments as a result of chemical reactions with clay minerals during the conversion of sediments to sedimentary rock. Thus, glauconite forms when the sedimentary rock forms, and a radiometric date indicates the time of the sedimentary rock's origin. Being a gas, however, the daughter product argon can easily escape from a mineral. Therefore, any date obtained from glauconite, or any other mineral containing the potassium 40 and argon 40 pair, must be considered a minimum age.

To obtain accurate radiometric dates, geologists must be sure that they are dealing with a *closed system*, meaning that neither parent nor daughter atoms have been added to or removed from the system since crystallization and that the ratio between them results only from radioactive decay. Otherwise, an inaccurate date will result. If daughter atoms have leaked out of the mineral being analyzed, the calculated age will be too young; if parent atoms have been removed, the calculated age will be too old.

Leakage may take place if the rock is heated or subjected to intense pressure as can sometimes occur during metamorphism. If this happens, some of the parent or daughter atoms may be driven from the mineral being analyzed, resulting in an inaccurate age determination. If the daughter product was completely removed, then one would be measuring the time since metamorphism (a useful measurement itself) and not the time since crystallization of the mineral (■ Figure 17.22). Because heat affects the parent–daughter ratio, metamorphic rocks are difficult to date accurately. Remember that although the parent–daughter ratio may be affected by heat, the decay rate of the parent element remains constant, regardless of any physical or chemical changes.

To obtain an accurate radiometric date, geologists must make sure that the sample is fresh and unweathered and that it has not been subjected to high temperatures or intense pressures after crystallization. Furthermore, it is sometimes possible to cross-check the radiometric date obtained by measuring the parent–daughter ratio of two different radioactive elements in the same mineral. For example, naturally occurring uranium consists of both uranium 235 and uranium 238 isotopes. Through various decay steps, uranium 235 decays to lead 207, whereas uranium 238 decays to lead 206 (Figure 17.19). If the minerals that contain both uranium isotopes have remained closed systems, the ages obtained from each parent–daughter ratio should agree closely and therefore should indicate the time of crystallization of the magma. If the ages do not closely agree, then other samples must be taken and ratios measured to see which, if either, date is correct.

Recent advances and the development of new techniques and instruments for measuring various isotope ratios have enabled geologists to analyze increasingly smaller samples and with a greater precision than ever before. Presently, the measurement error for many radiometric dates is typically less than 0.5% of the age, and in some cases is even better than 0.1%. Thus, for a rock 540 million years old (near the beginning of the Cambrian Period), the possible error could range from nearly 2.7 million years to less than 540,000 years.

Long-Lived Radioactive Isotope Pairs

Table 17.1 shows the five common, long-lived parent–daughter isotope pairs used in radiometric dating. Long-lived pairs have half-lives of millions or billions of years. All of these were present when Earth formed and are still present in measurable quantities. Other shorter-lived radioactive isotope pairs have decayed to the point that only small quantities near the limit of detection remain.

The most commonly used isotope pairs are the uranium-lead and thorium-lead series, which are used principally to date ancient igneous intrusives, lunar samples, and some meteorites. The rubidium-strontium pair is also used for very old samples and has been effective

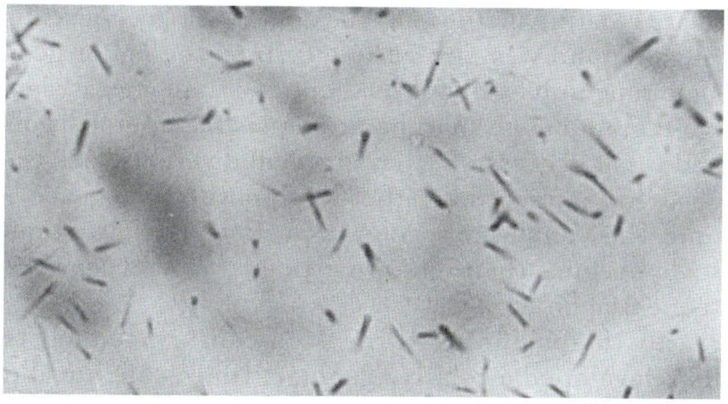

USGS

■ **Figure 17.23**

Each fission track (about 16 microns [= $\frac{16}{1000}$ mm] long) in this apatite crystal is the result of the radioactive decay of a uranium atom. The apatite crystal, which has been etched with hydrofluoric acid to make the fission tracks visible, comes from one of the dikes at Shiprock, New Mexico, and has a calculated age of 27 million years.

in dating the oldest rocks on Earth as well as meteorites. The potassium-argon method is typically used for dating fine-grained volcanic rocks from which individual crystals cannot be separated; hence the whole rock is analyzed. Because argon is a gas, great care must be taken to ensure that the sample has not been subjected to heat, which would allow argon to escape; such a sample would yield an age that is too young. Other long-lived radioactive isotope pairs exist, but they are rather rare and are used only in special situations.

Fission Track Dating

The emission of atomic particles that result from the spontaneous decay of uranium within a mineral damages its crystal structure. The damage appears as microscopic linear tracks that are visible only after etching the mineral with hydrofluoric acid. The age of a sample is determined from the number of fission tracks present and the amount of uranium the sample contains: the older the sample, the greater the number of tracks (■ Figure 17.23).

Fission track dating is of particular interest to geologists because the technique can be used to date samples ranging from only a few hundred to hundreds of millions of years old. It is most useful for dating samples from the time between about 40,000 and 1.5 million years ago, a period for which other dating techniques are not always suitable. One of the problems in fission track dating occurs when the rocks have later been subjected to high temperatures. If this happens, the damaged crystal structures are repaired by annealing, and consequently the tracks disappear. In such instances, the calculated age will be younger than the actual age.

Radiocarbon and Tree-Ring Dating Methods

Carbon is an important element in nature and is one of the basic elements found in all forms of life. It has three isotopes; two of these, carbon 12 and 13, are stable, whereas carbon 14 is radioactive (see Figure 3.3). Carbon 14 has a half-life of 5730 years plus or minus 30 years. The **carbon 14 dating technique** is based on the ratio of carbon 14 to carbon 12 and is generally used to date once-living material.

The short half-life of carbon 14 makes this dating technique practical only for specimens younger than about 70,000 years. Consequently, the carbon 14 dating method is especially useful in archaeology and has greatly helped unravel the events of the latter portion of the Pleistocene Epoch.

Carbon 14 is constantly formed in the upper atmosphere when cosmic rays, which are high-energy particles (mostly protons), strike the atoms of upper-atmospheric gases, splitting their nuclei into protons and neutrons. When a neutron strikes the nucleus of a nitrogen atom (atomic number 7, atomic mass number 14), it may be absorbed into the nucleus and a proton emitted. Thus, the atomic number of the atom decreases by 1, while the atomic mass number stays the same. Because the atomic number has changed, a new element, carbon 14 (atomic number 6, atomic mass number 14), is formed. The newly formed carbon 14 is rapidly assimilated into the carbon cycle and, along with carbon 12 and 13, is absorbed in a nearly constant ratio by all living organisms (■ Figure 17.24). When an organism dies, however, carbon 14 is not replenished, and the ratio of carbon 14 to carbon 12 decreases as carbon 14 decays back to nitrogen by a single beta decay step (Figure 17.24).

Currently, the ratio of carbon 14 to carbon 12 is remarkably constant in both the atmosphere and living organisms. There is good evidence, however, that the production of carbon 14, and thus the ratio of carbon 14 to carbon 12, has varied somewhat during the past several thousand years. This was determined by comparing ages established by carbon 14 dating of wood samples with ages established by counting annual tree rings in the same samples. As a result, carbon 14 ages have been corrected to reflect such variations in the past.

Tree-ring dating is another useful method for dating geologically recent events. The age of a tree can be determined by counting the growth rings in the lower part of the stem. Each ring represents one year's growth, and the pattern of wide and narrow rings can be compared among trees to establish the exact year in which the rings were formed. The procedure of matching ring patterns from numerous trees and wood fragments in a given area is called *cross-dating*. By correlating distinctive tree-ring sequences from living and nearby dead trees, scientists can construct a time scale that extends back about 14,000 years (■ Figure 17.25). When they match ring patterns to the composite ring scale, wood samples whose ages are not known can be accurately dated.

The applicability of tree-ring dating is somewhat limited because it can be used only where continuous tree records are found. It is therefore most useful in arid regions, particularly the southwestern United States.

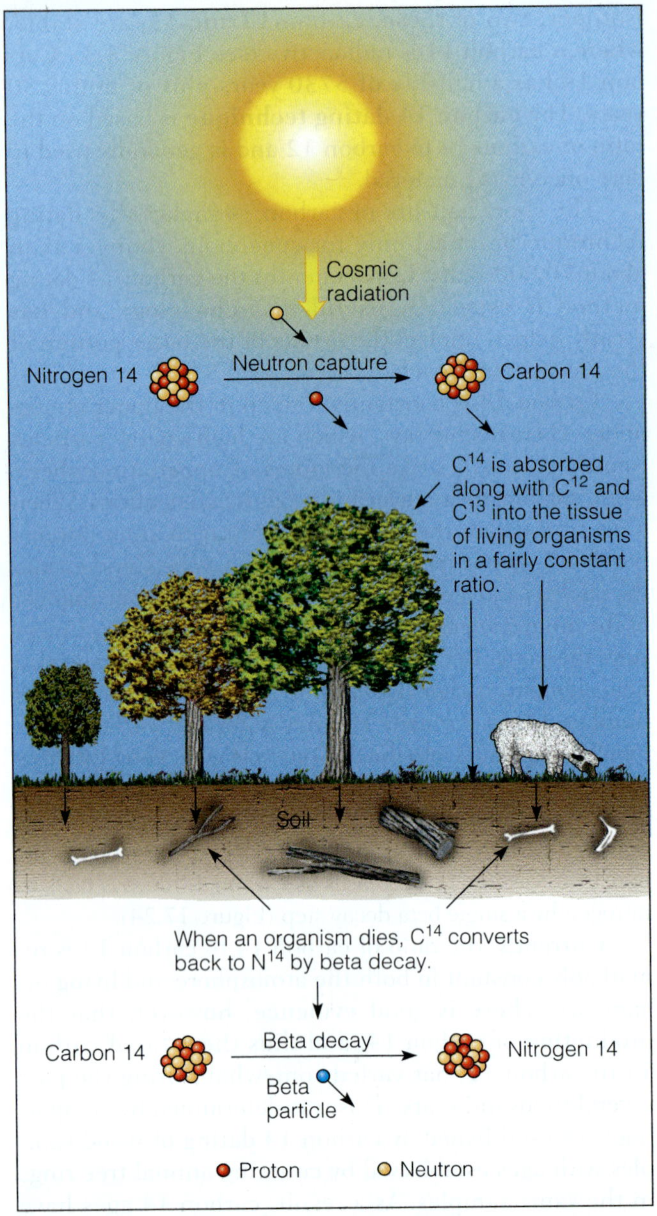

Figure 17.24

The carbon cycle showing the formation, dispersal, and decay of carbon 14.

HOW WAS THE GEOLOGIC TIME SCALE DEVELOPED?

The geologic time scale is a hierarchical scale in which the 4.6-billion-year history of Earth is divided into time units of varying duration (Figure 17.1). It was not developed by any one individual, but rather evolved, primarily during the 19th century, through the efforts of many people. By applying relative-dating methods to rock outcrops, geologists in England and western Europe defined the major geologic time units without the benefit of radiometric dating techniques. Using the principles of superposition and fossil succession, they correlated various rock exposures and pieced together a composite geologic section. This composite section is, in effect, a relative time scale because the rocks are arranged in their correct sequential order.

By the beginning of the 20th century, geologists had developed a relative geologic time scale but did not yet have any absolute dates for the various time-unit boundaries. Following the discovery of radioactivity near the end of the 19th century, radiometric dates were added to the relative geologic time scale (Figure 17.1).

Because sedimentary rocks, with rare exceptions, cannot be radiometrically dated, geologists have had to rely on interbedded volcanic rocks and igneous intrusions to apply absolute dates to the boundaries of the various subdivisions of the geologic time scale (■ Figure 17.26). An ash-fall or lava flow provides an excellent marker bed that is a time-equivalent surface, supplying a minimum age for the sedimentary rocks below and a maximum age for the rocks above. Ashfalls are particularly useful because they may fall over both marine and nonmarine sedimentary environments and can provide a connection between these different environments.

Thousands of absolute ages are now known for sedimentary rocks of known relative ages, and these absolute dates have been added to the relative time scale. In this way, geologists have been able to determine both the absolute ages of the various geologic periods and their durations (Figure 17.1).

STRATIGRAPHY AND STRATIGRAPHIC TERMINOLOGY

The recognition of a relative geologic time scale brought some order to *stratigraphy* (the study of the composition, origin, areal distribution, and age relationships of layered rocks). Problems remained, however, because many rock units are time transgressive; that is, a rock unit was deposited during one geologic period in a particular area and during another period elsewhere. To deal with both rocks and time, modern stratigraphic terminology includes two fundamentally different kinds of units: those defined by their content and those related to geologic time (Table 17.2).

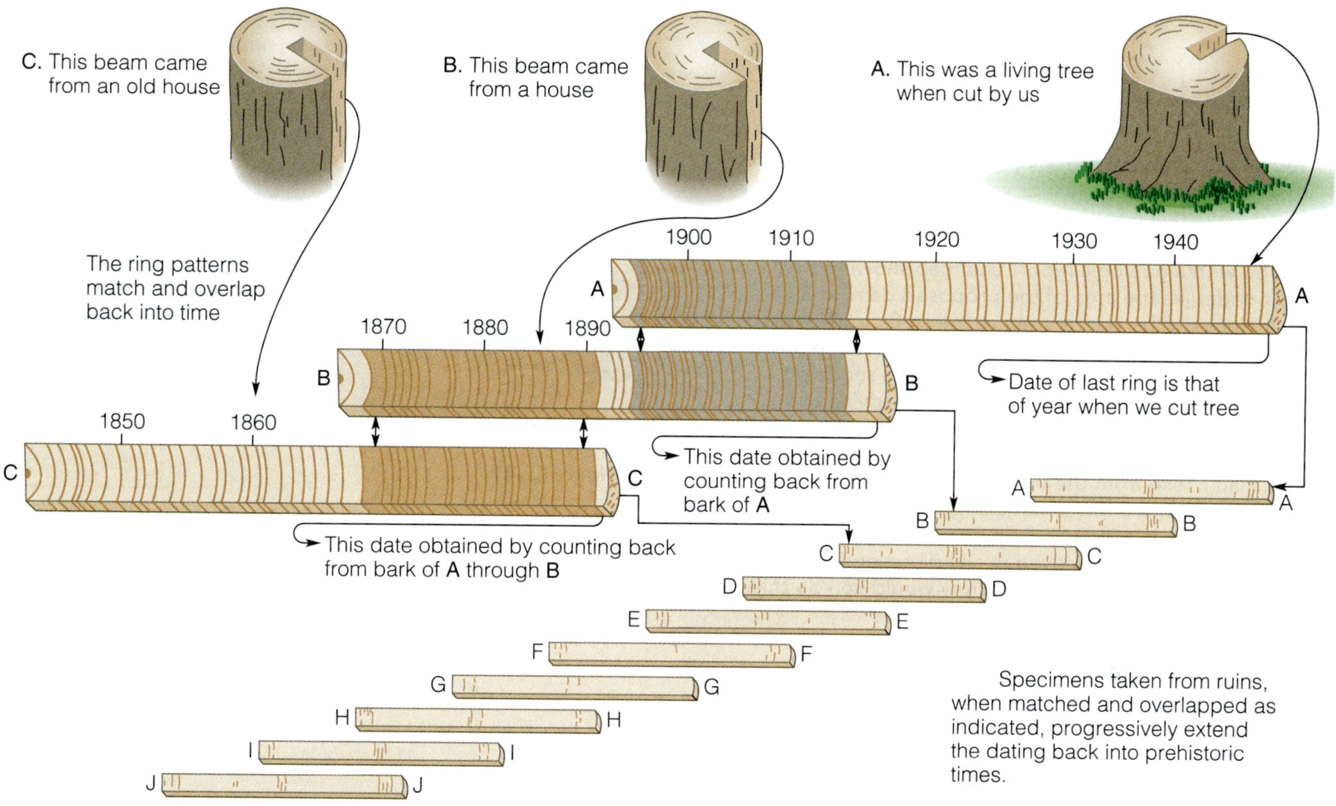

Figure 17.25

In the cross-dating method, tree-ring patterns from different woods are matched against each other to establish a ring-width chronology backward in time. Source: From *An Introduction to Tree-Ring Dating*, by Stokes and Smiley, 1968, p. 6. Reprinted by permission of University of Chicago Press.

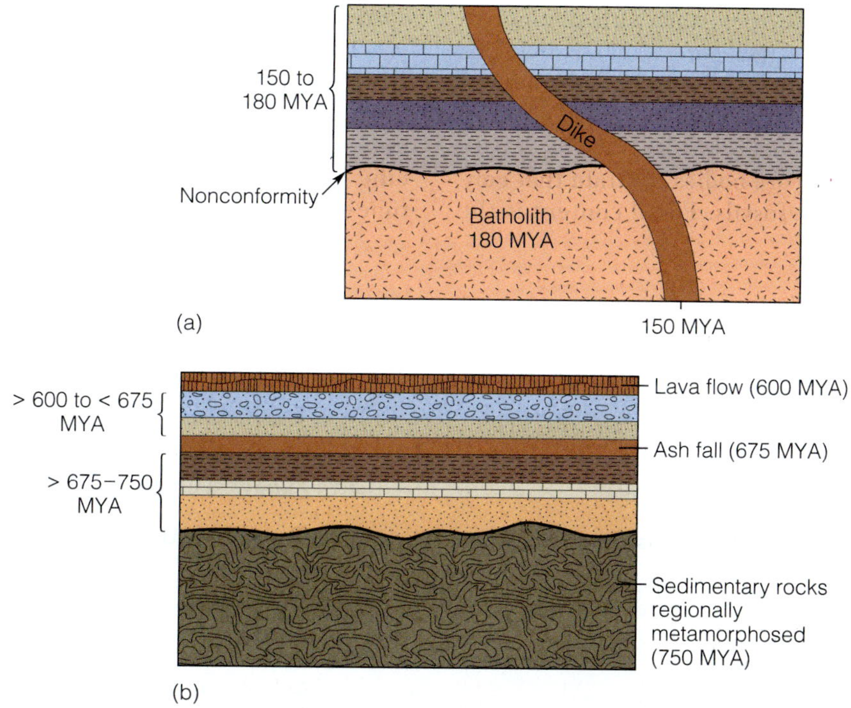

Figure 17.26

Absolute ages of sedimentary rocks can be determined by dating associated igneous rocks. In (a) and (b), sedimentary rocks are bracketed by rock bodies for which absolute ages have been determined.

What Would You Do?

You are a member of a regional planning commission that is considering a plan for constructing what is said to be a much-needed river dam that will create a recreational lake. Opponents of the dam project have come to you with a geologic report and map showing that a fault underlies the area of the proposed dam, and the fault trace can be clearly seen at the surface. The opponents say the fault may be active, and thus some day it will move suddenly, bursting the dam and sending a wall of water downstream. You seek the advice of a local geologist who has worked in the dam area, and she tells you she found a lava flow covering the fault, less than a mile from the proposed dam project site. Can you use this information along with a radiometric date from the lava flow to help convince the opponents that the fault has not moved in any direction (vertically or laterally) in the recent past? How would you do this, and what type of reasoning would you use?

Figure 17.27

The Madison Group in Montana consists of two formations, the Lodgepole Formation and the overlying Mission Canyon Formation. The Mission Canyon Formation is the rock unit exposed on the skyline. The underlying Lodgepole Formation is the rock covered on the slopes below.

Units defined by their content include **lithostratigraphic** and **biostratigraphic units.** Lithostratigraphic (*lith-* and *litho-* are prefixes meaning "stone" or "stonelike") units are defined by physical attributes of the rocks, such as rock type, with no consideration of time of origin. The basic lithostratigraphic unit is the *formation,* which is a mappable rock unit with distinctive upper and lower boundaries (■ Figure 17.27). Formations may be lumped together into larger units called *groups* or *supergroups* or divided into smaller units known as *members* and *beds* (Table 17.2).

Biostratigraphic units are bodies of strata that contain recognizably distinct fossils. Fossil content is the only criterion used to define them. Biostratrigraphic unit boundaries do not necessarily correspond to lithostratigraphic boundaries (■ Figure 17.28). The fundamental biostratigraphic unit is the *biozone.* Several types of biozones are recognized, one of which, the *concurrent range zone,* was discussed in the section on correlation.

The units related to geologic time include **time-stratigraphic units** (also known as chronstratigraphic units) and **time units** (Table 17.2). Time-stratigraphic units are units of rock that were deposited during a specific interval of time. The *system* is the fundamental time-stratigraphic unit. It is based on rocks in a partic-

Table 17.2

Classification of Stratigraphic Units

UNITS DEFINED BY CONTENT		UNITS EXPRESSING OR RELATED TO GEOLOGIC TIME	
Lithostratigraphic Units	**Biostratigraphic Units**	**Time-Stratigraphic Units**	**Time Units**
Supergroup	Biozones	Eonothem	Eon
Group		Erathem	Era
Formation		System	Period
Member		Series	Epoch
Bed		Stage	Age

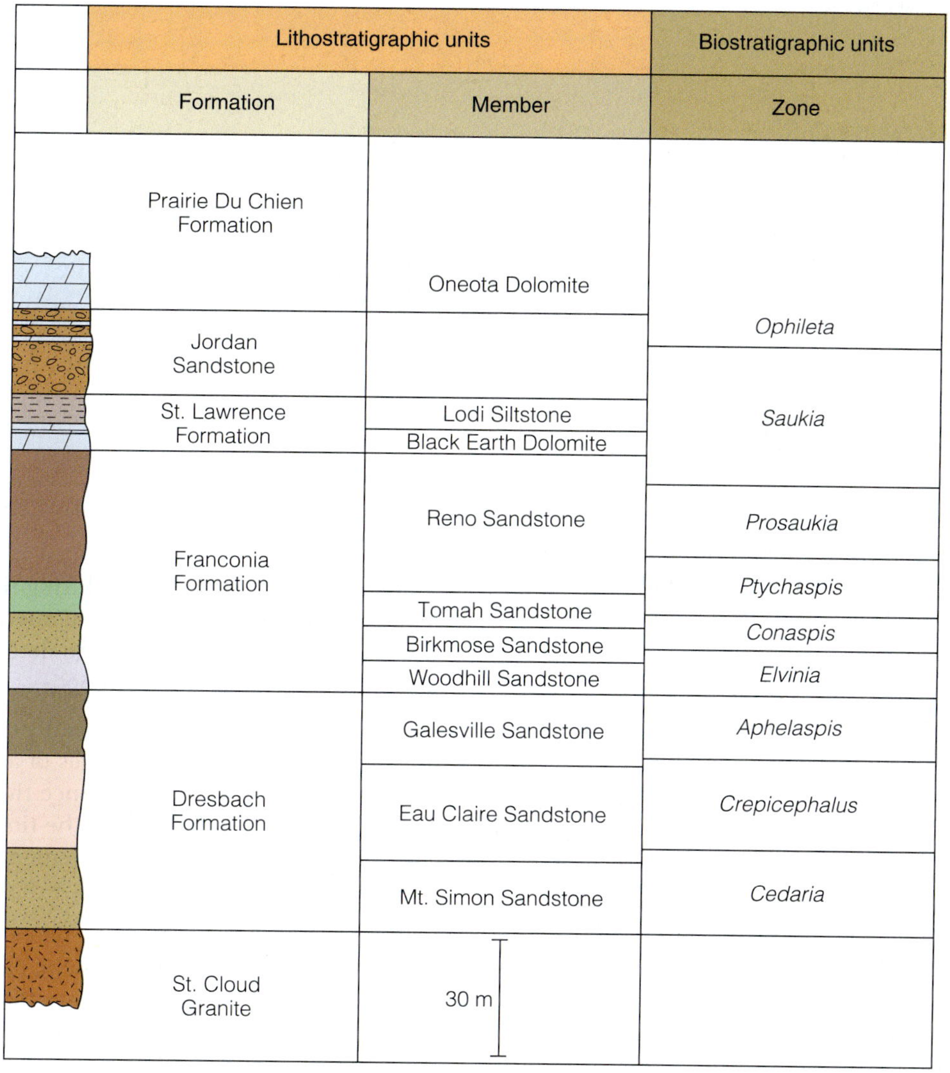

Figure 17.28

Relationships of biostratigraphic units to lithostratigraphic units in southeastern Minnesota. Notice that the biozone boundaries do not necessarily correspond to the lithostratigraphic boundaries.

ular area, the stratotype, and is recognized beyond the stratotype area primarily on the basis of fossil content.

Time units are simply units designating specific intervals of geologic time. The basic time unit is the *period,* but smaller units including epoch and age are also recognized. The time units period, epoch, and age correspond to the time-stratigraphic units system, series, and stage, respectively (Table 17.2). For example, the Cambrian Period is defined as the time during which strata of the Cambrian System were deposited. Time units of higher rank than period also exist. Eras include several periods, whereas eons include two or more eras.

17 GEO RECAP

Chapter Summary

- Relative dating involves placing geologic events in a sequential order as determined from their position in the geologic record. Absolute dating provides specific dates for events, expressed in years before the present.
- During the 18th and 19th centuries, attempts were made to determine Earth's age based on scientific evidence rather than revelation. Although some attempts were ingenious, they yielded a variety of ages that now are known to be much too young.
- James Hutton thought that present-day processes operating over long periods of time could explain all the geologic features of Earth. His observations were instrumental in establishing the principle of uniformitarianism.
- Uniformitarianism, as articulated by Charles Lyell, soon became the guiding principle of geology. It holds that the laws of nature have been constant through time and that the same processes operating today have operated in the past, although not necessarily at the same rates.
- Besides uniformitarianism, the principles of superposition, original horizontality, lateral continuity, cross-cutting relationships, inclusions, and fossil succession are basic for determining relative geologic ages and for interpreting Earth history.
- Surfaces of discontinuity encompassing significant amounts of geologic time are common in the geologic record. Such surfaces are unconformities and result from times of nondeposition, erosion, or both.
- Correlation demonstrates equivalency of units in different areas. Time equivalence is most commonly demonstrated by correlating strata that contain similar fossils.
- Radioactivity was discovered during the late 19th century, and soon afterward radiometric dating techniques enabled geologists to determine absolute ages for geologic events.
- Absolute dates for rocks are usually obtained by determining how many half-lives of a radioactive parent element have elapsed since the sample originally crystallized. A half-life is the time it takes for half of the radioactive parent element to decay to a stable daughter element.
- The most accurate radiometric dates are obtained from long-lived radioactive isotope pairs in igneous rocks. The most reliable dates are those obtained by using at least two different radioactive decay series in the same rock.
- Carbon 14 dating can be used only for organic matter such as wood, bones, and shells and is effective back to about 70,000 years ago. Unlike the long-lived isotopic pairs, the carbon 14 dating technique determines age by the ratio of radioactive carbon 14 to stable carbon 12.
- Through the efforts of many geologists applying the principles of relative dating, a relative geologic time scale was established by the mid-1800s.
- Most absolute ages of sedimentary rocks and their contained fossils are obtained indirectly by dating associated metamorphic or igneous rocks.
- Stratigraphic terminology includes two fundamentally different kinds of units: those based on content and those related to geologic time.

Important Terms

absolute dating (p. 472)
angular unconformity (p. 481)
biostratigraphic unit (p. 498)
carbon 14 dating technique (p. 495)
correlation (p. 487)
disconformity (p. 481)
fission track dating (p. 495)
guide fossil (p. 488)
half-life (p. 492)
lithostratigraphic unit (p. 498)
nonconformity (p. 484)
principle of cross-cutting relationships (p. 479)
principle of fossil succession (p. 481)
principle of inclusions (p. 479)
principle of lateral continuity (p. 479)
principle of original horizontality (p. 479)
principle of superposition (p. 479)
principle of uniformitarianism (p. 475)
radioactive decay (p. 491)
relative dating (p. 472)
time-stratigraphic unit (p. 498)
time unit (p. 498)
tree-ring dating (p. 495)
unconformity (p. 481)

Review Questions

1. Placing geologic events in sequential or chronological order as determined by their position in the geologic record is
a._____absolute dating; b._____correlation; c._____historical dating; d._____relative dating; e._____uniformitarianism.
2. Who is generally considered to be the founder of modern geology?
a._____Werner; b._____Lyell; c._____Steno; d._____Cuvier; e._____Hutton.
3. If a radioactive element has a half-life of 8 million years, the amount of parent material remaining after 24 million years of decay will be what fraction of the original amount?
a._____1/32; b._____1/16; c._____1/8; d._____1/4; e._____1/2.
4. If a feldspar grain within a sedimentary rock (such as a sandstone) is radiometrically dated, the date obtained will indicate when
a._____the feldspar crystal formed; b._____the sedimentary rock formed; c._____the parent radioactive isotope formed; d._____the daughter radioactive isotope(s) formed; e._____none of these.
5. In which type of radioactive decay does a neutron change to a proton in the nucleus by the emission of an electron?
a._____alpha decay; b._____beta decay; c._____electron capture; d._____fission track; e._____none of these.
6. Demonstrating time equivalency of rock units in different areas is called
a._____relative dating; b._____historical dating; c._____correlation; d._____absolute dating; e._____none of these.
7. How many half-lives are required to yield a mineral with 1250 atoms of U^{235} and 18,750 atoms of Pb^{206}?
a._____1; b._____2; c._____4; d._____8; e._____16.
8. In radiocarbon dating, what isotopic ratio decreases as carbon 14 decays back to nitrogen?
a._____nitrogen 14 to carbon 14; b._____carbon 14 to carbon 12; c._____carbon 13 to carbon 12; d._____nitrogen 14 to carbon 12; e._____none of these.
9. Considering the half-life of potassium 40, which is 1.3 billion years, what fraction of the original potassium 40 can be expected to remain within a given mineral crystal after 3.9 billion years?
a._____1/2; b._____1/4; c._____1/8; d._____1/16; e._____1/32.
10. Which of the following is the basic time-stratigraphic unit?
a._____period; b._____formation; c._____group; d._____system; e._____series.
11. What is the significance of an unconformity in correlation and relative dating?

12. In some places where disconformities are particularly difficult to discern from a physical point of view, using the principle of fossil succession helps us delineate such unconformities. How can using fossils help us find such hard-to-see disconformities?
13. If a rock or mineral were radiometrically dated using two or more radioactive isotope pairs (e.g., uranium 238 to lead 206 and rubidium 87 to thorium 87) and the analysis for those isotope pairs yielded distinctly different results, what possible explanation could be offered to explain how this happened? How can one rock have two correct ages?
14. How do time-stratigraphic units differ from lithostratigraphic units?
15. Where did Lord Kelvin go wrong? Explain his rationale for his age-of-Earth calculations. What discovery subsequently showed that Kelvin's calculations were, in fact, in error?
16. How does metamorphism affect the potential for accurate radiometric dating using any and all techniques discussed in this chapter? How are such radiometric dates affected by metamorphism, and why?
17. Describe the principle of uniformitarianism according to Hutton and Lyell. What is the significance of this principle?
18. What uncertainties are associated with trying to radiometrically date any sedimentary rock?
19. Describe how the principle of inclusions is important in recognizing a nonconformity.
20. Explain the concept of a concurrent range zone and how it relates to guide or index fossils.

World Wide Web Activities

Geology Now Assess your understanding of this chapter's topics with additional quizzing and comprehensive interactivities at

http://earthscience.brookscole.com/changingearth4e

as well as current and up-to-date weblinks, additional readings, and InfoTrac College Edition exercises.

Evolution—The Theory and Its Supporting Evidence

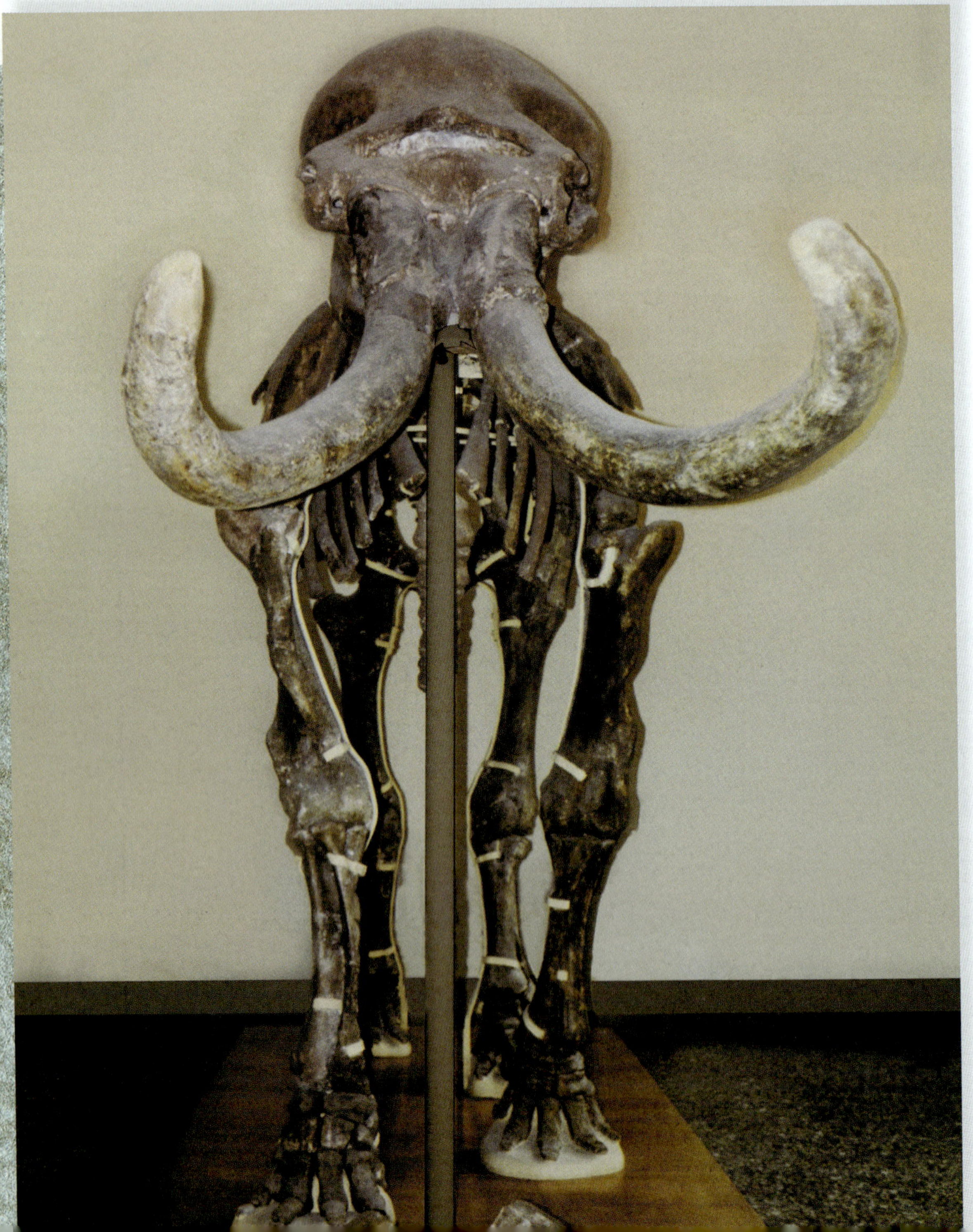

CHAPTER 18

OUTLINE

Geology⇌Now *This icon, which appears throughout the book, indicates an opportunity to explore interactive tutorials, animations, or practice problems available on the GeologyNow website at http://earthscience.brookscole.com/changingearth4e.*

OBJECTIVES

At the end of this chapter, you will have learned that:

- The central claim of the theory of evolution is that today's organisms descended, with modification, from ancestors that lived in the past.
- Jean-Baptiste de Lamarck in 1809 proposed the first widely accepted mechanism, inheritance of acquired characteristics, to account for evolution.
- In 1859, Charles Darwin and Alfred Wallace simultaneously published their views on evolution and proposed natural selection as a mechanism to bring about change.
- Experiments carried out by Gregor Mendel during the 1860s demonstrated that variations in populations are maintained rather than blended during inheritance, as previously thought.
- In the modern view of evolution, variation is accounted for mostly by sexual reproduction and by mutations in sex cells.
- A number of evolutionary trends, such as size increase and changing configuration of shells, teeth, or limbs, are well known for organisms for which sufficient fossil material is available.
- The theory of evolution is truly scientific because we can think of observations or experiments that would support it or render it incorrect.
- Fossils are important as evidence for evolution, but additional evidence comes from classification, biochemistry, molecular biology, genetics, and geographic distribution.

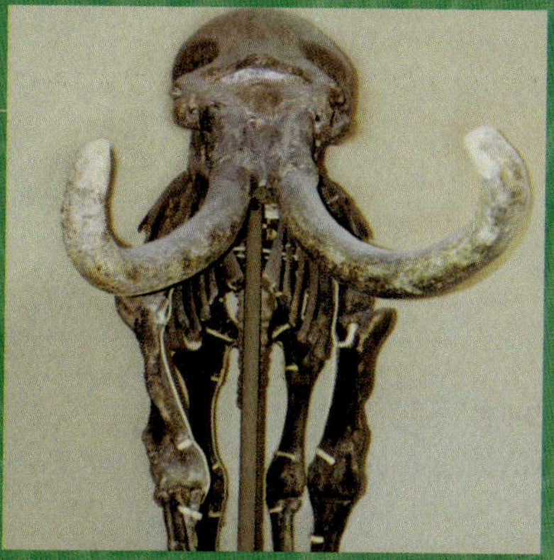

ome of the evidence for evolution is provided by fossils such as this arly Pleistocene mammoth known as Archidiskodon meridionalis *on splay in the Museum of Geology and Paleontology at the University of orence in Florence, Italy.* Source: Courtesy of Sue Monroe

Introduction

A rugged group of 13 large islands, 8 smaller ones, and 40 islets, all belonging to Ecuador, lies in the Pacific Ocean about 1000 km west of South America. Officially called the Archipelago de Colon after Christopher Columbus, the group is better known as the Galápagos Islands. During Charles Robert Darwin's five-year voyage (1831–1836) as an unpaid naturalist aboard the research vessel HMS *Beagle*, he visited the Galápagos Islands, where he made important observations that changed his ideas about the then widely held concept called *fixity of species*. According to this idea, all present-day species had been created in their present form and had changed little or not at all.

Darwin began his voyage not long after graduating from Christ's College of Cambridge University with a degree in theology, and although he was rather indifferent to religion, he fully accepted the biblical account of creation. During the voyage, though, his ideas began to change. For example, some of the fossil mammals he collected in South America were similar to present-day llamas, sloths, and armadillos yet also differed from them. He realized that these animals had descended with modification from ancestral species, and so he began to question the idea of fixity of species.

The finches and giant tortoises living on the Galápagos Islands particularly fascinated Darwin. He postulated that the 13 species of finches had evolved from a common ancestor species that somehow reached the islands as an accidental immigrant from South America. Indeed, their ancestor was very likely a single species resembling the blue-back grassquit finch now living along South America's Pacific Coast. The islands' scarcity of food accounts for the ancestral species evolving different physical characteristics, especially beak shape, to survive (■ Figure 18.1). And likewise the tortoises, that vary from island to island, still resemble tortoises living along South America's Pacific Coast, even though they differ in subtle ways.

Charles Darwin became convinced that organisms descended with modification from ancestors that lived during the past, which is the central claim of the **theory of evolution.** So why should you study evolution? For one thing, evolution involving inheritable changes in organisms through time is fundamental to biology and **paleontology,** the study of life history

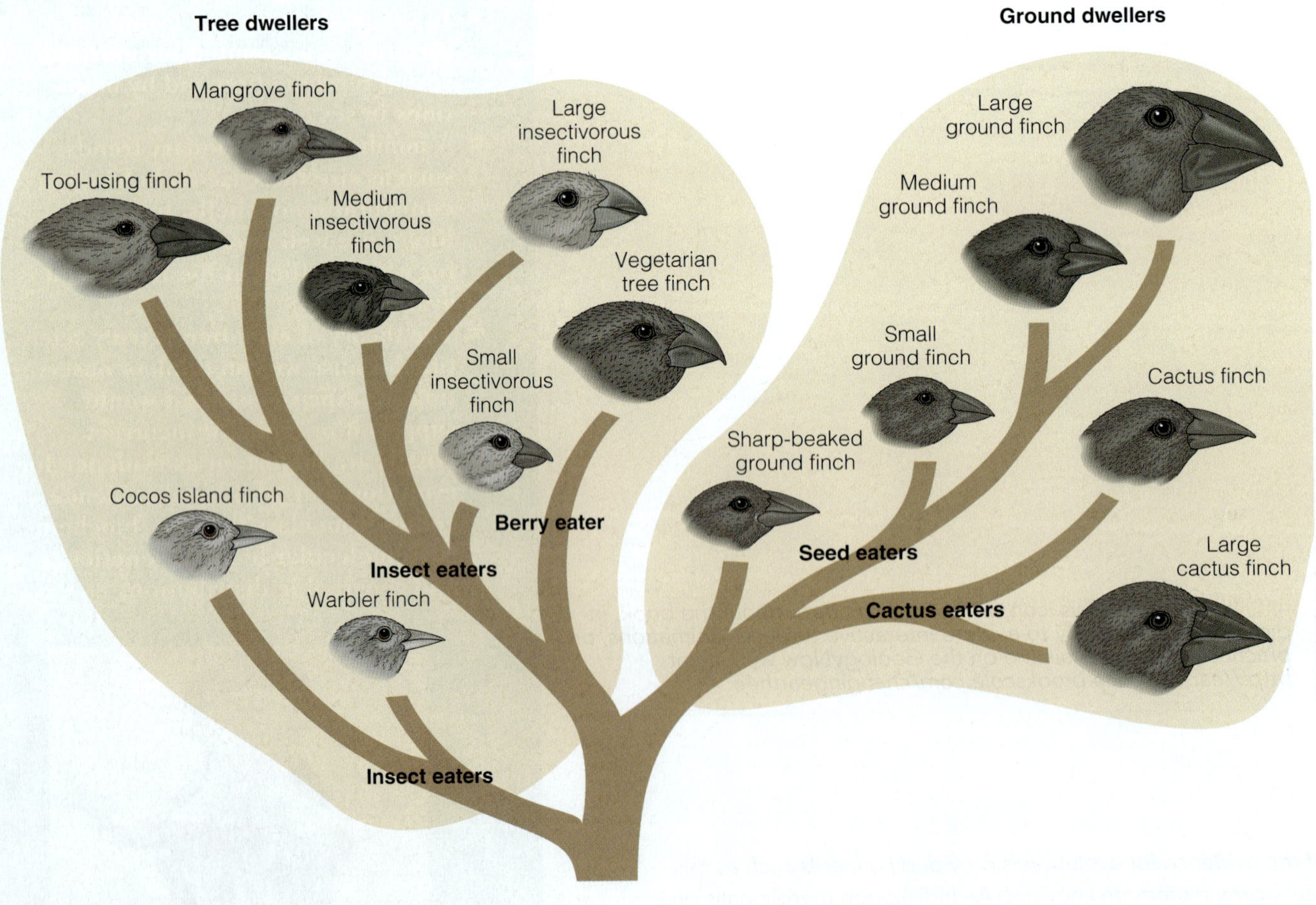

■ **Figure 18.1**

Darwin's finches from the Galápagos Islands arranged to show evolutionary relationships. Notice that the beak shape varies depending on diet.

as revealed by fossils. Furthermore, like plate tectonic theory, evolution is a unifying theory that explains an otherwise encyclopedic collection of facts. And finally, it serves as the framework for discussions of life history in the following chapters.

Unfortunately, many people have a poor understanding of the theory of evolution and hold a number of misconceptions. For example, many think that (1) evolution proceeds strictly by chance, (2) nothing less than fully developed structures such as eyes are of any use, (3) no transitional fossils (so-called missing links) connect ancestors and descendants, and (4) humans evolved from monkeys, so monkeys should no longer exist.

EVOLUTION: WHAT DOES IT MEAN?

We have stated the theory of evolution's central claim: Today's organisms have descended with modification from ancestors that lived during the past. This idea is usually attributed solely to Charles Darwin, but it was seriously considered long before he was born, even by some ancient Greeks and by philosophers and theologians during the Middle Ages. Nevertheless, among Europeans the prevailing belief well into the 1700s was that the works of the Greek philosopher Aristotle (384–322 B.C.) and the first two chapters of the book of Genesis contained all important knowledge. Literally interpreted, Genesis was taken as the final word on the origin and diversity of life as well as much of Earth history. To question any aspect of this interpretation was heresy.

The social and intellectual climate changed in 18th-century Europe, when the absolute authority of the church in all matters was questioned. Ironically, the very naturalists who were trying to find evidence supporting Genesis found more and more evidence that could not be reconciled with a literal reading of Scripture. For example, observations of sedimentary rocks previously attributed to a single worldwide flood led naturalists to conclude that they were deposited during a vast amount of time, in environments like those existing now. They reasoned that this evidence truly indicated how the rocks originated, for to infer otherwise implied a deception on the part of the Creator.

In this changing intellectual atmosphere, scientists gradually accepted the principle of uniformitarianism and Earth's great age, and the French zoologist Georges Cuvier demonstrated that many types of plants and animals had become extinct. In view of the accumulating fossil evidence as well as studies of living organisms, scientists became convinced that change from one species to another actually took place. However, they lacked a theoretical framework to explain evolution.

Jean-Baptiste de Lamarck and His Ideas on Evolution

Jean-Baptiste de Lamarck (1744–1829) was not the first to propose a mechanism to account for evolution, but in 1809 he was the first to be taken seriously. Lamarck contributed greatly to our understanding of the natural world, but unfortunately he is best remembered for his theory of **inheritance of acquired characteristics.** According to this idea, new traits arise in organisms because of their needs and somehow these characteristics are passed on to their descendants. In an ancestral population of short-necked giraffes, for instance, neck stretching to browse in trees gave them the capacity to have offspring with longer necks. In short, Lamarck thought that characteristics acquired during an individual's lifetime were inheritable.

Given the information available at the time, Lamarck's theory seemed logical and was widely accepted as a viable mechanism for evolution. Indeed, it was not totally refuted until decades later when scientists discovered that the units of heredity known as *genes* cannot be altered by any effort by an organism during its lifetime. Despite many attempts to demonstrate inheritance of acquired characteristics, none have been successful (see Geo-Focus 18.1).

The Contributions of Charles Robert Darwin and Alfred Russel Wallace

In 1859, Charles Robert Darwin (1809–1882) (■ Figure 18.2) published *On the Origin of Species,* in which he detailed his ideas on evolution and proposed a mechanism whereby evolution could take place. Although 1859 marks the beginning of modern evolutionary thought, Darwin actually formulated his ideas more than 20 years earlier but, being aware of the furor they would cause, was reluctant to publish them.

Darwin had concluded during his 1831–1836 voyage aboard the *Beagle* that species are not immutable and fixed (see the Introduction), but he had no idea what might bring about change in organisms through time.

What Would You Do?

Suppose that a powerful group in Congress mandated that all future genetic research had to conform to strict guidelines—specifically, that plants and animals should be exposed to particular environments so that they would acquire characteristics that would allow them to live in otherwise inhospitable areas. Furthermore, this group enacted legislation that prohibited any other type of genetic research. Why would it be unwise to implement this research program?

GEOFOCUS 18.1

The Tragic Lysenko Affair

When Jean-Baptiste de Lamarck (1744–1829) proposed his *theory of inheritance of acquired characteristics,* many scientists accepted it as a viable explanation to account for evolution. In fact, it was not fully refuted for several decades, and even Charles Darwin, at least for a time, accepted some kind of Lamarckian inheritance. Nevertheless, the notion that acquired characteristics could be inherited is now no more than an interesting footnote in the history of science. There was, however, one instance in recent times when the idea enjoyed some popularity, but with tragic consequences.

One of the most notable examples of adherence to a disproved scientific theory involved Trofim Denisovich Lysenko (1898–1976), who became president of the Soviet Academy of Agricultural Sciences in 1938. He lost this position in 1953 but in the same year was appointed director of the Institute of Genetics, a post he held until 1965. Lysenko endorsed Lamarck's theory of inheritance of acquired characteristics: He thought plants and animals could be changed in desirable ways by exposing them to a new environment. Lysenko reasoned that seeds exposed to dry conditions or to cold would acquire a resistance to drought or cold weather, and these traits would be inherited by future generations.

Lysenko accepted inheritance of acquired characteristics because he thought it was compatible with Maxist-Leninist philosophy. Accordingly, beginning in 1929 Soviet scientists were encouraged to develop concepts consistent with this philosophy. As president of the Academy of Agricultural Sciences and with the endorsement of the Central Committee of the Soviet Union, Lysenko did not allow any other research concerning inheritance. Those who publicly disagreed with him lost their jobs or were sent to labor camps.

Unfortunately for the Soviet people, inheritance of acquired characteristics had been discredited more than 50 years earlier. The results of Lysenko's belief in the political correctness of this theory and its implementation were widespread crop failures and famine and the thorough dismantling of Soviet genetic research in agriculture. In fact, it took decades for the Soviet genetic research programs to fully recover from this setback, after scientists finally became free to experiment with other theories of inheritance.

Lysenko's ideas on inheritance were not mandated in the Soviet Union because of their scientific merit but rather because they were deemed compatible with a belief system. The fact that Lysenko retained power and only his type of genetic research was permitted for more than 25 years is a testament to the absurdity of basing scientific theories on philosophic or political beliefs. In other words, a government mandate does not validate a scientific theory, nor does a popular vote or a decree by some ecclesiastic body. Legislating or decreeing that acquired traits are inheritable does not make it so.

However, his observations of selection practiced by plant and animal breeders and a chance reading of Thomas Malthus's essay on population gave him the ideas necessary to formulate his theory.

Plant and animal breeders practice **artificial selection** by selecting those traits they deem desirable and then breeding plants and animals with those traits, thereby bringing about a great amount of change. The fantastic variety of plants and animals so produced made Darwin wonder whether a process selecting among variant types in nature could also bring about change. He came to fully appreciate the power of selection when he read in Malthus's essay that far more animals are born than reach maturity, yet the adult populations remain rather constant. Malthus reasoned that competition for resources resulted in a high infant mortality rate, thus limiting population size.

In 1858, Darwin received a letter from Alfred Russel Wallace (1823–1913), a naturalist working in southern Asia, who had also read Malthus's essay and came to exactly the same conclusion that a natural process was selecting only a few individuals for sur-

America Museum of Natural History, #326697.

Figure 18.2

Charles Robert Darwin in 1840.

vival. Darwin and Wallace presented their idea, called *natural selection,* simultaneously in 1859 to the Linnaean Society in London.

Natural Selection—What Is Its Significance?

The salient points on **natural selection,** a mechanism that accounts for evolution, are summarized as follows:

1. Organisms in all populations possess heritable variations—size, speed, agility, visual acuity, digestive enzymes, color, and so forth.
2. Some variations are more favorable than others; that is, some variant types have a competitive edge in acquiring resources and/or avoiding predators.
3. Not all young survive to reproductive maturity.
4. Those with favorable variations are *more likely* to survive and pass on their favorable variations.

In colloquial usage natural selection is sometimes expressed as "survival of the fittest," which is misleading because it reduces to "the fittest are those that survive and are thus the fittest." But it actually involves differential rates of survival and reproduction. Therefore it is largely a matter of reproductive success. Having favorable variations does not guarantee survival for an individual, but in a population of perhaps thousands, those with favorable variations are more likely to survive and reproduce.

Natural selection works on the existing variation in a population, thus simply giving a competitive edge to some individuals, as in a population of giraffes. So evolution by natural selection and evolution by inheritance of acquired characteristics are both testable, but evidence supports the former, whereas attempts to verify the latter have failed (see Geo-Focus 18.1).

One misconception about natural selection is that among animals only the biggest, strongest, and fastest are likely to survive. Indeed, size and strength are important when male bighorn sheep compete for mates, but remember, females pass along their genes, too. Speed is certainly an advantage to some predators, but weasels and skunks are not very fast and survive quite well. In fact, natural selection may favor the smallest where resources are limited, as on small islands, or the most easily concealed, or those that adapt most readily to a new food source, or those that have the ability to detoxify some natural or human-made substance.

MENDEL AND THE BIRTH OF GENETICS

Critics of natural selection were quick to point out that Darwin and Wallace could not account for the origin of variations or explain how variations were maintained in populations. The critics reasoned that should a variant trait arise, it would blend with other traits and would be lost. Actually, the answers to these criticisms existed even then, but they remained in obscurity until 1900.

Mendel's Experiments

During the 1860s, Gregor Mendel, an Austrian monk, performed a series of controlled experiments with true-breeding strains of garden peas (strains that when self-fertilized always display the same trait, such as flower color). He concluded from these experiments that traits such as flower color are controlled by a pair of factors or what we now call **genes,** the basic units of inheritance, and that these factors (genes) controlling the same trait occur in alternative forms, or **alleles.** He further realized that one allele may be dominant over another and that offspring receive one allele of each pair from each parent (■ Figure 18.3).

We can summarize the most important aspects of Mendel's work as follows: The factors (genes) that control traits do not blend during inheritance, and even though traits may not be expressed in each generation,

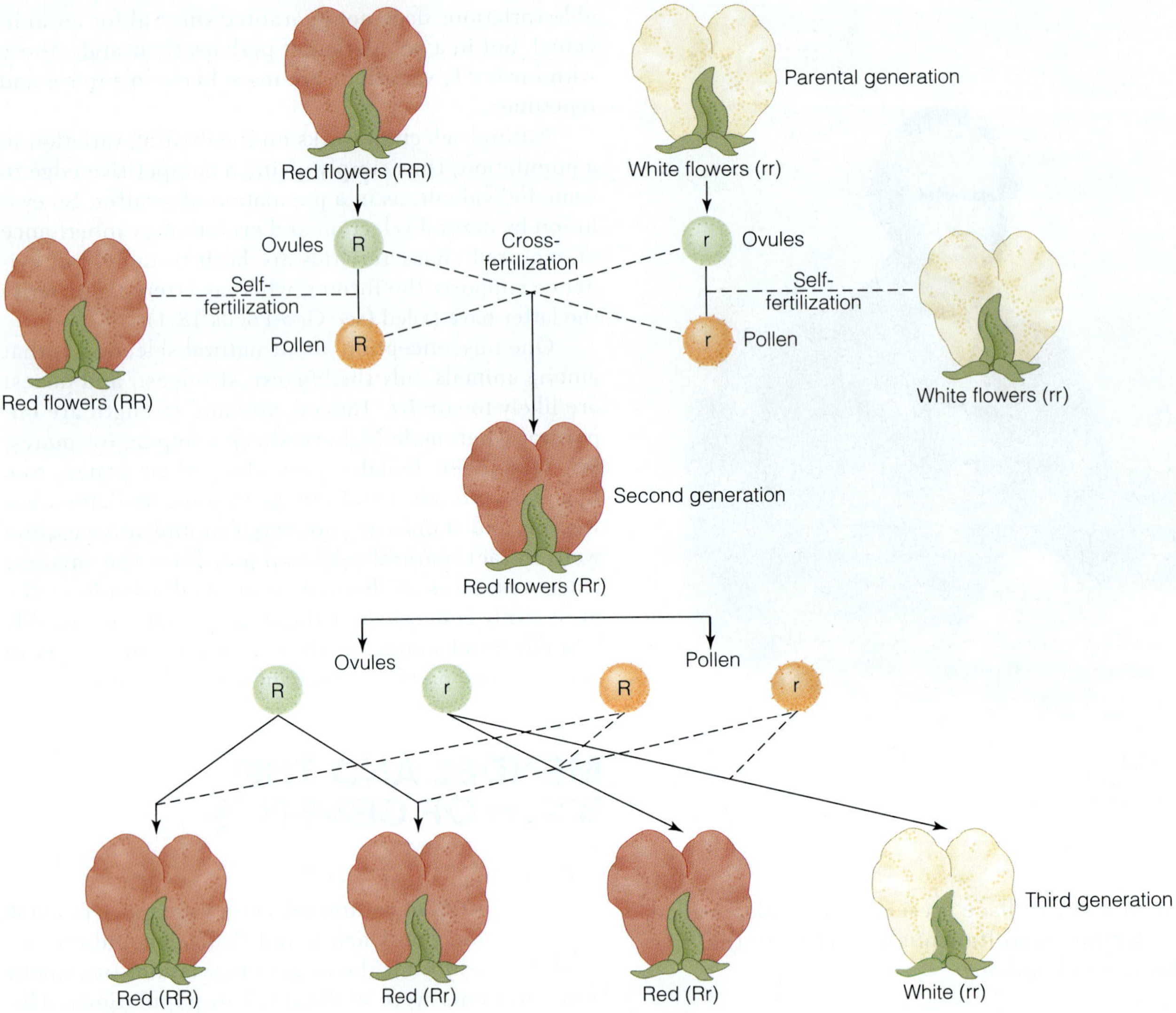

Figure 18.3

In his experiments with flower color in garden peas, Mendel used true-breeding stains. These plants, shown as the parental generation, when self-fertilized always yield offspring with the same trait as the parent. If the parental generation is cross-fertilized, though, all plants in the second generation receive the Rr combination of alleles; these plants will have red flowers because R is dominant over r. The second generation of plants produces ovules and pollen with the alleles shown and, when left to self-fertilize, produces a third generation with a ratio of three red-flowered plants to one white-flowered plant.

they are not lost (Figure 18.3). Therefore, alternative expressions of genes account for some variation in populations because traits do not blend, as previously thought, and variation can be maintained.

Mendelian genetics explains much about heredity, but we now know the situation is much more complex than he realized. Our discussion focused on a single gene controlling a trait (flower color in Figure 18.3), but in fact many genes control most traits, and some genes show incomplete dominance. Nevertheless, Mendel discovered the answers Darwin and Wallace needed, although they were published in an obscure journal and went unnoticed until three independent researchers rediscovered them in 1900.

Genes and Chromosomes

Complex, double-stranded, helical molecules of **deoxyri-bonucleic acid (DNA)** called **chromosomes** are found in the cells of all organisms (■ Figure 18.4a). Specific segments or regions of the DNA molecule are the basic hereditary units, the genes. The number of chromosomes is specific for a single species but varies among species. For instance, fruit flies have 8 chromosomes (4 pairs), humans have 46, and domestic horses have 64; chromosomes are always found in pairs carrying genes that control the same traits.

In sexually reproducing organisms, *sex cells* (pollen and ovules in plants and sperm and eggs in animals) are

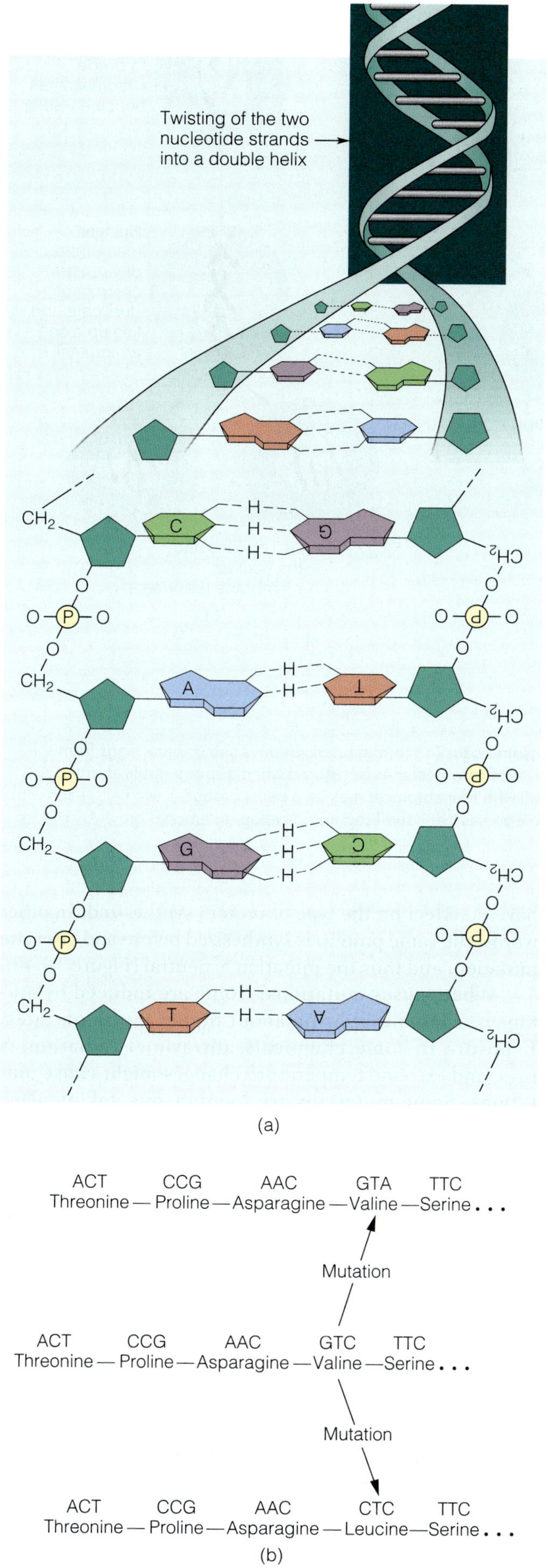

Figure 18.4

(a) Diagram of the double-stranded, helical DNA molecule. The molecule's two strands are joined by hydrogen bonds (H). Notice the bases A, G, C, and T, combinations of which code for amino acids during protein synthesis in cells. (b) A hypothetical protein showing its sequence of amino acids and two possible mutations, Notice that both GTC and GTA code for the amino acid valine, so the mutation shown at the top is neutral. However, a change from GTC to CTC results in the insertion of leucine for valine. This mutation might be harmful or beneficial.

produced when cells undergo a type of cell division known as **meiosis.** This process yields cells with only one chromosome of each pair, so all sex cells have only half the chromosome number of the parent cell (Figure 18.5a). During reproduction, a sperm fertilizes an egg (or pollen fertilizes an ovule), yielding an egg (or ovule) with the full set of chromosomes typical for that species (Figure 18.5b).

As Mendel deduced from his experiments, half of the genetic makeup of a fertilized egg comes from each parent. The fertilized egg, however, develops and grows by a cell division process called **mitosis** during which cells are simply duplicated; that is, there is no reduction in the chromosome number as in meiosis (Figure 18.5c).

THE MODERN VIEW OF EVOLUTION

During the 1930s and 1940s, the ideas developed by paleontologists, geneticists, population biologists, and others were merged to form a **modern synthesis** or neo-Darwinian view of evolution. The chromosome theory of inheritance was incorporated into evolutionary thinking; changes in genes (mutations) were seen as one source of variation in populations; Lamarck's idea of inheritance of acquired characteristics was completely rejected; and the importance of natural selection was reaffirmed. The modern synthesis also emphasized that evolution is gradual, a point that has been challenged.

What Brings About Variation?

Evolution by natural selection works on variations in populations, most of which are accounted for by the reshuffling of alleles from generation to generation during sexual reproduction. Given that each of thousands of genes might have several alleles, and that offspring receive half of their genes from each parent, the potential for variation is enormous. Any new variations arise by **mutations** involving some kind of change in the chromosomes or genes—that is, a change in the hereditary information.

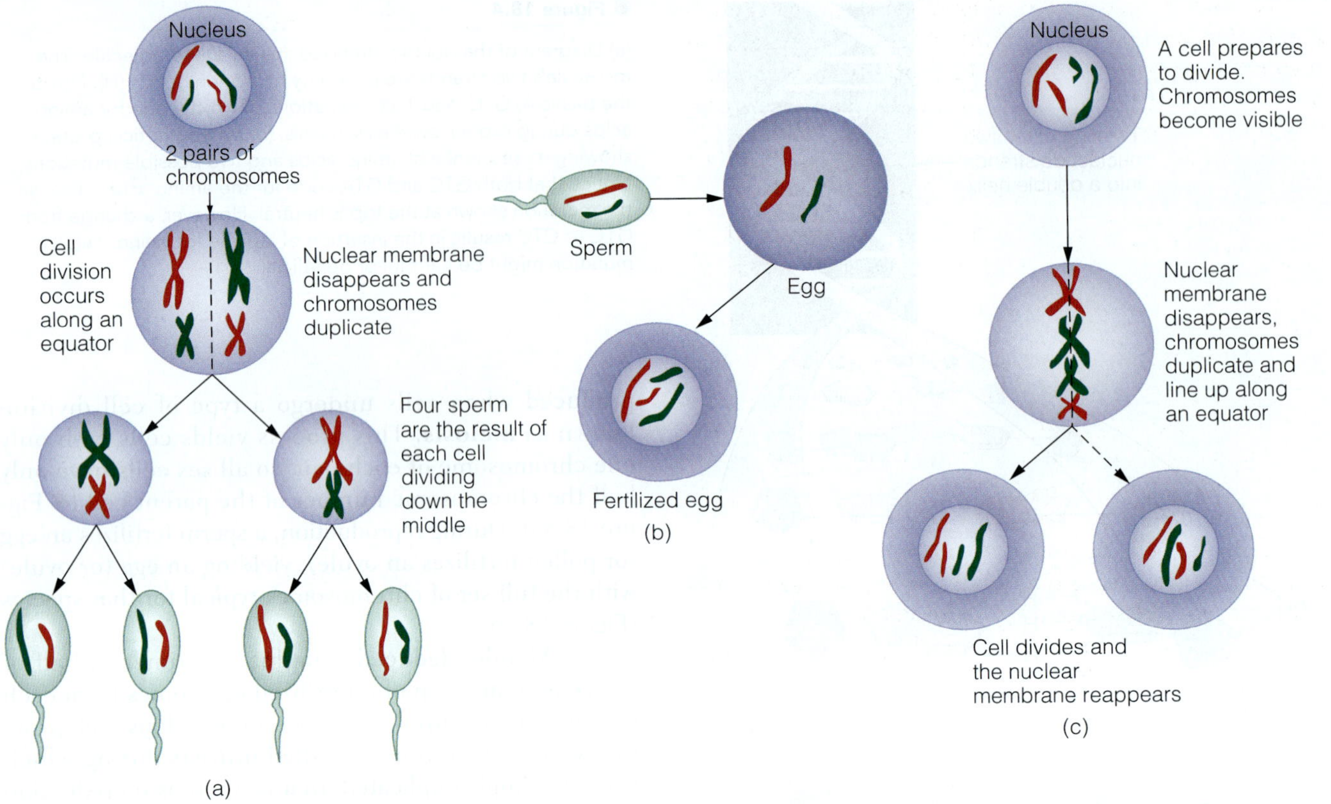

Geology Now ■ **Active Figure 18.5**

(a) During meiosis, sex cells form that contain one member of each chromosome pair. The formation of sperm is shown here; eggs form the same way, but only one of the four final eggs is functional. (b) The full number of chromosomes is restored when a sperm fertilizes an egg. (c) Mitosis results in the complete duplication of a cell. In this example, a cell with four chromosomes (two pairs) produces two cells, each with four chromosomes. Mitosis takes place in all cells except sex cells. Once an egg has been fertilized, the developing embryo grows by mitosis.

Whether a *chromosomal mutation* (affecting a large segment of a chromosome) or a *point mutation* (a change in a particular gene), as long as it takes place in a sex cell it is inheritable. To fully understand mutations we must explore them further. For one thing, they are random with respect to fitness, meaning they may be beneficial, neutral, or harmful.

If a species is well adapted to its environment, most mutations would not be particularly useful and might perhaps be harmful. But what was a harmful mutation can become a useful one if the environment changes. For instance, some plants have developed a resistance to contaminated soils around mines. Plants of the same species from the normal environment do poorly or die in contaminated soils, whereas contaminant-resistant plants do very poorly in the normal environment. Mutations for contaminant resistance probably occurred repeatedly in the population, but they were not beneficial until contaminated soils were present.

How can a mutation be neutral? In cells, information carried on chromosomes directs the formation of proteins by selecting the appropriate amino acids and arranging them into a specific sequence. However, some mutations have no effect on the type of protein synthesized. In other words, the same protein is synthesized before and after the mutation, and thus the mutation is neutral (Figure 18.4b).

What causes mutations? Some are induced by *mutagens,* agents that bring about higher mutation rates. Exposure to some chemicals, ultraviolet radiation, X rays, and extreme temperature changes might cause mutations. Some mutations are spontaneous, taking place in the absence of any known mutagen.

Speciation and the Rate of Evolution

Speciation, the phenomenon of a new species arising from an ancestral species, is well documented, but the rate and ways in which it takes place vary. First, though, let us be clear on what we mean by **species,** a biological term for a population of similar individuals that in nature interbreed and produce fertile offspring. Thus, a species is reproductively isolated from other species. This definition does not apply to organisms such as bacteria that reproduce asexually, but it is nevertheless useful for our discussion of plants, animals, fungi, and single-cell organisms called protistans.

Goats and sheep are distinguished by physical characteristics, and they do not interbreed in nature, thus they are separate species. Yet in captivity they can produce fertile offspring. Lions and tigers can also interbreed in captivity, although they do not interbreed in nature and their offspring are sterile, so they too are separate species. Domestic horses that have gone wild can interbreed with zebras to yield a *zebroid,* which is sterile, yet horses and zebras are separate species. It should be obvious from these examples that reproductive barriers are not complete in some species, indicating varying degrees of change from a common ancestral species.

The process of speciation involves a change in the genetic makeup of a population, which also may bring about changes in form and structure. According to the concept of **allopatric speciation,** species arise when a small part of a population becomes isolated from its parent population (■ Figure 18.6). Isolation may result from a marine transgression that effectively separates a once-interbreeding species, or a few individuals may somehow get to a remote area and no longer exchange genes with the parent population (Figure 18.1). Given these conditions and the fact that different selective pressures are likely, they may eventually give rise to a reproductively isolated species.

Although widespread agreement exists on allopatric speciation, scientists disagree on how rapidly a new species may evolve. According to Darwin and reaffirmed by the modern synthesis, the gradual accumulation of minor changes eventually brings about the origin of a new species, a phenomenon called **phyletic gradualism** (■ Figure 18.7a). Another view, known as **punctuated equilibrium,** holds that little or no change takes place in a species during most of its existence, and then evolution occurs rapidly, giving rise to a new species in perhaps as little as a few thousands of years (Figure 18.7b).

Proponents of punctuated equilibrium argue that few examples of gradual transitions from one species to another are found in the fossil record. Critics, however, point out that neither Darwin nor those who formulated the modern synthesis insisted that all evolutionary change was gradual and continuous, a view shared by many present-day biologists and paleontologists. Indeed, they allowed for times during which evolutionary change

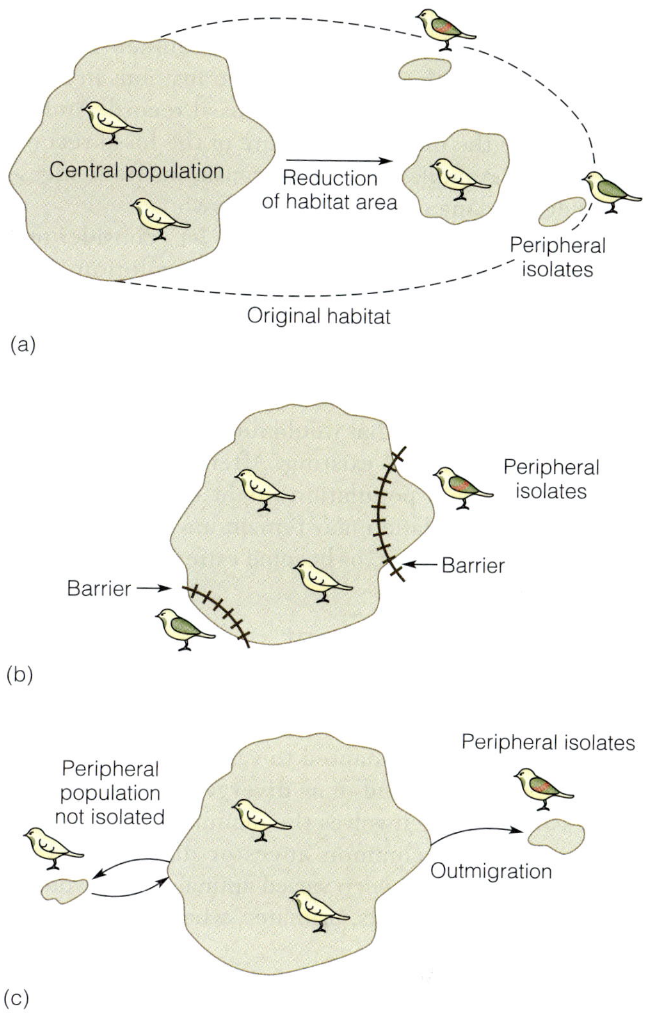

■ **Figure 18.6**

Allopatric speciation. (a) Reduction of the area occupied by a species may leave small, isolated populations at the periphery of the once more extensive range. In this example, members of both peripheral isolates evolved into new species, but the original population remained unchanged. (b) Geographic barriers form across parts of a central population's range, thereby isolating small populations. (c) Outmigration and the origin of a peripheral isolate.

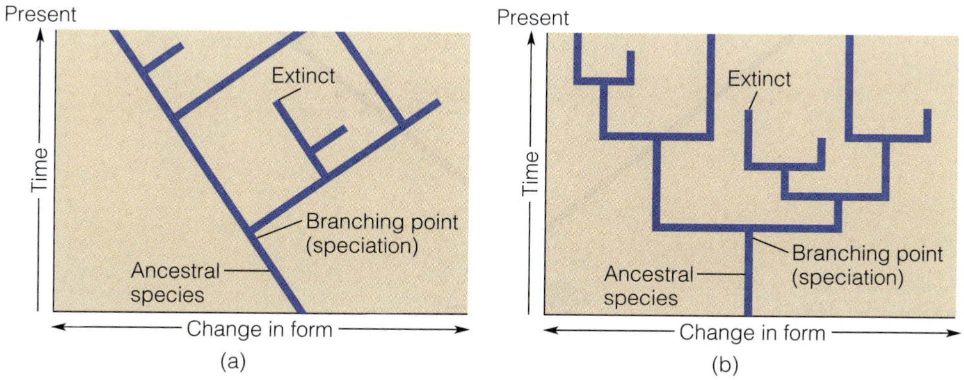

■ **Figure 18.7**

Comparison of two models for divergence from a common ancestor. (a) In phyletic gradualism, slow, continuous change takes place as a new species evolves from an ancestral one. (b) According to the punctuated equilibrium model, a species remains unchanged during most of its existence and new species evolve rapidly, perhaps in a few thousands of years.

in small populations could be quite rapid. Furthermore, deposition of sediments in most environments is not continuous; thus, the lack of gradual transitions in many cases is simply an artifact of the fossil record. And finally, despite the incomplete nature of the fossil record, a number of examples of gradual transitions from ancestral to descendant species are well known.

Before leaving speciation entirely, let's consider one of the common misconceptions about evolution. One antievolution argument is "If humans evolved from monkeys, why are there still monkeys?" Actually, there are two misconceptions here. First, no scientist has ever claimed that humans evolved from monkeys. And second, even if they had, that would not preclude the possibility of monkeys still existing. After all, in allopatric speciation a small population might evolve while the larger parent population may remain unchanged, evolve in some other direction, or become extinct.

Divergent, Convergent, and Parallel Evolution

The phenomenon of an ancestral species giving rise to diverse descendants adapted to various aspects of the environment is referred to as **divergent evolution.** An impressive example involves the mammals whose diversification from a common ancestor during the Late Mesozoic gave rise to such varied animals as platypuses, armadillos, rodents, bats, primates, whales, and rhinoceroses (■ Figure 18.8).

Divergent evolution leads to descendants that differ markedly from their ancestors. *Convergent evolution* and *parallel evolution* are processes whereby similar adaptations arise in different groups. Unfortunately, they differ in degree and are not always easy to distinguish. Nevertheless, both are common phenomena. **Convergent evolution** involves the development of similar characteristics in distantly related organisms, whereas similar characteristics arising in closely related organisms is **parallel evolution.** In both cases, similar characteristics develop independently because the organisms in question adapt to comparable environments. Perhaps the following examples will clarify the distinction between these two concepts.

During much of the Cenozoic, South America was an island continent with a unique mammalian fauna that evolved in isolation. Nevertheless, several mammals in South and North America adapted in similar ways so that they superficially resembled one another (■ Figure 18.9a)—a good example of convergent evolution. Likewise, convergence also accounts for the superficial similarities between Australian marsupial (pouched) mammals and placental mammals elsewhere—for example, catlike marsupial carnivores and true cats. Parallel evolution, in contrast, involves closely related organisms, such as jerboas and kangaroo rats, that independently evolved comparable features (Figure 18.9b).

Cladistics and Cladograms

Traditionally, scientists have depicted evolutionary relationships with *phylogenetic trees,* in which the horizontal axis represents anatomic differences and the vertical axis denotes time (■ Figure 18.10a). The patterns of ancestor–descendant relationships shown are based on a variety of characteristics, although the ones used are rarely specified. In contrast, a **cladogram** shows the relation-

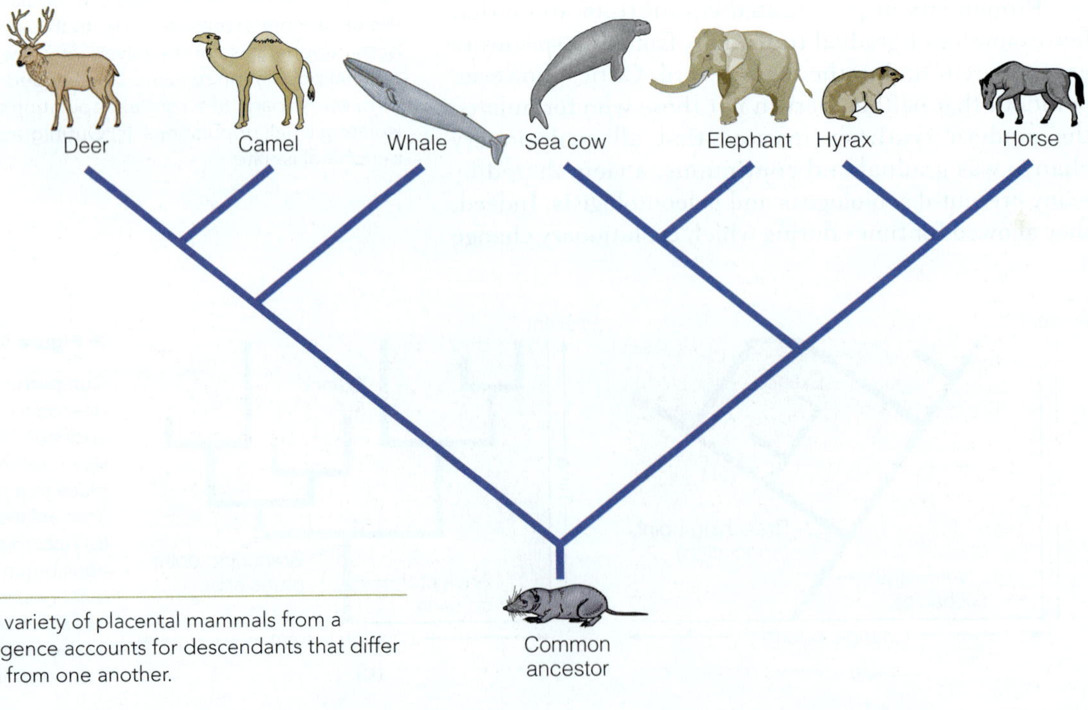

■ **Figure 18.8**

Divergent evolution of a variety of placental mammals from a common ancestor. Divergence accounts for descendants that differ from their ancestors and from one another.

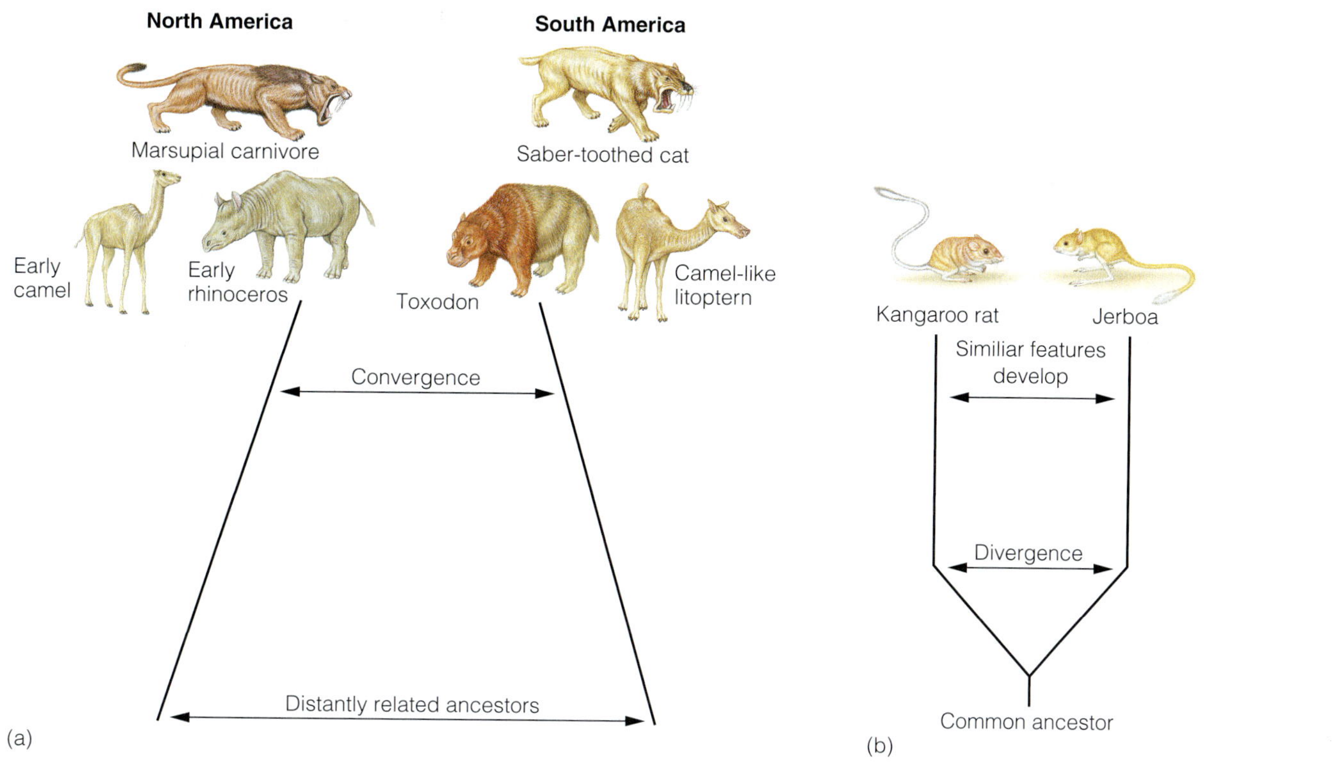

Figure 18.9

(a) Convergent evolution takes place when distantly related organisms give rise to species that resemble one another because they adapt in comparable ways. (b) Parallel evolution involves the independent origin of similar features in closely related organisms.

ships among members of a *clade,* a group of organisms including its most recent common ancestor (Figure 18.10b).

Cladistics is a type of biological analysis in which organisms are grouped together based on derived as opposed to primitive characteristics. For instance, all land-dwelling vertebrates have bones and paired limbs, so these characteristics are primitive and of little use in establishing relationships among them. However, hair and three middle ear bones are derived characteristics, sometimes called *evolutionary novelties,* because only one

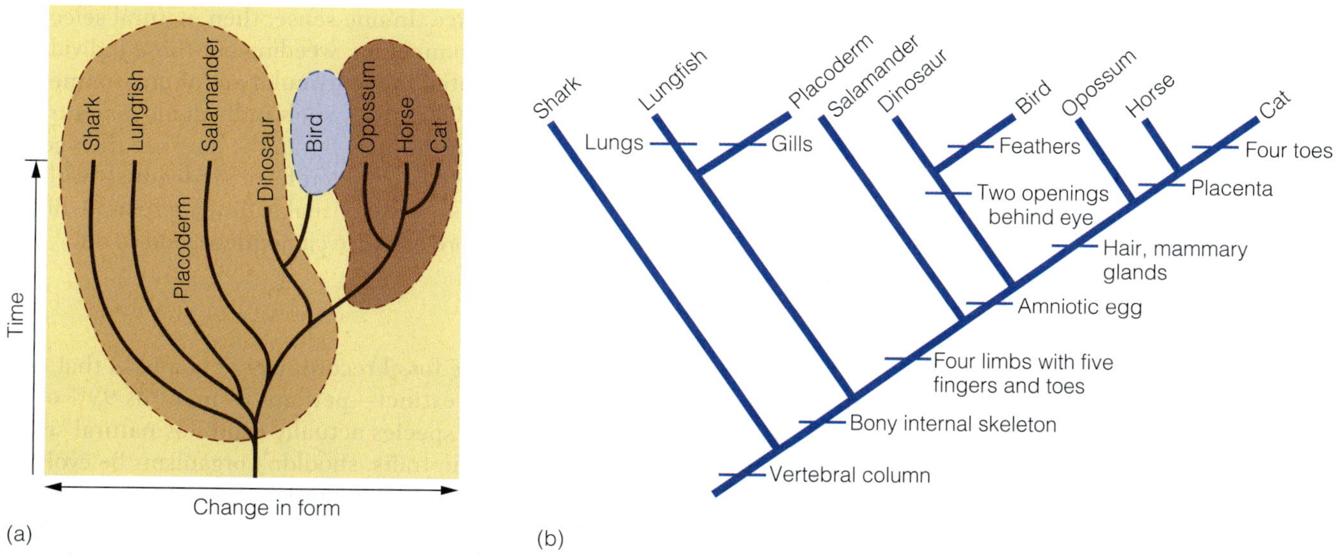

Figure 18.10

(a) A phylogenetic tree showing the relationships among various vertebrate animals. (b) A cladogram showing inferred relationships. Some of the characteristics used to construct this cladogram are indicated.

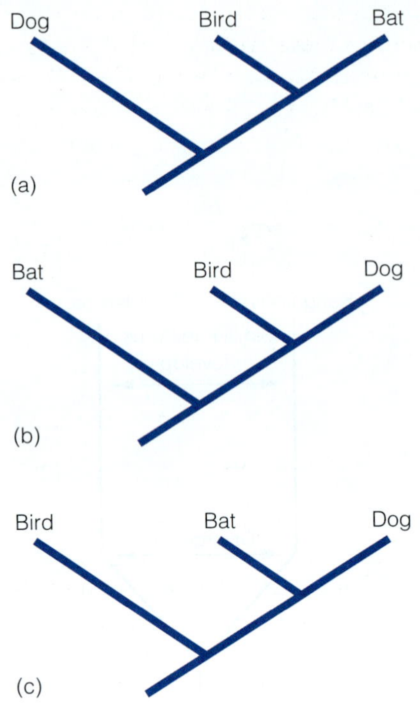

■ **Figure 18.11**

Cladograms showing three hypotheses for the relationships among birds, bats, and dogs. Derived characteristics such as hair and giving birth to live young indicate that dogs and bats are most closely related, as shown in (c).

subclade, the mammals, has them. If you consider only mammals, hair and middle ear bones are primitive characteristics, but live birth is a derived characteristic that serves to distinguish most mammals from the egg-laying mammals.

Any number of organisms can be depicted in a cladogram, but the more shown, the more complex and difficult it is to construct. Let's use an example of only three animals—bats, dogs, and birds. ■ Figure 18.11 shows three cladograms, each a different interpretation of their relationship. Bats and birds fly, so we might conclude they are more closely related to one another than to dogs (Figure 18.11a). But if we concentrate on evolutionary novelties, such as hair and giving birth to live young, we conclude that bats and dogs are more closely related than either is to birds (Figure 18.11c).

Cladistics and cladograms work well for living organisms, but when they are applied to fossils, care must be taken in determining what are primitive versus derived characteristics, especially in groups with poor fossil records. Furthermore, cladistic analysis depends solely on characteristics inherited from a common ancestor, so paleontologists must be especially careful of characteristics that result from convergent evolution. Nevertheless, cladistics is a powerful tool that has more clearly elucidated the relationships among many fossil lineages.

Evolutionary Trends

Evolutionary changes do not involve all aspects of an organism simultaneously because selection pressure is greater on some features than on others. As a result, a key feature we associate with a descendant group may appear before other features typical of that group. For example, the oldest known bird had feathers and the typical fused clavicles of birds, but it also retained many reptile characteristics (see Chapter 23). Accordingly, it represents the concept of **mosaic evolution,** meaning that organisms have recently evolved characteristics as well as some features of their ancestral group.

Paleontologists determine in some detail the *phylogeny,* or evolutionary history, and various *evolutionary trends* for groups of organisms if sufficient fossil material is available. Size increase is one of the most common evolutionary trends, but trends are extremely complex, they may be reversed, and several trends taking place may not all proceed at the same rate. One evolutionary trend in horses was larger size, but some now-extinct horses actually showed a size decrease. Other trends in horses include changes in their teeth and skull as well as lengthening of their legs and reduction in the number of toes, but these trends did not all take place at the same rate (see Chapter 23).

Isn't evolution by natural selection a random process? If so, how is it possible for a trend to continue long enough to account, just by chance, for such complex structures as eyes, wings, and hands? Actually, evolution by natural selection is a two-step process, and only the first step involves chance. First, variation must be present or arise in a population. Whether or not variations arising by mutations are favorable is indeed a matter of chance, but the second step involving natural selection is not, because only individuals with favorable variations are most likely to survive and reproduce. In one sense, then, natural selection is a process of elimination, weeding out those individuals not as well adapted to a particular set of environmental circumstances. Of course, such individuals may survive and reproduce just by luck, but in the long run their genes will be eliminated as selection favors individuals with favorable variations—better visual acuity, appropriate digestive enzymes, more effective camouflage, and so on.

Extinctions

Judging from the fossil record, most organisms that ever existed are now extinct—perhaps as many as 99% of all species. Now, if species actually evolve as natural selection favors certain traits, shouldn't organisms be evolving toward some kind of higher order of perfection or greater complexity? Certainly vertebrates are more complex, at least in overall organization, than are bacteria, but complexity does not necessarily mean they are superior in some survival sense—after all, bacteria have persisted for at least 3.5 billion years. Actually, natural selection does not yield some kind of perfect organism, but rather those adapted to a specific set of circumstances at a particular time. Thus a clam or lizard existing now is not somehow superior to those that lived millions of years ago.

The continual extinction of species is referred to as *background extinction,* to clearly differentiate it from a **mass extinction,** during which accelerated extinction rates sharply reduce Earth's biotic diversity. Extinction is a continuous occurrence, but so is the evolution of new species that usually, but not always, quickly exploit the opportunities created by another species' extinction. When the dinosaurs and their relatives died out, the mammals began a remarkable diversification as they began to occupy the niches left temporarily vacant.

Everyone is familiar with the mass extinction of dinosaurs and other animals at the end of the Mesozoic Era (see Chapter 22). The greatest extinction, though, during which perhaps more than 90% of all species died out, was at the end of the Paleozoic Era (see Chapter 21).

WHAT KINDS OF EVIDENCE SUPPORT EVOLUTIONARY THEORY?

When Charles Darwin proposed his theory of evolution, he cited supporting evidence such as classification, embryology, comparative anatomy, geographic distribution, and, to a limited extent, the fossil record. He had little knowledge of the mechanism of inheritance, and both biochemistry and molecular biology were unknown during his time. Studies in these areas, coupled with a more complete and much better understood fossil record, have convinced scientists that the theory is as well supported by evidence as any other major theory. Of course, scientists still disagree on many details, but the central claim of the theory is well established and widely accepted.

But is the theory of evolution truly scientific? That is, can testable predictive statements be made from it? First we must be clear on what a theory is and what we mean by "predictive." Scientists propose hypotheses to explain natural phenomena, test them, and in some cases raise them to the status of a theory—an explanation of some natural phenomenon well supported by evidence from experiments, observations, or both.

Almost everything in the sciences has some kind of theoretical underpinning—optics, the nature of matter, the present distribution of continents, diversity in the organic world, and so on. Of course, no theory is ever proven in some final sense, although it might be supported by substantial evidence; all are always open to question, to revision, and occasionally to replacement by a more comprehensive theory. In geology, plate tectonic theory has replaced geosyncline theory to explain the origin of mountains, but the new theory contains many elements of its predecessor.

As for prediction, the theory of evolution cannot predict the future. No one knows which existing species might become extinct, or what the descendants of horses or clams, if any, will look like 10 million years from now. Nevertheless, we can make a number of predictions about the present-day biologic world and many aspects of the fossil record that should be consistent with the theory if it is correct (Table 18.1).

Table 18.1

Some Predictions from the Theory of Evolution

1. If evolution has taken place, the oldest fossil-bearing rocks should have remains of organisms very different from those existing now, and more recent rocks should have fossils more similar to today's organisms.
2. If evolution by natural selection actually occurred, Earth must be very old, perhaps many millions of years old.
3. If today's organisms descended with modification from ones in the past, there should be fossils showing characteristics that connect orders, classes, and so on.
4. If evolution is true, closely related species should be similar not only in details of their anatomy but also in their biochemistry, genetics, and embryonic development, whereas distantly related species should show fewer similarities.
5. If the theory of evolution is correct—that is, living organisms descended from a common ancestor—classification of organisms should show a nested pattern or similarities.
6. If evolution actually took place, we would expect cave-dwelling plants and animals to most closely resemble those immediately outside their respective caves rather than being most similar to those in caves elsewhere.
7. If evolution actually took place, we would expect land-dwelling organisms on oceanic islands to most closely resemble those of nearby continents rather than those on other distant islands.
8. If evolution has taken place, a mechanism should exist that accounts for the evolution of one species to another.
9. If evolution occurred, we would expect mammals to appear in the fossil record long after the appearance of the first fish. Likewise, we would expect reptiles to appear before the first mammals or birds.
10. If we examine the fossil record of presumably related organisms such as horses and rhinoceroses, we should find that they were quite similar when they diverged from a common ancestor but became increasingly different as their divergence continued.

For example, closely related species such as wolves and coyotes should be similar not only in their overall anatomy but also in terms of their biochemistry, genetics, and embryonic development (point 4 in Table 18.1). Suppose they differed in their biochemical mechanisms as well as embryology. Obviously, our prediction would fail, and we would at least have to modify the theory. If other predictions also failed—for example, mammals appeared in the fossil record before fishes—we would have to abandon the theory and find a better explanation for our observations. Accordingly, the theory of evolution is truly scientific because it can at least in principle "be falsified"—that is, proven wrong.

Classification—A Nested Pattern of Similarities

Carolus Linnaeus (1707–1778) proposed a classification system in which organisms are given a two-part genus and species name; the coyote, for instance, is *Canis latrans.* Table 18.2 shows Linnaeus's classification scheme, which is a hierarchy of categories that becomes more inclusive as one proceeds up the list. The coyote (*Canis latrans*) and the wolf (*Canis lupus*) share numerous characteristics, so they are members of the same genus, whereas both share some but fewer characteristics with the red fox (*Vulpes fulva*), and all three are members of the family Canidae (■ Figure 18.12). All canids share some characteristics with cats, bears, and weasels and are grouped together in the order Carnivora, which is one of 18 living orders of the class Mammalia, all of whom are warm-blooded, possess fur or hair, and have mammary glands.

Linnaeus certainly recognized shared characteristics among organisms, but his intent was simply to categorize species he thought were specially created and immutable. Following the publication of Darwin's *On the Origin of Species* in 1859, however, biologists quickly realized that shared characteristics constituted a strong argument for evolution. After all, if present-day organisms actually descended from ancient species, we should expect a pattern of similarites between closely related species and fewer between more distantly related ones.

How Does Biological Evidence Support Evolution?

If all existing organisms actually descended with modification from ancestors that lived during the past, fundamental similarities should exist among all life-forms. As a matter of fact, all living things—be they bacteria, redwood trees, or whales—are composed mostly of carbon, nitrogen, hydrogen, and oxygen. Furthermore, their chromosomes consist of DNA and all cells synthesize proteins in essentially the same way.

Studies in biochemistry also provide evidence for evolutionary relationships. Blood proteins are similar among all mammals but also indicate that humans are most closely related to great apes, followed in order by Old World monkeys, New World monkeys, and lower primates such as lemurs. Biochemical tests support the idea that birds descended from reptiles, a conclusion supported by evidence in the fossil record.

The forelimbs of humans, whales, dogs, and birds are superficially dissimilar. Yet all are made up of the same bones; all have basically the same arrangement of muscles, nerves, and blood vessels; all are similarly arranged with respect to other structures; and all have a similar pattern of embryonic development (■ Figure 18.13). These **homologous structures,** as they are called, are simply basic vertebrate forelimbs modified for different functions; that is, they indicate derivation from a common ancestor. However, some similarities are unrelated to evo-

Table 18.2

Expanded Linnaen Classification Scheme (the animal classified in this example is the coyote, *Canis latrans*)

Kingdom	**Animalia**—Multicelled organisms; cells with nucleus (see Chapter 19); reproduce sexually; ingest preformed organic molecules for nutrients
Phylum	**Chordata**—Possess notochord, gill slits, dorsal hollow nerve cord at some time during life cycle (see Chapter 21)
Subphylum	**Vertebrata**—Those chordates with a segmented vertebral column (see Chapter 21)
Class	**Mammalia**—Warm-blooded vertebrates with hair or fur and mammary glands (see Chapter 22)
Order	**Carnivora**—Mammals with teeth specialized for a diet of meat (see Chapter 23)
Family	**Canidae**—The doglike carnivores (excludes hyenas, which are more closely related to cats)
Genus	***Canis***—Made up only of closely related species—coyotes and wolves (also includes domestic dogs)
Species	***latrans***—Consists of similar individuals that in nature can interbreed and produce fertile offspring

Figure 18.12

Classification of organisms based on shared characteristics. All members of the subphylum Vertebrata—fishes, amphibians, reptiles, birds, and mammals—have a segmented vertebral column Subphyla belong to more inclusive phyla (singular, *phylum*), and these in turn belong to kingdoms (Table 18.2). Among vertebrates only warm-blooded animals with hair or fur and mammary glands are mammals. Eighteen orders of mammals exist, including the order Carnivora shown here. The family Canidae includes only doglike carnivores, and the genus *Canis* includes only closely related species. The coyote, *Canis latrans*, stands alone as a species.

lutionary relationships. For instance, wings of insects and birds serve the same function but are dissimilar in both structure and development and are thus termed **analagous structures** (■ Figure 18.14).

Why do dogs have tiny, functionless toes on their back feet or forefeet or, in some cases, on all feet (■ Figure 18.15a)? These dewclaws are **vestigial structures**—that is, nonfunctional remnants of structures that were functional in their ancestors. Ancestral dogs had five toes on each foot, all of which contacted the ground. But as they evolved, they became toe walkers with only four digits on the ground and the big toes and thumbs were lost or reduced to their present state. Almost all species have vestigial structures—functionless toes in pigs and deer, remnants of toes in horses (Figure 18.15b), and whales and snakes with a pelvis but no rear limbs. No one would seriously argue that a dog's dewclaws are functional, nor would anyone convince most people that wisdom teeth are essential in humans (many people never have wisdom teeth, and most of those who do, have problems with them). Cave-dwelling salamanders and fish have vestigial eyes, whereas flightless beetles have totally useless but fully developed wings beneath fused wing covers.

Another type of evidence for evolution is observations of small-scale evolution in living organisms. We have already mentioned one example of small-scale evolution—the adaptations of some plants to contaminated soils. As a matter of fact, small-scale changes take place rapidly enough that new insecticides and pesticides must be developed continually because insects and rodents

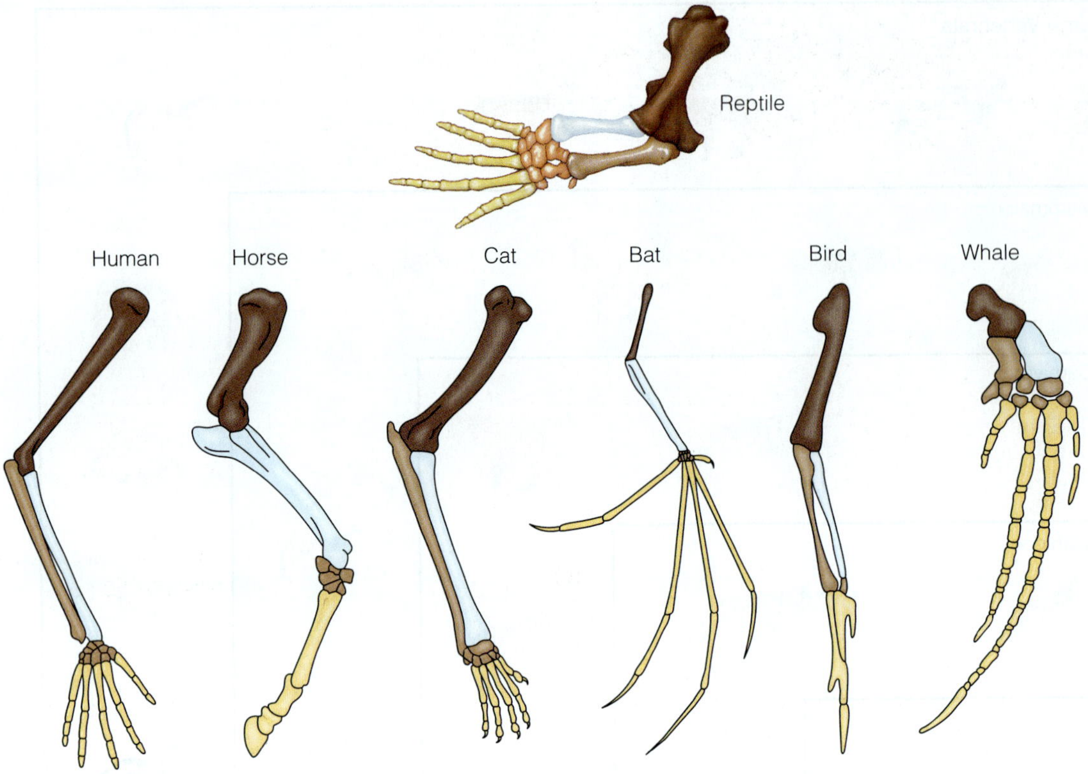

■ **Figure 18.13**

Homologous organs, the forelimbs of various mammals, compared with the forelimb of their ancestors among the reptiles. Homologous organs may serve different functions, but all are composed of the same bones, all have a similar arrangement with respect to other structures, and all undergo a similar embryonic development.

develop resistance to existing ones. And development of antibiotic-resistant strains of bacteria is a continuing problem in medicine. Whether the variation in these populations previously existed or was established by mutations is irrelevant. In either case, some variant types lived and reproduced, bringing about a genetic change.

Fossils: What Do We Learn From Them?

We have briefly discussed **fossils,** the remains or traces of organisms preserved in rocks, and noted that they are useful for determining environments of deposition (see Chapter 6) and are used for relative age determinations

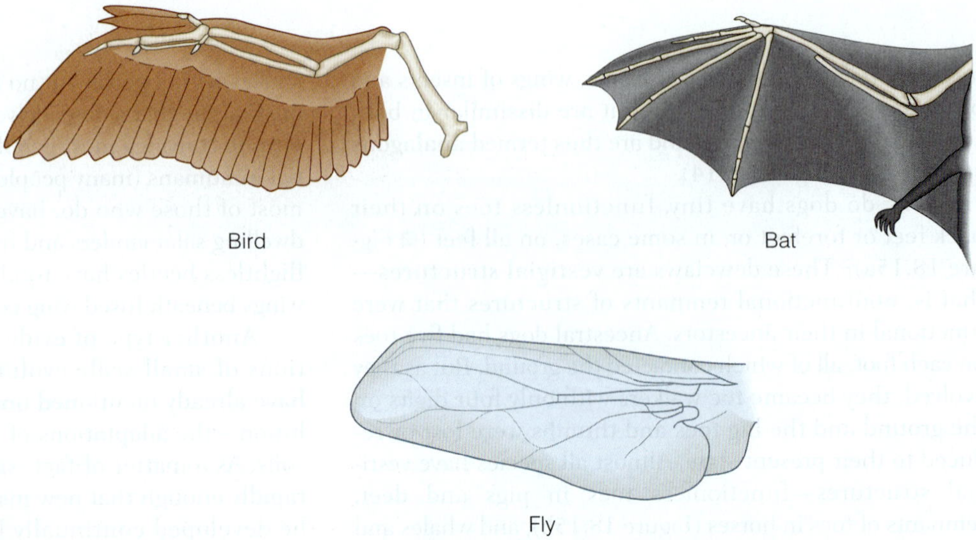

■ **Figure 18.14**

The fly's wings serve the same function as wings of birds and bats, so they are analogous, but they have a different structure and different embryologic development. Are any of these wings both analogous and homologous?

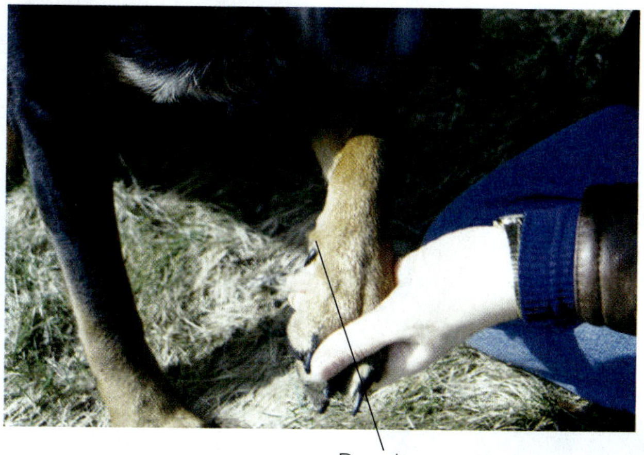

(a)

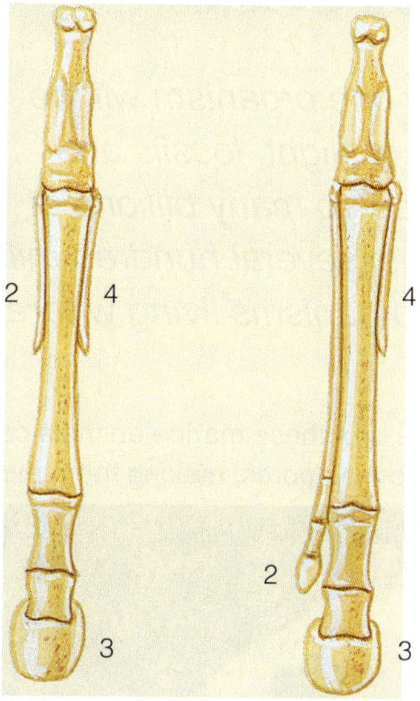

(b)

Figure 18.15

Vestigial structures. (a) Notice the dewclaw, a vestige of the thumb, on the forefoot of this dog. (b) Normal condition of a horse's back foot (left); the only functional toe is the third, but splints, small remnants of toes 2 and 4, remain as vestiges. Occasionally horses are born with one or both of these vestiges enlarged (right).

(see Chapter 17). The term **body fossil** applies to actual remains such as bones, teeth, and shells, whereas a **trace fossil** is an indication of organic activity, such as burrows and tracks. Fossilization takes place in several ways (Table 18.3 and see "Fossilization" on pages 522 and 523), and fossils in general are quite common, although the chances of some organisms being preserved are much better than for others. In fact, the fossil record is strongly biased toward organisms with durable skeletons that live in areas where burial is likely. Accordingly, marine invertebrates such as clams and corals are better represented in the fossil record than bats, birds, worms, and jellyfish.

Fossil marine invertebrates found far from the sea, and even high in mountains, led early naturalists to conclude that the fossil-bearing rocks were deposited during a worldwide flood. In 1508, Leonardo da Vinci realized that the fossil distribution was not what one would expect from a rising flood, but the flood explanation persisted and John Woodward (1665–1728) proposed a testable hypothesis. According to him, the density of organic remains determined the order in which they settled from floodwaters, so, logically, fossils in the oldest rocks should be denser than those in younger ones. Woodward's hypothesis was quickly rejected because observations did not support it; fossils of various densities are found throughout the fossil record.

The fossil record does show a sequence of different organisms but not one based on density, size, shape, or habitat. Rather, the sequence consists of first appearances

Table 18.3

Types of fossil preservation

Body fossils—unaltered remains	Original composition and structure retained
Freezing	Large Ice Age mammals in frozen ground (rare)
Mummification	Air drying and shriveling of soft tissues (rare)
Preservation in amber	Insects in hardened tree resin
Preservation in tar	Bones in asphaltlike substance at oil seeps (rare)
Body fossils—altered remains	Change in composition and/or structure of original
Permineralization	Addition of minerals to pores and cavities
Recrystallization	Change in crystal structure—for example, aragonite recrystallized as calcite
Replacement	One chemical compound replaces another; for example, pyrite replaces calcium carbonate; silicon dioxide replaces wood
Carbonization	Carbon film of leaf, insect, and so on, when volatile elements are lost
Trace fossils	Burrows, tracks, trails, nests, droppings (coprolites), or any other indication of organic activity
Molds and casts	Mold, a cavity having the shape of a bone or shell; cast, a mold filled by minerals or sediment

Fossilization

Although the chance that any one organism will be preserved in the fossil record is slight, fossils are nevertheless common because so many billions of organisms lived during the past several hundred million years. Hard skeletal parts of organisms living where burial was likely are most common.

The bones of this dinosaur and the shells of these marine animals called ammonites have had minerals added to their pores, making them more durable.

Sue Monroe

Sue Monroe

Sue Monroe

Sue Monroe

A tiny amphibian track (left) and a coprolite (fossilized feces, above) from a carnivorous mammal. The coprolite is about 5.5 cm long.

Trace fossils do not include actual remains—only tracks, burrows, nests, and droppings.

Sue Monroe

This object looks like a clam, but it is simply sediment that filled a space formed when a clam shell dissolved. The material filling the space, called a mold, is a cast.

Fossil insect replaced by silicon dioxide (SiO_2).

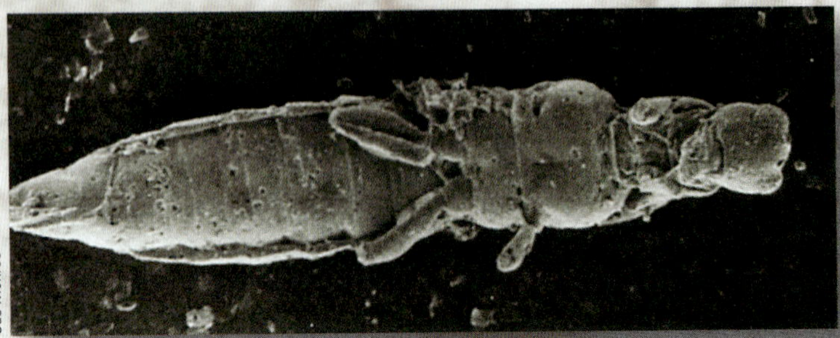

Sue Monroe

Wayne E. Moore

Fossilized tree stump at Florissant Fossil Beds National Monument, Colorado. The woody tissue has been replaced by silicon dioxide. Mudflows 3 to 6 m deep buried the lower parts of many trees at this site.

Sue Monroe

A palm frond and an insect preserved by carbonization.

Sovfoto/V.Khristoforov

This 6- or 7-month old frozen baby mammoth was found in Siberia in 1977. It measures 1.15 m long and 1 m high. Most of its hair has fallen out except around the feet. The carcass is about 40,000 years old.

J. Koivula/Science Source/Photo Researchers, Inc.

Insects preserved in amber, hardened resin secreted by coniferous trees.

Sue Monroe

This insect from the La Brea Tar Pits in Los Angeles is just one of hundreds of species of animals found in the asphaltlike substance in the pits.

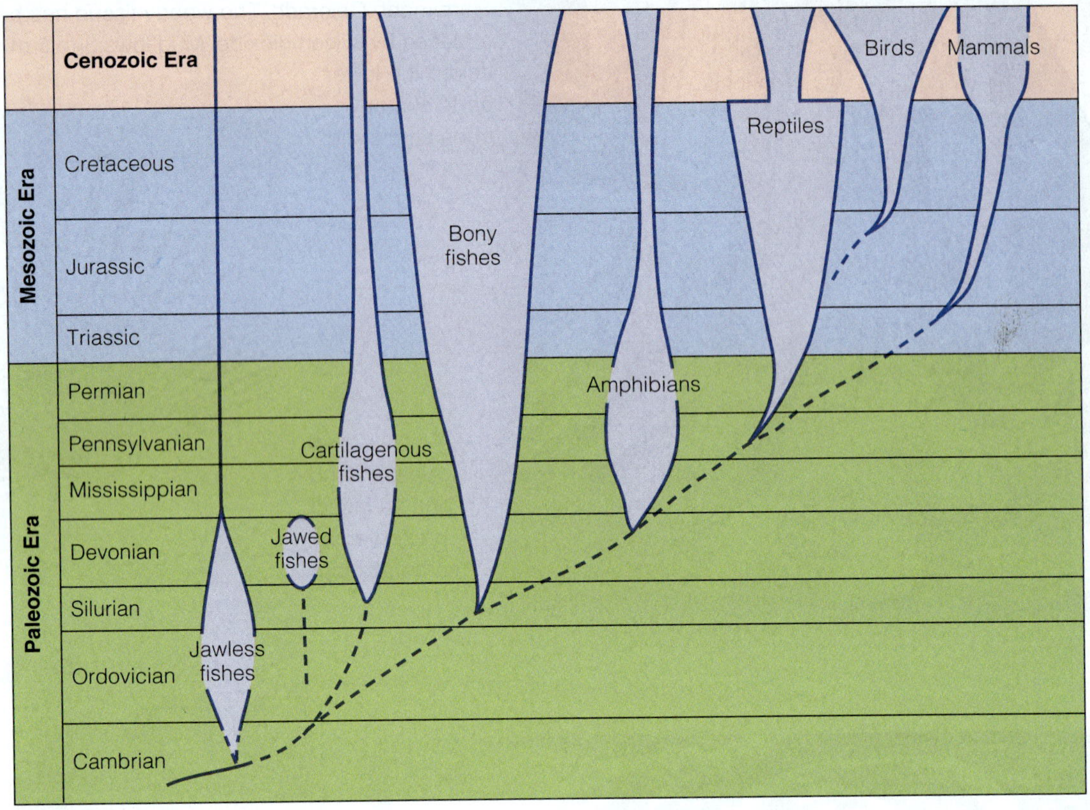

Figure 18.16

The times when the major groups of vertebrate animals appeared in the fossil record are shown by spindles, which also show the relative abundance of members of each group. Note the marked restriction of the reptile spindle at the end of the Mesozoic Era.

Source: Reproduced with permission from R. L. Carroll, *Patterns and Processes of Vertebrate Evolution*, p. 156 (Fig 7.4). 1988 Cambridge University Press.

of various groups of organisms through time (Figure 18.16). One-cell organisms appeared before multicelled ones, plants before animals, and invertebrates before vertebrates. Among vertebrates, fish appear first, followed in succession by amphibians, reptiles, mammals, and birds.

If fossils actually provide evidence to support the theory of evolution, we might ask, Where are the fossils, the so-called missing links, that connect descendants and ancestors? Paleontologists call these "missing links" *transitional fossils* to emphasize the fact that they show characteristics of different groups, and there are in fact many of them. For instance, Jurassic-aged fossils from Germany have anatomic features much like those of small carnivorous dinosaurs (dinosaur-like teeth, a long tail, braincase, and hind limbs), and yet they have feathers and the fused clavicles typical of birds. These creatures, known as *Archaeopteryx,* have precisely the features we would expect in the ancestors of birds (see Chapter 22).

Horses, rhinoceroses, and tapirs may seem an odd assortment of animals, but fossils and studies of living animals indicate that they, along with the extinct titanotheres and chalicotheres, share a common ancestor (Figure 18.17). If this is true, we can predict that as we trace these animals back in the fossil record, differentiating one from the other should become increasingly difficult. And, in fact, the earliest members of each are remarkably similar, differing mostly in size and details of their teeth. As diversification proceeded, though, the differences became more apparent.

Mesozoic mammal-like reptiles so closely resemble mammals that it is difficult to differentiate them from true mammals, and yet they clearly possess reptile characteristics (see Chapter 22). Until only a decade or two ago, the origin of whales was poorly understood. It turns out that the transition from land-dwelling ancestor to aquatic whales took place in southern Asia, a part of the world where the fossil record was poorly known. Now a number of fossils are available that show how and when whales evolved (see Chapter 23).

Little was known about the ancestry of dugongs and manatees, collectively called sea cows—until 1994. Then fossils were found of a Middle Eocene animal known as *Pezosiren portelli.* This sea cow had legs capable of supporting the animal on land, whereas sea cows today have limbs modified into paddles (Figure 18.18).

Pezosiren portelli is truly a transitional fossil, as are others that we will discuss in the following chapters.

Of course, we will never have enough fossils to document the evolutionary history of all living creatures, simply because the remains of some organisms are more likely to be preserved than those of others, and the accumulation of sediment varies in both time and space. Nevertheless, more and more fossils are found that clearly show ancestor–descendant relationships among many organisms or that help clarify evolutionary trends within specific lineages.

What Would You Do?

Suppose someone told you that evolution is "only a theory that has never been proven" and that "the fossil record shows a sequence of organisms in older to younger rocks that was determined by their density and habitat." Why is the first statement irrelevant to theories in general? What kinds of evidence could you cite to refute the second statement?

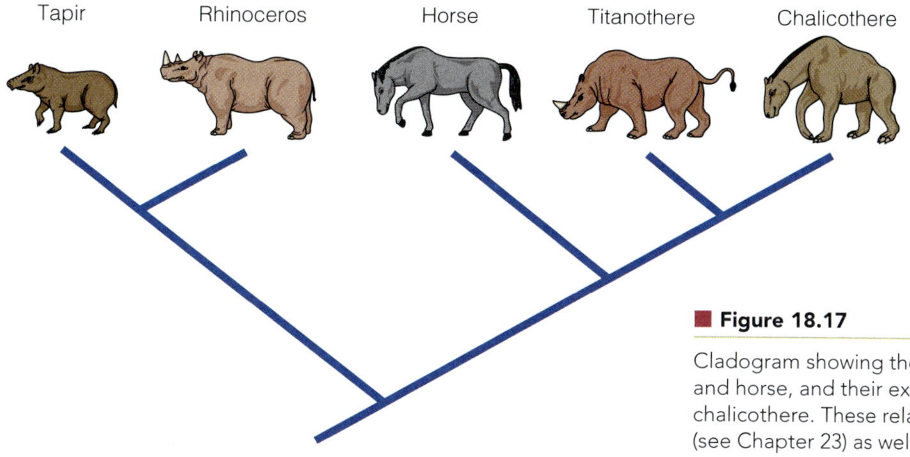

Figure 18.17

Cladogram showing the relationships among the tapir, rhinoceros, and horse, and their extinct relatives the titanothere and chalicothere. These relationships are well documented by fossils (see Chapter 23) as well as by studies of living animals.

Figure 18.18

(a) The Florida manatee or sea cow is a large, fully aquatic mammal living in the coastal waters of Florida and Georgia. (b) Skeleton of *Pezosiren portelli* found in Jamaica, an Eocene sea cow with four well-developed legs.

18 GEO RECAP

Chapter Summary

- Jean-Baptiste de Lamarck proposed the first formal theory of evolution to be taken seriously. Inheritance of acquired characteristics was his mechanism for evolution.
- In 1859, Charles Robert Darwin and Alfred Russel Wallace published their views on evolution and proposed natural selection as the mechanism for evolutionary change.
- Darwin's observations of variation in natural populations and artificial selection, as well as his reading of Thomas Malthus's essay on population, helped him formulate the idea that natural processes select favorable variants for survival.
- Gregor Mendel's breeding experiments with garden peas provided some of the answers regarding how variation is maintained and passed on.
- Genes are the hereditary determinants in all organisms. This genetic information is carried in the chromosomes of cells. Only the genes in the chromosomes of sex cells are inheritable.
- Sexual reproduction and mutations account for most variation in populations.
- Evolution by natural selection is a two-step process. First, variation must be produced and maintained in interbreeding populations, and second, favorable variants must be selected for survival.
- An important way in which new species evolve is by allopatric speciation. When a group is isolated from its parent population, gene flow is restricted or eliminated, and the isolated group is subjected to different selection pressures.
- Divergent evolution involves an ancestral stock giving rise to diverse species. The development of similar adaptive types in different groups of organisms results from parallel and convergent evolution.
- Scientists are increasingly using cladistic analyses to determine relationships among organisms.
- Extinctions take place continually, and times of mass extinctions resulting in marked decreases in Earth's biologic diversity have occurred several times.
- The theory of evolution is truly scientific because we can make observations that would support it or that could falsify it.
- Much of the evidence supporting the theory of evolution comes from classification, embryology, genetics, biochemistry, molecular biology, and present-day small-scale evolution.
- The fossil record also provides evidence for evolution in that it shows a sequence of different groups appearing through time, and some fossils show features we would expect in the ancestors of birds or mammals, and so on.

Important Terms

allele (p. 509)
allopatric speciation (p. 513)
analogous structure (p. 519)
artificial selection (p. 508)
body fossil (p. 521)
chromosome (p. 510)
cladistics (p. 515)
cladogram (p. 514)
convergent evolution (p. 514)
deoxyribonucleic acid (DNA) (p. 510)
divergent evolution (p. 514)
fossil (p. 520)
gene (p. 509)
homologous structure (p. 518)
inheritance of acquired characteristics (p. 507)
mass extinction (p. 517)
meiosis (p. 511)
mitosis (p. 511)
modern synthesis (p. 511)

mosaic evolution (p. 516)
mutation (p. 511)
natural selection (p. 509)
paleontology (p. 506)
parallel evolution (p. 514)
phyletic gradualism (p. 513)
punctuated equilibrium (p. 513)
species (p. 512)
theory of evolution (p. 506)
trace fossil (p. 521)
vestigial structure (p. 519)

Review Questions

1. Allopatric speciation involves a new species arising
a._____in less than 1000 years; b._____by interbreeding with another species; c._____from a small, isolated population; d._____by inheriting acquired characteristics; e._____when distantly related organisms adapt to comparable environments.
2. In Linnaeus's classification, a family consists of two or more related
a._____genera; b._____classes; c._____phyla; d._____orders; e._____clades.
3. When animals produce sex cells, they do so by a cell division process called
a._____abiotic synthesis; b._____allopatric speciation; c._____selection; d._____meiosis; e._____cross-fertilization.
4. The wings of butterflies and birds are _____structures.
a._____complementary; b._____analogous; c._____divergent; d._____punctuated; e._____allopatric.
5. According to the concept of punctuated equilibrium,
a._____species remain unchanged during most of their existence; b._____organisms from different phyla interbreed; c._____mitosis yields sex cells with half the chromosomes typical of a species; d._____natural selection favors the largest for survival; e._____acquired traits are inheritable.
6. Charles Darwin and _____proposed the theory of natural selection.
a._____William Smith; b._____Jean-Baptiste de Lemarck; c._____Alfred Wallace; d._____Charles Lyell; e._____Erasmus Darwin.
7. The diversification of a species into a number of descendant species is known as
a._____the modern synthesis; b._____divergent evolution; c._____parallel convergence; d._____morphologic selection; e._____punctuated equilibrium.
8. Chromosomes are complex helical molecules of
a._____potassium, aluminum, and silicon; b._____calcium carbonate; c._____homologous genes; d._____deoxyribonucleic acid; e._____bacteria and viruses.
9. One prediction of evolutionary theory is
a._____species are fixed and immutable; b._____mammals should appear in the fossil record after reptiles; c._____features acquired during an organism's lifetime are inheritable; d._____mitosis results in the production of sex cells; e._____humans evolved from monkeys.
10. Convergent evolution accounts for the fact that
a._____most species evolve in a few thousands of years; b._____the marine invertebrate fossil record is better than it is for birds; c._____phyletic gradualism takes place in small populations; d._____many Australian marsupials resemble placental mammals elsewhere; e._____cladistics is useful for analyzing evolutionary relationships.
11. What kinds of evidence should we find in the fossil record if the theory of evolution is correct?
12. Survival of the fittest implies that only the biggest and strongest live and reproduce. Is this really what natural selection is all about?
13. Compare the concepts of phyletic gradualism and punctuated equilibrium. Is there any evidence for either view? If so, what is it?
14. A cladogram shows that among carnivorous mammals, cats, hyenas, and mongooses are closely related. Yet hyenas look rather like dogs, whereas mongooses resemble weasels. What kinds of evidence from living animals and from fossils would lend credence to the idea that these animals constitute a closely related group?

15. Discuss the concept of allopatric speciation, and give two examples of how it might take place.
16. Draw three cladograms showing possible relationships among sharks, whales, and bears. Which cladogram best depicts the relationships among these animals, and what criteria did you use to make your decision?
17. How did the experiments carried out by Gregor Mendel help answer the questions that plagued Darwin and Wallace?
18. When we speak of classification, what do we mean by a nested pattern of similarities?
19. What is the difference(s) between analogous and homologous structures? Which ones provide evidence for evolution?
20. The theory of inheritance of acquired characteristics seems logical. So why doesn't it work?

World Wide Web Activities

Geology Now Assess your understanding of this chapter's topics with additional quizzing and comprehensive interactivities at

http://earthscience.brookscole.com/changingearth4e

as well as current and up-to-date weblinks, additional readings, and InfoTrac College Edition exercises.

Precambrian Earth and Life History

CHAPTER 19

OUTLINE

- Introduction
- What Happened During the Hadean?
- Continental Foundations—Shields, Platforms, and Cratons
- Archean Earth History
- Proterozoic Earth History

GEOLOGY IN UNEXPECTED PLACES: *Little Rock, Big Story*

- Origin and Evolution of the Atmosphere and Hydrosphere
- Life—Its Origin and Early History
- What Kinds of Resources Are Found in Precambrian Rocks?

GEO-FOCUS 19.1: *Banded Iron Formation—From Mine to Steel Mill*

- Geo-Recap

Geology Now *This icon, which appears throughout the book, indicates an opportunity to explore interactive tutorials, animations, or practice problems available on the GeologyNow website at http://earthscience.brookscole.com/changingearth4e.*

The Teton Range in Grand Teton National Park, Wyoming, is made up largely of Archean gneiss, schist, and granite. Younger rocks are also present in the range, but they are not visible in this view. This majestic mountain range resulted from uplift along a normal fault beginning about 10 million years ago. Erosion, especially by valley glaciers, accounts for the present rugged topography of the range.

Source: James S. Monroe

OBJECTIVES

At the end of this chapter, you will have learned that:

- The Precambrian encompasses more than 88% of all geologic time, yet we know less about its history than we do about more recent intervals of geologic time.
- During the earliest part of the Precambrian, called the Hadean, Earth differentiated and crust began to form.
- The Archean Eon was a time during which several small continental nuclei began to evolve, and the most common rocks were granite-gneiss complexes and greenstone belts.
- During the Proterozoic Eon, an essentially modern style of plate tectonics developed, and crust that formed during the Archean amalgamated into a large craton we now call Laurentia.
- Proterozoic rocks consist of sandstone-carbonate-shale assemblages, red beds, banded iron formations, glacial deposits, and a variety of others.
- The Precambrian atmosphere evolved from one that lacked free oxygen and an ozone layer, but even at the end of the Precambrian the atmosphere had much less free oxygen than it does now.
- Once Earth had cooled sufficiently, water vapor fell as precipitation, accounting for the origin of surface waters.
- During the Archean and Early Proterozoic, only single-celled bacteria existed.
- During the Proterozoic, organisms that reproduced sexually evolved and multicelled organisms appeared.
- Several natural resources are found in Precambrian rocks.

Introduction

Imagine a lifeless, waterless, hot planet with a noxious atmosphere. Storms form continually in the turbulent atmosphere, lightning flashes most of the time, but no rain falls. Even if organic matter were present, it would not burn because there is no free oxygen, and yet a continuous red glow comes from pools and streams of lava. Assuming we could somehow visit this hostile planet, we would have to be protected from high temperatures, intense ultraviolet radiation (no ozone layer existed), and meteorite and comet impacts. This description might seem like something from science fiction, but it is probably reasonably accurate for Earth shortly after it formed (■ Figure 19.1).

Although Earth's surface was barren at this time, our view of the Moon would have been spectacular because it was closer, and we would have experienced a much shorter day as Earth rotated in as little as 10 hours. This earliest episode in Earth history is known by the informal name *Precambrian*, which refers to all rocks beneath Cambrian-aged strata and to all time before the Cambrian Period. The Precambrian encompasses just over 88% of all geologic time, so if Earth's entire history were represented by a 24-hour clock, slightly more than 21 hours of it would be Precambrian (■ Figure 19.2). Unfortunately for geologists, no rocks are known for the first 640 million years of Earth history, so our narrative above is based only on our knowledge of the origin and evolution of terrestrial planets.

Geologists subdivide the Precambrian into two eons: the *Archean Eon* (4.0 to 2.5 billion years ago) and the *Proterozoic Eon* (2.5 billion to 545 million years ago) (Table 19.1). The interval from Earth's origin 4.6 billion years ago until the onset of the Archean is informally known as the *Hadean*. The Archean and Proterozoic are based solely on absolute ages, so they are geochronologic units rather than time-stratigraphic units. This departs from the standard practice in which systems based on stratotypes are the basic time-stratigraphic units (see Chapter 17); there are no Archean or Proterozoic stratotypes. There are, however,

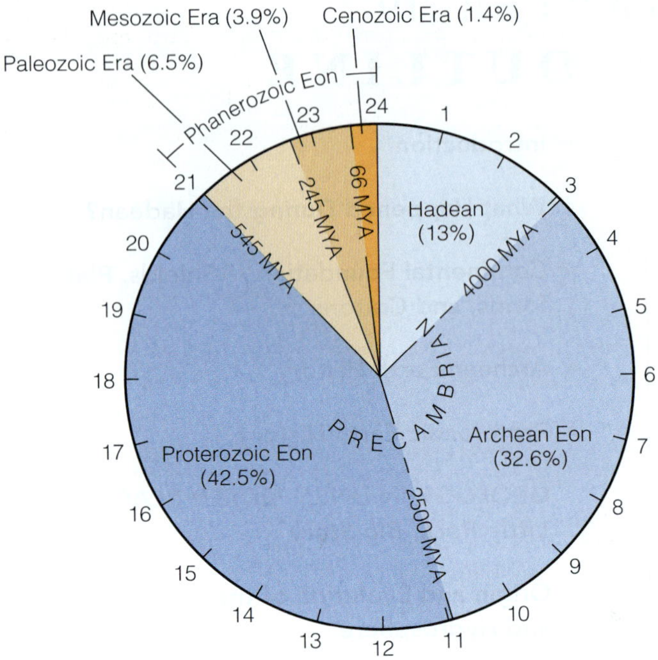

■ **Figure 19.2**

If all geologic time were represented by a 24-hour clock, the Precambrian would be more than 21 hours long and constitute more than 88% of all geologic time.

■ **Figure 19.1**

Earth as it probably appeared during the Hadean.

Archean- and Proterozoic-aged rocks known from many areas, but there are no rocks on Earth of Hadean age (except meteorites), so this term designates only a specific interval of geologic time.

Why should you study Precambrian Earth and life history? One reason is that it constitutes most of geologic time, and many events that occurred during the Precambrian set the stage for further evolution of the planet and life on it. In Chapter 1, we introduced the concept of *systems* and gave examples of systems' interactions when we discussed volcanism, plate tectonics, and various surface processes such as running water, glaciers, and wind. It was during the Precambrian that Earth's systems became operative, although not all at the same time or necessarily in their present form.

Earth did not differentiate into a core, mantle, and crust until millions of years after it initially formed. Once it did, though, internal heat was responsible for moving plates (plate tectonics); the crust began to evolve and it continues to do so. Also, early Earth's atmosphere evolved from a noxious, carbon dioxide–rich one to one with free oxygen and an ozone layer. Organisms appeared as long as 3.5 billion years ago, and they had a profound effect on changes in the atmosphere. Impacting meteorites and comets and gases derived from within Earth were responsible for the origin of the hydrosphere.

In short, when Earth first formed, it was vastly different than it is now, but during the Precambrian, it began to evolve, gradually becoming more like it is today. Evidence of these changes is found in the **geologic record**—that is, the record of prehistoric physical and biological events preserved in rocks. Our task in this and the following chapters is to investigate the geologic records for the intervals of geologic time designated Precambrian (Archean and Proterozoic), Paleozoic, Mesozoic, and Cenozoic.

WHAT HAPPENED DURING THE HADEAN?

Remember that *Hadean* is an informal term designating the first 640 million years of Earth history (Table 19.1). It was certainly an important time in the evolution of Earth, but one for which we have no geologic record. Indeed, the only rocks known of Hadean age are meteorites and some Moon rocks. Nevertheless, this was the time during which Earth accreted from planetesimals and later differentiated into a core and mantle. Recall from Chapter 1 that Earth and the other terrestrial planets had a similar early history involving accretion followed

Table 19.1

Terminology for Precambrian Rocks and Time in the United States

Time (millions of years ago)	U.S. Geological Survey, 1971		North American Commission on Stratigraphic Nomenclature, 1982*
500	Phanerozoic		Phanerozoic
	Precambrian Z	Proterozoic	Late Proterozoic
	800	Proterozoic	
1000		Proterozoic	900
	Precambrian Y	Proterozoic	Middle Proterozoic
1500	1600	Proterozoic	1600
2000	Precambrian X	Proterozoic	Early Proterozoic
2500	2500		2500
		Archean	Late Archean
3000		Archean	3000
	Precambrian W	Archean	Middle Archean
		Archean	3400
3500		Archean	Early Archean
4000			4000
4500			Hadean

*This scheme, which was proposed by Harrison and Peterman, is followed in this book.

by volcanism and an episode of intense meteorite impacts. But what do we know about Hadean crust?

Surely some crust existed more than 3.8 billion years ago because rocks that old are known from several areas, including Minnesota, Greenland, and South Africa. And many of these rocks are metamorphic, meaning that they formed from even older rocks. In fact, the Acasta Gneiss in Canada dates from 3.96 billion years ago, and being metamorphic, it proves that even older rocks were present. Furthermore, sedimentary rocks in Australia contain detrital minerals known as zircon ($ZrSiO_4$) dated at 4.2 billion years old, so older source rocks must have existed.

We can be certain that some crust was present by the latter part of the Hadean, but we can only speculate on what existed before that. For example, what were the composition and structure of the earliest crust, was it worldwide or more restricted, and were both oceanic and continental crust present?

Many geologists think that the earliest crust formed during an episode of partial melting that yielded magma that rose to form a thin, unstable crust of ultramafic rock—that is, igneous rock with less than 45% silica (see Chapter 4). Then followed a time of partial melting of ultramafic crust that yielded mafic magma (45–52% silica) that disrupted the ultramafic crust; because of its greater density, ultramafic crust was consumed at subduction zones—only more silica-rich crust is resistant to subduction.

Following this earliest episode of crust formation, a second stage of crustal evolution began. Once again partial melting took place, but this time it was mafic crust that partially melted, thus forming intermediate-composition magma (52–65% silica) and even felsic magma (>65% silica). These intermediate and felsic magmas were incorporated into developing volcanic island arcs that continued to evolve, acting as the nuclei for the accretion of true continents, a process that continued throughout the rest of the Precambrian and, in fact, continues even now.

Geologists are certain that early Earth possessed more residual heat from its origin and, more important, more heat generated by radioactive decay. As a result, Earth's internal heat was responsible for more rapid plate movements and for collisions of the evolving island arcs and protocontinents. Geologists are also certain that by the beginning of the Archean Eon, several granitic continental nuclei were present (■ Figure 19.3).

These earliest continents were made up mostly of volcanic and plutonic rocks, but once formed they were undoubtedly weathered and eroded, thus yielding sediment for sedimentary rocks. And, of course, continuing plate collisions and the heat from plutons resulted in metamorphism and metamorphic rocks (see the rock cycle in Figure 1.12). Unfortunately, so little data are available that we do not know the geographic extent of the early continents, nor do we know their positions with respect to one another. We can be sure, however, that they were barren and uninhabitable by Earth's present-day organisms, with the possible exception of bacteria.

CONTINENTAL FOUNDATIONS—SHIELDS, PLATFORMS, AND CRATONS

Continents are more than simply land areas above sea level. Indeed, they consist of rocks with an overall composition similar to granite, and continental crust is thicker and less dense than oceanic crust, which is made up of basalt and gabbro. Furthermore, a **shield** consisting of a vast area or areas of exposed ancient rocks is found on all continents. Continuing outward from the shields are broad **platforms** of buried Precambrian rocks that underlie much of each continent. Collectively, a shield and platform make up a **craton,** which we can think of as a continent's ancient nucleus.

The cratons are the foundations of continents, and along their margins more continental crust was added, a process called **continental accretion,** as they evolved to their present sizes and shapes. Both Archean and Proterozoic rocks are present in cratons, many of which indicate several episodes of deformation accompanied by metamorphism, igneous activity, and mountain building. However, the cratons have experienced remarkably little deformation since the Precambrian.

In North America, the exposed part of the craton is the **Canadian shield,** which occupies most of northeastern Canada, a large part of Greenland, the Adirondack Mountains of New York, and parts of the Lake Superior region in Minnesota, Wisconsin, and Michigan (■ Figure 19.4). Overall, the Canadian shield is an area of subdued topography, numerous lakes, and exposed Archean and Proterozoic rocks thinly covered in places by Pleistocene glacial deposits. The rocks themselves are plutonic, volcanic, and sedimentary, and metamorphic equivalents of all of these (■ Figure 19.5a).

Actually, the Canadian shield as well as the adjacent platform are made up of numerous units or smaller cratons that amalgamated along deformation belts during the Early Proterozoic. Absolute ages and structural trends differentiate these various smaller cratons from one another.

Drilling and geophysical evidence indicate that Precambrian rocks underlie much of North America, but beyond the Canadian shield they are exposed only in areas of deep erosion or uplift. For instance, Archean and Proterozoic rocks are present in the deeper parts of the Grand Canyon of Arizona as well as in the Appalachian Mountains and the Rocky Mountains (Figure 19.5b, c).

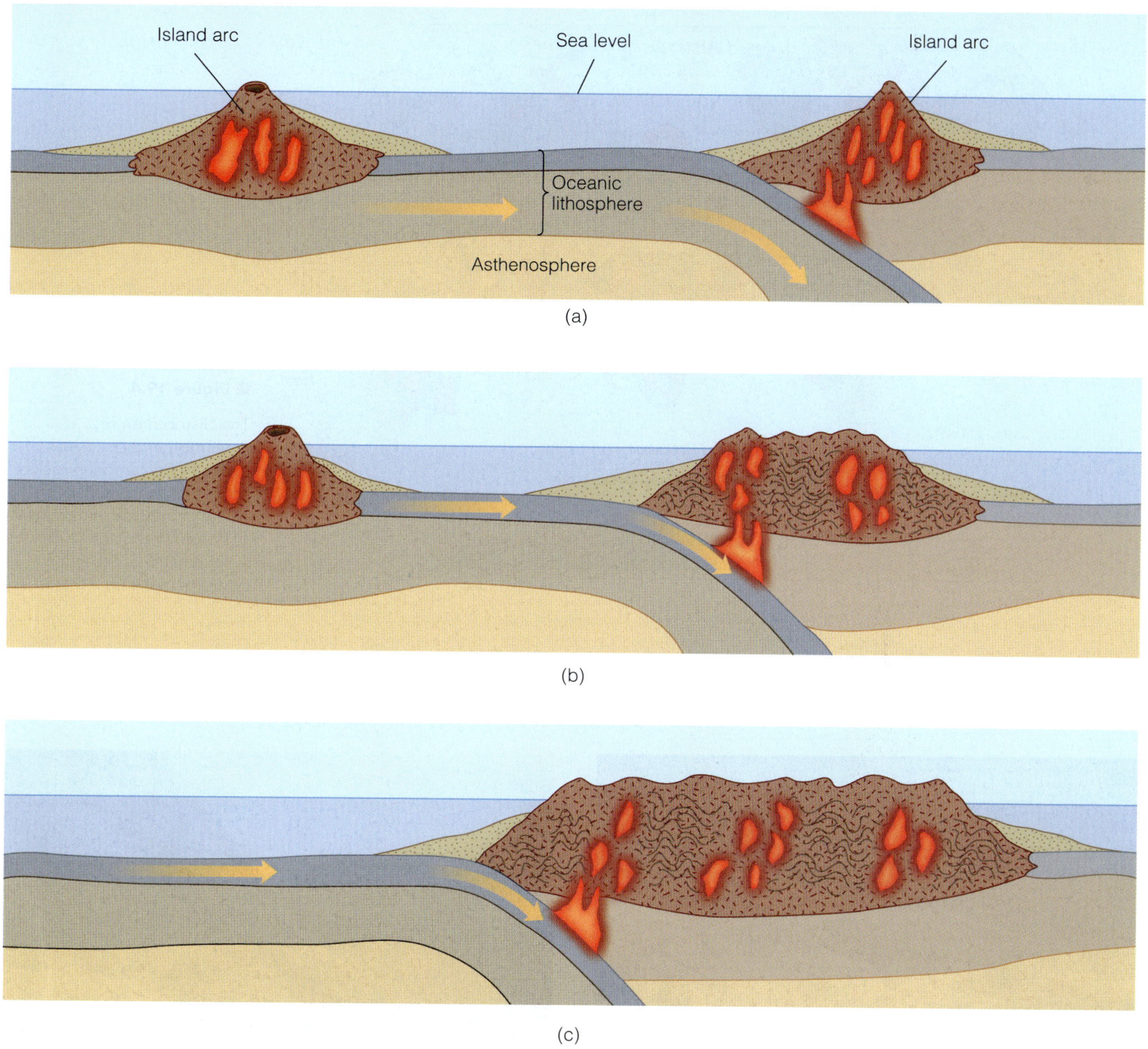

Geology Now ■ **Active Figure 19.3**

Three stages in the origin of granitic continental crust. Andesitic island arcs formed by the partial melting of basaltic oceanic crust are intruded by granitic magmas. As a result of plate movements, island arcs collide and form larger units or cratons. (a) Two island arcs on separate plates move toward each other. (b) The island arcs shown in (a) collide, forming a small craton, and another island arc approaches this craton. (c) The island arc shown in (b) collides with the craton.

ARCHEAN EARTH HISTORY

Geologists place the beginning of the Archean Eon at 4.0 billion years ago (Table 19.1), approximating the age of the oldest known rocks on Earth other than meteorites. Remember that the Acasta Gneiss in Canada is dated at 3.96 billion years old. The end of the Archean Eon, and the beginning of the Proterozoic Eon, at 2.5 billion years ago is arbitrary, but it does correspond to a time when the style of crustal evolution changed and when rock assemblages more like those of the present appeared in abundance.

The Archean Eon alone accounts for 32.6% of all geologic time, yet we review both its physical and biological history in just a few pages. One should not assume that it was an unimportant time in Earth history, though. The geologic record is more complete for more recent intervals of geologic time, and because Earth is so active, ancient rocks are more likely to have been eroded or changed by metamorphism. In short, older intervals of geologic time are represented by smaller volumes of rock,

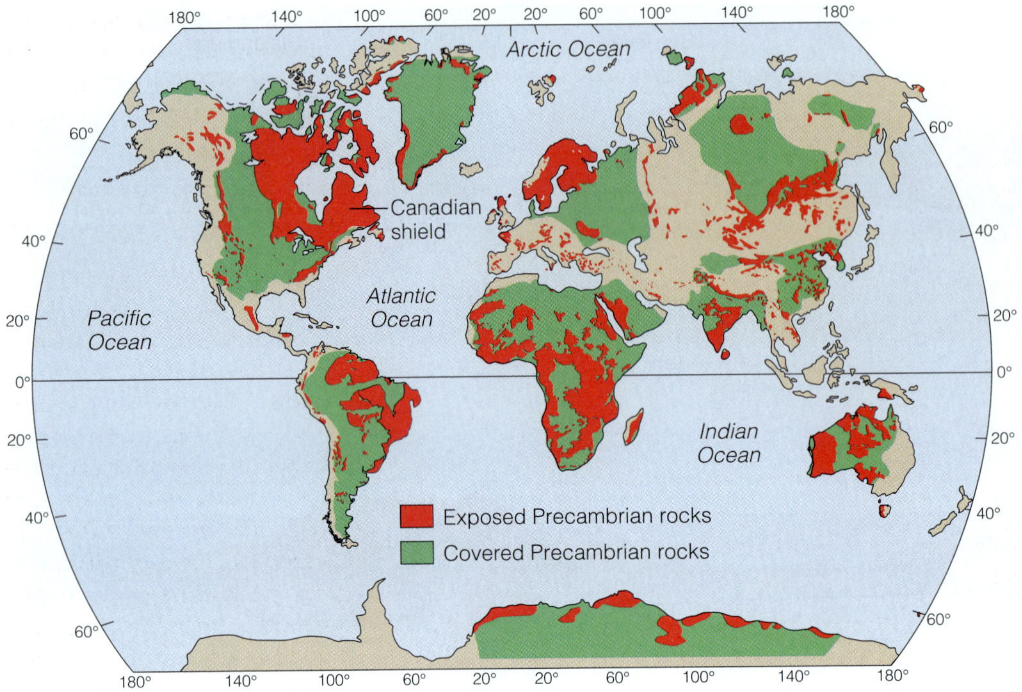

Figure 19.4

The distribution of Precambrian rocks. Areas of exposed Precambrian rocks constitute the shields, whereas the platforms consist of buried Precambrian rocks. A shield and its adjoining platform make up a craton.

James S. Monroe

(a)

James S. Monroe

(b)

Jeff Mayo/GeoPhoto Publishing

(c)

Figure 19.5

(a) Precambrian rocks of the Canadian shield. Basalt (dark) and granite (light) along the banks of the Chippewa River in Ontario, Canada. Beyond the Canadian shield, Precambrian rocks are found in areas of uplift, as in the Rocky Mountains in Colorado (b), and in areas of deep erosion, as in the Grand Canyon of Arizona (c).

Table 19.2

Chronologic Summary of Events Important in the Archean Development of Cratons*

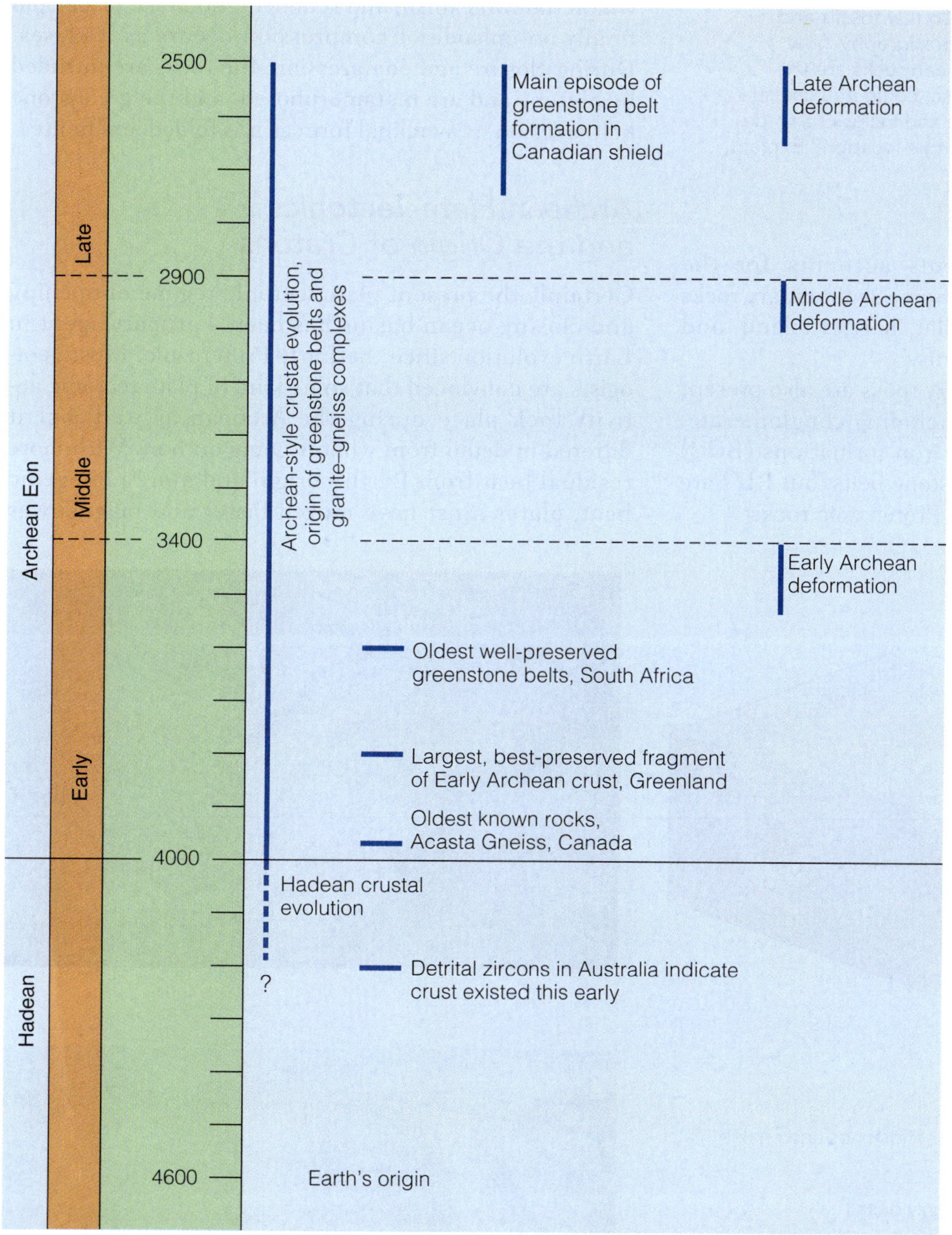

*Ages are in thousands of millions of years

especially sedimentary rock, and even if preserved, they are more difficult to find.

Archean Rocks

Archean rocks are found in many places. In North America, large parts of the Canadian shield consist of Archean rocks, and they are present in the Appalachian and Rocky Mountains (see the chapter opening photo) as well as in some areas of deep erosion such as the Grand Canyon (Figure 19.5b).

By far the most common Archean rocks are granite and gneiss, the latter being high-grade metamorphic rocks (Table 19.2). Although fewer have been found, rock successions known as **greenstone belts** are fairly common and tell something of Archean tectonic history. An ideal greenstone belt has three major rock units; the lower and middle units are mostly volcanic rocks, and the uppermost unit is made up chiefly of sedimentary rocks. The origin of the mineral chlorite during low-grade metamorphism gives these rocks a greenish color—hence the name *greenstone*. Most greenstone belts have a syncline-like structure and have been intruded by magma that cooled to form granitic plutons (■ Figure 19.6). Greenstone belts are also commonly complexly folded and cut by thrust faults.

Much of the volcanism responsible for the volcanic units in greenstone belts must have been subaqueous because pillow lavas are common, but some large volcanic centers built above sea level. Some of the most interesting volcanic rocks are ultramafic lava flows. In Chapter 4, we noted that eruptions producing these flows are rare in rocks younger than Archean because Earth's radiogenic heat production has decreased and near-surface temperatures are not high enough for ultramafic magma to reach the surface.

Sedimentary rocks are a minor component in the lower parts of greenstone belts, but they become increasingly abundant toward the top (Figure 19.6). Associations of graywacke, a sandstone with abundant clay and volcanic rock fragments, and argillite (slightly metamorphosed mudrock) are particularly common. Small-scale graded bedding and cross-bedding indicate that

What Would You Do?

Because Precambrian rocks have so few fossils and those present are of little use in stratigraphy, how would you demonstrate that Archean rocks are the same age as those elsewhere? Also, could you use superposition, original horizontality, and inclusions to decipher the geologic history at a single location? Explain.

deposition by turbidity currents accounts for the graywacke-argillite successions. Some sedimentary rocks were deposited in delta, tidal flat, barrier island, and shallow marine shelf environments.

A variety of other sedimentary rocks are also present in Archean greenstone belts, including conglomerate, carbonates, and chert. Banded iron formations (BIFs) are also found in Archean greenstone belts, but BIFs are much more common in areas of Proterozoic rocks.

According to one idea, greenstone belts developed in *back-arc marginal basins* that first opened and then closed (see Figure 10.21). An early stage of extension occurs, when the back-arc marginal basin opens, during which time volcanism and sedimentation take place, and finally an episode of compression occurs as it closes. During closure and compression, the rocks are intruded by plutons and are metamorphosed, and the greenstone belt takes on its synclinal form as it is folded and faulted.

Archean Plate Tectonics and the Origin of Cratons

Certainly the present plate tectonic regime of opening and closing ocean basins has been a primary agent in Earth evolution since the Early Proterozoic. Most geologists are convinced that some kind of plate tectonic activity took place during the Archean as well, but it differed in detail from what is going on now. With more residual heat from Earth's origin and more radiogenic heat, plates must have moved faster and magma was

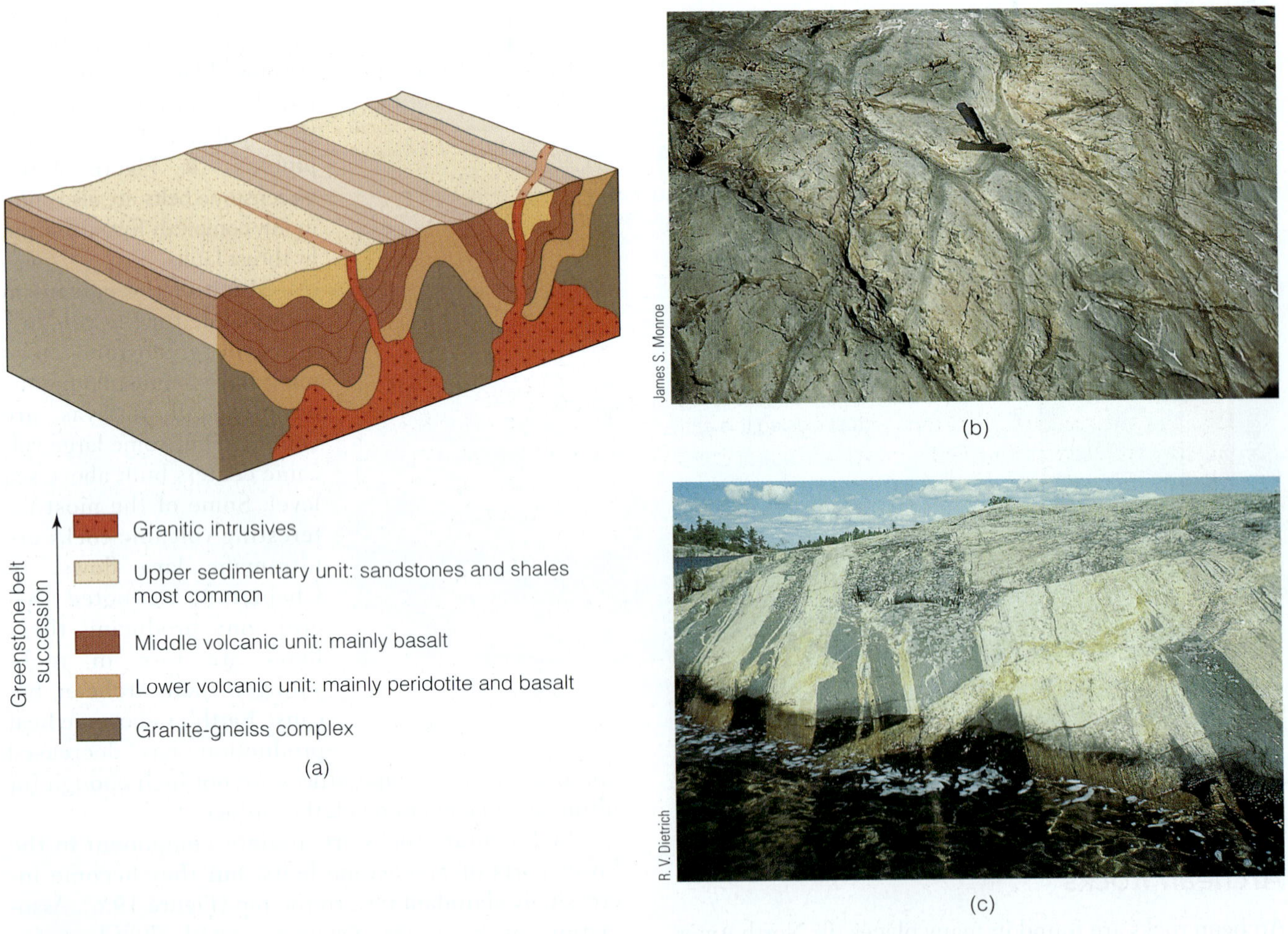

Figure 19.6

(a) Two adjacent greenstone belts. Older belts—those more than 2.8 billion years old—have an ultramafic unit overlain by a basaltic unit. In younger belts, the succession is from a basaltic lower unit to an andesite-rhyolite unit. In either case, the upper unit is mostly sedimentary rocks. (b) Pillow lava of the Ispheming greenstone belt in Michigan. (c) Gneiss from a granite-gneiss complex in Ontario, Canada.

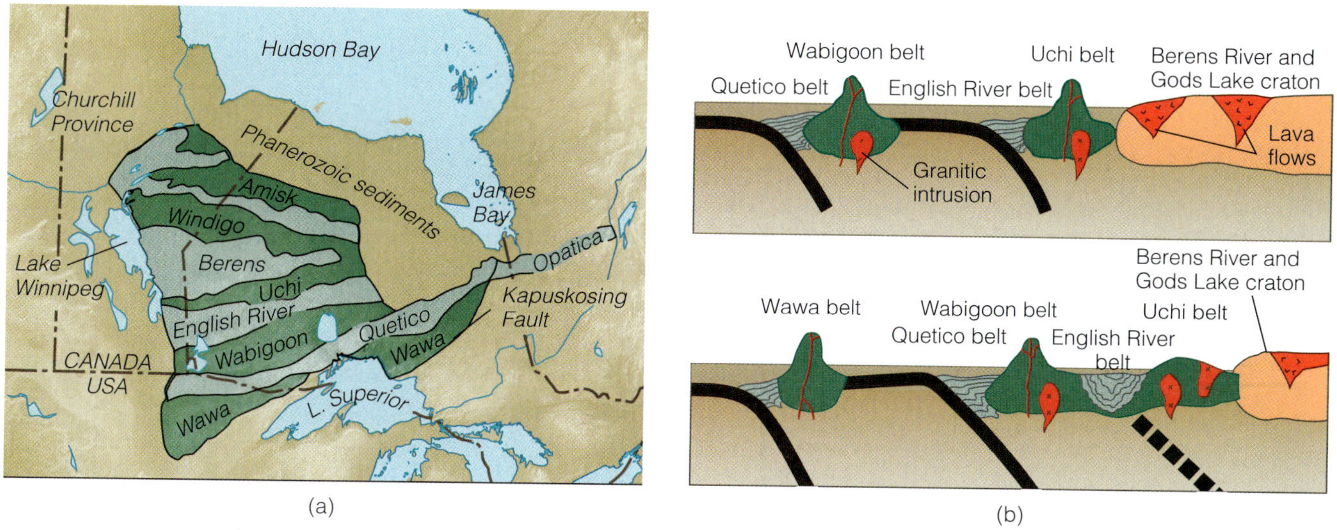

■ **Figure 19.7**

Origin of the southern Superior craton. (a) Geologic map showing greenstone belts (dark green areas) and granite gneiss subprovinces (light green areas). (b) Plate tectonic model for the development of the southern Superior craton. The figure represents a north-south cross section, and the upper diagram is an earlier stage of the lower diagram. Source: K.C. Condie, *Plate Tectonics and Crustal Evolution,* 2ed 1982, p. 218 (fig. 10.2). Reprinted with permission from the author.

generated more rapidly. As a result, continents no doubt grew more rapidly along their margins as plates collided with island arcs and other plates.

There were, however, marked differences between the Archean world and the one that followed. We have little evidence of Archean rocks deposited on broad, passive continental margins, but they become widespread in Proterozoic terrains. Deformation belts between colliding cratons indicate that Archean plate tectonics was active, but the *ophiolites* so typical of younger convergent plate boundaries are rare, although Late Archean-aged ophiolites are now known from several areas.

Certainly several small cratons existed by the beginning of the Archean and grew by continental accretion periodically during the rest of that eon, and they amalgamated into a larger unit during the Early Proterozoic. By the end of the Archean, perhaps 30% to 40% of the present volume of continental crust existed. A plate tectonic model for the Archean crustal evolution of the southern Superior craton of Canada relies on a series of events, including greenstone belt evolution, plutonism, and deformation (■ Figure 19.7). We can take this as a provisional model for Archean crustal evolution in general.

The events leading to the origin of the southern Superior craton (Figure 19.7) are part of a more extensive orogenic episode that took place near the end of the Archean. Deformation at this time was responsible for the formation of the Superior and Slave cratons and the origin of some Archean rocks in other parts of the Canadian shield as well as in Wyoming, Montana, and the Mississippi River Valley. By the time it was over, several sizable cratons existed that are now found in the older parts of the Canadian shield.

PROTEROZOIC EARTH HISTORY

The basic difference between Archean and Proterozoic Earth history is the style of crustal evolution. During the Archean, crust-forming processes generated greenstone belts and granite-gneiss complexes, that continued into the Proterozoic but at a considerably reduced rate. Most Archean rocks were altered by metamorphism, but many Proterozoic rocks have been little altered. And finally, widespread assemblages of sedimentary rocks deposited on passive continental margins are common in the Proterozoic but rare in the Archean.

We noted that Archean cratons assembled through a series of island-arc and minicontinent collisions (Figure 19.7). These provided the nuclei around which Proterozoic continental crust accreted, thereby forming much larger cratons. One large landmass so formed, called **Laurentia,** consisted mostly of North America and Greenland, parts of northwestern Scotland, and perhaps parts of the Baltic shield of Scandinavia. We emphasize the geologic evolution of Laurentia in the following sections (Table 19.3).

Proterozoic Evolution of Laurentia

The first major episode of the Proterozoic evolution of Laurentia took place during the Early Proterozoic, between 2.0 and 1.8 billion years ago (BYA). Several major **orogens** developed—zones of deformed rocks, many of

Table 19.3

Summary of the Proterozoic Geologic and Biologic Events Discussed in the Text

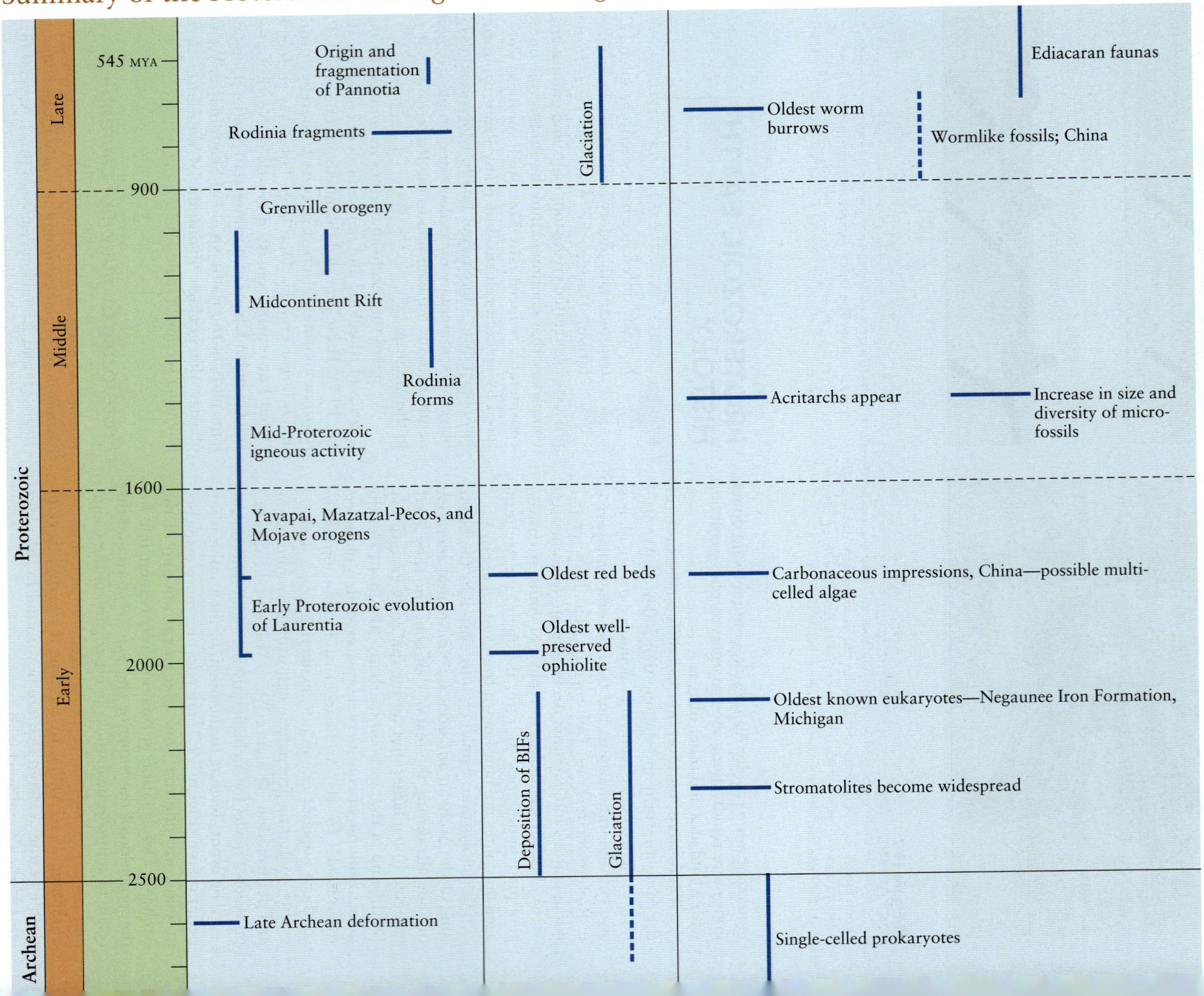

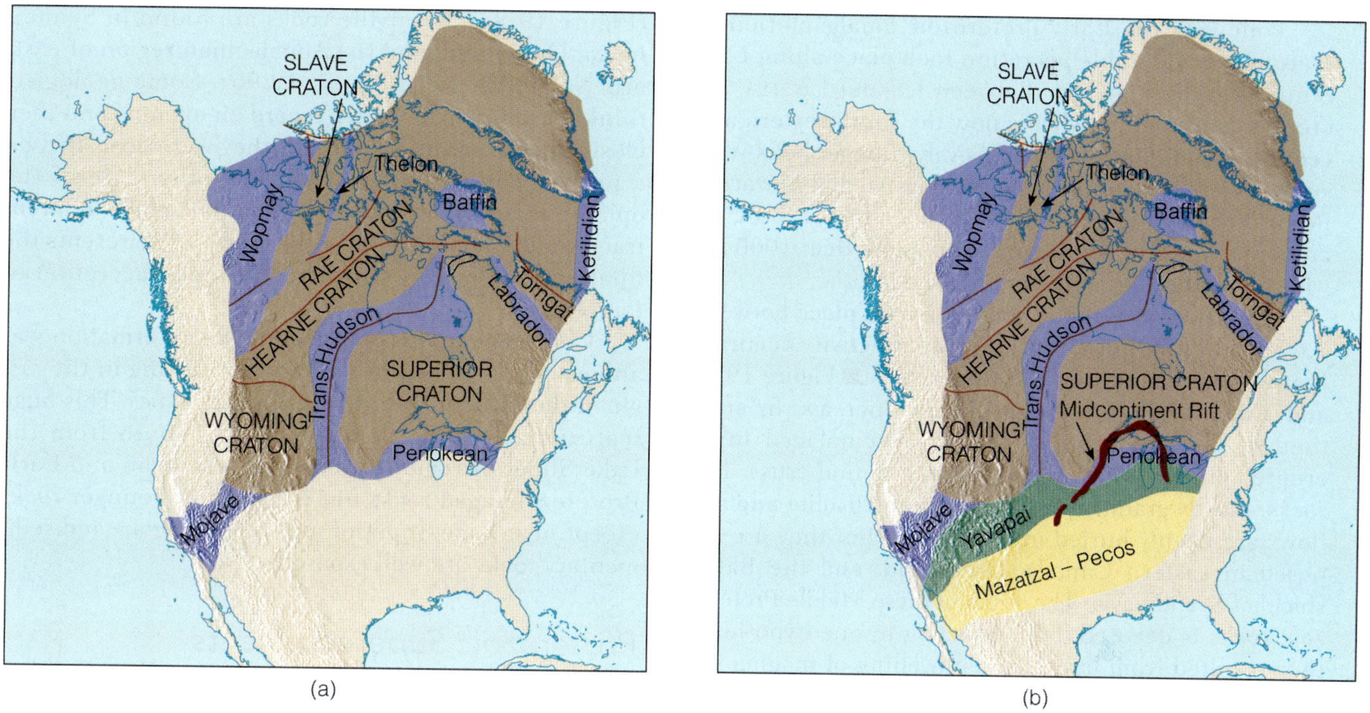

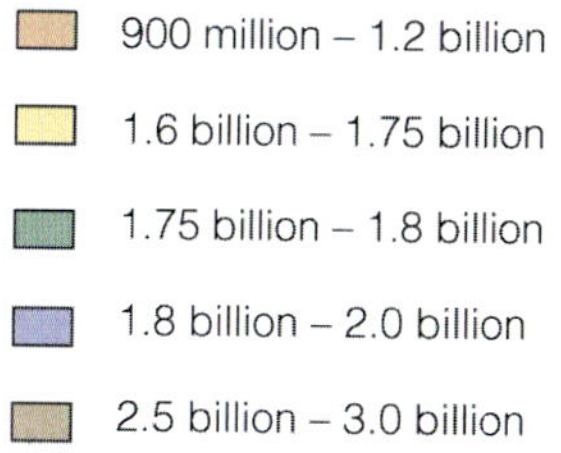

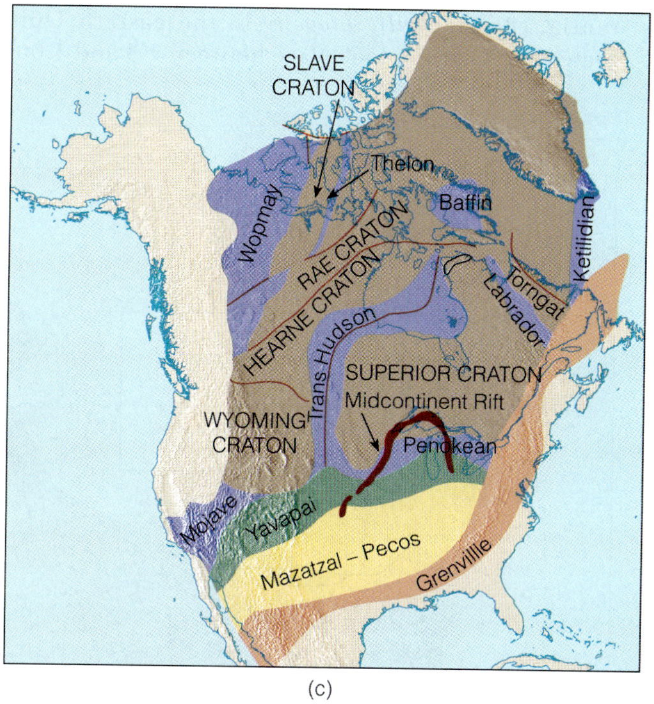

Figure 19.8

Proterozoic evolution of Laurentia. (a) During the Early Proterozoic, Archean cratons were sutured along deformation belts called *orogens.* (b) Laurentia grew along its southern margin by accretion of the Central Plains, Yavapai, and Mazatzal orogens. (c) A final episode of Protoerozoic accretion occurred during the Grenville orogeny. Source: Reprinted from K. C. Condie, *Plate Tectonics and Crustal Evolution*, 4th edition, p. 65 (Fig. 2.26), copyright © 1997 Butterworth-Heinemann. Reprinted with permission from Elsevier.

which were metamorphosed and intruded by plutons. So this was a time of continental accretion during which collisions between Archean-aged cratons formed a larger craton. Much of the North American craton had formed by 1.8 BYA (■ Figure 19.8a). For instance, the Slave and Rae cratons collided along the Thelon orogen, and rocks in the Trans-Hudson orogen provide a record of orogeny and suturing of the Superior, Hearne, and Wyoming cratons (Figure 19.8a).

Another notable Early Proterozoic event was the origin of the Wopmay orogen adjacent to the Slave craton in northwestern Canada. Rocks here record the oldest complete Wilson cycle—that is, opening and closing of an ocean basin. Furthermore, some rocks constitute a **sandstone-carbonate-shale assemblage,** a suite of rocks characteristic of passive continental margins that first became common and widespread during the Proterozoic.

Following the Early Proterozoic amalgamation of cratons, considerable accretion took place along Laurentia's southern margin. Between 1.8 and 1.6 BYA, accretion continued in what is now the southwestern and central United States as successively younger belts were sutured to the craton, forming the Yavapai and Mazatzal-Pecos orogens (Figure 19.8b). The net effect was the accretion of a belt of continental crust more than 1000 km wide along the southern margin of Laurentia.

No major continental accretion took place between 1.6 and 1.3 BYA, but extensive igneous activity occurred that was unrelated to orogenic activity (■ Figure 19.9a and Table 19.3). Laurentia did not increase in area, though, because magma was simply emplaced in or erupted onto previously existing continental crust. The rocks, mostly granitic plutons, and vast rhyolite and ash flows are deeply buried in most areas, but they are exposed in eastern Canada, Greenland, and the Baltic shield of Scandinavia. The origin of these Middle Proterozoic rocks is debated, but according to one hypothesis they resulted from large-scale upwelling of magma beneath a supercontinent.

Another important event in the evolution of Laurentia, the *Grenville orogeny* in the eastern United States and Canada, took place between 1.3 and 1.0 BYA (Figure 19.8c). Grenville rocks are found in Scandinavia, Greenland, and the Appalachian region of eastern North America (Figure 19.9b). Some geologists think these Grenville rocks record an opening and then closing ocean basin, so perhaps they were deposited on a passive continental margin. But others are of the opinion that Grenville deformation took place in an intracratonic basin. Whatever the cause, it represents the final episode of Proterozoic continental accretion of Laurentia.

Contemporaneous with Grenville deformation was an episode of rifting in Laurentia resulting in the origin of the **Midcontinent Rift** (Figure 19.8c). This huge feature has two branches extending south from the Lake Superior region. It cuts both Archean and Early Proterozoic-aged rocks but is buried by younger rocks except near Lake Superior, where its igneous and sedimentary rocks are well exposed.

Proterozoic Supercontinents

In previous chapters you learned that a *continent* is one of Earth's landmasses consisting of granitic crust with most of its surface above sea level. A *supercontinent*, in contrast, consists of all or at least much of the conti-

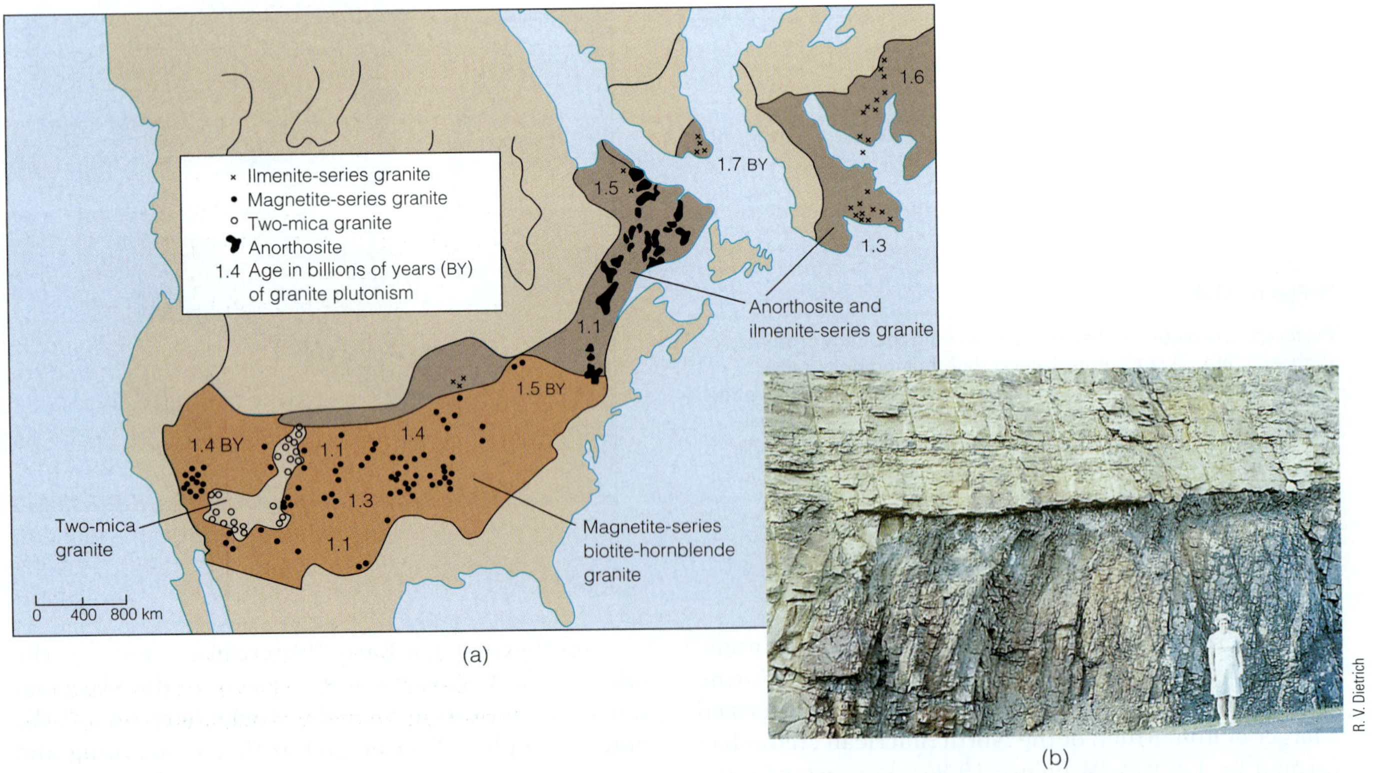

■ **Figure 19.9**

(a) Proterozoic igneous rocks unrelated to orogenic activity. Most of these rocks are Middle Proterozoic in age, and they are deeply buried over most of their extent. (b) These steeply dipping rock layers in New York were metamorphosed during the Grenville orogeny. They are now overlain by nearly horizontal layers of Late Cambrian-aged sedimentary rocks.

nents, so other than size it is the same as a continent. The supercontinent Pangaea that existed at the end of the Paleozoic Era is familiar, but few are aware of earlier supercontinents.

Before specifically addressing supercontinents, though, we must note that the present style of plate tectonics involving opening and then closing ocean basins had almost certainly been established by the Early Proterozoic. In fact, the oldest known complete *ophiolite* providing evidence for an ancient convergent plate boundary is the Jormua mafic-ultramafic complex in Finland. It is about 1.97 billion years old but nevertheless compares closely in detail with younger well-documented ophiolites.

Supercontinents may have existed as early as the Late Archean, but if so we have little evidence of them. The first supercontinent that geologists recognize with some certainty, known as *Rodinia* (■ Figure 19.10), assembled between 1.3 and 1.0 billion years ago and then began fragmenting 750 million years ago. Judging by the large-scale deformation, the *Pan-African orogeny,* that took place in what are now the Southern Hemisphere continents, geologists conclude that Rodinia's separate pieces reassembled about 650 million years ago and formed another supercontinent, this one known as *Pannotia.* And finally, by the latest Proterozoic, about 550 million years ago, fragmentation was under way again, giving rise to the continental configuration that existed at the onset of the Phanerozoic Eon (see Chapter 20).

What Would You Do?

Suppose you visit a planet that, like Earth, has continents and vast oceans. What evidence would indicate that this hypothetical planet's continents formed and evolved like those on Earth?

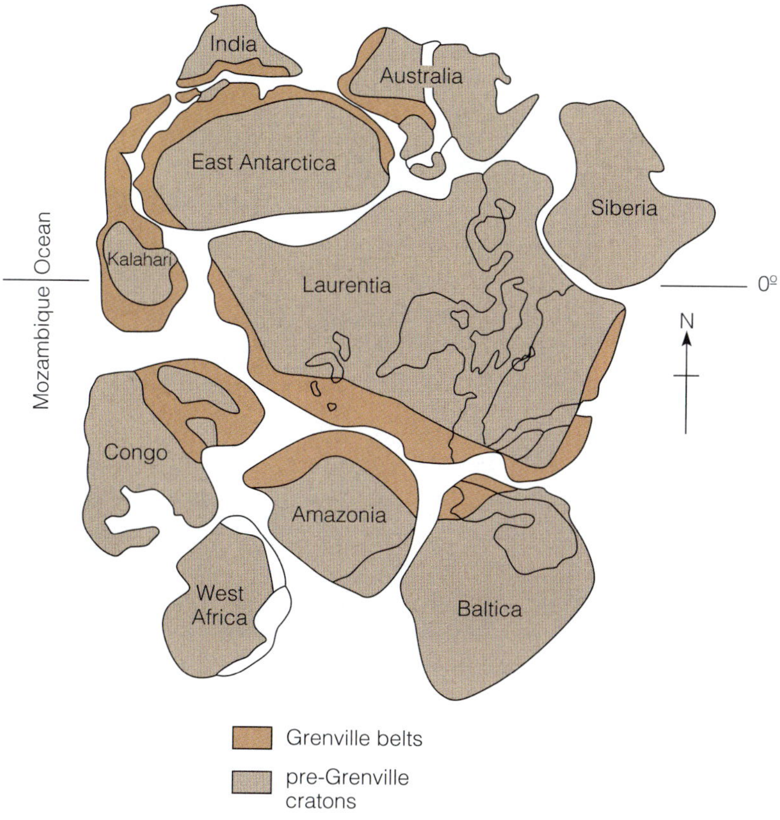

■ Figure 19.10

Possible configuration of the Late Proterozoic supercontinent Rodinia before it began fragmenting about 750 million years ago.

Proterozoic Rocks

Archean crust-forming processes generated granite-gneiss complexes and greenstone belts that were shaped into cratons. And although these same rock associations continued to form during the Proterozoic, they did so at a considerably reduced rate. In addition, many Archean rocks have been metamorphosed, although some are completely unaltered, whereas vast areas of Proterozoic rocks show little or no effects of metamorphism, and in many places Proterozoic rocks are separated from Archean rocks by unconformities.

Fully 60% of all Proterozoic rocks are sandstone-carbonate-shale assemblages that were deposited along rifted continental margins and in basins within cratons. Their widespread occurrence indicates that large, stable cratons were present with depositional environments much like those of the present. Early Proterozoic assemblages of these rocks are common in the Great Lakes region of the United States and Canada (■ Figure 19.11), and in the western part of the continent they are found in three Middle to Late Proterozoic basins (■ Figure 19.12). Wave-formed ripple marks and cross-bedding in the sandstones and stromatolites (structures formed by blue-green algae) in the carbonate rocks indicate that deposition took place in marginal marine and marine environments. Some of these rocks show evidence of deposition in continental environments, such as braided streams and alluvial fans.

Deposits along with their associated features indicate that two major episodes of continental glaciation took place during the Proterozoic. North America probably had an extensive ice sheet centered southwest of Hudson Bay during the Early Proterozoic (■ Figure 19.13a). Similar deposits of about the same age are present in the United States, Australia, and

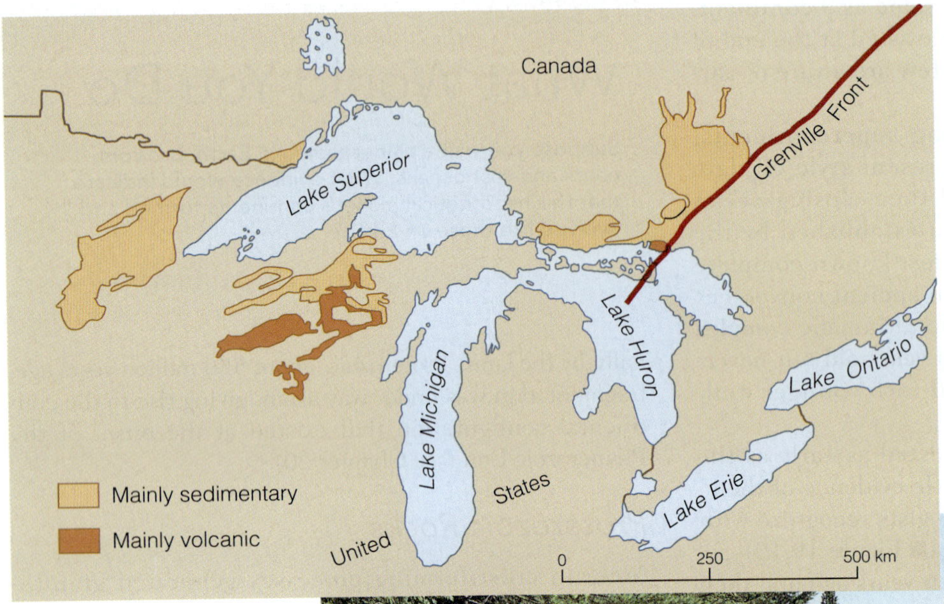

(a)

(b)

(c)

(d)

Figure 19.11

(a) Distribution of Early Proterozoic-aged rocks in the Great Lakes region. Outcrops in Michigan of the Sturgeon Quartzite (b), the Mesnard Quartzite (c), and the Kona Dolomite (d). The crests of the ripple marks in (c) point toward the observer, and the bulbous structures in (d) are *stromatolites* that resulted from the activities of cyanobacteria (blue-green algae). Source: G.M. Young, "Tectonic-Sedimentary History of Early Proterozic Rocks of the Northern Great Lakes Region," in *Early Proterozoic Rocks Geology of the Northern Great Lakes Region*, L. D. Medaris, Jr. (ed.), GSA Monitor 160, 1983, p. 16 (Fig 1), Geological Society of America.

(a)

(b)

(c)

(d)

Figure 19.12

(a) Middle to Late Proterozoic-aged sedimentary basins in the western United States and Canada. (b) Outcrop of red mudrock and (c) limestone with stromatolites in Glacier National Park, Montana. (d) Sandstone in the Grand Canyon, Arizona.

South Africa, but their ages are not precise enough to determine whether there was a single widespread period of glaciation or a number of glacial events in different areas at different times.

Another time of widespread glaciation is indicated by Late Proterozoic-aged glacial deposits on all continents except Antarctica (Table 19.3). Figure 19.13b shows the approximate distribution of these Late Proterozoic glaciers, but we must emphasize *approximate* because the geographic extent of glacial ice is unknown and these glaciers were not all present at the same time. Nevertheless, these glaciers were so widespread that they existed even in near-equatorial areas.

In Chapter 5, we briefly discussed **banded iron formations (BIFs).** These are sedimentary rocks that consist of alternating thin layers of silica (chert) and the iron oxide

What Would You Do?

As a working geologist, you encounter Proterozoic-aged conglomerate that you are convinced is a glacial deposit. Others, however, think it is simply stream-deposited gravel or perhaps an ancient landslide deposit. What kinds of evidence, if present, would verify your interpretation? That is, what attributes of the rock itself and its associated deposits might lead to your analysis?

minerals hematite (Fe_2O_3) and magnetite (Fe_3O_4) (■ Figure 19.14), although some BIFs contain other iron minerals. Archean BIFs are small, lens-shaped bodies measuring a few meters across and a few meters thick. In contrast, Proterozoic BIFs are much thicker, cover vast areas, and were probably deposited in shallow marine environments.

Iron is a highly reactive element that in the presence of oxygen forms rustlike oxides that do not readily dissolve in water. If free oxygen is absent, though, iron goes into solution and can accumulate in the oceans. Given that the Archean atmosphere probably had little free oxygen, little iron was dissolved in seawater. But about 2.3 billion years ago there was an increase in the abundance of photosynthesizing bacteria, a corresponding increase in the amount of free oxygen in the atmosphere and oceans, and the precipitation of dissolved iron and silica to form BIFs.

One type of cement binding detrital particles together is iron oxide (see Chapter 6). Many sedimentary rocks in the southwestern United States have small quantities of iron oxide cement and accordingly are reddish. These **red beds** of sandstone, siltstone, and shale do not appear in the geologic record until about 1.8 BYA (Figure 19.12b, d). Iron oxide cement forms under oxidizing conditions, which implies that Earth's atmosphere

(a)

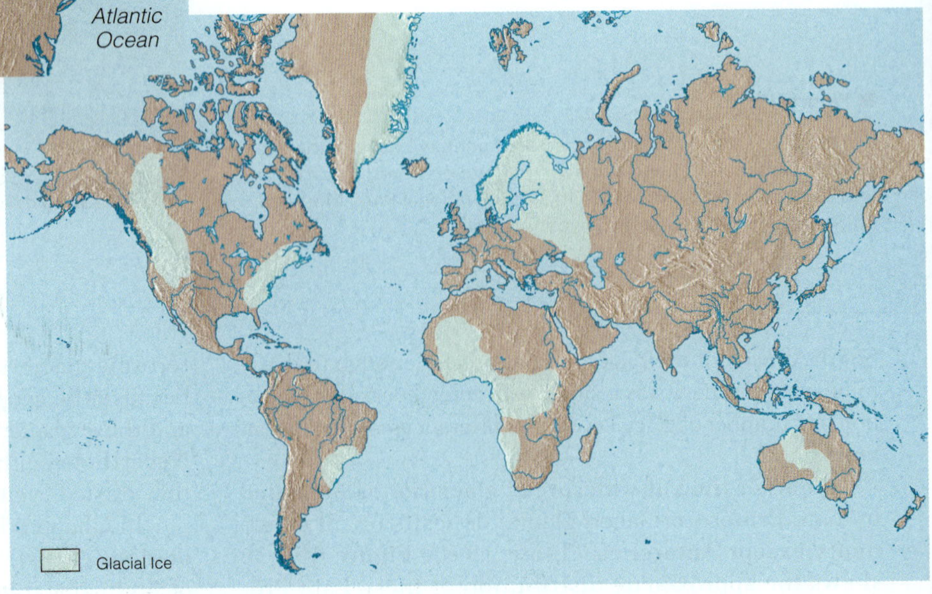

(b)

■ **Figure 19.13**

(a) Glacial deposits of about the same age in several areas indicate that an Early Proterozoic ice sheet was centered southwest of Hudson Bay. (b) Late Proterozoic glaciers shown on a map with the continents in their present positions. The extent of the ice is approximate. Source: Reprinted from L. A. Frakes, *Climates Throughout Geologic Time*, p. 39. Copyright © 1979 Elsevier Science Publishers.

GEOLOGY IN UNEXPECTED PLACES

Little Rock, Big Story

More than a million people visit Plymouth Rock at Pilgrim Memorial State Park in Plymouth, Massachusetts, each year, but most are surprised to find out that it is a rather small boulder (■ Figure 1). Legend holds that it is the landing place where the Pilgrims first set foot on the New World in 1620. In fact, the Pilgrims first landed near Provincetown on Cape Cod and then went on to the Plymouth area, but even then probably landed north of Plymouth, not at Plymouth Rock.

Actually the boulder that is Plymouth Rock was much larger when the Pilgrims landed, but an attempt to move the symbolic stone to the town square in 1774 split it in two. The lower half of the rock remained near the seashore. In 1880 the two pieces of the rock were reunited, and in 1921 the boulder was moved once again, this time to a stone canopy erected over the original site.

Plymouth Rock has great symbolic value but otherwise is a rather ordinary rock, although it is geologically interesting. The stone itself is from the 600+-million-year-old (Late Proterozoic) Dedham Granodiorite, an intrusive igneous rock similar to granite, which was carried to its present site and deposited by a glacier during the Ice Age (1.6 million to 10,000 years ago). The Dedham Granodiorite is part of an association of rocks in New England that geologists think represent a chain of volcanic islands similar to the Aleutian Islands. These volcanic islands were incorporated into North America when plates collided and caused a period of mountain building called the Taconic orogeny about 450 million years ago (see Chapter 20).

■ Figure 1

Plymouth Rock, where according to legend the Pilgrims first stepped ashore in the New World in 1620. They actually landed several miles away, but the rock nevertheless has great symbolic value.

Marcus Kazmierczak, www.mkaz.com

had some free oxygen by this time, although the amount was probably still only 1% or 2% of present levels.

ORIGIN AND EVOLUTION OF THE ATMOSPHERE AND HYDROSPHERE

We mentioned in the Introduction that early Earth was waterless, its atmosphere lacked free oxygen—that is, oxygen not combined with other elements—and no ozone layer was present so ultraviolet radiation was intense. The atmosphere now contains nearly 21% free oxygen and other important gases in trace amounts, especially carbon dioxide (CO_2) and water vapor. An ozone layer in the upper atmosphere blocks out most ultraviolet radiation, and now 71% of Earth's surface is covered by water. Both the atmosphere and hydrosphere have played an important role in the development of the biosphere. The obvious question is, What brought about such remarkable changes?

The Atmosphere

Before Earth had a differentiated core, it lacked a magnetic field and a *magnetosphere*, the area around the

(a)

(b)

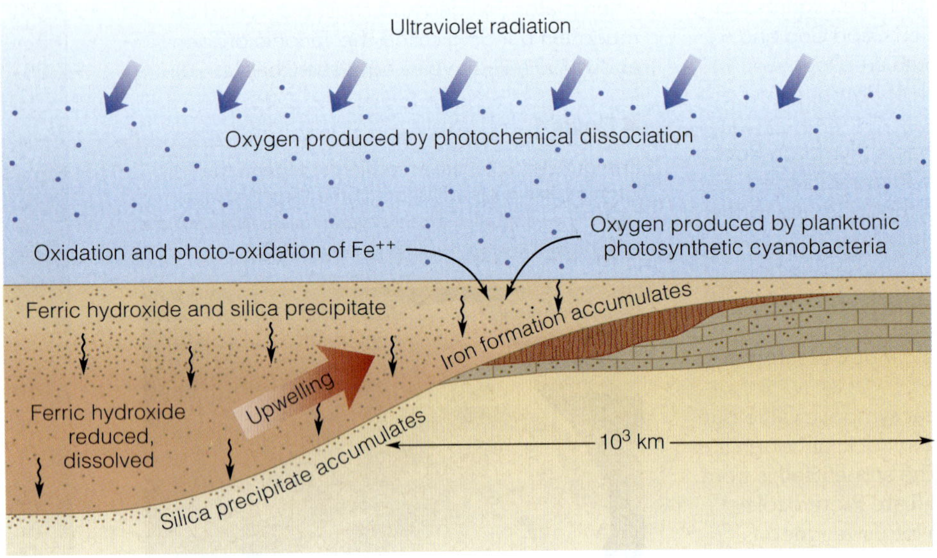

(c)

■ Figure 19.14

Early Proterozoic-aged banded iron formation (BIF). (a) At this outcrop in Ishpeming, Michigan, the rocks are brilliantly colored alternating layers of red chert and silver iron minerals. (b) A more typical outcrop of BIF near Negaunee, Michigan. (c) Depositional model for the origin of banded iron formations.

planet in which the magnetic field is confined. Accordingly, a strong solar wind, an outflow of ions from the Sun, would have swept away any gases that otherwise might have formed an atmosphere. Once a magnetosphere was established, though, gases from Earth's interior were released during volcanic eruptions and began to accumulate—a phenomenon known as **outgassing** (■ Figure 19.15).

We have every reason to think that Hadean and Archean volcanoes emitted the same gases as today's volcanoes do—mostly water vapor and lesser amounts of carbon dioxide, sulfur dioxide, carbon monoxide, nitrogen, hydrogen, and several others. These gases no doubt accumulated rapidly because Earth possessed more residual heat and more radiogenic heat, and as a result volcanism was ubiquitous. The atmosphere so formed was rich in carbon dioxide but contained little or no free oxygen, and with no free oxygen there was no ozone layer. Furthermore, as these volcanic gases reacted chemically, they very likely yielded atmospheric ammonia (NH_3) and methane (CH_4). In short, the atmosphere was quite unlike the one present now.

Archean sedimentary deposits with detrital minerals such as pyrite (FeS_2) and uraninite (UO_2) indicate an oxygen-deficient atmosphere because both oxidize rapidly when free oxygen is present. But oxidized iron is quite common in Proterozoic-aged rocks, so the atmosphere had at least some free oxygen by that time. Two processes account for introducing free oxygen into the atmosphere: photochemical dissociation and photosynthesis.

Photochemical dissociation is a process in which water (H_2O) vapor molecules are broken up by ultraviolet radiation in the upper atmosphere to yield hydrogen (H_2) and free oxygen (O_2) (■ Figure 19.16). But photochemical dissociation is a self-limiting process and probably accounts for no more than 2% of the present-day level of free oxygen. With this much free oxygen, ozone (O_3)

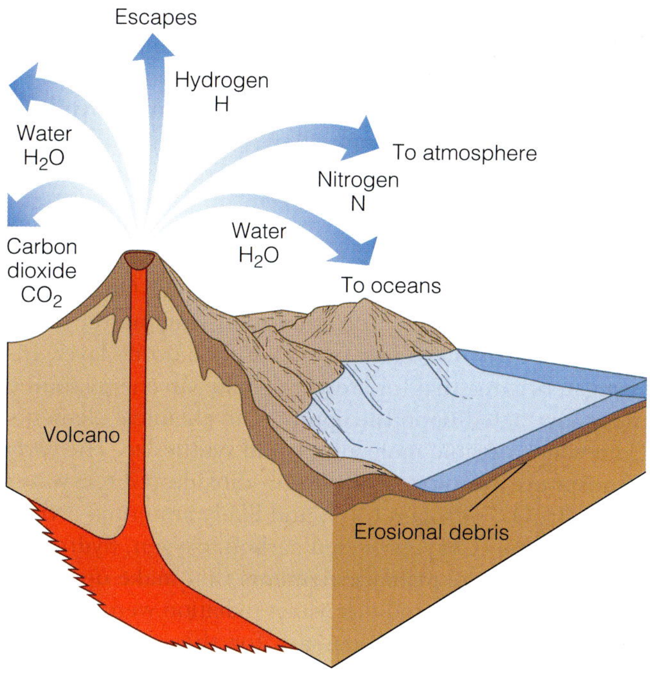

Figure 19.15

Outgassing and the origin of Earth's early atmosphere. Notice that although the atmosphere contains several gases, no free oxygen is present. Chemical reactions in this atmosphere also probably yielded methane and ammonia.

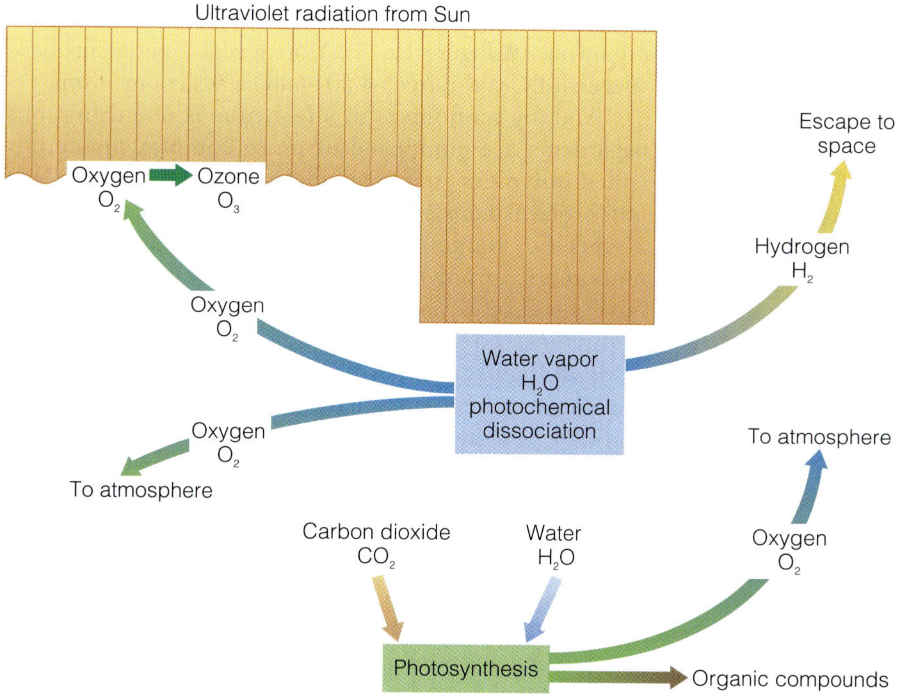

Figure 19.16

Photochemical dissociation and photosynthesis added free oxygen to the atmosphere. Once free oxygen was present, an ozone layer formed in the upper atmosphere and blocked most incoming ultraviolet radiation.

forms a barrier against incoming ultraviolet radiation, so another process must account for most of the atmosphere's free oxygen.

Photosynthesis is much more important in releasing free oxygen into the atmosphere, but obviously it could not take place until organisms practicing this metabolic process had evolved. Organisms that carry out photosynthesis combine carbon dioxide and water into the organic molecules they need for survival, and they release free oxygen as a waste product (Figure 19.16). We know from the fossil record that organisms that almost certainly practiced photosynthesis were present at least 3.5 billion years ago. Even so, at the end of the Archean Eon, the atmosphere may have had no more than 1% of its present free oxygen level.

During the Proterozoic Eon, the free oxygen level increased from 1% to perhaps 10% of its present level, but probably not until well into the Paleozoic Era, about 400 million years ago, did it reach its current concentration of 21% of the atmosphere. The presence of banded iron formations and red beds (Figure 19.14) provides compelling evidence that Earth's Proterozoic atmosphere was an oxidizing one.

The Hydrosphere

We already mentioned that when Earth first formed it was waterless and possessed a noxious atmosphere. We have outlined the history of the evolving atmosphere, and now we turn our attention to the hydrosphere, another of Earth's major systems. All water on Earth is part of the hydrosphere, but most of it—more than 97%—is in the oceans. Where did it come from, and how has it changed?

Certainly outgassing released water vapor from Earth's interior, and once the planet cooled sufficiently, water vapor condensed and surface waters began to accumulate, probably at some time during the Hadean. Another source of water vapor, and eventually liquid water, was meteorites and especially icy comets (Figure 19.1). Which of these, outgassing or meteorites and comets, was most important is unknown, but we do know that oceans were present by the Early Archean, although their volumes and geographic extent cannot be determined. Nevertheless, we can envision an early Earth with numerous erupting volcanoes and an early episode of intense meteorite and comet bombardment accounting for a rapid rate of surface water accumulation.

Volcanoes erupt today and meteorite and comet impacts still occur, so is the volume of the ocean still increasing? Probably so, but at a considerably reduced rate because (1) Earth's heat production and thus the amount of volcanism have decreased, and (2) the intense episode of impacts ended about 3.8 billion years ago. Accordingly, the rate of accumulation now is so slow that it adds insignificant amounts of surface water to the oceans.

Recall from Chapter 17 that one early attempt to determine Earth's age was to calculate how long it took for the oceans to reach their current salinity level—assuming, of course, that the oceans formed soon after Earth did, that they were freshwater to begin with, and that their salinity increased at a uniform rate. None of these assumptions is correct, so the ages determined were vastly different depending on which chemical element in seawater was analyzed. We now know that the very early oceans were salty, probably about as salty as they are now. That is, very early in their history the oceans reached chemical equilibrium and have remained in near-equilibrium conditions ever since.

LIFE—ITS ORIGIN AND EARLY HISTORY

Today, Earth's biosphere includes millions of species of animals, plants, one-celled organisms, fungi, and bacteria, whereas only bacteria are known from the Archean. In Chapter 18, we considered the evolutionary processes whereby life diversified, but here our concern is with how life originated and its earliest history. First we must be clear about what is living versus nonliving. Just how do today's plants and animals differ from nonliving things, and is the distinction always clear? Minimally, an organism must reproduce and practice some kind of metabolism. Reproduction ensures the long-term survival of a species as it perpetuates itself, whereas metabolism ensures the short-term survival of an individual as a chemical system.

If we use this reproduction–metabolism criterion, it would seem a simple matter to decide whether something is alive. Yet viruses, composed of a bit of DNA or RNA in a protein capsule, behave like living organisms when in an appropriate host cell, but outside a host cell they neither reproduce nor produce organic molecules. So, are they living or nonliving? Some biologists think they are nonliving; others think they represent another way of defining living. In either case, though, they illustrate that the distinction is not always clear.

Simple organic molecules known as microspheres form spontaneously and show greater organizational complexity than inorganic objects such as rocks. In fact, they grow and divide, but in ways that are more like random chemical processes. Consequently, they are not living, but they nevertheless show some characteristics of organisms. Furthermore, if life originated on Earth by a natural process, then it must have passed through some kind of prebiotic stage, perhaps similar to microspheres.

The Origin of Life

As early as 1924, the Russian biochemist A. I. Oparin postulated that life originated from nonliving matter when Earth had little or no free oxygen and no ozone layer. Investigators agree that for life to originate, an energy source must have acted upon the appropriate chemical elements from which organic molecules could synthesize. The early atmosphere composed of carbon dioxide (CO_2), water vapor (H_2O), nitrogen (N_2), and likely ammonia (NH_3) and methane (CH_4) provided carbon, oxygen, hydrogen, and nitrogen, the primary elements that make up all organisms. Lightning and ultraviolet radiation were two possible energy sources necessary for these elements to combine and form rather simple organic molecules known as **monomers,** such as amino acids.

Monomers are needed as basic building blocks of more complex organic molecules, but is it plausible that they formed in the manner postulated? Experimental evidence indicates it is. During the 1950s, Stanley Miller circulated gases approximating Earth's early atmosphere through a closed glass vessel and subjected the mixture to an electric spark to simulate lightning. Within a few days, the mixture became turbid and an analysis showed that Miller had synthesized several amino acids common in organisms (■ Figure 19.17). More recent experiments have successfully synthesized all 20 amino acids in organisms.

Making monomers in a test tube is one thing, but organisms are composed of more complex molecules called **polymers** such as nucleic acids (DNA and RNA) and proteins consisting of monomers linked together in a specific sequence. So how did this linking of monomers, or polymerization, take place? Researchers have successfully synthesized small molecules called *proteinoids,* consisting of more than 200 linked amino acids (■ Figure 19.18). In fact, when heated, dehydrated, and concentrated, amino acids spontaneously polymerize and form proteinoids.

We can call these artificially synthesized proteinoids *protobionts,* meaning they have characteristics between those of inorganic chemical compounds and living organisms. Suppose these protobionts came into existence in a manner similar to that outlined above. They would have been diluted and would have ceased to exist if they had not developed some kind of outer covering. In other words, they had to be self-contained chemical systems as today's cells are. In the experiments just mentioned, proteinoids have spontaneously aggregated into microspheres that have a cell-like outer covering and grow and divide somewhat like bacteria do (Figure 19.18).

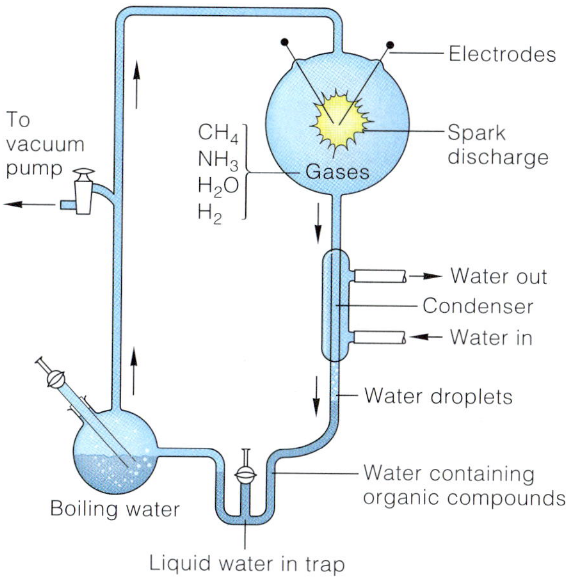

Figure 19.17

Experimental apparatus used by Stanley Miller. Several amino acids characteristic of organisms were artificially synthesized during Miller's experiments. Source: From S. L. Miller, "The Formation of Organic Compounds on the *Primitive Earth,*" in *Modern Ideas of Spontaneous Generation*, Nigrelli (ed.), Annuals of the New York Academy of Sciences, v 69, Art. 2, Aug, 30, 1957, p. 261 (Fig 1). Used with permission.

Perhaps the first steps leading to life took place as monomers formed in great abundance and polymerized, but little is known about how a reproductive mechanism came about. Microspheres divide but, as noted earlier, do so in a nonbiologic fashion. In fact, for some time researchers were baffled because in present-day organisms either DNA or RNA is necessary for reproduction, but these nucleic acids cannot replicate without protein enzymes, yet protein enzymes cannot be made without nucleic acids. Or so it seemed until a few years ago, when researchers discovered that some small RNA molecules can in fact reproduce without the aid of protein enzymes. So, the first replicating system might have been an RNA molecule. Just how these molecules were synthesized under conditions that existed on early Earth has not been resolved.

A common theme among investigators is that life originated when organic molecules were synthesized from atmospheric gases. But even this idea has been questioned by those who think the same sequence of events—that is, formation of monomers, then polymers—took place adjacent to hydrothermal vent systems (black smokers) on the seafloor.

Archean Organisms

Prior to the mid-1950s, scientists assumed that the fossils so abundant in Cambrian rocks had a long earlier history, but they had little direct knowledge of Precambrian life. During the early 1900s, Charles Walcott described layered moundlike structures from the Early Proterozoic-aged Gunflint Iron Formation of Canada that he proposed were constructed by algae. We now call these structures **stromatolites,** but not until 1954 did scientists demonstrate that these structures are actually the result of organic activity. In fact, stromatolites still form in a few areas where they originate by trapping sediment on sticky mats of photosynthesizing cyanobacteria, more commonly known as blue-green algae (Figure 19.19). Although widespread in Proterozoic rocks, they are now restricted to aquatic environments with especially high salinity where snails cannot live and graze on them.

Currently, the oldest known undisputed stromatolites are from 3.0-billion-year-old rocks in South Africa, but other probable ones have been discovered in 3.3- to

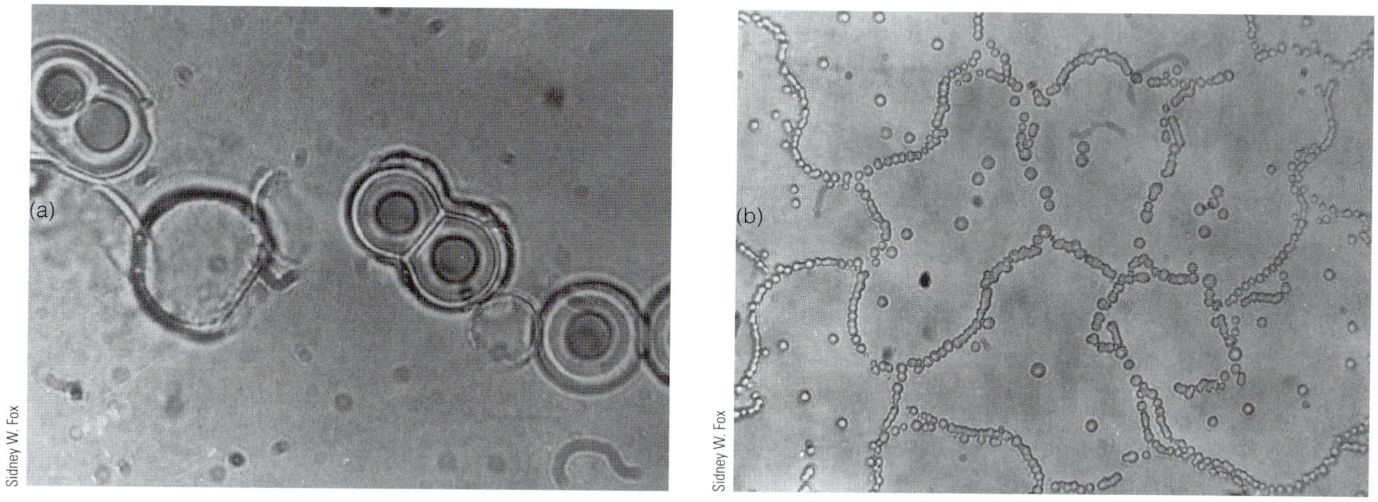

(a) (b)

Figure 19.18

(a) Bacteria-like proteinoid. (b) Proteinoid microspheres.

■ **Figure 19.19**

(a) Present-day stromatolites, Shark Bay, Australia. (b) Different types of stromatolites include irregular mats, columns, and columns linked by mats. Source: (b) R. L. Anstey and T. L. Chase, *Environments Through Time*, 1974, Macmillan Publishing Company.

3.5-billion-year-old rocks near North Pole, Australia (Table 19.3). Even more ancient evidence of life comes from small carbon spheres in 3.8-billion-year-old rocks in Greenland, but the evidence is not conclusive.

Cyanobacteria, the oldest known fossils, practice photosynthesis, a complex metabolic process that must have been preceded by a simpler process. So it seems reasonable that cyanobacteria were preceded by nonphotosynthesizing organisms for which we so far have no record. They must have resembled tiny bacteria, and because the atmosphere lacked free oxygen they must have been **anaerobic,** meaning they required no free oxygen. They also were likely **heterotrophic,** dependent on an external source of nutrients, rather than **autotrophic,** capable of manufacturing their own nutrients as in photosynthesis. And finally, they were **prokaryotic cells,** cells lacking a cell nucleus and other internal structures typical of more advanced *eukaryotic cells* (discussed in a later section).

We can characterize these earliest organisms as single-celled, anaerobic, heterotrophic prokaryotes (■ Figure 19.20). Furthermore, they reproduced asexually. Their energy source was likely adenosine triphosphate (ATP), which can be synthesized from simple gases and phosphate, so it was no doubt available in the early Earth environment. These cells may have simply acquired their ATP directly from their surroundings, but this situation could not have persisted for long because as more and more cells competed for the same resources, the supply should have diminished. Thus, a more sophisticated metabolic process developed, probably *fermentation,* an anaerobic process during which molecules such as sugars are split and release carbon dioxide, alcohol, and energy. As a matter of fact, most living prokaryotes practice fermentation.

Of course, the nature of the earliest cells is informed speculation, but we can say that the most significant event in Archean life history was the development of the autotrophic process of photosynthesis. These more advanced cells were still anaerobic and prokaryotic, but as autotrophs they no longer depended on an external source of preformed organic molecules for their nutrients. The Archean fossils in Figure 19.20 belong to the kingdom Monera, which is represented today by bacteria and cyanobacteria.

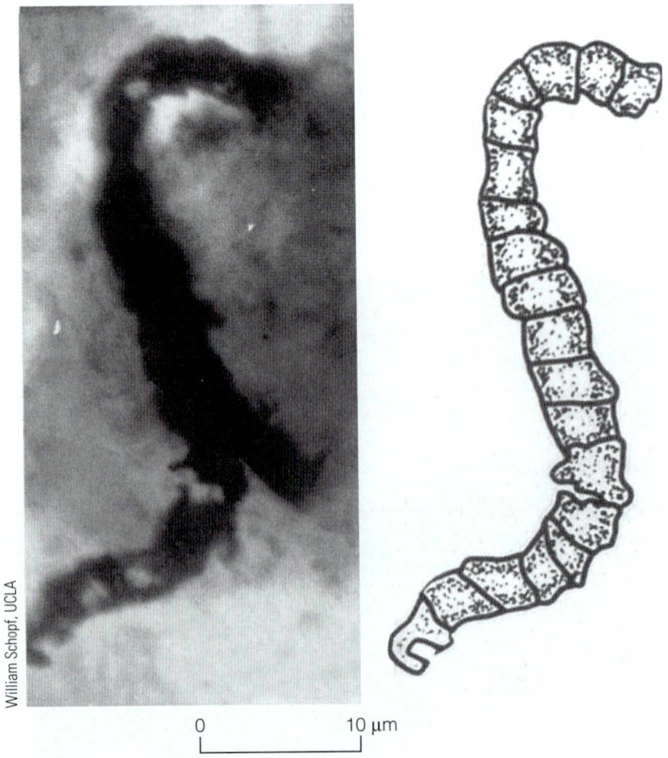

■ **Figure 19.20**

Photomicrograph and schematic restoration of fossil prokaryote from the 3.3- to 3.5-billion-year-old Warrawoona Group, Western Australia.

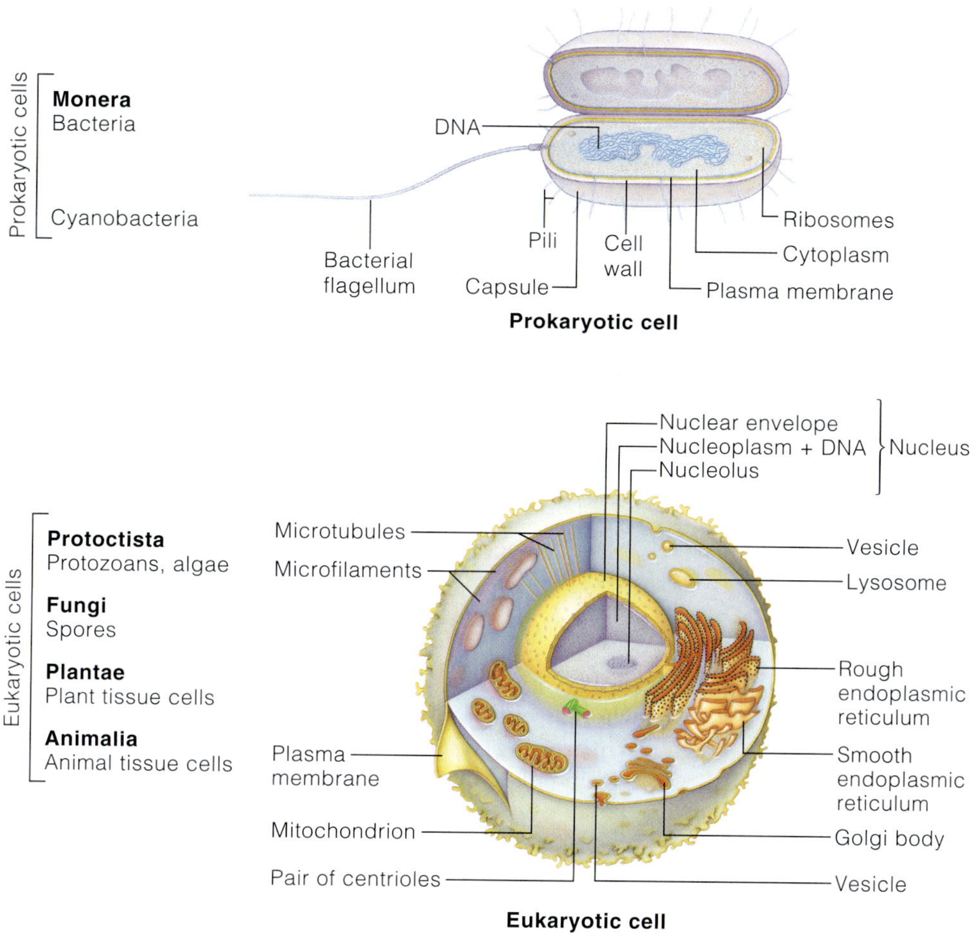

■ Figure 19.21

Prokaryotic and eukaryotic cells. Eukaryotic cells have a cell nucleus containing the genetic material and organelles such as mitochondria and plastids, as well as chloroplasts in plant cells. In contrast, prokaryotic cells are smaller and not nearly as complex as eukaryotic cells.

Life of the Proterozoic

The Early Proterozoic, like the Archean, is characterized mostly by a biota of single-celled, prokaryotic bacteria. No doubt thousands of varieties existed, but none of the more familiar organisms such as plants and animals were present. Before the appearance of cells capable of sexual reproduction, evolution was a comparatively slow process, accounting for the low organic diversity. But by 2.1 billion years ago, sexually reproducing cells appeared and the tempo of evolution picked up markedly thereafter.

A New Type of Cell Appears The origin of **eukaryotic cells** marks one of the most important events in life history (Table 19.3). These cells are much larger than prokaryotic cells; they have a membrane-bounded nucleus containing the genetic material, and most reproduce sexually (■ Figure 19.21). And most eukaryotes—that is, organisms made up of eukaryotic cells—are multicellular and aerobic, so they could not have existed until some free oxygen was present in the atmosphere.

No one doubts that eukaryotes were present by the Middle Proterozoic, and it is becoming increasingly clear that they existed even earlier. Rocks 1.4 to 1.7 billion years old in Montana and China have coiled, carbonaceous impressions of eukaryotes, probably algae, and 2.1-billion-year-old rocks in Michigan have similar impressions (■ Figure 19.22). Some rocks in Australia have

Bruce Runnegar

■ Figure 19.22

Fossil eukaryotic algae from the 2.1-billion-year-old Negaunee Iron Formation at Marquette, Michigan.

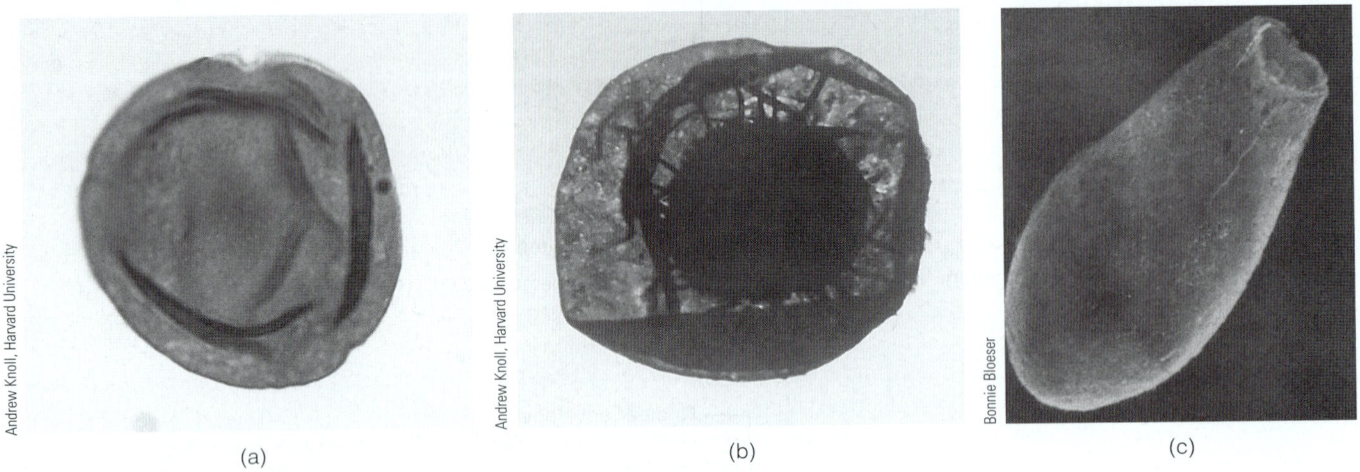

Figure 19.23

(a) and (b) Acritarchs are probably the cysts of algae. These common Late Proterozoic-aged microfossils are thought to represent eukaryotic organisms. (c) Vase-shaped microfossil, probably a cyst of some kind of algae.

fossils that show evidence of mitosis and meiosis, cell division processes used only by eukaryotes.

Cells larger than 60 microns appear in abundance at least 1.4 billion years ago, and many show an increased degree of organizational complexity compared to that of prokaryotic cells. An internal membrane-bounded cell nucleus is present in some, for instance. Furthermore, microscopic, hollow fossils known as *acritarchs* that probably represent cysts of planktonic algae become common during the Middle and Late Proterozoic (■ Figure 19.23a, b). And vase-shaped microfossils from rocks in the Grand Canyon have been tentatively identified as cysts of some kind of algae (Figure 19.23c).

Endosymbiosis and the Origin of Eukaryotic Cells

According to a widely accepted theory, eukaryotic cells formed from several prokaryotic cells that entered into a symbiotic relationship. Symbiosis, involving a prolonged association of two or more dissimilar organisms, is common today. In many cases, both symbionts benefit from the association, as in lichens, which were once thought to be plants but actually are symbiotic associations between fungi and algae.

In a symbiotic relationship, each symbiont must be capable of metabolism and reproduction, but the degree of dependence in some relationships is such that one symbiont cannot live independently. This may have been the case with Proterozoic symbiotic prokaryotes that became increasingly interdependent until the unit could exist only as a whole. In this relationship, though, one symbiont lived within the other, which is a special type of symbiosis called **endosymbiosis** (■ Figure 19.24).

Supporting evidence for endosymbiosis comes from studies of living eukaryotic cells containing internal structures called organelles, such as mitochondria and plastids, that have their own genetic material. In addition, prokaryotic cells synthesize proteins as a single system, whereas eukaryotic cells are a combination of protein-synthesizing systems. That is, some of the organelles within eukaryotic cells are capable of protein synthesis.

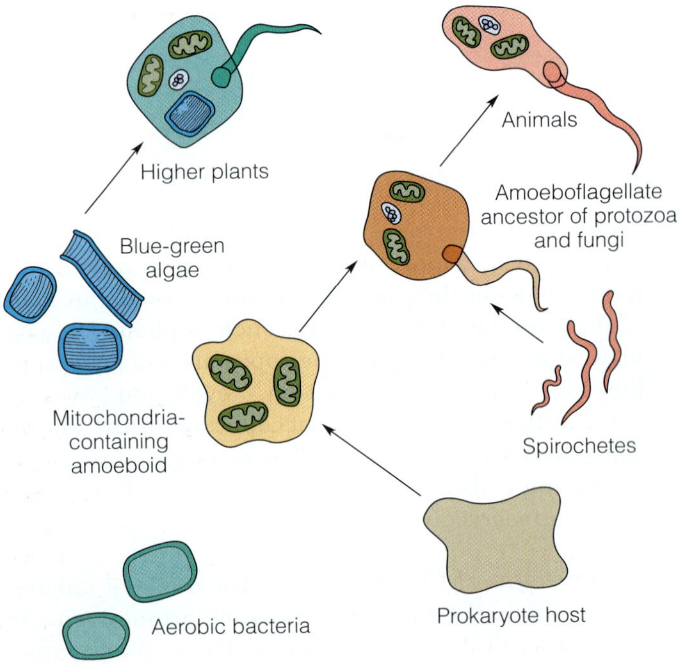

Figure 19.24

Endosymbiosis theory for the origin of eukaryotic cells. An aerobic bacterium and a larger host of the kingdom Monera united to form a mitochondria-containing amoeboid. An amoeboflagellate was formed by a union of the amoeboid and a bacterium of the spirochete group; this amoeboflagellate was the direct ancestor of two kingdoms—Fungi and Animalia. Another kingdom, Plantae, originated when this amoeboflagellate formed a union with blue-green algae (cyanobacteria) that became plastids.

These organelles with their own genetic material and protein-synthesizing capabilities are thought to have been free-living bacteria that entered into a symbiotic relationship, eventually giving rise to eukaryotic cells.

The Dawn of Multicelled Organisms **Multicelled organisms** are not only composed of many cells but also have cells specialized to perform specific functions such as reproduction and respiration. Unfortunately, the fossil record does not tell us how multicelled organisms arose from single-celled ancestors. However, studies of present-day organisms give some clues about how this transition might have taken place.

Suppose that a single-celled organism divided and formed a group of cells that did not disperse but remained together as a colony. The cells in some colonies may have become somewhat specialized similar to the situation in some living *colonial organisms.* Further specialization might have led to simple multicelled organisms such as sponges, consisting of cells that carry out functions such as reproduction, respiration, and food gathering. Carbonaceous impressions of Proterozoic multicelled algae are known from many areas (Figure 19.23), but the oldest are in the 2.1-billion-year-old Negaunee Iron Formation in Michigan (Figure 19.22).

The Ediacaran Fauna In 1947, the Australian geologist R. C. Sprigg discovered a unique assemblage of multicelled, soft-bodied animals preserved as molds and casts on the undersides of sandstone layers (■ Figure 19.25). Some investigators think that at least three present-day invertebrate phyla are represented: jellyfish and sea pens (phylum Cnidaria), segmented worms (phylum Annelida), and primitive members of the phylum Arthropoda. One wormlike Ediacaran fossil, *Spriggina,* has been cited as a possible ancestor of trilobites, and another *Tribachidium,* may be a primitive echinoderm (Figure 19.25). On the other hand, some think Ediacaran fossils represent an early evolutionary radiation distinct from the ancestry of existing invertebrates. These **Ediacaran faunas** existed between 670 and 570 million years ago and are now known on all continents except South America and Antarctica. The animals were widespread, but their fossils are not very common because all of them lacked durable skeletons.

Other Proterozoic Animal Fossils Although scarce, a few animal fossils older than those of the Ediacaran fauna are known. A jellyfish-like impression is present in rocks 2000 m below the Pound Quartzite, and in many areas burrows, presumably made by worms, are found in rocks at least 700 million years old. Wormlike fossils as well as fossil algae come from 700- to 900-million-year-old rocks in China, but the identity and age of these "fossils" have been questioned.

All known Proterozoic animals were soft-bodied, but there is some evidence that the earliest stages in the origin

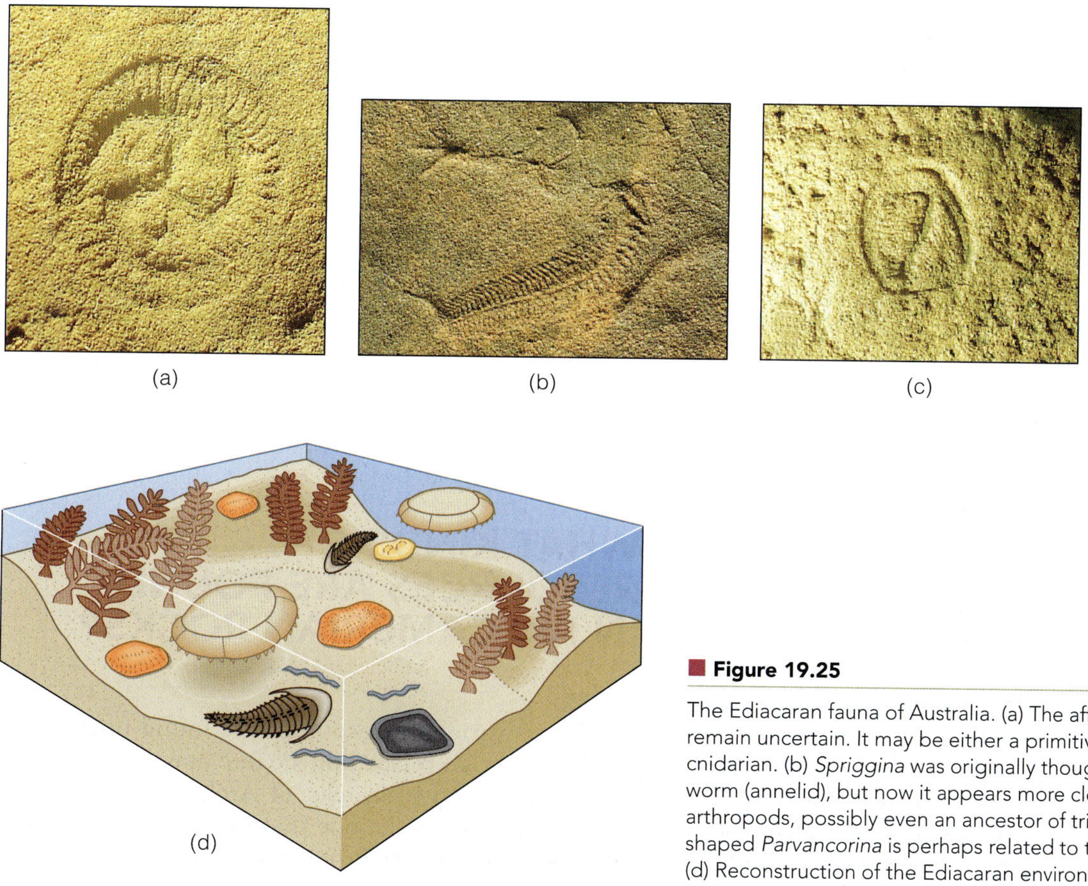
(a–c) South Australian Museum, Adelaide, South Australia

(a) (b) (c) (d)

■ **Figure 19.25**

The Ediacaran fauna of Australia. (a) The affinities of *Tribachidium* remain uncertain. It may be either a primitive echinoderm or a cnidarian. (b) *Spriggina* was originally thought to be a segmented worm (annelid), but now it appears more closely related to arthropods, possibly even an ancestor of trilobites. (c) Shield-shaped *Parvancorina* is perhaps related to the arthropods. (d) Reconstruction of the Ediacaran environment.

GEOFOCUS 19.1

Banded Iron Formation—From Mine to Steel Mill

The United States and Canada, both highly industrialized nations, owe a good deal of their economic success to abundant natural resources, although both nations, especially the United States, must import some essential commodities. But iron ore is present in great abundance in the Great Lakes region and in the Labrador Trough of eastern Canada. The giant iron-ore freighters that ply the Great Lakes sail from loading facilities in Ontario, Minnesota, and Michigan to iron-producing cities such as Hamilton, Ontario; Cleveland, Ohio; and Gary, Indiana.

Iron ore is produced from rocks of various ages, but by far most of it comes from Proterozoic *banded iron formation* (BIF) consisting of alternating layers of iron-bearing minerals (mostly hematite and magnetite) and chert. We discussed what is known about the origin of BIF, and remember that 92% of it was deposited during the Proterozoic Eon when Earth's atmosphere lacked free oxygen—that is, oxygen not combined with other elements. Most BIF is not particularly attractive, but some consists of alternating layers of brilliant, silvery, iron minerals and red chert, as at Jasper Knob in Ishpeming, Michigan (Figure 19.14a).

The richest iron ores in the Great Lakes region had been largely depleted by the time of World War II (1939–1945), but then a technologic innovation was developed that allowed mining of poorer-grade ores. In this process, the unusable rock is separated from the iron, which is then shaped into pellets before it is shipped to steel mills (■ Figure 1). In short, it was now possible to use what otherwise could not be recovered economically. Iron mining now accounts for a large part of the economies of several cities on or near the shores of Lake Superior and in eastern Canada.

The complete process from mine to steel mill consists of hauling the iron-bearing rock in huge dump trucks to a crusher, where it is ground to powder. The powdered ore is then processed to separate the iron from the unusable rock, either by floatation or by magnetic separation. In any case, the iron-rich powdered concentrate is dried and shaped into pellets in large rotating drums, and then finally baked at about 1300°C in huge rotary kilns. From the kilns, the pellets are either stored for later shipment or taken directly to iron-ore freighters at Duluth, Minnesota; Marquette, Michigan; or other ports. And then the iron-ore freighters make their way through

of skeletons were under way. Even some Ediacaran animals may have had a chitinous carapace, and others appear to have had areas of calcium carbonate. The odd creature known as *Kimberella* from the latest Proterozoic of Russia had a tough outer covering similar to that of some present-day marine invertebrates (■ Figure 19.26). Exactly what *Kimberella* was remains uncertain; some think it was a sluglike creature, whereas others think it was more like a mollusk.

Latest Proterozoic fossils of minute scraps of shell-like material as well as small toothlike denticles and spicules, presumably from sponges, indicate that several animals with skeletons or at least partial skeletons existed. But more durable skeletons of silica, calcium carbonate, and chitin (a complex organic substance) did not appear in abundance until the beginning of the Phanerozoic Eon 545 million years ago (see Chapter 21).

WHAT KINDS OF RESOURCES ARE FOUND IN PRECAMBRIAN ROCKS?

A variety of resources are found in Precambrian rocks, but the one most often associated with rocks of Archean age is gold, whereas iron ore is the one that immediately comes to mind when we discuss Proterozoic resources.

James S. Monroe

(a)

Sue Monroe

(b)

Figure 1

(a) The Empire Mine at Palmer, Michigan, where iron ore from the Early Proterozoic-aged Negaunee Iron Formation is mined and shaped into (b) pellets containing about 65% iron. The pellets measure about 1 cm in diameter.

the Great Lakes and deliver the pellets to the steel mills.

The availability of iron ore and a steel-making capacity is essential for large-scale industrialization. Accordingly, industrialized nations and developing nations must have domestic deposits of iron ore, or else they must import it. Japan is industrialized but has no iron-ore deposits and thus imports most of what it needs from Australia, which has vast deposits of Proterozoic BIF.

Currently, the nations that produce the most iron ore for export are Australia and Brazil, each accounting for about a third of the world's total exports, but China is actually the largest iron-ore producer, with more than 200 million metric tons per year. During 2001, the United States and Canada produced 60 million and 35 million metric tons of iron ore, respectively.

Archean Resources

Archean and Proterozoic rocks near Johannesburg, South Africa, have yielded more than 50% of the world's gold since 1886, but Archean gold is or has also been mined in the Superior craton in Canada and the Homestake Formation in South Dakota. Several Archean-aged massive sulfide deposits of zinc, copper, and nickel are known in Australia, Zimbabwe, and Canada, and similar deposits are currently forming adjacent to black smokers on the seafloor (see Chapter 9).

About one-fourth of the chrome reserves are in Archean rocks, especially in Zimbabwe; they probably formed along with crystals and settled in mafic and ultramafic plutons. The Stillwater Complex in Montana has low-grade chrome and platinum ores that have been mined and stockpiled during the past, but so far they have not been refined for their chrome or platinum.

Archean-aged pegmatites are mostly granitic and of little economic importance. However, some in Manitoba, Canada, and the Rhodesian Province in Africa contain valuable minerals. In addition to gem-quality minerals, a few Archean pegmatites are mined for lithium, beryllium, and rubidium.

Proterozoic Resources

Banded iron formations (BIFs), the world's major source of iron ores, are present on all continents, and 92% of all BIFs were deposited during the Late Proterozoic

Figure 19.26

All agree that *Kimberella*, from latest Proterozoic rocks in Russia, was an animal, but some think it was a sluglike creature, whereas others are of the opinion it was mollusklike.

(Figure 19.14). China, Brazil, and Australia, in that order, are the largest producers of iron ore. The United States must import about 30% of the iron ore it uses, mostly from Canada and Venezuela. Most of the iron ore mined in North America comes from mines in the Lake Superior region and in eastern Canada (see Geo-Focus 19.1 on the previous page).

The Sudbury mining district in Ontario, Canada, is an important area of nickel and platinum production. Nickel is essential for the manufacture of nickel alloys such as stainless steel and Monel metal (nickel plus copper), which are valued for their strength and resistance to corrosion and heat. The United States imports more than 50% of all the nickel it uses, most from the Sudbury mining district in Canada. Some platinum for jewelry, surgical instruments, and chemical and electrical equipment is also exported from Canada to the United States, but the largest exporter is South Africa. The United States also depends on South Africa for much of its chromite, the ore of chromium.

Economically recoverable oil and natural gas were discovered in Proterozoic rocks in China and Siberia, arousing interest in the Midcontinent Rift as a potential source of hydrocarbons. Some rocks within the rift are known to contain oil, but so far no oil or gas wells are operating.

19 GEO RECAP

Chapter Summary

- Geologists divide the Precambrian into an informal Hadean followed by two eons, the Archean and the Proterozoic.
- Each continent has an ancient, stable craton made up of a Precambrian shield and platform. The Canadian shield in North America is made up of several subunits.
- Archean rocks are mostly granite-gneiss complexes and subordinate greenstone belts. One model for the origin of greenstone belts holds that they formed in back-arc marginal basins.
- The amalgamation of Archean cratons and continental accretion along their margins account for the origin of a large landmass known as Laurentia.
- Many geologists think that Archean plates moved faster than plates do now because Earth possessed more radiogenic heat.
- Sandstone-carbonate-shale assemblages deposited on passive continental margins and in intracratonic basins are the most common Proterozoic-aged rocks.

- Widespread glaciers were present during the Early and Late Proterozoic.
- Earth's earliest atmosphere lacked free oxygen but was rich in carbon dioxide. It was derived by the release of gases during volcanism, a process called outgassing. Meteorite and comet impacts and outgassing yielded the hydrosphere.
- Deposition of widespread banded iron formations between 2.5 and 2.0 billion years ago and the first red beds about 1.8 billion years ago indicates that some free oxygen was present in the atmosphere.
- Energy such as lightning and ultraviolet radiation acting on chemical elements present on early Earth may have yielded the first living things. Some investigators think that RNA molecules were the first molecules capable of reproduction.
- All known Archean fossils represent prokaryotic bacteria. Stromatolites formed by photosynthesizing bacteria may date from 3.5 billion years ago.
- Endosymbiosis practiced by prokaryotic cells was likely responsible for the first eukaryotic cells.
- The oldest fossils of multicelled organisms are carbonaceous impressions, probably of algae, in rocks more than 2 billion years old.
- The Late Proterozoic Ediacaran faunas include the oldest well-documented animal fossils. None had durable skeletons, so their fossils are not common.

Important Terms

anaerobic (p. 552)
autotrophic (p. 552)
banded iron formation (BIF) (p. 545)
Canadian shield (p. 534)
continental accretion (p. 534)
craton (p. 534)
Ediacaran faunas (p. 555)
endosymbiosis (p. 554)
eukaryotic cell (p. 553)
geologic record (p. 533)
greenstone belt (p. 537)
heterotrophic (p. 552)
Laurentia (p. 539)
Midcontinent Rift (p. 542)
monomer (p. 550)
multicelled organism (p. 555)
orogen (p. 539)
outgassing (p. 548)
photochemical dissociation (p. 548)
photosynthesis (p. 549)
platform (p. 534)
polymer (p. 550)
prokaryotic cell (p. 552)
red bed (p. 546)
sandstone-carbonate-shale assemblage (p. 541)
shield (p. 534)
stromatolite (p. 551)

Review Questions

1. The uppermost unit in a typical greenstone belt is composed mostly of
   a._____sedimentary rocks; b._____ultramafic rocks; c._____andesite; d._____lava flows; e._____granite and gneiss.
2. The largest exposed area of the North American craton is the
   a._____Wyoming craton; b._____Grand Canyon; c._____Canadian shield; d._____American platform; e._____Minnesota lowlands.
3. One source of Earth's atmosphere was volcanic gases in a process known as
   a._____dewatering; b._____photosynthesis; c._____autotrophism; d._____fermentation; e._____outgassing.
4. A polymer is an organic molecule
   a._____capable of metabolism and reproduction; b._____made up of amino acids linked together; c._____characterized as heterotrophic and anaerobic; d._____that binds together loose sediment to make stromatolites; e._____found mostly in the lava flows of greenstone belts.
5. Most of the free oxygen was probably added to Earth's atmosphere by the process known as
   a._____photosynthesis; b._____intracontinental rifting; c._____cratonization; d._____polymerization; e._____photochemical dissociation.
6. Some scientists think the first self-replicating system might have been a(n)

a._____stromatolite; b._____RNA molecule; c._____bacterium; d._____ATP cell; e._____proteinoid.

7. Which one of the following rock associations is typical of passive continental margins?
a._____sandstone-granite-basalt; b._____basalt-andesite-ash fall; c._____granite-tillite-banded iron formation; d._____stromatolite-prokaryote-lava; e._____sandstone-carbonate-shale.

8. Most of the evidence for the origin of eukaryotic cells from prokaryotic cells comes from
a._____chemicals preserved in Archean rocks; b._____fossils in tillite; c._____studies of present-day organisms; d._____tracks and trails of trilobites; e._____Early Proterozoic carbonate rocks.

9. The Middle Proterozoic of Laurentia was a time of
a._____the origin of most of Earth's greenstone belts; b._____igneous activity, rifting, and the Grenville orogeny; c._____appearance of the first stromatolites; d._____origin of animals with durable skeletons of chitin, calcium carbonate, and silicon dioxide; e._____widespread glaciation and comet and meteorite impacts.

10. Why are Precambrian rocks so difficult to study compared with Phanerozoic rocks?

11. What is the evidence that some rocks were present before the beginning of the Archean Eon 4.0 billion years ago?

12. Why are ultramafic rocks so rare in rocks younger than Archean?

13. Summarize the experimental evidence that indicates both monomers and polymers could have formed by natural processes on early Earth.

14. Explain how photosynthesis and photochemical dissociation supplied oxygen to the Archean atmosphere.

15. How is it that geologists have few difficulties correlating Phanerozoic rocks of the same age, but find it hard or impossible to do the same with Archean rocks? Also, why are even radiometric dates of limited use in this endeavor?

16. What are stromatolites, and how do they form?

17. Explain how pre-Archean crust might have formed. Why is none of this ancient crust still present?
18. Explain how red beds, banded iron formations, and the earliest eukaryotic cells give us some idea of the composition of the Proterozoic atmosphere.
19. The Belt-Purcell Supergroup of the northwestern United States and adjacent parts of Canada is 4000 m thick and was deposited between 1.45 billion and 850 million years ago. Calculate the average rate of sediment accumulation in millimeters per year. Why is this figure unlikely to represent the real rate of sedimentation?
20. Discuss the endosymbiosis theory for the origin of eukaryotic cells.

World Wide Web Activities

GeologyNow Assess your understanding of this chapter's topics with additional quizzing and comprehensive interactivities at

http://earthscience.brookscole.com/changingearth4e

as well as current and up-to-date weblinks, additional readings, and InfoTrac College Edition exercises.

Paleozoic Earth History

CHAPTER 20

OUTLINE

Geology Now *This icon, which appears throughout the book, indicates an opportunity to explore interactive tutorials, animations, or practice problems available on the GeologyNow website at http://earthscience.brookscole.com/changingearth4e.*

OBJECTIVES

At the end of this chapter, you will have learned that:

- Six major continents were present at the beginning of the Paleozoic Era, and plate movement during the Paleozoic resulted in continental collisions leading to the formation of the supercontinent Pangaea at the end of the Paleozoic.
- The Paleozoic history of North America is subdivided into six cratonic sequences that represent major transgressive–regressive cycles.
- During the transgressive portions of each cycle, the North American craton was partially to completely covered by shallow seas in which a variety of detrital and carbonate sediments were deposited, resulting in widespread sandstone, shale, reef, and coal deposits.
- Mountain-building activity took place primarily along the eastern and southern margins (known as mobile belts) of the North American craton during the Paleozoic Era.
- In addition to large-scale plate interactions, microplate activity played an important role in the formation of Pangaea.
- Paleozoic-age rocks contain a variety of important mineral resources.

eproduction of William Smith's original geologic map of England. Published in 1815 and representing Smith's more than 20 years of meticulous study of the rocks and fossils of England, this hand-painted map was of immediate practical value in the search for rock and mineral resources.

ource: Courtesy of the Department of Earth Sciences, University of New Hampshire.

Introduction

August 1, 1815, is an important date in the history of geology. On that day William Smith, a canal builder, published the world's first true geologic map. More than 8 ft high and 6 ft wide, Smith's hand-painted geologic map of England represented more than 20 years of detailed study of the rocks and fossils of England.

England is a country rich in geologic history. Five of the six Paleozoic geologic systems (Cambrian, Ordovician, Silurian, Devonian, and Carboniferous) were described and named for rocks exposed in England. The Carboniferous coal beds of England helped fuel the Industrial Revolution, and the need to transport coal cheaply from where it was mined to where it was needed set off a flurry of canal building during the late 1700s and early 1800s. During this time William Smith, who was mapping coal mines, first began to notice how the same sequences of fossils are found in many areas. Smith then surveyed the English countryside for the most efficient canal routes to bring the coal to market. Much of his success was based on his ability to predict what rocks canal diggers would encounter. Realizing that his observations allowed him to unravel the geologic history of an area and correlate rocks from one region to another, William Smith set out to make the first geologic map of an entire country!

The story of how William Smith came to publish the world's first geologic map is a fascinating tale of determination and perseverance. Instead of gaining fame and success, however, Smith found himself, slightly less than four years later, in debtors' prison and, on his release after more than two months of imprisonment, homeless. If such a story can have a happy ending, William Smith at least lived long enough to finally be recognized and honored for the seminal contribution he made to the then-fledgling science of geology.

Just as William Smith applied basic geologic principles in deciphering the geology of England, we use these same principles to interpret the geology of the Paleozoic Era. In this chapter, we use the geologic principles and concepts discussed in earlier chapters to help explain how Earth's systems and associated geologic processes have interacted during the Paleozoic to lay the groundwork for the distribution of continental landmasses, ocean basins, and the topography we have today.

The Paleozoic history of most continents involves major mountain-building activity along their margins and numerous shallow-water marine transgressions and regressions over their interiors. These transgressions and regressions were caused by global changes in sea level that most probably were related to plate activity and glaciation. In this chapter, we examine the geologic history of North America in terms of major transgressions and regressions rather than a period-by-period chronology. Although the focus is on North America, we place those events in a global context.

CONTINENTAL ARCHITECTURE: CRATONS AND MOBILE BELTS

During the Precambrian, continental accretion and orogenic activity led to the formation of sizable continents. At least three large continents existed during the Late Proterozoic, and some geologists think that these landmasses later collided to form a single Pangaea-like supercontinent. This supercontinent began breaking apart sometime during the latest Proterozoic, and by the beginning of the Paleozoic Era, six major continents were present. Each continent can be divided into two major components: a craton and one or more mobile belts.

Cratons are the relatively stable and immobile parts of continents; they form the foundation upon which Phanerozoic sediments were deposited (■ Figure 20.1). Cratons typically consist of two parts: a shield and a platform.

Shields are the exposed portions of the crystalline basement rocks of a continent and are composed of Precambrian metamorphic and igneous rocks (see Figures 7.1 and 19.4) that reveal a history of extensive orogenic activity during the Precambrian. During the Phanerozoic, however, shields were extremely stable and formed the foundation of the continents.

Extending outward from the shields are buried Precambrian rocks that constitute a *platform*, another part of the craton. Overlying the platform are flat-lying or gently dipping Phanerozoic detrital and chemical sedimentary rocks that were deposited in widespread shallow seas that transgressed and regressed over the craton. These seas, called **epeiric seas,** were a common feature of most Paleozoic cratonic histories. Changes in sea level, caused primarily by continental glaciation as well as by plate movement, were responsible for the advance and retreat of the seas.

Whereas most of the Paleozoic platform rocks are still essentially flat lying, in some places they were gently folded into regional arches, domes, and basins (Figure 20.1). In many cases some of these structures stood out as low islands during the Paleozoic Era and supplied sediments to the surrounding epeiric seas.

Mobile belts are elongated areas of mountain-building activity. They are located along the margins of continents where sediments are deposited in the relatively shallow waters of the continental shelf and the deeper waters at the base of the continental slope. Dur-

Figure 20.1

The major cratonic structures and mobile belts of North America that formed during the Paleozoic Era.

ing plate convergence along these margins, the sediments are deformed and intruded by magma, creating mountain ranges.

Four mobile belts formed around the margin of the North American craton during the Paleozoic: the **Franklin, Cordilleran, Ouachita,** and **Appalachian mobile belts** (Figure 20.1). Each was the site of mountain building in response to compressional forces along a convergent plate boundary and formed such mountain ranges as the Appalachians and Ouachitas.

PALEOZOIC PALEOGEOGRAPHY

One result of plate tectonics is that Earth's geography is constantly changing. The present-day configuration of the continents and ocean basins is merely a snapshot in time. As the plates move about, the location of continents and ocean basins constantly changes. One of the goals of historical geology is to provide paleogeographic reconstructions of the world for the geologic past. By synthesizing all the pertinent paleoclimatic, paleomagnetic, paleontologic, sedimentologic, stratigraphic, and tectonic data available, geologists can construct paleogeographic maps. Such maps are simply interpretations of the geography of an area for a particular time in the geologic past.

The paleogeographic history of the Paleozoic Era, for example, is not as precisely known as for the Mesozoic and Cenozoic eras, in part because the magnetic anomaly patterns preserved in the oceanic crust were destroyed when much of the Paleozoic oceanic crust was subducted during the formation of Pangaea. Paleozoic paleogeographic reconstructions are therefore based primarily on structural relationships, climate-sensitive sediments such as red beds, evaporites, and coals, as well as the distribution of plants and animals.

At the beginning of the Paleozoic, six major continents were present. Besides these large landmasses, geologists have also identified numerous small microcontinents and island arcs associated with various microplates. We are primarily concerned, however, with the history of the six major continents and their relationships to one another. The six major Paleozoic continents are **Baltica** (Russia west of the Ural Mountains and the major part of northern Europe), **China** (a complex area consisting of at least three Paleozoic continents that were not widely separated and are here considered to include China, Indochina, and the Malay Peninsula), **Gondwana** (Africa, Antarctica, Australia, Florida, India, Madagascar, and parts of the Middle East and southern Europe), **Kazakhstania** (a triangular continent centered on Kazakhstan but considered by some to be an extension of the Paleozoic Siberian continent), **Laurentia** (most of present-day North America, Greenland, northwestern Ireland, and Scotland), and **Siberia** (Russia east of the Ural Mountains and Asia north of Kazakhstan and south of Mongolia). The paleogeographic reconstructions that follow (Figures 20.2, 20.3, and 20.4) are based on the methods used to determine and interpret the location, geographic features, and environmental conditions on the paleocontinents.

Early-Middle Paleozoic Global History

In contrast to today's global geography, the Cambrian world consisted of six major continents dispersed around the globe at low tropical latitudes (■ Figure 20.2a). Water circulated freely among ocean basins, and the polar regions were mostly ice-free. By the Late Cambrian, epeiric seas had covered areas of Laurentia, Baltica, Siberia, Kazakhstania, and China, whereas highlands were present in northeastern Gondwana, eastern Siberia, and central Kazakhstania.

During the Ordovician and Silurian periods, plate movement played a major role in the changing global geography (Figure 20.2b, c). Gondwana moved southward during the Ordovician and began to cross the South Pole, as indicated by Upper Ordovician tillites (sediments deposited by glaciers and later lithified) found today in the Sahara Desert. In contrast to the passive continental margin Laurentia exhibited during the Cambrian, an active convergent plate boundary formed along its eastern margin during the Ordovician, as indicated by the Late Ordovician *Taconic orogeny* that occurred in New England. During the Silurian, Baltica moved northwestward relative to Laurentia and collided with it to form the larger continent of **Laurasia.** This collision, which closed the northern Iapetus Ocean, is marked by the *Caledonian orogeny.* After this orogeny, the southern part of the Iapetus Ocean still remained open between Laurentia and Gondwana (Figure 20.2c). Siberia and Kazakhstania moved from a southern equatorial position during the Cambrian to north temperate latitudes by the end of the Silurian Period.

During the Devonian, as the southern Iapetus Ocean narrowed between Laurasia and Gondwana, mountain building continued along the eastern margin of Laurasia with the *Acadian orogeny* (■ Figure 20.3a). The erosion of the resulting highlands spread vast amounts of reddish fluvial sediments over large areas of northern Europe (Old Red Sandstone) and eastern North America (the Catskill Delta). Other Devonian tectonic events, probably related to the collision of Laurentia and Baltica, include the Cordilleran *Antler orogeny,* the *Ellesmere orogeny* along the northern margin of Laurentia (which may reflect the collision of Laurentia with Siberia), and the change from a passive continental margin to an active convergent plate boundary in the Uralian mobile belt of eastern Baltica. The distribution of reefs, evaporites, and red beds, as well as the existence of similar floras throughout the world, suggests a rather uniform global climate during the Devonian Period.

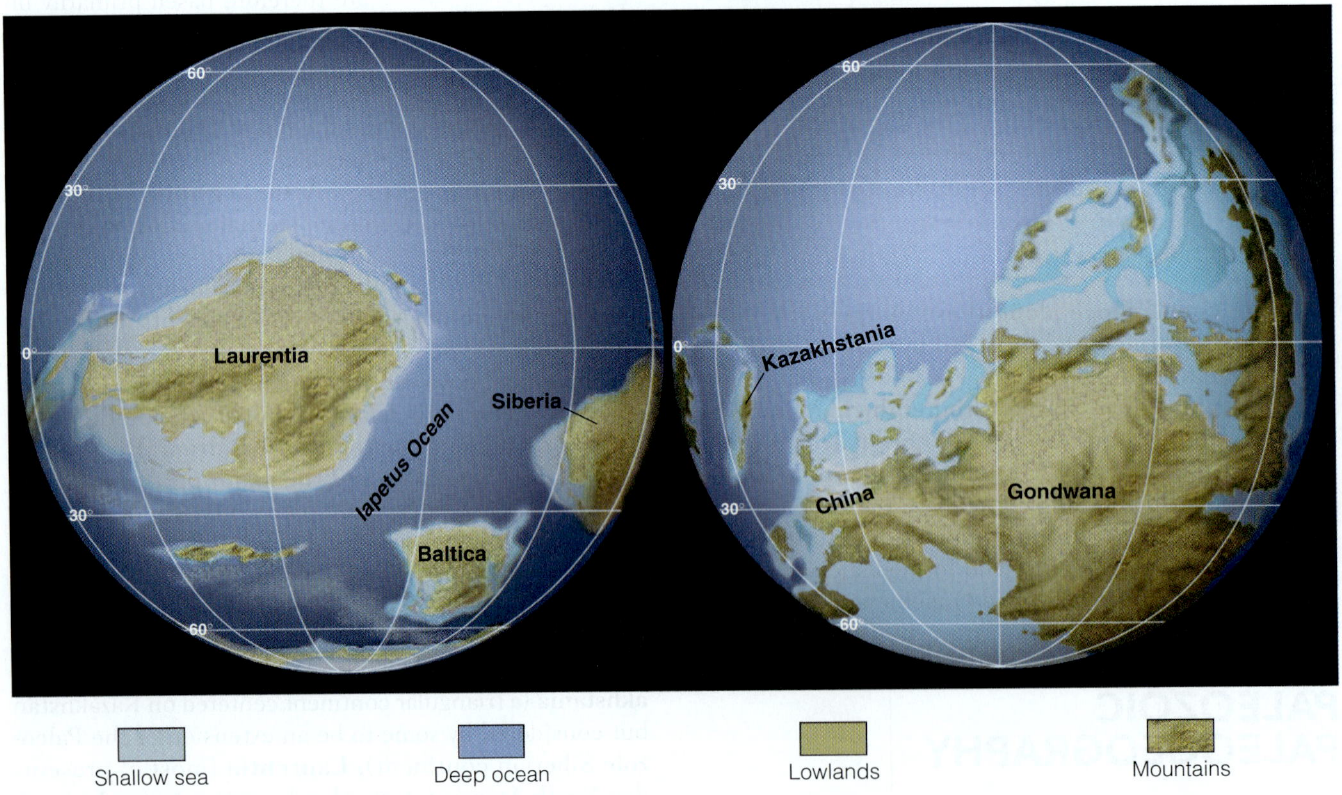

(a) Late Cambrian Period

Geology Now ■ **Active Figure 20.2**

Paleogeography of the world for the (a) Late Cambrian Period.

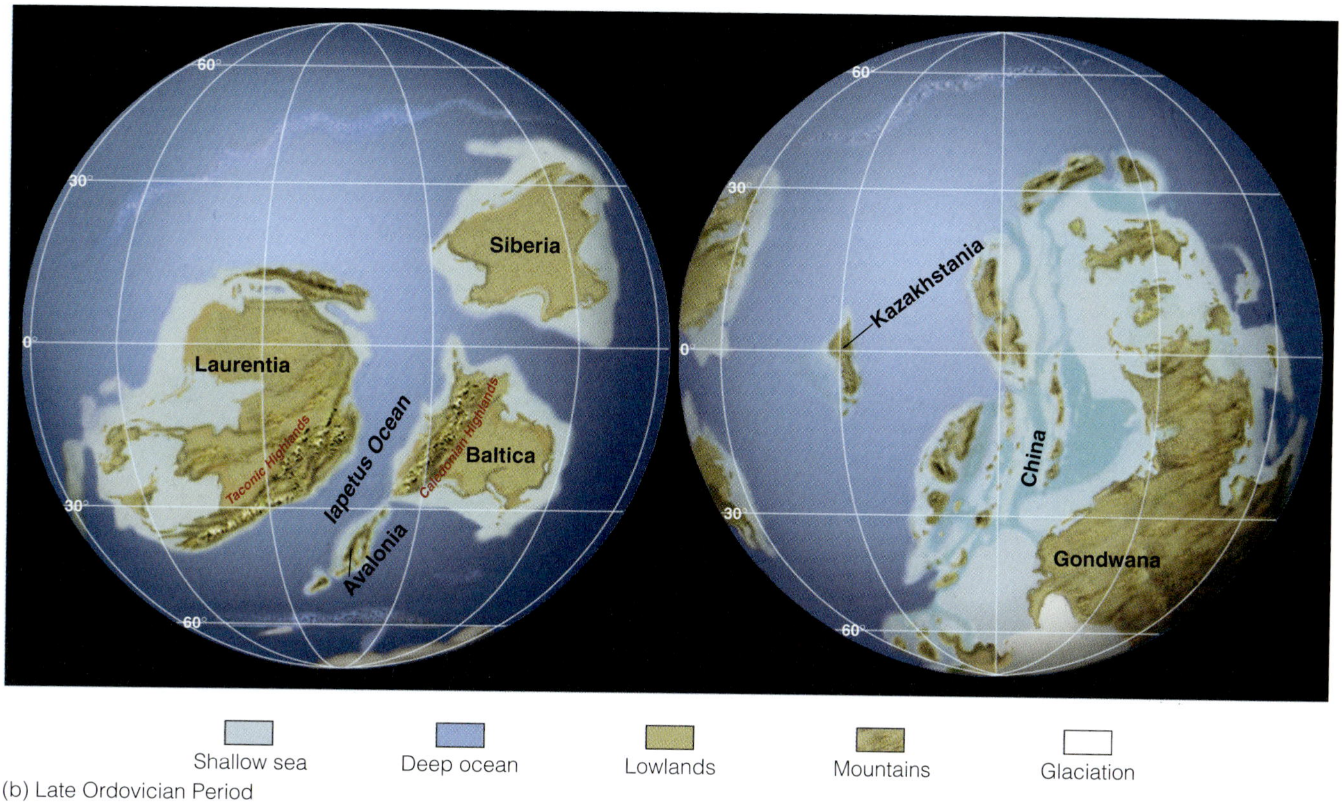

(b) Late Ordovician Period

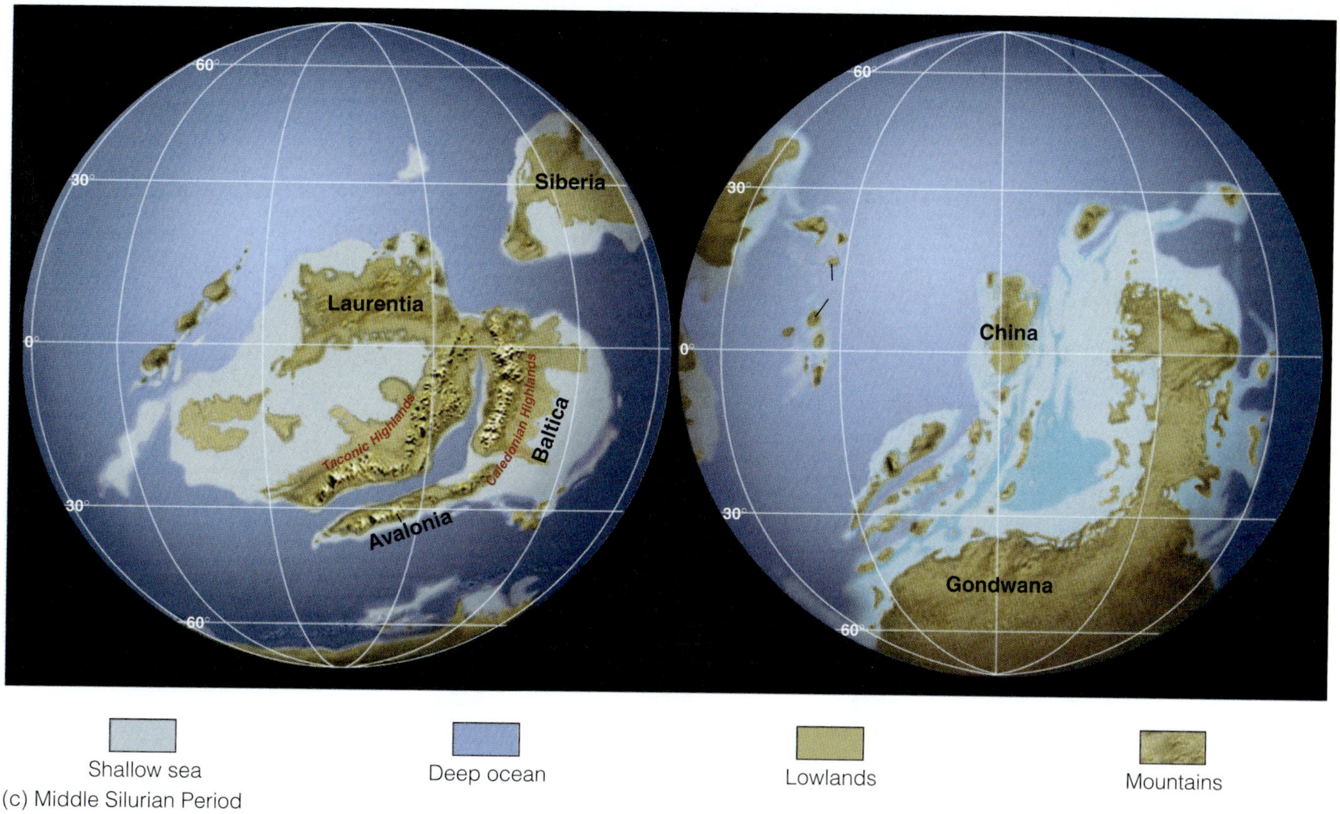

(c) Middle Silurian Period

GeologyNow ■ **Active Figure 20.2 *continued***

Paleogeography of the world for the (b) Late Ordovician Period and (c) Middle Silurian Period.

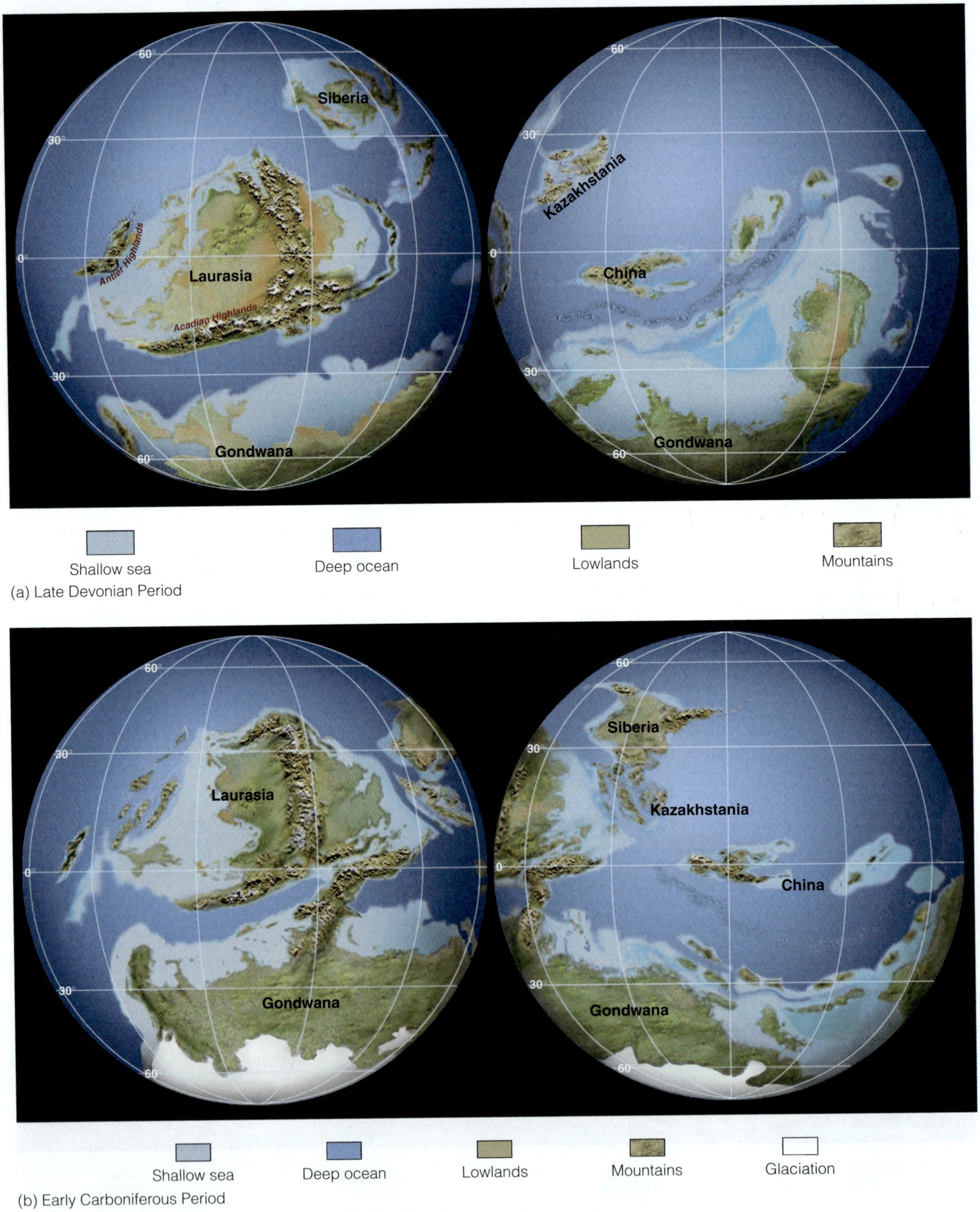

Figure 20.3

Paleogeography of the world for the (a) Late Devonian Period and (b) Early Carboniferous Period.

Late Paleozoic Global History

During the Carboniferous Period, southern Gondwana moved over the South Pole resulting in extensive continental glaciation (■ Figures 20.3b and 20.4a). The advance and retreat of these glaciers produced global changes in sea level that affected sedimentation patterns on the cratons. As Gondwana continued moving northward, it began colliding with Laurasia during the Early Carboniferous and continued suturing with it during the rest of the Carboniferous (Figures 20.3b and 20.4a). Because Gondwana rotated clockwise relative to Laurasia, deformation generally progressed in a northeast-to-southwest direction along the Hercynian, Appalachian, and Ouachita mobile belts of the two continents. The final phase of collision between Gondwana and Laurasia is indicated by the Ouachita Mountains of Oklahoma, formed by thrusting during the Late Carboniferous and Early Permian.

Elsewhere, Siberia collided with Kazakhstania and moved toward the Uralian margin of Laurasia (Baltica), colliding with it during the Early Permian. By the end of the Carboniferous, the various continental landmasses were fairly close together as Pangaea began taking shape.

The Carboniferous coal basins of eastern North America, western Europe, and the Donets Basin of the Ukraine all lay in the equatorial zone, where rainfall was high and temperatures were consistently warm. The absence of strong seasonal growth rings in fossil plants from these coal basins indicates such a climate. The fossil plants found in the coals of Siberia, however, show well-developed growth rings, signifying seasonal growth with abundant rainfall and distinct seasons such as in the temperate zones (latitudes 40 degrees to 60 degrees north).

Glacial conditions and the movement of large continental ice sheets in the high southern latitudes are indicated by widespread tillites and glacial striations in southern Gondwana (see Figure 2.6b). These ice sheets spread toward the equator and, at their maximum growth, extended well into the middle temperate latitudes.

The assembly of Pangaea was essentially completed during the Permian as a result of the many continental collisions that began during the Carboniferous (Figure 20.4b). Although geologists generally agree on the configuration and location of the western half of the supercontinent, there is no consensus on the number or configuration of the various terranes and continental blocks that composed the eastern half of Pangaea. Regardless of the exact configuration of the eastern portion, geologists know that the supercontinent was surrounded by various subduction zones and moved steadily northward during the Permian. Furthermore, an enormous single ocean, **Panthalassa,** surrounded Pangaea and spanned Earth from pole to pole (Figure 20.4b). Waters of this ocean probably circulated more freely than at present, resulting in more equable water temperatures.

What Would You Do?

You work for a travel agency and are putting together a raft trip down the Colorado River through the Grand Canyon. In addition to the usual information about such a trip, what kind of geologic information would you include in your brochure to make the trip appealing from an educational standpoint as well?

The formation of a single large landmass had climatic consequences for the terrestrial environment as well. Terrestrial Permian sediments indicate that arid and semiarid conditions were widespread over Pangaea. The mountain ranges produced by the *Hercynian, Alleghenian,* and *Ouachita orogenies* were high enough to create rainshadows that blocked the moist, subtropical, easterly winds—much as the southern Andes Mountains do in western South America today. This produced very dry conditions in North America and Europe, as evident from the extensive Permian red beds and evaporites found in western North America, central Europe, and parts of Russia. Permian coals, indicating abundant rainfall, were mostly limited to the northern temperate belts (latitude 40 degrees to 60 degrees north), whereas the last remnants of the Carboniferous ice sheets continued their recession.

PALEOZOIC EVOLUTION OF NORTH AMERICA

It is convenient to divide the geologic history of the North American craton into two parts: the first dealing with the relatively stable continental interior over which epeiric seas transgressed and regressed, and the other with the mobile belts where mountain building occurred.

In 1963, the American geologist Laurence Sloss subdivided the sedimentary rock record into six cratonic sequences. A **cratonic sequence** is a large-scale (greater than supergroup) lithostratigraphic unit representing a major transgressive–regressive cycle bounded by craton-wide unconformities (■ Figure 20.5). The transgressive phase, which is usually covered by younger sediments, commonly is well preserved, whereas the regressive phase of each sequence is marked by an unconformity. Where rocks of the appropriate age are preserved, each of the six unconformities can be shown to extend across the various sedimentary basins of the North American craton and into the mobile belts along the cratonic margin.

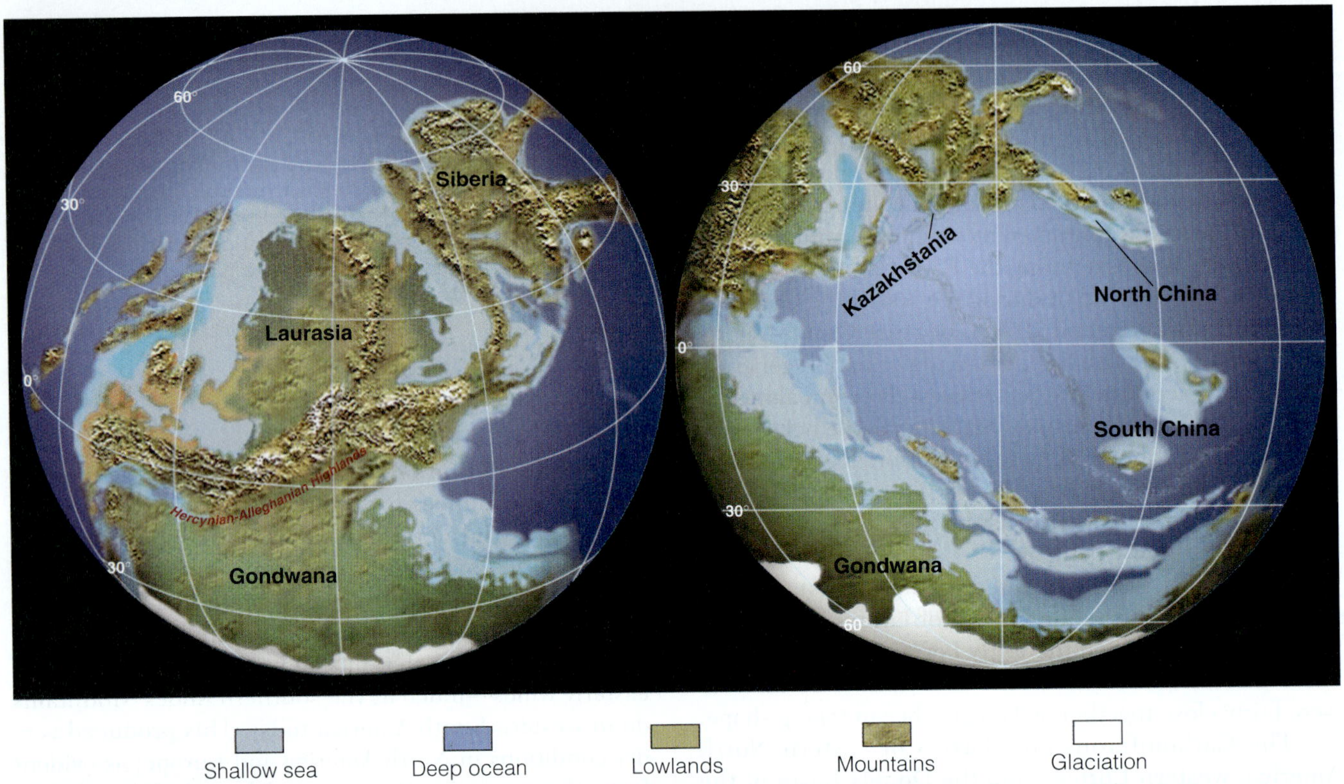

(a) Late Carboniferous Period

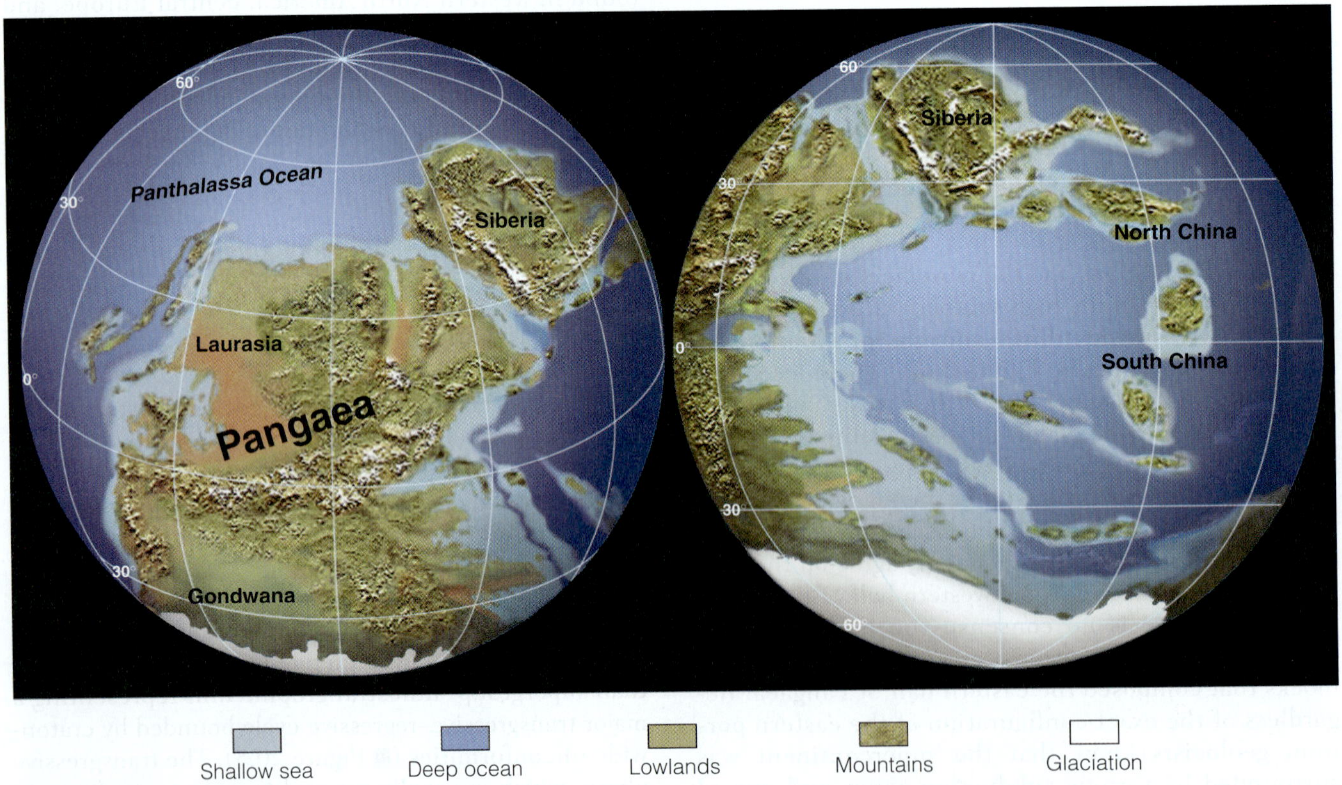

(b) Late Permian Period

Figure 20.4

Paleogeography of the world for the (a) Late Carboniferous Period and (b) Late Permian Period.

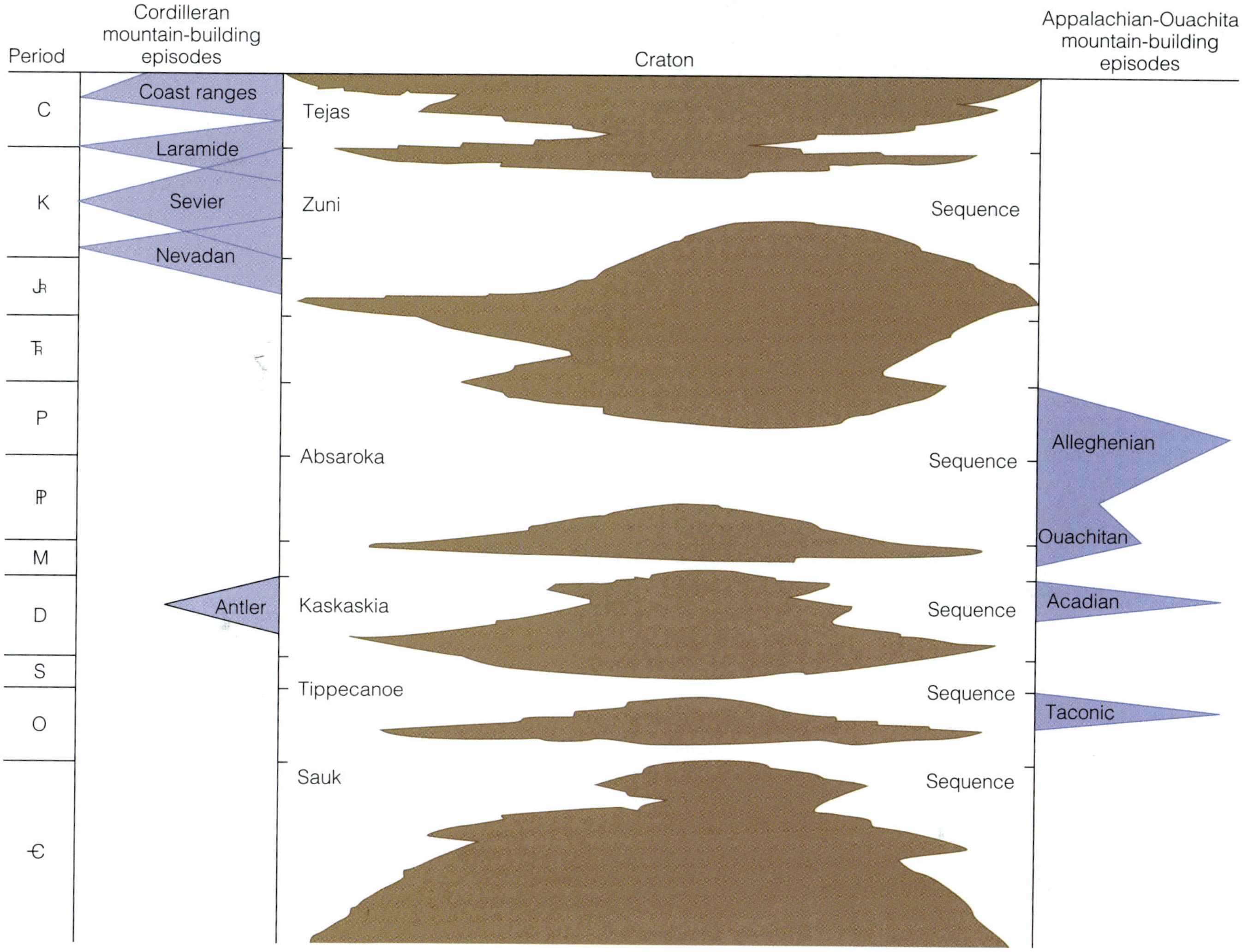

Figure 20.5

Cratonic sequences of North America. The white areas represent sequences of rocks separated by large-scale unconformities (brown areas). The major Cordilleran orogenies are shown on the left side, and the major Appalachian orogenies are on the right.

Geologists have also recognized major unconformity-bounded sequences in cratonic areas outside North America. Such global transgressive–regressive cycles of sea-level changes are thought to result from major tectonic and glacial events.

The realization that rock units can be divided into cratonic sequences and that these sequences can be further subdivided and correlated provides the foundation for an important concept in geology that allows high-resolution analysis of time and facies relationships within sedimentary rocks. **Sequence stratigraphy** is the study of rock relationships within a time-stratigraphic framework of related facies bounded by erosional or nondepositional surfaces. The basic unit of sequence stratigraphy is the *sequence*, which is a succession of rocks bounded by unconformities and their equivalent conformable strata. Sequence boundaries result from a relative drop in sea level. Sequence stratigraphy is an important tool in geology because it allows geologists to subdivide sedimentary rocks into related units that are bounded by time-stratigraphically significant boundaries. Geologists use sequence stratigraphy for high-resolution correlation and mapping as well as interpreting and predicting depositional environments.

THE SAUK SEQUENCE

Rocks of the **Sauk sequence** (Late Proterozoic–Early Ordovician) record the first major transgression onto the North American craton (Figure 20.5). During the Late Proterozoic and Early Cambrian, deposition of marine sediments was limited to the passive shelf areas of the Appalachian and Cordilleran borders of the craton. The craton itself was

GEOFOCUS 20.1

Pictured Rocks National Lakeshore

Exposed along the south shore of Lake Superior between Au Sable Point and Munising in Michigan's Upper Peninsula is the beautiful and imposing wavecut sandstone cliffs called Pictured Rocks (■ Figure 1). The rocks exposed in this area, part of which is designated a national lakeshore, make up the Upper Cambrian Munising Formation, which is divided into two members: the lower Chapel Rock Sandstone and the upper Miner's Castle Sandstone. The Munising Formation unconformably overlies the Upper Proterozoic Jacobsville Sandstone and is unconformably overlain by the Middle Ordovician Au Train Formation. The reddish brown, coarse-grained Jacobsville Sandstone was deposited in streams and lakes over an irregular erosion surface. Following deposition, the Jacobsville was slightly uplifted and tilted.

By the Late Cambrian, the transgressing Sauk Sea reached the Michigan area, and the Chapel Rock Sandstone was deposited. The principal source area for this unit was the Northern Michigan highlands, an area that corresponds to the present Upper Peninsula. Following deposition of the Chapel Rock Sandstone, the Sauk Sea retreated from the area.

During a second transgression of the Sauk Sea in this area, the Miner's Castle Sandstone was deposited. This second transgression covered most of the Upper Peninsula of Michigan and drowned the highlands that were the source for the older Chapel Rock Sandstone.

The source area for the Miner's Castle Sandstone was the Precambrian Canadian Shield area to the north and northeast. The Miner's Castle Sandstone contains rounder, better sorted, and more abundant quartz grains than the Chapel Rock Sandstone, indicating a different source area. A major unconformity separates the Miner's Castle Sandstone from the overlying Middle Ordovician Au Train Formation.

One of the most prominent features of Pictured Rocks National Lakeshore is Miner's Castle, a wave-cut projection along the shoreline (Figure 1b). The lower sandstone unit at water level is the Chapel Rock Sandstone, and the rest of the feature is composed of the Miner's Castle Sandstone. The two turrets of the castle formed as sea stacks during a time following the Pleistocene when the water level of Lake Superior was much higher.

above sea level and experiencing extensive weathering and erosion. Because North America was located in a tropical climate at this time and there is no evidence of any terrestrial vegetation, weathering and erosion of the exposed Precambrian basement rocks must have proceeded rapidly. During the Middle Cambrian, the transgressive phase of the Sauk began with epeiric seas encroaching over the craton (see Geo-Focus 20.1). By the Late Cambrian, the Sauk Sea had covered most of North America, leaving only a portion of the Canadian Shield and a few large islands above sea level (■ Figure 20.6). These islands, collectively named the **Transcontinental Arch,** extended from New Mexico to Minnesota and the Lake Superior region.

The sediments deposited both on the craton and along the shelf area of the craton margin show abundant evidence of shallow-water deposition. The only difference between the shelf and craton deposits is that the shelf deposits are thicker. In both areas, the sands are generally clean and well sorted, and commonly contain ripple marks and small-scale cross-bedding. Many of the carbonates are bioclastic (composed of fragments of organic remains), contain stromatolites, or have oolitic (small, spherical calcium carbonate grains) textures. Such sedimentary structures and textures indicate shallow-water deposition.

Recall from Chapter 6 that sediments become increasingly finer the farther away from land one goes. Therefore, in a stable environment where sea level remains the same, coarse detrital sediments are typically deposited in the nearshore environment, and finer-grained sediments are deposited in the offshore environ-

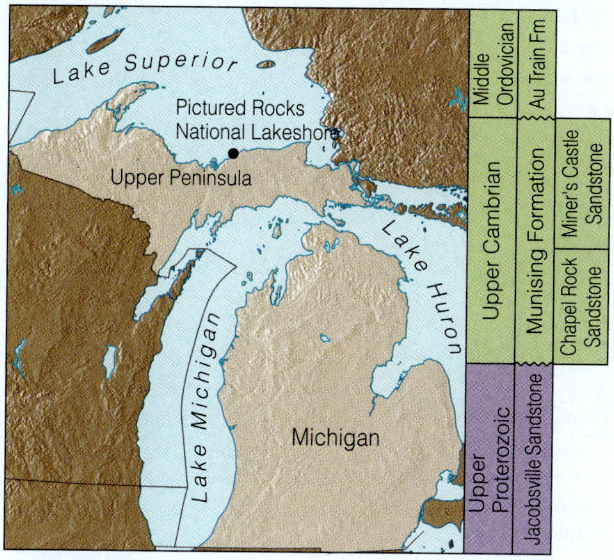

(a)

(b)

Figure 1

(a) Location of Pictured Rocks National Lakeshore and stratigraphy of rocks exposed in this area. (b) The photograph of the Munising Formation shows the shoreline projection called Miner's Castle. The two turrets at the top of Miner's Castle formed by wave action when the water level of Lake Superior was higher. The contact between the Miner's Castle and Chapel Rock sandstone members is located just above lake level.

ment. Carbonates form farthest from land in the area beyond the reach of detrital sediments. During a transgression, these facies (sediments that represent a particular environment) migrate in a landward direction (see Figure 6.21).

This time-transgressive facies relationship occurred not only in the Grand Canyon area of the craton during the Sauk sequence but elsewhere as well as the seas encroached from the Appalachian and Ouachita mobile belts onto the craton interior (■ Figure 20.7). Carbonate deposition dominated on the craton as the Sauk transgression continued during the Early Ordovician, and the islands of the Transcontinental Arch were soon covered by the advancing Sauk Sea. By the end of Sauk time, much of the craton was submerged beneath a warm, equatorial epeiric sea (Figure 20.2a).

THE TIPPECANOE SEQUENCE

As the Sauk Sea regressed from the craton during the Early Ordovician, it revealed a landscape of low relief. The rocks exposed were predominantly limestones and dolostones that experienced deep and extensive erosion because North America was still located in a tropical environment (■ Figure 20.8). The resulting cratonwide unconformity marks the boundary between the Sauk and Tippecanoe sequences.

Like the Sauk sequence, deposition of the **Tippecanoe sequence** (Middle Ordovician–Early Devonian)

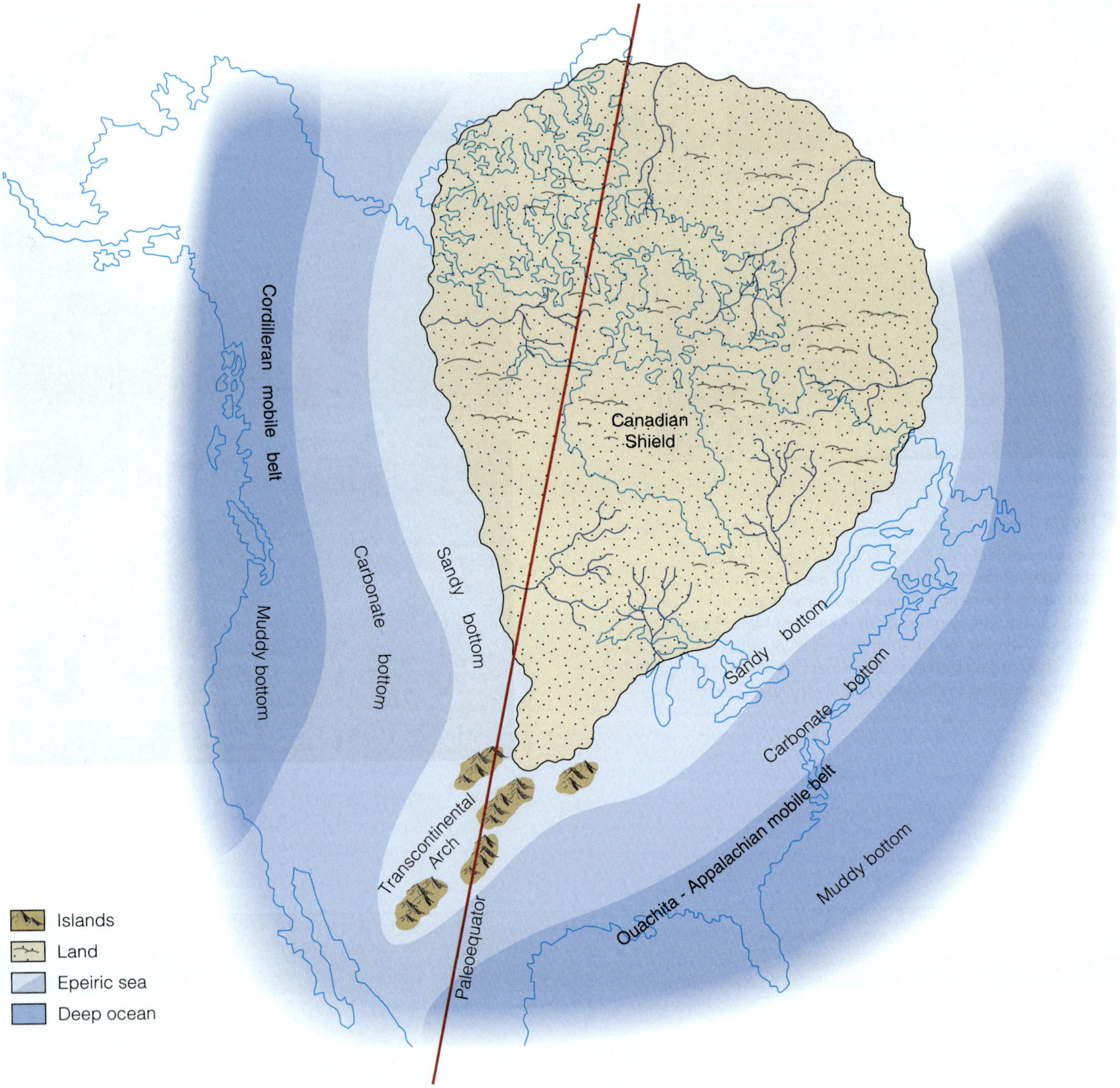

Figure 20.6

Paleogeography of North America during the Cambrian Period. Note the position of the Cambrian paleoequator. During this time North America straddled the equator as indicated in Figure 20.2a.

began with a major transgression onto the craton. This transgressing sea deposited clean, well-sorted quartz sands over most of the craton. The best known of the Tippecanoe basal sandstones is the St. Peter Sandstone, an almost pure quartz sandstone used in manufacturing glass. It occurs throughout much of the midcontinent and resulted from numerous cycles of weathering and erosion of Proterozoic and Cambrian sandstones deposited during the Sauk transgression.

The Tippecanoe basal sandstones were followed by widespread carbonate deposition (Figure 20.8). The limestones were generally the result of deposition by calcium carbonate–secreting organisms such as corals, brachiopods, stromatoporoids, and bryozoans. Besides the limestones, there were also many dolostones. Most of the dolostones formed as a result of magnesium replacing calcium in calcite, thus converting the limestones into dolostones.

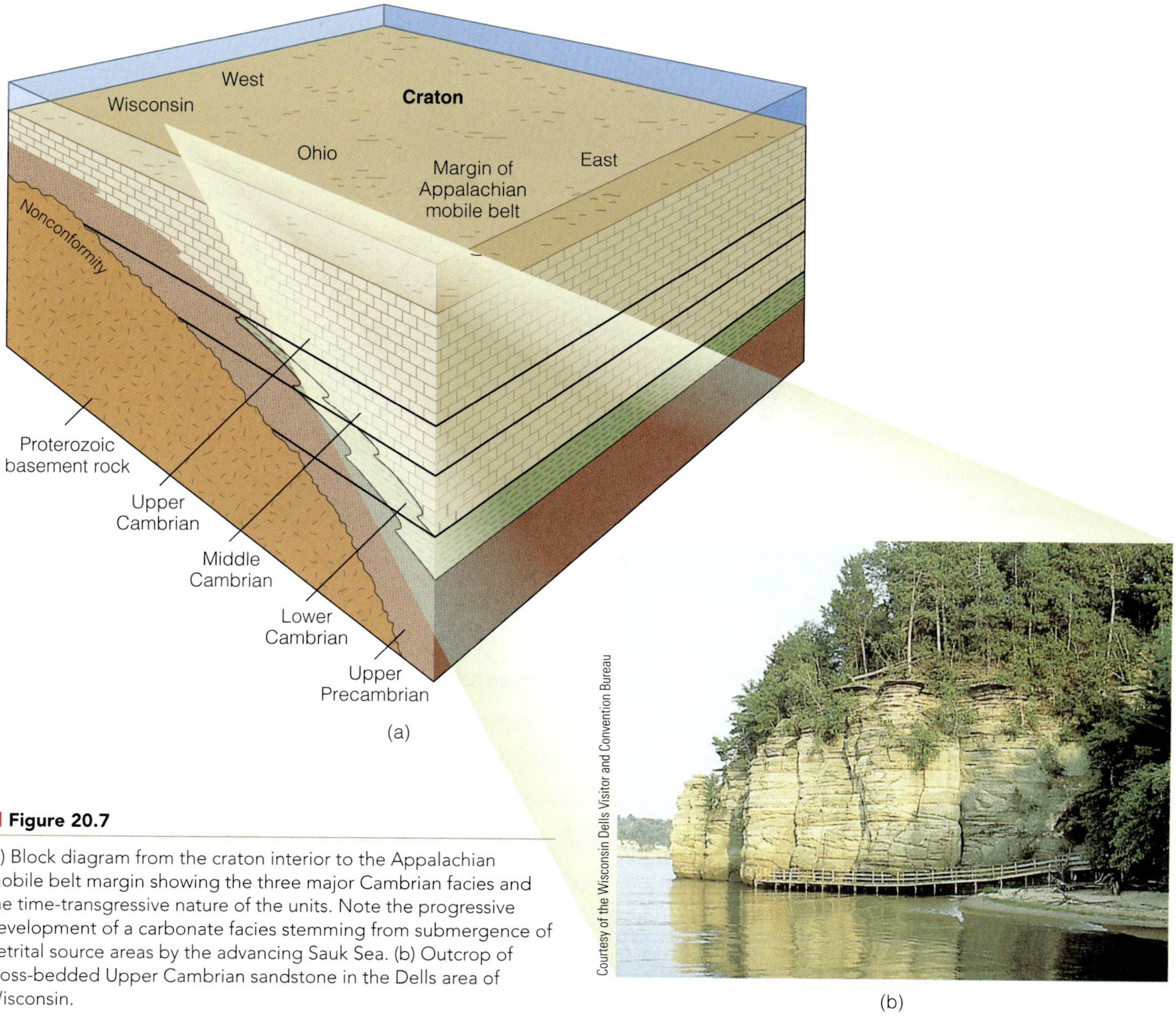

■ **Figure 20.7**

(a) Block diagram from the craton interior to the Appalachian mobile belt margin showing the three major Cambrian facies and the time-transgressive nature of the units. Note the progressive development of a carbonate facies stemming from submergence of detrital source areas by the advancing Sauk Sea. (b) Outcrop of cross-bedded Upper Cambrian sandstone in the Dells area of Wisconsin.

In the eastern portion of the craton, the carbonates grade laterally into shales. These shales mark the farthest extent of detrital sediments derived from weathering and erosion of the Taconic Highlands, a tectonic event we will discuss later.

Tippecanoe Reefs and Evaporites

Organic reefs are limestone structures constructed by living organisms, some of which contribute skeletal materials to the reef framework (■ Figure 20.9). Today corals and calcareous algae are the most prominent reef builders, but in the geologic past other organisms played a major role. Regardless of the organisms that dominate reef communities, reefs appear to have occupied the same ecological niche in the geologic past that they do today. Because of the ecological requirements of reef-building organisms, present-day reefs are confined to a narrow latitudinal belt between 30 degrees north and south of the equator. Corals, the major reef-building organisms today, require warm, clear, shallow water of normal salinity for optimal growth.

The size and shape of a reef are largely the result of the interactions among the reef-building organisms, the bottom topography, wind and wave action, and subsidence of the seafloor. Reefs also alter the area around them by forming barriers to water circulation or wave action.

Reefs have been common features since the Cambrian and have been built by a variety of organisms. The first skeletal builders of reeflike structures were archaeocyathids. These conical-shaped organisms lived during the Cambrian and had double, perforated, calcareous shell walls. Archaeocyathids built small mounds that

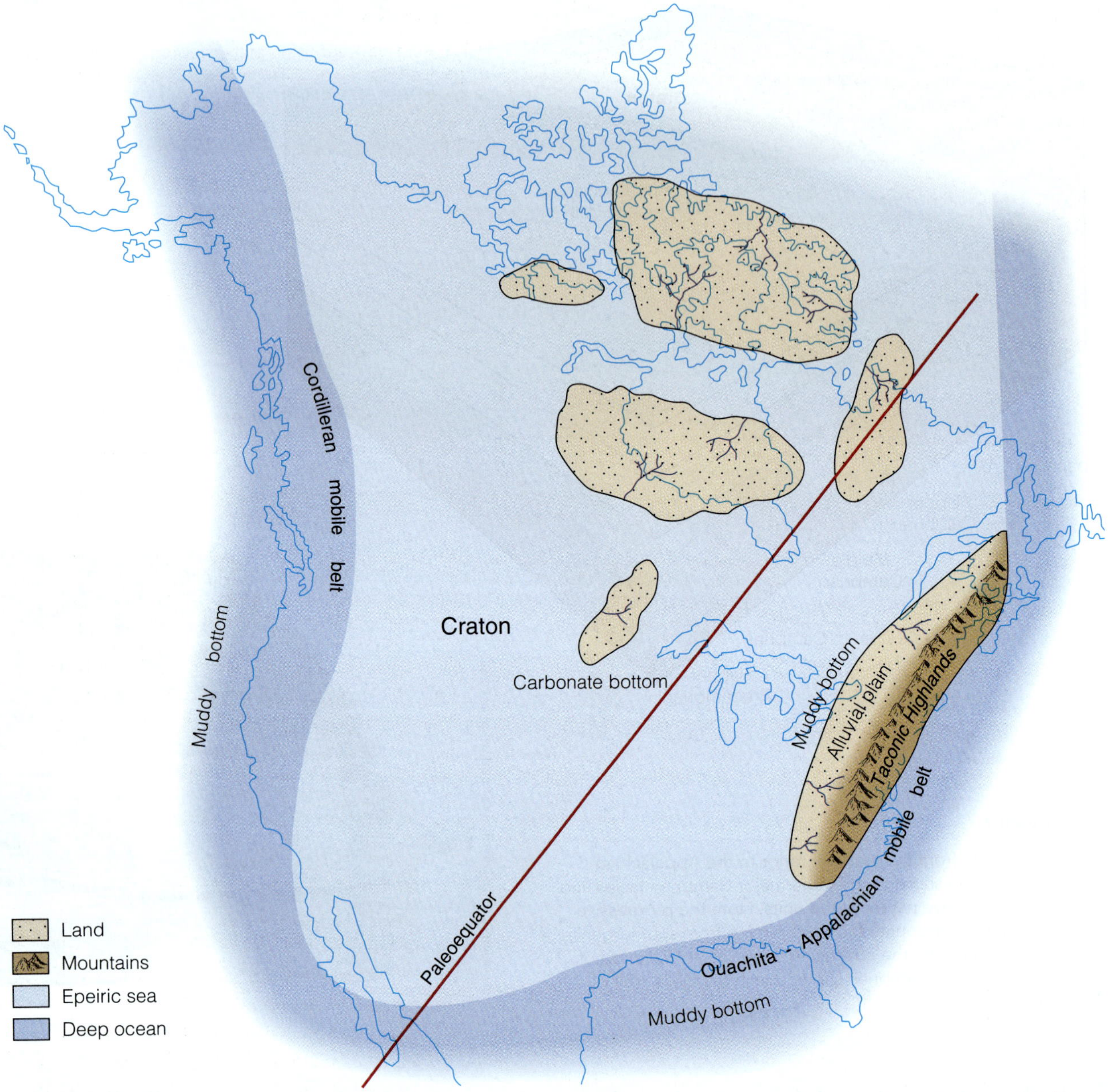

Figure 20.8

Paleogeography of North America during the Ordovician Period. Note that the position of the equator has shifted, indicating that North America was rotating counterclockwise.

have been found on all continents except South America (see Figure 21.5). Beginning in the Middle Ordovician, stromatoporoid-coral reefs became common in the low latitudes, and similar reefs remained so throughout the rest of the Phanerozoic Eon. The burst of reef building seen in the Late Ordovician through Devonian probably occurred in response to evolutionary changes triggered by the appearance of extensive carbonate seafloors and platforms beyond the influence of detrital sediments.

The Middle Silurian rocks (Tippecanoe sequence) of the present-day Great Lakes region are world-famous for their reef and evaporite deposits (Figure 20.10). The most famous structure in the region, the Michigan Basin, is a broad, circular basin surrounded by large barrier reefs. No doubt these reefs contributed to increasingly restricted circulation and the precipitation of Upper Silurian evaporites within the basin (Figure 20.11).

Within the rapidly subsiding interior of the basin, other types of reefs are found. *Pinnacle reefs* are tall, spindly structures up to 100 m high. They reflect the rapid upward growth needed to maintain themselves near sea level during subsidence of the basin (Figure 20.11a). Besides the pinnacle reefs, bedded carbonates

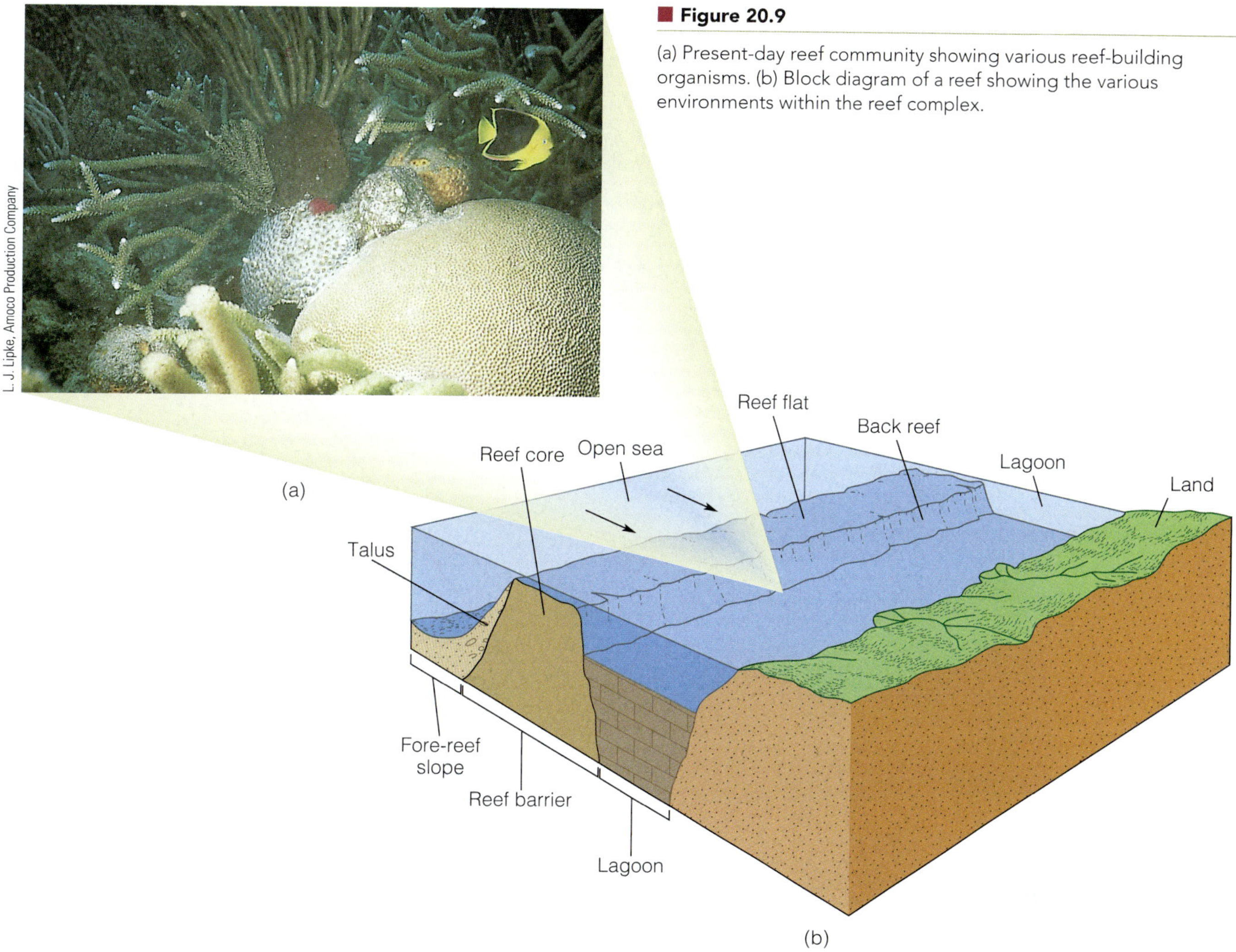

■ **Figure 20.9**

(a) Present-day reef community showing various reef-building organisms. (b) Block diagram of a reef showing the various environments within the reef complex.

and thick sequences of salt and anhydrite are also found in the Michigan Basin (Figure 20.11a, b, d).

As the Tippecanoe Sea gradually regressed from the craton during the Late Silurian, precipitation of evaporite minerals occurred in the Appalachian, Ohio, and Michigan Basins. In the Michigan Basin alone, approximately 1500 m of sediments were deposited, nearly half of which are halite and anhydrite. How did such thick sequences of evaporites accumulate? One possibility is that when sea level dropped, the tops of the barrier reefs were as high as or above sea level, thus preventing the influx of new seawater into the basin. Evaporation of the basinal seawater would result in the precipitation of salts. A second possibility is that the reefs grew upward so close to sea level that they formed a sill or barrier that eliminated interior circulation (■ Figure 20.12).

The End of the Tippecanoe Sequence

By the Early Devonian, the regressing Tippecanoe Sea had retreated to the craton margin, exposing an extensive low-land topography. During this regression, marine deposition was initially restricted to a few interconnected cratonic basins and, finally by the end of the Tippecanoe, to only the mobile belts surrounding the craton.

As the Tippecanoe Sea regressed during the Early Devonian, the craton experienced mild deformation, forming many domes, arches, and basins. These structures were mostly eroded during the time the craton was exposed, so they were eventually covered by deposits from the encroaching Kaskaskia Sea.

THE KASKASKIA SEQUENCE

The boundary between the Tippecanoe sequence and the overlying **Kaskaskia sequence** (Middle Devonian–Middle Mississippian) is marked by a major unconformity. As the Kaskaskia Sea transgressed over the low-relief landscape of the craton, most basal beds deposited consisted of clean, well-sorted quartz sandstones.

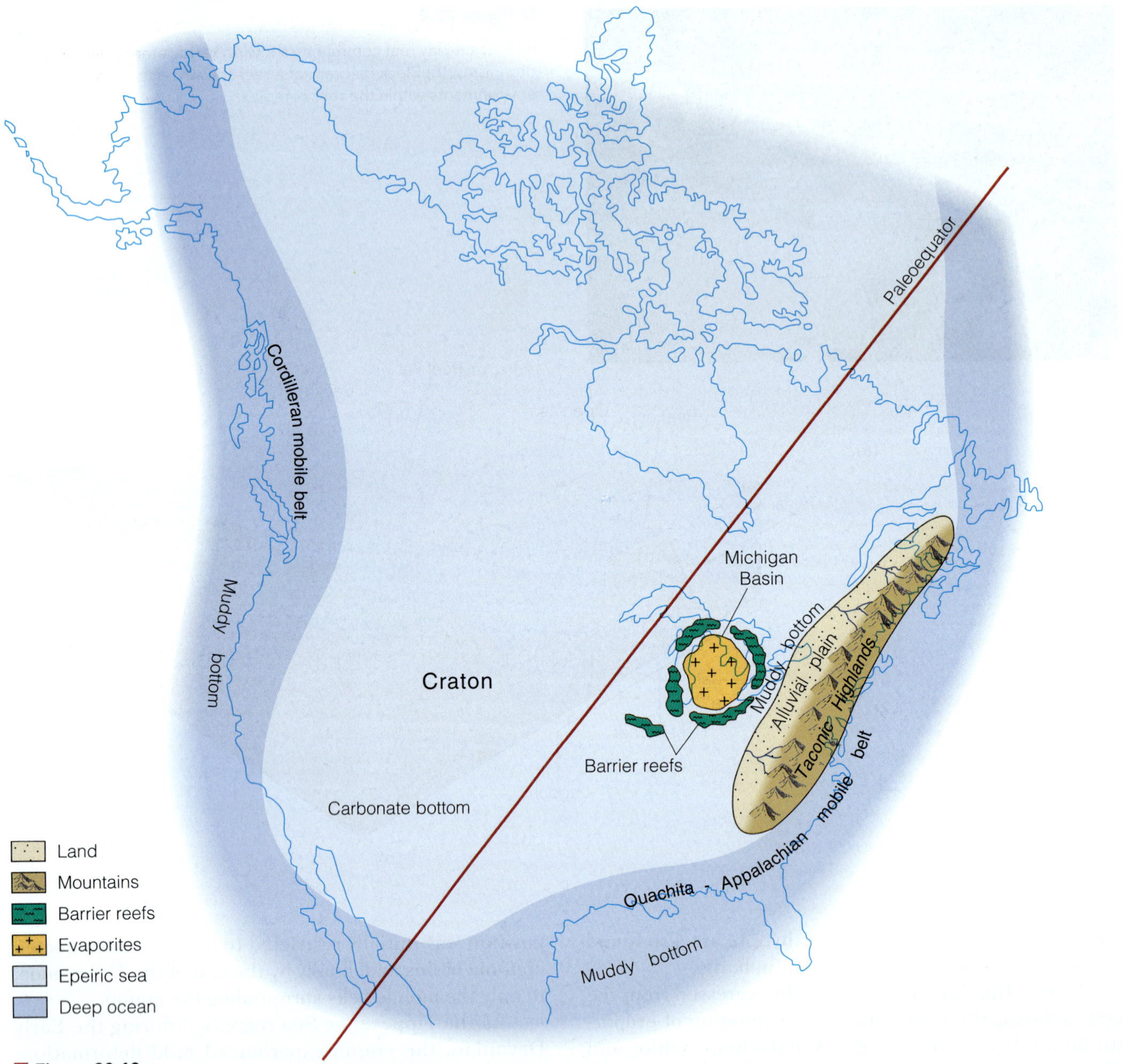

Figure 20.10

Paleogeography of North America during the Silurian Period. Note the development of reefs in the Michigan, Ohio, and Indiana-Illinois-Kentucky areas.

The source areas for the basal Kaskaskia sandstones were primarily the eroding highlands of the Appalachian mobile belt area (■ Figure 20.13), exhumed Cambrian and Ordovician sandstones cropping out along the flanks of the Ozark Dome, and exposures of the Canadian Shield in the Wisconsin area. The lack of similar sands in the Silurian carbonate beds below the Tippecanoe-Kaskaskia unconformity indicates that the source areas of the basal Kaskaskia detrital rocks were submerged when the Tippecanoe sequence was deposited. Stratigraphic studies indicate these source areas were uplifted and the Tippecanoe carbonates removed by erosion before the Kaskaskia transgression. Kaskaskian basal rocks elsewhere on the craton consist of carbonates that are frequently difficult to differentiate from the underlying Tippecanoe carbonates unless they are fossiliferous.

Except for widespread Upper Devonian and Lower Mississippian black shales, the majority of Kaskaskian rocks are carbonates, including reefs, and associated evaporite deposits. In many other parts of the world, such as southern England, Belgium, central Europe, Australia, and Russia, the Middle and early Late Devonian epochs were times of major reef building.

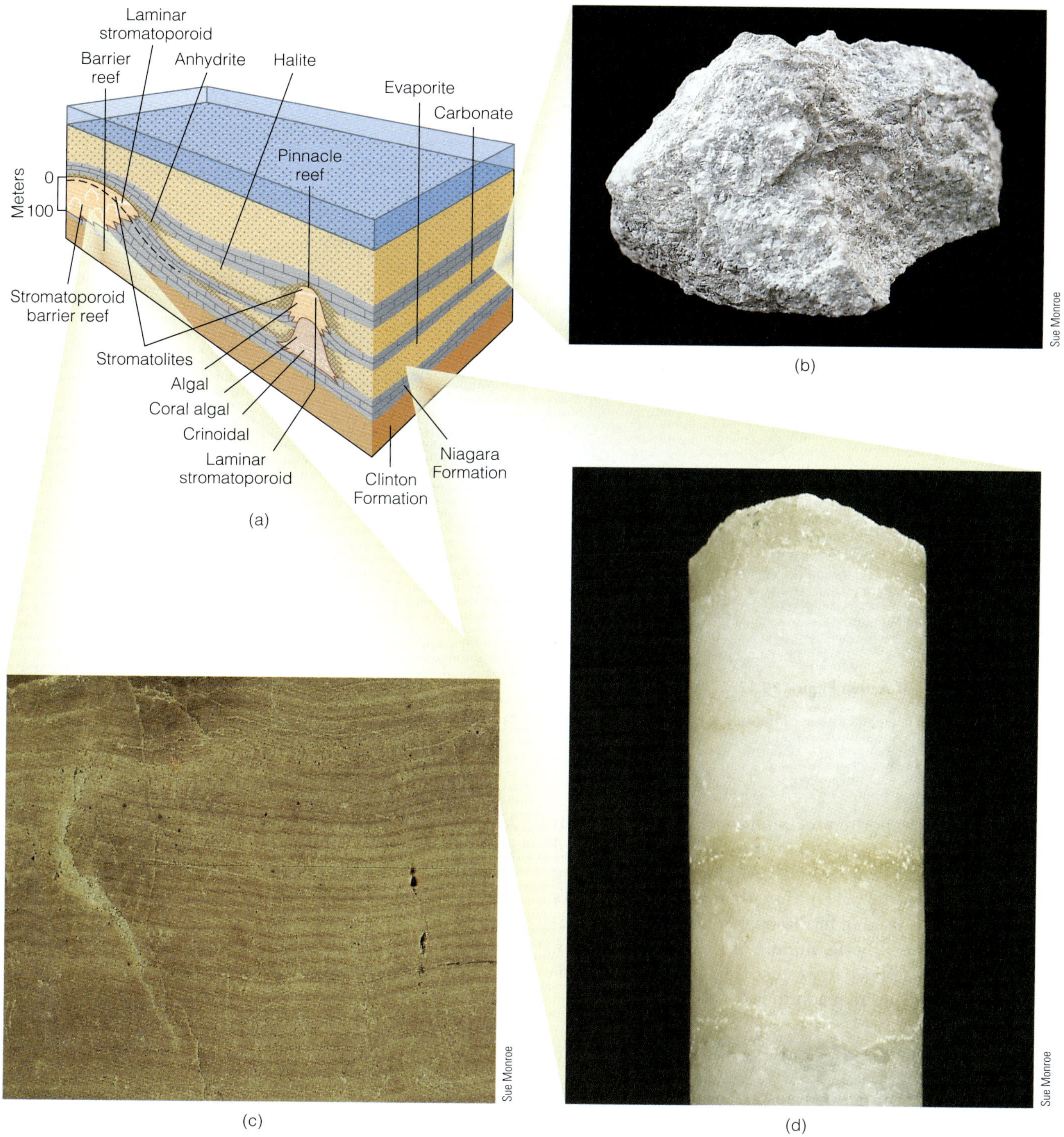

■ **Figure 20.11**

(a) Generalized block diagram of the northern Michigan Basin during the Silurian Period. (b) Limestone from the carbonate facies. (c) Cross section of a stromatoporoid colony from the stromatoporoid barrier reef facies. (d) Core of rock salt from the evaporite facies.

Source: (a) From K. J. Mesolella, J. D. Robinson, L. M. McCormick and A. R. Ormiston, "Cyclic Deposition of Silurian Carbonates and Evaporates in Michigan Basin," AAPG Bulletin, Vol. 58, No. 1, AAPG © 1958, Fig. 6, p. 40. Reprinted by permission of AAPG, whose permission is required for future use.

Reef Development in Western Canada

The Middle and Late Devonian reefs of western Canada contain large reserves of petroleum and have been widely studied from outcrops and in the subsurface (■ Figure 20.14). These reefs began forming as the Kaskaskia Sea transgressed southward into western Canada. By the end of the Middle Devonian, they had coalesced into a large barrier-reef system that restricted the flow of oceanic water into the back-reef platform, creating

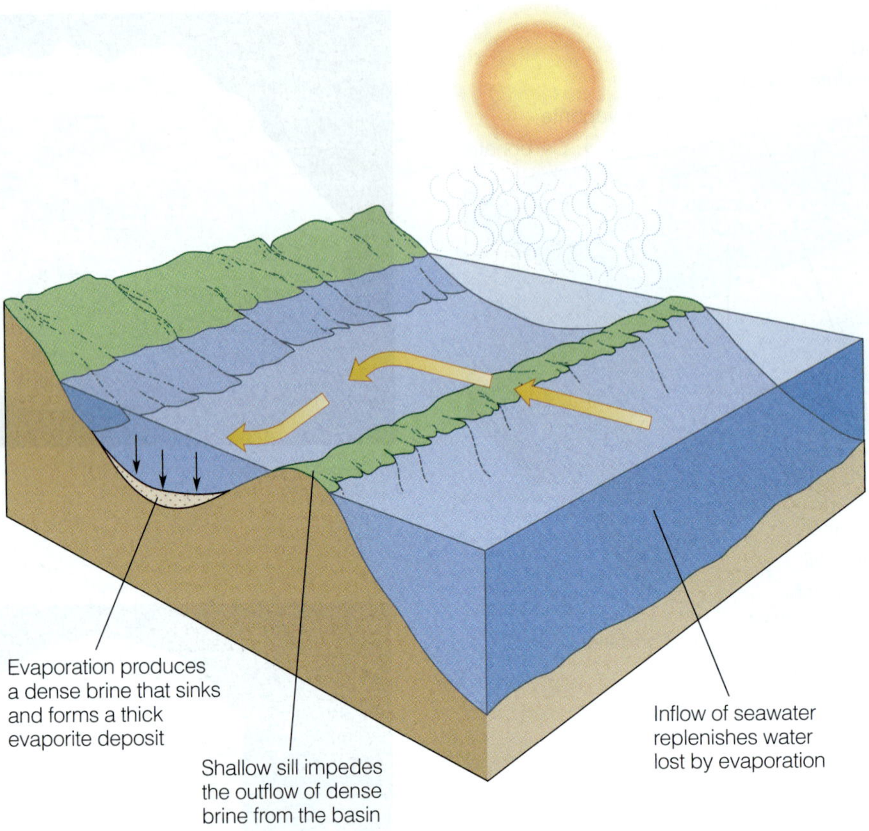

Geology⇌Now ■ **Active Figure 20.12**

Silled basin model for evaporite sedimentation by direct precipitation from seawater. Vertical scale is greatly exaggerated.

conditions for evaporite precipitation (Figure 20.14). In the back-reef area, up to 300 m of evaporites precipitated in much the same way as in the Michigan Basin during the Silurian (Figure 20.11). More than half of the world's potash, which is used in fertilizers, comes from these Devonian evaporites. By the middle of the Late Devonian, reef growth had stopped in western Canada, although nonreef carbonate deposition continued.

Black Shales

In North America, many areas of carbonate-evaporite deposition gave way to a greater proportion of shales and coarser detrital rocks beginning in the Middle Devonian and continuing into the Late Devonian. This change to detrital deposition resulted from the formation of new source areas brought on by the mountain-building activity associated with the Acadian orogeny in North America (Figure 20.13).

As the Devonian Period ended, a conspicuous change in sedimentation occurred over the North American craton with the appearance of widespread black shales. In the eastern United States, these black shales are commonly called the Chattanooga Shale, but they are known by a variety of local names elsewhere (for example, New Albany Shale and Antrim Shale). Although these black shales are best developed from the cratonic margins along the Appalachian mobile belt to the Mississippi Valley, correlative units can also be found in many western states and in western Canada (■ Figure 20.15a).

The Upper Devonian–Lower Mississippian black shales of North America are typically noncalcareous, thinly bedded, and less than 10 m thick (Figure 20.15b). Fossils are usually rare, but some Upper Devonian black shales do contain rich conodont faunas. Because most black shales lack body fossils, they are difficult to date and correlate. However, in places where conodonts (microscopic animals), acritarchs (microscopic algae), or plant spores are found, the lower beds are Late Devonian and the upper beds are Early Mississippian in age.

Although the origin of these extensive black shales is still being debated, the essential features required to produce them include undisturbed anaerobic bottom water, a reduced supply of coarser detrital sediment, and high organic productivity in the overlying oxygenated waters. High productivity in the surface waters leads to a shower of organic material, which decomposes on the

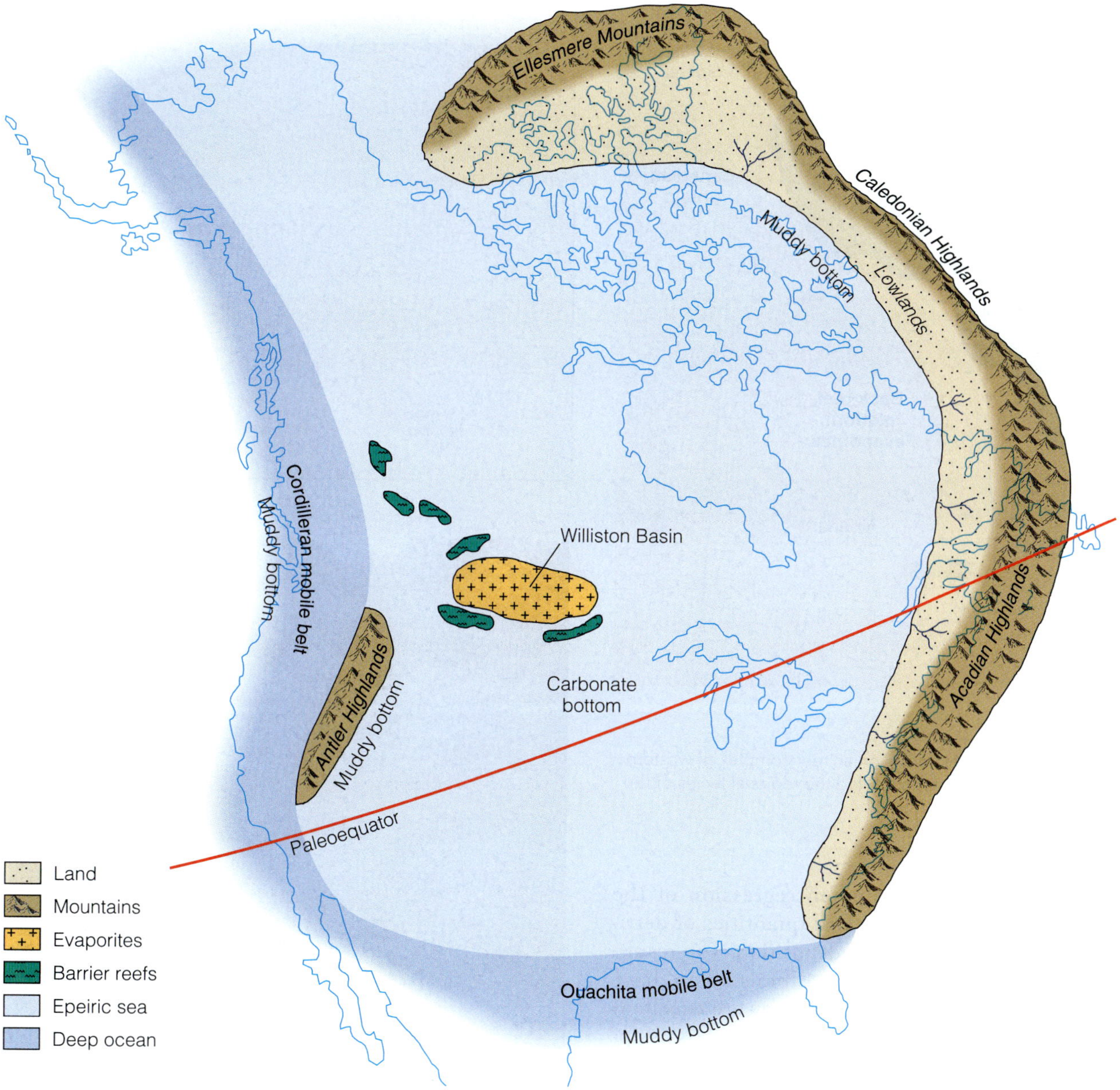

■ **Figure 20.13**

Paleogeography of North America during the Devonian Period.

undisturbed seafloor and depletes the dissolved oxygen at the sediment-water interface.

The wide extent of such apparently shallow-water black shales in North America remains puzzling. Nonetheless, these shales are rich in uranium and are an important source rock of oil and gas in the Appalachian region.

The Late Kaskaskia—A Return to Extensive Carbonate Deposition

Following deposition of the widespread Upper Devonian–Lower Mississippian black shales, carbonate sedimentation on the craton dominated the remainder of the Mississippian Period (■ Figure 20.16). During this time, a variety of carbonate sediments were deposited in the epeiric sea, as indicated by the extensive deposits of crinoidal limestones (rich in crinoid fragments), oolitic limestones, and various other limestones and dolostones (■ Figure 20.17). These Mississippian carbonates display cross-bedding, ripple marks, and well-sorted fossil fragments, all of which indicate a shallow-water environment. Analogous features can be observed on the present-day Bahama Banks. In addition, numerous small organic reefs occurred throughout the craton during the Mississippian. These were all much smaller than the large barrier-reef complexes that dominated the earlier Paleozoic seas.

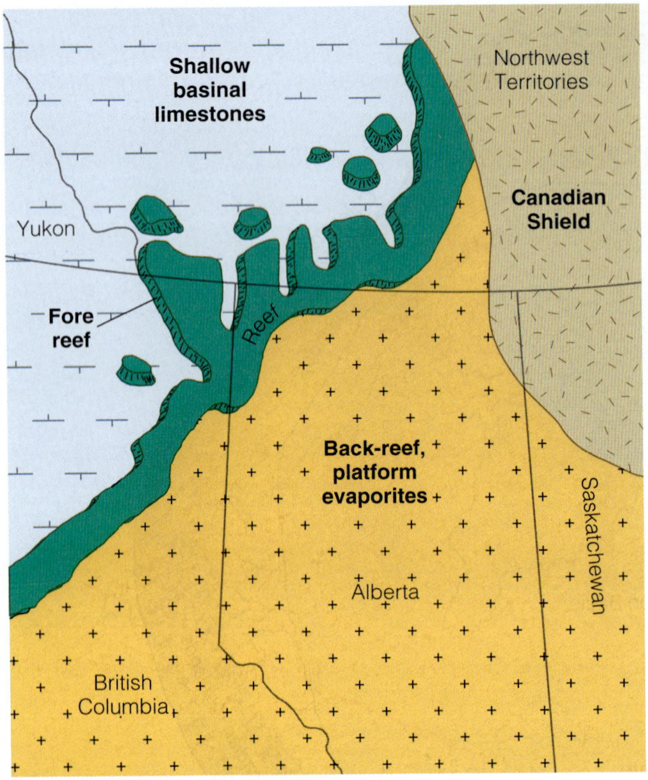

■ **Figure 20.14**

Reconstruction of the extensive Devonian reef complex of western Canada. These extensive reefs controlled the regional facies of the Devonian epeiric seas.

During the Late Mississippian regression of the Kaskaskia Sea from the craton, vast quantities of detrital sediments replaced carbonate deposition. The resulting sandstones, particularly in the Illinois Basin, have been studied in great detail because they are excellent petroleum reservoirs. Before the end of the Mississippian, the Kaskaskia Sea had retreated to the craton margin, once again exposing the craton to widespread weathering and erosion resulting in a cratonwide unconformity when the Absaroka Sea began transgressing back over the craton.

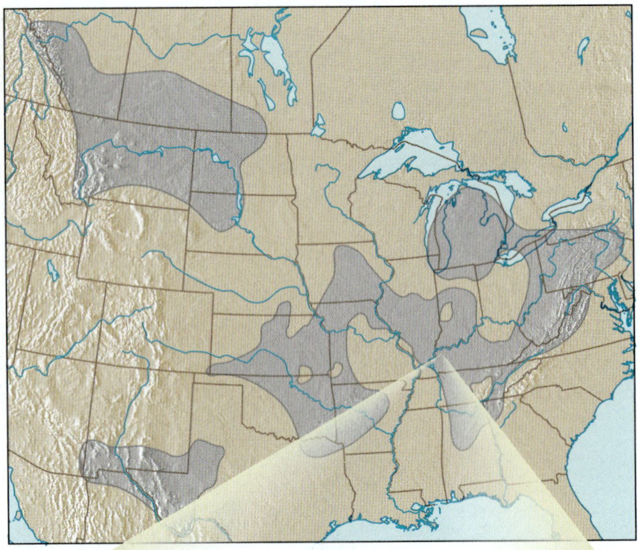

(a)

(b)

■ **Figure 20.15**

(a) The extent of the Upper Devonian to Lower Mississippian Chattanooga Shale and its equivalent units (such as the Antrim Shale and New Albany Shale) in North America. (b) Upper Devonian New Albany Shale, Button Mold Knob Quarry, Kentucky.

THE ABSAROKA SEQUENCE

The **Absaroka sequence** includes rocks deposited during the latest Mississippian through Early Jurassic. In this chapter, however, we are concerned only with the Paleozoic rocks of the Absaroka sequence. The extensive unconformity separating the Kaskaskia and Absaroka sequences essentially divides the strata into the North American Mississippian and Pennsylvanian systems. These two systems are equivalent to the European Lower and Upper Carboniferous systems, respectively. The rocks of the Absaroka sequence not only differ from those of the Kaskaskia sequence but also resulted from different tectonic regimes.

The lowermost sediments of the Absaroka sequence are confined to the margins of the craton. These deposits are generally thickest in the east and southeast, near the emerging highlands of the Appalachian and Ouachita mobile belts, and thin westward onto the craton. The rocks also reveal lateral changes from nonmarine detrital rocks and coals in the east, through transitional marine-nonmarine beds, to largely marine detrital rocks and limestones farther west (■ Figure 20.18).

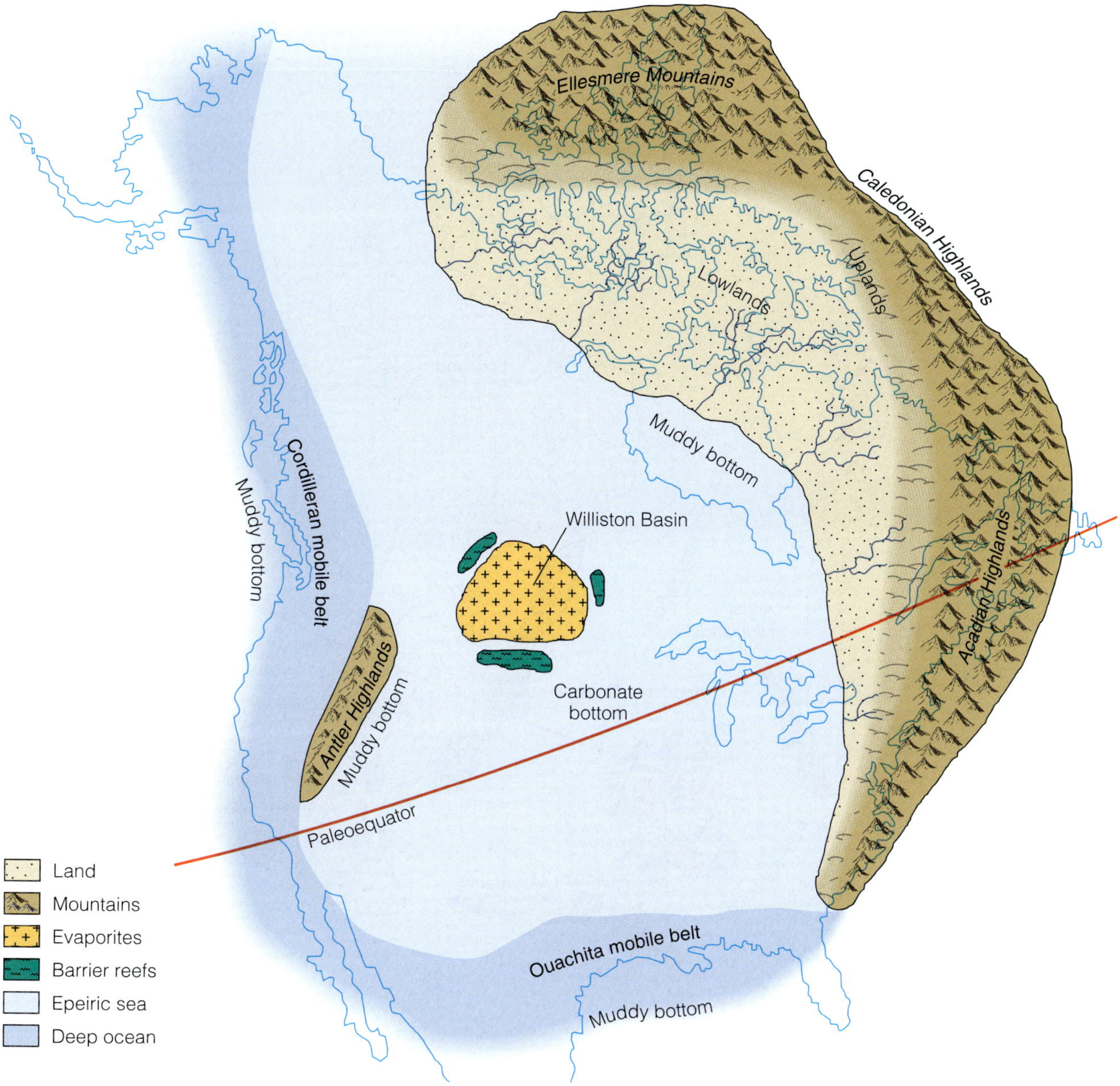

Figure 20.16

Paleogeography of North America during the Mississippian Period.

Figure 20.17

Mississippian limestones exposed near Bowling Green, Kentucky.

What Are Cyclothems, and Why Are They Important?

One characteristic feature of Pennsylvanian rocks is their cyclical pattern of alternating marine and nonmarine strata. Such rhythmically repetitive sedimentary sequences are known as **cyclothems.** They result from repeated alternations of marine and nonmarine

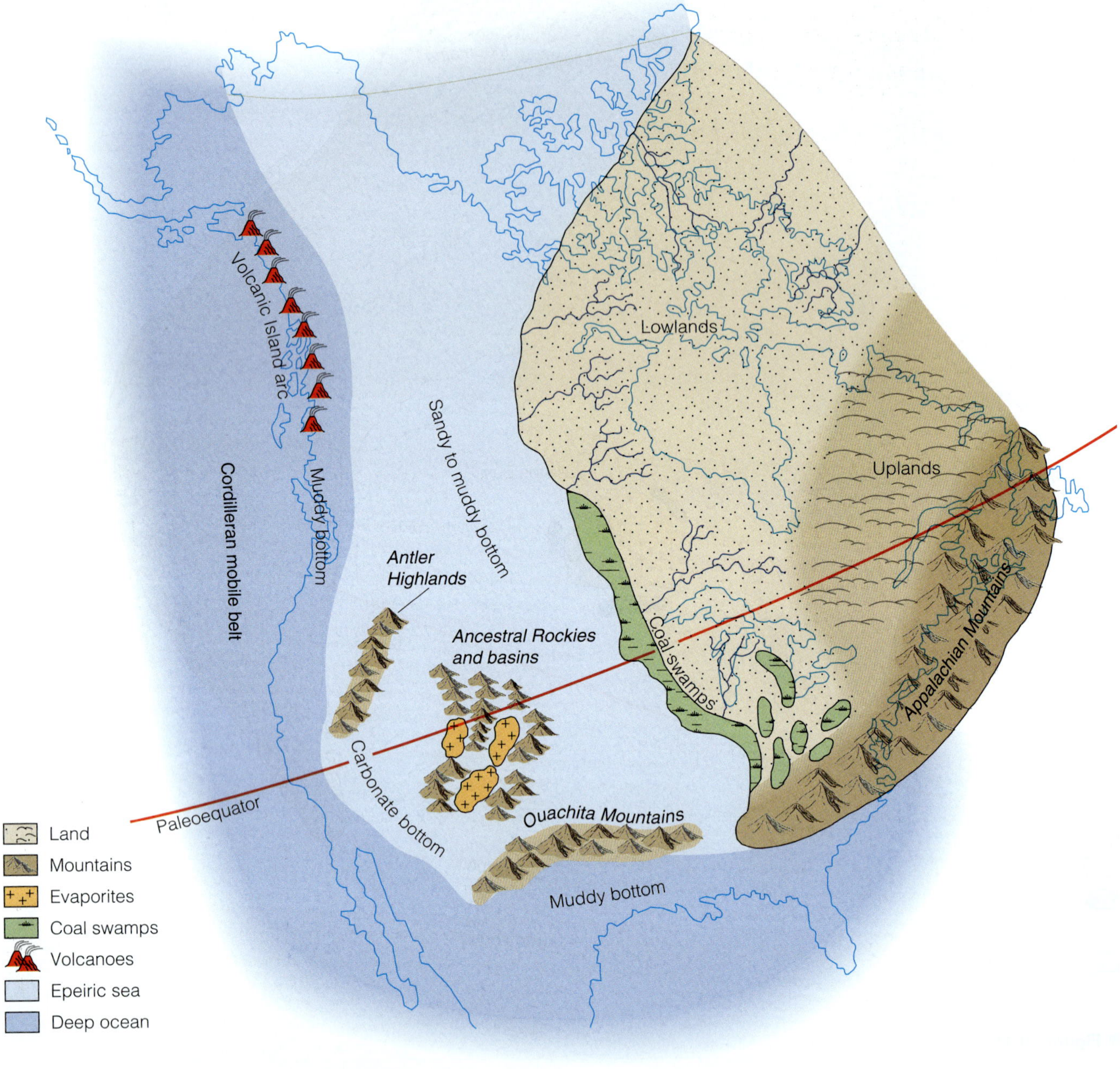

■ **Figure 20.18**

Paleogeography of North America during the Pennsylvanian Period.

environments, usually in areas of low relief. Although seemingly simple, cyclothems reflect a delicate interplay between nonmarine deltaic and shallow-marine interdeltaic and shelf environments.

For illustration, look at a typical coal-bearing cyclothem from the Illinois Basin (■ Figure 20.19a). Such a cyclothem contains nonmarine units, capped by a coal and overlain by marine units. Figure 20.19c shows the depositional environments that produced the cyclothem. The initial units represent deltaic and fluvial deposits. Above them is an underclay that frequently contains root casts from the plants and trees that make up the overlying coal. The coal bed results from accumulations of plant material and is overlain by marine units of alternating limestones and shales, usually with an abundant marine invertebrate fauna. The marine cycle ends with an erosion surface. A new cyclothem begins with a nonmarine deltaic sandstone. All the beds illustrated in the idealized cyclothem are not always preserved because of abrupt changes from marine to nonmarine conditions or removal of some units by erosion.

Cyclothems represent transgressive and regressive sequences with an erosional surface separating one cy-

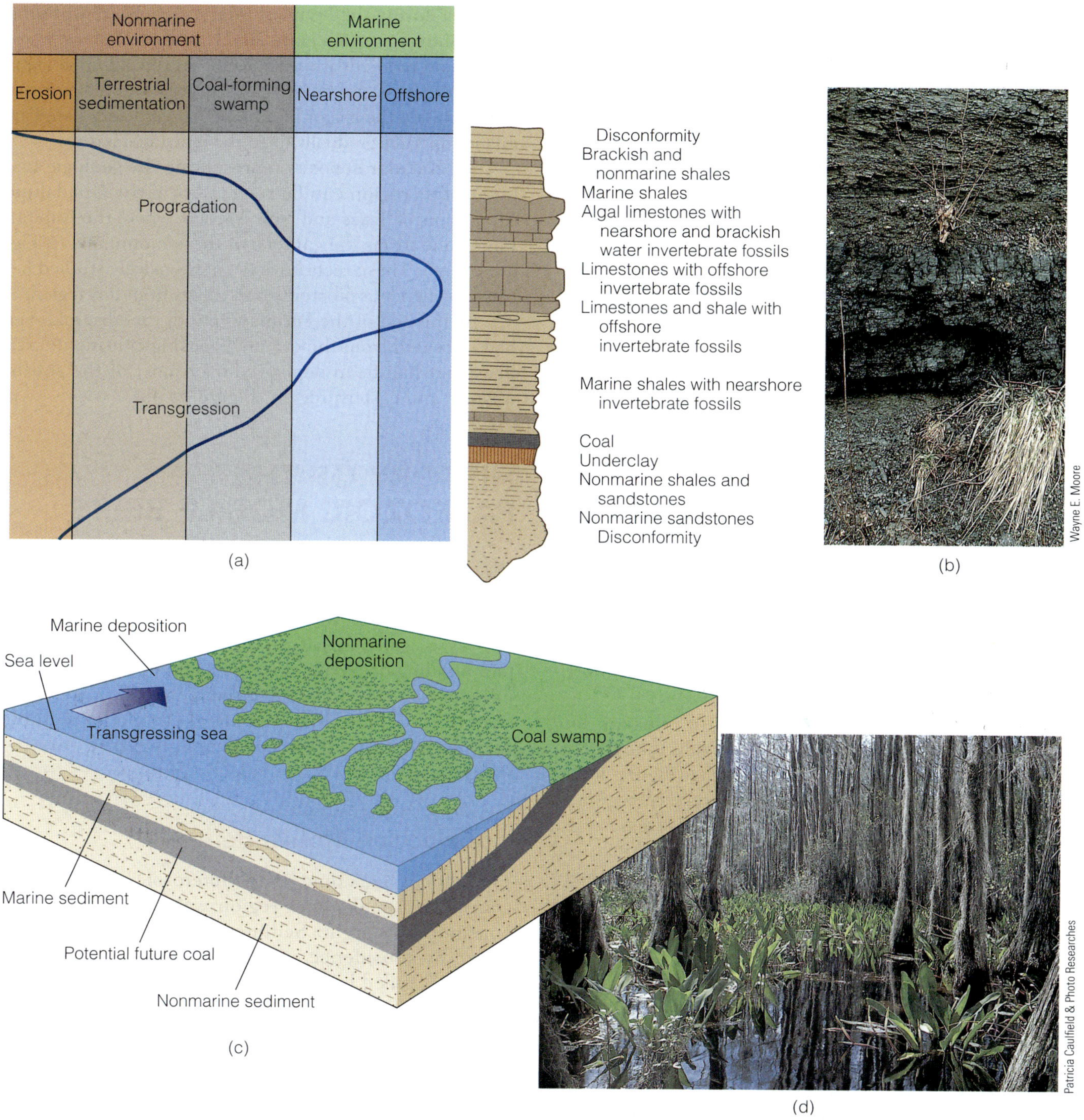

Figure 20.19

(a) Columnar section of a complete cyclothem. (b) Pennsylvanian coal bed, West Virginia. (c) Reconstruction of the environment of a Pennsylvanian coal-forming swamp. (d) The Okefenokee Swamp, Georgia, is a modern example of a coal-forming environment, similar to those occurring during the Pennsylvanian Period.

clothem from another. Thus, an idealized cyclothem passes upward from fluvial-deltaic deposits, through coals, to detrital shallow-water marine sediments, and finally to limestones typical of an open marine environment.

Such regularity and cyclicity in sedimentation over a large area require an explanation. The hypothesis currently favored by most geologists is a rise and fall of sea level related to advances and retreats of Gondwanan continental glaciers. When the Gondwanan ice sheets advanced, sea level dropped; when they melted, sea level rose. Late Paleozoic cyclothem activity on all the cratons closely corresponds to Gondwanan glacial–interglacial cycles.

Cratonic Uplift—The Ancestral Rockies

Recall that cratons are stable areas, and what deformation they do experience is usually mild. The Pennsylvanian Period, however, was a time of unusually severe cratonic deformation, resulting in uplifts of sufficient magnitude to expose Precambrian basement rocks. In addition to newly formed highlands and basins, many previously formed arches and domes, such as the Cincinnati Arch, Nashville Dome, and Ozark Dome, were also reactivated (Figure 20.1).

During the Late Absaroka (Pennsylvanian), the area of greatest deformation was in the southwestern part of the North American craton, where a series of fault-bounded uplifted blocks formed the **Ancestral Rockies** (■ Figure 20.20a). These mountain ranges had diverse geologic histories and were not all elevated at the same time. Uplift of these mountains, some of which were elevated more than 2 km along near-vertical faults, resulted in erosion of overlying Paleozoic sediments and exposure of the Precambrian igneous and metamorphic basement rocks (Figure 20.20b). As the mountains eroded, tremendous quantities of coarse, red arkosic sand and conglomerate were deposited in the surrounding basins. These sediments are preserved in many areas, including the rocks of the Garden of the Gods near Colorado Springs (Figure 20.20c) and at the Red Rocks Amphitheatre near Morrison, Colorado.

Intracratonic mountain ranges are unusual, and their cause has long been debated. It is now thought that the collision of Gondwana with Laurasia (Figure 20.4a) produced great stresses in the southwestern region of the North American craton. These crustal stresses were relieved by faulting that uplifted cratonic blocks and downwarped adjacent basins, forming a series of ranges and basins.

The Late Absaroka—More Evaporite Deposits and Reefs

While the various intracratonic basins were filling with sediment during the Late Pennsylvanian, the Absaroka Sea slowly began retreating from the craton. During the Early Permian, the Absaroka Sea occupied a narrow region from Nebraska through west Texas (■ Figure 20.21). By the Middle Permian, the sea had retreated to west Texas and southern New Mexico. The thick evaporite deposits in Kansas and Oklahoma show the restricted nature of the Absaroka Sea during the Early and Middle Permian and its southwestward retreat from the central craton.

During the Middle and Late Permian, the Absaroka Sea was restricted to west Texas and southern New Mexico, forming an interrelated complex of lagoonal, reef, and open-shelf environments (■ Figure 20.22). Three basins separated by two submerged platforms developed in this area during the Permian. Massive reefs grew around the basin margins (■ Figure 20.23), and limestones, evaporites, and red beds were deposited in the lagoonal areas behind the reefs. As the barrier reefs grew and the passageways between the basins became more restricted, Late Permian evaporites gradually filled the individual basins.

Spectacular deposits representing the geologic history of this region can be seen today in the Guadalupe Mountains of Texas and New Mexico, where the Capitan Limestone forms the caprock of these mountains (■ Figure 20.24). These reefs have been extensively studied because tremendous oil production comes from this region.

By the end of the Permian Period, the Absaroka Sea had retreated from the craton, exposing continental red beds that had been deposited over most of the southwestern and eastern region (Figure 20.4b).

HISTORY OF THE PALEOZOIC MOBILE BELTS

Having examined the Paleozoic history of the craton, we now turn our attention to the orogenic activity in the mobile belts. The mountain building occurring during this time had a profound influence on the climate and sedimentary history of the craton. In addition, it was part of the global tectonic regime that sutured the continents together, forming Pangaea by the end of the Paleozoic Era.

Appalachian Mobile Belt

Throughout Sauk time (Late Proterozoic–Early Ordovician), the Appalachian region was a broad, passive, continental margin. Sedimentation was closely balanced by subsidence as thick, shallow marine sands were succeeded by extensive carbonate deposits. During this time, the **Iapetus Ocean** was widening as a result of movement along a divergent plate boundary (■ Figure 20.25a).

Taconic Orogeny Beginning with the subduction of the Iapetus plate beneath Laurentia (an oceanic–continental convergent plate boundary), the Appalachian mobile belt was born (Figure 20.25b). The resulting **Taconic orogeny**—named after the present-day Taconic Mountains of eastern New York, central Massachusetts, and Vermont—was the first of several orogenies to affect the Appalachian region.

The Appalachian mobile belt can be divided into two depositional environments. The first is the extensive, shallow-water carbonate platform that formed the broad eastern continental shelf and stretched from Newfoundland to Alabama (Figure 20.25a). It formed during the

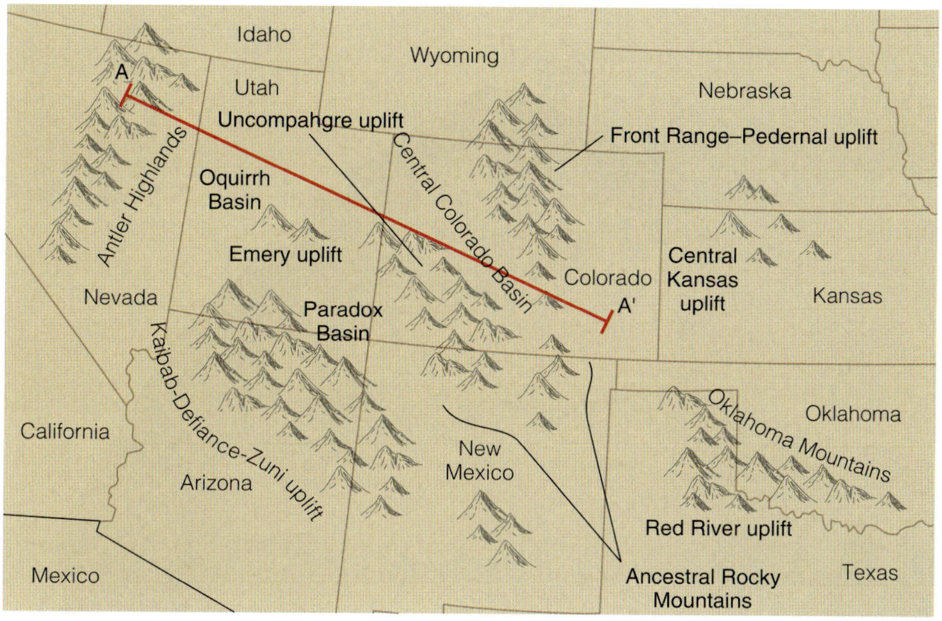

(a)

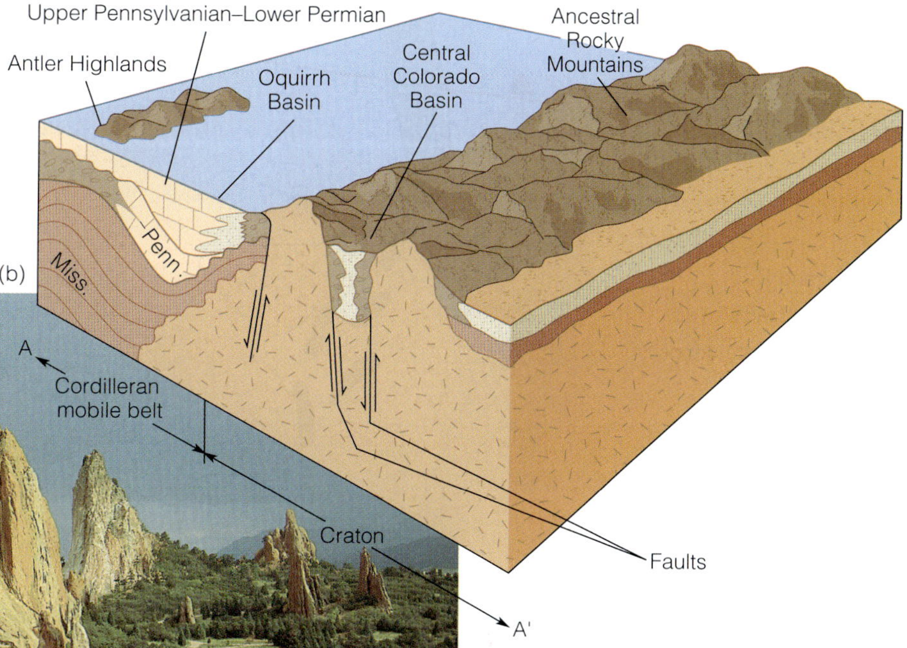

(b)

Tom Bean, 1993 /Tom and Susan bean. Inc.

(c)

■ **Figure 20.20**

(a) Location of the principal Pennsylvanian highland areas and basins of the southwestern part of the craton. (b) Block diagram of the Ancestral Rockies, elevated by faulting during the Pennsylvanian Period. Erosion of these mountains produced coarse, red sediments deposited in the adjacent basins. (c) Garden of the Gods, view from Near Hidden Inn, Colorado Springs, Colorado.

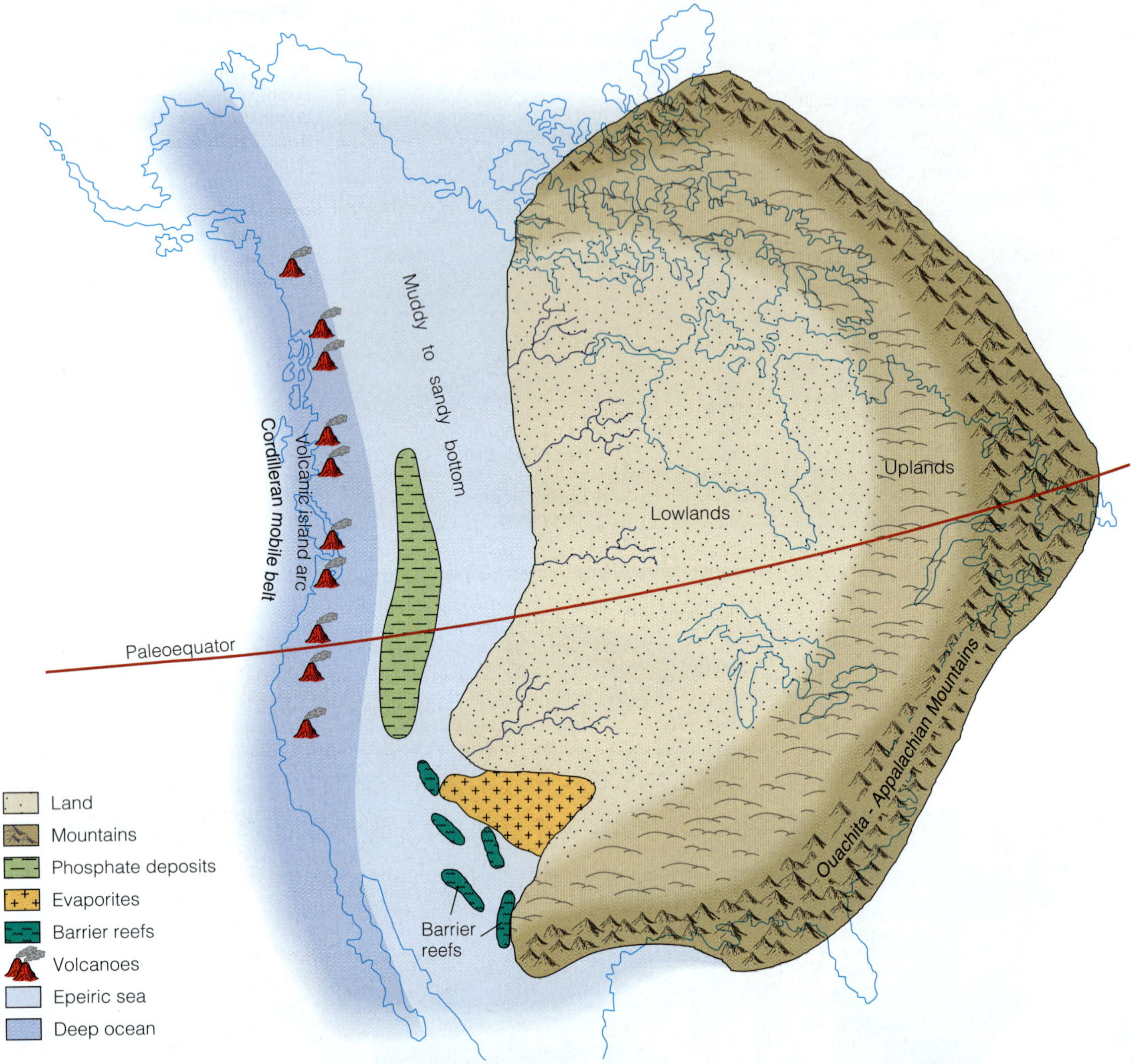

■ **Figure 20.21**

Paleogeography of North America during the Permian Period.

Sauk Sea transgression onto the craton when carbonates were deposited in a vast, shallow sea.

Carbonate deposition ceased along the east coast during the Middle Ordovician and was replaced by the second depositional environment, deepwater deposits characterized by thinly bedded black shales, graded beds, coarse sandstones, graywackes, and associated volcanics. This suite of sediments marks the onset of mountain building—in this case, the Taconic orogeny. The subduction of the Iapetus plate beneath Laurentia resulted in volcanism and downwarping of the carbonate platform (Figure 20.25b).

The final piece of evidence for the Taconic orogeny is the development of a large **clastic wedge,** an extensive accumulation of mostly detrital sediments deposited adjacent to an uplifted area. These deposits are thickest and coarsest nearest the highland area and become thinner and finer grained away from the source area, eventually grading into the carbonate cratonic facies (■ Figure 20.26). The clastic wedge resulting from the erosion of the Taconic Highlands is referred to as the **Queenston Delta.** Careful mapping and correlation of these deposits indicate that more than 600,000 km^3 of rock were eroded from the Taconic Highlands. Based on this figure, geologists estimate the Taconic Highlands were at least 4000 m high.

■ **Figure 20.22**

Location of the west Texas Permian basins and surrounding reefs.

■ **Figure 20.24**

The prominent Capitan Limestone forms the caprock of the Guadalupe Mountains. The Capitan Limestone is rich in fossil corals and associated reef organisms.

■ **Figure 20.23**

A reconstruction of the Middle Permian Capitan Limestone reef environment. Shown are brachiopods, corals, bryozoans, and large glass sponges.

Caledonian Orogeny The Caledonian mobile belt stretches along the western border of Baltica and includes the present-day countries of Scotland, Ireland, and Norway (Figure 20.2c). During the Middle Ordovician, subduction along the boundary between the Iapetus plate and Baltica (Europe) began, forming a mirror image of the convergent plate boundary off the east coast of Laurentia (North America).

The culmination of the **Caledonian orogeny** occurred during the Late Silurian and Early Devonian with the formation of a mountain range along the margin of Baltica. Red-colored sediments deposited along the front of the Caledonian Highlands formed a large clastic wedge known as the *Old Red Sandstone.*

Acadian Orogeny The third Paleozoic orogeny to affect Laurentia and Baltica began during the Late Silurian and concluded at the end of the Devonian Period. The **Acadian orogeny** affected the Appalachian mobile belt from Newfoundland to Pennsylvania as sedimentary rocks were folded and thrust against the craton.

As with the preceding Taconic and Caledonian orogenies, the Acadian orogeny occurred along an oceanic–continental convergent plate boundary. As the northern Iapetus Ocean continued to close during the Devonian, the plate carrying Baltica finally collided with Laurentia, forming a continental–continental convergent plate boundary along the zone of collision (Figure 20.3a).

Weathering and erosion of the Acadian Highlands produced the **Catskill Delta,** a thick clastic wedge

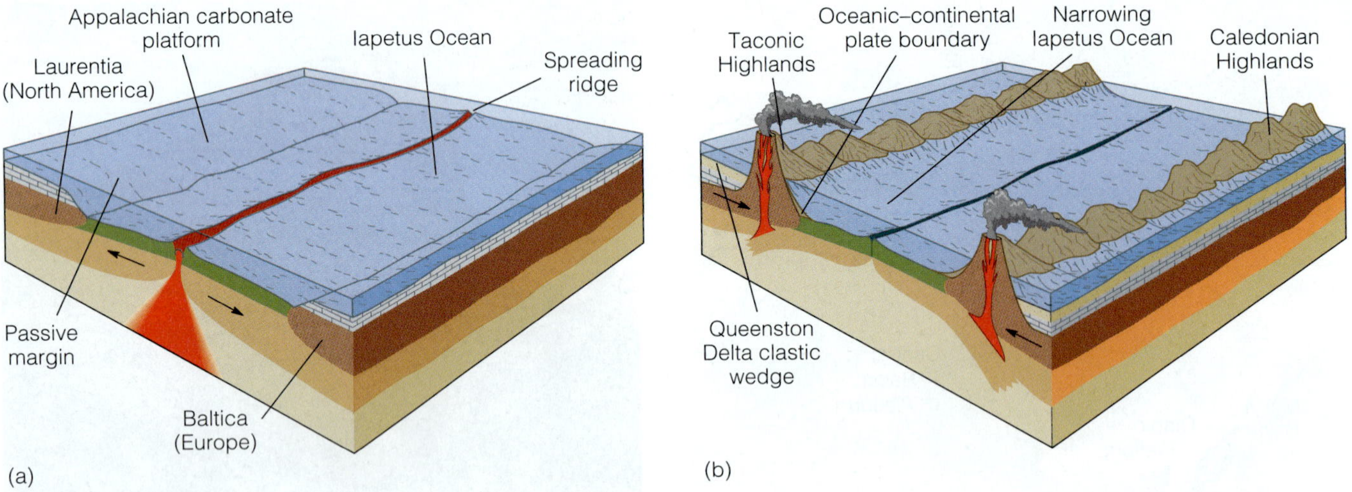

Figure 20.25

Evolution of the Appalachian mobile belt from the Late Proterozoic to the Late Ordovician. (a) During the Late Proterozoic to the Early Ordovician, the Iapetus Ocean was opening along a divergent plate boundary. Both the east coast of Laurentia and the west coast of Baltica were passive continental margins with large carbonate platforms. (b) Beginning in the Middle Ordovician, the passive margins of Laurentia and Baltica became oceanic–continental plate boundaries, resulting in orogenic activity (Figure 20.2b).

named for the Catskill Mountains in northern New York where it is well exposed. The Catskill Delta, composed of red, coarse conglomerates, sandstones, and shales, contains nearly three times as much sediment as the Queenston Delta.

The Devonian rocks of New York are among the most studied on the continent. A cross section of the Devonian strata clearly reflects an eastern source (Acadian Highlands) for the Catskill facies (Figure 20.27). These clastic rocks can be traced from eastern Pennsylvania, where the coarse clastics are approximately 3 km thick, to Ohio, where the deltaic facies are only about 100 m thick and consist of cratonic shales and carbonates.

The red beds of the Catskill Delta derive their color from the hematite in the sediments. Plant fossils and oxidation of the hematite indicate that the beds were deposited in a continental environment. Toward the west, the red beds grade laterally into gray sandstones and shales containing fossil tree trunks, which indicate a swamp or marsh environment.

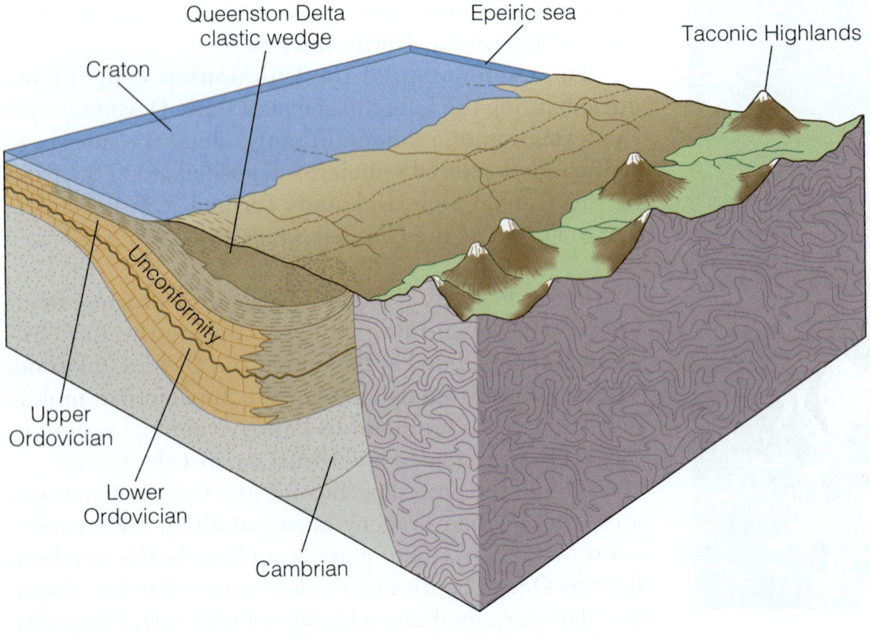

Figure 20.26

Reconstruction of the Taconic Highlands and Queenston Delta clastic wedge. The clastic wedge consists of thick, coarse-grained detrital sediments nearest the highlands and thins laterally into finer-grained sediments on the craton.

The Old Red Sandstone The red beds of the Catskill Delta have a European counterpart in the Devonian Old Red Sandstone of the British Isles (Figure 20.27). The Old Red Sandstone was a Devonian clastic wedge that grew eastward from the Caledonian Highlands onto the Baltica craton. The Old Red Sandstone, just like its North American Catskill counterpart, contains numerous fossils of freshwater fish, early amphibians, and land plants.

By the end of the Devonian Period, Baltica and Laurentia were sutured together, forming Laurasia (Figure 20.27). The red beds of the Catskill Delta can be traced north, through Canada and Greenland, to the Old Red Sandstone of the British Isles and into northern Europe. These beds were deposited in similar environments along the flanks of developing mountain chains formed at convergent plate boundaries.

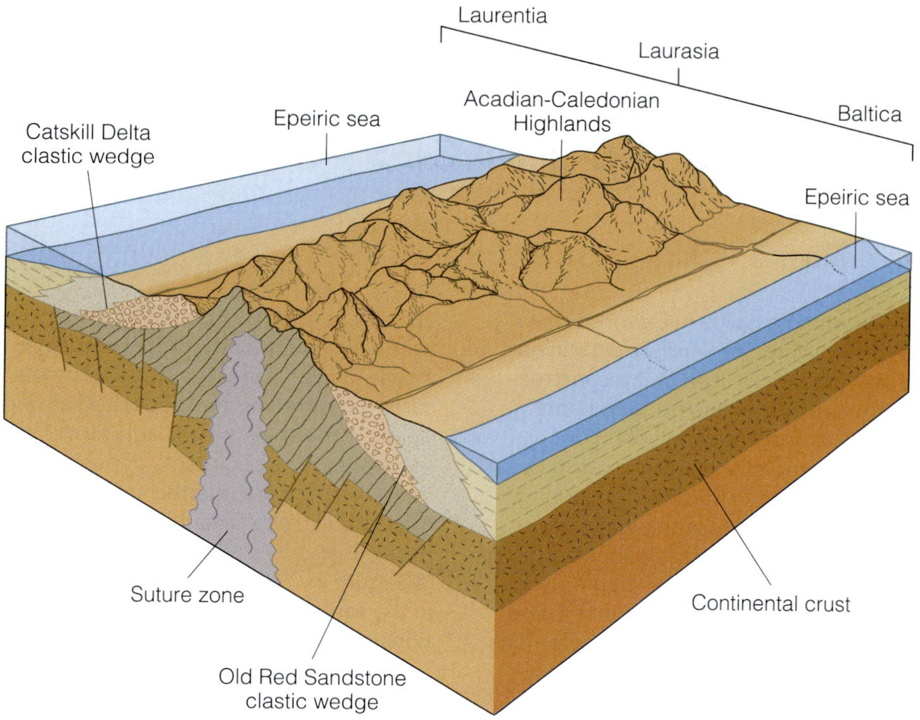

Figure 20.27

Block diagram showing the area of collision between Laurentia and Baltica. Note the bilateral symmetry of the Catskill Delta clastic wedge and the Old Red Sandstone and their relationship to the Acadian and Caledonian Highlands.

The Taconic, Caledonian, and Acadian orogenies were all part of the same major orogenic event related to the closing of the Iapetus Ocean (Figure 20.25 and 20.27). This event began with paired oceanic–continental convergent plate boundaries during the Taconic and Caledonian orogenies and culminated along a continental–continental convergent plate boundary during the Acadian orogeny as Laurentia and Baltica became sutured. After this, the Hercynian-Alleghenian orogeny began, followed by orogenic activity in the Ouachita mobile belt.

Hercynian-Alleghenian Orogeny The Hercynian mobile belt of southern Europe and the Appalachian and Ouachita mobile belts of North America mark the zone along which Europe (part of Laurasia) collided with Gondwana (Figure 20.3). While Gondwana and southern Laurasia collided during the Pennsylvanian and Permian periods in the area of the Ouachita mobile belt, eastern Laurasia (Europe and southeastern North America) joined together with Gondwana (Africa) as part of the **Hercynian-Alleghenian orogeny** (Figure 20.4a).

Initial contact between eastern Laurasia and Gondwana began during the Mississippian Period along the Hercynian mobile belt. The greatest deformation occurred during the Pennsylvanian and Permian periods and is referred to as the *Hercynian orogeny.* The central and southern parts of the Appalachian mobile belt (from New York to Alabama) were folded and thrust toward the craton as eastern Laurasia and Gondwana were sutured. This event in North America is referred to as the *Alleghenian orogeny.*

These three Late Paleozoic orogenies (Hercynian, Alleghenian, and Ouachita) represent the final joining of Laurasia and Gondwana into the supercontinent Pangaea during the Permian.

Cordilleran Mobile Belt

During the Late Proterozoic and Early Paleozoic, the Cordilleran area was a passive continental margin along which extensive continental shelf sediments were deposited. Thick sections of marine sediments graded laterally into thin cratonic units as the Sauk Sea transgressed onto the craton. Beginning in the Middle Paleozoic, an island arc formed off the western margin of the craton. This eastward-moving island arc and the western border of the craton collided during the Late Devonian and Early Mississippian, resulting in a highland area.

This orogenic event, the **Antler orogeny,** was caused by subduction and closed the narrow ocean basin that separated the island arc from the craton (■ Figure 20.28). Erosion of the resulting Antler Highlands produced large quantities of sediment that were deposited to the east in the epeiric sea covering the craton and to the west in the deep sea. The Antler orogeny was the first in a series of orogenic events to affect the Cordilleran mobile belt. During the Mesozoic and Cenozoic, this area was the site of major tectonic activity caused by

GEOLOGY IN UNEXPECTED PLACES

A Man's Home Is His Castle

Imagine visiting a castle in Devonshire, England, or Glasgow, Scotland, and noticing that the same rocks used in the construction of those castles can be seen in the Catskill Mountains of New York (■ Figure 1). Whereas most people who visit castles in Great Britain and elsewhere in Europe are learning about the history of the castles—when they were built, why they were built, who lived in them, and other historical facts—geologists frequently are looking at the rocks that make up the walls of a castle and trying to determine what type of rock it is, how old it is, and anything else they can learn about it.

During the Devonian Period (408 to 360 million years ago), the North American continent and what is now Europe were moving toward each other along an oceanic–continental convergent plate boundary. As movement along this boundary continued, the ocean basin separating these landmasses shrunk until the two continents collided in what is known as the Acadian orogeny. This mountain-building episode formed a large mountain range. Just as is happening today, those mountains began weathering and their eroded sediments were carried by streams and deposited as large deltas in the shallow seas adjacent to the mountains. The sediments deposited in what is now New York are referred to by geologists as the Catskill Delta. The counterpart in Great Britain is known as the Old Red Sandstone (■ Figure 2). These sediments were deposited in the same environment and reflect the conditions at the time of deposition.

Later during the Mesozoic Era (245 to 66 million years ago), as the supercontinent Pangaea broke apart along divergent plate boundaries, the Atlantic Ocean basin formed, separating North America from Europe. Even though the Devonian rocks of the present-day Catskill Mountains in New York are separated by several thousand kilometers from the Old Red Sandstone rocks of Great Britain, they were formed at the same time and deposited in the same environment hundreds of millions of years ago. That is why they have the same red color and composition and contain many of the same fossils.

So, the next time you see someone closely inspecting the rocks of a castle wall, there is a good chance that person is a geologist or someone, like yourself, who took a geology course.

■ Figure 1

Remains of Goodrich Castle in Herefordshire, England, dated to the period between 1160 and 1270. Goodrich Castle is a good example of a structure built from rocks quarried from the Old Red Sandstone, a Devonian-age formation that is the counterpart of the rocks found in the Catskill Mountains of New York.

Myth and Legends Photography

■ Figure 2

Closeup of cross-bedding in the Old Red Sandstone. This rock comes from a building in Glasgow, Scotland, and reflects that the sediments were deposited by a stream.

James S. Monroe

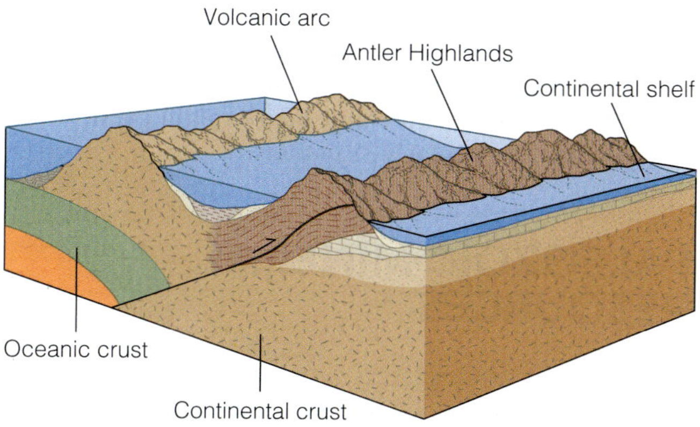

■ **Figure 20.28**

Reconstruction of the Cordilleran mobile belt during the Early Mississippian in which deep-water continental slope deposits were thrust eastward over shallow-water continental shelf carbonates, forming the Antler Highlands.

oceanic–continental convergence and accretion of various terranes.

Ouachita Mobile Belt

The Ouachita mobile belt extends for approximately 2100 km from the subsurface of Mississippi to the Marathon region of Texas. Approximately 80% of the former mobile belt is buried beneath a Mesozoic and Cenozoic sedimentary cover. The two major exposed areas in this region are the Ouachita Mountains of Oklahoma and Arkansas and the Marathon Mountains of Texas. Based on extensive study of the subsurface geology and the Ouachita and Marathon Mountains, geologists have learned that this region had a complex geologic history (■ Figure 20.29).

During the Late Proterozoic to Early Mississippian, shallow-water detrital and carbonate sediments were deposited on a broad continental shelf, and in the deeper-water portion of the adjoining mobile belt, bedded cherts and shales were accumulating (Figure 20.29a). Beginning in the Mississippian Period, the rate of sedimentation increased dramatically as the region changed from a passive continental margin to an active convergent plate boundary marking the beginning of the **Ouachita orogeny** (Figure 20.29b). Rapid deposition of sediments continued into the Pennsylvanian with the formation of a clastic wedge that thickened to the south. As much as 16,000 m of Mississippian- and Pennsylvanian-age rocks crop out in the Ouachita Mountains, attesting to the rapid rate of sedimentation during this time. The formation of a clastic wedge marks the beginning of uplift of the area and formation of a mountain range during the Ouachita orogeny.

Thrusting of sediments continued throughout the Pennsylvanian and Early Permian, driven by the compressive forces generated along the zone of subduction as Gondwana collided with Laurasia (Figure 20.29c). The collision of Gondwana and Laurasia is marked by the formation of a large mountain range, most of which was eroded during the Mesozoic Era. Only the rejuvenated

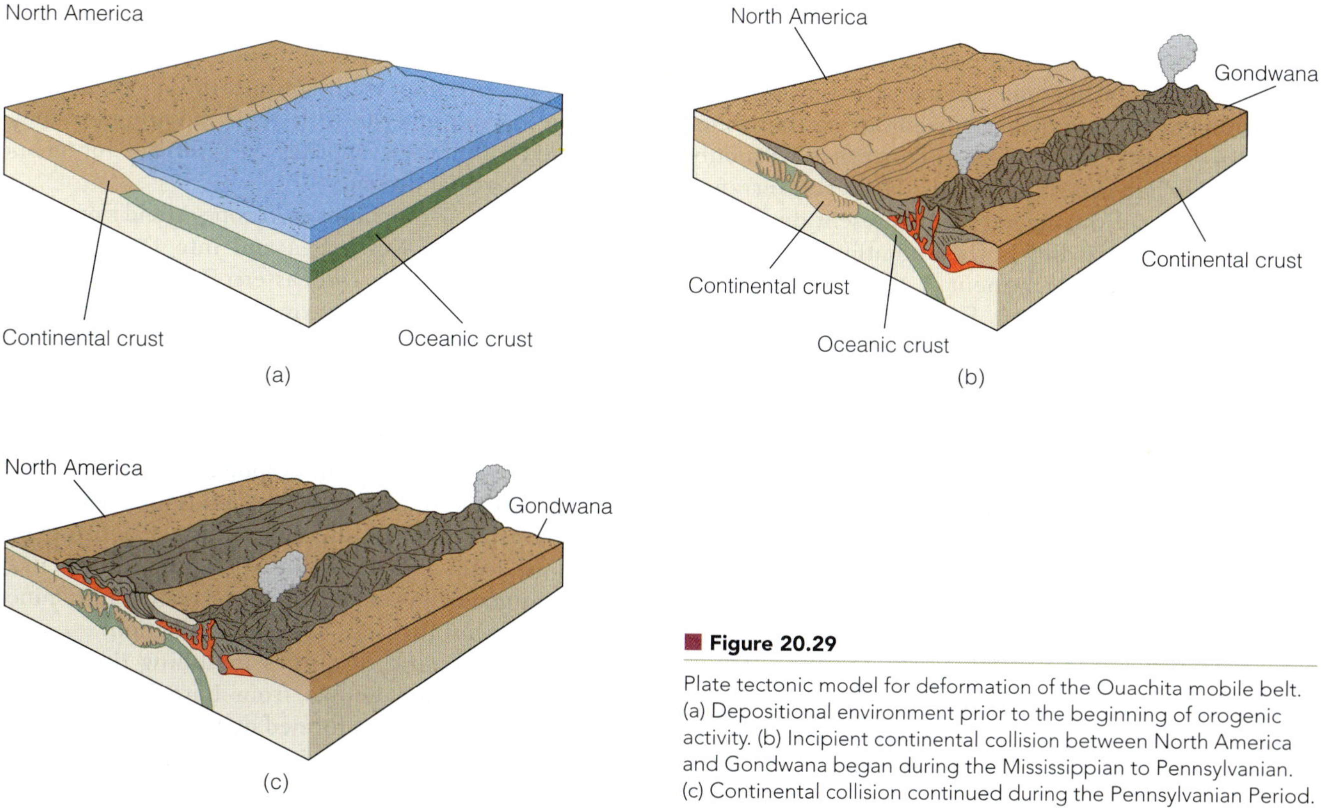

■ **Figure 20.29**

Plate tectonic model for deformation of the Ouachita mobile belt. (a) Depositional environment prior to the beginning of orogenic activity. (b) Incipient continental collision between North America and Gondwana began during the Mississippian to Pennsylvanian. (c) Continental collision continued during the Pennsylvanian Period.

Ouachita and Marathon Mountains remain of this once-lofty mountain range.

The Ouachita deformation was part of the general worldwide tectonic activity that occurred when Gondwana united with Laurasia. The Hercynian, Appalachian, and Ouachita mobile belts were continuous and marked the southern boundary of Laurasia (Figure 20.4). The tectonic activity that uplifted the Ouachita mobile belt was very complex and involved not only the collision of Laurasia and Gondwana but also several microplates and terranes between the continents that eventually became part of Central America. The compressive forces impinging on the Ouachita mobile belt also affected the craton by broadly uplifting the southwestern part of North America.

WHAT ROLE DID MICROPLATES AND TERRANES PLAY IN THE FORMATION OF PANGAEA?

We have presented the geologic history of the mobile belts bordering the Paleozoic continents in terms of subduction along convergent plate boundaries. It is becoming increasingly clear, however, that accretion along the continental margins is more complicated than the somewhat simple, large-scale plate interactions we have described. Geologists now recognize that numerous terranes or microplates existed during the Paleozoic and were involved in the orogenic events that occurred during that time.

In this chapter, we have been concerned with only the six major Paleozoic continents. However, terranes and microplates of varying sizes were present during the Paleozoic and participated in the formation of Pangaea. For example, the small continent of Avalonia consists of some coastal parts of New England, southern New Brunswick, much of Nova Scotia, the Avalon Peninsula of eastern Newfoundland, southeastern Ireland, Wales, England, and parts of Belgium and northern France. This terrane existed as a separate continent during the Ordovician and began to collide with Baltica during the Silurian and Laurentia (as part of Baltica) during the Devonian (Figures 20.2b, c and 20.3a).

Florida and parts of the eastern seaboard of North America make up the Piedmont terrane that was part of the larger Gondwana continent. This terrane became sutured to Laurasia during the Pennsylvanian Period. Numerous microplates and terranes occupied the region between Gondwana and Laurasia that eventually became part of Central America during the Pennsylvanian collision between these continents.

Thus, although the basic history of the formation of Pangaea during the Paleozoic remains the same, geologists now realize that microplates and terranes also played an important role and help explain some previously anomalous geologic situations.

PALEOZOIC MINERAL RESOURCES

Paleozoic-age rocks contain a variety of important mineral resources, including energy resources, metallic and nonmetallic mineral deposits, and sand and gravel for construction. Important sources of industrial or silica sand are the Upper Cambrian Jordan Sandstone of Minnesota and Wisconsin, the Middle Ordovician St. Peter Sandstone, the Lower Silurian Tuscarora Sandstone in Pennsylvania and Virginia, the Devonian Ridgeley Formation in West Virginia, Maryland, and Pennsylvania, and the Devonian Sylvania Sandstone in Michigan.

Silica sand has a variety of uses, including the manufacture of glass, refractory bricks for blast furnaces, and molds for casting iron, aluminum, and copper alloys. Some silica sands, called hydraulic fracturing sands, are pumped into wells to fracture oil- or gas-bearing rocks and provide permeable passageways for the oil or gas to migrate to the well.

Thick deposits of Silurian evaporites, mostly rock salt (NaCl) and rock gypsum ($CaSO_4 \cdot H_2O$) altered to rock anhydrite ($CaSO_4$), underlie parts of Michigan, Ohio, New York, and adjacent areas in Ontario, Canada. These rocks are important sources of various salts. In addition, barrier and pinnacle reefs in carbonate rocks associated with these evaporites are reservoirs for oil and gas in Michigan and Ohio.

The Zechstein evaporites of Europe extend from Great Britain across the North Sea and into Denmark, the Netherlands, Germany, and eastern Poland and Lithuania. Besides the evaporites themselves, Zechstein deposits form the caprock for the large reservoirs of the gas fields of the Netherlands and part of the North Sea region.

Other important evaporite mineral resources include those of the Permian Delaware Basin of west Texas and New Mexico, and Devonian evaporites in the Elk Point basin of Canada. In Michigan, gypsum is mined and used in the construction of sheetrock. Late Paleozoic–age limestones from many areas in North America are used in the manufacturing of cement. Limestone is also mined and used in blast furnaces for steel production.

Metallic mineral resources including tin, copper, gold, and silver are known from Late Paleozoic–age rocks, especially those deformed during mountain building. The host rocks for deposits of lead and zinc in southeast Missouri are Cambrian dolostones, although some

Ordovician rocks contain these metals as well. These deposits have been mined since 1720 but have been largely depleted. Now most lead and zinc mined in Missouri come from Mississippian-age sedimentary rocks.

The Silurian Clinton Formation crops out from Alabama north to New York, and equivalent rocks are found in Newfoundland. This formation has been mined for iron in many places. In the United States, the richest ores and most extensive mining occurred near Birmingham, Alabama, but only a small amount of ore is currently produced in that area.

Petroleum and natural gas are recovered in commercial quantities from rocks ranging in age from the Devonian through Permian. For example, Devonian rocks in the Michigan Basin, Illinois Basin, and the Williston Basin of Montana, South Dakota, and adjacent parts of Alberta, Canada, have yielded considerable amounts of hydrocarbons. Permian reefs and other strata in the western United States, particularly Texas, have also been important producers.

Although Permian-age coal beds are known from several areas, including Asia, Africa, and Australia, much of the coal in North America and Europe comes from Pennsylvanian (Late Carboniferous) deposits. Large areas in the Appalachian region and the midwestern United States are underlain by vast coal deposits (■ Figure 20.30). These coal deposits formed from the lush vegetation that flourished in Pennsylvanian coal-forming swamps (see Chapter 21).

What Would You Do?

You are the geology team leader of an international mining company. Your company holds the mineral rights on large acreages in various countries along the west coast of Africa. The leases on these mineral rights will shortly expire, and you've been given the task of evaluating which leases are the most promising. How do you think your knowledge of Paleozoic plate tectonics can help you in making these evaluations?

Much of this coal is bituminous coal, which contains about 80% carbon. It is a dense, black coal that has been so thoroughly altered that plant remains can be seen only rarely. Bituminous coal is used to make *coke,* a hard, gray substance made up of the fused ash of bituminous coal. Coke is used to fire blast furnances for the production of steel.

Some Pennsylvanian coal from North America is *anthracite,* a metamorphic type of coal containing up to 98% carbon. Most anthracite is in the Appalachian region (Figure 20.30). It is especially desirable because it burns with a smokeless flame and yields more heat per unit volume than other types of coal. Unfortunately, it is the least common type—much of the coal used in the United States is bituminous.

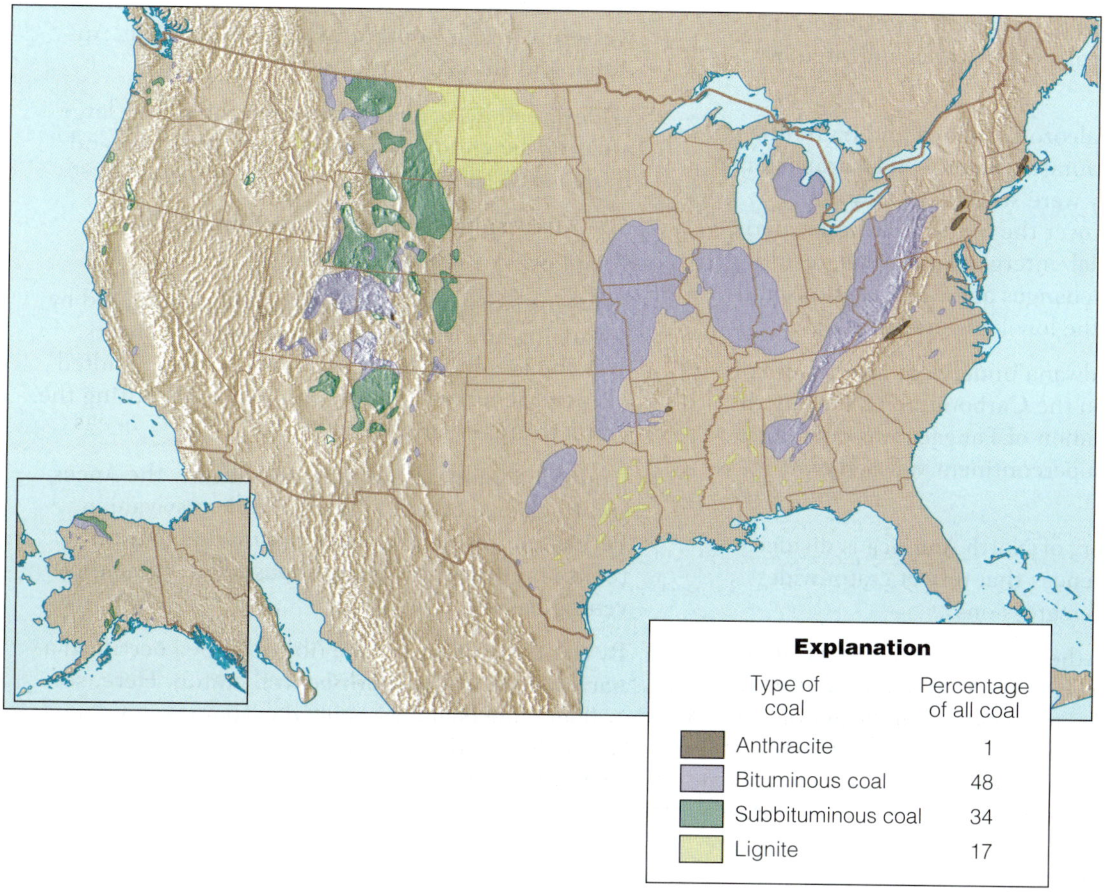

■ **Figure 20.30**

Distribution of coal deposits in the United States. The age of the coals in the midwestern states and the Appalachian region are mostly Pennsylvanian, whereas those in the West are mostly Cretaceous and Cenozoic.

20 GEO RECAP

Chapter Summary

Tables 20.1 (see p. 597) and 20.2 (see p. 598) provide summaries of the geologic history of the North American craton and mobile belts as well as global events and sea-level changes during the Paleozoic Era.

- Most continents consist of two major components: a relatively stable craton over which epeiric seas transgressed and regressed, surrounded by mobile belts in which mountain building took place.
- Six major continents existed at the beginning of the Paleozoic Era; four were located near the paleoequator.
- During the Early Paleozoic (Cambrian–Silurian), Laurentia was moving northward and Gondwana moved to a south polar location, as indicated by tillite deposits.
- During the Late Paleozoic, Baltica and Laurentia collided to form Laurasia. Siberia and Kazakhstania collided and finally were sutured to Laurasia. Gondwana moved over the South Pole and experienced several glacial–interglacial periods, resulting in global sea-level changes and transgressions and regressions along the low-lying craton margins.
- Laurasia and Gondwana underwent a series of collisions beginning in the Carboniferous. During the Permian, the formation of Pangaea was completed. Surrounding the supercontinent was a global ocean, Panthalassa.
- The geologic history of North America is divided into cratonic sequences that reflect cratonwide transgressions and regressions.
- The Sauk Sea was the first major transgression onto the craton. At its maximum, it covered the craton except for parts of the Canadian Shield and the Transcontinental Arch, a series of large, northeast–southwest trending islands.
- The Tippecanoe sequence began with deposition of an extensive sandstone over the exposed and eroded Sauk landscape. During Tippecanoe time, extensive carbonate deposition took place. In addition, large barrier reefs enclosed basins, resulting in evaporite deposition within these basins.
- The basal beds of the Kaskaskia sequence that were deposited on the exposed Tippecanoe surface consisted either of sandstones derived from the eroding Taconic Highlands or of carbonate rocks.
- Most of the Kaskaskia sequence is dominated by carbonates and associated evaporites. The Devonian Period was a time of major reef building in western Canada, southern England, Belgium, Australia, and Russia.
- Widespread black shales were deposited over large areas of the craton during the Late Devonian and Early Mississippian.
- The Mississippian Period was dominated for the most part by carbonate deposition.
- Transgressions and regressions, probably caused by advancing and retreating Gondwanan ice sheets, over the low-lying North American craton resulted in cyclothems and the formation of coals during the Pennsylvanian Period.
- Cratonic mountain building, specifically the Ancestral Rockies, occurred during the Pennsylvanian Period and resulted in thick nonmarine detrital rocks and evaporites being deposited in the intervening basins.
- By the Early Permian, the Absaroka Sea occupied a narrow zone of the south-central craton. Here, several large reefs and associated evaporites developed. By the end of the Permian Period, the Absaroka Sea had retreated from the craton.

Table 20.1

Summary of Early Paleozoic Geologic Events

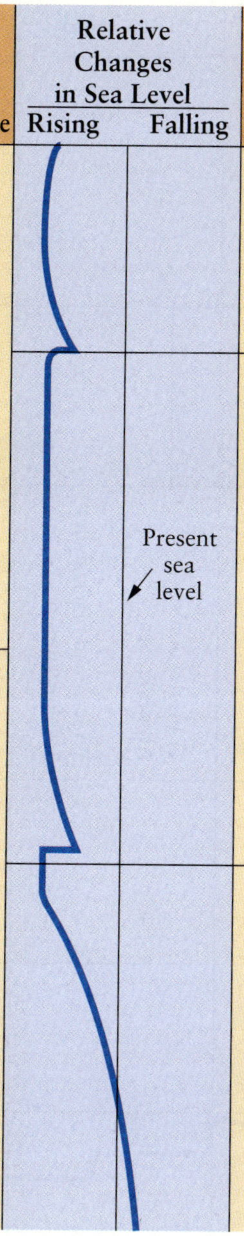

Age (millions of years)	Geologic Period	Sequence	Relative Changes in Sea Level (Rising / Falling)	Cordilleran Mobile Belt	Craton	Ouachita Mobile Belt	Appalachian Mobile Belt	Major Events Outside North America
408–438	Silurian	Tippecanoe			Extensive barrier reefs and evaporites common		Acadian orogeny	Caledonian orogeny
438–505	Ordovician	Tippecanoe; Sauk	Present sea level		Queenston Delta clastic wedge Transgression of Tippecanoe Sea Regression exposing large areas to erosion		Taconic orogeny	Continental glaciation in Southern Hemisphere
505–545	Cambrian	Sauk			Canadian Shield and Transcontinental Arch only areas above sea level Transgression of Sauk Sea			

- The eastern edge of North America was a stable carbonate platform during Sauk time. During Tippecanoe time, an oceanic–continental convergent plate boundary formed, resulting in the Taconic orogeny, the first of several orogenies to affect the Appalachian mobile belt.
- The newly formed Taconic Highlands shed sediments into the western epeiric sea, producing the Queenston Delta, a clastic wedge.
- The Caledonian, Acadian, Hercynian, and Alleghenian orogenies were all part of the global tectonic activity that assembled Pangaea.
- The Cordilleran mobile belt was the site of the Antler orogeny, a minor Devonian orogeny during which deep-water sediments were thrust eastward over shallow-water sediments.

Table 20.2

Summary of Late Paleozoic Events

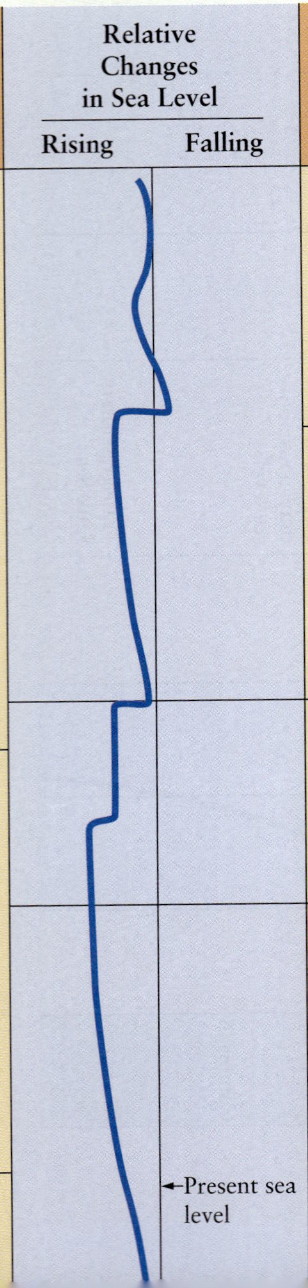

Age (millions of years)	Geologic Period	Sequence	Cordilleran Mobile Belt	Craton	Ouachita Mobile Belt	Appalachian Mobile Belt	Major Events Outside North America
245–286	Permian	Absaroka		Deserts, evaporites, and continental red beds in southwestern United States Extensive reefs in Texas area	Ouachita orogeny (late Permian onward)	Formation of Pangaea Allegheny orogeny	Formation of Pangaea Hercynian orogeny
286–320	Carboniferous: Pennsylvanian	Absaroka		Coal swamps common Formation of Ancestral Rockies	Ouachita orogeny	Allegheny orogeny	Hercynian orogeny Continental glaciation in Southern Hemisphere
320–360	Carboniferous: Mississippian	Absaroka / Kaskaskia		Transgression of Absaroka Sea Widespread black shales and limestones	Ouachita orogeny (early)		
360–	Devonian	Kaskaskia / Tippecanoe	Antler orogeny	Widespread black shales Catskill Delta clastic wedge Extensive barrier reef formation in western Canada Transgression of Kaskaskia Sea		Acadian orogeny	Old Red Sandstone clastic wedge in British Isles Caledonian orogeny

- Mountain building occurred in the Ouachita mobile belt during the Pennsylvanian and Early Permian. This tectonic activity was partly responsible for the cratonic uplift in the southwest, resulting in the Ancestral Rockies.
- During the Paleozoic Era, numerous terranes such as Avalonia existed and played an important role in forming Pangaea.
- Paleozoic-age rocks contain a variety of mineral resources, including building stone, limestone for cement, silica sand, evaporites, petroleum, coal, ores of iron, lead, and zinc, and other metallic deposits.

Important Terms

Absaroka sequence (p. 582)
Acadian orogeny (p. 589)
Ancestral Rockies (p. 586)
Antler orogeny (p. 591)
Appalachian mobile belt (p. 565)
Baltica (p. 565)
Caledonian orogeny (p. 589)
Catskill Delta (p. 589)
China (p. 565)
clastic wedge (p. 588)
Cordilleran mobile belt (p. 565)
cratonic sequence (p. 569)
cyclothem (p. 583)
epeiric sea (p. 564)
Franklin mobile belt (p. 565)
Gondwana (p. 565)
Hercynian-Alleghenian orogeny (p. 591)
Iapetus Ocean (p. 586)
Kaskaskia sequence (p. 577)
Kazakhstania (p. 565)
Laurasia (p. 566)
Laurentia (p. 565)
mobile belt (p. 564)
organic reef (p. 575)
Ouachita mobile belt (p. 565)
Ouachita orogeny (p. 593)
Panthalassa Ocean (p. 569)
Queenston Delta (p. 588)
Sauk sequence (p. 571)
sequence stratigraphy (p. 571)
Siberia (p. 565)
Taconic orogeny (p. 586)
Tippecanoe sequence (p. 573)
Transcontinental Arch (p. 572)

Review Questions

1. A major transgressive–regressive cycle bounded by cratonwide unconformities is a(n)
   a._____biostratigraphic unit; b._____cratonic sequence; c._____orogeny; d._____shallow sea; e._____cyclothem.
2. The Taconic orogeny resulted from what type of plate interaction?
   a._____oceanic–oceanic convergent; b._____oceanic–continental convergent; c._____continental–continental convergent; d._____divergent; e._____transform.
3. During which sequence were evaporites and reef carbonates the predominant cratonic rocks?
   a._____Kaskaskia; b._____Zuni; c._____Sauk; d._____Absaroka; e._____Tippecanoe.
4. During the Early Paleozoic Era, orogenic activity was confined to which mobile belt?
   a _____Franklin; b._____Cordilleran; c._____Ouachita; d._____Appalachian; e._____answers b and c.
5. During which sequence was the eastern margin of Laurentia a passive plate margin?
   a._____Zuni; b._____Tippecanoe; c._____Sauk; d._____Kaskaskia; e._____Absaroka.
6. Weathering of which highlands produced the Catskill Delta clastic wedge?
   a._____Transcontinental Arch; b._____Acadian; c._____Taconic; d._____Sevier; e._____Nevadan.
7. Uplift in the southwestern part of the craton during the Late Absaroka resulted in which mountainous region?
   a._____Antler Highlands; b._____Marathon; c._____Ancestral Rockies; d._____Appalachian; e._____none of these.
8. Rhythmically repetitive sedimentary sequences are
   a._____reefs; b._____orogenies; c._____cyclothems; d._____tillites; e._____evaporites.

9. Which was the first Paleozoic orogeny to occur in the Cordilleran mobile belt?

   a._____Acadian; b._____Alleghenian; c._____Antler; d._____Caledonian; e._____Ellesmere.

10. During which period did extensive continental glaciation of the Gondwana continent occur?

    a._____Cambrian; b._____Silurian; c._____Devonian; d._____Carboniferous; e._____Permian.

11. Discuss how plate movement during the Paleozoic Era affected worldwide weather patterns.

12. What are some of the methods geologists use to determine the locations of continents during the Paleozoic Era?

13. Discuss why cratonic sequences are a convenient way to study the geologic history of the Paleozoic Era.

14. According to estimates made from mapping and correlation, the Queenston Delta contains more than 600,000 km^3 of rock eroded from the Taconic Highlands. Based on this figure, geologists estimate the Taconic Highlands were at least 4000 m high. They also estimate that the Catskill Delta contains three times as much sediment as the Queenston Delta. From what you know about the geographic distribution of the Taconic Highlands and the Acadian Highlands, can you estimate how high the Acadian Highlands might have been?

15. What evidence indicates that the Iapetus Ocean began closing during the Middle Ordovician?

16. Based on the discussion of Milankovitch cycles and their role in causing glacial–interglacial cycles (see Chapter 14), could these cycles be partly responsible for the transgressive–regressive cycles that resulted in cyclothems during the Pennsylvanian Period?

17. What were the major differences between the Appalachian, Ouachita, and Cordilleran mobile belts during the Paleozoic Era?

18. How are the Caledonian, Acadian, Ouachita, Hercynian, and Alleghenian orogenies related to modern concepts of plate tectonics?

19. How does the origin of evaporite deposits of the Kaskaskia sequence compare with the origin of evaporites of the Tippecanoe sequence?

20. Discuss how sequence stratigraphy can be used to make global correlations and why it is so useful in reconstructing past events.

World Wide Web Activities

GeologyNow Assess your understanding of this chapter's topics with additional quizzing and comprehensive interactivities at

http://earthscience.brookscole.com/changingearth4e

as well as current and up-to-date weblinks, additional readings, and InfoTrac College Edition exercises.

Paleozoic Life History

CHAPTER 21
OUTLINE

GeologyNow *This icon, which appears throughout the book, indicates an opportunity to explore interactive tutorials, animations, or practice problems available on the GeologyNow website at http://earthscience.brookscole.com/changingearth4e.*

Diorama of the environment and biota of the Phyllopod bed of the Burgess Shale, British Columbia, Canada. In the background is a vertical wall of a submarine escarpment with algae growing on it. The large cylindrical ribbed organisms on the muddy bottom in the foreground are sponges. Source: Smithsonian Institution, transparency no. 86-13471A

OBJECTIVES

At the end of this chapter, you will have learned that:

- Animals with skeletons appeared abruptly at the beginning of the Paleozoic Era and experienced a short period of rapid evolutionary diversification.
- During the Paleozoic Era, the invertebrates experienced times of diversification followed by extinction, culminating in the greatest recorded mass extinction in Earth's history at the end of the Permian Period.
- Vertebrates first evolved during the Cambrian Period, and fish diversified rapidly during the Paleozoic Era.
- Amphibians first appear in the fossil record during the Devonian, having made the transition from water to land, and became extremely abundant during the Pennsylvanian Period when coal-forming swamps were widespread.
- The evolution of the amniote egg allowed reptiles to colonize all parts of the land beginning in the Late Mississippian.
- The pelycosaurs or finback reptiles were the dominant reptile group during the Permian and were the ancestors to the therapsids or mammal-like reptiles.
- The earliest land plants are known from the Ordovician Period, whereas the oldest known vascular land plants first appear in the Middle Silurian.
- Seedless vascular plants, such as ferns, were very abundant during the Pennsylvanian Period.
- With the onset of arid conditions during the Permian Period, the gymnosperms became the dominant element of the world's flora.

Introduction

On August 30 and 31, 1909, near the end of the summer field season, Charles D. Walcott, geologist and head of the Smithsonian Institution, was searching for fossils along a trail on Burgess Ridge between Mount Field and Mount Wapta, near Field, British Columbia, Canada. On the west slope of this ridge, he discovered the first soft-bodied fossils from the Burgess Shale, a discovery of immense importance in deciphering the early history of life. During the next week, Walcott and his collecting party split open numerous blocks of shale, many of which yielded the impressions of soft-bodied organisms beautifully preserved on bedding planes. Walcott returned to the site the following summer and located the shale stratum that was the source of his fossil-bearing rocks in the steep slope above the trail. He quarried the site and shipped back thousands of fossil specimens to the National Museum of Natural History, where he later cataloged and studied them.

The importance of Walcott's discovery is not that it was another collection of well-preserved Cambrian fossils, but rather that it provided geologists with a rare glimpse into a world previously almost unknown—that of the soft-bodied animals that lived some 530 million years ago. The beautifully preserved fossils from the Burgess Shale present a much more complete picture of a Middle Cambrian community than do deposits containing only fossils of the hard parts of organisms. In fact, 60% of the total fossil assemblage of more than 100 genera is composed of soft-bodied animals, a percentage comparable to present-day marine communities.

What conditions led to the remarkable preservation of the Burgess Shale fauna? The depositional site of the Burgess Shale lay at the base of a steep submarine escarpment. The animals whose exquisitely preserved fossil remains are found in the Burgess Shale lived in and on mud banks that formed along the top of this escarpment. Periodically, this unstable area would slump and slide down the escarpment as a turbidity current. At the base, the mud and animals carried with it were deposited in a deep-water anaerobic environment devoid of life. In such an environment, bacterial degradation did not destroy the buried animals, and thus they were compressed by the weight of the overlying sediments and eventually preserved as carbonaceous impressions.

In this chapter, we examine the history of Paleozoic life as a system of interconnected biologic and geologic events. The underlying processes of evolution and plate tectonics are the forces that drove this system. The opening and closing of ocean basins, transgressions and regressions of epeiric seas, the formation of mountain ranges, and the changing positions of the continents profoundly affected the evolution of the marine and terrestrial communities.

A time of tremendous biologic change began with the appearance of skeletonized animals near the Precambrian-Cambrian boundary. Following this event, marine invertebrates began a period of adaptive radiation and evolution. The earliest fossil records of vertebrates are fish, which evolved during the Cambrian. One group of fish was ancestral to the amphibians, which evolved during the Devonian. Reptiles evolved from amphibians during the Mississippian Period and were the dominant vertebrate animals on land by the end of the Paleozoic. Plants preceded animals onto land, and both plants and animals were confronted with the same basic problems in making the transition from water to land.

The end of the Paleozoic Era witnessed the greatest mass extinction in Earth's history. The marine invertebrate community was decimated, and many amphibians and reptiles also became extinct.

WHAT WAS THE CAMBRIAN EXPLOSION?

At the beginning of the Paleozoic Era, animals with skeletons appeared rather abruptly in the fossil record. In fact, their appearance is described as an explosive development of new types of animals and is referred to as the "Cambrian explosion" by most scientists. This sudden appearance of new animals in the fossil record is rapid, however, only in the context of geologic time, having taken place over millions of years during the Early Cambrian Period.

This seemingly sudden appearance of animals in the fossil record is not a recent discovery. Early geologists observed that the remains of skeletonized animals appeared rather abruptly in the fossil record. Charles Darwin addressed this problem in *On the Origin of Species* and observed that, without a convincing explanation, such an event was difficult to reconcile with his newly expounded evolutionary theory.

The sudden appearance of shelled animals during the Early Cambrian contrasts sharply with the biota that lived during the preceding Proterozoic Eon. Up until the evolution of the Ediacaran fauna, Earth was populated primarily by single-celled organisms. Recall from Chapter 19 that the Ediacaran fauna, which is found on all continents except Antarctica, consists primarily of multicelled soft-bodied organisms. Microscopic calcareous tubes, presumably housing wormlike suspension-feeding organisms, have also been found at some localities. In addition, trails and burrows, which represent the activities of worms and sluglike animals, are found associated with Ediacaran faunas throughout the world. The trails and burrows are similar to those made by present-day soft-bodied organisms.

Until recently, it appeared that there was a fairly long period of time between the extinction of the Ediacaran fauna and the first Cambrian fossils. That gap has been considerably narrowed in recent years with the discovery of new Proterozoic fossiliferous localities. Now, Proterozoic fossil assemblages continue right to the base of the Cambrian. Furthermore, recent work from Namibia indicates that Ediacaran-like fossils are even present above the first occurrence of Cambrian index fossils.

Nonetheless, the cause of the sudden appearance of so many different animal phyla during the Early Cambrian is still a hotly debated topic. Newly developed molecular techniques that allow evolutionary biologists to compare molecular sequences of the same gene from different species are being applied to the phylogeny of many organisms. In addition, new fossil sites and detailed stratigraphic studies are shedding light on the early history and ancestry of the various invertebrate phyla.

The Cambrian explosion probably had its roots firmly planted in the Proterozoic. However, the mechanism that triggered this event is still unknown and was likely a combination of factors, both biologic and geologic. For example, geologic evidence indicates that Earth was glaciated one or more times during the Proterozoic, followed by global warming during the Cambrian. These global environmental changes may have stimulated evolution and contributed to the Cambrian explosion. Whatever the ultimate cause of the Cambrian explosion, the appearance of a skeletonized fauna and the rapid diversification of that fauna during the Early Cambrian were major events in life history.

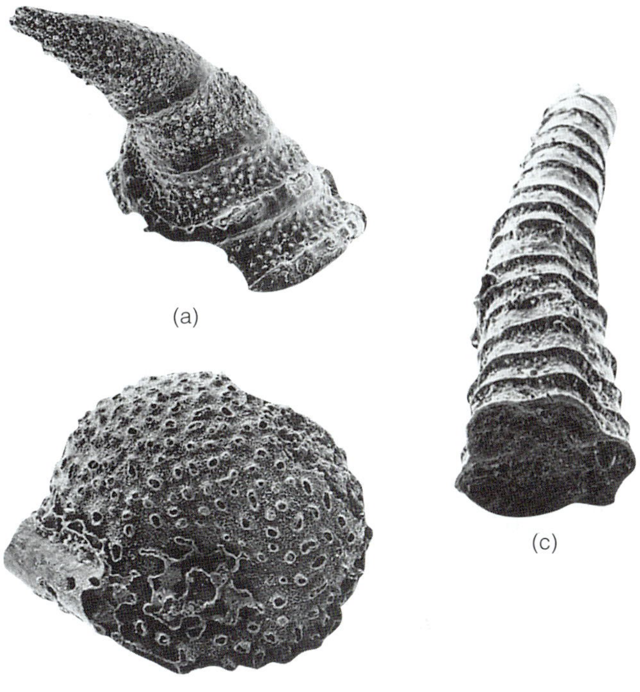
(a) (b) (c)

Figure 21.1

Three small (several millimeters in size) Lower Cambrian shelly fossils. (a) A conical sclerite (a piece of the armor covering) of *Lapworthella* from Australia. (b) *Archaeooides*, an enigmatic spherical fossil from the Mackenzie Mountains, Northwest Territories, Canada. (c) The tube of an anabaritid from the Mackenzie Mountains, Northwest Territories, Canada. Source: Courtesy of Simon Conway Morris and Stefan Bengston, University of Cambridge, UK.

THE EMERGENCE OF A SHELLY FAUNA

The earliest organisms with hard parts are Proterozoic calcareous tubes found associated with Ediacaran faunas from several locations throughout the world. These are followed by other microscopic skeletonized fossils from the Lower Cambrian (■ Figure 21.1) and the appearance of large skeletonized animals during the Cambrian explosion. Along with the question of why did animals appear so suddenly in the fossil record is the equally intriguing one of why they initially acquired skeletons and what selective advantage this provided. A variety of explanations about why marine organisms evolved skeletons have been proposed, but none are completely satisfactory or universally accepted.

The formation of an exoskeleton confers advantages on an organism: (1) It provides protection against ultraviolet radiation, allowing animals to move into shallower waters; (2) it helps prevent drying out in an intertidal environment; (3) whether an exo- or endoskeleton, it allows animals to increase their size and provides attachment sites for muscles; and (4) it provides protection against predators. Recent evidence of actual fossils of predators and specimens of damaged prey, as well as antipredatory adaptations in some animals, indicates the great impact of predation during the Cambrian. With predators playing an important role in the Cambrian marine ecosystem, any mechanism or feature that protected an animal would certainly be advantageous and confer an adaptive advantage to the organism.

Scientists currently have no clear answer about why marine organisms evolved mineralized skeletons during the Cambrian explosion and shortly thereafter. They undoubtedly evolved because of a variety of biologic and environmental factors. Whatever the reason, the acquisition of a mineralized skeleton was a major evolutionary innovation allowing invertebrates to successfully occupy a wide variety of marine habitats.

Table 21.1

The Major Invertebrate Groups and Their Stratigraphic Ranges

Phylum Protozoa	Cambrian–Recent	**Phylum Mollusca**	Cambrian–Recent
Class Sarcodina	Cambrian–Recent	Class Monoplacophora	Cambrian–Recent
Order Foraminifera	Cambrian–Recent	Class Gastropoda	Cambrian–Recent
Order Radiolaria	Cambrian–Recent	Class Bivalvia	Cambrian–Recent
Phylum Porifera	Cambrian–Recent	Class Cephalopoda	Cambrian–Recent
Class Demospongea	Cambrian–Recent	**Phylum Annelida**	Precambrian–Recent
Order Stromatoporoida	Cambrian–Oligocene	**Phylum Arthropoda**	Cambrian–Recent
Phylum Archaeocyatha	Cambrian	Class Trilobita	Cambrian–Permian
Phylum Cnidaria	Cambrian–Recent	Class Crustacea	Cambrian–Recent
Class Anthozoa	Ordovician–Recent	Class Insecta	Silurian–Recent
Order Tabulata	Ordovician–Permian	**Phylum Enchinodermata**	Cambrian–Recent
Order Rugosa	Ordovician–Permian	Class Blastoidea	Ordovician–Permian
Order Scleractinia	Triassic–Recent	Class Crinoidea	Cambrian–Recent
Phylum Bryozoa	Ordovician–Recent	Class Echinoidea	Ordovician–Recent
Phylum Brachiopoda	Cambrian–Recent	Class Asteroidea	Ordovician–Recent
Class Inarticulata	Cambrian–Recent	**Phylum Hemichordata**	Cambrian–Recent
Class Articulata	Cambrian–Recent	Class Graptolithina	Cambrian–Mississippian

PALEOZOIC INVERTEBRATE MARINE LIFE

Rather than focusing on the history of each invertebrate phylum (Table 21.1), we will survey the evolution of the Paleozoic marine invertebrate communities through time, concentrating on the major features and changes that took place. To do that, we need to briefly examine the nature and structure of living marine communities so that we can make a reasonable interpretation of the fossil record.

The Present Marine Ecosystem

In analyzing the present-day marine ecosystem, we must look at where organisms live, how they get around, and how they feed (■ Figure 21.2). Organisms that live in the water column above the seafloor are called *pelagic*. They are divided into two main groups: the floaters, or **plankton,** and the swimmers, or **nekton.**

Plankton are mostly passive and go where currents carry them. Plant plankton, such as diatoms, dinoflagellates, and various algae, are called *phytoplankton* and are mostly microscopic. Animal plankton are called *zooplankton* and are also mostly microscopic. Examples of zooplankton include foraminifera, radiolarians, and jellyfish. The nekton are swimmers and are mainly vertebrates such as fish; the invertebrate nekton include cephalopods.

Organisms that live on or in the seafloor make up the **benthos.** They are characterized as *epifauna* (animals) or *epiflora* (plants), for those that live on the seafloor, or as *infauna,* which are animals living in and moving through the sediments. The benthos are further divided into those organisms that stay in one place, called *sessile,* and those that move around on or in the seafloor, called *mobile.*

The feeding strategies of organisms are also important in terms of their relationships with other organisms in the marine ecosystem. There are basically four feeding groups: **suspension-feeding** animals remove or consume microscopic plants and animals as well as dissolved nutrients from the water; **herbivores** are plant eaters; **carnivore-scavengers** are meat eaters; and **sediment-deposit feeders** ingest sediment and extract the nutrients from it.

We can define an organism's place in the marine ecosystem by where it lives and how it eats. For example, a brachiopod is a benthonic, epifaunal suspension feeder, whereas a cephalopod is a nektonic carnivore.

An ecosystem includes several *trophic levels,* which are tiers of food production and consumption within a feeding hierarchy. The feeding hierarchy and hence energy flow in an ecosystem make up a food web of complex interrelationships among the producers, consumers, and decomposers (■ Figure 21.3). The **primary producers,** or *autotrophs,* are those organisms that manufacture their own food. Virtually all marine primary producers are phytoplankton. Feeding on the primary producers are the primary consumers, which are mostly suspension feeders. Secondary consumers feed on the primary consumers and thus are predators, whereas tertiary consumers, which are also predators, feed on the secondary consumers. Besides the producers and consumers, there are also transformers and decomposers. These are bacteria that break down the dead organisms that have not been consumed into organic compounds, which are then recycled.

When we look at the marine realm today, we see a complex organization of organisms interrelated by trophic interactions and affected by changes in the physical environment. When one part of the system changes, the whole structure changes, sometimes almost insignificantly and other times catastrophically.

As we examine the evolution of the Paleozoic marine ecosystem, keep in mind how geologic and evolutionary changes can have a significant impact on its composition and structure. For example, the major transgressions onto the craton opened up vast areas of shallow seas that could be inhabited. The movement of continents affected oceanic circulation patterns and caused environmental changes.

What Would You Do?

Congress has just voted a large cut in funding for the National Science Foundation (NSF). Because of reduced funding, some earth science areas will not be funded this year. One area slated for major funding reduction or no funding this year is paleontology. As a paleontology professor whose research depends on NSF funds, you will be testifying against cutting funding for paleontology. Your specialty is Paleozoic marine ecosystems. What arguments would you use to keep funding for such research? What possible outcomes from your research could you point to that might benefit society today?

Cambrian Marine Community

During the Cambrian Period, many new body plans evolved and animals moved into new niches. As might be expected, the Cambrian witnessed a higher percentage of such experiments than any other period of geologic history.

Although almost all the major invertebrate phyla evolved during the Cambrian Period (Table 21.1), many were represented by only a few species. Whereas trace fossils are common and echinoderms diverse, trilobites, brachiopods, and archaeocyathids made up the majority of Cambrian skeletonized life (■ Figure 21.4).

Trilobites were by far the most conspicuous members of the Cambrian marine invertebrate community and made up about half of the total fauna. Trilobites were benthonic, mobile, sediment-deposit feeders that crawled

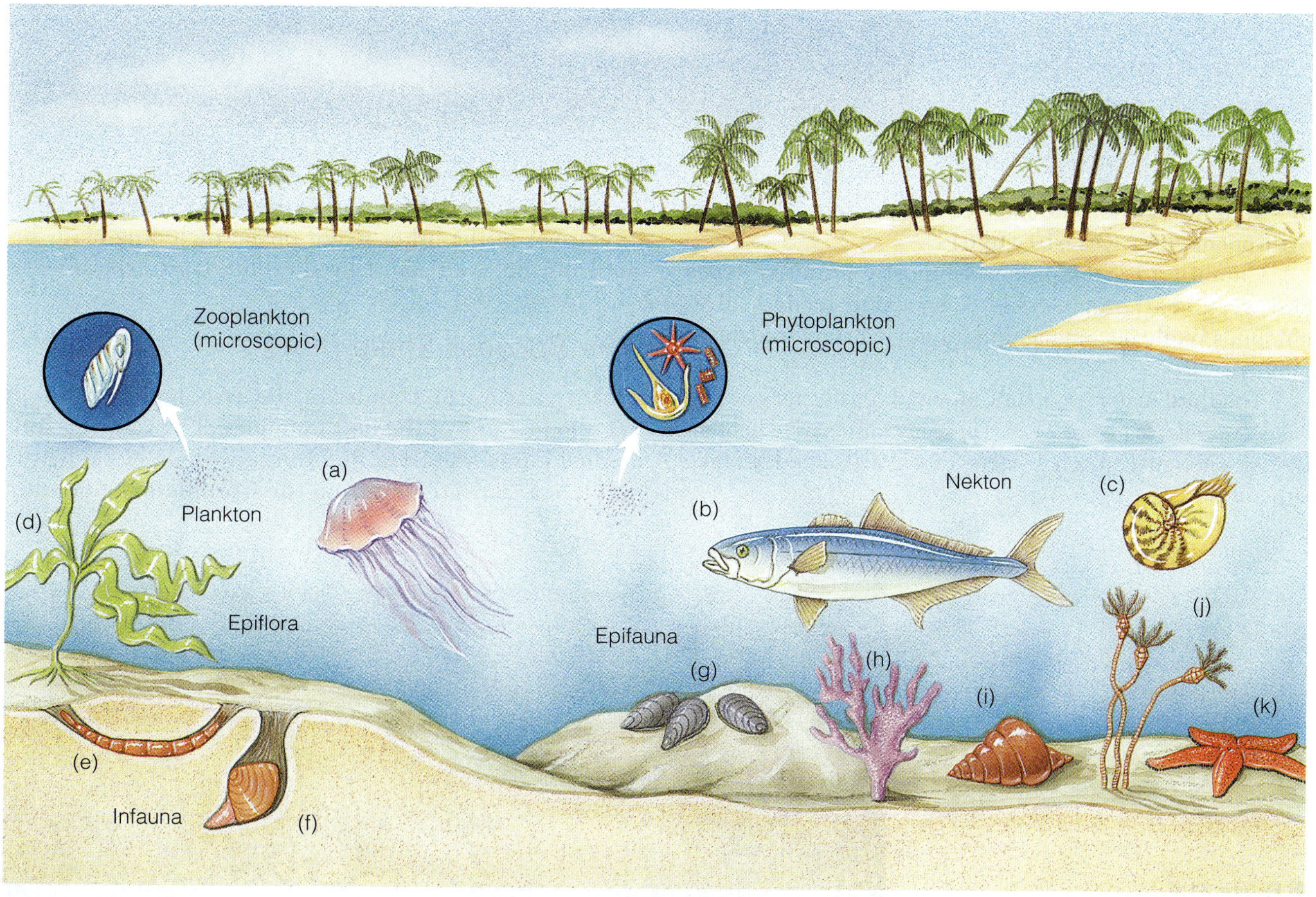

■ **Figure 21.2**

Where and how animals and plants live in the marine ecosystem. Plankton: (a) jellyfish. Nekton: (b) fish and (c) cephalopod. Benthos: (d) through (k). Sessile epiflora: (d) seaweed. Sessile epifauna: (g) bivalve, (h) coral, and (j) crinoid. Mobile epifauna: (k) starfish and (i) gastropod. Infauna: (e) worm and (f) bivalve. Suspension feeders: (g) bivalve, (h) coral, and (j) crinoid. Herbivores: (i) gastropod. Carnivore-scavengers: (k) starfish. Sediment-deposit feeders: (e) worm.

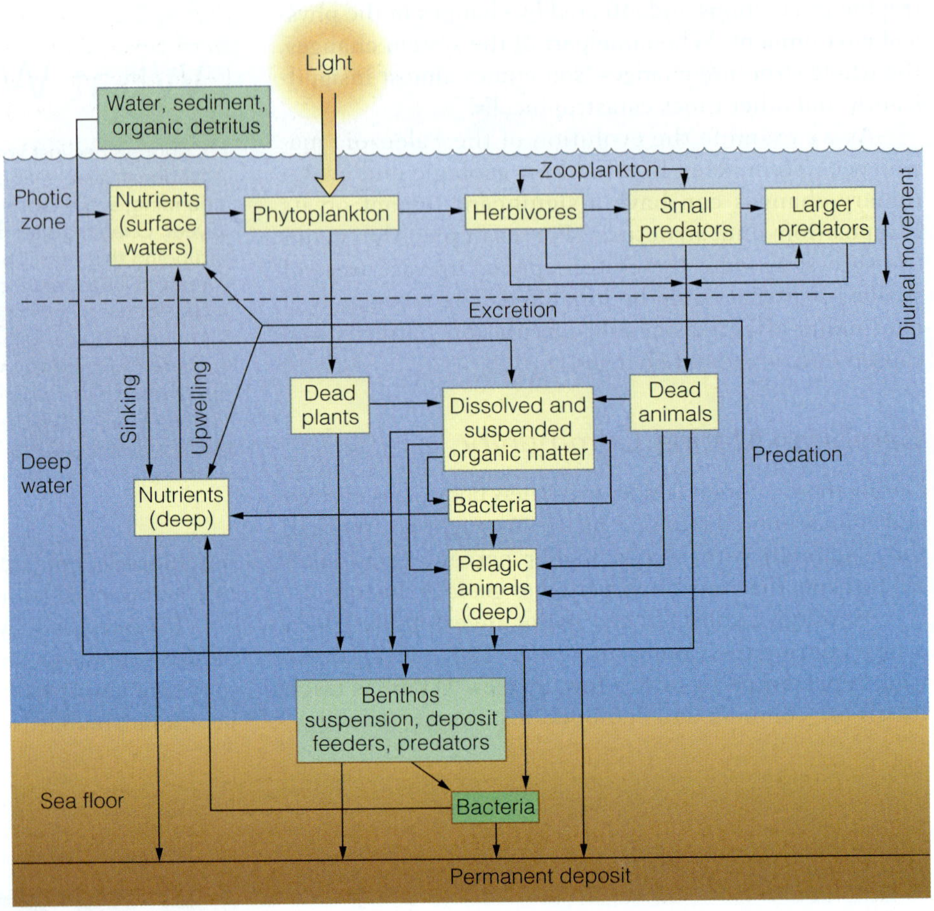

■ **Figure 21.3**

Marine food web showing the relationships among the producers, consumers, and decomposers.

or swam along the seafloor (see "Trilobites—Paleozoic Arthropods" on pages 610 and 611).

Cambrian *brachiopods* were mostly primitive types and were benthonic, sessile, suspension feeders. Brachiopods really became abundant during the Ordovician Period.

The third major group of Cambrian organisms were *archaeocyathids* (■ Figure 21.5). These organisms were benthonio, sessile, suspension feeders that constructed reeflike structures.

The rest of the Cambrian fauna consisted of representatives of the other major phyla, including many organisms that were short-lived evolutionary experiments.

The Burgess Shale Biota

No discussion of Cambrian life is complete without mentioning one of the best examples of a preserved soft-bodied fauna and flora, the Burgess Shale biota. As the Sauk Sea transgressed from the Cordilleran shelf onto

Carnegie Museum of Natural History

■ **Figure 21.4**

Reconstruction of a Cambrian marine community. Floating jellyfish, swimming arthropods, benthonic sponges, and scavenging trilobites are shown.

Figure 21.5

Restoration of a Cambrian reeflike structure built by archaeocyathids.

the western edge of the craton, Early Cambrian sands were covered by Middle Cambrian black muds that allowed a diverse soft-bodied benthonic community to be preserved. As we discussed in the Introduction, these fossils were discovered in 1909 by Charles D. Walcott near Field, British Columbia. They represent one of the most significant fossil finds of the 20th century because they consist of impressions of soft-bodied animals and plants (Figure 21.6), which are rarely preserved in the fossil record.

In recent years, the reconstruction, classification, and interpretation of many of the Burgess Shale fossils

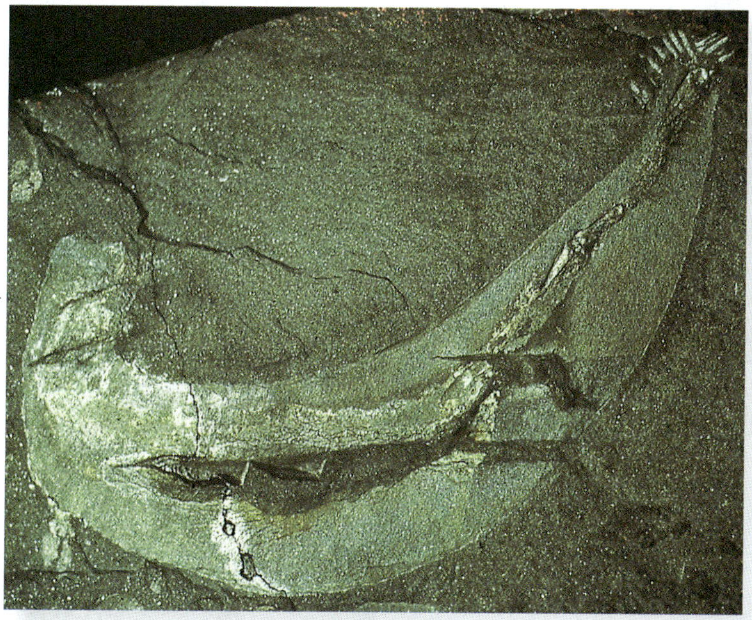

Douglas H. Erwin/National Museum of Natural History

(a) *Ottoia*

Douglas H. Erwin/National Museum of Natural History

(b) *Wiwaxia*

Douglas H. Erwin/National Museum of Natural History

(c) *Hallucigenia*

Douglas H. Erwin/National Museum of Natural History

(d) *Waptia*

Figure 21.6

Some of the fossil animals preserved in the Burgess Shale. (a) *Ottoia,* a carnivorous worm. (b) *Wiwaxia,* a scaly armored sluglike creature whose affinities remain controversial. (c) *Hallucigenia,* a velvet worm. (d) *Waptia,* an arthropod.

Trilobites—Paleozoic Arthropods

Trilobites, an extinct class of arthropods, are probably the favorite and most sought after of invertebrate fossils. They lived from the Early Cambrian until the end of the Permian and were most diverse during the Late Cambrian. More than 15,000 species of trilobites have been described, and they are currently grouped into nine orders. Trilobites had a worldwide distribution throughout the Paleozoic and they lived in all marine environments, from shallow, nearshore waters to deep oceanic settings. They occupied a wide variety of habitats. Most were bottom dwellers, crawling around the sea floor and scavenging organic detritus (see left) or feeding on microorganisms or algae. Others were free-swimming predators or filter feeders living throughout the water column.

The Manitoba Museum/Dr. Graham Young

The trilobite body is divided into three parts: A *cephalon* (head), containing the eyes, mouth, and sensory organs; the *thorax* (body), composed of individual segments; and the *pygidium* (tail). The name trilobite does not refer to its three main body parts, but means *three-lobed*, which corresponds to the three longitudinal lobes of the thorax — the axial lobe, and the two flanking pleural lobes. *Cedaria minor*, from the Cambrian-age Weeks Formation, Utah, illustrates the major body parts of a trilobite.

University of Utah-Geology and Geophysics

Peabody Museum of Natural History, Yale University, New Haven CT

Peabody Museum of Natural History, Yale University, New Haven CT

Model of dorsal (above) and ventral (right) anatomy of the Late Ordovician trilobite *Triarthrus eatoni.*

The appendages of trilobites are rarely preserved. However, from specimens in which the soft-part anatomy is preserved as an impression, paleontologists know that beneath each thoracic segment was a two-part appendage consisting of a gill-bearing outer branch used for respiration and an inner branch or walking leg composed of articulating limb segments. This model of the dorsal and ventral anatomy of *Triarthrus eatoni* is from the Late Ordovician-age Frankfort Shale in New York.

Trilobites range in size from a few millimeters in length, as found in many of the agnostid trilobites such as *Hypagnostus parvifrons* from the Cambrian-age Marjum Formation, Utah (far left), to more than 70 centimeters, as recorded by the world's largest trilobite *Isotellus rex* from the Late Ordovician-age Churchill River Group, Manitoba, Canada (left, below). Most trilobites, however, range between 3 and 10 cm long.

University of Utah-Geology and Geophysics

Hypagnostus parvifrons

The Manitoba Museum/Dr. Graham Young

Isotellus rex

Trilobites were either blind, as were the agnostid trilobites (*Hypagnostus parvifrons*), or possessed paired eyes. Sighted trilobites had one of two types of eyes. Holochroal eyes are characterized by generally hexagonal-shaped biconvex lenses that are packed closely together and covered by a single corneal layer that covers all the lenses. This type of eye is similar to a modern compound eye and results in mosaic vision, such as in the trilobite *Scutellum campaniferum* (below).

E.N.K. Clarkson

Scutellum campaniferum, above. (holochroal eyes)

Eophacops trapeziceps, below. (schizochroal eyes)

E.N.K. Clarkson

Schizochroal eyes are composed of relatively large, individual lenses, each of which is covered by its own cornea and separated from adjacent lenses by a thick dividing wall, such as in the trilobite *Eophacops trapeziceps* (above). Schizochroal lenses are arranged in rows and columns and most likely produced a visual field consisting of shadows.

Sinclair Stammers/SPL/Photo Researchers, Inc.

Some trilobites had the ability to enroll, presumably to protect their antennae, limbs, and soft ventral appendages. By doing so, only the hard exoskeleton would be exposed to potential enemies and the trilobite could still view its surroundings. This enrolled specimen is *Pliomera fisheri* (Ordovician age) from Putilowa, Poland.

have undergone a major change that has led to new theories and explanations of the Cambrian explosion of life. Recall that during the Late Proterozoic multicelled organisms evolved, and shortly thereafter animals with hard parts made their first appearance. These were followed by an explosion of invertebrate phyla during the Cambrian, some of which are now extinct. These Cambrian phyla represent the root stock and basic body plans from which all present-day invertebrates evolved. The question that paleontologists are still debating is how many phyla arose during the Cambrian, and at the center of that debate are the Burgess Shale fossils. For years, most paleontologists placed the bulk of the Burgess Shale organisms into existing phyla, with only a few assigned to phyla that are now extinct. Thus, the phyla of the Cambrian world were viewed as being essentially the same in number as the phyla of the present-day world, but with fewer species in each phylum. According to this view, the history of life has been simply a gradual increase in the diversity of species within each phylum through time. The number of basic body plans has therefore remained more or less constant since the initial radiation of multicelled organisms.

This view, however, has been challenged by other paleontologists who think that the initial explosion of varied life-forms in the Cambrian was promptly followed by a short period of experimentation and then extinction of many phyla. The richness and diversity of modern life-forms are the result of repeated variations of the basic body plans that survived the Cambrian extinctions. In other words, life was much more diverse in terms of phyla during the Cambrian than it is today. The reason members of the Burgess Shale biota look so strange to us is that no living organisms possess their basic body plan, and therefore many of them have been reassigned into new phyla.

Discoveries of new Cambrian fossils at localities such as Sirius Passet, Greenland, and Yunnan, China, have resulted in reassignment of some Burgess Shale specimens back into extant phyla. If these reassignments to known phyla prove correct, then no massive extinction followed the Cambrian explosion, and life has gradually increased in diversity through time. Currently, there is no clear answer to this debate, and the outcome will probably be decided as more fossil discoveries are made.

Ordovician Marine Community

A major transgression that began during the Middle Ordovician (Tippecanoe sequence) resulted in the most widespread inundation of the craton. This vast epeiric sea, which experienced a uniformly warm climate during this time, opened many new marine habitats that were soon filled by a variety of organisms.

Not only did sedimentation patterns change dramatically from the Cambrian to the Ordovician, but the fauna underwent equally striking changes. Whereas the Cambrian invertebrate community was dominated by trilobites, brachiopods, and archaeocyathids, the Ordovician was characterized by the continued diversification of brachiopods and by the adaptive radiation of many other animal phyla (such as bryozoans and corals), with a consequent dramatic increase in the diversity of the total shelly fauna (■ Figure 21.7).

During the Cambrian, archaeocyathids were the main builders of reeflike structures, but bryozoans, stromatoporoids, and tabulate and rugose corals assumed that role beginning in the Middle Ordovician. Many of these reefs were small patch reefs similar in size to those of the Cambrian but of a different composition, whereas others were quite large. Like present-day reefs, Ordovician reefs exhibited a high diversity of organisms and were dominated by suspension feeders.

The end of the Ordovician was a time of mass extinctions in the marine realm. More than 100 families

■ Figure 21.7

Recreation of a Middle Ordovician seafloor fauna. Cephalopods, crinoids, colonial corals, trilobites, and brachiopods are shown.

of marine invertebrates became extinct, and in North America alone, about half of the brachiopods and bryozoans died out. What caused such an event? Many geologists think these extinctions were the result of the extensive glaciation in Gondwana at the end of the Ordovician Period (see Chapter 20).

Silurian and Devonian Marine Communities

The mass extinction at the end of the Ordovician was followed by rediversification and recovery of many of the decimated groups. Brachiopods, bryozoans, gastropods, bivalves, corals, crinoids, and graptolites were just some of the groups that rediversified during the Silurian.

As discussed in Chapter 20, the Silurian and Devonian were times of major reef building. Whereas most of the Silurian radiations of invertebrates represented repopulating of niches, organic reef builders diversified in new ways, building massive reefs larger than any produced during the Cambrian or Ordovician. This repopulation was probably caused in part by renewed transgressions over the craton, and although a major drop in sea level occurred at the end of the Silurian, the Middle Paleozoic sea level was generally high (see Table 20.1).

The Silurian and Devonian reefs were dominated by tabulate and colonial rugose corals and stromatoporoids (■ Figure 21.8). Although the fauna of these Silurian and Devonian reefs was somewhat different from that of earlier reefs and reeflike structures, the general composition and structure are the same as in present-day reefs.

The Silurian and Devonian periods were also the time when *eurypterids* (arthropods with scorpionlike bodies and impressive pincers) were abundant, especially in brackish and freshwater habitats (■ Figure 21.9). Ammonoids, a subclass of the cephalopods, evolved from nautiloids during the Early Devonian and rapidly diversified. With their distinctive suture patterns, short stratigraphic ranges, and widespread distribution, ammonoids are excellent guide fossils for the Devonian through Cretaceous periods (Figure 21.8).

Another mass extinction occurred near the end of the Devonian and resulted in a worldwide near-total collapse of the massive reef communities. On land, however, the seedless vascular plants were seemingly unaffected, although the diversity of freshwater fish was greatly reduced.

The demise of the Middle Paleozoic reef communities highlights the geographic aspects of the Late Devonian mass extinction. The tropical groups were most severely affected; in contrast, the polar communities were seemingly little affected. Apparently, an episode of global cooling was largely responsible for the extinctions near the end of the Devonian. During such a cooling, the disappearance of tropical conditions would have had a severe effect on reef and other warm-water organisms. Cool-water species, in contrast, could have simply migrated toward the equator. Although cooling temperatures certainly played an important role in the Late Devonian extinctions, the closing of the Iapetus Ocean and the orogenic events of this time undoubtedly also played a role by reducing the area of shallow shelf environments where many marine invertebrates lived.

Carboniferous and Permian Marine Communities

The Carboniferous invertebrate marine community responded to the Late Devonian extinctions in much the same way the Silurian invertebrate marine community responded to the Late Ordovician extinctions—that is, by renewed adaptive radiation and rediversification. The brachiopods and ammonoids quickly recovered and again assumed important ecologic roles, whereas other groups, such as the lacy bryozoans and crinoids, reached their greatest diversity during the Carboniferous. With the decline of the stromatoporoids and the tabulate and rugose corals, large organic reefs such as those existing earlier in the Paleozoic virtually disappeared and were replaced by small patch reefs. These reefs were dominated by crinoids, blastoids, lacy bryozoans, brachiopods, and calcareous algae

■ **Figure 21.8**

Reconstruction of a Middle Devonian reef from the Great Lakes area. Shown are corals, cephalopods, trilobites, crinoids, and brachiopods.

and flourished during the Late Paleozoic (■ Figure 21.10).

The Permian invertebrate marine faunas resembled those of the Carboniferous but were not as widely distributed because of the restricted size of the shallow seas on the cratons and the reduced shelf space along the continental margins (see Figure 20.21). The spiny and oddshaped productids dominated the brachiopod assemblage and constituted an important part of the reef complexes that formed in the Texas region during the Permian (■ Figure 21.11). The fusulinids (spindle-shaped foraminifera), which first evolved during the Late Mississippian and greatly diversified during the Pennsylvanian, experienced a further diversification during the Permian. Because of their abundance, diversity, and worldwide occurrence, fusulinids are important guide fossils for the Pennsylvanian and Permian. Bryozoans, sponges, and some types of calcareous algae also were common members of the Permian invertebrate fauna.

■ **Figure 21.9**

Restoration of a Silurian brackish-marine bottom scene near Buffalo, New York. Shown are algae, eurypterids, gastropods, worms, and shrimp.

The Permian Marine Invertebrate Mass Extinction

The greatest recorded mass extinction to affect Earth occurred at the end of the Permian Period (■ Figure 21.12). Before the Permian ended, roughly 50% of all marine invertebrate families and about 90% of all marine invertebrate species became extinct. Fusulinids, rugose and tabulate corals, several bryozoan and brachiopod orders as well as trilobites and blastoids did not survive the end of the Permian. All these groups had been very successful during the Paleozoic Era. In addition, more than 65% of all amphibians and reptiles as well as nearly 33% of insects on land became extinct.

What caused such a crisis for both marine and land-dwelling organisms? Various hypotheses have been proposed, but no completely satisfactory answer has yet been found. Some scenarios put forth to explain the extinctions are: (1) a meteorite impact such as occurred at the end of the Cretaceous Period (see Chapter 22); (2) a widespread marine regression resulting from glacial conditions; (3) a reduction in shelf space caused by the formation of Pangaea; (4) climatic changes; and (5) oceanographic changes such as anoxia, salinity changes, and turnover of deep-ocean waters.

■ **Figure 21.10**

Reconstruction of marine life during the Mississippian, based on an Upper Mississippian fossil site at Crawfordville, Indiana. Invertebrate animals shown include crinoids, blastoids, lacy bryozoans, brachiopods, and small corals.

■ **Figure 21.11**

Reconstruction of a Permian patch-reef community from the Glass Mountains of west Texas. Shown are algae, productid brachiopods, cephalopods, sponges, and corals.

The Permian mass extinction took place during an 8-million-year interval at the end of the Permian Period, which seems to rule out a meteorite impact. However, recent findings indicate that a meteorite impact may have contributed to the Permian mass extinction. The second and third hypotheses can probably be eliminated because most collisions of the continents had already taken place by the end of the Permian and because the large-scale formation of glaciers occurred during the Pennsylvanian Period. In addition, current evidence indicates a time of sudden warming at the end of the Permian.

Currently, many scientists think a large-scale marine regression coupled with climatic changes in the form of global warming—caused by an increase in carbon dioxide levels—may have been responsible for the Permian mass extinctions. In this scenario, a widespread lowering of sea level occurred near the end of the Permian, greatly reducing the amount of shallow shelf space for marine organisms and exposing the shelf to erosion. Oxidation of the organic matter trapped in the sediments ensued, reducing atmospheric oxygen levels as well as releasing large quantities of carbon dioxide into the atmosphere, resulting in increased global warming. During this time, widespread volcanic eruptions also took place, releasing additional carbon dioxide into the atmosphere and contributing to increased climatic instability and ecologic collapse. At the end of the Permian, a rise in sea level flooded and destroyed the nearshore terrestrial habitats, causing the extinction of many terrestrial plants and animals.

Some scientists also suggest that a global turnover of deep-ocean waters would have increased the amount of carbon dioxide in the surface waters and released large amounts of carbon dioxide into the atmosphere. The end result would be the same: global warming and disruption of the marine and terrestrial ecosystems.

Regardless of the ultimate cause of the Permian mass extinctions, the fact is that Earth's biota was dramatically changed. The resulting Triassic marine faunas had low diversity, but the surviving species tended to be abundant and widely distributed around the world. This fauna provided the root stock from which the Mesozoic marine fauna evolved and repopulated the world's oceans (see Chapter 22).

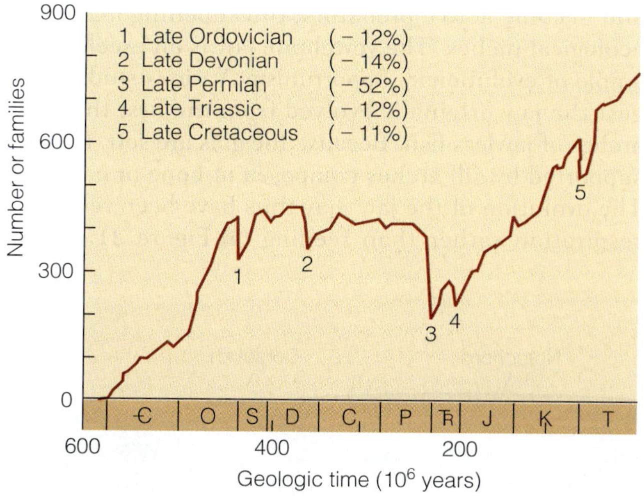

■ **Figure 21.12**

Phanerozoic diversity for marine invertebrate and vertebrate families. Note the three episodes of Paleozoic mass extinctions, with the greatest occurring at the end of the Permian Period.

Source: Reproduced with permission from "Mass Extinction in the Marine Fossil Record," by D. M. Raup, and J. J. Sepkoski, Science, Vol. 215, P. 1502 (figu 2). Copyrighted © 1982 American Association for the Advancement of Science.

VERTEBRATE EVOLUTION

A **chordate** (phylum Chordata) is an animal that has, at least during part of its life cycle, a notochord, a dorsal hollow nerve cord, and

What Would You Do?

One concern of environmentalists is that environmental degradation is leading to vast reductions in global biodiversity. As a paleontologist, you know that mass extinctions have taken place throughout Earth history. What facts and information can you provide from your geologic perspective that will help focus the debate on whether Earth's biota is being adversely affected by such human activities as industrialization, and what the possible outcome(s) might be if global biodiversity is severely reduced?

gill slits (■ Figure 21.13). **Vertebrates,** which are animals with backbones, are simply a subphylum of chordates.

The ancestors and early members of the phylum Chordata were soft-bodied organisms that left few fossils (■ Figure 21.14). Consequently, we know little about the early evolutionary history of the chordates or vertebrates. Surprisingly, a close relationship exists between echinoderms (Table 21.1) and chordates. They may even have shared a common ancestor because the development of the embryo is the same in both groups and differs completely from that of other invertebrates (■ Figure 21.15). Furthermore, the biochemistry of muscle activity and blood proteins and the larval stages are similar in both echinoderms and chordates.

The evolutionary pathway to vertebrates thus appears to have taken place much earlier and more rapidly than many scientists have long thought. Based on fossil evidence and recent advances in molecular biology, one scenario suggests that vertebrates evolved shortly after an ancestral chordate, probably resembling *Yunnanozoon,* acquired a second set of genes. According to this hypothesis, a random mutation produced a duplicate set of genes, letting the ancestral vertebrate animal evolve entirely new body structures that proved to be evolutionarily advantageous. Not all scientists accept this hypothesis, and the origin of vertebrates is still hotly debated.

FISH

The most primitive vertebrates are fish, and some of the oldest fish remains are found in the Upper Cambrian Deadwood Formation in northeastern Wyoming. Here phosphatic scales and plates of *Anatolepis,* a primitive member of the class Agnatha (jawless fish), have been recovered from marine sediments. All known Cambrian and Ordovician fossil fish have been found in shallow, nearshore marine deposits, whereas the earliest nonmarine fish remains have come from Silurian strata. This does not prove that fish originated in the oceans, but it does lend strong support to the idea.

As a group, fish range from the Late Cambrian to the present (■ Figure 21.16). The oldest and most primitive of the class Agnatha are the **ostracoderms,** whose name means "bony skin" (Table 21.2). These are armored jawless fish that first evolved during the Late Cambrian, reached their zenith during the Silurian and Devonian, and then became extinct.

The majority of ostracoderms lived on the seafloor. *Hemicyclaspis* is a good example of a bottom-dwelling ostracoderm (■ Figure 21.17a). Vertical scales allowed *Hemicyclaspis* to wiggle sideways, propelling itself along the seafloor, and the eyes on the top of its head enabled it to see such predators as cephalopods and jawed fish approaching from above. While moving along the seafloor, it probably sucked up small bits of food and sediments through its jawless mouth.

The evolution of jaws was a major advance among primitive vertebrates. Although their jawless ancestors could feed only on detritus, jawed fish could chew food and become active predators, thus opening many new ecological niches. The vertebrate jaw is an excellent example of evolutionary opportunism. Various studies suggest the jaw originally evolved from the first three gill arches of jawless fish. Because the gills are soft, they are supported by gill arches composed of bone or cartilage. The evolution of the jaw may thus have been related to respiration rather than feeding (■ Figure 21.18). By

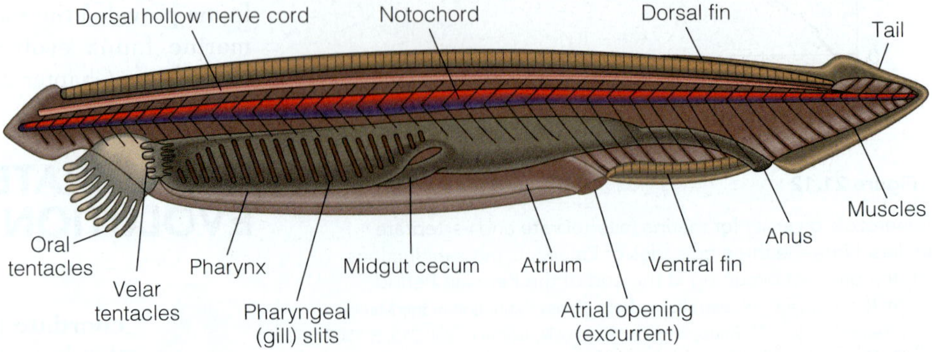

■ **Figure 21.13**

The structure of the lancelet *Amphioxus* illustrates the three characteristics of a chordate: a notochord, a dorsal hollow nerve cord, and gill slits. Source: Reproduced with permission from "The Lancelet" by M. Dale Stohes and N. D. Hallond. American Scientist 86: 552–560, fig. 2. Copyright © 1998 American Scientist.

■ **Figure 21.14**

Found in 525-million-year-old rocks in Yunnan province, China, *Yunnanozoon lividum*, a 5-cm-long animal is one of the oldest known chordates.

GeologyNow ■ **Active Figure 21.15**

(a) Arrangement of cells resulting from spiral cleavage. Cells in successive rows are nested between each other. Spiral cleavage is characteristic of all invertebrates except echinoderms.
(b) Arrangement of cells resulting from radial cleavage is characteristic of chordates and echinoderms. Cells are directly above each other.

evolving joints in the forward gill arches, jawless fish could open their mouths wider. Every time a fish opened and closed its mouth, it pumped more water past the gills, thereby increasing the oxygen intake. The modification from rigid to hinged forward gill arches let fish increase both their food consumption and oxygen intake, and the evolution of the jaw as a feeding structure rapidly followed.

The fossil remains of the first jawed fish are found in Lower Silurian rocks and belong to the **acanthodians**, a group of enigmatic fish characterized by large spines, scales covering much of the body, jaws, teeth, and reduced body armor (Figure 21.17c and Table 21.2). Although their relationship to other fish has not been well established, many scientists think the acanthodians

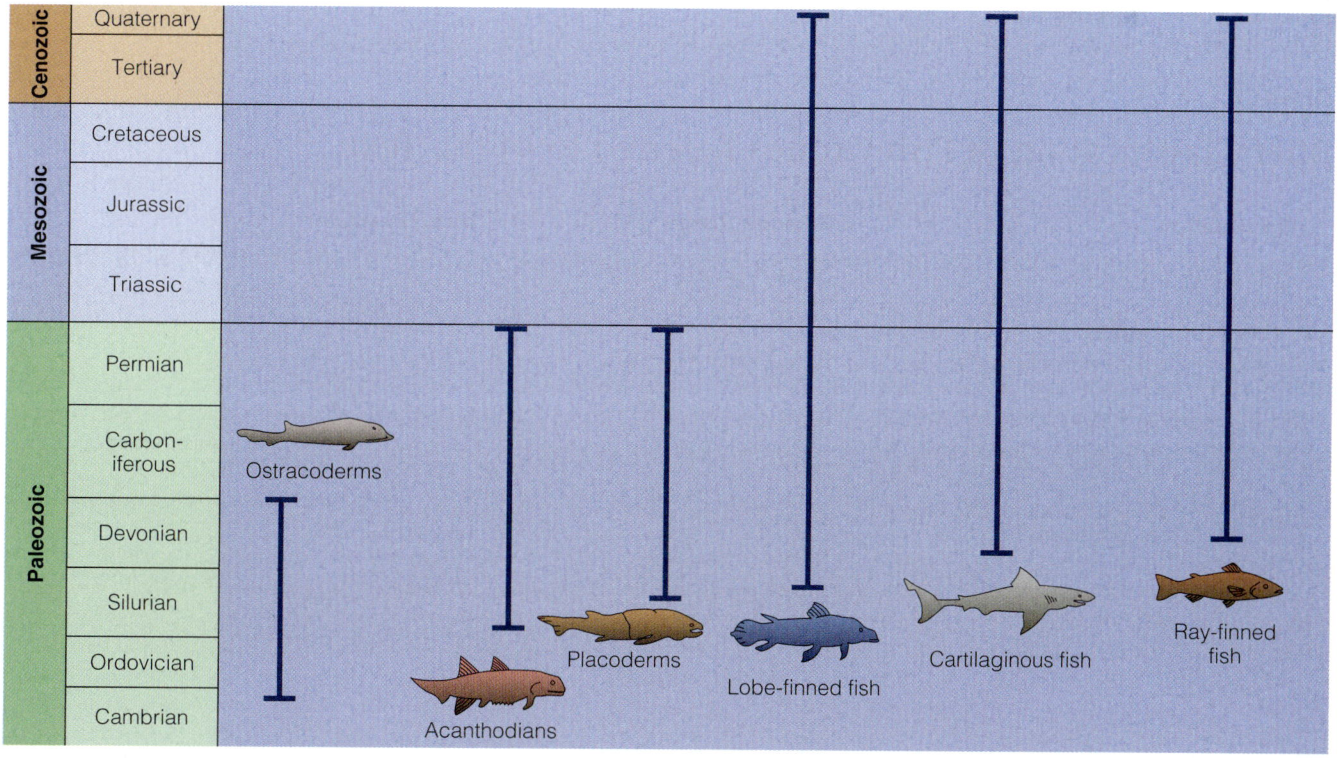

■ **Figure 21.16**

Geologic ranges of the major fish groups.

Figure 21.17

Recreation of a Devonian seafloor showing (a) an ostracoderm (*Hemicyclaspis*), (b) a placoderm (*Bothriolepis*), (c) an acanthodian (*Parexus*), and (d) a ray-finned fish (*Cheirolepis*).

Table 21.2

Brief Classification of Fish Groups Referred to in the Text

Classification	Geologic Range	Living Example
Class Agnatha (jawless fish) Early members of the class are called ostracoderms	Late Cambrian–Recent	Lamprey, hagfish No living ostracoderms
Class Acanthodii (the first fish with jaws)	Early Silurian–Permian	None
Class Placodermii (armored jawed fish)	Late Silurian–Permian	None
Class Chondrichthyes (cartilagenous fish)	Devonian–Recent	Sharks, rays, skates
Class Osteichyes (bony fish)	Devonian–Recent	Tuna, perch, bass, pike, carfish, trout, salmon, lungfish, *Latimeria*
Subclass Actinopterygii (ray-finned fish)	Devonian–Recent	Tuna, perch, bass, pike, catfish, trout, salmon
Subclass Sarcopterygii (lobe-finned fish)	Devonian–Recent	Lungfish, *Latimeria*
Order Dipnoi	Devonian–Recent	Lungfish
Order Crossopterygii	Devonian–Recent	*Latimeria*
Suborder Rhipidistia	Devonian–Permian	None

included the ancestors of the present-day bony and cartilaginous fish groups.

The other jawed fish that evolved during the Late Silurian were the **placoderms,** whose name means "plate-skinned" (Table 21.2). Placoderms were heavily armored, jawed fish that lived in both freshwater and the ocean. The placoderms showed considerable variety, including small bottom dwellers (Figure 21.17b) as well as large

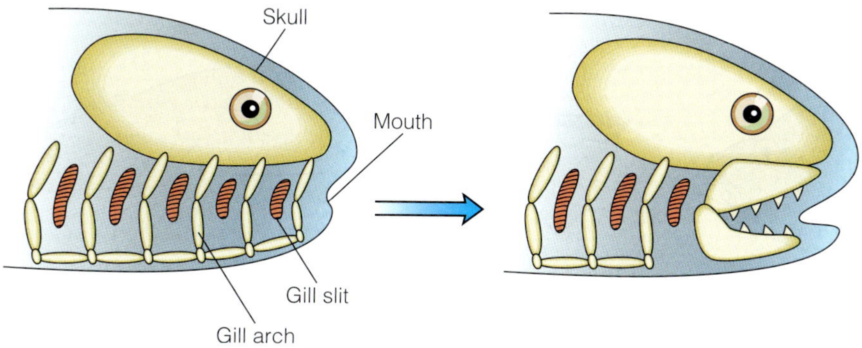

■ **Figure 21.18**

The evolution of the vertebrate jaw is thought to have begun from the modification of the first two or three anterior gill arches. This theory is based on the comparative anatomy of living vertebrates.

major predators such as *Dunkleosteus,* a Late Devonian fish that lived in the mid-continental North American epeiric seas (■ Figure 21.19a). It was by far the largest fish of the time, reaching a length of more than 12 m. It had a heavily armored head and shoulder region, a huge jaw lined with razor-sharp bony teeth, and a flexible tail, all features consistent with its status as a ferocious predator.

Besides the abundant acanthodians, placoderms, and ostracoderms, other fish groups, such as the cartilaginous and bony fish, also evolved during the Devonian Period. Small wonder, then, that the Devonian is informally called the "Age of Fish" because all major fish groups were present during this time period.

■ **Figure 21.19**

A Late Devonian marine scene from the midcontinent of North America. (a) The giant placoderm *Dunkleosteus* (length more than 12 m) is pursuing (b) the shark *Cladoselache* (length up to 1.2 m). Also shown are (c) the bottom-dwelling placoderm *Bothriolepsis* and (d) the swimming ray-finned fish *Cheirolepsis,* both of which reached a length of 40–50 cm.

Cartilaginous fish, class Chrondrichthyes (Table 21.2), represented today by sharks, rays, and skates, first evolved during the Middle Devonian, and by the Late Devonian, primitive marine sharks such as *Cladoselache* were abundant (Figure 21.19b). Cartilaginous fishes have never been as numerous or as diverse as their cousins, the bony fishes, but they were, and still are, important members of the marine vertebrate fauna.

Along with the cartilaginous fish, the **bony fish,** class Osteichthyes (Table 21.2), also first evolved during the Devonian. Because bony fish are the most varied and numerous of all the fishes, and because the amphibians evolved from them, their evolutionary history is particularly important. There are two groups of bony fish: the common *ray-finned fish* (Figure 21.19d) and the less familiar *lobe-finned fish* (Table 21.2).

The term *ray-finned* refers to the way the fins are supported by thin bones that spread away from the body (■ Figure 21.20a). From a modest freshwater beginning during the Devonian, ray-finned fish, which include most of the familiar fish such as trout, bass, perch, salmon, and tuna, rapidly diversified to dominate the Mesozoic and Cenozoic seas.

Present-day lobe-finned fish are characterized by muscular fins. The fins do not have radiating bones but rather articulating bones with the fin attached to the body by a fleshy shaft (Figure 21.20b). Two major groups of lobe-finned fish are recognized: *lungfish* and *crossopterygians* (Table 21.2).

Lungfish were fairly abundant during the Devonian, but today only three freshwater genera exist, one each in South America, Africa, and Australia.

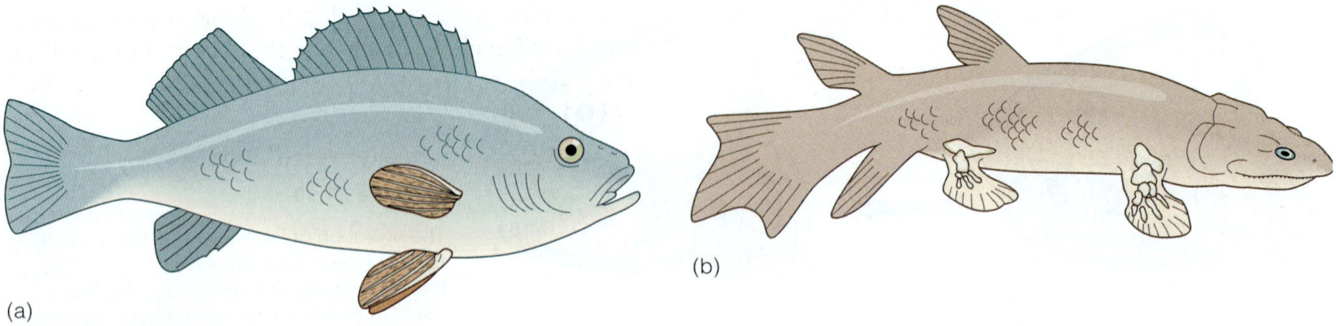

■ **Figure 21.20**

Arrangement of fin bones for (a) a typical ray-finned fish and (b) a lobe-finned fish. The muscles extend into the fin of the lobe-finned fish, allowing greater flexibility of movement than in the ray-finned fish.

Their present-day distribution presumably reflects the Mesozoic breakup of Gondwana (see Chapter 22).

The **crossopterygians** are an important group of lobe-finned fish because amphibians evolved from them. During the Devonian, two separate branches of crossopterygians evolved. One led to the amphibians, and the other invaded the sea. This latter group, the *coelacanths,* were thought to have become extinct at the end of the Cretaceous. In 1938, however, a fisherman caught a coelacanth in the deep waters off Madagascar, and since then several dozen more have been caught, both there and in Indonesia.

The group of crossopterygians that is ancestral to amphibians are *rhipidistians* (Table 21.2). These fish, reaching lengths greater than 2 m, were the dominant freshwater predators during the Late Paleozoic. *Eusthenopteron,* a good example of a rhipidistian crossopterygian, had an elongate body that let it move swiftly in the water, as well as paired muscular fins that could be used for moving on land (■ Figure 21.21). The structural similarity between crossopterygian fish and the earliest amphibians is striking and one of the better documented transitions from one major group to another (■ Figure 21.22).

■ **Figure 21.21**

Eusthenopteron, a member of the rhipidistian crossopterygians. The crossopterygians are the group from which the amphibians are thought to have evolved. *Eusthenopteron* had an elongate body and paired fins that could be used for moving about on land.

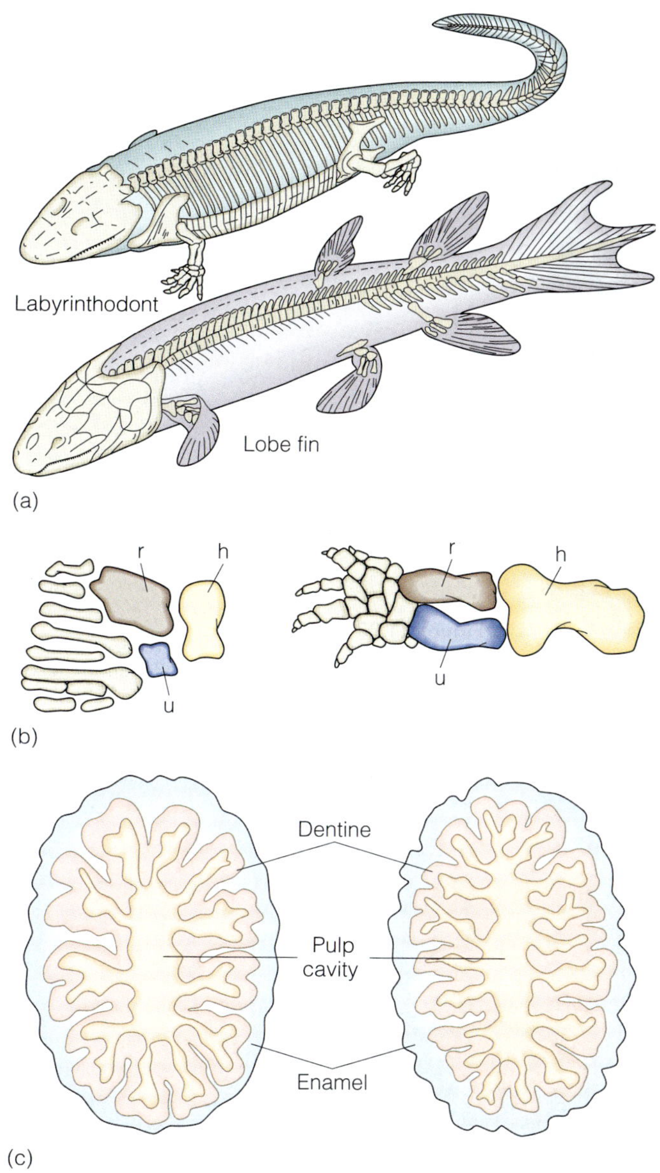

Figure 21.22

Similarities between the crossopterygian lobe-finned fish and the labyrinthodont amphibians. (a) Skeletal similarity. (b) Comparison of the limb bones of a crossopterygian (*left*) and amphibian (*right*); color identifies the bones (u = ulna, shown in blue, r = radius, mauve, h = humerus, gold) that the two groups have in common. (c) Comparison of tooth cross sections shows the complex and distinctive structure found in both the crossopterygians (*left*) and labyrinthodont amphibians (*right*).

AMPHIBIANS—VERTEBRATES INVADE THE LAND

Although amphibians were the first vertebrates to live on land, they were not the first land-living organisms. Land plants, which probably evolved from green algae, first evolved during the Ordovician. Furthermore, insects, millipedes, spiders, and even snails invaded the land before amphibians. Fossil evidence indicates that such land-dwelling arthropods as scorpions and flightless insects had evolved by at least the Devonian.

The transition from water to land required animals to surmount several barriers. The most critical were desiccation, reproduction, the effects of gravity, and the extraction of oxygen from the atmosphere by lungs rather than from water by gills. These problems were partly solved by the crossopterygians; they already had a backbone and limbs that could be used for walking and lungs that could extract oxygen (Figure 21.22).

Nevertheless, the question remains as to when animals made the transition from water to land and how and why it came about. The discovery in 1992 of fossilized tetrapod (from the Greek *tetra*, "four," and *podos*, "foot") footprints more than 365 million years old forced paleontologists to rethink how and when animals emerged onto land. The trackway Swiss geologist Iwan Stössel discovered that year on Valentia Island, off the southwest coast of Ireland, has shed light on the early evolution of tetrapods (■ Figure 21.23). From these footprints, geologists estimate that the creature was longer than 1 m and had fairly large back legs. Furthermore, instead of walking on dry land, this animal was probably walking or wading around in a shallow tropical stream filled with aquatic vegetation and predatory fish. This hypothesis is based on the fact the trackway showed no evidence of a tail being dragged behind it. Unfortunately, no bones are associated with the tracks to help reconstruct what this primitive tetrapod looked like.

One of the intriguing questions paleontologists ask is, Why did limbs evolve in the first place? Probably not for walking on land. In fact, many scientists think aquatic limbs made it easier for animals to move around in streams, lakes, or swamps that were choked with water plants or other debris. The scant fossil evidence also seems to support this hypothesis. Fossils of *Acanthostega*, a tetrapod found in 360-million-year-old rocks from Greenland, reveal an animal with limbs but one clearly unable to walk on land. Paleontologist Jenny Clack, who recovered hundreds of specimens of *Acanthostega*, points out that *Acanthostega*'s limbs were not strong enough to support its weight on land, and its rib cage was too small for the necessary muscles needed to hold its body off the ground. In addition, *Acanthostega* had gills and lungs, meaning it could survive on land, but it was more suited for the water. Clack thinks *Acanthostega* used its limbs to maneuver around in swampy, plant-filled waters, where swimming would be difficult and limbs were an advantage.

Fragmentary fossils from other tetrapods living at about the same time as *Acanthostega* suggest, however, that some of these early tetrapods may have spent more time on dry land than in the water. These oldest amphibians, many of which are also found in the Upper Devonian Old Red Sandstone of eastern Greenland and belong to genera such

■ **Figure 21.23**

Tetrapod trackway at Valentia Island, Ireland. These fossilized footprints, which are more than 365 million years old, are evidence of one of the earliest four-legged animals on land. Source: Photo courtesy of Ken Higgs, Department of Geology, University College–Cork, Ireland.

as *Ichthyostega,* had streamlined bodies, long tails, and fins along their backs. In addition to four legs, they had a strong backbone, rib cage, and pelvic and pectoral girdles, all of which were structural adaptations for walking on land (■ Figure 21.24). Many of these earliest amphibians seem to have inherited numerous characteristics from the crossopterygians with little modification (Figure 21.22).

The Late Paleozoic amphibians did not at all resemble the familiar frogs, toads, newts, and salamanders that make up the modern amphibian fauna. Rather, they displayed a broad spectrum of sizes, shapes, and modes of life (■ Figure 21.25). One group of amphibians were the **labyrinthodonts,** so named for the labyrinthine wrinkling and folding of the chewing surface of their teeth (Figure 21.22). Most labyrinthodonts were large animals, as long as 2 m. These typically sluggish creatures lived in swamps and streams, eating fish, vegetation, insects, and other small amphibians (Figure 21.25).

Labyrinthodonts were abundant during the Carboniferous, when swampy conditions were widespread (see Chapter 20), but they soon declined in abundance during the Permian, perhaps in response to changing climatic conditions. Only a few species survived into the Triassic.

■ **Figure 21.24**

A Late Devonian landscape in the eastern part of Greenland. Shown is *Ichthyostega*, an amphibian that grew to a length of about 1 m. The flora of the time was diverse, consisting of a variety of small and large seedless vascular plants.

■ **Figure 21.25**

Reconstruction of a Carboniferous coal swamp. The varied amphibian fauna of the time is shown, including the large labyrinthodont amphibian *Eryops (foreground)*, the larval *Branchiosaurus (center)*, and the serpentlike *Dolichosoma (background)*.

EVOLUTION OF THE REPTILES—THE LAND IS CONQUERED

Amphibians were limited in colonizing the land because they had to return to water to lay their gelatinous eggs. The evolution of the **amniote egg** (■ Figure 21.26) freed reptiles from this constraint. In such an egg, the developing embryo is surrounded by a liquid-filled sac called the *amnion* and provided with both a yolk, or food sac, and an allantois, or waste sac. The emerging reptile is in essence a miniature adult, bypassing the need for a larval stage in the water. The evolution of the amniote egg allowed vertebrates to colonize all parts of the land because they no longer had to return to the water as part of their reproductive cycle.

Many of the differences between amphibians and reptiles are physiologic and are not preserved in the fossil record. Nevertheless, amphibians and reptiles differ sufficiently in skull structure, jawbones, ear location, and limb and vertebral construction to suggest that reptiles evolved from labyrinthodont ancestors by the Late Mississippian.

Some of the oldest known reptiles are from the Lower Pennsylvania Joggins Formation in Nova Scotia, Canada. Here, remains of *Hylonomus* are found in the sediments filling in tree trunks. These earliest reptiles were small and agile and fed largely on grubs and insects. They are loosely grouped together as **protorothyrids**, whose members include the earliest reptiles (■ Figure 21.27). During the Permian Period, reptiles diversified and began to displace many amphibians. The reptiles succeeded partly because of their advanced method of reproduction and their more advanced jaws and teeth, as well as their ability to move rapidly on land.

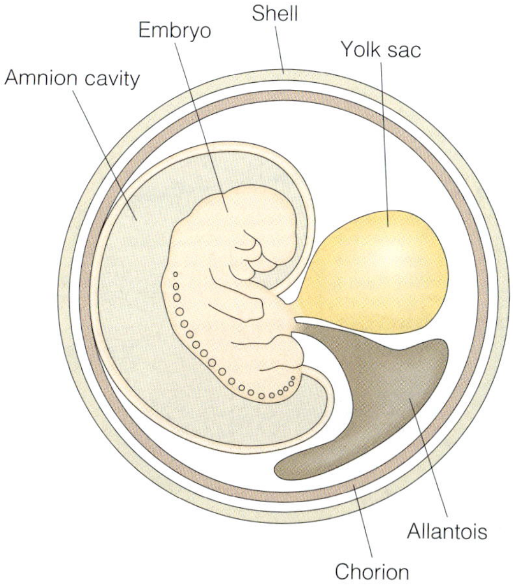

■ **Figure 21.26**

In an amniote egg, the embryo is surrounded by a liquid sac (amnion cavity) and provided with a food source (yolk sac) and waste sac (allantois). The evolution of the amniote egg freed reptiles from having to return to the water for reproduction and let them inhabit all parts of the land.

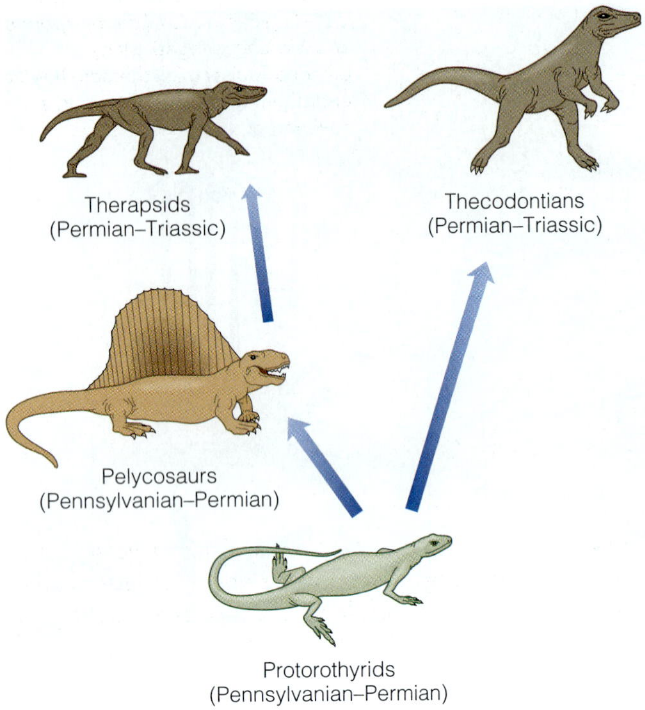

Figure 21.27

Evolutionary relationship among the Paleozoic reptiles.

The **pelycosaurs,** or finback reptiles, evolved from the protorothyrids during the Pennsylvanian and were the dominant reptile group by the Early Permian. They evolved into a diverse assemblage of herbivores, exemplified by *Edaphosaurus,* and carnivores such as *Dimetrodon* (Figure 21.28). An interesting feature of the pelycosaurs is their sail. It was formed by vertebral spines that, in life, were covered with skin. The sail has been variously explained as a sexual display, a means of protection, and a display to look more ferocious. The current consensus seems to be that the sail served as a thermoregulatory device, raising the reptile's temperature by catching the sun's rays or cooling it by facing the wind. Because pelycosaurs are considered to be the group from which therapsids evolved, it is interesting that they may have had some sort of body-temperature control.

The pelycosaurs became extinct during the Permian and were succeeded by the **therapsids,** mammal-like reptiles that evolved from the carnivorous pelycosaur lineage and rapidly diversified into herbivorous and carnivorous lineages (Figure 21.29). Therapsids were small- to medium-sized animals displaying the beginnings of many mammalian features: fewer bones in the skull because many of the small skull bones were fused; enlarged lower jawbone; differentiation of teeth for various functions such as nipping, tearing, and chewing food; and more vertically placed legs for greater flexibility, as opposed to the way the legs sprawled out to the side in primitive reptiles.

Furthermore, many paleontologists think therapsids were *endothermic,* or warm-blooded, enabling them to maintain a constant internal body temperature. This characteristic would have let them expand into a variety of habitats, and indeed the Permian rocks in which their fossil remains are found have a wide latitudinal distribution.

As the Paleozoic Era came to an end, the therapsids constituted about 90% of the known reptile genera and occupied a wide range of ecological niches. The mass extinctions that decimated the marine fauna at the close of the Paleozoic had an equally great effect on the terrestrial population. By the end of the Permian, about 90% of all marine invertebrate species were extinct compared with more than two-thirds of all amphibians and reptiles. Plants, in contrast, apparently did not experience as great a turnover as animals.

Figure 21.28

Most pelycosaurs or finback reptiles have a characteristic sail on their back. One hypothesis explains the sail as a thermoregulatory device. Other hypotheses are that it was a sexual display or a device to make the reptile look more intimidating. Shown here are (a) the carnivore *Dimetrodon* and (b) the herbivore *Edaphosaurus*.

Figure 21.29

A Late Permian scene in southern Africa showing various therapsids including *Dicynodon* (left foreground) and *Moschops* (right). Many paleontologists think therapsids were endothermic and may have had a covering of fur, as shown here.

PLANT EVOLUTION

When plants made the transition from water to land, they had to solve most of the same problems that animals did: desiccation, support, reproduction, and the effects of gravity. Plants did so by evolving a variety of structural adaptations that were fundamental to the subsequent radiations and diversification that occurred during the Silurian, Devonian, and later periods (see Geo-Focus 21.1) (Table 21.3). Most experts agree that the ancestors

Table 21.3

Major Events in the Evolution of Land Plants

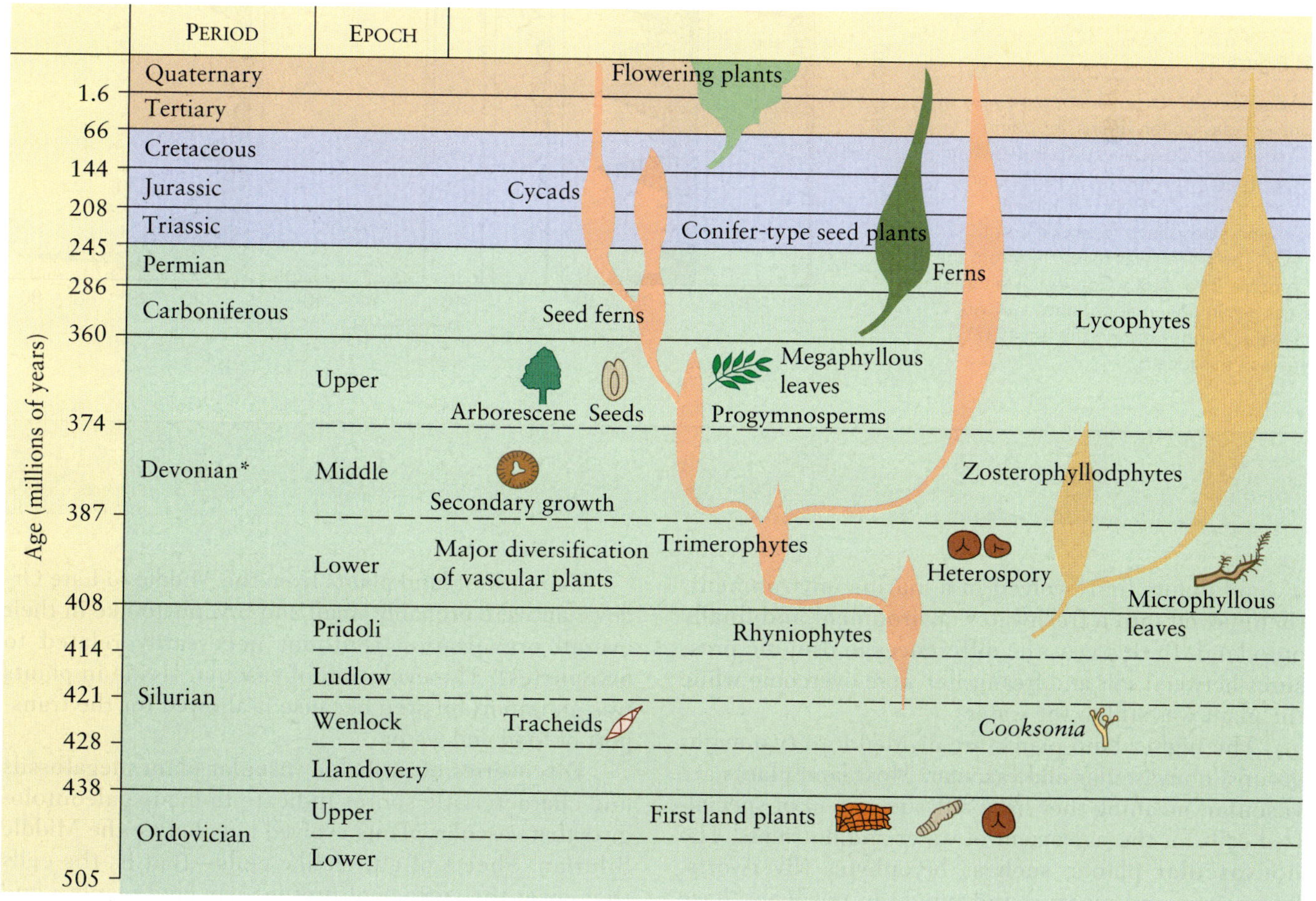

*The Devonian Period was a time of rapid evolution for the land plants. Major events were the appearance of leaves, heterospory, secondary growth, and the emergence of seeds.

GEOFOCUS 21.1

Palynology: A Link Between Geology and Biology

Palynology is the study of organic microfossils called *palynomorphs*. These include such familiar things as spores and pollen (both of which cause allergies for many people) (■ Figure 1a) and also such unfamiliar organisms as acritarchs (■ Figure 2), dinoflagellates (marine and freshwater single-celled phytoplankton, some species of which in high concentrations make shellfish toxic to humans) (see Figure

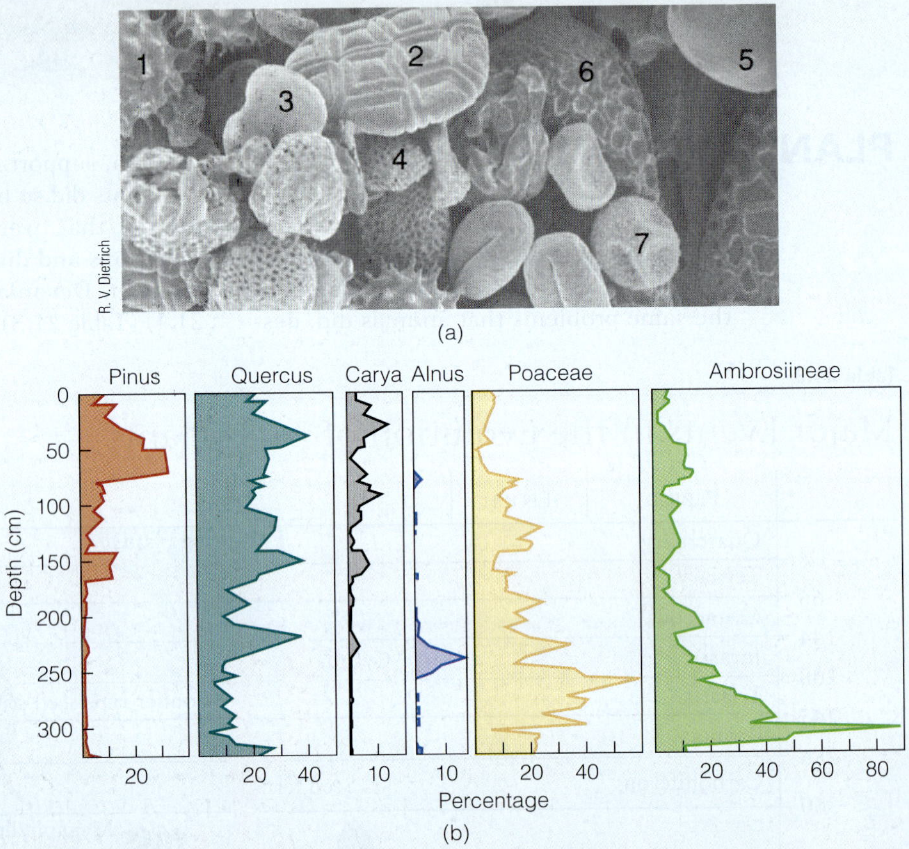

■ **Figure 1**

(a) Scanning electron microscope view of present-day pollen grains, including (1) sunflower, (2) acacia, (3) oak, (4) white mustard, (5) little walnut, (6) agave, and (7) ash juniper. (b) Pollen diagrams showing abundance for six different trees. The pollen was recovered from the Ferndale Bog, Atoka County, Oklahoma. Changes in pollen abundance indicate climate changes during the last 12,000 years at this locality.

of land plants first evolved in a marine environment, then moved into a freshwater environment and finally onto land. In this way, the differences in osmotic pressures between salt and freshwater were overcome while the plant was still in the water.

The higher land plants are divided into two major groups: nonvascular and vascular. Most land plants are **vascular,** meaning they have a tissue system of specialized cells for the movement of water and nutrients. The **nonvascular** plants, such as bryophytes (liverworts, hornworts, and mosses) and fungi, do not have these specialized cells, are typically small, and usually live in low, moist areas.

The earliest land plants from the Middle to Late Ordovician were probably small and bryophyte-like in their overall organization (but not necessarily related to bryophytes). The evolution of vascular tissue in plants was an important step because it allowed for the transport of food and water.

Discoveries of probable vascular plant megafossils and characteristic spores indicate to many paleontologists that vascular plants evolved long before the Middle Silurian. Sheets of cuticlelike cells—that is, the cells that cover the surface of present-day land plants—and tetrahedral clusters that closely resemble the spore tetrahedrals of primitive land plants have been reported from

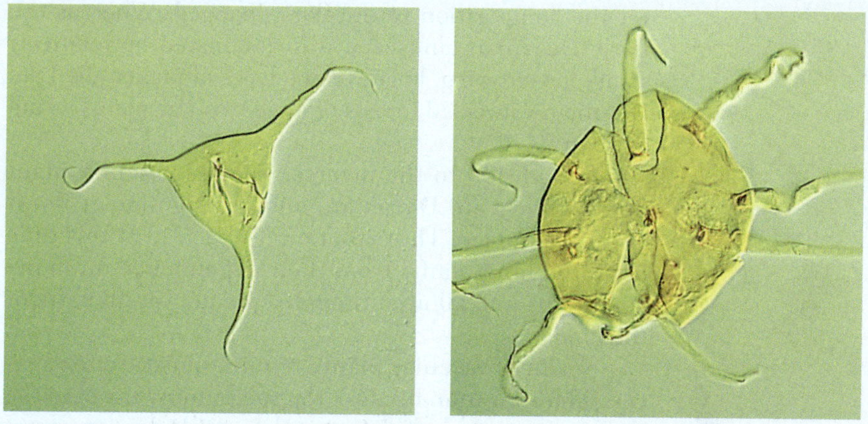

Figure 2

Acritarchs from the Upper Ordovician Sylvan Shale, Oklahoma. Acritarchs are organic-walled phytoplankton and were the primary food source for suspension feeders during the Paleozoic Era.

22.20c), chitinozoans (vase-shaped microfossils of unknown origin), scolecodonts (jaws of marine annelid worms), and microscopic colonial algae. Fossil palynomorphs are extremely resistant to decay and are extracted from sedimentary rocks by dissolving the rocks in various acids.

A specialty of palynology that attracts many biologists and geologists is the study of spores and pollen. By examining the fossil spores and pollen preserved in sedimentary rocks, palynologists can tell when plants colonized Earth's surface, which in turn influenced weathering and erosion rates, soil formation, and changes in the composition of atmospheric gases. Furthermore, because plants are not particularly common as fossils, the study of spores and pollen can frequently reveal the time and region for the origin and extinction of various plant groups.

Analyses of fossil spores and pollen are used to solve many geologic and biologic problems. One of the more important uses of fossil spores and pollen is determining the geologic age of sedimentary rocks. Because spores and pollen are microscopic, resistant to decay, deposited in both marine and terrestrial environments, extremely abundant, and part of the life cycle of plants (Figure 21.32), they are very useful for determining age. Many spore and pollen species have narrow time ranges that make them excellent guide fossils. Rocks considered lacking in fossils by paleontologists who were looking only for megafossils often actually contain thousands, even millions, of fossil spores or pollen grains that allow palynologists to date these so-called unfossiliferous rocks.

Fossil spores and pollen are also useful in determining the environment and climate in the past. Their presence in sedimentary rocks helps palynologists determine what plants and trees were living at the time, even if the fossils of those plants and trees are not preserved. Plants are very sensitive to climatic changes, and by plotting the abundance and types of vegetation present, based on their preserved spores and pollen, palynologists can determine past climates and changes in climates through time (Figure 1b).

The study of spores and pollen provides a tremendous amount of information about the vegetation in the past, its evolution, the type of climate, and changes in climate. In addition, spores and pollen are useful for dating rocks and correlating marine and terrestrial rocks, both regionally and globally.

Middle to Upper Ordovician rocks from western Libya and elsewhere.

The ancestor of terrestrial vascular plants was probably some type of green alga. Although no fossil record of the transition from green algae to terrestrial vascular plants exists, comparison of their physiology reveals a strong link. Primitive **seedless vascular plants** such as ferns resemble green algae in their pigmentation, important metabolic enzymes, and type of reproductive cycle. Furthermore, the green algae are one of the few plant groups to have made the transition from saltwater to freshwater. The evolution of terrestrial vascular plants from an aquatic, probable green algal ancestry was accompanied by various modifications that let them occupy this new harsh environment.

Silurian and Devonian Floras

The earliest known vascular land plants are small Y-shaped stems assigned to the genus *Cooksonia* from the Middle Silurian of Wales and Ireland. Together with Upper Silurian and Lower Devonian species from Scotland, New York State, and the Czech Republic, these earliest plants were small, simple, leafless stalks with a spore-producing structure at the tip (■ Figure 21.30); they are known as seedless vascular plants because they

Courtesy of Dianne Edwards, University College, Wales

■ Figure 21.30

The earliest known fertile land plant was *Cooksonia,* seen in this fossil from the Upper Silurian of South Wales. *Cooksonia* consisted of upright, branched stems terminating in sporangia (spore-producing structures). It also had a resistant cuticle and produced spores typical of a vascular plant. These plants probably lived in moist environments such as mud flats. This specimen is 1.49 cm long.

did not produce seeds. They also did not have a true root system. A *rhizome,* the underground part of the stem, transferred water from the soil to the plant and anchored the plant to the ground. The sedimentary rocks in which these plant fossils are found indicate that they lived in low, wet, marshy, freshwater environments.

An interesting parallel can be seen between seedless vascular plants and amphibians. When they made the transition from water to land, they had to overcome the problems such a transition involved. Both groups, though successful, nevertheless required a source of water in order to reproduce. In the case of amphibians, their gelatinous egg had to remain moist, and the seedless vascular plants required water for the sperm to travel through to reach the egg.

From this simple beginning, the seedless vascular plants evolved many of the major structural features characteristic of modern plants, such as leaves, roots, and secondary growth. These features did not all evolve simultaneously but rather at different times, a pattern known as *mosaic evolution.* This diversification and adaptive radiation took place during the Late Silurian and Early Devonian and resulted in a tremendous increase in diversity (■ Figure 21.31). During the Devonian, the number of plant genera remained about the same, yet the composition of the flora changed. Whereas the Early Devonian landscape was dominated by relatively small, low-growing, bog-dwelling types of plants, the Late Devonian witnessed forests of large tree-size plants as tall as 10 m.

In addition to the diverse seedless vascular plant flora of the Late Devonian, another significant floral event took place. The evolution of the seed at this time liberated land plants from their dependence on moist conditions and allowed them to spread over all parts of the land.

Seedless vascular plants require moisture for successful fertilization because the sperm must travel to the egg on the surface of the gamete-bearing plant (gametophyte) to produce a successful spore-generating plant (sporophyte). Without moisture, the sperm would dry out before reaching the egg (■ Figure 21.32a). In the seed method of reproduction, the spores are not released to the environment as they are in the seedless vascular plants, but are retained on the spore-bearing plant, where they grow into the male and female forms of the gamete-bearing generation. In the case of the **gymnosperms,** or flowerless seed plants, these are male and female cones (Figure 21.32b). The male cone produces pollen, which contains the sperm and has a waxy coating to prevent desiccation, and the egg, or embryonic seed, is contained in the female cone. After fertilization, the seed develops into a mature, cone-bearing plant. In this way the need for a moist environment for the gametophyte generation is solved. The significance of this development is that seed plants, like reptiles, were no longer restricted to wet areas but were free to migrate into previously unoccupied dry environments. Although the seedless vascular plants dominated the flora of the Carboniferous coal-forming swamps, the gymnosperms made up an important element of the Late Paleozoic flora, particularly in the nonswampy areas.

Late Carboniferous and Permian Floras

As discussed earlier, the rocks of the Pennsylvania Period (Late Carboniferous) are the major source of the world's coal. Coal results from the alteration of plant remains accumulating in low, swampy areas. The geologic and geographic conditions of the Pennsylvanian were ideal for the growth of seedless vascular plants, and consequently these coal swamps had a very diverse flora (■ Figure 21.33).

It is evident from the fossil record that whereas the Early Carboniferous flora was similar to its Late Devonian counterpart, a great deal of evolutionary experimentation was taking place that would lead to the highly successful Late Paleozoic flora of the coal swamps and adjacent habitats. Among the seedless vascular plants, the *lycopsids* and *sphenopsids* were the most important coal-forming groups of the Pennsylvanian Period.

The lycopsids were the dominant element of the coal swamps, achieving heights up to 30 m in such genera as *Lepidodendron* and *Sigillaria.* The Pennsylvanian

Figure 21.31

Reconstruction of an Early Devonian landscape showing some of the earliest land plants: (a) *Bucheria,* (b) *Protolepidodendron,* (c) *Dawsonites.*

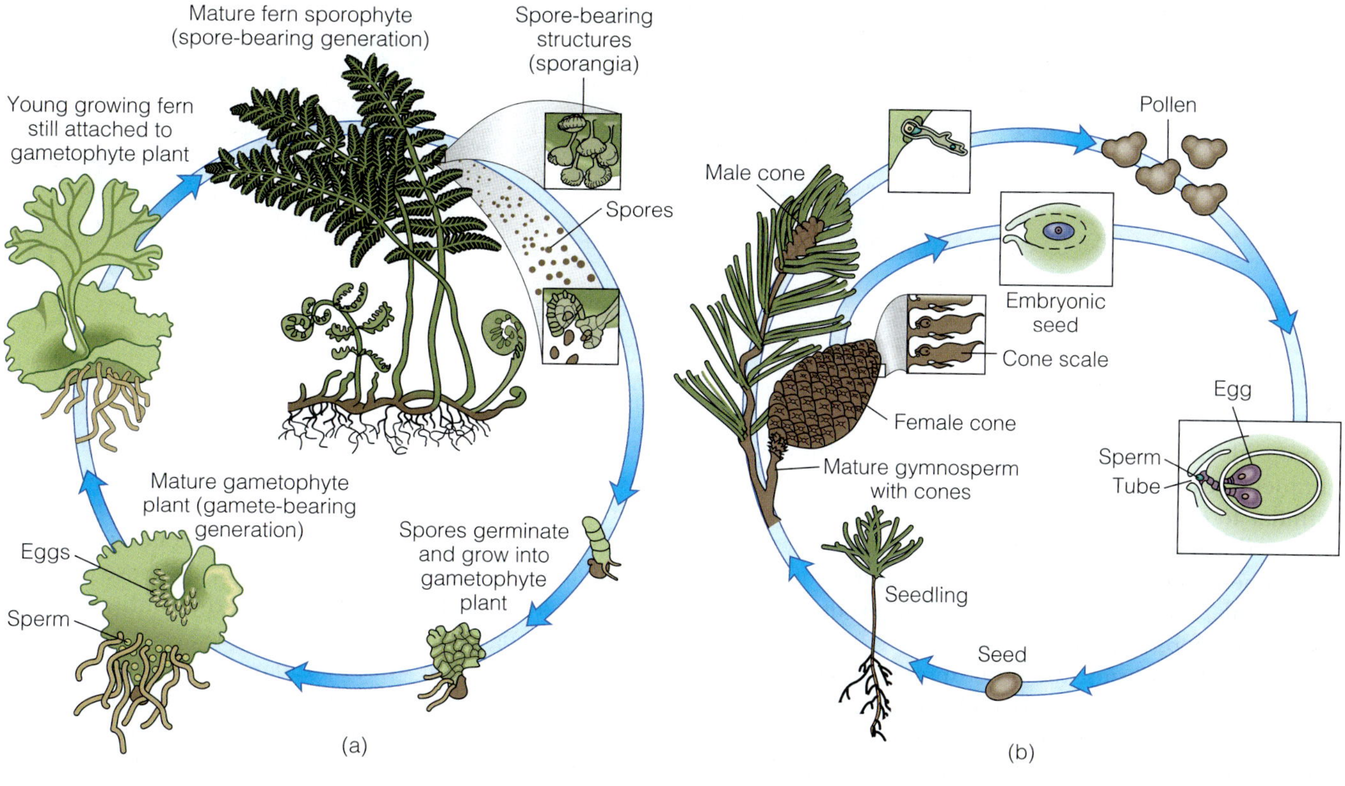

Geology⇌Now **Active Figure 21.32**

(a) Generalized life history of a seedless vascular plant. The mature sporophyte plant produces spores, which on germination grow into small gametophyte plants that produce sperm and eggs. The fertilized eggs grow into the spore-producing mature plant, and the sporophyte–gametophyte life cycle begins again. (b) Generalized life history of a gymnosperm plant. The mature plant bears both male cones that produce sperm-bearing pollen grains and female cones that contain embryonic seeds. Pollen grains are transported to the female cones by the wind. Fertilization occurs when the sperm moves through a moist tube growing from the pollen grain and unites with the embryonic seed, which then grows into a cone-bearing mature plant.

■ **Figure 21.33**

Reconstruction of a Pennsylvanian coal swamp with its characteristic vegetation. The amphibian is *Eogyrinus*.

lycopsid trees are interesting because they lacked branches except at their top, which had elongate leaves similar to the individual palm leaf of today. As the trees grew, the leaves were replaced from the top, leaving prominent and characteristic rows or spirals of scars on the trunk. Today the lycopsids are represented by small temperate-forest ground pines.

The sphenopsids, the other important coal-forming plant group, are characterized by joints and horizontal underground stem-bearing roots. Many of these plants, such as *Calamites*, average 5 to 6 m tall. Living sphenopsids include the horsetail (*Equisetum*) and scouring rushes (■ Figure 21.34). Small seedless vascular plants and seed ferns formed a thick undergrowth or ground cover beneath these treelike plants.

Not all plants were restricted to the coal-forming swamps. Among those plants occupying higher and drier ground were some of the *cordaites*, a group of tall gymnosperm trees that grew up to 50 m tall and probably formed vast forests (■ Figure 21.35). Another important nonswamp dweller was *Glossopteris*, the famous plant so abundant in Gondwana (see Figure 2.1), whose distribution is cited as critical evidence that the continents have moved through time.

The floras that were abundant during the Pennsylvanian persisted into the Permian, but because of climatic and geologic changes resulting from tectonic events they declined in abundance and importance. By the end of the Permian, the cordaites became extinct, and the lycopsids and sphenopsids were reduced to mostly small, creeping forms. Gymnosperms with lifestyles more suited to the warmer and drier Permian climates diversified and came to dominate the Permian, Triassic, and Jurassic landscapes.

■ **Figure 21.34**

Living sphenopsids include the horsetail *Equisetum*.

■ **Figure 21.35**

A cordaite forest from the Late Carboniferous. Cordaites were a group of gymnosperm trees that grew up to 50 m tall.

21 GEO RECAP

Chapter Summary

Table 21.4 (see page 632) summarizes the major evolutionary and geologic events of the Paleozoic Era and shows their relationships to one another.

- Multicelled organisms presumably had a long Precambrian history, during which they lacked hard parts. Invertebrates with hard parts suddenly appeared during the Early Cambrian in what is called the "Cambrian explosion." Skeletons provided such advantages as protection against predators and support for muscles, enabling organisms to grow large and increase their locomotor efficiency. Hard parts probably evolved as a result of various geologic and biologic factors rather than from a single cause.
- Marine organisms are classified as plankton if they are floaters, nekton if they swim, and benthos if they live on or in the seafloor.
- Marine organisms are divided into four basic feeding groups: suspension feeders, which consume microscopic plants and animals as well as dissolved nutrients from water; herbivores, which are plant

Table 21.4

Major Evolutionary and Geologic Events of the Paleozoic Era

Age (millions of years)	Geologic Period		Invertebrates	Vertebrates	Plants	Major Geologic Events
245–286	Permian		Largest mass extinction event to affect the invertebrates	Acanthodians, placoderms, and pelycosaurs become extinct Therapsids and pelycosaurs the most abundant reptiles	Gymnosperms diverse and abundant	Formation of Pangaea Alleghenian orogeny Hercynian orogeny
286–320	Carboniferous	Pennsylvanian	Fusulinids diversify	Amphibians abundant and diverse	Coal swamps with flora of seedless vascular plants and gymnosperms	Coal-forming swamps common Formation of Ancestral Rockies Continental glaciation in Gondwana
320–360	Carboniferous	Mississippian	Crinoids, lacy bryozans, blastoids become abundant Renewed adaptive radiation following extinctions of many reef builders	Reptiles evolve	Gymnosperms appear (may have evolved during Late Devonian)	Ouachita orogeny
360–408	Devonian		Extinctions of many reef-building invertebrates near end of Devonian Reef building continues Eurypterids abundant	Amphibians evolve All major groups of fish present—Age of Fish	First seeds evolve Seedless vascular plants diversify	Widespread deposition of black shale Acadian orogeny Antler orogeny
408–438	Silurian		Major reef building Diversity of invertebrates remains high	Ostracoderms common Acanthodians, the first jawed fish, evolve	Early land plants—seedless vascular plants	Caledonian orogeny Extensive barrier reefs and evaporites
438–505	Ordovician		Extinctions of a variety of marine invertebrates near end of Ordovician Major adaptive radiation of all invertebrate groups Suspension feeders dominant	Ostracoderms diversify	Plants move to land?	Continental glaciation in Gondwana Taconic orogeny
505–545	Cambrian		Many trilobites become extinct near end of Cambrian Trilobites, brachiopods, and archaeocyathids are most abundant	Earliest vertebrates—jawless fish called ostracoderms		First Phanerozoic transgression (Sauk) onto North American craton

eaters; carnivore-scavengers, which are meat eaters; and sediment-deposit feeders, which ingest sediment and extract nutrients from it.

- The marine ecosystem consists of various trophic levels of food production and consumption. At the base are primary producers, on which all other organisms are dependent. Feeding on the primary producers are the primary consumers, which in turn are fed on by higher levels of consumers. The decomposers are bacteria that break down the complex organic compounds of dead organisms and recycle them within the ecosystem.
- The Cambrian invertebrate community was dominated by three major groups: trilobites, brachiopods, and archaeocyathids. Little specialization existed among the invertebrates, and most phyla were represented by only a few species.
- The Middle Cambrian Burgess Shale contains one of the finest examples of a well-preserved soft-bodied biota in the world.
- The Ordovician marine invertebrate community marked the beginning of the dominance of the shelly fauna and the start of large-scale reef building. The end of the Ordovician Period was a time of major extinctions of many invertebrate phyla.
- The Silurian and Devonian periods were times of diverse faunas dominated by reef-building animals, whereas the Carboniferous and Permian periods saw a great decline in invertebrate diversity.
- Chordates are characterized by a notochord, dorsal hollow nerve cord, and gill slits. The earliest chordates were soft-bodied organisms that were rarely fossilized. Vertebrates are a subphylum of the chordates.
- Fish are the earliest known vertebrates, with their first fossil occurrence in Upper Cambrian rocks. They have had a long and varied history, including jawless and jawed armored forms (ostracoderms and placoderms), cartilaginous forms, and bony forms. Crossopterygians, a group of lobe-finned fish, gave rise to the amphibians.
- The link between crossopterygians and the earliest amphibians is convincing and includes a close similarity of bone and tooth structures. The transition from fish to amphibians occurred during the Devonian. During the Carboniferous, the labyrinthodont amphibians were the dominant terrestrial vertebrate animals.
- The earliest fossil record of reptiles is from the Late Mississippian. The evolution of an amniote egg was the critical factor in the reptiles' ability to colonize all parts of the land.
- Pelycosaurs were the dominant reptile group during the Early Permian, whereas therapsids dominated the landscape for the rest of the Permian Period.
- In making the transition from water to land, plants had to overcome the same basic problems as animals—namely, desiccation, reproduction, and gravity.
- The earliest fossil record of land plants is from Middle to Upper Ordovician rocks. These plants were probably small and bryophyte-like in their overall organization.
- The evolution of vascular tissue was an important event in plant evolution because it allowed food and water to be transported throughout the plant and provided the plant with additional support.
- The ancestor of terrestrial vascular plants was probably some type of green alga based on such similarities as pigmentation, metabolic enzymes, and the same type of reproductive cycle.
- The earliest seedless vascular plants were small, leafless stalks with spore-producing structures on their tips. From this simple beginning, plants evolved many of the major structural features characteristic of today's plants.
- By the end of the Devonian Period, forests with tree-sized plants up to 10 m tall had evolved. The Late Devonian also witnessed the evolution of flowerless seed plants (gymnosperms) whose reproductive style freed them from having to stay near water.
- The Carboniferous Period was a time of vast coal swamps, where conditions were ideal for the seedless vascular plants. With the onset of more arid conditions during the Permian, the gymnosperms became the dominant element of the world's forest.
- A major extinction occurred at the end of the Paleozoic Era, affecting the invertebrates as well as the vertebrates. Its cause is still being debated.

Important Terms

acanthodian (p. 617)
amniote egg (p. 623)
benthos (p. 606)
bony fish (p. 619)
carnivore-scavenger (p. 606)
cartilaginous fish (p. 619)
chordate (p. 615)
crossopterygian (p. 620)
gymnosperm (p. 628)
herbivore (p. 606)
labyrinthodont (p. 622)
nekton (p. 606)
nonvascular (p. 626)
ostracoderm (p. 616)
pelycosaur (p. 624)
placoderm (p. 618)
plankton (p. 606)
primary producer (p. 606)
protorothyrids (p. 623)
sediment-deposit feeder (p. 606)
seedless vascular plant (p. 627)
suspension feeder (p. 606)
therapsid (p. 624)
vascular (p. 626)
vertebrate (p. 616)

Review Questions

1. The greatest recorded mass extinction to affect Earth occurred at the end of which period?
a._____Cambrian; b._____Ordovician; c._____Devonian; d._____Permian; e._____Cretaceous.
2. The fossils of the Burgess Shale are significant because they provide a rare glimpse of
a._____the first shelled animals; b._____the soft-part anatomy of extinct groups; c._____soft-bodied animals; d._____answers a and b; e._____answers b and c.
3. What type of invertebrates dominated the Ordovician invertebrate community?
a._____epifaunal, benthonic, sessile suspension feeders; b._____infaunal, benthonic, sessile suspension feeders; c._____epifaunal, benthonic, mobile suspension feeders; d._____infaunal, nektonic carnivores; e._____epifloral, planktonic primary producers.
4. An exoskeleton is advantageous because it
a._____prevents drying out in an intertidal environment; b._____provides protection against ultraviolet radiation; c._____provides protection against predators; d._____provides attachment sites for development of strong muscles; e._____all of these.
5. Mass extinctions occurred at the end of which three periods?
a._____Cambrian, Ordovician, Permian;
b._____Ordovician, Silurian, Devonian;
c._____Ordovician, Devonian, Permian;
d._____Silurian, Pennsylvanian, Permian;
e._____Devonian, Pennsylvanian, Permian.
6. Amphibians evolved from which of the following groups?
a._____reptiles; b._____placoderms; c._____lungfish; d._____archaeocyathids; e._____lobe-finned fish.
7. Which algal group was the probable ancestor of vascular plants?
a._____green; b._____ blue-green; c._____brown; d _____red; e._____yellow.
8. The most significant evolutionary change that allowed reptiles to colonize all of the land was the evolution of
a._____a scaly covering; b._____limbs and a backbone capable of supporting the animals on land; c._____tear ducts; d._____an egg that contained food and waste sacs and surrounded the embryo in a fluid-filled sac; e._____the middle ear bones.
9. The first plant group that did not require a wet area for part of its life cycle was the
a._____flowering; b._____naked seedless; c._____seedless vascular; d._____angiosperms; e._____gymnosperms.
10. Which reptile group gave rise to mammals?
a._____labyrinthodonts; b._____acanthodians; c_____pelycosaurs; d._____protothyrids; e._____therapsids.
11. Discuss how changing geologic conditions affected the evolution of invertebrate life during the Paleozoic Era.
12. Draw a marine food web that shows the relationships among producers, consumers, and decomposers.

13. Discuss the significance of the appearance of the first shelled animals and possible causes for the acquisition of a mineralized exoskeleton.
14. Discuss how the incompleteness of the fossil record may play a role in such theories as the Cambrian explosion of life.
15. Discuss how changing geologic conditions affected the evolution of plants and vertebrates.
16. Describe the problems that had to be overcome before organisms could inhabit and completely colonize the land.
17. What are the major differences between the seedless vascular plants and the gymnosperms? Why are these differences significant in terms of exploiting the terrestrial environment?
18. Why is it likely that fish evolved in the seas and then migrated to freshwater environments? Could fish have evolved in freshwater and then migrated to the seas?
19. Outline the evolutionary history of fish.
20. Discuss some of the possible causes for the Permian mass extinction.

World Wide Web Activities

GeologyNow Assess your understanding of this chapter's topics with additional quizzing and comprehensive interactivities at

http://earthscience.brookscole.com/changingearth4e

as well as current and up-to-date weblinks, additional readings, and InfoTrac College Edition exercises.

Mesozoic Earth and Life History

CHAPTER 22

OUTLINE

- Introduction
- The Breakup of Pangaea
- Mesozoic History of North America

  GEOLOGY IN UNEXPECTED PLACES: *Ancient Seafloor in San Francisco*
- What Role Did Accretion of Terranes Play in the Growth of Western North America?
- Mesozoic Mineral Resources
- Life of the Mesozoic Era

  GEO-FOCUS 22.1: *Mary Anning's Contributions to Paleontology*
- Mesozoic Climates and Paleogeography
- Mass Extinctions—A Crisis in the History of Life
- Geo-Recap

GeologyNow *This icon, which appears throughout the book, indicates an opportunity to explore interactive tutorials, animations, or practice problems available on the GeologyNow website at http://earthscience.brookscole.com/changingearth4e.*

OBJECTIVES

At the end of this chapter, you will have learned that:

- The Mesozoic breakup of Pangaea profoundly affected geologic and biologic events.
- Most of North America was above sea level during much of the Mesozoic Era.
- A global rise in sea level during the Cretaceous Period resulted in an enormous interior seaway that divided North America into two large landmasses.
- Western North America was affected by four interrelated orogenies that took place at an oceanic–continental convergent plate boundary.
- Terrane accretion also affected the Mesozoic geologic history of western North America.
- Marine invertebrates that survived the Permian extinction event diversified and repopulated the Mesozoic seas.
- Land-plants changed markedly when flowering plants evolved during the Cretaceous.
- Dinosaurs, flying reptiles, and marine reptiles evolved during the Mesozoic.
- Birds evolved from small carnivorous dinosaurs, whereas mammals descended from mammal-like reptiles.
- Mesozoic extinctions resulted in the demise of dinosaurs and some of their relatives and several types of marine invertebrates.

Restoration of one of the oldest known dinosaurs, Staurikosaurus, *from the Late Triassic of Argentina and Brazil. It measures about 2 m long.*

Source: © 1990 Mark Hallett, All Rights Reserved

Introduction

Approximately 150 to 210 million years after the emplacement of massive plutons created the Sierra Nevada (Nevadan orogeny), gold was discovered at Sutter's Mill on the South Fork of the American River at Coloma, California. On January 24, 1848, James Marshall, a carpenter building a sawmill for John Sutter, found bits of the glittering metal in the mill's tailrace. Soon settlements throughout the state were abandoned as word of the chance for instant riches spread throughout California. Within a year after the news of the gold discovery reached the East Coast, the Sutter's Mill area was swarming with more than 80,000 prospectors, all hoping to make their fortune. At least 250,000 gold seekers prospected the Sutter's Mill area, and although most were Americans, they came from all over the world, even as far away as China. Most thought the gold was simply waiting to be taken and didn't realize that prospecting was hard work.

No one thought much about the consequences of so many people converging on the Sutter's Mill area, all intent on making easy money. In fact, life in the mining camps was extremely hard and expensive. The shop owners and traders frequently made more money than the prospectors. In reality, only a few prospectors ever hit it big or were even moderately successful. The rest barely eked out a living until they eventually abandoned their dream and went home.

The gold these prospectors sought was mostly in the form of placer deposits. Weathering of gold-bearing igneous rocks and mechanical separation of minerals by density during stream transport form placer deposits. Although many prospectors searched for the mother lode, all of the gold recovered during the gold rush came from placers (deposits of sand and gravel containing gold particles large enough to be recovered by panning). Panning is a common way of mining placer deposits. In this method, a miner dips a shallow pan into a streambed, swirls the material around, and pours off the lighter material. Gold, being about six times heavier than most sand grains and rock chips, concentrates on the bottom of the pan and can then be picked out (see "The Precious Metals," pages 74–75).

Although some prospectors dug $30,000 worth of gold dust a week out of a single claim and found gold practically on the surface of the ground, most of this easy gold was recovered very early during the gold rush. Most prospectors made only a living wage working their claims. Nevertheless, during the five years from 1848 to 1853 that constituted the gold rush proper, more than $200 million in gold was extracted.

The Mesozoic Era (245 to 66 million years ago) was an important time in Earth's history. The major geologic event was the breakup of Pangaea, which affected oceanic and climatic circulation patterns and influenced the evolution of the terrestrial and marine biotas. Other important Mesozoic geologic events resulting from plate movement include the origin of the Atlantic Ocean basin and the Rocky Mountains, the accumulation of vast salt deposits that eventually formed salt domes adjacent to which oil and natural gas were trapped, and the emplacement of huge batholiths that account for the origin of various mineral resources, including the gold that fueled the California gold rush of 1849.

THE BREAKUP OF PANGAEA

Just as the formation of Pangaea influenced geologic and biologic events during the Paleozoic, the breakup of this supercontinent profoundly affected geologic and biologic events during the Mesozoic. The movement of continents affected the global climatic and oceanic regimes as well as the climates of the individual continents. Populations became isolated or were brought into contact with other populations, leading to evolutionary changes in the biota. So great was the effect of this breakup on the world that it forms the central theme of this chapter.

Geologic, paleontologic, and paleomagnetic data indicate that the breakup of Pangaea took place in four general stages. The first stage began with rifting between Laurasia and Gondwana during the Late Triassic. By the end of the Triassic, the expanding Atlantic Ocean separated North America from Africa (■ Figure 22.1a). This change was followed by the rifting of North America from South America sometime during the Late Triassic and Early Jurassic.

Separation of the continents allowed water from the Tethys Sea to flow into the expanding central Atlantic Ocean, whereas Pacific Ocean waters flowed into the newly formed Gulf of Mexico, which at that time was little more than a restricted bay (■ Figure 22.2). During that time, these areas were in the low tropical latitudes where high temperatures and high rates of evaporation were ideal for the formation of thick evaporite deposits.

The second stage in Pangaea's breakup involved rifting and movement of the various Gondwana continents during the Late Triassic and Jurassic periods. As early as the Late Triassic, Antarctica and Australia,

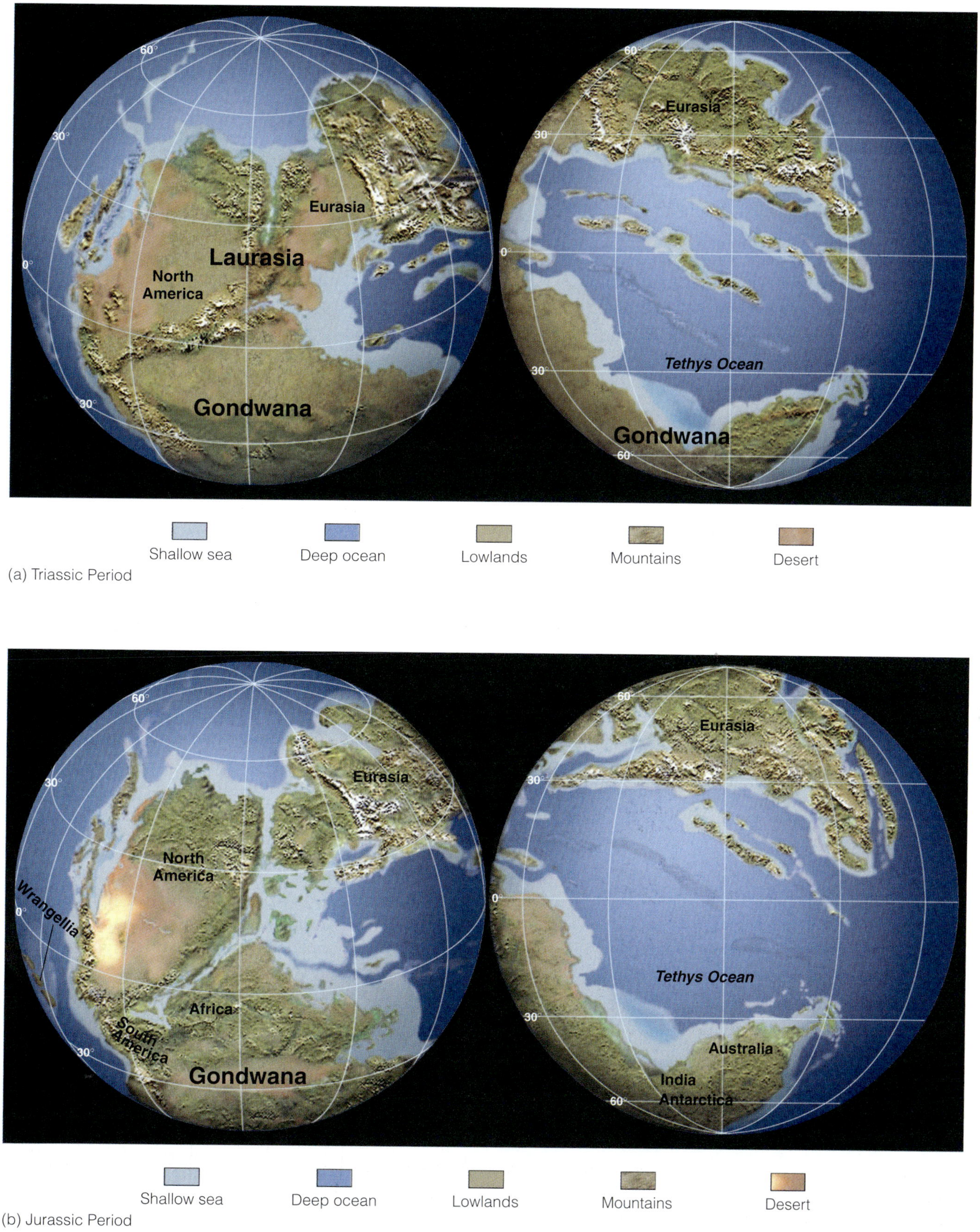

Figure 22.1

Paleogeography of the world during the (a) Triassic Period and (b) Jurassic Period of the Mesozoic Era. Source: From R. S. Dietz and J. C. Holden, Journal of Geophysical Research, Vol. 75, no. 26, pp. 4939–4956. Copyright © 1970 by the American Geophysical Union.

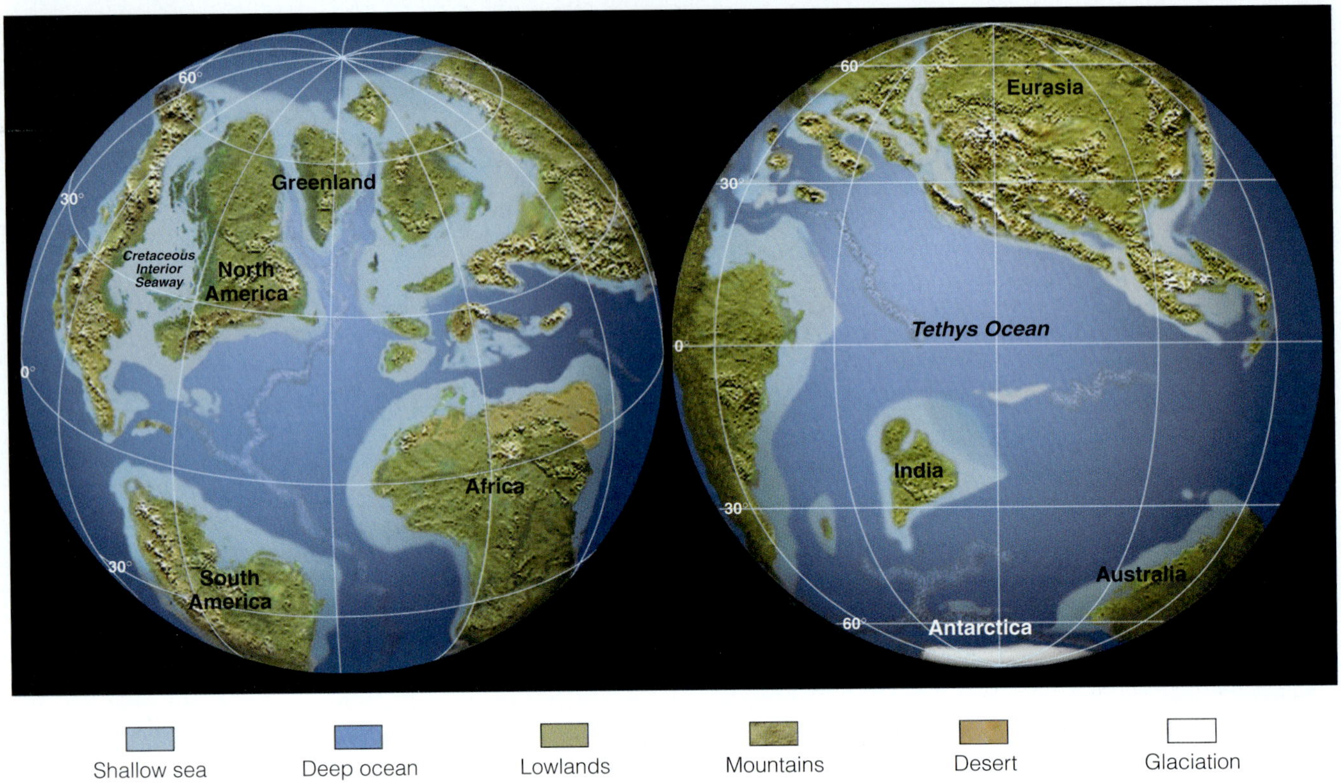

(c) Late Cretaceous Period

Figure 22.1 *Continued*

Paleogeography of the world during the (c) late Cretaceous Period of the Mesozoic Era.

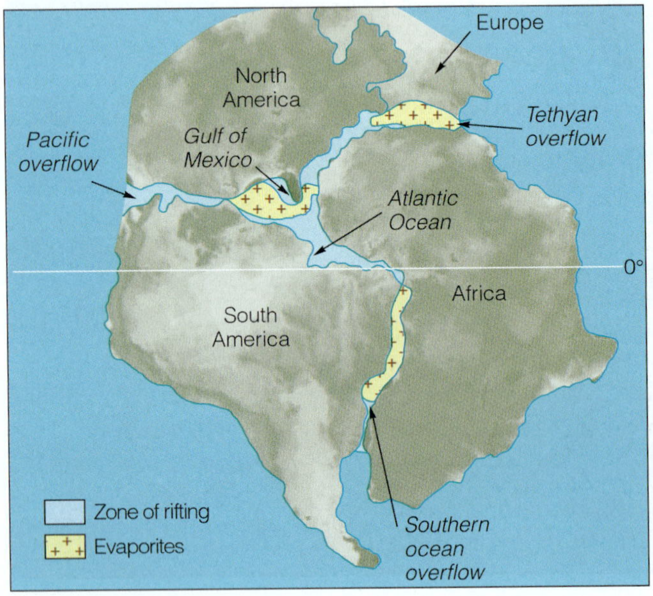

Figure 22.2

Evaporites accumulated in shallow basins as Pangaea broke apart during the Early Mesozoic. Water from the Tethys Sea flowed into the central Atlantic Ocean, and water from the Pacific Ocean flowed into the newly formed Gulf of Mexico. Marine water from the south flowed into the southern Atlantic Ocean.

which remained sutured together, separated from South America and Africa. Also during this time, India began to rift from the Gondwana continent and move northward (Figure 22.1a, b).

The third stage of breakup began during the Late Jurassic, when South America and Africa began separating (Figure 22.1b). The rifting and subsequent separation of these two continents formed a narrow basin where thick evaporite deposits accumulated from the evaporation of southern ocean waters (Figure 22.2). During this time, the eastern end of the Tethys Sea began to close as a result of the clockwise rotation of Laurasia and the northward movement of Africa. This narrow Late Jurassic and Cretaceous seaway between Africa and Europe was the forerunner of the present Mediterranean Sea.

By the end of the Cretaceous, Australia and Antarctica had separated, India had nearly reached the equator, South America and Africa were widely separated, and Greenland was essentially an independent landmass (Figure 22.1c).

The final stage in Pangaea's breakup occurred during the Cenozoic. Australia continued moving northward, and Greenland completely separated from Europe and North America and formed a separate landmass.

The Effects of the Breakup of Pangaea on Global Climates and Ocean Circulation Patterns

By the end of the Permian Period, Pangaea extended from pole to pole, covered about one-fourth of Earth's surface, and was surrounded by Panthalassa, a global ocean that encompassed about 300 degrees of longitude. Such a configuration exerted tremendous influence on the world's climate and resulted in generally arid conditions over large parts of Pangaea's interior.

The world's climates result from the complex interaction between wind and ocean currents and the location and topography of the continents. In general, dry climates occur on large landmasses in areas remote from sources of moisture and where barriers to moist air exist, such as mountain ranges. Wet climates occur near large bodies of water or where winds can carry moist air over land.

Past climatic conditions can be inferred from the distribution of climate-sensitive deposits. Evaporites are deposited where evaporation exceeds precipitation. Although desert dunes and red beds may form locally in humid regions, they are characteristic of arid regions. Coal forms in both warm and cool humid climates. Vegetation that is eventually converted into coal requires at least a good seasonal water supply; thus, coal deposits are indicative of humid conditions.

Widespread Triassic evaporites, red beds, and desert dunes in the low and middle latitudes of North and South America, Europe, and Africa indicate dry climates in those regions, whereas coal deposits are found mainly in the high latitudes, indicating humid conditions. These high-latitude coals are analogous to today's Scottish peat bogs or Canadian muskeg. The lands bordering the Tethys Sea were probably dominated by seasonal monsoon rains resulting from the warm, moist winds and warm oceanic currents impinging against the east-facing coast of Pangaea.

The temperature gradient between the tropics and the poles also affects oceanic and atmospheric circulation. The greater the temperature difference between the tropics and the poles, the steeper the temperature gradient, and the faster the circulation of the oceans and atmosphere. Oceans absorb about 90% of the solar radiation they receive, whereas continents absorb only about 50%, even less if they are snow covered. The rest of the solar radiation is reflected back into space. Therefore, areas dominated by seas are warmer than those dominated by continents. By knowing the distribution of continents and ocean basins, geologists can generally estimate the average annual temperature for any region on Earth, as well as determine a temperature gradient.

The breakup of Pangaea during the Late Triassic caused the global temperature gradient to increase because the Northern Hemisphere continents moved farther northward, displacing higher-latitude ocean waters. Because of the steeper global temperature gradient produced by a decrease in temperature in the high latitudes and the changing positions of the continents, oceanic and atmospheric circulation patterns greatly accelerated during the Mesozoic (■ Figure 22.3). Although the temperature gradient and seasonality on land were increasing during the Jurassic and Cretaceous, the middle- and higher-latitude oceans were still warm because warm waters from the Tethys Sea were circulating to the higher latitudes. The result was a relatively equable worldwide climate through the end of the Cretaceous.

MESOZOIC HISTORY OF NORTH AMERICA

The beginning of the Mesozoic Era was essentially the same in terms of tectonism and sedimentation as the preceding Permian Period in North America. Terrestrial sedimentation continued over much of the craton, and block faulting and igneous activity began in the Appalachian region as North America and Africa began separating (■ Figure 22.4). The newly forming Gulf of Mexico experienced extensive evaporite deposition during the Late Triassic and Jurassic as North America separated from South America (■ Figure 22.5).

A global rise in sea level during the Cretaceous resulted in worldwide transgressions onto the continents (■ Figure 22.6). These transgressions were caused by higher heat flow along the oceanic ridges due to increased rifting and the consequent expansion of oceanic crust. By the Middle Cretaceous, sea level probably was as high as at any time since the Ordovician, and approximately one-third of the present land area was inundated by epeiric seas.

Marine deposition was continuous over much of the North American Cordillera. A volcanic island arc system that formed off the western edge of the craton during the Permian was sutured to North America sometime later during the Permian or Triassic. This event is referred to as the *Sonoma orogeny* and will be discussed later in this chapter. During the Jurassic, the entire Cordilleran area was involved in a series of major mountain-building episodes resulting in the formation of the Sierra Nevada, the Rocky Mountains, and other lesser mountain ranges. Although each orogenic episode has its own name, the entire mountain-building event is simply called the *Cordilleran orogeny* (also discussed later in this chapter). With this simplified overview of the Mesozoic history of North America in mind, we will now examine the specific regions of the continent.

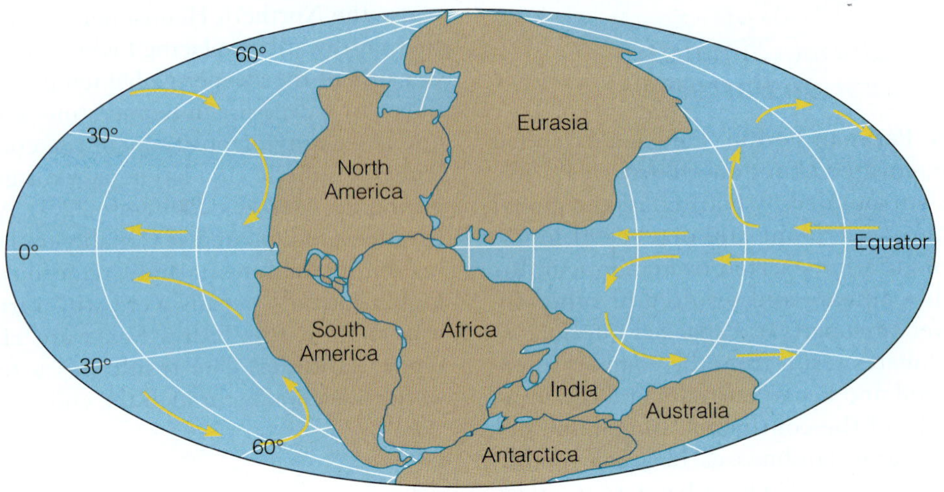

(a) Triassic Period

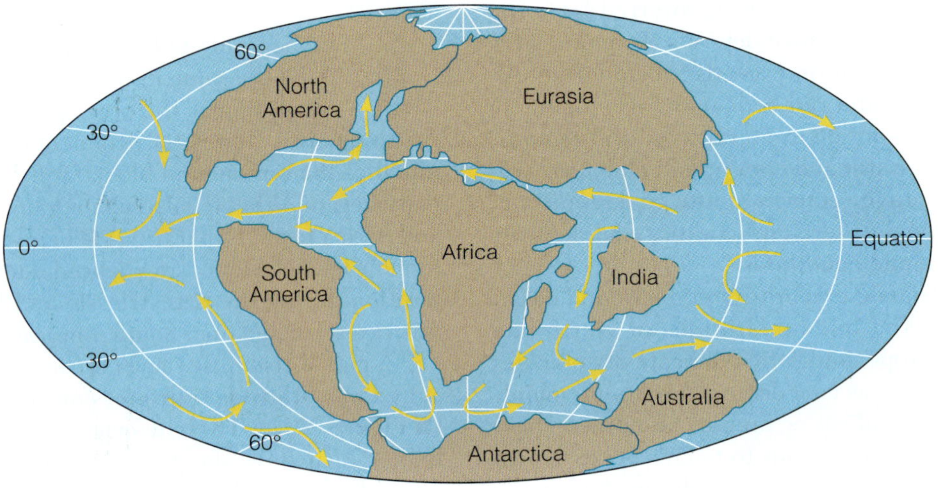

(b) Cretaceous Period

Figure 22.3

Oceanic circulation evolved from (a) a simple pattern in a single ocean (Panthalassa) with a single continent (Pangaea) to (b) a more complex pattern in the newly formed oceans of the Cretaceous Period.

Continental Interior

Recall that the history of the North American craton can be divided into unconformity-bound sequences reflecting advances and retreats of epeiric seas over the craton (see Figure 20.5). Although these transgressions and regressions played a major role in the Paleozoic geologic history of the continent, they were not as important during the Mesozoic. Most of the continental interior during the Mesozoic was well above sea level and was not inundated by epeiric seas. Consequently, the two Mesozoic cratonic sequences, the *Absaroka sequence* (Late Mississippian to Early Jurassic) and the *Zuni sequence* (Early Jurassic to Early Paleocene) (see Figure 20.5), are incorporated as part of the history of the three continental margin regions of North America.

Eastern Coastal Region

During the Early and Middle Triassic, coarse detrital sediments derived from erosion of the recently uplifted Appalachians (Alleghenian orogeny) filled the various intermontane basins and spread over the surrounding areas. As erosion continued during the Mesozoic, this once lofty mountain system was reduced to a low-lying plain. During the Late Triassic, the first stage in the breakup of Pangaea began with North America separating from Africa. Fault-block basins developed in response

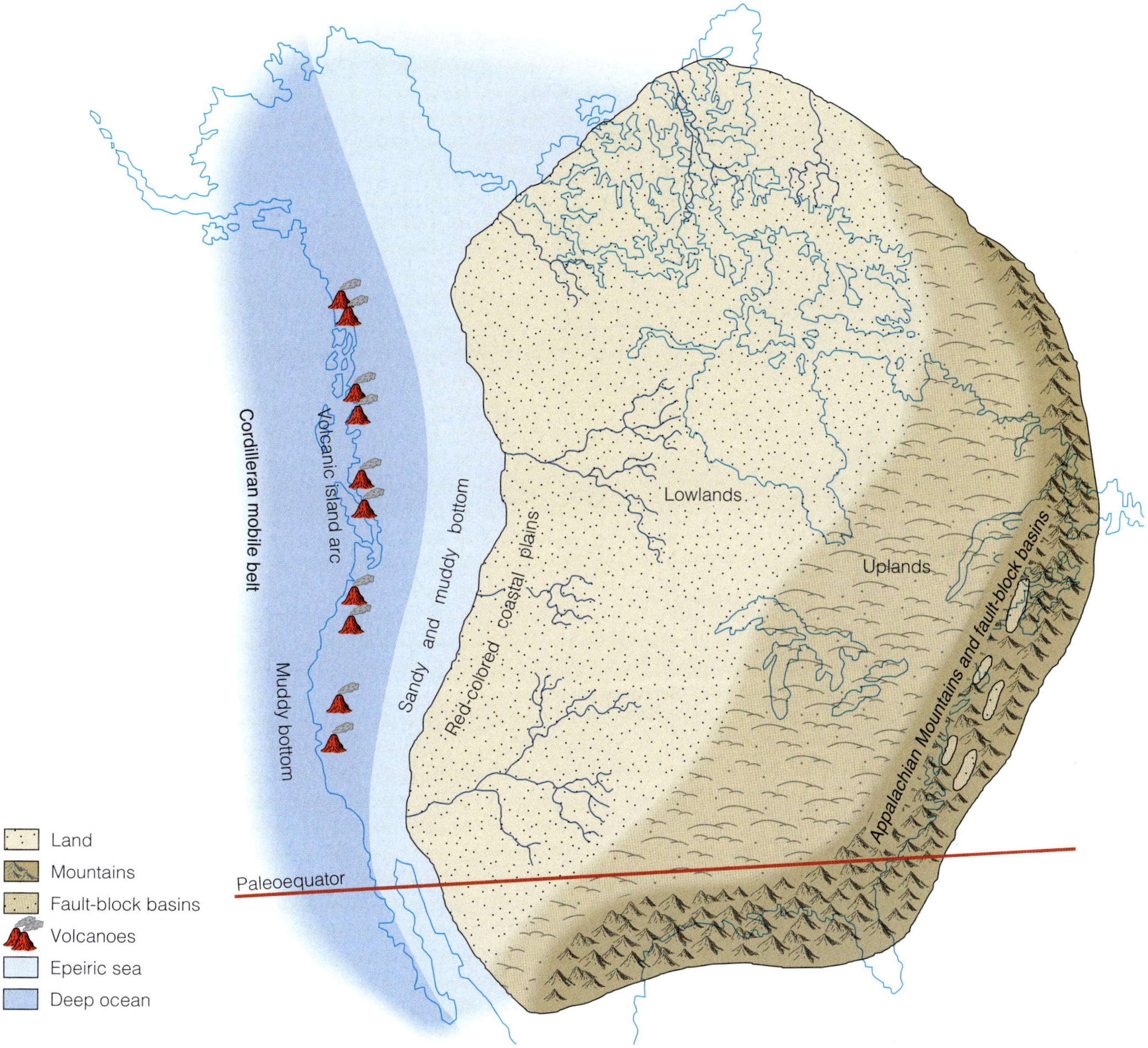

Figure 22.4

Paleogeography of North America during the Triassic Period.

to upwelling magma beneath Pangaea in a zone stretching from present-day Nova Scotia to North Carolina (Figure 22.7). Erosion of the adjacent fault-block mountains filled these basins with great quantities (up to 6000 m) of poorly sorted red nonmarine detrital sediments known as the *Newark Group*. Reptiles roamed along the margins of the lakes and streams that formed in these basins, leaving their footprints and trackways in the soft sediments. Although the Newark rocks contain numerous dinosaur footprints, they are almost completely devoid of dinosaur bones! The Newark Group is mostly Late Triassic in age, but in some areas deposition began in the Early Jurassic.

Concurrent with sedimentation in the fault-block basins were extensive lava flows that blanketed the basin floors as well as intrusions of numerous dikes and sills. The most famous intrusion is the prominent Palisades sill along the Hudson River in the New York–New Jersey area (Figure 22.7d).

As the Atlantic Ocean grew, rifting ceased along the eastern margin of North America, and this once-active plate margin became a passive, trailing continental margin. The fault-block mountains produced by this rifting continued to erode during the Jurassic and Early Cretaceous until all that was left was a large low-relief area. The

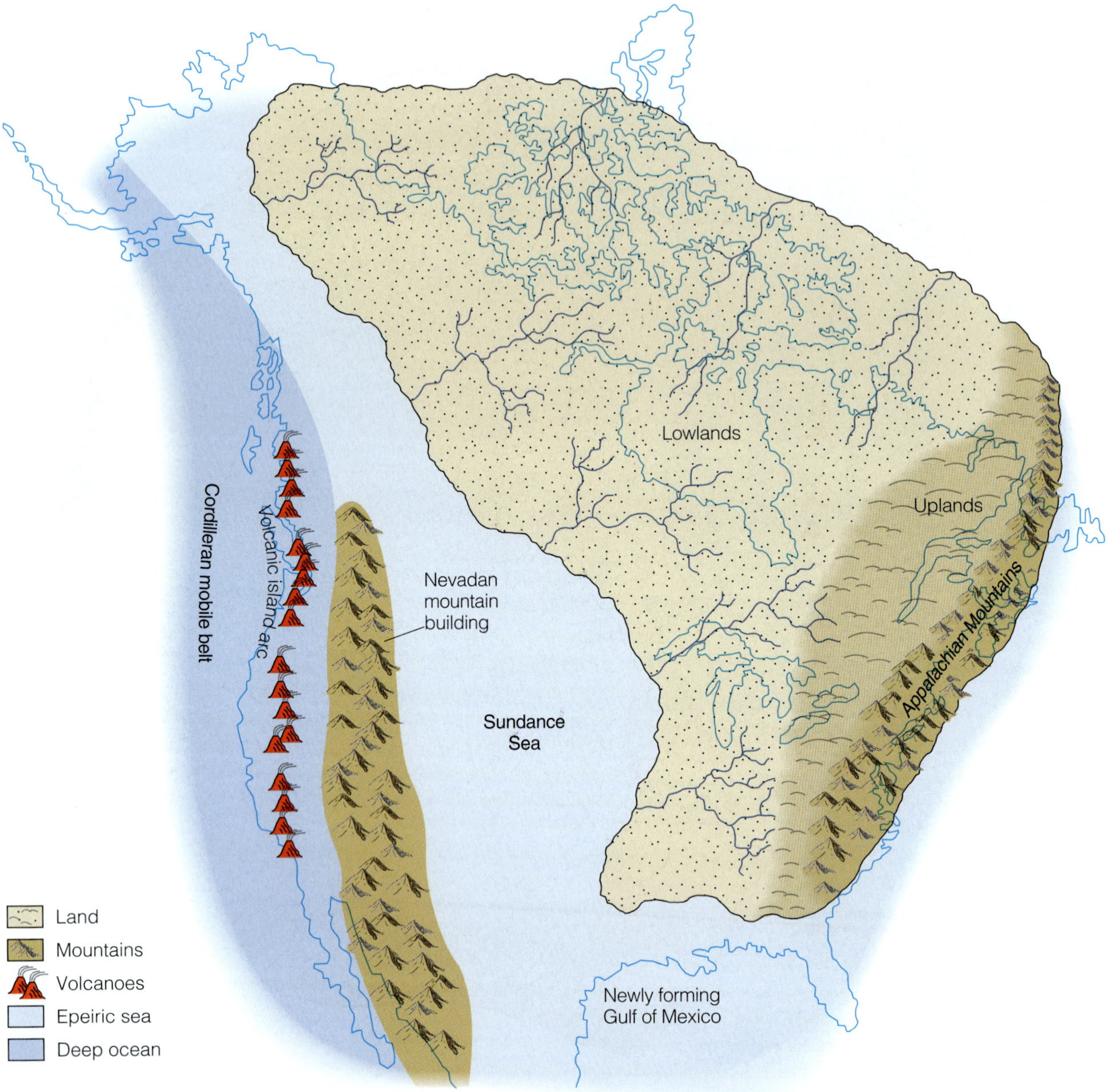

Figure 22.5

Paleogeography of North America during the Jurassic Period.

sediments produced by this erosion contributed to the growing eastern continental shelf. During the Cretaceous Period, the Appalachian region was reelevated and once again shed sediments onto the continental shelf, forming a gently dipping, seaward-thickening wedge of rocks up to 3000 m thick. These rocks are currently exposed in a belt extending from Long Island, New York, to Georgia.

Gulf Coastal Region

The Gulf Coastal region was above sea level until the Late Triassic (Figure 22.4). As North America separated from South America during the Late Triassic, the Gulf of Mexico began to form (Figure 22.5). With oceanic waters flowing into this newly formed, shallow, restricted basin, conditions were ideal for evaporite formation. More than 1000 m of evaporites were precipitated at this time, and most geologists think that these Jurassic evaporites are the source for the Tertiary salt domes found today in the Gulf of Mexico and southern Louisiana.

By the Late Jurassic, circulation in the Gulf of Mexico was less restricted, and evaporite deposition ended. Normal marine conditions returned to the area with al-

Figure 22.6

Paleogeography of North America during the Cretaceous Period.

ternating transgressing and regressing seas. The resulting sediments were covered and buried by thousands of meters of Cretaceous and Cenozoic sediments.

During the Cretaceous, the Gulf Coastal region, like the rest of the continental margin, was inundated by northward-transgressing seas (Figure 22.6). As a result, nearshore sandstones are overlain by finer sediments characteristic of deeper waters. Following an extensive regression at the end of the Early Cretaceous, a major transgression began during which a wide seaway extended from the Arctic Ocean to the Gulf of Mexico (Figure 22.6). Sediments deposited in the Gulf Coastal region during the Cretaceous formed a seaward-thickening wedge.

Reefs were also widespread in the Gulf Coastal region during the Cretaceous. Bivalves called *rudists* were the main constituent of many of these reefs (Figure 22.19c). Because of their high porosity and permeability, rudistoid reefs make excellent petroleum reservoirs.

Western Region

Mesozoic Tectonics

The Mesozoic geologic history of the North American Cordilleran mobile belt is very

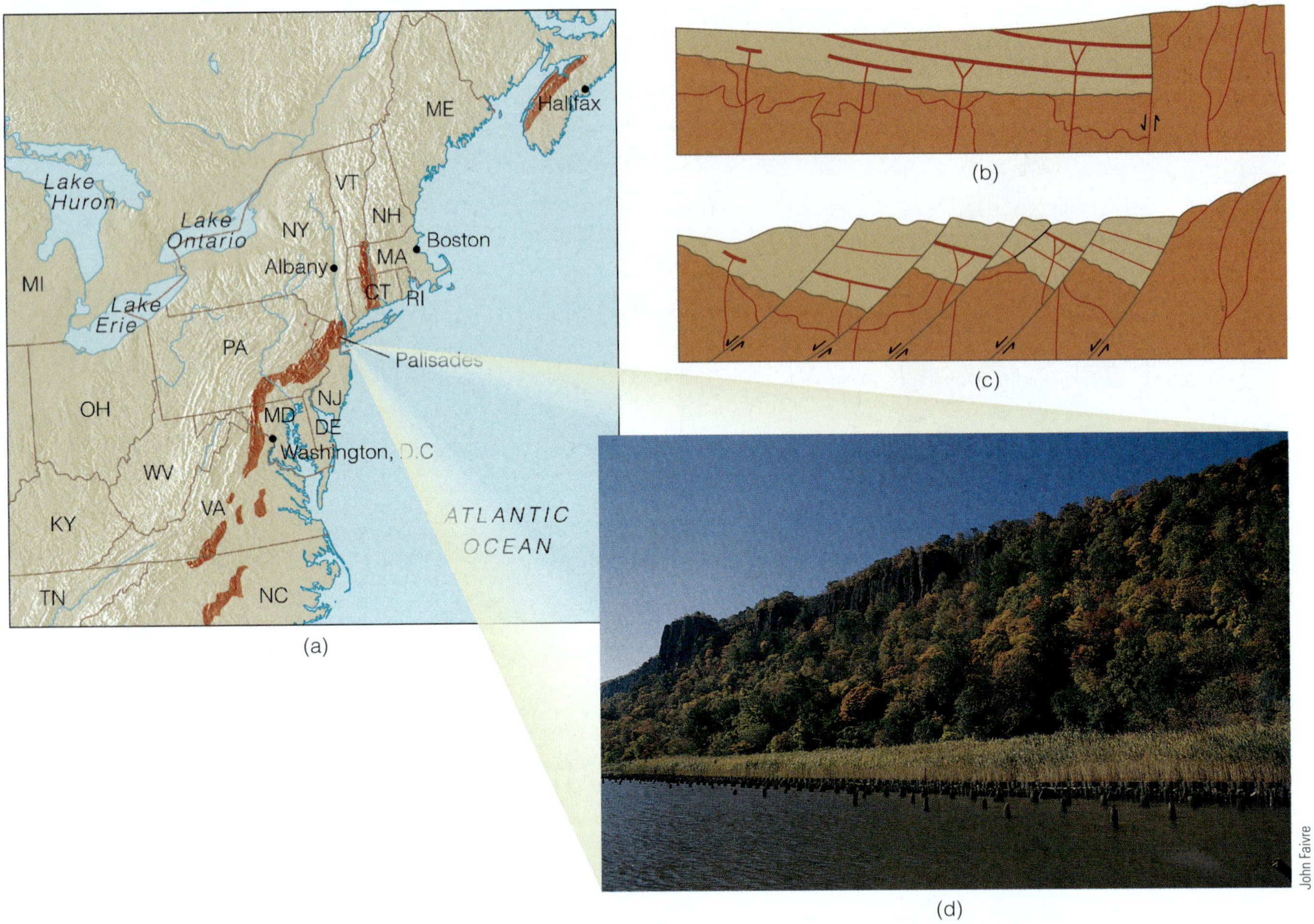

■ **Figure 22.7**

(a) Areas where Triassic fault-block basin deposits crop out in eastern North America. (b) After the Appalachians were eroded to a low-lying plain by the Middle Triassic, fault-block basins formed as a result of Late Triassic rifting between North America and Africa. (c) These valleys accumulated tremendous thickness of sediments and were themselves broken by a complex of normal faults during rifting. (d) Palisades of the Hudson River. This sill was one of many intruded into the Newark sediments during the Late Triassic rifting that marked the separation of North America from Africa.

complex, involving the eastward subduction of the oceanic Farallon plate under the continental North American plate. Activity along this oceanic–continental convergent plate boundary resulted in an eastward movement of deformation. This orogenic activity progressively affected the trench and continental slope, the continental shelf, and the cratonic margin, causing a thickening of the continental crust.

Except for the Late Devonian–Early Mississippian Antler orogeny (see Figure 20.28), the Cordilleran region of North America experienced little tectonism during the Paleozoic. During the Permian, however, an island arc and ocean basin formed off the western North American craton (Figure 22.4), followed by subduction of an oceanic plate beneath the island arc and the thrusting of oceanic and island arc rocks eastward against the craton margin (■ Figure 22.8). This event, known as the **Sonoma orogeny,** occurred at or near the Permian–Triassic boundary.

Following the Late Paleozoic–Early Mesozoic destruction of the volcanic island arc during the Sonoma orogeny, the western margin of North America became an oceanic–continental convergent plate boundary. During the Late Triassic, a steeply dipping subduction zone developed along the western margin of North America in response to the westward movement of North America over the Farallon plate. This newly created oceanic–continental plate boundary controlled Cordilleran tectonics for the rest of the Mesozoic and for most of the Cenozoic Era; this subduction zone marks the beginning of the modern circum-Pacific orogenic system.

Two subduction zones dipping in opposite directions formed off the west coast of North America during the

GEOLOGY IN UNEXPECTED PLACES

Ancient Seafloor in San Francisco

At 5:13 A.M. on April 18, 1906, a huge earthquake shook San Francisco, California, killing thousands and causing tremendous property damage. San Francisco is well known for past and potential earthquakes generated by movements on the San Andreas fault (see Chapter 8). But most people, including the city's residents, are unaware that rock exposures there provide excellent examples of ancient seafloor sediments and volcanic rocks that formed at or near spreading ridges. Indeed, the city is underlain by a succession of rocks scraped off against the continental margin as an oceanic plate was subducted beneath North America during the Jurassic and Cretaceous periods. Some of the rocks are called *mélange*, a geologic term for a chaotic mixture of rock fragments set in a matrix of clay formed by intense shearing during subduction.

Excellent exposures of some of these ancient rocks are on Alcatraz Island, which served as a maximum-security federal penitentiary from 1934 to 1963 and where such infamous criminals as Al Capone were once imprisoned. The rocks consist of thick-bedded graywacke, a type of sandstone with a large amount of rock fragments and clay (■ Figure 1). The original sediments were probably deposited in an oceanic trench much like the present-day Peru–Chili Trench along South America's west coast.

Other interesting rocks in and around San Francisco are pillow lava that originated by submarine volcanism and thinly bedded, highly deformed, bedded chert, another type of rock that originated on the deep seafloor (see Figure 1). So, in addition to outstanding scenery and its cultural atmosphere, San Francisco has an interesting geologic history.

■ Figure 1

The rocks exposed on Alcatraz Island in San Francisco Bay are thick-bedded sandstones that were deposited on the deep seafloor. The inset shows intensely folded, thinly bedded chert that also formed on the deep seafloor.

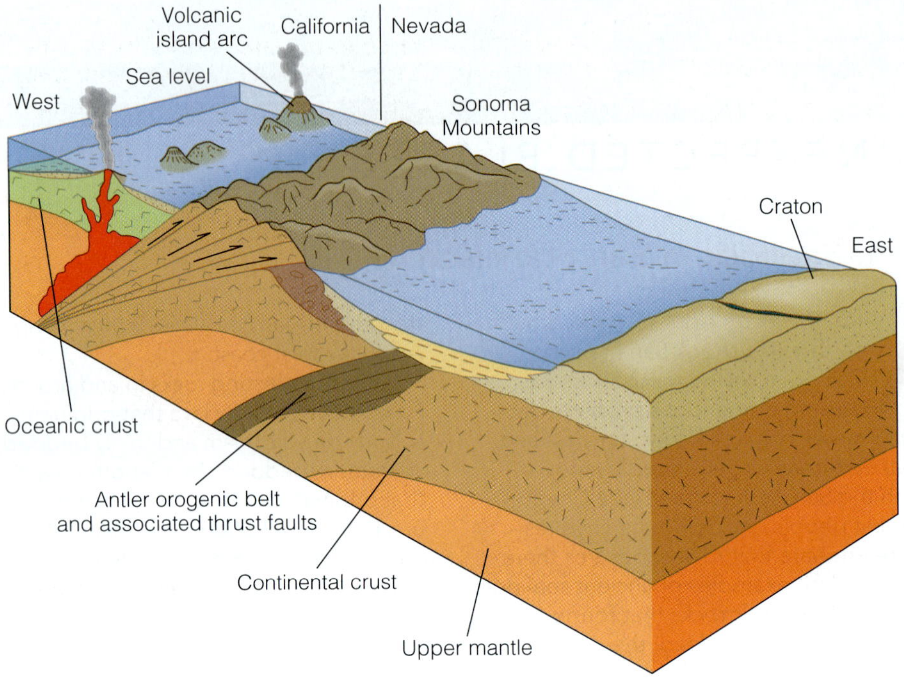

Figure 22.8

Tectonic activity that culminated in the Permian–Triassic Sonoma orogeny in western North America. The Sonoma orogeny was the result of a collision between the southwestern margin of North America and an island arc system.

Middle and early Late Jurassic (■ Figure 22.9). The more westerly subduction zone was eliminated by the westward-moving North American plate, which overrode the oceanic Farallon plate.

The *Franciscan Complex,* which is up to 7000 m thick, is an unusual rock unit consisting of a chaotic mixture of rocks that accumulated during the Late Jurassic and Cretaceous. The various rock types—graywacke, volcanic breccia, siltstone, black shale, chert, pillow basalt, and blueshist metamorphic rocks—suggest that continental shelf, slope, and deep-sea environments were brought together in a submarine trench when North America overrode the subducting Farallon plate (see Figure 7.20).

East of the Franciscan Complex and currently separated from it by a major thrust fault is the *Great Valley Group*. It consists of more than 16,000 m of Cretaceous conglomerates, sandstones, siltstones, and shales. These sediments were deposited on the continental shelf and slope at the same time the Francisan deposits were accumulating in the submarine trench (see Figure 7.20).

The general term **Cordilleran orogeny** is applied to the mountain-building activity that began during the Jurassic and continued into the Cenozoic (■ Figure 22.10). The Cordilleran orogeny consisted of a series of individual mountain-building events that occurred in different regions at different times. Most of this Cordilleran orogenic activity is related to the continued westward movement of the North American plate.

The first phase of the Cordilleran orogeny, the **Nevadan orogeny** (Figure 22.10), began during the Late Jurassic and continued into the Cretaceous as large volumes of granitic magma were generated at depth beneath the western edge of North America. These granitic masses ascended as huge batholiths that are now recognized as the Sierra Nevada, Southern California, Idaho, and Coast Range Batholiths (■ Figure 22.11).

By the Late Cretaceous, most of the volcanic and plutonic activity had migrated eastward into Nevada and Idaho. This migration was probably caused by a change from high-angle to low-angle subduction, resulting in the subducting oceanic plate reaching its melting depth farther east (■ Figure 22.12). Thrusting occurred progressively farther east, so that by the Late Cretaceous, it extended all the way to the Idaho–Washington border.

The second phase of the Cordilleran orogeny, the **Sevier orogeny,** was mostly a Cretaceous event (Figure 22.10). Subduction of the Farallon plate beneath the North American plate continued during this time, resulting in numerous overlapping, low-angle thrust faults in which blocks of older strata were thrust eastward on top of younger strata (■ Figure 22.13). This deformation produced generally north-south–trending mountain ranges that stretch from Montana to western Canada.

During the Late Cretaceous to Early Cenozoic, the final pulse of the Cordilleran orogeny occurred (Figure 22.10). The **Laramide orogeny** developed east of the Sevier orogenic belt in the present-day Rocky Mountain

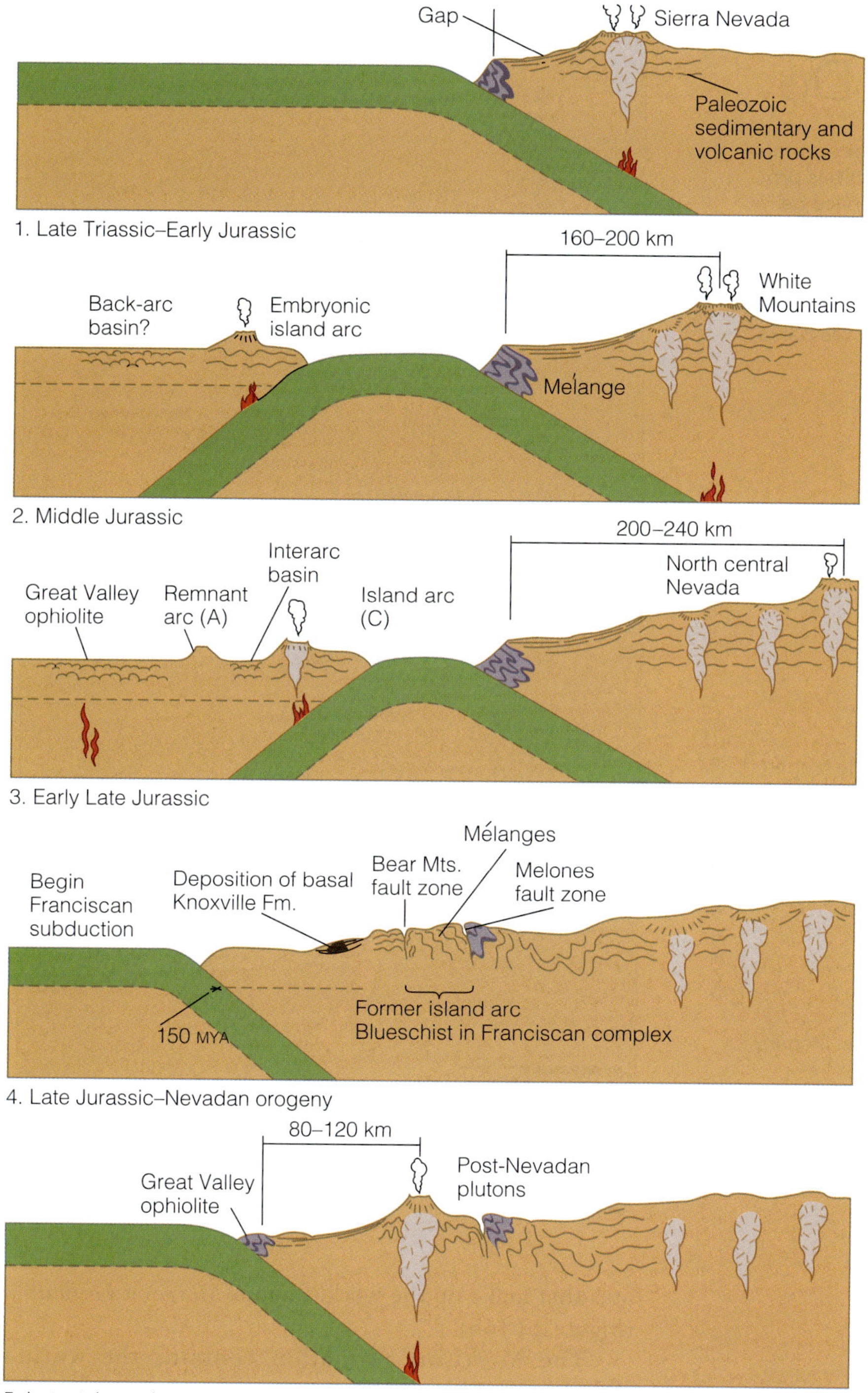

Figure 22.9

Interpretation of the tectonic evolution of the Sierra Nevada during the Mesozoic Era.
Source: Reprinted with permission from Geological Society of America, "Tectonic Evolution of the Western Sierra Nevada, California," in GSA Bulletin 86, 1975, p. 1334 (Fig. 3).

areas of New Mexico, Colorado, and Wyoming. Most features of the present-day Rocky Mountains resulted from the Cenozoic phase of the Laramide orogeny, and for that reason, we will discuss it in Chapter 23.

Mesozoic Sedimentation Concurrent with the tectonism in the Cordilleran mobile belt, Early Triassic sedimentation on the western continental shelf consisted of shallow-water marine sandstones, shales, and limestones. During the Middle and Late Triassic, the western shallow seas regressed farther west, exposing large areas of former seafloor to erosion. Marginal marine and nonmarine Triassic rocks, particularly red beds, contribute to the spectacular and colorful scenery of the region.

These rocks represent a variety of continental depositional environments. The Upper Triassic *Chinle Formation,* for example, is widely exposed throughout the Colorado Plateau region and is probably most famous for its petrified wood spectacularly exposed in Petrified Forest National Park, Arizona (Figure 22.14). This formation, as well as other Triassic formations in the Southwest, also contains the fossilized remains and tracks of amphibians and reptiles.

Early Jurassic deposits in a large part of the western region consist mostly of clean, cross-bedded sandstones indicative of wind-blown deposits. The thickest and most prominent of these is the *Navajo Sandstone,* a widespread cross-bedded sandstone that accumulated in a coastal dune environment along the southwestern margin of the craton. The sandstone's most distinguishing feature is its large-scale cross-beds, some of which are more than 25 m high (Figure 22.15).

Marine conditions returned to the region during the Middle Jurassic when a wide seaway called the **Sundance Sea** twice flooded the interior of western North America (Figure 22.5). The resulting deposits, the *Sundance Formation,* were produced from the erosion of tectonic highlands to the west that paralleled the shoreline. These highlands resulted from intrusive igneous activity and associated volcanism that began during the Triassic.

During the Late Jurassic, a mountain chain formed in Nevada, Utah, and Idaho as a result of the deformation produced by the Nevadan orogeny (Figure 22.16a). As the mountain chain grew and shed sediments

What Would You Do?

The U.S. economy has been in a recession for several years, and to reduce spending, the president has proposed that funding for the National Park Service be significantly reduced. As director of the National Park Service, you have been called to testify at the upcoming budget hearings. What arguments would you use to try and prevent the proposed reduction in funding? Would a knowledge of the geology and ecology of the parks be helpful? Explain.

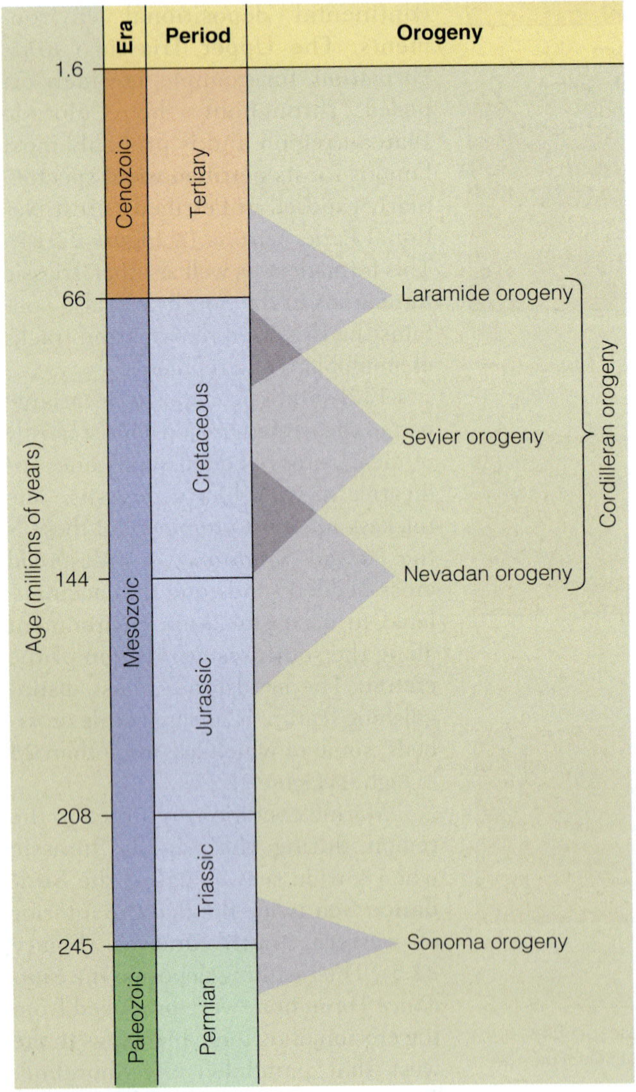

■ **Figure 22.10**

Mesozoic orogenies occurring in the Cordilleran mobile belt.

■ **Figure 22.11**

Location of Jurassic and Cretaceous batholiths in western North America.

eastward, the Sundance Sea began retreating northward. A large part of the area formerly occupied by the Sundance Sea was then covered by multicolored sandstones, mudstones, shales, and occasional lenses of conglomerates that make up the world-famous *Morrison Formation* (Figure 22.16b).

The Morrison Formation contains the world's richest assemblage of Jurassic dinosaur remains. Although most of the dinosaur skeletons are broken up, as many as 50 individuals have been found together in a small area.

Shortly before the end of the Early Cretaceous, Arctic waters spread southward over the craton, forming a large inland sea in the Cordilleran foreland basin area. Mid-Cretaceous transgressions also occurred on other continents, and all were part of the global mid-Cretaceous rise in sea level that resulted from accelerated seafloor spreading as Pangaea continued to fragment.

By the beginning of the Late Cretaceous, this incursion joined the northward-transgressing waters from the Gulf area to create an enormous **Cretaceous Interior**

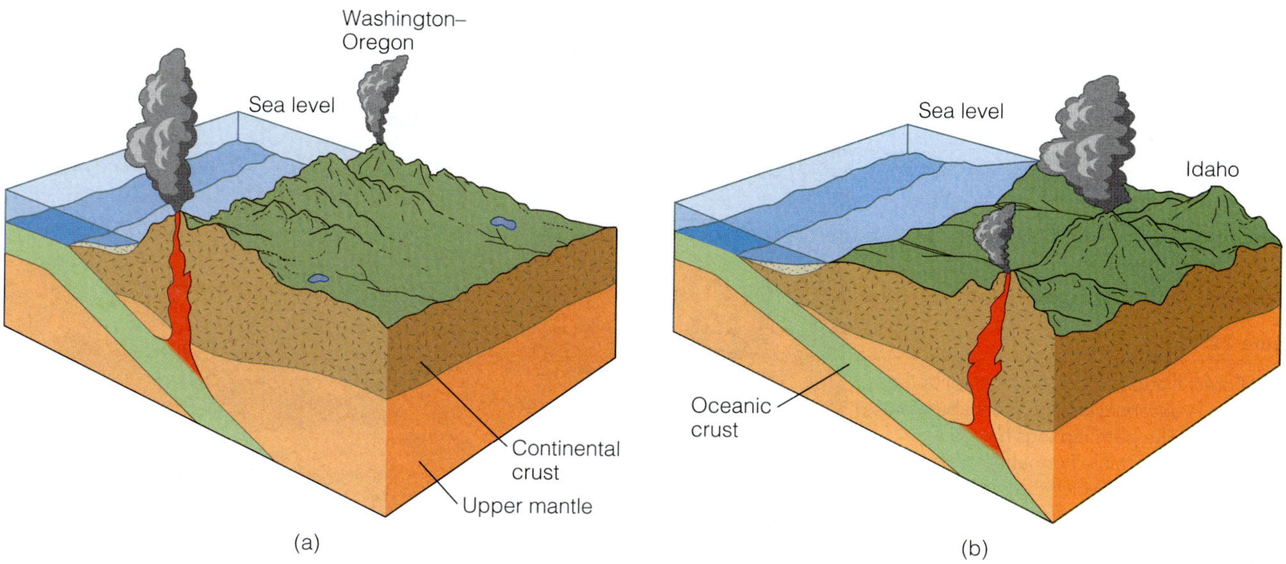

GeologyNow ■ **Active Figure 22.12**

A possible cause for the eastward migration of igneous activity in the Cordilleran region during the Cretaceous Period was a change from (a) high-angle to (b) low-angle subduction. As the subducting plate moved downward at a lower angle, the depth of melting moved farther east.

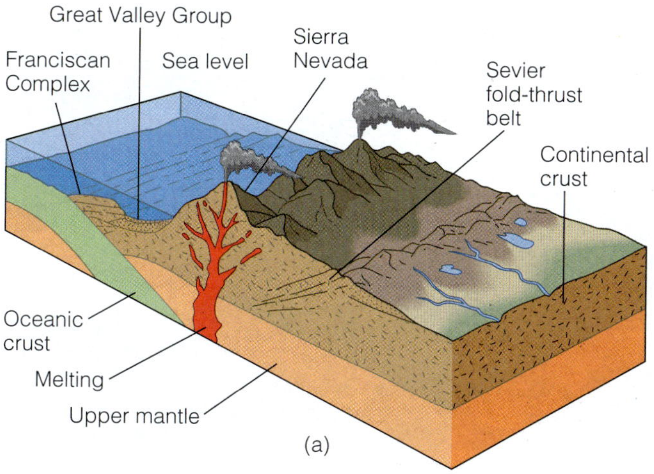

Seaway that occupied the area east of the Sevier orogenic belt. Extending from the Gulf of Mexico to the Arctic Ocean, and more than 1500 km wide at its maximum extent, this seaway effectively divided North America into two large landmasses until just before the end of the Late Cretaceous (Figure 22.6).

Cretaceous deposits less than 100 m thick indicate that the eastern margin of the Cretaceous Interior Seaway subsided slowly and received little sediment from the emergent, low-relief craton to the east. The western shoreline, however, shifted back and forth, primarily in response to fluctuations in the supply of sediment from the Cordilleran Sevier orogenic belt to the west. The facies relationships show lateral changes from conglomerate and coarse sandstone adjacent to the mountain belt through finer sandstones, siltstones, shales, and even limestones and chalks in the east (■ Figure 22.17). During times of particularly active mountain building, these coarse clastic wedges of gravel and sand prograded even farther east.

■ **Figure 22.13**

(a) The tectonic features of the Late Cretaceous Sevier orogeny were caused by subduction of the Farallon plate under the North American plate. (b) The Keystone thrust fault, a major fault in the Sevier overthrust belt, is exposed west of Las Vegas, Nevada. The sharp boundary between the light-colored Mesozoic rocks and the overlying dark-colored Paleozoic rocks marks the trace of the Keystone thrust fault.

© Stephen J. Kraseman/Photo Researchers, Inc.

Figure 22.14

Petrified Forest National Park, Arizona. All of the logs here are *Araucarioxylon*, the most abundant tree in the park. The petrified logs have been weathered from the Chinle Formation and are mostly in the position in which they were buried some 200 million years ago.

Alden C. Larson/JLM Visuals

Figure 22.15

Large cross-beds of the Jurassic Navajo Sandstone in Zion National Park, Utah.

As the Mesozoic Era ended, the Cretaceous Interior Seaway withdrew from the craton. During this regression, marine waters retreated to the north and south, and marginal marine and continental deposition formed widespread coal-bearing deposits on the coastal plain.

WHAT ROLE DID ACCRETION OF TERRANES PLAY IN THE GROWTH OF WESTERN NORTH AMERICA?

We have discussed orogenies along convergent plate boundaries resulting in continental accretion. Much of the material accreted to continents during such events is simply eroded older continental crust, but a significant amount of new material is added to continents as well—igneous rocks that formed as a consequence of subduction and partial melting, for example. Although subduction is the predominant influence on the tectonic history in many regions of orogenesis, other processes are also involved in mountain building and continental accretion, especially the accretion of terranes.

Geologists now know that portions of many mountain systems are composed of small accreted lithospheric blocks that are clearly of foreign origin. These *terranes* differ completely in their fossil content, stratigraphy, structural trends, and paleomagnetic properties from the rocks of the surrounding mountain system and adjacent craton. In fact, these terranes are so different from adjacent rocks that most geologists think they formed elsewhere and were carried great distances as parts of other

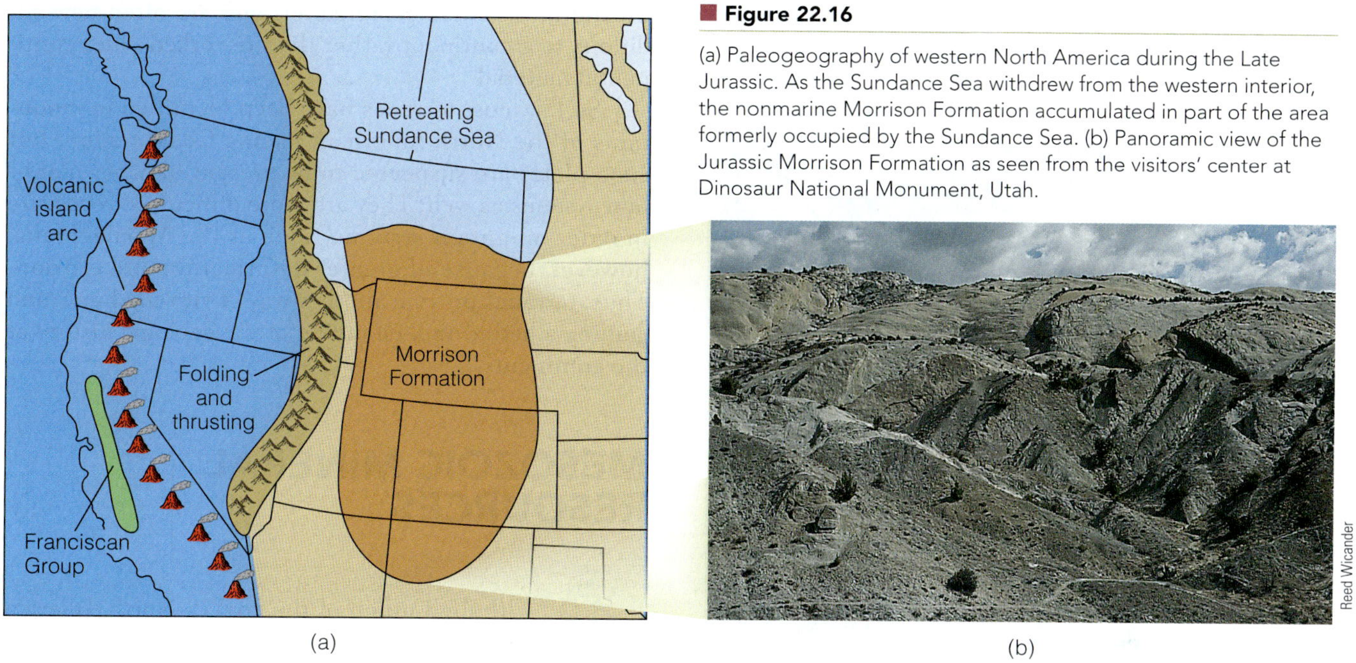

Figure 22.16

(a) Paleogeography of western North America during the Late Jurassic. As the Sundance Sea withdrew from the western interior, the nonmarine Morrison Formation accumulated in part of the area formerly occupied by the Sundance Sea. (b) Panoramic view of the Jurassic Morrison Formation as seen from the visitors' center at Dinosaur National Monument, Utah.

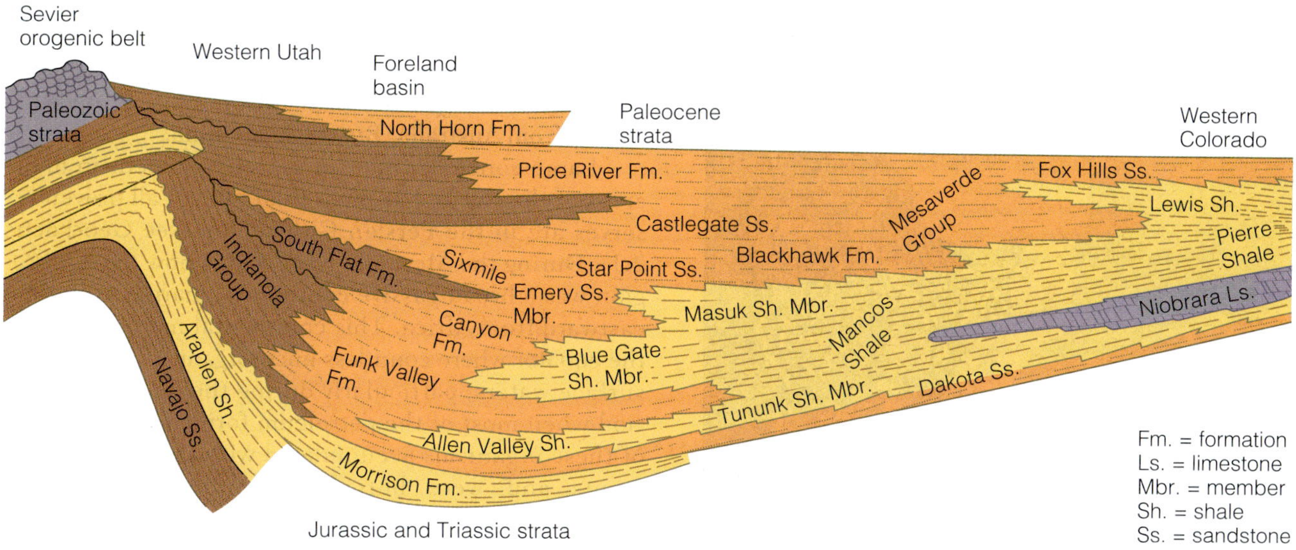

Figure 22.17

This restored west–east cross section of Cretaceous facies of the western Cretaceous Interior Seaway shows their relationship to the Sevier orogenic belt.

plates until they collided with other terranes or continents.

Geologic evidence indicates that more than 25% of the entire Pacific Coast from Alaska to Baja California consists of accreted terranes. The accreting terranes are composed of volcanic island arcs, oceanic ridges, seamounts, volcanic plateaus, hot-spot tracks, and small fragments of continents that were scraped off and accreted to the continent's margin as the oceanic plate on which they were carried was subducted under the continent. It is estimated that more than 100 different-sized terranes have been added to the western margin of North America during the last 200 million years (■ Figure 22.18). A good example of this are the Wrangellian terranes (Figure 22.1b) that have been accreted to North America's western margin (Figure 22.18).

The basic plate tectonic reconstruction of orogenies and continental accretion remains unchanged, but the details of such reconstructions are decidedly different in view of terrane tectonics. For example, growth along active continental margins is faster than along passive continental margins because of the accretion of terranes.

■ **Figure 22.18**

Some of the accreted lithospheric blocks called terranes that form the western margin of the North American Craton. The dark brown blocks probably originated as terranes and were accreted to North America. The light green blocks are possibly displaced parts of North America. The North American craton is shown in dark green. See Figure 22.1b for position of Wrangellian terranes during the Jurassic. Source: Reproduced with permission from "The Movement of Continents," by Zvi Ben-Avraham, American Scientist, Vol. 69, No. 3, p. 298 (Figure 9). Copyright © 1981 American Scientist.

Furthermore, these accreted terranes are often new additions to a continent rather than reworked older continental material.

So far, most terranes have been identified in mountains of the North American Pacific Coast region, but such plates are suspected to be present in other mountain systems as well. They are more difficult to recognize in older mountain systems, such as the Appalachians, however, because of greater deformation and erosion. Thus, terranes provide another way of viewing Earth and gaining a better understanding of the geologic history of the continents.

MESOZOIC MINERAL RESOURCES

Although much of the coal in North America is Pennsylvanian or Cenozoic in age, important Mesozoic coals occur in the Rocky Mountain states. These are mostly lignite and bituminous coals, but some local anthracites are present as well. Particularly widespread in western North America are coals of Cretaceous age. Mesozoic coals are also known from Alberta and British Columbia, Canada, as well as from Australia, Russia, and China.

Large concentrations of petroleum occur in many areas of the world, but more than 50% of all proven reserves are in the Persian Gulf region (see Geo-Focus 2.1). During the Mesozoic Era, what is now the Gulf region was a broad passive continental margin extending eastward from Africa. This continental margin lay near the equator where countless microorganisms lived in the surface waters, particularly during the Cretaceous Period when most of the petroleum formed. The remains of these organisms accumulated with the bottom sediments and were buried, beginning the complex processes of petroleum generation.

Similar conditions existed in what is now the Gulf Coast region of the United States and Central America. Here, petroleum and natural gas also formed on a broad shelf over which transgressions and regressions occurred. In this region, the hydrocarbons are largely in reservoir rocks that were deposited as distributary channels on deltas and as barrier-island and beach sands. Some of these hydrocarbons are associated with structures formed adjacent to rising salt domes. The salt, called the *Louann Salt*, initially formed in a long, narrow sea when North America separated from Europe and North Africa during the fragmentation of Pangaea (Figure 22.2).

The richest uranium ores in the United States are widespread in Mesozoic rocks of the Colorado Plateau area of Colorado and adjoining parts of Wyoming, Utah, Arizona, and New Mexico.

As noted in Chapter 19, Proterozoic banded iron formations are the main sources of iron ores. There are, however, some important exceptions. For example, the Jurassic-age "Minette" iron ores of Western Europe, composed of oolitic limonite and hematite, are important ores in France, Germany, Belgium, and Luxembourg. In Great Britain, low-grade iron ores of Jurassic age consist of oolitic siderite, which is an iron carbonate. And in Spain, Cretaceous rocks are the host rocks for iron minerals.

South Africa, the world's leading producer of gem-quality diamonds and among the leaders in industrial diamond production, mines these minerals from kimberlite pipes, conical igneous intrusions of dark gray or blue igneous rock. Diamonds, which form at great depth where pressure and temperature are high, are brought to the surface during the explosive volcanism that forms kimberlite pipes. Although kimberlite pipes have formed throughout geologic time, the most intense episode of such activity in South Africa and adjacent countries was during the Cretaceous Period. Triassic and Jurassic diamond-bearing kimberlites were also emplaced in Siberia.

In the Introduction we noted that the mother lode or source for the placer deposits mined during the California gold rush is in Jurassic-age intrusive rocks of the Sierra Nevada. Gold placers are also known in Cretaceous-age conglomerates of the Klamath Mountains of California and Oregon.

Porphyry copper was originally named for copper deposits in the western United States mined from porphyritic granodiorite, but the term now applies to large, low-grade copper deposits disseminated in a variety of rocks. These porphyry copper deposits are an excellent example of the relationship between convergent plate boundaries and the distribution, concentration, and exploitation of valuable metallic ores. Magma generated by partial melting of a subducting plate rises toward the surface, and as it cools, it precipitates and concentrates various metallic ores. The world's largest copper deposits were formed during the Mesozoic and Cenozoic in a belt along the western margins of North and South America (see Figure 2.29).

LIFE OF THE MESOZOIC ERA

Fossils and restorations of prehistoric floras and faunas—especially Mesozoic dinosaurs, marine reptiles, and flying reptiles—fascinate many people. Indeed, these animals have been popularized in books, television specials, and movies, including *Jurassic Park* (1993) and its two sequels, *The Lost World* (1997) and *Jurassic Park III* (2001), as well as *Dinosaur* (2000). Part of the popularity of these creatures is no doubt related to the tremendous increase in knowledge that scientists have gained about the Mesozoic biota during the last few decades.

The Mesozoic Era is commonly called the "Age of Reptiles," calling attention to the fact that reptiles predominated among land-dwelling vertebrate animals. Certainly the Mesozoic diversification of reptiles was an important event in life history, but other equally important, though not as well known, events also took place. Mammals, for instance, made their appearance during the Triassic, having evolved from mammal-like reptiles, and birds evolved probably from small carnivorous dinosaurs by Jurassic time. And, of course, there were a variety of reptiles in addition to dinosaurs—flying reptiles and marine reptiles being two of the most notable.

Important changes took place among plants when the first flowering plants (angiosperms) evolved and soon became the most common plants on land. Even though the major land plants from the Paleozoic Era persisted, and many still exist, they now make up less than 10% of all land plants. Marine invertebrates such as clams, snails, and cephalopods made a notable resurgence following the Permian mass extinctions.

Although Pangaea began fragmenting during the Triassic (Figure 22.2), the proximity of continents and mild Mesozoic climates made it possible for plants and animals to occupy extensive geographic ranges. But as the fragmentation continued, some continents became isolated, especially Australia and South America, and their faunas evolving in isolation became increasingly different from those elsewhere. Indeed, Australia has been a separate landmass since the Mesozoic and even now has a fauna unlike that of any other continent.

Marine Invertebrates and Phytoplankton

The Mesozoic was a time when marine invertebrates repopulated the seas. The Early Triassic invertebrate fauna was not very diverse, but by the Late Triassic the seas were once again swarming with invertebrates from planktonic foraminifera to cephalopods. The brachiopods that had been so abundant during the Paleozoic never completely recovered from their near extinction, and although they still exist, the bivalves have largely taken over their ecologic niche.

Mollusks such as cephalopods, bivalves, and gastropods were the most important members of the Mesozoic marine invertebrate fauna. Their rapid evolution and the fact that many cephalopods were nektonic make them excellent guide fossils (■ Figure 22.19a). The Ammonoidea, cephalopods with wrinkled sutures, constitute three groups: the goniatites, ceratites, and ammonites. The latter, though present during the entire Mesozoic, were most prolific during the Jurassic and Cretaceous.

Figure 22.19

(a) Cephalopods such as the Late Cretaceous ammonoids *Baculites* (foreground) and *Helioceros* (background) were important predators, and they are excellent guide fossils. (b) Bivalves, represented by two Cretaceous forms, were particularly diverse and abundant during the Mesozoic. (c) Two genera of reef-building bivalves called *rudists*. Rudists replaced corals as the main reef-building animals of the Mesozoic.

Most ammonites were coiled, some attaining diameters of 2 m, whereas others were uncoiled and led a near benthonic existence (Figure 22.19a). Ammonites became extinct at the end of the Cretaceous, but two related groups of cephalopods survived into the Cenozoic: the *nautiloids*, including the living pearly nautilus, and the *coleoids*, represented by the extinct belemnoids, which are good Jurassic and Cretaceous guide fossils, as well as by the living squid and octopus.

Mesozoic bivalves diversified to inhabit many epifaunal and infaunal niches. Oysters and clams (epifaunal suspension feeders) became particularly diverse and abundant, and despite a reduction in diversity at the end of the Cretaceous, remain important animals in the marine fauna today (Figure 22.19b). As is true now, where shallow marine waters were warm and clear, coral reefs proliferated. An important reef builder throughout the Mesozoic was a group of bivalves known as *rudists* (Figure 22.19c). Rudists are important because they displaced corals as the main reef builders during the later Mesozoic and are excellent guide fossils for the Late Jurassic and Cretaceous.

A new and familiar type of coral also appeared during the Triassic, the *scleractinians*. Whether scleractinians evolved from rugose corals or from an as yet unknown soft-bodied ancestor with no known fossil record is still unresolved. In addition, another invertebrate group that prospered during the Mesozoic was the echinoids. Echinoids were exclusively epifaunal during the Paleozoic but branched out into the infaunal habitat during the Mesozoic.

A major difference between Paleozoic and Mesozoic marine invertebrate faunas was the increased abundance and diversity of burrowing organisms. With few exceptions, Paleozoic burrowers were soft-bodied animals such as worms. The bivalves and echinoids, which were epifaunal animals during the Paleozoic, evolved various means of entering infaunal habitats. This trend toward an infaunal existence may have been an adaptive response to increasing predation from the rapidly evolving

■ **Figure 22.20**

(a) A Miocene coccolith from the Gulf of Mexico (left); a Pliocene–Miocene coccolith from the Gulf of Mexico (right). (b) Upper Miocene diatoms from Java (left and right). (c) Eocene dinoflagellates from Alabama (left) and the Gulf of Mexico (right). Source: (a) Merton E. Hill, (b) John Barron, USGS, (c) John H. Wrenn, Louisiana State University.

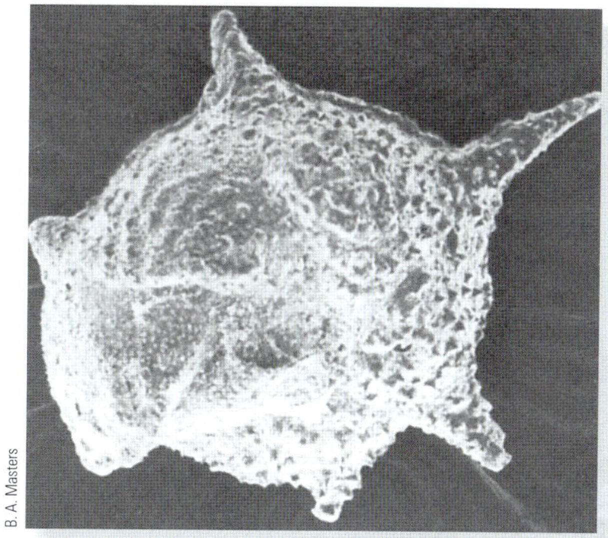

■ **Figure 22.21**

Planktonic foraminifera, represented by *Globotruncana calcarata* from the Cretaceous Pecan Gap Chalk of Texas, became diverse during the Jurassic and Cretaceous. Many died out at the end of the Cretaceous.

fish and cephalopods. Bivalves, for instance, expanded into the infaunal niche during the Mesozoic, and by burrowing they escaped predators.

The primary producers in the Mesozoic seas were various types of microorganisms. *Coccolithophores* are an important group of phytoplankton (■ Figure 22.20a) that first evolved during the Jurassic and became extremely common during the Cretaceous. *Diatoms* (Figure 22.20b), which build skeletons of silica, made their appearance during the Cretaceous, but they are more important as primary producers during the Cenozoic. Diatoms are presently most abundant in cooler oceanic waters, and some species inhabit freshwater lakes. *Dinoflagellates* were common during the Mesozoic and today are the major primary producers in warm water (Figure 22.20c).

The foraminifera (single-celled consumers) diversified rapidly during the Jurassic and Cretaceous and continued to be diverse and abundant to the present (■ Figure 22.21). The planktonic forms in particular diversified rapidly, but most genera became extinct at the end of the Cretaceous.

In general terms, we can think of the Mesozoic as a time of increasing complexity among the marine invertebrate fauna. At the beginning of the Triassic, diversity was low and food chains were short. Near the end of the Cretaceous, though, the marine invertebrate fauna was highly complex, with interrelated food chains. This evolutionary history reflects changing geologic conditions influenced by plate tectonic activity, discussed in Chapter 2.

Plants—Primary Producers on Land

Plants practice photosynthesis and thus lie at the base of the food chain on land, so we discuss them before considering land-dwelling animals. Just as during the Late Paleozoic, seedless vascular plants and gymnosperms dominated Triassic and Jurassic land-plant communities, and, in fact, representatives of both groups are still common. Among the gymnosperms, the large seed ferns became extinct by the end of the Triassic, but *ginkgos* remained abundant and still exist in isolated regions, and *conifers* continued to diversify and are now widespread in some terrestrial habitats, particularly at high elevations and high latitudes. A new group of gymnosperms known as *cycads* made its appearance during the Triassic. These palmlike plants became widespread and now exist in tropical and semitropical areas.

The long dominance of seedless plants and gymnosperms ended during the Early Cretaceous, perhaps the Late Jurassic, when many were replaced by **angiosperms,** or flowering plants (■ Figure 22.22). In 1998, scientists described the oldest known flowering plant from Early Cretaceous–aged rocks in China, and they report the root systems of probable flowering plants from Jurassic rocks.

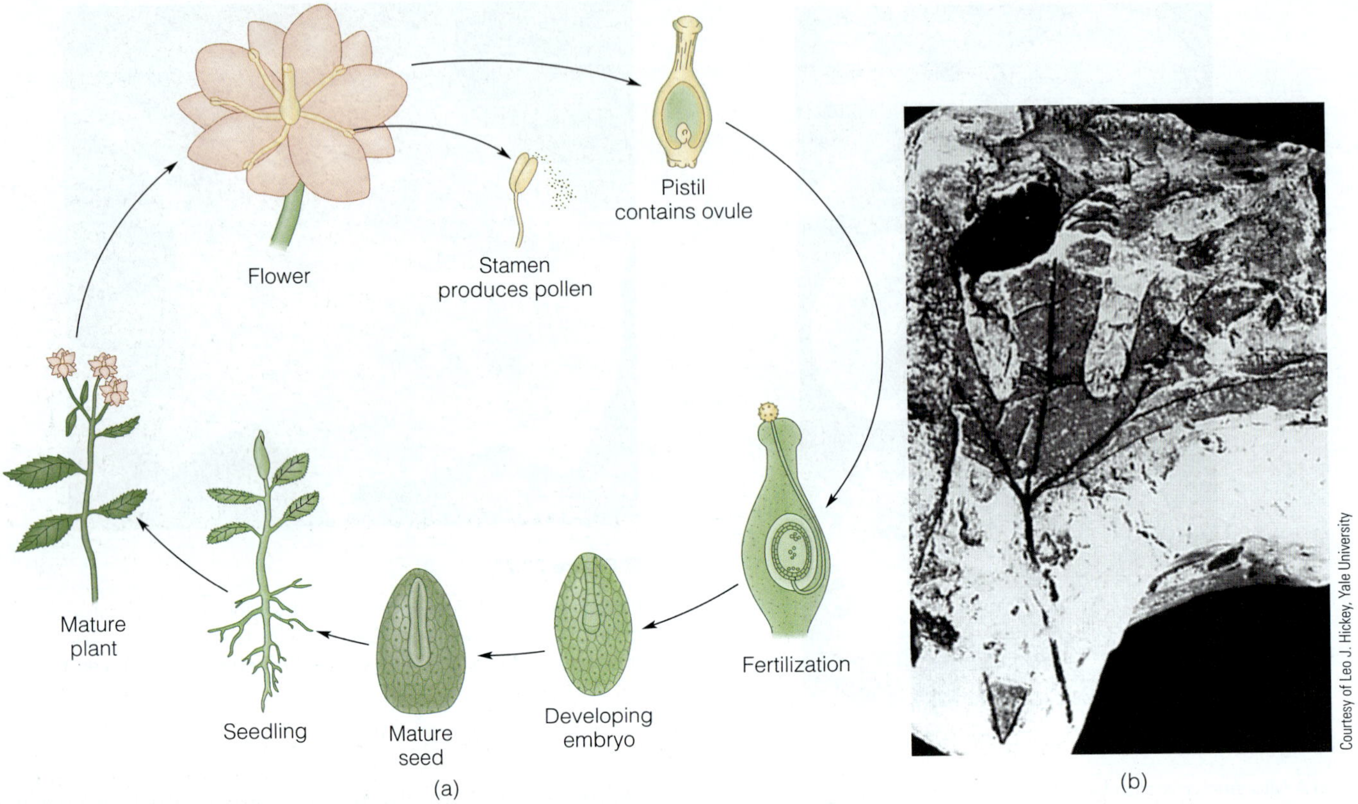

Figure 22.22

(a) The reproductive cycle in angiosperms. (b) Early Cretaceous angiosperm from Cecil County, Maryland.

Angiosperms probably evolved from specialized gymnosperms. Indeed, recent studies have identified both fossil and living gymnosperms that show close relationships to angiosperms. In any case, since they first evolved, angiosperms have adapted to nearly every terrestrial habitat from mountains to deserts, and some have even adapted to shallow coastal waters. Several factors account for their phenomenal success, but chief among them is their method of reproduction (Figure 22.22a). Two developments were particularly important: the evolution of flowers, which attract animal pollinators, especially insects; and the evolution of enclosed seeds.

Seedless vascular plants and gymnosperms are important and still flourish in many environments; in fact, many botanists regard ferns and conifers as emerging groups. Nevertheless, a measure of the angiosperms' success is that today with 250,000 to 300,000 species they account for more than 90% of all land-plant species, and they occupy some habitats in which other land plants do poorly or cannot exist.

The Diversification of Reptiles

Reptile diversification began during the Mississippian Period with the evolution of the protorothyrids, apparently the first animals to lay amniotic eggs (see Chapter 21). From this basic stock of so-called *stem reptiles,* all other reptiles as well as birds and mammals evolved (■ Figure 22.23). Recall from Chapter 21 that pelycosaurs were the dominant land vertebrates of the Pennsylvania and Permian.

Certainly reptiles diversified markedly during the Mesozoic Era, but we must ask some fundamental questions about reptiles in general and dinosaurs in particular. Several skeletal modifications and the fact that reptiles lay amniotic eggs clearly differentiate them from amphibians (see Chapter 21). Living reptiles such as lizards, snakes, crocodiles, and turtles lay amniotic eggs; all are cold-blooded, have a tough, scaly skin, and practice internal fertilization. Also, with the exception of turtles, all living reptiles and dinosaurs have two openings on the side of the skull in the temporal region. So dinosaurs have traditionally been classed as reptiles, but they also possessed several characteristics that set them apart.

Dinosaurs had teeth set in individual sockets, a reduced lower leg bone (fibula), a pelvis anchored to the vertebral column by three or more vertebrae, a ball-like head on their upper leg bone (femur), and elongate bones in the palate. In short, they differed from other reptiles in important ways. Another significant feature is that dinosaurs had a fully upright posture with their limbs directly beneath their bodies, rather than the

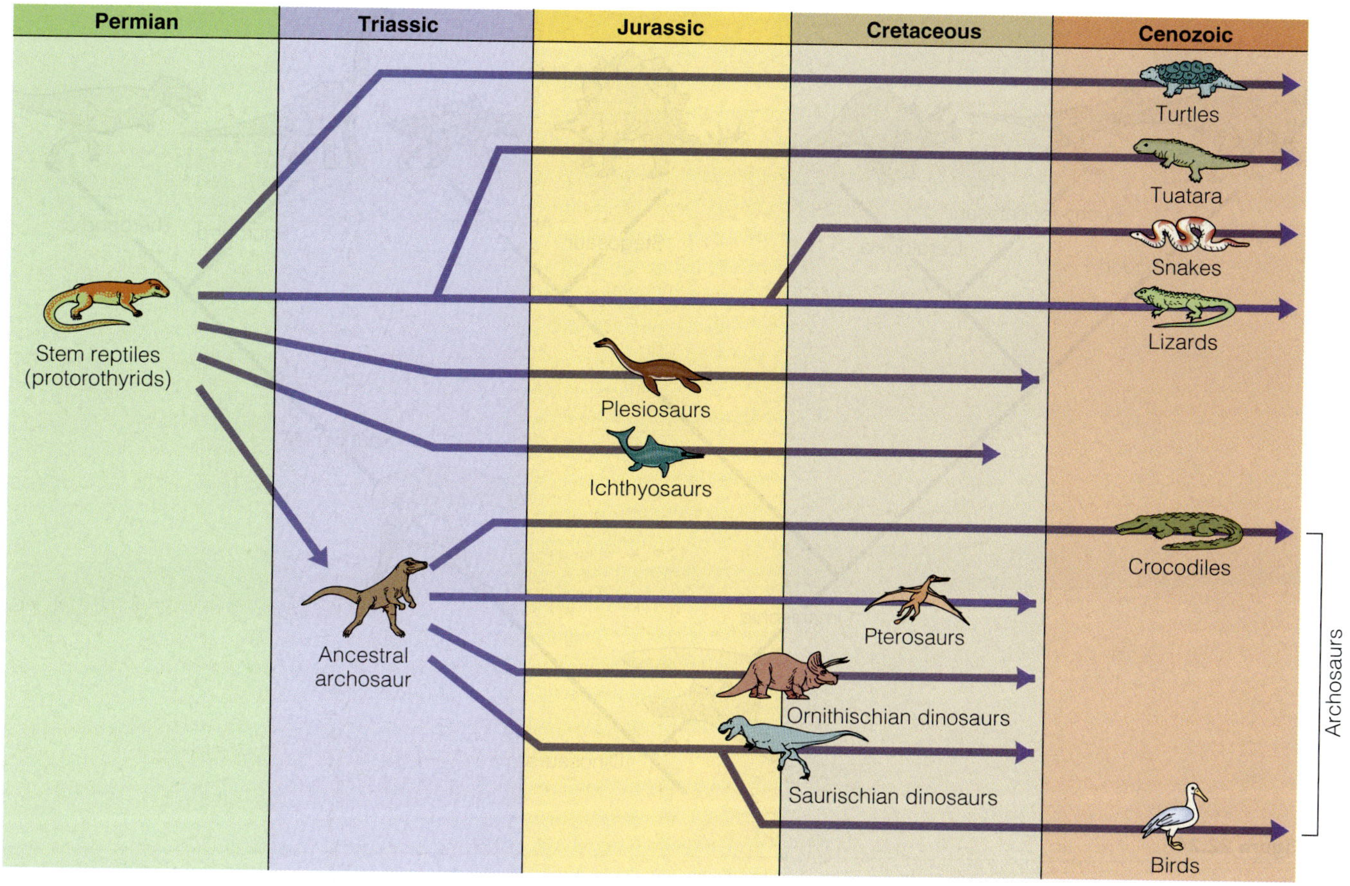

Figure 22.23

Relationships among fossil and living reptiles and birds.

sprawling stance of other reptiles (Figure 22.24). In fact, their upright posture coupled with modifications in their limbs accounts for more efficient locomotion than in other reptiles; this may have been one factor in their incredible success.

Contrary to popular belief, there were no flying dinosaurs or fully aquatic ones, although other reptiles occupied these niches. Nor were all dinosaurs large, although some certainly were. Note that dinosaurs lived only during the Mesozoic Era, unless we consider their evolutionary descendants, the birds.

Archosaurs and the Origin of Dinosaurs Reptiles known as **archosaurs** (*archo* meaning "ruling," and *sauros* meaning "lizard") include crocodiles, pterosaurs (flying reptiles), dinosaurs, and birds. Including such diverse animals in a single group implies that they share a common ancestor, and indeed they possess several characteristics that unite them. For instance, all have teeth set in individual sockets, except today's birds, but even the earliest birds had this feature. We now turn to a discussion of dinosaurs, but we will have more to say about pterosaurs and birds later in this chapter.

All dinosaurs share a number of characteristics yet differ enough for us to recognize two distinct orders: the **Saurischia** and **Ornithischia.** Each order has a distinctive pelvic structure: saurischian dinosaurs have a lizardlike pelvis and are thus called lizard-hipped dinosaurs; ornithischians have a birdlike pelvis and are called bird-hipped dinosaurs (Figure 22.24). For decades, paleontologists thought each order evolved independently during the Late Triassic, but it is now clear that they had a common ancestor much like archosaurs known from Middle Triassic rocks in Argentina (see chapter opening photo). These dinosaur ancestors were small (less than 1 m long), long-legged carnivores that walked and ran on their hind limbs, so they were **bipedal,** as opposed to **quadrupedal** animals that move on all four limbs.

Dinosaurs Sir Richard Owen proposed the term *dinosaur* in 1842 to mean "fearfully great lizard," although now "fearfully" has come to mean "terrible," thus the characterization of dinosaurs as "terrible lizards." But of course they were not terrible, or at least no more terrible than animals living today, and they were not lizards. Nevertheless, dinosaurs more than any other

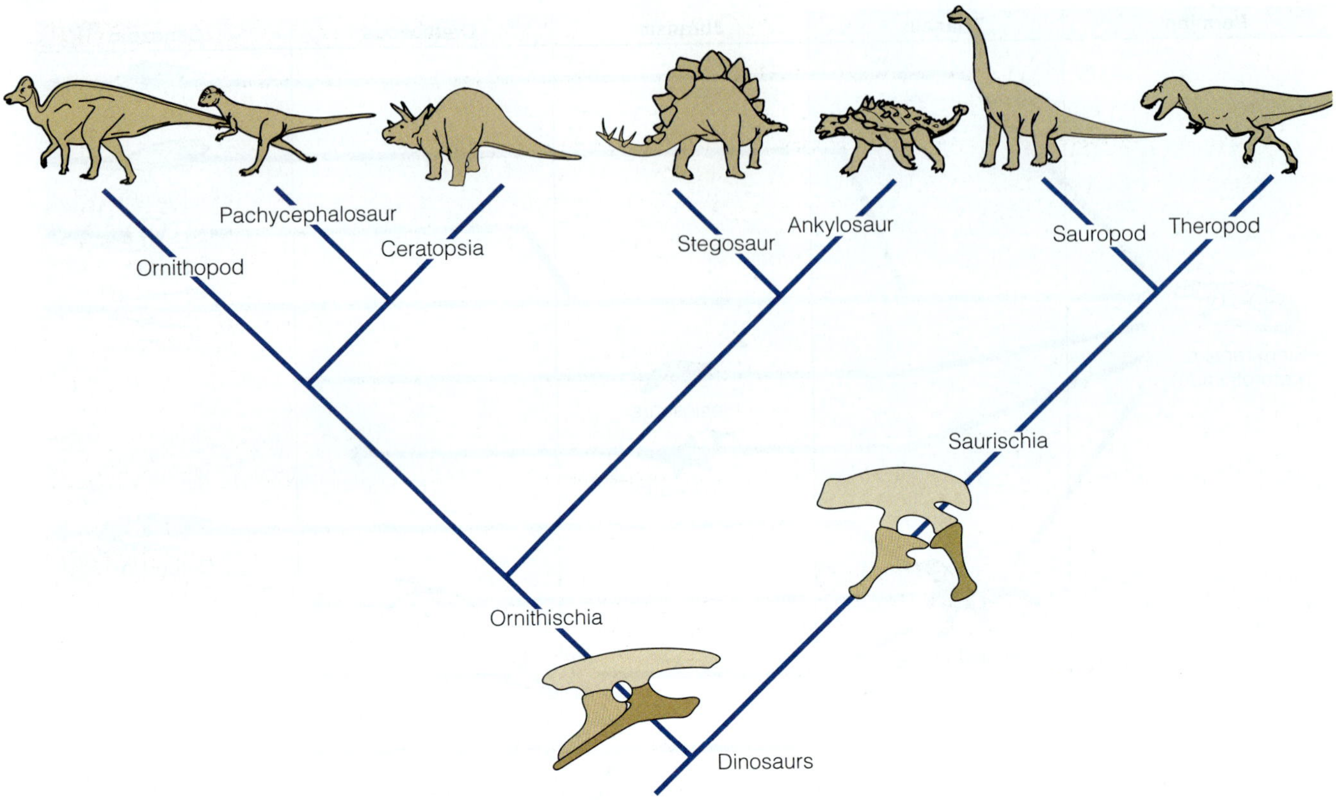

Figure 22.24

Cladogram showing dinosaur relationships. Pelvises of ornithischian and saurischian dinosaurs are shown for comparison. Among several subgroups of dinosaurs, theropods were carnivores and the others were herbivores.

animal have inspired awe and thoroughly captured the public imagination. Unfortunately, their popularization in cartoons, books, and movies has commonly been inaccurate and has contributed to misunderstandings. For instance, many people think that all dinosaurs were large, and because they are extinct they must have been poorly adapted. It is true that many were large—indeed, the largest animals ever to live on land. But not all were large. In fact, dinosaurs varied from giants weighing several tens of metric tons to those no larger than a chicken. And to consider them poorly adapted is to ignore the fact that dinosaurs were extremely diverse and widespread for more than 140 million years!

Although various media now portray dinosaurs as more active animals, the misconception that they were lethargic dim-witted beasts persists. Evidence now available indicates that some were quite active and perhaps even warm-blooded. It also appears that some species cared for their young long after hatching, a behavior more characteristic of birds and mammals. Although many questions remain unanswered about dinosaurs, their fossils and the rocks containing them are revealing more and more about their evolutionary relationships and behavior.

We noted that scientists recognize two distinct orders of dinosaurs: the Saurischia and Ornithischia (Figure 22.24 and Table 22.1). Among the saurischians, two suborders are defined: theropods and sauropods. All *theropods* were bipedal carnivores that ranged in size from tiny *Compsognathus* to comparative giants such as *Tyrannosaurus* and similar but even larger species from Africa and Argentina (see "Dinosaurs" on pages 662 and 663). The movie *Jurassic Park* and its sequel *The Lost World* popularized some of the smaller theropods such as *Velociraptor* with a large sickle-shaped claw on each hind foot. *Velociraptor* and its larger relative *Deinonychus* likely used these claws in a slashing type of attack. Some remarkable discoveries beginning in 1996 by Chinese paleontologists have yielded several species of small theropods with feathers. The significance of these fossils is discussed in later sections.

Included among the sauropods are the giant quadrupedal herbivores such as *Apatosaurus, Diplodocus,* and *Brachiosaurus,* the largest land animals of any kind (Table 22.1). According to one estimate, *Brachiosaurus*, a giant even by sauropod standards, weighed more than 75 metric tons! Partial remains discovered in Colorado, New Mexico, and elsewhere indicate that even larger

Table 22.1

Orders and Suborders of Dinosaurs

Order	Suborder	Familiar Genera	Comments*
Saurischia	Theropoda	*Allosaurus, Coelophysis, Compsognathus, Deinonychus, Tyrannosaurus,*** *Velociraptor*	Bipedal carnivores. Late Triassic to end of Cretaceous. Size from 0.6 to 15 m long, 2 or 3 kg to 7.3 metric tons. Some smaller genera may have hunted in packs.
	Sauropoda	*Apatosaurus, Brachiosaurus, Camarasaurus, Diplodocus, Titanosaurus*	Giant quadrupedal herbivores. Late Triassic to Cretaceous, but most common during Jurassic. Size up to 27 m long, 75 metric tons.† Trackways indicate sauropods lived in herds. Preceded in fossil record by the smaller prosauropods.
Ornithischia	Ornithopoda	*Anatosaurus, Camptosaurus, Hypsilophodon, Iguanodon, Parasaurolophus*	Some ornithopods, such as *Anatosaurus*, had a flattened bill-like mouth and are called duck-billed dinosaurs. Size from a few meters up to 13 m long and 3.6 metric tons. Especially diverse and common during the Cretaceous. Primarily bipedal herbivores but could also walk on all fours.
	Pachycephalosauria	*Stegoceras*	*Stegoceras* only 2 m long and 55 kg, but larger species known. Thick bones of skull cap might have been for butting contests for dominance and mates. Bipedal herbivores of Cretaceous.
	Ankylosauria	*Ankylosaurus*	*Ankylosaurus* more than 7 m long and about 2.5 metric tons. Heavily armored with bony plates on top of head, back, and sides. Quadrupedal herbivore.
	Stegosauria	*Stegosaurus*	A variety of stegosaurs are known, but *Stegosaurus* with bony plates on its back and a spiked tail is best known. Plates probably were for absorbing and dissipating heat. Quadrupedal herbivores that were most common during the Jurassic. *Stegosaurus* 9 m long, 1.8 metric tons.
	Ceratopsia	*Triceratops*	Numerous genera known. Some early ones were bipedal, but later large animals were quadrupedal herbivores. Much variation in size; *Triceratops* to 7.6 m long and 5.4 metric tons, with large bony frill over top of neck, three horns on skull, and beaklike mouth. Especially common during the Cretaceous.

* Lengths and weights are approximate from several sources.

**Until recently *Tyrannosaurus* at 4.5 metric tons was the largest known theropod, but now similar, larger animals are known from Argentina and Africa.

† Partial remains indicate even larger brachiosaurs existed, perhaps measuring 30 m long and weighing 135 metric tons.

sauropods existed, but just how large they were is unresolved. The sauropods were preceded in the fossil record by the smaller *prosauropods*.

Theropods and sauropods (saurischians) are no doubt the most familiar dinosaurs, but they were not nearly as diverse as the ornithischians. Dozens of species of ornithischians are included in five distinct suborders: ornithopods, pachycephalosaurs, ankylosaurs, stegosaurs, and ceratopsians (Table 22.1, see "Dinosaurs" on pages 662 and 663). All were herbivores and many were quadrupedal. Pachycephalosaurs, however, were bipeds, and many ornithopods were primarily bipedal but their well-developed forelimbs enabled them to walk on all fours as well.

Only about a dozen genera of pachycephalosaurs are known, all of which lived during the Cretaceous. Their most distinctive feature is thickened skull caps that may have been used in butting contests between males. Paleontologists have described at least 80 genera of ornithopods, which include the well-known duck-billed dinosaurs, some with elaborate head crests.

Dinosaurs

Dinosaurs are without a doubt the best known extinct animals. Hundreds of dinosaur species existed during the Late Triassic through the Cretaceous. Paleontologists now think that all dinosaurs descended from a common ancestor.

Sue Monroe

This skull of *Tyrannosaurus rex* is more than 1 m long. At about 12 m long and weighing 3 to 5 metric tons, *T. rex* was the largest theropod in North America.

Sue Monroe

Lifelike restoration of the Early Cretaceous 3-m-long theropod *Deinonychus* (terrible claw) in its probable attack posture. Notice the huge curved claws on the back feet and the well-developed forelimbs with sharp claws.

James Gurney/NGS Image Collection

Gigantosaurus was a huge theropod that lived during the Late Cretaceous in Argentina and Brazil. It was nearly 13 m long, weighed about 7.3 metric tons, and lived 25 million years before the more familiar *Tyrannosaurus rex*.

Carlyn Iverson

Scene from the Late Triassic showing *Brachiosaurus*, a giant even by dinosaur standards. It was about 25 m long and weighed perhaps 75 metric tons. Partial remains indicate that even larger sauropods existed. *Stegosaurus* in the foreground is noted for the rows of plates along its back and bony spikes at the end of its tail.

Eleanor M. Kish

This 6-m-long ankylosaur known as *Sauropelta* lived in North America during the Early Cretaceous. Some ankylosaurs had a large bony club at the end of the tail.

Skeleton of the rhinoceros-sized ceratopsian dinosaur *Triceratops*. It and other ceratopsian genera were quite common in North America during the Late Cretaceous.

Sue Monroe

Restoration of the Early Cretaceous feathered dinosaur *Caudipteryx*. This turkey-sized animal along with several other feathered dinosaurs comes from the Liaoning Province of northeastern China.

Archive Photos/Getty Images

During the Late Cretaceous, numerous ornithopods known as *Maiasaura* nested in colonies in northern Montana and tended their young long after hatching. In this scene a female leads her young to a feeding area.

Douglas Henderson

Carlyn Iverson

Some ornithopods had head crests that might have functioned as chambers to amplify bellowing. This scene from the Late Cretaceous shows *Corythosaurus* (back left) and *Parasaurolophus* (foreground).

Two Late Cretaceous pachycephalosaurs, *Stegoceras* (foreground) and *Pachycephalosaurus*. Notice the dome-shaped skulls that resulted from thickening of the bones. Perhaps these were used in butting contests.

Carlyn Iverson

The ankylosaurs were quadrupedal herbivores that relied on their heavy armor for protection from predators. Several stegosaurs are known, but *Stegosaurus* with bony plates on its back and a spiked tail is the most familiar. Although paleontologists debate the precise arrangement of the plates, no one doubts that the spiked tail was used for defense. Some of the earliest ceratopsians were bipeds, but all of the later ones were quadrupedal. Many of them had a large bony frill covering the top of the neck and a large horn or horn projecting from the skull.

Warm-Blooded Dinosaurs? Were dinosaurs **endotherms** (warm-blooded) like today's mammals and birds, or were they **ectotherms** (cold-blooded) like all of today's reptiles? Almost everyone now agrees some compelling evidence exists for dinosaur endothermy, but opinion is still divided among (1) those holding that all dinosaurs were endotherms, (2) those who think only some were endotherms, and (3) those proposing that dinosaur metabolism, and thus the ability to regulate body temperature, changed as they matured.

Bones of endotherms typically have numerous passageways that, when the animals are alive, contain blood vessels, but the bones of ectotherms have considerably fewer passageways. Proponents of dinosaur endothermy note that dinosaur bones are more similar to bones of living endotherms. Crocodiles and turtles have this so-called endothermic bone, but they are ectotherms, and some small mammals have bone more typical of ectotherms. Perhaps bone structure is related more to body size and growth patterns than to endothermy, so this evidence is not conclusive.

Endotherms must eat more than comparably sized ectotherms because their metabolic rates are so much higher. Consequently, endothermic predators require large prey populations and thus constitute a much smaller proportion of the total animal population than their prey, usually only a few percent. In contrast, the proportion of ectothermic predators to prey may be as high as 50%. Where data are sufficient to allow an estimate, dinosaur predators made up 3% to 5% of the total population. Nevertheless, uncertainties in the data make this argument less than convincing for many paleontologists.

A large brain in comparison to body size requires a rather constant body temperature and thus implies endothermy. And some dinosaurs were indeed brainy, especially the small- and medium-sized theropods, so brain size might be a convincing argument for these dinosaurs. Even more compelling evidence for theropod endothermy comes from their probable relationship to birds and from the rather recent discoveries in China of dinosaurs with feathers or a featherlike covering. Today, only endotherms have hair, fur, or feathers for insulation.

Some scientists point out that certain duck-billed dinosaurs grew and reached maturity much more quickly than would be expected for ectotherms and conclude that they must have been warm-blooded. Furthermore, a fossil ornithopod discovered in 1993 has a preserved four-chambered heart much like that of living mammals and birds. Three-dimensional imaging of this structure, now on display at the North Carolina Museum of Natural Sciences, has convinced many scientists that this animal was an endotherm.

Good arguments for endothermy exist for several types of dinosaurs, although the large sauropods were probably not endothermic but nevertheless were capable of maintaining a rather constant body temperature. Large animals heat up and cool down more slowly than smaller ones because they have a small surface area compared to their volume. With their comparatively smaller surface area for heat loss, sauropods probably retained heat more effectively than their smaller relatives.

In general, a fairly good case can be made for endothermy in many theropods and in some ornithopods. Nevertheless, disagreement exists, and for some dinosaurs the question is still open.

Dinosaur Behavior Various media typically depict dinosaurs as aggressive, dangerous beasts, but we have no reason to think they were any more aggressive or dangerous than land animals living now. Indeed, carnivorous dinosaurs, like today's predators, probably went for the easy kill—the young, old, or disabled—and avoided large dangerous prey, and the larger ones likely chased smaller predators away from their kill. No one doubts that carnivores dined on carrion if available and that they actively hunted, but did they hunt in packs as depicted in *Jurassic Park* and its sequels? Evidence for pack behavior is difficult to come by, but it does seem that diminutive *Coelophysis* and medium-sized *Deinonychus* probably were pack hunters.

Two kinds of evidence indicate that many herbivorous dinosaurs were gregarious, congregating in vast herds. One is fossil trackways showing that many dinosaurs of the same type moved together—various species of sauropods, for instance. Even more compelling evidence comes from *bone beds* in which only one species is found, indicating that large numbers of animals perished quickly from some kind of catastrophe. A bone bed in Canada with hundreds of ceratopsians probably formed after numerous animals died during a river crossing, and the estimated 10,000 hadrosaurs in northern Montana were no doubt killed when they were overwhelmed by volcanic gases or ash.

Many paleontologists are convinced that the crests of ornithopods were chambers that amplified bellowing, and horns, frills, and spikes in ceratopsians may have functioned in species recognition or displays to attract mates or establish dominance. Certainly bellowing, grunting, and posturing were used in defense and territorial disputes. Male pachycephalosaurs with thickened skull caps likely butted heads much like bighorn sheep do today.

Excavations in Montana reveal that three species of Late Cretaceous dinosaurs nested in colonies and used the same nesting area repeatedly just as some birds do today. One of these, *Maiasaura* ("good mother dinosaur"), laid eggs in nests spaced about 7 m apart, or about the length of an adult. Juveniles up to 1 m long were found in some nests, which is much longer than they were when they hatched. So the young must have remained in the nest area for some time, during which adults fed and protected them (see "Dinosaurs" on pages 662 and 663).

A 2.4-m-long ostrichlike dinosaur known as *Oviraptor* found in 1993 has its forelimbs encircling about 15 eggs and its back legs folded under its body. Whether this individual was incubating the eggs or protecting them from a predator cannot be resolved, but in either case it indicates a behavior we commonly attribute to birds and mammals. In 1998, paleontologists found a nesting ground in Argentina where thousands of Late Cretaceous titanosaurs (sauropods) built nests on a floodplain. Many of the softball-sized eggs were found with unhatched embryos, but it seems these dinosaurs simply laid their eggs and then abandoned them.

Dinosaurs as sluggish, dim-witted beasts—the old view—have been replaced in recent decades by a new view in which these animals were more active and exhibited behaviors more like those of mammals and birds. As far as intellect is concerned, some dinosaurs, particularly the small- and medium-sized carnivores, had large brains, at least in comparison to body size (see the preceding section on Warm-Blooded Dinosaurs?).

Flying Reptiles Paleozoic insects were the first animals to achieve flight, but the first among vertebrates were *pterosaurs*, or flying reptiles, which were common in the skies from the Late Triassic until their extinction at the end of the Cretaceous (■ Figure 22.25). Adaptations for flight include a wing membrane supported by an elongated fourth finger, light hollow bones, and development of those parts of the brain associated with muscular coordination and sight. Because at least one pterosaur species had a coat of hair or hairlike feathers, possibly it, and perhaps all pterosaurs, were endotherms.

Pterosaurs are generally depicted in movies as large creatures, but some were no bigger than today's sparrows, robins, and crows. A few species, however, had wingspans of several meters, and one Cretaceous pterosaur found in Texas had a wingspan of at least 12 m! Nevertheless, even the largest species probably weighed no more than a few tens of kilograms.

Experiments and studies of fossils indicate that the bones of large pterosaurs such as *Pteranodon* (Figure 22.25b) were too weak for sustained wing flapping. These comparatively large animals probably took advantage of thermal updrafts to stay airborne, mostly by soaring but occasionally flapping their wings for maneuvering. In contrast, smaller pterosaurs probably stayed aloft by vigorously flapping their wings just as present-day small birds do.

(a)

(b)

■ Figure 22.25

(a) This long-tailed pterosaur from the Jurassic of Europe had a wingspan of about 0.6 m. (b) *Pteranodon* was a Cretaceous pterosaur with a wingspan of more than 6 m.

GEOFOCUS 22.1

Mary Anning's Contributions to Paleontology

Paleontologists use fossils to study life of the past, so part of their efforts are spent in finding and collecting fossils. Certainly Western European men dominated the early history of this field, but this situation no longer prevails. Indeed, men and women from many countries are now making significant contributions. Perhaps the most notable early exception is Mary Anning (1799–1847), who began a remarkable career as a fossil collector when she was only 11 years old.

Mary Anning was born in Lyme Regis on England 's southern coast. When only 15 months old, she survived a lightning strike that, according to one report, killed three girls, and according to another, killed a nurse tending her. In 1810, Mary's father, a cabinet maker who also sold fossils part time, died leaving the family nearly destitute. Mary Anning (■ Figure 1) expanded the fossil business and became a professional fossil collector known to the paleontologists of her time, some of whom visited her shop to buy fossils or gather information. She collected fossils from the Dorset coast near Lyme Regis and is reported to have been the inspiration for the tongue twister, "She sells sea shells sitting on the seashore."

Soon after her father's death, Mary Anning made her first important discovery, a nearly complete skeleton of a Jurassic ichthyosaur, which was described in 1814 by Sir Everard Home. The sale of this fossil specimen provided considerable financial relief for her family. In 1821, she made a second major discovery and excavated the remains of a plesiosaur. And in 1828, she found the first pterosaur in England, which was sent to the eminent geologist William Buckland at Oxford University.

By 1830, Mary Anning's fortunes began declining as collectors and museums had fewer funds with which to buy fossils. Indeed, she may once again have become destitute were it not for her geologist

Marine Reptiles The most familiar Mesozoic marine reptiles are the porpoise-like *ichthyosaurs* (■ Figure 22.26a). Most of these fully aquatic animals were about 3 m long, but one species reached about 12 m. All ichthyosaurs had a streamlined body, a powerful tail for propulsion, and flipperlike forelimbs for maneuvering. Their numerous sharp teeth indicate that they were carnivores, and preserved stomach contents reveal a diet of fish, cephalopods, and other marine organisms. Ichthyosaurs were so completely aquatic that is doubtful they could come onto land, so females probably retained eggs within their bodies and gave birth to live young. A few fossils with small ichthyosaurs within the appropriate part of the body cavity support this interpretation.

An interesting side note in the history of paleontology is the story of Mary Anning (1799–1847). When only 11 years old, she discovered and directed the excavation of a nearly complete ichthyosaur in southern England (see Geo-Focus 22.1).

Another well-known group of Mesozoic marine reptiles, the *plesiosaurs*, belonged to one of two groups: short-necked and long-necked (Figure 22.26b). Most were modest-sized animals 3.6 to 6 m long, but one species found in Antarctica measures 15 m. Short-necked plesiosaurs might have been bottom feeders, but their long-necked cousins probably used their necks in a snakelike fashion to capture fish with their numerous sharp teeth. These animals probably came ashore to lay their eggs.

Mosasaurs were Late Cretaceous marine lizards related to the present-day Komodo dragon or monitor lizard. Some species measured no more than 2.5 m long, but a few such as *Tylosaurus* were giants, measuring up to 9 m long (Figure 22.26c). Mosasaur limbs resemble paddles and were probably used mostly for maneuvering while the long tail provided propulsion. All were predators, and preserved stomach contents indicate that they ate fish, birds, smaller mosasaurs, and a variety of invertebrates including ammonoids.

friend Henry Thomas de la Beche, also a resident of Lyme Regis. De la Beche drew a fanciful scene called *Duria antiquior,* meaning "An earlier Dorset," in which he brought to life the fossils Mary Anning had collected. The scene was made into a lithograph that was printed and sold widely, and its proceeds went directly to Mary Anning.

Mary Anning died of cancer in 1847, and although only 48 years old, she had a fossil-collecting career that spanned 36 years. Her contributions to paleontology are now widely recognized but, unfortunately, soon after her death she was mostly forgotten. Apparently, the people who purchased her fossils were credited with finding them. "It didn't occur to them to credit a woman from the lower classes with such astonishing work. So an uneducated little girl, with a quick mind and an accurate eye, played a key role in setting the course of the 19th century geologic revolution. Then—we simply forgot about her."*

Figure 1

Mary Anning, who lived in Lyme Regis on England's southern coast, began collecting and selling fossils when she was only 11 years old.

*John Lienhard, University of Houston.

From Reptiles to Birds

Several fossils from the Jurassic Solnhofen Limestone of Germany have many of the physical features of small theropods, and yet they also have feathers and the fused clavicles typical of birds. This creature known as *Archaeopteryx,* which is classified as a bird, retained dinosaur-like teeth as well as a long tail, hind limb structure, and a brain case much like that of dinosaurs (■ Figure 22.27).

Most paleontologists now agree that some kind of small theropod was the ancestor of birds. Even the wishbone (fused clavicles) so typical of birds is found in a number of theropods. Discoveries in China in recent years of theropods with a feathery covering provide more evidence for a close relationship between theropods and birds (see "Dinosaurs" on pages 662 and 663). However, another possible bird ancestor, a small reptile known as *Longisquama,* was discovered during the 1970s in Kyrgystan. *Longisquama* is probably an archosaur, and it too appears to have feathers, but some paleontologists think that these are actually elongated scales.

Another Mesozoic bird from China is slightly younger than *Archaeopteryx* and possesses both primitive and advanced features. For example, it retains abdominal ribs similar to those of *Archaeopteryx* and theropods, but it has a reduced tail more like that of birds today. One of the oldest known toothless birds comes from the Late Jurassic or Early Cretaceous of Asia.

Unfortunately, the fossil record for *Archaeopteryx* is insufficient to resolve whether it is the ancestor of today's birds or an animal that died out with no evolutionary descendants. Of course, that in no way diminishes its having both reptile and bird characteristics or its status as a transitional fossil. Its place in the evolutionary scheme was challenged when two crow-sized fossils known as *Protoavis* were discovered. According to at least one paleontologist, *Protoavis* from the Late Triassic is older than *Archaeopteryx* and has hollow bones and

■ **Figure 22.26**

Mesozoic marine reptiles. (a) Ichthyosaurs. (b) A long-necked plesiosaur. (c) *Tylosaurus*, a Late Cretaceous mosasaur, measured up to 9 m long.

the breastbone structure of birds. But because these specimens are fragmentary and no feather impressions were found, many paleontologists think they are fossils of small theropods.

One of the most vocal early critics of Darwin's *On the Origin of Species* was the cleric St. George Jackson Mivart (1827–1900). Mivart claimed in his 1871 book *The Genesis of Species* that complex structures such as bird wings must have been created in their present form because anything less would be of no use. Yet, today young partridges have tiny wings that are of no use for flying but are still useful. The partridges flap their wings when they run up tree trunks to escape predators, a technique also commonly used by adult partridges. If their wings are taped down, they cannot climb trees. Only later when their wings have developed further are they capable for using them for flight.

Origin and Early History of Mammals

We briefly described *therapsids,* or the advanced mammal-like reptiles, in Chapter 21. They diversified into numerous species of herbivores and carnivores and during the Permian were the most diverse and numerous terrestrial vertebrates. One particular group of carnivorous therapsids called *cynodonts* was the most mammal-like of all and by the Late Triassic gave rise to mammals.

The transition from **cynodonts** to mammals is well documented by fossils and is so gradational that classification of some fossils as either reptile or mammal is difficult. We can easily recognize living mammals as warm-blooded animals with hair or fur that have mammary glands and, except for the platypus and spiny anteater, give birth to live young.

Obviously these criteria for recognizing living mammals are inadequate for classifying fossils. For them, we must use skeletal structure only. Several skeletal modifications characterize the transition from mammal-like reptiles to mammals, but distinctions between the two groups are based largely on details of the middle ear, the lower jaw, and the teeth (Table 22.2).

Reptiles have only one small bone in the middle ear—the stapes—whereas mammals have three—the incus, the malleus, and the stapes. Also, the lower jaw of a mammal is composed of a single bone called the *den-*

Tom McHugh/Photo Researchers

■ **Figure 22.27**

Fossil of *Archaeopteryx,* from the Jurassic-aged Solnhofen Limestone of Germany. Notice the feather impressions on the wings and the long tail. This animal had feathers and a wishbone, making it a bird, but in most details of its anatomy it resembled small theropod dinosaurs. For example, it had reptilian teeth, claws on its wings, and a long tail—none of which are found on birds today.

What Would You Do?

While visiting one of our national parks, you observe limestone layers with fossil trilobites and brachiopods dipping at a 50-degree angle, but the overlying sandstone beds with dinosaur fossils dip at only 15 degrees. Furthermore, a basalt dike cuts through all the strata. How would you explain the sequence of events accounting for your observations? What basic geologic principles did you use to make your interpretation, and is it possible to determine an absolute age for any of the events? If so, explain.

tary, but a reptile's jaw is composed of several bones (■ Figure 22.28). In addition, a reptile's jaw is hinged to the skull at a contact between the articular and quadrate bones, whereas in mammals the dentary contacts the squamosal bone of the skull (Figure 22.28).

During the transition from cynodonts to mammals, the quadrate and articular bones that had formed the joint between the jaw and skull in reptiles were modified into the incus and malleus of the mammalian middle ear (Figure 22.28). Fossils clearly document the progressive enlargement of the dentary until it became the only element in the mammalian jaw. Likewise, a progressive change from the reptile to mammal jaw joint is documented by

Table 22.2

Characteristics and How They Changed During the Transition from Reptiles to Mammals

Characteristic	Typical Reptile	Cynodont	Mammal
Lower jaw	Dentary and several other bones	Dentary enlarged, other bones reduced	Dentary bone only, except in earliest mammals
Jaw–skull joint	Articular–quadrate	Articular–quadrate; some advanced cynodonts had both the reptile jaw–skull joint and the mammal jaw–skull joint	Dentary–squamosal
Middle ear bones	Stapes	Stapes	Stapes, incus, malleus
Secondary palate	Absent	Partly developed	Well developed
Teeth	No differentiation	Some differentiation	Fully differentiated into incisors, canines, and chewing teeth
Tooth replacement	Teeth replaced continuously	Only two sets of teeth in some advanced cynodonts	Two sets of teeth
Occlusion (chewing teeth meet surface to surface to allow grinding)	No occlusion	Occlusion in some advanced cynodonts	Occlusion
Endothermic vs. Ectothermic	Ectothermic	Probably endothermic	Endothermic
Body covering	Scales	One fossil shows skin similar to that of mammals	Skin with hair or fur

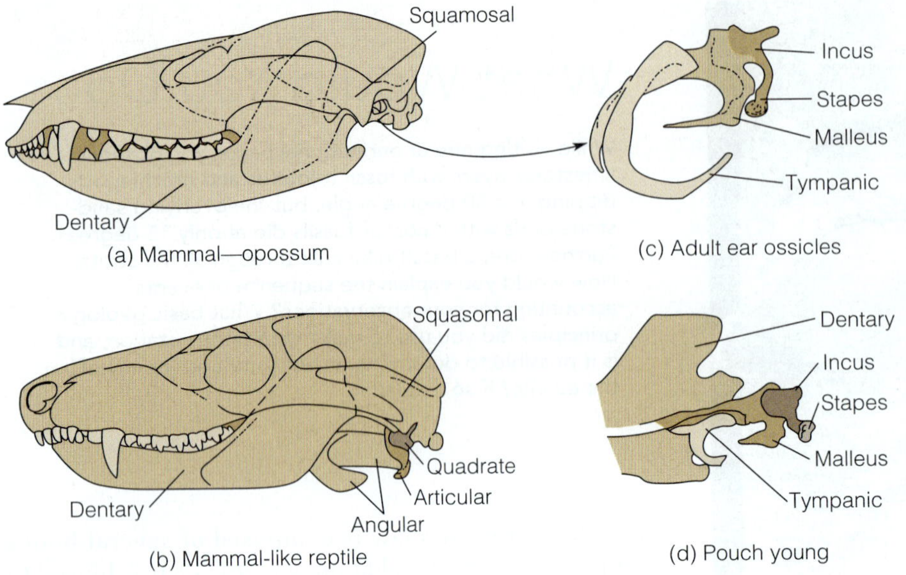

Figure 22.28

(a) The skull of an opossum showing the typical mammalian dentary–squamosal jaw joint. (b) The skull of a cynodont shows the articular–quadrate jaw joint of reptiles. (c) Enlarged view of an adult opossum's middle ear bones. (d) View of the inside of a young opossum's jaw showing that the elements of the middle ear are attached to the dentary during early development. This is the same arrangement of bones that is found in the adults of the ancestral mammals. Source: Reproduced with permission from J. S. Hobson, "The Mammal-Like Reptiles: A Study of Transitional Fossils," *The American Biology Teacher*, v. 49, no. 1, 1987, pp. 18–22. National Association of Biology Teachers, Reston, Virginia.

fossil evidence. In fact, some of the most advanced cynodonts were truly transitional because they had a compound jaw joint consisting of (1) the articular and quadrate bones typical of reptiles and (2) the dentary and squamosal bones as in mammals (Table 22.2).

In Chapter 18, we noted that the study of embryos provides evidence for evolution. Opossum embryos clearly show that the middle ear bones of mammals were originally part of the jaw. In fact, even when opossums are born, the middle ear elements are still attached to the dentary (Figure 22.28), but as they develop further, these elements migrate to the middle ear and a typical mammal jaw joint develops.

Mammals have teeth fully differentiated into incisors, canines, and chewing teeth to perform specific functions—nipping, tearing, and grinding. In typical reptiles, all the teeth look much the same, except in cynodonts whose teeth do in fact show some differentiation. In addition, reptiles have teeth replaced throughout their lives, whereas mammals and cynodonts have only two sets of teeth during their lifetimes—a set of baby teeth and the permanent adult teeth. Some of the more advanced cynodonts also show occlusion between the upper and lower chewing teeth; that is, the chewing teeth meet surface to surface to allow grinding, another mammal characteristic (Table 22.2).

Mammals possess a secondary palate, a bony shelf that separates the nasal passages from the mouth cavity, which is an adaptation for chewing and breathing at the same time, a necessary requirement for endotherms with their high demands for oxygen. This feature is also partly developed in cynodonts, especially the more advanced ones (■ Figure 22.29). This evidence along with a preserved cynodont skin similar to that of mammals indicates that these ancestral animals were endotherms.

True mammals were present by Late Triassic time, but their diversity remained low during the rest of the Mesozoic. Many of these earliest mammals were truly transitional animals in that they possessed both reptile and mammal characteristics. For instance, the Triassic mammals known as triconodonts had both the mammal

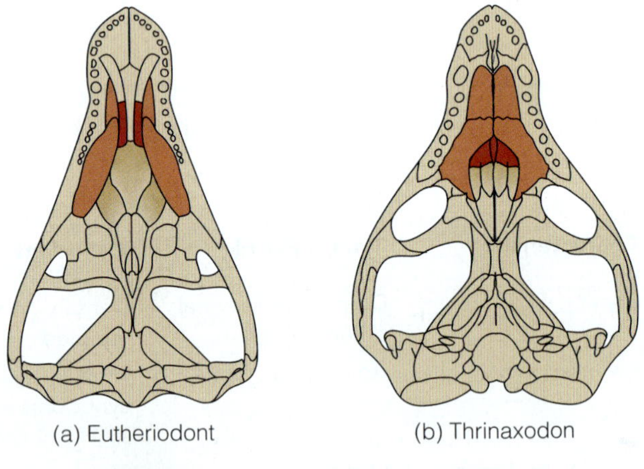

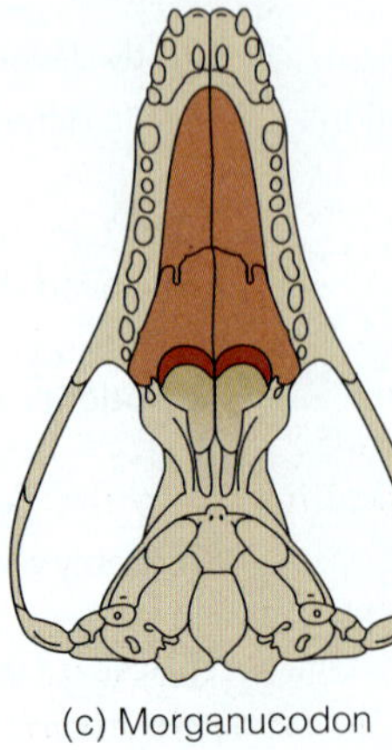

Figure 22.29

Views of the bottoms of skulls of (a) an early therapsid, (b) a cynodont, and (c) an early mammal showing the progressive development of the bony secondary palate (brown). Source: Reproduced with permission from J. S. Hobson, "The Mammal-Like Reptiles: A Study of Transitional Fossils," *The American Biology Teacher*, v. 49, no. 1, 1987, pp. 18–22. National Association of Biology Teachers, Reston, Virginia.

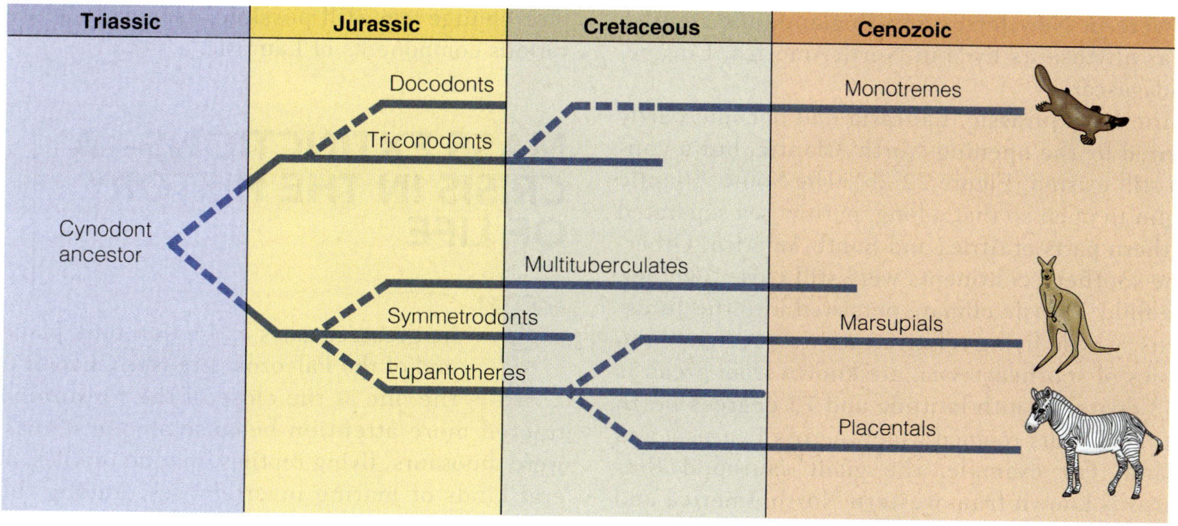

Figure 22.30

Relationships among the various recognized groups of early mammals and their descendants. Mammal evolution proceeded along two branches: one leading to today's monotremes, or egg-laying mammals, and the other to marsupial and placental mammals.

and the reptile jaw–skull joint. In short, some mammalian features evolved before others, thus once again illustrating the concept of *mosaic evolution.*

After mammals had evolved, they diverged into two distinct branches. One is the triconodonts and their probable descendants, the **monotremes,** or egg-laying mammals, which includes today's platypus and spiny anteater of the Australian region. The second branch includes the **marsupial** (pouched) **mammals** and the **placental mammals.** Both marsupials and placentals give birth to live young, but in marsupials the young are immature and develop further in the mother's pouch. Placental mammals nourish their developing young through a specialized membrane, the *placenta,* and the newborns are more mature. The divergence of marsupials and placentals may have taken place by the Early Cretaceous, but undoubtedly both were present by the Late Cretaceous (Figure 22.30 and 22.31).

Figure 22.31

The first known placental mammals were members of the order Insectivora, such as these in a scene from the Late Cretaceous. These animals measured only a few centimeters long. Insectivores today are represented by shrews, moles, and hedgehogs. Source: Adapted with permission from J. Benes, *Prehistoric Animals and Plants,* 1979, pp. 174–175, illustration by Zdenek Burian, published by Hippocrene Books, Inc.

MESOZOIC CLIMATES AND PALEOGEOGRAPHY

Fragmentation of the supercontinent Pangaea began by the Late Triassic and continues to the present, but during much of the Mesozoic, close connections existed between the various landmasses. The proximity of these landmasses, however, is not sufficient to explain Mesozoic biogeographic distributions because climates are also effective barriers to wide dispersal. During much of the Mesozoic, though, climates were more equable and lacked the strong north and south zonation characteristic of the present. In short, Mesozoic plants and animals had greater opportunities to occupy much more extensive geographic ranges.

Pangaea persisted as a single unit throughout most of the Triassic (Figure 22.2a), and the Triassic climate was warm-temperate to tropical, although some areas, such as the present southwestern United States, were arid. Mild temperatures extended 50 degrees north and south of the equator, and even the polar regions may have been temperate. The fauna had a truly worldwide distribution. Some dinosaurs had continuous ranges across Laurasia and Gondwana, the peculiar gliding lizards were in New Jersey and England, and reptiles known as phytosaurs lived in North America, Europe, and Madagascar.

By the Late Jurassic, Laurasia had become partly fragmented by the opening North Atlantic, but a connection still existed (Figure 22.2b). The South Atlantic had begun to open so that a long, narrow sea separated the southern parts of Africa and South America. Otherwise, the southern continents were still close together.

The mild Triassic climate persisted into the Jurassic. Ferns, whose living relatives are now restricted to the tropics of southeast Asia, are known from areas as far as 63 degrees south latitude and 75 degrees north latitude. Dinosaurs roamed widely across Laurasia and Gondwana. For example, the giant sauropod *Brachiosaurus* is known from western North America and eastern Africa. Stegosaurs and some families of carnivorous dinosaurs lived throughout Laurasia and in Africa.

By the Late Cretaceous, the North Atlantic had opened further, and Africa and South America were completely separated (Figure 22.2c). South America remained an island continent until late in the Cenozoic and its fauna, evolving in isolation, became increasingly different from faunas elsewhere. Marsupial mammals reached Australia from South America via Antarctica, but the South American connection was eventually severed. Placentals, other than bats and a few rodents, never reached Australia, thus explaining why marsupials continue to dominate the continent's fauna even today.

Cretaceous climates were more strongly zoned by latitude, but they remained warm and equable until the close of that period. Climates then became more seasonal and cooler, a trend that persisted into the Cenozoic. Dinosaur and mammal fossils demonstrate that interchange was still possible, especially between the various components of Laurasia.

MASS EXTINCTIONS—A CRISIS IN THE HISTORY OF LIFE

The greatest mass extinction took place at the end of the Paleozoic Era (see Chapter 21), but the one at the close of the Mesozoic has attracted more attention because among its casualties were dinosaurs, flying reptiles, marine reptiles, and several kinds of marine invertebrates. Among the latter were ammonites, which had been so abundant throughout the Mesozoic, rudistid bivalves, and some planktonic organisms.

A hypothesis explaining Mesozoic extinctions that has become popular since 1980 is based on a discovery at the Cretaceous–Paleogene boundary in Italy—a 2.5-cm-thick clay layer with a remarkably high concentration of the platinum group element iridium (■ Figure 22.32a). High iridium concentrations have now been identified at many other Cretaceous–Paleogene boundary sites.

The significance of this *iridium anomaly* is that iridium is rare in crustal rocks but is found in much higher concentrations in some meteorites. Accordingly, some investigators propose a meteorite impact to explain the iridium anomaly and further postulate that the impact of a meteorite perhaps 10 km in diameter set in motion a chain of events leading to extinctions. Some Cretaceous–Paleogene boundary sites also contain soot and

(a)

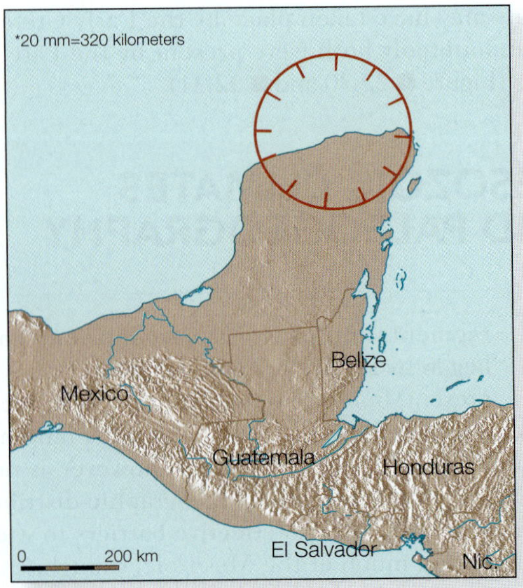

(b)

Geology Now ■ **Active Figure 22.32**

(a) Closeup view of the iridium-rich Cretaceous–Paleogene boundary clay in the Raton Basin, New Mexico. (b) Proposed meteorite impact crater centered on Chicxulub on the Yucatán Peninsula of Mexico.

shock-metamorphosed quartz grains, both cited as additional evidence of an impact.

According to the impact hypothesis, about 60 times the mass of the meteorite was blasted from the crust high into the atmosphere, and the heat generated at impact started raging forest fires that added more particulate matter to the atmosphere. Sunlight was blocked for several months, causing a temporary cessation of photosynthesis, food chains collapsed, and extinctions followed. Furthermore, with sunlight greatly diminished, Earth's surface temperatures were drastically reduced, adding to the biologic stress. Another proposed consequence of an impact is that sulfuric acid (H_2SO_4) and nitric acid (HNO_3) resulted from vaporized rock and atmospheric gases. Both would have contributed to strongly acid rain that might have had devastating effects on vegetation and marine organisms (see Tables 22.3 and 22.4 on pages 674 and 675).

The iridium anomaly is real, but its origin and significance are debatable. We know very little about the distribution of iridium in crustal rocks or how it may be distributed and concentrated. Some geologists suggest that the iridium was derived from within Earth by volcanism, but this idea is not well supported by evidence.

Some now claim that a probable impact site centered on the town of Chicxulub on the Yucatán Peninsula of Mexico has been found (Figure 22.32b). The structure lies beneath layers of sedimentary rock, measures 180 km in diameter, and appears to be the right age. Evidence supporting the conclusion that the Chicxulub structure is an impact crater includes shocked quartz, what appears to be the deposits of huge waves, and tektites, which are small pieces of rock that were melted during the proposed impact and hurled into the atmosphere.

Even if a meteorite did hit Earth, did it lead to these extinctions? If so, both terrestrial and marine extinctions must have occurred at the same time. To date, strict time equivalence between terrestrial and marine extinctions has not been demonstrated. And recent investigations by Gerta Keller and her students at Princeton University lead them to think that the extinctions actually took place about 300,000 years after the Chicxulub meteorite impact.

The selective nature of the extinctions is also a problem. In the terrestrial realm, large animals were the most drastically affected, but not all dinosaurs were large, and crocodiles, close relatives of dinosaurs, were not affected. Some paleontologists think that dinosaurs, some marine invertebrates, and many plants were already on the decline and headed for extinction before the end of the Cretaceous. A meteorite impact may have simply hastened the process.

In the final analysis, Mesozoic extinctions have not been explained to everyone's satisfaction. Most geologists now concede that a large meteorite impact occurred, but we also know that vast outpourings of lava were taking place in what is now India. Perhaps these brought about detrimental atmospheric changes. Furthermore, the vast shallow seas that covered large parts of the continents had mostly withdrawn by the end of the Cretaceous, and the mild equable Mesozoic climates became harsher and more seasonal by the end of that era. But the fact remains that these extinctions were very selective, and no single explanation accounts for all aspects of this crisis in life history.

22 GEO RECAP

Chapter Summary

Tables 22.3 and 22.4 (see pages 674 and 675) provide summaries of Mesozoic geologic and biologic events, respectively.

- The breakup of Pangaea can be summarized as follows:
  1. The first stage involved the separation of North America from Africa during the Late Triassic, followed by the rifting of North America from South America.
  2. The second stage involved the separation of Antarctica, India, and Australia from South America and Africa during the Jurassic. During this time, India broke away from the still-united

Table 22.3

Mesozoic Geologic Events

Age (millions of years)	Geologic Period	Sequence	Relative Changes in Sea Level (Rising / Falling)	Cordilleran Mobile Belt	Gulf Coastal Region	Eastern Coastal Region	Global Plate Tectonic Events
66–144	Cretaceous	Zuni	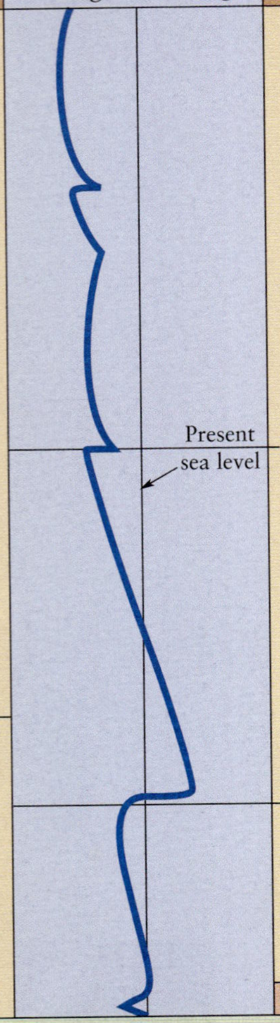	Jurassic and Cretaceous tectonism controlled by eastward subduction of the Pacific plate beneath North America and accretion of microplates; Cordilleran orogeny: Laramide orogeny; Sevier orogeny; Nevadan orogeny	Major Late Cretaceous transgression. Reefs particularly abundant. Regression at end of Early Cretaceous. Early Cretaceous transgression and marine sedimentation	Appalachian region uplifted. Erosion of fault-block mountains formed during the Late Triassic to Early Jurassic	South America and Africa are widely separated. Greenland begins separating from Europe
144–208	Jurassic	Zuni; Absaroka		Nevadan orogeny; Jurassic and Cretaceous tectonism controlled by eastward subduction of the Pacific plate beneath North America and accretion of microplates	Sandstones, shales, and limestones are deposited in transgressing and regressing seas. Thick evaporites are deposited in newly formed Gulf of Mexico	Erosion of fault-block mountains formed during the Late Triassic to Early Jurassic. Fault-block mountains and basins develop in eastern North America	South America and Africa begin separating in the Late Jurassic
208–245	Triassic	Absaroka		Subduction zone develops as a result of westward movement of North America. Sonoma orogeny	Gulf of Mexico begins forming during Late Triassic	Deposition of Newark Group; lava flows, sills, and dikes	Breakup of Pangaea begins with rifting between Laurasia and Gondwana. Supercontinent Pangaea still in existence

Table 22.4

Mesozoic Biologic Events

Age (millions of years)	Geologic Period	Invertebrates	Vertebrates	Plants	Climate	Plate Tectonics
66–144	Cretaceous	Extinction of ammonites, rudists, and most planktonic foraminifera at end of Cretaceous Continued diversification of ammonites and belemnoids Rudists become major reef builders	Extinctions of dinosaurs, flying reptiles, marine reptiles, and some marine invertebrates Placental and marsupial mammals diverge	Angiosperms evolve and diversify rapidly Seedless plants and gymnosperms still common but less varied and abundant	North-south zonation of climates more marked but remains equable Climate becomes more seasonal and cooler at end of Cretaceous	Further fragmentation of Pangaea South America and Africa have separated Australia separated from South America but remains connected to Antarctica North Atlantic continues to open
144–208	Jurassic	Ammonites and belemnoid cephalopods increase in diversity Scleractinian coral reefs common Appearance of rudist bivalves	GREATEST DIVERSITY OF DINOSAURS First birds (may have evolved in Late Triassic) Time of giant sauropod dinosaurs	Seedless vascular plants and gymnosperms only	Much like Triassic Ferns with living relatives restricted to tropics live at high latitudes, indicating mild climates	Fragmentation of Pangaea continues, but close connections exist among all continents
208–245	Triassic	The seas are repopulated by invertebrates that survived the Permian extinction event Bivalves and echinoids expand into the infaunal niche	Mammals evolve from cynodonts Cynodonts become extinct Ancestral archosaur gives rise to dinosaurs Flying reptiles and marine reptiles evolve	Land flora of seedless vascular plants and gymnosperms as in Late Paleozoic	Warm-temperate to tropical Mild temperatures extend to high latitudes; polar regions may have been temperate Local areas of aridity	Fragmentation of Pangaea begins in Late Triassic

Antarctica and Australia landmass and began moving northward.

3. During the third stage, which began in the Late Jurassic, South America and Europe began separating and Europe and Africa began converging.

4. The final stage in Pangaea's breakup occurred during the Cenozoic, when Greenland separated from Europe and North America.

- The breakup of Pangaea influenced global climatic and atmospheric circulation patterns. Although the temperature gradient from the tropics to the poles gradually increased during the Mesozoic, overall global temperatures remained equable.
- Except for incursions along the continental margin and two major transgressions (the Sundance Sea and the Cretaceous Interior Seaway), the North American craton was above sea level during the Mesozoic Era.
- The Eastern Coastal Plain was the initial site of the separation of North America from Africa that began during the Late Triassic. During the Cretaceous Period, it was inundated by marine transgressions.
- The Gulf Coastal region was the site of major evaporite accumulation during the Jurassic as North America rifted from South America. During the Cretaceous, it was inundated by a transgressing sea, which, at its maximum, connected with a sea transgressing from the north to create the Cretaceous Interior Seaway.
- Mesozoic rocks of the western region of North America were deposited in a variety of continental and marine environments. One of the major controls of sediment distribution patterns was tectonism.
- Western North America was affected by four interrelated orogenies: the Sonoma, Nevadan, Sevier, and Laramide. Each involved igneous intrusions as well as eastward thrust faulting and folding.
- The cause of the Sonoma, Nevadan, Sevier, and Laramide orogenies was the changing angle of subduction of the oceanic Farallon plate under the continental North American plate. The timing, rate, and, to some degree, the direction of plate movement were related to seafloor spreading and the opening of the Atlantic Ocean.
- Orogenic activity associated with the oceanic–continental convergent plate boundary in the Cordilleran mobile belt explains the structural features of the western margin of North America. It is thought, however, that more than 25% of the North American western margin originated from the accretion of terranes.
- Mesozoic rocks contain a variety of mineral resources, including coal, petroleum, uranium, gold, and copper.
- Among the marine invertebrates, survivors of the Permian extinction diversified and gave rise to increasingly complex Mesozoic marine invertebrate communities.
- Triassic and Jurassic land-plant communities were composed of seedless vascular plants and gymnosperms. Angiosperms evolved during the Early Cretaceous, diversified rapidly, and soon became the dominant land plants.
- Dinosaurs evolved during the Late Triassic but were most abundant and diverse during the Jurassic and Cretaceous. The two distinct orders of dinosaurs, based on pelvic structure, are Saurischia (lizard-hipped) and Ornithischia (bird-hipped).
- Small pterosaurs were probably active, wing-flapping fliers, whereas large ones may have depended on thermal updrafts and soaring to stay aloft. At least one pterosaur species had hair or feathers, so it was likely endothermic.
- The fish-eating, porpoise-like ichthyosaurs were thoroughly adapted to an aquatic life. Plesiosaurs were heavy-bodied marine reptiles that probably came ashore to lay eggs, whereas mosasaurs were marine lizards.
- Birds probably evolved from small carnivorous dinosaurs. The oldest known bird, *Archaeopteryx*, appeared during the Jurassic, but few other Mesozoic birds are known.
- The earliest mammals evolved during the Late Triassic, but they are difficult to distinguish from advanced cynodonts. Details of the teeth, middle ear, and lower jaw are used to distinguish the two.
- Several types of Mesozoic mammals existed, but all were small and their diversity was low. A group of Mesozoic mammals gave rise to both marsupials and placentals during the Cretaceous.
- Because the continents were close together during much of the Mesozoic and climates were mild even at high latitudes, animals and plants dispersed widely.
- Mesozoic mass extinctions account for the disappearance of dinosaurs, several other groups of reptiles, and some marine invertebrates. One hypothesis holds that the extinctions were caused by the impact of a large meteorite.

Important Terms

angiosperm (p. 657)
archosaur (p. 659)
bipedal (p. 659)
Cordilleran orogeny (p. 648)
Cretaceous Interior Seaway (p. 650)
cynodont (p. 668)
ectotherm (p. 664)
endotherm (p. 664)
Laramide orogeny (p. 648)
marsupial mammal (p. 671)
monotreme (p. 671)
Nevadan orogeny (p. 648)
Ornithischia (p. 659)
placental mammal (p. 671)
quadrupedal (p. 659)
Saurischia (p. 659)
Sevier orogeny (p. 648)
Sonoma orogeny (p. 646)
Sundance Sea (p. 649)

Review Questions

1. What is the evidence for the breakup of Pangaea?
   a._____rift valleys; b._____dikes; c._____great quantities of poorly sorted nonmarine detrital sediments; d._____sills; e._____ all of these.
2. The time of greatest post-Paleozoic inundation of the craton occurred during which geologic period?
   a._____Triassic; b._____Jurassic; c._____Cretaceous; d._____Paleogene; e._____Neogene.
3. A possible cause for the eastward migration of igneous activity in the Cordilleran region during the Cretaceous was a change from
   a._____oceanic–oceanic convergence to oceanic–continental convergence; b._____high-angle to low-angle subduction; c._____divergent to convergent plate margin activity; d._____divergent plate margin activity to subduction; e._____subduction to divergent plate margin activity.
4. The Jurassic formation or complex famous for dinosaur fossils is the
   a._____Franciscan; b._____Sundance; c._____Chinle; d._____Navajo; e._____Morrison.
5. The orogeny responsible for the present-day Rocky Mountains is the
   a._____Laramide; b._____Sevier; c._____Sonoma; d._____Antler; e._____Nevadan.
6. All carnivorous dinosaurs belong to the group known as
   a. _____theropods; b. _____ankylosaurs; c. _____mosasaurs; d. _____ammonites; e. _____gymnosperms.
7. Which of the following pairs of animals were marine reptiles?
   a._____angiosperm–ginkgo; b._____ichthyosaur–plesiosaur; c._____cephalopod–rudist; d._____cynodont–marsupial; e._____monotreme–endotherm.
8. Which important group of marine phytoplankton first evolved during the Jurassic and remain numerous today?
   a._____oysters; b._____burrowing worms; c._____coccolithophores; d._____belemnoids; e._____teleosts.
9. Dinosaurs, crocodiles, pterosaurs, and birds are collectively known as
   a._____placentals; b._____archosaurs; c._____mosasaurs; d._____invertebrates; e._____stem reptiles.
10. An important Mesozoic event in the history of land plants was the
    a._____origin of ferns and horsetail rushes; b._____extinction of cycads and conifers; c._____first appearance and diversification of angiosperms; d._____dominance of seedless vascular plants; e._____prevalence of ginkgos and ceratopsians.
11. Briefly summarize the evidence for and against endothermy in dinosaurs.
12. Discuss the similarities and differences between the orogenic activity that occurred in the Appalachian mobile belt during the Paleozoic Era and that which occurred in the Cordilleran mobile belt during the Mesozoic Era.
13. How do the Mesozoic marine invertebrate communities differ from those of the Paleozoic?

14. The breakup of Pangaea influenced the distribution of continental landmasses, ocean basins, and oceanic and atmospheric circulation patterns, which in turn affected the distribution of natural resources, landforms, and evolution of the world's biota. Reconstruct a hypothetical history of the world for a different breakup of Pangaea, one in which the continents separate in a different order or rift apart in a different configuration. How would such a scenario affect the distribution of natural resources? Would the distribution of coal and petroleum reserves be the same? How might evolution be affected? Would human history be different?
15. Discuss the modifications that occurred for flight in pterosaurs and for an aquatic life in ichthyosaurs.
16. How did the breakup of Pangaea affect oceanic and climatic circulation patterns?
17. How did the land-plant communities of the Triassic and Jurassic differ from those of the Cretaceous?
18. How does terrane accretion change our interpretations of the geologic history of the western margin of North America? How does it relate to the Mesozoic orogenies that took place in that area?
19. Explain and diagram how increased seafloor spreading can cause a rise in sea level along the continental margins.
20. Briefly summarize the evidence for and against the hypothesis that a meteorite impact caused Mesozoic mass extinctions.

World Wide Web Activities

Geology⇌Now Assess your understanding of this chapter's topics with additional quizzing and comprehensive interactivities at

http://earthscience.brookscole.com/changingearth4e

as well as current and up-to-date weblinks, additional readings, and InfoTrac College Edition exercises.

Cenozoic Earth and Life History

CHAPTER 23

OUTLINE

Geology⇌Now *This icon, which appears throughout the book, indicates an opportunity to explore interactive tutorials, animations, or practice problems available on the GeologyNow website at http://earthscience.brookscole.com/changingearth4e.*

OBJECTIVES

At the end of this chapter, you will have learned that:

- The Mesozoic breakup of Pangaea continued during the Cenozoic, accounting for orogenies in two major belts.
- The North American Cordillera experienced deformation, the origin of mountains, volcanism, uplift, and deep erosion.
- An epeiric sea briefly occupied North America's interior lowlands.
- Thick sedimentary deposits accumulated along the Gulf and Atlantic Coastal Plains.
- Cenozoic uplift and erosion account for the present-day Appalachian Mountains.
- Pleistocene continental glaciers covered vast areas of the Northern Hemisphere continents.
- Cenozoic rocks contain several resources such as oil and gold.
- Marine invertebrates such as foraminifera and mollusks were abundant during the Cenozoic.
- Mammals diversified and eventually gave rise to today's familiar mammals.
- A variety of large mammals and birds existed during the Pleistocene.
- Many large land-dwelling mammals became extinct at the end of the Pleistocene.
- Primates were present throughout the Cenozoic, but human evolution took place during the Pliocene and Pleistocene.

Oligocene- and Miocene-aged sedmentary rocks, mostly siltstone and sandstone, and volcanic ash layers exposed at Scott's Bluff National Monument, Nebraska. Source: James S. Monroe

Introduction

One reason to study Cenozoic Earth and life history is that the present distribution of land and sea, oceanic and atmospheric circulation patterns, and Earth's present distinctive topography and biota evolved during that time. For instance, the Appalachian Mountains began their overall evolution during the Proterozoic Eon, but their present distinctive topography resulted largely from Cenozoic uplift and erosion. Likewise, the Sierra Nevada of the western United States, the Himalayas in Asia, and the Andes in South America owe their existing characteristics to Cenozoic events. Indeed, some of these mountain systems remain tectonically active.

Geologists divide the Cenozoic into three periods of unequal duration—Paleogene, Neogene, and Quaternary—and each period is further subdivided into epochs (■ Figure 23.1). The Paleogene and Neogene have recently replaced the Tertiary Period, so you will see many publications with the Cenozoic consisting of only the Tertiary and Quaternary Periods. At 65.5 million years long, the Cenozoic Era is only 1.4% of all geologic time, or just 20 minutes on our hypothetical 24-hour clock (see Figure 19.2). Although brief in the context of geologic time, by any other measure 65.5 million years is longer than we can even imagine, certainly long enough for significant evolution of Earth and its biota.

Obviously the Quaternary makes up only a small part of the Cenozoic Era, yet it is an important time in Earth and life history. Notice that the Quaternary Period consists of two epochs, the Pleistocene and Holocene or Recent, the former beginning 1.6 million years ago and ending 10,000 years ago (Figure 23.1). The Pleistocene, or what is popularly called the Ice Age, was a time when continental glaciers covered vast areas, especially on the Northern Hemisphere continents.

Recall from Chapter 19 that when Earth formed, it was hot, barren, and waterless, the atmosphere was noxious, and all known organisms were single-celled bacteria or their precursors. By Cenozoic time, not only was Earth taking on its present physical appearance, but its biota continued to evolve as plants and animals appeared that were more and more familiar. In this chapter, we emphasize the continuing evolution of mammals. Remember, mammals first appeared during the Late Triassic, but during much of the Mesozoic they retained several characteristics of their ancestors, the mammal-like reptiles; they were small and not particularly diverse. By Cenozoic time, though, mammals had clearly differentiated from their progenitors.

In addition to the evolution of mammals, equally important events took place among plants. The angiosperms continued to dominate land-plant communities, as they had during the Cretaceous, and now constitute more than 90% of all land plants. Although birds evolved during the Jurassic, the families now common appeared during the Paleogene and Neogene, reached their maximum diversity during the Quaternary, and have declined slightly since then. Following the extinctions at the end of the Mesozoic, marine invertebrates diversified, giving rise to the present-day familiar fauna of the seas.

In this chapter, we also examine the primates—in particular, the origin and evolution of humans. We should point out that new discoveries and new techniques for scientific analyses are leading to new hypotheses about human ancestry. In previous editions of this book, we stated that the earliest fossil evidence of hominids (humans and their extinct ancestors) is from 4.4-million-year-old rocks in Africa. Since then, discoveries have pushed the age back to almost 7 million years. And by the time you read this chapter, new discoveries may have changed some of the conclusions based on what we currently know.

Era	Period		Epoch	Duration, millions of years (approx.)	Millions of years ago (approx.)
Cenozoic	Quaternary	Quaternary	Holocene/Recent		0.01
			Pleistocene	1.59	1.6
	Tertiary	Neogene	Pliocene	3.7	5.3
			Miocene	18.7	24
		Paleogene	Oligocene	13	37
			Eocene	21	58
			Paleocene	8	66

■ **Figure 23.1**

The geologic time scale for the Cenozoic Era. In this book we use Paleogene and Neogene rather than Tertiary, although the latter is still widely used.

CENOZOIC PLATE TECTONICS AND OROGENY—AN OVERVIEW

The Late Triassic fragmentation of the supercontinent Pangaea (see Figure 22.1a) began an episode of plate motions that continues even now. As a result, Cenozoic orogenic activity has been concentrated in two major zones or belts: the *Alpine–Himalayan belt* and the *circum-Pacific belt* (see Figure 10.20). The Alpine–Himalayan belt includes the mountainous regions of southern Europe and north Africa and extends eastward through the Middle East and India and into southeast Asia, whereas the circum-Pacific belt, as its name implies, nearly encircles the Pacific Ocean basin.

Within the Alpine–Himalayan orogenic belt, the *Alpine orogeny* began during the Mesozoic, but major deformation also took place from the Eocene to Late Miocene as the African and Arabian plates moved northward against Eurasia. Deformation resulting from plate convergence formed the Pyrenees Mountains between Spain and France, the Alps of mainland Europe, the Apennines of Italy, and the Atlas Mountains of North Africa (see Figure 10.20). Active volcanoes in Italy and Greece and seismic activity in much of southern Europe and the Middle East indicate that this orogenic belt remains geologically active.

Farther east in the Alpine–Himalayan orogenic belt, the Himalayan orogen resulted from the collision of India with Asia (see Figure 10.23). The exact time of this collision is uncertain, but sometime during the Eocene India's northward drift rate decreased abruptly, indicating the probable time of collision. In any event, two continental plates became sutured, accounting for the location of the present-day Himalayas far inland rather than near a plate boundary.

Plate subduction in the circum-Pacific orogenic belt occurred throughout the Cenozoic, giving rise to orogens in the Aleutians, the Philippines, Japan, and along the west coasts of North, Central, and South America. For example, the Andes Mountains in western South America formed as a result of convergence of the Nazca and South American plates (see Figure 10.22). Spreading at the East Pacific Rise and subduction of the Cocos and Nazca plates beneath Central and South America, respectively, account for continuing seismic, volcanic, and orogenic activity in these regions.

PALEOGENE AND NEOGENE EVOLUTION OF NORTH AMERICA

In the Introduction we noted that many of Earth's features have long histories but that they owe their present distinctive aspects to events that took place during the Cenozoic Era. Of course, the Paleogene and Neogene make up nearly 98% of the Cenozoic (Figure 23.1), so it was during this time that much of the evolution of North America took place.

The North American Cordillera

The **North American Cordillera** is a complex mountainous segment of the circum-Pacific orogenic belt extending from Alaska, through Canada and the United States, and into central Mexico (■ Figure 23.2). It was more or less continuously deformed during the Late Jurassic to Early Paleogene as the Nevadan, Sevier, and Laramide orogenies progressively affected areas from west to east (see Figure 22.10). But the final episode of deformation, the Late Cretaceous to Eocene **Laramide orogeny**, differed from the previous orogenies. It took place much farther inland than is typical; deformation was mostly in the form of vertical uplifts as opposed to compression-induced folding and faulting, and little volcanism and intrusion of plutons took place during deformation.

Orogenies resulting from oceanic–continental convergence usually take place very near a continental plate margin (see Figure 10.22). To account for the Laramide orogeny far from a plate boundary, geologists have proposed a model involving shallow subduction and cessation of igneous activity. During the Late Cretaceous, a subduction zone existed along the western margin of North America where the Farralon plate was subducted at about a 50-degree angle, and igneous activity took place 150 to 200 km inland from the trench (■ Figure 23.3). By the Early Paleogene, the westward moving North American plate had overridden part of the Farallon plate. As a result of nearly horizontal subduction, igneous activity migrated farther inland and eventually ceased because the descending Farallon plate no longer penetrated to the mantle.

Another consequence of the decreased angle of subduction was a change in the style of deformation. The fold-thrust tectonism of the Sevier orogeny gave way to large-scale buckling and fracturing, which produced fault-bounded uplifts with intervening intermontane basins. These basins were the depositional sites for Paleogene and Neogene sediments eroded from the uplifts.

The Laramide orogen is centered in the middle and southern Rockies in Wyoming and Colorado, but deformation also took place far to the north and south. In Montana and Alberta, Canada, large slabs of strata were transported to the east along overthrust faults (see "Types of Faults" on pages 272 and 273). Laramide deformation also took place in Alaska, and a major fold-thrust belt developed in northern Mexico. Laramide structural trends can be traced south into the area of Mexico City. Cessation of Laramide deformation apparently coincided with an increasing angle of descent of the Farallon plate (Figure 23.3).

Figure 23.2

The North American Cordillera, which is a segment of the circum-Pacific organic belt, and the major provinces of North America.

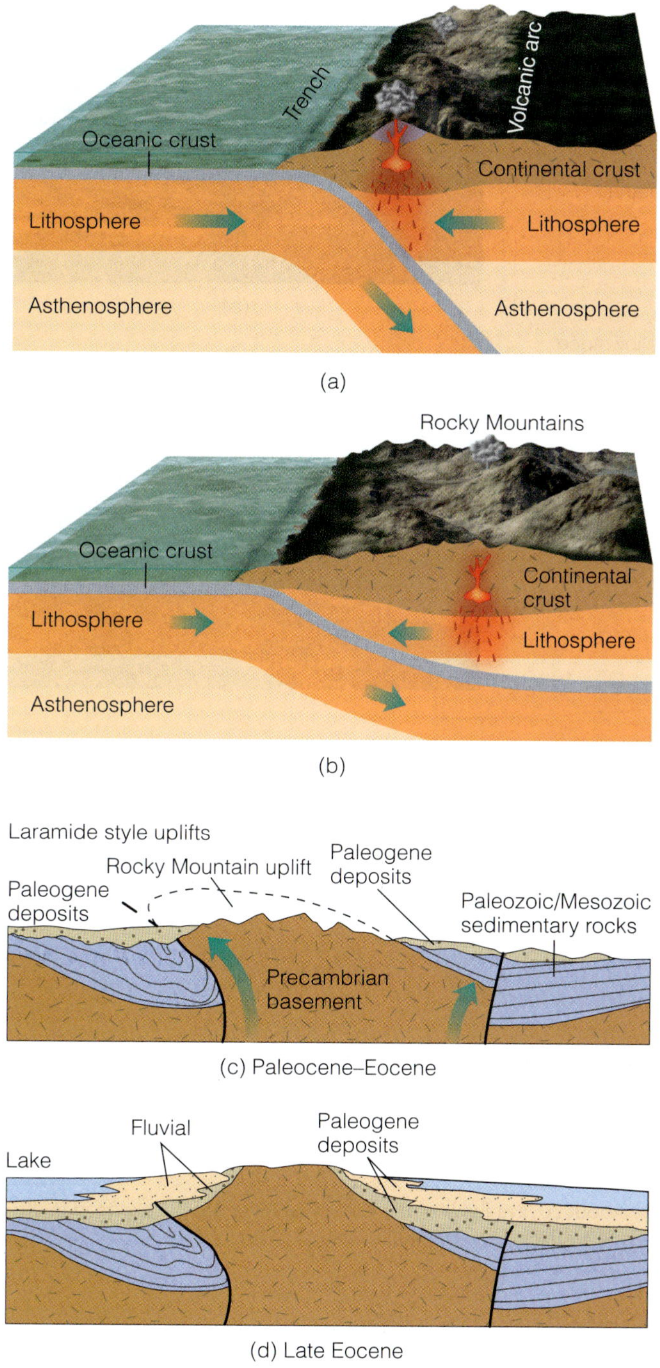

Figure 23.3

(a) During orogenies at oceanic–continental plate boundaries, the subducting plate descended at a steep angle, perhaps 50 degrees, and a volcanic arc forms 100–150 km inland. (b) During the Laramide orogeny, the subducting plate apparently moved nearly horizontally beneath the continent, resulting in mountain building further inland and minimal igneous activity. (c–d) This Laramide style of deformation yielded vertical uplifts bounded by faults rather than the compression-induced structures typical of earlier stages of the Cordilleran orogeny.

Although the vast batholiths of western North America were emplaced mostly during the Mesozoic, intrusive activity continued into the Cenozoic. Numerous small plutons were emplaced, including copper and molybdenum-bearing stocks in Utah, Nevada, Arizona, and New Mexico. In addition, Devil's Tower, Wyoming (see Geo-Focus 4.1), and a number of other volcanic necks were emplaced during the Cenozoic.

The Cordillera was also the site of considerable Cenozoic volcanism although it varied in eruptive style and location (■ Figure 23.4). A large area in Washington and adjacent parts of Oregon and Idaho is underlain by about 200,000 km^3 of Miocene basalt lava flows known as the Columbia River basalts (see Figure 5.13). The flows issued from long fissures and are now well exposed in the deep gorges cut by the Snake and Columbia Rivers. The Snake River Plain in Idaho (Figure 23.4c) is actually a depression filled mostly by Pliocene and younger basalt lava flows. Bordering the Snake River Plain on the northeast is the Yellowstone Plateau of Wyoming, an area of Late Pliocene and Quaternary rhyolitic and some basaltic volcanism. The volcanoes of the Cascade Range were built by andesitic volcanism during the Pliocene, Pleistocene, and Recent epochs.

A large area in the Cordillera known as the **Basin and Range Province** (Figure 23.2) has experienced crustal extension and the development of north-south–oriented normal faults since the Late Miocene. Differential movement on these faults has produced a basin-and-range structure, consisting of mountain ranges separated by broad valleys (■ Figure 23.5).

The Colorado Plateau (Figure 23.2) is a vast area of deep canyons, broad mesas, volcanic mountains, and brilliantly colored rocks (■ Figure 23.6). During the Early Paleogene this area, which was formerly near sea level, was uplifted and deformed into broad anticlines and arches and basins. A number of large normal faults also cut the area, but overall deformation was far less intense than it was elsewhere in the Cordillera. Neogene uplift elevated the region to the 1200–1800-m elevations seen today. As uplift proceeded, deposition ceased and erosion of the deep canyons began.

The present plate tectonic elements of the Pacific Coast developed as a result of the westward drift of the North American plate, the partial consumption of the Farallon plate, and the collision of North America with the Pacific–Farallon Ridge (■ Figure 23.7). Most of the Farallon plate was consumed at a subduction zone along the West Coast of North America, and now only two small remnants exist—the Juan de Fuca and Cocos plates. Westward drift of the North American plate also resulted in its collision with the Pacific–Farallon Ridge and the origin of the Queen Charlotte and San Andreas transform faults along the west coasts of British Columbia, Canada, and California, respectively (Figure 23.7).

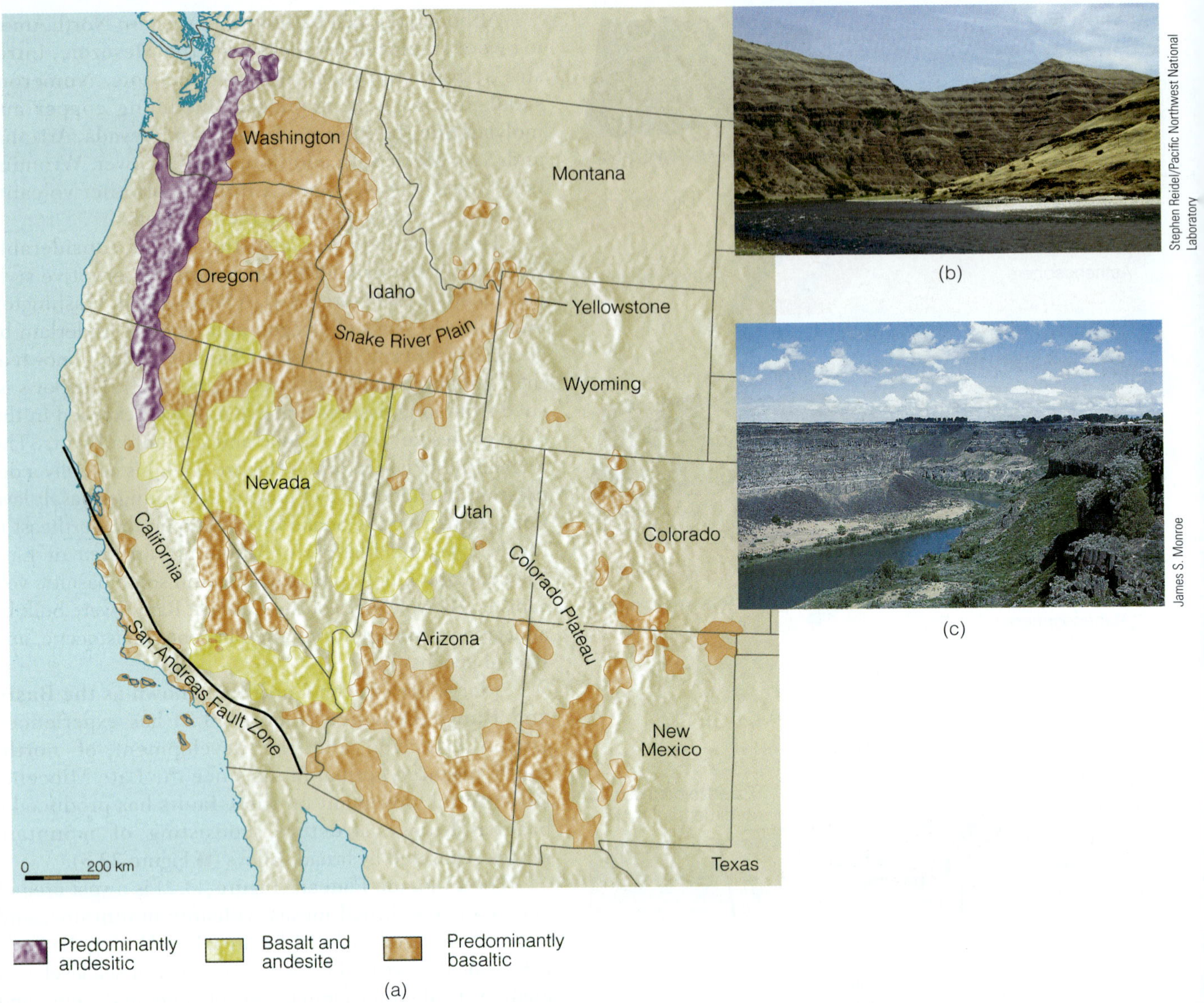

Figure 23.4

(a) Distribution of Cenozoic volcanic rocks in the western United States. (b) About 20 lava flows of the Columbia River basalts exposed in the canyon of the Grand Ronde River in Washington. (c) Lava flows of the Snake River Plain near Twin Falls, Idaho.

Much of the Cordilleran deformation and volcanism occurred during the Paleogene and Neogene periods, but these activities continued through the Quaternary to the present. For example, many of California's coastal oil- and gas-producing fold structures formed during the Pleistocene. Pleistocene tectonism also affected the Sierra Nevada in California, the Teton Range of Wyoming, and parts of the central and northern Rocky Mountains. Continued subduction of the Juan de Fuca and Cocos plates accounts for Pleistocene and Recent seismic activity and volcanism in the Pacific Northwest and Mexico, respectively.

The Interior Lowlands

During the Cretaceous, most of the western part of the Interior Lowlands (Figure 23.2) was covered by the Zuni epeiric sea. By Early Paleogene time, the Zuni Sea had withdrawn from most of North America, but a sizable remnant arm of the sea was still present in North Dakota during the Paleocene Epoch. Sediments derived from Laramide highlands to the west and southwest were transported to this remnant sea where they were deposited in transitional and marine environments. Elsewhere within the western Interior Lowlands, sedi-

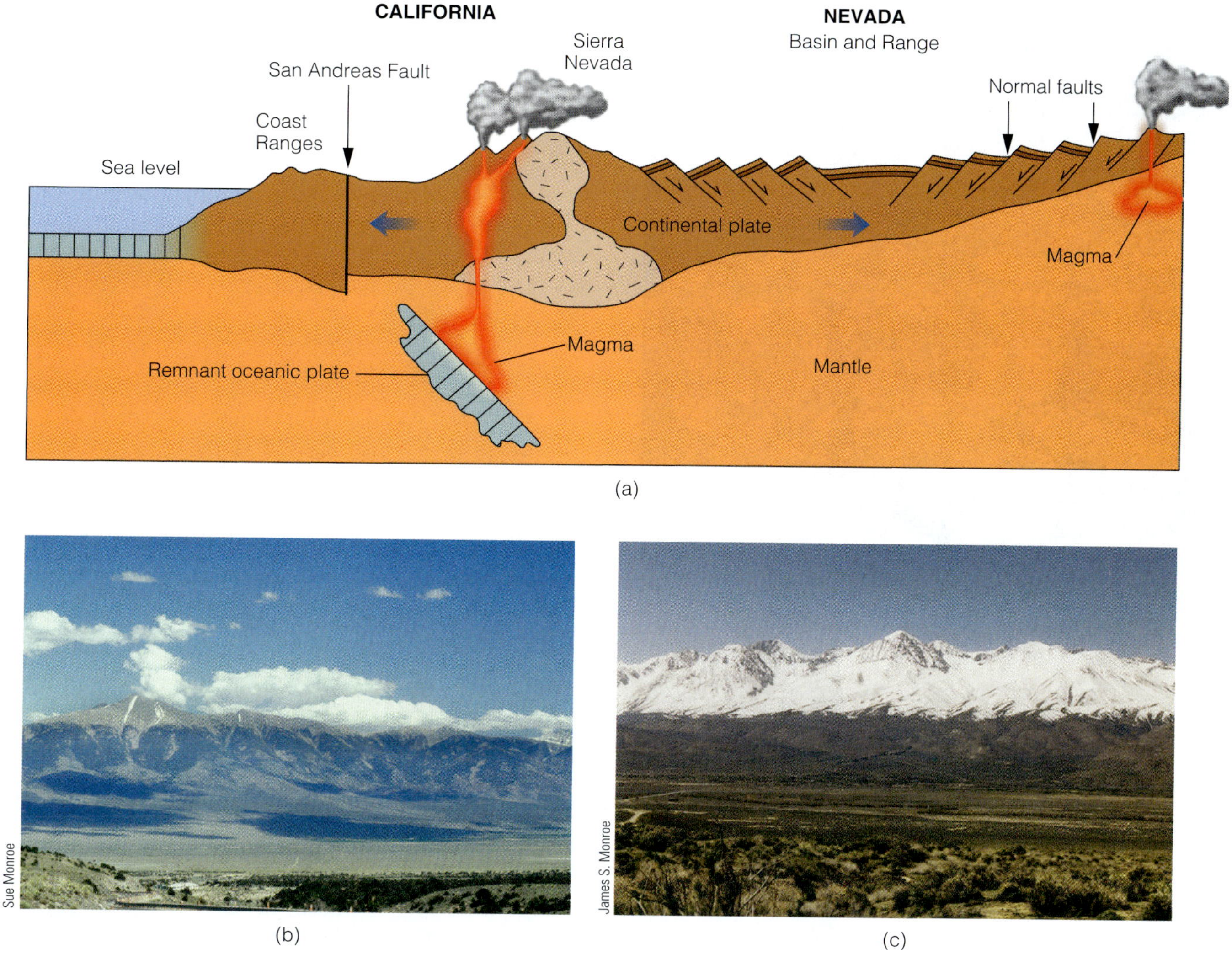

Figure 23.5

(a) Generalized cross section of the Basin and Range Province in Nevada. Notice that the ranges are bounded by normal faults. (b) The Snake Range in eastern Nevada, with one of the basins in the foreground. (c) View of the Sierra Nevada, which marks the western boundary of the Basin and Range Province. These mountains are bounded by normal faults and rise more than 3000 m above the basins to the east.

ment was transported eastward from the Cordillera and deposited in continental environments. It formed large, eastward-thinning wedges that now underlie the Great Plains (see the chapter opening photo).

Igneous activity was not widespread in the Interior Lowlands, but in some local areas it was significant. In northeastern New Mexico, for example, Neogene extrusive volcanism produced volcanoes and numerous lava flows. Small intrusive bodies were emplaced in Colorado, Montana, South Dakota, and Wyoming.

Eastward, beyond the Great Plains section of the Interior Lowlands, Cenozoic deposits, other than those of Pleistocene glaciers, are uncommon. Much of the Interior Lowlands was subjected to erosion during the Cenozoic. Of course, the eroded material had to be deposited somewhere, and that was on the Gulf Coastal Plain (Figure 23.2).

The Gulf Coastal Plain

Following the final withdrawal of the Cretaceous Zuni Sea, the Cenozoic **Tejas epeiric sea** made a brief appearance on the continent. But even at its maximum extent, it was largely restricted to the Atlantic and Gulf Coastal Plains. In fact, its greatest incursion onto North America was in the area of the Mississippi Valley, where it extended as far north as southern Illinois.

Sedimentary facies development on the Gulf Coastal Plain was controlled largely by a regression of the Cenozoic Tejas epeiric sea. Its regression, however,

Reed Wicander

(a)

Sue Monroe

(b)

■ **Figure 23.6**

The Colorado Plateau is underlain by a variety of rocks of various ages as in the Grand Canyon of Arizona (a) and Valley of the Gods in Utah (b).

was periodically reversed by minor transgressions; eight transgressive-regressive episodes are recorded in Gulf Coastal Plain sedimentary rocks.

The Gulf Coast sedimentation pattern was established during the Jurassic and persisted through the Cenozoic. Sediments were derived from the eastern Cordillera, western Appalachians, and Interior Lowlands and were transported toward the Gulf of Mexico. In general, the sediments form seaward-thickening wedges that grade from terrestrial facies in the north to progressively more offshore marine facies in the south (■ Figure 23.8).

Much of the Gulf Coastal Plain was dominated by detrital sediment deposition during the Cenozoic. In the Florida section of the coastal plain and the Gulf Coast of Mexico, though, thick sequences of carbonate rocks were deposited. A carbonate platform was established in Florida during the Cretaceous, and shallow-water carbonate deposition continued through the Early Paleogene. Carbonate deposition continues in Florida Bay, the Florida Keys, and the Great Bahama Bank.

Eastern North America

The eastern seaboard has been a passive continental margin since Late Triassic rifting separated North America from North Africa and Europe. Some seismic activity still occurs there (see Figure 8.7), but overall

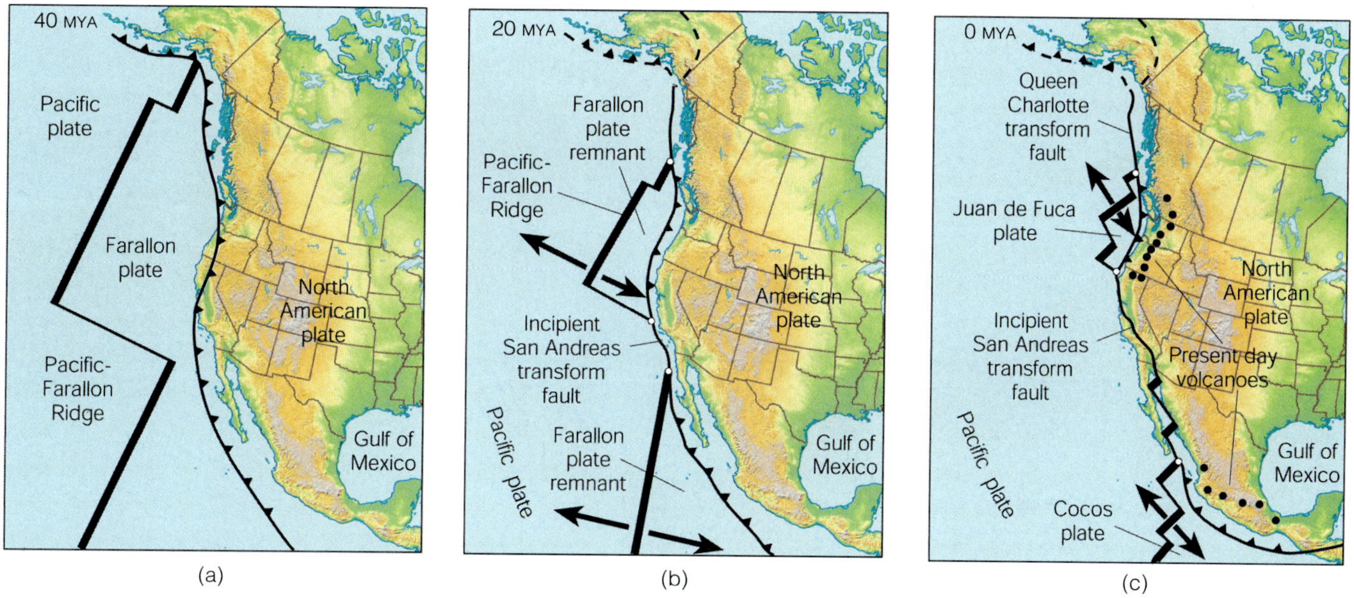

Figure 23.7

Three stages in the westward drift of North America, its collision with the Pacific–Farallon Ridge, and the origin of the Queen Charlotte and San Andreas faults. As the North American plate overrode the ridge, its margin became bounded by transform faults rather than a subduction zone. Source: Reprinted with permission from W. R. Dickinson, "Cenozoic Plate Tectonic Setting of the Cordilleran Region in the Western United States," *Cenozoic Paleogeography of the Western United States*. Pacific Coast Symposium 3, 1979, p. 2 (Fig 1).

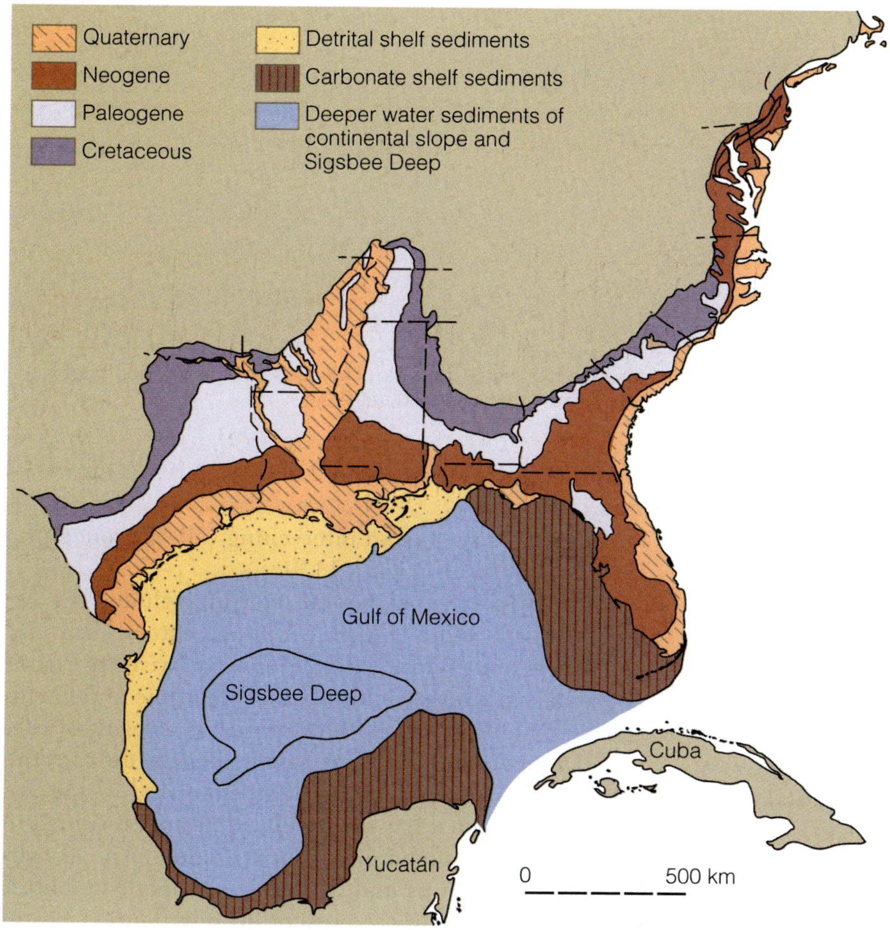

Figure 23.8

Cenozoic deposition on the Gulf Coastal Plain. Depositional provinces in the Gulf of Mexico and surface geology of the northwestern Gulf Coastal Plain. Source: Reprinted with permission from Walker and Coleman, "Atlantic and Gulf Coastal Province," in *DNAG, Centennial Special*, Vol 2, Geomorphic Systems of North America, 1987, p. 52 (Fig 1).

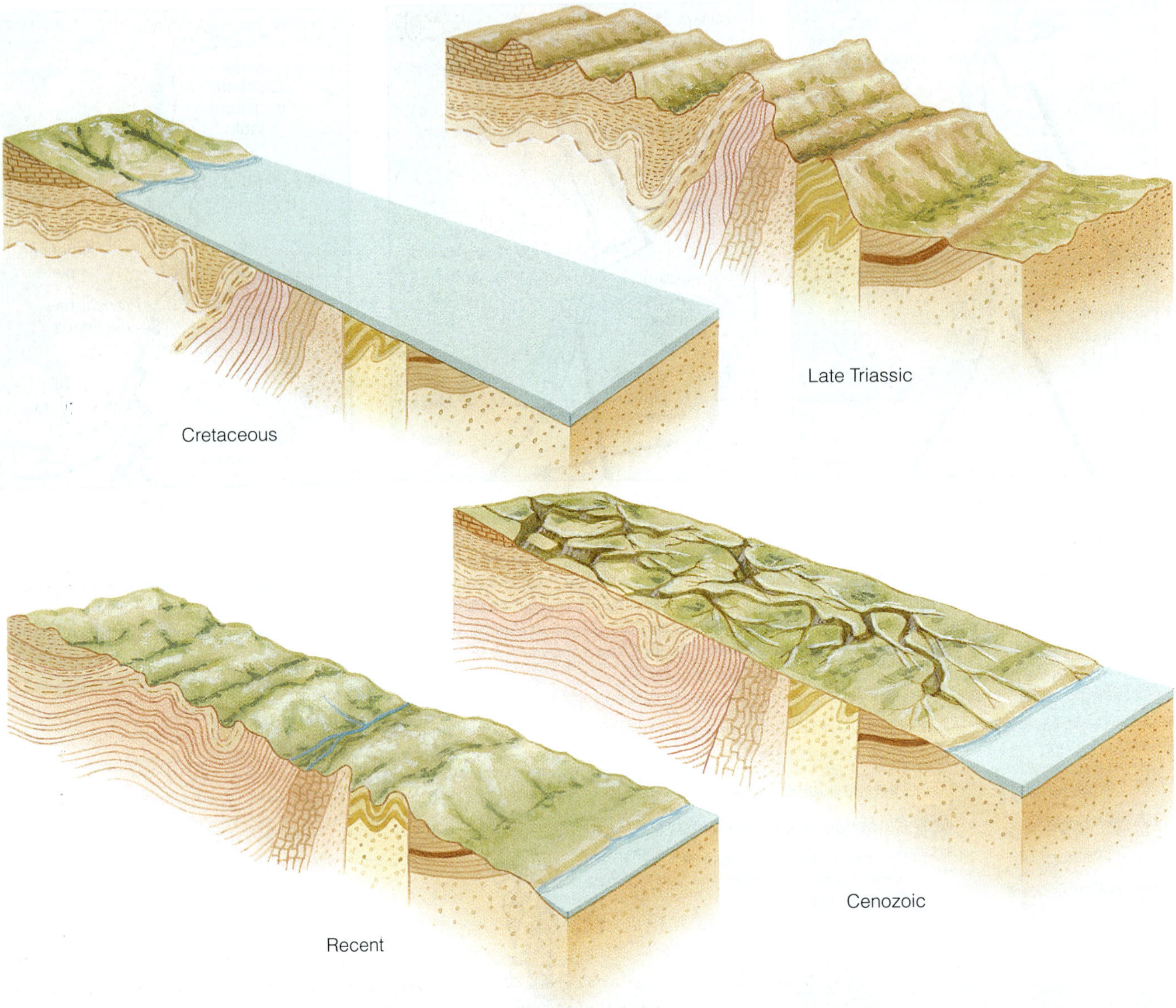

Figure 23.9

The origin of the present topography of the Appalachian Mountains. Erosion in response to Cenozoic uplift accounts for this topography.
Source: Reprinted with permission from D. Johnson, Stream Sculpture on the Atlantic Slope, 1931, Columbia University Press.

the region lacks the geologic activity characteristic of active continental margins.

The present distinctive topography of the Appalachian Mountains is the product of Cenozoic uplift and erosion. By the end of the Mesozoic, the Appalachian Mountains had been eroded to a plain (Figure 23.9). Cenozoic uplift rejuvenated the streams, which responded by renewed downcutting. As the streams eroded downward, they were superposed on resistant strata and cut large canyons across these rocks. For example, the distinctive topography of the Valley and Ridge Province is the product of Cenozoic erosion and preexisting geologic structures. It consists of northeast-southwest–trending ridges of resistant upturned strata and intervening valleys eroded into less resistant strata (Figure 23.9).

The Atlantic continental margin includes the Atlantic Coastal Plain (Figure 23.2) and extends seaward across the continental shelf, slope, and rise (Figure 23.10a). It possesses a number of Mesozoic and Cenozoic sedimentary basins that formed as a result of rifting. Deposition in these basins began during the Jurassic, and even though sediments of this age are known only from a few deep wells, they are presumed to underlie the entire margin of the continent. The distribution of Cretaceous and Cenozoic sediments is better known because both are present in the Atlantic Coastal

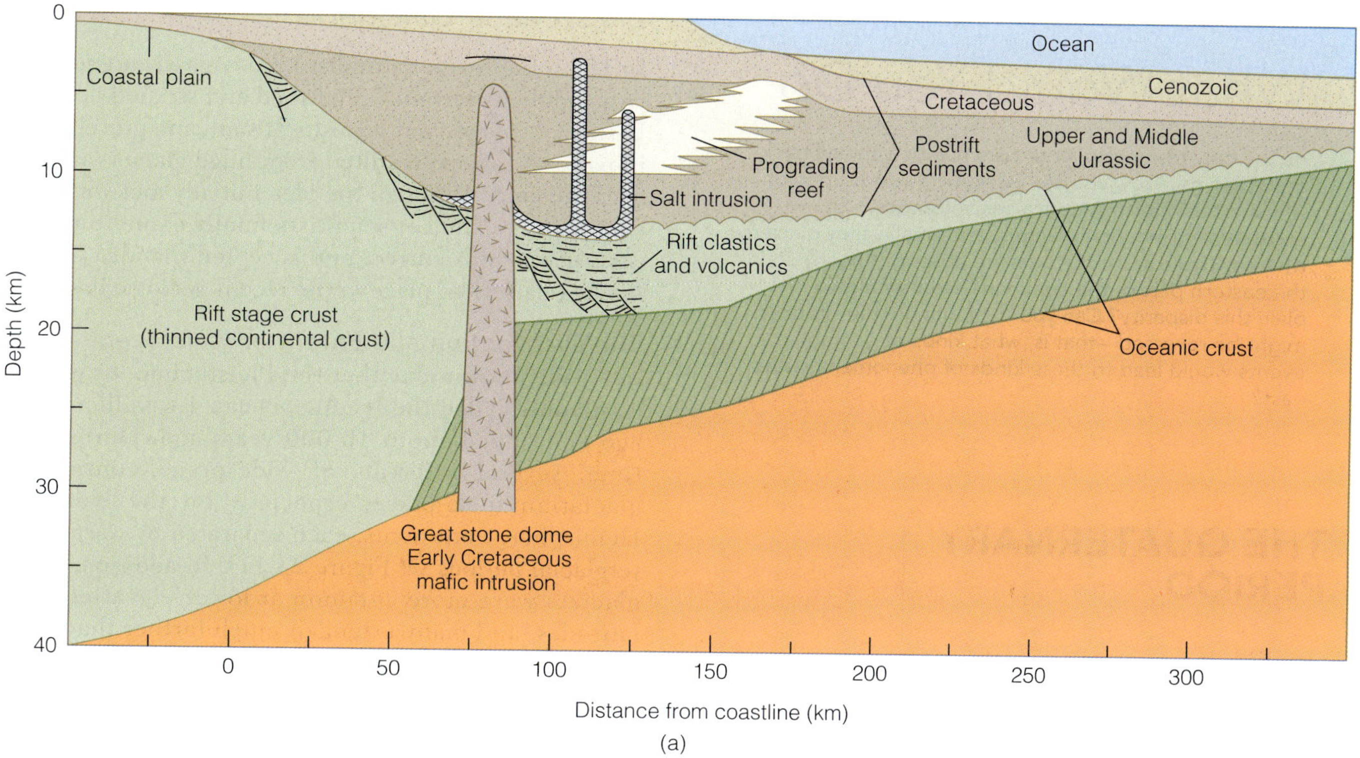

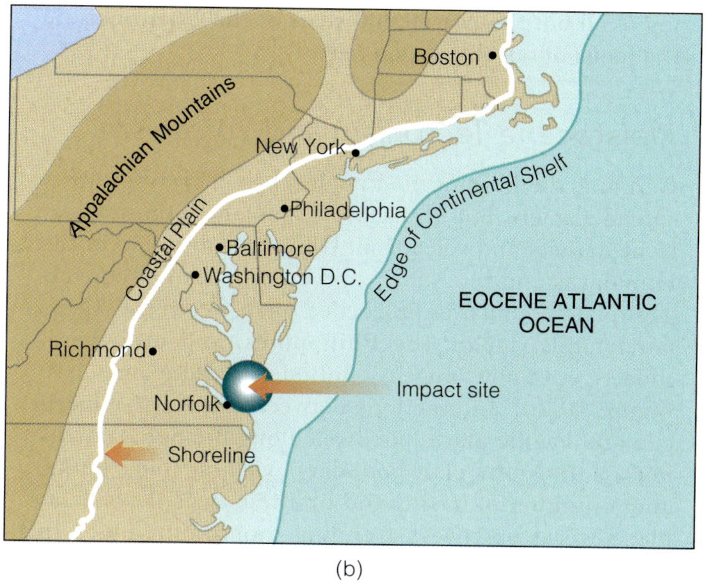

Figure 23.10

(a) The continental margin in eastern North America. The coastal plain and the continental shelf are covered mostly by Cenozoic sandstones and shales, which are underlain by Cretaceous and probably Jurassic sedimentary rocks. (b) Site of the Eocene Chesapeake Bay asteroid or comet impact.

Plain, and both have been penetrated by wells on the continental shelf.

In general, the sedimentary rocks of the Atlantic Coastal Plain are part of a seaward-thinning wedge that dips gently seaward. In some places, such as off the coast of New Jersey, these deposits are up to 14 km thick. The best-studied wedge of these sedimentary rocks is in the Baltimore Canyon Trough, an area that also exhibits the structures typical of a passive continental margin (Figure 23.10a).

Although the Atlantic Coastal Plain formed in the manner discussed above, an event took place during the Eocene (about 35 million years ago) that not only had a profound influence on the continental margin but still affects the area today. Geologic evidence is accumulating that an asteroid or comet 3 to 5 km in diameter hit this area, leaving an 85-km-diameter crater, which is now buried beneath 300 to 500 m of sediment (Figure 23.10b). Some of the lingering effects of this event are ground subsidence and disruption of aquifers in the region.

What Would You Do?

As the only resident of your community with any background in geology, you are asked by a curious person why western North America has volcanoes, earthquakes, mountain ranges, and small valley glaciers, whereas these same features are absent or nearly so in the eastern part of the continent. How would you explain this disparity? Can you think of how the situation might be reversed—that is, what kinds of geologic events would lead to these kinds of phenomena in the east?

THE QUATERNARY PERIOD

The most recent 1.6 million years of geologic time, the *Quaternary Period*, is divided into the *Pleistocene Epoch* (Ice Age) and the *Holocene* or *Recent Epoch* (Figure 23.1). the Pleistocene, consisting of all but the last 10,000 years of the Quaternary, is the focus of our discussion here.

Pleistocene Tectonism and Volcanism

Certainly the Pleistocene Epoch is best known for widespread glaciers, but it was also a time of continuing tectonic activity and volcanism. Orogeny continued in the Himalayas and the Andes, and deformation at convergent plate boundaries proceeded unabated in the Aleutian Islands, Japan, the Philippines, and elsewhere. Indeed, even now orogeny continues in these areas (see Figure 10.20). Interactions between the North American and Pacific plates persisted along the San Andreas fault (a transform plate boundary), yielding folds, faults, and a number of basins and uplifted areas. For example, several east-west–trending mountain ranges in southern California owe their existence to stresses created along a bend in the San Andreas fault.

Ongoing subduction of remnants of the Farallon plate beneath Central America and the Pacific Northwest accounts for Quaternary and present-day volcanism in these two regions. Although the Cascade Range began evolving far back in the Paleocene, the large composite volcanoes such as Mount Shasta and Mount Rainier as well as the lava dome known as Lassen Peak are mostly Pleistocene and Recent (see "Cascade Range Volcanoes" on pages 694 and 695). Huge calderas formed in what is now Yellowstone National Park in Wyoming following colossal eruptions that began during the Late Neogene and continuing into the Pleistocene (see Geo-Focus 23.1).

Pleistocene Glaciation

In 1837, the Swiss naturalist Louis Agassiz argued that large boulders (erratics), polished and striated bedrock, U-shaped valleys, and deposits of sand and gravel found in parts of Europe resulted from huge glaciers moving over the land. Although the idea initially met with considerable resistance, scientists finally came to realize that Agassiz was correct and accepted the idea that an Ice Age had taken place in the recent geologic past.

The Distribution and Extent of Pleistocene Glaciers

We know today that the Pleistocene, or what is commonly called the Ice Age, began 1.6 million years ago and ended about 10,000 years ago. During this time, several intervals of widespread continental glaciation took place, especially on the Northern Hemisphere continents, each separated by warmer interglacial periods (■ Figure 23.11). In addition, valley glaciers were more common at lower elevations and latitudes, and many extended much farther than their shrunken remnants do today (see Geo-Focus 14.1). Unfortunately, scientists do not know whether we are still in an interglacial period or entering another cooler glacial interval.

As one would expect, the climatic effects responsible for Pleistocene glaciers were worldwide. Nevertheless, Earth was not as frigid as portrayed in movies and cartoons, nor was the onset of the climatic conditions leading to glaciation very rapid. Indeed, evidence from several types of investigations indicates that the climate gradually cooled from the Eocene through the Pleistocene. Furthermore, evidence from oxygen isotope data (the ratio of O^{18} to O^{16}) from deep-sea cores shows that 20 major warm-cold cycles have occurred during the last 2 million years.

The Effects of Glaciation

Glaciation has had many direct and indirect effects. Glaciers moving over Earth's surface have produced distinctive landscapes in much of Canada, the northern tier of states, and the mountains of the West (see Chapter 14). Sea level has risen and fallen with the formation and melting of glaciers, and these changes in turn have affected the margins of continents. Glaciers have also altered the world's climate, causing cooler and wetter conditions in some areas that are arid to semiarid today. In addition to the usual evidence of glacial activity, one of the largest floods in history was caused by the collapse of an ice dam in eastern Washington.

More than 70 million km^3 of snow and ice blanketed the continents during the maximum glacial coverage of the Pleistocene. The storage of ocean waters in glaciers lowered sea level 130 m and exposed large areas of the present-day continental shelves, which were soon covered by vegetation. Lowering of sea level also affected the base level of rivers and streams. When sea level

(a)

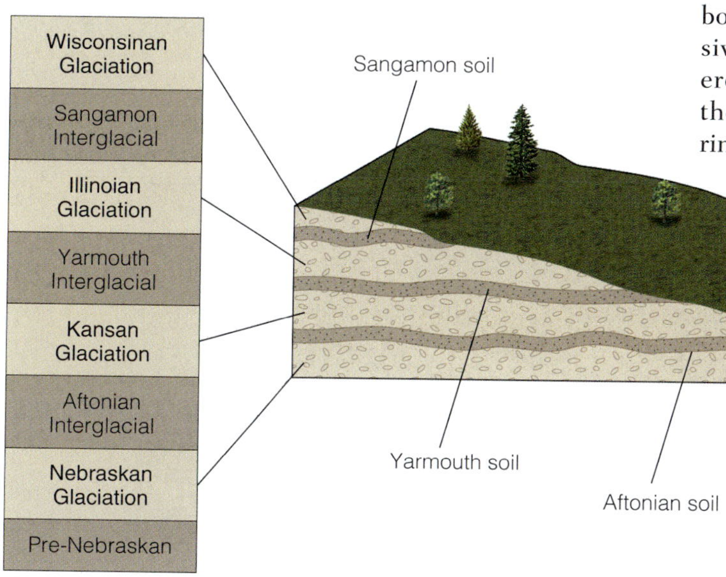

(b)

GeologyNow ■ Active Figure 23.11

(a) Centers of ice accumulation and maximum extent of Pleistocene glaciers in North America. (b) Standard terminology for Pleistocene glacial and interglacial stages in North America.

dropped, streams eroded downward as they sought to adjust to a lower base level.

From such glacial features as terminal moraines, erratics, and drumlins (see Chapter 14), it seems that at their greatest extent Pleistocene glaciers covered about three times as much of Earth's surface as they do now (Figure 23.11a). That is, they covered more than 40 million km^2, and like the vast ice sheets now present in Greenland and Antarctica they were probably 3 km thick. Geologists have identified four major Pleistocene glacial episodes that took place in North America—the *Nebraskan, Kansan, Illinoian,* and *Wisconsinan,* each named for the states in which the most southerly glacial deposits are well exposed. The three interglacial stages are named for localities of well-exposed soils and other deposits (Figure 23.11b). In Europe, six or seven major glacial advances and retreats are recognized.

Recent studies show that there were an as yet undetermined number of pre-Illinoian glacial advances and retreats in North America, and that the glacial history of this continent is more complex than previously thought. In view of this evidence, the traditional four-part subdivision of the North American Pleistocene will have to be modified.

Stream channels in coastal areas were extended and deepened along the emergent continental shelves. When sea level rose with the melting of the glaciers, the lower ends of stream valleys along the East Coast of North America were flooded and are now important harbors, whereas just off the West Coast, they form impressive submarine canyons. Great amounts of sediment eroded by the glaciers were transported by streams to the sea and thus contributed to the growth of submarine fans along the base of the continental slope.

We noted in Chapter 10 that as the Pleistocene ice sheets formed and increased in size, the weight of the ice caused the crust to slowly subside deeper into the mantle. In some places, Earth's surface was depressed as much as 300 m below the preglacial elevations. As the ice sheets retreated by melting, the downwarped areas gradually rebounded to their former positions.

During the Wisconsinan glacial stage, many large lakes existed in what are now dry basins in the southwestern United States. These lakes formed as a result of greater precipitation and overall cooler temperatures (especially during the summer), which lowered the evaporation rate. At the same time, increased precipitation and runoff helped maintain high water levels. Lakes that formed during those times are *pluvial lakes,* and they correspond to the expansion of glaciers elsewhere. The largest of these lakes was Lake Bonneville, which attained a maximum size of 50,000 km^2 and a depth of at least 335 m (■ Figure 23.12 on page 699). The vast salt deposits of the Bonneville Salt Flats west of Salt Lake City, Utah, formed as parts of this ancient lake dried up; Great Salt Lake is simply the remnant of this once-vast lake. Another large pluvial lake (Lake Manly) existed in Death Valley, California, which is now the hottest, driest place in North America.

In contrast to pluvial lakes, which form far from glaciers, *proglacial lakes* are formed by the meltwater

Cascade Range Volcanoes

Several large volcanoes and hundreds of smaller volcanic vents are present in the Cascade Range of the Pacific Northwest. Medicine Lake Volcano and Newberry Volcano are shield volcanoes that lie just east of the main trend of the Cascade Range. Volcanism began in the range during the Oligocene, but the most recent episode started about 5 million years ago, and of course, it continues to the present.

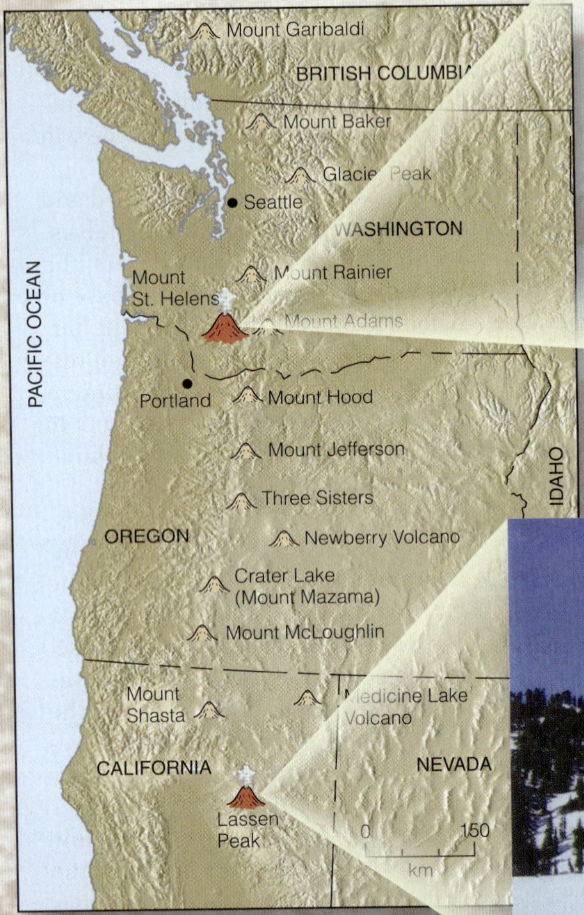

Mount St. Helens, a composite volcano, as it appeared from the east in 1978.

Lassen Peak today. This most southerly peak in the Cascade Range is made up of 2 km^3 of material including the bulbous masses of rock visible in this image.

Sue Monroe

Of these volcanoes, only Lassen Peak and Mount St. Helens erupted during the 1900s. Mount St. Helens began erupting again in October, 2004.

Lassen Peak is a lava dome that formed about 27,000 years ago on the northeast flank of a deeply eroded composite volcano known as Mount Tehama.

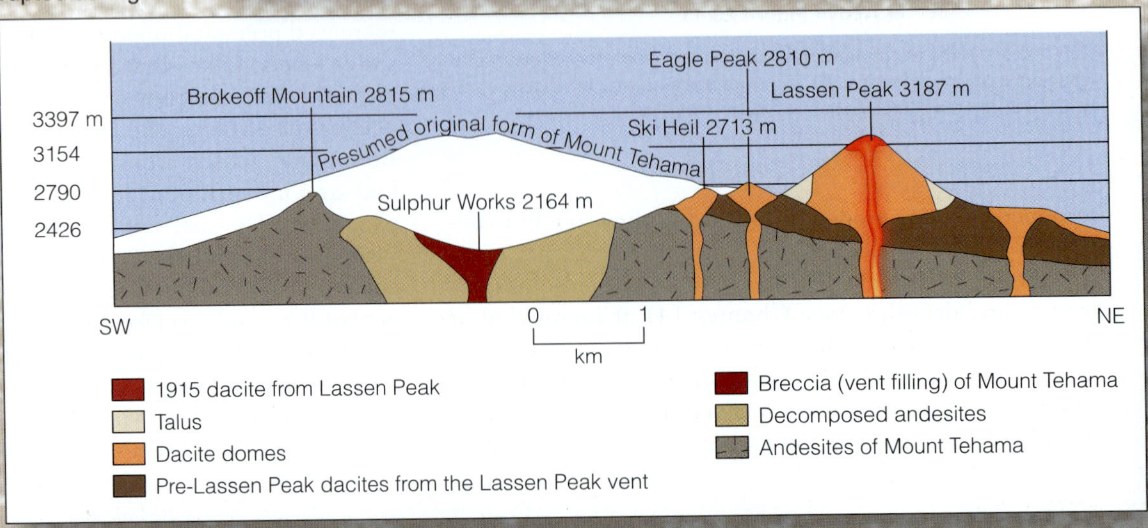

Courtesy of Keith Ronnholm

Mount St. Helens' lateral blast on May 18, 1980, took place when a bulge on the volcano's north face collapsed, reducing the pressure on gas-charged magma.

PhotoDisc/ Getty Images

Shortly after the lateral blast on May 18, 1980, this 19-km-high ash and steam cloud erupted from Mount St. Helens.

Lynn Topinka/ USGS

Mount St. Helens' lateral blast killed about 63 people and leveled 600 km^2 of forest. These trees in Smith Creek Valley are in part of the blast zone. For scale, see the two men within the red circle.

Lassen Peak erupted numerous times from 1914 to 1917. This eruption took place in 1915.

Sue Monroe

A huge steam explosion called the Great Hot Blast leveled the area in the foreground in 1915. In the 88 years since the eruption, trees are becoming reestablished in this area known as the Devastated Area.

GEOFOCUS 23.1

Supervolcanoes

Geologists have no formal definition for the term *supervolcano*, but we can take it to mean an eruption that results in the explosive ejection of hundreds of cubic kilometers of pyroclastic materials and the origin of a huge caldera. Fortunately, such voluminous eruptions are rare, and none have occurred during historic time. One supervolcano of particular interest in North America is the one that yielded the Yellowstone caldera in Wyoming. Continuing earthquakes and hydrothermal activity (hot springs and geysers) remind us that renewed volcanism in this area is possible.

The Yellowstone caldera lies within the confines of Yellowstone National Park, the first area in the United States set aside as a national park. Yellowstone National Park is noted for its scenery, wildlife, boiling mud pots, hot springs, and geysers, especially Old Faithful, but few tourists are aware of the region's volcanic history. The fact that a large body of magma is present beneath the surface is accepted among geologists, many of whom are convinced that the Yellowstone caldera remains active.

On three separate occasions supervolcano eruptions followed the accumulation of rhyolitic magma beneath the surface of the Yellowstone region. Each eruption yielded a widespread blanket of volcanic ash and pumice as well as collapse of the surface and origin of a large caldera. We can summarize Yellowstone's volcanic history by noting that supervolcano eruptions took place 2 million years ago, 1.3 million years ago, and 600,000 years ago. It was during this last huge eruption that the present-day Yellowstone caldera originated, which is actually part of a larger composite caldera resulting from the three cataclysmic events noted above (■ Figure 1a).

The magnitude of these supervolcano eruptions is difficult to imagine. In fact, the pyroclastic flows and airborne ash cover not only the areas within and adjacent to the Yellowstone region but also a large part of the western United States and northern Mexico. Nothing in the immediate areas of the eruptions could have survived the intense heat and choking clouds of volcanic ash.

Since the last supervolcano eruption, rhyolitic magma has continued to accumulate beneath the caldera, thereby elevating parts of its floor in what are known as *resurgent domes*. Precise leveling shows that the caldera floor continues to rise; more than 80 cm of uplift has taken place since 1923. And between 150,000 and 75,000 years ago, an additional 1000 km^3 of pyroclastic materials was erupted within the caldera. An excellent example is the Yellowstone Tuff into which the picturesque Grand Canyon of the Yellowstone River is incised (Figure 1b).

Many geologists are convinced that a *mantle plume*, a cylindrical mass of magma rising from the mantle, underlies the area. As this rising magma nears the surface, it triggers volcanic eruptions, and because the magma is rhyolitic the eruptions are particularly explosive. The Yellowstone hot spot, as it is called, is one of only a few dozen hot spots recognized on Earth.

that accumulates along the margins of glaciers. In fact, in many proglacial lakes one shoreline is the ice front itself, while the other shorelines consist of moraines. Lake Agassiz was a large proglacial lake covering about 250,000 km^2 of North Dakota, Manitoba, Saskatchewan, and Ontario. It persisted until the glacial ice along its northern margin melted, at which time the lake was able to drain northward into Hudson Bay.

CENOZOIC MINERAL RESOURCES

The United States is the third largest producer of petroleum, accounting for about 17% of the world's total. Much of this production comes

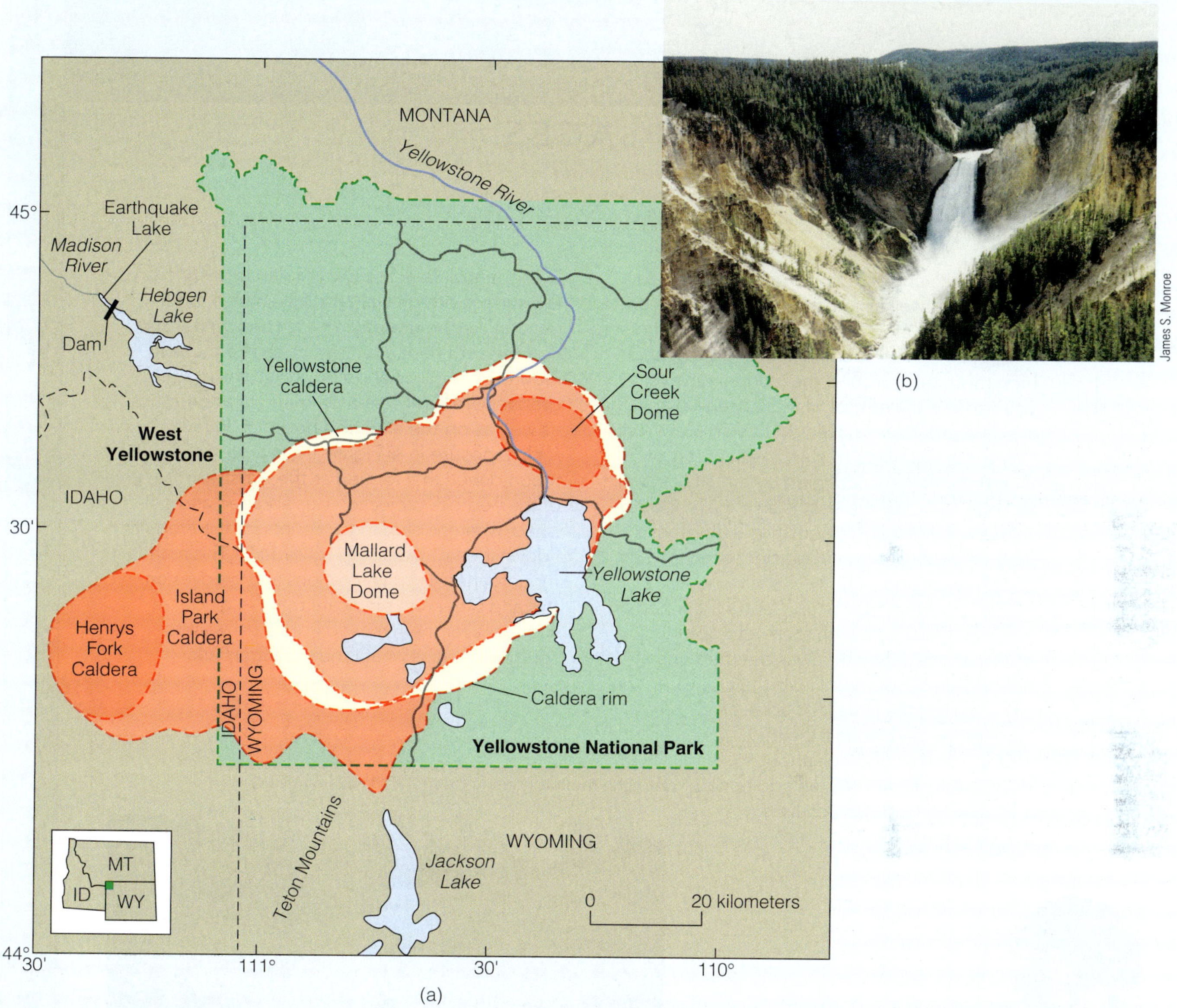

Figure 1

(a) The Yellowstone caldera formed following voluminous eruptions about 600,000 years ago. It is part of a larger composite caldera. (b) The walls of the Grand Canyon of the Yellowstone River are made up of hydrothermally altered Yellowstone Tuff that partly fills the Yellowstone Caldera.

from Cenozoic reservoirs of the Gulf Coastal Plain and the adjacent continental shelf. On the Gulf Coastal Plain, most of the petroleum is in structural traps related to salt domes and other structures. Several basins in southern California are also important areas of petroleum production. The Green River Formation of Wyoming, Utah, and Colorado has huge reserves of oil shale and evaporites.

Diatomite, is a sedimentary rock composed of the microscopic shells of diatoms, which are single-celled marine and freshwater plants that secrete skeletons of silica (SiO_2). This rock, also called diatomaceous earth, is used chiefly in gas purification and to filter liquids such as molasses, fruit juices, water, and sewage. The United States is the leader in diatomite production, mostly from Cenozoic deposits in California, Oregon, and Washington.

GEOLOGY IN UNEXPECTED PLACES

Evidence of Glaciation in New York City

Our discussion of glaciers here as well as in Chapter 14 has concentrated on Antarctica, Greenland, the Canadian Arctic islands, and high mountains around the world. However, you can see the effects of Pleistocene glaciers in some urban areas. Central Park in New York City is a good place to start. As a matter of fact, the area now occupied by the city as well as Long Island to the east is an excellent place to see erosional and depositional landforms created by a continental glacier as recently as 20,000 years ago.

The area that makes up Central Park had an irregular landscape with rock outcroppings, bluffs, and marshes that was deemed unsuitable for development. Nevertheless, when the city acquired the initial 700 acres of land for the park during the mid-1800s, it had to displace about 1600 poor residents. In 1863, an additional 143 acres of land was acquired, bringing the park to its present size.

In Central Park many of the rocks resting on the surface are glacial erratics—that is, boulders unlike the underlying rock that were carried far from their sources, much like those in Figure 14.14. Glacial polish, striations, and grooves are easily seen on many of the erratics as well as on the exposed bedrock. In fact, the rock exposure in ■ Figure 1 is a large roche mountonnée that formed as the "upstream" side of the bedrock projection was abraded and polished, whereas plucking accounts for the irregular "downstream" side (see Figure 14.9). This roche mountonnée and the striations and grooves on this and other rocks indicate that the glacier responsible for them moved from northwest to southeast across what is now Central Park.

■ **Figure 1**

A roche mountonnée in Central Park in New York City.

D. D. Trent

Huge deposits of low-grade lignite and subbituminous coals in the northern Great Plains are becoming increasingly important resources. These coal deposits are Late Cretaceous to Early Paleogene in age and are most extensive in the Williston and Powder River basins of Montana, Wyoming, and North and South Dakota. In addition to having a low sulfur content, some of these coals are in beds 30 to 60 m thick!

Gold production from the Pacific Coast, particularly California, comes mostly from Cenozoic gravels. The gold is found in placer deposits, which formed as concentrations of minerals separated from weathered debris by fluvial processes.

A variety of other mineral deposits are also important. For example, the United States must import almost all manganese used in the manufacture of steel. The

Figure 23.12

Pleistocene lakes in the western United States. Lake Missoula was a proglacial lake, but the others were pluvial lakes.

largest manganese deposits are in Cenozoic rocks in Russia. One molybdenum deposit in Colorado accounts for much of the world production of this element. Cenozoic sand and gravel, as well as evaporites, building stone, and clay deposits, are quarried from areas around the world.

Sand and gravel deposits resulting from glacial activity are a valuable Quaternary resource in many formerly glaciated areas. Most Pleistocene sands and gravels originated as floodplain deposits, outwash sediment, or esker deposits. The bulk of the sand and gravel in the United States and Canada is used in construction and as roadbase and fill for highway and railway construction.

The periodic evaporation of pluvial lakes in the Death Valley region of California during the Pleistocene led to the concentration of many evaporite minerals such as borax. During the 1880s, borax was transported from Death Valley by the famous 20-mule team wagon trains.

Another Quaternary resource is peat, a vast potential energy resource that has been developed in Canada and Ireland. Peatlands formed from plant assemblages as the result of particular climate conditions.

PALEOGENE AND NEOGENE LIFE HISTORY

Earth's flora and fauna continued to evolve during the Cenozoic Era as more and more familiar kinds of plants and animals made their appearance. And even though we emphasize the evolution of mammals in this chapter, you should be aware of other important life events. Flowering plants continued to dominate land-plant communities, the present-day groups of birds evolved during the Paleogene, and some marine invertebrates continued to diversify, eventually giving rise to today's marine fauna.

Mammals coexisted with dinosaurs for more than 100 million years, yet their Mesozoic fossil record indicates that they were not abundant, diverse, or very large. Extinctions at the end of the Mesozoic eliminated dinosaurs and some of their relatives, thereby creating the adaptive opportunities that mammals quickly exploited. The "Age of Mammals," as the Cenozoic Era is commonly called, had begun.

Marine Invertebrates and Phytoplankton

The Cenozoic marine ecosystem was populated by those plants, animals, and single-celled organisms that survived the Mesozoic extinctions. Especially prolific Cenozoic invertebrate groups were foraminifera, radiolarians, corals, bryozoans, mollusks, and echinoids. The marine invertebrate community in general became more provincial during the Cenozoic because of changing ocean currents and temperature differences with latitude.

Only a few species in each major group of phytoplankton survived into the Cenozoic. The coccolithophores, diatoms, and dinoflagellates all recovered from their Late Cretaceous reduction in numbers. The diatoms were particularly abundant during the Miocene, probably because of increased volcanism during this time. Volcanic ash provided increased dissolved silica in seawater and was used by the diatoms to construct their skeletons.

The foraminifera were a major component of the Cenozoic marine invertebrate community. Though dominated by relatively small forms (■ Figure 23.13), it included some exceptionally large forms that lived in the warm waters of the Cenozoic Tethys Sea. Shells of these larger foraminifera accumulated to form thick limestones, some of which were used by the ancient Egyptians to construct the Pyramids of Giza.

The corals, having relinquished their reef-building role to the rudists during the mid-Cretaceous, again became the dominant reef builders during the Cenozoic. Other suspension feeders such as bryozoans and crinoids were also abundant and successful during the Cenozoic. Perhaps the least important of the Cenozoic marine invertebrates were the brachiopods, with fewer than 60 genera surviving today.

Just as during the Mesozoic, bivalves and gastropods were two of the major groups of marine invertebrates, and they had a markedly modern appearance. Following the extinction of the ammonites and belemnites at the end of the Cretaceous, the Cenozoic cephalopod fauna consisted of nautiloids and cephalopods lacking shells, such as squids and octopuses.

The echinoids continued their expansion in the infaunal habitat and were particularly prolific during the Cenozoic. New forms such as sand dollars evolved during this time from biscuit-shaped ancestors.

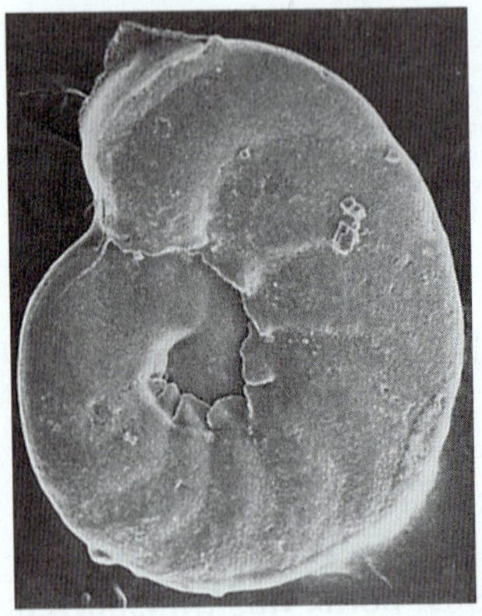
(a)

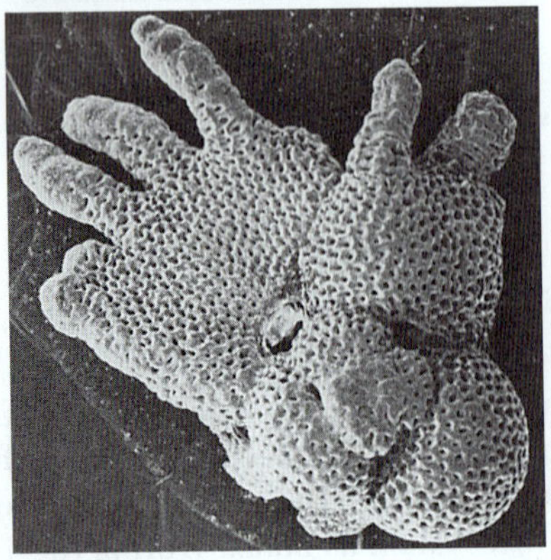
(b)

B. A. Masters

■ **Figure 23.13**

Foraminifera of the Cenozoic Era. (a) *Cibicides americanus*, a benthonic form from the Early Miocene of California. (b) A planktonic form, *Globigerinoides fistulosus*, from the Pleistocene, South Pacific Ocean.

Paleogene and Neogene Birds

The first members of many of the living orders of birds, including owls, hawks, ducks, penguins, and vultures, evolved during the Early Paleogene. Beginning during the Miocene Epoch, a marked increase in the variety of songbirds took place, and by 5 to 10 million years ago many of the existing genera of birds were present. Even hummingbirds much like those today had evolved by Oligocene time. Birds adapted to numerous habitats and continued to diversify during the Pleistocene, but since then their diversity has decreased slightly.

Several varieties of large flightless birds existed during the Pleistocene (discussed later) and today, but some remarkable predatory birds lived during the Paleogene and Neogene. For instance, *Diatryma*, a heavily built, flightless bird that stood about 2 m high, lived during the Paleocene and Eocene in North America and Europe (■ Figure 23.14). About 25 species of flightless birds up to 3 m tall are known from South America, where they were

James S. Monroe

■ **Figure 23.14**

Restoration of *Diatryma,* a large, flightless, predatory bird that stood more than 2 m tall. It lived during the Paleocene and Eocene in North America and Europe.

the dominant predators until they were replaced by big cats and dogs that migrated from North America.

Diversification of Mammals

Among living mammals **monotremes,** such as the platypus, lay eggs, whereas **marsupials** and **placentals** give birth to live young. Marsupial mammals are born in an immature, almost embryonic condition, and then undergo further development in their mother's pouch. In placental mammals, on the other hand, the amnion of the amniote egg (see Figure 21.26) has fused with the walls of the uterus, forming a *placenta.* Nutrients and oxygen carried from the mother to the embryo through the placenta permit the young to develop much more fully before birth.

A measure of the success of placental mammals is that more than 90% of all mammals, fossil and living, are placental. In contrast, judging from the fossil record, monotremes have never been very common, and the only living ones are platypuses and spiny anteaters of the Australian region. Marsupials have been more successful, at least in terms of number of species and geographic distribution, but even they have been largely restricted to South America and the Australian region.

What Would You Do?

You are a science teacher who through remarkably good fortune receives numerous unlabeled mammal and plant fossils from a generous benefactor. All of the fossils came from Oligocene- and Miocene-aged rocks. You're not too concerned with identifying genera and species, but you do want to show your students various mammal adaptations for diet and speed. What features of the skulls, teeth, and bones would allow you to infer which animals were herbivores (grazers versus browsers) and carnivores, and which ones were speedy runners? Also, could you use the fossil leaves to make any inference about ancient climates?

Although mammals first appeared during the Triassic, major diversification began during the Paleocene and continued throughout the Cenozoic. Several groups of Paleocene mammals are considered *archaic,* meaning that they were holdovers from the Mesozoic Era or they did not give rise to any of today's mammals (■ Figure 23.15). Also among these mammals were the first rodents, rabbits, primates, carnivores, and hoofed mammals. However, even these had not yet become clearly differentiated from their ancestors, and the differences between herbivores and carnivores were slight. Most were small; large mammals were not present until the Late Paleocene, and the first giant terrestrial mammals did not appear until the Eocene (■ Figure 23.16).

Diversification continued during the Eocene when several more types of mammals appeared, but if we could go back and visit this time, we would probably not recognize many of these animals. Some would be vaguely familiar, but the ancestors of horses, camels, rhinoceroses, and elephants would bear little resemblance to their living descendants. By Oligocene time, all the basic groups of existing mammals—that is, the orders—were present, but diversification continued as more familiar families and genera appeared. Miocene and Pliocene mammals were mostly mammals that we could readily identify, although a few types were unusual (■ Figure 23.17).

Cenozoic Mammals

Mammals arose from mammal-like reptiles known as cynodonts during the Late Triassic, so two-thirds of their evolution was during the Mesozoic Era (see Chapter 22). However, following the Mesozoic extinctions, mammals began to diversify and soon became the most abundant land-dwelling vertebrates. Now, more than 4000 species exist, ranging from tiny shrews to giants such as elephants and whales.

When people consider mammals, they think mostly of larger species such as elephants, horses, deer, dogs, and

■ **Figure 23.15**

The archaic mammals of the Paleocene Epoch included such animals as (1) *Protictis,* an early carnivore; (2) insectivores; (3) *Ptilodus;* and (4) *Pantolambda,* which stood about 1 m tall.

The Field Museum, Chicago, Neg. #CK46T

■ **Figure 23.16**

The uintatheres were Eocene rhinoceros-sized mammals with three pairs of bony protuberances on the skull and saberlike upper canine teeth.

Figure 23.17

Mural showing Pliocene mammals of the western North America grasslands. The animals shown include (1) *Amebeledon*, a shovel-tusked mastodon; (2) *Teleoceras*, a short-legged rhinoceros; (3) *Cranioceras*, a horned, hoofed mammal; (4) a rodent; (5) a rabbit; (6) *Merycodus*, an extinct pronghorn; (7) *Synthetoceras*, a hoofed mammal with a horn on its snout; and (8) *Pliohippus*, a one-toed grazing horse.

cats, but they fail to realize that most mammals are small, weighing less than 1 kg. In fact, with few exceptions, rodents, insectivores, rabbits, and bats fall into this category and they constitute fully 75% of all mammal species. These animals adapted to the microhabitats unavailable to larger mammals, or in the case of bats, became the only flying mammals. With this in mind, we now turn our attention to some of the larger ones, especially hoofed mammals, carnivores, elephants, and whales.

Hoofed Mammals **Ungulate** is a general term referring to several types of hoofed mammals but especially the orders Artiodactyla and Perissodactyla. About 170 living species of antelope, camels, giraffes, deer, goats, peccaries, pigs, and several others are even-toed hoofed mammals, or **artiodactyls,** the most common living ungulates. In marked contrast, only 16 species of horses, rhinoceroses, and tapirs are **perissodactyls,** or odd-toed hoofed mammals. As even- and odd-toed imply, artiodactyls have two or four toes while perissodactyls have one or three (■ Figure 23.18b).

All ungulates are herbivores, but some are **grazers,** meaning they feed on grasses, and others are **browsers,** feeding on the tender shoots, twigs, and leaves of trees and

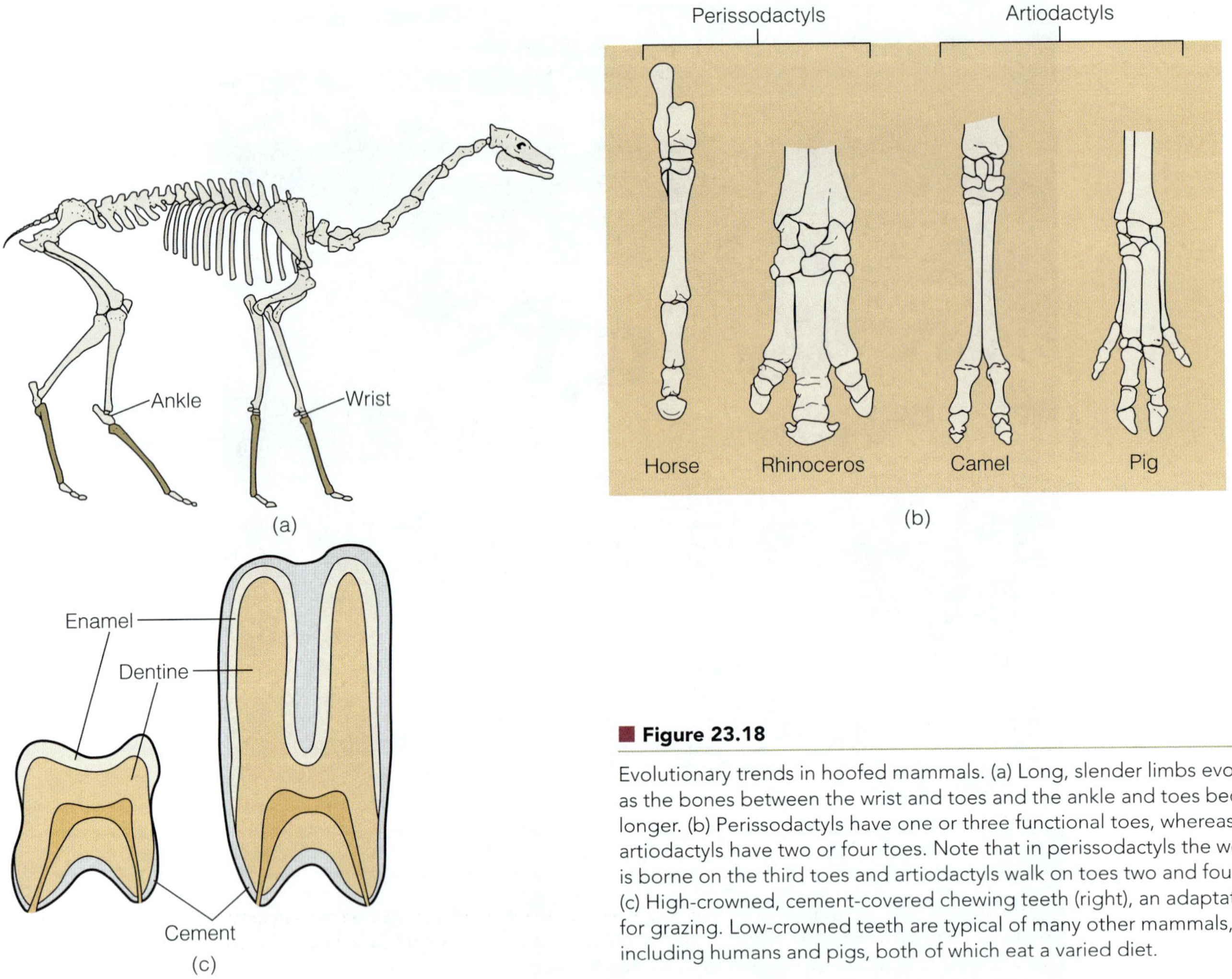

■ **Figure 23.18**

Evolutionary trends in hoofed mammals. (a) Long, slender limbs evolved as the bones between the wrist and toes and the ankle and toes became longer. (b) Perissodactyls have one or three functional toes, whereas artiodactyls have two or four toes. Note that in perissodactyls the weight is borne on the third toes and artiodactyls walk on toes two and four. (c) High-crowned, cement-covered chewing teeth (right), an adaptation for grazing. Low-crowned teeth are typical of many other mammals, including humans and pigs, both of which eat a varied diet.

bushes. When grasses grow through soil, they pick up tiny pieces of sand that are quite abrasive to teeth, so the grazing ungulates developed high-crowned chewing teeth resistant to abrasion (Figure 23.18c). Browsers, on the other hand, never developed these kinds of chewing teeth.

Some ungulates are small and depend on concealment to avoid predators; others, such as rhinoceroses, are so large that size alone is enough to discourage predators, at least for adults. But many of the more modest-sized ungulates are speedy runners. Adaptations for running include elongation of some of the limb bones as well as reduction in the number of bony limb elements, especially toes. Accordingly, the limbs of speedy ungulates are long and slender (Figure 23.18a).

Rabbit-size ancestral artiodactyls of the Early Eocene differed little from their ancestors but gave rise to numerous families, several of which are now extinct (■ Figure 23.19). Small four-toed camels, for instance, appeared early in this diversification and were common in North America well into the Pleistocene. In fact, most of their evolution took place on this continent and only during the Pliocene did they migrate to South America and Asia where they now exist.

Certainly the most common existing artiodactyls are the *bovids*, represented by cattle, goats, bison, and many species of antelope. They first appeared in the Miocene fossil record and continued to diversify throughout the rest of the Cenozoic. Most bovid evolution took place in Europe and northern Asia, but these creatures have since migrated to southern Asia and Africa where they are most common today.

The perissodactyls—horses, rhinoceroses, tapirs—and the extinct titanotheres and chalicotheres are united by several shared characteristics (see Figure 18.17). Furthermore, the fossil record shows they all evolved from a common ancestor during the Eocene. Their diversity increased through the Oligocene, but since then they have declined markedly and now constitute less than 10% of the world's hoofed mammal fauna.

Horses have a particularly good fossil record, which shows that present-day *Equus* evolved from a tiny Eocene ancestor (■ Figure 23.20). Most of their evolu-

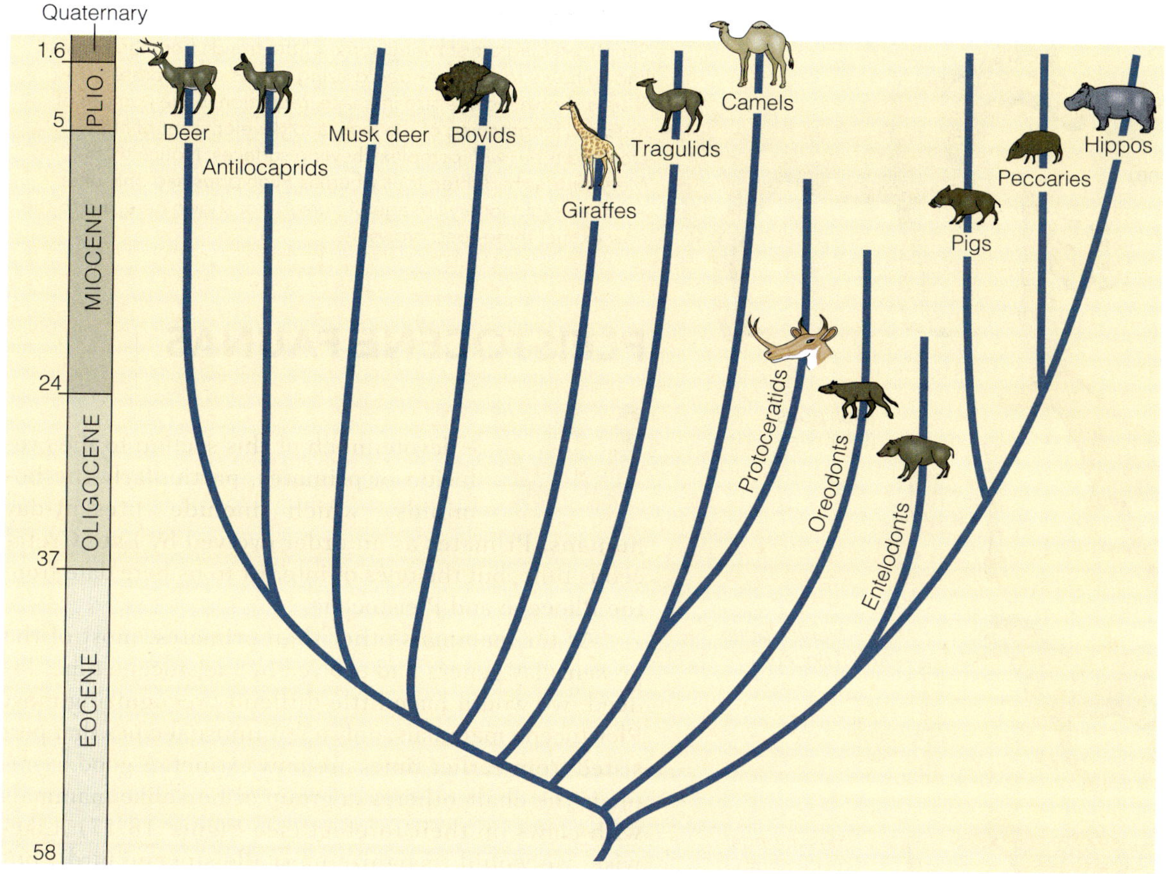

■ **Figure 23.19**

Relationships among the living artiodactyls and some of those extinct. Most artiodactyls are ruminants—that is, cud-chewing animals—the major exception being the pigs and their relatives the peccaries and hippos. The bovids, consisting of dozens of species of sheep, goats, antelope, and bison, are by far the most diverse and abundant artiodactyls.

tion took place in North America. Fossils show several trends such as increased size, lengthening of the limbs, and development of high-crowned chewing teeth as horses became speedy grazing animals (Table 23.1). There was, however, another branch of horses leading to three-toed browsers that became extinct during the Pleistocene.

Other Mammals—Carnivores, Elephants, and Whales During the Paleocene, the *miacids,* small carnivorous mammals with short heavy limbs, made their appearance. These small creatures were ancestors to all later members of the order Carnivora, which includes, among others, today's dogs, cats, hyenas, bears, weasels, and seals. All have well-developed canine teeth for slashing and tearing, and most also developed a pair of large shearing teeth (■ Figure 23.21a). Some of the better-known fossil carnivores are the saber-toothed cats, or what are more commonly called saber-toothed tigers (Figure 23.21b).

Elephants (order Proboscidea), the largest land mammals, evolved from pig-sized ancestors during the Eocene. And by Oligocene time, they clearly showed the trend toward large size, a long snout (proboscis), and large tusks. Mastodons with teeth adapted for browsing were present by the Miocene, and during the Pliocene the present-day elephants and mammoths diverged from a common ancestor. During most of the Cenozoic, elephants were widespread on the northern continents, but now only two species exist in southern Asia and Africa.

We briefly mentioned whales in Chapter 18 in Fossils: What Do We Learn from Them?, noting that until recently little was known about their transition from land-dwelling ancestor to fully aquatic whales. Although a number of questions remain unanswered, the fossils now available indicate that whales appeared during the Early Eocene and by the Late Eocene had become diverse and widespread. Eocene whales still possessed vestigial rear limbs, their teeth resembled those of their land-dwelling ancestors, their nostrils (blowhole) were not on top of the head, and they were proportioned quite differently from living whales (■ Figure 23.22). By Oligocene time, both groups of living whales—the toothed whales and the baleen whales—had evolved.

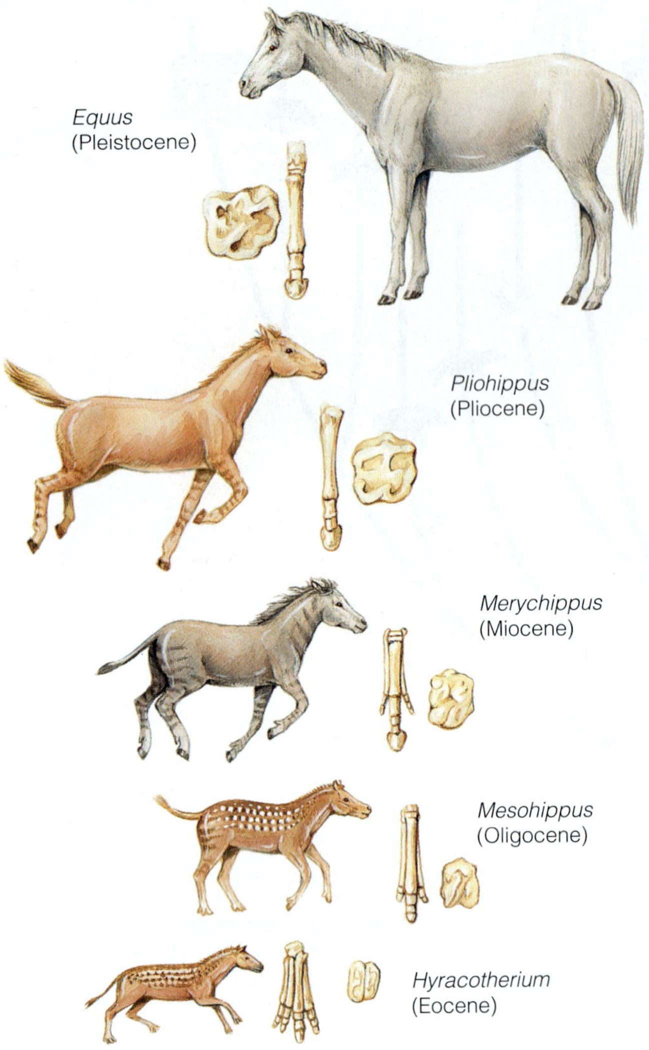

Figure 23.20

Simplified diagram showing some of the trends from the earliest known horse to the one-toed grazing horses of the present. Trends shown include size increase, reduction in the number of toes and lengthening of the legs, and development of high-crowned teeth with complex chewing surfaces. Notice that *Merychippus* had three toes whereas *Pliohippus* had only one. Another evolutionary lineage of horses, not shown here, led to the now extinct three-toed browsers.

PLEISTOCENE FAUNAS

We devote much of this section to the evolution of primates, particularly the hominids, which include present-day humans. Primates as an order evolved by Late Cretaceous time, but the ones of interest to us here date from the Pliocene and Pleistocene.

As for mammals other than primates, most of the present-day genera had evolved by Pleistocene time. Indeed, we would have little difficulty recognizing most Pleistocene mammals; only a few unusual types that persisted from earlier times are now extinct. A good example is the chalicotheres, a group of horselike mammals with claws on their forefeet (see Figure 18.17). Likewise, we would recognize most Pleistocene birds, but some large ground-dwelling species are now extinct.

Mammals and Birds

One of the most remarkable aspects of the Pleistocene mammalian fauna is that so many large species existed.

Table 23.1

Overall Trends in the Cenozoic Evolution of the Present-Day Horse *Equus*

1. Size increases.
2. Legs and feet become longer, an adaptation for running.
3. Lateral toes are reduced to vestiges. Only toe three remains functional in *Equus*.
4. The back straightens and stiffens.
5. Incisor teeth become wider.
6. Molarization of premolars yields a continuous row of teeth for grinding vegetation.
7. The chewing teeth, molars and premolars, become high-crowned and cement-covered for grinding abrasive grasses.
8. Chewing surfaces of premolars and molars become more complex, also an adaptation for grinding abrasive grasses.
9. Front part of skull and lower jaw become deeper to accommodate high-crowned premolars and molars.
10. Face in front of eye becomes longer to accommodate high-crowned teeth.
11. Brain becomes larger and more complex.

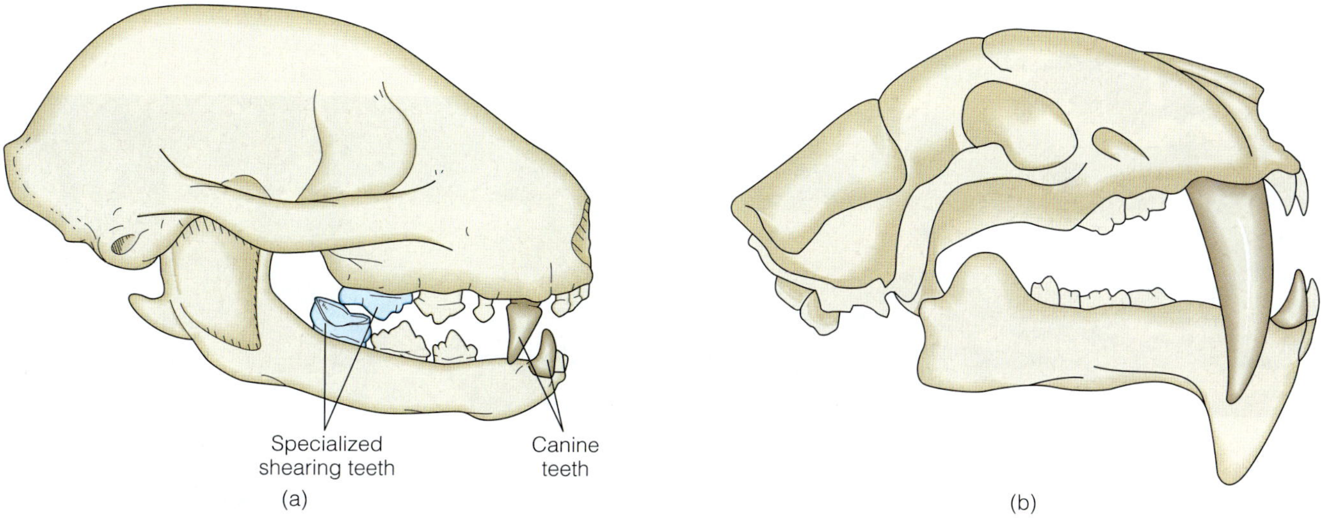

Figure 23.21

(a) This present-day cat skull and jaw show the specialized sharp-crested shearing teeth present in carnivorous mammals. Note also the enlarged canines. (b) A number of Cenozoic saber-toothed cats had huge canine teeth. This one is the Oligocene saber-tooth *Euismilus*.

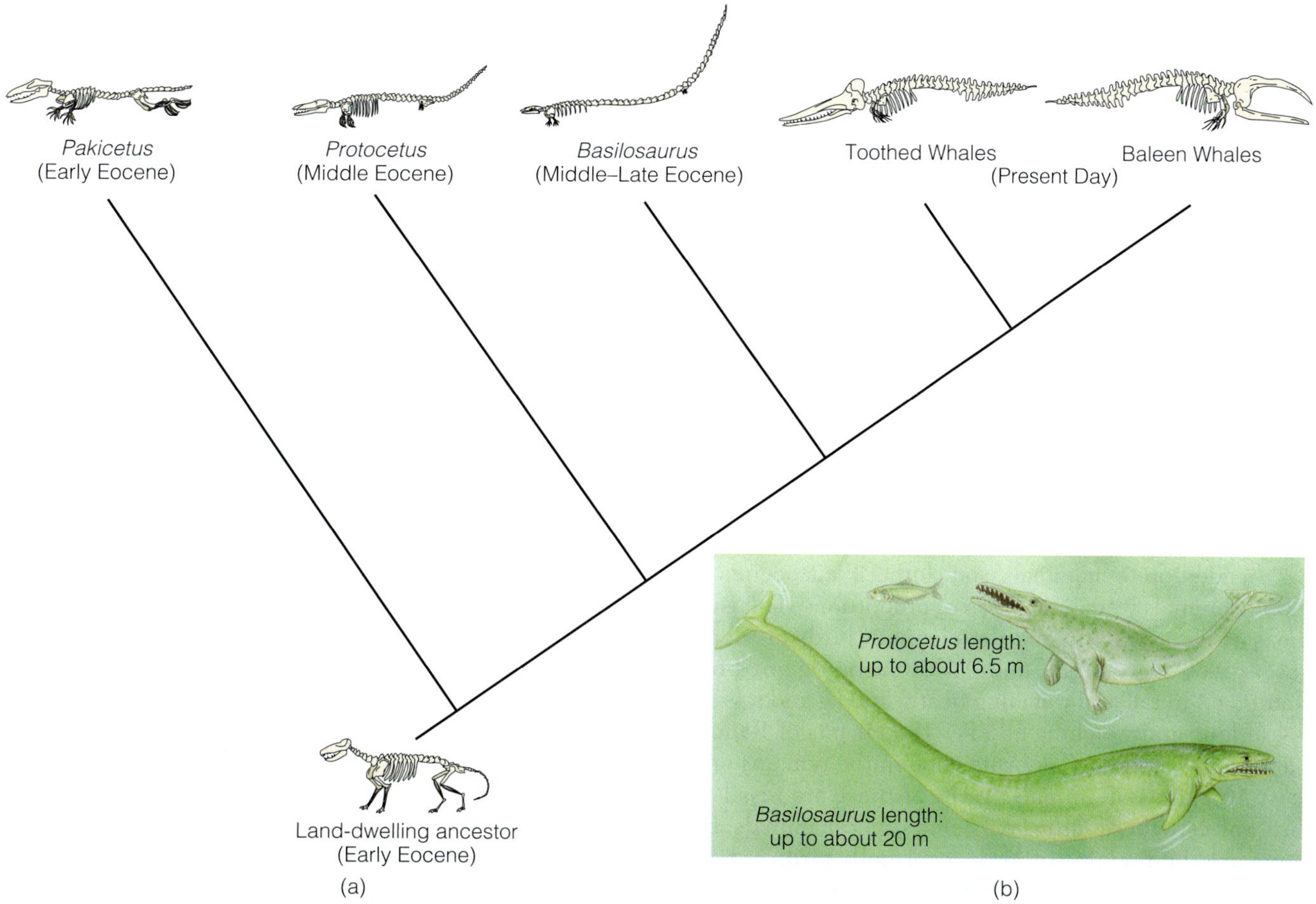

Figure 23.22

(a) Cladogram showing the relationships among some fossil and living whales and their land-dwelling ancestors. Note that *Pakicetus* had well-developed hind limbs, but only vestiges remain in *Protocetus* and *Basilosaurus*. (b) Restoration of *Protocetus* and *Basilosaurus*. Although *Basilosaurus* was a fully aquatic whale, it differed considerably from today's whales. Source: (a) Reprinted with permission from Mammalian Evolution in Major Features of Vertebrate Evolution, Short Course on Paleontolgoy, no. 7, Dr. R. Prothero and R. M. Schoch (eds.).

George C. Page Museum

(a)

Erika Simons/Florida Museum of Natural History

(b)

■ Figure 23.23

(a) A mural by Charles R. Knight showing Late Pleistocene mammals at Rancho La Brea, California. A saber-toothed cat and giant vultures are in the foregound. Notice the herd of mammoths in the distance. (b) Among the diverse Pliocene and Pleistocene mammals of Florida were 6-m-long giant ground sloths and armored mammals known as glyptodonts that weighed more than 2 metric tons. Horses, camels, elephants, carnivores, and a variety of rodents were also common.

In North America, for example, there were mastodons and mammoths, giant bison, huge ground sloths, giant camels, and beavers nearly 2 m tall at the shoulder (■ Figure 23.23). Kangaroos standing 3 m tall, wombats the size of rhinoceroses, leopard-sized marsupial lions, and large platypuses characterized the Pleistocene fauna of Australia. In Europe and parts of Asia lived cave bears, elephants, and the giant deer commonly called the Irish elk with an antler spread of 3.35 m. The evolutionary trend toward large body size was perhaps an adaptation to the cooler temperatures of the Pleistocene. Large animals have proportionately less surface area compared to their volume and thus retain heat more effectively than do smaller animals.

In addition to mammals, some other Pleistocene vertebrate animals were of impressive proportions. The giant moas of New Zealand and the elephant birds of Madagascar were very large, and Australia had giant birds standing 3 m tall and weighing nearly 500 kg and a lizard 6.4 m long and weighing 585 kg. The tar pits of Rancho La Brea in southern California contain the remains of at least 200 kinds of animals. Many of these are fossils of dire wolves, saber-toothed cats, and other mammals, but some are the remains of birds, especially birds of prey, and a giant vulture with a wingspan of 3.6 m.

Primate Evolution

Primates are difficult to characterize as an order because they lack the strong specializations found in most other mammalian orders. We can, however, point to several trends in their evolution that help define primates and are related to their *arboreal,* or tree-dwelling, ancestry. These include changes in the skeleton and mode of locomotion; an increase in brain size; a shift toward smaller, fewer, and less specialized teeth; and the evolution of stereoscopic vision and a grasping hand with opposable thumb. Not all of these trends took place in every primate group, nor did they evolve at the same rate in each group. In fact, some primates have retained certain primitive features, whereas others show all or most of these trends.

The primate order is divided into two suborders (Table 23.2). The *prosimians,* or lower primates, include the lemurs, lorises, tarsiers, and tree shrews; they are the oldest primate lineage, with a fossil record extending

Table 23.2

Classification of the Primates

Order Primates: Lemurs, lorises, tarsiers, monkeys, apes, humans
- Suborder Prosimii: Lemurs, lorises, tarsiers (lower primates)
- Suborder Anthropoidea: Monkeys, apes, humans (higher primates)
  - Superfamily Hominoidea: Apes, humans
    - Family Hominidae: Humans

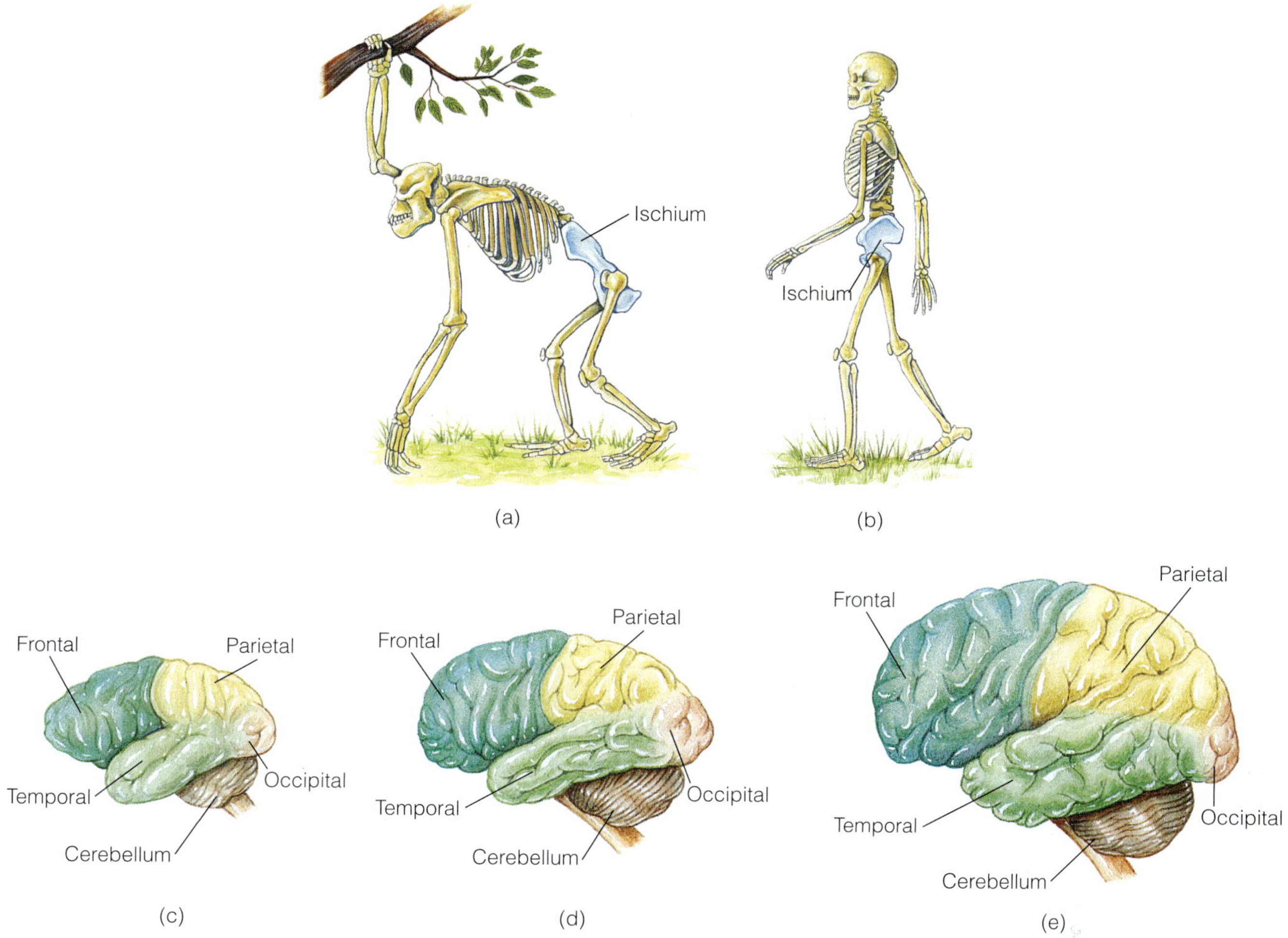

Geology Now ■ **Active Figure 23.24**

Comparison between quadrupedal and bipedal locomotion in gorillas and humans. (a) In gorillas the ischium bone is long, and the entire pelvis is tilted toward the horizontal. (b) In humans the ischium bone is much shorter, and the pelvis is vertical. (c–e) An increase in brain size and organization is apparent in comparing the brains of (c) a New World monkey, (d) a great ape, and (e) a present-day human.

back to the Paleocene. Sometime during the Late Eocene, the *anthropoids*, or higher primates that include monkeys, apes, and humans, evolved from a prosimian lineage. By the Oligocene, the anthropoids were a well-established group with both Old World monkeys (Africa, Asia) and New World monkeys (Central and South America) having evolved during this epoch. The *hominoids*, the group containing apes and humans, diverged from Old World monkeys sometime before the Miocene, but exactly when is still being debated. It is generally accepted, however, that hominoids evolved in Africa.

Hominids

The **hominids** (family Hominidae), the primate family that includes present-day humans and their extinct ancestors (Table 23.2), have a fossil record extending back almost 7 million years. Several features distinguish them from other hominoids. Hominids are bipedal; that is, they have an upright posture, which is indicated by several modifications in their skeleton (■ Figure 23.24a, b). In addition, they show a trend toward a large and internally reorganized brain (Figure 23.24c–e). Other features include a reduced face and reduced canine teeth, omnivorous feeding, increased manual dexterity, and the use of sophisticated tools.

Many anthropologists think that these hominid features evolved in response to major climatic changes that began during the Miocene and continued into the Pliocene. During this time, vast savannas replaced the African tropical rain forests where the lower primates and Old World monkeys had been so abundant. As the savannas and grasslands continued to expand, the hominids made the transition from true forest dwelling to life in an environment of mixed forests and grasslands.

At present, there is no clear consensus on the evolutionary history of the hominid lineage. This is because of

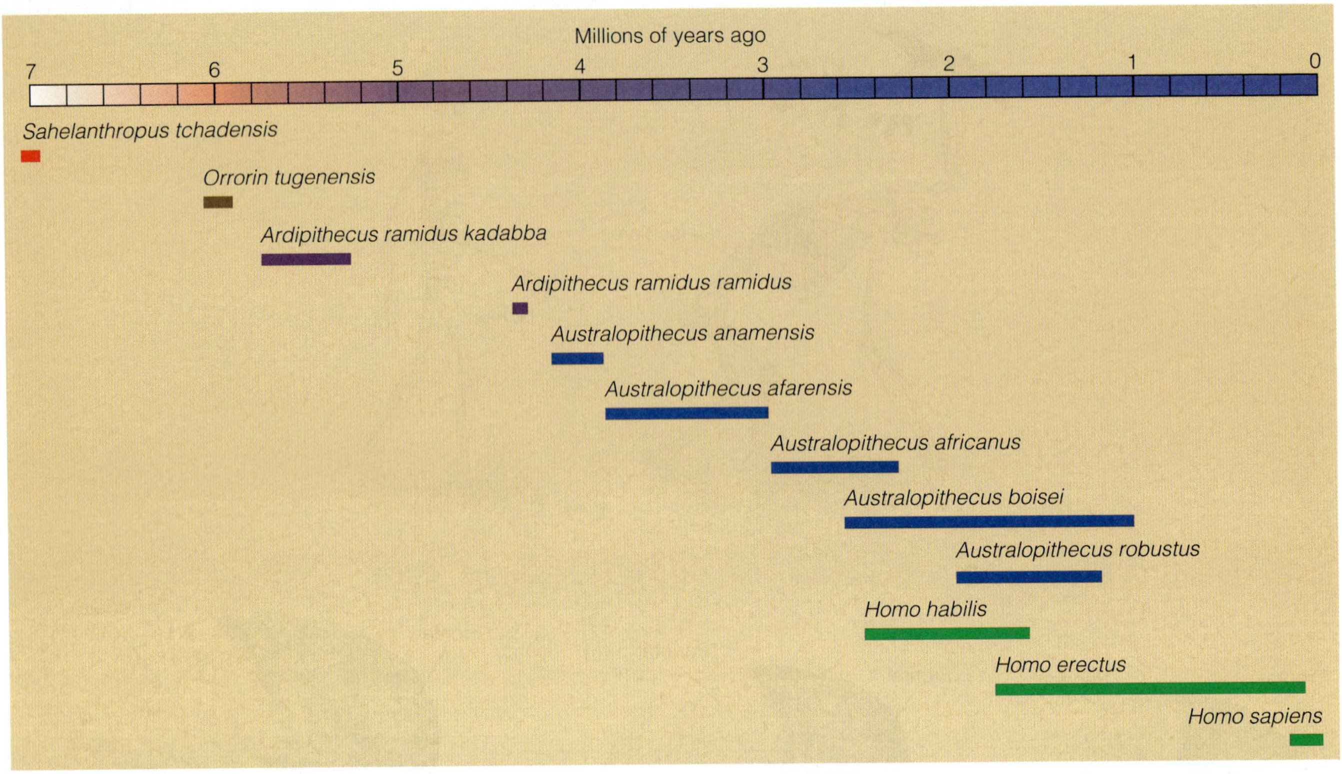

Figure 23.25

The geologic ranges for the commonly accepted species of hominids.

the incomplete fossil record of hominids as well as new discoveries, and also because some species are known only from partial specimens or fragments of bone. There is even disagreement on the total number of hominid species. A complete discussion of all the proposed hominid species and the various competing schemes of hominid evolution is beyond the scope of this chapter. However, we will briefly discuss the generally accepted species (■ Figure 23.25) and present some of the current theories of hominid evolution. Remember that although the fossil record of hominid evolution is not complete, what there is, is well documented. Furthermore, the interpretation of that fossil record precipitates the often vigorous and sometimes acrimonious debates concerning our evolutionary history.

Discovered in northern Chad's Djurab Desert in July 2002, the nearly 7-million-year-old skull and dental remains of *Sahelanthropus tchadensis* make it the oldest known hominid yet unearthed and very close to the time when humans diverged from our closest living relative, the chimpanzee (■ Figure 23.26). It is presently hypothesized that *Sahelanthropus tchadensis* was bipedal in its walking habits, but until bones from its legs and feet are found, that remains conjecture. The next oldest hominid is *Orrorin tugenensis,* whose fossils have been dated at 6 million years old and consist of bits of jaw, isolated teeth, finger, arm, and partial upper leg bones (Figure 23.25). At this time, there is still debate as to exactly where *Orrorin tugenensis* fits in the hominid lineage.

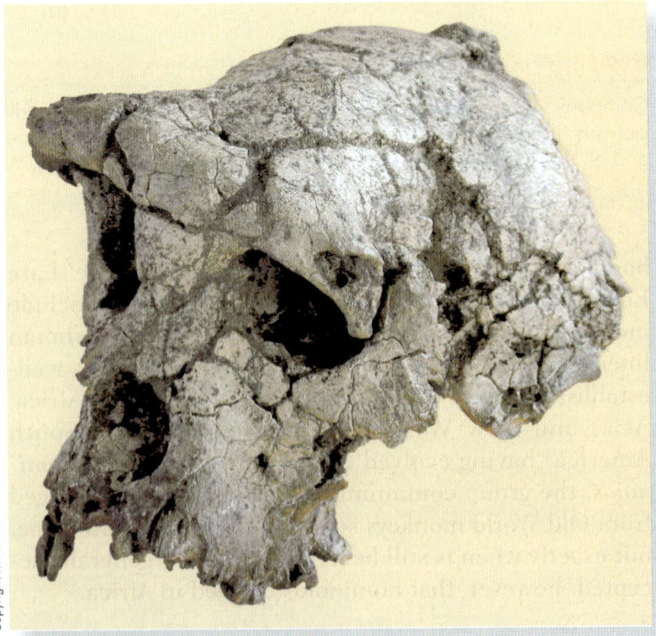

Figure 23.26

Discovered in Chad in 2002 and dated at nearly 7 million years old, this skull of *Sahelanthropus tchadensis* is presently the oldest known hominid (the branch of primates that includes present-day humans and their, extinct ancestors).

■ **Figure 23.27**

Recreation of a Pliocene landscape showing members of *Australopithecus afarensis* gathering and eating various fruits and seeds.

Sometime between 5.8 and 5.2 million years ago, another hominid, *Ardipithecus ramidus kadabba,* was in eastern Africa. Although many paleoanthropologists think both *Orrorin tugenensis* and *Ardipithecus ramidus kadabba* were habitual bipedal walkers and thus on a direct evolutionary line to humans, others are not as impressed with the fossil evidence and are reserving judgment. Until more fossil evidence is found and analyzed, supporting any single evolutionary scheme of hominid evolution presented here would be premature.

Australopithecines **Australopithecine** is a collective term for all Pliocene and Pleistocene members of the genus *Australopithecus.* Currently five species are recognized: *A. anamensis, A. afarensis, A. africanus, A. robustus,* and *A. boisei.* Many paleontologists accept the evolutionary scheme in which *A. anamensis,* the oldest known australopithecine, is ancestral to *A. afarensis,* who in turn is ancestral to *A. africanus* and the genus *Homo,* as well as the side branch of australopithecines represented by *A. robustus* and *A. boisei.*

The oldest known australopithicine is *Australopithecus anamensis.* Discovered at Kanapoi, a site near Lake Turkana, Kenya, by Meave Leakey of the National Museums of Kenya and her colleagues, this 4.2-million-year-old bipedal species has many features in common with its younger relative, *Australopithecus afarensis,* yet is more primitive in other characteristics, such as its teeth and skull.

Australopithecus afarensis (■ Figure 23.27), which lived 3.9–3.0 million years ago, was fully bipedal and

■ **Figure 23.28**

Recreation of a Pleistocene setting in Europe in which members of *Homo erectus* are using fire and stone tools.

exhibited great variability in size and weight. Members of this species ranged from just over 1 m to about 1.5 m tall and weighed between 29 and 45 kg. They had a brain size of 380–450 cubic centimeters (cc), larger than the 300–400 cc of a chimpanzee but much smaller than that of present-day humans (1350 cc average).

The skull of *A. afarensis* retained many apelike features, including massive brow ridges and a forward-jutting jaw, but its teeth were intermediate between those of apes and humans. The heavily enameled molars were probably an adaptation to chewing fruits, seeds, and roots (Figure 23.27).

A. afarensis was succeeded by *Australopithecus africanus,* which lived 3.0–2.3 million years ago. The differences between the two species are relatively minor, although *A. africanus* was slightly larger and had a flatter face and a somewhat larger brain.

Both *A. afarensis* and *A. africanus* differ markedly from the so-called robust species *A. boisei* (2.6–1.0 million years ago) and *A. robustus* (2.0–1.2 million years ago), neither of which had any evolutionary descendants.

Most scientists accept the idea that the robust australopithecines form a separate lineage from the other australopithecines that went extinct 1 million years ago.

The Human Lineage The earliest member of our own genus **Homo** is *Homo habilis,* which lived 2.5–1.6 million years ago. *H. habilis* evolved from the *A. afarensis* and *A. africanus* lineage and coexisted with *A. africanus* for about 200,000 years (Figure 23.25). *H. habilis* had a larger brain (700 cc average) than its australopithicine ancestors but smaller teeth. The evolutionary transition from *H. habilis* to *H. erectus* appears to have occurred in a short time, between 1.8 and 1.6 million years ago.

In contrast to the australopithecines and *H. habilis,* which are unknown outside Africa, *Homo erectus* was a widely distributed species, having migrated from Africa during the Pleistocene (■ Figure 23.28). Although *H. erectus* developed regional variations in form, the species differed from modern humans in several ways. Its brain size of 800–1300 cc, although much larger than that of *H. habilis,* was still less than the average for *Homo sapiens* (1350 cc). The skull of *H. erectus* was thick-walled, its face was massive, it had prominent brow ridges, and its teeth were slightly larger than those of present-day humans. *H. erectus* was comparable in size to modern humans, standing between 1.6 and 1.8 m tall and weighing between 53 and 63 kg.

The archaeological record indicates that *H. erectus* was a tool maker. Furthermore, some sites show evidence that its members used fire and lived in caves, an advantage for those living in more northerly climates (Figure 23.28).

Currently, a heated debate surrounds the transition from *Homo erectus* to our own species, *Homo sapiens.* Pale-

■ **Figure 23.29**

Archaeological evidence indicates that Neanderthals lived in caves and participated in ritual burials, as depicted in this painting of a burial ceremony such as occurred approximately 60,000 years ago at Shanidar Cave, Iraq.

oanthropologists are split into two camps. On the one side are those who support the "out of Africa" view, in which our ancestors, *Homo erectus,* migrated from Africa perhaps as recently as 100,000 years ago and populated Europe and Asia, driving the earlier hominid population to extinction.

The alternative explanation, the "multiregional" hypothesis, maintains that early modern humans did not have an isolated origin in Africa but rather established separate populations throughout Eurasia. Occasional contact and interbreeding between these populations enabled our species to maintain its overall cohesiveness while still preserving the regional differences in people we see today. Regardless of which theory turns out to be correct, our species, *H. sapiens,* most certainly evolved from *H. erectus.*

Perhaps the most famous of all fossil humans are the **Neanderthals,** who inhabited Europe and the Near East from about 200,000 to 30,000 years ago (■ Figure 23.29). The most notable difference between Neanderthals and present-day humans is in the skull. Neanderthal skulls were long and low with heavy brow ridges, a projecting mouth, and a weak, receding chin. Their brain was slightly larger on average than our own and somewhat differently shaped. The Neanderthal body was somewhat more massive and heavily muscled than ours, with rather short limbs, much like those of other cold-adapted people of today.

Based on specimens from more than 100 sites, we now know that Neanderthals were not much different from us, only more robust. Europe's Neanderthals were the first humans to move into truly cold climates, enduring miserably long winters and short summers as they pushed north into tundra country. Their remains are found chiefly in caves and hutlike rock shelters, which also contain a variety of specialized stone tools and weapons. Furthermore, archaeological evidence indicates that Neanderthals commonly took care of their injured and buried their dead, frequently with such grave items as tools, food, and perhaps even flowers (Figure 23.29).

About 30,000 years ago, humans closely resembling modern Europeans moved into the region inhabited by Neanderthals and completely replaced them. **Cro-Magnons,** the name given to the successors of Neanderthals in France, lived from about 35,000 to 10,000 years ago; during this period the development of art and technology far exceeded anything the world had seen before. Using paints made from manganese and iron oxides, Cro-Magnon people painted hundreds of scenes on the ceilings and walls of caves in France and Spain, where many of them are still preserved today.

With the appearance of Cro-Magnons, human evolution has become almost entirely cultural rather than biologic. Since the evolution of the Neanderthals about 200,000 years ago, humans have gone from a stone culture to a technology that has allowed us to visit other planets with space probes and land men on the Moon. It remains to be seen how we will use this technology in the future and whether we will continue as a species, evolve into another species, or become extinct as many groups have before us.

Pleistocene Extinctions

Extinctions have occurred continually during life history, but at times of mass extinctions Earth's biotic diversity sharply declined, as at the ends of the Paleozoic and Mesozoic eras. In marked contrast, the Pleistocene extinctions were rather modest, but they did have a profound effect on genera of large terrestrial mammals, those weighing more than 40 kg, and some large flightless birds. Particularly hard hit were Australia and the Americas. In Australia, 15 of the continent's 16 genera of large mammals died out, North America lost 33 of 45 genera of large mammals, and in South America 46 of 58 such genera went extinct. In contrast, Europe lost only 7 of 23 such genera, and in Africa south of the Sahara only 2 of 44 died out.

These facts lead to three questions: (1) What caused Pleistocene extinctions? (2) Why did these extinctions eliminate mostly large mammals? (3) Why were extinctions most severe in Australia and the Americas? No explanation is completely satisfactory, but two competing hypotheses are currently being debated. One holds that rapid climatic changes at the end of the Pleistocene caused extinctions, whereas another, called *prehistoric overkill,* holds that human hunters were responsible.

Rapid changes in climate and vegetation occurred over much of Earth's surface during the late Pleistocene as glaciers began retreating. The North American and northern Eurasian open-steppe tundras were replaced by conifer and broadleaf forests as warmer and wetter conditions prevailed. The Arctic region flora changed from a productive herbaceous one that supported a variety of large mammals to a relatively barren water-logged tundra that supported a far sparser fauna. The southwestern U.S. region also changed from a moist area with numerous lakes, where saber-tooth cats, giant ground sloths, and mammoths roamed, to a semiarid environment unable to support a diverse fauna of large mammals.

Rapid changes in climate and vegetation can certainly affect animal populations, but the climate hypothesis presents problems. First, why didn't the large mammals migrate to more suitable habitats as the climate and vegetation changed? After all, many other animal species did. For example, reindeer and the Arctic fox lived in southern France during the last glaciation and migrated to the Arctic when the climate became warmer. The second problem is the apparent lack of correlation between extinctions and earlier glacial advances and retreats throughout the Pleistocene Epoch. Previous changes in climate were not marked by episodes of mass extinctions.

Proponents of the prehistoric overkill hypothesis argue that the mass extinctions in North and South America and Australia coincided closely with the arrival of humans. Perhaps hunters had a tremendous impact on the faunas of North and South America about 11,000 years ago because the animals had no previous experience with humans. The same thing happened much earlier in Australia soon after people arrived about 40,000 years ago. No large-scale extinctions occurred in Africa and most of Europe because animals in those regions had long been familiar with humans.

One problem with the prehistoric overkill hypothesis is that archaeological evidence indicates that early human inhabitants of North and South America, as well as Australia, probably lived in small, scattered communities, gathering food and hunting. How could a few hunters decimate so many species of large mammals? However, humans have caused major extinctions on oceanic islands. For example, in a period of about 600 years after arriving in New Zealand, humans exterminated several species of the large, flightless birds called moas.

A second problem is that present-day hunters concentrate on smaller, abundant, and less dangerous animals. The remains of horses, reindeer, and other small animals are found in many prehistoric sites in Europe, whereas mammoth and woolly rhinoceros remains are scarce.

Finally, few human artifacts are found among the remains of extinct animals in North and South America, and there is little evidence that the animals were hunted. Countering this argument is the assertion that the impact on the previously unhunted fauna was so swift as to leave little evidence.

The reason for the extinctions of large Pleistocene mammals is still unresolved and probably will be for some time. It may turn out that the extinctions resulted from a combination of different circumstances. Populations that were already under stress from climatic changes were perhaps more vulnerable to hunting, especially if small females and young animals were the preferred targets.

23 GEO RECAP

Chapter Summary

- Cenozoic tectonism was concentrated in the Alpine–Himalayan and circum-Pacific belts. Each belt is composed of smaller areas of deformation called orogens.
- The Cenozoic evolution of the North American Cordillera included deformation during the Laramide orogeny, extensional tectonics that yielded basin-and-range structures, extensive intrusive and extrusive igneous activity, and uplift and erosion.
- One model for the Laramide orogeny involves near horizontal subduction of the Farallon plate beneath North America, resulting in fault-bounded uplifts in the area of the present-day Rocky Mountains.
- As the North American plate drifted westward, it collided with the Pacific–Farallon Ridge, at which time subduction ceased and the continent became bounded by large transform faults, except in the Pacific Northwest where subduction continues.
- Sediments eroded from Laramide uplifts were deposited in intermontane basins of the Great Plains, whereas a wedge of sediments pierced by salt domes is found on the Gulf Coastal Plain.
- Cenozoic uplift and erosion were responsible for the present topography of the Appalachian Mountains. As the Appalachians eroded, much of the sediment was deposited on the Atlantic Coastal Plain.
- Vast glaciers covered about 30% of Earth's land surface during the Pleistocene. About 20 warm-cold Pleistocene climatic cycles are recognized from evidence found in deep-sea cores.
- Large parts of North America and other Northern Hemisphere continents were covered by continental glaciers several times during the Pleistocene.
- Cenozoic mineral resources include sand and gravel, placer deposits of gold, some evaporite minerals such as borax, and oil and natural gas.
- Marine invertebrate groups that survived the extinctions at the end of the Mesozoic continued to diversify, giving rise to the present-day marine fauna.
- The Paleocene mammalian fauna was made up of Mesozoic holdovers and several new orders. During this time of mammal diversification, several orders became extinct, but by Eocene time most living mammal orders were present.
- Shrewlike placental mammals that evolved during the Late Cretaceous were ancestors for the orders of placental mammals that evolved during the Cenozoic.
- Among the hoofed mammals (artiodactyls and perissodactyls) adaptations include modifications of the teeth for grinding vegetation and changes in their limbs for speed.
- The evolutionary history of horses is particularly well documented by fossils, but scientists also know much about the evolution of other hoofed mammals as well as elephants, whales, and some carnivores.
- The primates evolved during the Paleocene. Several trends help characterize primates and differentiate them from other mammalian orders, including a change in overall skeletal structure and mode of locomotion, an increase in brain size, stereoscopic vision, and evolution of a grasping hand with opposable thumb.
- The primates are divided into two suborders: prosimians and anthropoids. The prosimians are the oldest primate lineage and include lemurs, lorises, tarsiers, and tree shrews. The anthropoids include the New and Old World monkeys, apes, and hominids, which are humans and their extinct ancestors.
- The oldest known hominid is *Sahelanthropus tchadensis,* dated at nearly 7 million years old. It was followed by *Orrorin tugenensis* at 6 million years old and then two subspecies of *Ardipithecus*

at 5.8 and 4.4 million years old, respectively. These early hominids were succeeded by the australopithecines, a fully bipedal group that evolved in Africa 4.2 million years ago. Currently, five australopithecine species are known: *Australopithecus anamensis, A. afarensis, A. africanus, A. robustus,* and *A. boisei.*

- The human lineage began about 2.5 million years ago in Africa with the evolution of *Homo habilis,* which survived as a species until about 1.6 million years ago.
- *Homo erectus* evolved from *H. habilis* about 1.8 million years ago and was the first hominid to migrate out of Africa. The transition from *H. erectus* to *H. sapiens* is still unresolved. There currently is not enough evidence to determine which of the two competing hypotheses—the "out of Africa" hypothesis or the "multiregional" hypothesis—is correct.
- Neanderthals inhabited Europe and the Near East between 200,000 and 30,000 years ago and were not much different from us, only more robust and with differently shaped skulls. They made specialized tools and weapons, apparently took care of their injured, and buried their dead.
- The Cro-Magnons were the successors of the Neanderthals and lived from about 35,000 to 10,000 years ago. They were highly skilled nomadic hunters, formed living groups of various sizes, and were also skilled cave painters.
- Modern humans succeeded the Cro-Magnons about 10,000 years ago and have spread throughout the world.

Important Terms

artiodactyl (p. 703)
australopithecine (p. 711)
Basin and Range Province (p. 685)
browser (p. 703)
Cro-Magnon (p. 713)
grazer (p. 703)
hominid (p. 709)
***Homo* (p. 712)**
Laramide orogeny (p. 683)
marsupial mammal (p. 701)
monotreme (p. 701)
Neanderthal (p. 713)
North American Cordillera (p. 683)
perissodactyl (p. 703)
placental mammal (p. 701)
primate (p. 708)
Tejas epeiric sea (p. 689)
ungulate (p. 703)

Review Questions

1. The Cenozoic history of the Appalachian Mountains involves mostly
a._____movement along a large transform fault; b._____uplift and erosion; c._____metamorphism and igneous intrusions; d._____vertical uplifts during the Laramide orogeny; e._____ compression-induced overturned folds and thrust faults.

2. Ongoing volcanism takes place in the Cascade Range because of
a._____deposition of a wide coastal plain;
b._____the origin of banded iron formations;
c._____the presence of the Tejas epeiric sea;
d._____subduction of the Juan de Fuca plate;
e._____rifting along the San Andreas fault.

3. The oldest known primate lineage is the
a._____prosimians; b._____anthropoids;
c._____hominids; d._____pelycosaurs;
e._____insectivores.

4. A vast area of overlapping lava flows in the northwestern United States is the
a._____Basin and Range Province;
b._____Tejas epeiric sea; c._____Columbia River basalts; d._____Colorado Plateau;
e._____Gulf Coastal Plain.

5. Hoofed mammals with teeth adapted for a diet of grass are known as
a._____browsers; b._____monotremes;
c._____miacids; d._____grazers; e._____carnivores.

6. Which were the first hominids to migrate out of Africa and from which present-day humans evolved?
a._____*Australopithecus robustus;*
b. _____*Ardipithecus ramidus;* c._____*Homo habilis;* d._____*Homo erectus;* e._____*Homo sapiens.*

7. The Cenozoic history of the Basin and Range Province involved
   a._____ large-scale block-faulting; b._____broad uplift with little deformation; c._____carbonate deposition in a shallow, tropical sea; d._____deep erosion by continental glaciers; e._____meteorite impact and volcanism.
8. Large body size in mammals during the Pleistocene may have been an adaptation for
   a._____cooler temperatures; b._____increased predation; c._____longer summers; d._____lower elevations; e._____dwelling on grass-covered plains.
9. Which of the following features distinguish hominids from other hominoids?
   a._____bipedalism; b._____a large and internally reorganized brain; c._____a reduced face and reduced canine teeth; d._____use of sophisticated tools; e._____all of these.
10. Although there were probably many Pleistocene glacial stages in North America, the four now recognized are the
   a._____ Nebraskan-Kansan-Illinoin-Wisconsinan; b._____ Athabascan-Ontarioan-Iowan-Pennsylvanian; c._____Mississippian-Cambrian-Albanian-Yarmouthian; d._____Aftonian-Ohioan-Kentuckyan-Lahontan; e._____Kewatin-Labradoran-Canadian-Missoulan.
11. Explain how the Laramide orogeny differs from most orogenies at convergent plate boundaries.
12. What kinds of evidence indicate that widespread glaciers were present during the Pleistocene?
13. What are the three basic groups of mammals, and how do they differ from one another?
14. During the Paleozoic, eastern North America experienced considerable deformation, whereas during the Mesozoic and Cenozoic, most deformation took place in the western part of the continent. What accounts for this changing pattern of deformation?
15. Summarize the evidence for and against the prehistoric overkill hypothesis.
16. Briefly outline the Cenozoic geologic history of the Appalachian Mountains.
17. How have the teeth of some ungulates been modified for grazing?
18. What kinds of sediments accumulated in the Interior Lowlands and on the Gulf Coastal Plain, and what was the source of these sediments?
19. How does the fossil record demonstrate that animals as different as horses and rhinoceroses evolved from a common ancestor? (*Hint:* Also see Chapter 18.)
20. What factors do you think will influence the future course of human evolution? Can we as a species control the direction evolution takes?

World Wide Web Activities

GeologyNow Assess your understanding of this chapter's topics with additional quizzing and comprehensive interactivities at

http://earthscience.brookscole.com/changingearth4e

as well as current and up-to-date weblinks, additional readings, and InfoTrac College Edition exercises.

Physical and Historical Geology in Perspective

CHAPTER 24

In this book we have investigated the two major subdivisions of geology: physical geology and historical geology. Chapters 1–17 covered (1) Earth materials—that is, minerals and rocks as well as the processes by which they form and change; (2) Earth's internal processes, such as the magnetic field, and internal heat, which is responsible for moving plates, mountain building, and seismic activity; and (3) various surface processes including mass wasting, running water, glaciers, and groundwater, all of which yield distinctive landscapes. In Chapters 18–23, we reviewed Earth and life history as revealed by the geologic record. But few of you will become geologists, so how will you benefit from the information and insights you gained on these topics?

Quite simply, many of you will enter professions in which at least part of your work will involve geologic considerations. You may be a city planner, a city council member, a county commission, or a member of a planning board for siting a sanitary landfill, for ensuring an adequate supply of groundwater for your community, or for developing zoning regulations for construction on floodplains or in a seismically active area. Perhaps you will become a developer or a contractor and have to contend with slope stability, expansive soils, and areas prone to flooding. Or maybe you will be an engineer involved in planning and constructing large structures such as bridges, dams, power plants, and highways in tectonically active regions (■ Figure 24.1).

Geology Now *This icon, which appears throughout the book, indicates an opportunity to explore interactive tutorials, animations, or practice problems available on the GeologyNow website at http://earthscience.brookscole.com/changingearth4e.*

About 60 million years ago, basalt lava flows covered a landscape of forests where rivers flowed over chalk (limestone) bedrock in what is now Northern Ireland. Then for 2 million years, weathering yielded laterite, a deep red soil, seen about halfway up the slope. Once again lava flows covered the area, and as before they flowed into valleys where they cooled slowly and formed columns (see Figure 5.6). Since then, glaciers, running water, waves, and mass wasting have modified the area. In short, the present-day landscape resulted from interactions among Earth's systems during the last several tens of millions of years. Source: James S. Monroe

James S. Monroe

(a)

James S. Monroe

(b)

Figure 24.1

(a) This freeway crosses a small valley a short distance from the San Andreas fault in California. Engineers had to take into account the near certainty that this structure will be badly shaken during an earthquake. (b) A bridge crossing part of a reservoir in a mountainous region. Rock type and geologic structures were important considerations when planning and erecting this structure.

Various federal and state agencies plan for and mitigate the effects of *geologic hazards* such as volcanic eruptions, earthquakes, floods, and landslides. These are the more spectacular, and sometimes catastrophic, geologic hazards, but others are more subtle—soil creep and radon gas, for example. As a matter of fact, soil creep and expansive soils account for more property damage than any other geologic hazard, but of course they do not cause injuries or fatalities. Radon gas, however, poses a long-term health risk in some areas.

Geologic hazards account for thousands of fatalities and billions of dollars in damages each year. Although the incidence of hazardous events has not increased, fatalities and damages have because more and more people live in disaster-prone areas. Population density and poor construction are obvious elements in disasters such as earthquakes and coastal flooding in the developing world. As we noted in Chapter 1, most scientists agree that the single greatest environmental problem is overpopulation.

Of course, we cannot eliminate geologic hazards and other natural hazards such as hurricanes and freezing temperatures, but we can better understand these phenomena, enact zoning and land-use regulations, and at the very least decrease the amount of human suffering and damage. Unfortunately, geologic information that is readily available is often ignored. A case in point is the Turnagain Heights subdivision in Anchorage, Alaska, that was so heavily damaged when the soil beneath it liquefied during the 1964 earthquake (see Figure 11.19). Not only were reports on soil stability ignored or overlooked before homes were built there, but since 1964 new homes have been built on part of the same site! In some areas, development continues near shorelines where erosion will surely destroy homes and other structures (see Figure 16.12b).

In Chapter 1, we defined a *system* as a combination of related parts that operate in an organized fashion (see Figure 1.1), and we gave examples of interacting systems throughout the text. For instance, in Chapter 5 we mentioned volcanism and its impact on the atmosphere, hydrosphere, and biosphere, and in Chapter 2 we discussed moving plates and how they affect Earth's systems. We also considered interacting systems in the broader context of geologic time in Chapter 17.

Remember from previous discussions that during its earliest history, Earth's surface was hot and dry, volcanism was ubiquitous, and the atmosphere was likely composed mostly of carbon dioxide. Meteorites and comets flashed through this primitive atmosphere, and because no ozone layer existed, cosmic radiation was intense. Gases derived from comets and from within Earth formed the surface waters, but not until organisms were present, perhaps 3.5 billion years ago, did the atmosphere begin to accumulate free oxygen as a waste product of photosynthesis. Once oxygen was present in sufficient quantities, an ozone layer formed and oxygen-dependent organisms evolved. In short, Earth's systems became established; our planet began to evolve and gradually became like it is today.

The Moon, Mercury, and Mars show much less internal activity and their surface processes are minimal compared to those on Earth. Why, then, is Earth so active? And will it eventually be as inactive as these other celestial bodies? Earth is, of course, much larger than the other bodies mentioned, and its greater mass accounts for greater gravitational attraction, thus enabling it to retain an atmosphere and liquid surface water (hydrosphere); its distance from the Sun is also important. Among the planets and moons of the solar system, Earth is the only body known to support life (biosphere). Earth remains an internally active planet because it possesses sufficient internal heat for volcanism, seismic activity, and plate movements, but scientists are convinced that Earth's internally generated heat will gradually diminish and in the far distant future these activities will cease. In the meantime, though, Earth remains a dynamic planet (■ Figure 24.2).

We mentioned in Chapter 17 that the concept of time measured in millions and billions of years sets geology apart from most of the other sciences, although astronomers are certainly concerned with time of this magnitude. Indeed, geology is a four-dimensional science: vast amounts of time (geologic time) are an essential part of the science. The latter chapters in this book considered the physical and biologic evolution of Earth and the concept of geologic time in some detail.

Perhaps your chosen profession will not rely on geologic considerations, but as a concerned citizen you cannot ignore important environmental issues. Certainly global warming, acid runoff and acid rain, ozone depletion, and waste disposal will have an impact on our standard of living and continued existence on this planet. What are the positive and negative aspects of storing nuclear wastes at Yucca Mountain, Nevada? After all, the wastes already exist, so even if we were to shut down every nuclear power-generating plant today, the problem of disposal of existing wastes would remain. Perhaps you approve or disapprove of constructing a dam for flood control and generating electricity. In either case, your appeal is better when it is based on an understanding of the science involved rather than on only emotion.

In several chapters we emphasized the importance of resources such as ores of iron and copper, hydrocarbons, phosphate, and others. These resources affect our standing of living, and they figure in decisions made by business, industry, and government. Indeed, as we noted in Chapter 1, the distribution of resources and our access to them are factors that shape foreign policy. Furthermore, dependence on politically unstable nations for essential commodities is risky, and the fact that we must import so much is not good for our balance of payments. As of April 2003, the United States imported almost 57% of the petroleum it needs and all or large proportions of many other mineral commodities (see Figure 3.19).

On another level, geology may enrich your appreciation of the scenic wonders of our national parks and monuments, as well as the very nature of the planet wherever you might travel. Perhaps this course in geology will enhance your understanding of geologic processes and how they operate through time. Ten or twenty years from now, you may not remember terms such as *anticline, aréte, barchan,* or *tombolo,* but chances are if you see these features, you will have a good idea of how they formed. You may not remember the precise phrase *principle of superposition,* but should you visit the Grand Canyon in Arizona, you will no doubt figure out the relative ages of the exposed rock layers. And likewise, you may have forgotten the details of the Taconic orogeny, but if you visit the Appalachian Mountains, you probably will remember that they have a long complex history of plate collisions, uplift, and erosion.

Most of this book has necessarily been descriptive, although we discussed many principles and concepts, too. But one objective of this book, and much of your secondary education, is to develop your skill as a critical thinker. As opposed to simple disagreement, critical thinking involves evaluating the support or evidence for a particular point of view. Although your exposure to geology at this time is limited, you do have the fundamentals needed to appraise why geologists accept plate tectonic theory and to logically evaluate an opposing hypothesis should one be formulated and presented.

In conclusion, the most important lesson you can learn from the study of physical and historical geology is that Earth is an extremely complex planet in which interactions among its systems have resulted in changes in the atmosphere, hydrosphere, biosphere, and the planet itself. We tend to view our planet from the perspective of our own lives, or at most from the perspective of recorded history, but we commonly overlook the fact that Earth has changed and continues to do so—though not always on a scale we can readily appreciate. Geology therefore is not a static science, but one, like the dynamic Earth it seeks to understand, that constantly evolves as new information and methods of investigation become available.

Figure 24.2

Some of Earth's interacting systems. (a) Plate divergence in the Red Sea (lower right) results in northward movement of the Arabian plate (upper right) against the Eurasian plate. (b) Although several processes account for this spectacular scenery in Yosemite National Park, California, erosion by glaciers and running water, both parts of the hydrosphere, was responsible for most of it. (c) This huge wind-blown sand dune near Fallon, Nevada, is 180 m high and about 4 km long. (d) Layering in these ancient rocks in Michigan resulted from the activities of cyanobacteria or blue-green algae. Not only have these organisms left their mark in the rocks, but they also added oxygen to early Earth's atmosphere during photosynthesis.

Atlantic Southeast Region

Geology of the Atlantic Southeast Region

John R. Huntsman

University of North Carolina Wilmington

Looking east from Chimney Rock, North Carolina with a view of Lake Lure in the foreground. Resistant granitic gneiss underlies the rounded hills behind Lake Lure as well as Chimney Rock. Resistant quartz-rich metamorphic rocks of Kings Mountain and the South Mountains underlie the resistant ridges along the far distant skyline that mark the exposed eastern boundary of the Piedmont zone.

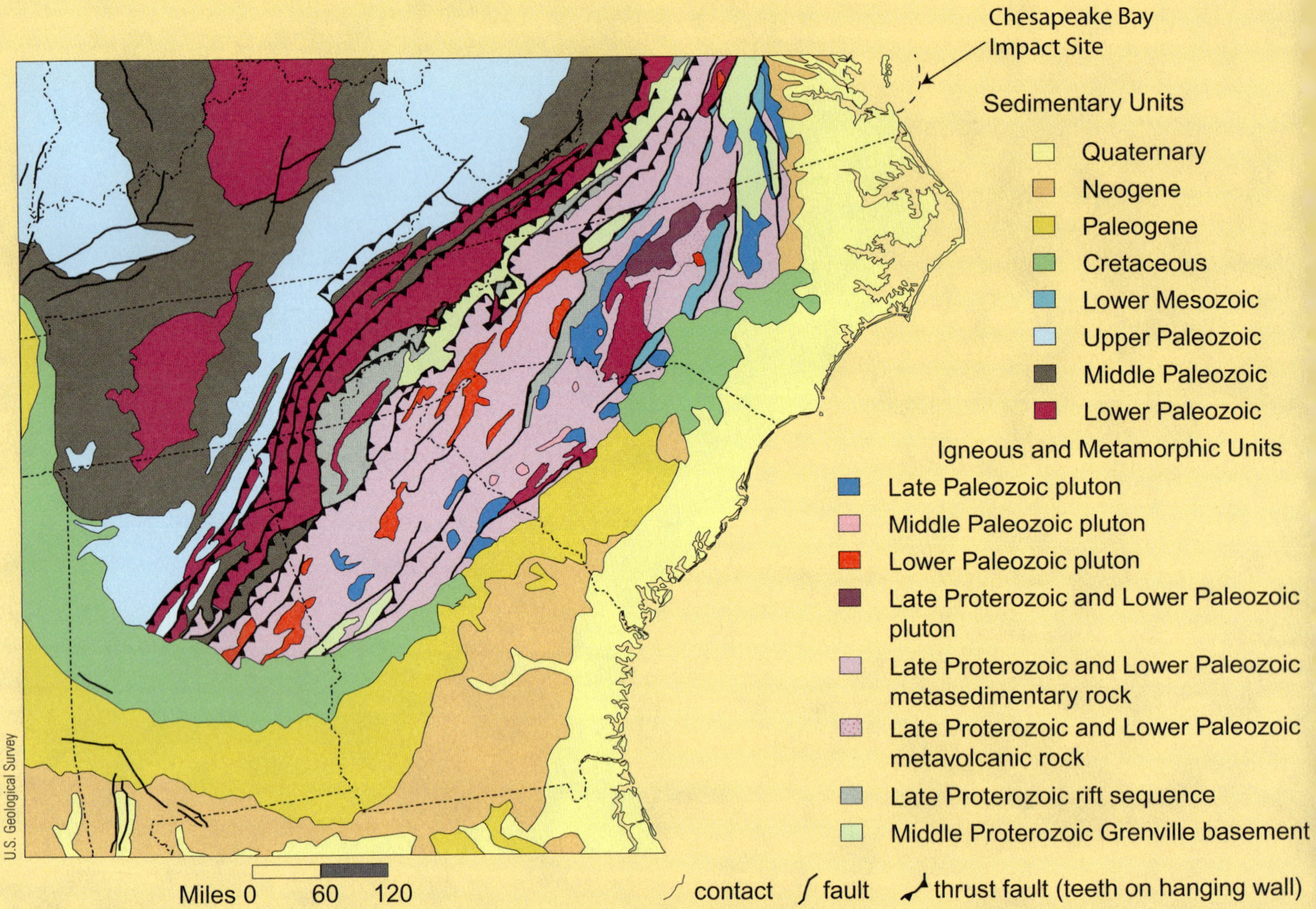

Geologic Map of the Atlantic Southeast Region.

ESSENTIAL QUESTIONS TO ASK

Atlantic Southeast.1 Introduction

- *Why aren't active volcanoes present within the Atlantic Southeast region despite a record of seismic activity?*
- *How has the geology of the Atlantic Southeast region contributed to the economic history of the United States?*

Atlantic Southeast.2 Proterozoic Geology and Tectonics

- *What events led to the formation of Rodinia?*
- *What evidence exists for interpreting a continental rifting event along the margin of Laurentia in the Late Proterozoic?*

Atlantic Southeast.3 Late Proterozoic Climates, Paleogeography, and Life

- *What is the significance of glacial climates in both Laurentia and peri-Gondwanan exotic terranes during the Late Proterozoic?*

Atlantic Southeast.4 Early Paleozoic History of the Laurentian Continental Margin

- *What was Iapetus, and where was it located?*
- *How do the Early Paleozoic strata of the Laurentian continental margin differ from the underlying Late Proterozoic stratified rocks?*

Atlantic Southeast.5 Proterozoic and Early Paleozoic History of the Terranes Comprising the Piedmont

- *How does the geologic history of an exotic terrane that existed within the Iapetus Ocean marginal to Gondwana differ from that of Laurentia?*
- *What evidence exists that demonstrates the size of Iapetus?*

Atlantic Southeast.6 Middle Ordovician through Late Paleozoic Geology and Tectonics

- *What events comprise the Taconic Orogeny?*
- *What events comprise the Acadian Orogeny?*
- *What events comprise the Alleghanian Orogeny?*

Atlantic Southeast.7 Mesozoic and Cenozoic Geology and Tectonics

- *What is a rift-to-drift sequence?*
- *Why do some rift basins have strata dipping northwest while others have strata dipping southeast?*

Atlantic Southeast.8 Natural and Anthropogenic Hazards of the Atlantic Southeast Region

- *What current natural hazards facing the population of this region today pose the greatest threats to property and life?*
- *What are the chances of Myrtle Beach, South Carolina, and Virginia Beach, Virginia, being destroyed by a catastrophic tsunami?*

 The Adaptable Courseware Program consists of products and additions to existing Brooks/Cole products that are produced from camera-ready copy. Peer review, class testing, and accuracy are primarily the responsibility of the author(s). Geology of the Atlantic Southeast Region/John R. Huntsman - First Edition ISBN-13: 978-0-7593-9080-5, ISBN-10: 0-7593-9080-0.
Printed in Canada

Atlantic Southeast.1
Introduction

The Atlantic Southeast region (Figure 1) is a fundamental region not only for learning and understanding the history and growth of the United States but also because many classical ideas of geology originated from studies of rocks, soils, rivers, beaches, and other processes in this region. The eroded southern remnants of the Paleozoic **Appalachian orogenic belt** dominate the western two thirds of the region, and the overlapping, essentially horizontal strata and fringing barrier islands of the Coastal Plain occupy the eastern third. The extensive fluvial and estuary systems of the Coastal Plain provided transportation routes and protected harbors for the ships bringing early explorers and colonists to the region during the seventeenth century. The extensive beaches of the Coastal region continue to play an important role for tourism in addition to the rich history and natural beauty found there. The natural beauty and attractions of the current geography are better understood as the products of balanced geologic processes that have occurred repeatedly through time.

The Appalachian region provided the central framework for the concept of geosynclines as proposed by James Hall and J. D. Dana in the 1800s. Hall noticed that Paleozoic limestone strata in the central region of North America were thicker in the Appalachians even though all of the limestone layers originally accumulated in shallow waters. The correct explanation for this paradox is for the floor of the basin to gradually subside at a pace that kept the uppermost sedimentary layer within shallow water. This realization of a mobile crust fostered new thinking that linked various observations of geologic processes into a logical explanation for linear basins containing vast thicknesses of sediments. Hall and many of those who followed used Appalachian stratigraphy to link sedimentation, sedimentary facies, volcanism and plutonism, folding, faulting, metamorphism, and even crustal growth by outward accretion of geosynclines to ancient cratons. Plate tectonics supplanted geosynclinal theory in the late 1960s, but the basic concept of a mobile crust is a fundamental root of the newer tectonic concepts. Whereas geosynclines stressed vertical motions within the crust, plate tectonic theory demonstrates the equal, if not more dominant, role of horizontal displacements of continental and oceanic crust and underlying lithosphere in explaining the way that the Earth works. The Appalachian region was one of the first mountain systems to be explained in terms of plate tectonic theory. The Atlantic Southeast region provides an excellent opportunity to study the complete historical cycle of the Appalachians from beginning to end (see Frontispiece Geologic Map). The region also permits a chance to evaluate more recent effects of Cenozoic climate change on coastal margins and may provide guiding insights into the future chapters of geologic history.

▶ Figure Atlantic SE.1 Physiographic provinces in the Southern Appalachians. Boundaries are approximate. Base digital elevation map: U.S. Geological Survey.

Current Tectonic Setting

The Atlantic Southeast region currently is classified as a passive continental margin. When a large continent splits, smaller fragments begin drifting apart. **Passive margins** occur along the trailing edge of drifting continents and border ocean basins that expand with increasing separation of the continents. Passive margins lack any of the major earthquake and magmatic activity that is so prevalent at the leading edges of drifting continents at convergent plate boundaries. Earthquake activity does occur in passive margins, but the frequency of major events is very low and magnitudes are not catastrophic. The seismic activity results from interactions between consistent horizontal compressive stresses and older structures such as faults, **shear zones**, or plate boundaries. Over time, the older structure moves slightly to relieve the accumulating pressure from the applied stresses. During the evening of August 31, 1886, near Charleston, South Carolina, an earthquake with an estimated Richter magnitude exceeding 7.0 created widespread damage in the Charleston area, including the deaths of some 60 people. This is the only known major earthquake to have affected the southeast region in recorded history. A few moderate earthquakes have occurred in an area extending through eastern Tennessee and into southeastern West Virginia with Richter magnitudes less than 5.5, much lower than the Charleston earthquake. Even so, the U.S. Geological Survey (USGS) predicts that part of the Atlantic Southeast region may experience moderately low to moderate seismic activity within the next 50 years (Figure 2).

Present-Day Natural Hazards

The significant threats to life and property in the Atlantic Southeast region stem from natural hazards related to tropical storms. The region borders the southeastern Atlantic

▶ **Figure Atlantic SE.2** Potential for seismic hazards within the next 50 years. Peak acceleration values listed in percentage of gravitational acceleration (% g).

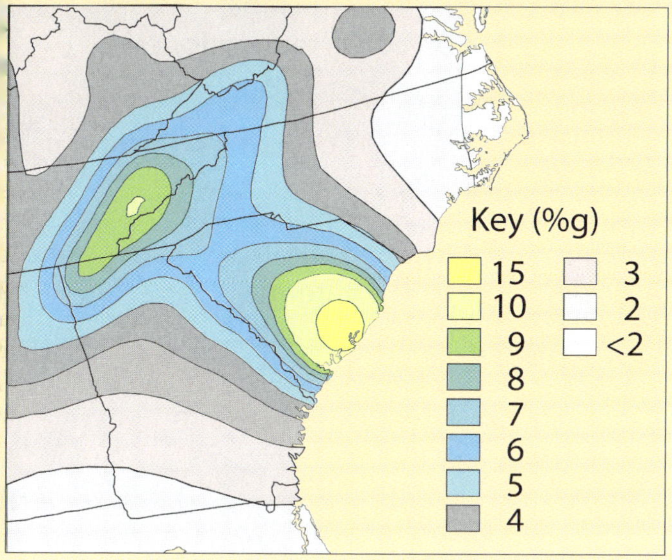

Basin, where many historical events of major hurricanes and nor'easters abound.

Coastal storms and hazards recur at frequent intervals and cause much destruction of property from severe coastal erosion, high winds, extensive flooding, and even significant mass wasting in mountainous regions located far from the direct landfall of the storm.

Resource Potential

The Atlantic Southeast region has a long history of production of ore, gemstone, and industrial minerals that have had significant economic significance to the United States. All of these resources originated with appropriate tectonic events comprising the geologic history of the southern Appalachians. Discovery of gold in 1799 in the central Piedmont of North Carolina led to exploration, discovery, and eventual production of different metallic resources including gold, silver, copper, lead, and zinc. Current interests for discovering additional metallic mineral deposits in the Piedmont region of the Carolinas will probably discover more deposits, but high volumes of ore are not expected. In addition to base and precious metallic ores, iron-rich sedimentary rocks within the Valley and Ridge and Blue Ridge became valued commodities because of their ready access to nearby rivers for ready smelting and transportation. A few forges remain as interesting historical markers to this history. These old iron-works dating back to early colonial history provided raw materials for use in local manufacturing of finished products including farm and household equipment and tools, rims for wagon wheels, and gun barrels and other weapons.

Nonmetallic resources are produced in much greater quantities than metallic mineral resources in the region. The Blue Ridge and Piedmont areas are renowned for their gem minerals including emeralds, rubies, sapphires, and diamonds. The most significant resources, though, are dimension stone and other industrial rocks and minerals. The largest open-faced **granite** quarry in the world can be seen at Mount Airy, North Carolina. Stone quarried today from the quarry perfectly matches stone extracted 100 years ago because of the prized aspect of Mount Airy granite's homogeneous fabric and texture. This stone is in high demand for use in monuments and other less glamorous applications such as curbstones in cities including Boston, New York, and Washington, DC. North Carolina also ranks as the number one source of olivine, a mineral critical for manufacturing high-temperature furnaces for smelting other metals and making alloys.

Geologic processes of weathering, erosion, and transportation of sediment provided necessary steps in forming several natural resources within the Coastal Plain, including sand and gravel deposits (valued for high-grade silica extraction and use in construction), phosphate (important ingredient for fertilizers), and even placer deposits of heavy minerals (important sources of metals used in producing alloys, as well as gold and minerals used for making abrasives).

Energy resources available in the southeast region include lignite, coal, oil, and natural gas. Bituminous coal fields occur throughout the later Paleozoic sedimentary rocks of the Appalachians. Peat and lignite deposits of the Coastal Plain are hydrocarbon resources for energy and other uses. The world's first successful well drilled to find oil was in the Appalachians, and the Paleozoic strata of the southern Appalachians continue to have significant production of both oil and natural gas. Limited studies and exploratory investigations suggest the potential of hydrocarbon resources offshore in the coastal waters of the continental shelf. Existing legislation bans these tracts from detailed exploration and prevents any significant evaluation of their real potential.

Our understanding of the geology and geologic history of the Appalachians helps us understand the interrelations among geologic processes, mineral and rock deposits, and even hydrologic resources. We can make realistic predictions about the future availabilities of resources that are vital to the health and welfare of our nation. Geologic efforts also enable us to find better solutions to environmental concerns demanding assessment and evaluation. The best way to plan for the future of this historically important area is to study the geologic past and understand its changes through time.

A Note on Physiographic Subdivisions

Physiography is the study of landforms and other physical features of the Earth's surface. Processes of physical and chemical weathering, erosion, and deposition modify landscapes that are very rugged with high relief to those that are very subdued with low relief. Internal geologic characteristics of rocks in differing geographic areas produce distinctive features that lead to natural groupings of similar landforms, or physiographic subdivisions. In the Appalachians, physiographic subdivisions have always provided a useful, convenient scheme for discussing the geology, a tradition continued here (see Figure 1). From west to east, the region consists of folded and faulted sedimentary rocks of the Plateaus and Valley and Ridge, the igneous and metamorphic rocks of the Blue Ridge and Piedmont, and the eastward-thickening wedge of younger subhorizontal strata of the Coastal Plain.

Atlantic Southeast.2 Proterozoic Geology and Tectonics

The geologic history of the Atlantic Southeast region begins during an ancient interval of geologic time known as the Proterozoic Eon. The compilation of the geologic time scale itself is an excellent example of using the logic of science to piece together many scattered tidbits of information into a useful tool. Major or mega blocks of time lasting several hundred to several hundreds of millions of years are called eons. Successively smaller intervals of times are eras, periods, and epochs, respectively. In studying rocks that are older than approximately 600 million years, geologists rely heavily on using isotope-dating techniques for deciphering a time sequence in these metamorphic and igneous rocks. The Proterozoic Eon spanned approximately 2 billion years of Earth history before the dawn of the present Phanerozoic Eon and its fossil-bearing rocks. Many geologists recognize that an important and fundamental transition in the way that the Earth works occurred during the Proterozoic. Active and relatively rapid geologic processes operating with thin crusts, small continents, and small ocean basins characterize geologic workings of the earlier Archean Eon. The more stable and slower geologic processes of the current Phanerozoic Eon reflect the increasing stability that progressively accumulated during the Proterozoic through cooling and thickening of the continental crust within a tectonic framework similar to that of today. A stable continental crust is thick enough to slow down the transfer of heat from the Earth's mantle the Earth's surface. As a thicker crust reduces the flow of heat, then less thermal energy is available to create quick and large uplifts with rapid tectonic activity in the continental interiors. Stability allows continents to increase their surface area exposed to surface processes over long periods. In turn, more effective weathering and efficient erosion results in deposition of mature sediments within large, established sedimentary basins. Although the Proterozoic contains tracks of immature sedimentary rocks mixed with extensive volumes of volcanic rocks that characterize typical rocks of the Archean Eon, such assemblages are less extensive and are separated by broad regions of relatively undisturbed, quiet water sediments unaffected by significant magmatic or tectonic activity. The marked increase in stability of Proterozoic continental crust enables us to use present-day geologic characteristics and processes associated with plate margins as guides for interpreting the geologic history of the oldest rocks existing within the Atlantic Southeast region.

▶ **Figure Atlantic SE.3** Sketch map of Rodinia with composite continents.

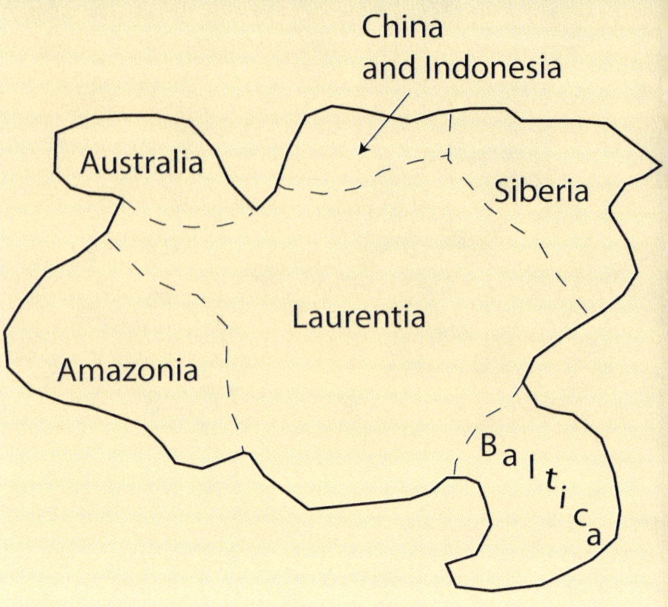

Metamorphic rocks exposed in northeastern Tennessee and northwestern North Carolina are approximately 1.8 billion year old, forming originally as sedimentary and igneous rocks associated with a large and stable continent known as **Laurentia**. Similar and slightly younger rocks comprise an axial zone within the Blue Ridge that compares favorably with rocks along the eastern and southeastern margin of Laurentia known as the **Grenville** Province. Geologists interpret the Grenville rocks as the products of a series of tectonic events associated with collisions between eastern Laurentia and ancient smaller continents that now probably exist within South America and Australia. The collisions spanned an interval of approximately 200 million years (1.2–1.0 billion years) and were part of a process that assembled the supercontinent **Rodinia** (Figure 3). The Atlantic Southeast provides some insights into this construction.

Geologists know virtually nothing about the intervening ocean basin consumed by the collisions between eastern Laurentia and the approaching continents. Exposures of Grenville-aged rocks along the Blue Ridge (Figure 4) support an interpretation of colliding continents driven over the edge of Laurentia in the same manner that Asia currently overrides northern India in the Himalaya Mountains. The larger size and greater thickness of Proterozoic continents generated large, compressive stresses during the collisions that extended to great depths. The combination of large stresses and higher temperatures at depth along the edges of the colliding continents resulted in deforming rocks at elevated temperatures, a major requirement for high-temperature and high-pressure regional metamorphism.

▶ **Figure Atlantic SE.4** Exposure of Grenville basement rock near Mars Hill, North Carolina. Darker layers of biotite-hornblende gneiss and lighter layers of biotitic granite gneiss.

Part of the horizontally directed compressive stress cut through the rocks, creating tabular, thick masses along zones of ductile shearing (mylonite zones). Eventually, blocks of lower crustal rocks thrust forward and upward along the mylonite zones to shallower depths within the crust. The progressive deformation squeezed the continental crusts involved in the collision into narrower intervals by telescoping the lower and upper parts of the crusts into thicker, vertically stacked layers of higher-grade metamorphosed rocks on top of lower-grade, shallow crustal rocks, including sedimentary rocks characteristic of stable continental margins. The collision process generated heat through friction and decompression that resulted in partial melting and subsequent widespread intrusive activity that welded the zone of collisions to Laurentia as part of the formation of the much larger Rodinian supercontinent.

Grenville-aged rocks in the Atlantic Southeast region occur either as elongated masses of metamorphosed plutonic rocks within the axial region of the Blue Ridge or as isolated, domelike masses surrounded by younger rocks within the Piedmont. Although these rocks possess distinctive remnant evidence of their Grenville history, their current geologic and geographic positions record a complex history of younger, overprinting, non-Grenville tectonic events. Grenville-aged rocks of the Blue Ridge and Piedmont are fragments of Rodinia that rifted apart from the supercontinent during the late Proterozoic. Similar fragments are found in other continents that drifted apart with the disaggregating of Rodinia. Some of the larger fragments now comprise a group of Laurentian continents; the remainder constitutes the Gondwanan continents. Smaller fragments of Rodinia existed as independent land masses (**microcontinents**) fringing the peripheral margins of both Laurentian and Gondwanan continents and produced similar, but not identical, geologic histories. Ironically, much younger geologic and tectonic events consolidated the rifted fragments back into Laurentia during the assembly of the next supercontinent, **Pangaea**.

Late Proterozoic Geologic History

Continental rifting typically follows the final consolidation of a supercontinent. Metamorphosed sedimentary and related igneous rocks of the Blue Ridge in Virginia, North Carolina, and Tennessee are relics of active continental rifting within Rodinia approximately 750 million years ago during the Late Proterozoic. Linear basins bordered by normal faults opened within the crust and began filling with a very thick volume of rapidly deposited sedimentary rocks. Later Paleozoic orogenic events obliterated much of the contiguous regional record for rifting of Rodinia, but careful research on selected areas in the Atlantic Southeast region reveals some aspects of this important stage in the geologic history of the Appalachians.

Rifting of a supercontinent begins with a broad circular or domelike uplift of the crust and lithosphere situated over a rising plume of asthenosphere (Figure 5). Heating of the lithosphere causes it to buoyantly rise upward. The continental crust situated at the top of the lithosphere stretches over the rising dome and begins to rupture along a three-armed pattern of major fracture zones that radiate outward from the central area of uplift. Each arm hosts a growing system of normal faults and narrow, linear, sedimentary basins within the down-dropped blocks (graben). As more separation occurs along the faults, the basins become wider and deeper. High topographic relief between the uplifted horsts of continental crust and the floors of the sedimentary basins in graben drives a rapid process of erosion and deposition that pours very coarse-grained and poorly sorted clastic sediments into the basins. Alluvial fans accumulate at the base of the steep slopes of adjacent horsts. Fluvial environments characterize the interior of the basins. Deep lakes also can form within the lowest parts of the basins. The linearity of the basins requires rapid transitions between sedimentary facies over a narrow distance. Sandstones represent the finer-grained clastic sediments of the fluvial systems and locally contain significant amounts of feldspar, mica, and other types of sedimentary and igneous rock. The fluvial facies change into mud-rich sediments of deep lakes. Interior drainage within the basins favors the formation of playa lakes and deposition of layers containing evaporite minerals (salts) interbedded within the pile of clastic sediments. Continued extension and rifting results in the floors of the basins and lakes extending to depths well below sea level. However, only two of the arms and basin systems

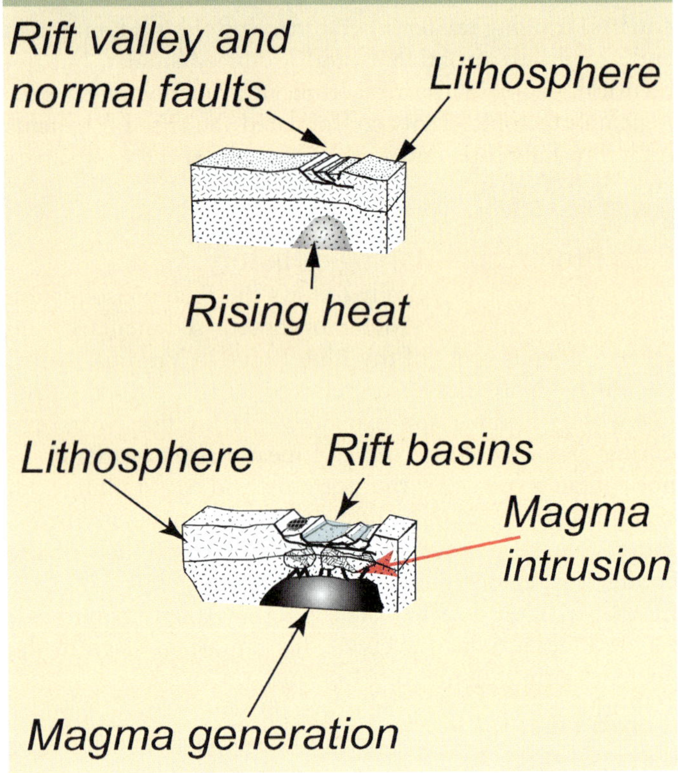

► **Figure Atlantic SE.5** Processes of continental rifting. Heat of rising plume lifts overlying continental crust and thins lithosphere. Extensional normal faults accommodate uplift and thinning. Partial melting of underlying material leads to basalt magmas that rise and intrude along channels created by fault system.

continue to grow and expand, eventually merging into a juvenile ocean basin (Figure 6). The third arm ceases to expand, but its aborted rift history can be preserved in the rock record.

The rising plume affecting the lithosphere also generates magmas that intrude the crust along rift-related normal faults. Volcanoes form with repeated eruptions, leaving a record of alternating flows of **basalt** within various levels of sediments deposited in the basins. Basalt dikes and sills also penetrate the sedimentary pile as satellite intrusions branching from the feeder system for the magma. Small plutons of alkali-enriched granitic originate within the thinned continental crust and rise to shallow depths underlying the **rift basin**. As the two favored systems of rift basins begin to merge as one, an incipient ocean basin with juvenile oceanic crust grows from the increased volume of injected basaltic magma. Through time and growth of the oceanic crust, the rift basins fringing the edge of the ocean basin gradually cool, become tectonically inactive, and subside under the growing weight of transgressing marine sediments. The buried rift valleys and their cover of fine-grained marine sediments comprise a characteristic stratigraphic package on the trailing edge of the drifting continent known as a passive margin sequence. As long as the ocean basin continues to grow by spreading from its central ridge system, the passive margin exists without major interruptions for very long time intervals of geologic time.

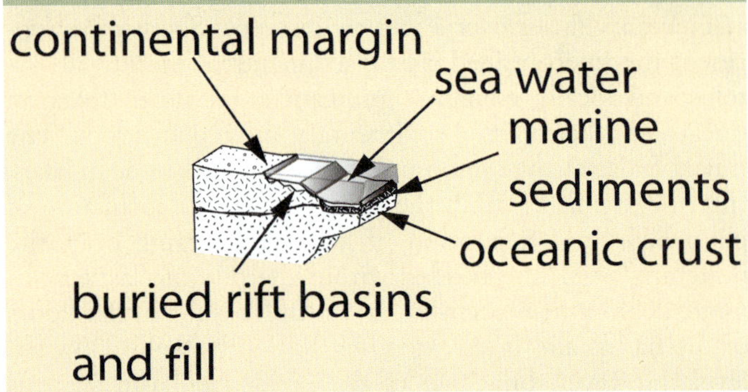

► **Figure Atlantic SE.6** Establishing the passive continental margin. Continued rifting results in merging of two arms of original rift. Thinned and subsided crust of rift basin becomes flooded by surrounding sea waters. Continued subsidence occurs with cooling and progressive movement of continental margin from juvenile ocean basin.

Late Proterozoic stratified rocks unconformably overlie Grenville basement rocks within linear basins locally bound by faults at several centers within the Blue Ridge. One extremely thick sequence, the Great Smoky Group, underlies the Great Smoky Mountains of western North Carolina and eastern Tennessee (Figure 7). Characteristic rocks of

► **Figure Atlantic SE.7** Folded metasandstone and schist exposed in Great Smoky Mountains National Park at Woolyback Overlook, North Carolina.

the metamorphosed sequence include very coarse conglomerates grading into feldspathic sandstones, mudrocks and interstratified turbidite deposits, and shales deposited under less-energetic conditions. At other locations in the Blue Ridge, dikes and sills of fine-grained gabbro (known locally in North Carolina as the Bakersville) intrude Late Proterozoic rocks and represent feeder systems for layers of basalt and rhyolite found within the various successions (Grandfather Mountain and Mount Rogers Formations, for example). Small stocks of alkaline-rich granite (Crossnore group of plutons along the Blue Ridge), whose absolute ages coincide with the relative stratigraphic positions of Late Proterozoic rocks, intrude both the strata and the underlying Grenville basement along a trend that is traceable northward within the Blue Ridge from North Carolina through Virginia.

The geologic events responsible for continental rifting and associated magmatic activity often work to concentrate metallic elements into resources. Within the Blue Ridge, many old prospects and mines in both the crystalline rocks of the basement, stratified deposits, and cross-cutting plutonic rocks have yielded significant amounts of copper and iron and even precious metals including gold and silver.

Atlantic Southeast.3 Late Proterozoic Climates, Paleogeography, and Life

Late Proterozoic stratified rocks of the Blue Ridge in the Mount Rogers area of southwestern Virginia and similar rocks in the Grandfather Mountain area of northwestern North Carolina provide some interesting clues supporting the presence of continental glaciers at this time. Thick, massive units consisting of poorly sorted (boulder- to silt-sized) clasts within the uppermost parts of the Late Proterozoic strata grade into areas of mudrock consisting of alternating thin lamina of silt and clay. Some of the thinly laminated clay mudrock units contain large, somewhat angular clasts of exotic rocks including granite and rhyolite. The clasts penetrate and deform underlying lamina of clay and silt. However, the laminar stratification also envelopes the upper parts of the exotic clasts, indicating that sedimentation continued long after the introduction of the fragments in the sediment. Possible interpretations for explaining the exotic clasts as dropstones include falling ejecta from local volcanoes (for the rhyolite) and rafting of continental sediments by icebergs into meltwater lakes (for the granite). If the laminated mudrocks represent varved deposits of glacial lakes, then the massive conglomeratic units could represent glacial tillites. The paucity of additional exposures exhibiting these features leaves an open interpretation for the Blue Ridge strata as glacial deposits. No unambiguous exposures of polished and striated basement exist below the postulated glacial deposits in the Blue Ridge. Indirect support for widespread continental glaciation in the Late Proterozoic includes sedimentologic and stratigraphic interpretations from both surrounding and younger overlying strata, but the interpretation remains inconclusive. Other regions in the world yield relatively conclusive evidence of Late Proterozoic glaciation, but their stratigraphic correlation with similar-aged strata in the Blue Ridge is not well-defined and certain.

Reported discoveries of fossils in the Late Proterozoic stratified succession include visible impressions of probable multicellular organisms and microscopic fossils identified as acritarchs. The poor preservation of the indistinct impressions prevents a positive identification of these fossils to any particular type of Late Proterozoic soft-bodied organism, but little doubt exists regarding the age of the impressions. The range of time during which acritarchs lived extends from the Late Proterozoic into the Early Paleozoic. Their local occurrences within strata that correlate with known uppermost units of the Late Proterozoic succession help to constrain the age of these rocks in the Blue Ridge. Structural complexities inhibit conclusive correlations between exposures containing other reported fossil discoveries and known fossil-bearing strata in either Late Proterozoic or Early Paleozoic strata within the Blue Ridge area. Widespread, pervasive deformation and metamorphism during younger Paleozoic orogenic events, as well as erosion of potentially fossiliferous rocks obliterated most chances of discovering evidence of life in the Late Proterozoic succession of the Blue Ridge.

Atlantic Southeast.4 Early Paleozoic History of the Laurentian Continental Margin

Gondwana and its peripheral microcontinents continued to drift away from the passive eastern margin of Laurentia during the Early Paleozoic (Figure 8). Clastic sediments transported by rivers draining Laurentia entered the flooded continental shelf where currents worked to grade and sort them during their transition from a fluvial to shallow marine environment. Quiet, shallow marine waters permitted settling of suspended mud and precipitation of dissolved ions in the streams as layers of terrigenous mud and chemical sediments. The collective assemblage of marine sandstone, shale, and carbonate begins to transgress across the subsiding continental shelf as a **drift sequence** of strata and marks the complete transition from rifting to drifting (Figure 9).

Fluvial sandstones and conglomerates grade upward into finer-grained and relatively thin marine shales and carbonates and mark the base of the Cambrian stratigraphic section in the southern Appalachians. Locally, relatively thin amygdaloidal basalt flows appear in an alternating sequence within the lowest parts of the coarse clastic strata. At some

▶ Figure Atlantic SE.8 Sketch map of southeastern Laurentia during the Late Proterozoic and juvenile Iapetus Ocean.

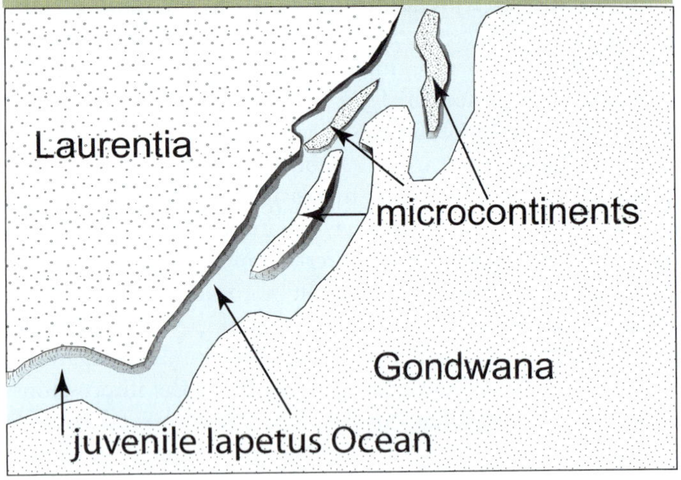

▶ Figure Atlantic SE.9 Suggested profile of Laurentian passive margin during the latest Proterozoic and earliest Paleozoic.

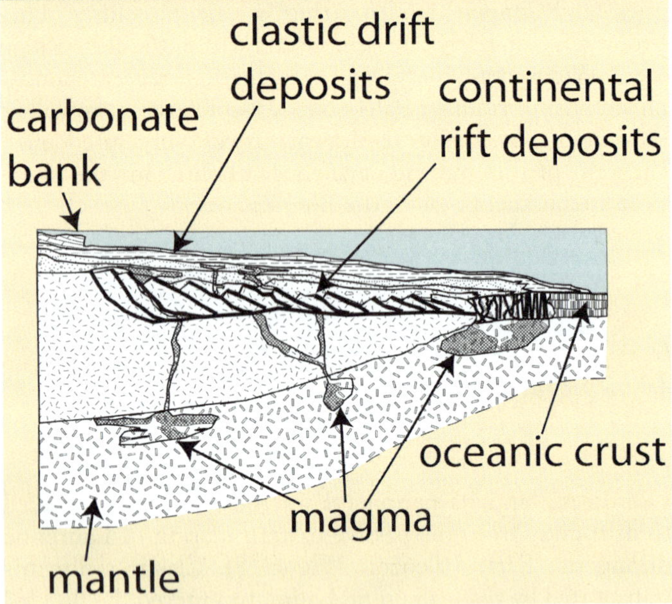

localities in the Blue Ridge, no distinctive break occurs between the uppermost Proterozoic stratified sequence and the lowermost Cambrian conglomerates and sandstones. However, basal Cambrian strata overlie older rock units with a marked **unconformity** at many other localities. All of this reveals that the Cambrian strata compose the lower part of an onlapping, or transgressive, package of marine sedimentary rocks representing the drift sequence of the early **Iapetus** Ocean basin. An interesting aspect of the basal Cambrian conglomerates and sandstones is the relative abundance of large, somewhat angular clasts of fresh potassium feldspar. Weathering processes usually degrade feldspar minerals into clay on exposure to air and water, but the clasts in the onlapping Cambrian clastic sediments argue for very little weathering before deposition. Warm ambient temperatures and the presence of sufficient water favor a fast rate in the chemical decomposition of minerals during weathering. Although direct evidence supporting widespread glaciation during the Late Proterozoic in the Blue Ridge is lacking, essentially unweathered clastic feldspar grains in conglomerates of the lowermost Cambrian argues for rapid deposition under near-glacial temperatures. Basal Cambrian conglomerates from other localities in North America have similar characteristics and confirm that cold and dry climates existed across the continent at this time. No other significant rock fragments are present as clastic grains in the basal conglomerates and sandstones, a purity that suggests a stable source area of granitic composition with low relief. An extensively glaciated continent matches these criteria.

The basal sequence of fluvial and marine clastic strata grades upward into marine shales and thicker carbonate units. Dolomite and chert comprise the bulk of the carbonate rocks reflecting very shallow marine waters associated with a carbonate bank. Carbonate deposition dominated the passive margin of eastern Laurentia for all of the Cambrian and into the Ordovician. A slight **regression** occurred within the Late Cambrian as the Rome trough opened as a rift located to the west of the passive margin within a more interior area of Laurentia (Figure 10). Rifting uplifted the margins of the trough, which then shed clastic sediments eastward toward the passive margin. After the uplift and subsequent erosion, renewed carbonate deposition on the passive margin continued without interruption into the early Middle Ordovician (Figure 11).

Carbonate deposition on the passive margin during the Late Cambrian and Ordovician exhibits cyclic repetitions of deeper facies of limestone and dolomite grading upward into very shallow carbonates with algal mats. Ordovician carbonate sequences in this region consist predominantly of tidal-dominated, very thinly laminated dolomitic limestones.

▶ Figure Atlantic SE.10 Sketch map of geography of Laurentian margin during the Cambrian.

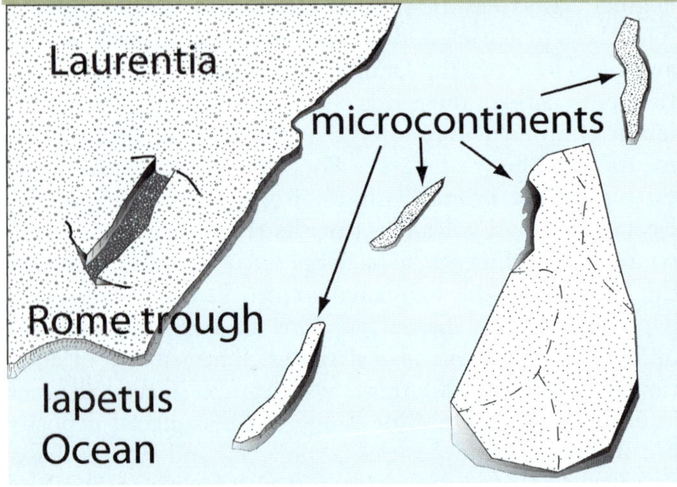

▶ **Figure Atlantic SE.11** Details of the Cambrian-Ordovician carbonate bank of eastern Laurentia.

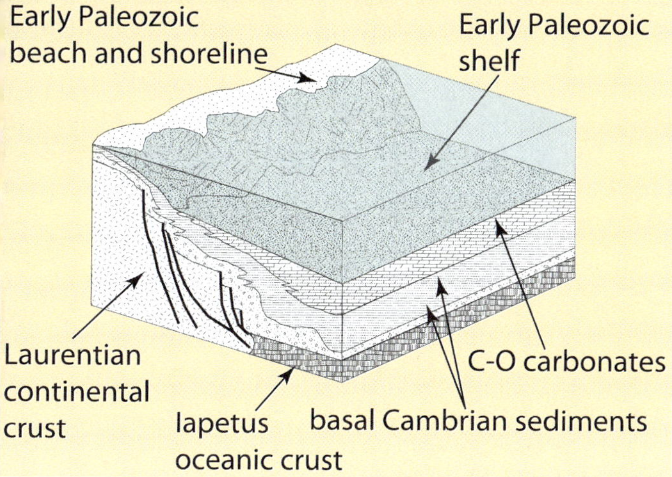

Exposures of these very shallow carbonates above sea level enhanced the growth of solution-related karstic features along a horizon that later marked a prominent **disconformity**. The pits, closed depressions, collapse breccias, and sinkholes characterize a weathered surface within carbonate rocks that became favorable locations for economic mineral deposits including manganese, copper, lead, and zinc.

Evidence of continental slope and deeper sediments of Cambrian and Early Ordovician age, where preserved in areas that lay east of the carbonate bank, consists of thin and somewhat laterally discontinuous sandstone interfingering with mudrock and limestone. These strata generally reflect limited contributions of sediments from the continent shelf lying to the west. Our principle clues for reconstructing the paleogeographic setting of this time occur in deformed rocks that were transported great distances from their original position during the major tectonic events of younger Paleozoic age. Before establishing the remaining history of Paleozoic sedimentary rocks of eastern Laurentia, let us shift our focus to the Late Proterozoic and early Paleozoic geology of the Iapetus Ocean basin that bordered the passive margin.

Atlantic Southeast.5 Proterozoic and Early Paleozoic History of the Terranes Comprising the Piedmont

Most of our knowledge regarding the early geologic history of the Iapetus Ocean basin comes from careful studies and analyses of exotic and **suspect terranes** currently recognized within the Blue Ridge and Piedmont provinces. A **terrane** consists of an area of rocks possessing similar characteristics and structural histories. Usually, the size of a terrane is at least several tens of kilometers. However, later Paleozoic orogenic events in the Appalachians greatly reduced terrane dimensions by thinning and extending, stretching and pulling, compressing, folding, refolding, and shearing. Recognized terranes within the Appalachians possess continental or oceanic characteristics. Mixtures of relatively low-grade metasedimentary and metaigneous rocks characterize terranes that formed in the uppermost levels of the crust. Later tectonic events have not obliterated or greatly modified primary features of the rocks such as bedding and original stratigraphy. Rocks that formed at deeper levels in the terranes generally have higher degrees of metamorphism and exhibit multiple, overprinting phases of dominantly **ductile deformation** that obliterate all primary structures.

Terranes generally represent a variety of origins including fragments of continents, remnants of oceanic islands, intact or fragmented oceanic plateaus, and other possibilities. Paleomagnetic studies of rocks and/or fossils discovered within a terrane assist in determining its original geographic location, particularly if the terrane is relatively young and has not been overprinted by too many later events. An **exotic terrane** is one whose original location is reasonably known. The original geographic setting of a suspect terrane is not known with reasonable certainty. Recognized terranes in the Atlantic Southeast region ultimately originated with the Proterozoic rifting of Rodinia into smaller continents (in this case, Laurentia and Gondwana), continental fragments, and microcontinents. Many of the terranes recognized within the Blue Ridge are remnants of Laurentia that drifted far enough away from the mother continent to cultivate their own variation of a drift-related stratigraphic sequence. Isotopic analyses of other Blue Ridge terranes suggest that some could be fragments of Gondwana that either drifted closer to Laurentia over time or were simply left behind as their mother continent drifted farther away from Laurentia.

Fault zones (either brittle or ductile) delineate terrane boundaries. The incorporation of a terrane within an **orogen** is the result of either convergent or transform tectonics related to plate collision along active boundaries. Identification of a terrane boundary helps in unraveling the relative sequence of geologic and tectonic events leading to the collision. Terranes can experience tectonic activity before their involvement in a later collision. The process of one colliding terrane merging into another terrane or continent is called **docking**. The **suture zone** marks the docking event and includes the terrane boundary and **stitching plutons** that weld the masses together. Stitching plutons intrude rocks on both sides of a terrane boundary and have ages that are similar to the terrane boundaries.

Prominent fault zones, identification of ophiolite complexes, and recognition of stitching plutons provide evidence of terrane boundaries and suture zones in the Blue Ridge and Piedmont. Ophiolite complexes are segments of oceanic lithosphere and crust characterized by a distinctive association of rocks: basal ultramafic rocks overlain by

Figure Atlantic SE.12 Exposure of deformed and partially melted layers or blocks of amphibolite gneiss in a matrix of finer-grained feldspar-quartz rock. Note hat and tubular borings for blasting as references of scale. Several, but not all, of the amphibolites are highlighted. The amphibolites are interpreted as remnants of part of the ophiolite sequence as suggested by the linking arrows. U.S. 23-441 at Savannah Church, North Carolina.

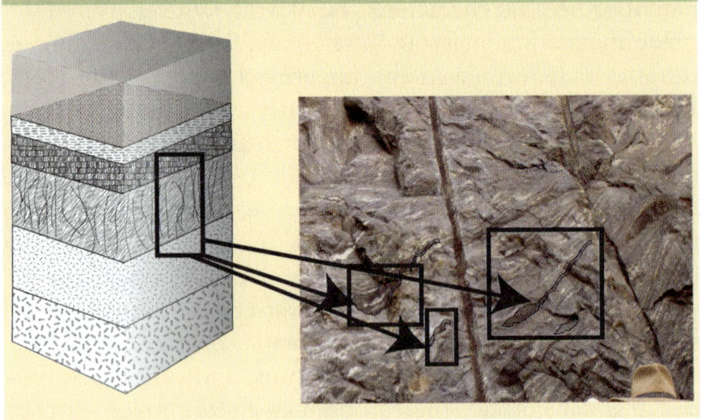

massive layers of gabbro, and an overlying complex of sheeted basaltic dikes, pillow lavas, and basaltic oceanic crust. Deep-sea muds and oozes comprise a thin sedimentary veneer capping the uppermost oceanic crust. Ophiolites in the Blue Ridge and Piedmont are recognized by their ultramafic and mafic compositions either as mappable contiguous bodies or as discontinuous, stretched, and thinned layers of centimeter-sized fragments within gneiss (Figure 12). Later deformation and metamorphism obliterated the originally continuous sequence from the ultramafic mantle through the oceanic crust comprising the ideal ophiolite. Nonetheless, the juxtaposition of ultramafic and mafic rocks even as discontinuous layers within gneiss faulted against Grenville basement gneisses or other bedrocks of continental origin delineates a terrane boundary. Paleozoic orogenic events overprinting earlier Paleozoic and late Proterozoic suture zones have made interpretations of the early Iapetus Ocean basin difficult, but not impossible.

Metamorphic rock types in the Piedmont Province range from slates to gneisses, greenstones to amphibolites, and quartzites to marbles. Protoliths include ophiolite complexes; mafic and felsic plutons associated with a **magmatic arc** or arcs; volcanic ash beds, tuffs, flows, and volcanic-related sedimentary rocks; quartz sandstones; mudrocks; and limestones and dolomites. Reported metamorphic grades range from lowest greenschist facies into upper amphibolite facies. Pressure types also vary from low- to high-pressure metamorphism. Compilations of mineral and whole-rock ages of Piedmont rocks date the origins of these rocks between the Late Proterozoic up through the Late Paleozoic. Shear zones and thrust faults dissect the Piedmont into smaller areas, each containing an internal coherent assemblage of characteristic rocks typical of exotic and suspect terranes.

The eastern segment of the Piedmont contains remnant magmatic arcs that formed during docking and collision events of terranes whose ages are older than the Iapetus Ocean. Plutonic rocks of the terranes fix the basement underlying these arcs as either oceanic or hybrid. Hybrid crust contains elements of both oceanic and continental settings and suggests a **subduction zone** origin involving a small continental fragment or microcontinent. Rodinia's dismemberment scattered small fragments of continental crust including microcontinents across Iapetus. Limited fossil data within terranes of the eastern Piedmont place their original position around the periphery of Iapetus in the vicinity of distal Gondwana, suggesting that Iapetus was a large ocean basin.

Atlantic Southeast.6 Middle Ordovician through Late Paleozoic Geology and Tectonics

Taconic and Post-Taconic Paleozoic Geologic History of the Laurentian Continental Margin

Uplift and a regional regression during the Middle Ordovician resulted in extensive weathering and erosion of the Lower Paleozoic passive margin sequence of Laurentia. A regional **transgression** deposited a layer of younger carbonate sediments over weathered remnants of the earlier Paleozoic carbonate bank. Major subsidence adjacent to the shelf created a deep marine basin that began filling with submarine fan deposits including turbidite sequences and deep-water clastic sediments derived from eastern source areas. The drastic shift from the underlying passive margin sequence to this thick mass of mixed coarse- to fine-grained, immature clastic sediments including volcanic rock fragments reflects the tectonic origin for the strata deposited within a deep-sea, subduction-related trench. As the trench became completely filled, its sediments began drowning carbonate deposits with clastic material.

A wedge of Late Ordovician and Early Silurian clastic sediments overlies remnants of the subduction-related clastic sediments that overwhelmed the former Laurentian shelf margin during the Middle Ordovician (Figure 13). The very mature conglomerates, sands, and reddish muds reflect a mixture of beach, deltaic, and westward-flowing fluvial systems as part of a broad coastal plain along the flank of eroding highlands. The Lower Silurian red beds yield iron ore when the concentration of iron in certain thick layers is unusually high. Many beds of the Lower Silurian Clinton Group supplied ore for the iron and steel industry of the United States in the nineteenth and twentieth centuries. The Silurian clastic sequence graded westward into marine muds and carbonate sediments. Although erosion has removed part of the stratigraphic record of the later Silurian and Early Devonian in the Atlantic Southeast, deposition continued uninterrupted through part of the Early Devonian. Depositional environments of the

▶ **Figure Atlantic SE.13** Folded and thrust-faulted Upper Ordovician and Lower Silurian clastic sediments in southwestern Virginia. Red coloration of sandstones on the left results from iron oxides. Deformation of the strata resulted from Alleghanian Orogeny in Late Paleozoic.

Devonian ranged from deep basin with accumulations of reduced organic-rich mud (deprived of oxygen) upward into shallower waters containing coarser-grained sandy and carbonate sediments. Devonian black shales of the deeper marine facies serve as source beds for petroleum and natural gas that later migrated and accumulated within overlying reservoir rocks, many of which are still in production. Carbonate rocks of the Lower Devonian yield a prolific quantity of diverse marine fossils. The unusually pure layers of Devonian marine sandstones provide excellent sources of sand for making high-quality glass and glassware. Within the lower interval of the black shale, layers of altered volcanic ash, derived from an eastern source, are locally capped by an interval of chert. Rapid filling of the Appalachian basin began during the middle Devonian. Fluvial and deltaic sediments associated with a westward-advancing fluvial system encroached into the region as the Catskill Delta sequence. Catskill strata occur in more northerly parts of the Appalachian Basin, but intervals of brown and red shale with greenish gray sandstone and black shale are their distant equivalents present in the southern parts of the Appalachian Basin. Lower Mississippian fluvial sediments overlie the Devonian and grade upward into a marine mud and carbonate sequence.

Cyclic deposits characterize Lower Pennsylvanian strata, beginning with a basal sequence of nonmarine fluvial sandstones and associated floodplain deposits. The fluvial systems flowed through a broad area containing lakes and swamps. Relative submergence of this large coastal plain resulted in the progressive deposition of marine shale and limestone. Relative sea level began to lower as reflected by deposition of brackish marine shale over the limestone. Continued regression leads to the start of the cycle again with fluvial channels cutting into the upper parts of the brackish marine shale. Within the Appalachians, most of these cyclic deposits appear within the Appalachian Plateaus, but small areas of the section of the Pennsylvanian locally occur in the western Valley and Ridge. Organic sediments of the swamp deposits eventually became peat, which transformed into the abundant bituminous coal deposits of the Appalachians. Literally hundreds of cycles can be counted throughout the region, demonstrating the tremendous volume of the economic coal resources present. Above the coal-bearing intervals, a sequence of fluvial and floodplain deposits of conglomerate, sandstone, and reddish-colored mudrocks persist through the remainder of the exposed section of Pennsylvanian strata. This uppermost part of the section comprises a westward-advancing clastic wedge deposited in front of a highland region.

Bituminous coal is one of the progressive rankings of coal based on the amount of carbon present. Lignite is the category for low-rank coals, bituminous for middle ranks, and anthracite has the highest amount of carbon present within the coal. Lower amounts of carbon within the coal permit greater amounts of noncarbon materials to be present including sulfide minerals such as pyrite. Waste rock excavated during mining of bituminous coal exposes large amounts of disseminated pyrite to weathering. Iron sulfide decomposes into various iron and sulfur compounds including sulfuric acid, which enters streams draining the mined area. Acid waters have significant and deleterious environmental effects within local watersheds. Laws and regulations for reclaiming old and recently depleted mines attempt to minimize the environmental damage from acid waters through requirements that isolate and seal the pyrite-bearing material from further exposure to the atmosphere and surface waters.

Major industrial centers originally located in the central and southern Appalachians because of the close proximity to bituminous coal resources. In particular, the close proximity of coal, limestone, iron ore, and major rivers for transportation were major factors in the success of the early iron and steel industry in the central and southern Appalachians. Coal was fused and partially burned to make coke, which burned at higher temperatures in a blast furnace. Limestone acted as a flux when mixed with iron ore in the furnace to smelt the raw iron, or pig iron, from the ore. The molten flux floated on top of the liquid iron to form porous, volcanic-looking material called slag. Piles of slag could be seen as tall hills in many of the old steel mills that populated this part of the Appalachian Mountains.

Paleozoic Geology and Tectonic History of the Piedmont

Subduction zones mark the descent of one plate under another during their collision along a **convergent plate boundary**. Horizontally directed compressive forces and stresses generate a consistent sequence of topographic characteristics at subduction zones including a trench, **accretionary complex**, **forearc** basin, and a magmatic or volcanic arc. As opposed to the extensional rifting and normal faulting of divergent plate boundaries, compressive folding and reverse faulting deform rocks caught within the vise of convergent boundaries. Rifting at divergent boundaries expresses the upward rise of a convecting plume or mass of asthenosphere and culminates with lateral drifting of continents as the new lithosphere and crust cool. Subduction zones form because the convective motion begins a downward path into the mantle. The downward turn effectively drags the overlying oceanic crust, creating a deep trench in the sea floor. The deepest known trenches reach depths as much as 35,000 feet. As subduction proceeds, the trench deepens. Any sediment or weakly consolidated material on the sea floor is scraped off by the overlying plate, accumulating into a wedge-shaped mass of squeezed and thrusted sediments and rocks known as an accretionary complex. The complex grows progressively wider and taller as subduction continues, eventually reaching the trench. The subducted lithosphere reached depths within the mantle where water and other volatiles can escape from the descending rocks into the overlying mantle leading to the generation of magmas and volcanoes of the volcanic arc. Instability of the growing accretionary complex leads to its partial collapse, opening a basin on top of the accretionary complex and in front of the volcanic arc (the forearc basin). Submarine fans and turbidite flows emanating from the elevated volcanic arc soon begin to accumulate in the basin as their escape toward the trench is now slowed. Parts of the descending oceanic crust may be pried off during subduction below the accretionary wedge and become incorporated into its base. Compression within the wedge results in folding, and thrust faulting. Thrust faults climb upward from the base of the accretionary complex and outward toward the trench and may eventually affect parts of the overlying forearc basin. A chaotic mixture (mélange) forms and consists of clastic sediment, deep-sea oozes and muds, immature sands, breccias, conglomerates, volcanic rock fragments, and even pieces of the ocean floor and crust (ophiolite material).

Scraping of the descending slab may shear off a large flake or fragment of oceanic lithosphere that becomes embedded within the deforming accretionary complex as either a partial (dismembered) or intact ophiolite. The dimensions of a subduction zone are capable of incorporating ophiolites or large slabs of other intact rock masses that have areas reaching several hundreds of meters.

► **Figure Atlantic SE.14** Sketch map of western Iapetus and the eastern Laurentian margin during the middle Ordovician. A subduction zone and trench formed in front of the advancing composite terrane of the Piedmont. Eventually, the eastern margin of Laurentia partially descended in this trench resulting in the Taconic Orogeny.

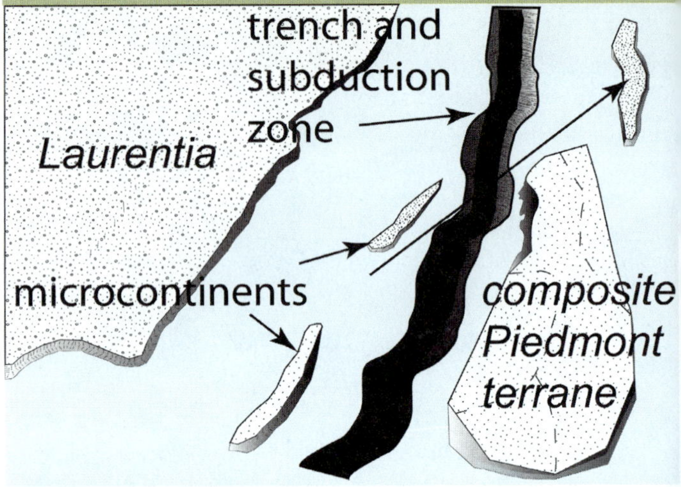

Subduction around Iapetus began during the Cambrian as marked by docking of older terranes containing Late Proterozoic sedimentary rocks docking with other small terranes into larger masses (a **composite terrane**). An eastward-dipping subduction zone formed in front of approaching terranes of the eastern Blue Ridge and western Piedmont, partially dragging the eastern margin of Laurentia downward, an event that triggered the Ordovician **Taconic** Orogeny (Figure 14). The oblique approach of this progressive collision resulted in terranes of the western Piedmont and eastern Blue Ridge docking to Laurentia before the arrival of the eastern Piedmont terranes. Late Ordovician plutons stitched a suture whose trace lies somewhere at depth and within a major though-going shear, the Central Piedmont Shear Zone. Uplift of terranes above the subduction zone and/or back-thrusting them toward Laurentia accounts for uplift, erosion, and subsidence of the remaining passive margin during the Late Ordovician before its flooding with clastic sediments during the Early Silurian.

Iapetus continued to close after the events of the Taconic Orogeny. Post-Taconic granitic plutons intruded through shear zones that juxtaposed shallow and deep segments of metamorphosed terranes. During the Devonian, the closure of Iapetus changed from a mode dominated by subduction to one characterized by right-handed (dextral) strike-slip motion combined with compression, a type of motion known as **transpression** (Figure 15). The Brevard Zone, one of several major shear zones reactivated at this time, parallels the regional strike of the Blue Ridge and generally lies at the base of the Blue Ridge Front in western North Carolina. Many of the Taconic thrust faults associated with subduction zones became reactivated as Devonian dextral shear zones. The timing and size of these oblique, regional shear zones reflects an origin rooted with a large-scale docking event, possibly a large microcontinent, with Laurentia.

▸ **Figure Atlantic SE.15** Diagram demonstrating dextral strike-slip tectonics of the Acadian Orogeny. Oblique convergence of the buoyant composite terranes of Piedmont and Carolina with Laurentia resulted in dextral shearing. The collision reactivated older Taconic faults, which also experienced dextral shearing. Note that the closest part of the Piedmont terrane was involved with shearing long before more distant parts on the trailing edge of the terrane. Although absolute ages of deformed rocks in both parts of the larger terrane can indicate a significant discrepancy in times of shearing, the nature of oblique collisions correctly reveals that the two ages are end points of a progressive series and not two separate events.

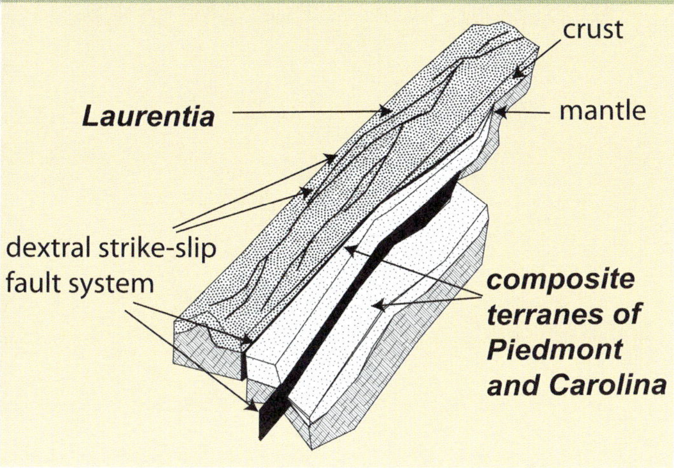

▸ **Figure Atlantic SE.16** Schematic profile of the Alleghanian Orogeny. Gondwana collided with Laurentia and thrust the rocks and structures formed during the Acadian and Taconic events westward along a detachment zone, which can be traced below the Appalachian orogen eastward below the present Coastal Plain. VR represents the Valley and Ridge. BR is the Blue Ridge.

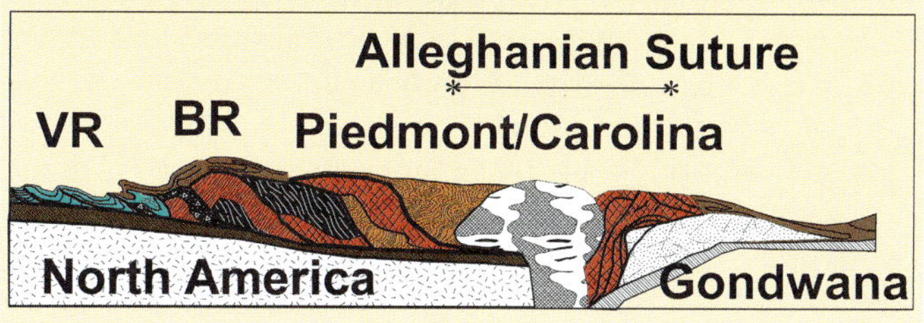

The San Andreas Fault Zone, a classic **strike-slip fault**, experienced a similar history of transition from a subduction to a transpressive mode of collision. Collectively, the dextral shearing, granitic magmas, and accompanying low-grade metamorphic processes comprise the **Acadian** Orogeny in the Southern Appalachians.

Devonian magmas in the Piedmont and Blue Ridge are mostly granitic but are associated with significant economic mineral deposits. Over an approximate area of 500 km^2, the Spruce Pine district in the Blue Ridge of North Carolina contains several thousand pegmatite bodies and very light-colored granitoid intrusions known as alaskites. Economic deposits of minerals present within the Spruce Pine district include kyanite, chromite, beryl, potassium feldspar, muscovite, kaolin, and quartz (of extremely high quality for fiber optics, optical glass, and microchips for electronics). Gemstones produced include emeralds, aquamarine, and golden beryl. Uranium minerals are also present in the pegmatite bodies. Within the Piedmont, the Acadian granites are larger in area, display more effects of metamorphism, and have created significant mineralization in contact with host schistose and metavolcanic rocks including the famous spodumene deposits within the Kings Mountain district.

The final events in the closure of Iapetus occurred during the Late Paleozoic **Alleghanian** Orogeny (Figure 16). Metamorphism and deformation of this event was strong enough to overprint many effects of the Taconic and Acadian events through reactivation as structures in the Alleghanian sequence of compression. The compression resulted in large-scale lateral overthrusting of the Valley and Ridge, Blue Ridge, and Piedmont along detachment horizons beginning near the base of the passive margin sequence. Compression also reactivated Acadian shear zones as thrust faults. Widespread granitic plutons in the Piedmont comprise the largest orogenic-related magmas in the Atlantic Southeast and are prized as sources of building and dimension stone for many cities and monuments across the United States. Alleghanian regional metamorphism reached upper amphibolite facies, which is lower than the peak Taconic metamorphic zones in the Blue Ridge. However, Alleghanian metamorphic recrystallization was pervasive and obliterated characteristic features of both the Taconic and Acadian orogenic events. Metamorphism accompanied granite magmatism, but many of the Alleghanian plutons such as the Rolesville **batholith** near Raleigh, North Carolina outlasted the peak metamorphism. Some Alleghanian thrust faults are clearly pre-Alleghanian metamorphism, others are contemporaneous with the regional recrystallization, while others moved well after metamorphism. Sometime during the Early Pennsylvanian within the Piedmont, Alleghanian thrusting transformed into Alleghanian dextral shear zones, some of which may record strike separations of up to 200 km.

Detachment horizons or faults occur when a relatively incompetent or weak layer within a horizontally layered pile of sedimentary rocks is squeezed and pushed from the side. Strata overlying the weaker layer move as a unit in the direction of the applied stress while the strata below remain relatively stationary. The detachment surface progressively moves through the weak layer until a lens or interbedded layer of stronger rock forces the detachment to climb upward at a steeper angle. Eventually, the detachment surface enters into an overlying weaker layer where that passage follows a nearly horizontal direction once again. The Paleozoic stratigraphy of the Appalachians favored using weaker shale intervals for subhorizontal motion along detachments (Figure 17), whereas sandstones act as more competent units with higher angled thrust faults. In a similar, but not identical, manner, horizontal detachments formed in crystalline rocks of the Piedmont and Blue Ridge. Movement along the detachments created long, linear trains of subparallel plunging anticlines and synclines in the Paleozoic sedimentary rocks now existing as the Valley and Ridge. Stacks of crystalline thrust sheets formed in both the Piedmont and Blue Ridge as a consequence of Alleghanian compression. Thrusting began during the early Carboniferous within Piedmont crystalline thrust sheets and became progressively younger toward the west with the emplacement of the youngest thrust sheets in the Valley and Ridge in the Permian. The main event triggering the Alleghanian Orogeny was the oblique collision of Laurentia with a comparable-sized Gondwanan continent such as Africa. Interestingly, the change from thrusting to dextral strike-slip motions in the Piedmont resulted in Early Pennsylvanian uplift and erosion of marginal strata in the Valley and Ridge. Continued transpression in the Piedmont created push-up ridges that quickly eroded releasing pulses of coarse-grained clastic wedges across the wide coastal plain in front of the uplifted hinterland.

The Alleghanian Orogeny is part of a global system of colliding continents that welded together during the Late Paleozoic as the supercontinent Pangaea. The Atlantic Southeast region is fortunate to have rocks that recorded the time and history of events that transpired between the formation of the most recent supercontinents, Rodinia and Pangaea. Just as its predecessor broke apart into smaller continental masses that drifted away, Pangaea began to rift and separate into separate continents that we now recognize as North America, South America, Africa, and Europe. We pick up their geologic histories as we progress forward in time into the Mesozoic and Cenozoic.

▶ **Figure Atlantic SE.17** Hunter Valley thrust near Duffield, Virginia. Cambrian carbonate (Conasauga Group, Middle Cambrian) of hanging wall section in contact with Devonian shale. Brunton pocket transit along the trace of fault toward the right of the photograph. The Devonian shale is ductile along the fault, whereas the overlying carbonate deformed in a more brittle manner. HV stands for the Hunter Valley thrust fault.

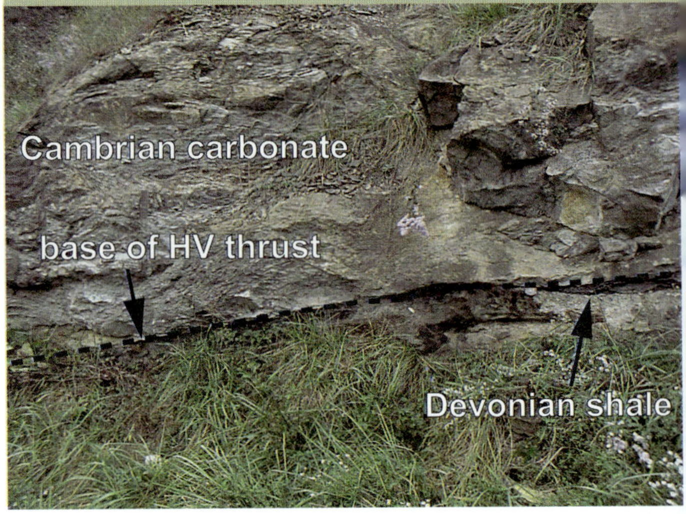

Atlantic Southeast.7 Mesozoic and Cenozoic Geology and Tectonics

Early Mesozoic Rift Basins

Pangaea began rifting during latest Paleozoic time in the southern Appalachians. Linear trends of brittle fractures, discordant zones of silicified fault breccia, and thick veins of milky quartz crosscut older Alleghanian features of Piedmont rocks at low, oblique angles. In Virginia and North Carolina, Late Triassic rift basins unconformably overlie Piedmont rocks, including these zones of brittle fractures. The relative age and regional trends of the brittle fracture zones suggest that tensile forces and stress fields responsible for their formation progressively initiated continental uplift and extension necessary for the younger rift basins.

Steeply dipping normal fault systems border the Late Triassic rift basins. In some basins, the main border fault dips toward the northwest, whereas in others, the dip is toward the southeast. Active faulting occurred during deposition of sediments within the basins. Strata dip toward the southeast in basins bordered by a northwest-dipping border fault. The opposite is true for basins bordered by southeast-dipping faults. Deep drilling records and geophysical studies reveal the presence of additional early Mesozoic rift basins deeply buried below the Coastal Plain and the present continental shelf. Thick sequences of younger strata overlie these buried basins. Somewhat later in the Triassic and early Jurassic, dikes and sills of diabase and flows of basalt intruded and flowed above the sediments filling the basins, as well as the surrounding rocks of the Piedmont (Figure 18). The parent magmas for these mafic igneous rocks ultimately created the embryonic crust underlying the North Atlantic Ocean. The history of the early Mesozoic rift basins is similar to those associated with the Late Proterozoic breakup of Rodinia.

The stratigraphy of the basins follows a characteristic and predictable threefold sequence: (1) a basal conglomerate, containing clasts of both granitic rocks and gneiss, and

▶ **Figure Atlantic SE.18** Jurassic dike exhibiting spheroidal weathering in the Piedmont of North Carolina.

John R. Huntsman

► Figure Atlantic SE.19 Inclined strata of Cow Branch Formation in a quarry near Eden, North Carolina. This locality is within the Dan River Basin, a Mesozoic rift basin.

coarse-grained arkosic sandstone; (2) an overlying sequence of interbedded dark gray to black fossiliferous siltstone (Figure 19), organic-rich shale, and thin coal layers; (3) and reddish brown siltstone and mudstone, arkosic to pebbly sandstone, and conglomerate. The strata reflect fluvial and lake environments within the interior of the basins bordered by a combination of fluvial/alluvial fan environments adjacent to the border faults. Deep drilling in buried basins under the present continental shelf reveals thick sequences of evaporite minerals and carbonate rocks as additional characteristic depositional facies. Fossils document a diverse record of life including plants (pollen, spores, leafs, twigs, roots, and wood), invertebrates, and vertebrates (fish and tetrapods including tracks and footprints) (Figure 20).

► Figure Atlantic SE.20 Vertebrate footprint preserved in Cow Branch Formation, Dan River Basin, Eden, North Carolina.

John R. Huntsman

Rift-to-Drift Transition

Mesozoic and younger strata exposed along the continental margins of eastern North America, North Africa, and Western Europe document the progressive changes whereby continental rift facies become drift facies of a passive margin sequence. A seaward-thickening wedge of both nonmarine and marine strata comprises the exposed Mesozoic drift sequence in the Coastal Plain. A prominent unconformity separates Early Cretaceous basal strata of the Coastal Plain sequence from rocks of the underlying Piedmont including the Late Triassic rift basins and younger basalts. Significant weathering and erosion of the Piedmont during an interval spanning tens of millions of years established a land surface of relatively low relief. However, a thick series of Jurassic carbonates and evaporite deposits overlie the buried rift basins. An interval of coarse-grained clastic sedimentary rocks of Cretaceous age overlies the carbonate sequence and is similar to thinner strata exposed in the Coastal Plain. The steady lateral migration of this clastic wedge eventually transgressed over the continental margin. By Late Cretaceous time, the transgression had reached the Fall Line, which delineates the western edge of the exposed Coastal Plain.

The Fall Line refers to a remarkably consistent trend of a narrow zone of small waterfalls common to all streams flowing from the Piedmont into the Coastal Plain. This "line of falls" significantly impeded exploration and settlement to early colonists dependent on rivers and streams for major transportation. The Fall Line results from the higher elevations left by resistant weathering of quartz-rich crystalline rocks of the Piedmont relative to lower elevations of the less resistant semiconsolidated strata of the Coastal Plain. During the erosion of the Piedmont before the Late Cretaceous, a significant topographic break existed at the Fall Line that effectively stopped the westward marine transgression of the Cretaceous clastic wedge across an eroded continental shelf inclined gently to the east. This imparted a very low angle of primary dip away from Piedmont toward the outer Coastal Plain and continental shelf in successive Coastal Plain strata.

Lower Cretaceous units of the exposed Coastal Plain section consist of a succession of fluvial and deltaic strata that laterally grade eastward into coarse-grained clastic sediments deposited in open marine shelf, lagoonal, and carbonate bank/reef environments. A thick sequence of coarse-grained conglomerates and arkosic sandstones grades upward from the base of the Lower Cretaceous units into clay beds and finer-grained sandstones. At some localities south of North Carolina, the clay layers become thick and consist of very pure economic deposits of kaolin. Sedimentary structures such as desiccation features, slump structures, and cross bedding, and remnants of woody plant material and root casts attest to the fluvial origin of the exposed Lower Cretaceous strata. Decreased grain sizes and better maturity of overlying sand, mud, and marl marks the completion of the rift-to-drift transition Lower Cretaceous.

The Late Cretaceous time interval reflects complex shifting of deltas, prodeltas, delta plains, and marine shelf facies across the continental shelf and margin. Upper Cretaceous thin sands, dark clays, cross-bedded micaceous sands, and gravels represent deposits within either a lower delta plain or a delta front environment. Burrows and channels disrupt bedding. Fragments of petrified wood coated with pyrite occur in other strata that also contain rounded clasts of pure kaolin and thicker beds of pure clay. Some layers consist of a mixture of glauconite mixed with clay, and thin layers of quartz sand containing phosphate pebbles and abraded fragments of bone and shells. The clastic layers change facies laterally into abundantly fossiliferous marine limestone beds containing gigantic fossil oysters. Some of the Upper Cretaceous units present in the coastal plain of North Carolina are either absent in northern South Carolina or present only as very thin beds. This distribution resulted from the presence of the emerging Cape Fear Arch. This structure greatly influenced the patterns of deposition and the distribution of sedimentary facies. Thinner sections of Upper Cretaceous strata coincide with the crest of the arch, whereas thicker and more complete sections occur as deeper embayed basins along its flanks. The gentle dips of strata away from the crest of the crest suggest the arch originated as a rising fold, possibly a domelike feature related to regional warping of the continental shelf at this time.

Tertiary and Quaternary Geology

Cenozoic strata preserve a record of repeated intervals of shallow marine transgressions and regressions of marine waters in the Coastal Plain. Cyclic changes in climate during an overall cooling from the much warmer global temperatures in the Cretaceous provided the greatest factors driving these changes in global, or eustatic, sea level. Local variations in patterns of sedimentary facies reflect additional tectonic influences on deposition. Cenozoic strata generally become younger and lower in elevation outward from the inner Coastal Plain toward the present-day shoreline and outlying continental shelf.

Recall that a relative rise in sea level results in a transgression, whereas a relative drop in sea level results in a regression. Transgressions, then, generally deposit new sediment, whereas regressions erode existing sediments and older strata exposed by the falling sea. Sea level constitutes the regional base level for streams, which, by definition, is the elevation below which no fluvial erosion occurs. Streams maintain a dynamic balance among erosion, transportation, and deposition of their sediment load. Significant lowering of base level drives the stream to actively erode down through its channel. Strong and rapid downward erosion creates entrenched stream meanders. Cenozoic regressions eroded older Coastal Plain strata leaving partial remnants in entrenched stream channels and tributaries. Thick aprons of coarse-grained, nonmarine sediments are typical deposits of a regression. Thinner and finer-grained marine strata result from a transgression. Subsequent rises in sea level (base level) deposit finer-grained and texturally mature sediments above eroded and dissected surfaces that contain remnants of coarser-grained and less mature regressive deposits. Transgressions initially enter and fill entrenched stream channels before eventually merging as laterally extensive sheets or layers of sediments as sea level rises progressively. A basic understanding of transgression and regression effects on deposition is necessary for understanding all of the subtle changes in Coastal Plain deposits during the Cenozoic.

The oldest Cenozoic strata are Paleocene in age and consist of sands, clays, and muds deposited in open marine waters of the continental shelf, or shallower and more restricted environments associated with barrier islands and the back-barrier lagoons, beaches and estuaries, and river environments. A significant transgression during the Eocene deposited lime-rich mud rocks and fossiliferous limestones on a broad but shallow continental shelf bordered by a narrow coastal plain. Fossils within the carbonate strata provide excellent correlations with similar transgressive cycles in other parts of the world. Miocene marine limestones unconformably overlie Eocene strata in the exposed Coastal Plain, but seismic studies on the outer continental margin demonstrate a more complete Miocene transgression occurred with a lower sequence of clastic strata underlying the limestones. The landward limit of the Miocene transgression lay west of the exposed Miocene strata in the Coastal Plain. Local remnants of a marine limestone unit, possibly of Miocene age, directly overlie Piedmont crystalline rocks west of the Fall Line in central North Carolina.

Pliocene and younger transgressions in the Cenozoic were not as extensive as those of the Eocene and Miocene. Older Pliocene deposits are regressive and generally consist of sands and gravels containing pebbles of phosphate, carbonized branches and twigs of plants, rip-up clasts of older sediments, and calcareous sands. Finer-grained, fossiliferous quartz sands, and sandy and clay-rich muds embody younger Pliocene transgressive deposits. Fossils in the sands contain a rich assemblage of clams and snails, very large oysters, and teeth and bones of vertebrate fossils. Lower Pleistocene deposits contain very fossiliferous marine sands and muds sometimes exposed as a hard, cemented coquina ("beachrock") at isolated locations along the coastline. Younger Quaternary deposits include beach and barrier island sands and dune deposits, and lagoonal silts and muds mixed with fluvial sands. Further inland from the coastal counties, gravels, clay-rich silts, and sands of fluvial origin mixed with dune deposits, alluvium, and colluvial material of probable Pliocene and Pleistocene ages unconformably overlie highly weathered older Coastal Plain strata and Piedmont rocks. Some of the sand deposits are valued for their purity as high-silica resources (Figure 21). Recent erosion has isolated these inland or upland deposits from their equivalent marine deposits present in the outer coastal plain.

Active sedimentation in the present-day coastal zone is part of a regional transgression induced by melting of Pleistocene glaciers over the past 10,000 to 20,000 years. The chain of barrier islands that lie seaward of bays and estuaries along the coast offer insights to the processes assisting the landward migration of the barrier islands. Littoral currents formed by breaking waves along the beach establish a seasonal direction of sand movement parallel to the length of the islands. Overwash of sand from the seaward face of the islands into small lobate deltas document the role of wave action in storms pushing sediment across the islands. Aeolian dunes (Figure 22) overlapping older marsh, back-barrier, and estuarine deposits on the lee side of the islands signify the important role of steady breezes in moving the sand. Tidal inlets form as breaks in the line of islands that permit the movement of daily tidal currents to move sediment. As inlets migrate or shift over time, so do the tidal channels and deltas, and overall expansion and contraction of the islands results. The time scale for these dynamic changes is measured in years and tens of years, resulting in constant threats to human structures willfully placed within this dynamic zone. Regional problems and

Figure Atlantic SE.21 Pliocene Citronelle Formation ("Lilesville Gravels") near Lilesville, North Carolina. Weathering of very minor iron oxide minerals has stained the surface of this unit into its distinctive red, yellow, and orange colors. Note the channels filled with the gravels cut into older weathered strata below.

John R. Huntsman

Figure Atlantic SE.22 Dune deposits exposed by action of tropical storm, southern New Hanover County, North Carolina. Note cross bedding at top of eroded bluff.

John R. Huntsman

expenses created by coastal dynamics and geologic processes provide plenty of fuel for controversy in the fields of science, engineering, urban and cultural development, politics, and economics.

The Cenozoic Era contains a repeated history of climatic changes accompanied by rising and falling sea levels. Pleistocene glaciers advanced over North America at least four times. Each advance is associated with a major regression, and each interval of warmer climates between the glacial advances contains a period of transgression. Global records indicate that the Cenozoic records a gradual global cooling from a Cretaceous maximum temperature, which far exceeded the average global temperature of today. The subject of global warming constitutes the center of discussions in many environmental, economic, and political circles. The current debates question the role of human activity in the form of releasing greenhouse gases into the atmosphere as major factors in global warming despite the obvious geologic observations that present-day Earth is not nearly as warm as it was during previous times of global warming. It is entirely possible that even given the probable increase in the rate of global warming from automobile and industrial emissions of greenhouse gases, the peak of global temperatures may not be as catastrophic as some contend. However, the documented dynamic changes within the coastal zone over the past hundred years suggest that even limited increases in sea levels produce significant problems for humans. To error greatly on the side of caution in releasing additional gases that can contribute to more rapid rates of global warming is a viable and perhaps practical choice. We have a moral and ethical responsibility to leave a habitat for successive generations to experience the wonderful aspects of geology.

Carolina Bays are interesting phenomena of geology present widely within the Coastal Plain of the Atlantic Southeast region, but a majority of geologists and earth scientists has yet to completely agree on their origin or origins. More than 500,000 Carolina Bays occur as circular to elliptical depressions from Delaware and New Jersey to Florida, but they are particularly concentrated in eastern regions of the Carolinas. Some of the bays are occupied by lakes, whereas others may be wetlands populated by Bay trees, which is the root of the term *Carolina Bay.* Most of the bays are elliptical in plan with an elongation direction trending NNW-SSE. The bays have variable lengths and widths from a few meters to several hundreds of meters. Small ridges of sand generally occur on the southeastern margins of the bays. At some locales, the sand ridges merge outward into fields and arrays of curved and linear dunes (parabolic and longitudinal). Carolina Bays formed between 100,000 and 200,000 years ago based on the age of underlying strata or unconsolidated surface deposits. Various hypotheses exist to explain the bays, including natural geologic processes and catastrophic encounters with extraterrestrial objects. A suggested model for the origin of Carolina Bays focuses on prevailing northwest winds blowing southeast during arid to semiarid conditions of a Pleistocene glacial episode. Water subsequently accumulated in deflation basins leading to wind- and wave-driven sand deposits along the southeastern margins of the depressions. Because Carolina Bays are unique geologic features, their exact origin will continue to be controversial and unsettled.

Chesapeake Bay Impact

A comet or an asteroid with a diameter of 3 to 5 km burst into the shallow submerged continental shelf in the area of southern Chesapeake Bay about 35 million years ago. The impact created an enormous crater measuring 85 km in diameter and more than 1 km in depth that is now buried under about 500 m of Cenozoic strata. Rock fragments ripped from strata underlying the continental shelf, including the deeper rocks of the basement, blew upward during the impact, and then resettled into the crater as a chaotic mass of slump blocks and brecciated rubble. The cataclysmic event created a low area on the continental shelf that has affected younger geologic processes including diversion of streams and watersheds, subsidence of land areas over the crater, faulting, and interference with groundwater systems. Some of these effects still pose concerns for today's population in the southeastern Virginia and the eastern shore of Maryland. Apart from localized effects in the area created during the impact, significant disruptions to contemporaneous geologic processes throughout the Atlantic Southeast are not yet recognized and confirmed.

Atlantic Southeast.8 Natural and Anthropogenic Hazards of the Atlantic Southeast Region

Recent Seismic Activity and Cenozoic Tectonics

The history of seismic events throughout the southeastern region indicates that seismic activity is not randomly distributed but occurs as episodic events within clustered areas. Seismologists and geologists recognize this pattern as the result of earthquakes generated within the North American plate and not as the product of interactions at a convergent or **transform plate boundary**. Intraplate earthquakes result when dominantly compressive, NNE-trending horizontal stresses currently existing in this part of the North American plate interact with local structures such as buried border faults of Mesozoic basins or reactivated shear zones and terrane boundaries in older basement rocks. Such interactions are unlikely to produce frequent, high-magnitude earthquakes in this region.

The Charleston earthquake discussed in the introduction had an estimated magnitude of greater than 7.0 and documented intensity of X. Widespread damage to structures occurred within an epicentral region of approximately 30 × 50 km. Felt reports of the earthquake came from areas as far away as central Ohio and western West Virginia. No fault scarp of the earthquake is known as a result of the unconsolidated nature of the Coastal Plain sediments throughout the affected area. Geomorphologic and other indirect studies suggest that the epicenter of the earthquake overlays the junction of two faults: a high-angle, northwesterly-trending fault and the southern segment of a regionally extensive, northeast-trending fault zone. Analyses of stream terraces show narrow zones of anomalous channel characteristics that defy explanation by normal processes of erosion and deposition. The epicentral area of the earthquake abounded with examples of liquefaction related to seismic shaking. Liquefaction requires saturated conditions in unconsolidated sediments to form clastic dikes, sand blows, and lateral spreading of the ground triggered by seismic shaking. In the Charleston area, evidence of seismic-induced liquefaction occurs in sediments that are buried to maximum depths of less than several meters and the age of which is younger than 250,000 years. Recent discoveries of several prehistoric liquefaction events observed in the walls of excavated trenches suggest a recurrence interval of approximately 1,000 years for damaging earthquakes (magnitude > 5.5) in the Charleston area.

Seismic hazard maps indicate that the Charleston area has the potential for significant ground acceleration (in excess of 10% g, or 1 m/sec/sec) at least once within the next 50 years. Similar peak values are projected for the central area of the eastern Tennessee seismic region that extends and overlaps with northeastern Alabama, northwestern Georgia, western North Carolina, southwestern Virginia, and southeastern West Virginia. The Giles County area, situated along the border of Virginia and West Virginia, is another area of recent moderate earthquake activity in the southeast. Seismic events in these regions are relatively small but occur with noticeable recurrence intervals of several years' duration. Minor structural damage occurs, but larger damages can be widespread with seismic magnitudes approaching 5.5. Much of the Atlantic Southeast can expect even lower peak ground accelerations within the next 50 years and, therefore, lesser chances of felt earthquakes. The expected values of acceleration, though, should prompt citizens in the areas of greater potential to ensure that foundations of houses and other buildings are adequately prepared to withstand the estimated potential motions. The least damage will occur to buildings underlain by consolidated rock. However, existing structures underlain by unconsolidated to semiconsolidated sediments, weathered rock, and thick soils should be reinforced as necessary to meet local building codes that have been established to meet the predicted seismic hazards.

Groundwater Issues (Hydrogeology)

Groundwater supplies at least 25% of the water for drinking in the Atlantic Southeast region. All of the groundwater in this region originates from precipitation including rain and snow. Average annual precipitation values range from as low as 30 inches/year to more than 80 inches/year. Groundwater is stored in rocks and unconsolidated deposits that comprise **aquifers**, or systems of aquifers, and **confining units**. A saturated rock or deposit that has enough space to hold water between the solid particles or minerals (called pore space) and that permits the free flowage of the water through the rock or deposit (often called permeability) defines an aquifer. Any saturated rock or deposit that essentially limits the flow or passage of water through it at extremely slow rates is a confining unit. Groundwater systems within a physiographic province consist either as layers of connected and mutually dependent aquifers or as a sequence of separate, unconnected, nearly identical aquifers that act as a single unit. Up to six aquifers, including a surface aquifer, are present within the Coastal Plain. Three aquifers within consolidated rocks, and an unconfined aquifer in overlying, unconsolidated surface deposits, define the groundwater system in the Piedmont and Blue Ridge. The Valley and Ridge and Plateaus contain two systems within consolidated strata, as well as a local, overlying surface aquifer.

Aquifer systems are at risk for contamination from two general sources: direct infiltration by contaminated surface and near-surface waters; and direct recharge of contaminants into the aquifer by injection, leaky wells, and other faulty structures. Very slow velocities and nonturbulent flow of groundwater allow narrow plumes of contamination from spreading laterally and undergoing natural purification. In areas of low population density, practical methods of remediation do not require much more than time and a new well. However, urbanization and industrialization in concentrated areas has created more complex problems requiring more significant efforts at remediation including excavation and transportation of contaminant soils. Actions and regulations may call for actual "cleaning" of the groundwater by a combination of pumping, filtration, extraction, or introduction of catalysts, which can cause a rapid breakdown of the contaminating agents within the groundwater. Practical solutions may also call for installation of municipal sewer systems and water supplies in place of individual septic systems and wells. Sources of contaminants in the Atlantic Southeast region include improperly functioning septic systems; leaking sewer pipes, mechanical breakdowns of municipal sewer systems, or both; industrial and chemical wastes leaking from storage facilities; pesticides and fertilizers applied to crops; and excessive withdrawals of groundwater leading to saltwater invasion of freshwater aquifers in coastal regions.

Climate Changes and Eustatic Sea Level

Exposures of peat and mud typical of back-barrier island deposits, tree roots and stumps, and many other occurrences of similar oddities indicate that sea level is currently rising. Tidal records of elevations from around the world demonstrate a consistent fractional increase in sea level each year for at least the past 100 years. The obvious cause for this rise is melting of continental glaciers in the past 10,000 years as the Earth's climate has warmed. Pleistocene glaciation repeatedly affected the globe with at least 20 cycles of cooling and warming as recorded by oxygen isotope and carbon dioxide data from terrestrial and marine sources. Preserved marine terraces and transgressive cycles recorded a history of cyclic changes of **eustatic sea level** throughout the Tertiary in Coastal Plain strata. Although the record of multiple cycles of warming and cooling is clear, explanations for the variation of climate are not in agreement. However, for our purposes, the main concern of climate change is its effect within the coastal zone. Regardless of causes leading to warming and predictions of how much warming will occur, processes in the coastal zone are greatly modified even with relatively small changes in sea level. As sea level rises, shorelines and barrier islands move progressively landward. The gentle angle of slope on the continental shelf results in a larger amount of lateral migration per small elevation in sea level; that is, the rise of sea level is balanced by a very large run of the shoreline over land areas. Barrier islands migrate by the action of littoral currents, overwash, and inlet migration. This means that sand on the beaches must shift position and fill inlets as the island slowly creeps landward. Anything that lies in the pathway of migration is affected, including the actions of humans. Our society can follow a responsible path of adjustments to the migration of barrier islands and beaches that minimizes expenses and inconveniences presented by this natural process. Disagreements on possible causes and catalysts of global warming only postpone our ability to establish practical adjustments for a large population exposed to the risks of rising sea level. Inhabitants of coastal zones must remain diligent to short- and long-term changes in sea level. Human activity must also institute and maintain actions and plans that minimize our obstructions to coastal zone processes that are responding to changing sea levels.

The coastal areas of the Atlantic Southeast region currently experience high levels of interest for tourism, commercial development, and urbanization. Many activities prove to be at odds with the concentrated energy of dynamic coastal processes, and this leads to many conflicting interests and policies. "Starving" the beaches from receiving sand includes damming of rivers or creating obstacles and obstructions to the littoral and barrier systems. Attempts to control the movement of littoral sand and sediment often have led to a net loss of sediment in the system, which, in turn, leads to more erosion. Large tropical storms can move much sediment from the shoreline, but subsequent daily processes can restore this sand onto the beach. Seasonal storms such as nor'easters can create as much erosion, if not more, and the system takes longer to repair itself. Until all of the local populations are fully aware of the hazards created by development in coastal areas—the problems of erosion of sand coupled with migration of barrier islands onshore as an effect of rising sea level—policies that promote more accommodations to the natural dynamics of the system will never gain acceptance.

Tropical Storms, Floods, and Mass Movements

The Atlantic Southeast region routinely experiences the effects of tropical storms generated within the Atlantic Basin, including the Gulf of Mexico and Caribbean Sea. High magnitude, but relatively infrequent, storms produce catastrophic destruction over large areas by devastating winds and flooding. High amounts of rainfall associated with large, slow-moving tropical systems effectively drown principal streams of major drainage basin, which results in backing up and overflowing of tributary streams within the basin. Floodwaters in the lower reaches of a drainage basin can last for weeks before flowage lowers the levels below flood stage. Storm surge associated with a given storm affecting a coastal region may exceed the maximum elevation of a protective dune field, thereby flooding low-lying areas behind the barrier. Over the past century in the southeastern United States, improved construction, planning and zoning practices, and improved analyses and predictions of tropical storms have resulted in fewer deaths from tropical storms. Drastic increases in costs related to physical damages to structures from storm-driven wind and rains over the same time interval, however, reflect an increasing population in the coastal zone, and given the decadal or longer episodic cycle of high-magnitude storms, prompts officials to levels of emergency planning and response that exceed historical levels.

Regions within the southeast outside of the Coastal Zone also experience the effects of tropical storm systems. Although hurricane force winds rapidly diminish as storms track inland, the systems are still capable of spawning tornadoes and producing copious amounts of rain. Widespread flooding can occur directly from the precipitation, as well as the damming effect on principal drainage streams when the entire basin is overwhelmed. Flooding in mountainous areas such as the Blue Ridge and Valley and Ridge is particularly hazardous because of flash-flood events and mass movements. Thick sections of unconsolidated regolith on steep slopes become saturated with heavy rains. Their growing instability leads to rapid downslope movements of the regolith as landslides or mudslides. Debris slides in western North Carolina associated with tropical storms in 2004 and 2005 made national news for the magnitude of their destruction. Widespread damage from mass movements in the mountains, though, is historical as written accounts in the early part of the twentieth century document debris flows and landslides attributed to slow-moving remnants of hurricanes and tropical storms. In addition to rainfall exceeding several inches, mass movements in the mountains are influenced by other factors including soils already saturated, human activities that provoke or initiate slope instability (clear cutting, for example), and the underlying geologic framework. Increased risks for widespread and frequent mass movements can only result as more of the mountainous areas within the southeast are developed as urban and recreation areas. Methods of careful planning and action plans should be followed to minimize the risks and adjustments to this natural hazard.

Tsunami Threats

The Banda Aceh earthquake and tsunami of December 2004 reminded the world of the severe levels of death and destruction associated with a tsunami. What is the potential for such levels if a tsunami struck the southeast region? The answer lies with understanding that a tsunami occurs in response to a sudden displacement of seawater related to rapid uplift or subsidence of the sea floor by faulting or from a submarine landslide pushing displaced water laterally in front of the moving debris. Destructive tsunamis correspond to major earthquakes whose Richter magnitudes greatly exceed 7.0, and by massive submarine landslides that involve tremendous volumes of debris. The relative dearth of large fault zones capable of suddenly raising or lowering the sea floor adjacent to the southeastern region of the United States precludes a destructive tsunami related to earthquakes. Within the Atlantic Ocean Basin, the Caribbean offers potential for larger earthquakes and faults capable of generating destructive tsunami affecting the Atlantic Southeast region, but the separation distance could limit the size of the waves reaching this region. Another threat for tsunami in the southeast is catastrophic collapse of volcanic islands in the Cape Verde chain. Models suggest that tsunami of several meters in height could travel and strike the East Coast from such a catastrophic collapse and landslide. Images of crowns, tracks, and toes of submarine landslides along our own continental margin and slope raise concerns about tsunami forming in this area. However, no information is available as to the actual time(s) of formation of these structures, which constrains predictions of events. Also, any tsunami created by landslides on our continental margin would travel eastward, which limits the impact of the event on the East Coast. An additional possible source for an East Coast tsunami is by a large meteorite or bolide impact in the Atlantic Ocean Basin. Variations in submarine topography may prevent catastrophic damage from tsunami in most East Coast settings by obstructing, diverting, and interfering with incoming waves. Chances are excellent that we will never experience a major tsunami along the southeastern coast in our lifetimes.

Environmental Concerns

Four general areas of concern raise issues related to environmental degradation within the Atlantic Southeast region: coastal resources, hydrologic resources, waste disposal and septic issues, and urbanization. Although all of these areas overlap with each other, none is completely evaluated unless the underlying geology is known and understood.

Urbanization is the overriding umbrella for most environmental problems in the region. The region hosts major cities and ports of the eastern seaboard, as well as alluring beaches and other natural areas that invite the attention of locals and visitors. The popularity of the region to many people places an ever-increasing demand for real estate, transportation, utilities, and other resources, each of which carries a complex and burgeoning infrastructure. Deciding the best location of a new road, intersection, or bridge is a critical task, but equally important are issues concerning the locations of materials needed for their construction, potential loss of other resources by paving over them, and the potential contamination effects of storm water runoff into adjacent soils, wetlands, streams, and estuaries. The location of the coastal zone and plain is really an important transition between the naturally dynamic environments of the marine and terrestrial realms. We now understand that the subsurface below many areas of the land are records of changes and transitions between environments through time. Their vertical succession downward is a useful guide to lateral changes or disruptions in buried layers. Just as our effects on land are significant, so, too, are our efforts below the land surface. We must always ensure that we have the best information available for both the surface and subsurface materials for any project that is proposed. Not to do so is unethical and poor stewardship for successive generations.

Hydrologic resources are vital to urbanization. Hydrology includes both surface water and groundwater systems in this case. Rivers continue to provide a major source of freshwater for consumption by people, industry, and agriculture, and groundwater resources are becoming significant additional resources. A natural system of cleansing or purifying both surface and groundwater resources lies within soils and wetlands adjacent to streams and overlying recharge areas of aquifers. The term *wetland* actually refers to particular relations among soil type, soil moisture, and the water table. Urbanization must remain vigilant in correctly identifying and protecting wetland areas for protecting water resources, as well as natural living environments and ecosystems that promote the pleasing aesthetics of the region. Protection of wetlands and soils overlying unconfined and **confined aquifers** is critical because remediation of contaminated aquifers is expensive, time-intensive, and difficult. As surface water becomes unusable for any number of reasons, some urban areas are turning to aquifers for additional water to meet their needs. This need places heavy demands on aquifers, and care must be taken to ensure that natural recharge can keep pace with the volume of water mined or extracted. If excessive water is withdrawn from an aquifer, the porosity and conductivity can be lost and never recovered. Similar hydrologic issues occur in more consolidated bedrock in the uplands and mountainous areas of the southeastern region. One significant difference in aquifer systems of crystalline rocks is the thickness of regolith overlying the aquifer. The regolith effectively exists as an unconfined aquifer that recharges the fracture porosity of the underlying crystalline rocks. Disturbance of the regolith can significantly affect the health of the aquifer. Aquifer systems in the Valley and Ridge and Plateaus include geologic frameworks that are similar to the preceding discussions. Any plans to use these precious resources to meet increased needs of urbanization, commerce, and agriculture must be balanced by wise decisions, timely methods, and ethical concerns for prolonging the useful life of the water resource.

People cannot live without generating wastes. Waste disposal issues are pervasive in urban areas. Environmental problems result if the conditions existing within disposal sites cannot decompose the wastes completely before entering the environment. Landfills are engineered excavations where solid waste is buried daily in specific cells or small areas of the excavation. Rainfall and surface moisture can infiltrate into the cell despite preventative measures including a cap of relatively impermeable soil over the cell. These waters percolate downward, partially leaching decomposing wastes, and enter the ground as leachate. Insufficient thicknesses of soil between the base of the landfill and the water table leachate to enter an aquifer system. Lining the base of landfills with thick impermeable media such as plastic sheeting has become a practical means of reducing leachate for new landfills, but remediation of leachate contamination at older sites often include shutting down the entire operation. If potable water supplies have been damaged by contamination, then new supplies are found after significant effort and expense. Contamination of surface and groundwater supplies by septic tank systems is similar to the leachate problems with landfills, although the scale is different. One solution for failing septic tanks and contamination of groundwater is to install a municipal water and sewer system. This is not a foolproof solution, though, because numerous examples of breaks in sewer lines dumping millions of gallons of raw sewage into stream, estuaries, or city streets can be found in the archives of many urban newspapers.

Review Workbook

SUMMARY

Atlantic Southeast.1 Introduction

■ *Why aren't active volcanoes present within the Atlantic Southeast region despite a record of seismic activity?*

The current tectonic setting of the Atlantic Southeast region is that of a passive margin. Passive margins do not involve either active rifting or subduction, both of which trigger partial melting of the mantle and crust, a critical component in the formation of magmas. Moderate earthquake activity within the region results from intraplate stresses interacting with older tectonic features.

■ *How has the geology of the Atlantic Southeast region contributed to the economic history of the United States?*

The first gold rush occurred in central North Carolina in the earliest 1800s when the country was barely 20 years old. Iron ores, coal, and limestone in the Valley and Ridge and Plateaus were the essential ingredients for the steel industry of the United Other resources in the various parts of the Appalachians hav provided nonmetallic mineral and rock deposits, raw materia and sources of energy for many industries across the nation.

Atlantic Southeast.2 Proterozoic Geology and Tect

■ *What events led to the formation of Rodinia?*

Rodinia was an early supercontinent that formed by contine continent collisions of all the world's continents during the Proterozoic. The Grenville Orogeny is one of the processes to Rodinia and represents thrusting and stacking of thrust pl involving segments of the lower continental crust, a process that of the Himalaya Mountains.

■ *What evidence exists for interpreting a continental rifting e along the margin of Laurentia in the Late Proterozoic?*

Metamorphosed, thick, stratified sequences of conglomerate, arkosic sandstone, quartz sandstone, mudrock, and basalt dikes, sills, and flows unconformably overlie Grenville basement rocks and are bordered by a fault system. Younger Paleozoic strata overlie the Late Proterozoic stratified rocks. The stratified sequences are discontinuous along strike and tend to be narrow.

Atlantic Southeast.3 Late Proterozoic Climates, Paleogeography, and Life

■ *What is the significance of glacial climates in both Laurentia and peri-Gondwanan exotic terranes during the Late Proterozoic?*
The peri-Gondwanan terranes were separated from each other and Laurentia by the Iapetus Ocean. For these separated areas to have experienced identical glacial events requires an extremely cold global climate that supported continental glaciation.

Atlantic Southeast.4 Early Paleozoic History of the Laurentian Continental Margin

■ *What was Iapetus, and where was it located?*
Iapetus was an ocean that formed during the Late Proterozoic when Rodinia broke apart into Laurentia, Gondwana, and various chips and microcontinents. Iapetus was the ocean that existed between Laurentia and Gondwana.

■ *How do the Early Paleozoic strata of the Laurentian continental margin differ from the underlying Late Proterozoic stratified rocks?*
The Early Paleozoic strata of the Laurentian continental margin consist of a basal section of coarse-grained fluvial and transitional clastic sediments overlain by a thinner sequence of shale and mature quartz sandstones. A thicker interval of carbonate rocks caps the sequence. A few basalt flows occur within the lowermost parts of the basal clastic interval. Overall, the Early Paleozoic strata are much thinner and cleaner marine sediments compared with the significantly thicker and immature clastic rocks comprising the Late Proterozoic stratified rocks.

Atlantic Southeast.5 Proterozoic and Early Paleozoic History of the Terranes Comprising the Piedmont

■ *How does the geologic history of an exotic terrane that existed within the Iapetus Ocean marginal to Gondwana differ from that of Laurentia?*
The exotic terranes of the Iapetus Ocean basin reveal a history of docking events that occurred no later than the Cambrian. During this same time, the Appalachian side, Laurentia, was a drifting passive continental margin.

■ *What evidence exists that demonstrates the size of Iapetus?*
Rare fossil occurrences in some of the eastern Piedmont terranes indicate affinities with fauna that are found in terranes and continents known to have existed on the other side of Iapetus from Laurentia. Some of the terranes have isotope ratios in their basement rocks that group with the basement rocks and isotopes of Gondwanan continents. The composite nature of many Piedmont terranes documents a Proterozoic and Early Paleozoic history of docking and collision events that are not recorded along the margin of Laurentia, which is another argument in favor of a large ocean basin.

Atlantic Southeast.6 Middle Ordovician through Late Paleozoic Geology and Tectonics

■ *What events comprise the Taconic Orogeny?*
The Taconic Orogeny resulted from the attempted subduction of eastern Laurentia below a large composite terrane (the Piedmont). Rapid erosion and deposition of clastic sediments from the approaching Piedmont terrane filled up the trench and overflowed on the passive continental shelf of Laurentia. Together with a preceding uplift of the margin, the flooding of the continental shelf below the clastic wedge from the east buried the carbonate bank of reefs that had dominated the continental shelf throughout most of the Cambrian and Early Ordovician. The subduction event also accreted terranes with continental basements to Laurentia together with thrusting and shearing, intense metamorphism, and magmatism.

■ *What events comprise the Acadian Orogeny?*
The Acadian Orogeny in the southern Appalachians was a time of pervasive deformation related to dextral shearing along a transpressive boundary. Subduction processes were essentially absent. Regional metamorphism was relatively low. Devonian plutons of granitic rock are present as relatively small but abundant bodies within some geographic areas.

■ *What events comprise the Alleghanian Orogeny?*
The Alleghanian Orogeny consists of the events related to the ultimate closure of Iapetus when Gondwana collided with Laurentia. The collision resulted in widespread thrusting of deeper parts of the accreted Piedmont terranes over shallower parts of the Laurentian crust. The folding and thrusting of the Valley and Ridge formed at this time. Widespread metamorphism and granitic intrusions occurred throughout many of the composite Piedmont terranes.

Atlantic Southeast.7 Mesozoic and Cenozoic Geology and Tectonics

■ *What is a rift-to-drift sequence?*
A rift-to-drift sequence begins with a base of very coarse clastic sedimentary rocks and basaltic igneous rocks of continental rift basins. Overlying this base is a sequence of marine clastic rocks and overlying carbonates. The marine strata become part of a transgressive sequence that crosses the continental shelf onto a subsiding continental margin.

■ *Why do some rift basins have strata dipping northwest while others have strata dipping southeast?*
The strata accumulate in the basin as faulting progressively opens the basin. Border faults of the basins occur on either the northwest or southeast side. As northwest border fault dips southeast, then the strata will dip northwest toward the border fault. The opposite is true for southeast border faults.

Atlantic Southeast.8 Natural and Anthropogenic Hazards of the Atlantic Southeast Region

■ *What current natural hazards facing the population of this region today pose the greatest threats to property and life?*
The greatest threats are tropical storms and their associated high winds, storm surge, tornadoes, mass wasting in areas of higher relief, and extensive flooding.

■ *What are the chances of Myrtle Beach, South Carolina, and Virginia Beach, Virginia, being destroyed by a catastrophic tsunami?*
Very little because the tectonics that result in catastrophic tsunamis are not active around the East Coast of North America. We are so far distant from any active sources that natural attenuation of the tsunamis should prevent them from striking with catastrophic effects. Exceptions, of course, are meteorite or bolide impacts in the adjacent waters of the Atlantic, but none is expected within the near future.

ESSENTIAL TERMS TO KNOW

Acadian – a Devonian-aged orogeny in the southern Appalachians characterized by dextral strike-slip faulting, low-grade regional metamorphism, and intrusion of granitic rock.

accretionary complex – a wedge-shaped pile of mixed rock types that accumulates in front of the plate overlying a subduction zone by off-scraping and underplating of rocks from the descending slab to the base of the upper plate.

Alleghanian – a Late Paleozoic orogeny in the southern Appalachians characterized by thrust faulting, folding, regional metamorphism of upper amphibolite facies, late-stage dextral strike-slip faulting, and widespread granitic batholiths and smaller intrusions.

Appalachian orogenic belt – a linear range of mountains, valleys, plateaus, and piedmont uplands, and the rocks contained within, that extends from Alabama through Newfoundland. The Appalachians formed from a series of tectonic events in the Paleozoic Era.

aquifer – a porous and permeable rock that is saturated with groundwater and is capable of producing sufficient quantities for use.

basalt – the most common type of mafic volcanic rock. Oceanic crust consists of basalt. Continental rift basins also have basalt as lava as well as dikes and sills (properly called diabase).

batholith – a very large pluton composed of granite. Granite forms at convergent plate boundaries during orogenic events.

composite terrane – an exotic or suspect terrane that has formed by the docking of smaller terranes.

confined aquifer – an aquifer overlain by an impermeable layer of rock.

confining unit – the name of the impermeable layer of rock overlying an aquifer.

convergent plate boundary – the boundary between two colliding lithospheric plates where the directed motion is oriented at a moderate to high angle to the boundary.

disconformity – a type of unconformity between two sedimentary rock units sharing a common primary stratification; the unconformity is more or less parallel to the primary stratification.

docking – a term for the welding of two tectonic terranes, or a terrane and a continent, into one object.

drift sequence – refers to the mature marine strata that overlie and bury continental strata in rift basins; part of a passive continental margin.

ductile deformation – minerals and rocks lose their size and shape becoming distorted as the result of excess straining of the rock to applied stresses. Ductile deformation commonly occurs when the rock is warm enough to deform without shattering into fragments.

eustatic sea level – another name for global sea level; a reference level or datum for worldwide correlations of strata.

exotic terrane – a terrane whose original location is known.

forearc – a general term referring to rocks and accumulating immature sediments that occur in front of the volcanic arc region and behind the accretionary complex of a subduction zone in a convergent plate margin.

Grenville – a Proterozoic orogenic event affecting the eastern side of Laurentia.

Gondwana – the name of the large continent consisting of Africa, South America, Antarctica, and Australia; rifted apart from Laurentia during the breakup of Rodinia in the Late Proterozoic.

granite – a felsic plutonic rocks consisting of an equal mixture of potassium feldspar, plagioclase feldspar, and quartz. Granite is a common pluton found in orogenic belts and is the most common felsic plutonic rock in continental crust.

Iapetus – the ocean that formed in the Late Proterozoic between Laurentia and Gondwana. The closing of Iapetus occurred during the Paleozoic and is responsible for the orogenic activity in the Appalachians.

Laurentia – a former continent contained within the present continent of North America. Laurentia was part of the supercontinent Rodinia that broke apart in the Late Proterozoic. The Appalachian orogenic belt formed along the eastern margin of Laurentia.

magmatic arc (volcanic arc, plutonic arc) – a curved trend of magmatic activity that parallels and commonly sits behind subduction zones and deep-sea trenches at convergent plate boundaries.

microcontinent – a relatively small landmass or large island with a basement of continental crust. Microcontinents originate in several ways including fragments created by rifting of larger continents.

orogen – a linear zone of rocks that exhibits a history of crustal mobility and active tectonics. Orogens typically consist of a core containing high-grade metamorphic rocks, granites, and shear zones. Linear zones of lesser deformed and metamorphosed rocks commonly occur between the core of the orogen and its perimeter.

Pangaea – the supercontinent that formed during the Late Paleozoic. The collision of Gondwana with Laurentia that resulted in the Alleghanian Orogeny was one of the global collisions that formed Pangaea.

passive margin – an area of relatively quiet tectonic activity that occurs along the trailing edge of a drifting continent. Continental weathering, erosion, and deposition by rivers and streams characterize the clastic sediments of a passive margin. Carbonate sedimentation results from the chemical precipitation of dissolved ions in the marine waters overlying the passive margin.

regression – the process of lowering of relative sea level. Regressions occur on continental margins as relative sea level falls, or as the adjacent land mass rises, or both.

rift basin – a linear basin that is bordered by a normal fault on one side. Rift basins form from the thinning and extension of continental crust by rising heat from below associated with a mantle plume. Rift basins have relatively high margins and steep slopes where alluvial fans accumulate. Interior drainages in the central areas of rift basins contain lakes where evaporite deposits may form around the edges.

Rodinia – the supercontinent that formed in the Middle Proterozoic and broke apart in the Late Proterozoic. Rodinia's breakup provided the setting for Paleozoic orogenic events in the Appalachians.

shear zone – a general name for a zone of similar faults. Commonly, shear zone is associated with ductile deformation.

stitching pluton – an igneous intrusion along a terrane boundary that penetrates through both terranes and has the same relative age as the terrane boundary.

strike-slip fault – a type of fault motion characterized by separation parallel to the trend (strike) of the fault.

subduction zone – an inclined zone at a convergent plate boundary along which the lithosphere is pulled and pushed downward toward the mantle under an overriding continent or terrane. Magmas can form above subduction zones and rise upward through the crust to feed plutonic or volcanic arcs. Deep-sea trenches coincide with the upper end of subduction zones. Accretionary complexes and forearc basins occur behind the trench and in front of the volcanic arc in subduction zones.

suspect terrane – a terrane whose original location is not known for certain.

suture zone – a terrane boundary that has been stitched together.

Taconic – the name of the middle Ordovician Orogeny in the Appalachian Orogenic belt. The eastern margin of Laurentia was partially subducted below a composite terrane represented by the Piedmont.

terrane – an area within an orogenic belt that is bounded completely by faults and has a unique assemblage of rocks with distinctive characteristics that reflect an origin in a distant or different location external to the orogen.

transform plate boundary – a plate boundary characterized by directed motion more or less parallel to its length.

transgression – the process of rising of relative sea level. Transgressions occur on continental margins as relative sea level rises or as the adjacent land mass subsides, or both.

transpression – a combination of reverse faulting and strike-slip faulting. Transpression commonly is associated with transform plate boundaries.

unconformity – a type of depositional contact between rock units along which is a significant missing interval of time as the result of erosion, nondeposition, or both.

Geology of the Atlantic Southeast Region is written to augment introductory courses in physical and historical geology taught within the area encompassing the southern Appalachians. For over 150 years, many classical geologic and tectonic concepts originated and were successfully tested in this region extending from northern Virginia through Georgia, and from eastern Tennessee to the Atlantic.

Geology of the Atlantic Southeast Region presents a synopsis of nearly 2 billion years of geologic history documenting the birth and destruction of continents and ocean basins. After a brief introduction to the physiographic provinces of the region, the chapter follows a standard historical sequence from the Proterozoic to the present. Each section relates general packages of rocks to corresponding geologic environments such as continental rifts, passive continental margins, volcanic arcs, terranes, and collisions. The text ties the rock successions and environments to relevant tectonic and orogenic events and teaches the student the process of interpreting historical and regional geology from the rock record. Where appropriate, the chapter also highlights important mineral and energy resources of the region produced from different rock sequences. The final section of the text focuses on environmental issues of interest and concern including coastal erosion, seismic activity, hydrology, and tropical storms.

Also Available

People often decide on their residence or business location based on a desire to live or work in scenic environments without understanding the hazards around them. This series of reprinted chapters from Donald and David Hyndman's *Natural Hazards and Disasters* has been carefully selected to provide students with the opportunity to see how specific natural hazards are relevant to their regional area.

Each volume has been reprinted in its entirety, including end-of-chapter problems and "Case In Point" case studies, for easy integration into a variety of introductory courses where discussion of natural hazards can add relevant context to the regular coursework.

Available Volumes, with select case studies and examples:

Tsunami: The Great Wave ISBN 0534615228	**Floods and Human Interactions** ISBN 0495161500
Waves, Beaches, and Coastal Erosion: Rivers of Sand ISBN 0495161489	**Thunderstorms and Tornadoes** ISBN 0495161497
Hurricanes and Nor'Easter: The Big Winds & Hurricane Katrina a Case Study of the Costliest Disaster in U.S. History ISBN 0495153214	

To learn more about *Natural Hazards and Disaster* by Donald and David Hyndman, please visit: academic.cengage.com/earthscience/hyndman.

John R. Huntsman

CENGAGE Learning

For your lifelong learning solutions, visit custom.cengage.com

Visit our corporate website at cengage.com

ISBN 13: 978-0-7593-9080-5
ISBN 10: 0-7593-9080-0
9 780759 390805

Appendix A
English–Metric Conversion Chart

	English Unit	Conversion Factor	Metric Unit	Conversion Factor	English Unit
Length	Inches (in.)	2.54	Centimeters (cm)	0.39	Inches (in.)
	Feet (ft)	0.305	Meters (m)	3.28	Feet (ft)
	Miles (mi)	1.61	Kilometers (km)	0.62	Miles (mi)
Area	Square inches (in.^{2})	6.45	Square centimeters (cm^2)	0.16	Square inches (in.^{2})
	Square feet (ft^2)	0.093	Square meters (m^2)	10.8	Square feet (ft^2)
	Square miles (mi^2)	2.59	Square kilometers (km^2)	0.39	Square miles (mi^2)
Volume	Cubic inches (in.^{3})	16.4	Cubic centimeters (cm^3)	0.061	Cubic inches (in.^{3})
	Cubic feet (ft^3)	0.028	Cubic meters (m^3)	35.3	Cubic feet (ft^3)
	Cubic miles (mi^3)	4.17	Cubic kilometers (km^3)	0.24	Cubic miles (mi^3)
Weight	Ounces (oz)	28.3	Grams (g)	0.035	Ounces (oz)
	Pounds (lb)	0.45	Kilograms (kg)	2.20	Pounds (lb)
	Short tons (st)	0.91	Metric tons (t)	1.10	Short tons (st)
Temperature	Degrees Fahrenheit (°F)	$-32° \times 0.56$	Degrees centigrade (Celsius)(°C)	$\times 1.80 + 32°$	Degrees Fahrenheit (°F)

Examples:

10 inches = 25.4 centimeters; 10 centimeters = 3.9 inches

100 square feet = 9.3 square meters; 100 square meters = 1080 square feet

50°F = 10.1°C; 50°C = 122°F

Appendix B
Periodic Table of the Elements

47	Atomic Number
Ag	Symbol of Element
silver	Name of Element
107.9	Atomic Mass Number (rounded to three significant figures)

Representative Elements | Transition Elements | Inner-Transition Elements | Noble Gases

Period	(1)* I A	(2) II A	(3) III B	(4) IV B	(5) V B	(6) VI B	(7) VII B	(8) VIII B	(9)
1	1 H hydrogen 1.008								
2	3 Li lithium 6.941	4 Be beryllium 9.012							
3	11 Na sodium 22.99	12 Mg magnesium 24.31							
4	19 K potassium 39.10	20 Ca calcium 40.08	21 Sc scandium 44.96	22 Ti titanium 47.90	23 V vanadium 50.94	24 Cr chromium 52.00	25 Mn manganese 54.94	26 Fe iron 55.85	27 Co cobalt 58.93
5	37 Rb rubidium 85.47	38 Sr strontium 87.62	39 Y yttrium 88.91	40 Zr zirconium 91.22	41 Nb niobium 92.91	42 Mo molybdenum 95.94	43 Tc technetium 98.91	44 Ru ruthenium 101.1	45 Rh rhodium 102.9
6	55 Cs cesium 132.9	56 Ba barium 137.3	57 La lanthanum 138.9	72 Hf hafnium 178.5	73 Ta tantalum 180.9	74 W tungsten 183.9	75 Re rhenium 186.2	76 Os osmium 190.2	77 Ir iridium 192.2
7	87 Fr francium (223)	88 Ra radium 226.0	89 Ac actinium (227)	104 Rf rutherfordium (261)	105 Db dubnium (262)	106 Sg seaborgium (263)	107 Bh bohrium (262)	108 Hs hassium (265)	109 Mt meitnerium (266)

Lanthanides	58 Ce cerium 140.1	59 Pr praseodymium 140.9	60 Nd neodymium 144.2	61 Pm promethium (147)	62 Sm samarium 150.4
Actinides	90 Th thorium 232.0	91 Pa protactinium 231.0	92 U uranium 238.0	93 Np neptunium 237.0	94 Pu plutonium (244)

( ) Indicates mass number of isotope with longest known half-life.

* Number in ( ) heading each column represents the group designation recommended by the American Chemical Society Committee on Nomenclature.

(10)	(11) I B	(12) II B	(13) III A	(14) IV A	(15) V A	(16) VI A	(17) VII A	(18) Noble Gases
								2 **He** helium 4.003
			5 **B** boron 10.81	6 **C** carbon 12.01	7 **N** nitrogen 14.01	8 **O** oxygen 16.00	9 **F** fluorine 19.00	10 **Ne** neon 20.18
			13 **Al** aluminum 26.98	14 **Si** silicon 28.09	15 **P** phosphorus 30.97	16 **S** sulfur 32.06	17 **Cl** chlorine 35.45	18 **Ar** argon 39.95
28 **Ni** nickel 58.71	29 **Cu** copper 63.55	30 **Zn** zinc 65.37	31 **Ga** gallium 69.72	32 **Ge** germanium 72.59	33 **As** arsenic 74.92	34 **Se** selenium 78.96	35 **Br** bromine 79.90	36 **Kr** krypton 83.80
46 **Pd** palladium 106.4	47 **Ag** silver 107.9	48 **Cd** cadmium 112.4	49 **In** indium 114.8	50 **Sn** tin 118.7	51 **Sb** antimony 121.8	52 **Te** tellurium 127.6	53 **I** iodine 126.9	54 **Xe** xenon 131.3
78 **Pt** platinum 195.1	79 **Au** gold 197.0	80 **Hg** mercury 200.6	81 **Tl** thallium 204.4	82 **Pb** lead 207.2	83 **Bi** bismuth 209.0	84 **Po** polonium (210)	85 **At** astatine (210)	86 **Rn** radon (222)
110 **Uun** ununnilium (269)	111 **Uuu** unununium (272)	112 **Uub** ununbium (277)	113	114 **Uuq** ununquadium (289)	115	116 **Uuh** ununhexium (289)	117	118 **Uuo** ununoctium (293)

63 **Eu** europium 152.0	64 **Gd** gadolinium 157.3	65 **Tb** terbium 158.9	66 **Dy** dysprosium 162.5	67 **Ho** holmium 164.9	68 **Er** erbium 167.3	69 **Tm** thulium 168.9	70 **Yb** ytterbium 173.0	71 **Lu** lutetium 175.0
95 **Am** americium (243)	96 **Cm** curium (247)	97 **Bk** berkelium (247)	98 **Cf** californium (251)	99 **Es** einsteinium (254)	100 **Fm** fermium (257)	101 **Md** mendelevium (258)	102 **No** nobelium (255)	103 **Lr** lawrencium (256)

Answers
Multiple-Choice Review Questions

Chapter 1
1. d; 2. c; 3. e; 4. d; 5. b; 6. d; 7. c; 8. c; 9. a; 10. b; 11. e

Chapter 2
1. b; 2. e; 3. b; 4. c; 5. c; 6. c; 7. e; 8. a; 9. c; 10. a

Chapter 3
1. b; 2. e; 3. a; 4. c; 5. b; 6. a; 7. d; 8. e; 9. c; 10. b

Chapter 4
1. c; 2. e; 3. a; 4. b; 5. a; 6. d; 7. c; 8. b; 9. a; 10. a

Chapter 5
1. c; 2. b; 3. a; 4. e; 5. d; 6. c; 7. b; 8. c; 9. a; 10. b

Chapter 6
1. b; 2. a; 3. c; 4. b; 5. e; 6. b; 7. e; 8. c; 9. b; 10. a; 11. b; 12. e

Chapter 7
1. c; 2. e; 3. d; 4. d; 5. d; 6. b; 7. e; 8. e; 9. c; 10. c

Chapter 8
1. d; 2. c; 3. d; 4. c; 5. a; 6. c; 7. b; 8. d; 9. c; 10. e

Chapter 9
1. a; 2. c; 3. e; 4. b; 5. d; 6. a; 7. c; 8. a; 9. c; 10. c

Chapter 10
1. b; 2. d; 3. a; 4. e; 5. e; 6. c; 7. b; 8. b; 9. d; 10. a

Chapter 11
1. b; 2. b; 3. b; 4. c; 5. b; 6. c; 7. e; 8. c; 9. d; 10. e

Chapter 12
1. a; 2. c; 3. e; 4. a; 5. d; 6. a; 7. c; 8. b; 9. b; 10. c

Chapter 13
1. d; 2. b; 3. c; 4. d; 5. b; 6. a; 7. d; 8. c; 9. d

Chapter 14
1. e; 2. c; 3. a; 4. c; 5. d; 6. b; 7. d; 8. b; 9. a; 10. c

Chapter 15
1. c; 2. b; 3. d; 4. d; 5. d; 6. e; 7. d; 8. c; 9. d; 10. b

Chapter 16
1. a; 2. d; 3. b; 4. e; 5. c; 6. a; 7. b; 8. a; 9. d; 10. d

Chapter 17
1. d; 2. e; 3. c; 4. a; 5. b; 6. c; 7. c; 8. b; 9. c; 10. d

Chapter 18
1. c; 2. a; 3. d; 4. b; 5. a; 6. c; 7. b; 8. d; 9. b; 10. d

Chapter 19
1. a; 2. c; 3. e; 4. b; 5. a; 6. b; 7. e; 8. c; 9. b

Chapter 20
1. b; 2. b; 3. e; 4. d; 5. c; 6. b; 7. c; 8. c; 9. c; 10. d

Chapter 21
1. d; 2. e; 3. a; 4. e; 5. c; 6. e; 7. a; 8. d; 9. e; 10. e

Chapter 22
1. e; 2. c; 3. b; 4. e; 5. a; 6. a; 7. b; 8. c; 9. b; 10. c

Chapter 23
1. b; 2. d; 3. a; 4. c; 5. d; 6. d; 7. a; 8. a; 9. e; 10. a

Glossary

aa Lava flow with a surface of rough, angular blocks and fragments.

abrasion The process whereby rock is worn smooth by the impact of sediment transported by running water, glaciers, waves, or wind.

Absaroka sequence Widespread Upper Mississippian to Lower Jurassic sedimentary rocks bounded above and below by unconformities; deposited during a transgressive–regressive cycle of the Absaroka Sea.

absolute dating Using various radioactive decay dating techniques to assign ages in years before the present to rocks. (*See also* relative dating)

abyssal plain Vast flat area on the seafloor adjacent to the continental rise of a passive continental margin.

Acadian orogeny An episode of Devonian deformation in the northern Appalachian mobile belt resulting from the collision of Baltica with Laurentia.

acanthodian Any Early Silurian to Devonian member of the class Acanthodii, the first fish with jaws or a jawlike structure.

active continental margin A continental margin with volcanism and seismicity at the leading edge of a continental plate where oceanic lithosphere is subducted. (*See also* passive continental margin)

allele Alternative form of a single gene controlling different versions of the same trait.

allopatric speciation Model for the evolution of a new species from a small, geographically isolated part of a larger parent population.

alluvial fan A cone-shaped accumulation of mostly sand and gravel deposited where a stream flows from a mountain valley onto an adjacent lowland.

alluvium A collective term for all detrital sediment transported and deposited by running water.

amniote egg Egg in which the embryo develops in a fluid-filled cavity called the amnion. The egg also contains a yolk sac and a waste sac.

anaerobic Refers to organisms that do not need oxygen for respiration.

analogous structure Body part, such as wings of insects and birds, that serves the same function but differs in structure and development. (*See also* homologous structure)

Ancestral Rockies Late Paleozoic uplift in the southwestern part of the North American craton.

angiosperm Vascular plants that have flowers and seeds; the flowering plants.

angular unconformity An unconformity below which older rocks dip at a different angle (usually steeper) than overlying strata. (*See also* disconformity *and* nonconformity)

anticline A convex upward fold in which the oldest exposed rocks coincide with the fold axis and all strata dip away from the axis.

Antler orogeny Late Devonian to Mississippian deformation that affected the Cordilleran mobile belt from Nevada to Alberta, Canada.

aphanitic texture A texture in igneous rocks in which individual mineral grains are too small to be seen without magnification; results from rapid cooling of magma and generally indicates an extrusive origin.

Appalachian mobile belt A mobile belt along the eastern margin of North America extending from Newfoundland to Georgia.

archosaur A term referring to the ruling reptiles—dinosaurs, flying reptiles (pterosaurs), crocodiles, and birds.

arête A narrow, serrated ridge between two glacial valleys or adjacent cirques.

artesian system A confined groundwater system with high hydrostatic pressure that causes water to rise above the level of the aquifer.

artificial selection The practice of selectively breeding plants and animals for desirable traits.

artiodactyl The even-toed hoofed mammals, such as goats, cattle, antelope, and swine; members of the order Artiodactyla.

aseismic ridge A ridge or broad area rising above the seafloor that lacks seismic activity.

ash Pyroclastic materials that measure less than 2 mm.

assimilation A process whereby magma changes composition as it reacts with country rock.

asthenosphere The part of the mantle that lies below the lithosphere; it behaves plastically and flows slowly.

atom The smallest unit of matter that retains the characteristics of an element.

atomic mass number The number of protons plus neutrons in the nucleus of an atom.

atomic number The number of protons in the nucleus of an atom.

aureole A zone surrounding a pluton in which contact metamorphism took place.

australopithecine A term referring to several extinct species of the genus *Australopithecus* that existed during the Pliocene and Pleistocene epochs.

autotrophic Describes organisms that synthesize organic molecules from inorganic raw materials, as in photosynthesis. (*See also* heterotrophic)

backshore That part of a beach that is usually dry, being water covered only by storm waves or especially high tides.

Baltica One of six major Paleozoic continents; composed of Russia west of the Ural Mountains, Scandinavia, Poland, and northern Germany.

banded iron formation (BIF) Sedimentary rocks made up of thin alternating bands of silica (chert) and iron minerals (mostly hematite and magnetite).

barchan dune A crescent-shaped sand dune with its tips pointing downwind.

barrier island A long, narrow island of sand parallel to a shoreline but separated from the mainland by a lagoon.

basal slip Movement involving a glacier sliding over its underlying surface.

basalt plateau A plateau built up by horizontal or nearly horizontal overlapping lava flows that erupted from fissures.

base level The level below which a stream or river cannot erode; sea level is ultimate base level.

basin A circular fold in which all strata dip inward toward a central point and the youngest exposed strata are in the center.

Basin and Range Province An area of Cenozoic block-faulting centered on Nevada but extending into adjacent states and northern Mexico.

batholith An irregularly shaped, discordant pluton with at least 100 km^2 of surface area.

baymouth bar A spit that has grown until it closes off a bay from the open sea.

beach Any deposit of sediment extending landward from low tide to a change in topography or where permanent vegetation begins.

beach face The sloping area of a beach that is exposed to wave swash.

bed An individual layer of rock, especially sediment or sedimentary rock. (*See also* strata)

bed load That part of a stream's sediment load, mostly sand and gravel, transported along its bed.

benthos Any organism that lives on or in the bottom sediments of seas or lakes.

berm A platform of sediment with a steeply sloping seaward face deposited by waves; some beaches have no berm, whereas others may have several.

Big Bang A model for the evolution of the universe in which a dense, hot state was followed by expansion, cooling, and a less dense state.

biochemical sedimentary rock Any sedimentary rock produced by the chemical activities of organisms. (*See also* chemical sedimentary rock)

biostratigraphic unit An association of sedimentary rocks defined by its fossil content.

bipedal Walking on two legs as a means of locomotion, as in humans, birds, and some dinosaurs. (*See also* quadrupedal)

black smoker A type of submarine hydrothermal vent that emits a black plume of hot water colored by dissolved minerals.

body fossil The actual remains of an organism, such as bones, teeth, shells, and, rarely, the soft parts. (*See also* trace fossil)

bonding The process whereby atoms join to other atoms.

bony fish Fish with an internal skeleton of bone; the class Osteichthyes.

Bowen's reaction series A series of minerals that form in a specific sequence in cooling magma or lava; originally proposed to explain the origin of intermediate and felsic magma from mafic magma.
braided stream A stream with multiple dividing and rejoining channels.
breaker A wave that steepens as it enters shallow water until its crest plunges forward.
browser Any animal that eats tender shoots, twigs, and leaves. (*See also* grazer)
butte An isolated, steep-sided, pinnacle-like hill formed when resistant cap rock is breached allowing erosion of less resistant underlying rocks.

caldera A large, steep-sided, oval to circular depression usually formed when a volcano's summit collapses into a partially drained underlying magma chamber.
Caledonian orogeny A Silurian-Devonian episode of deformation along the northwestern margin of Baltica as it collided with Laurentia.
Canadian Shield The exposed part of the North American craton.
carbon 14 dating technique Absolute dating technique relying on the ratio of C^{14} to C^{12} in an organic substance; useful back to about 70,000 years ago.
carbonate mineral A mineral with the carbonate radical $(CO_3)^{-2}$, as in calcite ($CaCO_3$) and dolomite [$CaMg(CO_3)_2$].
carbonate rock Any rock, such as limestone and dolostone, made up mostly of carbonate minerals.
carnivore-scavenger Any animal that depends on other animals, living or dead, for nutrients.
cartilaginous fish Sharks, rays, and skates and their extinct relatives that have an internal skeleton of cartilage.
Cascade Range A mountain range with several active volcanoes in northern California, Oregon, Washington, and southern British Columbia, Canada.
Catskill Delta A Devonian clastic wedge deposited adjacent to highlands that formed during the Acadian orogeny.
cave A natural subsurface opening generally connected to the surface and large enough for a person to enter.
cementation The process whereby minerals crystallize in the pore spaces of sediment and bind the loose particles together.
chemical sedimentary rock Sedimentary rock made up of minerals that were dissolved during chemical weathering and later precipitated from seawater, more rarely lake water, or extracted from solution by organisms. (*See also* biochemical sedimentary rock)
chemical weathering The decomposition of rocks by chemical alteration of parent material.
China One of six major Paleozoic continents; composed of southeast Asia, including China, Indochina, part of Thailand, and the Malay Peninsula.
chordate Animals of the phylum Chordata; all have a notochord, dorsal hollow nerve cord, and gill slits at some time during their life cycle.
chromosome Double-stranded, helical molecule of deoxyribonucleic acid (DNA); specific segments of chromosomes are genes.
cinder cone A small, steep-sided volcano made up of pyroclastic materials resembling cinders that accumulate around a vent.
circum-Pacific belt A zone of seismic and volcanic activity and mountain building that nearly encircles the Pacific Ocean basin.
cirque A steep-walled, bowl-shaped depression on a mountainside at the upper end of a glacial valley.
cladistics A type of analysis of organisms in which they are grouped together on the basis of derived as opposed to primitive characteristics.
cladogram A diagram that shows the probable evolutionary relationships among members of a clade, a group of organisms including their most recent common ancestor.
clastic wedge An accumulation of detrital sediments eroded from and deposited adjacent to an uplifted area (e.g., the Catskill Delta).
cleavage Breakage along internal planes of weakness in mineral crystals.
columnar jointing The phenomenon of forming columns bounded by fractures in some igneous rocks as they cooled and contracted.
compaction Reduction in the volume of a sedimentary deposit that results from its own weight and the weight of any additional sediment deposited on top of it.
complex movement A combination of different types of mass movements in which no single type is dominant; usually involves sliding and flowing.
composite volcano (stratovolcano) A volcano composed of lava flows and pyroclastic layers, typically of intermediate composition, and mudflows.
compound Any substance resulting from the bonding of two or more different elements (e.g., water, H_2O, and quartz, SiO_2).
compression Stress resulting when rocks are squeezed by external forces directed toward one another.
concordant pluton Intrusive igneous body whose boundaries parallel the layering in the country rock. (*See also* discordant pluton)
cone of depression A cone-shaped depression around a well where water is pumped from an aquifer faster than it can be replaced.
contact metamorphism Metamorphism of country rock adjacent to a pluton.
continental accretion The process in which the continents grow by additions of Earth materials along their margins.
continental–continental plate boundary A convergent plate boundary along which two continental lithospheric plates collide.
continental drift The theory that the continents were joined into a single landmass that broke apart with the various fragments (continents) moving with respect to one another.
continental glacier A glacier that covers a vast area (at least 50,000 km^2) and is not confined by topography; also called an ice sheet.
continental margin The area separating the part of a continent above sea level from the deep seafloor.
continental rise The gently sloping part of the continental margin between the continental slope and the abyssal plain.
continental shelf The very gently sloping part of the continental margin between the shoreline and the continental slope.
continental slope The relatively steeply inclined part of the continental margin between the continental shelf and the continental rise or between the continental shelf and an oceanic trench.
convergent evolution The development of similar features in two or more distantly related groups of organisms resulting from adapting to a comparable lifestyle. (*See also* parallel evolution)
convergent plate boundary The boundary between two plates that move toward each other.
Cordilleran mobile belt A large region of deformation along the western margin of North America bounded on the west by the Pacific Ocean and on the east by the Great Plains.
Cordilleran orogeny An episode of deformation affecting the western margin of North America from Jurassic to Early Cenozoic time; divided into three separate phases called the Nevadan, Sevier, and Laramide orogenies.
core The interior part of Earth beginning at a depth of 2900 km that probably consists of iron and nickel.
Coriolis effect The apparent deflection of a moving object from its anticipated course because of Earth's rotation. Winds and oceanic currents are deflected clockwise in the Northern Hemisphere and counterclockwise in the Southern Hemisphere.
correlation Demonstration of the physical continuity of rock units or biostratigraphic units, or demonstration of time equivalence as in time-stratigraphic correlation.
country rock Any preexisting rock that has been intruded by a pluton or altered by metamorphism.
covalent bond A chemical bond formed by the sharing of electrons between atoms.
crater An oval to circular depression at the summit of a volcano resulting from the eruption of lava, pyroclastic materials, and gases.
craton The stable nucleus of a continent consisting of a shield of Precambrian rocks and a platform of buried ancient rocks.
cratonic sequence A widespread sequence of sedimentary rocks bounded above and below by unconformities that was deposited during a transgressive–regressive cycle of an epeiric sea.
creep A widespread type of mass wasting in which soil or rock moves slowly downslope.
crest The highest part of a wave.
Cretaceous Interior Seaway A Late Cretaceous arm of the sea that effectively divided North America into two large landmasses.
Cro-Magnon A race of *Homo sapiens* that lived mostly in Europe from 35,000 to 10,000 years ago.
cross-bedding A type of bedding in which layers are deposited at an angle to the surface on which they accumulate, as in sand dunes.
crossopterygian A specific type of lobe-finned fish that had lungs; the ancestors of amphibians were among crossopterygians.
crust Earth's outermost layer; the upper part of the lithosphere that is separated from the mantle by the Moho; divided into continental and oceanic crust.

crystal A naturally occurring solid of an element or compound with a specific internal structure that is manifested externally by planar faces, sharp corners, and straight edges.
crystal settling The physical separation and concentration of minerals in the lower part of a magma chamber or pluton by crystallization and gravitational settling.
crystalline solid A solid in which the constituent atoms are arranged in a regular, three-dimensional framework.
Curie point The temperature at which iron-bearing minerals in cooling magma or lava attain their magnetism.
cyclothem A sequence of cyclically repeating sedimentary rocks resulting from alternating periods of marine and nonmarine deposition; commonly contain a coal bed.
cynodont A type of carnivorous therapsid (advanced mammal-like reptile); ancestors of mammals are among cynodonts.

debris flow A type of mass wasting that involves a viscous mass of soil, rock fragments, and water that moves downslope; debris flow has larger particles than mudflows and contain less water.
deflation The removal of sediment and soil by wind.
deformation A general term for any change in shape or volume or both of rocks in response to stress; involves folding and fracturing.
delta An alluvial deposit formed where a stream or river flows into the sea or a lake.
density The mass of an object per unit volume; usually expressed in grams per cubic centimeter (g/cm^3).
deoxyribonucleic acid (DNA) The chemical substance of which chromosomes are composed; the genetic material in all organisms.
depositional environment Any site such as a floodplain or beach where physical, biologic, and chemical processes yield a distinctive kind of sedimentary deposit.
desert Any area that receives less than 25 cm of rain per year and that has a high evaporation rate.
desert pavement A surface mosaic of close-fitting pebbles, cobbles, and boulders found in many dry regions; results from wind erosion of sand and smaller particles.
desertification The expansion of deserts into formerly productive lands.
detrital sedimentary rock Sedimentary rock made up of the solid particles (detritus) of preexisting rocks.
differential pressure Pressure that is not applied equally to all sides of a rock body.
differential weathering Weathering that occurs at different rates on rocks, thereby yielding an uneven surface.
dike A tabular or sheetlike discordant pluton.
dip A measure of the maximum angular deviation of an inclined plane from horizontal.
dip-slip fault A fault on which all movement is parallel with the dip of the fault plain. (*See also* normal fault *and* reverse fault)
discharge The volume of water in a stream or river moving past a specific point in a given interval of time; expressed in cubic meters per second (m^3/sec) or cubic feet per second (ft^3/sec).
disconformity An unconformity above and below which the rock layers are parallel. (*See also* angular unconformity *and* nonconformity)
discontinuity A boundary across which seismic wave velocity or direction of travel changes abruptly, such as the mantle–core boundary.
discordant pluton Pluton with boundaries that cut across the layering in the country rock. (*See also* concordant pluton)
dissolved load The part of a stream's load consisting of ions in solution.
divergent evolution The diversification of a species into two or more descendant species.
divergent plate boundary The boundary between two plates that are moving apart.
divide A topographically high area that separates adjacent drainage basins.
dome A rather circular geologic structure in which all rock layers dip away from a central point and the oldest exposed rocks are at the dome's center.
drainage basin The surface area drained by a stream or river and its tributaries.
drainage pattern The regional arrangement of channels in a drainage system.
drumlin An elongate hill of till formed by the movement of a continental glacier or by floods.
dune A mound or ridge of wind-deposited sand.
dynamic metamorphism Metamorphism in fault zones where rocks are subjected to high differential pressure.

earthflow A mass wasting process involving the downslope movement of water-saturated soil.
earthquake Vibrations caused by the sudden release of energy, usually as a result of displacement of rocks along faults.
ectotherm Any of the cold-blooded vertebrate animals such as amphibians and reptiles; an animal that depends on external heat to regulate body temperature. (*See also* endotherm)
Ediacaran faunas Name for all Late Proterozoic faunas with animal fossils similar to those of the Ediacara fauna of Australia.
elastic rebound theory An explanation for the sudden release of energy that causes earthquakes when deformed rocks fracture and rebound to their original undeformed condition.
elastic strain A type of deformation in which the material returns to its original shape when stress is relaxed.
electron A negatively charged particle of very little mass that encircles the nucleus of an atom.
electron shell Electrons orbit an atom's nucleus at specific distances in electron shells.
element A substance composed of atoms that all have the same properties; atoms of one element can change to atoms of another element by radioactive decay, but otherwise they cannot be changed by ordinary chemical means.
emergent coast A coast where the land has risen with respect to sea level.
end moraine A pile or ridge of rubble deposited at the terminus of a glacier. (*See also* terminal moraine *and* recessional moraine)
endosymbiosis A symbiotic relationship in which one symbiont lives permanently within another organism. Organelles in cells such as plastids and mitochondria are thought to have resulted from endosymbiosis.
endotherm Any of the warm-blooded vertebrate animals, such as birds and mammals, that maintain their body temperature within narrow limits by internal processes. (*See* ectotherm)
epeiric sea A broad, shallow sea that covers part of a continent, such as the Sauk Sea.
epicenter The point on Earth's surface directly above the focus of an earthquake.
erosion The removal of weathered materials from their source area by running water, wind, glaciers, and waves.
erratic A rock fragment carried some distance from its source and deposited on bedrock of a different composition.
esker A long, sinuous ridge of stratified drift deposited by running water in a tunnel beneath stagnant ice.
eukaryotic cell A cell with internal structures such as mitochondria and an internal membrane-bounded nucleus containing chromosomes. (*See also* prokaryotic cell)
evaporite Any sedimentary rock, such as rock salt, formed by inorganic chemical precipitation of minerals from evaporating water.
Exclusive Economic Zone (EEZ) An area extending 371 km seaward from the coast of the United States and its possessions in which the United States claims rights to all resources.
exfoliation dome A large, rounded dome of rock resulting when concentric layers of rock are stripped from the surface of a rock mass.

fault A fracture along which rocks on opposite sides of the fracture have moved parallel with the fracture surface.
fault plane A fault surface that is more or less planar.
felsic magma Magma with more than 65% silica and considerable sodium, potassium, and aluminum, but little calcium, iron, and magnesium. (*See also* intermediate magma *and* mafic magma)
ferromagnesian silicate Any silicate mineral that contains iron, magnesium, or both. (*See also* nonferromagnesian silicate)
fetch The distance the wind blows over a continuous water surface.
fiord An arm of the sea extending into a glacial trough eroded below sea level.
firn Granular snow formed by partial melting and refreezing of snow; material transitional between snow and glacial ice.
fission track dating The absolute dating process in which small linear tracks (fission tracks) resulting from alpha decay are counted in mineral crystals.
fissure eruption A volcanic eruption in which lava or pyroclastic materials issue from a long, narrow fissure (crack) or group of fissures.
floodplain A low-lying, flat area adjacent to a channel that is partly or completely water-covered when a stream or river overflows its banks.
fluid activity An agent of metamorphism in which water and carbon dioxide promote metamorphism by increasing the rate of chemical reactions.

focus The site within Earth where an earthquake originates and energy is released.
fold A type of geologic structure in which planar features in rock layers such as bedding and foliation have been bent.
foliated texture A texture in metamorphic rocks in which platy and elongate minerals are aligned in a parallel fashion.
footwall block The block of rock that lies beneath a fault plane. (*See also* hanging wall block)
foreshore That part of a beach covered by water at high tide but exposed during low tide.
fossil Remains or traces of prehistoric organisms preserved in rocks. (*See also* body fossil *and* trace fossil)
fracture A break in rock resulting from intense applied pressure.
Franklin mobile belt The most northerly belt of deformation in North America; mostly in Greenland and the Canadian Arctic islands.
frost action The disaggregation of rocks by repeated freezing and thawing of water in cracks and crevasses.

gene A specific segment of a chromosome constituting the basic unit of heredity. (*See also* allele)
geologic record The record of prehistoric physical and biological events preserved in rocks.
geologic structure Any feature in rocks that results from deformation, such as folds, joints, and faults.
geologic time scale A chart arranged so that the designation for the earliest part of geologic time appears at the bottom followed upward by progressively younger time designations.
geology The science concerned with the study of Earth materials (minerals and rocks), surface and internal processes, and Earth history.
geothermal energy Energy that comes from steam and hot water trapped within Earth's crust.
geothermal gradient Earth's temperature increase with depth; it averages 25°C/km near the surface but varies from area to area.
geyser A hot spring that periodically ejects hot water and steam.
glacial budget The balance between expansion and contraction of a glacier in response to accumulation versus wastage.
glacial drift A collective term for all sediment deposited directly by glacial ice (till) and by meltwater streams (outwash).
glacial ice Water in the solid state within a glacier; forms as snow partially melts and refreezes and compacts so that it is transformed first to firn and then to glacial ice.
glacial polish A smooth, glistening rock surface formed by the movement of sediment-laden ice over bedrock.
glacial surge A time of greatly accelerated flow in a glacier; commonly results in displacement of a glacier's terminus by several kilometers.
glacier A mass of ice on land that moves by plastic flow and basal slip.
***Glossopteris* flora** A Late Paleozoic association of plants found only on the Southern Hemisphere continents and India; named for its best-known genus, *Glossopteris*.
Gondwana A major Paleozoic continent composed of South America, Africa, Australia, India, and parts of southern Europe, Arabia, and Florida.
graded bedding Sedimentary layer in which a single bed shows a decrease in grain size from bottom to top.
graded stream A stream that has an equilibrium profile in which a delicate balance exists among gradient, discharge, flow velocity, channel characteristics, and sediment load so that neither significant deposition nor erosion takes place within its channel.
gradient The slope over which a stream or river flows expressed in m/km or ft/mi.
gravity anomaly A departure from the expected force of gravity; anomalies may be positive or negative.
grazer An animal that eats low-growing vegetation, especially grasses. (*See also* browser)
greenstone belt A linear or podlike association of igneous and sedimentary rocks particularly common in Archean terrains.
ground moraine The layer of sediment released from melting ice as a glacier's terminus retreats.
groundwater Underground water stored in the pore spaces of soil, sediment, and rock.
guide fossil Any easily identified fossil with an extensive geographic distribution and short geologic range useful for determining the relative ages of rocks in different areas.
guyot A flat-topped seamount of volcanic origin rising more than 1 km above the seafloor.
gymnosperm A flowerless, seed-bearing plant.

half-life The time necessary for half of the original number of radioactive atoms of an element to decay to a stable daughter product; for example, the half-life for potassium 40 is 1.3 billion years.
hanging valley A tributary glacial valley whose floor is at a higher level than that of the main glacial valley.
hanging wall block The block of rock that overlies a fault plane. (*See also* footwall block)
hardness A term used to express the resistance of a mineral to abrasion.
headland Part of a shoreline commonly bounded by cliffs that extends out into the sea or a lake.
heat An agent of metamorphism.
herbivore An animal that depends directly on plants as a source of nutrients.
Hercynian-Alleghenian orogeny Pennsylvanian to Permian orogenic event during which the Appalachian mobile belt in eastern North America and the Hercynian mobile belt of southern Europe were deformed.
heterotrophic Any organism that depends on preformed organic molecules from its environment for nutrients. (*See also* autotrophic)
hominid Abbreviated form of Hominidae, the family of bipedal primates that includes *Australophithecus* and *Homo*.
Homo The genus of hominids consisting of *Homo sapiens* and their ancestors *Homo erectus* and *Homo habilis*.
homologous structure Body part in different organisms that has a similar structure, similar relationships to other structures, and similar development but does not necessarily serve the same function; for example, forelimbs in whales, dogs, and bats. (*See also* analogous structure)
horn A steep-walled, pyramid-shaped peak formed by the headward erosion of at least three cirques.
hot spot Localized zone of melting below the lithosphere; detected at the surface by volcanism.
hot spring A spring in which the water temperature is warmer than the temperature of the human body (37°C).
hydraulic action The removal of loose particles by the power of moving water.
hydrologic cycle The continuous recycling of water from the oceans, through the atmosphere, to the continents, and back to the oceans, or from the oceans, through the atmosphere, and back to the oceans.
hydrolysis The chemical reaction between hydrogen (H^+) ions and hydroxyl (OH^-) ions of water and a mineral's ions.
hydrothermal A term referring to hot water as in hot springs and geysers.
hypothesis A provisional explanation for observations that is subject to continual testing. If well supported by evidence, a hypothesis may be called a theory.

Iapetus Ocean A Paleozoic ocean between North America and Europe that closed when the continents collided during the Late Paleozoic.
ice cap A dome-shaped mass of glacial ice that covers less than 50,000 km^2.
igneous rock Any rock formed by cooling and crystallization of magma or lava or the consolidation of pyroclastic materials.
incised meander A deep, meandering canyon cut into bedrock by a stream or river.
index mineral A mineral that forms within specific temperature and pressure ranges during metamorphism.
infiltration capacity The maximum rate at which soil or sediment absorbs water.
inheritance of acquired characteristics Jean-Baptiste de Lamarck's mechanism for evolution; holds that characteristics acquired during an individual's lifetime can be inherited by descendants.
inselberg An isolated, steep-sided erosional remnant rising above a desert plain.
intensity The subjective measure of the kind of damage done by an earthquake as well as people's reaction to it.
intermediate magma Magma with a silica content between 53% and 65% and an overall composition intermediate between mafic and felsic magma.
ion An electrically charged atom produced by adding or removing electrons from the outermost electron shell.
ionic bond A chemical bond resulting from the attraction between positively and negatively charged ions.
isostasy *See* principle of isostasy.
isostatic rebound The phenomenon in which unloading of the crust causes it to rise until it attains equilibrium.

joint A fracture along which no movement has occurred or where movement is perpendicular to the fracture surface.

Jovian planet Any of the four planets (Jupiter, Saturn, Uranus, and Neptune) that resemble Jupiter. All are large and have low mean densities, indicating that they are composed mostly of lightweight gases, such as hydrogen and helium, and frozen compounds, such as ammonia and methane.

kame Conical hill of stratified drift originally deposited in a depression on a glacier's surface.
karst topography Landscape consisting of numerous caves, sinkholes, and solution valleys formed by groundwater solution of rocks such as limestone and dolostone.
Kaskaskia sequence A widespread sequence of Devonian and Mississippian sedimentary rocks bounded above and below by unconformities that was deposited during a transgressive–regressive cycle of the Kaskaskia Sea.
Kazakhstania One of six major Paleozoic continents; a triangular-shaped continent centered on Kazakhstan.

labyrinthodont Any of the Devonian to Triassic amphibians characterized by teeth with complexly folded enamel.
laccolith A concordant pluton with a mushroomlike geometry.
lahar A mudflow composed of pyroclastic materials such as ash.
Laramide orogeny Late Cretaceous to Early Cenozoic phase of the Cordilleran orogeny; responsible for many structural features of the present-day Rocky Mountains.
lateral moraine Ridge of sediment deposited along the margin of a valley glacier.
laterite A red soil, rich in iron or aluminum or both, resulting from intense chemical weathering in the tropics.
Laurasia A Late Paleozoic Northern Hemisphere continent made up of North America, Greenland, Europe, and Asia.
Laurentia A Proterozoic and Early-Middle Paleozoic continent composed of North America, Greenland, parts of Scotland, and perhaps part of the Baltic shield of Scandinavia.
lava Magma at Earth's surface.
lava dome A bulbous, steep-sided mountain formed by viscous magma moving upward through a volcanic conduit.
lava flow A stream of magma flowing over Earth's surface.
lava tube A tunnel beneath the solidified surface of a lava flow through which lava flows; also, the hollow space left when the lava within a tube drains away.
lithification The process of converting sediment into sedimentary rock by compaction and cementation.
lithosphere Earth's outer, rigid part consisting of the upper mantle, oceanic crust, and continental crust.
lithostatic pressure Pressure exerted on rocks by the weight of overlying rocks.
lithostratigraphic unit A body of rock, such as a formation, defined solely by its physical attributes.
loess Wind-blown deposit of silt and clay.
longitudinal dune A long ridge of sand generally parallel to the direction of the prevailing wind.
longshore current A current resulting from wave refraction found between the breaker zone and a beach that flows parallel to the shoreline.
Love wave (L-wave) A surface wave in which the individual particles of material move only back and forth in a horizontal plane perpendicular to the direction of wave travel.
luster The appearance of a mineral in reflected light. Luster is metallic or nonmetallic, although the latter has several subcategories.

mafic magma Magma with between 45% and 52% silica and proportionately more calcium, iron, and magnesium than intermediate and felsic magma.
magma Molten rock material generated within Earth.
magma chamber A reservoir of magma within Earth's upper mantle or lower crust.
magma mixing The process whereby magmas of different composition mix together to yield a modified version of the parent magmas.
magnetic anomaly Any deviation, such as a change in average strength, in Earth's magnetic field.
magnetic reversal The phenomenon involving the complete reversal of the north and south magnetic poles.
magnitude The total amount of energy released by an earthquake at its source. (*See also* Richter Magnitude Scale)
mantle The thick layer between Earth's crust and core.
marine regression The withdrawal of the sea from a continent or coastal area, resulting in the emergence of the land as sea level falls or the land rises with respect to sea level.
marine terrace A wave-cut platform now above sea level.
marine transgression The invasion of a coastal area or a continent by the sea, resulting from a rise in sea level or subsidence of the land.
marsupial mammal The pouched mammals such as kangaroos and wombats that give birth to their young in a very immature state.
mass extinction Greatly accelerated extinction rate that results in a marked decrease in biodiversity, such as the mass extinction at the end of the Cretaceous Period.
mass wasting The downslope movement of Earth materials under the influence of gravity.
meandering stream A stream that has a single, sinuous channel with broadly looping curves.
mechanical weathering Disaggregation of rocks by physical processes that yields smaller pieces that retain the composition of the parent material.
medial moraine A moraine carried on the central surface of a glacier; formed where two lateral moraines merge.
Mediterranean belt A zone of seismic and volcanic activity extending through the Mediterranean region of southern Europe and eastward to Indonesia.
meiosis Cell division that yields sex cells, sperm and eggs in animals and pollen and ovules in plants, in which the number of chromosomes is reduced by half. (*See also* mitosis)
mesa A broad, flat-topped erosional remnant bounded on all sides by steep slopes.
metamorphic facies A group of metamorphic rocks characterized by particular minerals that formed under the same broad temperature and pressure conditions.
metamorphic rock Any rock that has been changed from its original condition by heat, pressure, and the chemical activity of fluids, as in marble and slate.
metamorphic zone The region between lines of equal metamorphic intensity known as isograds.
Midcontinent Rift A Late Proterozoic rift in Laurentia in which volcanic and sedimentary rocks accumulated.
Milankovitch theory An explanation for the cyclic variations in climate and the onset of ice ages as a result of irregularities in Earth's rotation and orbit.
mineral A naturally occurring, inorganic, crystalline solid that has characteristic physical properties and a narrowly defined chemical composition.
mitosis Cell division that results in two cells that have the same number of chromosomes as the parent cell; takes place in all cells except sex cells. (*See also* meiosis)
mobile belt An elongate area of deformation generally at the margins of a craton, such as the Appalachian mobile belt.
modern synthesis A combination of ideas of scientists regarding evolution that includes chromosome theory of inheritance, mutations as a source of variation, and gradualism, but rejects the idea of inheritance of acquired characteristics.
Modified Mercalli Intensity Scale A scale with values from I to XII used to characterize earthquakes based on damage.
Mohorovičić discontinuity (Moho) The boundary between Earth's crust and mantle.
monocline A bend or flexure in otherwise horizontal or uniformly dipping rock layers.
monomer A comparatively simple organic molecule, such as an amino acid, that can link with other monomers to form more complex polymers such as proteins.
monotreme The egg-laying mammals; includes only the platypus and spiny anteater of the Australian region.
mosaic evolution The concept that not all parts of an organism evolve at the same rate, thus yielding organisms with features retained from the ancestral condition as well as more recently evolved features.
mud crack A crack in clay-rich sediment that forms in response to drying and shrinkage.
mudflow A flow consisting mostly of clay- and silt-sized particles and up to 30% water that moves downslope under the influence of gravity.
multicelled organism Any organism made up of many cells as opposed to a single cell; possesses cells specialized to perform specific functions.
mutation A change in the genes of organisms; yields some of the variation on which natural selection acts.

native element A mineral composed of a single element, such as gold.
natural levee A ridge of sandy alluvium deposited along the margins of a channel during floods.
natural selection A mechanism that accounts for differential survival and reproduction among members of a species; the

mechanism proposed by Darwin and Wallace to account for evolution.

Neanderthal A type of human that lived in the Near East and Europe from 200,000 to 30,000 years ago; may be a subspecies of *Homo* (*Homo sapiens neanderthalensis*) or a separate species (*Homo neanderthalensis*).

nearshore sediment budget The balance between additions and losses of sediment in the nearshore zone.

nekton Actively swimming organisms, such as fish, whales, and squid. (*See also* plankton)

neutron An electrically neutral particle found in the nucleus of an atom.

Nevadan orogeny Late Jurassic to Cretaceous phase of the Cordilleran orogeny; most strongly affected the western part of the Cordilleran mobile belt.

nonconformity An unconformity in which stratified sedimentary rocks overlie an erosion surface cut into igneous or metamorphic rocks. (*See also* angular unconformity *and* disconformity)

nonferromagnesian silicate A silicate mineral that has no iron or magnesium. (*See also* ferromagnesian silicate)

nonfoliated texture A metamorphic texture in which there is no discernable preferred orientation of minerals.

nonvascular Plants that lack specialized tissues for transporting fluids.

normal fault A dip-slip fault on which the hanging wall block has moved downward relative to the footwall block. (*See also* reverse fault)

North American Cordillera A complex mountainous region in western North America extending from Alaska into central Mexico.

nucleus The central part of an atom consisting of protons and neutrons.

nuée ardente A fast moving, dense cloud of hot pyroclastic materials and gases ejected from a volcano.

oblique-slip fault A fault showing both dip-slip and strike-slip movement.

oceanic–continental plate boundary A convergent plate boundary along which oceanic lithosphere is subducted beneath continental lithosphere.

oceanic–oceanic plate boundary A convergent plate boundary along which two oceanic plates collide and one is subducted beneath the other.

oceanic ridge A submarine mountain system composed mostly of basalt found in all ocean basins.

oceanic trench A long, narrow feature restricted to active continental margins and along which subduction occurs.

ooze Deep-sea sediment composed mostly of shells of marine animals and plants.

ophiolite A sequence of igneous rocks representing a fragment of oceanic lithosphere; composed of peridotite overlain successively by gabbro, sheeted basalt dikes, and pillow lava.

organic evolution *See* theory of evolution.

organic reef A wave-resistant limestone structure with a framework of animal skeletons.

Ornithischia One of the two orders of dinosaurs, characterized by a birdlike pelvis; includes ornithopods, stegosaurs, ankylosaurs, pachycephalosaurs, and ceratopsians. (*See also* Saurischia)

orogen A linear part of Earth's crust that was or is being deformed during an orogeny; part of an orogenic belt.

orogeny An episode of mountain building involving deformation, usually accompanied by igneous activity, and crustal thickening.

ostracoderm The "bony-skinned" fish characterized by bony armor but no jaws or teeth; appeared during the Late Cambrian, making them the oldest known vertebrates.

Ouachita mobile belt An area of deformation along the southern margin of the North American craton.

Ouachita orogeny A period of mountain building that took place in the Ouachita mobile belt during the Pennsylvanian Period.

outgassing The process whereby gases released from Earth's interior by volcanism formed an atmosphere.

outwash plain The sediment deposited by meltwater discharging from a continental glacier's terminus.

oxbow lake A cutoff meander filled with water.

oxidation The reaction of oxygen with other atoms to form oxides or, if water is present, hydroxides.

pahoehoe A type of lava flow with a smooth ropy surface.

paleomagnetism Residual magnetism in rocks, studied to determine the intensity and direction of Earth's past magnetic field.

paleontology The use of fossils to study life history and relationships among organisms.

Pangaea The name Alfred Wegener proposed for a supercontinent consisting of all Earth's landmasses at the end of the Paleozoic Era.

Panthalassa Ocean A Late Paleozoic ocean that surrounded Pangaea.

parabolic dune A crescent-shaped dune with its tips pointing upwind.

parallel evolution The development of similar features in two or more closely related but separate lines of descent as a consequence of similar adaptations. (*See also* convergent evolution)

parent material The material that is chemically and mechanically weathered to yield sediment and soil.

passive continental margin A continental margin within a tectonic plate as in the East Coast of North America where little seismic activity and no volcanism occur; characterized by a broad continental shelf and a continental slope and rise.

pedalfer Soil formed in humid regions with an organic-rich A horizon and aluminum-rich clays and iron oxides in horizon B.

pediment An erosion surface of low relief gently sloping away from the base of a mountain range.

pedocal Soil characteristic of arid and semiarid regions with a thin A horizon and a calcium carbonate–rich B horizon.

pelagic clay Brown or red deep-sea sediment composed of clay-sized particles.

pelycosaur Pennsylvanian to Permian reptiles that had some mammal characteristics; many species had large fins on the back.

perissodactyl The order of odd-toed hoofed mammals, which includes present-day horses, rhinoceroses, and tapirs.

permafrost Ground that remains permanently frozen.

permeability A material's capacity to transmit fluids.

phaneritic texture Igneous rock texture in which minerals are easily visible without magnification.

photochemical dissociation A process by which water molecules in the upper atmosphere are disrupted by ultraviolet radiation to yield free oxygen (O_2) and hydrogen (H_2).

photosynthesis The metabolic process in which organic molecules are synthesized from water and carbon dioxide, using the radiant energy of the Sun captured by chlorophyll-containing cells.

phyletic gradualism The concept that a species evolves gradually and continuously as it gives rise to new species. (*See also* punctuated equilibrium)

pillow lava Bulbous masses of basalt, resembling pillows, formed when lava is rapidly chilled under water.

placental mammal All mammals with a placenta to nourish the developing embryo, as opposed to egg-laying mammals (monotremes) and pouched mammals (marsupials).

placoderm Late Silurian through Permian "plate-skinned" fish with jaws and bony armor, especially in the head and shoulder area.

plankton Aquatic organisms that float passively, such as phytoplankton (plants) and zooplankton (animals). (*See also* nekton)

plastic flow The flow that takes place in response to pressure and causes deformation with no fracturing.

plastic strain Permanent deformation of a solid with no failure by fracturing.

plate An individual segment of the lithosphere that moves over the asthenosphere.

plate tectonic theory The theory holding that large segments of Earth's outer part (lithospheric plates) move relative to one another.

platform The part of a craton that lies buried beneath flat-lying or mildly deformed sedimentary rocks.

playa A dry lakebed found in deserts.

pluton An intrusive igneous body that forms when magma cools and crystallizes within the crust, such as a batholith or sill.

plutonic (intrusive igneous) rock Igneous rock that formed from magma intruded into or formed in place within the crust.

point bar The sediment body deposited on the gently sloping side of a meander loop.

polymer A comparatively complex organic molecule, such as nucleic acids and proteins, formed by monomers linking together. (*See also* monomer)

porosity The percentage of a material's total volume that is pore space.

porphyritic texture An igneous texture with minerals of markedly different sizes.

pressure release A mechanical weathering process in which rocks that formed under pressure expand on being exposed at the surface.

primary producer Organism in a food chain, such as a bacterium or green plant, that manufactures its own organic molecules, and on which other members of the food chain depend for sustenance. (*See also* autotrophic)

primate Any of the mammals that belong to the order Primates; includes prosimians (lemurs and tarsiers), monkeys, apes, and humans.

principle of cross-cutting relationships A principle holding that an igneous intrusion or fault must be younger than the rocks it intrudes or cuts across.

principle of fossil succession A principle holding that fossils, and especially groups or assemblages of fossils, succeed one another through time in a regular and predictable order.

principle of inclusions A principle holding that inclusions or fragments in a rock unit are older than the rock unit itself; for example, granite inclusions in sandstone are older than the sandstone.

principle of isostasy The theoretical concept of Earth's crust "floating" on a dense underlying layer. (*See* isostatic rebound)

principle of lateral continuity A principle holding that rock layers extend outward in all directions until they terminate.

principle of original horizontality According to this principle, sediments are deposited in horizontal or nearly horizontal layers.

principle of superposition A principle holding that in a vertical sequence of undeformed sedimentary rocks, the relative ages of the rocks can be determined by their position in the sequence—oldest at the bottom followed by successively younger layers.

principle of uniformitarianism A principle holding that we can interpret past events by understanding present-day processes, based on the idea that natural processes have always operated in the same way.

prokaryotic cell A cell that lacks a nucleus and organelles such as mitochondria and plastids; the cells of bacteria and cyanobacteria. (*See also* eukaryotic cell)

proton A positively charged particle found in the nucleus of an atom.

protorothyrid A loosely grouped category of small lizardlike reptiles.

punctuated equilibrium A concept holding that a new species evolves rapidly, in perhaps a few thousand years, and then remains much the same during its several million years of existence. (*See also* phyletic gradualism)

P-wave A compressional, or push-pull, wave; the fastest seismic wave and one that can travel through solids, liquids, and gases; also called a primary wave.

P-wave shadow zone An area between 103 and 143 degrees from an earthquake focus where little P-wave energy is recorded by seismographs.

pyroclastic (fragmental) texture A fragmental texture characteristic of igneous rocks composed of pyroclastic materials.

pyroclastic materials Fragmental substances, such as ash, explosively ejected from a volcano.

pyroclastic sheet deposit Vast, sheetlike deposit of felsic pyroclastic materials erupted from fissures.

quadrupedal A term referring to locomotion on all four limbs, as in dogs and horses. (*See also* bipedal)

Queenston Delta A clastic wedge resulting from erosion of the highlands formed during the Taconic orogeny.

quick clay A clay deposit that spontaneously liquefies and flows like water when disturbed.

radioactive decay The spontaneous change of an atom to an atom of a different element by emission of a particle from its nucleus (alpha and beta decay) or by electron capture.

rainshadow desert A desert found on the lee side of a mountain range because precipitation falls mostly on the windward side of the range.

rapid mass movement Any kind of mass wasting that involves a visible downslope displacement of material.

Rayleigh wave (R-wave) A surface wave in which individual particles of material move in an elliptical path within a vertical plane oriented in the direction of wave movement.

recessional moraine An end moraine that forms when a glacier's terminus retreats, then stabilizes, and a ridge or mound of till is deposited. (*See also* end moraine *and* terminal moraine)

red beds Sedimentary rocks, mostly sandstone and shale, with red color due to the presence of iron oxides.

reef A moundlike, wave-resistant structure composed of the skeletons of organisms.

reflection The return to the surface of some of a seismic wave's energy when it encounters a boundary separating materials of different density or elasticity.

refraction The change in direction and velocity of a seismic wave when it travels from one material into another of different density or elasticity.

regional metamorphism Metamorphism that occurs over a large area, resulting from high temperatures, tremendous pressure, and the chemical activity of fluids within the crust.

regolith The layer of unconsolidated rock and mineral fragments and soil that covers most of the land surface.

relative dating The process of determining the age of an event as compared to other events; involves placing geologic events in their correct chronologic order but does not involve a consideration of when the events occurred in number of years ago. (*See also* absolute dating)

reserve The part of the resource base that can be extracted economically.

resource A concentration of naturally occurring solid, liquid, or gaseous material in or on Earth's crust in such form and amount that economic extraction of a commodity from the concentration is currently or potentially feasible.

reverse fault A dip-slip fault on which the hanging wall block has moved upward relative to the footwall block. (*See also* normal fault)

Richter Magnitude Scale An open-ended scale that measures the amount of energy released during an earthquake.

rip current A narrow surface current that flows out to sea through the breaker zone.

ripple mark Wavelike (undulating) structure produced in granular sediment, especially sand, by unidirectional wind and water currents or by oscillating wave currents.

rock A solid aggregate of one or more minerals, as in limestone and granite, or a consolidated aggregate of rock fragments, as in conglomerate, or masses of rocklike materials, such as coal and obsidian.

rock cycle A group of processes through which Earth materials may pass as they are transformed from one major rock type to another.

rockfall A type of extremely fast mass wasting in which rocks fall through the air.

rock-forming mineral Any mineral common in rocks that is important in their identification and classification.

rock slide Rapid mass wasting in which rocks move downslope along a more or less planar surface.

runoff The surface flow in streams and rivers.

salt crystal growth A mechanical weathering process in which salt crystals growing in cracks and pores disaggregate rocks.

sandstone-carbonate-shale assemblage An association of sedimentary rocks typically found on passive continental margins.

Sauk sequence A widespread association of sedimentary rocks bounded above and below by unconformities that was deposited during a latest Proterozoic to Early Ordovician transgressive–regressive cycle of the Sauk Sea.

Saurischia An order of dinosaurs characterized by a lizardlike pelvis; includes theropods, prosauropods, and sauropods. (*See also* Ornithischia)

scientific method A logical, orderly approach that involves gathering data, formulating and testing hypotheses, and proposing theories.

seafloor spreading The theory that the seafloor moves away from spreading ridges and is eventually consumed at subduction zones.

seamount A submarine volcanic mountain rising at least 1 km above the seafloor.

sediment Loose aggregate of solids derived by weathering from preexisting rocks, or solids precipitated from solution by inorganic chemical processes or extracted from solution by organisms.

sediment-deposit feeder Animal that ingests sediment and extracts nutrients from it.

sedimentary facies Any aspect of a sedimentary rock unit that makes it recognizably different from adjacent sedimentary rocks of the same or approximately the same age.

sedimentary rock Any rock composed of sediment such as limestone and sandstone.

sedimentary structure Any feature in sedimentary rock that formed at or shortly after the time of deposition, such as cross-bedding, animal burrows, and mud cracks.

seedless vascular plant Plant with specialized tissues for transporting fluids and nutrients that reproduces by spores rather than seeds, as in ferns and horsetail rushes.

seismic profiling A method in which strong waves generated at an energy source penetrate the layers beneath the seafloor. Some of the energy is reflected back from various layers to the surface, making it possible to determine the nature of the layers.

seismograph An instrument that detects, records, and measures the various waves produced by earthquakes.

seismology The study of earthquakes.
sequence stratigraphy The study of rock relationships within a time-stratigraphic framework of related facies bounded by widespread unconformities.
Sevier orogeny Cretaceous phase of the Cordilleran orogeny that affected the continental shelf and slope areas of the Cordilleran mobile belt.
shear strength The resisting forces that help maintain a slope's stability.
shear stress The result of forces acting parallel to one another but in opposite directions; results in deformation by displacement of adjacent layers along closely spaced planes.
shield A vast area of exposed ancient rocks on a continent; the exposed part of a craton.
shield volcano A dome-shaped volcano with a low, rounded profile built up mostly by overlapping basalt lava flows.
shoreline The area between mean low tide and the highest level on land affected by storm waves.
Siberia One of six major Paleozoic continents; composed of Russia east of the Ural Mountains, and Asia north of Kazakhstan and south of Mongolia.
silica A compound of silicon and oxygen.
silica tetrahedron The basic building block of all silicate minerals; consists of one silicon atom and four oxygen atoms.
silicate A mineral that contains silica, such as quartz (SiO_2).
sill A tabular or sheetlike concordant pluton.
sinkhole A depression in the ground that forms by the solution of the underlying carbonate rocks or by the collapse of a cave roof.
slide Mass wasting involving movement of material along one or more surfaces of failure.
slow mass movement Mass movement that advances at an imperceptible rate and is usually detectable only by the effects of its movement.
slump Mass wasting that takes place along a curved surface of failure and results in the backward rotation of the slump mass.
soil Regolith consisting of weathered materials, water, air, and humus that can support vegetation.
soil degradation Any process leading to a loss of soil productivity; may involve erosion, chemical pollution, or compaction.
soil horizon A distinct soil layer that differs from other soil layers in texture, structure, composition, and color.
solar nebula theory A theory for the evolution of the solar system from a rotating cloud of gas.
solifluction Mass wasting involving the slow downslope movement of water-saturated surface materials; especially the flow at high elevations or high latitudes where the flow is underlain by frozen soil.
solution A reaction in which the ions of a substance become dissociated in a liquid and the solid substance dissolves.
Sonoma orogeny A Permian-Triassic orogeny caused by the collision of an island arc with the southwestern margin of North America.
species A population of similar individuals that in nature can interbreed and produce fertile offspring.
specific gravity The ratio of a substance's weight, especially a mineral, to an equal volume of water at 4°C.
spheroidal weathering A type of chemical weathering in which corners and sharp edges of rocks weather more rapidly than flat surfaces, thus yielding spherical shapes.
spit A fingerlike projection of a beach into a body of water such as a bay.
spring A place where groundwater flows or seeps out of the ground.
stock An irregularly shaped discordant pluton with a surface area smaller than 100 km^2.
stoping A process in which rising magma detaches and engulfs pieces of the country rock.
strain Deformation caused by stress.
strata (*singular,* stratum) Refers to layering in sedimentary rocks. (*See also* bed)
stratified drift Glacial deposits that show both stratification and sorting.
stream terrace An erosional remnant of a floodplain that formed when a stream was flowing at a higher level.
stress The force per unit area applied to a material such as rock.
strike The direction of a line formed by the intersection of an inclined plane and a horizontal plane.
strike-slip fault A fault involving horizontal movement of blocks of rock on opposite sides of a fault plane. (*See also* dip-slip fault)
stromatolite A biogenic sedimentary structure, especially in limestone, produced by the entrapment of sediment on sticky mats of photosynthesizing bacteria.
submarine canyon A steep-walled canyon best developed on the continental slope, but some extend well up onto the continental shelf.
submarine fan A cone-shaped sedimentary deposit that accumulates on the continental slope and rise.
submarine hydrothermal vent A crack or fissure in the seafloor through which superheated water issues. (*See also* black smoker)
submergent coast A coast along which sea level rises with respect to the land or the land subsides.
Sundance Sea A wide seaway that existed in western North America during the Middle Jurassic Period.
superposed stream A stream that once flowed on a higher surface and eroded downward into resistant rocks while maintaining its course.
suspended load The smallest particles (silt and clay) carried by running water, which are kept suspended by fluid turbulence.
suspension feeder Animal that consumes microscopic plants and animals or dissolved nutrients from water.
S-wave A shear wave that moves material perpendicular to the direction of travel, thereby producing shear stresses in the material it moves through; also known as a secondary wave; S-waves travel only through solids.
S-wave shadow zone Those areas more than 103 degrees from an earthquake focus where no S-waves are recorded.
syncline A down-arched fold in which the youngest exposed rocks coincide with the fold axis and all strata dip toward the axis.
system A combination of related parts that interact in an organized fashion; Earth systems include the atmosphere, hydrosphere, biosphere, and solid Earth.

Taconic orogeny An Ordovician episode of mountain building that resulted in the deformation of the Appalachian mobile belt.
talus Accumulation of coarse, angular rock fragments at the base of a slope.
Tejas epeiric sea A Cenozoic sea largely restricted to the Gulf and Atlantic Coastal Plains, coastal California, and the Mississippi Valley.
tension A type of stress in which forces act in opposite directions but along the same line, thus tending to stretch an object.
terminal moraine An end moraine consisting of a ridge or mounds of rubble marking the farthest extent of a glacier. (*See also* end moraine *and* recessional moraine)
terrane A block of rock with characteristics different from those of surrounding rocks. Terranes probably represent seamounts, oceanic rises, and other seafloor features that accreted to continents during orogenies.
terrestrial planet Any of the four innermost planets (Mercury, Venus, Earth, and Mars). They are all small and have high mean densities, indicating that they are composed of rock and metallic elements.
theory An explanation for some natural phenomenon that has a large body of supporting evidence. To be scientific, a theory must be testable—for example, plate tectonic theory.
theory of evolution The theory holding that all organisms are related and that they descended with modification from organisms that lived during the past.
therapsid Permian to Triassic mammal-like reptiles; the ancestors of mammals are among a group of terapsids known as cynodonts.
thermal convection cell A type of circulation of material in the asthenosphere during which hot material rises, moves laterally, cools and sinks, and is reheated and continues the cycle.
thermal expansion and contraction A type of mechanical weathering in which the volume of rocks changes in response to heating and cooling.
thrust fault A type of reverse fault in which a fault plane dips less than 45 degrees.
tide The regular fluctuation of the sea's surface in response to the gravitational attraction of the Moon and Sun.
till All sediment deposited directly by glacial ice.
time-stratigraphic unit A body of strata that was deposited during a specific interval of geologic time; for example, the Devonian System (a time-stratigraphic unit) was deposited during the Devonian Period.
time unit Any of the units such as eon, era, period, epoch, and age used to refer to specific intervals of geologic time.
Tippecanoe sequence A widespread body of sedimentary rocks bounded above and below by unconformities that was deposited during an Ordovician to Early Devonian transgressive–regressive cycle of the Tippecanoe Sea.
tombolo A type of spit that extends out from the shoreline and connects the mainland with an island.
trace fossil Any indication of prehistoric organic activity such as tracks, trails, and

nests preserved in rocks. (*See also* body fossil)

Transcontinental Arch Area extending from Minnesota to New Mexico that stood above sea level as several large islands during the Cambrian transgression of the Sauk Sea.

transform fault A fault along which one type of motion is transformed into another; commonly displaces oceanic ridges; on land recognized as a strike-slip fault, such as the San Andreas fault.

transform plate boundary Plate boundary along which plates slide past one another and crust is neither produced nor destroyed.

transverse dune A ridge of sand with its long axis perpendicular to the wind direction.

tree-ring dating The process of determining the age of a tree or wood in a structure by counting the number of annual growth rings.

trough The lowest point between wave crests.

tsunami A large sea wave that is usually produced by an earthquake but can also result from submarine landslides and volcanic eruptions.

turbidity current A sediment–water mixture, denser than normal seawater, that flows downslope to the deep seafloor.

unconformity A break in the geologic record represented by an erosional surface separating younger strata from older rocks. (*See also* nonconformity, disconformity, *and* angular unconformity)

ungulate An informal term referring to the hoofed mammals, especially the orders Artiodactyla and Perissodactyla.

U-shaped glacial trough A valley with steep or vertical walls and a broad, rather flat floor formed by the movement of a glacier through a stream valley.

valley A linear depression bounded by higher areas such as ridges or mountains.

valley glacier A glacier confined to a mountain valley or an interconnected system of mountain valleys.

valley train A long, narrow deposit of stratified drift confined within a glacial valley.

vascular Land plants that have specialized tissues for transporting fluids.

velocity A measure of distance traveled per unit of time, as in the flow velocity in a stream or river.

ventifact A stone with a surface polished, pitted, grooved, or faceted by wind abrasion.

vertebrate Any animal that has a segmented vertebral column, as in fish, amphibians, reptiles, birds, and mammals.

vesicle A small hole or cavity formed by gas trapped in cooling lava.

vestigial structure Structure in an organism that no longer serves any or only a limited function, such as dewclaws in dogs and wisdom teeth in humans.

viscosity A fluid's resistance to flow.

volcanic explosivity index (VEI) A semiquantitative scale for the size of a volcanic eruption based on evaluation of criteria such as volume of material explosively erupted and height of eruption cloud.

volcanic (extrusive igneous) rock An igneous rock formed when magma is extruded onto Earth's surface where it cools and crystallizes, or when pyroclastic materials become consolidated.

volcanic neck An erosional remnant of the material that solidified in a volcanic pipe.

volcanic pipe The conduit connecting the crater of a volcano with an underlying magma chamber.

volcanic tremor Ground motion lasting from minutes to hours resulting from magma moving beneath the surface, as opposed to the sudden jolts produced by most earthquakes.

volcanism The processes whereby magma and its associated gases rise through the crust and are extruded onto the surface or into the atmosphere.

volcano A hill or mountain formed around a vent as a result of the eruption of lava and pyroclastic materials.

water table The surface that separates the zone of aeration from the underlying zone of saturation.

water well A well made by digging or drilling into the zone of saturation.

wave An undulation on the surface of a body of water, resulting in the water surface rising and falling.

wave base The depth corresponding to about one-half wavelength below which water is unaffected by surface waves.

wave-cut platform A beveled surface that slopes gently seaward; formed by the erosion and retreat of a sea cliff.

wave refraction The bending of waves so that they more nearly parallel the shoreline.

weathering The physical breakdown and chemical alteration of rocks and minerals at or near Earth's surface.

zone of accumulation The part of a glacier where additions exceed losses and the glacier's surface is perennially covered with snow. Also refers to horizon B in soil where soluble material leached from horizon A accumulates as irregular masses.

zone of aeration The zone above the water table that contains both air and water within the pore spaces of soil, sediment, or rock.

zone of saturation The area below the water table in which all pore spaces are filled with water.

zone of wastage The part of a glacier where losses from melting, sublimation, and calving of icebergs exceed the rate of accumulation.

Index

Italic page numbers refer to illustrations.